*Jörg Kärger,
Douglas M. Ruthven, and
Doros N. Theodorou*

**Diffusion in Nanoporous
Materials**

Related Titles

Cejka, Jiri / Corma, Avelino / Zones, Stacey (eds.)

Zeolites and Catalysis

882 Pages, 2 Volumes, Hardcover

318 Fig. (42 Colored Fig.), 75 Tab.

2010

ISBN: 978-3-527-32514-6

Kulprathipanja, Santi (ed.)

Zeolites in Industrial Separation and Catalysis

594 Pages, Hardcover

310 Fig. (10 Colored Fig.)

2010

ISBN: 978-3-527-32505-4

Hirscher, Michael (ed.)

Handbook of Hydrogen Storage

353 Pages, Hardcover

158 Fig. (6 Colored Fig.), 24 Tab.

2010

ISBN: 978-3-527-32273-2

Farrusseng, David (ed.)

Metal-Organic Frameworks

392 Pages, Hardcover

185 Fig. (26 Colored Fig.), 28 Tab.

2011

ISBN: 978-3-527-32870-3

Bruce, Duncan W. / Walton, Richard I. / O'Hare, Dermot (eds.)

Porous Materials

350 Pages, Hardcover

2010

ISBN: 978-0-470-99749-9

Bräuchle, Christoph / Lamb, Don Carroll / Michaelis, Jens (eds.)

Single particle tracking and single molecule energy transfer

343 Pages, Hardcover

112 Fig. (46 Colored Fig.), 3 Tab.

2010

ISBN: 978-3-527-32296-1

Klages, Rainer / Radons, Günter / Sokolov, Igor M. (eds.)

Anomalous Transport Foundations and Applications

584 Pages, Hardcover

163 Fig. (11 Colored Fig)

2008

ISBN: 978-3-527-40722-4

Deutschmann, Olaf (ed.)

Modeling and Simulation of Heterogeneous Catalytic Reactions From the Molecular Process to the Technical System

354 Pages, Hardcover.

155 Fig. (25 Colored Fig.), 20 Tab.

2011

ISBN: 978-3-527-32120-9

Reichl, Linda E.

A Modern Course in Statistical Physics 3rd revised and updated edition

411 Pages, Softcover

2009

ISBN: 978-3-527-40782-8

Rao, B. L. S. Prakasa

Statistical Inference for Fractional Diffusion Processes

Series: Wiley Series in Probability and Statistics

280 Pages, Hardcover

2010

ISBN: 978-0-470-66568-8

*Jörg Kärger, Douglas M. Ruthven,
and Doros N. Theodorou*

Diffusion in Nanoporous Materials

Volume 1

WILEY-VCH Verlag GmbH & Co. KGaA

The Authors

Prof. Dr. Jörg Kärger
Universität Leipzig
Abteilung Grenzflächenphysik
Linnestr. 5
04103 Leipzig

Prof. Dr. Douglas M. Ruthven
Dept. of Chemical Engineering
University of Maine
Orono, ME 04469-5737
USA

Prof. Doros N. Theodorou
Nat. Techn. Univ. of Athens
School of Chemical Engineering
9 Heroon Polytechniou Street
15780 Athens
Griechenland

All books published by **Wiley-VCH** are carefully produced. Nevertheless, authors, editors, and publisher do not warrant the information contained in these books, including this book, to be free of errors. Readers are advised to keep in mind that statements, data, illustrations, procedural details or other items may inadvertently be inaccurate.

Library of Congress Card No.: applied for

British Library Cataloguing-in-Publication Data
A catalogue record for this book is available from the British Library.

Bibliographic information published by the Deutsche Nationalbibliothek
The Deutsche Nationalbibliothek lists this publication in the Deutsche Nationalbibliografie; detailed bibliographic data are available on the Internet at http://dnb.d-nb.de.

© 2012 Wiley-VCH Verlag & Co. KGaA, Boschstr. 12, 69469 Weinheim, Germany

All rights reserved (including those of translation into other languages). No part of this book may be reproduced in any form – by photoprinting, microfilm, or any other means – nor transmitted or translated into a machine language without written permission from the publishers. Registered names, trademarks, etc. used in this book, even when not specifically marked as such, are not to be considered unprotected by law.

Cover Design Grafik-Design Schulz, Fußgönheim
Typesetting Thomson Digital, Noida, India
Printing and Binding betz-druck GmbH, Darmstad

Printed in the Federal Republic of Germany
Printed on acid-free paper

Print ISBN: 978-3-527-31024-1
ePDF ISBN: 978-3-527-65130-6
ePub ISBN: 978-3-527-65129-0
mobi ISBN: 978-3-527-65128-3
oBook ISBN: 978-3-527-65127-6

To Birge, Patricia and Fani

Contents

Preface *XXV*
Acknowledgments *XXIX*

Content of Volume 1

Part I	**Introduction** *1*	
1	**Elementary Principles of Diffusion** *3*	
1.1	Fundamental Definitions *4*	
1.1.1	Transfer of Matter by Diffusion *4*	
1.1.2	Random Walk *6*	
1.1.3	Transport Diffusion and Self-Diffusion *7*	
1.1.4	Frames of Reference *9*	
1.1.5	Diffusion in Anisotropic Media *10*	
1.2	Driving Force for Diffusion *12*	
1.2.1	Gradient of Chemical Potential *12*	
1.2.2	Experimental Evidence *15*	
1.2.3	Relationship between Transport and Self-diffusivities *16*	
1.3	Diffusional Resistances in Nanoporous Media *17*	
1.3.1	Internal Diffusional Resistances *17*	
1.3.2	Surface Resistance *19*	
1.3.3	External Resistance to Mass Transfer *19*	
1.4	Experimental Methods *21*	
	References *24*	
Part II	**Theory** *25*	
2	**Diffusion as a Random Walk** *27*	
2.1	Random Walk Model *27*	
2.1.1	Mean Square Displacement *27*	
2.1.2	The Propagator *29*	
2.1.3	Correspondence with Fick's Equations *32*	
2.2	Correlation Effects *33*	
2.2.1	Vacancy Correlations *33*	

2.2.2	Correlated Anisotropy	35
2.3	Boundary Conditions	35
2.3.1	Absorbing and Reflecting Boundaries	35
2.3.2	Partially Reflecting Boundary	38
2.3.3	Matching Conditions	39
2.3.4	Combined Impact of Diffusion and Permeation	40
2.4	Macroscopic and Microscopic Diffusivities	42
2.5	Correlating Self-Diffusion and Diffusion with a Simple Jump Model	44
2.6	Anomalous Diffusion	47
2.6.1	Probability Distribution Functions of Residence Time and Jump Length	47
2.6.2	Fractal Geometry	49
2.6.3	Diffusion in a Fractal System	53
2.6.4	Renormalization	54
2.6.5	Deviations from Normal Diffusion in Nanoporous Materials: A Retrospective	55
	References	57
3	**Diffusion and Non-equilibrium Thermodynamics**	**59**
3.1	Generalized Forces and Fluxes	59
3.1.1	Mechanical Example	59
3.1.2	Thermodynamic Forces and Fluxes	60
3.1.3	Rate of Generation of Entropy	61
3.1.4	Isothermal Approximation	63
3.1.5	Diffusion in a Binary Adsorbed Phase	64
3.2	Self-Diffusion and Diffusive Transport	65
3.3	Generalized Maxwell–Stefan Equations	67
3.3.1	General Formulation	67
3.3.2	Diffusion in an Adsorbed Phase	68
3.3.3	Relation between Self- and Transport Diffusivities	71
3.4	Application of the Maxwell–Stefan Model	72
3.4.1	Parameter Estimation	72
3.4.2	Membrane Permeation	73
3.4.3	Diffusion in Macro- and Mesopores	74
3.5	Loading Dependence of Self- and Transport Diffusivities	75
3.5.1	Self-Diffusivities	75
3.5.2	Transport Diffusivities	77
3.5.3	Molecular Simulation	78
3.5.4	Effect of Structural Defects	78
3.6	Diffusion at High Loadings and in Liquid-Filled Pores	80
	References	81
4	**Diffusion Mechanisms**	**85**
4.1	Diffusion Regimes	85
4.1.1	Size-Selective Molecular Sieving	85

4.2	Diffusion in Macro- and Mesopores	*87*
4.2.1	Diffusion in a Straight Cylindrical Pore	*87*
4.2.1.1	Knudsen Mechanism	*88*
4.2.1.2	Viscous Flow	*90*
4.2.1.3	Molecular Diffusion	*91*
4.2.1.4	Transition Region	*91*
4.2.1.5	Self-Diffusion/Tracer Diffusion	*92*
4.2.1.6	Relative Importance of Different Mechanisms	*92*
4.2.1.7	Surface Diffusion	*92*
4.2.1.8	Combination of Diffusional Resistances	*93*
4.2.2	Diffusion in a Pore Network	*94*
4.2.2.1	Dusty Gas Model	*94*
4.2.2.2	Effective Medium Approximation	*95*
4.2.2.3	Tortuosity Factor	*95*
4.2.2.4	Parallel Pore Model	*96*
4.2.2.5	Random Pore Model	*97*
4.2.2.6	Capillary Condensation	*97*
4.3	Activated Diffusion	*99*
4.3.1	Diffusion in Solids	*99*
4.3.2	Surface Diffusion Mechanisms	*100*
4.3.2.1	Vacancy Diffusion	*100*
4.3.2.2	Reed–Ehrlich Model	*101*
4.3.3	Diffusion by Cage-to-Cage Jumps	*104*
4.4	Diffusion in More Open Micropore Systems	*106*
4.4.1	Mobile Phase Model	*106*
4.4.2	Diffusion in a Sinusoidal Field	*107*
4.4.3	Free-Volume Approach for Estimating the Loading Dependence	*108*
	References	*108*
5	**Single-File Diffusion**	*111*
5.1	Infinitely Extended Single-File Systems	*112*
5.1.1	Random Walk Considerations	*112*
5.1.2	Molecular Dynamics	*116*
5.2	Finite Single-File Systems	*119*
5.2.1	Mean Square Displacement	*119*
5.2.2	Tracer Exchange	*120*
5.2.3	Catalytic Reactions	*123*
5.2.3.1	Dynamic Monte Carlo Simulations and Adapting the Analysis for Normal Diffusion	*123*
5.2.3.2	Rigorous Treatment	*125*
5.2.3.3	Molecular Traffic Control	*130*
5.3	Experimental Evidence	*132*
5.3.1	Ideal versus Real Structure of Single-File Host Systems	*132*
5.3.2	Experimental Findings Referred to Single-File Diffusion	*135*
5.3.2.1	Pulsed Field Gradient NMR	*135*

5.3.2.2	Quasi-elastic Neutron Scattering *136*
5.3.2.3	Tracer Exchange and Transient Sorption Experiments *137*
5.3.2.4	Catalysis *138*
	References *139*

6	**Sorption Kinetics** *143*
6.1	Resistances to Mass and Heat Transfer *143*
6.2	Mathematical Modeling of Sorption Kinetics *145*
6.2.1	Isothermal Linear Single-Resistance Systems *145*
6.2.1.1	External Fluid Film or Surface Resistance Control *145*
6.2.1.2	Micropore Diffusion Control *146*
6.2.1.3	Effect of Particle Shape *149*
6.2.1.4	Macropore Diffusion Control *149*
6.2.2	Isothermal, Linear Dual-Resistance Systems *151*
6.2.2.1	Surface Resistance plus Internal Diffusion *151*
6.2.2.2	Transient Concentration Profiles *152*
6.2.2.3	Two Diffusional Resistances (Biporous Solid) *156*
6.2.3	Isothermal Nonlinear Systems *160*
6.2.3.1	Micropore Diffusion Control *160*
6.2.3.2	Macropore Diffusion Control *160*
6.2.3.3	Semi-infinite Medium *161*
6.2.3.4	Adsorption Profiles *162*
6.2.3.5	Desorption Profiles *165*
6.2.3.6	Experimental Uptake Rate Data *167*
6.2.3.7	Adsorption/Desorption Rates and Effective Diffusivities *170*
6.2.3.8	Approximate Analytic Representations *172*
6.2.3.9	Linear Driving Force Approximation *172*
6.2.3.10	Shrinking Core Model *173*
6.2.3.11	Surface Resistance Control – Nonlinear Systems *176*
6.2.4	Non-isothermal Systems *179*
6.2.4.1	Intraparticle Diffusion Control *181*
6.2.4.2	Experimental Non-isothermal Uptake Curves *183*
6.3	Sorption Kinetics for Binary Mixtures *185*
6.3.1.1	Counter-Diffusion *186*
6.3.1.2	Co-diffusion *187*
	References *188*

Part III	**Molecular Modeling** *191*
7	**Constructing Molecular Models and Sampling Equilibrium Probability Distributions** *193*
7.1	Models and Force Fields for Zeolite–Sorbate Systems *194*
7.1.1	Molecular Model and Potential Energy Function *194*
7.1.2	*Ab Initio* Molecular Dynamics of Zeolite-Sorbate Systems *203*

7.1.3	Computer Reconstruction of Meso- and Macroporous Structure	*204*
7.2	Monte Carlo Simulation Methods	*206*
7.2.1	Metropolis Monte Carlo Algorithm	*206*
7.2.2	Canonical Monte Carlo	*207*
7.2.3	Grand Canonical Monte Carlo	*210*
7.2.4	Gibbs Ensemble and Gage Cell Monte Carlo	*215*
7.3	Free Energy Methods for Sorption Equilibria	*217*
7.4	Coarse-Graining and Potentials of Mean Force	*222*
	References *224*	
8	**Molecular Dynamics Simulations**	*227*
8.1	Statistical Mechanics of Diffusion	*227*
8.1.1	Self-diffusivity	*227*
8.1.2	Mass Fluxes in a Microporous Medium	*229*
8.1.3	Transport in the Presence of Concentration Gradients in Pure and Mixed Sorbates	*232*
8.2	Equilibrium Molecular Dynamics Simulations	*235*
8.2.1	Integrating the Equations of Motion: the Velocity Verlet Algorithm	*235*
8.2.2	Multiple Time Step Algorithms: rRESPA	*238*
8.2.3	Domain Decomposition	*241*
8.2.4	Molecular Dynamics of Rigid Linear Molecules	*241*
8.2.5	Molecular Dynamics of Rigid Nonlinear Molecules: Quaternions	*245*
8.2.6	Constraint Dynamics in Cartesian Coordinates	*249*
8.2.7	Extended Ensemble Molecular Dynamics	*253*
8.2.8	Example Application of Equilibrium MD to Pure Sorbates	*257*
8.2.9	Example Applications of MD to Mixed Sorbates	*259*
8.3	Non-equilibrium Molecular Dynamics Simulations	*265*
8.3.1	Gradient Relaxation Method	*265*
8.3.2	External Field NEMD	*267*
8.3.3	Dual Control Volume Grand Canonical Molecular Dynamics	*269*
	References *272*	
9	**Infrequent Event Techniques for Simulating Diffusion in Microporous Solids**	*275*
9.1	Statistical Mechanics of Infrequent Events	*276*
9.1.1	Time Scale Separation and General Expression for the Rate Constant	*276*
9.1.2	Transition State Theory Approximation and Dynamical Correction Factor	*281*
9.1.3	Multidimensional Transition State Theory	*283*
9.1.4	Infrequent Event Theory of Multistate Multidimensional Systems	*290*
9.1.5	Numerical Methods for Infrequent Event Analysis	*291*

9.2	Tracking Temporal Evolution in a Network of States *292*
9.2.1	Master Equation *292*
9.2.2	Kinetic Monte Carlo Simulation *293*
9.2.3	Analytical Solution of the Master Equation *295*
9.3	Example Applications of Infrequent Event Analysis and Kinetic Monte Carlo for the Prediction of Diffusivities in Zeolites *296*
9.3.1	Self-diffusivity at Low Occupancies *296*
9.3.2	Self-diffusivity at High Occupancy *300*
	References *301*

Part IV Measurement Methods *303*

10 Measurement of Elementary Diffusion Processes *305*

10.1	NMR Spectroscopy *306*
10.1.1	Basic Principles: Behavior of Isolated Spins *306*
10.1.1.1	Classical Treatment *306*
10.1.1.2	Quantum Mechanical Treatment *308*
10.1.2	Behavior of Nuclear Spins in Compact Material *309*
10.1.2.1	Fundamentals of the NMR Line Shape *311*
10.1.2.2	Effect of Molecular Motion on Line Shape *311*
10.1.2.3	Fundamentals of Pulse Measurements *314*
10.1.3	Resonance Shifts by Different Surroundings *318*
10.1.3.1	Spins in Different Chemical Surroundings *318*
10.1.3.2	Anisotropy of the Chemical Shift *320*
10.1.3.3	Quadrupole NMR *320*
10.1.4	Impact of Nuclear Magnetic Relaxation *323*
10.2	Diffusion Measurements by Neutron Scattering *326*
10.2.1	Principle of the Method *326*
10.2.1.1	Experimental Procedure *326*
10.2.2	Theory of Neutron Scattering *327*
10.2.3	Scattering Patterns and Molecular Motion *330*
10.2.3.1	Fundamental Relations *330*
10.2.3.2	Thermodynamic Factor *331*
10.2.3.3	Evidence on the Elementary Steps of Diffusion *334*
10.2.3.4	Intermediate Scattering Function *334*
10.2.3.5	Range of Measurement *335*
10.3	Diffusion Measurements by Light Scattering *337*
10.3.1	Theory *337*
10.3.2	Experimental Issues for Observing Light Scattering Phenomena: Index Matching *339*
10.3.2.1	Index Matching *339*
10.3.2.2	Optical Mixing Techniques *340*
10.3.2.3	Filter Techniques *341*
	References *343*

11	**Diffusion Measurement by Monitoring Molecular Displacement** *347*
11.1	Pulsed Field Gradient (PFG) NMR: Principle of Measurement *348*
11.1.1	PFG NMR Fundamentals *348*
11.1.2	Basic Experiment *351*
11.2	The Complete Evidence of PFG NMR *352*
11.2.1	Concept of the Mean Propagator *352*
11.2.2	Evidence of the Mean Propagator *353*
11.2.3	Concept of Effective Diffusivity *354*
11.3	Experimental Conditions, Limitations, and Options for PFG NMR Diffusion Measurement *355*
11.3.1	Sample Preparation *355*
11.3.2	Basic Data Analysis and Ranges of Observation *356*
11.3.3	Pitfalls *358*
11.3.3.1	Gradient Pulse Mismatch *358*
11.3.3.2	Mechanical Instabilities *358*
11.3.3.3	Benefit of Extra-large Stray-Field Gradients *359*
11.3.3.4	Impedance by Internal Gradients *359*
11.3.3.5	Impedance by Contaminants *360*
11.3.4	Measurement of Self-diffusion in Multicomponent Systems *361*
11.3.4.1	Fourier-Transform PFG NMR *361*
11.3.4.2	Alternative Approaches *362*
11.3.5	Diffusion Anisotropy *362*
11.3.5.1	"Single-Crystal" Measurements *362*
11.3.5.2	Powder Measurement *363*
11.3.5.3	Evidence on Host Structure *364*
11.4	Different Regimes of PFG NMR Diffusion Measurement *364*
11.4.1	Effect of Finite Crystal Size on the Measurement of Intracrystalline Diffusion *365*
11.4.1.1	Correlating the True and Effective Diffusivities in the Short-Time Limit *365*
11.4.1.2	Correlating the Effective Diffusivities with the Crystal Dimensions in the Long-Time Limit *367*
11.4.2	Long-Range Diffusion *368*
11.4.2.1	Tortuosity Factor and the Mechanism of Diffusion *369*
11.4.3	Covering the Complete Range of Observation Times, the "NMR-Tracer Desorption" Technique *371*
11.4.3.1	Two-Dimensional Monte-Carlo Simulation *371*
11.4.3.2	Two-Region Model *374*
11.4.3.3	NMR Tracer Desorption (= Tracer-Exchange) Technique *376*
11.4.3.4	Observation of Surface Barriers *377*
11.5	Experimental Tests of Consistency *379*
11.5.1	Variation of Crystal Size *379*
11.5.2	Diffusion Measurements with Different Nuclei *379*
11.5.3	Influence of External Magnetic Field *380*

11.5.4	Comparison of Self-diffusion and Tracer Desorption Measurements	*380*
11.5.5	Blocking of the Extracrystalline Space	*380*
11.5.6	Determination of Crystal Size for Restricted Diffusion Measurements	*381*
11.5.7	Long-Range Diffusion	*382*
11.5.8	Tracer Exchange Measurements	*382*
11.6	Single-Molecule Observation	*383*
11.6.1	Basic Principles of Fluorescence Microscopy	*384*
11.6.2	Time Averaging Versus Ensemble Averaging	*385*
11.6.3	Correlating Structure and Mass Transfer in Nanoporous Materials	*388*
	References	*389*
12	**Imaging of Transient Concentration Profiles**	*395*
12.1	Different Options of Observation	*396*
12.1.1	Positron Emission Tomography (PET)	*397*
12.1.2	X-Ray Monitoring	*397*
12.1.3	Optical Microscopy	*398*
12.1.4	IR Microscopy	*399*
12.1.5	Magnetic Resonance Imaging (MRI)	*400*
12.2	Monitoring Intracrystalline Concentration Profiles by IR and Interference Microscopy	*403*
12.2.1	Principles of Measurement	*403*
12.2.2	Data Analysis	*406*
12.3	New Options for Experimental Studies	*408*
12.3.1	Monitoring Intracrystalline Concentration Profiles	*408*
12.3.1.1	New Options of Correlating Transport Diffusion and Self-diffusion	*412*
12.3.1.2	Visualizing Guest Profiles in Transient Sorption Experiments	*412*
12.3.1.3	Testing the Relevance of Surface Resistances by Correlating Fractional Uptake with Boundary Concentration	*413*
12.3.1.4	Applying Boltzmann's Integration Method for Analyzing Intracrystalline Concentration Profiles	*414*
12.3.2	Direct Measurement of Surface Barriers	*415*
12.3.2.1	Concentration Dependence of Surface Permeability	*415*
12.3.2.2	Correlating the Surface Permeabilities through Different Crystal Faces	*417*
12.3.3	Adapting the Concept of Sticking Probabilities to Nanoporous Particles	*421*
	References	*423*
13	**Direct Macroscopic Measurement of Sorption and Tracer Exchange Rates**	*427*
13.1	Gravimetric Methods	*427*
13.1.1	Experimental System	*428*

13.1.2	Analysis of the Response Curves 429
13.1.2.1	Intrusion of Heat Effects 429
13.1.2.2	Criterion for Negligible Thermal Effects 430
13.1.2.3	Intrusion of Bed Diffusional Resistance 431
13.1.2.4	Experimental Checks 432
13.1.3	The TEOM and the Quartz Crystal Balance 432
13.2	Piezometric Method 433
13.2.1	Mathematical Model 434
13.2.2	Single-Step Frequency Response 436
13.2.3	Single-Step Temperature Response 436
13.3	Macro FTIR Sorption Rate Measurements 437
13.4	Rapid Recirculation Systems 440
13.4.1	Liquid Phase Systems 441
13.5	Differential Adsorption Bed 441
13.6	Analysis of Transient Uptake Rate Data 443
13.6.1	Time Domain Matching 443
13.6.2	Method of Moments 444
13.7	Tracer Exchange Measurements 445
13.7.1	Detectors for Radioisotopes 446
13.7.2	Experimental Procedure 447
13.8	Frequency Response Measurements 447
13.8.1	Theoretical Model 448
13.8.2	Temperature Frequency Response 451
13.8.3	Measurement Limits 451
13.8.4	Experimental Systems 452
13.8.5	Results 454
13.8.6	Frequency Response Measurements in a Flow System 455
	References 456
14	**Chromatographic and Permeation Methods of Measuring Intraparticle Diffusion** 459
14.1	Chromatographic Method 460
14.1.1	Mathematical Model for a Chromatographic Column 460
14.1.1.1	Time Domain Solutions 462
14.1.1.2	Form of Response Curves 463
14.1.2	Moments Analysis 464
14.1.2.1	First and Second Moments 464
14.1.2.2	Use of Higher Moments 466
14.1.2.3	HETP and the van Deemter Equation 466
14.2	Deviations from the Simple Theory 468
14.2.1	"Long-Column" Approximation 468
14.2.1.1	Pressure Drop 468
14.2.1.2	Nonlinear Equilibrium 469
14.2.1.3	Heat Transfer Resistance 470
14.3	Experimental Systems for Chromatographic Measurements 470

14.3.1	Dead Volume *471*	
14.3.2	Experimental Conditions *471*	
14.4	Analysis of Experimental Data *472*	
14.4.1	Liquid Systems *473*	
14.4.2	Vapor Phase Systems *473*	
14.4.3	Intrusion of Axial Dispersion *477*	
14.5	Variants of the Chromatographic Method *479*	
14.5.1	Step Response *479*	
14.5.2	Limited Penetration Regime *480*	
14.5.3	Wall-Coated Column *480*	
14.5.4	Direct Measurement of the Concentration Profile in the Column *480*	
14.6	Chromatography with Two Adsorbable Components *481*	
14.7	Zero-Length Column (ZLC) Method *483*	
14.7.1	General Principle of the Technique *483*	
14.7.2	Theory *485*	
14.7.2.1	Basic Theory for Intraparticle Diffusion Control *485*	
14.7.2.2	Short-Time Behavior *486*	
14.7.2.3	Diffusion in Macroporous Particles *488*	
14.7.3	Deviations from the Simple Theory *489*	
14.7.3.1	Surface Resistance and/or Fluid Film Resistance *489*	
14.7.3.2	Measurement of Surface Resistance by ZLC *489*	
14.7.3.3	Fluid Phase Hold-Up *490*	
14.7.3.4	Effect of Isotherm Nonlinearity *491*	
14.7.3.5	Heat Effects *492*	
14.7.4	Practical Considerations *493*	
14.7.4.1	Choice of Operating Conditions and Extraction of the Time Constant *493*	
14.7.4.2	Preliminary Checks *494*	
14.7.4.3	Equilibration Time *494*	
14.7.4.4	Partial Loading Experiment *495*	
14.7.5	Extensions of the ZLC Technique *497*	
14.7.5.1	Tracer ZLC *497*	
14.7.5.2	Counter-current ZLC (CCZLC) *498*	
14.7.5.3	Liquid Phase Measurements *498*	
14.8	TAP System *500*	
14.9	Membrane Permeation Measurements *501*	
14.9.1	Steady-State Permeability Measurements *501*	
14.9.2	Wicke–Kallenbach Steady-State Method *503*	
14.9.2.1	Macro/Mesopore Diffusion Measurements *503*	
14.9.2.2	Intracrystalline Diffusion *503*	
14.9.3	Time Lag Measurements *504*	
14.9.3.1	Linear Systems (Constant D) *504*	
14.9.3.2	Single-Crystal Zeolite Membrane *505*	
14.9.3.3	Nonlinear Systems *506*	
14.9.3.4	Evaluation of Single-Crystal Membrane Technique *507*	

14.9.4	Transient Wicke–Kallenbach Experiment 508
14.9.5	Comparison of Transient and Steady-State Measurements 510
	References 510

Content of Volume 2

Part V   Diffusion in Selected Systems 515

15   Amorphous Materials and Extracrystalline (Meso/Macro) Pores 517
15.1   Diffusion in Amorphous Microporous Materials 518
15.1.1   Diffusion in Microporous Glass 518
15.1.2   Diffusion in Microporous Carbon 520
15.2   Effective Diffusivity 524
15.2.1   Direct Measurement of Tortuosity in a Porous Catalyst Particle 525
15.2.2   Determination of Macropore Diffusivity in a Catalyst Particle under Reaction Conditions 526
15.3   Diffusion in Ordered Mesopores 527
15.4   Diffusion through Mesoporous Membranes 530
15.4.1   Mesoporous Vycor Glass 530
15.4.2   Mesoporous Silica 531
15.4.2.1   Permeance Measurements 532
15.4.2.2   Modified Mesoporous Membranes 533
15.5   Surface Diffusion 534
15.5.1   Determination of Surface Diffusivities 534
15.5.2   Concentration Dependence 535
15.6   Diffusion in Liquid-Filled Pores 538
15.7   Diffusion in Hierarchical Pore Systems 539
15.7.1   Ordered Mesoporous Material SBA-15 539
15.7.2   Activated Carbon with Interpenetrating Micro- and Mesopores 541
15.7.3   Diffusion in "Mesoporous" Zeolites 544
15.8   Diffusion in Beds of Particles and Composite Particles 545
15.8.1   Approaches by Mathematical Modeling 546
15.8.2   Temperature Dependence 548
15.8.3   Multicomponent Systems 550
15.9   More Complex Behavior: Presence of a Condensed Phase 552
15.9.1   Diffusion in Porous Vycor Glass: Hysteresis Effects 553
15.9.2   Supercritical Transition in an Adsorbed Phase 555
15.9.3   Diffusion in a Composite Particle of MCM-41 556
           References 557

16   Eight-Ring Zeolites 561
16.1   Eight-Ring Zeolite Structures 561
16.1.1   LTA Structure 562
16.1.2   Cation Hydration 563
16.1.3   Structure of CHA 564

16.1.3.1	Structure of DDR (ZSM-58) 565
16.1.4	Concentration and Temperature Dependence of Diffusivity 566
16.2	Diffusion in Cation-Free Eight-Ring Structures 567
16.2.1	Effect of Window Dimensions 567
16.2.2	Diffusion of CO_2 and CH_4 in DDR 569
16.3	Diffusion in 4A Zeolite 571
16.4	Diffusion in 5A Zeolite 573
16.4.1	Uptake Rate Measurements 576
16.4.2	Measurements with Different Crystal Sizes 579
16.4.3	PFG NMR Measurements over Different Length Scales 580
16.4.4	Intra-Cage Jumps 581
16.4.5	Comparison of Different Zeolite Samples and Different Measurement Techniques 581
16.5	General Patterns of Behavior in Type A Zeolites 582
16.5.1	Activation Energies 584
16.5.2	Pre-exponential Factor 585
16.6	Window Blocking 587
16.6.1	Sorption Cut-Off 587
16.6.2	Monte Carlo Simulations 588
16.6.3	Effective Medium Approximation 590
16.6.4	Diffusion in NaCaA Zeolites 591
16.6.5	3A Zeolites 592
16.7	Variation of Diffusivity with Carbon Number 593
16.7.1	Loading Dependence of Self-diffusivity 595
16.8	Diffusion of Water Vapor in LTA Zeolites 595
16.9	Deactivation, Regeneration, and Hydrothermal Effects 596
16.9.1	Effect of Water Vapor 596
16.9.2	Kinetic Behavior of Commercial 5A: Evidence from Uptake Rates 597
16.9.3	Effects of Pelletization and other Aspects of Technological Performance: Evidence from NMR Studies 599
16.10	Anisotropic Diffusion in CHA 602
16.11	Concluding Remarks 603
	References 604
17	**Large Pore (12-Ring) Zeolites** 607
17.1	Structure of X and Y Zeolites 607
17.2	Diffusion of Saturated Hydrocarbons 609
17.2.1	Evidence from NMR 609
17.2.2	Comparison of NMR and ZLC Data for NaX 615
17.2.3	Isoparaffins and Cyclohexane 616
17.2.4	n-Octane Diffusion in NaY, USY and NaX 618
17.2.5	Diffusion in NaCaX: Effect of Ca^{2+} Cations 619
17.2.6	Diffusion Measurements as Evidence of Structural Imperfection 621
17.3	Diffusion of Unsaturated and Aromatic Hydrocarbons In NaX 623
17.3.1	Light Olefins 623

17.3.2	C_8 Aromatics	624
17.3.2.1	Macroscopic Measurements	624
17.3.2.2	Comparison with Microscopic Measurements	625
17.3.3	Benzene	627
17.3.3.1	Comparison of Macroscopic and Microscopic Measurements	627
17.3.3.2	Diffusion Mechanism	628
17.3.3.3	Hysteresis	631
17.3.4	Discrepancy between Macroscopic and Microscopic Diffusivity Measurements for Aromatics in NaX	632
17.4	Other Systems	633
17.4.1	Water in NaX and NaY	633
17.4.2	Methanol	635
17.4.3	Triethylamine	635
17.4.4	Hydrogen	636
17.5	PFG NMR Diffusion Measurements with Different Probe Nuclei	636
17.5.1	Fluorine Compounds	637
17.5.2	Xenon and Carbon Monoxide and Dioxide	638
17.5.3	Nitrogen	639
17.6	Self-diffusion in Multicomponent Systems	640
17.6.1	Hydrocarbons	640
17.6.1.1	*n*-Heptane–Benzene	640
17.6.1.2	Benzene–Perfluorobenzene	641
17.6.1.3	*n*-Butane–Perfluoromethane	641
17.6.1.4	Ethane–Ethene	642
17.6.2	Self-diffusion under the Influence of Co-adsorption and Carrier Gases	643
17.6.2.1	Effect of Moisture on Self-diffusion	643
17.6.2.2	Water and Ammonia	643
17.6.2.3	Effect of an Inert Carrier Gas	644
17.6.3	Self-diffusion in Multicomponent Systems Evolving during Catalytic Conversion	645
17.6.3.1	Cyclopropane into Propene	645
17.6.3.2	Isopropanol into Acetone and Propene	647
	References	648
18	**Medium-Pore (Ten-Ring) Zeolites**	**653**
18.1	MFI Crystal Structure	654
18.1.1	Saturation Capacity	655
18.1.2	Molecular Sieve Behavior	659
18.2	Diffusion of Saturated Hydrocarbons	659
18.2.1	Linear Alkanes	659
18.2.1.1	Microscale Measurements	661
18.2.1.2	Macroscale Measurements	663
18.2.1.3	Molecular Simulations	664
18.2.1.4	Loading Dependence of Diffusivity	665

18.2.1.5	Microdynamic Behavior	665
18.2.2	Diffusion of Branched and Cyclic Paraffins	666
18.2.2.1	Isobutane at Low Loadings	667
18.2.2.2	Cyclohexane and Alkyl Cyclohexanes	669
18.2.2.3	2,2-Dimethylbutane	670
18.2.2.4	Comparison of Diffusivities for C_6 Isomers	670
18.2.2.5	Comparison of Diffusivities for Linear and Branched Hydrocarbons	673
18.2.2.6	Diffusion of Isobutane at High Loadings	674
18.3	Diffusion of Aromatic Hydrocarbons	676
18.3.1	Diffusion of Benzene	676
18.3.1.1	Self- and Transport Diffusion	680
18.3.1.2	Benzene Microdynamics	680
18.3.2	Diffusion of C_8 Aromatics	681
18.3.2.1	Uptake Curves	681
18.3.2.2	ZLC and TZLC Measurements	683
18.3.2.3	Frequency Response Data for *p*-Xylene	684
18.3.2.4	Evidence from Membrane Permeation Studies	685
18.3.2.5	Diffusion of *o*-Xylene and *m*-Xylene	686
18.4	Adsorption from the Liquid Phase	686
18.5	Microscale Studies of other Guest Molecules	688
18.5.1	Tetrafluoromethane	688
18.5.2	Water and Methanol	689
18.5.3	Ammonia	689
18.5.4	Hydrogen	691
18.6	Surface Resistance and Internal Barriers	692
18.6.1	Surface Resistance	692
18.6.1.1	Macroscopic Rate Measurements	693
18.6.1.2	Surface Etching	694
18.6.1.3	Measurement of Transient Concentration Profiles	695
18.6.1.4	Other Surface Effects	696
18.6.2	Intracrystalline Barriers	696
18.6.3	Sub-structure of MFI Crystals	697
18.6.4	Assessing Transport Resistances at Internal and External Boundaries by Micro-imaging	699
18.6.5	Evidence from PFG NMR Diffusion Studies	701
18.6.6	Differences in Diffusional Behavior of Linear and Branched Hydrocarbons	702
18.7	Diffusion Anisotropy	703
18.7.1	Correlation Rule for Structure-Directed Diffusion Anisotropy	703
18.7.2	Comparison of Measured Profiles	704
18.7.3	Evidence by PFG NMR Measurements	705
18.7.4	Limits of the Correlation Rule	707
18.7.5	Anisotropic Diffusion in a Binary Adsorbed Phase	710
18.7.6	Anisotropy of Real Crystals	710

18.8	Diffusion in a Mixed Adsorbed Phase	712
18.8.1	Blocking Effects by a Co-adsorbed Second Guest Species	712
18.8.1.1	Methane in the Presence of Benzene	712
18.8.1.2	Methane in the Presence of Pyridine and Ammonia	713
18.8.1.3	n-Butane in the Presence of Isobutane	714
18.8.2	Two-Component Diffusion	716
18.8.2.1	Methane and Tetrafluoromethane	716
18.8.2.2	Methane and Xenon	716
18.8.2.3	Methane and Ammonia	717
18.8.2.4	Permeation Properties of Nitrogen and Carbon Dioxide	717
18.8.2.5	Counter-current Desorption of p-Xylene–Benzene	717
18.8.2.6	Co- and Counter-diffusion of Benzene and Toluene	718
18.8.2.7	Counter-diffusion of Isobutane and n-Butane	720
18.9	Guest Diffusion in Ferrierite	722
	References	723

19 Metal Organic Frameworks (MOFs) 729

19.1	A New Class of Porous Solids	730
19.2	MOF-5 and HKUST-1: Diffusion in Pore Spaces with the Architecture of Zeolite LTA	732
19.2.1	Guest Diffusion in MOF-5	733
19.2.2	Guest Diffusion in CuBTC	736
19.3	Zeolitic Imidazolate Framework 8 (ZIF-8)	739
19.3.1	An Ideal Case where Experimental Self- and Transport Diffusivities are interrelated from First Principles	740
19.3.2	Membrane-Based Gas Separation Following the Predictions of Diffusion Measurements	744
19.4	Pore Segments in Single-File Arrangement: Zn(tbip)	747
19.5	Breathing Effects: Diffusion in MIL-53	751
19.6	Surface Resistance	754
19.6.1	Experimental Observations	754
19.6.2	Conceptual Model for a Surface Barrier in a One-Dimensional System: Blockage of Most of the Pore Entrances	757
19.6.2.1	Simulation Results	758
19.6.2.2	An Analytic Relationship between Surface Permeability and Intracrystalline Diffusivity	759
19.6.2.3	Surface Resistance with Three-Dimensional Pore Networks	760
19.6.2.4	Estimating the Fraction of Unblocked Entrance Windows	760
19.6.3	Generalization of the Model	761
19.6.3.1	Intracrystalline Barriers	761
19.6.3.2	Activation Energies	761
19.7	Concluding Remarks	762
	References	764

Part VI Selected Applications 769

20 Zeolite Membranes 771
20.1 Zeolite Membrane Synthesis 772
20.2 Single-Component Permeation 773
20.2.1 Selectivity and Separation Factor 774
20.2.2 Modeling of Permeation 776
20.3 Separation of Gas Mixtures 779
20.3.1 Size-Selective Molecular Sieving 779
20.3.2 Diffusion-Controlled Permeation 781
20.3.3 Equilibrium-Controlled Permeation 783
20.4 Modeling Permeation of Binary Mixtures 784
20.4.1 Maxwell–Stefan Model 784
20.4.1.1 Concentration Profile through the Membrane 785
20.4.1.2 Importance of Mutual Diffusion 789
20.4.2 More Complex Systems 789
20.4.2.1 Membrane Thickness 791
20.4.2.2 Support Resistance 791
20.5 Membrane Characterization 792
20.5.1 Bypass Flow 792
20.5.2 Perm-Porosimetry 793
20.5.3 Isotherm Determination 795
20.5.4 Analysis of Transient Response 797
20.6 Membrane Separation Processes 797
20.6.1 Pervaporation Process for Dehydration of Alcohols 797
20.6.2 CO_2–CH_4 Separation 798
20.6.3 Separation of Butene Isomers 799
20.6.4 MOF (Metal Organic Framework) Membranes for H_2 Separation 799
20.6.5 Amorphous Silica Membranes 800
20.6.6 Stuffed Membranes 801
20.6.7 Membrane Modules 801
20.6.8 Membrane Reactors 802
20.6.9 Barriers to Commercialization 802
References 803

21 Diffusional Effects in Zeolite Catalysts 807
21.1 Diffusion and Reaction in a Catalyst Particle 807
21.1.1 The Effectiveness Factor 808
21.1.2 External Mass Transfer Resistance 811
21.1.3 Temperature Dependence 812
21.1.4 Reaction Order 814
21.1.5 Pressure Dependence 814
21.1.6 Non-isothermal Systems 815
21.1.7 Relative Importance of Internal and External Resistances 816

21.2	Determination of Intracrystalline Diffusivity from Measurements of Reaction Rate *817*	
21.2.1	Temperature Dependence of "Effective Diffusivity" *819*	
21.3	Direct Measurement of Concentration Profiles during a Diffusion-Controlled Catalytic Reaction *819*	
21.3.1	Reaction of Furfuryl Alcohol on HZSM-5 *820*	
21.3.2	Reaction in Mesoporous MCM-41 *821*	
21.4	Diffusional Restrictions in Zeolite Catalytic Processes *822*	
21.4.1	Size Exclusion *822*	
21.4.2	Catalytic Cracking over Zeolite Y *823*	
21.4.3	Catalytic Cracking Over HZSM-5 *824*	
21.4.4	Catalytic Cracking over other Zeolites *825*	
21.4.5	Activation Energies *825*	
21.4.6	Xylene Isomerization *826*	
21.4.7	Selective Disproportionation of Toluene *828*	
21.4.8	MTG Reaction *830*	
21.4.9	MTO Process *831*	
21.5	Coking of Zeolite Catalysts *833*	
21.5.1	Information from PFG NMR *834*	
21.5.2	Information from Fluorescence Microscopy *835*	
	References *835*	

Notation *839*

Index *851*

Preface

Diffusion at the atomic or molecular level is a universal phenomenon, occurring in all states of matter on time scales that vary over many orders of magnitude, and indeed controlling the overall rates of many physical, chemical, and biochemical processes. The wide variety of different systems controlled by diffusion is well illustrated by the range of the topics covered in the *Diffusion Fundamentals Conference* series (http://www.uni-leipzig.de/diffusion/). For both fundamental and practical reasons diffusion is therefore important to both scientists and engineers in several different disciplines. This book is concerned primarily with diffusion in microporous solids such as zeolites but, since the first edition was published (in 1992 under the title *Diffusion in Zeolites and Other Microporous Solids*), several important new micro- and mesoporous materials such as metal organic frameworks (MOFs) and mesoporous silicas (e.g., MCM-41 and SBA-15) have been developed. In recognition of these important developments the scope of this new edition has been broadened to include new chapters devoted to mesoporous silicas and MOFs and the title has been modified to reflect these major changes.

In addition to the important developments in the area of new materials, over the past 20 years, there have been equally important advances both in our understanding of the basic physics and in the development of new theoretical and experimental approaches for studying diffusion in micro- and mesoporous solids. Perhaps the most important of these advances is the remarkable development of molecular modeling based on numerical simulations. Building on the rapid advances in computer technology, Monte Carlo (MC) and molecular dynamic (MD) simulations of adsorption equilibrium and kinetics have become almost routine (although kinetic simulations must still be treated with caution unless confirmed by experimental data). In recognition of the importance of these developments the new edition contains three authoritative new chapters, written mainly by Doros Theodorou, dealing with the principles of molecular simulations and their application to the study of diffusion in porous materials.

With respect to experimental techniques, over the past 20 years, neutron scattering has advanced from a scientific curiosity to a viable and valuable technique for studying diffusion at short length scales. Interference microscopy (IFM) has also become a practically viable technique, providing unprecedented insights into diffusional behavior by allowing direct measurement of the internal concentration profiles during transient adsorption or desorption processes. In contrast, the early promise of light scattering techniques has not yet been fulfilled as the practical difficulties have, so far, proved insurmountable. We are, however, witnessing impressive advances in our understanding of a wide variety of systems through the application of single-molecule visualization techniques. As a highlight of this development the book includes experimental confirmation of the celebrated ergodic theorem.

As with the first edition our intention in writing this book has been to present a coherent summary and review of both the basic theory of diffusion in porous solids and the major experimental and theoretical techniques that have been developed for studying and simulating the behavior of such systems. The theoretical foundations of the subject and indeed some of the experimental approaches borrow heavily from classical theories of diffusion in solids, liquids, and gases. We have therefore attempted to include sufficient background material to allow the book to be read without frequent reference to other sources.

The book is divided into six parts, of which the first four, dealing with basic theory, molecular simulations, and experimental methods are included in Volume I. The "experimental" chapters cover both macroscopic measurements, in which adsorption/desorption rates are followed in an assemblage of adsorbent particles, and microscopic methods (mainly PFG NMR and QENS) in which the movement of the molecules themselves is followed, as well as the new imaging techniques such as IFM and IRM in which concentration profiles or fluxes within a single crystal are measured. Parts Five and Six, in Volume II, deal with diffusion in selected systems and with the practical application of zeolites as membranes and catalysts.

The first edition contained considerable discussion of the discrepancies between microscopic and macroscopic measurements. These discrepancies have now been largely resolved, but it turns out that in many zeolite crystals structural defects are much more important than was originally thought. As a result, in such systems, the measured diffusivity is indeed dependent on the length scale of the measurement and the diffusivity as a structurally perfect crystal is often approached only at the very short length scales probed by neutron scattering. Another important feature that has become apparent only through the application of detailed IFM measurements is the prevalence of surface resistance. In many zeolite and MOF crystals the resistance to transport at the crystal surface is significant and has been shown to result from the blockage of a large fraction of the pore openings. Such detailed insights, which depend on the application of new experimental techniques, have become possible only recently.

Throughout the text and in the major tables we have generally used SI units although our adherence to that system has not been slavish and, particularly with respect to pressure, we have generally retained the original units.

The selection of the material for a text of this kind inevitably reflects the biases and interests of the authors. In reviewing the literature of the subject we have made no attempt to be comprehensive but we hope that we have succeeded in covering, or at least mentioning, most of the more important developments.

Jörg Kärger, Leipzig, Germany
Douglas M. Ruthven, Orono, Maine, USA
Doros N. Theodorou, Athens, Greece

Acknowledgments

A book of this kind is inevitably a collaborative project involving not only the authors but their research students, colleagues, and associates, many of whom have contributed, both directly and indirectly, over a period of many years. Our early collaboration, in the days of the GDR, would not have been possible without the support and encouragement of two well-known pioneers of zeolite research, Professor Wolfgang Schirmer (Academy of Sciences of the GDR) and Professor Harry Pfeifer (University of Leipzig). Much of the early experimental work was carried out by Dr Jürgen Caro (now Professor of Physical Chemistry at the University of Hanover), using large zeolite crystals provided by Professor Zhdanov (University of Leningrad) and the home-made PFG NMR spectrometer that was constructed and maintained by Dr Wilfried Heink.

Since German re-unification both the formal and financial difficulties of research collaboration have been greatly reduced and the list of collaborators, many of whom are mentioned in the cited references, has become too long to name individuals. For the historical record it is, however, appropriate to mention the contributions of a few key people who were involved in the development of the new experimental and molecular modeling techniques that were used to obtain most of the information presented in this new edition. Jeffrey Hufton (now with Air Products Inc.) and Stefano Brandani (now Professor of Chemical Engineering at Edinburgh University) were mainly responsible for the development of "tracer ZLC," which allowed the first *direct* comparisons of "macroscopic" and "microscopic" measurements of self-diffusion in zeolites. The successful development of interference microscopy to allow direct visualization of the transient intracrystalline concentration profiles was largely due to the efforts of Ulf Schemmert and Sergey Vasenkov (now professors at the University of Applied Sciences in Leipzig and the University of Florida, respectively) and the parallel development of infrared microscopy to allow the visualization of the profiles of individual species in a multicomponent system was largely due to Dr Christian Chmelik (University of Leipzig). The development of molecular simulation techniques for studying sorption and diffusion in zeolites owes much to Larry June (now with Shell Oil), Randy Snurr, and Ed Maginn (now professors at Northwestern University and the University of Notre Dame, respectively), Professor Alexis Bell (University of California, Berkeley), and Dr George Papadopoulos (NTU Athens). We should also mention the work of Hervé Jobic (CNRS, Villeurbanne), who has developed neutron scattering as a viable experimental technique for studying intracrystalline diffusion over very short time and distances, comparable to those accessible by molecular dynamics simulations.

We are grateful to numerous funding agencies, especially the National Research Foundations of Germany, Canada, and the United States, the Alexander von Humboldt Foundation, DECHEMA and the Fonds der Chemischen Industrie, the European Community, and several companies, notably, ExxonMobil who have provided research support as well as valuable technical assistance over many years.

Finally, we would also like to thank Wiley-VCH and especially our editor Bernadette Gmeiner for her efficient collaboration in the preparation and editing of the manuscript and also our wives, Birge, Patricia, and Fani, for all their help and support throughout the course of this project.

Jörg Kärger
Douglas M. Ruthven
Doros N. Theodorou
5 December 2011

Part I
Introduction

1
Elementary Principles of Diffusion

The tendency of matter to migrate in such a way as to eliminate spatial variations in composition, thereby approaching a uniform equilibrium state, is well known. Such behavior, which is a universal property of matter at all temperatures above absolute zero, is called diffusion and is simply a manifestation of the tendency towards maximum entropy or maximum randomness. The rate at which diffusion occurs varies widely, from a time scale of seconds for gases to millennia for crystalline solids at ordinary temperatures. The practical significance therefore depends on the time scale of interest in any particular situation.

Diffusion in gases, liquids, and solids has been widely studied for more than a century [1–3]. In this book we are concerned with the specific problem of diffusion in porous solids. Such materials find widespread application as catalysts or adsorbents, which is a subject of considerable practical importance in the petroleum and chemical process industries and have recently attracted even more attention due to their potential as functional materials with a broad range of applications ranging from optical sensing to drug delivery [4]. To achieve the necessary surface area required for high capacity and activity, such materials generally have very fine pores. Transport through these pores occurs mainly by diffusion and often affects or even controls the overall rate of the process. A detailed understanding of the complexities of diffusional behavior in porous media is therefore essential for the development, design, and optimization of catalytic and adsorption processes and for technological exploitation of porous materials in general. Moreover, systematic diffusion studies in such systems lead to a better understanding of such fundamental questions as the interaction between molecules and solid surfaces [5] and the behavior of molecular systems of reduced dimensionality [6–8].

One class of microporous materials that is of special interest from both practical and theoretical points of view is the zeolites, where this term is used in its broad sense to include both microporous crystalline aluminosilicates and their structural analogs such as the titanosilicates and aluminophosphates. These materials form the basis of many practical adsorbents and catalysts. They combine the advantages of high specific area and uniform micropore size and, as a result, they offer unique properties such as size selective adsorption that can be exploited to achieve

Diffusion in Nanoporous Materials. Jörg Kärger, Douglas M. Ruthven, and Doros N. Theodorou.
© 2012 Wiley-VCH Verlag GmbH & Co. KGaA. Published 2012 by Wiley-VCH Verlag GmbH & Co. KGaA.

practically useful separations and to improve the efficiency of catalytic processes. The regularity of the pore structure, which is determined by the crystal structure rather than by the mode of preparation or pretreatment, offers the important advantage that it is possible, in such systems, to investigate the effect of pore size on the transport properties. In more conventional adsorbents, which have a very much wider distribution of pore size, such effects are more difficult to isolate. In the earlier chapters of this book diffusion in nanoporous solids is treated from a general perspective, but the later chapters focus on zeolitic adsorbents; because of their practical importance, these materials have been studied in much greater detail than amorphous materials.

Since the first edition of this book was published [9], an important new class of nanoporous materials based on metal–organic frameworks (MOFs) has been discovered and studied in considerable detail. Although their composition is quite different, MOFs are structurally similar to the zeolites and show many similarities in their diffusional behavior. Some of the recent studies of these materials are reviewed in Chapter 19.

1.1
Fundamental Definitions

1.1.1
Transfer of Matter by Diffusion

The quantitative study of diffusion dates from the early work of two pioneers, Thomas Graham and Adolf Fick (for a detailed historical review, see, for example, Reference [10]), during the period 1850–1855. Graham's initial experiments, which led to Graham's law of diffusion, involved measuring the rate of interdiffusion of two gases, at constant pressure, through a porous plug [11, 12]. He concluded that:

> The diffusion or spontaneous inter-mixture of two gases in contact is, in the case of each gas, inversely proportional to the density of the gas.

In later experiments with salt solutions he, in effect, verified the proportionality between the diffusive flux and the concentration gradient, although the results were not reported in precisely those terms. He also established the very large difference in the orders of magnitude of gas and liquid diffusion rates.

Fick's contribution was to recognize that Graham's observations could be understood if the diffusion of matter obeys a law of the same general form as Fourier's law of heat conduction, an analogy that remains useful to this day. On this basis he formulated what is now generally known as Fick's first law of diffusion, which is in fact no more than a definition of the "diffusivity" (D):

$$J = -D\frac{\partial c}{\partial z} \quad (1.1)$$

or, more generally:

$$J = -D \operatorname{grad} c$$

He showed that for diffusion in a parallel-sided duct with a constant diffusivity, this leads to the conservation equation:

$$\frac{\partial c}{\partial t} = D \frac{\partial^2 c}{\partial z^2} \tag{1.2}$$

or:

$$\frac{\partial c}{\partial t} = D \operatorname{div}(\operatorname{grad} c)$$

which is commonly known as Fick's second law of diffusion. He then proceeded to verify these conclusions by a series of ingenious experiments involving the measurement of concentration profiles, under quasi-steady state conditions, in conical and cylindrical vessels in which uniform concentrations were maintained at the ends [10, 13, 14].

These experiments were carried out with dilute solutions in which the diffusivity is substantially independent of composition. The definition of Eq. (1.1) makes no such assumption and is equally valid when the diffusivity varies with concentration. The additional assumption that the diffusivity does not depend on concentration is, however, introduced in the derivation of Eq. (1.2). The more general form of the conservation equation, allowing for concentration dependence of the diffusivity, is:

$$\frac{\partial c}{\partial t} = \frac{\partial}{\partial z}\left[D(c)\frac{\partial c}{\partial z}\right] \tag{1.3}$$

or:

$$\frac{\partial c}{\partial t} = \operatorname{div}\left[D(c) \operatorname{grad} c\right]$$

which reverts to Eq. (1.2) when D is constant.

In an isothermal binary system, Eq. (1.1) may also be written, equivalently, in terms of the gradient of mole fraction or (for gases) the partial pressure:

$$J = -D\frac{\partial c}{\partial z} = -cD\frac{\partial y_A}{\partial z} = -\frac{D}{RT}\frac{\partial P_A}{\partial z} \tag{1.4}$$

but these formulations are no longer equivalent in a non-isothermal system. Momentum transfer arguments lead to the conclusion that for diffusion in a gas mixture the gradient of partial pressure should be regarded as the fundamental "driving force," since that formulation remains valid even under non-isothermal conditions. A more detailed discussion of this point has been given by Haynes [15]. However, in this book problems of diffusion under non-isothermal conditions are not addressed in any substantial way and so the equivalence of Eq. (1.4) can generally be assumed.

The mathematical theory of diffusion, which has been elaborated in detail by Crank [16], depends on obtaining solutions to Eq. (1.1) [or Eqs. (1.2) and (1.3)] for the appropriate initial and boundary conditions. A number of such solutions are summarized in Chapter 6 for some of the situations commonly encountered in the measurement of diffusivities. In this chapter we present only the solution for one simple case that is useful for elaboration of the analogy between diffusion and a "random walk."

1.1.2
Random Walk

In the late 1820s, that is, about 20 years before the experiments of Graham and Fick, the Scottish botanist Robert Brown gave a detailed description of another phenomenon that turned out to be closely related to diffusion [10, 17]. On observing a suspension of pollen grains with the aid of the then new achromatic microscope he noticed that the individual particles undergo a sequence of rapid and apparently random movements. Today we know this behavior results from the continuously changing interaction between small particles and the molecules of the surrounding fluid. Although this microdynamic explanation was only suggested much later [18] this phenomenon is generally referred to as Brownian motion. The close relationship between Brownian motion and diffusion was first elaborated by Einstein [19] and, eventually, turned out to be nothing less than the ultimate proof of nature's atomic structure [10]. An experimentally accessible quantity that describes Brownian motion is the time dependence of the concentration distribution of the Brownian particles (diffusants) that were initially located within a given element of space. To apply Fick's equations [Eqs. (1.1) and (1.2)] to this process the particles initially within this space element must be considered to be distinguishable from the other particles, that is, they must be regarded as "labeled." The concentration distribution of these labeled particles will obey Eq. (1.2), which, in this situation, holds exactly since the total concentration of particles (and therefore their mobility or diffusivity) remains constant throughout the region under consideration.

It is easy to show by differentiation that for a constant diffusivity system:

$$c = \frac{A}{\sqrt{t}} e^{-z^2/4Dt} \tag{1.5}$$

(in which A is an arbitrary constant) is a general solution of Eq. (1.2). The total quantity of diffusing substance (M), assuming a parallel-sided container of unit cross-sectional area and infinite length in the z direction, is given by:

$$M = \int_{-\infty}^{+\infty} c \, dz \tag{1.6}$$

and, on writing $\xi^2 = z^2/4Dt$, we see that:

$$M = 2A\sqrt{D} \int_{-\infty}^{+\infty} e^{-\xi^2} d\xi = 2A\sqrt{\pi D} \tag{1.7}$$

Substitution in Eq. (1.5) shows that if this quantity of solute is initially confined to the plane at $z=0$, the distribution of solute at all later times will be given by:

$$\frac{c}{M} = \frac{e^{-z^2/4Dt}}{\sqrt{4\pi Dt}} \tag{1.8}$$

The corresponding solution for isotropic diffusion from a point source in three-dimensional space may be derived in a similar way:

$$\frac{c}{M} = \frac{e^{-r^2/4Dt}}{(4\pi Dt)^{3/2}} \tag{1.9}$$

where r represents the position vector from the origin. Equations (1.8) and (1.9) thus give the probability of finding, at position r, a particle (or molecule) that was located at the origin at time zero. This quantity is termed the "propagator" and, as it is a Gaussian function, it is completely defined by the mean square half-width or the "mean square displacement" of the diffusants, which may be found directly from Eqs. (1.8) or (1.9) by integration:

$$\langle z^2(t) \rangle = \int_{-\infty}^{+\infty} z^2 \frac{e^{-z^2/4Dt}}{\sqrt{4\pi Dt}} dz = 2Dt$$

$$\langle r^2(t) \rangle = \int r^2 \frac{e^{-r^2/4Dt}}{(4\pi Dt)^{3/2}} dz = 6Dt \tag{1.10}$$

These equations are generally known as Einstein's relations [19] and provide a direct correlation between the diffusivity, as defined by Fick's first equation, and the time dependence of the mean square displacement, which is the most easily observable quantitative feature of Brownian motion.

Chapter 2 explores the equivalence between a "random walk" and diffusion in greater detail. Starting from the assumption that the random walkers may step with equal probability in any direction, it is shown that the distribution and mean square displacement for a large number of random walkers, released from the origin at time zero, are given by Eqs. (1.8–1.10). From the perspective of the random walk one may therefore elect to consider Eq. (1.10) as defining the diffusivity and, provided the diffusivity is independent of concentration, this definition is exactly equivalent to the Fickian definition based on Eq. (1.1).

1.1.3
Transport Diffusion and Self-Diffusion

Two different diffusion phenomena may be distinguished: mass transfer (or transport diffusion) resulting from a concentration gradient (Figure 1.1a) and Brownian molecular motion (self-diffusion), which may be followed either by tagging a certain fraction of the diffusants (Figure 1.1b) or by following the trajectories of a large number of individual diffusants and determining their mean square displacement

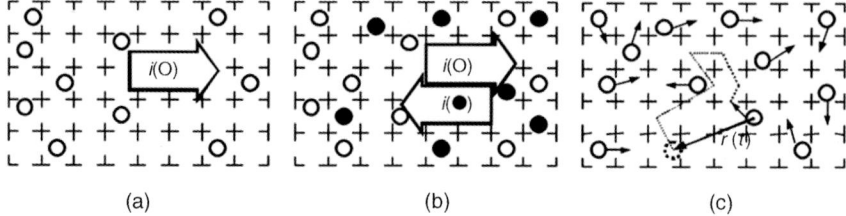

Figure 1.1 Microscopic situation during the measurement of transport diffusion (a) and self-diffusion by following the flux of labeled molecules (○) counterbalanced by that of the unlabeled molecules (●, b) or by recording the displacement of the individual molecules (c).

(Figure 1.1c). Because of the difference in the microphysical situations represented by Figures 1.1a–c the diffusivities in these two situations are not necessarily the same.

Following general convention we call the diffusivity corresponding to the situation represented by Figure 1.1a (in which there is a concentration gradient rather than merely a gradient in the fraction of marked molecules) the transport diffusivity (D), since this coefficient is related directly to the macroscopic transport of matter. Completely synonymously, the terms collective or chemical diffusion are sometimes also used [3].

The quantity describing the rate of Brownian migration under conditions of macroscopic equilibrium (Figure 1.1b and c) is referred to as the tracer or self-diffusivity ($\mathcal{D}$). A formal definition of the self-diffusivity may be given in two ways based on either Eqs. (1.1) or (1.10):

$$J^* = -\mathcal{D} \frac{\partial c^*}{\partial z}\bigg|_{c=\text{const}} \tag{1.11}$$

or:

$$\langle r^2(t) \rangle = 6\mathcal{D}t \tag{1.12}$$

but, as noted above, these definitions are in fact equivalent. Note that the self-diffusivity may vary with the total concentration, but it does not vary with the fraction of marked molecules.

Although both diffusion and self-diffusion occur by essentially the same micro-dynamic mechanism, namely, the irregular (thermal) motion of the molecules, the coefficients of transport diffusion and self-diffusion are generally not the same. Their relationship is discussed briefly in Section 1.2.3 and in greater detail in Section 3.3.3 as well as in Chapters 2 and 4 on the basis of various model assumptions for molecular propagation.

Mass transfer phenomena following Eqs. (1.11) and (1.12) are referred to as normal diffusion. It is shown in Section 2.1.2 that this describes the common situation in porous materials. Anomalous diffusion [7, 8, 20] leads to a deviation from the linear interdependence between the mean square displacement and the observation time as predicted by Eq. (1.12), which may formally be taken account of by considering the self-diffusivity as a parameter that depends on either the observation time in Eq. (1.12) or on the system size in Eq. (1.11) [21]. Such deviations, however, necessitate a highly correlated motion with a long "memory" of the diffusants, which occurs under only very exceptional conditions such as in single-file systems (Chapter 5). Anomalous diffusion is generally therefore of no technological relevance for mass transfer in nanoporous materials.

1.1.4
Frames of Reference

The situation shown in Figure 1.1a is only physically reasonable in a microporous solid where the framework of the solid permits the existence of an overall gradient of concentration under isothermal and isobaric conditions. Furthermore, in such systems the solid framework provides a convenient and unambiguous frame of reference with respect to which the diffusive flux may be measured. In the more general case of diffusion in a fluid phase the frame of reference must be specified to complete the definition of the diffusivity according to Eq. (1.1). For the interdiffusion of two components A and B we may write:

$$J_A = -D_A \frac{\partial c_A}{\partial z}, \quad J_B = -D_B \frac{\partial c_B}{\partial z} \tag{1.13}$$

If the partial molar volumes of A and B are different ($V_A \neq V_B$), the interdiffusion of the two species will lead to a net (convective) flow relative to a fixed coordinate system. The total volumetric flux is given by:

$$J_V = V_A D_A \frac{\partial c_A}{\partial z} + V_B D_B \frac{\partial c_B}{\partial z} \tag{1.14}$$

and the plane across which there is no net transfer of volume is given by $J_V = 0$. If there is no volume change on mixing:

$$V_A c_A + V_B c_B = \text{constant} \tag{1.15}$$

$$V_A \frac{\partial c_A}{\partial z} + V_B \frac{\partial c_B}{\partial z} = 0 \tag{1.16}$$

For both Eqs. (1.14) and (1.16) to be satisfied with $J_V = 0$ and V_A and V_B finite, it follows that $D_A = D_B$. The interdiffusion process is therefore described by a single diffusivity provided that the fluxes, and therefore the diffusivity, are defined relative to the plane of no net volumetric flow. The same result can be shown to hold even when there is a volume change on mixing, provided that the fluxes are defined relative to the plane of no net mass flow. In general the interdiffusion of two components can always be described by a single diffusivity but the frame of reference required to achieve this simplification depends on the nature of the system.

To understand the definition of the diffusivity for an adsorbed phase we must first consider the more general case of diffusion in a convective flow. The diffusive flux (relative to the plane of no net molal flow) is conventionally denoted by J and the total flux, relative to a fixed frame of reference, by N. For a binary system (A, B) we have:

$$N_A = J_A + y_A N = J_A + y_A (N_A + N_B) \tag{1.17}$$

If component B is stationary ($N_B = 0$) then:

$$J_A = N_A (1 - y_A) \tag{1.18}$$

which thus defines the relationship between the fluxes N_A and J_A. Diffusion of a mobile species within a porous solid may be regarded as a special case of binary diffusion in which one component (the solid) is immobile. The flux, and therefore the diffusivity, is normally defined with respect to the fixed coordinates of the solid rather than with respect to the plane of no net molal flux. There is no convective flow, so:

$$N_A = J'_A = \frac{J_A}{1 - y_A} = -D'_A \frac{\partial c_A}{\partial z}; \quad D'_A = \frac{D_A}{(1 - y_A)} \tag{1.19}$$

but J'_A and D'_A are now defined in the fixed frame of reference. In discussing diffusion in an adsorbed phase the distinction between J' and J and between D' and D is generally not explicit. The symbols J and D are commonly applied to fluxes in both fluid and adsorbed phases but it is important to understand that their meanings are not identical. This is especially important when applying results derived for diffusion in a homogeneous fluid to diffusion in a porous adsorbent.

1.1.5
Diffusion in Anisotropic Media

Extension of the unidimensional diffusion equations to diffusion in two or three dimensions [e.g., Eqs. (1.2), (1.3) or (1.8) and (1.9)] follows in a straightforward manner for an isotropic system in which the diffusivity in any direction is the same. In most macroporous adsorbents, randomness of the pore structure ensures that the diffusional properties should be at least approximately isotropic. For intracrystalline diffusion the situation is more complicated. When the crystal structure is cubic, intracrystalline diffusion should be isotropic since the micropore structure must then be identical in all three principal directions. However, when the crystal symmetry is anything other than cubic, the pore geometry will generally be different in the different principal directions, so anisotropic diffusion is to be expected. Perhaps the most important practical example is diffusion in ZSM-5/silicalite, which is discussed in Chapter 18.

In an isotropic medium the direction of the diffusive flux at any point is always perpendicular to the surface of constant concentration through that point, but this is not true in a nonisotropic system. This means that, for a nonisotropic system, Eq. (1.1) must be replaced by:

$$-J_x = D_{xx}\frac{\partial c}{\partial x} + D_{xy}\frac{\partial c}{\partial y} + D_{xz}\frac{\partial c}{\partial z}$$

$$-J_y = D_{yx}\frac{\partial c}{\partial x} + D_{yy}\frac{\partial c}{\partial y} + D_{yz}\frac{\partial c}{\partial z} \quad (1.20)$$

$$-J_z = D_{zx}\frac{\partial c}{\partial x} + D_{zy}\frac{\partial c}{\partial y} + D_{zz}\frac{\partial c}{\partial z}$$

In this notation the coefficients D_{ij} (with $i, j = x, y, z$) represent the contribution to the flux in the i direction from a concentration gradient in the j direction. The set:

$$\begin{bmatrix} D_{xx} & D_{xy} & D_{xz} \\ D_{yx} & D_{yy} & D_{yz} \\ D_{zx} & D_{zy} & D_{zz} \end{bmatrix}$$

is commonly called the diffusion tensor.

The equivalent of Eq. (1.2) for a (constant diffusivity) non-isotropic system is:

$$\frac{\partial c}{\partial t} = D_{xx}\frac{\partial^2 c}{\partial x^2} + D_{yy}\frac{\partial^2 c}{\partial y^2} + D_{zz}\frac{\partial^2 c}{\partial z^2} + (D_{yz} + D_{zy})\frac{\partial^2 c}{\partial y\,\partial z}$$
$$+ (D_{zx} + D_{xz})\frac{\partial^2 c}{\partial z\,\partial x} + (D_{xy} + D_{yx})\frac{\partial^2 c}{\partial x\,\partial y} = 0 \quad (1.21)$$

but it may be shown that a transformation to the rectangular coordinates ξ, η, ζ can always be found, which reduces this to the form:

$$\frac{\partial c}{\partial t} = D_1\frac{\partial^2 c}{\partial \xi^2} + D_2\frac{\partial^2 c}{\partial \eta^2} + D_3\frac{\partial^2 c}{\partial \zeta^2} \quad (1.22)$$

If we make the further substitutions:

$$\xi_1 = \xi\sqrt{D/D_1}, \quad \eta_1 = \eta\sqrt{D/D_2}, \quad \zeta_1 = \zeta\sqrt{D/D_3}$$

in which D may be arbitrarily chosen, Eq. (1.22) reduces to:

$$\frac{\partial c}{\partial t} = D\left(\frac{\partial^2 c}{\partial \xi_1^2} + \frac{\partial^2 c}{\partial \eta_1^2} + \frac{\partial^2 c}{\partial \zeta_1^2}\right) \quad (1.23)$$

which is formally identical with the diffusion equation for an isotropic system. In this way many of the problems of diffusion in nonisotropic systems can be reduced to the corresponding isotropic diffusion problems. Whether this is possible in any given situation depends on the boundary conditions, but where these are simple (e.g., step change in concentration at $t = 0$) this reduction is usually possible. The practical consequence of this is that in such cases one may expect the diffusional behavior to be similar to an isotropic system so that measurable features such as the transient uptake curves will be of the same form. However, the apparent diffusivity derived by matching such curves to the isotropic solution will be a complex average of D_1, D_2, and D_3, the diffusivities in the three principal directions ξ, η, and ζ. It is in general not possible to extract the individual values of D_1, D_2, and D_3 with satisfactory accuracy, although given the values of the principal coefficients (e.g., from an a priori prediction) it would be possible to proceed in the reverse direction and calculate the value of the apparent diffusivity.

1.2
Driving Force for Diffusion

1.2.1
Gradient of Chemical Potential

Fick's first law [Eq. (1.1)] and the equivalent definition of the diffusivity according to Eq. (1.10) both carry the implication that the driving force for diffusion is the gradient of concentration. However, since diffusion is simply the macroscopic manifestation of the tendency to approach equilibrium, it is clear that the true driving force must be the gradient of chemical potential (μ). This seems to have been explicitly recognized first by Einstein [22]. If the diffusive flux is considered as a flow driven by the gradient of chemical potential and opposed by frictional forces, the steady-state energy balance for a differential element is simply:

$$f u_A = -\frac{d\mu_A}{dz} \tag{1.24}$$

where u_A is the flow velocity of component A and f is a friction coefficient. The flux (J_A) is given by $u_A c_A$. To relate the chemical potential to the concentration we may consider the equilibrium vapor phase in which, neglecting deviations from the ideal gas law, the activity may be identified with the partial pressure:

$$\mu_A = \mu_A^0 + RT \ln p_A \tag{1.25}$$

The expression for the flux may then be written:

$$J_A = u_A c_A = -\frac{RT}{f} \frac{d \ln p_A}{d \ln c_A} \frac{dc_A}{dz} \tag{1.26}$$

Comparison with Eq. (1.1) shows that the transport diffusivity is given by:

$$D_A = \frac{RT}{f} \frac{d \ln p_A}{d \ln c_A} = D_0 \frac{d \ln p_A}{d \ln c_A} \tag{1.27}$$

where $d \ln p_A / d \ln c_A$ represents simply the gradient of the equilibrium isotherm in logarithmic coordinates. This term [the "thermodynamic (correction) factor"] may vary substantially with concentration and, in general, approaches a constant value of 1 only at low concentrations within the Henry's law region.

The principle of the chemical potential driving force is also implicit in the Stefan–Maxwell formulation [23, 24] (presented in Section 3.3) which, for a binary system, may be written in the form:

$$-\frac{\partial}{\partial z}\left(\frac{\mu_A}{RT}\right) = \frac{y_B}{\mathcal{D}_{AB}}(u_A - u_B) \quad (1.28)$$

where y_B denotes the mole fraction of component B, $\mathcal{D}_{AB}$ is the Stefan–Maxwell diffusivity, and u_A, u_B are the diffusive velocities. For an isothermal system with no net flux, Eq. (1.28) reduces to:

$$J_A = -\mathcal{D}_{AB}\frac{d\ln p_A}{d\ln c_A}\cdot\frac{dc_A}{dz} \quad (1.29)$$

which is equivalent to Eq. (1.26).

An alternative and equivalent form may be obtained by introducing the activity coefficient γ_A (defined by $f_A \approx p_A = \gamma_A c_A$ where f_A is the fugacity):

$$J_A = -\mathcal{D}_{AB}\left(\frac{\partial \ln \gamma_A}{\partial \ln c_A} + 1\right)\frac{dc_A}{dz} \quad (1.30)$$

This form of expression was applied by Darken [25] in his study of interdiffusion in binary metal alloys. The use of "thermodynamically corrected" diffusion coefficients is therefore sometimes attributed to Darken. However, it is apparent from the preceding discussion that the idea actually predates Darken's work by many years and is probably more correctly attributed to Maxwell and Stefan or Einstein.

The same formulation can obviously be used to represent diffusion of a single component (A) in a porous adsorbent (B). In this situation $u_B = 0$ and $\mathcal{D}_{AB}$ is the diffusivity for component A relative to the fixed coordinates of the pore system. Furthermore, in a microporous adsorbent there is no clear distinction between molecules on the surface and molecules in the "gas" phase in the central region of the pore. It is therefore convenient to consider only the total "intracrystalline" concentration (q). Assuming an ideal vapor phase, the transport equation is then written in the form:

$$J = -D\frac{dq}{dz}, \quad D = D_0\frac{d\ln p}{d\ln q} \quad (1.31)$$

D_0, defined in this way, is generally referred to as the "corrected diffusivity" and $d\ln p/d\ln q$ ($\equiv \Gamma$) as the "thermodynamic factor." Comparison with Eq. (1.29) shows that, under the specified conditions, D_0 is identical to the Stefan–Maxwell diffusivity $\mathcal{D}_{AB}$.

If the system is thermodynamically ideal ($p \propto q$) $d\ln p/d\ln q \to 1.0$ and the Fickian and corrected diffusivities become identical. However, in the more general case of a thermodynamically nonideal system, the Fickian transport diffusivity is seen to be the product of a mobility coefficient (D_0) and the thermodynamic correction factor $d \ln p/d \ln q$, which arises from nonlinearity of the relationship between activity and concentration. Thermodynamic ideality is generally approached only in dilute systems (gases, dilute liquid or solid solutions) and, under these conditions, one may also expect negligible interaction between the diffusing molecules, leading to a diffusivity that is independent of concentration. Since diffusion is commonly first encountered under these near ideal conditions, the idea that the diffusivity should be constant and that departures from such behavior are in some sense abnormal, has become widely accepted. In fact, except in dilute systems, the Fickian diffusivity is generally found to be concentration dependent. Equation (1.31) shows that this concentration dependence may arise from the concentration dependence of either D_0 or $d \ln p/d \ln q$.

In liquid-phase systems both these effects are often of comparable magnitude [26] and one may therefore argue that there is little practical advantage to be gained from using the corrected diffusivity (D_0) rather than the Fickian transport diffusivity (D). The situation is different in adsorption systems. In the saturation region the equilibrium isotherm becomes almost horizontal so that $d \ln p/d \ln q \to \infty$ whereas in the low-concentration (Henry's law) region $d \ln p/d \ln q \to 1.0$. The concentration dependence of this factor and, as a result, the concentration dependence of the Fickian diffusivity is therefore generally much more pronounced than the concentration dependence of the corrected diffusivity. Indeed, for many systems the corrected diffusivity has been found experimentally to be almost independent of concentration. Correlation of transport data for adsorption systems in terms of the corrected diffusivity is therefore to be preferred for practical reasons since it generally provides a simpler description.

In addition to these practical considerations there is a strong theoretical argument in favor of using corrected diffusivities. According to Eq. (1.31), and as will become clearer from the statistical mechanical considerations presented in Section 8.1.3, the transport diffusivity is seen to be a hybrid quantity, being the product of a mobility coefficient and a factor related to the driving force. In attempting to understand transport behavior at the molecular level it is clearly desirable to separate these two effects. Two systems with the same transport diffusivity may, as a result of large differences in the correction factor, have very different molecular mobilities. In any fundamental analysis the "corrected" diffusivity is therefore clearly the more useful quantity.

Beyond the Henry's law region the simple Langmuir model is often used to represent the behavior of adsorption systems in an approximate way. For a single adsorbed component:

$$\theta = \frac{q}{q_s} = \frac{bp}{1+bp}; \quad \frac{d \ln p}{d \ln q} = \frac{1}{1-q/q_s} = \frac{1}{1-\theta} \tag{1.32}$$

where θ is referred to as the fractional loading and b is the adsorption equilibrium constant (per site). This expression has the correct asymptotic behavior ($p \to 0$, $q \to Kp$ where $K = bq_s$ and $p \to \infty$, $q \to q_s$) and, although it provides an accurate representation of the isotherms for only a few systems, it provides a useful approximate representation for many systems. The extension to a binary system is:

$$\theta_A = \frac{q_A}{q_{As}} = \frac{b_A p_A}{1 + b_A p_A + b_B p_B} \tag{1.33}$$

The partial derivatives required for the analysis of diffusion in a binary system (Section 3.3.2) follow directly:

$$\frac{\partial \ln p_A}{\partial \ln q_A} = \frac{1-\theta_B}{1-\theta_A-\theta_B}; \quad \frac{\partial \ln p_A}{\partial \ln q_B} = \frac{\theta_A}{1-\theta_A-\theta_B} \tag{1.34}$$

1.2.2
Experimental Evidence

Direct experimental proof that the driving force for diffusive transport is the gradient of chemical potential, rather than the gradient of concentration, is provided by the experiments of Haase and Siry [27, 28] who studied diffusion in binary liquid mixtures near the consolute point. At the consolute point the chemical potential, and therefore the partial pressures, are independent of composition so that, according to Eq. (1.29), the transport diffusivity should be zero. The consolute point for the system n-hexane–nitrobenzene occurs at 20 °C at a mole fraction 0.422 of nitrobenzene. The system shows complete miscibility above this temperature but splits into two separate phases at lower temperatures. The opposite behavior is shown by the system water–triethylamine, for which the consolute temperature occurs at 18 °C at a mole fraction of triethylamine of 0.087. The mixture is completely miscible at lower temperatures but separates into two phases at higher temperature. Figure 1.2 shows the results of diffusion measurements. In both systems the Fickian diffusivity approaches zero as the consolute temperature is approached, as required by Eq. (1.29). The behavior of the water–triethylamine system is especially noteworthy since the diffusivity actually decreases with increasing temperature as the upper consolute point (18 °C) is approached. Such behavior, which follows naturally from the assumption that chemical potential is the driving force, cannot be easily accounted for in terms of a strictly Fickian model.

Despite the compelling evidence provided by Haase and Siry's experiments, the contrary view has been expressed that diffusive transport is a stochastic process for which the true driving force must be the gradient of concentration [29]. This argument is based on the random walk model with the implicit assumption that molecular propagation is a purely random process that occurs with equal a-priori probability in any direction. In fact when the relationship between activity and concentration is nonlinear, the propagation probabilities in the presence of a chemical potential gradient are not the same in all directions. To reconcile the random walk argument with the implications of Eq. (1.31) requires only the additional assumption that the a-priori jump probability varies in proportion to the local gradient of chemical potential.

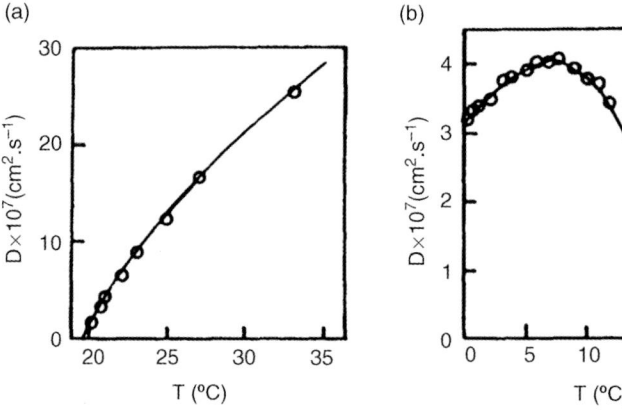

Figure 1.2 Variation of Fickian diffusivity with temperature for liquid mixtures of the critical composition, close to the consolute point: (a) n-hexane–nitrobenzene, mole fraction of nitrobenzene = 0.422, consolute temperature = 20 °C; (b) water–triethylamine, mole fraction triethylamine = 0.087, consolute temperature = 18 °C. Reprinted from Turner [28], with permission.

1.2.3
Relationship between Transport and Self-diffusivities

A first approximation to the relationship between the self- and transport diffusivities may be obtained by considering Eq. (1.26). In a mixture of two identical species, distinguishable only by their labeling (Figure 1.1b), the relation between p_A and c_A is clearly linear, and so the self-diffusivity is given simply by $\mathcal{D} = RT/f$. The expression for transport diffusivity [Eq. (1.27)] may therefore be written in the form:

$$D = \mathcal{D} \frac{d \ln p}{d \ln q} \tag{1.35}$$

implying the self-diffusivity can be equated with the corrected transport diffusivity ($\mathcal{D} = D_0$). In conformity with this equation it has been shown experimentally that in a dilute binary liquid solution the mutual or transport diffusivity approaches the self-diffusivity of the solute [30, 31]. It has therefore been generally assumed that in an adsorption system the transport and self-diffusivities should coincide in the low concentration limit where the nonlinearity correction vanishes and encounters between diffusing molecules occur only infrequently. Satisfactory agreement between transport and self-diffusivities has indeed been demonstrated experimentally for several adsorption systems. However, the argument leading to Eq. (1.35) contains the hidden assumption that the "friction coefficient" is the same for both self-diffusion (where there is no concentration gradient) and for transport diffusion (where there is a concentration gradient). Such an assumption is only valid if the adsorbent can be regarded as an inert framework that is not affected in any significant way by the presence of the sorbate.

A series of informative examples of systems following Eq. (1.35) are given in Section 19.3.1. They include cases where the thermodynamic factor yields values both larger and smaller than 1, thus giving rise to self-diffusivities both smaller (the usual case for nanoporous host–guest systems) and larger than the transport diffusivities. Note in particular Figure 19.12, which illustrates the correlation of the thermodynamic factor with the shape of the adsorption isotherm.

1.3
Diffusional Resistances in Nanoporous Media

Nowadays materials with pore diameters in the range 1–100 nm (10–1000 Å) are commonly referred to as "nanoporous" but, according to the IUPAC classification [32], pores are classified in three different categories based on their diameter:

$$\text{micropores } d < 20 \text{ Å}; \quad \text{mesopores } 20 \text{ Å} < d < 500 \text{ Å}; \quad \text{macropores } 500 \text{ Å} < d$$

This division, although somewhat arbitrary, is based on the difference in the types of forces that control adsorption behavior in the different size ranges. In the micropore range, surface forces are dominant and an adsorbed molecule never escapes from the force field of the surface even when at the center of the pore. In mesopores, capillary forces become important, while the macropores actually contribute very little to the adsorption capacity, although of course they play an important role in the transport properties. This classification is appropriate where small gaseous sorbates are considered, but for larger molecules the micropore regime may be shifted to substantially large pore sizes.

1.3.1
Internal Diffusional Resistances

Different mechanisms of diffusion control the transport in different regions of porosity. Diffusion in micropores is dominated by interactions between the diffusing molecule and the pore wall. Steric effects are important and diffusion is an activated process, proceeding by a sequence of jumps between regions of relatively low potential energy (sites). Since the diffusing molecules never escape from the force field of the pore walls it is logical to consider the fluid within the pore as a single "adsorbed" phase. Diffusion within this regime is known variously as "configurational" diffusion, "intra-crystalline" diffusion, or simply "micropore" diffusion but these terms are essentially synonymous.

Within the macropore range the role of the surface is relatively minor. Diffusion occurs mainly by the bulk or molecular diffusion mechanism, since collisions between diffusing molecules occur more frequently than collisions between a diffusing molecule and the pore wall, although of course this depends on the pressure. Within the mesopore range Knudsen diffusion is generally more important but there may also be significant contributions from surface diffusion and capillarity effects. Chapter 4 gives a more detailed discussion.

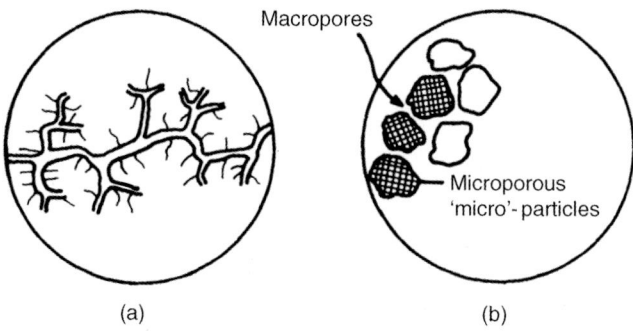

Figure 1.3 Two common types of microporous adsorbent: (a) homogeneous particle with a wide range of pore size distribution and (b) composite pellet formed from microporous microparticles giving rise to a well-defined bimodal distribution of pore size.

Uptake rate measurements with sufficiently large zeolite crystals can generally be interpreted according to a simple single (micropore) diffusion resistance model but with small commercial crystals the situation is not so straightforward. The assemblage of crystals in the measuring device can act like a macroporous adsorbent since the diffusion rate may be significantly affected, indeed controlled, by transport within the intercrystalline space. To interpret kinetic data in these circumstances it may be necessary to use a more complicated model including both "micropore" and "macropore" diffusional resistances.

The situation is even more complicated in commercial pelleted adsorbents. Two common types are shown schematically in Figure 1.3. In materials such as silica or alumina (Figure 1.3a) there is generally a wide distribution of pore size with no clear distinction between micropores and meso/macropores. In such adsorbents it is experimentally possible to measure only an average diffusivity and the relative contribution from pores of different size is difficult to assess. The situation is somewhat simpler in many zeolite and carbon molecular sieve adsorbents since these materials generally consist of small microporous particles (of zeolite or carbon sieve) aggregated together, often with the aid of a binder, to form a macroporous pellet of convenient size (Figure 1.3b). In such adsorbents there is a well-defined bimodal distribution of pore size so that the distinction between the micropores and the meso/macropores is clear.

Depending on the particular system and the conditions, either macropore or micropore diffusion resistances may control the transport behavior or both resistances may be significant. In the former case a simple single-resistance diffusion model is generally adequate to interpret the kinetic behavior but in the latter case a more complex dual resistance model that takes account of both micropore and macropore diffusion may be needed. Some of these more complex situations are discussed in Chapter 6. In any particular case the nature of the controlling regime may generally be established by varying experimental conditions such as the particle size.

1.3.2
Surface Resistance

Mass transfer through the surface of a zeolite crystal (or other nanoporous adsorbent particle) can be impeded by various mechanisms, including the collapse of the genuine pore structure close to the particle boundary and/or the deposition of strongly adsorbed species on the external surface of the particle. This may result in either total blockage of a fraction of the pore mouths or pore-mouth narrowing as well as the possibility that the surface may be covered by a layer of dramatically reduced permeability for the guest species under consideration. In all these cases the flux though the particle boundary can be represented by a surface rate coefficient (k_s) defined by:

$$J = k_s(q^* - q) \tag{1.36}$$

where ($q^* - q$) represents the difference between the equilibrium concentration of the adsorbed phase and the actual boundary concentration within the particle. If the surface resistance is brought about by a homogeneous layer of thickness δ with dramatically reduced diffusivity D_s, the surface rate coefficient is easily seen to be given by $k_s = R_s D_s/\delta$, with R_s denoting the ratio of the guest solubilities in the surface layer and in the genuine particle pore space.

1.3.3
External Resistance to Mass Transfer

In addition to any surface resistance and the internal diffusional resistances discussed above, whenever there is more than one component present in the fluid phase, there is a possibility of external resistance to mass transfer. This arises because, regardless of the hydrodynamic conditions, the surface of an adsorbent or catalyst particle will always be surrounded by a laminar boundary layer through which transport can occur only by molecular diffusion. Whether or not the diffusional resistance of this external fluid "film" is significant depends on the thickness of the boundary layer, which in turn depends on the hydrodynamic conditions. In general, for porous particles, this external resistance to mass transfer is smaller than the internal pore diffusional resistance but it may still be large enough to have a significant effect.

External resistance is generally correlated in terms of a mass transfer coefficient (k_f), defined in the usual manner according to a linearized rate expression of similar form to that used to represent surface resistance [Eq. (1.36)]:

$$J = k_f(c - c_s^*) \tag{1.37}$$

in which c is the sorbate concentration in the (well-mixed) bulk phase and c^* is the fluid phase concentration that would be at equilibrium with the adsorbed phase concentration at the particle surface. The capacity of the fluid film is small compared with that of the adsorbent particle and so there is very little accumulation of sorbate within the film. This implies a constant flux and a linear concentration gradient through the film. The time required to approach the steady state profile in the film will be small so that, even in a transient situation, in which the adsorbed phase concentration changes with time, the profile through the film will be of the same form, although the slope will decrease as equilibrium is approached and the rate of mass transfer declines. This is shown schematically in Figure 1.4.

The concentration gradient through the film is given by $(c - c^*)/\delta$ where δ is the film thickness and, comparing Eqs. (1.1) and (1.36), it is evident that $k_f = D/\delta$. However, since δ is generally unknown and can be expected to vary with the hydrodynamic conditions, this formulation offers no real advantage over a direct correlation in terms of the mass transfer coefficient. It is mainly for this reason that external fluid film resistance is generally correlated in terms of a mass transfer coefficient while internal resistances are correlated in terms of a diffusivity. For an isolated spherical adsorbent particle in a stagnant fluid it may be easily shown (by analogy with heat conduction) that:

$$k_f \approx 2D/d; \quad Sh = k_f d/D \approx 2 \tag{1.38}$$

Under flow conditions the Sherwood number (Sh) may be much greater than 2.0. The relevant dimensionless parameters that characterize the hydrodynamics are the

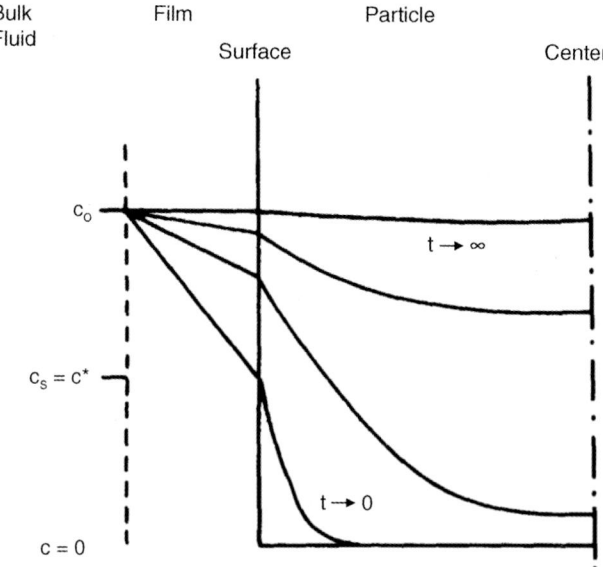

Figure 1.4 Schematic diagram showing the form of concentration profile for an initially sorbate-free particle exposed at time zero to a steady external fluid phase concentration or sorbate under conditions of combined external fluid film and internal diffusion control.

Schmidt number ($Sc \equiv \eta/\varrho D$) and the Reynolds number ($Re \equiv \varrho \varepsilon v d/\eta$). Correlations of the form:

$$Sh = f(Re, Sc) \tag{1.39}$$

have been presented for various well-defined fluid–solid contacting patterns. For example, for flow through a packed bed [33, 34]:

$$Sh = 2.0 + 1.1\,Re^{2/3}Sc^{-0.3} \tag{1.40}$$

This correlation has been shown to be valid for both gases and liquids over a wide range of flow conditions.

1.4
Experimental Methods

There are three distinct but related aspects to the study of diffusion: the investigation of the elementary process at the molecular level, the study of tracer or self-diffusion, and the measurement of transport diffusion. The first two involve measurements under equilibrium conditions while the third type of study necessarily requires measurements under non-equilibrium conditions. A wide variety of different experimental techniques have been applied to all three classes of measurement; a short summary is given in Table 1.1, in which the various techniques are classified according to the scale of the measurement. Some of these methods are discussed in detail in Part Four (Chapters 10–14).

The study of the elementary steps of diffusion requires measurement of the movement of individual molecules and this can be accomplished only by spectroscopic methods. Nuclear magnetic resonance (NMR) and neutron scattering have been successfully applied. NMR phenomena are governed by the interaction of the magnetic dipole (or for nuclei with spin $I > \frac{1}{2}$, the electric quadrupole) moments of the nuclei with their surroundings. Information can therefore be obtained concerning the spatial arrangement of individual molecules and the rate at which the positions and orientations of the molecules are changing. In scattering experiments with neutrons, molecular motion may be traced over distances of a few Ångströms up to the nanometer range. Such distances are, however, short compared with the length scales required for the study of the overall diffusion process.

Diffusion in the strict sense can be studied only over distances substantially greater than the dimensions of the diffusing molecules. Such measurements fall into two broad classes: self-diffusion measurements that are made by following the movement of labeled molecules under equilibrium conditions and transport diffusion measurements that are made by measuring the flux of molecules under a known gradient of concentration. The "microscopic" equilibrium techniques (incoherent QENS and PFG NMR) measure self-diffusion directly by determining the mean square molecular displacement in a known time interval. The "macroscopic" techniques generally measure transport diffusion at the length scale of the individual

Table 1.1 Classification of experimental methods for measuring intracrystalline diffusion in nanoporous solids. After References [35, 36].

	Non-equilibrium		Equilibrium
	Transient	Stationary	
Macroscopic	Sorption/desorption Frequency response (FR) Zero length column (ZLC) IR-FR Positron emission profiling (PEP) Temporal product analysis (TAP) IR spectroscopy	Membrane permeation Effectiveness factor in chemical reactions	Tracer sorption/desorption Tracer ZLC (zero length column)
Mesoscopic	IR microscopy (IRM)	Single-crystal-permeation	Tracer IR microscopy
Microscopic	Interference microscopy (IFM) IR micro-imaging (IRM)		Tracer IR micro-imaging Pulsed field gradient NMR (PFG NMR) Quasi-elastic neutron scattering (QENS)

crystal by following either adsorption/desorption kinetics, under transient conditions, or flow through a zeolite membrane, generally under steady-state conditions. Such techniques can be adapted to measure self-diffusion by using an isotopically labeled tracer. Single-crystal FTIR and single crystal permeance measurements can be regarded as intermediate (mesoscopic) techniques since measurements are made on an individual crystal. Recently, microscopic measurement of transport diffusion has also become possible. In coherent QENS, the relevant information is extracted from density fluctuations analogous to the situation in light scattering. In interference microscopy (IFM) this information is acquired by monitoring the evolution of intracrystalline concentration profiles.

In recent years it has become clear that the length scale at which intracrystalline diffusion measurements are made can be important since the effect of structural defects becomes increasingly important when the measurement scale spans many unit cells. As a result, the diffusivity values derived from macroscopic measurements may be much smaller than those from microscopic measurements which approximate more closely the behavior for an ideal zeolite crystal. Uniquely among the techniques considered, PFG NMR offers the possibility of varying the length scale from sub-micron to several microns (or even mm under favorable conditions), thus allowing a direct quantitative assessment of the impact of structural defects. NMR

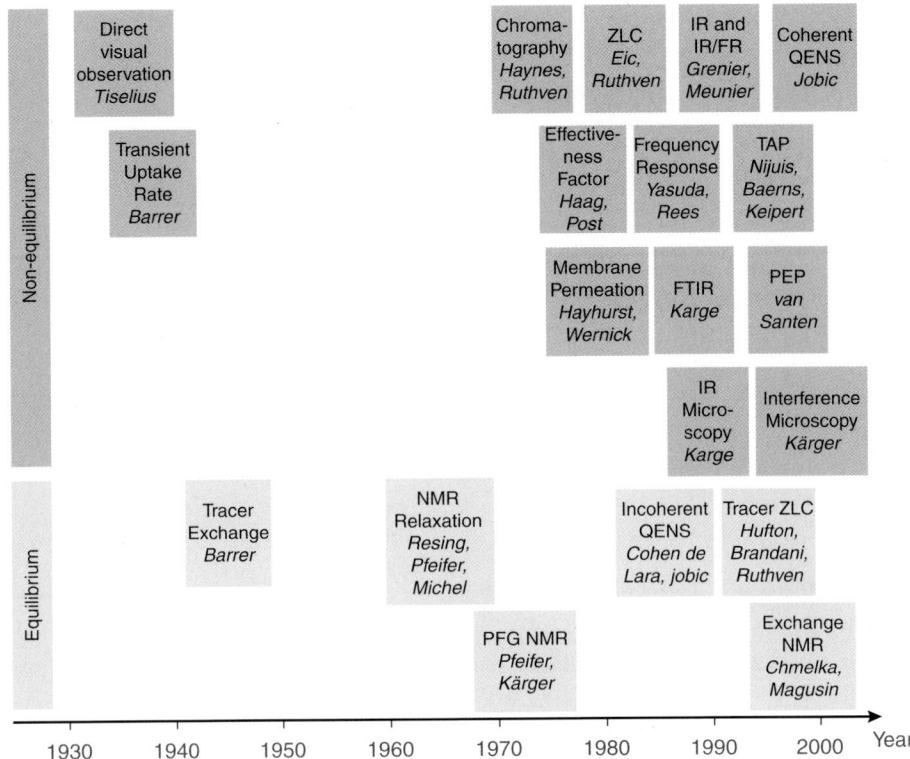

Figure 1.5 Measurement of zeolitic diffusion: historical development [37].

methods are also applicable to the measurement of long-range (intercrystalline or macropore) diffusion in an adsorbent particle. NMR labeling may also be applied to tracer diffusion measurements, thus providing essentially the same information that can be obtained from measurements with isotopically labeled molecules.

Figure 1.5 depicts the historical development of the study of intracrystalline diffusion and shows how the stimulus provided by the earliest PFG NMR measurements led to the introduction of a large spectrum of new experimental techniques.

With the impressive increase of computer power, over the last few decades molecular modeling and computer simulation have become powerful tools that complement the direct measurement of diffusion. The unique option to "play" with system parameters that, in reality, are invariable provides insights into the diffusion mechanisms that are often inaccessible from "real" experiments. With the widespread availability of very fast computers this approach, which is discussed in Part Three (Chapters 7–9), has become increasingly popular. In assessing results derived from molecular simulations it is, however, important not to lose sight of the limitations of this approach. Our knowledge of repulsive forces remains rudimentary and, for hindered diffusion, the impact of such forces is often dominant. As a result minor errors in the assumed force field can lead to very large errors in the predicted diffusivities, especially for diffusion in small pores. Direct experimental validation therefore remains critically important.

References

1 Jost, W. (1960) *Diffusion in Solids, Liquids and Gases*, Academic Press, New York.
2 Cussler, E.L. (1984) *Diffusion*, Cambridge University Press, Cambridge.
3 Heitjans, P. and Kärger, J. (2005) *Diffusion in Condensed Matter: Methods, Materials, Models*, Springer, Berlin, Heidelberg.
4 Laeri, F., Schüth, F., Simon, U., and Wark, M. (2003) *Host-Guest Systems Based on Nanoporous Crystals*, Wiley-VCH Verlag GmbH, Weinheim.
5 Ertl, G. (2008) *Angew. Chem. Int. Ed.*, **47**, 3524.
6 Drake, J.M. and Klafter, J. (1990) *Phys. Today*, **43**, 46.
7 Ben-Avraham, D. and Havlin, S. (2000) *Diffusion and Reaction in Fractals and Disordered Systems*, Cambridge University Press, Cambridge.
8 Klages, R., Radons, G., and Sokolov, I.M. (eds) (2008) *Anomalous Transport*, Wiley-VCH Verlag GmbH, Weinheim.
9 Kärger, J. and Ruthven, D.M. (1992) *Diffusion in Zeolites and Other Microporous Solids*, John Wiley & Sons, Inc., New York.
10 Philibert, J. (2010) in *Leipzig, Einstein, Diffusion* (ed J. Kärger), Leipziger Universitätsverlag, Leipzig, p. 41.
11 Graham, T. (1850) *Philos. Mag.*, **2**, 175, 222 and 357.
12 Graham, T. (1850) *Phil Trans. Roy. Soc. London*, **140**, 1.
13 Fick, A.E. (1855) *Ann. Phys.*, **94**, 59.
14 Fick, A.E. (1855) *Phil. Mag.*, **10**, 30.
15 Haynes, H.W. (1986) *Chem. Eng. Educ.*, **20**, 22.
16 Crank, J. (1975) *Mathematics of Diffusion*, Oxford University Press, London.
17 Brown, R. (1828) *Phil. Mag.*, **4**, 161; (1830) **8**, 41.
18 Gouy, G. (1880) *Comptes Rendus*, **90**, 307.
19 Einstein, A. (1905) *Ann. Phys.*, **17**, 349.
20 Klafter, J. and Sokolov, I.M. (August, 29 2005) *Phys. World*, **18**, 29.
21 Russ, S., Zschiegner, S., Bunde, A., and Kärger, J. (2005) *Phys. Rev. E*, **72**, 030101-1-4.
22 Einstein, A. (1906) *Ann. Phys.*, **19**, 371.
23 Maxwell, J.C. (1860) *Phil. Mag.*, **19**, 19, 20, 21; See also Niven, W.D. (ed.) (1952) *Scientific Papers of J. C. Maxwell*, Dover, New York, p. 629.
24 Stefan, J. (1872) *Wien. Ber.*, **65**, 323.
25 Darken, L.S. (1948) *Trans. AIME*, **175**, 184.
26 Ghai, R.K., Ertl, H., and Dullien, F.A.L. (1974) *AIChE J.*, **19**, 881; (1975) **20**, 1.
27 Haase, R. and Siry, M. (1968) *Z. Phys. Chem. Frankfurt*, **57**, 56.
28 Turner, J.C.R. (1975) *Chem. Eng. Sd.*, **30**, 1304.
29 Danckwerts, P.V. (1971) in *Diffusion Processes*, vol. 2 (eds J.N. Sherwood, A.V. Chadwick, W.M. Muir, and F.L. Swinton), Gordon and Breach, London, p. 45.
30 van Geet, A.L. and Adamson, A.W. (1964) *J. Phys. Chem.*, **68**, 238.
31 Adamson, A.W. (1957) *Angew. Chem.*, **96**, 675.
32 Schüth, F., Sing, K.S.W., and Weitkamp, J. (eds) (2002) *Handbook of Porous Solids*, Wiley-VCH Verlag GmbH, Weinheim.
33 Wakao, N. and Funazkri, T. (1978) *Chem. Eng. Sd.*, **33**, 1375.
34 Wakao, N. and Kaguei, S. (1982) *Heat and Mass Transfer in Packed Beds*, Gordon and Breach, London, ch. 4.
35 Kärger, J. and Ruthven, D.M. (2002) in *Handbook of Porous Solids* (eds F. Schüth, K.S.W. Sing, and J. Weitkamp), Wiley-VCH Verlag GmbH, Weinheim, p. 2087.
36 Chmelik, C. and Kärger, J. (2010) *Chem. Soc. Rev.*, **39**, 4864.
37 Kärger, J. (2002) *Ind. Eng. Chem. Res.*, **41**, 3335.

Part II
Theory

2
Diffusion as a Random Walk

Diffusion arises from the random thermal motion of the ultimate particles (atoms or molecules) of which the system under consideration is composed. Many of the important features of diffusion can be understood and interpreted, both qualitatively and quantitatively, by considering the process as a random walk in which the particles move by a sequence of steps of equal length, taken at constant frequency, and with equal probability in any direction. This simple model can be extended to more complex systems by relaxing these restrictions to allow for variations in step size and frequency and for the possibility of non-random correlations between the directions of successive steps. The concept of the simple random walk model was introduced in Section 1.1.2. In this chapter the implications of this model are explored in greater detail and the formal mathematical theory is developed to show the correspondence with the classical Fickian laws of diffusion. The modeling approach is extended to the case of diffusion in a composite material and used to examine the relationship between transport diffusion and self-diffusion. The extension of the principle of the random walk to disordered structures and the possibility of fractal behavior, including the occurrence of "anomalous" diffusion, are also considered.

2.1
Random Walk Model

2.1.1
Mean Square Displacement

The simplest situation to consider is a one-dimensional system in which steps of $\pm l$ are taken with equal probability in the z direction. The time between jumps (τ) and therefore the jump frequency ($1/\tau$) are assumed to be constant and we assume further that the directions of successive jumps are completely uncorrelated, so that the probability of a positive step is always $1/2$. Such a model would describe the motion of a man, constrained to walk along a straight line, who throws a coin at each step to decide whether to proceed forward or backwards.

Diffusion in Nanoporous Materials. Jörg Kärger, Douglas M. Ruthven, and Doros N. Theodorou.
© 2012 Wiley-VCH Verlag GmbH & Co. KGaA. Published 2012 by Wiley-VCH Verlag GmbH & Co. KGaA.

The mean square distance covered by our random walker after n steps may be written as:

$$\langle z^2(n) \rangle = \left\langle \left(\sum_{i=1}^{n} l_i \right)^2 \right\rangle \equiv \sum_{i=1}^{n} \langle l_i^2 \rangle + \sum_{i \neq j} \langle l_i l_j \rangle \qquad (2.1)$$

in which $l_i \, (= \pm l)$ denotes the shift in position brought about by the i-th step. Since the steps are uncorrelated one has $\langle l_i l_j \rangle = \langle l_i \rangle \langle l_j \rangle$. With $\langle l_i \rangle = 0$ and $(|l_i| = l)$ Eq. (2.1) thus becomes:

$$\langle z^2(n) \rangle = n l^2 \qquad (2.2)$$

The number of steps is related to the time by $n = t/\tau$ so the mean square displacement finally becomes:

$$\langle z^2(t) \rangle = 2 \mathcal{D} t \qquad (2.3)$$

with:

$$\mathcal{D} = l^2 / 2\tau \qquad (2.4)$$

It is shown below that $\mathcal{D}$, defined in this way, is in fact the self-diffusivity, as defined in Eq. (1.11), that is, the proportionality factor between the concentration gradient and the flux of labeled molecules (corresponding to a set of a multitude of simultaneously roaming random walkers). Thus, Eq. (2.3) turns out to be a simple form of the famous Einstein equation (also referred to as the Einstein–Smoluchowski equation) predicting proportionality between the mean square displacement of particles and the time over which these displacements are recorded [1].

These considerations may be easily extended to two- or three-dimensional space with variable step sizes. To do this the scalar quantities in Eqs. (2.1) and (2.2) are simply replaced by the corresponding vectors. Provided that the steps are uncorrelated and there is no preferential direction for particle propagation ($\langle \mathbf{l}_i \rangle = 0$) the cross term in Eq. (2.1) is still zero and the square of the step length is simply replaced by the average value:

$$\langle \mathbf{l}_i^2 \rangle = \sum_{i=1}^{n} \langle \mathbf{l}_i^2 \rangle / n = l^2 \qquad (2.5)$$

Equation 2.2 therefore remains valid with the difference that l^2 now represents the mean square step size. The relationship between the mean square displacement and the diffusivity, however, depends on the dimensionality of the system. In two- and three-dimensional systems only one-half and one-third (respectively) of the jumps occur in a given direction and so, to maintain consistency with the diffusion equation, we have corresponding to Eqs. (2.3) and (2.4):

$$\langle r^2(t) \rangle = 2k \mathcal{D} t \qquad (2.6)$$

$$\mathcal{D} = l^2 / 2k\tau \qquad (2.7)$$

where k (=1, 2, or 3) is the dimensionality and $\langle r^2 \rangle$ represents the mean square displacement. A more formal derivation of these relations is given below.

2.1.2
The Propagator

To determine the mean square displacement of a random walker one may follow the propagation of a large number of such walkers under identical conditions. In addition to the mean square displacement such a measurement would also provide the complete distribution curve of the positions of random walkers at any specified observation time. This may be expressed as a probability density function $P(\mathbf{r}, t)$ where $P(\mathbf{r}, t)d\mathbf{r}$ represents the probability of finding at time t a random walker in the volume element $d\mathbf{r}$, at a distance $\mathbf{r}$ from the starting point. This function is termed the "propagator" and contains the maximum information that can be obtained from the chaotic movement of an ensemble of random walkers. Over certain space and time scales, the propagator is accessible by experimental observation, with PFG NMR ([2], see Section 11.2) and QENS ([3], see Section 10.2).

We will now sketch how, under the conditions of "normal diffusion" as considered by the simple model of the previous section, the propagator may be expressed analytically. A more detailed description may be found in Reference [4] and the original reference by Chandrasekhar [5]. We start by introducing the probability $P(m, n)$ that, after n steps, the net shift of the random walker will be m steps in the positive direction. With n_+ and n_- ($= n - n_+$) denoting the numbers of steps in positive and negative directions, one has:

$$m = n_+ - n_- = 2n_+ - n \quad \text{or} \quad n_+ = \frac{m+n}{2} \tag{2.8}$$

The probability $P(n, n_+)$ that in a random sequence of n steps, n_+ steps will be in the positive direction and consequently n_- steps in the negative direction is given by:

$$P(n, n_+) = \left(\frac{1}{2}\right)^n \frac{n!}{(n-n_+)! n_+!} = \left(\frac{1}{2}\right)^n \frac{n!}{[(n-m)/2]![(n+m)/2]!} \tag{2.9}$$

This is the product of two factors, the first representing the probability of any particular sequence of positive and negative steps in a sequence of n steps, and the second representing the number of possible different sequences with n_+ positive steps and n_- negative steps. For large numbers of steps Eq. (2.9) may be transformed into an analytic expression from which, with Eq. (2.8), one eventually obtains:

$$P(m, n) = \sqrt{\frac{2}{\pi n}} e^{-m^2/2n} \tag{2.10}$$

The probability $P(z, t) \, dz$ of finding a random walker in the interval z to $z + dz$ (where $z = ml$) after time t (or $n = t/\tau$ steps) is therefore:

$$P(z,t)dz = P\left(\frac{z}{l}, \frac{t}{\tau}\right)\frac{dz}{2l} = \sqrt{\frac{2}{(\pi t/\tau)}} \exp\left(\frac{-z^2}{2l^2 t/\tau}\right)\frac{dz}{2l} \tag{2.11}$$

The factor $dz/2l$ gives the number of possible final points within the interval dz. These points are separated by a distance $2l$ since, for a given number of steps (n) corresponding to a fixed time, one more step in the positive direction automatically implies one less step in the negative direction. From Eqs. (2.4) and (2.11) comes:

$$P(z,t)dz = \frac{1}{\sqrt{4\pi \mathcal{D} t}} e^{-z^2/4\mathcal{D} t} dz \tag{2.12}$$

from which it may be seen that the propagator function is given by:

$$P(z,t) = \frac{1}{\sqrt{4\pi \mathcal{D} t}} e^{-z^2/4\mathcal{D} t} \tag{2.13}$$

Equation (2.13) emerges as a special case of the central limit theorem (CLT) of statistics [6, 7] which predicts that the superposition of elementary "steps" with identical probability distributions will always lead to a probability distribution given by a Gaussian. This relation is identical with the distribution of labeled molecules diffusing from a plane source into unlabeled but otherwise identical surroundings [Eq. (1.8)] calculated from Fick's second law. The mean square displacement may be obtained directly by integration:

$$\langle z^2(t) \rangle = \int_{z=-\infty}^{+\infty} z^2 P(z,t) dz = 2\mathcal{D} t \tag{2.14}$$

which is evidently identical to Eq. (2.3).

The generalization to a three-dimensional isotropic system is straightforward. Since movement in the three principal directions (x, y, z) is assumed to be uncorrelated, the three-dimensional propagator may be written as the product of the three one-dimensional functions:

$$P(\mathbf{r}, t) = P(x, t) P(y, t) P(z, t) \tag{2.15}$$

which, with Eq. (2.13), leads immediately to:

$$P(\mathbf{r}, t) = \frac{1}{(4\mathcal{D}\pi t)^{3/2}} e^{-r^2/4\mathcal{D} t} \tag{2.16}$$

This is identical to Eq. (1.9). Integration gives for the mean square displacement:

$$\langle r^2(t) \rangle = 6\mathcal{D} t \tag{2.17}$$

in accordance with Eq. (2.6).

The quantity $\mathcal{D}$ in Eqs. (2.16) and (2.17) has been introduced in Eq. (2.6), for a one-dimensional system, in which τ denotes the mean time between jumps in the

z direction. For diffusion in two or three equivalent dimensions it is clearly more reasonable to take τ as the mean time between successive jumps, regardless of direction, so that:

$$\tau_{1\text{dim.}} = 2\tau_{2\text{dim.}} = 3\tau_{3\text{dim.}} \tag{2.18}$$

leading to Eq. (2.7).

Irrespective of the simple model considerations under which we have shown the validity of Eqs. (2.16) and (2.17), these equations can be shown to hold much more generally, as a consequence of the central limit theorem of statistics [6, 7]. For this, it is sufficient to require that (i) the system under study may be subdivided into a large number of volume elements that, at least in a statistical sense, may be considered to be identical and (ii) the total time of the diffusion measurement (the "observation time") may be subdivided into time intervals during which the molecules "lose their memory," that is, time intervals large enough so that molecular propagation at the end of the interval is not affected by the situation given at the beginning of the interval. In diffusion studies with nanoporous materials the sizes of the porous solids under study generally exceed the pore diameters by orders of magnitude and, obeying requirement (ii), the measurements are performed over time spans much larger than the mean time of molecular exchange between adjacent pores. As a rule, therefore, molecular diffusion may be assumed to follow the laws of "normal" diffusion as reflected by the Gaussian propagator [Eq. (2.13)] and by numerous experimental studies, as exemplified by Figure 2.1. Some important cases giving rise to deviations are considered in Section 2.6.

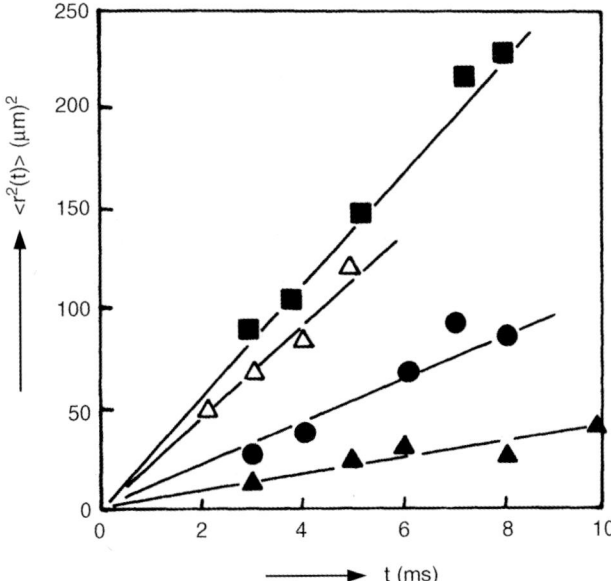

Figure 2.1 Time dependence of mean square displacement of Xe in zeolite NaX (■), NaCaA (•), H-ZSM-5 (▲), and of CH$_4$ in H-ZSM-5 (△) at 293 K, as measured by pulsed field gradient NMR. Reprinted from Heink et al. [8] with permission.

2.1.3
Correspondence with Fick's Equations

The propagator can be thought of equivalently either as defining the probability of finding, in a given spatial element at time t, a random walker (marked molecule) released from the origin at time zero or as the ratio of the number of random walkers (marked molecules) within the specified element to the number released (from the origin at time zero). A Gaussian distribution is completely defined by its mean square width, which is directly related to the elapsed time through Eq. (2.17), in which the parameter $\mathcal{D}$ corresponds to the self-diffusivity. The same relations (with $\mathcal{D}$ replaced by D) apply also to transport diffusion provided that the diffusing molecules do not interfere with one another. This requirement is obviously fulfilled at sufficiently low sorbate concentrations when molecular propagation is controlled by the interaction with the pore wall rather than with other molecules. This corresponds with the Henry's law region of proportionality between the sorbate concentration and the surrounding pressure.

The correspondence between Fickian diffusion and the random walk model that was demonstrated in relation to Eqs. (2.16) and (2.17) may be shown to apply also to the flux equations. The calculation procedure is illustrated by Figure 2.2 (see, for example, Reference [4]). In the case of tracer- (or self-) diffusion, Fick's law [Eq. (1.11)] and Einstein's relation [Eq. (1.12)] are thus found to be completely equivalent, provided that sufficiently large space and time scales [items (i) and (ii) above] are considered. The same reasoning may be transferred to transport diffusion, provided that the individual molecules may be assumed to diffuse independently of each other, that is, in the range of low concentrations where the rates of diffusion and self-diffusion are identical.

In the context of Figure 1.1, by the same reasoning, the self-diffusivities and transport diffusivities may be shown to coincide. Analogously, one may also anticipate that these quantities will differ when molecular interactions become relevant. In most nanoporous host–guest systems the host–host (i.e., intermolecular) interaction is greatly exceeded by the host–guest interaction (i.e., by the attraction of the molecules by the pore walls). Hence, under the conditions of a macroscopic concentration gradient (Figure 1.1a), a molecular random walk should not be expected to remain isotropic. In fact, owing to the larger free space on the right-hand side (range of low concentrations), each individual molecule is more likely to be shifted to the right. Hence, the flux arising from the concentration gradient (as observed in tracer diffusion experiments simply due to the gradient of labeled molecules) is enhanced, that is, the transport diffusivities exceed the self-diffusivities. If the molecular interaction exceeds the interaction with the pore wall, that is, if the molecules "prefer" to stick to each other (in Reference [9] referred to as the "clustering effect"), this reasoning has to be reversed. Now, under the influence of a concentration gradient, molecular equilibration over the sample will be reduced and the self-diffusivity (i.e., the redistribution rate under a uniform concentration) will be greater than the transport diffusivity. In Sections 3.2 and 3.3 the formalism of irreversible thermodynamics is used to quantify this reasoning,

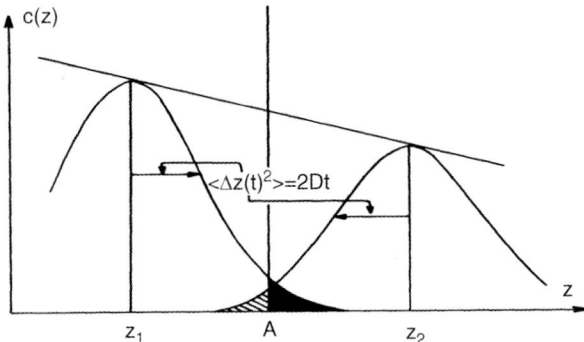

Figure 2.2 Determination of the flux through a cross-section A from the time dependence of the propagator function. The differently shaded areas represent the numbers of molecules that, during time t, have moved from z_1 (z_2) into the space beyond A. Integration over all starting points z, on the left and right of A, yields, by difference, the total number of molecules that have diffused through section A (from left to right) in this time interval and thus the net flux. The flux calculated in this way is identical to that calculated from Eq. (1.1) with the diffusivity calculated from the mean square width of the propagator.

which, in Section 19.3.1, is shown to be convincingly confirmed by recent experimental evidence.

2.2 Correlation Effects

2.2.1 Vacancy Correlations

Let us consider the diffusion path of a constituent (atom, ion) of a solid. When the particle jumps from a certain lattice point, a vacancy is left at the starting point so that the next jump is more likely to occur in the reverse direction than in any other direction. Deviating from the situation of Section 2.1, subsequent displacements can no longer be considered as uncorrelated. Correspondingly, the cross terms $\langle l_i, l_j \rangle$, in Eq. (2.1) are no longer zero and one may anticipate that these cross terms will lead to mean square displacements that are smaller than for an uncorrelated situation [for which Eq. (2.2) holds]. This phenomenon has been referred to as the "correlation effect" of diffusion in solids [10–15]. To distinguish this particular type of correlation in a particle's random walk, the term "vacancy correlation" has been coined [16].

Under such conditions, the self-diffusivity may be represented in the form:

$$\mathcal{D}(\theta) = \mathcal{D}(0)(1-\theta)f(\theta) \tag{2.19}$$

Here, $\mathcal{D}(0)$ is the diffusivity that would result in a series of uncorrelated jumps as represented by Eq. (2.7). The factor $(1-\theta)$ represents the mean field approximation

Table 2.1 Correlation factors for vacancy diffusion.

Lattice type	Correlation factor
Diamond	0.5
BCC	0.7272
FCC	0.7815
Hexagonal (two-dimensions)	0.3333
Triangular (two-dimensions)	0.5601

for the diffusivity reduction by site occupancy. It denotes the probability that an arbitrarily selected site is vacant and takes into account that jumps are only possible if they are directed to a vacancy (with θ denoting the site occupancy); $f(\Theta)$ (<1) is referred to as the correlation factor and expresses the additional reduction in the diffusivities due to the effect of vacancy correlations, as discussed above. In a typical solid the vacancies are isolated from each other and the correlation factor becomes independent of their density. The correlation factor is then a function only of the lattice architecture. Examples are summarized in Table 2.1.

For vacancy-mediated self-diffusion in a cubic lattice, the correlation factors as provided by Table 2.1 are found to be nicely approached by the following rule of thumb [15]:

$$f = 1 - \frac{2}{z} \tag{2.20}$$

with z denoting the lattice coordination number. This relation may be rationalized by taking into account that a vacated site will be re-occupied with a probability close to $1/z$ by the particle that has just left it. Since a backward jump cancels the effect of the preceding one, the "efficiency" of any jump is diminished by twice this probability in comparison with that of uncorrelated jumps.

In a "one-dimensional lattice," that is, in a chain of sites, each of which can be occupied by only one particle, the effect of vacancy correlation is to prohibit any type of normal diffusion. A molecular random walk under such conditions is referred to as single-file diffusion [17, 18] and is considered in greater detail in Chapter 5. This fundamental difference in the impact of vacancy correlation on diffusion in multidimensional lattices and in single-file systems is correlated with the particle "memory." Particles diffusing in a lattice will progressively "lose" their memory the further they have moved away from their starting point. Eventually, when the distance covered significantly exceeds the individual step length, step directions become independent of the starting point. This tendency may be rationalized both intuitively and by considering the probabilities of the respective series of jumps [14, 15] and ensures that one may find a space scale over which subsequent displacements are independent of each other, which is a prerequisite for the occurrence of ordinary diffusion.

In a single-file system, however, the shifted particles "retain their memory" over infinitely long intervals of time. A particle may only be displaced in one direction if all particles "in front" of it happen to be shifted in the same direction. This in turn would mean a free space reduction "in front" of and enhancement "behind" the particle. Consequently, particle displacement is more likely to occur in the backward direction rather than in the same direction as the previous jump. In an infinitely extended single-file system this "anti-correlation" is preserved regardless of the displacement and therefore impedes normal diffusion. This result is found to be nicely reproduced by Eq. (2.20). With $z = 2$, the correlation factor is found to be zero and hence, with Eq. (2.19), so is the self-diffusivity.

2.2.2
Correlated Anisotropy

Guest diffusion in nanoporous materials is subject to the given pore space topology. Correspondingly, diffusion in anisotropic materials must be generally described by the three principal elements D_i of a diffusion tensor [Eq. (1.20)] rather than by a single diffusivity. In the directions of the principal axes [Eq. (1.22)], both Fick's equations [Eqs. (1.1) and (1.2)] and the Einstein equation [Eq. (2.3)] may be applied in the one-dimensional notation, with the respective principal element of the diffusion tensor. On their way through the host system the diffusing molecules must obviously follow the spatial constraints imposed by the pore network of the adsorbent. This means in particular that molecular displacements in the direction of one principal axis may be automatically accompanied by displacements in the directions of the other axes, leading to a correlation between the principal values of the diffusion tensor.

To quantify these correlation rules one may exploit the fact that in some nanoporous host–guest systems, including zeolites of structure type CHA (Section 16.1.3 and Reference [19]) and MFI (Section 18.1 and Reference [20]), molecular migration may be adequately described by a sequence of jumps on a lattice, with nodes and connecting lines. They correspond, respectively, in MFI to the channel intersections and channel segments and in CHA to the cavities and connecting windows. Inherent in such models is the assumption that the particle "memory" is much shorter than the time of passage between adjacent nodes. In Sections 16.10 and 18.7.1 we will refer to this geometry-related type of correlation in more detail.

2.3
Boundary Conditions

2.3.1
Absorbing and Reflecting Boundaries

In our discussion so far both classical diffusion according to Fick's equations and molecular propagation according to the random walk model have been considered

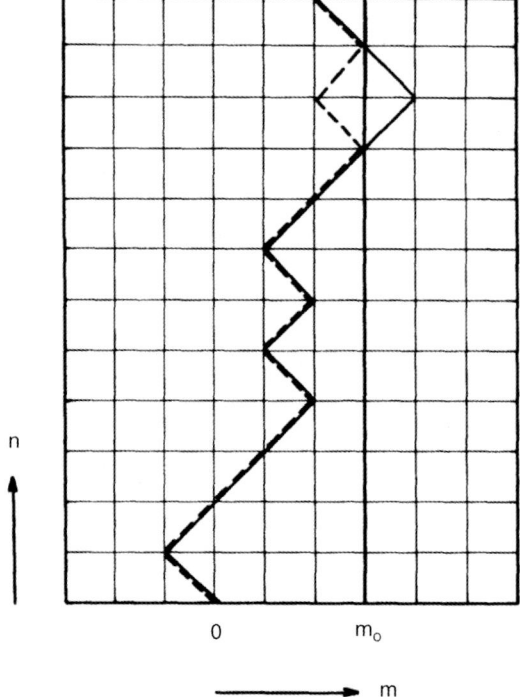

Figure 2.3 Diffusion path for a random walker in unconfined space (solid line) and in the presence of a reflecting barrier at m_0 (broken line) (n is the step number and m represents the position for a random walk in one dimension).

only in a homogeneous region of infinite extent. It is also necessary to consider the ways in which the development of the concentration profile may be modified by boundaries.

Following our initial considerations we refer to the step sequence for one-dimensional diffusion discussed in Section 2.1 (see References [4, 5]). In Figure 2.3 such a diffusion path is drawn, but we now assume that free molecular propagation in the direction considered is possible only up to a certain value of m denoted by m_0. In principle the confinement of the molecule by the boundary may be considered to be between two physically limiting situations: (i) any molecule encountering the position m_0 is reflected with absolute certainty or (ii) any molecule encountering this position is absorbed with absolute certainty, that is, it is removed from the region of consideration. Considering all possible jump sequences in the unconfined space we note that, with a boundary at $m = m_0$, certain of these sequences become impossible (i.e., all sequences involving points $m > m_0$). For any of these sequences we may construct an acceptable path by reflecting the position of the path for which $m > m_0$ with respect to the line $m = m_0$ (Figure 2.3). Obviously, for the limiting case of a perfectly reflecting boundary all these reflected sequences will occur in addition to those jump sequences for which the molecules have remained entirely within the unconfined region $m < m_0$. From the perspective of a random walk one can say that at any encounter with the reflecting boundary at m_0 the next jump will move the molecule to position $m_0 - 1$ with absolute certainty, rather than with probability 0.5 as in an unconfined situation. Thus, to obtain the total number of step sequences ending at some position $m < m_0$, we have simply to add the number of step sequences that, for unrestricted diffusion, would end at position m and at the position symmetric to m with respect to m_0, that is, $2m_0 - m$. The desired probability is thus:

$$P_r(n, m, m_0) = P(n, m) + P(n, 2m_0 - m) \tag{2.21}$$

where $P_r(n, m, m_0)$ denotes the probability that a molecule (or random walker) will have arrived at position m, if propagation is confined by a perfectly reflecting boundary at position m_0.

In the other extreme limit of a perfectly absorbing boundary at position m_0, any molecule that would have traveled to a position beyond m_0 is removed so that all step sequences that involve points beyond m_0 must be eliminated. By the same logic we must also eliminate all sequences that, although ending at $m < m_0$, pass at some point into the region $m > m_0$ (in the unconfined situation). According to our previous consideration the number of such diffusion paths is the same as the number of those sequences that would lead, in the case of unrestricted diffusion, to the position symmetric with respect to m_0, that is, to $2m_0 - m$. Thus the probability that, after n steps, our random walker will have arrived at position m, with an absorbing boundary at m_0 is:

$$P_a(n, m, m_0) = P(n, m) - P(n, 2m_0 - m) \tag{2.22}$$

From Eq. (2.13) this may be expressed as:

$$\begin{aligned} P_r(z, z_0, t) &= P(z, t) + P(2z_0 - z, t) \\ &= \frac{1}{\sqrt{4\pi Dt}} \left[\exp\left(-\frac{z^2}{4Dt}\right) + \exp\left(-\frac{(2z_0 - z)^2}{4Dt}\right) \right] \end{aligned} \tag{2.23}$$

$$\begin{aligned} P_a(z, z_0, t) &= P(z, t) - P(2z_0 - z, t) \\ &= \frac{1}{\sqrt{4\pi Dt}} \left[\exp\left(-\frac{z^2}{4Dt}\right) - \exp\left(-\frac{(2z_0 - z)^2}{4Dt}\right) \right] \end{aligned} \tag{2.24}$$

These equations give the probability densities that, during interval t, a molecule starting at the origin will suffer a displacement z when there is a reflecting or absorbing boundary at z_0. As indicated in Figure 2.4 the probability distribution functions for the cases of molecular confinement are simply obtained by adding or subtracting the reflection of the distribution curves for the unconfined situation.

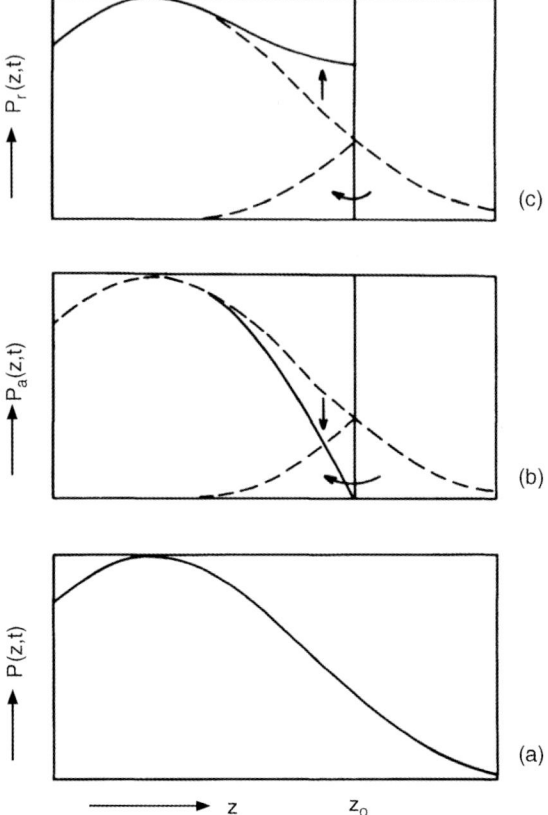

Figure 2.4 Construction of the propagator under the influence of a reflecting barrier (c) and an absorbing barrier (b) from the propagator for unconstrained diffusion (a).

Figure 2.4 also indicates that the probability distributions constructed in this way have the following properties:

$$\frac{d}{dz} P_r(z, z_0, t)\big|_{z=z_0} = 0 \tag{2.25}$$

$$P_a(z_0, z_0, t) = 0 \tag{2.26}$$

which may be easily seen from Eqs. (2.23) and (2.24). Translated into the language of diffusion these statements imply that at a reflecting boundary the gradient of concentration (and therefore the flux in the perpendicular direction) is zero while at an absorbing boundary the concentration itself is zero.

2.3.2
Partially Reflecting Boundary

In the event that the boundary behaves in a manner intermediate between the two extreme cases so far considered we note that the number of molecules absorbed must be related to the number of molecules that are transported to the boundary in the relevant time interval. Suppose that a molecule reaching position m_0 is absorbed with probability P_a and reflected with probability $1 - P_a$. The net number of molecules jumping from position $m - 1$ to position m (per unit cross sectional area A and in time dt) is given by:

$$j = [c(m-1)Al - c(m)Al]\frac{dt/2\tau}{A\,dt} \tag{2.27}$$

where $c(m)$ denotes the concentration of lattice position m. With the self-diffusion coefficient given by $\mathcal{D} = l^2/2\tau$ [Eq. (2.4)] and realizing that:

$$\frac{c(m-1) - c(m)}{l} = -\frac{dc}{dz} \tag{2.28}$$

equation (2.27) is immediately seen to be equivalent to the familiar expression for the flux perpendicular to the boundary:

$$J = -\mathcal{D}\frac{dc}{dz} \tag{2.29}$$

Since this flux must be equal to the rate at which molecules are absorbed at the boundary:

$$J = -\mathcal{D}\frac{dc_A}{dz} = c(m)\frac{P_a l}{2\tau} = c(m)\frac{P_a \mathcal{D}}{l} \tag{2.30}$$

which yields the important result that, in the limit $l \to 0$, for a fixed absorption probability P_a the limit P_a/l goes to infinity. Since the left-hand side represents a real flux density that must remain finite as $l \to 0$ it follows that $c(m)$ must go to zero. This means that for any finite absorption probability one reaches the boundary condition for a completely absorbing boundary. To come to some intermediate value in the small time limit P_a must decrease with l. Therefore, we introduce an absorption parameter, $a = P_a \mathcal{D}/l$, with which Eq. (2.30) may be written in the following form:

$$J = -\mathcal{D}\frac{dc}{dz}\Big|_{z=z_0} = ac(z_0) \tag{2.31}$$

a assumes values of 0 and ∞ in the limiting cases of perfectly reflecting and absorbing boundaries.

2.3.3
Matching Conditions

As the diffusion equation is a parabolic partial differential equation (second order in the space variable), to fully define conditions at a boundary between two regions, characterized, for example by different diffusivities, two matching conditions are needed. The first and most obvious condition is continuity of the flux across the boundary, which for a non-absorbing boundary requires:

$$J_1 = -D_1 \frac{\partial c_1}{\partial z}\bigg|_{\text{Boundary},1} = +D_2 \frac{\partial c_2}{\partial z}\bigg|_{\text{Boundary},2} = -J_2 \quad (2.32)$$

The second matching condition may be understood as a relation (often an equilibrium relation) between the concentrations on either side of the interface. If there is mass transfer resistance at the interface:

$$J_1 = k_f [c_1 - c_1^*] \quad (2.33)$$

where c_1^* is in equilibrium with c_2. The parameter k_f is referred to as the interface permeance or the fluid side mass transfer coefficient. With Eq. (2.32) this gives:

$$-D\frac{\partial c}{\partial z}\bigg|_{\text{surface}} - k_f [c_1 \; c_1^*] \quad (2.34)$$

In some cases, under equilibrium conditions, the concentrations on both sides of the boundary will be the same. However, deviations from this simple situation will occur when there is a change in the free space density across the boundary or, for adsorbed phases, whenever there is a change in adsorption equilibrium. This concerns, in particular, the surface of the nanoporous particles, that is, the interface between the (internal) crystal bulk phase and the surrounding (liquid or gaseous) fluid. In this case, k_f is replaced by the surface permeance (also called the solid side mass transfer coefficient or, in the context of its experimental measurement, surface permeability – see Sections 12.3.2 and 19.6), which is commonly denoted by the symbol α (or k_s).

Importantly, the surface permeance, that is, the transport resistance on the crystal surface, is not correlated with the change in free energy (i.e., the differences in energy and entropy) between the intra- and intercrystalline spaces. It is true that the increase in potential energy from inside to outside reduces the escape rate of the molecules out of crystals, just as a decrease in entropy from outside to inside (caused by a decrease in the number of degrees of freedom due to pore confinement) decreases the entrance rate. However, these effects are already accounted for in the change of the equilibrium concentrations so that any decrease in the escape or entrance rates is exactly compensated by a corresponding increase in the respective populations and, hence, in the "attempt" rates to escape from or enter the crystal.

2.3.4
Combined Impact of Diffusion and Permeation

So far we have implicitly assumed that the porous materials under investigation are homogeneous and structurally ideal, that is, over millions of unit cells the pore space is assumed to be statistically uniform and of infinite extent in all crystallographic directions. However, entropic considerations dictate that such an ideal situation is highly improbable. As displayed in Figure 2.5, application of high-resolution transmission electron microscopy reveals, for example, stacking faults of mirror-twin type

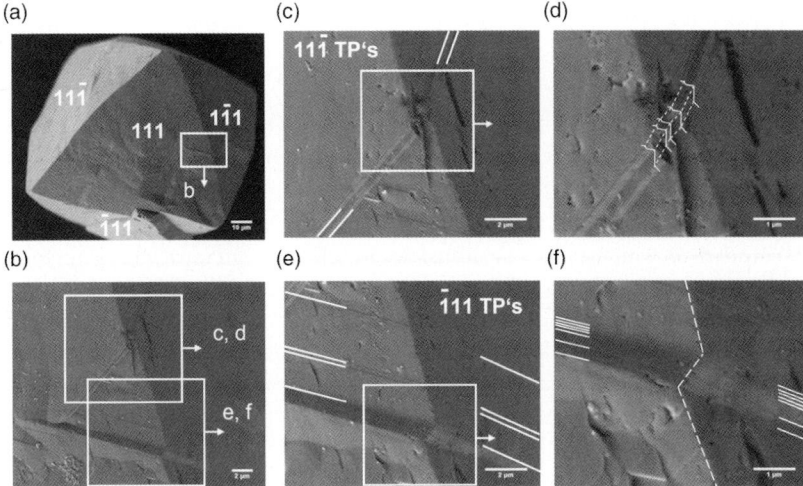

Figure 2.5 Secondary electron micrographs of a typical NaX faujasite powder particle: (a) regularly shaped cuboctahedron exhibiting (111)-type surfaces; (b) triangular meeting zone of (111) and (1$\bar{1}$1) twin planes (TPs); (c, d) and (e, f) close-up of (1$\bar{1}$1) TPs. Reprinted from Reference [21] with permission.

on the (111)-planes of the cubic framework of zeolite NaX [21]. These faults have been interpreted as planes of dramatically reduced permeation, leading to a notable reduction of the diffusivities with increasing diffusion path lengths as observed on comparing the results of quasi-elastic neutron scattering (QENS, Section 10.2) and pulsed field gradient NMR (PFG NMR, Section 11.1). Similar trends in the diffusivities were observed in numerous studies [22–26], providing strong evidence for the existence of transport barriers in the intracrystalline space.

For diffusion path lengths that exceed the separation between such transport resistances, the effective diffusivities resulting from a formal application of either Fick's first law [Eq. (1.1)] or the Einstein relation [Eq. (1.12)] must reflect the influences of both the finite permeation rate through these resistances and the genuine bulk diffusivity D in the ideal structure. In the simplest case of parallel transport resistances of equal mutual distance l and permeability α [used synonymously with the mass transfer coefficient (or permeance) k_f in Eq. (2.34)], one obtains:

$$\frac{1}{D_{\text{eff}}} = \frac{1}{D} + \frac{1}{\alpha l} \tag{2.35}$$

This relation may easily be shown to hold for both transport diffusion and tracer (i.e., self-) diffusion.

As a special case of diffusion barriers one may consider impermeable planes with circular "holes." Assuming holes of radius r in a square-lattice arrangement of mutual distance L ($\gg r$) [27, 28] the permeability is found to be:

$$\alpha = \frac{2Dr}{L^2} \tag{2.36}$$

An interesting consequence of this expression is that the permeability through an impermeable membrane with holes should be proportional to the diffusivity D in the surrounding medium! By inserting Eq. (2.36) into Eq. (2.35), the effective diffusivity is also found to obey this proportionality, yielding:

$$D_{\text{eff}} = \frac{D}{1 + L^2/(2rl)} \tag{2.37}$$

In comparison with the genuine diffusivity, the effective diffusivity turns out to be reduced merely by a geometrical factor, involving the separation l of the resistances and the separation L and diameters $2r$ of the "holes" in these resistances. Postulating the occurrence of such transport resistances in the interior of zeolite crystallites, Eq. (2.37) evidently provides a straightforward explanation as to why the diffusivities determined in macroscopic diffusion experiments (i.e., with mean diffusion paths $\sqrt{\langle z^2 \rangle}$ notably larger than the separation l of these internal resistances) and in microscopic experiments (with $\sqrt{\langle z^2 \rangle} < l$) have often been found to reflect similar trends in their dependence on structural features of the nanoporous host–guest system, although differing by orders of magnitude in their absolute values [29–32] (see also Section 19.6.3).

Sections 19.6.1 and 19.6.2 provide an example revealing a close correlation between intracrystalline diffusion and surface permeation in experimental measurement. This correlation will be shown to be based on a modified version of Eq. (2.36).

It is worth emphasizing that Eq. (2.36) attains its simple form only in the limit of sufficiently small holes ($r \ll L$). For axial diffusion in tubes with periodic partition by impenetrable planes with central, circular holes, the influence of the finite hole size may be approximated analytically by the expression [33]:

$$\alpha = \frac{2D}{\pi R} \nu f(\nu), \quad \text{with} \quad f(\nu) = \frac{1 + 1.37\nu - 0.37\nu^4}{(1-\nu^2)^2} \tag{2.38}$$

where $\nu = r/R$ is the ratio between the radii of the holes and of the tube. For $r \ll R$, that is, $\nu \to 0$ and $f(\nu) \to 1$, the permeability again results as the bulk diffusivity times the hole diameter divided by the plane area per hole, in complete agreement with Eq. (2.36).

2.4
Macroscopic and Microscopic Diffusivities

It is explained in Section 2.1 that the propagation of a diffusing species, within a region of uniform properties, is characterized by a Gaussian distribution, provided that the time interval considered is long enough to contain a large number of elementary steps, yet short enough to preclude the escape of any significant quantity of the diffusing species at the boundaries. Accordingly, molecular migration in an adsorbent–adsorbate system may be characterized on at least three different scales. First we may consider the elementary diffusion steps that may proceed by many different mechanisms, depending on the internal structure of the molecule and the interactions between the diffusing molecule and the adsorbent or between the diffusing molecules themselves. The process may occur either by distinct jumps or by a more or less continuous creeping mechanism (segmental diffusion) [34, 35]. Quite clearly, the elementary steps do not obey the laws of diffusion since these are probabilistic laws that apply only in a statistical sense when the number of steps is large. There are several experimental and computational techniques by which the elementary diffusion steps can be investigated and this topic is pursued in Chapters 8–10.

The sequence of elementary steps within a homogeneous region of space (e.g., within a single microporous zeolite crystal) gives rise to a diffusion process on a scale commensurate with the size of the zeolite crystals or, more generally, the size of quasi-homogeneous regions. Since these dimensions are typically of the order of a few micrometers, we refer to this process as microscopic, intracrystalline, or simply micropore diffusion. Sometimes, in this context also the term configurational diffusion has been used [36].

If, on the other hand, one considers diffusion through an assemblage of microporous particles, for example, in a bed of zeolite crystals or in a composite pellet of the type shown in Figure 1.3b, one may again define an appropriate diffusion coefficient that characterizes the overall long-range migration through the composite medium consisting of the microporous particles and the interparticle void spaces. In many cases, particularly in gas–solid adsorption systems, the molecular density in the adsorbed phase is very much greater than in the gas phase, while the molecular mobility is very much higher in the gas than in the adsorbed phase. Under these conditions transport occurs almost entirely through the gas phase while the adsorbed phase provides a large reservoir of molecules that exchange continuously with those in the gas phase and the long-range or macro-diffusivity is given by:

$$\mathcal{D}_{lr} = \frac{\varepsilon_p}{\varepsilon_p + (1-\varepsilon_p)(q/c)} \mathcal{D}_{inter} = p_{inter} \mathcal{D}_{inter} \tag{2.39}$$

where p_{inter} denotes the fraction of molecules and $\mathcal{D}_{inter}$ the diffusivity in the intercrystalline space.

From the perspective of the random walk model this expression may be justified as follows. The total displacement of a diffusing molecule may be represented as the sum of the displacements within the individual particles and in the interparticle void space, which, to a reasonable approximation, can be considered uncorrelated, using Eq. (2.17):

$$\mathcal{D}_{lr} = \frac{\langle r^2_{intra}(t) \rangle}{6t} + \frac{\langle r^2_{inter}(t) \rangle}{6t} \tag{2.40}$$

Since the ratio of the mean lifetimes in the intraparticle and interparticle regions (τ_{intra}, τ_{inter}) must coincide with the ratio of the relative populations in these regions (p_{intra}, p_{inter}) and since, in the case of the gas phase, p_{inter} will be much smaller than

p_{intra}, it follows that $\tau_{inter} \ll \tau_{intra}$. The second term in Eq. (2.40) may therefore be expressed as follows:

$$\frac{\langle r_{inter}^2(t)\rangle}{6t} = \mathcal{D}_{inter}\frac{t_{inter}}{t} = \mathcal{D}_{inter}\left(\frac{\tau_{inter}}{\tau_{intra}+\tau_{inter}}\right) \qquad (2.41)$$
$$= p_{inter}\mathcal{D}_{inter}$$

Equations (2.40) and (2.41) differ by the term $(r_{intra}^2)/6t$. However, this term is negligibly small, as may be seen from the following argument. The total mean displacement cannot exceed the displacement that would be observed within an individual crystal of infinite size during the same time interval:

$$\langle r_{intra}^2(t)\rangle \leq 6\mathcal{D}_{intra}t_{intra} \approx 6\mathcal{D}_{intra}t \qquad (2.42)$$

Therefore:

$$\frac{\langle r_{intra}^2(t)\rangle}{6t} \leq \mathcal{D}_{intra} \qquad (2.43)$$

Thus, provided that:

$$\mathcal{D}_{intra} \ll p_{inter}\mathcal{D}_{inter} \qquad (2.44)$$

the first term in Eq. (2.40) may be neglected. In the other limit where:

$$\mathcal{D}_{intra} \gg p_{inter}\mathcal{D}_{inter} \qquad (2.45)$$

the molecules are essentially confined within an individual crystallite (see Section 11.4.1 and situation shown in Figure 11.2b), so that the high intracrystalline mobility does not contribute to long-range diffusion. The second term in Eq. (2.40) is again dominant, leading to Eq. (2.39). Section 15.8 presents the results of rigorous calculations of the long-range diffusivities in model systems where an analytical treatment is possible. In all cases considered, the rigorously calculated long-range diffusivities are found to be reasonably approached by Eq. (2.39). In Section 11.4.3.1, Eq. (2.39) is also shown to be confirmed by kinetic Monte Carlo simulations.

2.5
Correlating Self-Diffusion and Diffusion with a Simple Jump Model

Although the processes of transport diffusion and self-diffusion occur by the same microdynamic mechanism, the coefficients of transport diffusion and self-diffusion are in general different, as noted in Section 1.1. Their precise relationship can be determined only from a detailed microdynamic model although some general information on the form of this relationship may be deduced from more general thermodynamic considerations (Sections 3.2 and 3.3). The quantitative treatment of jump sequences, as outlined in the present chapter, has been shown to provide a useful means of correlating the macroscopically observable features of diffusion with the elementary processes. This approach is now used to establish the form of the relationship between the self- and transport diffusivities on the basis of a specific microdynamic model.

Let us assume that the molecular propagation occurs by a sequence of activated jumps between adjacent adsorption sites such as, for example, between the adjacent cavities in the crystals of zeolite A or MOF ZIF-8 (Chapter 16 and Section 19.3). Within the Stefan–Maxwell formalism (Section 3.3) this assumption is equivalent with the implication of an infinitely large mutual Stefan–Maxwell diffusivity $Ð_{ij}$, that is, with the assumption that the diffusive fluxes are counterbalanced exclusively by the drag exerted by the pore network, while any intermolecular drag effects are negligible. This means that any molecule accommodated by a given cage is able to leave into any of the adjacent cages with equal probability, with a lifetime $\tau = \tau(q)$, which is a function only of the local molecular density. The self-diffusivity in such a system [Eq. (2.7)] is given simply by:

$$\mathcal{D} = \frac{\lambda^2}{6\tau(q)} \tag{2.46}$$

where λ is the distance between the centers of adjacent cages (or sites).

Let us now consider a plane of area A separating two adjacent layers of supercages. In the case of self-diffusion the total concentration q is the same on both sides of this plane and the mean lifetimes (τ) are therefore also the same. Assuming, for simplicity, a cubic arrangement of cages so that one-sixth of all jumps pass through the area A, the total number of marked molecules crossing the plane in time dt from left to right will be given by:

$$N^*_{l \to r} = q^*(z) \frac{\lambda A}{6} \frac{dt}{\tau(q)} \tag{2.47}$$

and in the reverse direction:

$$N^*_{r \to 1} = q^*(z+\lambda) \frac{\lambda A}{6} \frac{dt}{\tau(q)} \approx \left[q^*(z) + \frac{dq^*}{dz}\lambda\right] \frac{\lambda A}{6} \frac{dt}{\tau(q)} \tag{2.48}$$

The net diffusion flux is therefore given by:

$$J^* = \frac{N^*_{1 \to r} - N^*_{r \to 1}}{A\,dt} = \frac{-\lambda^2}{6\tau(q)} \frac{dq^*}{dz} \tag{2.49}$$

from which the self-diffusion coefficient can be directly identified as $\lambda^2/6\tau(q)$ in conformity with Eq. (2.46).

The diffusion coefficient for the analogous transport diffusion situation may be derived in a similar way but there is one important difference: since there is now a difference in total concentration, the lifetimes $\tau(q)$ on both sides of the plane will be different. We therefore have:

$$N_{1 \to r} = q(z) \frac{\lambda A}{6} \frac{dt}{\tau[q(z)]} \tag{2.50}$$

$$N_{r\to 1} = q(z+\lambda)\frac{\lambda A}{6}\frac{dt}{\tau[q(z+\lambda)]} = \frac{\lambda A}{6}dt\left[\frac{q(z)}{\tau[q(z)]} + \frac{d}{dq}\left(\frac{q}{\tau}\right)\frac{dq}{dz}\lambda\right] \quad (2.51)$$

$$J = -\left[\frac{\lambda^2}{6}\frac{d}{dq}\left(\frac{q}{\tau}\right)\right]\frac{dq}{dz}, \quad D = \frac{\lambda^2}{6}\frac{d}{dq}\left(\frac{q}{\tau}\right) \quad (2.52)$$

If τ is constant Eq. (2.52) indicates that the Fickian diffusivity will be independent of concentration, as in the low-concentration limit. In general, however, Eq. (2.52) implies that the Fickian diffusivity will be concentration dependent, the form of the concentration dependence being determined by the function $\tau(q)$. It is thus easy to see, from the microdynamic perspective, how the strong concentration dependence predicted for a nonlinear equilibrium system from the thermodynamic driving force model [Eq. (1.31)] may arise. For example, if we assume:

$$\frac{q}{\tau(q)} = \frac{q_s}{\tau_0}\ln\left(\frac{1}{1-q/q_s}\right) \quad (2.53)$$

we obtain from Eq. (2.52):

$$D = \frac{D_0}{1-q/q_s} \quad (2.54)$$

which is the form predicted from Eq. (1.31) for a system that follows the Langmuir equilibrium isotherm. Comparing the expressions for $\mathcal{D}$ and D [Eqs. (2.46) and (2.52)] we see that:

$$D(q) = \frac{d}{dq}[\mathcal{D}(q)q] \quad (2.55)$$

Equations (2.52) and (2.55) allow some interesting general conclusions to be drawn. If the self-diffusivity is independent of concentration $D(q) = \mathcal{D}(q)$ and we see that the tracer and transport diffusivities must be identical, as already deduced for non-interacting molecules from physical reasoning.

Clearly, the same conclusion holds true if the transport diffusivity is independent of concentration. This is evidently the situation in the low concentration limit where τ is independent of concentration. Conversely, if either D or $\mathcal{D}$ is concentration dependent τ must also be concentration dependent and the two coefficients cannot be equal.

If the self-diffusivity varies inversely with concentration:

$$\mathcal{D}(q) \propto 1/q \quad (2.56)$$

it follows that the transport diffusivity will be zero. This means that, even in the presence of a concentration gradient, there is no diffusive flux. This corresponds in the thermodynamic sense to the condition $d\mu/dq = 0$, as discussed in Section 1.2. This situation may be understood by realizing that, taken together, Eqs. (2.46) and (2.56) imply that the mean lifetime of a molecule in an individual cage is directly proportional to the number of molecules within that cage, which is precisely the condition for dynamic equilibrium between the different populations.

It follows that if $\mathcal{D}$ varies with a higher inverse power of concentration the transport diffusivity becomes negative. This implies that the net molecular flux is directed in the direction of increasing molecular concentration. This situation may again be understood by considering Eq. (2.46), from which it is evident that the mean molecular lifetime must increase more than linearly with concentration. The microdynamic consequence is that the population of cages with higher molecular populations will increase at the expense of cages with lower molecular populations. This situation obviously corresponds to a negative diffusivity. According to the thermodynamic model this implies $d\mu/dq < 0$. Such a situation can exist for example in a non-isothermal system or in phase separation from a metastable system in which mixing on the molecular scale has been forced beyond the point of equilibrium.

It should be emphasized that these conclusions are true only for the particular microdynamic model. In Section 4.3.3 the theory of absolute reaction rates will be shown to provide an example where such relations should be applicable. Furthermore, it may be shown (see Reference [4]) that the validity of Eq. (2.52) is in no way affected by the simplifying assumption (constant cage occupation perpendicular to the occupation gradient) made in the present derivation. Thus, this simple model may be regarded as an informative example to illustrate the way in which the features of the probability distribution of molecular propagation may be shifted under non-equilibrium conditions.

2.6
Anomalous Diffusion

2.6.1
Probability Distribution Functions of Residence Time and Jump Length

In Section 2.1 any random particle movement was shown to give rise to well-defined patterns of particle transport. Completely equivalently, these patterns appear in Fick's first law [proportionality between particle flux and concentration gradient, see Figure 1.1b and Eq. (1.11)] and the Einstein relation [proportionality between mean squared particle displacements and observation time, see Figure 1.1c and Eq. (1.12)]. Mass transfer following these laws is referred to as normal diffusion. As the only prerequisite, the relevant time and space scales of particle displacements must be large enough so that the overall process of migration can be considered as a sum of many displacements occurring independently of each other. Hence, the particle "memory" must be assumed to be much shorter than the observation time.

One option for considering mass transfer under conditions that do not allow these simplifying assumptions is based on the CTRW (continuous time random walk [6, 37, 38]) model. In this model, mass transfer is considered to occur in a sequence of jumps, which is characterized by a probability distribution function

(PDF) for both the residence times between two consecutive jumps and the jump lengths. Deviations from normal diffusion may be expected to occur for PDFs that give rise to a diverging mean value of the *residence times*. In this case, referred to as "subdiffusion," the mean square displacement:

$$\langle z^2(t) \rangle \sim t^\kappa \quad (\kappa < 1) \tag{2.57}$$

increases less than linearly with time.

We show that this divergence in the mean residence time is correlated with the existence of a particle memory by rationalizing that for particles "with no memory" only finite mean waiting times can be expected. For such particles, each instant of time during their rest between succeeding jumps is equivalent. This means in particular that the jump probabilities dt/τ during each time interval dt are identical (where $1/\tau$, for now, is introduced as nothing more than a factor of proportionality). The difference in the corresponding PDFs at two subsequent instants of time may be noted as:

$$p(t+dt) - p(t) = -p(t) \times dt/\tau \tag{2.58}$$

whence one obtains:

$$p(t) = \frac{1}{\tau} \exp\left(-\frac{t}{\tau}\right) \tag{2.59}$$

resulting in a finite mean waiting time $\tau \equiv \int tp(t)dt$, which thus yields an easily comprehensible meaning for the proportionality factor, $1/\tau$. Equations (2.58) and (2.59) reflect a situation typical of guest diffusion in homogeneous nanoporous media. Hence, deviations from normal diffusion as a result of residence-time-related divergencies may be expected to occur in rather exceptional situations and only when the system under study has not yet attained dynamic equilibrium [39].

Divergences in the *mean jump lengths* may result in the opposite behavior, namely, in mean square displacements increasing faster than linearly with time, which is referred to as "super-diffusion." In nanoporous materials, this type of molecular random walk can become relevant when, in the Knudsen regime of diffusion (see Section 4.2.1.1), the pore space allows molecular "flights" over correspondingly long distances. This is the case, for example, in arrays of sufficiently long parallel tubes [40–43]. This type of molecular random walk is referred to as Lévy flights or, in its generalization, Lévy walks [38, 44].

Implying constant flight velocities and a power law distribution:

$$p(r) \propto 1/r^{1+\beta} \tag{2.60}$$

of the flight lengths r, the time dependence of the mean square displacement $\langle r^2(t) \rangle$ is found to follow well-defined patterns, with large differences for different values of β [38]. For $\beta > 2$, that is, for distribution functions decaying sufficiently fast with increasing jump lengths, $\langle r^2(t) \rangle$ is proportional to t in complete accordance with the requirements of normal diffusion. For $0 < \beta < 1$, on the other hand, the resulting mean square displacements are found to scale as $\langle r^2(t) \rangle \sim t^2$, following the pattern

$r(t) \sim t$ of a uniform movement with constant velocity. Intermediate values of β yield "intermediate" time dependencies of $\langle r^2(t) \rangle$, that is, scaling with $t^2/\ln t$ for $\beta = 1$, with $t^{3-\beta}$ for $1 < \beta < 2$, and $t \ln t$ for $\beta = 2$.

Simulating Knudsen diffusion in infinitely extended two-dimensional model channels has been shown to lead to the latter dependency and, hence, to a situation that can no longer be treated in terms of normal diffusion [43]. Notably, this problem disappears in three-dimensional pores. Here, parallelism with the pore axis as a prerequisite for extremely large flights depends on two rather than on only one angle. As a consequence, $p(r)$ is found to scale with r^{-4}, that is, with $\beta = 3$, complying with the requirement $\beta > 2$ for normal diffusion [45, 46]. Under these conditions, transport diffusion and self-diffusion were shown to coincide (as expected quite generally for Knudsen diffusion) and to be equally retarded by increasing surface roughness [41, 47]. In the usual case of diffusion measurements over space scales notably exceeding the mean flight lengths, flight length divergencies may therefore be excluded as a possible reason for anomalous diffusion.

Divergences in the residence times and/or in the jump lengths of guest molecules in porous materials are thus found to give rise to anomalous diffusion only under very exceptional conditions, which, currently, do not appear to be of much practical relevance. In the following we show that deviations from normal diffusion may, however, be dictated by the pore structure itself. We have illustrated in Section 2.1 that, as a necessary and sufficient prerequisite for normal diffusion, it must be possible to subdivide the relevant observation time into intervals such that the displacements of each individual particle during these intervals are independent of each other. Among the systems for which this requirement cannot be fulfilled, the so-called single-file systems have attracted most attention [48]. They consist of one-dimensional pores filled with guest molecules that are too bulky to pass each other. Hence, molecules that have been shifted in one direction are, in the subsequent interval of time, more likely to move in the opposite direction. For infinitely long pores, this tendency (i.e., the particle "memory") is preserved over arbitrarily long times. Chapter 5 is devoted exclusively to the treatment of this phenomenon.

In the following we consider extremely inhomogeneous host systems that are the opposite of the homogeneous, infinitely extended one-dimensional pores required for single-file diffusion. These inhomogeneities, however, are assumed to exhibit certain features of self-similarity. This self-similarity permits the introduction of "fractal" dimensions and leads to transport phenomena that may deviate notably from normal diffusion. The discussion is limited to an introduction to the topic; further information may be found in the extensive literature of this field [38, 49–53].

2.6.2
Fractal Geometry

Let us consider a part of a geometrical object obtained by reducing the space scale by a certain factor, a. In many cases the part, defined in this way, turns out to be an exact replica of the original object. A trivial example is a cube (edge length L), a square or a straight line for which such a procedure yields a similar cube, square or straight line of dimension L/a, that is, a reduced replica of the original object. Denoting the mass or volume of the original object and its reduced scale replica by M_0 and $M_{1/a}$, respectively, we may introduce the generalized dimension d_f, defined by:

$$\frac{M_0}{M_{1/a}} = a^{d_f} \tag{2.61}$$

Obviously, in the cases of a cube, a square, or a straight line the generalized dimension defined in this way corresponds with the topological dimension of these objects (3, 2, and 1, respectively). However, this concept of self-similarity, and the possibility of applying Eq. (2.61), may actually be encountered for a wide variety of different objects. The generalized dimension defined according to Eq. (2.61) does not necessarily assume an integer value; when it is non-integral the object is called a "fractal."

Figure 2.6 shows three classical examples of self-similar objects [49, 50, 54]. To obtain objects that are extended in three-dimensional space, the representations in Figure 2.6a and b may be considered to be cross-sections through a solid object that is uniform in the direction perpendicular to the plane of the paper. From Eq. (2.61) it follows that in this case the fractal dimension is simply increased by one, that is, 2.262 for the three-dimensional extension of the Koch curve and 2.585 for the corresponding extension of the Sierpinski gasket. All these dimensions, including that of the Menger sponge, lie between 2 and 3. This may be intuitively interpreted as an indication of the space filling character of the surfaces of these objects, which approaches 100% as the fractal dimension approaches 3.

As a result of this geometric feature the internal area measured by adsorption (i.e., the product of the area per molecule and the number of molecules in a monolayer) will increase as the size of the probe molecule is decreased [50, 55]. This simply reflects the fact that, as the size of the probe molecule decreases, smaller cracks and holes become accessible. This may be used as the basis for an alternative definition of the fractal dimension. As shown in Figure 2.7 for the Koch curve, a reduction in the diameter of the (spherical) molecules by a factor a (in this example 3) increases the number of molecules adsorbed by a factor equal to the number of images of the original object produced by the scale reduction (in this example 4):

$$\frac{N_{1/a}}{N_0} = \frac{M_0}{M_{1/a}} = a^{d_f} \tag{2.62}$$

where N_0 and $N_{1/a}$ denote the respective numbers of molecules forming a "monolayer." The second equality follows from Eq. (2.61). Again it may be seen that in the limiting cases of a straight line, a plane, and a completely filled space (corresponding to a cube) d_f assumes the integer values 1, 2, and 3, respectively.

So far in the example of Figure 2.6 we have considered self-similarity only in the strict sense, that is, for systems with an exactly repeating structure at all scales.

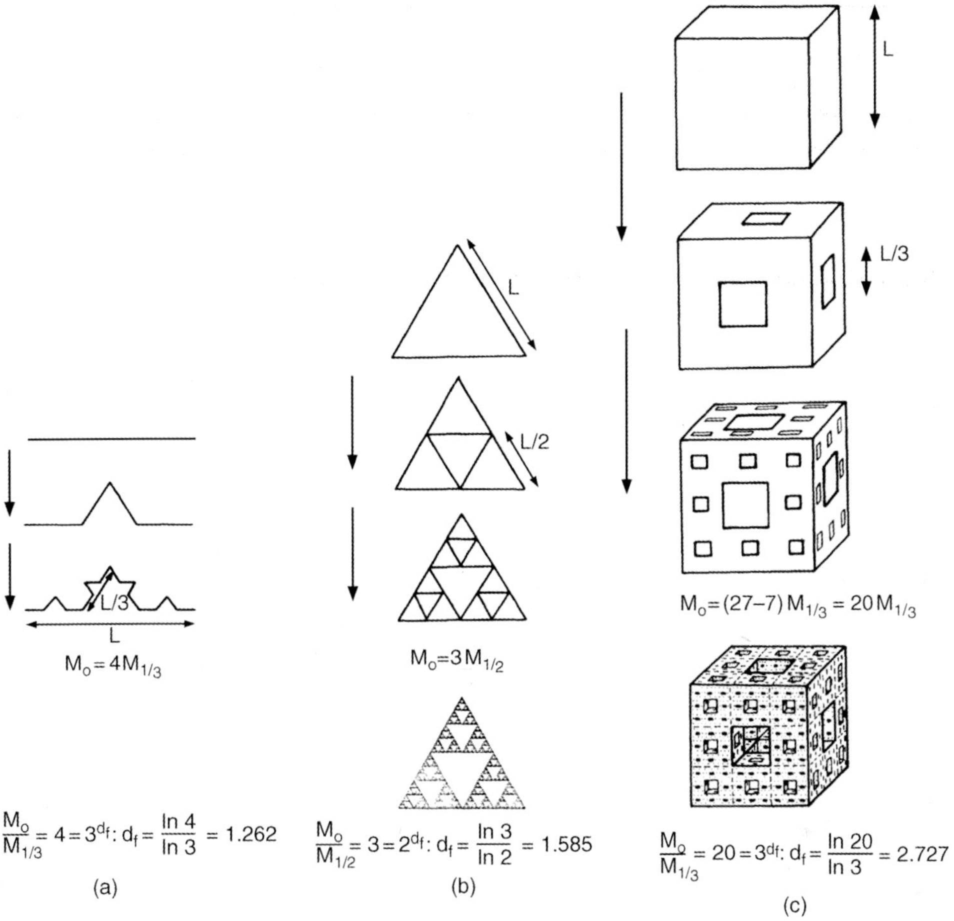

2.6 Anomalous Diffusion | 51

Figure 2.6 Examples of fractal objects showing the manner of construction and their fractal dimensions, (a) Koch curve, (b) Sierpinski gasket, and (c) Menger sponge.

$\dfrac{M_o}{M_{1/3}} = 4 = 3^{d_f}: d_f = \dfrac{\ln 4}{\ln 3} = 1.262$ (a)

$\dfrac{M_o}{M_{1/2}} = 3 = 2^{d_f}: d_f = \dfrac{\ln 3}{\ln 2} = 1.585$ (b)

$M_o = (27-7)\,M_{1/3} = 20\,M_{1/3}$

$\dfrac{M_o}{M_{1/3}} = 20 = 3^{d_f}: d_f = \dfrac{\ln 20}{\ln 3} = 2.727$ (c)

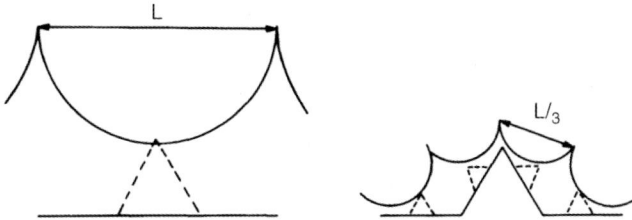

Figure 2.7 Monolayers of adsorbate on a fractal surface for different molecular diameters (L and $L/3$). The fine structure delineated by the broken lines remains "invisible" to molecules of the size considered but would become "visible" to a smaller adsorbate molecule.

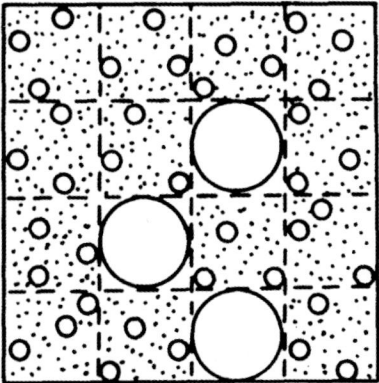

Figure 2.8 Fractal model for a system of parallel cylindrical pores. To generate the structure three of the sixteen squares in the 4 × 4 grid are filled with one large pore (of the same diameter) at each iteration.

The definitions of Eqs. (2.61) and (2.62), however, apply also if self-similarity is satisfied in a statistical sense. Obviously, in real adsorbents fractality is most likely to occur in this way. As an example, Figure 2.8 shows a fractal model of a cylindrical pore system [55]. It is constructed by dividing a square cross-section into 4×4 (=16) smaller squares, n (=3) of which, selected in a random manner, are filled by cylindrical pores. The remaining ($a^2 - n = 13$) squares are again subdivided and the procedure is repeated. Figure 2.8 shows such a pore distribution obtained by three such iterations. If we consider the total area of the circles in the cross section we obtain, in complete analogy with the Menger sponge:

$$M_0 = (a^2 - n) M_{1/a} \tag{2.63}$$

and:

$$\frac{M_0}{M_{1/a}} = (a^2 - n) = a^{d_f} \tag{2.64}$$

That is:

$$d_f = \frac{\ln(a^2 - n)}{\ln a} \tag{2.65}$$

Again, considering the extension perpendicular to the plane of the paper, 1 must be added to this dimension. In the example considered above with $a = 4$, $n = 3$ one therefore obtains $d_f = \ln 13 / \ln 4 + 1 = 2.85$.

The limits beyond which self-similarity no longer applies are called the inner (lower) and outer (upper) cut-offs of fractal behavior. In any case an absolute lower limit is imposed by the atomic dimensions and an upper limit by the overall macroscopic dimensions. With some microporous carbons, the range of fractality was found to cover several orders of magnitude (from 1 nm to 1 μm) [55, 56].

2.6.3
Diffusion in a Fractal System

In any pore structure the presence of constrictions and blind pores leads to a lengthening of the diffusion path and thus to a reduction in the rate of diffusive propagation. However, in a non-fractal system the statistically homogeneous nature of the pore network ensures that the relative influence of such effects is independent of the scale. The diffusing molecules always encounter the same distribution (for example of pore size) so the linear relationship between the mean square displacement and the observation time is not altered by changes in the scale of observation. By contrast, in a fractal system these effects are no longer independent of scale. As the scale is increased, the relative influence of blind pores and other constrictions that lead to a lengthening of the diffusion path becomes progressively more important. As a result, the mean square displacement in a fractal system increases less than linearly with the observation time:

$$\langle r^2(t) \rangle \propto t^\kappa, \quad (\kappa < 1.0) \tag{2.66}$$

A quantitative estimate of this behavior is most straightforward for diffusion within an array of entangled channels whose arrangement may be assumed to coincide with the diffusion path of a random walker. We define a curvilinear coordinate [s(t)] that follows the axis of the channel and diffusion with respect to this coordinate is assumed to follow the ordinary laws, leading to a mean square diffusion path that increases in proportion to the elapsed time:

$$\langle s^2(t) \rangle \propto t \tag{2.67}$$

However, as a result of the entanglement of the channels, the mean square separation $\langle r^2(t) \rangle$ between the starting point and the point reached after time t is related to the curvilinear path length $s(t)$ by:

$$\langle r^2(t) \rangle \propto s(t) \tag{2.68}$$

Combining relations (2.67) and (2.68) yields:

$$\langle r^2(t) \rangle \propto t^{1/2} \tag{2.69}$$

that is, in this example $\kappa = 0.5$.

Relation (2.66) may be applied to yield a quite general determination of the fractal dimension of the random walk path. Recalling that (i) the "mass" M is proportional to the elapsed time and (ii) the factor by which the space scale is reduced (a) is inversely dependent on $\langle r^2(t) \rangle^{1/2}$, one obtains:

$$\frac{M_0}{M_{1/a}} = \frac{t_0}{t_{1/a}} = \frac{\langle r^2(t_0) \rangle^{1/\kappa}}{\langle r^2(t_{1/a}) \rangle^{1/\kappa}} = a^{2/\kappa} \tag{2.70}$$

The second equality follows from relation (2.66) and the third equality results by representing the scale reduction factor a as the ratio of the respective root mean square displacements $\langle r^2(t_0)\rangle^{1/2}/\langle r^2(t_{1/a})\rangle^{1/2}$. Comparing relations (2.61) and (2.70) yields the following relation between the anomalous diffusion exponent κ and the fractal dimension d_ω of the random walker's path:

$$\kappa = 2/d_\omega \qquad (2.71)$$

Notably, d_ω and d_f are independent quantities since the same values of d_f may be obtained for quite different structures, which may lead to large differences in transport behavior and correspondingly different values of d_ω. With $\kappa = 1$, from Eq. (2.71) the fractal dimension of the trajectory of a random walker in isotropic (non-fractal) space is found to be 2. Note that this result holds for normal diffusion in both two- and three-dimensional spaces!

2.6.4
Renormalization

A useful method of determining the exponent κ for fractal diffusion is provided by the concept of "renormalization." The basic idea of this approach is that the relationship between space and time as given by relation (2.62) must be preserved when space and time scales are simultaneously changed. We may illustrate this procedure by considering the simple case of one-dimensional diffusion with a fixed jump length, as discussed in Section 2.1. Instead of steps of length l_1 at time intervals τ_1 we consider effective steps of length $2l_1$ and seek the corresponding time interval τ_2 during which the diffusing species will have reached the next but one position (Figure 2.9). Obviously this position can be reached only after an even number of jumps. We have therefore to find the probability P_n that a displacement $2l_1$ is achieved after n steps and the corresponding value of $\tau_2(n)$. The mean time τ_2 then follows from the summation:

$$\tau_2 = \sum P_n \tau_2(n) \qquad (2.72)$$

Clearly, the time $\tau_2(n)$ for n double steps is $2n\tau$. Pairs of steps of length l_1 lead, with equal probabilities $1/2$, to displacements 0 or $2l_1$. So we obtain:

$$P(n) = \left(\frac{1}{2}\right)^n \qquad (2.73)$$

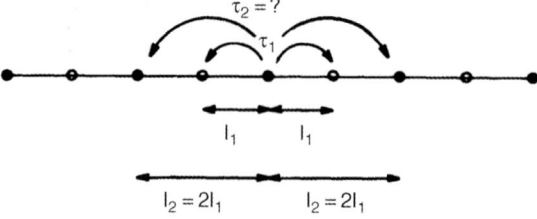

Figure 2.9 Renormalization scheme for a one-dimensional random walk with constant step size.

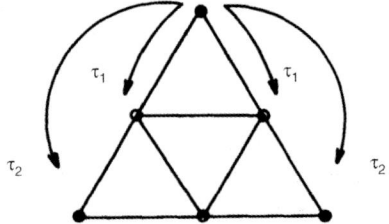

Figure 2.10 Renormalization scheme for random walk through the vertices of a Sierpinski gasket.

and:

$$\tau_2 = \sum_{n=1}^{\infty} \left(\frac{1}{2}\right)^n (2n\tau_1) = 2\tau_1 \sum_{n=1}^{\infty} \left(\frac{1}{2}\right)^n n = 4\tau_1 \qquad (2.74)$$

Since relation (2.66) must apply for any time and length it follows that:

$$\left(\frac{l_1}{l_2}\right)^2 = \left(\frac{\tau_1}{\tau_2}\right)^{\kappa} \qquad (2.75)$$

from which, since $l_2 = 2l_1$ and $\tau_2 = 4\tau_1$, one obtains the familiar result $\kappa = 1$.

For diffusion in a real fractal system, the equivalent procedure is more complicated and we therefore restrict the discussion to the case of the Sierpinski gasket (Figure 2.6b), for which the renormalization scheme for two-dimensional diffusion is given in Figure 2.10. It may be shown [38, 57] that the mean times of migration (τ_2) between the marked positions (filled circles) and the step times for jumps between the original positions obey the relation $\tau_2 = 5\tau_1$. With $l_2 = 2l_1$ and Eq. (2.75) this yields $\kappa = \ln 4/\ln 5 = 0.86$ for the anomalous exponent for a random walk across the vertices of a Sierpinski gasket.

2.6.5
Deviations from Normal Diffusion in Nanoporous Materials: A Retrospective

For any sequence of displacements with equal probability distributions, the final probability distribution function of molecular displacements resulting from the sum of the individual displacements is known to approach a Gaussian, owing to the central limit theorem of statistics. In Section 2.1, such a probability distribution has been shown to lead to Fick's laws [Eqs. (1.1) and (1.2)] and the Einstein relation [Eq. (2.6)] and to lead, therefore, to normal diffusion. Deviations from normal diffusion have to be associated, therefore, with mechanisms that give rise to a particle "memory," which ensures that the probability distribution of future displacements is correlated with the sequence of previous displacements.

Let us first assume that the particle memory operates in such a way that large displacements give rise to higher probabilities for displacements in the same direction. This situation would apply to Knudsen diffusion in the perspective of Levy flights since, during the flight, the molecules will clearly continue their movement in the same direction. This region of diffusion anomaly obviously disappears when the displacements that are considered span distances that are large in comparison with the flight lengths. Hence, for correspondingly large observation times, anomalous diffusion of this type (referred to as super-diffusion) may be excluded.

The reverse situation, namely, that molecular displacements are more likely to be followed by displacements in the opposite direction, is encountered under the influence of internal resistances. For displacements much smaller than the mutual separation of these resistances, their influence is still negligibly small, giving rise to mean square displacements that increase with time. With further increasing displacements, however, the influence of these resistances will become increasingly important, leading to a deviation of the mean square displacements towards smaller values. Correspondingly, the slope of the log–log plot of $\langle r^2(t) \rangle$ versus t would decrease continuously from its initial value 1.

The proportionality $\langle r^2(t) \rangle \sim t^\kappa$ with an exponent $\kappa \neq 1$, independent of the observation time t, is associated with the occurrence of "anomalous" diffusion. The constancy of the exponent κ with increasing time has to be ensured by some self-similarity in the pore structure, which may be made quantitative by using the concept of fractal dimensions. The origin of the less than linear increase in the mean square displacement with the observation time may be illustrated by considering the Sierpinski gasket shown in Figure 2.6b as a typical representative of a self-similar structure. Obviously, with increasing displacements, the molecules will get to sites that are adjacent to larger and larger "empty" triangles. Since we assume (see also Figure 2.10) that mass transfer proceeds by jumps between adjacent vertices, these "empty" triangles represent impenetrable regions for particle propagation, leading to its progressive retardation and, hence, to a less than linear increase of the mean square displacement with the observation time.

For experimental observation of such dependencies one has to require that the observed displacements are in fact within those spatial extents that can be described by a fractal geometry. This means that they have to be between the lower and upper "cut-off" of the fractal structure. Further, molecular jumps between adjacent vertices as implied in the theoretical model must also be required to approach the situation in reality. This would probably be true if molecular propagation is controlled by surface diffusion. However, when diffusion is significantly affected by transport in the gas phase, the situation may be reversed. Since the relative contributions of surface and bulk diffusion to overall mass transport (Section 4.2) depend strongly on the temperature, the same system may behave in quite different ways at different temperatures [58]. Thus, in a bed of zeolite crystals of type NaX with n-butane as a probe molecule the molecular mean square displacement has been found to increase both faster (for high temperatures) and slower (for low temperatures) than linearly with the observation time [59].

Deviating from the cases considered so far, the anomaly of single-file diffusion is brought about by the very specific particle–particle interaction in such systems, rather than by any hierarchy in the pore space. Again, however, experimental observation necessitates that, over the displacements under study, molecular diffusion proceeds under the conditions required by their theoretical treatment. Given the most recent results which show that real nanoporous materials tend to deviate significantly in their diffusion patterns from the behavior expected for ideal systems, the detection of well-defined structure–mobility relations under the conditions of anomalous diffusion remains one of the challenging tasks of current diffusion research.

References

1. Kärger, J. (2010) *Leipzig, Einstein, Diffusion*, Leipziger Universitätsverlag, Leipzig.
2. Kärger, J. and Heink, W. (1983) *J. Magn. Reson.*, **51**, 1.
3. Jobic, H. (2008) in *Adsorption and Diffusion* (eds H.G. Karge and J. Weitkamp), Springer, Berlin, Heidelberg, p. 207.
4. Kärger, J. and Ruthven, D.M. (1992) *Diffusion in Zeolites and Other Microporous Solids*, John Wiley & Sons, Inc., New York.
5. Chandrasekhar, S. (1943) *Rev. Mod. Phys.*, **15**, 1.
6. Weiss, G.H. (1994) *Aspects and Applications of Random Walks*, North-Holland, Amsterdam.
7. Feller, W. (1970) *An Introduction to Probability Theory and its Applications*, John Wiley & Sons, Inc., New York.
8. Heink, W., Kärger, J., Pfeifer, H., and Stallmach, F. (1990) *J. Am. Chem. Soc.*, **112**, 2175.
9. Krishna, R. and van Baten, J. (2010) *Langmuir*, **26**, 10854.
10. Manning, J.R. (1968) *Diffusion Kinetics for Atoms in Crystals*, van Nostrand, Princeton, NJ.
11. Bardeen, J. and Herring, C. (1951) in *Atom Movements* (ed J.H. Holloman), CRC Press, Cleveland, OH, p. 87.
12. Le Claire, A.D. and Lidiard, A.B. (1956) *Philos. Mag.*, **1**, 518.
13. Compaan, K. and Haven, Y. (1956) *Trans. Faraday Soc.*, **52**, 786.
14. Allnatt, A.R. and Lidiard, A.B. (1993) *Atomic Transport in Solids*, Cambridge University Press, Cambridge.
15. Mehrer, H. (2007) *Diffusion in Solids*, Springer, Berlin.
16. Kärger, J., Vasenkov, S., and Auerbach, S.M. (2003) in *Handbook of Zeolite Science and Technology* (eds S.M. Auerbach, K.A. Carrado, and P.K. Dutta), Marcel Dekker, New York, p. 341.
17. Rickert, H. (1964) *Z. Phys. Chem. NF*, **43**, 129.
18. Kärger, J., Petzold, M., Pfeifer, H., Ernst, S., and Weitkamp, J. (1992) *J. Catal.*, **136**, 283.
19. Bär, N.K., Kärger, J., Pfeifer, H., Schäfer, H., and Schmitz, W. (1998) *Microporous Mesoporous Mater.*, **22**, 289.
20. Kärger, J. (1991) *J. Phys. Chem.*, **95**, 5558.
21. Feldhoff, A., Caro, J., Jobic, H., Ollivier, J., Krause, C.B., Galvosas, P., and Kärger, J. (2009) *ChemPhysChem.*, **10**, 2429.
22. Takaba, H., Yamamoto, A., Hayamizu, K., Oumi, Y., Sano, T., Akiba, E., and Nakao, S. (2004) *Chem. Phys. Lett.*, **393**, 87.
23. Takaba, H., Yamamoto, A., Hayamizu, K., and Nakao, S. (2005) *J. Phys. Chem. B*, **109**, 13871.
24. Vasenkov, S., Böhlmann, W., Galvosas, P., Geier, O., Liu, H., and Kärger, J. (2001) *J. Phys. Chem. B*, **105**, 5922.
25. Vasenkov, S. and Kärger, J. (2002) *Microporous Mesoporous Mater.*, **55**, 139.
26. Adem, Z., Guenneau, F., Springuel-Huet, M.A., and Gedeon, A. (2008) *Microporous Mesoporous Mater.*, **114**, 337.
27. Dudko, O.K., Berezhkovskii, A.M., and Weiss, G.H. (2004) *J. Chem. Phys.*, **121**, 11283.
28. Dudko, O.K., Berezhkovskii, A.M., and Weiss, G.H. (2005) *J. Phys. Chem. B*, **109**, 21296.
29. Ruthven, D.M. and Post, M.F.M. (2001) in *Introduction to Zeolite Science and Practice* (eds H. van Bekkum, E.M. Flanigen, and J.C. Jansen), Elsevier, Amsterdam, p. 525.
30. Kärger, J. and Ruthven, D.M. (1981) *J. Chem. Soc., Faraday Trans. I*, **77**, 1485.
31. Kärger, J. and Ruthven, D.M. (1989) *Zeolites*, **9**, 267.

32 Kärger, J. and Ruthven, D.M. (1997) in *Progress in Zeolite and Microporous Materials* (eds H. Chon, S.K. Ihm, and Y.S. Uh), Elsevier, Amsterdam, p. 1843.

33 Berezhkovskii, A.M., Monine, M.I., Muratov, C.B., and Shvartsman, S.Y. (2006) *J. Chem. Phys.*, **124**, 36103.

34 Barrer, R.M. (1980) in *The Properties and Applications of Zeolites* (ed R. Townsend), The Chemical Society, London, p. 3.

35 Barrer, R.M. (1978) presented at Symposium on the Characterisation of Porous Solids, Neuchatel, 9–13 July 1978.

36 Weisz, P.B. (1973) *Chem. Tech.*, **3**, 498.

37 Klafter, J. and Sokolov, I.M.(August, 29 2005) *Phys. World*, **18**, 29.

38 Ben-Avraham, D. and Havlin, S. (2000) *Diffusion and Reaction in Fractals and Disordered Systems*, Cambridge University Press, Cambridge.

39 Lubelski, A., Sokolov, I.M., and Klafter, J. (2008) *Phys. Rev. Lett.*, **100**.

40 Kärger, J., Valiullin, R., and Vasenkov, S. (2005) *New J. Phys.*, **7**, 15.

41 Zschiegner, S., Russ, S., Valiullin, R., Coppens, M.O., Dammers, A.J., Bunde, A., and Kärger, J. (2008) *Eur. Phys. J.*, 109.

42 Valiullin, R. and Khokhlov, A. (2006) *Phys. Rev. E*, **73**, 051605.

43 Russ, S. (2009) *Phys. Rev. E*, **80**, 61133.

44 Shlesinger, M.F. and Klafter, J. (1986) in *On Growth and Form* (eds H.E. Stanley and N. Ostrowsky), Nijhoff, Dordrecht.

45 Russ, S., Zschiegner, S., Bunde, A., and Kärger, J. (2005) *Phys. Rev. E*, **72**, 030101-1-4.

46 Bunde, A., Kärger, J., Russ, S., and Zschiegner, S. (2005) in *Diffusion Fundamentals* (eds J. Kärger, F. Grinberg, and P. Heitjans), Leipziger Universitätsverlag, Leipzig, p. 68.

47 Zschiegner, S., Russ, S., Bunde, A., and Karger, J. (2007) *EPL*, **78**, 20001.

48 Kärger, J. (2008) in *Adsorption and Diffusion* (eds H.G. Karge and J. Weitkamp), Springer, Berlin, Heidelberg, p. 329.

49 Mandelbrot, B.B. (1982) *The Fractal Geometry of Nature*, Freeman, San Francisco, CA.

50 Pfeifer, P. (1985) *Chimia*, **39**, 120.

51 Bunde, A. and Havlin, S. (1996) *Fractals and Disordered Systems*, Springer, Berlin.

52 Coppens, M.O. (2003) *Nature Inspired Chemical Engineering (Inaugural Lecture)*, Delft University Press, Delft.

53 Coppens, M.O. (2001) *Colloids Surf. A*, **257**, 187–188.

54 Havlin, S. and Ben-Avraham, D. (1987) *Adv. Phys.*, **36**, 695.

55 Spindler, H. and Vojta, G. (1988) *Z. Chem.*, **28**, 421.

56 Spindler, H., Ackermann, W.-G., and Kraft, M. (1988) *Z. Phys. Chem. (Leipzig)*, **269**, 1233.

57 Given, J.A. and Mandelbrot, B.B. (1983) *J. Phys. A*, **16**, L565.

58 Dvoyashkin, M., Valiullin, M., and Kärger, J. (2007) *Phys. Rev. E*, **75**, 41202.

59 Kärger, J. and Spindler, H. (1991) *J. Am. Chem. Soc.*, **113**, 7571.

3
Diffusion and Non-equilibrium Thermodynamics

Classical thermodynamics deals only with equilibrium states and any transition between different states must be considered to proceed through a sequence of equilibrium states. Clearly, such a sequence may be passed in either direction and it is therefore commonly called a "reversible path." Examples of such reversible processes include freezing/melting or adsorption/desorption under conditions of phase equilibrium between the solid/liquid or adsorbed/gaseous phases. By contrast the processes involved in diffusion phenomena are obviously irreversible and occur in only one direction. Such processes cannot be described within the framework of classical thermodynamics. The science of irreversible thermodynamics has developed from the early attempts of Onsager and others to apply thermodynamic reasoning to non-equilibrium processes.

3.1
Generalized Forces and Fluxes

The novel feature that has to be accounted for in any attempt to extend thermodynamic reasoning to non-equilibrium processes is the time dependence of the intensive variables such as temperature and concentration. The basic postulate of irreversible thermodynamics is that the generalized fluxes (of heat and matter) are linearly related to the relevant thermodynamic driving forces. These terms are drawn from the language of classical mechanics and, for a better understanding of their meaning, it is useful to consider first a simple mechanical example.

3.1.1
Mechanical Example

We consider a body moving in a viscous medium under the influence of gravity. We may regard the potential energy as the fundamental quantity that determines the development of the system since, in the absence of additional external forces, any movement must be such as to reduce the potential energy. The stable state of the system is therefore characterized by minimum potential energy. Neglecting, for

Diffusion in Nanoporous Materials. Jörg Kärger, Douglas M. Ruthven, and Doros N. Theodorou.
© 2012 Wiley-VCH Verlag GmbH & Co. KGaA. Published 2012 by Wiley-VCH Verlag GmbH & Co. KGaA.

simplicity, the density of the fluid in comparison with that of the solid body, the potential energy is given by:

$$u_p(z) = u_o + F_g z \tag{3.1}$$

in which z represents the coordinate in the vertical direction and u_o is a constant representing the potential energy at the reference position ($z=0$). The gravitational force F_g is given by the product of the mass and the gravitational acceleration g.

The body will accelerate under the influence of the gravitational force until it reaches a limiting velocity (v_∞) at which this force is precisely balanced by the frictional (viscous) drag exerted on the body by the fluid. According to Stokes' law this limiting velocity is directly proportional to the force F_g:

$$v_\infty \propto F_g \tag{3.2}$$

A second relationship between v_∞ and F_g may also be derived directly by differentiation of Eq. (3.1). In the long time limit:

$$\frac{du_p}{dt} = -F_g v_\infty \tag{3.3}$$

where we have put $\lim_{t\to\infty} dz/dt = -v_\infty$. The negative sign arises since the height (z) decreases with time.

3.1.2
Thermodynamic Forces and Fluxes

Equation (3.3) shows that the decrease in potential energy is equal to the product of the velocity and the driving force, while from Stokes' law the velocity (or for a set of particles, the flux) may be considered as a consequence of this same driving force. In non-equilibrium thermodynamics [1–3] this concept is transferred to the time development of a thermodynamic system that is considered to consist of such a large number of particles or subsystems that it can be treated as quasi-homogeneous and therefore describable by variables of state such as temperature and pressure. According to the second law of thermodynamics, the quantity that governs the time dependence of an isolated thermodynamic system is the entropy and this quantity is therefore analogous to potential energy in the mechanical system just considered. Only processes that lead to an increase in entropy are possible, so the necessary and sufficient condition for a stable state (in an isolated system) is that the entropy has reached its maximum value. By analogy with the mechanical example [Eq. (3.3)] it is postulated that entropy production due to internal processes may be represented in the form:

$$\frac{ds}{dt} = -\sum_{k=1}^{N} J_k X_k \tag{3.4}$$

where the quantities J and X are considered to be generalized fluxes and forces between which there exists a linear relationship of the type:

$$J_k = -\sum_{j=1}^{N} L_{kj} X_j \tag{3.5}$$

in which the coefficients L are the Onsager phenomenological coefficients. Equation (3.5) is the multidimensional equivalent of Stokes' law in the mechanical analogy. It is shown below that the generalized fluxes may be either the mass flows of different components or the flow of heat.

As it stands, Eq. (3.5) does not contain any specific information since it merely represents the simplest way in which the changes within a system may be related to the physical forces responsible for these changes. Familiar examples of relationships of this type are Fourier's law of heat conduction, Ohm's law, and Fick's first law. The special feature of Eq. (3.5) is that it includes cross-coefficients that take account of interaction effects such as the effect of a temperature gradient on the flux of matter or the effect of a concentration gradient of one component on the flux of another.

The matrix of the phenomenological coefficients turns out to be symmetric, which implies that:

$$L_{kj} = L_{jk} \tag{3.6}$$

This is known as the Onsager reciprocity relation and is derived by making use of the principle of microscopic reversibility. It has been confirmed by investigating the cross correlations between mass, heat, and charge transfer under conditions such that all generalized forces and fluxes can be measured experimentally.

3.1.3
Rate of Generation of Entropy

To obtain the set of generalized forces and fluxes, as defined by Eqs. (3.4) or (3.5), the time dependence of the entropy must be considered. To do this a second basic assumption is introduced, namely, that, for states not too far displaced from equilibrium, the entropy and the other extensive thermodynamic functions depend on the state variables as if the whole system were at equilibrium. Accordingly, even under non-equilibrium conditions, the change in entropy within a small differential volume element containing different molecular species may be written, according to the second law of thermodynamics, in the form:

$$dS = \frac{dU}{T} + \frac{p\,dV}{T} - \sum_j \frac{\mu_j}{T} dn_j \tag{3.7}$$

The quantities U, V, T, ϱ, μ_j, and n_j have their usual thermodynamic significance, denoting, respectively, the internal energy, volume, temperature, pressure, chemical potential, and numbers of molecules of the various components. For an adsorption system, the volume of the system, which is determined by the specific volume of the porous solid, may be considered to be constant, so Eq. (3.7) reduces to:

$$ds = \frac{du}{T} - \sum \frac{\mu_j}{T} dc_j \tag{3.8}$$

in which s and u denote specific values of S and U per unit volume of the system.

Integrating over the system volume, the total change in entropy per unit time is given by:

$$\sigma = \int_v \frac{ds}{dt} dV \equiv \int_v \left[\frac{1}{T} \frac{du}{dt} - \sum_j \frac{\mu_j}{T} \frac{dc_j}{dt} \right] dV \tag{3.9}$$

The implications of this expression can be seen most easily for a unidimensional system, where the integration extends over only one coordinate. The same logic can, however, be applied in vector form to yield the general result for a three-dimensional system.

In one dimension the conservation equations for matter and heat are:

$$\frac{dc_i}{dt} = -\frac{dJ_i}{dz}, \quad \frac{du}{dt} = -\frac{dQ}{dz} \tag{3.10}$$

in which J_i represents the fluxes (or more strictly the flux densities) of the various components and Q represents the heat flux in the z direction:

$$\sigma = \int_v \left[-\frac{1}{T} \frac{dQ}{dz} + \sum_i \frac{\mu_i}{T} \frac{dJ_i}{dz} \right] dz \tag{3.11}$$

But:

$$\frac{d}{dz}\left(\frac{Q}{T}\right) = Q \frac{d}{dz}\left(\frac{1}{T}\right) + \frac{1}{T}\frac{dQ}{dz} \tag{3.12}$$

and:

$$\frac{d}{dz}\left(\frac{\mu J}{T}\right) = \left(\frac{\mu}{T}\right)\frac{dJ}{dz} + J \frac{d}{dz}\left(\frac{\mu}{T}\right) \tag{3.13}$$

Equation (3.11) can therefore be written as:

$$\sigma = -\int_v \frac{d}{dz}\left[\frac{Q}{T} - \sum_i \frac{\mu_i J_i}{T} \right] dz + \int_v \left[Q \frac{d}{dz}\left(\frac{1}{T}\right) - \sum_i J_i \frac{d}{dz}\left(\frac{\mu_i}{T}\right) \right] dz \tag{3.14}$$

The first term on the right-hand side represents the difference of the term within the square brackets between the (z) planes, defining the extent of the system in the direction of the flux. This represents simply the change in entropy arising from the fluxes of heat and matter in and out of the system while the second term represents the entropy production by internal processes:

$$\sigma_{internal} = \int \left(\frac{ds}{dt}\right)_{internal} dz = \int_v \left[Q \frac{d}{dz}\left(\frac{1}{T}\right) - \sum_i J_i \frac{d}{dz}\left(\frac{\mu_i}{T}\right) \right] dz \tag{3.15}$$

$$\left(\frac{ds}{dt}\right)_{\text{internal}} = Q\frac{d}{dz}\left(\frac{1}{T}\right) - \sum_i J_i \frac{d}{dz}\left(\frac{\mu_i}{T}\right) \quad (3.16)$$

The general form is:

$$\left(\frac{ds}{dt}\right)_{\text{internal}} = Q\,\text{grad}\left(\frac{1}{T}\right) - \sum_i J_i \,\text{grad}\left(\frac{\mu_i}{T}\right) \quad (3.17)$$

Comparison with Eq. (3.4) yields for the generalized thermodynamic forces of heat and mass transfer:

$$\begin{aligned} X_Q &= -\text{grad}\left(\frac{1}{T}\right) = +\frac{1}{T^2}\text{grad }T \\ X_{J_i} &= +\text{grad}\left(\frac{\mu_i}{T}\right) = -\frac{\mu_i}{T^2}\text{grad }T + \frac{1}{T}\text{grad }\mu_i \end{aligned} \quad (3.18)$$

Thus, for single-component diffusion in an adsorbent under conditions of non-uniform temperature we would have:

$$\begin{aligned} J &= -L_{\text{matter}}\,\text{grad}\left(\frac{\mu}{T}\right) - L_{\text{cross}}\frac{1}{T^2}\text{grad }T \\ Q &= -L_{\text{cross}}\,\text{grad}\left(\frac{\mu}{T}\right) - L_{\text{heat}}\frac{1}{T^2}\text{grad }T \end{aligned} \quad (3.19)$$

in which, because of the reciprocity relation, we have retained only one cross coefficient.

These equations show the coupling between the mass and heat fluxes in a non-isothermal system. They are the starting point for the analysis of "thermal diffusion" in which a mass flux is driven by a temperature gradient. This effect has been utilized for isotope separation, for example, in the Clusius–Dickel column [4]. However, the thermal diffusion coefficient, which is directly proportional to the cross-coefficient, is very small, and, consequently, large temperature gradients with many theoretical stages are needed to achieve a useful separation.

3.1.4
Isothermal Approximation

Because of this coupling the analysis of diffusion, even of a single component, in a non-isothermal system is clearly complicated. However, we are primarily concerned with diffusion in a porous solid. If the diffusion of heat is substantially faster than the diffusion of matter the adsorbent may be considered to be isothermal with substantial simplification of the flux equations. Such an approximation appears to be generally valid for zeolite adsorbents for which the thermal diffusivity ($\lambda/\varrho C_p$) is of order

$10^{-7}\,\mathrm{m^2\,s^{-1}}$ [5], compared with intracrystalline diffusivities, which are generally smaller than $10^{-9}\,\mathrm{m^2\,s^{-1}}$.

Under isothermal conditions the flux for each component is given by:

$$J_i = -\sum_k L_{ik}\,\mathrm{grad}\,\mu_k \qquad (3.20)$$

in which the factor $1/T$ has been incorporated into the phenomenological coefficients. For diffusion of a single component in a porous adsorbent:

$$J = -L\frac{\partial \mu}{\partial z} \qquad (3.21)$$

where $L = D_o q/RT$. This corresponds to Eq. (1.24), which reduces, as shown in Section 1.2 to:

$$J = -D\frac{\partial q}{\partial z}; \quad D = D_o\frac{d\ln p}{d\ln q} \qquad (1.31)$$

Extension to a multicomponent (isothermal) system follows naturally:

$$J_i = -\sum_j D_{ij}\,\mathrm{grad}\,q_j; \quad D_{ij} = RT\sum_k L_{ik}\frac{\partial \ln p_k}{\partial q_j}(q_i\ldots q_n) \qquad (3.22)$$

This is discussed further in Section 3.3. In contrast to the phenomenological coefficients, the coefficients D_{ij} do not necessarily form a symmetric matrix.

3.1.5
Diffusion in a Binary Adsorbed Phase

For a binary adsorbed phase (components A and B):

$$\begin{aligned}J_A &= -L_{AA}\,\mathrm{grad}\,\mu_A - L_{AB}\,\mathrm{grad}\,\mu_B \\ J_B &= -L_{BB}\,\mathrm{grad}\,\mu_B - L_{AB}\,\mathrm{grad}\,\mu_A\end{aligned} \qquad (3.23)$$

In direct analogy with Eq. (3.19) these equations account for the effect of component B on the flux of component A (and vice versa) and, because of the reciprocity relation, they contain only one cross coefficient. The equations are coupled directly through the cross terms and indirectly through the effect of component B on the chemical potential of A. The latter effect is generally more important.

Krishna and van Baten [19] have shown that the ratio $L_{AB}/\sqrt{(L_{AA}\,L_{BB})}$ is bounded between 0 and 1. The zero value corresponding to negligible cross terms (the Habgood model) [23], which is generally a good approximation for cage-type zeolites such as LTA, CHA, ERI, and DDR, reduces the expression for the diffusivities [Eq. (3.22)] to:

$$D_{ij} = RTL_i \cdot \frac{\partial \ln p_i}{\partial q_j} \qquad (3.24)$$

The other limit $[L_{AB}/\sqrt{(L_{AA}\,L_{BB})} \to 1]$ represents a system in which correlation effects are dominant, such as one-dimensional transport in carbon nanotubes.

3.2
Self-Diffusion and Diffusive Transport

To obtain a relationship between the coefficients of transport and self-diffusion we may apply the formalism of irreversible thermodynamics to the situation represented in Figure 1.1 [6, 7]. For self-diffusion (Figure 1.1b) we consider the inter-diffusion of two identical species (A^* and A, representing the marked and unmarked molecules having identical mobilities). The phenomenological equations have the form of Eq. (3.23) with B replaced by A^*. The total concentration is uniform:

$$(c = c_A + c_{A^*}), \quad J_A = -J_{A^*}, \quad \text{grad } c_A = -\text{grad } c_{A^*}, \quad \text{grad } \mu_i = \frac{RT}{c_i} \text{ grad } c_i$$

so that:

$$J_A = RT \left\{ \frac{L_x}{c_{A^*}} - \frac{L(c_A, c_{A^*})}{c_A} \right\} \text{grad } c_A$$

$$J_{A^*} = RT \left\{ \frac{L_x}{c_A} - \frac{L(c_{A^*}, c_A)}{c_{A^*}} \right\} \text{grad } c_{A^*}$$

(3.25)

and, by comparison with Eq. (1.11), we see that the self-diffusivity is given by:

$$\mathcal{D}_A = RT \left\{ \frac{L(c_A, c_{A^*})}{c_A} - \frac{L_x}{c_{A^*}} \right\}$$

$$\mathcal{D}_{A^*} = RT \left\{ \frac{L(c_{A^*}, c_A)}{c_{A^*}} - \frac{L_x}{c_A} \right\}$$

(3.26)

These expressions must be identical since the diffusivity cannot depend on the labeling. From this identity we can immediately derive the relationship between the coefficients:

$$L_x = \frac{L(c_A, c_{A^*}) c_{A^*} - L(c_{A^*}, c_A) c_A}{c_A - c_{A^*}}$$

(3.27)

For transport diffusion (Figure 1.1a) the flux depends only on the chemical potential gradient of the species under consideration, and so we may write:

$$\mathbf{J} = -L(c, 0) RT \frac{d \ln p}{dc} \text{ grad } c$$

(3.28)

where $L(c,0)$ denotes the straight coefficient for the A–A^* system in which the concentration of A^* is reduced to zero. Comparison with Eq. (1.1) yields for the transport diffusivity:

$$D = RT L(c, 0) \frac{d \ln p}{dc} = RT \frac{L(c, 0)}{c} \frac{d \ln p}{d \ln c}$$

(3.29)

Comparison between the coefficients of diffusion and self-diffusion is complicated by the existence of at least three different phenomenological coefficients in Eqs. (3.26) and (3.29). However, these coefficients are not all independent. A relationship between them may be established by again considering a system of identical marked and unmarked molecules but, this time, in the presence of a gradient of total concentration:

$$\text{grad } c = \text{grad } c_A + \text{grad } c_{A^*} \tag{3.30}$$

Following this approach Kärger [7] has shown that the relationship between the self and corrected transport diffusivities must have the general form:

$$\frac{D_0}{\mathcal{D}} = \frac{D\,d\ln c}{\mathcal{D}\,d\ln p} = 1 + c\beta(c)/\alpha(c) \tag{3.31}$$

where D_0 is the corrected diffusivity [defined by Eq. (1.31)] and α and β are functions only of total concentration.

This expression has the correct asymptotic form since it shows that, in the low concentration limit ($c \to 0$), $D_0 \to \mathcal{D}$ as expected. Since $\beta(c)$ is proportional to the cross coefficient while $\alpha(c)$ is proportional to the straight coefficient it also shows that the approximation $D_0 \to \mathcal{D}$ will hold even at higher concentrations when the cross coefficient is small.

There is one further comment on the form of Eq. (3.31) that is possible on the basis of irreversible thermodynamics. Since the entropy production by internal processes must always be positive we have from Eq. (3.4):

$$\sum_{k=1}^{N} J_k X_k < 0 \tag{3.32}$$

and with Eq. (3.5), for the two components under consideration, this leads to:

$$L_{AA} X_A^2 + L_{AA^*} X_A X_{A^*} + L_{A^*A} X_A X_{A^*} + L_{A^*A^*} X_{A^*}^2 > 0 \tag{3.33}$$

which must evidently be valid for *any* choice of the generalized forces. Putting alternately X_A and X_{A^*} equal to zero it follows that both L_{AA} and $L_{A^*}L_{A^*}$ must be greater than zero. The necessary and sufficient condition for inequality (3.33) to be satisfied is:

$$(L_{AA^*} + L_{A^*A})^2 < 4 L_{AA} L_{A^*A^*} \tag{3.34}$$

which may be expressed in terms of the functions α and β as:

$$4\beta^2 c_A^2 c_{A^*}^2 < 4(\alpha c_A + \beta c_A^2)(\alpha c_{A^*} + \beta c_{A^*}^2) \tag{3.35}$$

or:

$$0 < 1 + \frac{\beta c}{\alpha} \tag{3.36}$$

As may be seen from Eq. (3.31) this relation expresses only the fact that the transport diffusivity (D) must always be a positive quantity as long as $d\ln p/d\ln c$ is positive. By the same reasoning, the corrected diffusivity (D_0) is seen to be always positive since it follows from Einstein's relation [Eq. (1.10)] that the self-diffusivity is always positive. The possibility of a negative diffusivity arises when $d\ln p/d\ln c$ is negative, as in the situation when phase separation occurs. This result corresponds to the situation discussed in Section 2.5. Within the framework of irreversible thermodynamics, any further specification of the term $\beta(c)/\alpha(c)$ is impossible and further conclusions can therefore be drawn only from experiment or from an appropriate microdynamic model.

With the approximation of negligible cross coefficients the diffusivities in a multicomponent system may be related directly to the self-diffusivities of the individual components. The argument needed to derive the relationship follows closely the derivation of Eq. (3.34). In the case of a multicomponent system the transport diffusivities are given by Eq. (3.24), while the self-diffusivities are given by $\mathcal{D}_i = RTL_i/q_i$, which is the equivalent of Eq. (3.26) with the cross coefficient term neglected. The required relationship between the self- and transport diffusivities, for any component in the mixture, is therefore:

$$\frac{D_{ij}}{\mathcal{D}_i} = \frac{q_i \, \partial \ln p_i}{q_j \, \partial \ln q_j} \tag{3.37}$$

For a binary Langmuirian system this yields:

$$\frac{D_{AA}}{\mathcal{D}_A} = \frac{1-\theta_B}{1-\theta_A-\theta_B}, \quad \frac{D_{AB}}{\mathcal{D}_A} = \frac{\theta_A}{1-\theta_A-\theta_B} \tag{3.38}$$

with the corresponding expressions for D_{BB} and D_{BA}. It may be noted that the "cross" diffusivities D_{AB} and D_{BA} are not identical but this is to be expected since the Onsager reciprocity relation applies only between the phenomenological coefficients and not between the resulting diffusivities.

Equation (3.38) yields the expected result that for sufficiently small concentrations the "straight" diffusivities approach the self-diffusivities while the "cross" diffusivities become vanishingly small. Subtracting the two expressions it may be seen that the difference between the straight and cross diffusivities is simply the self-diffusivity.

3.3
Generalized Maxwell–Stefan Equations

3.3.1
General Formulation

In recent years, largely as a result of the work of Krishna and his associates [8–22], the Maxwell–Stefan equations have attracted attention as an alternative to the traditional irreversible thermodynamic formulation. Both approaches are formally equivalent but the Maxwell–Stefan formulation has the important advantage that the rate parameters are simply related to directly measurable quantities, thus making this approach more suitable for use in the correlation of experimental data and predictive models. Reference [22] provides a useful review of the application of the Maxwell–Stefan model to diffusion in an adsorbed phase and detailed comparisons with the simple Habgood model [23].

The Maxwell–Stefan model, which was originally developed to describe diffusion in a homogeneous gas or liquid phase, considers the diffusion coefficients as inverse drag coefficients representing the interchange of momentum between the different types of molecules. The general form of the equation for diffusion in a multicomponent mixture is:

$$\frac{c_i}{RT}\frac{\partial \mu_i}{\partial z} = \sum_{j \neq i} \frac{c_i J_j - c_j J_i}{c \, \text{Ð}_{ij}} \qquad (3.39)$$

in which $J_i = c_i u_i$ denotes the diffusive flux of the i-th component and Ð_{ij} represents the Stefan–Maxwell diffusivities and c the total concentration. For an ideal vapor (or liquid) phase the left-hand side of this equation reduces to $\partial c_i/\partial z$, so for equimolar counter-diffusion in a binary system:

$$N_i = J_i = -J_j = -\text{Ð}_{ij}\frac{\partial c_i}{\partial z} \qquad (3.40)$$

which is identical with the usual Fickian formulation with $\text{Ð}_{ij} = D_{ij}$. The corresponding expression for a non-ideal fluid phase (thermodynamic activity a_i) is:

$$J_i = -\text{Ð}_{ij}\frac{d \ln a_i}{d \ln c_i}\frac{\partial c_i}{\partial z} \qquad (3.41)$$

which corresponds to Eq. (1.29).

3.3.2
Diffusion in an Adsorbed Phase

As noted in Section 1.1, in a microporous adsorbent there is no convective flow and the diffusive flux is measured relative to the fixed frame of reference defined by the pore structure rather than with respect to the plane of no net volumetric flow. With this definition $J_i = N_i$. To extend the Stefan–Maxwell model to diffusion in an adsorbed phase Krishna considers the "vacancies" (or the free space within the pores) as an additional component. It is also convenient to express the concentrations in terms of fractional occupancies (θ_i) defined by $\theta_i = q_i/q_s$.

For diffusion of a single adsorbed species (A) through a stationary adsorbent (denoted by v – vacancies) Eq. (3.39) becomes:

$$\frac{-q_A}{RT}\frac{\partial \mu_A}{\partial z} = \frac{\theta_v N_A - \theta_A N_v}{\text{Ð}_{Av}} \qquad (3.42)$$

Since $N_v = 0$ this expression is equivalent to the simple thermodynamic transport equation [see Eqs. 1.24–1.27] with $D_{0A} = \text{Ð}_{Av}/\theta_v = \text{Ð}_{Av}/(1-\theta_A)$.

For diffusion in a binary adsorbed phase we have three components (A, B, and v) so with $J_i = N_i$, $\text{Ð}_{iv} = \theta_v D_{0i}$ and $N_v = 0$, Eq. (3.39) becomes:

$$\frac{-q_A}{RT}\frac{\partial \mu_A}{\partial z} = \frac{\theta_B N_A - \theta_A N_B}{\text{Ð}_{AB}} + \frac{N_A}{D_{0A}}$$

$$\frac{-q_B}{RT}\frac{\partial \mu_B}{\partial z} = \frac{\theta_A N_B - \theta_B N_A}{\text{Ð}_{AB}} + \frac{N_B}{D_{0B}} \qquad (3.43)$$

The general form for a multicomponent system is:

$$-\frac{q_i}{RT}\nabla\mu_i = \sum_{j=1, j\neq i}^{n} \frac{\theta_j N_i - \theta_i N_j}{\mathcal{D}_{ij}} + \frac{N_i}{D_{oi}} \quad (3.44)$$

Evidently, these expressions contain two different types of diffusivities; D_{oA} and D_{oB} represent the hindrance due to interaction with the pore walls and $\mathcal{D}_{AB}$ represents the drag due to interaction between the diffusing molecules. $\mathcal{D}_{AB}$ captures the same effect as L_{cross} in the Onsager formulation. If $\mathcal{D}_{AB} \to \infty$ Eq. (3.43) reverts to the simple thermodynamic model with zero cross coefficients [Eq. (1.24)].

With some algebraic rearrangement Eq. (3.43) may be put in the form of Eq. (3.23):

$$N_A = J_A = -\frac{q_A}{RT}\frac{D_{oA}(\mathcal{D}_{AB} + \theta_A D_{oB})}{[\mathcal{D}_{AB} + \theta_A D_{oB} + \theta_B D_{oA}]}\mathrm{grad}\,\mu_A - \frac{q_B}{RT}\frac{\theta_A D_{oA} D_{oB}}{[\mathcal{D}_{AB} + \theta_A D_{oB} + \theta_B D_{oA}]}\mathrm{grad}\,\mu_B$$

$$N_B = J_B = -\frac{q_B}{RT}\frac{D_{oB}(\mathcal{D}_{AB} + \theta_B D_{oA})}{[\mathcal{D}_{AB} + \theta_A D_{oB} + \theta_B D_{oA}]}\mathrm{grad}\,\mu_B - \frac{q_A}{RT}\frac{\theta_B D_{oA} D_{oB}}{[\mathcal{D}_{AB} + \theta_A D_{oB} + \theta_B D_{oA}]}\mathrm{grad}\,\mu_B$$

$$(3.45)$$

thus giving the formal relationship between the phenomenological coefficients and the Stefan–Maxwell diffusivities:

$$\frac{RTL_{AA}}{q_A} = \frac{D_{oA}(\mathcal{D}_{AB} + \theta_A D_{oB})}{\mathcal{D}_{AB} + \theta_A D_{oB} + \theta_B D_{oA}}$$

$$\frac{RTL_{BB}}{q_B} = \frac{D_{oB}(\mathcal{D}_{AB} + \theta_B D_{oA})}{\mathcal{D}_{AB} + \theta_A D_{oB} + \theta_B D_{oA}}$$

$$\frac{RTL_{AB}}{q_B} = \frac{\theta_A D_{oA} D_{oB}}{\mathcal{D}_{AB} + \theta_A D_{oB} + \theta_B D_{oA}}$$

$$\frac{RTL_{BA}}{q_A} = \frac{\theta_B D_{oA} D_{oB}}{\mathcal{D}_{AB} + \theta_A D_{oB} + \theta_B D_{oA}}$$

$$(3.46)$$

Note that, in accordance with the Onsager reciprocity principle, $L_{AB} = L_{BA}$. In this formulation we have assumed $\mathcal{D}_{AB} = \mathcal{D}_{BA}$, which is generally true for a fluid phase and for an adsorbed phase when the saturation capacity is constant ($q_{sA} = q_{sB}$). A more general form of the reciprocity relation that is valid for unequal saturation capacities is given by Eq. (8.25) (Section 8.1.3). The corresponding formulation in terms of the Fickian diffusivities follows directly from the combination of Eq. (3.46) with Eq. (3.22).

For an ideal Langmuir system the thermodynamic factors are given by:

$$\theta_A\frac{\partial \ln p_A}{\partial \theta_A} = \frac{1-\theta_B}{1-\theta_A-\theta_B},\quad \theta_A\frac{\partial \ln p_A}{\partial \theta_B} = \frac{\theta_A}{1-\theta_A-\theta_B}$$

$$\theta_B\frac{\partial \ln p_B}{\partial \theta_A} = \frac{\theta_B}{1-\theta_A-\theta_B},\quad \theta_B\frac{\partial \ln p_A}{\partial \theta_B} = \frac{1-\theta_A}{1-\theta_A-\theta_B}$$

$$(3.47)$$

yielding for the flux equations [9, 10]:

$$\frac{N_A}{q_s} = -\frac{D_{oA}}{(1-\theta_A-\theta_B)} \cdot \frac{[(1-\theta_B)Đ_{AB}+\theta_A D_{oB}]\frac{\partial \theta_A}{\partial z} + \theta_A[Đ_{AB}+D_{oB}]\frac{\partial \theta_B}{\partial z}}{(Đ_{AB}+\theta_A D_{oA}+\theta_B D_{oB})}$$

$$\frac{N_B}{q_s} = -\frac{D_{oB}}{(1-\theta_A-\theta_B)} \cdot \frac{[(1-\theta_A)Đ_{AB}+\theta_B D_{oA}]\frac{\partial \theta_B}{\partial z} + \theta_B[Đ_{AB}+D_{oA}]\frac{\partial \theta_A}{\partial z}}{(Đ_{AB}+\theta_A D_{oA}+\theta_B D_{oB})}$$

(3.48)

If $Đ_{AB} \to \infty$ these equations reduce to the simplified form originally suggested by Habgood in 1958 [23]:

$$N_A = -\frac{D_{oA} q_s}{(1-\theta_A-\theta_B)}\left[(1-\theta_B)\frac{\partial \theta_A}{\partial z} + \theta_A \frac{\partial \theta_B}{\partial z}\right]$$

$$N_B = -\frac{D_{oB} q_s}{(1-\theta_A-\theta_B)}\left[(1-\theta_A)\frac{\partial \theta_B}{\partial z} + \theta_B \frac{\partial \theta_A}{\partial z}\right]$$

(3.49)

Figure 3.1 shows the variation with loading of the thermodynamic factors and the Fick and Maxwell–Stefan diffusivities, calculated from Eqs. (3.24), (3.46), and (3.47) for a Langmuirian system in which the corrected diffusivities are assumed to be independent of loading and $Đ_{ii} = D_{oi}$. Evidently, for such a system, the Onsager coefficients and the Fick diffusivities are strongly concentration dependent.

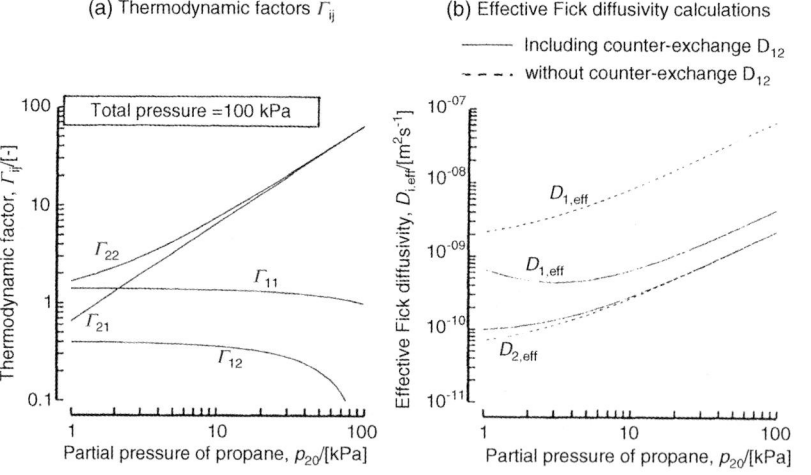

Figure 3.1 Variation of the thermodynamic factors and Fickian diffusivities with composition calculated according to Eqs. (3.24), (3.46), and (3.47) for a Langmuirian system in which the corrected diffusivities are constant. The system considered is methane (1)–propane (2) in silicalite at a total hydrocarbon pressure of one atmosphere at 303 K, as studied by Kapteijn, van de Graaf and Moulijn [24, 25]. The Langmuir parameters are $b_1 = 4 \times 10^{-6}\, Pa^{-1}$, $b_2 = 6.5 \times 10^{-4}\, Pa^{-1}$, and $q_s = 2.84\, mmol\, cm^{-3}$ (1.58 mmol g^{-1}), and the corrected single-component diffusivities are $D_{oA} \equiv Đ_1 = 1.04 \times 10^{-9}\, m^2\, s^{-1}$ and $D_{oB} \equiv Đ_2 = 3.4 \times 10^{-11}\, m^2\, s^{-1}$. The thermodynamic factors are defined by $\Gamma_{ij} = (q_i/q_j)(\partial \ln p_i/\partial \ln q_j)$. Reprinted from Keil *et al.* [9] with permission.

3.3.3
Relation between Self- and Transport Diffusivities

In the special case of tracer (or self-) diffusion we consider the two species A and B as identical in physical properties (A and A*) so $D_{0A} = D_{0A^*}$, $\partial \theta_A/\partial z = -\partial \theta_{A^*}/\partial z$, $N_A = -N_{A^*}$, $Ð_{AA} = Ð_{AA^*} = Ð_{A^*A^*}$ and the total loading is constant ($\theta = \theta_A + \theta_{A^*}$). Equation (3.43) thus reduces to:

$$N_A = -\frac{q_s \, \partial \theta_A/\partial z}{(1/D_{oA} + \theta/Ð_{AA})} = -N_{A^*} \tag{3.50}$$

or:

$$\frac{1}{\mathcal{D}_A} = \frac{1}{D_{0A}} + \frac{\theta}{Ð_{AA}} \tag{3.51}$$

which provides the relationship between the self-diffusivity and the corrected transport diffusivity as derived by Krishna and Paschek [12]. Like Eq. (3.21) this expression predicts that in general $\mathcal{D}_A < D_{0A}$ but $\mathcal{D}_A \to D_{0A}$ when $Ð_{AA} \to \infty$ (rapid self-exchange) or in the Henry's law region ($\theta \to 0$). Equation (3.51) may also be regarded as a definition of $Ð_{AA}$ and the basis for its experimental determination. Both $\mathcal{D}$ and D_0 are in general concentration dependent. Since the transport diffusivity (D_A) is related to D_{0A} through Eq. (1.31) one may expect that in general:

$$D_A > D_{0A} > \mathcal{D}_A \tag{3.52}$$

For diffusion of deuterium in the open pore system of NaX zeolite it is possible to determine both transport and self-diffusivity as well as the thermodynamic correction factor from neutron scattering measurements. Figure 3.2 presents the results of

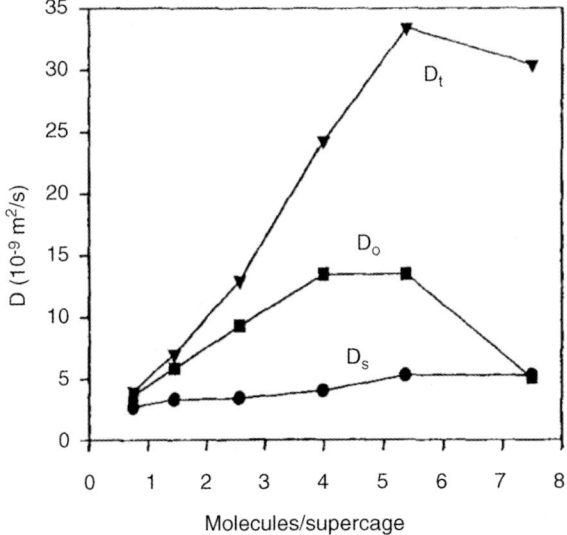

Figure 3.2 Loading dependence of Fickian diffusivity (D_t), corrected diffusivity (D_0) and self-diffusivity ($D_s \equiv \mathcal{D}$) for D_2 in NaX zeolite at 100 K measured by neutron scattering. Reprinted from Jobic, Kärger, and Bee [26] with permission.

such experiments, showing the loading dependence of D, D_0, and $\mathcal{D}$ [26]. The inequality expressed by Eq. (3.52) is evidently fulfilled at all loadings. For this system the self-diffusivity ($\mathcal{D}$) is almost independent of loading and the corrected diffusivity (D_0) passes through a maximum at intermediate loadings, approaching $\mathcal{D}$ at both low and high loadings. The variation of the Fickian diffusivity (D) is much stronger due to the strong loading dependence of the thermodynamic correction factor ($\partial \ln p / \partial \ln q$).

3.4
Application of the Maxwell–Stefan Model

3.4.1
Parameter Estimation

The Maxwell–Stefan model provides a useful theoretical framework for the analysis and interpretation of binary and multicomponent diffusion data but, more importantly, it allows an approximate prediction of binary or multicomponent kinetics from single-component data. The Maxwell–Stefan diffusivities for the individual components in a mixture (D_{0i}) are assumed to be the same as the single-component values measured at the same temperature and total fractional loading as for the mixture ($\theta = \Sigma \theta_i$). This assumption appears intuitively reasonable and its validity has been confirmed experimentally for several systems. However, it is not universally valid and, by molecular simulations, it has been shown to fail in systems such as CO_2–CH_4 in DDR where strong segregation effects lead to preferential occupation of the eight-ring windows by CO_2, thus obstructing the diffusion of CH_4 far more strongly than is predicted by the mutual diffusion effect [17, 30].

If binary kinetic data are available the mutual diffusivities ($\mathcal{D}_{ij}$) may be determined directly by fitting the measured fluxes to Eq. (3.43) using the values of D_{0A} and D_{0B} from the single-component data. Predicting the mutual diffusivities without recourse to binary measurements is less straightforward but still possible provided that both transport and self-diffusivity data are available for the single components. The self-exchange diffusivities ($\mathcal{D}_{ii}$) may be found from Eq. (3.51). To estimate the mutual diffusivities Skoulidas et al. [14] have suggested using a geometric mean approximation such as in the Vignes correlation [27] for diffusion in liquids:

$$\mathcal{D}_{AB} = (\mathcal{D}_{AA})^{\frac{\theta_A}{\theta_A + \theta_B}} \cdot (\mathcal{D}_{BB})^{\frac{\theta_B}{\theta_A + \theta_B}} = (\mathcal{D}_{AA})^{\frac{q_A}{q_A + q_B}} \cdot (\mathcal{D}_{BB})^{\frac{q_B}{q_A + q_B}} \quad (3.53)$$

or, when $q_{sA} \neq q_{sB}$:

$$q_{sA} \mathcal{D}_{BA} = q_{sB} \mathcal{D}_{AB} = [q_{sB} \mathcal{D}_{AA}]^{\frac{q_A}{q_A + q_B}} \cdot [q_{sA} \mathcal{D}_{BB}]^{\frac{q_B}{q_A + q_B}} \quad (3.54)$$

The validity of this approximation has been confirmed, for several systems, by MD simulations but not, as yet, by experiment. Using the values of $\mathcal{D}_{AA}$ and $\mathcal{D}_{BB}$ estimated from Eq. (3.51) the values of $\mathcal{D}_{AB}$ and $\mathcal{D}_{BA}$ may be estimated from Eq. (3.53) [or Eq. (3.54)].

In earlier studies it was assumed that these interpolation formulae could be applied with $Ð_{AA}$ and $Ð_{BB}$ replaced by the single-component corrected diffusivities (D_{0A}, D_{0B}) [7–9] but molecular simulations suggest that this is a good approximation only for small pore structures such as CHA, LTA, and DDR, which consist of cages separated by small (eight-ring) windows. For more open structures $Ð_{AA}$ may be much smaller than D_{0A} [19], implying that the resistance due to mutual diffusion can be comparable with or even larger than the resistance due to molecule–pore interactions.

More recently, Krishna and van Baten [21] have recommended a slightly different approach that offers several practical advantages. The Maxwell–Stefan equation is written in terms of the concentrations, expressed in terms of the accessible pore volume, and mole fractions in the adsorbed phase (q_A, q_B, x_A, x_B) rather than the fractional occupancies (θ_A, θ_B), yielding in place of Eq. (3.43):

$$\frac{-q_A}{RT}\frac{\partial \mu_A}{\partial z} = \frac{x_B N_A - x_A N_B}{Ð^p_{AB}} + \frac{N_A}{D_{0A}} \tag{3.55}$$

with a similar expression for component B. The mutual diffusivities in the two formulations are therefore related by:

$$Ð_{AB}\, q_s = Ð^P_{AB}(q_A + q_B) = Ð^P_{BA}(q_A + q_B) = Ð_{BA} q_s \tag{3.56}$$

A simple linear average based on fluid phase mole fractions is used to estimate the mutual diffusivity ($Ð^P_{AB}$) from single-component diffusivity data [Eq. (3.51) with $Ð_{ii}$ replaced by $\theta_i Ð^p_{ii}$]:

$$Ð^P_{AB} = Ð^P_{BA} = x_A Ð^P_{AA} + x_B Ð^P_{BB} \tag{3.57}$$

This approach has the advantage that it requires only single-component self- and corrected diffusivities (at the required loading) plus the binary equilibrium isotherm to predict the fluxes of both components in the binary system. It does not require knowledge of the saturation limit.

From a series of MD simulations Krishna and van Baten have shown that the mutual diffusivities ($Ð^P_{ij}$) are simply related to the corrected binary diffusivities in the free fluid at the same total concentration within the pores. The ratio $Ð^P_{A/B}/Ð_{AB\,fluid}$ is linearly related to the "degree of confinement," defined as the ratio of the kinetic diameter to the pore diameter. Thus, in the absence of sufficiently extensive transport and self-diffusivity data for the adsorbed species, an approximate estimate of the mutual diffusivities may be obtained from liquid phase self-diffusivity data using the recommended (structure dependent) factors [21]. This approach is especially useful for mesoporous materials since, for larger pores, $Ð^P_{AB} \approx Ð_{AB\,fluid}$.

3.4.2
Membrane Permeation

The Maxwell–Stefan model has been widely applied to membrane permeation where a suitable model is needed to predict the performance of a binary or multicomponent system from single-component measurements. Specific examples are discussed in

Chapter 20 but, based on past experience, a few general comments may be made here. The usefulness of this model depends on the complexity of the diffusional behavior of the relevant system. When mutual diffusion effects are negligible, as is often the case in small pore zeolite systems, and when the loading is sufficiently small that the corrected diffusivity can be taken as constant the Maxwell–Stefan model reduces to the Habgood model [or to a modified Habgood model in which the binary Langmuir isotherm is replaced by a more realistic model such as the ideal adsorbed solution theory (IAST)]. For such systems a reasonably accurate prediction of the binary performance can be obtained from the single-component diffusivities and the main challenge is to provide a sufficiently accurate representation of the binary equilibria, since uncertainty in the thermodynamic factors is often the major source of error. A modest increase in complexity arises when the loading level is higher so that the concentration dependence of the corrected diffusivity cannot be neglected (but the other approximations remain valid). A further modification of the Habgood model in which the concentration dependence of the corrected diffusivity is accounted for in terms of the Reed–Ehrlich model [28] has been shown to provide a good representation of the behavior of the CO_2–CH_4/SAPO-34 system [29] – see Figure 20.13.

For light hydrocarbons in silicalite mutual diffusion effects are significant but the equilibrium isotherm can be reasonably well represented by the binary Langmuir model and the corrected diffusivities are essentially constant. The permeation of ethane–methane and propane–methane mixtures through a silicalite membrane was successfully modeled using Eq. (3.48) with the mutual diffusivities estimated from the single component values according to the Vignes correlation [Eq. (3.53) with $\mathcal{D}_{AA}$ and $\mathcal{D}_{BB}$ replaced by the single component corrected diffusivities] [25] – see Figure 20.9. The effect of mutual diffusion is to reduce the flux of the faster species and to increase (slightly) the flux of the slower species in comparison with a non-interacting system. The effect on the slower species is, however, minor and so the flux of this species may be estimated reasonably well from the simple Habgood model.

Recent molecular simulations have cast some doubt on the generality of the Maxwell–Stefan approach. For example, the permeation of CO_2–CH_4 mixtures through a DDR membrane could not be properly modeled, even if the mutual diffusivities were taken as fitting parameters. This has been shown to be due to the preferential location of the CO_2 molecules in the eight-ring windows [17, 30]. The resulting reduction in the methane permeation in the mixture is not captured by the Maxwell–Stefan model. Nevertheless, from a practical perspective this model still provides the only reasonable approach to the prediction of diffusion in multicomponent systems from single-component data.

3.4.3
Diffusion in Macro- and Mesopores

For pore diameters greater than about 2 nm (i.e., for meso- and macropores) the diffusivities within the pore are essentially equal to those for the free fluid. The coefficients D_{0A} and D_{0B} represent the thermodynamically corrected single-component diffusivities of components A and B in the pore system, evaluated at the same loading as the total loading for the mixture. The formal definitions are therefore the same as for diffusion in a microporous solid although the diffusion mechanisms are quite different.

In macro- and mesopores the dominant mechanisms are Knudsen diffusion and viscous flow, with possibly some contribution from surface diffusion for significantly adsorbed species. The relative importance of these mechanisms depends on the pore diameter. When surface diffusion is insignificant a reasonable estimate of D_{0A} and D_{0B} can be made simply from the sum of the Knudsen and viscous flow contributions. Thus, provided that the thermodynamic and physical properties of the system are available it is possible to estimate all the diffusivities in Eq. (3.55) and hence to obtain a reasonable prediction of the fluxes at any defined composition. Notably, of course, the diffusivities estimated in this way refer to a straight cylindrical pore so they must be multiplied by the appropriate combination of porosity and tortuosity factor to obtain the effective values – see Section 4.2.

3.5
Loading Dependence of Self- and Transport Diffusivities

The complexity of the loading dependence of the diffusivities determines, to a large extent, the usefulness of the Maxwell–Stefan model for mixture diffusion. As this topic is addressed in detail in Chapters 15–19, only a few general comments are made here.

3.5.1
Self-Diffusivities

For mobile systems involving hydrogen-containing sorbates the PFG NMR technique provides a convenient and reliable way of measuring self-diffusion in an adsorbed phase and this approach has been used to establish the loading dependence of the self-diffusivity for many systems (see, for example, References [31–35]). Tracer exchange techniques, notably TZLC (tracer zero-length column), have also been widely applied [36–39]. A wide range of different patterns has been observed (Figure 3.3). The patterns shown in Figure 3.3a–e are commonly referred to as types I–V.

The data shown in Figure 3.4 for methanol–NaX [36] show both a well-defined maximum in self-diffusivity at intermediate loadings and excellent agreement between PFG NMR and TZLC data. However, for some other systems such as light alkanes in silicalite and NaX and aromatics in NaX the self-diffusivities determined by these techniques show substantial disagreement in both magnitude and trend [37, 38].

The general pattern appears to be that for small-pore systems in which mobility is restricted mainly by the steric hindrance of the windows the self-diffusivity shows a modestly increasing trend, whereas in more open larger-pore systems the opposite trend is observed. For highly polar molecules in cationic zeolites, for which sorbate–

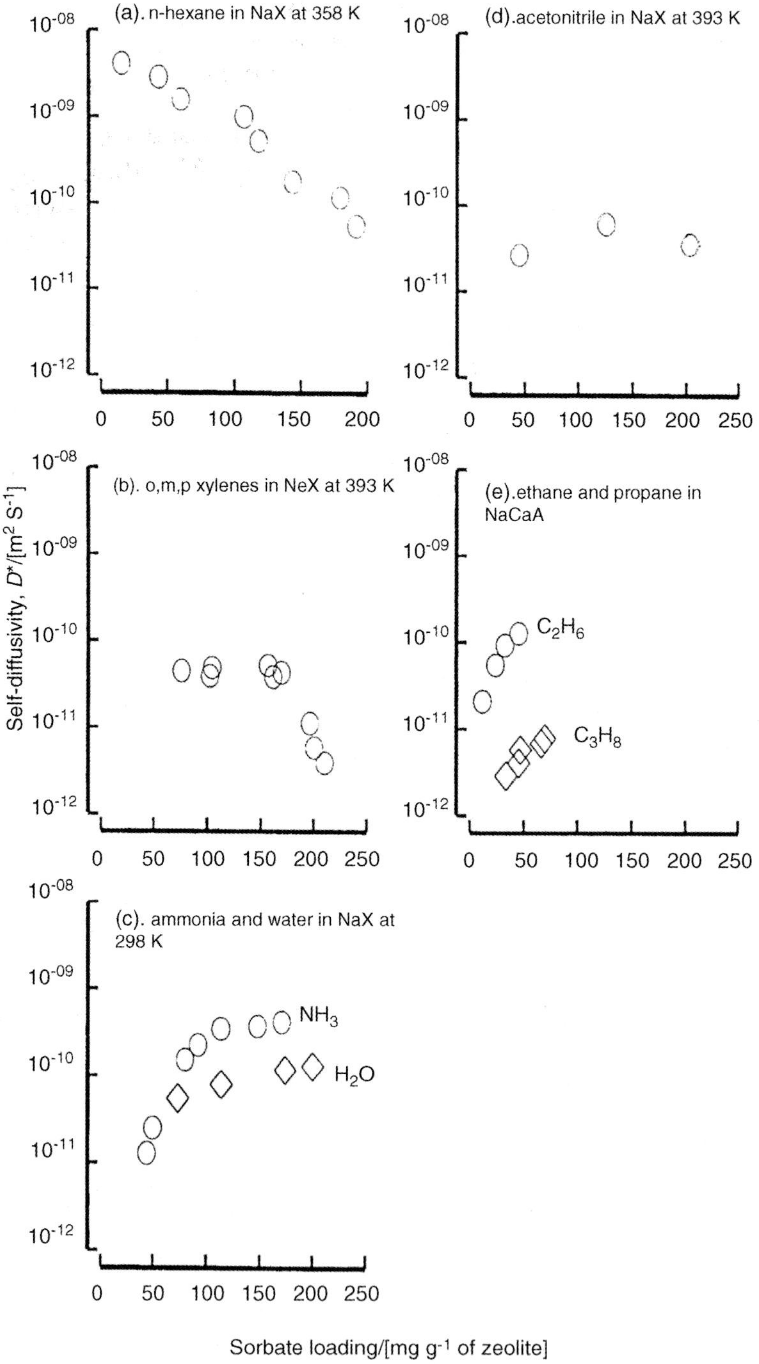

Figure 3.3 Loading dependence of self-diffusivities measured by PFG NMR showing the variety of different trends observed. Data of Kärger and Pfeifer [34]. From Keil *et al.* [9] with permission.

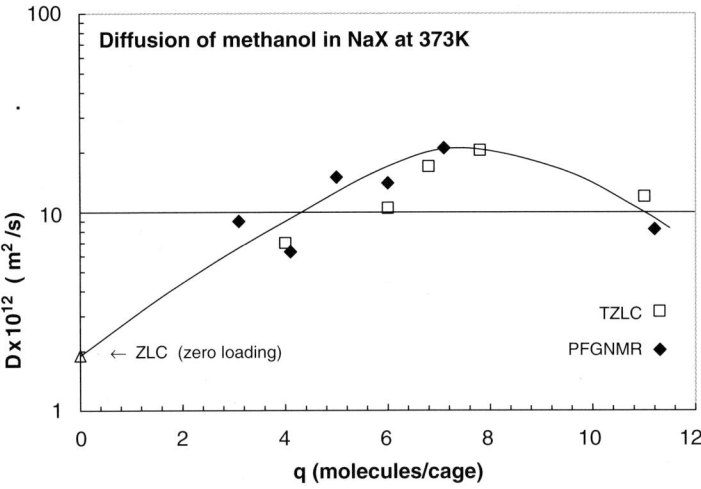

Figure 3.4 Self-diffusivity ($\mathcal{D}$) of methanol in NaX crystals at 100 °C; comparison of PFG NMR and tracer ZLC data. Note the extrapolation to the ZLC value of D_o at low loading. Plotted from data of Brandani et al. [36].

sorbent interactions are strong, the self-diffusivity increases at low loadings and may pass through a maximum, as observed for methanol–NaX (see Figure 3.4).

3.5.2
Transport Diffusivities

In general, outside the Henry's law region, the transport diffusivities increase strongly with loading due mainly to the dominant effect of the thermodynamic correction factor, which, for a favorable isotherm, increases strongly as saturation is approached. Of greater interest are the corrected diffusivities, calculated from the measured transport diffusivities in accordance with Eq. (1.31). The Reed–Ehrlich model (Section 4.3.2.2) suggests that, except in the saturation region, the loading dependence of the corrected diffusivity is much weaker than that of the Fickian diffusivity and, for a wide range of different systems (e.g., linear alkanes–5A [40–42], light gases–4A [40, 43], benzene–silicalite [44–47], and aromatics–NaX [39, 48–51]) the corrected diffusivities have been found experimentally to be approximately independent of loading. The system benzene–silicalite (Section 18.3.1) is particularly notable since the isotherm shows a well-defined inflexion, leading to a maximum in the thermodynamic factor and a corresponding maximum in the Fickian diffusivity at intermediate loadings, but the corrected diffusivity is still approximately constant (Figure 18.19) [45]. However, determination of the thermodynamic factor requires very accurate equilibrium isotherms and, especially at high loadings, this factor is subject to considerable uncertainty. A loading dependent corrected diffusivity will therefore be observed experimentally only when the variation is relatively strong.

For several systems (light alkanes–5A, TEA–NaX, and benzene–silicalite) [51–53] the self- and corrected transport diffusivities appear to be almost the same. Equation (3.51) implies that, for such systems, the mutual diffusivities ($Ð_{ij}$) must be relatively large so that the effect of mutual diffusion is not important. In such cases prediction of the diffusional behavior is greatly simplified, as noted above.

3.5.3
Molecular Simulation

Because of the difficulties inherent in the experimental measurements molecular simulations have been widely used to establish both the magnitudes and trends of self-diffusivities, corrected (transport) diffusivities, and mutual diffusivities ($\mathcal{D}$, D_0, and $Ð_{ij}$). This approach has the advantage that, once the force field and the interaction parameters are established, both the equilibrium isotherms (and hence the thermodynamic factors) and diffusivities may be accurately predicted. The mutual diffusivities are especially difficult to determine experimentally and the experimental values are subject to large uncertainties so this approach is inherently attractive. However, any error in the assumed force field or in the molecular parameters (especially the effective radius) can lead to very large errors in the predicted diffusivities and their trends with loading. It has recently been shown that the standard 6-12 potential used in most molecular simulations does not correctly capture the repulsive interaction, with the result that the predicted diffusivities, especially in small-pore systems, may be substantially larger than the true values [30] – see Section 16.2.2.

Figure 3.5 shows some examples of the trends predicted by MD simulations [18]. In general the small-pore systems at low loadings show approximately constant or modestly increasing trends of D_o and $\mathcal{D}$ with loading while the more open structures show decreasing trends in qualitative similarity with the available experimental data. At high loadings the calculated values of the corrected diffusivities always decrease dramatically, in accordance with the Reed–Ehrlich model (Section 4.3.2.2) [28]. This is generally not observed experimentally, possibly because the measurements seldom extend to sufficiently high fractional loadings. In general the predicted concentration dependences are stronger than those observed experimentally and approximately constant values of $\mathcal{D}$ and D_0, which are often observed experimentally, are predicted for only a few systems. However, in view of the inherent limitations of the MD calculations, such results, especially for small-pore zeolites, should be treated with caution, unless supported by experimental data.

3.5.4
Effect of Structural Defects

There is increasing evidence that experimentally measured intracrystalline diffusivities, especially in more open pore systems, often depend on the scale

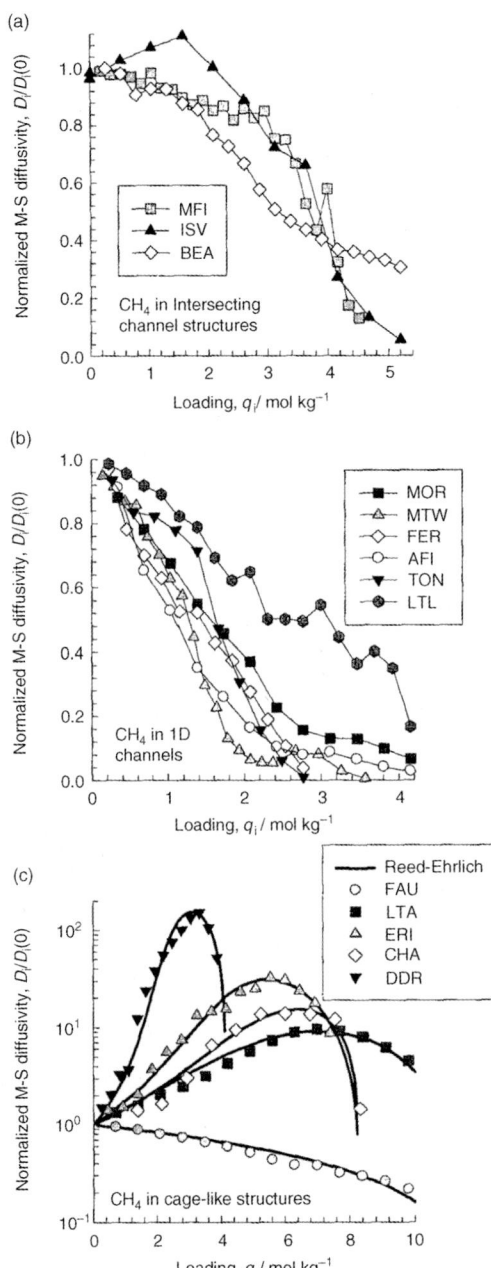

Figure 3.5 MD simulation results showing the loading dependence of the corrected diffusivities (D_o) for CH_4 in several different zeolite structures: (a) intersecting channels, (b) one-dimensional channels, and (c) cages separated by eight-ring windows. The lines in (c) are calculated according to the Reed–Ehrlich model [28] – see Figure 4.7. Reprinted from Krishna *et al.* [18] with permission.

of the measurement, as is discussed in greater detail in Chapters 17 and 18. This has been attributed to the effect of structural defects, which, especially in an open pore system, may reduce the diffusivities substantially below the values for an ideal crystal. Even if the problems associated with existing force fields can be overcome a molecular simulation will always reflect the transport behavior for an ideal structure. If, as now appears probable, structural defects have a major effect in real crystals, the practical relevance of such predictions becomes questionable.

3.6
Diffusion at High Loadings and in Liquid-Filled Pores

The modeling of pore diffusion at high loadings and in liquid systems presents particular difficulties. As an initial approach it would appear that the Maxwell–Stefan model or the simpler Habgood model [Eq. (3.49)] can be extended simply by setting the equilibrium pressures for all species equal to their saturation vapor pressures. However, it is well known that for most real systems the Langmuir model becomes increasingly inaccurate as saturation is approached. A more fundamental objection to the use of such models under conditions approaching full saturation is revealed by a recent study of Lettat et al. [54], which shows that, except when the molecular volumes of all components are the same, both the generalized Maxwell–Stefan model, as developed by Krishna, and the simpler Habgood model are thermodynamically incompatible with the assumption that the adsorbent is always fully saturated. To correctly represent the behavior of mixtures of different molecules under saturation conditions it is necessary to account for the convective flows within the micropores resulting from the exchange of molecules of different size.

Despite these fundamental objections a useful approximate treatment has been developed by Krishna and van Baten [22]. To see the simplicity of their approximation it is convenient to rewrite Eq. (3.55) in matrix form. For a binary (A,B) fluid:

$$N_A = -\Delta_{AA}\left(\frac{q_A}{RT}\right)\nabla\mu_A - \Delta_{AB}\left(\frac{q_B}{RT}\right)\nabla\mu_B \tag{3.58}$$

with an equivalent expression for N_B, where:

$$[\Delta] \equiv \begin{bmatrix} \Delta_{AA} & \Delta_{AB} \\ \Delta_{BA} & \Delta_{BB} \end{bmatrix} = \begin{bmatrix} \frac{1}{D_{0A}} + \frac{x_B}{\mathcal{D}_{AB}} & -\frac{x_A}{\mathcal{D}_{AB}} \\ -\frac{x_B}{\mathcal{D}_{AB}} & \frac{1}{D_{0B}} + \frac{x_A}{\mathcal{D}_{AB}} \end{bmatrix}^{-1} \tag{3.59}$$

In small pore systems at low loading, mutual diffusion effects are small ($Ð_{0i}/Ð_{ij} \to 0$) so $[\Delta]$ reduces to:

$$[\Delta] = \begin{bmatrix} \dfrac{1}{D_{0A}} & 0 \\ 0 & \dfrac{1}{D_{0B}} \end{bmatrix}^{-1} \tag{3.60}$$

which is the simple Habgood model [23] [Eq. (3.49)]. In the high loading limit where mutual diffusion effects are dominant Krishna and van Baten [22] show that:

$$\Delta_{AA} = \Delta_{AB} = \frac{1}{D_{0A}} + \frac{q_B}{q_A D_{0B}} \tag{3.61}$$

(with equivalent expressions for Δ_{BB}) so that the matrix becomes:

$$[\Delta] = \begin{bmatrix} \dfrac{1}{D_{0A}} + \dfrac{q_B}{q_A D_{0B}} & \dfrac{1}{D_{0A}} + \dfrac{q_B}{q_A D_{0B}} \\ \dfrac{q_A}{q_B D_{0A}} + \dfrac{1}{D_{0B}} & \dfrac{q_A}{q_B D_{0A}} + \dfrac{1}{D_{0B}} \end{bmatrix}^{-1} \tag{3.62}$$

and the flux expressions reduce to:

$$\begin{aligned} -N_A \left(\frac{1}{D_{0A}} + \frac{q_B}{q_A D_{0B}} \right) &= \left(\frac{q_A}{RT} \right) \nabla \mu_A + \left(\frac{q_B}{RT} \right) \nabla \mu_B \\ -N_B \left(\frac{q_A}{q_B D_{0A}} + \frac{1}{D_{0B}} \right) &= \left(\frac{q_A}{RT} \right) \nabla \mu_A + \left(\frac{q_B}{RT} \right) \nabla \mu_B \end{aligned} \tag{3.63}$$

These expressions contain only the thermodynamically corrected (Maxwell–Stefan) diffusivities for the single components (D_{0A} and D_{0B}) at the same fractional pore filling as for the mixture, thus eliminating the requirement to estimate the mutual diffusivities. Since D_{0A} and D_{0B} are amenable to direct experimental measurement this approach provides, at least in principle, a useful approximate approach to the prediction of pore diffusion at high loadings. However, notably, the justification for this simplified model rests only on molecular simulations and, as yet, it has not been verified experimentally.

References

1 Onsager, L. (1931) *Phys. Rev.*, **37**, 405; (1931) **38**, 2265.
2 De Groot, S.R. and Mazur, P. (1962) *Non-Equilibrium Thermodynamics*, Elsevier, Amsterdam.
3 Prigogine, I. (1967) *Thermodynamics of Irreversible Processes*, John Wiley & Sons, Inc., New York.
4 Clusius, K. and Dickel, G. (1939) *Z. Phys. Chem. B*, **44**, 397–450 and 451–473.
5 Meunier, F., Sun, L.M., Abdallah, K., and Grenier, Ph. (1991) in *Fundamentals of Adsorption* (eds A.B. Mersmann and S.E. Scholl), Engineering Foundation, New York, pp. 573–584.

6 Ash, R. and Barrer, R.M. (1967) *Surf. Sci.*, **8**, 461.
7 Kärger, J. (1973) *Surf. Sci.*, **36**, 797.
8 Krishna, R. and Wesselingh, J.A. (1990) *Chem. Eng. Sci.*, **45**, 1779–1791 and (1997) **52**, 861–911.
9 Keil, F.J., Krishna, R., and Coppens, M.-O. (2000) *Rev. Chem. Eng.*, **16**, 71–197.
10 Kapteijn, F., Moulijn, J.A., and Krishna, R. (2011) *Chem. Eng. Sci.*, **55**, 2923–2930.
11 Krishna, R. (2000) *Chem. Phys. Lett.*, **326**, 477–484.
12 Paschek, D. and Krishna, R. (2001) *Chem. Phys. Lett.*, **333**, 278–284.
13 Krishna, R. and Baur, B. (2003) *Sep. Purif. Technol.*, **33**, 213–254.
14 Skoulidas, A.J., Sholl, D.S., and Krishna, R. (2003) *Langmuir*, **19**, 7973–7988.
15 Krishna, R. and van Baten, J.M. (2005) *J. Phys. Chem. B*, **109**, 6386–6396.
16 Krishna, R. and van Baten, J.M. (2007) *Chem. Eng. Technol.*, **30**, 1235–1241.
17 Krishna, R. and van Baten, J.M. (2007) *Chem. Phys. Lett.*, **446**, 344–349.
18 Krishna, R., van Baten, J.M., Garcia-Perez, E., and Calero, S. (2007) *Ind. Eng. Chem. Res.*, **46**, 2974–2986.
19 Krishna, R. and van Baten, J.M. (2008) *Chem. Eng. Sci.*, **63**, 3120–3140.
20 Krishna, R. and van Baten, J.M. (2008) *Microporous Mesoporous Mater.*, **109**, 91–108.
21 Krishna, R. and van Baten, J.M. (2009) *Chem. Eng. Sci.*, **64**, 3159–3178.
22 Krishna, R. and van Baten, J.M. (2010) *J. Phys. Chem. C*, **114**, 11557–11563.
23 Habgood, H.W. et al. (1958) *Can. J. Chem.*, **36**, 1384–1397; Habgood, H.W. et al. (1966) *Sep. Sci.*, **1**, 219.
24 Kapteijn, F., van de Graaf, J., and Moulijn, J. (1988) *J. Mol. Catal. A: Chem.*, **134**, 201.
25 van de Graaf, J., Kapteijn, F., and Moulijn, J. (1999) *A.I.Ch.E. J.*, **45**, 497.
26 Jobic, H., Kärger, J., and Bee, M. (1999) *Phys. Rev. Letters*, **82**, 4260.
27 Vignes, A. (1986) *Ind. Eng. Chem. Fund.*, **5**, 189.
28 Reed, D.A. and Ehrlich, G. (1981) *Surf. Sci.*, **102**, 588–609.
29 Li, S., Falconer, J.L., Noble, R.D., and Krishna, R. (2007) *J. Phys. Chem. C*, **111**, 5075–5082.
30 Sholl, D.S. and Jee, S.E. (2009) *J. Am. Chem. Soc.*, **131**, 7896–7904.
31 Germanus, A., Kärger, J., and Pfeifer, H. (1984) *Zeolites*, **4**, 188.
32 Germanus, A., Kärger, J., Pfeifer, H., Samulevic, N.N., and Shdanov, S.P. (1985) *Zeolites*, **5**, 91.
33 Caro, J., Bülow, M., Schirmer, W., Kärger, J., Heink, W., and Pfeifer, H. (1985) *J. Chem. Soc., Faraday Trans. I*, **81**, 2541–2550.
34 Kärger, J. and Pfeifer, H. (1987) *Zeolites*, **7**, 90.
35 Heink, W., Kärger, J., Pfeifer, H., Selverdu, P., Datema, K., and Nowak, A. (1992) *J. Chem. Soc., Faraday Trans. I*, **88**, 515–519.
36 Brandani, S., Ruthven, D.M., and Kärger, J. (1995) *Zeolites*, **15**, 494–495.
37 Brandani, S., Xu, Z., and Ruthven, D.M. (1996) *Microporous Mater.*, **7**, 323–331.
38 Brandani, S., Hufton, J., and Ruthven, D.M. (1995) *Zeolites*, **15**, 624–631.
39 Goddard, M. and Ruthven, D.M. (1986) *Zeolites*, **6**, 445–448.
40 Yucel, H. and Ruthven, D.M. (1980) *J. Chem. Soc., Faraday Trans. I*, **76**, 60–70, and 71–79.
41 Doetsch, I.H., Ruthven, D.M., and Loughlin, K.F. (1974) *Can. J. Chem.*, **52**, 2717.
42 Vavlitis, A.P., Ruthven, D.M., and Loughlin, K.F. (1981) *J. Colloid Interface Sci.*, **84**, 526.
43 Yucel, H. and Ruthven, D.M. (1980) *J. Colloid Interface Sci.*, **74**, 186.
44 Zikanova, A., Bülow, M., and Schlodder, H. (1987) *Zeolites*, **7**, 115.
45 Shah, D.B., Hayhurst, D.T., Evanina, G., and Gao, J. (1988) *A.I.Ch.E. J.*, **34**, 1713.
46 Shen, D. and Rees, L.V.C. (1991) *Zeolites*, **11**, 667–671.
47 Ruthven, D.M. (2007) *Adsorption*, **13**, 225–230.
48 Goddard, M. and Ruthven, D.M. (1986) *Zeolites*, **6**, 275.
49 Goddard, M. and Ruthven, D.M. (1986) *Zeolites*, **6**, 283.

50 Goddard, M. and Ruthven, D.M. (1986) in *Proceedings 7th International Zeolite Conference Tokyo* (eds Y. Murakani, A. Lijina, and J. Ward), Kodansha, Tokyo, p. 467.
51 Kärger, J. and Ruthven, D.M. (1989) *Zeolites*, **9**, 267.
52 Kärger, J. and Ruthven, D.M. (1981) *J. Chem. Soc., Faraday Trans. I*, **77**, 1485–1496.
53 Brandani, S., Jama, M., and Ruthven, D.M. (2000) *Microporous Mesoporous Mater.*, **35–36**, 283–300.
54 Lettat, K., Jolimaitre, E., Tayakout, M., and Tondeur, D., *A.I.ChE J.* – in press (doi: 10.1002/aic.12679).

4
Diffusion Mechanisms

In Chapters 2 and 3 we examined various ways of describing diffusion phenomena, but the mechanisms by which diffusion may proceed were not considered. Clearly, the mechanism will in general depend on the nature of the diffusing molecules and their interaction with the surroundings. Depending on the particular system and the conditions, diffusion in a nanoporous solid may show features common to diffusion in vapor, liquid, or solid states. The models used to represent diffusion in small pores or in an adsorbed phase borrow heavily from theories of diffusion in homogeneous systems but, since several excellent treatises on these subjects are available [1–5], such systems are not considered here.

4.1
Diffusion Regimes

The diffusion mechanism depends on the pore diameter relative to the effective diameter of the diffusing molecules. The main regimes are indicated in Figure 4.1. In large pores transport is dominated by molecular diffusion and viscous flow, with an increasing contribution from Knudsen diffusion as the pore size decreases. When the pore diameter becomes comparable with the molecular diameter the molecule never escapes from the force field at the pore wall. Steric hindrance becomes dominant and we see a very rapid decrease in diffusivity with increasing molecular size. This regime is commonly referred to as *intracrystalline diffusion, zeolitic diffusion,* or *configurational diffusion*. It is of great practical interest since, under appropriately selected conditions, a very efficient size-selective separation becomes possible.

4.1.1
Size-Selective Molecular Sieving

Size-selective molecular sieve separations can operate by two quite different mechanisms. If the sizes of the guest molecules relative to the pore size are such that only one species (or one class of species) in a mixture can enter the pores, then clearly only that species will be admitted, leading to a highly efficient size-selective separation either

Diffusion in Nanoporous Materials. Jörg Kärger, Douglas M. Ruthven, and Doros N. Theodorou.
© 2012 Wiley-VCH Verlag GmbH & Co. KGaA. Published 2012 by Wiley-VCH Verlag GmbH & Co. KGaA.

4 Diffusion Mechanisms

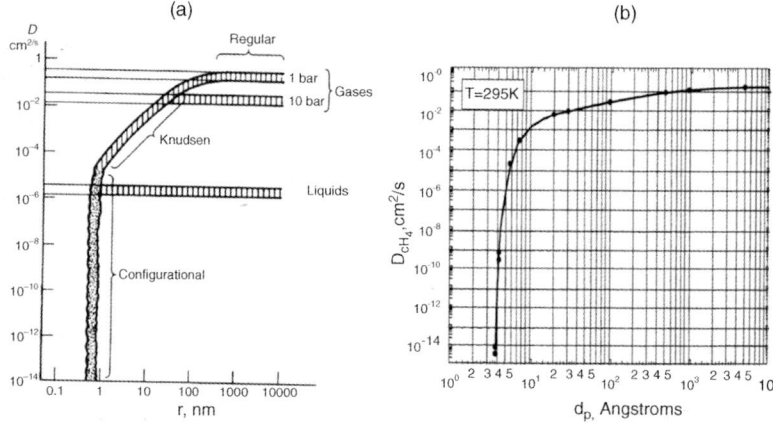

Figure 4.1 (a) Schematic diagram showing the variation of diffusivity with pore radius. The different regimes are indicated. (b) Diffusivity of methane in a straight cylindrical pore (at 295 K) as a function of pore diameter from molecular simulation. From Rao and Sircar [8] with permission.

by selective adsorption or by selective membrane permeation, as illustrated schematically in Figure 4.2a. There is, however, a second possibility, sometimes called "selective surface flow" (Figure 4.2b). In general the larger molecules will be more strongly adsorbed and, provided that they are not sterically excluded, they therefore

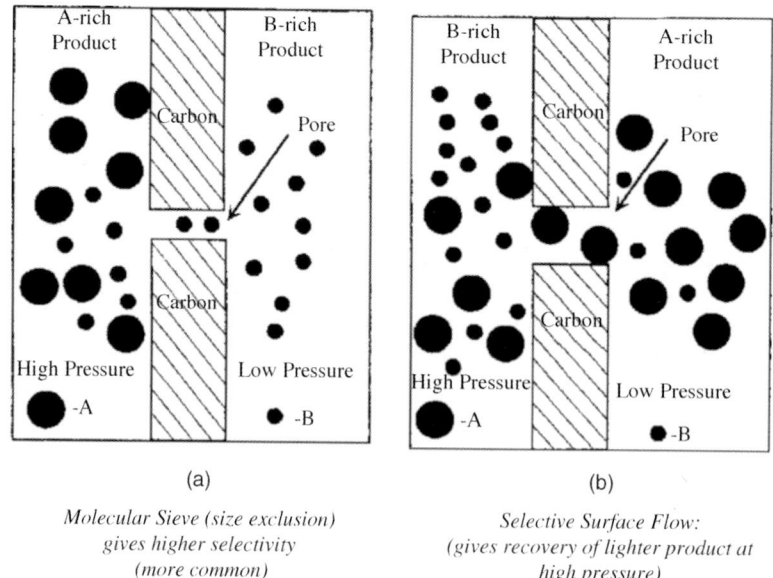

Figure 4.2 The two different mechanisms of molecular sieving. Reprinted from Rao and Sircar [8] with permission.

tend to occupy the pore space preferentially, thus excluding the smaller molecules from the pores by competitive adsorption. This leads to preferential adsorption or permeation of the *larger* molecules.

This idea was suggested and verified experimentally by Barrer and his coworkers many years ago [6, 7] and it was re-investigated from a commercial perspective by Rao and Sircar in the 1990s [8]. The main problem seems to be that the smaller molecules can never be totally excluded from the pores (from entropy considerations) so the selectivity is generally not sufficiently high to be of commercial interest. Most size-selective separations therefore operate on the simple size-exclusion principle. Nevertheless, the idea that it is possible to adsorb or permeate preferentially either the larger or the smaller species by careful control of the pore size is certainly intriguing.

4.2
Diffusion in Macro- and Mesopores

Beyond the molecular sieving region a range of different diffusion mechanisms become dominant, depending largely on the pore size relative to the size of the diffusing molecules. In this section we focus on diffusion in relatively large macro- and mesopores, in which non-activated diffusion mechanisms are dominant. Diffusion in macro- and mesopores is of intrinsic interest but it is also important because macro- and mesoporosity is commonly incorporated into microporous adsorbents and catalysts to improve the mass transfer characteristics [9, 10]. As the pore size decreases, surface diffusion, which occurs by an activated process, becomes increasingly important, eventually becoming the dominant mechanism in the intracrystalline pores of zeolites and related materials. Activated diffusion mechanisms are discussed in subsequent sections.

4.2.1
Diffusion in a Straight Cylindrical Pore

Although a real pore structure generally consists of a more or less random network of interconnecting pores of varying diameter and orientation, it is helpful to consider first the mechanisms by which diffusion can occur by reference to a straight cylindrical pore. This allows the discussion of diffusion mechanisms to be separated from the effects of geometry and pore size distribution, which complicate the behavior of real porous solids. To simplify the discussion further we assume that the pore diameter is large relative to the diameter of the diffusing molecule, so that zeolitic (configurational) diffusion and steric effects are excluded. Simultaneously, the pore is assumed to be sufficiently long so that finite-size effects in the axial direction [11] are excluded. We consider first the transport of a single component through such a pore system. The discussion is then extended to binary and multicomponent systems, in which the effect of intermolecular collisions must be considered.

4.2.1.1 Knudsen Mechanism

In small pores or at low pressures, the mean free path may become comparable with or even greater than the pore diameter, so that collisions between a molecule and the pore wall occur more frequently than intermolecular collisions. A molecule hitting the wall exchanges energy with the atoms or molecules of the surface, with the result that it is reflected diffusely, that is, the velocity of the molecule leaving the surface bears no relation to the velocity of the incident molecule and its direction is purely random. The stochastic nature of this process gives rise to a Fickian expression for the flux in which the diffusivity depends only on the pore size and the mean molecular velocity. This is known as Knudsen diffusion.

In the Knudsen regime the rate at which momentum is transferred to the pore walls greatly exceeds the transfer of momentum between diffusing molecules, so that the latter can be neglected. An expression for the Knudsen diffusivity in a straight cylindrical pore can therefore be obtained directly from kinetic theory. The rate at which molecules collide with a unit area of the pore wall is given by $c\bar{v}/4$ where c is the molecular concentration and $\bar{v}$ the mean molecular velocity. If the average velocity in the z direction is $\bar{v}_z$, the momentum flux to an element of the wall of a cylindrical pore is given by $(c\bar{v}/4)(m\bar{v}_z)(2\pi r\,dz)$. A force balance therefore yields:

$$-\pi r^2\,dp = \frac{c\bar{v}}{4} m\bar{v}_z 2\pi r\,dz \tag{4.1}$$

or:

$$\bar{v}_z = \frac{-2r}{mc\bar{v}}\frac{dp}{dz}$$

The flux is given by:

$$J = c\bar{v}_z = -D_K \frac{\partial c}{\partial z} \tag{4.2}$$

where D_K is the Knudsen diffusivity. Since $p = ck_BT$ and from kinetic theory the mean velocity is given by:

$$\tilde{v} = \left(\frac{8k_BT}{\pi m}\right)^{\frac{1}{2}} \tag{4.3}$$

it follows from Eqs. (4.1) and (4.2) that:

$$D_K = \frac{2rk_BT}{m\bar{v}} = r\sqrt{\left(\frac{\pi k_BT}{2m}\right)} \tag{4.4}$$

More rigorous application of kinetic theory, taking proper account of the distribution of molecular velocities, yields the commonly quoted result:

$$D_K = r\sqrt{\frac{32k_BT}{9\pi m}} = \frac{2}{3}r\bar{v} = 97r\sqrt{T/M} \tag{4.5}$$

where in the last expression r is in meters, T in Kelvin, M is the gram molecular weight and D_K is in m^2 s^{-1}. This differs from Eq. (4.4) by the factor $3\pi/8$ or about

18%. The Knudsen diffusivity varies only weakly with temperature and it is independent of pressure, since the mechanism does not depend on intermolecular collisions. The inverse square root dependence on molecular weight is the same as for molecular diffusion.

The validity of the Knudsen model has been confirmed experimentally, for weakly adsorbed species, by Grüner and Huber [12], who measured diffusion of helium and argon, over a wide range of temperature, in a single crystal of mesoporous silicon with cylindrical pores of approximately 12-nm diameter. Theoretical studies by Nicholson, Bhatia et al. [13–20] and Krishna [21, 22] suggest that for more strongly adsorbed species in smaller pores the Knudsen model may over-predict the diffusivity by as much as an order of magnitude. According to this theory the force field near the pore wall causes curvature of the trajectory of the diffusing molecule, thus lengthening the path and reducing the diffusivity relative to the Knudsen value. However, this has not been confirmed experimentally. Indeed the available experimental data appear to confirm the validity of the simple Knudsen model even under conditions of significant adsorption. For example, Figure 4.3 shows the experimental pore diffusivities measured by Reyes et al. [23] for N_2, Xe, and isobutane (iC_4) in mesoporous silica adsorbent (mean pore radius 3.5 nm) plotted against the Knudsen diffusivities calculated from Eq. (4.5). Although isobutane is quite strongly adsorbed, while N_2 is only very weakly adsorbed, the pore diffusivities all conform to Eq. (4.5) with a tortuosity factor of about 4.2.

The experimental permeance data for mesoporous glass and silica membranes (Section 15.4) also appear to be consistent with the simple Knudsen model. However, such measurements are complicated by the effects of surface diffusion, pore size

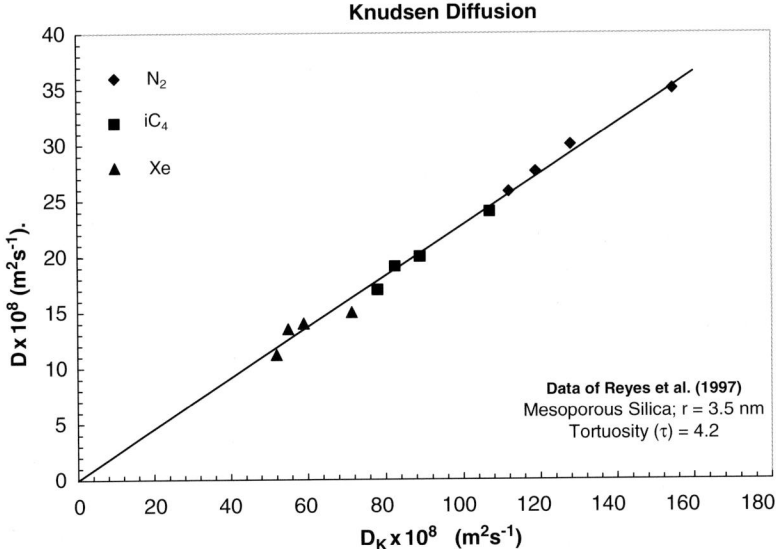

Figure 4.3 Experimental pore diffusivities for nitrogen, xenon, and isobutane (iC_4) in mesoporous silica, showing conformity with the Knudsen model [Eq. (4.5)]. Data of Reyes et al. [23].

distribution, and support resistance, so these results are less definitive. Although the simple Knudsen model appears to be widely applicable, it has been found that, in sufficiently small pores, increasing surface roughness may lead to a substantial reduction in diffusivity. In such a situation the coefficients of self- and transport diffusivity are affected in exactly the same way, in accordance with the general principle that, for non-interacting particles, the rates of self- and transport diffusion should always be the same [24, 25] and in marked contrast to the behavior in the transitional regime where self-diffusion occurs more slowly than diffusive transport, as discussed below.

4.2.1.2 Viscous Flow

If there is a difference in total pressure between the ends of a capillary, there will be bulk (laminar) flow in accordance with Poiseuille's equation. This is formally equivalent to a Fickian diffusivity given by:

$$D_{vis} = Pr^2/8\eta \tag{4.6}$$

where P is the absolute pressure and η is the viscosity. Since any such forced flow operates in parallel with the diffusive flux, it is reasonable to assume, at least as a first approximation, that this contribution is additive. The total pore diffusivity (in a single-component system) is therefore given (in SI units) by:

$$D = D_K + D_{vis} = 97r(T/M)^{\frac{1}{2}} + Pr^2/8\eta \tag{4.7}$$

At ordinary pressures the viscosity is approximately independent of pressure, so Eq. (4.7) predicts a linear increase of diffusivity with pressure, with an intercept (at zero pressure) corresponding to the Knudsen diffusivity. However, at very low pressures the viscosity becomes pressure dependent and the "no-slip" condition at the pore wall, which is assumed in the derivation of Poiseuille's equation, also breaks down, with the result that the pressure dependence of the diffusivity generally follows the form sketched in Figure 4.4. A semi-empirical expression to account for this

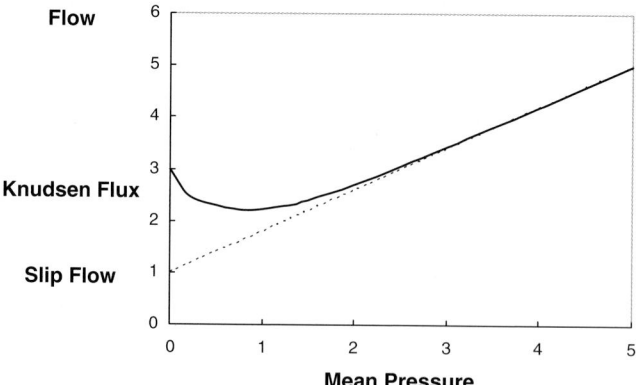

Figure 4.4 Variation of flux with mean pressure in the low-pressure region, showing the transition from Knudsen to viscous flow.

behavior was originally developed by Knudsen [26] and further refined by Cunningham and Williams [27] –see Do [28].

4.2.1.3 Molecular Diffusion

When the diffusing gas contains two or more different molecular species, the effect of intermolecular collisions must be considered. Whether this is significant depends on the ratio of the pore diameter to the mean free path (which determines the relative frequency of intermolecular collisions and molecule–wall collisions). When the pore diameter is sufficiently large, the effect of molecule–wall collisions becomes negligible and the diffusivity (for a straight cylindrical pore) approaches the molecular diffusivity. This limit will always apply for liquid systems.

4.2.1.4 Transition Region

When the mean free path is comparable with the pore diameter, the transfer of momentum between diffusing molecules and between molecules and the pore wall are both significant and diffusion occurs by the combined effects of both the molecular and Knudsen mechanisms. We now consider a binary gas mixture (A and B). From Eqs. (4.1) and (4.4), the pressure drop arising from transfer of momentum from component A to the pore wall is given by:

$$-\frac{1}{k_B T}\left(\frac{dp_A}{dz}\right)_{wall} = \frac{N_A}{D_K} \qquad (4.8)$$

while the rate of transfer of momentum from A to B is given by:

$$\frac{1}{k_B T}\left(\frac{dp_A}{dz}\right)_{gas} = \frac{N_A - y_A N}{D_{AB}} = \frac{N_A - y_A(N_A + N_B)}{D_{AB}} \qquad (4.9)$$

Assuming these two effects to be additive:

$$\frac{N_A}{D} = -\frac{1}{k_B T}\left(\frac{dp_A}{dz}\right)_{total} = \frac{N_A}{D_K} + \frac{N_A - y_A(N_A + N_B)}{D_{AB}} \qquad (4.10)$$

The combined transport diffusivity is therefore given by:

$$\frac{1}{D} = \frac{1}{D_K} + \frac{1}{D_{AB}}[1 - y_A(1 + N_B/N_A)] \qquad (4.11)$$

as derived by Scott and Dullien [29] and others [30, 31].

The factor $1 - y_A(1 + N_B/N_A)$ represents simply the ratio N_A/J_A, that is, the ratio of the diffusive flux to the total flux. For equimolar counter-diffusion $N_B = -N_A$, and Eq. (4.11) reduces to the simple reciprocal addition rule:

$$\frac{1}{D} = \frac{1}{D_K} + \frac{1}{D_{AB}} \qquad (4.12)$$

Evidently, this is always a good approximation when y_A is small. However, when neither of these conditions is fulfilled, such an approximation should be treated with caution. Since $D_K \propto r$ and $D_{AB} \propto 1/P$ this expression shows clearly the transition from Knudsen diffusion in small pores at low pressure to molecular diffusion in larger pores and at higher pressure.

For transport diffusion of a single component $N_B = 0$ and $y_A = 1$, so the molecular diffusivity has no effect and $D = D_K$. This is to be expected from the momentum balance argument, since the transfer of momentum by collisions between molecules of the same type leads to no net loss of momentum and therefore has no effect on the gradient of partial pressure.

4.2.1.5 Self-Diffusion/Tracer Diffusion

For tracer or self-diffusion component B is the labeled species and under equilibrium conditions $N_B/N_A = -1$ so Eq. (4.12) should always be valid with $D_{AB} = D_{AA}$, the molecular self-diffusivity. We may therefore conclude that for diffusion of an isotopically labeled mixture in the transitional regime, under non-equilibrium conditions, the overall concentration profile (the development of which is governed by D_{KA}) will approach equilibrium more rapidly than the species profile that is governed by the combined self-diffusivity given by Eq. (4.12) with $D_{AB} = D_{AA}$.

4.2.1.6 Relative Importance of Different Mechanisms

A rigorous analysis of the combined effects of molecular diffusion, Knudsen diffusion, and viscous flow is mathematically difficult. A simplified analysis based on the "dusty gas" model (Section 4.2.2.1) suggests that, at least under isothermal conditions, the addition rule [Eq. (4.7)] is still approximately valid with D_K replaced by D, calculated from Eq. (4.11). A simple order-of-magnitude calculation, the results of which are summarized in Table 4.1, shows that, for air at atmospheric pressure and room temperature, the contribution from viscous flow becomes important only in relatively large pores. However, the viscous contribution is more important at higher pressures and, at 10 atm, amounts to nearly 40% even in pores of only 200 Å diameter.

4.2.1.7 Surface Diffusion

The three transport mechanisms considered so far all involve diffusion through the fluid phase in the central region of the pore. If there is significant adsorption

Table 4.1 Relative importance of molecular, Knudsen diffusion, and Poiseuille flow for air at 20 °C in a straight cylindrical pore.

p (atm)	D_m (cm² s⁻¹)	r (cm)	D_K (cm² s⁻¹)	D (cm² s⁻¹)ᵃ⁾	D_{vis} (cm² s⁻¹)	D_{total} (cm² s⁻¹)	$\dfrac{D_{vis}}{D_{total}}$
1.0	0.2	10⁻⁶	0.03	0.027	0.0007	0.027	0.026
		10⁻⁵	0.3	0.121	0.07	0.19	0.37
		10⁻⁴	3.0	0.19	7.0	7.2	0.97
10	0.02	10⁻⁶	0.03	0.012	0.007	0.019	0.37
		10⁻⁵	0.3	0.019	0.7	0.719	0.97
		10⁻⁴	3.0	0.020	70	70	1.0

a) D is assumed to be given by Eq. (4.12).

on the pore wall, there is the possibility of an additional flux due to diffusion through the adsorbed phase or "surface diffusion." Physically adsorbed molecules are relatively mobile and, although the mobility is substantially smaller than in the vapor phase, if adsorption equilibrium is favorable, the molecular density in the adsorbed layer may be relatively high. The fluxes through the gas phase and the adsorbed phase are to a first approximation independent and therefore additive, so that the diffusivity will be given by the sum of the pore and surface contributions. If the diffusivity is defined on the basis of the overall guest molecule concentration in the pore space (as is usual in a PFG NMR measurement) the combined diffusivity (D) is given simply by [32]:

$$D = p_p D_p + p_s D_s \tag{4.13a}$$

where D_s is the surface diffusivity, D_p denotes the contributions from Knudsen and molecular diffusion, as well as from Poiseuille flow (if significant), and p_p and p_s denote the fractions of molecules in the pore volume and on the pore surface, respectively. However, if the combined diffusivity is referred to the concentration gradient in the gas phase (within the pore), as is usual in macroscopic measurements, it is necessary to allow for the difference in molecular densities between the two phases. The corresponding expression is then:

$$D = D_p + K' D_s \tag{4.13b}$$

where K' is the ratio p_s/p_p which corresponds to the dimensionless adsorption equilibrium constant expressed in terms of pore volume. Considering a pore within a solid matrix, it is more usual to express the equilibrium constant in terms of the volume of the (microporous) solid, so that in place of Eq. (4.13b) we have:

$$D = D_p + K\left(\frac{1-\varepsilon_p}{\varepsilon_p}\right) D_s \tag{4.14}$$

where ε_p represents the porosity and K is the dimensionless Henry constant based on solid volume.

It is clear from Eq. (4.14) that the relative importance of surface diffusion depends on the ratio KD_s/D_p. Surface diffusion is an activated process (Section 4.3) but the diffusional activation energy is generally smaller than the heat of adsorption, so that the product KD_s increases with decreasing temperature. Surface diffusion is therefore generally insignificant at temperatures that are high relative to the normal boiling point of the sorbate.

4.2.1.8 Combination of Diffusional Resistances
The Knudsen and molecular diffusion processes behave like resistances in series, whereas viscous flow and surface diffusion occur through independent parallel

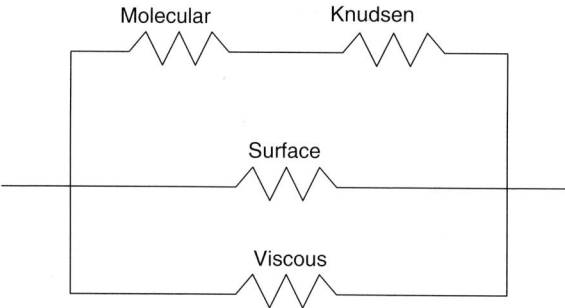

Figure 4.5 Combination of diffusional resistances according to a resistor network model.

paths. We may therefore consider the combination of diffusional resistances in a pore in terms of the equivalent resistor network (Figure 4.5).

4.2.2
Diffusion in a Pore Network

4.2.2.1 Dusty Gas Model

The molecular velocity varies in a complex way over the cross section of a pore and this makes rigorous analysis of the transport behavior very difficult. To avoid this difficulty Maxwell [33] introduced a simplified model ("dusty gas"), in which the solid matrix is replaced by a dummy species of very massive molecules constrained by unspecified external forces to have zero drift velocity. The system is thus considered as a mixture of the n true gaseous species plus the dummy species (the dust). The Chapman–Enskog kinetic theory is then applied to this $(n + 1)$ component mixture. Interaction of the gas molecules with the pore wall is simulated by the interaction with the "dust," but, since the dust is dispersed uniformly through the system, the difficult problem of accounting for the variations of flux and composition over the pore cross section is avoided. By varying the proportion of "dust" it is possible to span the transition from the molecular regime, in which momentum transfer occurs primarily by molecular collisions, to the Knudsen regime in which momentum transfer occurs entirely by collision with the "dust."

The "dusty gas" model has been investigated in detail by Evans *et al.* [34]. For isothermal diffusion this analysis, in essence, confirms the validity of Eq. (4.11) as a representation of the combined effects of Knudsen and molecular diffusion. A major advantage of this model is that it provides a rational basis for the investigation of diffusion under non-isothermal conditions, from which conclusions may be drawn concerning the relative importance of thermal diffusion and the conditions under which such effects may be neglected. Detailed accounts of this approach have been given by Jackson [35] and by Cunningham and Williams [27].

The "dusty gas" model provides, in principle, a theoretical framework from which the transport behavior of a random pore structure may be predicted, but it has serious limitations. For example, viscous flow is not included and the model does not contain any parameters relating to the key features of a real pore system (porosity, pore shape, and pore size distribution). This greatly reduces its practical value.

There have been numerous more recent attempts to develop more sophisticated molecular models as well as numerical simulations, including multiscale simulations [36, 37], to describe diffusion in a random network of macro/mesopores. But the goal of developing a simple and tractable procedure to allow the diffusional resistance of a random pore system to be reliably predicted from the basic models for diffusion in straight cylindrical pores remains elusive.

4.2.2.2 Effective Medium Approximation

The problem of accounting for pore geometry and connectivity in a random pore lattice was considered by Burganos et al. [38–40], who developed the effective medium approximation, which allows the main effects of the pore structure to be represented in a comparatively straightforward way using average macroscopic properties. However, when the pore size distribution is broad, different mechanisms are likely to be dominant in the different pore sizes. This makes statistical averaging very difficult and, indeed, impossible for many real systems under conditions of practical interest.

4.2.2.3 Tortuosity Factor

The procedure that has been generally adopted in practice is to assume that the diffusivity in a porous solid can be related to the diffusivity, under comparable physical conditions, in a straight cylindrical pore (diameter equal to the mean pore diameter) by a simple numerical "tortuosity factor" (τ):

$$D = D_p/\tau \tag{4.15a}$$

This form is appropriate for PFG NMR measurements where the measured diffusivities refer directly to the actual guest concentration within the pores of the solid [41]. However, for macroscopic measurements the porosity (ε_p) must be included in the numerator to take account of the fact that transport occurs only through the pores and not through the solid matrix:

$$D = \varepsilon_p D_p/\tau \tag{4.15b}$$

The tortuosity factor, which is assumed to be a property of the adsorbent and independent of the diffusing species, is assumed to account for all other effects, including the increased path length, the effects of connectivity, and variation of the pore diameter. Notably, this definition of the tortuosity factor is widely but not universally accepted. Some authors (for example, Do [28]) define the tortuosity factor as τ^2, leading to possible confusion for the reader.

Since the porosity is typically about 0.3 and the tortuosity factor is typically 3, the combined effect of these factors is to reduce the diffusivity in the pore system by about an order of magnitude relative to diffusion in a straight cylindrical pore under similar conditions. Considering only the orientation effect, that is, modeling the pore structure as a random assemblage of uniform cylindrical capillaries oriented with equal probability in all directions in three dimensions, yields $\tau = \langle \cos^2 \theta \rangle^{-1} = 3$, where θ is the polar angle in a spherical polar coordinate system [42, 43]. The same value was also derived by Dullien [44] for diffusion in a cubic pore structure with a uniform repeating unit. Similar logic applied to diffusion in one dimension, as in a membrane, yields $\tau = 2$ so one may expect that, in general, the tortuosity factor for membrane diffusion will be lower than that for diffusion into a particle.

Experimental tortuosity factors generally fall in the range $2 < \tau < 5$. There is a general correlation between tortuosity and porosity, the tortuosity increasing as porosity decreases, so that high tortuosity factors are generally associated with highly compressed low-porosity pellets. A similar correlation may be established on considering tortuosity and disorder of the pore network [45]. Dullien [46] has shown that, if the pore structure is characterized in sufficient detail, a reasonably accurate prediction of the tortuosity factor can be made. However, this requires a detailed knowledge of the pore shape as well as the pore size distribution. In practice it is generally simpler to treat the tortuosity as an empirical constant.

The basic assumption that the behavior of a random pore network can be described in this simple way with a species-independent tortuosity factor is clearly an oversimplification. For example, it has been shown both experimentally [47] and from modeling studies [48–50] that the tortuosity factors (for the same adsorbent) may be different for Knudsen and molecular diffusion.

The idea that fluid within the pore may be considered as two distinct phases (an adsorbed layer with a central vapor phase region) becomes increasingly unrealistic as the pore size is reduced. The tortuosity factor was originally introduced to describe diffusion in macroporous solids in which transport occurs mainly by molecular diffusion, Knudsen diffusion, and viscous flow with minimal contributions from surface diffusion. The experimental evidence suggests that for such systems this simplification generally provides a reasonable approximation. Extension of this concept to mesopores and complex macro/mesopore networks is clearly questionable but, in the absence of any tractable alternative, such an approach has been widely adopted.

4.2.2.4 Parallel Pore Model

In sufficiently large pores, diffusion on the microscopic scale proceeds at the same rate as in the bulk fluid. In such a situation, provided that the pore size distribution is narrow, Eq. 4.15 provides an unambiguous definition of the tortuosity factor. However, except in the bulk diffusion regime, the pore diffusivity depends on the pore radius so, to complete the definition, it is necessary to specify the radius at which D_p is evaluated. The simplest approach is to use the average pore radius as suggested by Satterfield [51]. However, when the pore size distribution is broad it is more logical to define D_p as the pore volume average value:

$$\langle \varepsilon_p D_p \rangle = \int_0^\infty D(r) f(r) \mathrm{d}r; \quad D = \frac{\langle \varepsilon_p D_p \rangle}{\tau} \qquad (4.16)$$

where $f(r)\mathrm{d}r$ represents the fraction of pore volume in the radius interval r to $r + \mathrm{d}r$. This model, first introduced by Feng and Stewart [52], is commonly referred to as the "parallel pore model," since it is physically equivalent to considering the pore structure as an assemblage of pores of different sizes connected in parallel. Wang and Smith [53] showed that this model yields constant values for the tortuosity of a given adsorbent, independent of the sorbate or temperature, whereas the values based on the average pore radius did not show this consistency. Evidently, the values of τ based on the mean pore size will in general be smaller than the values based on the volume average [Eq. (4.16)] so, when comparing tortuosity factors reported by different authors, it is important to clarify which definition is being used.

4.2.2.5 Random Pore Model

The parallel pore concept is clearly appropriate only for certain pore structures. One can also visualize structures in which small and large pores are connected in series or in series–parallel arrangements. The "random pore" model was developed in an attempt to provide a simple approximate model for such systems. It is assumed that the pore structure can be regarded as built up from stacked layers of microporous particles (porosity ε'_p) with macroporosity (ε_p) between the particles. Considering the four possible paths in a two-layer assemblage of this kind leads to the following expression for the effective diffusivity [54]:

$$D_e = \varepsilon_p^2 D_p + (\varepsilon'_p)^2 \frac{(1+3\varepsilon_p)}{1-\varepsilon_p} D'_p \qquad (4.17)$$

where D_p and D'_p are the macro- and micro-pore diffusivity, respectively. In the limit such that either $\varepsilon'_p \to 0$ or $D'_p \to 0$ Eq. (4.17) reduces simply to $D_e \approx \varepsilon_p^2 D_p$. Comparison with Eq. 4.15 shows that in terms of this model $\tau \approx 1/\varepsilon_p$. Since ε_p is typically ∼0.3 this is roughly consistent with the commonly observed behavior ($\tau \approx 3.0$). Equation (4.17) has the advantage that it provides explicit recognition of the observation that the tortuosity generally increases as the porosity decreases.

4.2.2.6 Capillary Condensation

As a result of the lowering of the equilibrium vapor pressure, arising from the effect of surface tension, condensation occurs in a small capillary at a vapor pressure well below the saturation vapor pressure for the free liquid. The reduction in vapor pressure is given by the familiar Kelvin equation:

$$\frac{P}{P_0} = \exp\left(\frac{-2\sigma V_m \cos\theta}{rRT}\right) \qquad (4.18)$$

where σ, θ, and V_m denote, respectively, the surface tension, the contact angle, and the molar volume. This expression can also be thought of as defining the critical pore radius at which condensation will occur under a given equilibrium vapor pressure.

Diffusion in the capillary condensation regime is complicated to follow, as illustrated by the examples given in Section 15.9. As soon as a pore fills with condensate, the vapor flux through that pore is cut off and transport then depends on surface diffusion together with any bulk flow induced by the capillary forces. Intuitively one might therefore expect the apparent diffusivity to be greatly reduced. This has been confirmed by NMR studies combined with a lattice model simulation in which it was observed that the establishment of equilibrium following an increase of pressure occurred at rates substantially slower than expected from the molecular diffusivities [55]. This retardation was attributed to the presence of substantial local maxima in the Gibbs free energy surface. These maxima must be overcome by local fluctuations before the system can proceed to a lower free energy state. Eventually, the time constants for this process may become very large and the differences in free energy between the various possible molecular arrangements very small. Under these conditions, on any practical time scale the system may approach an array of different quasi-equilibrium states rather than a unique state of true equilibrium [56].

However, it is also possible, when there is a sufficiently broad distribution of pore size, that the flux, and therefore the apparent diffusivity, may be greatly enhanced. This counterintuitive result was first pointed out by Carman and Haul [57] and a more detailed analysis has been given by Weisz [58]. One may consider the condensate within a pore to be subject to an equivalent pressure $-2\sigma \cos(\theta/r)$ resulting from the surface tension forces. Considering a capillary of the critical radius in a pressure gradient (or a gradient of sorbate partial pressure), we will have condensation at the upstream end of the capillary and evaporation at the downstream end, where the pressure is lower. The condensing sorbate will flow through the capillary in viscous flow under the combined effects of the pressure gradient and the surface tension forces. This can be regarded as an additional flux replacing the gas phase flux. A quantitative estimate of this flux may be obtained by equating the surface tension force to the viscous drag. Comparing the capillary flux estimated in this way, for reasonable estimates of the physical parameters, with the Knudsen flux (estimated for pores of the same diameter in the absence of any condensation) one obtains:

$$\frac{J_{\text{capillary}}}{J_{\text{Knudsen}}} \sim \frac{3-10^3}{\Delta p} \tag{4.19}$$

where Δp is the pressure difference (in atmospheres) between the two ends of the capillary. Since Δp is generally small, this ratio is large. Clearly, any capillary that is filled with condensate will in effect create a "short circuit," leading to a reduction in the length of the diffusion path and a corresponding increase in the apparent diffusivity. Provided that only a relatively small fraction (α) of the pores is filled with condensate:

$$\frac{D_{\text{app}}}{D} \sim \frac{1}{1-\alpha} \tag{4.20}$$

The conditions under which capillary condensation occurs are also those under which significant surface diffusion may be expected. The study of this phenomenon is therefore complicated by the difficulty of separating the contributions from the various transport mechanisms.

4.3
Activated Diffusion

4.3.1
Diffusion in Solids

Solids differ from liquids or gases in that the atoms or molecules are localized and vibrate about well-defined positions in the crystal lattice. Occasionally a vibrating molecule will have sufficient energy to leave its site and jump to a neighboring position. Since such a transition requires the molecule to surmount a substantial energy barrier this is an activated diffusion process in which the diffusivity varies with temperature according to an equation of the Arrhenius form:

$$\mathcal{D} = \mathcal{D}_\infty \, e^{-E/RT} \qquad (4.21)$$

with the apparent activation energy (E) corresponding to the sum of the energy barrier that has to be overcome in an individual diffusion step towards a neighboring vacant site and the energy consumed for the formation of the vacancies. Depending on the given lattice type and hopping mechanism, the parameters D_∞ and E may follow quite different dependencies. Their specification is given in the literature [59–61]. In the present context, for simplicity, we confine ourselves to a face-centered cubic metal. Here, the activation energy (E) is found to be about two-thirds of the latent heat of vaporization of the metal (L) [1]. Since both diffusion and vaporization require removal of the metal atoms from their equilibrium position in the lattice, such a relationship is physically understandable.

Figure 4.6 shows schematically a plausible model for diffusion in a face-centered cubic metal lattice. Diffusion in the x direction is considered and we regard this as occurring by an activated jump of the atom, originally located at the center of plane 2, to the corresponding position in plane 4, which is assumed to be vacant. During transit the atoms in plane 3 are displaced from their equilibrium positions (symbolized in the

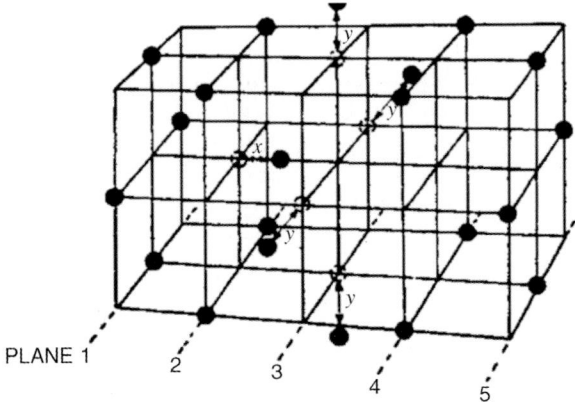

Figure 4.6 Activated diffusion in a face-centered cubic crystal.

figure by arrows over distances y) and this represents the major contribution to the activation energy of the process. A simple quantitative estimate based on the semi-empirical 6-9 Mie potential function suggests that the observed ratio $E/L \approx 0.67$ is physically reasonable [62].

According to the random walk model for diffusion in one dimension $\mathcal{D} = l^2/2\tau$; an estimate of the jump distance (l) may be obtained from the lattice parameter while the jump frequency ($1/\tau$) may be estimated from the natural vibration frequency (ν) multiplied by a Boltzmann factor ($e^{-E/RT}$) representing the probability that the vibrating atom has an energy greater than E:

$$\frac{1}{\tau} = 2\nu\, e^{-E/RT}, \quad \mathcal{D} = l^2\nu\, e^{-E/RT}, \quad \mathcal{D}_\infty = l^2\nu \tag{4.22}$$

4.3.2
Surface Diffusion Mechanisms

Diffusion within the adsorbed layer on a solid surface can occur by several different mechanisms, some of which are similar to diffusion in the solid phase.

4.3.2.1 Vacancy Diffusion

Let us suppose that the adsorbed molecules are held at localized sites, thus forming a molecular array [63, 64]. We consider only transport in the z direction. The jump rate for all molecules in this direction is ν but the jump will succeed only if it is directed towards a vacant site. The net flow, per unit time, from plane 1 to plane 2 is then given by:

$$\dot{n} = n_1\nu(1-n_2/m) - n_2\nu(1-n_1/m) = \nu(n_1-n_2) \tag{4.23}$$

where n_i is the number of molecules in the i-th layer, the capacity of which is m. With A and δ denoting the area and separation of the layers, respectively, the expression for the flux is obtained as:

$$J = \frac{\dot{n}}{A} = \nu\delta\frac{(n_1-n_2)}{\delta A} \tag{4.24}$$

Since $n/\delta A$ is simply the local concentration and $(n_1-n_2)/\delta A$ is just the concentration difference between two layers separated by distance δ:

$$J = -\nu\delta^2(dc/dz) \tag{4.25}$$

from which, by comparison with Fick's equation:

$$D = \nu\delta^2 \tag{4.26}$$

Although the probability of a successful jump decreases with concentration, the diffusivity, according to this simple mechanism, remains constant. This may be understood since, according to this mechanism, the diffusion process involves an exchange of molecules and vacancies and the sum of these two quantities must always remain constant (equal to the saturation capacity). This situation is formally similar to counter-diffusion in a binary system, relative to the plane of no net flux [Eqs. (1.13)–(1.16)], and by the same logic it follows that the diffusivity of the molecules and vacancies must be the same. Since any concentration dependence would be expected to affect the molecular and vacancy diffusivities in opposite ways, this requirement can reasonably be satisfied only if the diffusivity is independent of concentration.

Considering molecular self-diffusion, the rate of successful jumps is given by $v(1-\theta)$ with $\theta = q/q_s$, the fractional site occupation. Thus, with Eq. (2.4), the expression for the diffusivity becomes:

$$\mathcal{D} = v\delta^2(1-\theta) \tag{4.27}$$

and combining this with Eq. (4.26):

$$\frac{D}{\mathcal{D}} = \frac{1}{(1-\theta)} \tag{4.28}$$

in conformity with the Darken equation [Eq. (1.35)] for a Langmuir isotherm [for which $d\ln p/d\ln q = 1/(1-\theta)$]. The possibility that either v or δ may vary with concentration has not been considered so, even within the framework of this simple model, more complex forms of concentration dependence are possible.

This model has been extended to multicmponent diffusion by Qureshi and Wei [65]. For a binary adsorbed phase (A, B):

$$J_A = -v_A\delta^2(1-\theta_B)\frac{dq_A}{dz} - v_A\delta^2\theta_A\frac{dq_B}{dz} \tag{4.29}$$

The mass balance ($dq/dt = -dJ/dz$) yields a set of partial differential equations that can be solved to obtain expressions for the transient sorption curves obtained for any specified initial and boundary conditions. This model has been used to interpret co-diffusion and counter-diffusion experiments with mixtures of benzene and toluene on silicalite (Section 18.8.2.6). For either $\theta_B = 0$ or $\theta_A + \theta_B = $ constant and $v_A = v_B$, Eq. (4.29) reduces to Eqs. (4.26) or (4.28) as the relevant expression for the transport and tracer diffusivities in a single-component adsorbed phase. It may be noted that the form of Eq. (4.29) is quite different from the corresponding Maxwell–Stefan expression [Eq. (3.43)] or the simplified Habgood form [Eq. (3.49)], which assume that chemical potential gradient is the driving force.

4.3.2.2 Reed–Ehrlich Model

A useful model for surface diffusion has been developed by Reed and Ehrlich [66], who treated the adsorbed phase as a lattice gas. In the ideal case where there is no interaction between molecules this leads to the Langmuir isotherm. If it is assumed that the probability of a successful jump is proportional to the fraction of unoccupied sites $(1-\theta)$, the diffusivity will be independent of loading since this factor exactly cancels out the thermodynamic correction factor $[1/(1-\theta)]$ so that $D(\theta) = D_0/(1-\theta) = v(0)\delta^2$, in accordance with Eq. (4.26). In the more realistic case where there is significant interaction between the diffusing molecules the behavior becomes more complex, since it is necessary to allow for the effect of molecular interactions on the jump rate. Reed and Ehrlich assume that the jump distance remains constant and they estimate the change in the jump rate due to molecular interactions (considering both the apparent activation energy and the pre-exponential factor) by calculating the variation in the standard chemical potential with loading from the "quasi-chemical approximation," as developed by Lacher [67] and Fowler and Guggenheim [68], which yields for the chemical potential:

$$\mu = \mu^\circ + RT \ln\left(\frac{\theta}{1-\theta}\right) + \frac{zRT}{2} \ln\left[\frac{(\beta-1+2\theta)(1-\theta)}{(\beta+1-2\theta)\theta}\right] \tag{4.30}$$

where z is the coordination number. The corresponding expression for the isotherm is:

$$bp = \frac{\theta}{1-\theta}\left(\frac{2-2\theta}{\beta+1-2\theta}\right)^z \tag{4.31}$$

and for the thermodynamic correction factor:

$$\left(\frac{\partial \ln p}{\partial \ln \theta}\right)_T = \frac{1}{1-\theta}\left[1 + \frac{z}{2}\left(\frac{1-\beta}{\beta}\right)\right] \tag{4.32}$$

where:

$$\beta = [1-4\theta(1-\theta)(1-\eta)]^{\frac{1}{2}}; \quad \eta = e^{-w/k_B T} \tag{4.33}$$

where w is the interaction energy between two molecules occupying adjacent sites on the lattice. The jump rate is given by:

$$\frac{\nu(\theta)}{\nu(0)} = \frac{(1+\varepsilon)^{z-1}}{(1+\eta\varepsilon)^z} \tag{4.34}$$

where:

$$\varepsilon = (\beta-1+2\theta)/[2\eta(1-\theta)] \tag{4.35}$$

This yields for the corrected diffusivity:

$$\frac{D_0(\theta)}{D_0(0)} = \frac{(1+\varepsilon)^{z-1}}{(1+\eta\varepsilon)^z} \tag{4.36}$$

where $D_0(0) = D(0) = \nu(0)\delta^2$ and for the Fickian diffusivity:

$$\frac{D(\theta)}{D(0)} = \frac{(1+\varepsilon)^{z-1}}{(1-\theta)(1+\eta\varepsilon)^z}\left[1 + \frac{z}{2}\left(\frac{1-\beta}{\beta}\right)\right] \tag{4.37}$$

If $w \to 0$, $\beta \to 1$, $\eta \to 1$, $\varepsilon \to \theta/(1-\theta)$, Eq. (4.37) reduces to:

$$D(\theta) = D(0) = \nu(0)\delta^2 \tag{4.38}$$

as for an ideal Langmuir system.

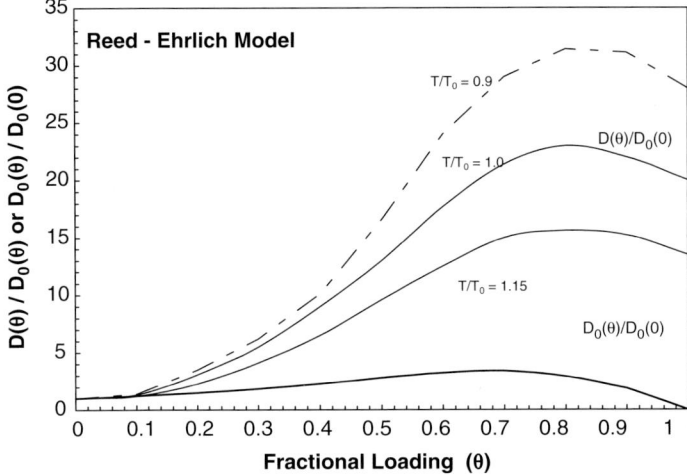

Figure 4.7 Variation of normalized transport diffusivity $D(\theta)/D_0(0)$ for various values of the parameter w $(= k_B T_0)$ and corrected diffusivity $D_0(\theta)/D_0(0)$ at $T = T_0$ calculated from the Reed–Ehrlich model [66] [Eqs. (4.36) and (4.37)].

In terms of transition state theory (Section 4.3.3) this approach is equivalent to assuming that the free energy of the transition state is independent of loading but the energy and entropy of activation vary since the free energy of the adsorbed molecules is affected by molecular interactions.

Through Eqs. (4.36)–(4.38), the parameters β, η, and ε can all be expressed as functions of the loading (θ), the coordination number (z), and the interaction energy ($w = k_B T_0$), so the model is in effect a two-parameter model. Figure 4.7 shows the variation of $D(\theta)/D_0(0)$ and $D_0(\theta)/D_0(0)$ with loading, calculated from Eqs. (4.36) and (4.37) for a fixed value of z ($= 4$) and selected values of w. The curves have qualitatively similar shapes but the range of variation (the height of the maximum) is an order of magnitude smaller for D_0 than for D, thus supporting the view that, apart from any theoretical argument, there is a practical advantage to be gained from correlating the diffusivity data for adsorbed phases in terms of the corrected diffusivity rather than the Fickian diffusivity. At loadings less than 70% of saturation the model predicts a modestly increasing trend of D_0 with loading followed by a strong decline as saturation is approached.

As the Reed–Ehrlich model is based on a lattice model, it predicts that at the saturation limit the corrected diffusivity should fall to zero. But, since the thermodynamic factor tends to infinity, the Fickian diffusivity, which corresponds to the product of the jump rate and the thermodynamic factor, approaches a finite limit given by $D(\theta \to 1)/D_0(0) = \exp[(z-1)w/k_B T]$. When there is no interaction $w \to 0$ and the Fickian diffusivity is independent of loading. Although experimental data at high loadings are limited a sharp decline in D_0 is commonly observed, especially for open pore systems, but the observed decline is generally not as severe as that predicted by the model. In reality one might expect that, at saturation, the diffusivity

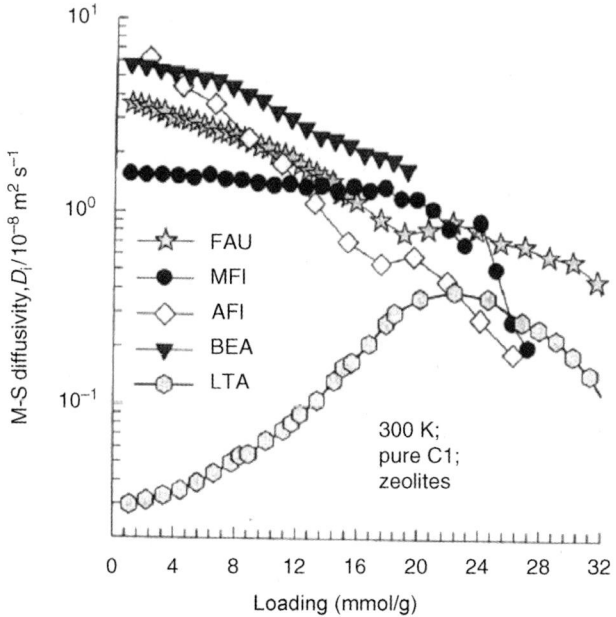

Figure 4.8 Variation of corrected diffusivity (D_0 = Maxwell–Stefan diffusivity) for methane in 12-ring zeolites from molecular simulations. The curve for the eight-ring zeolite LTA is shown for comparison, with permission from Krishna and van Baten [69].

in an open pore system would fall to some fraction of the liquid phase diffusivity at comparable conditions (reduced by an effective tortuosity factor), rather than to zero. In small-pore systems a less severe reduction may be expected.

Figure 4.8 shows the results of molecular simulations showing the variation of the corrected diffusivity of methane with loading for several 12-ring and one eight-ring zeolite (LTA). The behavior of some other cage structures is shown in Figure 3.5c. Evidently, the patterns of loading dependence are quite different for the large and small pore materials. In the more open structures the corrected diffusivity declines only slightly with loading, at least until about 70% of saturation but the behavior in zeolite A shows the opposite trend at low loadings, with a pronounced maximum. Both these trends can be fitted by the Reed–Ehrlich model with appropriate parameters (see Figure 3.5c)! Notably, the loading levels shown in such plots derived from molecular simulations extend well above the levels that can be achieved in practice, so only the range up to about 50–60% loading is accessible to experimental verification.

4.3.3
Diffusion by Cage-to-Cage Jumps

For many host–guest systems, the diameter of the "windows" between adjacent cavities is close to the size of the guest molecules. Examples include n-alkanes in the various "narrow-pore" eight-ring zeolites considered in Chapter 16. Under such conditions, molecular passage from one cage to an adjacent one becomes the rate-determining step of molecular propagation so that the diffusivities can be determined by simply estimating the rate of passage through the windows between adjacent cages. Calculations of this type represent a special and particularly simple case among the "infrequent event techniques" of diffusion simulation, which are discussed in greater detail in Chapter 9.

By considering molecular displacements notably exceeding the separation (l) between the centres of adjacent cages, with Eqs. (2.6) and (2.7) the self-diffusivity is easily seen to be given by:

$$\mathcal{D} = l^2/6\tau \tag{4.39}$$

Using the formalism of the transition state theory [TST [70–73], see Section 9.1 and Eq. (9.34)], the mean life time τ within one cage may be estimated by the relation:

$$1/\tau = \kappa \frac{kT}{h} \frac{f^+}{f_A} \exp(-E_0/k_B T) \tag{4.40}$$

where f_A and f^+ denote, respectively, the (molecular) partition function in the host cage (molecular ground state) and the "reduced" partition function for a molecule in the window (the transition state). Translational motion in the direction of the axis through the window is factored out and gives rise to the frequency factor $k_B T/h$, with h denoting Planck's constant. E_0 is the difference between the two energy levels. The transmission coefficient κ is a correction factor that takes into account that occupation of the transition state does not necessarily lead to diffusional jumps into the neighboring cage. It can be estimated from MD simulations (Chapter 9) and will, for simplicity, be assumed to be of the order of 1 in the following discussion. The cumbersome calculation of f_A may be circumvented by introducing, instead, the partition function f_g per unit volume in free gas phase, which is related to f_A through the equilibrium isotherm, which may be written in terms of the partition functions as:

$$n_A(p) = \frac{f_A}{f_g} \frac{p}{k_B T} \exp(U_o/k_B T) \tag{4.41}$$

where U_o denotes the difference in the zero energy levels between the free gas and the adsorbed state and $n_A(p)$ is the number of molecules per cage at pressure p [71]. Combining Eqs. (4.39)–(4.41) yields:

$$\mathcal{D} = \frac{l^2}{h} \frac{p}{n_A} \exp\left(-\frac{E_o - U_o}{k_B T}\right) = n_c \frac{l^2 f^+}{h f_g} \frac{p}{q} \exp\left(-\frac{E_o - U_o}{k_B T}\right) \tag{4.42}$$

where in the second equation, by introducing the number of cavities per volume (n_c), the particle number per cage is replaced by the particle concentration q. Eqs. (2.55) and (4.42) immediately yield the transport diffusivity:

$$D = n_c \frac{l^2 f^+}{h f_g} \frac{dp}{dq} \exp\left(-\frac{E_o - U_o}{k_B T}\right) \tag{4.43}$$

From Eqs. (4.42) and (4.43), the interrelation between self- and transport diffusion is easily seen to yield Eq. (1.35). Totally equivalently, the self-diffusivity is thus found to coincide with the corrected diffusivity. This is an immediate consequence of the model in which, by implying that molecular passages through the windows are very infrequent events, any cross-correlation between the diffusing molecules is neglected: it is exactly this correlation that, in Sections 3.2 and 3.3.3, is shown to reduce the self-diffusivity in comparison with the transport diffusivity. Since all other factors do not vary significantly with the guest concentration, it follows from Eqs. (4.42) and (4.43) that the concentration dependence is given mainly by the terms p/q and dp/dq. For Langmuir-type isotherms, the derivative dp/dq is easily seen to increase notably faster with increasing loading than the mere ratio p/q. Therefore, for narrow-pore systems, the self- or corrected diffusivity is often found to vary only insignificantly with increasing loading, while the transport diffusivity increases dramatically.

Their simple structure, with the prefactor:

$$n_c \frac{l^2}{h} \frac{f^+}{f_g}$$

as the only free parameter makes Eqs. (4.42) and (4.43) very useful as a basis for the discussion of experimental diffusivity data in narrow-pore host materials, in particular for rationalizing the dependence of the diffusivity on temperature and loading. These options are exploited in Chapters 16 and 19. More complex channel structures clearly require more complicated approaches. In particular this is true if molecular exchange between adjacent cavities is only possible from particular sites, rather than from anywhere within a cavity. This situation is considered in the "relevant site model" [74, 75], which combines the concept of cage-to-cage jumps with the Maxwell–Stefan model. Although the combination of concepts based on different mechanisms requires several arbitrary assumptions, this approach has be shown to be applicable to an impressive variety of different systems [76, 77]. However, in contrast to the simple approach provided by Eqs. (4.42) and (4.43), which contain only one unknown parameter, the more complex model contains four parameters that must be estimated from the experimental data.

4.4
Diffusion in More Open Micropore Systems

4.4.1
Mobile Phase Model

In the previous discussion of intracrystalline diffusion we have assumed that the adsorbed species can be considered as a single phase. An alternative possibility that may provide a more realistic model for some systems is to assume the existence of two adsorbed states, mobile and immobile [78]. The immobile molecules are assumed to be localized on the adsorption sites and to diffuse only slowly between sites. By contrast the mobile phase is assumed to diffuse according to a random walk with a self-diffusivity D_{mobile}. In the first instance we assume rapid equilibrium between the two states so that the fraction of the molecules in the mobile state is given by the equilibrium isotherm, $c_{\text{mobile}} = f(c_a)$. The overall diffusivity in a self-diffusion experiment will be given by:

$$\mathcal{D} = \frac{c_{\text{mobile}}}{c_{\text{total}}} \mathcal{D}_{\text{mobile}} \tag{4.44}$$

Considering the flux through the mobile phase resulting from a gradient in overall concentration the transport diffusivity is given by:

$$D = \mathcal{D}_{\text{mobile}} \frac{dc_{\text{mobile}}}{dc_{\text{total}}} \tag{4.45}$$

and it is evident that the combination of Eqs. (4.44) and (4.45) again leads to the simple relation, Eq. (1.35), correlating self- and transport diffusion.

4.4.2
Diffusion in a Sinusoidal Field

In contrast to the situation considered in Section 4.3, in some nanoporous host–guest systems diffusion takes place in a more uniform potential energy field. As a limiting case for such systems a first approximation to the expected behavior may be obtained by considering the field as a three-dimensional sinusoidal potential:

$$U = U_0 + \frac{1}{2} V_0 \left(1 - \cos\frac{2\pi x}{l}\right) + \frac{1}{2} V_0 \left(1 - \cos\frac{2\pi y}{l}\right) + \frac{1}{2} V_0 \left(1 - \cos\frac{2\pi z}{l}\right) \tag{4.46}$$

as suggested by T.L. Hill [79]. U_0 is the potential energy of the minima (sites or cages) and V_0 is the energy barrier between adjacent sites or cages. For fast energy exchange between the guest molecules and the host lattice, from Eq. (4.22) with the natural oscillation frequency $\nu = (V_0/2ml^2)^{1/2}$, the self-diffusivity is found to be given by [73, 80]:

$$\mathcal{D} = \frac{1}{\sqrt{2}} \left(\frac{V_0}{m}\right)^{\frac{1}{2}} \exp(-V_0/k_B T) \tag{4.47}$$

As demonstrated in References [73, 80], the diffusivity estimated from Eq. (4.47) varies by less than a factor of 2 between the extreme cases of infinitely fast and infinitely slow energy exchange. Although Eq. (4.46) provides only a crude representation of the potential energy field within a microporous adsorbent it provides an easy means for an order-of-magnitude check of the consistency of kinetic and equilibrium data. The first comparison that can be made is between the activation energy for self-diffusion (V_0) and the pre-exponential factor, both of which are experimentally accessible from diffusion measurements over a range of temperature:

$$\mathcal{D}_\infty = \frac{l}{\sqrt{2}} \left(\frac{V_0}{m}\right)^{\frac{1}{2}} \tag{4.48}$$

A second possibility is provided by a comparison of the experimental activation energy with the pre-exponential factor of the Henry's law equilibrium constant. By using the same model for the adsorbed phase under equilibrium conditions the following relationship is obtained [80, 81]:

$$K_\infty = \left[\exp\left(-\frac{V_0}{k_B T}\right) I_0\left(\frac{V_0}{2k_B T}\right)\right]^3 (V_0/k_B T) \qquad (4.49)$$

where I_0 is the modified Bessel function of the first kind. Some comparisons between experimental equilibrium and kinetic data on the basis of these expressions are given in Section 17.2.1

4.4.3
Free-Volume Approach for Estimating the Loading Dependence

In the derivation of Eqs. (4.47)–(4.49), any interaction between the guest molecules is neglected. Such an assumption is justified for sufficiently small loadings. With increasing loading, mutual molecular collisions will lead to jump lengths decreasing with increasing loading and, hence, to decreasing diffusivities. To a first-order approximation, this effect may be taken account of by equating the jump length with the cube root of the mean free volume per molecule. Analytical expressions for the concentration dependence of the diffusivity attainable in this way by adopting the well-established formalism of the free-volume theory [82] may be found in the literature [73, 83].

References

1 Mehrer, H. (2007) *Diffusion in Solids*, Springer Series in Solid State Science, vol. 155, Springer, Berlin.
2 Tyrrel, H.J.V. and Harris, K.R. (1984) *Diffusion in Liquids*, Butterworth, Guildford.
3 Kirkaldy, J.S. and Young, D.J. (1987) *Diffusion in the Condensed State*, Institute of Metals.
4 Cussler, E.L. (2009) *Diffusion: Mass Transfer in Fluid Systems*, Cambridge University Press.
5 Heitjans, P. and Kärger, J. (2005) *Diffusion in Condensed Matter: Methods, Materials, Models*, Springer, Berlin, Heidelberg.
6 Ash, R., Barrer, R.M., and Pope, C.G. (1963) *Proc. Roy. Soc. A*, **271**, 1–18.
7 Ash, R., Barrer, R.M., and Lowson, R.T. (1973) *J. Chem. Soc., Faraday Trans.1*, **69**, 2166–2178.
8 Rao, M.B. and Sircar, S. (1993) *J. Membrane Sci.* **85**, 253–264; (1996) **110**, 109–118 and (1993) *Gas Sep. Purif.*, **7**, 279–284.
9 Choi, M., Cho, H.S., Srivastava, R., Venkatesan, C., Choi, D.H., and Ryoo, R. (2006) *Nat. Mater.*, **5**, 718.
10 Wang, G., Johannessen, E., Kleijn, C.R., de Leeuwa, S.W., and Coppens, M.O. (2007) *Chem. Eng. Sci.*, **62**, 5110–5116.
11 Comets, F., Popov, S., Schütz, G.M., and Vachkovskaia, M. (2010) *J. Stat. Phys.*, **140**, 948.
12 Gruener, S. and Huber, P. (2008) *Phys. Rev. Lett.*, **100**, 064502.
13 Nicholson, D., Petrou, J., and Petropoulos, J.H. (1979) *J. Colloid Interface. Sci.*, **71**, 570–579.
14 Nicholson, D. and Petropoulos, J.H. (1981) *J. Colloid Interface. Sci.*, **83**, 420–427.
15 Nicholson, D. and Petropoulos, J.H. (1985) *J. Colloid Interface. Sci.*, **106**, 538–546.
16 Bhatia, S., Jepps, O., and Nicholson, D. (2004) *J. Chem. Phys.*, **120**, 4472–4485.

17 Jepps, O.G., Bhatia, S.K., and Searles, D.J. (2003) *J. Chem. Phys.*, **119**, 1719–1730.
18 Jepps, O.G., Bhatia, S.K., and Searles, D.J. (2003) *Phys. Rev. Lett.*, **91**, 126102.
19 Bhatia, S.K. and Nicholson, D. (2006) *AIChE J.*, **52**, 29–38.
20 Bhatia, S.K. and Nicholson, D. (2011) *Chem. Eng. Sci.*, **66**, 284.
21 Krishna, R. and van Baten, J.M. (2009) *Chem. Eng. Sci.*, **64**, 870–882.
22 Krishna, R.R. and van Baten, J.M. (2011) *J. Membr. Sci.*, **369**, 545.
23 Reyes, S.C., Sinfelt, J.H., DeMartin, G.J., and Ernst, R.H. (1997) *J. Phys. Chem. B*, **101**, 614–622.
24 Russ, S., Zschiegner, S., Bunde, A., and Kärger, J. (2005) *Phys. Rev. E*, **72**, 030101.
25 Zschiegner, S., Russ, S., Valiullin, R., Coppens, M.-O., Dammers, A.J., Bunde, A., and Kärger, J. (2008) *Eur. Phys. J.*, 109–120.
26 Kennard, E.H. (1938) *Kinetic Theory of Gases*, McGraw-Hill, New York.
27 Cunningham, R.E. and Williams, R.J.J. (1980) *Diffusion in Gases and Porous Media*, Plenum Press, New York.
28 Do, D.D. (1998) *Adsorption Analysis: Equilibria and Kinetics*, Imperial College Press, London, p. 383.
29 Scott, D.S. and Dullien, F.A.L. (1962) *AIChE J.*, **8**, 113.
30 Evans, R.B., Watson, G.M., and Mason, E.A. (1961) *Chem. Phys.*, **35**, 2076.
31 Rothfield, L.B. (1963) *AIChE J.*, **9**, 19.
32 Dvoyashkin, M., Valiullin, R., and Kärger, J. (2007) *Phys. Rev. E*, **75**, 041202.
33 Maxwell, J.C. (1860) *Philos. Mag.*, **20**, 21.
34 Evans, B., Watson, R.G.M., and Mason, E.A. (1962) *J. Chem. Phys.*, **36**, 1894; Evans, B., Watson, R.G.M., and Mason, E.A. (1963) **38**, 1808.
35 Jackson, R. (1977) *Transport in Porous Catalysts*, Elsevier, Amsterdam, ch. 3.
36 Albo, S., Broadbelt, L.J., and Snurr, R.Q. (2006) *AIChE J.*, **52**, 3679.
37 Albo, S., Broadbelt, L.J., and Snurr, R.Q. (2007) *Chem. Eng. Sci.*, **62**, 6843.
38 Burganos, V.N. and Sotirchos, S.V. (1987) *AIChE J.*, **33**, 1678–1689.
39 Burganos, V.N. and Sotirchos, S.V. (1989) *Chem. Eng. Sci.*, **44**, 2541–2562.
40 Burganos, V.N. and Payatakes, A.C. (1992) *Chem. Eng. Sci.*, **47**, 1383–1400.
41 Vasenkov, S. and Kortunov, P. (2005) *Diffusion Fundam.*, **1**, 2.1–2.11.
42 Johnson, M.F.L. and Stewart, W.E. (1965) *J. Catal.*, **4**, 248.
43 Haring, R.E. and Greenkorn, R.A. (1970) *AIChE J.*, **16**, 477.
44 Dullien, F.A.L. (1975) *AIChE J.*, **21**, 299.
45 Russ, S. (2009) *Phys. Rev. E*, **80**, 61133.
46 Dullien, F.A.L. (1979) *Porous Media, Fluid Transport and Pore Structure*, Academic Press, New York, ch. 3.
47 Vasenkov, S., Geir, O., and Kärger, J. (2003) *Eur. Phys. J.*, **12**, 35–38.
48 Burganos, V.N. (1998) *J. Chem. Phys.*, **109**, 6772–6779.
49 Papadopoulos, G.K., Theodorou, D.N., Vasenkov, S., and Kärger, J. (2007) *J. Chem. Phys.*, **126**, 094702.
50 Zalc, J.M., Reyes, S.C., and Iglesia, E. (2004) *Chem. Eng. Sci.*, **59**, 2947.
51 Satterfield, C.N. (1970) *Mass Transfer in Heterogeneous Catalysis*, MIT Press, Cambridge, MA.
52 Feng, C. and Stewart, W.E. (1973) *Ind. Eng. Chem. Fundam.*, **12**, 143.
53 Wang, C.T. and Smith, J.M. (1983) *AIChE J.*, **29**, 132.
54 Wakao, N. and Smith, J.M. (1962) *Chem. Eng. Sci.*, **17**, 825.
55 Valiullin, R., Naumov, S., Galvosas, P., Kärger, J., Woo, H.-J., Porcheron, F., and Monson, P.A. (2006) *Nature*, **430**, 965–968.
56 Naumov, S., Valiullin, R., Monson, P., and Kärger, J. (2008) *Langmuir*, **24**, 6429–6432.
57 Carman, P.C. and Haul, R.A.W. (1951) *Proc. R. Soc. London, Ser. A*, **209**, 38.
58 Weisz, P.B. (1975) *Ber. Bunsenges. Phys. Chem.*, **79**, 798.
59 Mehrer, H. and Stolwijk, N.A. (2009) in *Diffusion Fundamentals III* (eds C. Chmelik, N.K. Kanellopoulos, J. Kärger, and D. Theodorou), Leipzig University Press, Leipzig, pp. 21–52.
60 Mehrer, H. (2007) *Diffusion in Solids*, Springer, Berlin.
61 Allnatt, A.R. (1993) in *Atomic Transport in Solids* (ed. A.B. Lidiard), Cambridge University Press.
62 Moelwyn-Hughes, E.A. (1961) *Physical Chemistry*, Pergamon Press, Oxford, p. 658.

63 Riekert, L. (1970) *Adv. Catal.*, **21**, 281.
64 Riekert, L. (1971) *AIChE J.*, **17**, 446.
65 Qureshi, W.R. and Wei, J. (1990) *J. Catal.*, **126**, 126.
66 Reed, D.A. and Ehrlich, G. (1981) *Surf. Sci.*, **102**, 588–609.
67 Lacher, J.R. (1937) *Proc. Cambridge. Philos. Soc.*, **33**, 518; Lacher, J.R. (1937) *Proc. R. Soc. London, Ser. A*, **161**, 525.
68 Fowler, R.H. and Guggenheim, E.A. (1939) *Statistical Thermodynamics*, Cambridge University Press, p. 441.
69 Krishna, R. and van Baten, J.M. (2009) *Chem. Eng. Sci.*, **64**, 3159–3178.
70 Laidler, K.J. (1965) *Chemical Kinetics*, McGraw Hill, London.
71 Ruthven, D.M. and Derrah, R.I. (1972) *J. Chem. Soc., Faraday Trans. I*, **68**, 2332.
72 Kärger, J., Pfeifer, H., and Haberlandt, R. (1980) *J. Chem. Soc., Faraday Trans. I*, **76**, 1569.
73 Kärger, J. and Ruthven, D.M. (1992) *Diffusion in Zeolites and Other Microporous Solids*, John Wiley & Sons, Inc., New York.
74 van den Bergh, J., Ban, S., Vlugt, T.J.H., and Kapteijn, F. (2010) *Sep. Purif. Technol.*, **73**, 151.
75 van den Bergh, J., Ban, S., Vlugt, T.J.H., and Kapteijn, F. (2009) *J. Phys. Chem. C*, **113**, 17840.
76 van den Bergh, J., Ban, S., Vlugt, T.J.H., and Kapteijn, F. (2009) *J. Phys. Chem. C*, **113**, 21856.
77 van den Bergh, J., Gascon, J., and Kapteijn, F. (2010) in *Zeolites and Catalysis: Synthesis, Reactions and Applications* (eds J. Cejka, A. Corma, and S. Zones), Wiley-VCH Verlag GmbH, Weinheim.
78 Boddenberg, B., Haul, R., and Oppermann, G. (1970) *Surf. Sci.*, **22**, 79.
79 Hill, T.L. (1960) *Introduction to Statistical Thermodynamics*, Addison Wesley, Reading, MA.
80 Ruthven, D.M. and Doetsch, I.H. (1976) *J. Chem. Soc., Faraday Trans I*, **72**, 1043.
81 Kärger, J., Bülow, M., and Haberlandt, R. (1977) *J. Colloid Interface Sci.*, 60.
82 Cohen, M.H. and Turnbull, D.J. (1959) *J. Chem. Phys.*, **31**, 1164.
83 Kärger, J., Pfeifer, H., Rauscher, M., and Walter, A. (1980) *J. Chem. Soc., Faraday Trans. I*, **76**, 717.

5
Single-File Diffusion

Among the features that determine molecular transport in nanoporous materials, deviations from the ideal textbook structure are particularly important. For example, transport resistances ("barriers"), either on the surface (Sections 12.3.2 and 19.6) or within the bulk phase of the host particle (see Section 18.6.5), may retard mass transfer while transport rates may be enhanced by the presence of intracrystalline cracks and mesopores (see Section 15.7.3 and References [1, 2]). In this chapter we consider a phenomenon that is particularly sensitive to any irregularities in the host structure and which also serves as a textbook example of a situation in which molecular transport, even in an ideal structure, does not obey the laws of ordinary diffusion.

We assume that the host framework compels the diffusants to remain in the same order, as for example when the structure consists of an array of parallel nonintersecting channels with diameters small enough so that adjacent sorbate molecules cannot pass one another. At the beginning of Section 2.1, a random walk was described as "the motion of a man, constrained to walk along a straight line, who throws a coin at each step to decide whether to proceed forward or backwards." Now we have to consider quite a number of men on such a line, subject to the additional constraint that each man is allowed to move only if the step leads to an unoccupied position. Figure 5.1 demonstrates how such a situation may be visualized in the tiers of a lecture hall.

Systems of this type are referred to as single-file systems. Besides their relevance as a model case for studying highly-correlated motion, molecules in single-file arrangement are of obvious technological interest, including their potential use as traps for reducing the cold-start hydrogen emission from automotive exhaust [4] and their negative impact on separation by zeolite membranes at high loadings (see Chapter 20 for details). Single-file systems may be compared with a string of pearls, where one pearl may only be shifted in a given direction if there is sufficient free space, that is, if all pearls originally situated in this range have also been shifted in the same direction [5]. Since, therefore, the concentration of molecules "in front" of the shifted molecule tends to exceed the concentration behind it, subsequent displacements are more likely to be directed backward than forward, deviating from the requirement of ordinary diffusion (Section 2.1) that the direction of a molecular jump should be independent of past history.

Diffusion in Nanoporous Materials. Jörg Kärger, Douglas M. Ruthven, and Doros N. Theodorou.
© 2012 Wiley-VCH Verlag GmbH & Co. KGaA. Published 2012 by Wiley-VCH Verlag GmbH & Co. KGaA.

Figure 5.1 Unconventional way of illustrating the constraints leading to single-file diffusion in the lecture hall for experimental physics at Leipzig University. Reprinted from Reference [3] with permission.

Long before the term single-file diffusion was introduced into the zeolite community [6] it was used quite generally for the description of any type of one-dimensional transport of hard-core particles, including such diverse phenomena as transport in the ion channels of biological membranes and along dislocation lines in crystals [7, 8]. The abundance of intriguing questions associated with such systems has fascinated mathematicians and theoretical physicists over many years [9–13]. Reference [5] demonstrates the dramatic effect of single-file diffusion on the effectiveness of catalytic reactions in zeolites. The subsequent growing interest in investigating single-file phenomena in zeolites [14–21] has contributed greatly to increasing efforts to develop a comprehensive general theoretical treatment of this phenomenon [22–27]. A useful review is given in Reference [28], to which we shall refer repeatedly in this chapter.

We start by considering the phenomenon of single-file diffusion in an infinitely extended system. In practical applications, however, one generally has to take into account that mass transfer occurs in finite host crystals or particles. The influence of single-file diffusion under such conditions will be considered in the subsequent section. Finally, we shall see that in many of the reported experimental studies the results obtained reflect the fact that the experimental system was finite and do not reflect the behavior of an ideal infinite system.

5.1
Infinitely Extended Single-File Systems

5.1.1
Random Walk Considerations

Generally, the elementary steps of molecular propagation, as traced, for example, by the experimental techniques presented in Chapter 10 or as predicted by molecular modeling, will be much more complex than considered in Chapter 2. However, on considering sufficiently large displacements, under the conditions of normal diffusion these deviations have been found to be of no relevance: all diffusion phenomena are adequately described by Fick's equations [Eqs. (1.1) and (1.2)], together with the appropriate boundary and matching conditions. A similar procedure will now be applied to molecular transport in single-file systems. We shall again profit from the fact that, even under the conditions of anomalous diffusion, all relevant features of mass transfer are adequately reflected by this simple picture. Moreover, with several nanoporous host systems [including, for example, the MOFs of type Zn(tbip) considered in Section 19.4] the pore space is found to be formed by "chains" of equally spaced cavities that may accommodate just one guest molecule. For such systems, the simple jump model of single-file diffusion provides an exact description of the elementary steps of propagation.

Diffusion is assumed to occur along a linear chain of sites of distance λ, with τ denoting the mean time between subsequent jump attempts. Jump attempts are performed with equal probability in either direction, but are only successful if they are directed to a vacant site. In this case, the mean square displacement in the file direction is found to be [11–14, 25, 29, 30]:

$$\left\langle (\Delta z)^2 \right\rangle = \lambda^2 \frac{1-\theta}{\theta} \sqrt{\frac{2}{\pi}} \sqrt{\frac{t}{\tau}} \tag{5.1}$$

where θ denotes the site occupancy. As the most important finding, molecular mean square displacement is thus found to scale with the square root of the time rather than with the time itself as predicted by the Einstein relation [Eqs. (1.12) and (2.6)] for ordinary diffusion. Similarly as observed, for example, for anomalous diffusion in fractal networks (Figure 2.6 and Eq. (2.66)], the rate of molecular propagation is thus found to decrease with increasing observation time and, correspondingly, with increasing mean displacements. In fractal networks (Section 2.6) this retardation can be attributed to the fact that the diffusants are forced to make increasingly lengthy detours with increasing distance from the origin. In single-file systems, the retarding mechanism is an effect of the whole ensemble of diffusants. Increasing displacements lead to increasing mean concentrations "in front" of the diffusant under study, which hinder further propagation.

It is worthwhile recollecting there is no "memory" in the system as a whole. Equilibrium considerations therefore require that in both single-file systems and fractal networks the evolution of the guest system will not depend on its past. The future of the system depends exclusively on the given state, that is, on the positions of the individual molecules, no matter how they have got there. The situation becomes different, however, when we consider the behavior of the individual guest molecules. Both mechanisms described above lead to an imbalance in the number of accessible sites that a shifted molecule may reach. The "past" of each individual diffusant (namely the distance over which it has traveled) is thus "recorded" by its surroundings: with increasing particle concentrations "in front of the diffusant" in the case of single-file diffusion and with decreasing numbers of sites to jump to in fractal networks. Either effect leads to an increasing probability for backward-directed displacements and, hence, to a sub-diffusive molecular propagation. Thus, the existence of a "particle memory" can in fact be shown to give rise to deviation from ordinary diffusion as explained in Section 2.1.

Adopting the notation of the Einstein relation [Eqs. (1.12) and (2.6)], Eq. (5.1) may be written as:

$$\langle (\Delta z)^2 \rangle = 2F\sqrt{t} \tag{5.2}$$

with the so-called mobility factor of single-file diffusion [5, 22]:

$$F = \lambda^2 \frac{1-\theta}{\theta} \frac{1}{\sqrt{2\pi\tau}} \tag{5.3}$$

The derivation of Eq. (5.1) is complicated by the fact that the displacement of a particular molecule cannot be considered without considering the displacements of all other molecules. It is true that also under the conditions of ordinary diffusion molecular propagation is affected by the existence of other molecules. This appears, for example, in the different patterns of concentration dependence of the diffusivities (Figure 3.3). In all these cases, however, the molecules can be assumed to have lost their memories after sufficiently large observation times and displacements, respectively. In Reference [14] this complication is circumvented by considering the particle shift as a consequence of the movement of the vacancies within the single-file systems. This movement can be considered to be uncorrelated, so that a derivation of Eq. (5.1) by elementary arguments becomes possible. Moreover, in this way also all higher moments $\langle (\Delta z)^\nu \rangle$ of molecular displacements ($\nu = 2, 4, 6...$) may be calculated. This knowledge permits the determination of the probability distribution of the molecular displacements, the so-called propagator. As in the case of normal diffusion [cf. Eq. (2.13)], the propagator is found to be a Gaussian [15]:

$$P(\Delta z, t) = \left(2\pi \langle (\Delta z)^2 \rangle\right)^{-\frac{1}{2}} \exp\left[-(\Delta z)^2 / 2 \langle (\Delta z)^2 \rangle\right] \tag{5.4}$$

where now, clearly, the mean square displacement follows Eq. (5.2) rather than Eq. (2.6).

The mean square displacement in single-file systems may be shown, quite generally, to be related to the movement of an individual molecule by the expression [29, 30]:

$$\langle (\Delta z)^2 \rangle = l \langle |\Delta s| \rangle \tag{5.5}$$

where $\langle |\Delta s| \rangle$ denotes the mean value of molecular displacement of a lone particle (i.e., of a diffusant without any interaction with other particles), and l stands for the mean free distance (clearance) between adjacent molecules. Under the supposition that an isolated diffusant in the single-file system is subjected to normal diffusion, with Eq. (2.13) the mean displacement of an isolated particle may be calculated to be:

$$\langle |\Delta s| \rangle = \sqrt{\frac{4\mathcal{D}}{\pi}} \sqrt{t} \tag{5.6}$$

Combining Eqs. (5.2), (5.5), and (5.6), the mobility factor F of a single-file system and the diffusivity of a sole molecule in this system may thus be shown to be related to each other by the expression:

$$\mathcal{D} = \pi F^2/l^2 \tag{5.7}$$

As required, Eq. (5.1) results as a special case of Eq. (5.5) by using Eq. (5.6) with the relation $\mathcal{D} = \lambda^2/(2\tau)$ for the self-diffusivity of an isolated particle [Eq. (2.4)] and by using the relation:

$$l = \lambda \frac{1-\theta}{\theta} \tag{5.8}$$

where the step length λ has been set equal to the particle diameter σ.

Note that the validity of Eqs. (5.2) and (5.4) has been shown to hold quite generally for any type of diffusive motion under single-file confinement, including any interaction potential between the diffusants [22–25].

The single-file expressions Eqs. (5.1) and (5.5) are only valid for sufficiently large observation times. In the limit of short observation times, that is, in a time regime where the individual molecules have not yet become "aware" of their neighbors, molecular displacement may be described by a diffusion-like motion with a diffusivity:

$$\mathcal{D}(\theta) = \frac{\lambda^2}{2\tau}(1-\theta) \tag{5.9}$$

where, as a mean-field approach [31], the additional factor $(1-\theta)$ takes into account that a jump attempt is only successful with the probability $1-\theta$. Equating the resulting mean square displacements with the respective coefficients for ordinary diffusion and single-file diffusion, the cross-over time between the regimes of normal diffusion and of single-file diffusion is seen to be given by:

$$t_{normal \leftrightarrow sf} = \frac{2\tau}{\pi\theta^2} \tag{5.10}$$

Since normal and single-file diffusion are described by the same propagator [Eq. (5.4)], owing to the analogy between Eq. (5.2) with the Einstein relation [Eq. (2.6)], the propagation pattern of a given particle in a single-file system coincides with that of normal diffusion if the self-diffusivity $\mathcal{D}$ is substituted by F and the time t by $\sqrt{t}$. This analogy, however, is of only limited value for the treatment of practical problems, since in the case of normal diffusion there is no counterpart for the correlation in the movement of distant particles, which is an essential feature of single-file systems. This correlation prohibits extending this analogy to the treatment of molecular ensembles subject to single-file diffusion with given initial and boundary conditions [32–34].

5.1.2
Molecular Dynamics

In this section, rather than aiming at the investigation of real systems (as considered in Chapters 8 and 9), we exploit the options of molecular dynamics (MD) simulations in order to visualize the physics of single-file diffusion and some of its peculiarities. Figure 5.2 shows the results of simulations [35] under idealized conditions: spherical molecules of diameter σ propagating in an unstructured (i.e., smooth) tube that is narrow enough to ensure single-file confinement. The molecule–molecule and molecule–wall interactions are given by Lennard-Jones type potentials and an infinite tube is simulated by considering finite tube elements with periodic boundary conditions.

As a first peculiar feature, the simulated displacements appear to be limited by some maximum value. This apparently strange behavior may be easily shown to be inherent to the chosen simulation conditions. Since there is no net force acting from "outside" on the particle system in the axial direction, the position of the center-of-mass in each tube element is preserved. The maximum mean square displacement attainable under such conditions may be shown to be [35]:

$$\left\langle (\Delta z)^2 \right\rangle_\infty = \frac{1}{6}(1-\theta)^2 \frac{n\sigma^2}{\theta^2} = \frac{1}{6}\frac{(1-\theta)^2}{\theta} L\sigma \tag{5.11}$$

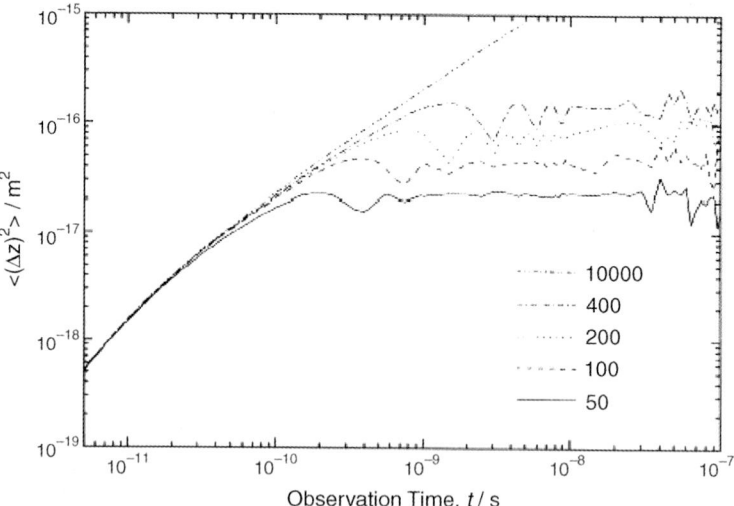

Figure 5.2 Time dependence of the mean square displacement $\langle(\Delta z)^2\rangle$ for MD simulations in a smooth narrow tube at a relative occupancy $\Theta = 0.2$ for increasing lengths of the tube elements and, hence, for increasing particles numbers (n) (as indicated in the legend) considered in the simulations; $\langle z^2 \rangle$ tends to a limiting value $\langle(\Delta z)^2\rangle_\infty$. For comparison, the result for 10 000 particles is also given. Reprinted from Reference [35] with permission.

where $L \equiv n\sigma/\theta$ stands for the length of the considered channel element. These limiting values are found to agree well with the simulation results shown in Figure 5.2. It turns out that one has to consider as many as 10 000 particles to avoid unwanted limitation effects during the considered time interval. This type of restriction does not exist in the case of normal diffusion, where the molecules are able to change their positions relative to each other. Obviously, in such a situation, arbitrarily large molecular displacements are possible, without the necessity of a shift of the center-of-mass.

Secondly, the proportionality between $\langle (\Delta z)^2 \rangle$ and the observation time t as revealed by Figure 5.2 appears to be in striking contrast to the $\sqrt{t}$ behavior to be expected on the basis of Eq. (5.2) for single-file systems. This apparent conflict, however, is immediately explained by Eq. (5.5). Since a lone particle in an unstructured tube behaves completely deterministically with a displacement $\langle |\Delta s| \rangle$ increasing in proportion with the observation time, Eq. (5.5) requires that, in a smooth tube as considered in these simulations, the mean square displacement $\langle (\Delta z)^2 \rangle$ must be proportional to t rather than to $\sqrt{t}$. In turn, proportionality between the mean square displacement and the square root of observation time can only be expected if the movement of the isolated molecule is a random walk. In Reference [35] this condition was fulfilled in the simplest way by either introducing an additional force, stochastically acting on the individual molecules, or by considering tubes with periodically varying diameters. In both cases, molecular displacements in single-file systems were found to follow Eq. (5.2) over observation times, covering two orders of magnitude. In this way, the $\sqrt{t}$ behavior of single-file diffusion as resulting from random walk considerations was confirmed by MD simulations. Moreover, the MD simulations were found to reflect satisfactorily even the concentration dependence predicted from the simple jump model [Eq. (5.1)] provided that the occupancy θ is defined with respect to a chain-like arrangement of the molecules along the tube axis.

The exclusion of mutual passage of the molecules within the zeolite channels is crucial for the occurrence of single-file diffusion. First attempts to perform such discriminations by MD simulations with methane and ethane in $AlPO_4$-5 have been presented in References [36–38]. In these studies both the methane and ethane molecules are found to be readily able to pass each other. Since such simulations are extremely sensitive to the potentials used [39–41], the evidence from such results is still under discussion.

Figure 5.3 shows the results of MD simulations with varying channel diameters. The simulations have been carried out to provide an overview of the different time regimes of molecular propagation possible in single-file systems [42]. For sufficiently narrow channels, after an initial "ballistic" period with $(\Delta z)^2 \propto t^2$, the mean square displacements follow the expected single-file dependence $(\Delta z)^2 \propto \sqrt{t}$. The regime of normal diffusion (Eq. 5.9), intermediate between the ballistic regime and normal diffusion, becomes scarcely visible. With increasing channel diameters deviations from single-file diffusion to the time dependence $(\Delta z)^2 \propto t$ of normal diffusion occur for shorter and shorter observation times. For sufficiently large observation times, the molecular displacements are completely determined by molecules passing one another, even though such events occur only rarely. Adopting Eqs. (2.3) and (2.4),

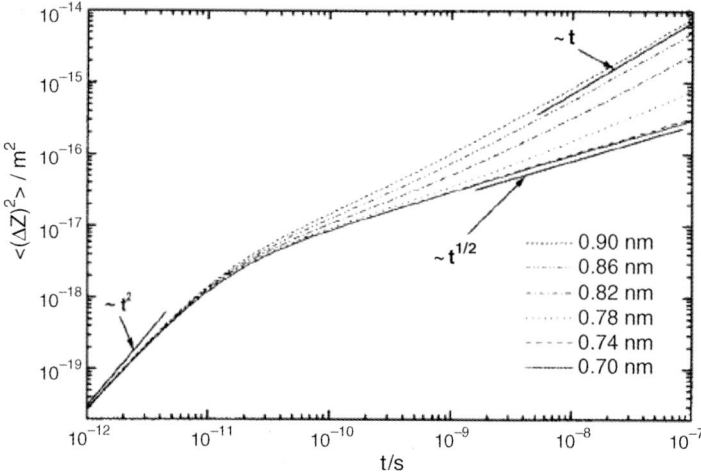

Figure 5.3 Various time regimes of molecular propagation in single-file systems as resulting from MD simulations. The legend indicates the channel diameters considered in the different simulations. In all cases, the diameter of the diffusants was assumed to be 0.383 nm. Reprinted from Reference [42] with permission.

by simple random-walk arguments the mean square displacement may then be shown to obey the relation:

$$\left\langle (\Delta z)^2 \right\rangle = a^2 t/\tau_{ex} \qquad (5.12)$$

with τ_{ex} and a denoting, respectively, the mean exchange time of a molecule with one of its neighbors and the mean distance between adjacent molecules. Following the time dependence of ordinary diffusion, $\langle (\Delta z)^2 \rangle$ now increases in proportion to the observation time. The mean square displacements calculated on the basis of Eq. (5.12) are found to be in satisfactory agreement with the simulated displacements. Since the calculation of $\langle (\Delta z)^2 \rangle$ requires much larger simulation times than τ_{ex}, the use of Eq. (5.12) saves simulation time when considering the long-time regime in zeolite channels in which the mutual passing of adjacent molecules eventually becomes dominant.

In MD simulations, Skoulidas and Sholl [43] determined the mean square displacement of CF_4 in ZSM-12 at 298 K (at a loading of one molecule per unit cell) as a function of time. The observed patterns nicely followed the dependency shown in Figure 5.3 for the larger channel cross sections, revealing a clear distinction between the regimes of ballistic motion, single-file diffusion, and, once the time is long enough so that the molecules were able to pass each other sufficiently often, normal diffusion.

The concept of single-file diffusion has been successfully applied for MD simulations in carbon nanotubes [44–50], yielding both the square-root time dependence of the molecular mean square displacement and a remarkably high mobility of the individual, isolated diffusants. In References [51–54], by MD simulations, the astonishingly high single-particle mobility in single-file systems has been attributed to a concerted motion of clusters of the adsorbed molecules.

5.2
Finite Single-File Systems

In the preceding section, the proportionality of the molecular displacements with the square root of observation time rather than with the time itself has been shown to be the preeminent feature of diffusion in infinitely extended single-file systems. We now demonstrate that the situation changes dramatically on considering single-file systems of finite extent. This, however, is exactly the situation with which one generally has to deal in practice. We shall see that, under such conditions, the mean square displacement soon follows the laws of ordinary diffusion, albeit with a diffusivity dramatically reduced in comparison with the diffusivity of an isolated, single molecule in the file. Subsequently, we will discuss the consequences of this confinement on the rates by which the guest molecules are able to enter or leave the nanoporous particles. This includes the case of counter-diffusion during chemical reactions within the files. It is important to emphasize that the particular features of single-file confinement do not appear in mere uptake or release experiments since it is of no relevance for the rate of these processes whether the molecules are able to mutually exchange their positions or not.

5.2.1
Mean Square Displacement

In contrast to the behavior in an infinitely extended file, molecular propagation within a finite single-file system with rapid molecular exchange between the file boundaries and the surroundings is soon found to follow the laws of normal diffusion. This behavior may be understood intuitively by realizing that, in an appropriately selected time interval, a molecule may enter the single-file system on one side while another molecule may leave the single-file system on the other side. Such a coincidence of events shifts the total chain of molecules and hence any particle by one position. This type of molecular shift, however, is uncorrelated with the next one, caused by the same sequence of events. We have shown in Chapter 2 that such behavior, characteristic of a Markovian process, leads to normal diffusion. Considering the elementary steps quantitatively, the mean square displacement of this process may be shown to obey the relation [21, 42, 55, 56]:

$$\left\langle (\Delta z)^2 \right\rangle = 2\mathcal{D}\frac{1-\theta}{n}t = 2\mathcal{D}\frac{1-\theta}{\theta}\frac{\lambda}{L}t \tag{5.13}$$

with L denoting the file length; $\mathcal{D} = \lambda^2/2\tau$ is the diffusivity of an isolated particle in the single-file system. Through Einstein's relation [Eqs. (1.12) and (2.6)], from Eq. (5.13) the effective diffusivity of a particle in the chain turns out to be:

$$\mathcal{D}_{\text{eff}} = \mathcal{D}\frac{1-\theta}{n} = \mathcal{D}\frac{1-\theta}{\theta}\frac{\lambda}{L} \tag{5.14}$$

and is thus found to be smaller than the diffusivity $\mathcal{D}(\theta) = \mathcal{D}(1-\theta)$ in the mean field approach for normal diffusion [see Eq. (5.9)] by exactly the number of molecules in the file (n) and smaller than the diffusivity $\mathcal{D}$ of an isolated particle by the factor $n/(1-\theta)$. This effective diffusivity is sometimes referred to as the "center-of-mass" diffusivity. Irrespective of the small diffusivity, the proportionality with t signifies that the mean square displacement in finite single-file systems with open ends is soon governed by a diffusion-type dependence, Eq. (5.13), rather than by the "single-file" relation, Eq. (5.1). Equating Eqs. (5.13) and (5.1), the crossover time from single-file diffusion to normal diffusion in finite single-file systems with open ends is found to be:

$$t_{\text{sf}\leftrightarrow\text{cmdiff}} = L^2/\pi\mathcal{D} \tag{5.15}$$

with $\mathcal{D} = \lambda^2/2\tau$. Inserting Eq. (5.15) into Eq. (5.13) [or Eq. (5.1)] yields:

$$\langle\Delta z^2\rangle_{\text{sf}\leftrightarrow\text{cmdiff}} = \frac{2}{\pi}\frac{1-\theta}{\theta}\lambda L \tag{5.16}$$

Most importantly, the distance, starting from which particle displacements are determined by normal diffusion [namely, by the "center-of-mass" diffusivity as given by Eq. (5.14)], is thus found to be much smaller than the total file length, being only of the order of the geometric means of the file length and the length of the individual diffusion steps!

In contrast to the case of "open" files just considered, reflecting boundary conditions will lead to deviations towards shorter rather than larger displacements, in comparison with an infinitely extended file. With increasing observation time, the mean square displacements may be shown to approach a limiting value given by [42]:

$$\langle(\Delta z)^2\rangle_\infty = \frac{1}{3}\frac{(1-\theta)^2}{\theta}L\sigma \tag{5.17}$$

Experiments with single-file systems of finite extension are thus found to depend sensitively on the boundary conditions. To study single-file phenomena in infinite files, as considered in Section 5.1, it is therefore important to make sure that the molecular displacements are far below the limiting values given by Eqs. (5.16) and (5.17). On considering the conditions for unperturbed measurement of intracrystalline diffusion by PFG NMR in Section 11.4.1 we shall see that, in the case of ordinary diffusion, the requirements are much less stringent.

5.2.2
Tracer Exchange

Molecular uptake and release in single-file systems follows the patterns derived in Chapter 6 from the Fick's laws. Both processes are governed by transport diffusion which is unaffected by the confinement condition of single-file diffusion: the exclusion of mutual exchange between neighboring particles leaves the flux rates unaffected [57]. However, when the movement of different species must be distinguished, the mutual correlation of the molecules in single-file systems notably complicates the predictions of the evolution of the particle distribution.

As an example, Figure 5.4 shows the initial stage of the process of tracer exchange resulting from dynamic Monte Carlo simulations [58]. By representing the product of

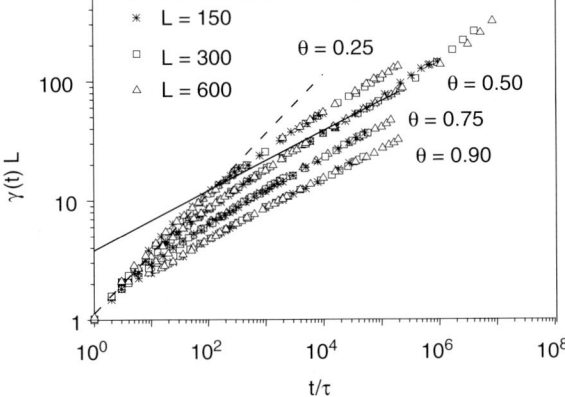

Figure 5.4 Total amount of tracer molecules during tracer exchange in a single-file system as a function of the number of simulation steps obtained by dynamic Monte Carlo simulations for various file lengths (site number L in the figure) and loadings (θ). The dashed and solid lines show the best fit lines for $\theta = 0.5$ with the slope of $1/2$ and $1/4$, as expected when normal and single-file diffusion, respectively, are dominant. Reprinted from Reference [58] with permission.

the file length L (represented as the total number of sites) and the relative number $\gamma(t) = m(t)/m(\infty)$ of exchanged molecules, Figure 5.4 shows the fraction of molecules exchanged. Since all file lengths yield coinciding values, there is no overlap in the profiles of the labeled molecules entering from the file ends.

By realising that, in first-order approximation, the total amount of exchanged molecules increases linearly with the square root of the molecular displacements, $\langle (\Delta z)^2 \rangle$, the observed time dependence reflects the three time regimes relevant for diffusion in single-file systems: (i) normal diffusion, which changes to (ii) single-file diffusion as soon as the encounters between different molecules become relevant, until (iii) molecular shifts are controlled by the center-of-mass diffusion. In the time regime of single-file diffusion, Figure 5.4 reveals a distinct parallel shift towards larger values with decreasing loadings, which reflects increasing single-file mobility factors as required by Eq. (5.3). The transition to the regime of center-of-mass diffusion is indicated for two file lengths with a loading of $\theta = 0.5$ where the simulation times have been chosen to be sufficiently large.

Equation (5.16) shows that even molecular displacements much smaller than the file length L are controlled by normal diffusion with a diffusivity given by Eq. (5.14). This result is reflected by the simulation data for tracer exchange shown in Figure 5.5. It is found that over the, by far, largest part of the file the concentration profiles during the exchange process are surprisingly well represented by the corresponding solution [Eq. (6.29)] of the classical diffusion equation, that is, of Fick's second Law, Eq. (1.2). In the approach that has been followed, a certain portion at the two file margins has been assumed to attain the equilibrium concentration instantaneously, so that the exchange had to be followed over N^* rather than over N sites. Table 5.1 indicates the values of the ratio N^*/N and of the ratio $\mathcal{D}_{\text{sim}}/\mathcal{D}_{\text{eff}}$ between the diffusivity used in

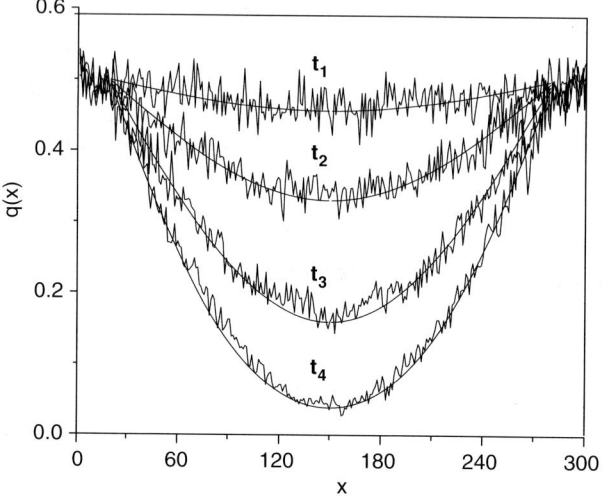

Figure 5.5 Concentration profiles of labeled molecules during tracer exchange in single-file systems of length L obtained by dynamic Monte Carlo simulations (oscillating lines) with those resulting for normal diffusion [Eq. (6.29)] with appropriately chosen values for the diffusivity (D_{sim}) and file length (N^*) as given in Table 5.1 (solid lines) at times $t_1 = 0.93 \times 10^6 \tau$, $t_2 = 2.1 \times 10^6 \tau$, $t_3 = 3.7 \times 10^6$, and $t_4 = 7.6 \times 10^6 \tau$ (τ is the duration of the elementary diffusion step). Reprinted from Reference [58] with permission.

the analytical simulations and the effective file diffusivity [as given by Eq. (5.14)] that yield the best agreement with the simulation data. Not unexpectedly, the site number N^* relevant for these simulations approaches the total number with increasing file length. Interestingly, $\mathcal{D}_{sim}$, though being of the order of $\mathcal{D}_{eff}$, does not seem to approach this value.

Inserting the effective diffusivity [Eq. (5.14)] into the corresponding expression (the first statistical moment for sorption limited by 1D diffusion, $\tau_{intra,diff} = L^2/(12\mathcal{D})$ – see Section 13.6, Table 13.1, and Reference [59]) yields:

$$\tau_{intra} = \frac{\theta}{1-\theta} \frac{L^3}{12\mathcal{D}\lambda} \tag{5.18}$$

Under single-file confinement, the intracrystalline molecular life time is thus seen to scale with the third power of the crystal size.

Table 5.1 Comparison of the values of $\mathcal{D}_{sim}$ and N^* that yield the best fit between the calculated and simulated profiles in Figure 5.5 with the center-of-mass diffusivity $\mathcal{D}_{eff}$ and site number N [58].

N	$\mathcal{D}_{sim}/\mathcal{D}_{eff}$	N^*/N
150	1.51	0.83
300	1.49	0.87
600	1.80	0.92

5.2.3
Catalytic Reactions

The combined influence of molecular transport and catalytic reaction in adsorbate–adsorbent systems is generally described by adding a reaction term to the Fickian differential equations of mass transfer and solving the resulting equation for the relevant initial and boundary conditions. Details of this procedure are described in Section 21.1. In the present section we discuss some of the peculiarities that arise if mass transfer in the catalyst is subject to single-file rather than to ordinary diffusion.

Transport inhibition under single-file conditions is intuitively expected to be much more severe, since a product molecule can only get out of the file if all the other molecules in front of it have been shifted in the same direction. Since the probability of such a combined process is very low, the rate of exchange of the product molecules with the surrounding atmosphere must be dramatically smaller than in the case of ordinary diffusion, where the movement of the molecules is uncorrelated. This situation is reflected by the different scaling patterns of the molecular exchange time with increasing crystal size in the cases of normal and single-file diffusion, as discussed at the end of the previous section.

5.2.3.1 Dynamic Monte Carlo Simulations and Adapting the Analysis for Normal Diffusion

As in the case of tracer exchange, quantitative information about the correlated effect of transport and catalytic reactions in single-file systems may be obtained by Monte Carlo simulations. Figure 5.6 illustrates the situation arising from the combined

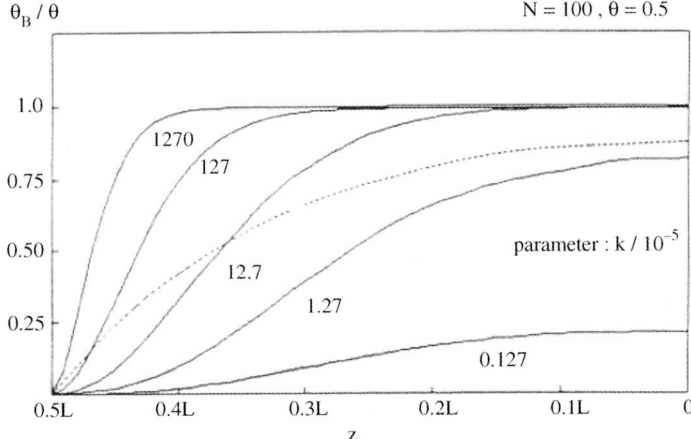

Figure 5.6 Concentration profiles of molecules of species B within a single-file system for the first-order reaction A $\rightarrow$ B under stationary conditions and comparison with the dependence to be expected for ordinary diffusion [broken line, Eq. (5.23)]. The quantity $2L(k/D)^{1/2}$ (the Thiele modulus ϕ) in Eq. (5.23) has been chosen to coincide with the generalized Thiele modulus [cf. Eq. (5.24)] of the single-file reaction for $\kappa = 1.27 \times 10^{-5}$ ($\phi = 2.77$); z denotes the distance from the middle of the file and $L = N\lambda$ its length. Reprinted from Reference [5] with permission.

effects of diffusion and catalytic reaction in a single-file system for the case of a monomolecular first-order reaction A → B [5]. For simplicity, it is assumed that the molecular species A and B are completely equivalent in their microdynamic properties. Moreover, it is assumed that in the gas phase A is in large excess so that only molecules of type A are captured by the marginal sites of the file. Figure 5.6 shows the concentration profile of the reaction product B within the single-file system under stationary conditions. The parameter of this representation is the probability κ that during the mean time between two jump attempts (τ) a molecule of type A is converted into B. It is related to the intrinsic reactivity k by the equation:

$$\kappa = k \cdot \tau \tag{5.19}$$

For normal diffusion, in Chapter 21 [Eq. (21.9)] the corresponding concentration profile is derived as:

$$\frac{\theta_B}{\theta} = 1 - \frac{\cosh\left(z\sqrt{k/D}\right)}{\cosh\left(L/2\sqrt{k/D}\right)} \equiv 1 - \frac{\cosh\left(\frac{2z}{L}\phi\right)}{\cosh\phi} \tag{5.20}$$

with:

$$\phi = \frac{L}{2}\sqrt{\frac{k}{D}} \tag{5.21}$$

denoting the Thiele modulus.

For comparison, Figure 5.6 also displays the concentration profile of species B for the case of normal diffusion as given by Eq. (21.9), with the particular choice of the parameter $\phi = 2.77$.

Figure 5.7 shows the effectiveness factor [Eq. (21.6)], which represents the ratio between the actual (mean) reaction rate in the catalyst particle and the rate in the absence of diffusion limitation:

$$\eta = \frac{k^*}{k} = \frac{\bar\theta_A}{\theta} \tag{5.22}$$

For normal Fickian diffusion the effectiveness factor depends only on the Thiele modulus (Section 21.1). The relevant expression for a one-dimensional system is given by Eq. (21.10):

$$\eta = \frac{\tanh\phi}{\phi} \tag{5.23}$$

This dependence is represented by the dotted line in Figure 5.7. The Thiele concept may be adapted to any transport mechanism [5] by expressing the Thiele modulus in terms of the intracrystalline mean life time τ_{intra}, rather than the intracrystalline diffusivity. With the relation $\tau_{intra,diff} = L^2/(12\mathcal{D})$ used above for deriving Eq. (5.18), Eq. (5.21) may be converted into:

$$\phi = \sqrt{3k\tau_{intra}} \tag{5.24}$$

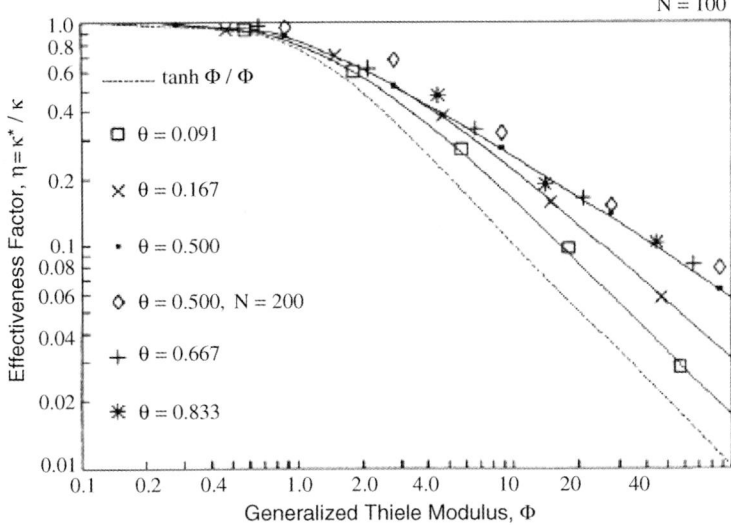

Figure 5.7 Effectiveness factor $\eta = \kappa^*/\kappa$ of a single-file reaction plotted as a function of the generalized Thiele modulus. Reprinted from Reference [5] with permission.

which, in contrast to Eq. (5.21), is not restricted to normal diffusion and includes the case of single-file diffusion. In this way, reactivities under the conditions of normal and single-file diffusion may be directly related to each other. The Thiele modulus of the example of normal diffusion in Figure 5.6 was thus chosen to coincide with the generalized modulus for the single-file case with $\kappa = 1.27 \times 10^{-5}$. Figure 5.7 demonstrates that the dependence of the effectiveness factor in single-file systems on the generalized Thiele modulus is at least qualitatively reflected by the classical dependence of the Thiele concept, as indicated by Eq. (5.23). One should not be confused by the slightly larger values in the case of single-file systems for comparable generalized Thiele moduli, which are indicated by the smaller area below the single-file profile curve for $\kappa = 1.27 \times 10^{-5}$ in Figure 5.6 in comparison with the area under the corresponding profile for normal diffusion with the same Thiele modulus. In the comparison between single-file diffusion and ordinary diffusion the dominant effect is clearly the dramatic enhancement of τ_{intra} and hence of the Thiele modulus, leading to a correspondingly dramatic reduction of the effectiveness factor.

5.2.3.2 Rigorous Treatment
As a consequence of the high degree of correlation, a rigorous treatment of the phenomena occurring in single-file systems should be based on joint probabilities covering the occupancy and further suitable quantities with respect to each individual site. These joint probabilities may be subject to master equations, correlating the file populations after the individual steps of diffusion or reaction. Examples of such studies may be found in References [60–64]. They are summarized in Reference [28], to which we refer in the following discussion.

The description of the state and the evolution of a single-file system during a first-order reaction A → B with uniform molecular dynamic properties may be based on three different sets of probabilities [60]:

1) The configuration probability $\theta^{\sigma_1 \sigma_2 \ldots \sigma_N}$ (with $\sigma_i = 0$ or 1) denoting the probability of a particular occupation pattern, where $\sigma_i = 0$ (1) means that the i-th site is vacant (occupied).
2) The residence time distribution function $\varphi_i^{\sigma_1 \ldots \sigma_{i-1} * \sigma_{i+1} \ldots \sigma_N}(\tau)$ providing the probability (density) that the particle configuration is given by the parameter set $\sigma_1 \ldots \sigma_{i-1} 1 \sigma_{i+1} \ldots \sigma_N$, and that the particle on site i has entered the file a time interval τ ago.
3) The reactant concentration profile $\eta_i^{\sigma_1 \ldots \sigma_{i-1} * \sigma_{i+1} \ldots \sigma_N}$ representing the probability that site i is occupied by a reactant molecule and that the occupation of the other sites (either by reactant or product molecules) is described by the set of parameters $\sigma_1 \ldots \sigma_{i-1}, \sigma_{i+1} \ldots \sigma_N$. Note that $\eta_i^{\sigma_1 \ldots \sigma_{i-1} * \sigma_{i+1} \ldots \sigma_N}$ corresponds to the effectiveness factor defined by Eqs. (5.22) and (21.6), but now referred to only a particular site and a certain file occupation pattern.

All these probabilities are time-invariant under stationary conditions. This means in particular that the probability distribution for the residence time of a particle on a given site (and for a given configuration) is invariant with time. After residing over a time τ within the single-file system, a reactant molecule A will not have been converted into a product molecule B with the probability $\exp(-k\tau)$. Therefore, the reactant concentration profile follows easily by multiplying the residence time distribution function by the probability that for a particular residence time there was no conversion, and then integrating over all residence times τ, yielding:

$$\eta_i^{\sigma_1 \ldots \sigma_{i-1} * \sigma_{i+1} \ldots \sigma_N} = \int_0^\infty e^{-k\tau} \varphi_i^{\sigma_1 \ldots \sigma_{i-1} * \sigma_{i+1} \ldots \sigma_N}(\tau) \, d\tau \qquad (5.25)$$

Thus, for a first-order reaction, the reactant concentration profile is given by the Laplace transform of the residence time distribution (a well-known theorem in chemical reaction engineering).

All these probability functions can be calculated by solving the master equations and transferred into experimentally relevant quantities by summing over the two possible states (vacant and occupied) of all sites of the system. For the relative occupancy of site i one thus obtains:

$$\theta_i = \sum_{\sigma_1=0}^{1} \cdots \sum_{\sigma_{i-1}=0}^{1} \sum_{\sigma_{i+1}=0}^{1} \cdots \sum_{\sigma_N=0}^{1} \theta^{\sigma_1 \ldots \sigma_{i-1} 1 \sigma_{i+1} \ldots \sigma_N} \qquad (5.26)$$

In a completely equivalent way one may determine the residence time distribution $\varphi_i(\tau)$ of site i and the probability η_i that site i is occupied by a reactant molecule. The intracrystalline residence time τ_i is determined by the residence time distribution

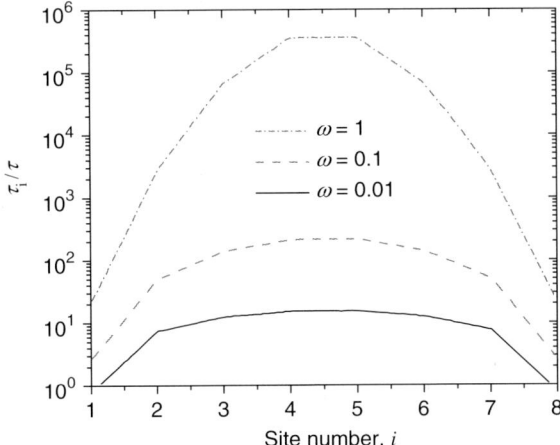

Figure 5.8 Average residence time profile (in units of the time τ between two jump attempts) of the particles in the single-file system for different values of the particle–particle interaction parameter ω. Reprinted from Reference [60] with permission.

function $\varphi_i(\tau)$ via:

$$\tau_i = \int_0^\infty \tau\, \varphi_i(\tau)\, d\tau \qquad (5.27)$$

and denotes the average time that a particle, which is found at site i, has already spent in the channel.

Although the master equations clearly represent analytical expressions for the intrinsic dynamics in single-file systems, the calculations become rather time consuming with increasing file lengths. As a reasonable compromise between computation time and information gained, in Reference [60] single-file systems with eight sites have been considered. As an example, Figure 5.8 shows the average intracrystalline residence times for the particles residing on each of the eight sites considered in this study [60]. In these calculations the assumption of a simple hard core interaction used so far has even been dropped by introducing a particle–particle interaction parameter ω. This is a measure of the reduction in the hopping rates if a jump starts from a position adjacent to another molecule, in comparison with the jump rate of an isolated molecule (with $\omega = 1$ denoting no interaction). As expected, particle mean life times decay from the file center towards the margins and increase with increasing particle–particle interaction.

The mean life time τ_{intra} of an arbitrarily selected particle follows as a mean over all values τ_i. Note that Figure 5.8 displays the average time particles have spent in the whole channel rather than on just the given site.

Summing over all values $\sigma_i = 0$ and 1 and, subsequently, over all sites i, Eq. (5.25) yields:

$$\eta(k) = \int_0^\infty e^{-k\tau} \varphi(\tau)\, d\tau \qquad (5.28)$$

where $\varphi(\tau)$ denotes the residence time distribution function of the molecules in the system and $\eta(k)$ is the effectiveness factor [Eqs. (5.22) and (21.6)]. Adopting the reasoning leading to Eq. (5.21), Eq. (5.28) is found to hold quite generally for any system with an intrinsic reactivity k and a stationary residence time distribution. The residence time distribution curve is not only a function of theoretical relevance; it is also experimentally accessible as the response curve of a tracer ZLC (zero-length column) experiment (Section 14.7) [65–69].

Interestingly, the residence time distribution may also be used as a generating function for the tracer exchange curve as considered in Sections 5.2.2 and 13.7. The relative amount of tracer exchange at a certain time t is simply the sum over all molecules that have left the system between time zero and the present instant and which, therefore, have been replaced by the labeled ones. This means, in other words, that one has to integrate over all molecules with residence times between zero and t:

$$\gamma(t) = \int_0^t \varphi(\tau)\, d\tau \tag{5.29}$$

Reversing the order, the probability distribution of life times results from the curve of tracer exchange via:

$$\varphi(\tau) = \frac{d}{dt}\gamma(t)\Big|_{t=\tau} \tag{5.30}$$

Equations (5.28) and (5.29) hold quite generally for systems with particles with residence times subject to a stationary probability distribution. The interrelation of the "characterizing" functions $\eta(k)$, $\gamma(t)$, and $\varphi(\tau)$ and their wide field of application for stationary population is described in great detail in References [70, 71].

Figure 5.9 displays the probability distribution function $\varphi(\tau)$ and the effectiveness factor $\eta(k)$, which have been calculated via Eqs. (5.30) and (5.28) from the tracer exchange curves in the limiting cases of single-file diffusion, normal diffusion, and barrier confinement. The fact that in all cases the residence time distribution function is found to decrease monotonically may be rationalized as a general property. Owing to the assumed stationary nature of the residence time distribution function, the number of molecules with a residence time τ is clearly the same at any instant of time. The number of molecules with a residence time $\tau + \Delta\tau$ may therefore be considered as the number of molecules with a residence time τ minus the number of molecules that will leave the system in the subsequent time interval $\Delta\tau$. Therefore, $\varphi(\tau)$ must quite generally be a monotonically decaying function.

Figure 5.9a reflects the fact that in the long-time range all curves are simple exponentials. It is therefore impossible to determine the dominant transport mechanism from this part of the distribution curve. In complete agreement with the simulation results given in Figure 5.7, for a given value of $k\tau_{intra}$ and, hence, of the Thiele modulus, the analytically determined effectiveness factors for the typical single-file case (i.e., for large concentrations) as shown in Figure 5.9b are slightly

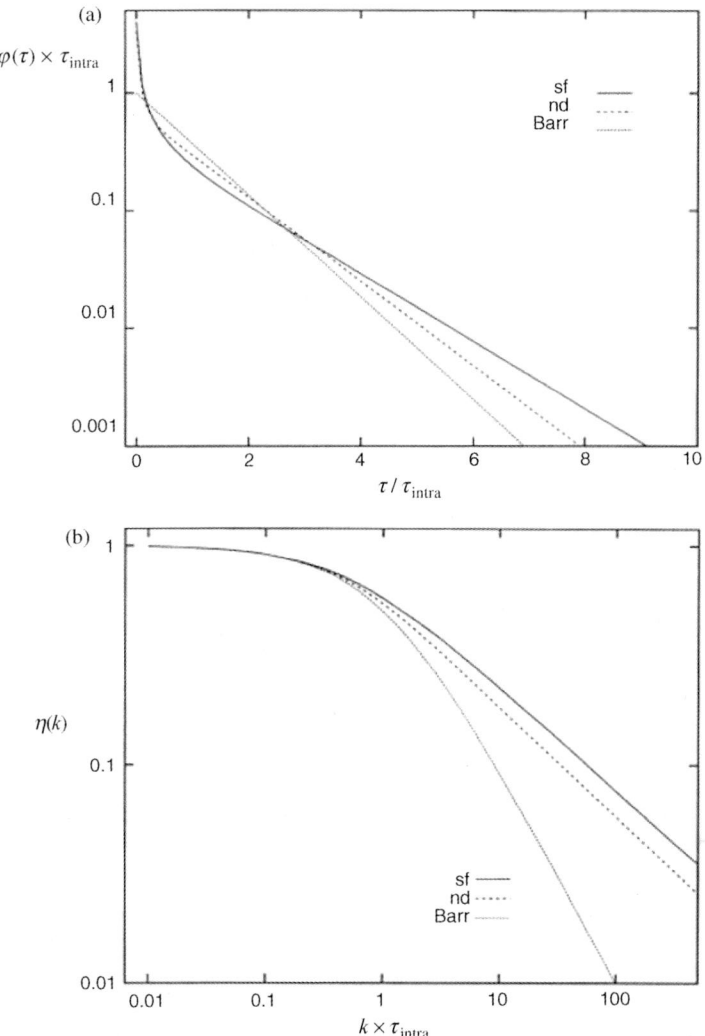

Figure 5.9 Probability distribution function $\varphi(\tau)$ (a) and effectiveness factor $\eta(k)$ (b) corresponding to the tracer exchange curves in the limiting cases of dominating single-file diffusion (sf), normal diffusion (nd), and surface barriers (Barr) as a function of the quotient of τ and τ_{intra}. Reprinted from Reference [72] with permission.

larger than in the case of normal-diffusion limitation. In addition, the effectiveness factor in the case of barrier limitation is found to be even smaller. Interestingly, this sequence also appears in the concentration profiles, which are most pronounced in the case of single-file diffusion (cf. Figure 5.6) and which degenerate to a horizontal line reflecting uniform concentration over the whole sample in the case of transport limitation by surface barriers.

5.2.3.3 Molecular Traffic Control

The concept of "molecular traffic control" (MTC) was proposed by Derouane and Gabelica to explain the apparent lack of transport limitations in some catalytic reactions occurring within zeolites with intersecting channel systems [73]. It refers to

> "a situation in which reactant molecules preferentially enter the catalyst through a given channel system while the products diffuse out by another, thereby preventing the occurrence of important counter-diffusion limitations in the catalytic conversion."

Reactivity enhancement by optimizing the intracrystalline transport of the molecules involved has been demonstrated to occur in networks of interconnecting single-file systems with different accommodation probabilities for the molecules participating in chemical reactions [74, 75]. In the given case, reactivity enhancement by "molecular traffic control" is caused by preferential adsorption of the reactant and product molecules along different diffusion paths (channels) in the interior of the catalyst particle and a corresponding reduction of their mutual interference [76–80].

This tendency can be understood by realizing that the mean lifetime in a single-file system [Eq. (5.18)] scales with the third power of the file length L. We use this proportionality to estimate the mean life time required for the reactant and product molecules to diffuse from one channel intersection to an adjacent one, with L being proportional to the number of sites between the intersections. This type of reasoning clearly applies only for a system of uniform overall concentration, that is, if the sum of the reactant and product concentrations is constant throughout the system. Without the difference in the adsorption preferences of the reactant and product molecules this would also be true for the single-file network. Since the components are attributed to the individual channels with different probabilities, however, the concentration gradients of the individual components inherent to chemical reactions (i.e., falling concentrations from outside to inside for the reactant molecules and increasing concentrations for the product molecules) would also appear to some extent in the channels. Under the influence of such concentration gradients, molecular transport in single-file systems proceeds under the conditions of normal diffusion with the mean lifetime, scaling with L^2 rather than with L^3. Hence, for sufficiently large distances L between the channel intersections, the MTC conditions will ensure a smaller degree of transport inhibition and thus a higher effective reactivity.

Differences in the occupation probabilities of different channels (namely, the straight and sinusoidal ones in MFI-type zeolites) as the main prerequisite of MTC have been shown to exist for CH_4 and CF_4 by two-component molecular simulations (Sections 18.1 and 18.8.2.1) [81]. Particularly detailed studies of this type [82] have been performed with boggsite (BOG), a zeolite possessing two sets of channels of different diameters: ten-ring (roughly 5.1×5.3 Å) and 12-ring (roughly 6.9 Å in diameter), intersecting each other at right angles. Under the conditions considered in the simulations (300 K, equimolar mixture with a total loading of 13.76 molecules per unit cell) the large (SF_6) molecules were found to locate themselves preferentially in the wide pores, while the small (Xe) molecules preferred the narrow pores. More specifically, the site occupancy of SF_6 was determined as 63.7% in the wide pores, 28.6% in the narrow pores, and 7.7% in the intersections, while the site occupancy of Xe was found to be 24.4% in the wide pores, 66.9% in the narrow pores, and 8.7% in the intersections.

The following conditions were found to be fulfilled by the simulated mixture at equilibrium:

1) In the narrow pore, the small molecule has a higher diffusivity than the large molecule.
2) In the wide pore, the large molecule has a higher diffusivity than the small molecule.
3) The large molecule has a higher diffusivity in the wide pore than in the narrow pore.
4) The small molecule has a higher diffusivity in the narrow pore than in the wide pore.

In boggsite, the narrow pores are quite corrugated, while the wide pores are relatively smooth. Condition 1 results mainly from trapping of the large molecules in wide regions of the narrow pores by the corrugations. Condition 2, on the other hand, results from the fact that a large molecule in the narrow pore tends to move along the centerline of the pore, rather than clinging to the pore walls. As the narrow pore is relatively smooth, diffusion is accelerated relative to the small molecule. This unexpected increase of diffusivity as the diameter of confined molecules approaches the diameter of a confining, relatively smooth, pore has been anticipated by Derouane et al. [83] and termed the "floating molecule" effect. The effect has been investigated by MD simulations in great detail [84–87] and is now generally referred to as the "levitation effect" (Section 17.2.3). Furthermore, the presence of large molecules reduces the concentration of small molecules in the wide pores, driving them to a regime where the activation energy for their diffusional motion is higher [85]. Condition 3 results from the fact that the large molecule is quite confined in the narrow pore. Condition 4, on the other hand, is due to the segregation of the large, pore-blocking molecules out of the narrow pores, which allows small molecules to move more freely in the narrow pores. Conditions 1–4 are quite sensitive to the exact dimensions of the sorbed molecules and to the geometry and topology of the pore network. Small changes in the sorbed molecule diameters may cause some of these conditions to disappear.

To confirm that conditions 1–4 above lead to an MTC effect under non-equilibrium conditions (in the presence of concentration gradients), Clark et al. devised an ingeniously simple MD relaxation experiment [82]. They first equilibrated a mixture of 2085 small and large molecules at the occupancy and temperature stated above within the primary simulation box, which was part of an infinite boggsite crystal, using periodic boundary conditions. They then removed all periodic images of the sorbed molecules, creating an initial configuration with the concentration of both species being high inside the primary simulation box (at the levels stated above) and zero everywhere else within the infinite zeolite crystal. They proceeded with an

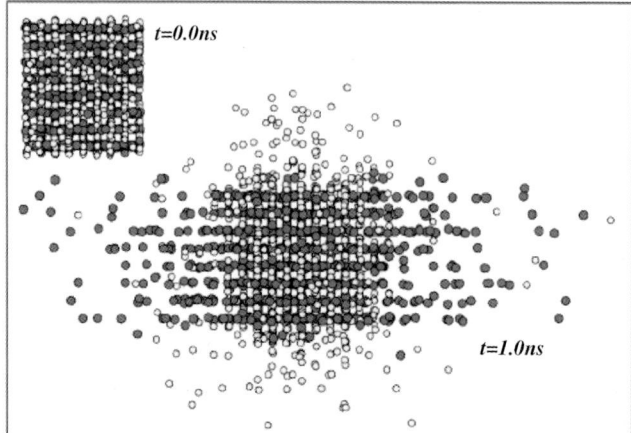

Figure 5.10 Initial ($t=0$) and final ($t=1$ ns) configurations from a transient relaxation simulation of xenon (small molecules, light spheres) and sulfur hexafluoride (large molecules, dark spheres) in an infinite crystal of boggsite, designed to check molecular traffic control effects. Initially, the mixture of molecules is equilibrated within the finite domain of the zeolite shown, using periodic boundary conditions. At $t=0$ all periodic images are removed and the mixture allowed to relax isothermally within the infinite crystal. Large SF_6 molecules are seen to escape the initial domain preferentially along the wide, 12-ring pores (horizontal direction). Conversely, the smaller Xe molecules are seen to escape the initial domain preferentially along the narrow, ten-ring pores (vertical direction). The infinite, periodic boggsite lattice has been omitted from the figure, for clarity. From Reference [82] with permission.

isothermal MD simulation of the system for 1 ns. As a result of the sharp concentration gradients present in the initial configuration, both types of molecules diffused out of the primary simulation box into the surrounding crystal. Remarkably, this diffusion occurred preferentially along the wide channels for the large molecules and along the narrow channels for the small molecules (Figure 5.10). The presence of MTC was thus confirmed under the conditions of the simulation.

5.3
Experimental Evidence

5.3.1
Ideal versus Real Structure of Single-File Host Systems

Numerous nanoporous materials, including zeolites of type ZSM-12, -22, -23, and -48, $AlPO_4$-5, -8, and -11, mordenite and VPI-5, and ordered mesoporous materials of type MCM-41 and SBA-15, are known to consist of parallel channel pores. Since ordered mesoporous materials are, in general, not formed as the result of a crystallization process, only in very exceptional cases [88] are the host particles found to contain a uniform array of channel pores. Moreover, also in these cases (Section 15.3) the

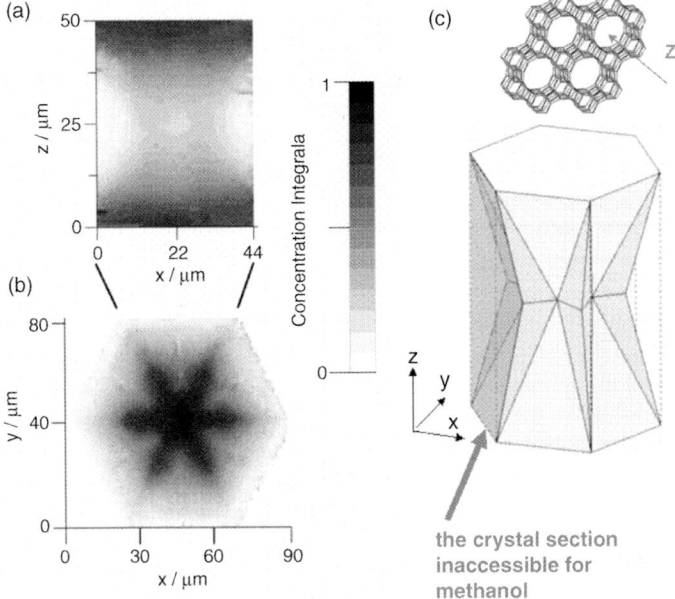

Figure 5.11 Profiles of the integrals of intracrystalline concentration of methanol in a CrAPO-5 crystal in the plane parallel to the crystal axes (a) and in the plane perpendicular to the crystal axes (b), together with the suggested internal structure of CrAPO-5 crystals (c). The profiles were measured under equilibrium with 1 mbar of methanol in the gas phase. Darker regions in parts (a) and (b) correspond to larger concentration integrals. In part (c) only the crystal component shown in different shades of gray is accessible for methanol molecules. x, y, and z are the crystallographic directions (the channel direction is z). Reprinted from Reference [99] with permission.

channel walls turn out to be penetrable, giving rise to substantial deviations from the ideal single-file confinement.

Structural irregularities are, however, also found in crystalline material. As an example, Figure 5.11 shows the projection of the equilibrium concentration of methanol in a zeolite CrAPO-5 [89, 90] onto planes parallel (Figure 5.11a) and perpendicular (Figure 5.11b) to the crystal axis [91, 92] as observed by interference microscopy (Section 12.2.1). Like AlPO$_4$-5, zeolite CrAPO-5 is a member of the family of AFI-type zeolites. The structural model deduced from these data (Figure 5.11c) shows that the crystals are far from the nanoscopic "bundles of macaroni" that one might assume on the basis of the structures as revealed by XRD. The CrAPO-5 crystals rather appear to contain channels, which are accessible only from one side. Only the central channels are accessible from both sides. Similarly, crystals of type SAPO-5, another representative of the AFI family, revealed a substructure that appeared in guest distributions to follow the form of a dumb-bell [91, 93]. These structural peculiarities are correlated with features of the crystallization process [94, 95] and have been confirmed by investigating the template-removal process via

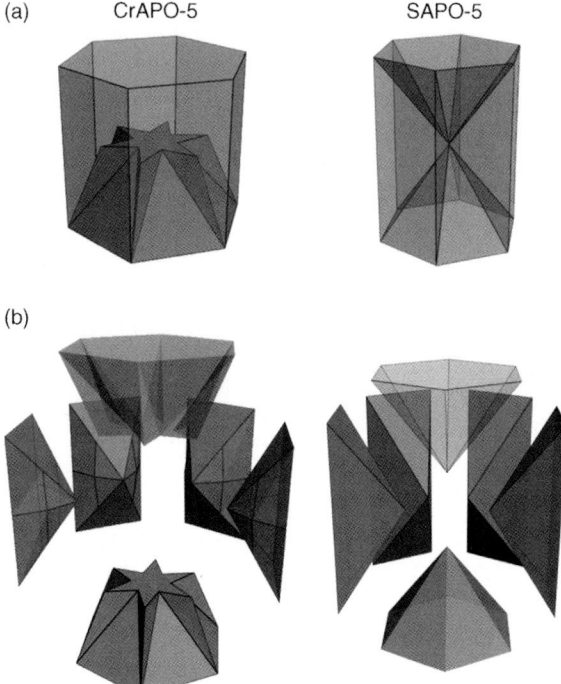

Figure 5.12 Three-dimensional representation of the reconstructed intergrowth structures (based on confocal fluorescence microscopy measurements) of zeolite CrAPO-5 (upper subunit not shown) and SAPO-5 (a) and "opened" representations without front subunits (b). Reprinted from Reference [97] with permission.

optical and fluorescence micro-spectroscopy [96]. From these measurements it has become possible to develop the structure models shown in Figure 5.12 [97, 98].

In Reference [93] the profiles during molecular uptake were found to be symmetric toward the center of the crystal. The profiles could be reproduced by kinetic Monte Carlo simulation but the mean jump rate had to be assumed to decrease during the uptake process.

In some nanoporous host–guest systems even this symmetry is lost. As an example, Figure 5.13 displays the evolution of the concentration profiles of isobutane during uptake by AlPO$_4$-5 [100]. Interestingly, this asymmetry becomes even more pronounced in tracer exchange experiments. If we assume that the asymmetry is caused by a non-uniform distribution of structural defects over the crystal, which happens to lead to a difference in the effective file lengths on either side of the crystal, this tendency can be explained by the different dependencies of the equilibration times on the file lengths. According to Eq. (5.18), the equilibration time under single-file confinement scales with the third power of the file length, while, under diffusion limitation, it scales with only the second power. Remembering that single-file confinement appears in the evolution of concentration profiles only under the

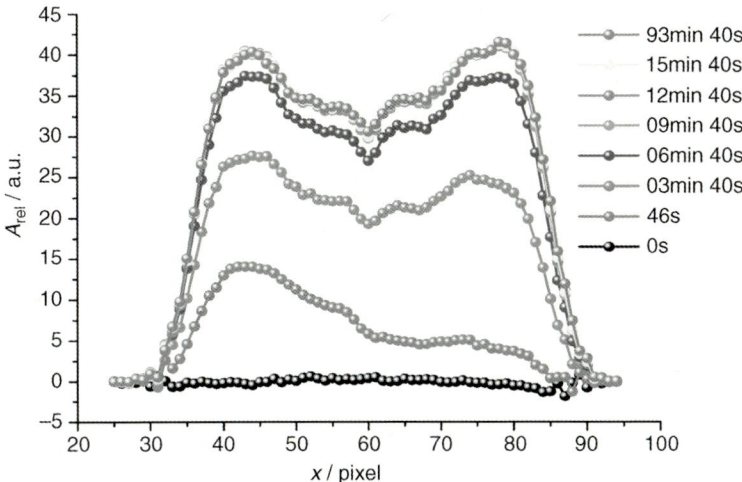

Figure 5.13 Evolution of the intracrystalline concentration of isobutane in AlPO$_4$-5 after a pressure step from zero to 10.26 mbar in the surrounding atmosphere. From Reference [100] with permission.

conditions of tracer exchange (i.e., counter-diffusion) it is clear that the effect of differences in the file lengths will be amplified in tracer exchange.

The given examples illustrate that, in contrast to one-dimensional arrangements of colloidal particles [23, 24, 101], nanoporous materials generally do not have the perfect structure of single-file systems. Having in mind that the nanoporous objects under study are typically composed of something like 10^9 unit cells, the observed deviations from ideality become easily understandable, in particular if one takes into account that the effect of structural defects is much more pronounced in 1D-channel systems than in 3D-pore networks.

5.3.2
Experimental Findings Referred to Single-File Diffusion

The preceding section showed that, as a consequence of inherent structural defects, a nanoporous crystal of single-file structure generally cannot be considered as a unique array of nanoscopic channels. Rather, certain sub-regions within these crystals may be expected to show the characteristic features of single-file diffusion. Some examples are considered in this section.

5.3.2.1 Pulsed Field Gradient NMR
This technique (Sections 11.1–11.4) offers the unique option to follow mean molecular diffusion paths over orders of magnitude, with 1 μm as a typical lower limit. In infinitely long single-file systems, molecular mean square displacements [Eq. (5.2)] are known to scale with the square root of the observation time rather than with the time itself [Eqs. (1.12) and (2.6)] as for ordinary diffusion. We have shown (Section 5.2.1), however, that molecular displacements in single-file systems are

dramatically affected by boundary effects that [by, for example, Eq. (5.16)] are found to become relevant for displacements of only a few micrometers, that is, close to the lower limit of measurability by NMR, for crystals as large as 100 μm. Deviations from structural ideality have moreover been shown to make the length of the unperturbed channel segments substantially smaller than the crystal dimensions. Remarkably, even under these restrictions well-documented NMR diffusion studies in zeolite AlPO$_4$-5 [16, 18, 19] yielded the time dependence $\langle(\Delta z)^2\rangle \propto \sqrt{t}$ of single-file diffusion in infinitely extended systems.

In Section 5.2.1 the two limiting cases of open and closed ends were shown to lead, respectively, to an enhancement and a reduction of the mean square displacement in comparison with an infinite single-file system. Therefore, under the influence of appropriate, intermediate boundary conditions, even in finite single-file systems, molecular propagation may be found to follow the time law [Eq. (5.2)] characteristic of single-file diffusion in an infinite system. This provides a possible explanation of the reported $\langle(\Delta z)^2\rangle \propto \sqrt{t}$ dependency.

5.3.2.2 Quasi-elastic Neutron Scattering

Like pulsed field gradient NMR, quasi-elastic neutron scattering (Section 10.2) is able to trace molecular displacements. These displacements, however, have to be nanometers rather than micrometers. As a consequence, the range of measurement is affected by the transition from the regime of normal diffusion to single-file diffusion [Eq. (5.10)] rather than from single-file diffusion to (normal) center-of-mass diffusion [Eq. (5.15)] which is relevant for the NMR measurements.

Deviating from the NMR studies [16, 18, 19], for both methane and ethane in AlPO$_4$-5 the QENS data could be fitted much better to a model based on unidirectional normal diffusion than to single-file diffusion [102]. Since the minimum scattering vectors of the QENS studies correspond to maximum displacements of the order of 3 nm, it appears that both the methane and ethane molecules must be able to pass each other in the channels. With respect to methane, such a conclusion would be in agreement with Reference [103] but would deviate from the predictions made in References [16, 18, 19]. For ethane, the QENS data did not support the conclusions reached in Reference [16].

For cyclopropane in AlPO$_4$-5 and methane in ZSM-48, however, the QENS data also show the typical scattering pattern of single-file diffusion [102, 104]. For methane in ZSM-48 one may even observe the transition from normal diffusion (at low concentrations, $\mathcal{D} \approx 2.5 \times 10^{-9}\,\mathrm{m^2\,s^{-1}}$) to single-file diffusion (at higher concentrations, $F \approx 2 \times 10^{-12}\,\mathrm{m^2\,s^{-1/2}}$, data for 155 K). Such behavior is in complete agreement with intuition. A quantitative treatment on the basis of Eq. (5.7), however, fails to confirm this agreement. The single-file mobility factors F measured at high concentrations would correspond to much larger single-particle diffusivities than those measured for low concentrations.

The inconsistencies in the results, both within one and the same set of experiments and between different techniques of measurement, illustrate the difficulties of discriminating unambiguously between ordinary and single-file diffusion by following molecular mean square displacements in nanoporous materials. New perspectives are opened by the recent application of the neutron spin echo technique (Section 10.2), which has been shown to allow a better distinction between (one-dimensional) normal and single-file diffusion [105].

5.3.2.3 Tracer Exchange and Transient Sorption Experiments

A second means for identifying single-file diffusion is provided by the dramatic difference in the dependence of the equilibration time on the crystal/file length. While, for tracer exchange under single-file conditions, the equilibration time [i.e., the first moment of the tracer exchange curve – see Eq. (13.8)] scales with the third power of the file length [Eq. (5.18)], under diffusion limitation (i.e., for uptake and release quite in general and for tracer exchange when the molecules are able to mutually exchange their positions) the equilibration (i.e., uptake, release or exchange) time scales with only the square of the file length (Table 13.1). In this context, one also has to realize that the first moment of the curve under limitation by surface barriers (Table 13.1) scales with only the file length itself. Such a situation has recently been observed with well-formed crystals of sodium mordenite, which may be considered as a model system for single-file diffusion. Monitoring the evolution of the guest distribution during n-hexane uptake via IR micro-imaging (Section 12.2) revealed totally flat profiles, which indicate the dominant role of surface resistances in the overall uptake process [106]. Comparable influences of single-file diffusion and barrier limitation [107–109] may thus be expected to lead to a scaling behavior similar to diffusion limitation.

On exploiting the different dependences of the uptake/release or exchange times on the crystal dimension to explore the governing mechanisms of molecular transport, one must also be aware of a second complication, since one cannot necessarily assume coinciding structural properties. Differences in the composition of the individual crystals and in their structural properties in the surface layer and, to a lesser extent, in the bulk phase may cause notable differences in both the surface permeabilities (Section 12.3.2) and diffusivities and/or single-file mobilities between different crystals!

The options of interference and IR microscopy for imaging the evolution of intracrystalline concentration profiles in individually selected crystals (Section 12.2) and, hence, for a microscopic determination of diffusivities and surface permeabilities provide a promising new approach to the study of single-file phenomena during transient sorption experiments. An example of these possibilities is given in Section 19.4, where comparative studies of uptake and tracer exchange are exploited to estimate the genuine single-file length in a nanoporous host material, Zn(tbip), with a one-dimensional pore system.

In a (classical) tracer exchange experiment with propane in a bed of crystals of type $AlPO_4$-5, at a loading of about 0.7 molecules per unit cell the intracrystalline mean life time of propane at room temperature was determined to 1.3×10^4 s, while molecular uptake was observed to proceed with a time constant of less than 100 s [28, 110]. This time constant may be considered as an estimate of the mean life time of an isolated molecule without single-file interaction [first column, second line in Table 13.1, with the diffusivity given by Eq. (2.7)]. Since tracer exchange and molecular uptake have been studied with the identical host, following the discussion in Section 5.2.2, the differences in the observed time constants may be attributed to the differences in the relevant (effective) diffusivities, namely, Eqs. (2.7) and (5.14). In Section 5.2.1, their ratio has been shown to be of the order of the number of molecules in the file. The resulting estimate of more than 100 approaches the order of magnitude expected in the experiments.

In the first tracer ZLC measurements [111] with propane, in AlPO-5 tracer exchange was found to be rather fast and there was no indication of single-file behavior. Moreover, it appears that the (transport) diffusivities resulting from ZLC measurements are in complete agreement with the concentration dependence of the (self-) diffusivities resulting from the correspondingly interpreted tracer ZLC response curves. In the case of single-file diffusion, one should rather have expected substantial differences. As a possible explanation of the difference from the PFG NMR measurements and the NMR tracer exchange studies one has to realize that the sorbate concentrations considered in the tracer ZLC studies are much smaller than those in the PFG NMR measurements. It is therefore possible that the molecular transport monitored in the tracer ZLC studies represents only the behavior of that a fraction of the channels in which imperfections allow molecular exchange with the surroundings.

5.3.2.4 Catalysis

In Section 5.2.3 single-file confinement was shown to dramatically reduce the effectiveness factor of catalytic reactions. This reduction is a consequence of the significant enhancement of the intracrystalline mean life time and, hence, of the generalized Thiele modulus [see Section 21.1 and Eq. (5.24)]. Interestingly, nevertheless, over many years, this peculiarity of molecular transport was largely ignored in the context of catalysis. However, it is also certainly true that in several host–guest systems offering single-file properties, including aromatics in mordenite, molecular transport turned out to have negligible influence on the reaction [112–114]. This is not necessarily in conflict with the above considerations, since catalytic reactions are generally run at high temperatures with correspondingly low sorbate concentrations and with the catalytic component applied as very small crystallites. Owing to the much stronger dependence of the intracrystalline mean life time on the crystal size [Eq. (5.18)] this latter aspect is particularly important for systems with single-file confinement. Finally, in Section 5.3.1 the real structure of nanoporous materials has been shown to deviate notably from the ideal structure so that defects may allow a mutual passage of the reactant and product molecules even in one-dimensional channel systems.

The first studies of catalytic processes that confirm the relevance of single-file diffusion in zeolite catalysis were carried out by Sachtler and coworkers. Considering the palladium-catalyzed conversion of neopentane in supports of different topology, characteristic differences were observed [115]. With zeolite HY, the process proceeded at substantially lower apparent activation energies than in SiO_2-supported catalysts. By contrast, equivalent experiments with zeolite L showed an enhancement of the activation energies. The Pd/HY data, obviously, follow the usual pattern where, under diffusion control, the apparent activation energy is known to decrease with increasing temperature (Section 21.1.3 and Figure 21.3). Transport inhibition of the single-file type may be expected in the unidirectional channels of zeolite L, but, surprisingly, the apparent activation energy was found to be increased! By way of Eq. (5.18), this effect may be attributed to a reduction in the intracrystalline mean life time with decreasing loading, provoked by the temperature increase. Extensive model calculations following the formalism presented in Section 5.2.3 [116, 117] support this supposition. In Reference [20] this effect was confirmed by comparative studies of neopentane reactions over Pt/H-mordenite and Pt/SiO$_2$ catalysts, including H/D exchange, isomerization, and hydrogenolysis. At 150 °C the Arrhenius plot for neopentane conversion over Pt/H-mordenite shows a break, indicating the crossover from the single-file controlled to the chemically controlled regime. In Reference [118], the transport inhibition in single-file systems was demonstrated by a characteristic change in the probability distribution of the number of exchanged atoms during the H/D exchange of cyclopentane on Pd/mordenites. When there is no transport resistance, most likely zero, five, or ten hydrogen atoms are exchanged. This is an immediate consequence of the elementary step of exchange, which, with high probability, involves all five hydrogen atoms on the side of the molecule that is in contact with the metal. Such a distribution has been observed for transition metals on amorphous supports and in zeolite Y [119]. As a consequence of the large residence time under single-file conditions, however, before desorbing into the gas phase the molecules react many times so that the initial "double-U-shaped" probability distribution of the number of exchanged hydrogen atoms is lost. Following these studies, in Reference [120] the hydroisomerization of n-hexane over Pt/H-mordenites was successfully described by including the loading dependence of the effective diffusivities as given by the single-file expression, Eq. (5.14).

References

1 Kocirik, M., Kornatowski, J., Masarik, V., Novak, P., Zikanova, A., and Maixner, J. (1998) *Microporous Mesoporous Mater.*, **23**, 295.

2 Kortunov, P., Vasenkov, S., Chmelik, C., Kärger, J., Ruthven, D.M., and Wloch, J. (2004) *Chem. Mater.*, **16**, 3552.

3 Kärger, J., Hahn, K., Kukla, V., and Rödenbeck, C. (1998) *Physik. Blätter*, **54**, 811.

4 Czaplewski, K.F., Reitz, T.L., Kim, Y.J., and Snurr, R.Q. (2002) *Microporous Mesoporous Mater.*, **56**, 55.

5 Kärger, J., Petzold, M., Pfeifer, H., Ernst, S., and Weitkamp, J. (1992) *J. Catal.*, **136**, 283.

6 Riekert, L. (1970) *Adv. Catal.*, **21**, 281.

7 Rickert, H. (1964) *Z. Phys. Chem. NF*, **43**, 129.

8 Hodkin, A.L. and Keynes, R.D. (1955) *J. Physiol.*, **128**, 61.

9 Jepsen, D. (1965) *J. Math. Phys. (N.Y.)*, **6**, 405.

10 Harris, T.E. (1965) *J. Appl. Probab.*, **2**, 323.

11 Fedders, P.A. (1978) *Phys. Rev. B*, **17**, 40.

12 van Bejeren, H., Kehr, K.W., and Kutner, R. (1983) *Phys. Rev. B*, **28**, 5711.

13 Kehr, K.W., Mussawisade, K., Schütz, G.M., and Wichmann, T. (2005) in *Diffusion in Condensed Matter: Methods, Materials, Models* (eds P. Heitjans and J. Kärger), Springer, Berlin, Heidelberg, p. 745.

14 Kärger F J. (1992) *Phys. Rev. A*, **45**, 4173.

15 Kärger, J. (1993) *Phys. Rev. E*, **47**, 1427.

16 Gupta, V., Nivarthi, S.S., McCormick, A.V., and Davis, H.T. (1995) *Chem. Phys. Lett.*, **247**, 596.

17 Brandani, S. (1996) *J. Catal.*, **160**, 326.

18 Hahn, K., Kärger, J., and Kukla, V. (1996) *Phys. Rev. Lett.*, **76**, 2762.
19 Kukla, V., Kornatowski, J., Demuth, D., Girnus, I., Pfeifer, H., Rees, L.V.C., Schunk, S., Unger, K.K., and Kärger, J. (1996) *Science.*, **272**, 702.
20 Lei, G.D., Carvill, B.T., and Sachtler, W.M.H. (1996) *Appl. Catal., A*, **142**, 347.
21 Nelson, P.H. and Auerbach, S.M. (1999) *J. Chem. Phys.*, **110**, 9235.
22 Burada, P.S., Hänggi, P., Marchesoni, F., Schmid, G., and Talkner, P. (2009) *ChemPhysChem*, 45.
23 Lutz, C., Kollmann, M., Leiderer, P., and Bechinger, C. (2004) *J. Phys.: Condens. Mat.*, **16**, S4075–S4083.
24 Lutz, C., Kollmann, M., and Bechinger, C. (2004) *Phys. Rev. Lett.*, **93**, 26001.
25 Kollmann, M. (2003) *Phys. Rev. Lett.*, **90**, 180602.
26 Marchesoni, F. and Taloni, A. (2007) *Chaos*, **17**, 043112.
27 Taloni, A. and Marchesoni, F. (2006) *Phys. Rev. Lett.*, **96**, 020601.
28 Kärger, J. (2008) in *Adsorption and Diffusion* (eds H.G. Karge and J. Weitkamp), Springer, Berlin, Heidelberg, p. 329.
29 Levitt, D.G. (1973) *Phys. Rev. A*, **8**, 3050.
30 Hahn, K. and Kärger, J. (1995) *J. Phys. A-Math. Gen.*, **28**, 3061.
31 Ben-Avraham, D. and Havlin, S. (2000) *Diffusion and Reaction in Fractals and Disordered Systems*, University Press, Cambridge.
32 Roque-Malherbe, R. and Ivanov, V. (2001) *Microporous Mesoporous Mater.*, **47**, 25.
33 Kärger, J. (2002) *Microporous Mesoporous Mater.*, **56**, 321.
34 Roque-Malherbe, R.M. (2002) *Microporous Mesoporous Mater.*, **56**, 322.
35 Hahn, K. and Kärger, J. (1996) *J. Phys. Chem.*, **100**, 316.
36 Keffer, D., McCormick, A.V., and Davis, H.T. (1996) *Mol. Phys.*, **87**, 367.
37 Bhide, S.Y. and Yashonath, S. (2002) *J. Phys. Chem. A*, **106**, 7130.
38 Demontis, P., Gonzalez, J.G., Suffritti, G.B., and Tilocca, A. (2001) *J. Am. Chem. Soc.*, **123**, 5069.
39 Demontis, P., Suffritti, G.B., Bordiga, S., and Buzzoni, R. (1995) *J. Chem. Soc., Faraday Trans.*, **91**, 525.
40 Fritzsche, S., Haberlandt, R., Hofmann, G., Kärger, J., Heinzinger, K., and Wolfsberg, M. (1997) *Chem. Phys. Lett.*, **265**, 253.
41 Fan, J.F., Xia, Q.Y., Gong, X.D., and Xiao, H.M. (2001) *Chin. J. Chem.*, **19**, 251.
42 Hahn, K. and Kärger, J. (1998) *J. Phys. Chem.*, **102**, 5766.
43 Skoulidas, A.I. and Sholl, D.S. (2003) *J. Phys. Chem. A.*, **107**, 10132.
44 MacElroy, J.M.D. and Suh, S.H. (2001) *Microporous Mesoporous Mater.*, **48**, 195.
45 Jakobtorweihen, S., Verbeek, M.G., Lowe, C.P., Keil, F.J., and Smit, B. (2005) *Phys. Rev. Lett.*, **95**, 044501.
46 Krishna, R. and van Baten, J.M. (2006) *Ind. Eng. Chem. Res.*, **45**, 2084.
47 Jakobtorweihen, S., Lowe, C.P., Keil, F.J., and Smit, B. (2006) *J. Chem. Phys.*, **124**, 154706(1)–154706(13).
48 Jakobtorweihen, S., Verbeek, M.G., Lowe, C.P., Keil, F.J., and Smit, B. (2007) *J. Chem. Phys.*, **127**, 024904(1)–024904(11).
49 Liu, Y.C., Shen, J.W., Gubbins, K.E., Moore, J.D., Wu, T., and Wang, Q. (2008) *Phys. Rev. B*, **77**, 125438(1)–125438(7).
50 Liu, Y.C., Moore, J.D., Chen, Q., Roussel, J., Wang, Q., and Gubbins, K.E. (2009) in *Diffusion Fundamentals III* (eds C. Chmelik, N.K. Kanellopoulos, J. Kärger, and D. Theodorou), Leipziger Universitätsverlag, Leipzig, p. 164.
51 Sholl, D.S. and Fichthorn, K.A. (1997) *Phys. Rev. Lett.*, **79**, 3569.
52 Sholl, D.S. and Fichthorn, K.A. (1997) *Phys. Rev. A*, **55**, 7753.
53 Sholl, D.S. and Fichthorn, K.A. (1997) *J. Chem. Phys.*, **107**, 4384.
54 Sholl, D.S. and Lee, C.K. (2000) *J. Chem. Phys.*, **112**, 817.
55 Nelson, P.H. and Auerbach, S.M. (1999) *Chem. Eng. J.*, **74**, 43.
56 Rödenbeck, C. and Kärger, J. (1999) *J. Chem. Phys.*, **110**, 3970.
57 Kutner, R. (1981) *Phys. Lett.*, **81**, 239.
58 Vasenkov, S. and Kärger, J. (2002) *Phys. Rev. E*, **66**, 052601-1.
59 Barrer, R.M. (1978) *Zeolites and Clay Minerals as Sorbents and Molecular Sieves*, Academic Press, London.
60 Rödenbeck, C., Kärger, J., and Hahn, K. (1997) *Phys. Rev.*, **55**, 5697.

61. Rödenbeck, C., Kärger, J., and Hahn, K. (1998) *Phys. Rev. E.*, **57**, 4382.
62. Nedea, S.V., Jansen, A.P.J., Lukkien, J.J., and Hilbers, P.A.J. (2002) *Phys. Rev. E.*, **65**, 6701.
63. Nedea, S.V., Jansen, A.P.J., Lukkien, J.J., and Hilbers, P.A.J. (2002) *Phys. Rev. E*, **66**, 66705.
64. Nedea, S.V., Jansen, A.P.J., Lukkien, J.J., and Hilbers, P.A.J. (2003) *Phys. Rev. E*, **67**, 46707.
65. Rödenbeck, C., Kärger, J., and Hahn, K. (1997) *Collect. Czech. Chem. C*, **62**, 995.
66. Brandani, S., Hufton, J., and Ruthven, D. (1995) *Zeolites*, **15**, 624.
67. Hufton, J.R., Brandani, S., and Ruthven, D.M. (1994) in *Zeolites and Related Microporous Materials: State of the Art 1994* (eds J. Weitkamp, H.G. Karge, H. Pfeifer, and W. Hölderich), Elsevier, Amsterdam, p. 1323.
68. Ruthven, D.M. and Post, M.F.M. (2001) in *Introduction to Zeolite Science and Practice* (eds H. van Bekkum, E.M. Flanigen, and J.C. Jansen), Elsevier, Amsterdam, p. 525.
69. Ruthven, D.M. and Brandani, S. (2000) in *Recent Advances in Gas Separation by Microporous Ceramic Membranes* (ed. N.K. Kanellopoulos), Elsevier, Amsterdam, p. 187.
70. Rödenbeck F C., Kärger, J., Schmidt, H., Rother, T., and Rödenbeck, M. (1999) *Phys. Rev. E*, **60**, 2737.
71. Rödenbeck, C., Kärger, J., and Hahn, K. (1998) *Ber. Bunsen-Ges. Phys. Chem. Chem. Phys.*, **102**, 929.
72. Rödenbeck, C. (1997) Mathematische Modellierung von Transport- und Umwandlungsprozessen in Single-file Systemen. PhD thesis Leipzig University.
73. Derouane, E.G. and Gabelica, Z. (1980) *J. Catal.*, **65**, 486.
74. Kärger, J., Bräuer, P., and Pfeifer, H. (2000) *Phys. Chem. Chem. Phys.*, **214**, 1707.
75. Neugebauer, N., Bräuer, P., and Kärger, J. (2000) *J. Catal.*, **194**, 1.
76. Bräuer, P., Brzank, A., and Kärger, J. (2003) *J. Phys. Chem. B*, **107**, 1821.
77. Brzank, A., Schutz, G.M., Brauer, P., and Kärger, J. (2004) *Phys. Rev. E*, **69**, 031102.
78. Brzank, A. and Schütz, G.M. (2005) *Appl. Catal., A*, **288**, 194.
79. Bräuer, P., Brzank, A., and Kärger, J. (2006) *J. Chem. Phys.*, **124**, 034713.
80. Brzank, A. and Schutz, G.M. (2006) *J. Chem. Phys.*, **124**, 214701.
81. Snurr, R.Q. and Kärger, J. (1997) *J. Phys. Chem. B*, **101**, 6469.
82. Clark, L.A., Ye, G.T., and Snurr, R.Q. (2000) *Phys. Rev. Lett.*, **84**, 2893.
83. Derouane, E.G., André, J.M., and Lucas, A.A. (1988) *J. Catal.*, **110**, 486.
84. Yashonath, S. and Bandyopadhyay, S. (1994) *Chem. Phys. Lett.*, **228**, 284.
85. Bandyopadhyay, S. and Yashonath, S. (1995) *J. Phys. Chem.*, **99**, 4286.
86. Yashonath, S. and Ghorai, P.K. (2008) *J. Phys. Chem. B*, **112**, 665.
87. Bhide, S.Y. and Yashonath, S. (2004) *Mol. Phys.*, **102**, 1057.
88. Stallmach, F., Kärger, J., Krause, C., Jeschke, M., and Oberhagemann, U. (2000) *J. Am. Chem. Soc.*, **122**, 9237.
89. Kornatowski, J., Zadrozna, G., Wloch, J., and Rozwadowski, M. (1999) *Langmuir*, **15**, 5863.
90. Kornatowski, J., Zadrozna, G., Rozwadowski, M., Zibrowius, B., Marlow, F., and Lercher, J.A. (2001) *Chem. Mater.*, **13**, 4447.
91. Lehmann, E., Chmelik, C., Scheidt, H., Vasenkov, S., Staudte, B., Kärger, J., Kremer, F., Zadrozna, G., and Kornatowski, J. (2002) *J. Am. Chem. Soc.*, **124**, 8690.
92. Lehmann, E., Vasenkov, S., Kärger, J., Zadrozna, G., Kornatowski, J., Weiss, Ö., and Schüth, F. (2003) *J. Phys. Chem. B*, **107**, 4685.
93. Lehmann, E., Vasenkov, S., Kärger, J., Zadrozna, G., and Kornatowski, J. (2003) *J. Chem. Phys.*, **118**, 6129.
94. Schunk, S.A., Demuth, D.G., SchulzDobrick, B., Unger, K.K., and Schuth, F. (1996) *Microporous Mater.*, **6**, 273.
95. Jacobs, W., Demuth, D.G., Schunk, S.A., and Schuth, F. (1997) *Microporous Mater.*, **10**, 95.
96. Karwacki, L., Stavitski, E., Kox, M.H.F., Kornatowski, J., and Weckhuysen, B.M. (2007) *Angew. Chem. Int. Ed.*, **46**, 7228.
97. Karwacki, L. (2010) Internal architecture and molecular transport barriers of large zeolite crystals. Thesis, University of Utrecht.

98 Karwacki, L., van der Bij, H.E., Kornatowski, J., Cubillas, P., Drury, M.R., de Winter, D.A.M., Anderson, M.W., and Weckhuysen, B.M. (2010) *Angew. Chem. Int. Ed.*, **49**, 6790.

99 Kärger, J., Valiullin, R., and Vasenkov, S. (2005) *New J. Phys.*, **7**, 15.

100 Titze, T. (2009) Untersuchung des Stofftransportes in nanoporösen Materialien mittels IR-Microimaging Diploma Thesis, Leipzig University.

101 Wei, Q.H., Bechinger, C., and Leiderer, P. (2000) *Science*, **287**, 625.

102 Jobic, H., Hahn, K., Kärger, J., Bée, M., Noack, M., Grinus, I., Tuel, A., and Kearley, G.J. (1997) *J. Phys. Chem.*, **101**, 5834.

103 Nivarthi, S.S., McCormick, A.V., and Davis, H.T. (1994) *Chem. Phys. Lett.*, **229**, 297.

104 Hahn, K., Jobic, H., and Kärger, J. (1999) *Phys. Rev. E*, **59**, 6662.

105 Jobic, H. and Farago, B. (2008) *J. Chem. Phys.*, **129**, 171102.

106 Zhang, L., Chmelik, C., van Laak, A.N.C., Kärger, J., de Jongh, P.E., and de Jong, K.P. (2009) *Chem. Commun.*, 6424.

107 Gulin-Gonzalez, J., Schuring, A., Fritzsche, S., Kärger, J., and Vasenkov, S. (2006) *Chem. Phys. Lett.*, **430**, 60.

108 Schuring, A., Vasenkov, S., and Fritzsche, S. (2005) *J. Phys. Chem. B*, **109**, 16711.

109 Vasenkov, S., Schuring, A., and Fritzsche, S. (2006) *Langmuir*, **22**, 5728.

110 Kukla, V. (1997) NMR-Untersuchungen zur Moleküldiffusion in Zeolithen mit eindimensionaler Porenstruktur PhD Thesis, Leipzig University.

111 Brandani, S., Ruthven, D.M., and Kärger, J. (1997) *Microporous Mater.*, **8**, 193.

112 Karge, H.G., Ladebeck, J., Sarbak, Z., and Hatada, K. (1982) *Zeolites*, **2**, 94.

113 Karge, H.G. and Weitkamp, J. (1986) *Chem.-Ing.-Tech.*, **58**, 946.

114 Weitkamp, J. and Ernst, S. (1994) *Catal. Today*, **19**, 107.

115 Kapinski, Z., Ghandi, S.N., and Sachtler, W.M.H. (1993) *J. Catal.*, **141**, 337.

116 Rödenbeck, C., Kärger, J., and Hahn, K. (1998) *J. Catal.*, **176**, 513.

117 Rödenbeck, C., Kärger, J., Hahn, K., and Sachtler, W. (1999) *J. Catal.*, **183**, 409.

118 Lei, G.D. and Sachtler, W.M.H. (1993) *J. Catal.*, **140**, 601.

119 Vanbroekhoven, E.H., Schoonhoven, J., and Ponec, V. (1985) *Surf. Sci.*, **156**, 899.

120 de Gauw, F.J.M.M., van Grondelle, J., and van Santen, R.A. (2001) *J. Catal.*, **204**, 53.

6
Sorption Kinetics

In previous chapters we have considered the fundamental principles of diffusion and some of the mechanisms by which diffusion occurs in an adsorbed phase. However, the relationship between the diffusional behavior and the macroscopic adsorption/desorption kinetics has not yet been addressed. This is the focus of the present chapter. An understanding of this relationship is essential for the practical application of microporous materials as catalysts and selective adsorbents. It is also critically important for the measurement of diffusivities, which is considered in Part Four of this book. The fluid–solid contactors used in several large-scale adsorption processes, such as water purification with powdered activated carbon, are simply scaled up versions of the laboratory systems used for sorption kinetic studies. The theory presented here therefore serves directly as the basis for modeling, design, and analysis of such processes. Much of this theory is drawn from the books by Crank [1] and Carslaw and Jaeger [2], in which numerous solutions of the "diffusion equation" and the analogous "heat conduction equation" are summarized.

6.1
Resistances to Mass and Heat Transfer

Physical adsorption is an extremely rapid process so that in a porous catalyst or adsorbent the overall rate of adsorption or desorption is almost always controlled by mass or heat transfer resistances rather than by the intrinsic rate of sorption at the active surface.

A biporous adsorbent of the type sketched in Figures 1.3b and 6.1 offers at least two and possibly three distinct resistances to mass transfer: the diffusional resistance of the micropores within the microparticles, the diffusional resistance of the meso- and macropores, and a possible barrier to mass transfer at the external surface of the microparticle. In addition, when adsorption occurs from a binary or multicomponent fluid phase there may be an external resistance associated with diffusion through the surrounding laminar fluid film.

The relative magnitude of these resistances varies widely, depending on the particular system and the conditions. Figure 6.1 shows the forms assumed by the

Diffusion in Nanoporous Materials. Jörg Kärger, Douglas M. Ruthven, and Doros N. Theodorou.
© 2012 Wiley-VCH Verlag GmbH & Co. KGaA. Published 2012 by Wiley-VCH Verlag GmbH & Co. KGaA.

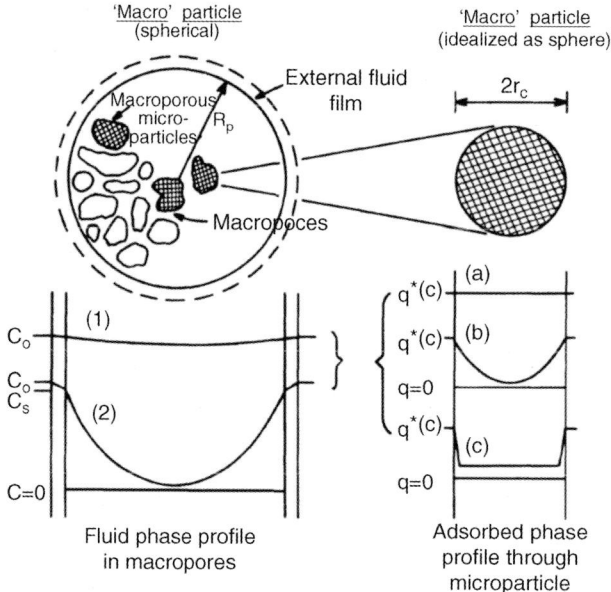

Figure 6.1 Schematic diagram showing the form of the transient concentration profile through a biporous adsorbent particle. External film resistance is assumed to be small. Profiles (1) and (2) correspond to small and large macropore diffusion resistance, respectively. The microparticle profiles correspond to (a) negligible microparticle resistance, (b) significant micropore diffusional resistance, and (c) mass transfer resistance at the microparticle surface.

intraparticle concentration profile in some of the more commonly encountered regimes. An order-of-magnitude analysis suggests that external fluid film resistance (Section 1.3.3) is seldom dominant although it is often large enough to exert some effect on the sorption rate. The relative importance of macro- and micro-resistances depends on the ratio of the diffusional time constants $[(D_{\mathrm{micro}}/r_c^2)]/(D_{\mathrm{macro}}/R_p^2)$. This can vary over several orders of magnitude so both the extreme limits of micro- or macro-control, as well as intermediate cases in which both these resistances are significant, are relatively common. Varying the sizes of the macro- and microparticles provides a straightforward experimental test to distinguish between these resistances.

Since adsorption is in general an exothermic process the temperature of the adsorbent will remain constant only when the rate of heat dissipation by conduction, convection, and radiation is high relative to the adsorption rate. In parallel with the mass transfer resistances one may distinguish three distinct resistances to heat transfer: conduction through micro- and macro-particles and convection/radiation from the external surface. The latter is generally the most important, as can be demonstrated by a simple order-of-magnitude analysis (Section 21.1.6). The conditions under which thermal effects may be neglected and the system treated as isothermal are also considered in Section 21.1.6.

6.2
Mathematical Modeling of Sorption Kinetics

Mass transfer rates are conveniently represented by the diffusion model [Eqs. (1.1)–(1.3)]. At sufficiently low concentrations or for small differential steps (linear systems) the diffusivity can be considered as independent of concentration, thus greatly simplifying the mathematical model. However, this simplification is generally not appropriate over large ranges of concentration since the equilibrium relationships then become nonlinear, leading to a significant concentration dependence of the diffusivity. A further complication arises when the isothermal approximation breaks down since it is then necessary to consider mass and heat transfer as coupled processes. In the discussion of sorption kinetics we therefore first consider linear isothermal (constant diffusivity) systems and then extend the discussion to nonlinear and non-isothermal systems.

The simplest situation is adsorption or desorption in a single zeolite crystal (or other homogeneous particle) since, for such a system, the intrusion of extracrystalline resistances to heat and mass transfer is generally minimal. We therefore consider first the behavior of a single adsorbent particle and then extend the discussion to composite particles or assemblages of crystals, such as are commonly used in processes or indeed in most kinetic measurements.

6.2.1
Isothermal Linear Single-Resistance Systems

6.2.1.1 External Fluid Film or Surface Resistance Control

If diffusion within the adsorbent particle is very rapid or if the adsorbent is nonporous, so that adsorption can occur only on the external surface, then the rate of sorption will be controlled by diffusion through the laminar fluid film surrounding the particle. Under these conditions there is no gradient of concentration through the particle and the adsorbed phase concentration throughout the particle is at equilibrium with the fluid phase concentration at the surface. The sorption rate is then given by

$$\frac{d\bar{q}}{dt} = k_f a (C - C^*) \tag{6.1}$$

where k_f is the film mass transfer coefficient and a is the external surface area per unit volume. For a parallel sided slab of thickness $2l$, $a = 1/l$ and, for a spherical particle of radius R_p, $a = 3/R_p$. If the equilibrium relationship is linear ($q^* = KC$), which will generally be true when the concentration change is sufficiently small, and the bulk fluid phase concentration C is maintained constant the rate expression may be written equivalently in the form:

$$\frac{d\bar{q}}{dt} = \frac{k_f a}{K}(q^* - \bar{q}) \tag{6.2}$$

where q^* is the final equilibrium adsorbed phase concentration. The same form of rate expression applies (with k_f replaced by the "solid film" coefficient $k_s = D/\delta$) when the rate is controlled by diffusional resistance through a thin film (thickness δ) at the surface of the solid due, for example, to pore blockage or constriction of the pore mouths. The general rate expression for combined surface and external film resistance follows from the principle of additivity of resistances:

$$\frac{d\bar{q}}{dt} = k\,a(q^* - \bar{q}) \tag{6.3}$$

where the overall coefficient (k) is given by:

$$\frac{1}{k} = \frac{K}{k_f} + \frac{1}{k_s} \tag{6.4}$$

Note that the symbol α (rather than k) is also commonly used for this coefficient (e.g., in Crank's text book [1]). For a step change in concentration at time zero:

$$\begin{aligned} t < 0, \quad & C = q = 0 \\ t > 0, \quad & C = C_\infty = q_\infty/K \end{aligned} \tag{6.5}$$

Eq. (6.3) may be integrated directly to yield:

$$\frac{\bar{q}}{q_\infty} = 1 - e^{-kat} \tag{6.6}$$

which corresponds to a simple exponential approach to equilibrium.

Since diffusion within the pores of the particle is almost always slower than diffusion through the external fluid film, external film resistance usually makes only a minor contribution, except in the case of non-porous particles. Surface resistance is, however, very common and sometimes completely rate controlling, especially for zeolite crystals. The physical significance of the surface rate coefficient may be understood by considering it as the quotient of the effective diffusivity and thickness of the solid surface layer ($k_s = D_s/\delta$).

6.2.1.2 Micropore Diffusion Control

When diffusion within the microparticle is rate controlling, the uptake rate is given by the appropriate solution of the transient diffusion equation, which, for a spherical adsorbent particle, may be written in the form:

$$\frac{\partial q}{\partial t} = \frac{1}{r^2}\frac{\partial}{\partial r}\left(r^2 D_c \frac{\partial q}{\partial r}\right) \tag{6.7}$$

If the uptake occurs over a small change in the adsorbed phase concentration we may assume a constant diffusivity, thus simplifying Eq. (6.7) to:

$$\frac{\partial q}{\partial t} = D_c\left(\frac{\partial^2 q}{\partial r^2} + \frac{2}{r}\frac{\partial q}{\partial r}\right) \tag{6.8}$$

For a step change in concentration at time zero the relevant initial and boundary conditions are:

$$t < 0, \quad C = C_0, \quad q = q_0 \text{(independent of } r \text{ and } t\text{)} \tag{6.9a}$$

$$t \geq 0, \quad C = C_\infty, \quad q(r_c, t) \to q_\infty \tag{6.9b}$$

$$t \to \infty, \quad C = C_\infty, \quad q(r, t) \to q_\infty \tag{6.9c}$$

$$\left.\frac{\partial q}{\partial t}\right|_{r=0} = 0 \text{ for all } t \tag{6.9d}$$

and we obtain the familiar solution for the transient sorption curve:

$$\frac{m_t}{m_\infty} = \frac{\bar{q}-q_0}{q_\infty-q_0} = 1 - \frac{6}{\pi^2} \sum_{n=1}^{\infty} \frac{1}{n^2} \exp\left(-\frac{n^2\pi^2 D_c t}{r_c^2}\right) \tag{6.10}$$

which may also be written in the equivalent form:

$$\frac{m_t}{m_\infty} = \frac{\bar{q}-q_0}{q_\infty-q_0} = 6\left(\frac{D_c t}{r_c^2}\right)^{1/2}\left[\frac{1}{\sqrt{\pi}} + 2\sum_{n=1}^{\infty} \text{ierfc}\left(\frac{nr_c}{\sqrt{D_c t}}\right)\right] - \frac{3D_c t}{r_c^2} \tag{6.11}$$

This form is more convenient for short times, since the summation in Eq. (6.10) converges only very slowly when t is small. A simplified expression for the initial region of the uptake curve may be obtained by neglecting the higher-order terms in Eq. (6.11):

$$\frac{m_t}{m_\infty} \approx \frac{6}{\sqrt{\pi}}\sqrt{\frac{D_c t}{r_c^2}} \tag{6.12}$$

This corresponds physically to the situation where the concentration front has not yet reached the center of the particle so that diffusion occurs as if in a semi-infinite medium. The corresponding solution for particles of any arbitrary shape can then be written in the form:

$$\frac{m_t}{m_\infty} \approx \frac{2A}{V}\left(\frac{D_c t}{\pi}\right)^{1/2} \tag{6.13}$$

where $A/V = a$ is the ratio of external surface area to particle volume.

In the long-time region, all terms except the first in the series of exponential terms in Eq. (6.10) become negligible so that the uptake curve approaches the asymptotic form:

$$\frac{m_t}{m_\infty} = 1 - \frac{6}{\pi^2}\exp\left[-\frac{\pi^2 D_c t}{r_c^2}\right] \tag{6.14}$$

from which it is apparent that a plot of $\ln(1 - m_t/m_\infty)$ versus t will approach a straight line with slope $-\pi^2 D_c/r_c^2$ and intercept $\ln(6/\pi^2)$ at $t = 0$. The difference from Eq. (6.6), for which such a plot passes through the origin, is clearly apparent, and provides one way of distinguishing between intraparticle diffusion and a surface resistance to mass transfer. In practice this limiting form is somewhat less useful than the expression for the initial rate since any distribution of particle size leads to curvature of the semi-logarithmic plot. Furthermore, the long-time region of the uptake curve is more sensitive than the initial region to intrusion of thermal effects.

The assumption of a constant boundary condition [Eq. (6.9a)], which is implicit in the derivation of these solutions, is valid when the uptake within the adsorbent sample is very small compared with the capacity of the system. For measurements with strongly adsorbed species, however, this situation may require an impractically large system volume. Furthermore, in measuring uptake rates from the liquid phase it is generally more convenient to monitor the fluid phase concentration (rather than the weight of the particle or the adsorbed phase concentration) and to do this it is necessary to operate with a relatively small system volume, so that the change in the fluid concentration following the initial step will be large and easily measurable. The boundary conditions given by Eq. (6.9a) must then be replaced by:

$$t < 0, \quad C = C_0, \quad q = q_0$$
$$t \geq 0, \quad C = C'_0, \quad q(r_c, t) = q^*(t)$$
$$-V_f \frac{dC}{dt} = \frac{3V_s}{r_c} D_c \frac{\partial q}{\partial r}\bigg|_{r_c} \tag{6.15}$$

The solution for the uptake curve then becomes:

$$\frac{m_t}{m_\infty} = \frac{\bar{q} - q_0}{q_\infty - q_0} = 1 - 6 \sum_{n=1}^{\infty} \frac{\exp(-D_c p_n^2 t / r_c^2)}{9\Lambda/(1-\Lambda) + (1-\Lambda)p_n^2} \tag{6.16}$$

where p_n is given by the nonzero roots of:

$$\tan p_n = \frac{3p_n}{3 + (1/\Lambda - 1)p_n^2} \tag{6.17}$$

and:

$$\Lambda = \frac{C'_0 - C_\infty}{C'_0 - C_0} = \frac{V_s}{V_f}\left(\frac{q_\infty - q_0}{C'_0 - C_0}\right) \tag{6.18}$$

where Λ is the fraction of the sorbate added in the step that is finally taken up by the adsorbent.

A set of theoretical uptake curves calculated from this expression is shown in Figure 6.2a, from which it is evident that, when Λ is greater than about 0.1, the assumption of a constant boundary condition [Eq. (6.10)] will lead to a significant error in the derived diffusivity value.

Notably, no assumption concerning the form of the equilibrium relationship has been introduced, so that the validity of these expressions is quite general as long as the system is isothermal and the concentration change is small enough to permit the assumption of a constant diffusivity. Under these conditions the equations for adsorption and desorption are symmetric, so the concentration profiles and sorption curves for desorption are identical when expressed in terms of the fractional approach to equilibrium.

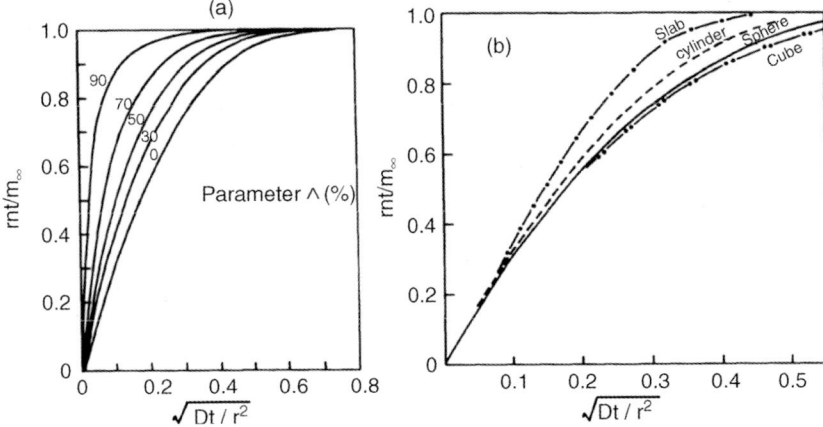

Figure 6.2 Theoretical uptake curves showing: (a) approach to equilibrium of a spherical adsorbent particle in response to a step change in the sorbate concentration at the external surface, calculated according to Eqs. (6.16) and (6.17); the parameter is Λ [Eq. (6.18)]. (b) Comparison of theoretical uptake curves ($\Lambda = 0$) for a sphere, a cube, an infinite cylinder, and a parallel-sided slab with time constants based on the equivalent spherical radius (Table 6.1).

6.2.1.3 Effect of Particle Shape

In many instances adsorbent particles are not uniform spheres and it is therefore pertinent to consider the extent to which the solution of the diffusion equation may be affected by particle shape. Solutions corresponding to Eq. (6.10) may be derived for any regular particle shape; Table 6.1 gives the expressions for a sphere, a cube, a parallel-sided slab, and an infinite cylinder. These expressions are less convenient than the simple form for spherical particles and it is therefore common practice to use the spherical particle expressions with an equivalent spherical particle radius r, defined as the radius of the sphere having the same external surface to volume ratio. The validity of this approximation is shown in Figure 6.2b. The approximation is clearly excellent in the initial region of the uptake curve but becomes less satisfactory in the long-time region. Nevertheless, for a cube and an infinite cylinder the deviation (in terms of dimensionless time) is only about 15 and 23%, respectively, at 90% approach to equilibrium. This is within the margin of error of most uptake rate measurements, so the approximation may be regarded as acceptable. The deviation is somewhat greater for a parallel-sided slab, suggesting that, where there is extreme departure from a compact particle shape, the mean equivalent radius approximation should be treated with caution, particularly in the long-time region.

6.2.1.4 Macropore Diffusion Control

If micropore diffusion is rapid, the concentration profile through a microparticle will be essentially uniform and the uptake rate will be controlled by diffusion through the macropores of the gross particle. A similar situation can arise when a bed of small adsorbent particles, such as zeolite crystals, is used to measure the uptake rate, for

Table 6.1 Comparison of solutions for the uptake curve with particles of different shape.

Shape	Solution
Sphere	$\dfrac{m_t}{m_\infty} = 1 - \dfrac{6}{\pi^2} \sum_{n=1}^{\infty} \dfrac{1}{n^2} \exp(-n^2 \pi^2 Dt/r^2)$ r = radius of sphere
Cube	$\dfrac{m_t}{m_\infty} = 1 - \dfrac{512}{\pi^6} \sum_{l=1}^{\infty}\sum_{m=1}^{\infty}\sum_{n=1}^{\infty} \dfrac{\exp\left[-\dfrac{\pi^2 Dt}{4a^2}\left((2l-1)^2 + (2m-1)^2 + (2n-1)^2\right)\right]}{(2l-1)^2(2m-1)^2(2n-1)^2}$ $2a$ = cube edge, $\bar{r} = a$
Parallel-sided slab	$\dfrac{m_t}{m_\infty} = 1 - \dfrac{8}{\pi^2} \sum_{n=0}^{\infty} \dfrac{1}{(2n+1)^2} \exp\left[-D(2n+1)^2\pi^2 t/4l^2\right]$ $2l$ = slab thickness, $\bar{r} = 3l$
Infinite cylinder	$\dfrac{m_t}{m_\infty} = 1 - 4 \sum_{m=1}^{\infty} \dfrac{1}{\alpha_n^2} \exp(-\alpha_n^2 Dt/r^2)$ $J_0(\alpha_n) = 0$ J_0 is zero order Bessel function of the first kind r = cylinder radius, $\bar{r} = 3r/2$

example, in a gravimetric system. If intracrystalline diffusion is rapid, the rate of uptake may be controlled by diffusion through the interparticle voids, just as in macropore-controlled uptake in a composite particle. These situations are formally similar, but not identical, to the micropore control situation. The key difference is that the accumulation of sorbate generally occurs mainly in the micropores so the uptake rate depends on the micropore capacity as well as on the macropore diffusivity.

The differential mass balance for a spherical shell element is:

$$(1-\varepsilon_p)\frac{\partial q}{\partial t} + \varepsilon_p \frac{\partial c}{\partial t} = \varepsilon_p D_p \left(\frac{\partial^2 c}{\partial R^2} + \frac{2}{R}\frac{\partial c}{\partial R}\right) \quad (6.19)$$

in which we have assumed the pore diffusivity to be independent of concentration. This is generally valid for a differential concentration change and in sorption from the vapor phase under conditions of Knudsen diffusion it is valid even when the concentration change is large. If the equilibrium relationship is linear ($q^* = Kc$) this equation may be written as:

$$\frac{\partial c}{\partial t} = \frac{\varepsilon_p D_p}{\varepsilon_p + (1-\varepsilon_p)K}\left[\frac{\partial^2 c}{\partial R^2} + \frac{2}{R}\frac{\partial c}{\partial R}\right] \quad (6.20)$$

which is formally identical with Eq. (6.8), except that D_c is replaced by $D_e = \varepsilon_p D_p / [\varepsilon_p + (1-\varepsilon_p)K]$, which is evidently the effective macropore diffusivity.

The relevant initial and boundary conditions for a step change in concentration at the external surface of the particle at time zero are:

$$c(R, 0) = c_0, \quad q(R, 0) = q_0$$
$$c(R_p, t) = c_\infty, \quad q(R_p, t) = q_\infty \tag{6.21}$$
$$\left.\frac{\partial c}{\partial R}\right|_{R=0} = \left.\frac{\partial q}{\partial R}\right|_{R=0} = 0$$

The solution for the uptake curve is therefore identical to Eqs. (6.10) or (6.11) with D_c/r_c^2 replaced by $(D_p/R_p^2)/[1 + K(1-\varepsilon_p)/\varepsilon_p]$, and the corresponding solution for the finite reservoir case is obtained by the same substitution. The equilibrium constant that appears in the expression for the effective diffusivity will vary with temperature according to an expression of the form $K = K_\infty \exp(-\Delta U_0/RT)$. Even though the temperature dependence of the macropore diffusivity (D_p) is generally modest, the effective diffusivity in the macro-controlled regime will therefore vary strongly with temperature in a manner reminiscent of an activated diffusion process ($D = D_\infty e^{-E/RT}$) but with an apparent activation energy roughly equal to the heat of adsorption. Since, for adsorption from the gas phase, the equilibrium constant is generally large, the effective diffusivity will be very much smaller than the pore diffusivity and no evidence concerning the nature of the controlling resistance can be deduced simply from the magnitude of the effective diffusivity. Furthermore, if measurements are made differentially in a system in which the isotherm has the normal type 1 form, the value of K, which represents the slope of the isotherm over the range of the differential step, will decrease with concentration, leading to an increasing trend of the effective diffusivity with concentration similar to that which is commonly observed for micropore-controlled systems.

The regime of macropore diffusion control corresponds with the measurement of long-range diffusion in PFG NMR (Section 11.4.2), which is further considered in Section 15.8.

6.2.2
Isothermal, Linear Dual-Resistance Systems

Subject to the same restrictions of an isothermal system and a differential concentration step (over which a linear equilibrium relationship and constant diffusivity may be assumed), it is relatively easy to obtain expressions for the uptake curve for some more complex cases involving more than one diffusional resistance.

6.2.2.1 Surface Resistance plus Internal Diffusion
For the combination of micropore diffusion within the particle and surface resistance the relevant boundary condition, replacing Eq. (6.9b), is:

$$D_c \left(\frac{\partial q}{\partial r}\right)_{r_c} = k(q_\infty - q|_{r_c}) \tag{6.22}$$

where $q|_{r_c}$ represents the concentration at the particle surface. For a step change in bulk concentration from C_0 to C_∞ at $t=0$ the solution for the uptake curve is:

$$\frac{m_t}{m_\infty} = \frac{\bar{q}-q_0}{q_\infty-q_0} = 1 - \sum_{n=1}^{\infty} \frac{6L^2 \exp(-\beta_n^2 D_c t/r_c^2)}{\beta_n^2[\beta_n^2 + L(L-1)]} \quad (6.23)$$

where $L = kr_c/D_c$ and β_n represents the roots of the equation:

$$\beta_n \cot \beta_n + L - 1 = 0 \quad (6.24)$$

For $L \to \infty$ (large k, small D_c) $\beta_n \to n\pi$ and Eq. (6.23) reverts to Eq. (6.10), the solution for intraparticle diffusion control. In the other limit for $L \to 0$, β is small so we may use only the first term in the series expansion:

$$\beta \cot \beta - 1 \approx -\beta^2/3, \quad \beta^2 - 3k_f r_c/KD_c = 3L \quad (6.25)$$

Equation (6.23) then reverts to the solution for surface resistance control [Eq. (6.6)]. For a parallel-sided slab the corresponding expressions [replacing Eqs. (6.23) and (6.24)] are:

$$\frac{m_t}{m_\infty} = 1 - 2L^2 \sum_{n=1}^{\infty} \frac{e^{-\beta_n^2 Dt/\ell^2}}{[\beta_n^2 + L(L+1)]\beta_n^2} \quad (6.26)$$

where:

$$L = k\ell/D_c = \beta_n \tan \beta_n \quad (6.27)$$

When macropore and film resistances are dominant, the situation is similar. In place of Eq. (6.22) we have:

$$\varepsilon_p D_p \frac{\partial c}{\partial R}\bigg|_{R_p} = k_f(C_\infty - c_s) \quad (6.28)$$

leading to Eq. (6.23) with $L = k_f R_p/\varepsilon_p D_p$. In general $Sh \geq 2.0$ ($Sh =$ Sherwood number) so $k_f \geq D_m/R_p$ [see Eq. (1.38)]. Except when there is a significant surface diffusion contribution, $D_p \leq D_m/\tau$ [see Eq. (4.15)]. Clearly, therefore, in a porous particle L will normally be greater than unity. In the absence of surface resistance the situation $L \ll 1.0$, corresponding to external fluid film control, can be reached only if there is a substantial contribution from surface diffusion within the macropores. Figure 6.3 illustrates the effect of surface resistance in modifying the form of the transient sorption curve.

6.2.2.2 Transient Concentration Profiles

The expression for the transient concentration profile for a one-dimensional system (parallel-sided slab of thickness 2ℓ) in which the sorption rate is controlled by the combined effects of intracrystalline diffusion and surface resistance is [2]:

$$\frac{q-q_0}{q_\infty-q_0} = 1 - 2L \sum_{n=1}^{\infty} \frac{e^{-\beta_n^2 Dt/\ell^2}}{[\beta_n^2 + L(L+1)]} \cdot \frac{\cos(\beta_n x/\ell)}{\cos \beta_n} \quad (6.29)$$

where β_n is given by Eq. (6.27).

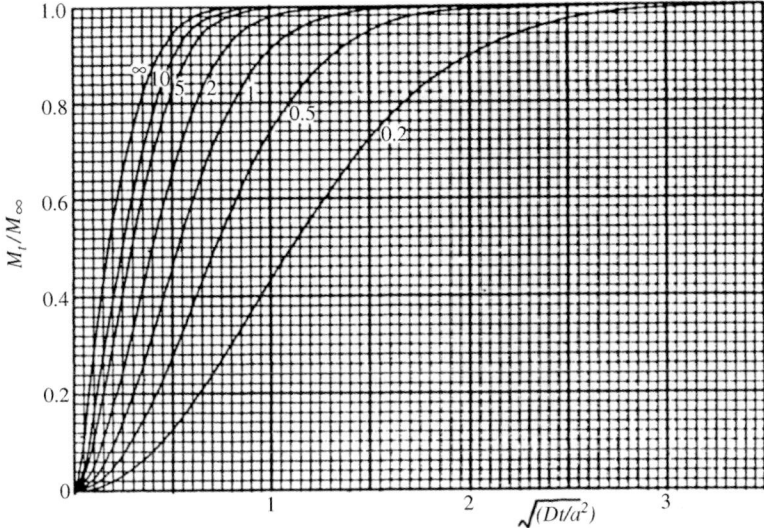

Figure 6.3 Effect of surface resistance to mass transfer on the form of the sorption (or desorption) curve for a spherical adsorbent particle of radius *a* subjected to a step change in concentration in the ambient fluid. Curves are calculated from Eq. (6.23), parameter L. Reprinted from Crank [1] with permission.

The corresponding expressions for a spherical particle have been given by Crank [1].

If $L \to 0$ (surface resistance control) the concentration within the particle becomes uniform, varying with time in accordance with Eq. (6.6). When L is large ($L \to \infty$) Eq. (6.29) reduces to the expression for internal diffusion control:

$$\frac{q-q_0}{q_\infty-q_0} = 1 - \frac{4}{\pi}\sum_{n=0}^{\infty}\frac{(-1)^n}{2n+1}\cos[(2n+1)\pi x/2\ell].\exp\left[-(2n+1)^2 D\pi^2 t/4\ell^2\right]$$

(6.30)

Representative experimental data showing the form of the transient profiles for adsorption and desorption of methanol in ferrierite (a one-dimensional system since only the eight-ring pores are accessible from the exterior – for more details see Section 12.3.1) are shown in Figure 6.4, together with the theoretical curves calculated from Eq. (6.29), which evidently provide a good representation of the observed behavior. In this pressure range the isotherm is essentially linear, so the adsorption and desorption curves are mirror images.

At the surface ($x = \ell$) Eq. (6.29) reduces to:

$$\frac{q_{\text{surf}}-q_0}{q_\infty-q_0} = 1 - 2L\sum_{n=1}^{\infty}\frac{e^{-\beta_n^2 Dt/\ell^2}}{[\beta_n^2 + L(L+1)]}$$

(6.31)

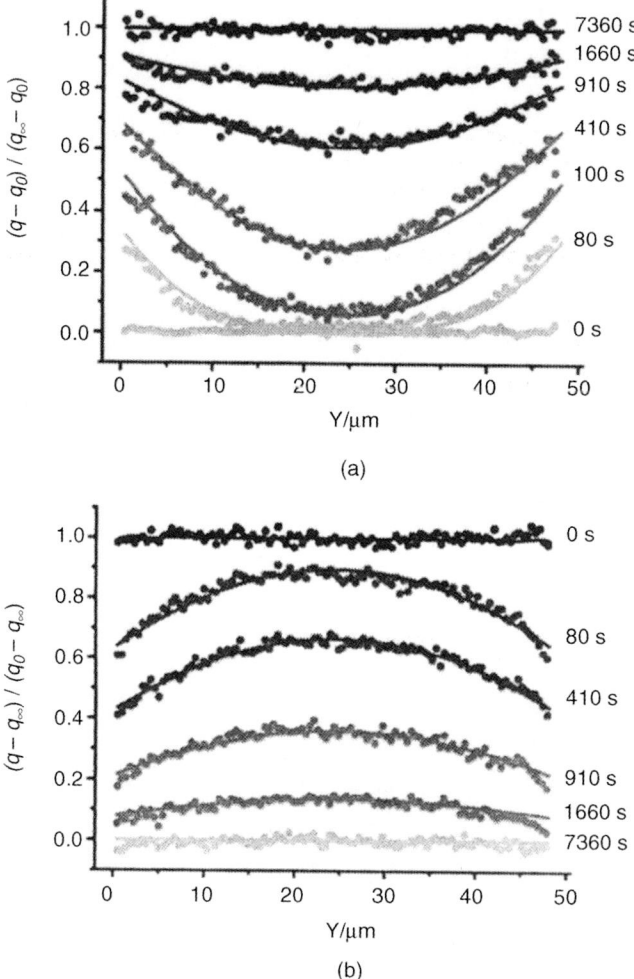

Figure 6.4 Experimental adsorption (a) and desorption (b) concentration profiles measured by interference microscopy for methanol in ferrierite (pressure steps 5–10 and 10–5 mbar). The lines show the best fit of the profiles to the dual-resistance model [Eq. (6.29)]. Reprinted from Kortunov et al. [3] with permission.

where q_{surf} represents the concentration in the adsorbed phase just inside the high resistance surface "skin." The expression for the uptake curve is given by Eq. (6.26). In the long-time region only the first terms of the summations are significant so, by comparing the two equations, it follows that:

$$\frac{q_{surf}-q_0}{q_\infty-q_0} = 1 - \frac{\beta_1^2}{L}\left(1-\frac{m_t}{m_\infty}\right) = \frac{\beta_1^2}{L}\left(\frac{m_t}{m_\infty}\right) + \left(1-\frac{\beta_1^2}{L}\right) \quad (6.32)$$

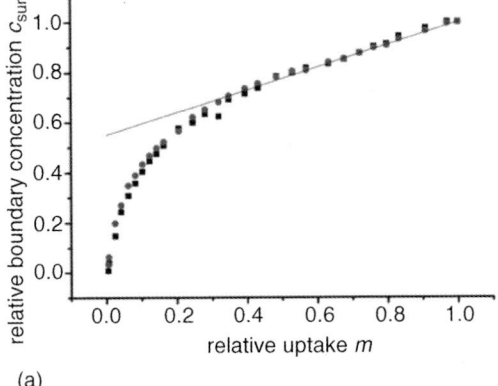

(a)

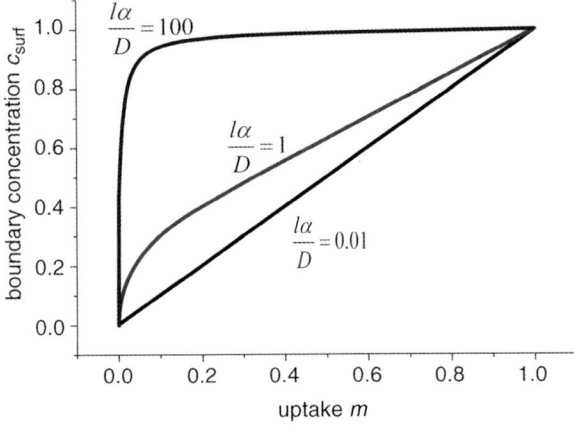

(b)

Figure 6.5 (a) Variation of reduced surface concentration with fractional uptake for methanol adsorption in ferrierite from the experimental data shown in Figure 6.4; (b) theoretical plot showing the variation of reduced surface concentration with fractional loading calculated from Eqs. (6.26) and (6.31) for selected values of the parameter L. Reprinted from Reference [57] with permission.

A plot of the dimensionless surface concentration $(q_{surf}-q_o)/(q_\infty-q_o)$ versus the fractional uptake m_t/m_∞ will therefore have a slope equal to β_1^2/L and an intercept on the ordinate equal to $(1-\beta_1^2/L)$ as shown in Figure 6.5a. Since β_1^2 and L are related through Eq. (6.27) such plots provide a straightforward way of estimating the parameter L (which is simply the ratio of the time constants for internal diffusion and surface resistance) from experimental profiles.

The variation of surface concentration with fractional uptake may be calculated, for any value of the parameter L, from Eqs. (6.26) and (6.31). This relationship is shown for selected values of L in Figure 6.5b. When L is small [k (denoted by α on the figure) small, D large] the uptake rate is controlled by surface resistance, leading to a linear variation of surface concentration with fractional uptake. When L is large the rate is controlled by internal diffusion. The surface concentration rapidly attains its equilibrium value and then remains constant. A plot of surface concentration versus fractional uptake therefore provides a sensitive test for the intrusion of surface resistance.

6.2.2.3 Two Diffusional Resistances (Biporous Solid)

An expression for the uptake curve for a system with two coupled intraparticle diffusional resistances appears to have been derived first by Ruckenstein et al. [4] The system they considered was a composite macroporous particle consisting of an assemblage of smaller microporous microparticles, as in Figure 1.3b. An analytic expression for the uptake curve in response to a step change in sorbate pressure was derived subject to the usual assumptions of an isothermal system and a linear equilibrium isotherm. Essentially the same problem was considered by Ma and Lee [5] and by Lee [6], who obtained a somewhat more tractable expression and extended the analysis to a finite volume system in which the fluid phase concentration varies with time [as in the derivation of Eq. (6.16)]. This extension is important as it permits the application of this model to uptake rate measurements by the piezometric method (Chapter 13) and in liquid systems. The model equations and the analytic solution obtained by Lee [6] are summarized in Table 6.2 and the form of the uptake curves is shown in Figure 6.6. The limiting cases corresponding to macropore or micropore control [Eqs. (6.16) and (6.17)] are recovered when $\beta \to \infty$ or $\beta \to 0$. For intermediate values of β the curves have a different shape that cannot be adequately represented by a single-resistance diffusion model. The problem of diffusion in a biporous adsorbent has also been investigated by Zolotarev and Ulin [7, 8] using moments analysis (Section 13.6).

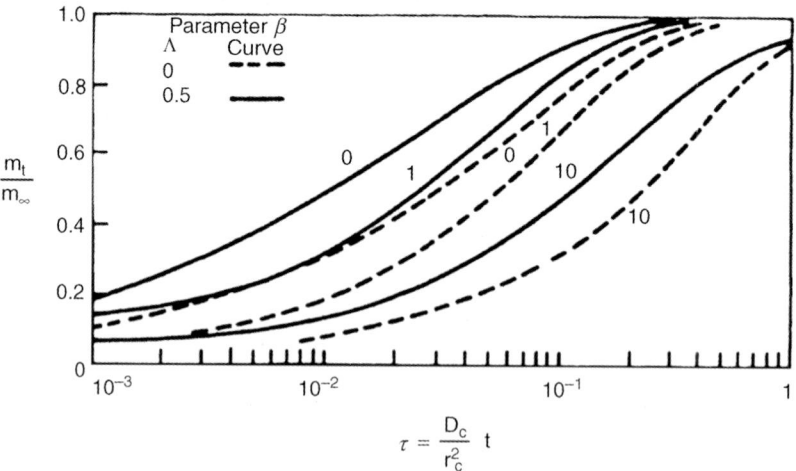

Figure 6.6 Theoretical uptake curves for a biporous adsorbent calculated according to the analysis of Lee [6] (Table 6.2), showing the transition from micropore to macropore diffusion control and the difference in the form of the uptake curve for intermediate values of the parameter β. Reprinted with permission from Reference [6].

Table 6.2 Dual diffusion resistance model for a biporous particle.

Intracrystalline diffusion in microparticle

$$\frac{\partial q}{\partial t} = \frac{1}{r^2}\frac{\partial}{\partial r}\left(r^2 D_c \frac{\partial q}{\partial r}\right)$$

$$\frac{\partial q}{\partial r}(0,t) = 0, \quad q(r_c,t) = Kc(R,t)$$

$$\bar{q}(R,t) = \frac{3}{r_c^3}\int_0^{r_c} qr^2 dr$$

Macropore diffusion

$$\frac{1}{R^2}\frac{\partial}{\partial R}\left(R^2 D_p \frac{\partial c}{\partial R}\right) = \frac{\partial c}{\partial t} + \left(\frac{1-\varepsilon_p}{\varepsilon_p}\right)\frac{\partial \bar{q}}{\partial t}$$

$$\frac{\partial c}{\partial R}(0,t) = 0, \quad c(R_p,t) = C(t), \quad q(r,0) = c(R,0) = 0$$

$$\frac{m_t}{m_\infty} = \frac{3}{R_p^3}\int_0^{R_p}\frac{[(1-\varepsilon_p)\bar{q} + \varepsilon_p c]R^2\, dR}{[(1-\varepsilon_p)\bar{q}_\infty + \varepsilon_p c_\infty]}$$

General solution (Λ finite)

$$\frac{m_t}{m_\infty} = 1 - \frac{1}{\Lambda}\sum_{m=1}^{\infty}\sum_{n=1}^{\infty}\frac{\exp(-p_{n,m}^2 D_c t/r_c^2)}{1 + \frac{1}{4}\left[9\gamma/\beta - p_{n,m}^2(1+\beta p_{n,m}^2/9\gamma)/u_n^2\right]L_{n,m}}$$

where:

$$L_{n,m} = \left(2\alpha + \frac{\beta}{1-\Lambda}\right) - \left(\frac{u_n^2 + \alpha p_{n,m}^2}{p_{n,m}^2}\right)\left[1 + \frac{(1-\Lambda)(u_n^2 + \alpha p_{n,m}^2)}{\beta}\right]$$

with $p_{n,m}$ given by the roots of:

$$u_n^2 = \frac{\beta}{1-\Lambda}(p_{n,m}\cot p_{n,m} - 1) - \alpha p_{n,m}^2$$

and u_n given by the roots of:

$$u_n \coth u_n = 1 + \frac{p_{n,m}^2 \beta}{9\gamma}, \quad \Lambda = \frac{C_0 - C_\infty}{C_0}, \quad \gamma = \frac{\beta\Lambda}{\beta + 3\alpha(1-\Lambda)}$$

$$\alpha = (D_c/r_c^2)/(D_p/R_p^2), \quad \beta = 3\alpha(1-\varepsilon_p)q_\infty/\varepsilon_p C_0, \quad \tau = D_c t/r_c^2$$

Solution for $\Lambda \to 0$ (infinite vessel)

$$\frac{m_t}{m_\infty} = 1 - \frac{18}{\beta + 3\alpha}\sum_{m=1}^{\infty}\sum_{n=1}^{\infty}\left(\frac{n^2\pi^2}{p_{n,m}^4}\right)\frac{e^{-p_{n,m}^2 D_c t/r_c^2}}{\left\{\alpha + \frac{\beta}{2}\left[1 + \frac{\cot p_{n,m}}{p_{n,m}}(p_{n,m}\cot p_{n,m} - 1)\right]\right\}}$$

where $p_{n,m}$ is given by the solution of the equation:

$$\alpha p_{n,m}^2 - n^2\pi^2 = \beta(p_{n,m}\cot p_{n,m} - 1)$$

Approximate solution for $\beta \gg \alpha$ (negligible macropore accumulation)

$$\frac{m_t}{m_\infty} = 1 - \sum_{m=1}^{\infty}\sum_{n=1}^{\infty}\frac{36 n^2 \pi^2 \exp(-p_{n,m}^2 D_c t/r_c^2)}{\beta^2 p_{n,m}^4\left[1 + \frac{\cot p_{n,m}}{p_{n,m}}(p_{n,m}\cot p_{n,m} - 1)\right]}$$

where $p_{n,m}$ is given by the roots of:

$$\beta(p_{n,m}\cot p_{n,m} - 1) = -n^2\pi^2$$

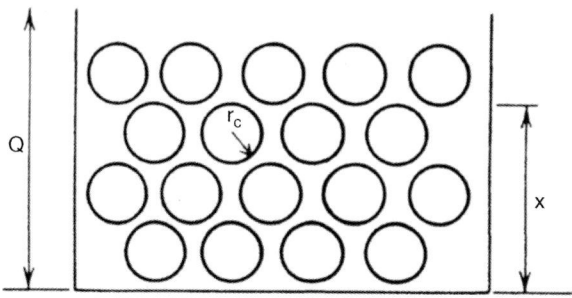

Figure 6.7 Idealized bed of uniform spherical microporous adsorbent particles.

A problem that is formally similar to diffusion in a biporous adsorbent particle also arises when diffusion into a sample bed occurs at a rate comparable with or slower than diffusion into the adsorbent particles within the bed. The situation is as sketched in Figure 6.7 and the solution for the uptake curve, which was obtained by the same procedure as for the biporous particle, [9] is summarized in Table 6.3.

In the limiting situation where $\beta \to \infty$, $p \to 0$, and $p \cot p - 1 \to -p^2/3$ the last two equations in Table 6.3 reduce to:

$$\frac{m_t}{m_\infty} = 1 - \sum_{n=0}^{\infty} \frac{2}{(n+\frac{1}{2})^2 \pi^2} \exp\left[-\left(n+\frac{1}{2}\right)^2 \pi^2 \frac{D_e t}{l^2}\right] \tag{6.33}$$

$$\beta p_{n,m}^2 = 3(n+\tfrac{1}{2})^2 \pi^2$$

$$\frac{D_e t}{l^2} = \frac{\varepsilon D_p C_0 t}{(1-\varepsilon) q_0 l^2} = \frac{3\tau}{\beta} \tag{6.34}$$

This corresponds to the situation in which diffusion within the particles is rapid so that the uptake rate is controlled entirely by diffusion into the adsorbent sample (bed diffusion). Equation (6.31) is equivalent to the expression given by Crank [1] for uptake into a parallel-sided adsorbent slab. In the more general case of Λ finite, the expression for the limiting situation corresponding to bed diffusion control ($\beta \to \infty$) is obtained as:

$$\frac{m_t}{m_\infty} = 1 - \frac{1}{\Lambda} \sum_{n=1}^{\infty} \frac{2 e^{-p_n^2 \tau}}{[1/(1-\Lambda)] + p_n^2 \beta / 3\Lambda} \tag{6.35}$$

where p_n and q_n are the roots of:

$$p_n^2 = 3(1-\Lambda) q_n^2 / \beta, \quad \tan q_n / q_n = -\Lambda/(\Lambda - 1) \tag{6.36}$$

These expressions provide the basis for the analysis of transient uptake curves under conditions corresponding to bed diffusion control and allow the conditions

Table 6.3 Diffusion in a bed of porous adsorbent particles.

Micropore diffusion

$$\frac{1}{r^2}\frac{\partial}{\partial r}\left(r^2 D_c \frac{\partial q}{\partial r}\right) = \frac{\partial q}{\partial t}$$

$$\frac{\partial q}{\partial r}(0,t) = 0, \quad q(r_c, t) = Kc(x,t)$$

$$\bar{q}(x,t) = \frac{3}{r_c^3}\int_0^{r_c} qr^2 dr$$

Bed diffusion

$$D_P \frac{\partial^2 c}{\partial z^2} = \frac{\partial c}{\partial t} + \left(\frac{1-\varepsilon}{\varepsilon}\right)\frac{\partial \bar{q}}{\partial t}$$

$$\frac{\partial c}{\partial z}(0,t) = 0, \quad c(l,t) = C(t), \quad q(r,0) = Kc(z,0) = 0$$

where l is the bed depth and ε is the bed voidage. The fractional uptake is given by:

$$\frac{m_t}{m_\infty} = \frac{1}{1+\left(\frac{3\alpha}{\beta}\right)(1-\Lambda)}\int_0^t \left[\frac{\bar{q}}{q_\infty} + \left(\frac{3\alpha}{\beta}\right)\frac{c}{C_0}\right]\frac{dz}{l}$$

where $\alpha = (D_c/r_c^2)/(D_p/l^2), \beta = 3\alpha(1-\varepsilon)q_\infty/\varepsilon C_0, \tau = D_c t/r_c^2$

General solution (Λ finite)

$$1 - \frac{m_t}{m_\infty} = \frac{1}{\Lambda}\sum_{n=1}^{\infty}\sum_{m=1}^{\infty}\frac{e^{-p_{n,m}^2 \tau}}{1+LM}$$

$$L = \left\{\alpha + \frac{\beta}{(1-\Lambda)}\left[1 + \frac{\cot p_{n,m}}{p_{n,m}}(p_{n,m}\cot p_{n,m} - 1)\right]\right\}$$

$$M = \left\{\frac{-p_{n,m}^2}{2u_n \coth u_n} + \frac{3\Lambda}{\beta + 3\alpha(1-\Lambda)}\left(\frac{1}{2} + \frac{1}{2u_n \coth u_n}\right)\right\}$$

where u_n and $p_{n,m}$ are given by roots of:

$$\frac{\coth u_n}{u_n} = \frac{3\Lambda}{p_{n,m}^2[\beta + 3\alpha(1-\Lambda)]}$$

$$u_n^2 = \frac{\beta}{1-\Lambda}(p_{n,m}\cot p_{n,m} - 1) - \alpha p_{n,m}^2$$

Notably, p is always real but u may be either real or imaginary. When u is imaginary ($u = iv$), coth (u_n/u_n) is replaced by $-\cot v/v$

Solution for $\Lambda \to 0$ [infinite vessel, $u \to i(n+\frac{1}{2})\pi$]

$$1 - \frac{m_t}{m_\infty} = \frac{6}{(\beta + 3\alpha)}\sum_{m=1}^{\infty}\sum_{n=1}^{\infty}\frac{(n+\frac{1}{2})\pi^2 e^{-p_{n,m}^2 \tau}}{p_{n,m}^4\left\{\alpha + \frac{\beta}{2}\left[1 + \frac{\cot p_{n,m}}{p_{n,m}}(p_{n,m}\cot p_{n,m} - 1)\right]\right\}}$$

where $p_{n,m}$ is given by the roots of:

$$\beta(p_{n,m}\cot p_{n,m} - 1) - \alpha p_{n,m}^2 = -\left(n+\tfrac{1}{2}\right)^2 \pi^2$$

Approximate solution for negligible accumulation in the voids ($\alpha/\beta \to 0$)

$$1 - \frac{m_t}{m_\infty} = \frac{12}{\beta^2}\sum_{n=1}^{\infty}\sum_{m=1}^{\infty}\frac{(n+\frac{1}{2})^2\pi^2 e^{-p_{n,m}^2 \tau}}{p_{n,m}^4\left[1 + \frac{\cot p_{n,m}}{p_{n,m}}(p_{n,m}\cot p_{n,m} - 1)\right]}$$

with $p_{n,m}$ given by:

$$\beta(p_{n,m}\cot p_{n,m} - 1) = -\left(n+\tfrac{1}{2}\right)^2 \pi^2$$

under which bed diffusion limitations become significant to be estimated from the system parameters. In the other limit when $\beta \to 0$, corresponding to intraparticle diffusion control, $p \to m\pi$ and the expressions given in Table 6.3 reduce to Eqs. (6.16) and (6.17) or, for $\Lambda \to 0$ to Eq. (6.10).

6.2.3
Isothermal Nonlinear Systems

In the preceding analysis we considered the diffusivity to be constant, which is a good approximation when the change in concentration is small. However, if the uptake curve is measured over a large concentration step, the assumption of a constant diffusivity may be a poor approximation for both micropore- and macropore-controlled systems.

6.2.3.1 Micropore Diffusion Control
In many zeolitic systems the concentration dependence of the micropore diffusivity is given approximately by Eq. (1.31) with D_0 constant. If the equilibrium isotherm can be represented by the Langmuir expression, this gives a diffusivity that increases strongly with loading, in accordance with Eq. (1.32). The appropriate form for the diffusion equation (for spherical particles) becomes:

$$\frac{\partial q}{\partial t} = \frac{D_0}{r^2} \frac{\partial}{\partial r}\left(\frac{r^2}{1-q/q_s}\frac{\partial q}{\partial r}\right) \tag{6.37}$$

6.2.3.2 Macropore Diffusion Control
The situation in the case of macropore diffusion control (i.e., diffusion in a composite pellet or a bed of adsorbent particles) is similar and solutions for the uptake curve have been obtained numerically by Kocirik et al. [10] and Youngquist [11]. For a nonlinear Langmuir system:

$$\frac{dq^*}{dc} = \frac{bq_s}{(1+bc)^2} = bq_s(1-q/q_s)^2 \tag{6.38}$$

where q^* denotes the equilibrium value of q. If we consider adsorption from the gas phase $q \gg c$ and since $\partial c/\partial R = (\partial q/\partial R)(dc/dq^*)$ we have, in place of Eq. (6.19):

$$\frac{\partial q}{\partial t} = \frac{\varepsilon_p D_p}{(1-\varepsilon_p)bq_s}\frac{1}{R^2}\frac{\partial}{\partial R}\left(\frac{R^2}{(1-q/q_s)^2}\frac{\partial q}{\partial R}\right) \tag{6.39}$$

which is equivalent to a concentration dependent diffusivity $D = D_0(1-q/q_s)^{-2}$.

Such systems provide important examples of the general class of systems for which the diffusivity increases strongly with loading. When the diffusivity is concentration dependent the adsorption and desorption processes become asymmetric (see, for example, Figure 12.8). From intuitive reasoning it may be deduced that, when the diffusivity increases with loading, adsorption will be faster than desorption and vice versa. In general, when the diffusivity is concentration dependent it is not possible to

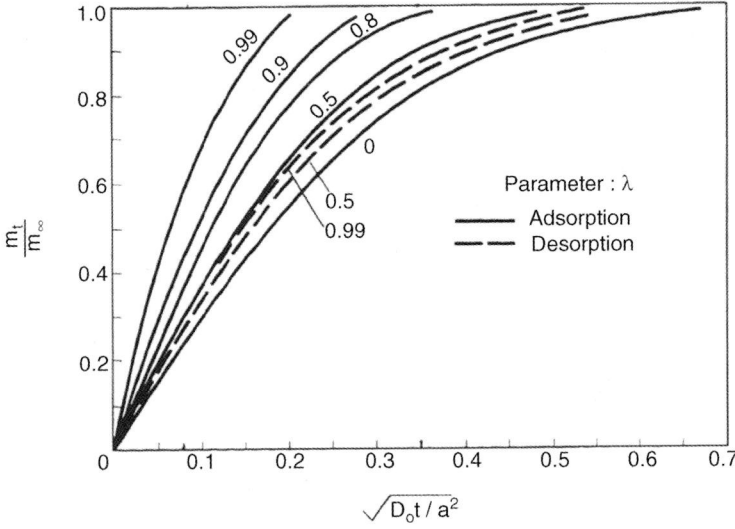

Figure 6.8 Numerically calculated transient adsorption/desorption curves for sorption in a spherical particle with $D/D_o = 1/(1-\lambda Q)$, $\lambda = r/(r-1)$. Reprinted from Garg and Ruthven [12] with permission.

obtain a formal analytic solution to the transient diffusion equation. Numerical solutions for both the above forms of concentration dependence have been obtained by Garg and Ruthven [12]. The theoretical uptake curves for adsorption and desorption in a spherical particle, calculated form Eq. (6.37), are shown in Figure 6.8, from which the large difference between adsorption and desorption rates is clearly apparent.

6.2.3.3 Semi-infinite Medium

For diffusion in a semi-infinite medium with a concentration dependent diffusivity an analytic solution for the uptake curve can sometimes be obtained by using the Boltzmann substitution [1] and, in some cases, such solutions provide a useful approximation for diffusion in a finite body.

As an example we consider diffusion in a parallel-sided slab (thickness 2ℓ) subjected at time zero to a step change in the surface concentration of sorbate from 0 to q_o (ads.) or from q_o to 0 (des.) [13]. Assuming internal diffusion control the adsorption/desorption kinetics are governed by the partial differential equation:

$$\frac{\partial Q}{\partial \tau} = \frac{\partial}{\partial X}\left[f(Q) \cdot \frac{\partial Q}{\partial X}\right] \qquad (6.40)$$

where $Q = q/q_o$ and the concentration dependence of the diffusivity is given by $D/D_o = f(Q)$, $X = x/\ell$, and $\tau = D_o t/\ell^2$, with the initial and boundary conditions:

$$\tau < 0, Q = 0 (\text{ads.}) \text{ or } Q = 1 (\text{des.}) \quad \text{for all } x$$
$$\tau \geq 0, Q = (0, \tau) = 1 (\text{ads.}) \text{ or } Q(0, \tau) = 0 (\text{des.}) \quad \text{for all } \tau \qquad (6.41)$$

$$\left.\frac{\partial q}{\partial x}\right|_{x=\ell} = \left.\frac{\partial Q}{\partial X}\right|X = 1 = 0 \tag{6.42}$$

In the initial region the concentration front has not penetrated to the center of the slab so the system behaves as a semi-infinite medium. In this situation the last boundary condition of Eq. (6.42) may be replaced by:

$$\left.\frac{\partial q}{\partial x}\right|_{x\to\infty} = 0; \quad Q|_{x\to\infty} = 0(\text{Ads.}) \text{ or } 1.0(\text{Des.}) \tag{6.43}$$

This allows the Boltzmann transformation $(y = x/2\sqrt{D_o t})$ to be used to reduce Eq. (6.40) to the ordinary differential equation:

$$-2y\frac{dQ}{dy} = \frac{d}{dy}\left[f(Q)\frac{dQ}{dy}\right] \tag{6.44}$$

(note that distance x is measured from the external surface at which $x=0$). With the initial and boundary conditions:

$$\begin{aligned}t < 0: &\quad Q(y) = 0(\text{ads.}) \text{ or } 1.0(\text{des.}) \\ t \geq 0: &\quad Q(0) = 1.0(\text{ads.}) \text{ or } 0(\text{des.}) \text{ for all } t \\ &\quad Q(y\to\infty) = 0(\text{ads.}) \text{ or } 1.0(\text{des.})\end{aligned} \tag{6.45}$$

$$\left.\frac{dQ}{dy}\right|_{y\to\infty} = 0 \tag{6.46}$$

Integration of Eq. (6.44) yields:

$$\int_{y=0}^{\infty} y\,dQ = \frac{1}{2}\left[f(Q)\frac{dQ}{dy}\right]_{y=0} - \frac{1}{2}\left[f(Q)\cdot\frac{dQ}{dy}\right]_{y\to\infty} \tag{6.47}$$

Since the gradient of concentration is zero for large y this simplifies to:

$$A = -\int_{Q=0}^{1} y\,dQ = -\frac{r}{2}\left.\frac{dQ}{dy}\right|_{y=0} (\text{ads.}) = \frac{1}{2}\left.\frac{dQ}{dy}\right|_{y=0} (\text{des.}) \tag{6.48}$$

where A is the area under the Q versus y profile, corresponding to the amount of material adsorbed or desorbed. Equation (6.48) provides a convenient check on the validity of numerical (or analytic) approximations to the solution for the concentration profile.

6.2.3.4 Adsorption Profiles

For a Langmuirian system for which $f(Q) = (1-\lambda Q)^{-1}$, a formal analytic solution has been given, in parametric form, by Fujita [14] (summarized by Crank [1]):

$$Q = \frac{r}{r-1}\left[1-e^{-2I(\phi)}\right] \approx 1-e^{-2I(\phi)} \text{ (large } r\text{)}; \quad y = \frac{1}{\sqrt{2\mu}}[f(\phi)-\phi]e^{I(\phi)} \tag{6.49}$$

where:

$$I(\phi) = \int_0^q \frac{d\phi}{f(\phi)}; \quad f(\phi) = \sqrt{\phi^2 - 2\mu \ln \phi}; \quad r = \frac{1}{1-\lambda} \quad (6.50)$$

The parameter μ is defined by the relationship:

$$I(1) = \int_0^1 \frac{d\phi}{f(\phi)} = \frac{1}{2} \ln r \quad (6.51)$$

Figure 6.9 shows the variation of the parameter μ with r calculated by numerical integration of Eq. (6.51).

When μ is small:

$$f(\phi) - \phi \approx -\frac{\mu \ln \phi}{\phi}; \quad I(\phi) \approx I(1) + \ln \phi = \ln(\phi \sqrt{r}) \quad (6.52)$$

With these approximations the limiting expression for the concentration profile (for r large) may be obtained in explicit form:

$$Q = 1 - \frac{1}{r} \exp\left[2\sqrt{\frac{2}{\mu r}} \cdot y\right] \quad (6.53)$$

The limiting slope is given by:

$$-r \frac{dQ}{dy}\bigg|_{y=0} = 2\sqrt{\frac{2}{\mu r}} \quad (6.54)$$

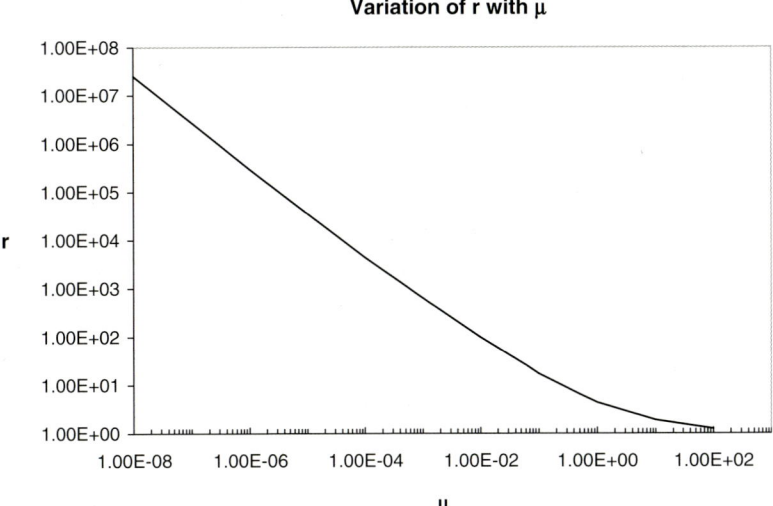

Figure 6.9 Variation of diffusivity ratio r with parameter μ calculated from Eqs. (6.50) and (6.51).

This approximation is valid only for the region where $Q \to 1.0$ but the range of validity increases when r is large.

Another limiting explicit approximation that is valid as $Q \to 0$ may be obtained by noting that (for $\phi \to 0$):

$$I(\phi) \approx \frac{1}{\sqrt{2\mu}} \int_0^\phi \frac{d\phi}{(-\ln \phi)^{\frac{1}{2}}} \approx 2\sqrt{\frac{\pi}{\mu}} \text{erfc}(u) \tag{6.55}$$

where $u^2 = -\ln \phi$. We thus obtain:

$$y \approx 2\sqrt{\frac{\pi}{\mu}} \text{erfc}(u); \quad Q \approx 2\sqrt{\frac{\pi}{\mu}} \text{erfc}(y) \tag{6.56}$$

as the asymptotic expressions for the limit $\phi \to 0, u \to \infty$.

Figure 6.10 shows concentration profiles for various values of r, calculated from Eq. (6.49). Also shown are the limiting profiles calculated from the asymptotic expressions [Eqs. (6.50) and (6.53)] for selected values of r. When r is large Eq. (6.53) provides a good approximation over most of the range but Eq. (6.56) is valid only at very low values of Q. Since $y = x/(2\sqrt{D_o t}) = X/2\sqrt{\tau}$ one may plot directly the profiles of Q versus X for a finite slab (for times less than the time at which the penetrating wave reaches the center of the slab). Such a plot is shown in Figure 6.11, from which the form of the penetrating wave is apparent. Note that, in contrast to the shockwave type of behavior observed for a macropore diffusion controlled system

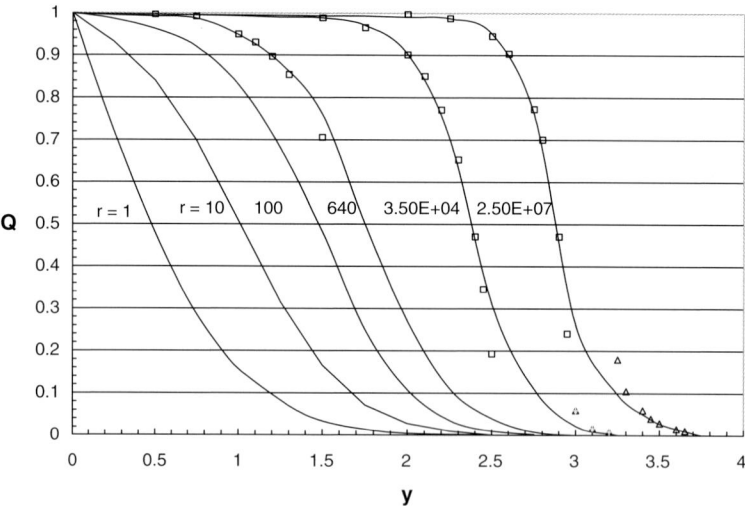

Figure 6.10 Concentration profiles for adsorption calculated from Eqs. (6.49) and (6.50) showing how the form of the concentration wave changes with r. Note that when r is large the asymptotic expression [Eq. (6.53)□] provides a good approximation, except when Q is small. For small values of Q the profile approaches Eq. 6.56 (Δ).

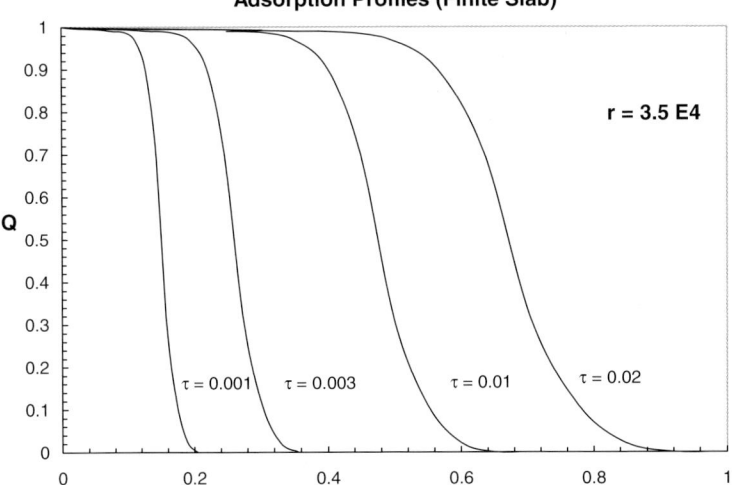

Figure 6.11 Profiles of Figure 6.10 re-plotted in the coordinates Q versus X (for a finite slab) for various values of τ. Note the increasing spread of the wave form as τ increases.

with irreversible adsorption [15], the concentration wave in the present system disperses as it penetrates.

6.2.3.5 Desorption Profiles

Fujita presented only the solution for adsorption, but by following his procedure the corresponding desorption problem may be solved to yield, for the transient profile:

$$Q = \left(\frac{r}{r-1}\right)[1-\exp\{2[I(\phi)-I(1)]\}]$$
$$y = \sqrt{\frac{r}{2\mu}}[f(\phi)-\phi]\exp[I(1)-I(\phi)] \tag{6.57}$$

where the symbols and integrals have the same meanings as for the adsorption case. When $r \to 1$, Eqs. (6.49) and (6.57) both reduce to the well-known error-function forms for a constant diffusivity system [13].

Figure 6.12 shows concentration profiles calculated from Eq. (6.57). As is to be expected, and in contrast to a linear system, the profiles for desorption and adsorption are quite different. The desorption profiles show no inflexion. The initial slope increases only slightly with increasing r, but the curvature increases so that, when r is large, the gradient at some distance from the surface is small. Physically this form is to be expected since, when r is large, the desorption rate is controlled by diffusion through the surface region in which $D \to D_o$. The limiting slope (for $y \to 0$) is given by $\sqrt{(8/\mu r)}$.

Experimentally measured profiles for adsorption and desorption of water vapor in a cylindrical pellet of NaX zeolite with only the ends exposed (so that the system is one-dimensional) are shown in Figure 6.13. This system is macropore diffusion

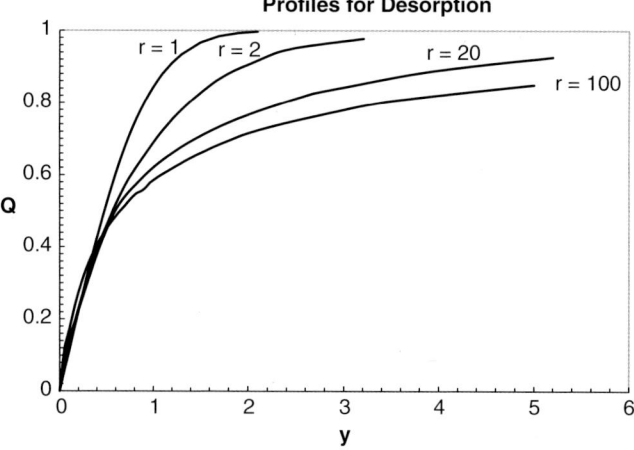

Figure 6.12 Profiles for desorption from a semi-infinite medium for selected values of r, calculated from Eq. (6.57).

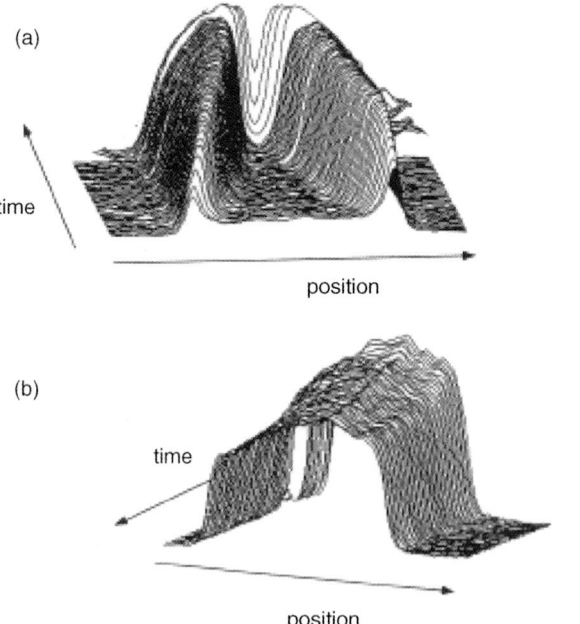

Figure 6.13 Experimental concentration profiles for (a) adsorption and (b) desorption of water vapor in a pellet of NaX zeolite measured by MRI. Reprinted from Bär *et al.* [15] with permission.

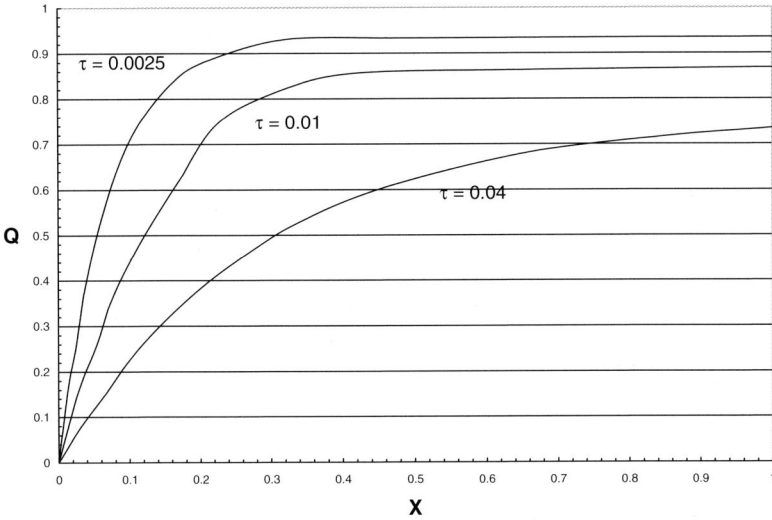

Figure 6.14 Transient profiles for desorption from a parallel sided slab with $r = 1000$, calculated from Eq. (6.58).

controlled but the diffusivity increases very strongly with loading. The profiles clearly have the forms predicted by the mathematical model: an advancing wave with a sharp front during adsorption and a monotonic profile with a high gradient near the surface during desorption.

Whereas for adsorption the semi-infinite medium boundary condition [Eq. (6.43)] is satisfied except in the very last stages of the transient uptake, this is not true for desorption, as may be seen from Figure 6.12. When r is large the desorption rate is controlled by diffusion through the surface layer in which the diffusivity has a very low value relative to the value in the central region where, as a result of the high diffusivity, the concentration profile is quite flat. As a simple model we therefore approximate a finite body (parallel sided slab) as a semi-infinite medium in which the concentration is essentially uniform, except in the surface region, but decreases with time. This leads to the approximate expression for the transient profile during desorption:

$$Q = e^{-\sqrt{2\tau}} \left(1 - e^{-X/\sqrt{2\tau}}\right) \tag{6.58}$$

Representative profiles calculated from this expression are shown in Figure 6.14.

6.2.3.6 Experimental Uptake Rate Data

Experimental uptake rate data obtained by Kondis and Dranoff [16, 17] for ethane in 4A zeolite were analyzed according to the isothermal diffusion model and the results are summarized in Figure 6.15. The isotherms conform approximately to the Langmuir expression. For small concentration steps the adsorption and desorption

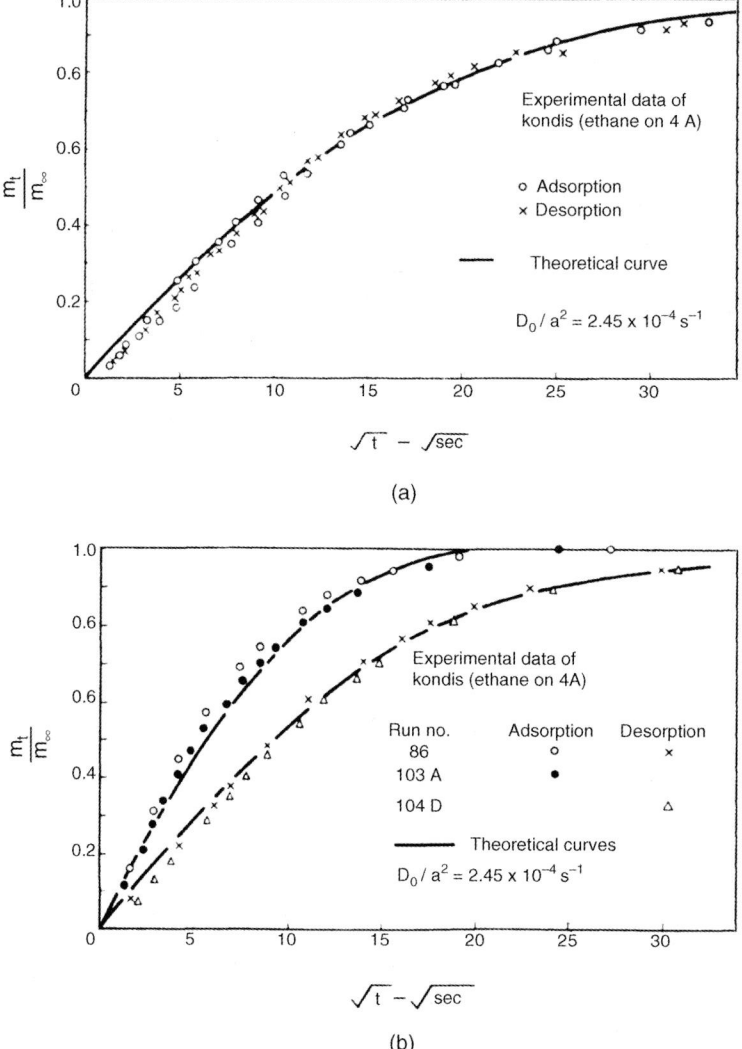

Figure 6.15 Comparison of theoretical and experimental uptake curves for ethane in 4A zeolite: (a) uptake curves measured differentially at low concentrations, showing conformity with the constant diffusivity model [Eq. (6.10)]; (b) integral uptake curve, measured over larger concentration steps, showing conformity with theoretical curves calculated from Eq. (6.37) with D_0 derived from the low concentration data. Note that the difference between adsorption and desorption curves is fully accounted for by the nonlinearity of the system. Reprinted from Garg and Ruthven [12] with permission.

curves were essentially mirror images and could be well represented by Eq. (6.10) (Figure 6.15a). For larger integral concentration steps, however, adsorption becomes much faster than desorption (Figure 6.15b) but the curves predicted from the model [Eq. (6.37)] using the value of D_0 from the low concentration measurements with λ

from the equilibrium data provide an excellent representation of the experimental curves. Clearly, the observed difference in sorption rates between adsorption and desorption is due entirely to the effect of nonlinearity and there is no evidence of any difference in intrinsic molecular mobility.

Experimental studies of sorption of several hydrocarbons in pelleted zeolite under conditions of macropore control have been reported by Ruthven and Derrah [18] and by Youngquist et al. [11]. The experimentally observed behavior is in accordance with the theory outlined above. Sorption curves measured differentially over small concentration steps show no difference between adsorption and desorption. As a result of the decreasing slope of the isotherm, the effective diffusivity increases rapidly with concentration in approximate conformity with Eq. (6.39). Tortuosity factors (5–6) calculated from the pore diffusivities according to Eq. (4.15) are consistent for the three sorbates and essentially independent of temperature.

Integral measurements made over large concentration steps show the expected large difference between adsorption and desorption rates, as shown in Figure 6.16, but the experimental curves are well approximated by the theoretical curves derived

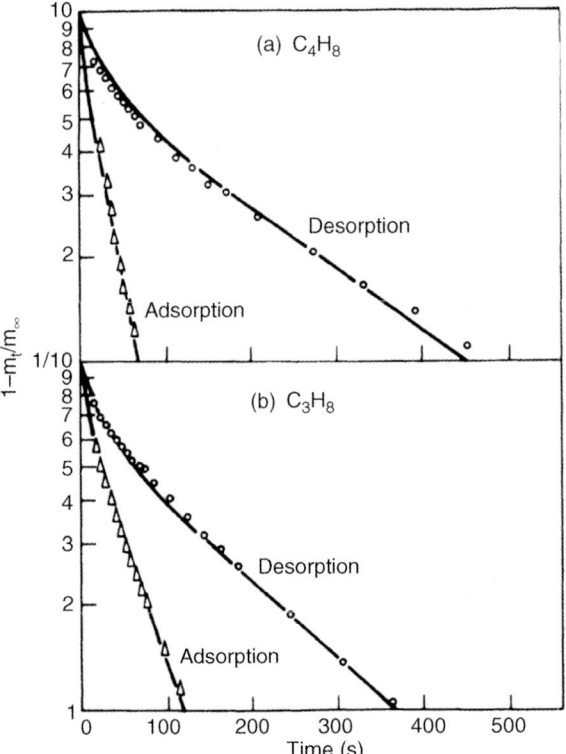

Figure 6.16 Integral sorption curves for (a) butylene and (b) propylene in Davison C-421 molecular sieve pellets with theoretical curves calculated numerically from Eq. (6.39). Relevant parameters are given in Table 6.4 Reprinted from Ruthven and Derrah [18] with pemission.

Table 6.4 Analysis of sorption curves for large concentration steps (Figure 6.16).

Sorbate		Pressure (Torr)		Concentration (mmol g^{-1})				Diffusivity (cm^2 s^{-1})	
		p_0	p_∞	q_0	q_∞	q_s	λ	$D_e \times 10^{5 a)}$	$D_{e0} \times 10^{5 b)}$
C$_4$H$_8$	Ads	1.1	86	1.35	1.68	1.71	0.92	14.1	0.64
	Des	86	1.1	1.35	1.68	1.71	0.92	2.2	0.51
C$_3$H$_8$	Ads	29	105	0.80	1.29	1.56	0.65	8.0	1.78$^{b)}$
	Des	100	12	0.51	1.27	1.56	0.72	2.7	1.15

a) D_e = effective diffusivity from sorption curve.
b) D_{e0} = limiting effective diffusivity as $q \to 0$; value for propane from differential measurements is about 2×10^{-5} cm^2 s^{-1}.

from the numerical solution of Eq. (6.39) with the diffusivity values from the differential measurements at low concentrations (Table 6.4).

6.2.3.7 Adsorption/Desorption Rates and Effective Diffusivities

Numerically calculated transient adsorption and desorption curves for a spherical adsorbent particle in which the diffusivity varies according to $D/D_o = 1/(1 - \lambda Q)$ are shown in Figure 6.8. The effective diffusivity may be obtained directly from the limiting slope (for $t \to 0$). Effective diffusivities (for adsorption) may also be derived from Fujita's solution. The adsorption rate is given by:

$$\frac{dm}{dt} = -rD_0 \frac{dq}{dx}\bigg|_{x=0} = -\frac{Dq_0}{2\sqrt{D_0 t}} \cdot \frac{dQ}{dy}\bigg|_{y=0} \tag{6.59}$$

and on integration:

$$m_t = -r\sqrt{D_0 t}\, q_0 \frac{dQ}{dy}\bigg|_{y=0} \tag{6.60}$$

When the diffusivity ratio is large ($\lambda \to 1.0$, $r \to \infty$) the concentration profile assumes the form of a penetrating wave (Figures 6.10, 6.11 and 6.13a), so the semi-infinite medium approximation will be valid throughout most of the uptake. In this regime the fractional approach to equilibrium ($m_\infty = q_0 \ell$) will be given by:

$$\frac{m_t}{m_\infty} = -r\sqrt{\frac{D_0 t}{\ell^2}} \cdot \frac{dQ}{dy}\bigg|_{y=0} = -r\sqrt{\tau}\frac{dQ}{dy}\bigg|_{y=0} \tag{6.61}$$

This expression shows that the "$\sqrt{t}$ law" is still valid even when the diffusivity is strongly concentration dependent.

Prediction of the uptake rate depends on estimating the concentration gradient at the surface:

$$\left[\frac{dQ}{dy}\bigg|_{y=0} = \kappa\right]$$

When r is large we may use Eq. (6.54), which yields:

$$\frac{m_t}{m_\infty} \approx 2\sqrt{\frac{2\tau}{\mu r}} \tag{6.62}$$

By comparison with a linear system for which the approach to equilibrium is given by:

$$\frac{m_t}{m_\infty} \approx 2\sqrt{\frac{D_0 t}{\pi \ell^2}} = 2\sqrt{\frac{\tau}{\pi}} \tag{6.63}$$

we see that the "effective" diffusivity (D_e) for adsorption is given by:

$$\frac{D_e}{D_0} = \frac{2\pi}{\mu r} \tag{6.64}$$

The product μr decreases with increasing r so the effective diffusivity will increase continuously (rather than approaching an asymptotic limit) as the isotherm approaches the rectangular form.

Effective diffusivities for desorption follow directly from our simplified model:

$$-\ell \frac{dq_0}{dt} = D_0 \frac{dq}{dx}\bigg|_{x=0} = \frac{q_0}{2}\sqrt{\frac{D_0}{t}} \frac{dQ}{dy}\bigg|_{y=0} \tag{6.65}$$

We assume that:

$$\frac{dQ}{dy}\bigg|_{y=0} = \kappa$$

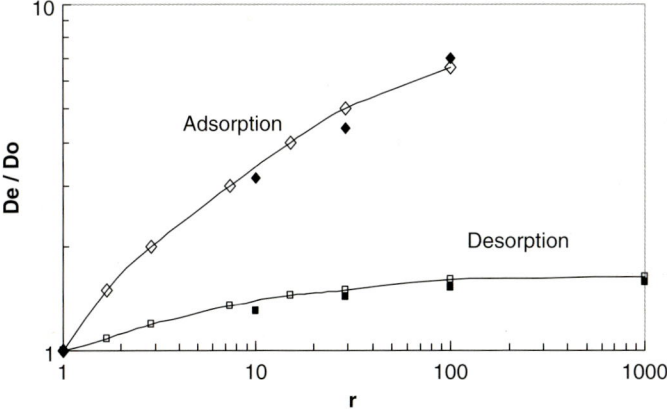

Figure 6.17 Variation of effective integral diffusivity with diffusivity ratio (r) for a Langmuirian system. Open symbols are from the numerical solution for a spherical particle (Figure 6.8), filled symbols are from Eqs. (6.64) and (6.66).

remains approximately constant (true for large r). With this approximation, after some algebra we find that the effective diffusivity is given, approximately, by:

$$\frac{D_e}{D_o} = \frac{\kappa^2 \pi}{4} \tag{6.66}$$

and for large r, $\kappa^2 \to 2$, so $D_e/D_o \to \pi/2$.

Effective diffusivity ratios for adsorption and desorption for a Langmuirian system, calculated from the numerical solution of the diffusion equation [Eq. (6.39)] and from Eqs. (6.64) and (6.66) for a wide range of r values are shown in Figure 6.17. The diffusivity ratio for adsorption is much higher than that for desorption and there is evidently good agreement between the numerically calculated values and the approximate expressions derived above.

6.2.3.8 Approximate Analytic Representations

Crank has pointed out two useful analytic approximations that provide surprisingly accurate representations of the adsorption/desorption curves for concentration dependent systems. For adsorption a weighted mean diffusivity defined by:

$$D_e = \frac{5}{3q_o^{5/3}} \int_0^{q_o} q^{2/3} D(q) \, dq \tag{6.67}$$

has been shown to provide a good prediction of the uptake rate over an integral concentration step, regardless of the precise form of the concentration dependence of the diffusivity. The corresponding expression for desorption is:

$$D_e = \frac{1.85}{q_o^{1.85}} \int_0^{q_o} (q_o - q)^{0.85} D(q) \, dq \tag{6.68}$$

6.2.3.9 Linear Driving Force Approximation

Except in the initial stages of sorption the concentration profile through the adsorbent particle is of approximately parabolic form:

$$q - q_o = a(R_P^2 - R^2) \tag{6.69}$$

where q_o is the final equilibrium loading and a is a constant. The concentration gradient at the surface and the average concentration through the particle are given by:

$$\left.\frac{\partial q}{\partial R}\right|_{R=R_P} = -2aR_P; \quad \bar{q} - q_o = \frac{2}{5}aR_P^2 \tag{6.70}$$

We may therefore express the gradient at the surface in terms of the rate of change of the average concentration, thus obtaining the correspondence with the diffusion model:

$$4\pi R_P^2 D \left.\frac{\partial q}{\partial R}\right|_{R=R_P} = -\frac{4}{3}\pi R_P^3 k(\bar{q} - q_o) \tag{6.71}$$

whence we see that, for the same flux:

$$k = \frac{15D}{R_P^2}; \quad \frac{\partial q}{\partial t} \approx k(\bar{q} - q_o) \tag{6.72}$$

This provides a simple explicit "linear driving force approximation" for the rate expression, as originally suggested by Glueckauf [20, 21]. This approximation is widely used in numerical simulations of adsorption processes where the use of a diffusion model would make the model too unwieldy. For processes that operate close to equilibrium this simple model works well but it breaks down for kinetically controlled processes that operate far from equilibrium since it severely underestimates the initial uptake rate.

6.2.3.10 Shrinking Core Model

For a large concentration step a highly favorable type I isotherm (e.g., a Langmuir isotherm) approaches the rectangular or irreversible form:

$$c = 0, \quad q^* = 0, \quad c > 0, \quad q^* = q_s \tag{6.73}$$

The rectangular model therefore provides a useful and mathematically tractable model for analysis of the sorption curves for strongly adsorbed species.

We consider a composite macroporous pellet of the type sketched in Figure 1.3b where most of the sorption capacity lies in the microparticles (zeolite crystals) while the dominant resistance to mass transfer is diffusion through the channels or pores between the microparticles. Intracrystalline diffusion is assumed to be sufficiently rapid to maintain the sorbate concentration practically uniform through a microparticle and at equilibrium with the sorbate concentration in the micropore just outside the microparticle. Under these conditions the concentration profile in the adsorbed phase penetrates as a shock front separating the shrinking core region to which sorbate has not yet penetrated, from the saturated outer shell, as sketched in Figure 6.18.

Adsorption occurs entirely at the shock front ($R = R_f$). Over the region $R_p > R > R_f$ the flow of sorbate through the pores is therefore constant:

$$4\pi R^2 \varepsilon_p D_p \frac{\partial c}{\partial R} = k(t) \tag{6.74}$$

Integration at constant t yields:

$$c = \frac{-k}{4\pi R \varepsilon_p D_p} + \text{const.} \tag{6.75}$$

from which, using the boundary conditions at the particle surface and at:

$$R = R_f (R = R_p, c = c_0; R = R_f, c = 0) \tag{6.76}$$

we obtain:

$$R^2 \frac{\partial c}{\partial R} = \frac{k}{4\pi \varepsilon_p D_p} = \frac{c_0}{1/R_f - 1/R} \tag{6.77}$$

A mass balance at $R = R_f$ gives for the flow across the control surface:

$$4\pi \varepsilon_p D_p \left(R^2 \frac{\partial c}{\partial R} \right) = \frac{4\pi \varepsilon_p D_p c_0}{1/R_f - 1/R} = -4\pi R_f^2 q_s \frac{dR_f}{dt} \tag{6.78}$$

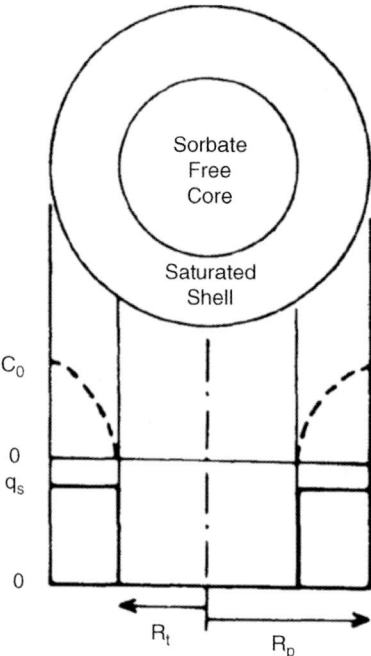

Figure 6.18 Schematic diagram showing the form of the concentration profile through an adsorbent particle for sorption with a rectangular isotherm under conditions of macropore diffusion control.

which may be integrated directly to obtain:

$$\frac{\varepsilon_p D_p C_0 t}{R^2 q_s} = -\frac{1}{2}\left(\frac{R_f}{R_p}\right)^2 + \frac{1}{3}\left(\frac{R_f}{R_p}\right)^3 + \frac{1}{6} \tag{6.79}$$

and since $1 - m_t/m_\infty = (R_f/R_p)^3$, the uptake curve is given by:

$$6\tau = 1 + 2\left(1 - \frac{m_t}{m}\right) - 3\left(1 - \frac{m_t}{m}\right)^{\frac{2}{3}} \tag{6.80}$$

where:

$$\tau = \frac{\varepsilon_p D_p c_0}{R_p^2 \; q_s} t \tag{6.81}$$

The stoichiometric time is given by $\tau = 1/6$. Equation (6.80) may also be written in explicit form [22]:

$$1 - \frac{m_t}{m_\infty} = \left\{\frac{1}{2} + \cos\left[\frac{\pi}{3} + \frac{1}{3}\cos^{-1}(1 - 12\tau)\right]\right\}^3 \left[0 < \tau < \frac{1}{6}\right] \tag{6.82}$$

These expressions appear to have been first obtained by Weisz and Goodwin [23] in relation to the burn-off of carbon from coked catalysts and first applied in the context of adsorption by Dedrick and Beekman [24] and Timofeev [25]. Although the algebraic forms of Eqs. (6.10) and (6.80) or (6.82) are very different the shapes of the corresponding uptake curves are in fact very similar.

Equations (6.80) or (6.82) are applicable only to constant pressure (infinite volume) systems since a constant boundary condition was assumed in the derivation. The corresponding expression for a finite volume system has also been derived [26] where:

$$\tau = I_2 - I_1 \tag{6.83}$$

$$I_1 = \int_1^\eta \frac{\eta \, d\eta}{1-\Lambda-\Lambda\eta^3} = \frac{1}{3\Lambda\lambda} \left\{ \frac{1}{2} \ln \left[\frac{(1-\Lambda-\Lambda\eta^3)(1+\lambda)^3}{(\eta+\lambda)^3} \right] \right\} \tag{6.84}$$

$$+ \sqrt{3} \tan^{-1} \left\{ \frac{2\eta-\lambda}{\lambda\sqrt{3}} - \sqrt{3} \tan^{-1} \left[\frac{2-\lambda}{\lambda\sqrt{3}} \right] \right\}$$

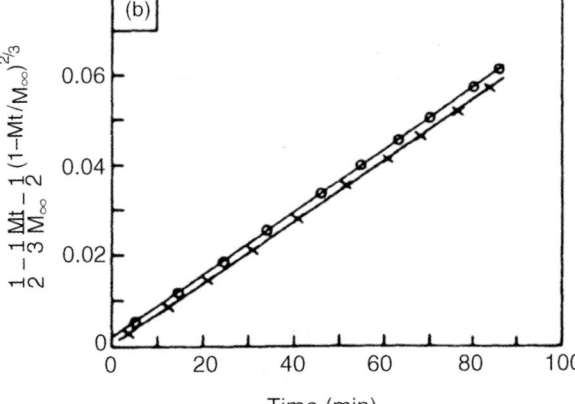

Figure 6.19 Experimental uptake curves for moisture on 4A zeolite pellets, showing conformity with Eq. (6.80)

Table 6.5 Details of experimental uptake curves for H_2O-4A shown in Figure 6.19.

Experimental conditions	Derived parameters	
	Parameter	Value
Adsorbent: Zeosorb 4A (Laporte)	$\varepsilon_p D_p$ (cm^{2}s^{-1})	0.0135
$R_p = 2.93$ mm	D_p (cm^{2}s^{-1})	0.039
$c_0 = 1.5 \times 10^{-5}$ g cm^{-3} (gas)	D_K^a (cm^{2}s^{-1})	4.0
$p_{H_2O} = 0.02$ atm	D_m^a (cm^{2}s^{-1})	0.26
$T = 303$ K	Tortuosity factor	5.0
$\varepsilon_p = 0.34$		
Mean macropore diameter 1 μm		

$$I_2 = \int_1^\eta \frac{\eta^2 \, d\eta}{1-\Lambda-\Lambda\eta^3} = \frac{1}{3\Lambda}\ln[1-\Lambda-\eta^3\Lambda] \quad (6.85)$$

with:

$$\lambda^3 = \left(\frac{1-\Lambda}{\Lambda}\right) \quad \text{and} \quad \eta^3 = 1-\bar{q}/q_s = 1-m_t/m_\infty \quad (6.86)$$

If external film resistance is significant Eq. (6.83) is replaced by:

$$\tau = \left[1+\frac{\varepsilon_p D_p}{k_f R_p}\right]I_2(\Lambda,\eta)-I_1(\Lambda,\eta) \quad (6.87)$$

Application of the irreversible adsorption model is illustrated in Figure 6.19, which shows the experimental uptake curves obtained by Kyte [27] for sorption of moisture on 4A zeolite pellets. The measurements were carried out gravimetrically with a single particle of adsorbent suspended in a flowing air stream containing a small partial pressure of moisture. Table 6.5 gives brief details of the experimental conditions and the derived parameters. The macropore diameter is large and the total pressure is relatively high (atmospheric). Under these conditions macropore diffusion occurs almost entirely by the molecular mechanism. Application of this model to liquid phase adsorption is discussed by Teo and Ruthven. [26].

6.2.3.11 Surface Resistance Control – Nonlinear Systems [28]

Surface resistance control under linear conditions leads to a linear concentration profile through the surface layer and a simple exponential expression for the transient uptake curve [Eq. (6.6)]. For larger concentration changes the diffusivity (in the surface layer) becomes concentration dependent, leading to more complex kinetic behavior. For example, if the diffusivity in the surface layer increases with concentration in accordance with Eqs. (1.31) and (1.32) (for a Langmuirian system with a

6.2 Mathematical Modeling of Sorption Kinetics

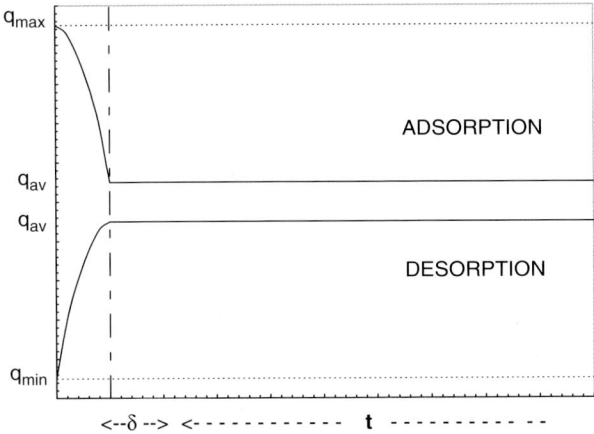

Figure 6.20 Sketch showing the form of the transient concentration profiles for a surface resistance controlled system in which the diffusivity increases strongly with loading.

chemical potential gradient as the driving force) the concentration profile through the surface layer will have the form shown in Figure 6.20.

Since the thickness of the surface layer (δ) is small compared with the half thickness (or radius) of the adsorbent particle there will be negligible accumulation within the surface layer. The flux through the surface layer (J) will therefore be constant (at any given time):

$$J = -D(q)\frac{\partial q}{\partial z} = -\frac{D_o}{1-q/q_s}\frac{\partial q}{\partial z} \tag{6.88}$$

and integrating across the surface layer (for adsorption):

$$J = -\frac{D_o q_s}{\delta}\ln\left(\frac{q_s-\bar{q}}{q_s-q_{max}}\right) = \ell\frac{d\bar{q}}{dt} \tag{6.89}$$

Integrating with respect to time:

$$\frac{D_o t}{\delta \ell} = \int_{q_o\bar{q}}^{q} \frac{dq}{q_s \ln\left(\frac{q_s-\bar{q}}{q_s-q_{max}}\right)} = (1-\lambda)\int_{u}^{u_o}\frac{du}{\ln u} = (1-\lambda)[I(u_o)-I(u)] \tag{6.90}$$

where: $u = \left(\frac{q_s-\bar{q}}{q_s-q_{max}}\right)$, $u_o = \left(\frac{q_s-q_o}{q_s-q_{max}}\right)$ and $\lambda = q_{max}/q_s$. The variable u is simply related to the fractional approach to equilibrium (m_t/m_∞). When $q_o = 0$ this relationship takes the simple form:

$$\frac{m_t}{m_\infty} = \frac{1-u(1-\lambda)}{\lambda} \tag{6.91}$$

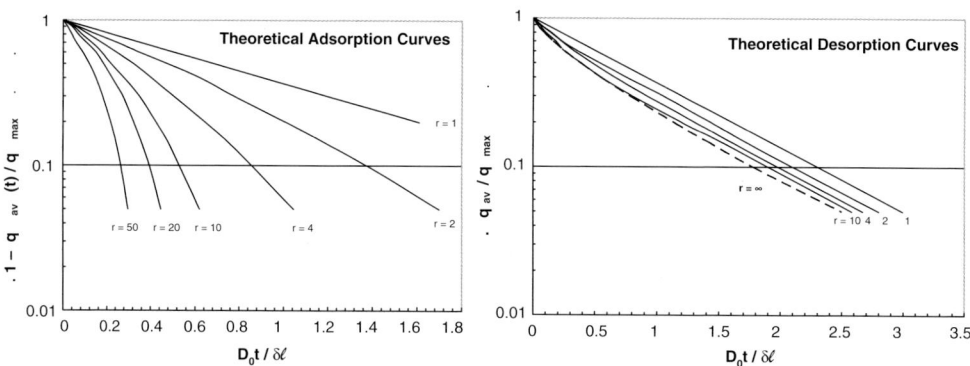

Figure 6.21 Theoretical adsorption and desorption curves for a Langmuirian system showing the variation with the diffusivity ratio $[r = D_{max}/D_0 = 1/(1-\lambda)]$. Note that for large r the desorption curves approach the asymptotic form of Eq. (6.94).

The corresponding expressions for desorption are:

$$\frac{D_0 t}{\delta \ell} = \int_{q_0}^{\bar{q}} \frac{d\bar{q}}{q_s \ln(1-\bar{q}/q_s)} = -\int_{v_0}^{v} \frac{dv}{\ln v} = I(v_0) - I(v) \qquad (6.92)$$

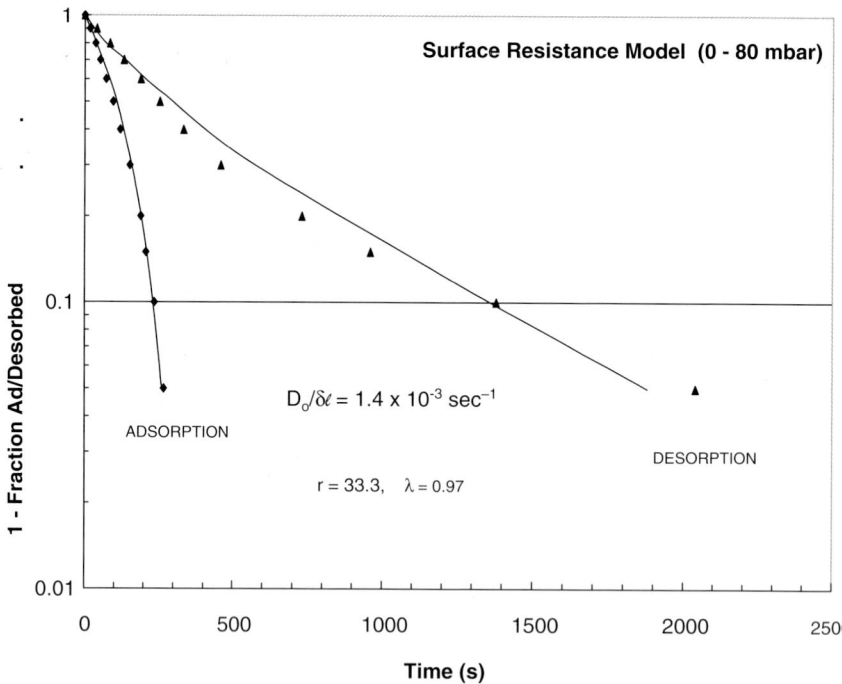

Figure 6.22 Comparison of experimental (points) and theoretical transient adsorption and desorption curves for methanol–ferrierite for pressure steps 0–80 and 80–0 mbar [29].

where $v = 1 - \bar{q}/q_s$, $v_o = 1 - q_o/q_s = 1 - \lambda$ and for $q_\infty = 0$, and:

$$\frac{m_t}{m_\infty} = 1 - (1-v)/\lambda \tag{6.93}$$

As $\lambda \to 1$, $v_o \to 0$, $I(v_o) \to 0$ and $v \to m_t/m_\infty$ so the desorption curves approach an asymptotic form given by:

$$D_0 t/\delta\ell \approx -I(v) = -I(m_t/m_\infty) \tag{6.94}$$

The form of the adsorption and desorption curves for various values of the ratio D_{max}/D_0 [$= 1/(1-\lambda)$], calculated from Eqs. (6.87) and (6.89) is shown in Figure 6.21. As an example, Figure 6.22 shows the fit of the theoretical model to the experimental adsorption and desorption curves for methanol–ferrierite for an integral concentration step (0–80 or 80–0 mbar). For this system under these conditions detailed IFM measurements show that the kinetics are controlled by surface resistance [29]. This relatively simple model provides a useful analytic representation of the behavior of surface resistance controlled kinetics that is commonly encountered for aged or surface-coked adsorbents.

6.2.4
Non-isothermal Systems

Since the heat effects associated with adsorption are often comparatively large, the assumption of isothermal behavior is a valid approximation only when sorption rates are relatively slow or in counter-diffusion systems (e.g., tracer exchange) where the heat effects cancel. The problem of non-isothermal sorption under conditions typical of a gravimetric uptake rate measurement has been considered by several authors. We present here a summary of the treatment given by Lee and Ruthven [30–32]. Table 6.6 summarizes some of the other available solutions.

A simple order-of-magnitude analysis [30] suggests that the dominant heat transfer resistance is generally the external film resistance and this has been confirmed experimentally by measuring intra- and extra-particle temperatures during an uptake experiment (Figure 6.23) [34, 35]. The dominant resistance to mass transfer may be intracrystalline (micropore) diffusion, intra-pellet (macropore) diffusion, or diffusion within the interstices of the sample bed. Macropore diffusion within a composite pellet or diffusion within an adsorbent bed are both formally similar since an assemblage of particles within the adsorbent sample is in effect equivalent to a composite pellet. The two situations of greatest practical importance in relation to the analysis of transient uptake curves are therefore the combination of external heat transfer resistance with either intraparticle (micropore) diffusion resistance or extraparticle (bed) diffusion resistance.

In a non-isothermal system there are two distinct effects: the temperature dependence of the equilibrium adsorbed phase concentration at the surface of the

Table 6.6 Non-isothermal models for transient uptake curves.

	Mass transfer resistances			Heat transfer resistances		
External film	Macropore (or bed) diffusion	Micropore (intraparticle) diffusion	External	Internal (conduction)	Analytic, numerical or moments	Reference[d]
X			X		A	Suzuki [34]
		X	X		A	Haul [35]
		X	X	X	A	Haul [36]
		X	X		A	Ruthven [31, 32]
	X		X		A	Ruthven [32]
		X	X	X	A[e]	Meunier [37]
X			X		M	Kocirik [38–40]
X			X		A	Sircar [41]
X			X		A, N[c]	Do [42]
X			X		N	Ma [43]
X				X	N	Hlavacek [44]
X			X	X	N	Hlavacek [45]
X			X		N	Hlavacek [46]
X			X		N	Anderson [47]

Adapted from Haynes *Catal. Rev. Sci. Eng.* **30**, 563 (1988).

a) For convenience in tracking further work references are cited by principal investigator, rather than first author.
b) Chemical potential driving force.
c) Analytic approximations are investigated.

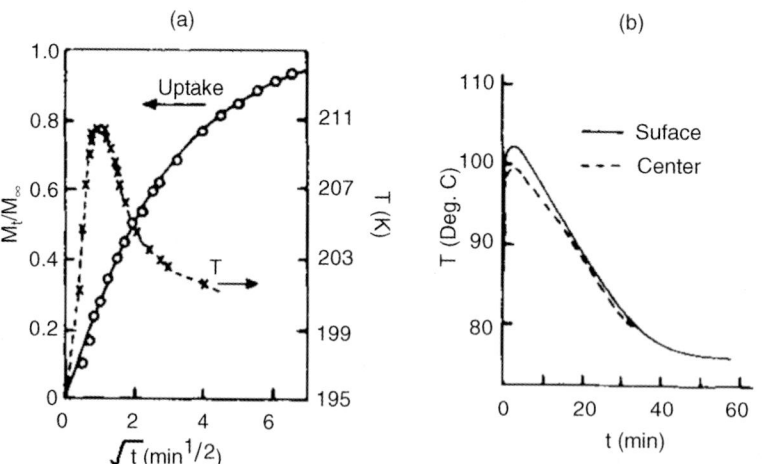

Figure 6.23 Experimental uptake curves and corresponding temperature response for an adsorbent particle during non-isothermal sorption. (a) Adsorption of N_2 at 195 K on 4A zeolite [47]; (b) adsorption of *n*-heptane at 300 K on 5A zeolite pellets [48]. Reprinted from Eagan *et al.* [47] and Ilavsky *et al.* [48] with permission.

adsorbent and the temperature dependence of the diffusivity. The latter effect may be eliminated by reducing the magnitude of the concentration step over which the uptake curve is measured, but the former effect is independent of the step size. Invariance of an experimental uptake curve to the size of the concentration step therefore provides no evidence that the system is effectively isothermal. This was first pointed out by Chihara et al. [34].

6.2.4.1 Intraparticle Diffusion Control

Assuming intraparticle diffusion control with a constant (micropore) diffusivity, the response of a microporous particle or zeolite crystal to a small differential step change in sorbate concentration at the external surface is described by the following set of equations:

$$\frac{\partial q}{\partial t} = \frac{1}{r^2} \frac{\partial}{\partial r} \left(r^2 D_c \frac{\partial q}{\partial r} \right) \tag{6.95}$$

$$\bar{q} = \frac{3}{r_c^3} \int_0^{r_c} qr^2 \, dr \tag{6.96}$$

$$(-\Delta H) \frac{d\bar{q}}{dt} = C_s \frac{dT}{dt} + ha(T - T_0) \tag{6.97}$$

$$q(r, 0) = 0; \quad \frac{\partial q}{\partial r}(0, t) = 0 \tag{6.98}$$

The equilibrium relationship at crystal surface is assumed to be linear:

$$\frac{q' - q'_0}{q_0 - q'_0} = 1 + \left(\frac{\partial q^*}{\partial T} \right)_p \left(\frac{T - T_0}{q_0 - q'_0} \right) \tag{6.99}$$

where $q'(r)$ represents the (time dependent) adsorbed phase concentration at the external surface of the particle and $(\partial q^*/\partial T)_p$ is the slope of the equilibrium isobar that is assumed constant over the relevant range; C_s is the heat capacity per unit particle volume. The expression for the uptake curve is:

$$\frac{m_t}{m_\infty} = 1 - \sum_{n=1}^{\infty} \frac{9[(p_n \cot p_n - 1)/p_n^2]^2 \exp(-p_n^2 D_c t/r_c^2)}{\frac{1}{\beta'} + \frac{3}{2}[p_n \cot p_n(p_n \cot p_n - 1)/p_n^2 + 1]} \tag{6.100}$$

where p_n is given by the roots of the equation:

$$3\beta'(p_n \cot p_n - 1) = p_n^2 - \alpha' \tag{6.101}$$

and the parameters α' and β' are defined by:

$$\alpha' = \frac{ha \, r_c^2}{C_s D_c}, \quad \beta' = \frac{\Delta H}{C_s} \left(\frac{\partial q^*}{\partial T} \right)_p \tag{6.102}$$

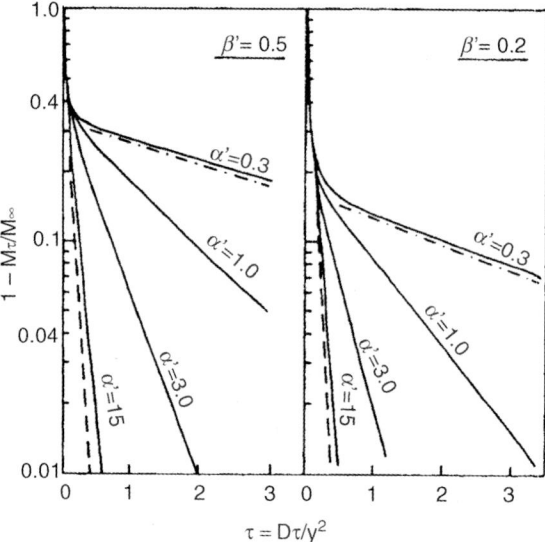

Figure 6.24 Theoretical uptake curves calculated from the non-isothermal model [Eqs. (6.100) and (6.101)] showing the effect of heat transfer resistance. As $\alpha' \to \infty$ the curves approach the limiting isothermal solution [– –, Eq. (6.10)] while for $\alpha' \to 0$ the curves approach the limiting form for heat transfer control [–·–·, Eq. (6.106)]. Reprinted from Ruthven et al. [32] with permission.

The corresponding expression for the temperature history is:

$$\frac{(T-T_0)}{(q_0-q'_0)}\left(\frac{\partial q^*}{\partial T}\right)_p = \sum_{n=1}^{\infty} \frac{-3[(p_n \cot p_n - 1)/p_n^2]\exp(-p_n^2 D_c t/r_c^2)}{\frac{1}{\beta'} + \frac{3}{2}[p_n \cot p_n(p_n \cot p_n - 1)/p_n^2 + 1]} \quad (6.103)$$

If macropore diffusion rather than intracrystalline diffusion is dominant the same equations apply but with D_c/r_c^2 replaced by D_e/R_p^2, where D_e is given by $\varepsilon_p D_p/[\varepsilon_p + (1-\varepsilon_p)K]$.

The full solution for the more exact model including both intraparticle heat conduction and external heat transfer resistance has been given by Haul and Stremming [35, 36] and by Sun and Meunier [37] but for most purposes the simplified solution outlined above is fully adequate. Figure 6.24 shows the general features of the uptake curves. The limiting case of isothermal behavior [Eq. (6.10)] is recovered when either $\alpha' \to \infty$ (infinitely rapid heat transfer) or $\beta' \to 0$ (infinite heat capacity). By comparing the solutions for a range of parameter values it was shown that the ratio α'/β' provides a useful criterion for assessing the deviation from isothermal behavior: when this ratio is greater than about 60 the error in the value of D_c/r_c^2 obtained by matching the uptake curve to the isothermal solution does not exceed 15% over the range 0–85% fractional approach to equilibrium [31].

When diffusion is rapid (α' small) the kinetics of sorption are controlled entirely by heat transfer. The limiting behavior may be derived by considering the asymptotic form of $(p_n \cot p_n - 1)$, which, for small values of p, may be replaced by the series expansion:

$$p_n \cot p_a - 1 \approx -(p^2/3 + p^4/45 + \cdots) \quad (6.104)$$

The first root of Eq. (6.101) is then given by:

$$p_1^2 = \frac{\alpha'}{1+\beta'} \quad (6.105)$$

and the expression for the uptake curve reduces to:

$$\frac{m_t}{m_\infty} = 1 - \left(\frac{\beta'}{1+\beta'}\right) \exp\left[-\frac{ha}{C_s}\frac{t}{(1+\beta')}\right] \quad (6.106)$$

which is equivalent to the expression derived by King and Cassie [49]. This limiting form may also be derived directly from the heat balance by assuming that equilibrium between the adsorbed and fluid phase concentrations is maintained at all times.

6.2.4.2 Experimental Non-isothermal Uptake Curves

Figure 6.25 shows representative uptake curves for pentane in 5A zeolite. The curves are clearly of the same general form as the theoretical curves for non-isothermal sorption shown in Figure 6.24 and the increasing intrusion of heat transfer resistance with

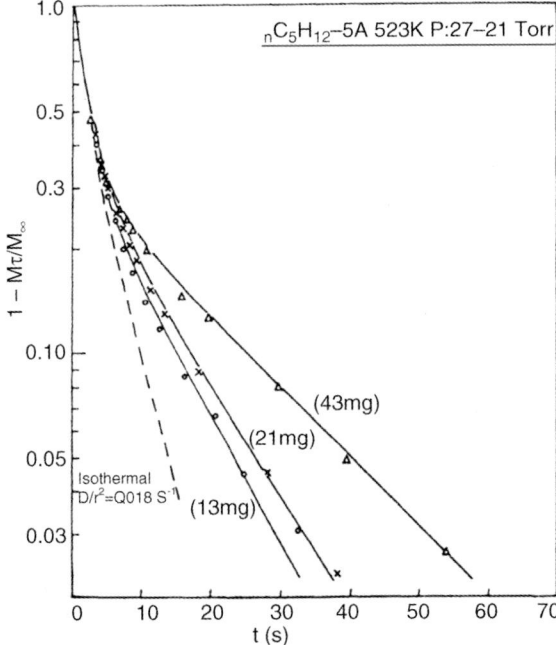

Figure 6.25 Experimental uptake curves for n-pentane in Linde 5A crystals (3.6 μm), showing increasing intrusion of heat transfer resistance as the quantity of sample is increased. The dashed line corresponds to the isothermal solution [Eq. (6.10)] with $D_c/r_c^2 = 0.018$ s^{-1}. Reprinted from Lee and Ruthven [30] with permission.

sample size is clearly apparent. This is to be expected since the external area per unit sample volume decreases with sample size. When heat transfer resistance is significant it affects mainly the long-time region of the uptake curve; the initial region remains close to the ideal isothermal curve until the heat transfer resistance becomes completely dominant. It is thus possible to obtain a reasonable estimate of the diffusivity from the initial part of the uptake curve, even under moderately non-isothermal conditions. Varying the configuration (external area/volume ratio) and/or the quantity of the adsorbent provides a simple experimental test not only for the intrusion of extracrystalline diffusional resistances but also for the intrusion of heat transfer resistance. Only if the experimental uptake curve is substantially invariant to such changes is it correct to interpret the data in terms of intraparticle diffusion control.

The extreme limit of heat transfer control [Eq. (6.106)] is illustrated in Figure 6.26, which shows data for sorption of CO_2 in small 5A zeolite crystals, in which diffusion is rapid. The uptake curves are essentially independent of crystal size but vary with the

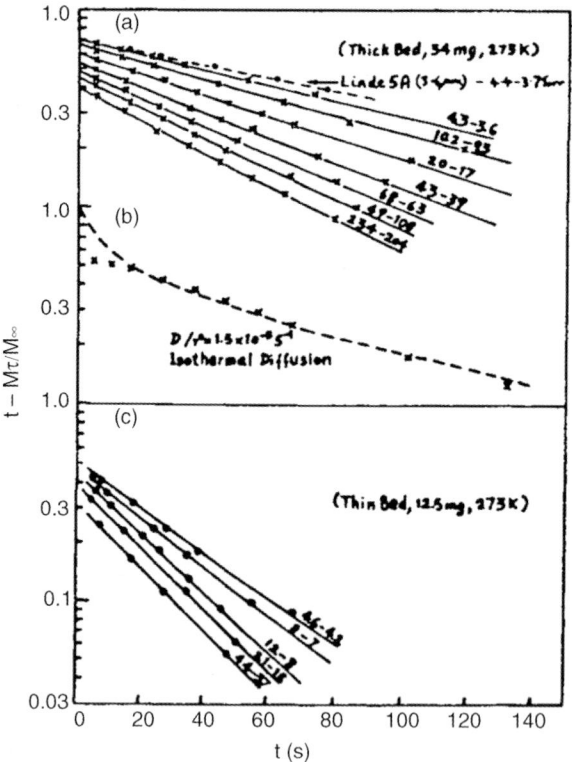

Figure 6.26 Experimental uptake curves for CO_2 in 5A zeolite crystals at 273 K, showing the limiting situation corresponding to heat transfer control [Eq. (6.106)]. Note the faster approach to equilibrium in the "thin bed" as a result of the larger external area per unit sample volume. Curve (b) shows a case where the heat transfer limited uptake curve lies fortuitously close to the ideal curve for isothermal diffusion control, except in the short time region. From Ruthven and Lee [32] with permission.

size of the sample (the bed thickness) as a result of the variation in the surface area/ volume ratio and the effective heat capacity. The initial diffusion-controlled uptake is extremely fast so that the measurable part of the curve lies entirely in the heat transfer controlled regime. Under such conditions no useful diffusion data can be extracted.

6.3
Sorption Kinetics for Binary Mixtures

The analysis of macropore diffusion in binary or multicomponent systems presents no especial problems since the transport properties of one component are not strongly affected by the other components in the system. In an adsorbed phase, however, the situation is more complex since in addition to a possible effect on the mobility the driving force for each component (the gradient of chemical potential) is modified, through the multicomponent adsorption equilibrium isotherm, by the concentrations of all other components in the system. Even if the mobilities remain constant the diffusion equations for each component are therefore directly coupled. A detailed discussion of diffusion in a binary adsorbed phase, from the perspective of the Maxwell–Stefan model, is given in Chapters 3 and 8, and the implications for membrane systems are discussed in Chapter 20. Here we consider only the basic theory of adsorption/desorption kinetics in a binary adsorption system.

A theoretical study of sorption kinetics in a binary Langmuirian adsorbed phase was presented by Round *et al.* [50] and a very similar analysis was reported by Kärger and Bülow [51]. Starting from the irreversible thermodynamic formulation (Section 3.1) and neglecting the cross coefficient terms, the fluxes of the two components are given by:

$$J_A = -D_{0A} \left[\frac{\partial \ln p_A}{\partial \ln q_A} \right] \frac{\partial q_A}{\partial z}, \quad J_B = -D_{0B} \left[\frac{\partial \ln p_B}{\partial \ln q_B} \right] \frac{\partial q_B}{\partial z} \tag{6.107}$$

where $p_A(q_A, q_B)$ and $p_B(q_A, q_B)$ are the vapor pressures in equilibrium with a binary adsorbed phase of composition (q_A, q_B). If it is assumed that the corrected diffusivities (D_{0A} and D_{0B}) are independent of the composition of the adsorbed phase and that the equilibrium relationship obeys the binary Langmuir expression:

$$\theta_A = \frac{q_A^*}{q_s} = \frac{b_A p_A}{1 + b_A p_A + b_B p_B}, \quad \theta_B = \frac{b_B p_B}{1 + b_A p_A + b_B p_B} \tag{6.108}$$

a differential mass balance for a shell element then yields for the relevant form of the diffusion equation:

$$\frac{\partial \theta_A}{\partial t} = \frac{D_{0A}}{(1-\theta_A-\theta_B)} \left\{ (1-\theta_B)\left(\frac{\partial^2 \theta_A}{\partial r^2} + \frac{2}{r}\frac{\partial \theta_A}{\partial r}\right) + \theta_A \left(\frac{\partial^2 \theta_B}{\partial r^2} + \frac{2}{r}\frac{\partial \theta_B}{\partial r}\right) \right\}$$

$$+ \frac{D_{0A}}{(1-\theta_A-\theta_B)^2} \left\{ (1-\theta_B)\frac{\partial \theta_A}{\partial r} + \theta_A \frac{\partial \theta_B}{\partial r} \right\} \left\{ \frac{\partial \theta_A}{\partial r} + \frac{\partial \theta_B}{\partial r} \right\}$$

$$\tag{6.109}$$

with a similar expression for component B.

To describe the binary diffusion we evidently need two coefficients, D_{0A} and D_{0B}, and the process cannot, in general, be represented as a simple diffusion process characterized by a single exchange diffusivity.

6.3.1.1 Counter-Diffusion

Such a representation is, however, possible in the special case of equimolar counter-diffusion, which requires $D_{0A} = D_{0B}$, $b_A = b_B$ and $\theta_{A\infty} - \theta_{AO} = -(\theta_{B\infty} - \theta_{BO})$. Under these conditions Eq. (6.106) reduces to:

$$\frac{\partial \theta_A}{\partial t} = -\frac{\partial \theta_B}{\partial t} = D_{AO}\left\{\frac{\partial^2 \theta_A}{\partial r^2} + \frac{2}{r}\frac{\partial \theta_A}{\partial r}\right\} = -D_{BO}\left\{\frac{\partial^2 \theta_B}{\partial r^2} + \frac{2}{r}\frac{\partial \theta_B}{\partial r}\right\} \quad (6.110)$$

which is the same form as the single-component diffusion equation with $D_{0A} = D_{0B} = D_0$.

Figure 27 shows theoretical counter-diffusion curves calculated from Eq. (6.110). "A" is the adsorbing species and the initial and final loadings of both components are indicated. In Figure 27a the curves for the two components are quite different, showing that the exchange process cannot be represented by a single effective diffusivity. Even though $D_{0A} = D_{0B}$ and $b_A = b_B$, the adsorption of A is substantially faster than the desorption of B as a result of the difference in the concentrations of the two components. The adsorption of A is not greatly affected by the presence of B since the uptake curve for A is close to the diffusion curve calculated from Eq. (6.10) with $D_c = D_{0A}$. The desorption curve for B is, however, of a different shape and it does not follow the form of Eq. (6.10). Attempts to match the desorption curve for B to the single-component diffusion curve lead to calculated diffusivity values that vary with time. The rate of desorption of B is evidently substantially reduced by the presence of even a relatively small amount of A as a consequence of the effect on the binary isotherm, which determines the driving force.

At higher loading (Figure 6.27b), the behavior approaches more closely the simple case of equimolar counter-diffusion. The adsorption curve for A and the desorption curve for B are now quite similar and their form approximates the simple diffusion curve with an effective diffusivity given by $D_{0A} = D_{0B}$. If $D_{0A} \ll D_{0B}$ the behavior is broadly similar but the effective diffusivity approximates to D_{0A} while if $D_{0B} \ll D_{0A}$ the effective diffusivity approximates to D_{0B}. In these calculations only differential changes in sorbate concentration have been considered. The situation is obviously more complicated for integral concentration changes although, in principle, the uptake curves may still be derived by numerical solution of Eq. (6.109).

Satterfield, Katzer, and coworkers [53–55] have measured counter-diffusion of several liquid hydrocarbons in small crystals of NaY zeolite. Measurements were made over relatively large concentration steps and the validity of representing the exchange behavior in terms of a single effective diffusion coefficient was not considered. Desorption was found to be substantially slower than adsorption in a

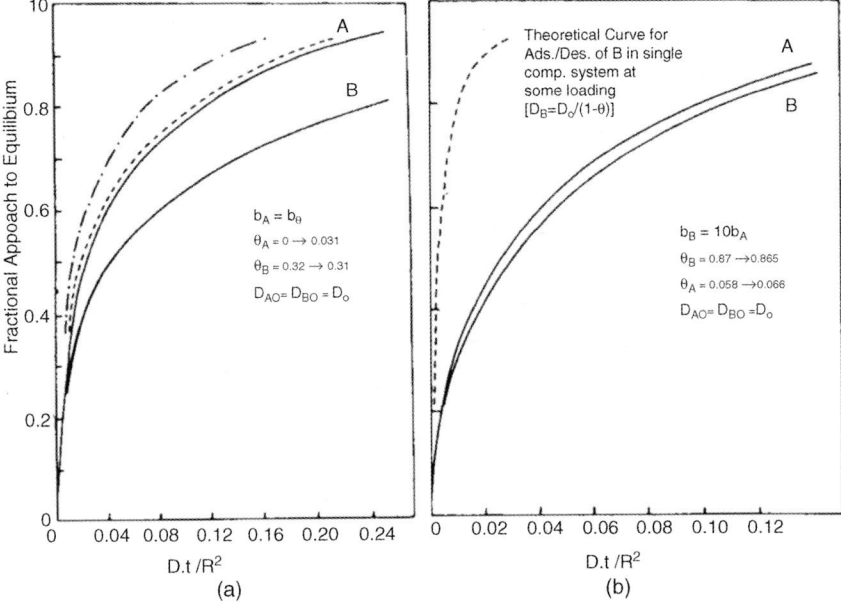

Figure 6.27 Theoretical uptake curves for counter-diffusion in a binary Langmuir system, calculated from Eq. (6.109). A is adsorbing, B desorbing and $D_{OA} = D_{OB} = D_0$. The continuous lines show the theoretical adsorption and desorption curves. In (a) the theoretical curve calculated from the single-component diffusion equation [Eq. (6.10)] with $D = D_0$ is shown and this is close to the curve for A derived from the binary expression. The curve for $D = D_0/(1-\theta)$ is also shown. In (b) the adsorption and desorption curves for both components are close to the theoretical curve derived from Eq. (6.10) with $D = D_0$ and much slower than the curve for $D = D_0/(1-\theta)$ Reprinted from Taylor [52].

single-component system and the effective diffusivities calculated by matching the exchange curves to Eq. (6.10) were found to vary strongly with fractional uptake, decreasing as equilibrium was approached. This is precisely the pattern of behavior to be expected, according to the above analysis, when the loading of the desorbing component is high and that of the adsorbing component is low (as in Figure 6.27a).

6.3.1.2 Co-diffusion

In a co-diffusion experiment, in which the concentration of both components is increased, the uptake curves generally have the form shown in Figure 6.28. The faster diffusing species is adsorbed initially to a concentration level that exceeds the final equilibrium position. It then desorbs to the equilibrium level as the slower diffusing species penetrates. Such behavior has been observed experimentally for n-heptane–benzene on NaX zeolite [51] and for N_2–CH_4 on 4A zeolite [56]. The form of the

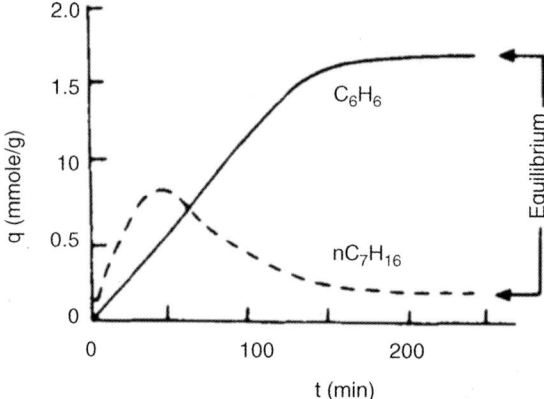

Figure 6.28 Experimental uptake curves for sorption of a heptane–benzene mixture on NaX zeolite crystals at 359 K ($p_{benzene} = 24.3$ Torr, $p_{heptane} = 50.6$ Torr). Heptane, the faster diffusing species, shows an overshoot. Reprinted from Kärger and Bülow [51] with permission.

concentration response curves is qualitatively predicted from the solution of Eq. (6.109). Quantitative agreement, however, is not to be expected since the isotherms do not conform precisely to the Langmuir model.

References

1 Crank, J. (1956) *Mathematics of Diffusion*, Oxford University Press, London.
2 Carslaw, H.S. and Jaeger, J.C. (1959) *Conduction of Heat in Solids*, Clarendon Press, Oxford.
3 Kortunov, P., Chmelik, C., Kärger, J., Rakoczy, R.A., Ruthven, D.M., Traa, Y., Vasenkov, S., and Weitkamp, J. (2005) *Adsorption*, **11**, 235–244.
4 Ruckenstein, E., Vaidyanathan, A.S., and Youngquist, G.R. (1971) *Chem. Eng. Sci.*, **26**, 1306.
5 Ma, Y.H. and Lee, T.Y. (1976) *AIChE J.*, **22**, 147.
6 Lee, L.-K. (1978) *AIChE J.*, **24**, 531.
7 Zolotarev, P.P. and Ulin, V.I. (1975) *Izv. Akad. Nauk SSSR7 Ser. Khim.*, 2367; Zolotarev, P.P. and Ulin, V.I. (1977) *Izv. Akad. Nauk SSSR7 Ser. Khim.*, 505.
8 Voloshchuk, A.M., Zolotarev, P.P., and Ulin, V.I. (1974) *Izv. Akad. Nauk SSSR, Ser. Khim.*, 1250.
9 Ruthven, D.M. (1984) *Principles of Adsorption and Adsorption Processes*, John Wiley & Sons, Inc., New York, p. 187.
10 Kocirik, M., Zikanova, A., and Dubsky, J. (1973) *Ind. Eng. Chem. Fundam.*, **12**, 440.
11 Youngquist, G.R., Allen, J.L., and Eisenberg, J. (1971) *Ind. Eng. Chem. Prod. Res. Dev.*, **10**, 308.
12 Garg, D.R. and Ruthven, D.M. (1972) *Chem. Eng. Sci.*, **27**, 417.
13 Ruthven, D.M. (2007) *Diffusion Fundamentals II* (eds S. Brandani, C. Chmelik, J. Kärger, and R. Volpe), Leipzig University Press, Leipzig, pp. 262–272.
14 Fujita, H. (1952) *Textile Res. J.*, **22**, 757–823.
15 Bär, N.-K., Balcom, B.J., and Ruthven, D.M. (2002) *Ind. Eng. Chem. Res.*, **41**, 2320–2329.
16 Kondis, E.F. and Dranoff, J.S. (1970) *Adv. Chem.*, **102**, 171.
17 Kondis, E.F. and Dranoff, S.S. (1971) *Ind. Eng. Chem. Process Design Dev.*, **10**, 108.
18 Ruthven, D.M. and Derrah, R.I. (1972) *Can. J. Chem. Eng.*, **50**, 743.
19 Ruthven, D.M. (2004) *Chem. Eng. Sci.*, **59**, 4531–4545.

20 Glueckauf, E. and Coates, J.E. (1947) *J. Chem. Soc.*, 1315.
21 Glueckauf, E. (1955) *Trans. Faraday Soc.*, **51**, 1540.
22 Brauch, V. and Schlunder, E.V. (1975) *Chem. Eng. Sci.*, **30**, 540.
23 Weisz, P.B. and Goodwin, R.D. (1963) *J. Catal.*, **2**, 397.
24 Dedrick, R.L. and Beekman, R.B. (1967) *Chem. Eng. Prog. Symp. Ser.*, **63** (74), 68.
25 Timofeev, D.P. (1971) *Adv. Chem.*, **102**, 247.
26 Teo, W.K. and Ruthven, D.M. (1986) *Ind. Eng. Chem. Process Design Dev.*, **25**, 17.
27 Kyte, W.S. (1970) PhD Thesis, University of Cambridge.
28 Ruthven, D.M., Heinke, L., and Kärger, J., (2010) *Microporous Mesoporous Mats*, **132**, 94.
29 Kortunov, P., Heinke, L., Vasenkov, S., Chmelik, C., Shah, D.B., Kärger, J., Rakoczy, R.A., Traa, Y., and Weitkamp, J. (2006) *J. Phys. Chem. B*, **110**, 23821.
30 Lee, L.K. and Ruthven, D.M. (1979) *J. Chem. Soc., Faraday Trans. I*, **75**, 2406.
31 Ruthven, D.M., Lee, L.-K., and Yucel, H. (1980) *AIChE J.*, **26**, 16.
32 Ruthven, D.M. and Lee, L.-K. (1981) *AIChE J.*, **27**, 654.
33 Haynes, H. W. (1988) *Catal. Rev. Sci. Eng.*, **30**, 563
34 Chihara, K., Suzuki, M., and Kawazoe, K. (1976) *Chem. Eng. Sci.*, **31**, 505.
35 Haul, R., Heintz, W., and Stremming, H. (1980) *Properties and Applications of Zeolites* (ed. R.P. Townsend), Special Publication No. 33, The Chemical Society, London, p. 27.
36 Haul, R. and Stremming, H. (1984) *J. Colloid Interface Sci.*, **97**, 348.
37 Sun, L.M. and Meunier, F. (1987) *Chem. Eng. Sci.*, **42**, 1585.
38 Kocirik, M., Kärger, J., and Zikanova, A. (1979) *J. Chem. Technol. Biotechnol.*, **29**, 339.
39 Kocirik, M., Smutek, M., Bezus, A., and Zikanova, A. (1980) *Collect. Czech. Chem. Commun.*, **45**, 3392.
40 Bezus, A., Zikanova, A., Smutek, M., and Kocirik, M. (1981) *Collect. Czech. Chem. Commun.*, **46**, 678.
41 Sircar, S. and Kumar, R. (1983) *ACS Symp. Ser.*, **223**, 171; see also: (1984) *J. Chem. Soc., Faraday Trans. I*, **80**, 2489.
42 Bhaskar, G.V. and Do, D.D. (1989) *Chem. Eng. Sci.*, **44**, 1215.
43 Kmiotek, S.J., Wu, P., and Ma, Y.H. (1982) *AIChE Symp. Ser.*, **78** (219), 83.
44 Brunovska, A., Hlavacek, V., Ilavsky, J., and Valtyin, J. (1978) *Chem. Eng. Sci.*, **33**, 1385.
45 Brunovska, A., Hlavacek, V., Ilavsky, J., and Valtyin, J. (1980) *Chem. Eng. Sci.*, **35**, 757.
46 Brunovska, A., Ilavsky, J., and Hlavacek, V. (1981) *Chem. Eng. Sci.*, **36**, 123.
47 Eagan, J.D., Kindl, B., and Anderson, R.B. (1971) *Adv. Chem.*, **102**, 165.
48 Ilavsky, J., Brunovska, A., and Hlavacek, V. (1980) *Chem. Eng. Sci.*, **35**, 2475.
49 King, G. and Cassie, A.B.D. (1940) *Trans. Faraday Soc.*, **26**, 445.
50 Round, G.F., Habgood, H.W., and Newton, R. (1966) *Sep. Sci.*, **1**, 219.
51 Kärger, J. and Bülow, M. (1975) *Chem. Eng. Sd.*, **30**, 893.
52 Taylor, R.A. (1979) PhD Thesis, University of New Brunswick, Fredericton.
53 Satterfield, C.N. and Katzer, J.R. (1971) *Adv. Chem.*, **102**, 193.
54 Satterfield, C.N. and Cheng, C.S. (1972) *AIChE J.*, **18**, 724.
55 Moore, R. and Katzer, J.R. (1972) *AIChE J.*, **18**, 816.
56 Habgood, H.W. (1958) *Can. J. Chem.*, **36**, 1384.
57 Heinke, L., Kortunov, P., Tzoulaki, D., and Kärger, J. (2007) The options of interference microscopy to explore the significance of intracrystalline diffusion and surface permeation for overall mass transfer on nanoporous materials. *Adsorption*, **13**, 215–223.

Part III
Molecular Modeling

7
Constructing Molecular Models and Sampling Equilibrium Probability Distributions

In the last 30 years an additional tool has emerged for the investigation of diffusion phenomena in zeolites: molecular modeling and computer simulation. This starts from detailed quantitative models of atomic-level structure and interactions and proceeds to derive thermodynamic and transport properties based on the principles of dynamics and statistical mechanics. Molecular simulations of materials in general and of zeolite–sorbate systems in particular have witnessed unprecedented growth in recent years. One factor responsible for this is that the computer power available at a given cost has been expanding exponentially. The number of transistors on an integrated circuit has been doubling, if not every 18 months as stated by the celebrated Moore's law [1], certainly every two years. Even more important to the growth of computational modeling of materials has been the development of new methods and algorithms that allow problems to be solved more efficiently with given computational resources. It is estimated that between 1970 and 2000 a nine-order of magnitude improvement was realized in the performance of computational approaches for solving some statistical mechanical problems, of which only four orders of magnitude are attributable to advances in computer hardware.

Molecular modeling and simulation is an ideal complement to experimental investigations of sorption and diffusion in zeolites because (i) it can provide detailed information on the molecular mechanisms and thereby help uncover the microscopic processes underlying phenomena observed in the laboratory; (ii) it can be used to explore "what if" questions concerning the effect of changes in chemical constitution, crystal structure, and material morphology on sorption and transport properties, and thus suggest how the basic mechanisms can be harnessed to develop materials with desired performance characteristics for specific applications; (iii) when carried out at the appropriate level, it can predict experimental observables, such as vibrational spectra, X-ray and neutron diffractograms, and quasi-elastic neutron scattering functions, in addition to self- and transport diffusivities, activation energies, sorption isotherms, and heats of sorption, and thereby help to interpret these observables, resolve possible discrepancies between different measurement methods, and provide estimates under conditions that are difficult to access experimentally. Conversely, experiments are valuable in validating the models and methods invoked by modeling and in realizing the design principles arrived at

Diffusion in Nanoporous Materials. Jörg Kärger, Douglas M. Ruthven, and Doros N. Theodorou.
© 2012 Wiley-VCH Verlag GmbH & Co. KGaA. Published 2012 by Wiley-VCH Verlag GmbH & Co. KGaA.

through modeling. Progress in understanding and designing zeolitic materials is best achieved through coordinated efforts involving experimental measurements, theoretical analyses, and computer simulations [2].

Challenges faced by modeling stem from:

1) Our incomplete knowledge of interaction potentials; progress in electronic structure calculations, quantum mechanics/molecular mechanics (QM/MM) and *ab initio* molecular dynamics methods is of strategic importance in meeting this challenge;
2) our limited ability to represent defective structures, surfaces and interfaces, and complex morphologies encountered in real-life zeolitic materials;
3) the high computational demands of simulation methods, which limit the length scales and time scales that can be addressed – atomistic molecular dynamics (MD) simulation on a conventional computer workstation, for example, can simulate systems of dimension on the order of 100 nm for times on the order of 1 μs, which are too short for many phenomena and processes occurring in zeolites. The development of multi-scale modeling approaches consisting of several levels, each level addressing phenomena over a specific window of time and length scales, receiving input from more fundamental levels and providing input to more coarse-grained ones, is a promising avenue for meeting this challenge.

In this part of the book (Chapters 7–9) we outline some molecular simulation methods useful in understanding sorption and diffusion phenomena in zeolites and other nanoporous solids. The exposure will necessarily be brief, with emphasis on the calculation of experimental observables and on the elucidation of siting, conformation, and modes of motion in the pores. For a more detailed exposition of methods, the reader is referred to textbooks on molecular simulation [2, 3]. This chapter will focus on constructing molecular models and predicting sorption equilibria through Monte Carlo (MC) simulations. Chapter 8 addresses the prediction of intracrystalline dynamics and diffusion through molecular dynamics simulations. Chapter 9 focuses on the analysis and simulation of infrequent events, which is a valuable tool for accessing long-time dynamical properties.

7.1
Models and Force Fields for Zeolite–Sorbate Systems

7.1.1
Molecular Model and Potential Energy Function

To study a zeolite–sorbate system computationally, one must first choose the model representation to be used. Typically, the model system consists of a nanometer-sized domain of zeolitic material and some molecules sorbed in it. The variables in terms of which one describes the configuration of the model system are the configurational "degrees of freedom." Typically, these are chosen as the position vectors $r_1, r_2, \ldots, r_N$ of all atoms constituting the zeolite and the sorbate molecules. One may, however, choose to use multi-atom moieties instead of individual atoms (e.g., methyl and methylene segments to describe a normal alkane in a so-called "united-atom" representation, which is computationally more economical than a fully atomistic model). We will use the term "interaction sites" to denote the N entities whose positions must be specified to fix the microscopic state of the system. Instead of position coordinates, one may choose to use generalized coordinates as degrees of freedom. For example, one may choose to describe a sorbed nitrogen molecule in terms of its center-of-mass $r_{cm} = (r_1 + r_2)/2$, the polar:

$$\psi_1 = \cos^{-1}\left[\hat{z} \cdot \frac{r_2 - r_1}{|r_2 - r_1|}\right]$$

and azimuthal:

$$\psi_2 = \pm\cos^{-1}\left(\hat{x} \cdot \frac{r_2 - r_1}{\left\{(r_2 - r_1)^2 - [(r_2 - r_1) \cdot \hat{z}]^2\right\}^{1/2}}\right)$$

angles, specifying its orientation, and the bond length $l = |r_2 - r_1|$ in place of the Cartesian coordinates of its two atoms, r_1 and r_2 (Figure 7.1). Both the vector of generalized coordinates $q = (r_{cm}, \psi_1, \psi_2, l)$ and the vector of Cartesian coordinates $r = (r_1, r_2)$ are six-dimensional and there is a Jacobian of transformation relating elementary volumes between the two descriptions.

Of central importance in simulations is the force field, which specifies the total potential energy $\mathcal{V}(q)$ of the model system as a function of the microscopic degrees of freedom.

In a classical description, the potential energy function $\mathcal{V}(q)$ determines the equations of motion governing the temporal evolution of all microscopic degrees of freedom (compare Chapter 8). Symbolizing by $\dot{q}$ the time derivative of the vector of (generalized) coordinates q and defining the Lagrangian [4] $\mathcal{L}(q, \dot{q}) = \mathcal{K} - \mathcal{V}$ as the

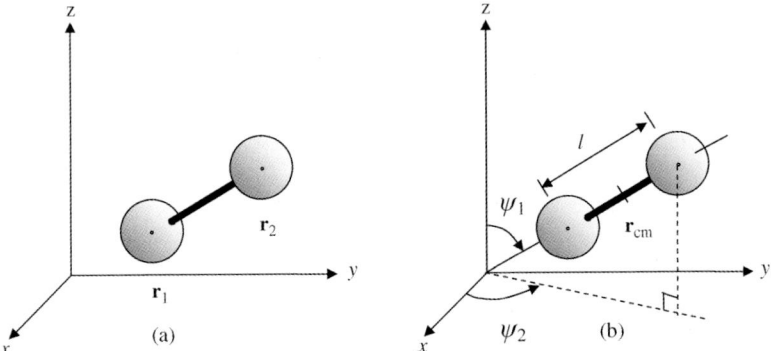

Figure 7.1 Description of the configuration of a diatomic molecule in space: (a) in terms of the Cartesian coordinates of the atoms, (r_1, r_2); (b) in terms of the generalized coordinates $(r_{cm}, \psi_1, \psi_2, l)$. Unit vectors along the x and z axes are denoted as $\hat{x}$ and $\hat{z}$, respectively.

difference between the kinetic energy $\mathcal{K}$ and the potential energy $\mathcal{V}$ of the system, the equations of motion can be written as:

$$\frac{d}{dt}\left(\frac{\partial \mathcal{L}}{\partial \dot{q}}\right) = \frac{\partial \mathcal{L}}{\partial q} \tag{7.1}$$

Lagrange's Eq. (7.1) is equivalent to Hamilton's variational principle, namely, that the motion of the system from time t_1 to time t_2 is such that the line integral $I = \int_{t_1}^{t_2} \mathcal{L}\, dt$ (action) has a stationary value for the correct path of motion [4]. One may change variables from $(q, \dot{q})$ to (q, p), with $p = \partial \mathcal{L}/\partial \dot{q}$ being the vector of generalized momenta, and invoke an alternative description of the dynamics in terms of the Hamiltonian $\mathcal{H}(q, p) = \dot{q} \cdot p - \mathcal{L}(q, \dot{q})$, which is the Legendre transform of $-\mathcal{L}$ with respect to $\dot{q}$. Then, Hamilton's equations of motion [4] provide an equivalent description to Eq. (7.1):

$$\begin{aligned} \dot{q} &= \frac{\partial \mathcal{H}}{\partial p} \\ \dot{p} &= -\frac{\partial \mathcal{H}}{\partial q} \end{aligned} \tag{7.2}$$

The potential energy function also plays a central role in shaping the thermodynamic properties of a system. Let us consider our system at equilibrium under the constraints of constant total number of interaction sites N, volume V, and temperature T. These macroscopic constraints, corresponding to a closed system surrounded by rigid diathermal walls, define the canonical ensemble of equilibrium statistical mechanics. Let us use the $3N$-dimensional vector of positions of all interaction sites, $r \equiv (r_1, r_2, \ldots, r_N)$, to describe the atomistic configuration. In other words, let us choose r as q. Then, the vector of generalized momenta p consists simply of the $3N$ components of the particle momenta. The kinetic energy function has the form:

$$\mathcal{K}(p) = \sum_{i=1}^{N} \frac{p_i^2}{2m_i}$$

with m_i being the mass of site i. The potential energy function $\mathcal{V}(r)$, on the other hand, depends only on the atomic coordinates. In this representation, the second Hamilton's Eq. (7.2) reduces to Newton's second law of motion.

In a classical description, the probability density in the $6N$-dimensional phase space (r, p) of atomistic microstates is of the form:

$$\varrho^{NVT}(r, p) = \frac{1}{Q(N, V, T)} \frac{1}{\prod_\alpha N_\alpha! h^{3N}} \exp[-\beta \mathcal{H}(r, p)] \tag{7.3}$$

where:

β equals $1/(k_B T)$ with $k_B = R/N_{\text{Avo}}$ being the Boltzmann constant,
R is the ideal gas constant,
and N_{Avo} is Avogadro's number.

The Hamiltonian in this case is simply $\mathcal{H}(r, p) = \mathcal{K}(p) + \mathcal{V}(r)$. N_α stands for the number of molecules of type α, and the product of factorials in the denominator of Eq. (7.3) runs over all distinguishable types of molecules present in the system. This product of factorials and the factor h^{3N}, with h being Planck's constant, ensures consistency with the correct quantum mechanical formulation; according to the latter, molecules of the same type are indistinguishable and the simultaneous specification of each coordinate of each site along with the corresponding component of momentum is subject to Heisenberg's uncertainty principle. The normalizing factor $Q(N,V,T)$ in Eq. (7.3) is the canonical partition function:

$$Q(N, V, T) = \frac{1}{\prod_\alpha N_\alpha! h^{3N}} \int \exp[-\beta \mathcal{H}(\mathbf{r},\mathbf{p})] \, \mathrm{d}^{3N}p \, \mathrm{d}^{3N}r \qquad (7.4)$$

By virtue of the separability of kinetic and potential energy contributions in this classical treatment, the integration over momenta in Eq. (7.4) reduces to the calculation of a product of Gaussian integrals. As a consequence, the canonical partition function can be rewritten as:

$$Q(N, V, T) = \frac{1}{\prod_\alpha N_\alpha! \prod_{i=1}^{N} \Lambda_i^3} Z(N, V, T) \qquad (7.5)$$

where $\Lambda_i = [\beta h^2/(2\pi m_i)]^{\frac{1}{2}}$ is the thermal wavelength of site i and $Z(N,V,T)$ is the configurational integral:

$$Z(N, V, T) = \int \exp[-\beta \mathcal{V}(\mathbf{r})] \, \mathrm{d}^{3N}r \qquad (7.6)$$

Integrating Eq. (7.3) over all momentum space, we obtain the configuration-space probability density of an equilibrium system in the canonical ensemble as a marginal distribution:

$$\varrho^{NVT}(\mathbf{r}) = \frac{1}{Z(N, V, T)} \exp[-\beta \mathcal{V}(\mathbf{r})] \qquad (7.7)$$

From the canonical partition function one immediately obtains a fundamental equation for the system in the Helmholtz energy representation:

$$A(N, V, T) = -k_B T \ln Q(N, V, T) \qquad (7.8)$$

Equation (7.8) constitutes an important link between atomic-level structure and interactions and macroscopic thermodynamic properties.

From the above it is evident that the potential energy function $\mathcal{V}(\mathbf{q})$ plays a central role in determining both dynamics and thermodynamics. In zeolite–sorbate systems, the potential energy function is usually assumed to consist of additive contributions from the zeolite framework, from the sorbate molecules, and from zeolite–sorbate interactions:

$$\mathcal{V} = \mathcal{V}_z + \mathcal{V}_s + \mathcal{V}_{zs} \qquad (7.9)$$

Contribution $\mathcal{V}_s$ is typically modeled as a sum of bonded and nonbonded intramolecular energy terms contributed by individual sorbate molecules i (which may be of the same or different kinds) plus a sum of intermolecular interactions contributed by pairs of sorbate molecules:

$$\mathcal{V}_s = \sum_{i=1}^{N_s} \mathcal{V}_{si}^{bonded} + \sum_{i=1}^{N_s} \mathcal{V}_{si}^{nonbonded} + \sum_{i=1}^{N_s-1} \sum_{j=i+1}^{N_s} \mathcal{V}_{sij}^{nonbonded} \quad (7.10)$$

with N_s being the total number of sorbate molecules. As a general rule, the choice of $\mathcal{V}_s$ is based on simulations of pure sorbates and sorbate mixtures in the fluid state, the requirement being that it provides an accurate representation of thermodynamic properties (e.g., liquid density as a function of temperature, enthalpies of vaporization and mixing), fluid phase equilibria (e.g., binodal curve for vapor–liquid equilibrium), and, if possible, transport properties (e.g., self-diffusivity in the liquid state). Many sophisticated force fields exist for this purpose, thanks to painstaking efforts in fitting electronic structure calculations and experimental measurements. An example is provided by the condensed-phase optimized molecular potentials for atomistic simulation studies (COMPASS) force field of Sun [5], which has been adopted by commercial simulation packages. The form postulated for $\mathcal{V}_s$ in the COMPASS force field is shown in Eq. (7.11):

$$\begin{aligned}
\mathcal{V}_s = & \sum_l \left[k_2^l(l-l_0)^2 + k_3^l(l-l_0)^3 + k_4^l(l-l_0)^4 \right] + \sum_\theta \left[k_2^\theta(\theta-\theta_0)^2 + k_3^\theta(\theta-\theta_0)^3 + k_4^\theta(\theta-\theta_0)^4 \right] \\
& + \sum_\phi [k_1^\phi(1-\cos\phi) + k_2^\phi(1-\cos2\phi) + k_3^\phi(1-\cos3\phi)] + \sum_\chi k_2^\chi \chi^2 + \sum_{l,l'} k^{ll'}(l-l_0)(l'-l'_0) \\
& + \sum_{l,\theta} k^{l\theta}(l-l_0)(\theta-\theta_0) + \sum_{l,\phi}(l-l_0)\left[k_1^{l\phi}\cos\phi + k_2^{l\phi}\cos2\phi + k_3^{l\phi}\cos3\phi\right] \\
& + \sum_{\theta,\phi}(\theta-\theta_0)\left[k_1^{\theta\phi}\cos\phi + k_2^{\theta\phi}\cos2\phi + k_3^{\theta\phi}\cos3\phi\right] + \sum_{\theta,\theta'} k^{\theta\theta'}(\theta-\theta_0)(\theta'-\theta'_0) \\
& + \sum_{\theta,\theta',\phi} k^{\theta\theta'\phi}(\theta-\theta_0)(\theta'-\theta'_0)\cos\phi \\
& + \frac{1}{4\pi\varepsilon_0}\sum_{i,j}\frac{q_iq_j}{r_{ij}} + \sum_{i,j}\varepsilon_{ij}\left[2\left(\frac{r_{ij}^0}{r_{ij}}\right)^9 - 3\left(\frac{r_{ij}^0}{r_{ij}}\right)^6\right]
\end{aligned}$$

$$(7.11)$$

The two last terms in Eq. (7.11) are nonbonded interaction terms, depending on the distances r_{ij} between pairs of atoms that are separated by three or more bonds or belong to different molecules. The r_{ij}^{-9} and r_{ij}^{-6} contributions represent repulsive (excluded volume) and dispersion interactions. The Coulombic contributions, dependent on partial charges q_i on atoms, are an important component of interactions between polar groups. All terms in Eq. (7.11) before the last two are valence terms. They depend on bond lengths (l), bond angles (θ), torsion angles (ϕ) or out-of-plane angles (χ) formed by pairs, triplets, and quadruplets of atoms bonded together, respectively. The cross-coupling terms (dependent on more than one l, θ, ϕ, or χ in Eq. (7.11) are necessary for the accurate prediction of vibrational frequencies and variations in bonded geometry associated with conformational changes.

In practice, simpler force fields have been found to provide satisfactory predictions for the thermodynamics and dynamics of fluids and zeolite–sorbate systems. In these simple force fields, $\mathcal{V}_{si}^{bonded}$ is typically represented as a sum of bond stretching, bond angle bending, and torsional terms:

$$\mathcal{V}_{si}^{bonded} = \mathcal{V}_{si}^{bond} + \mathcal{V}_{si}^{angle} + \mathcal{V}_{si}^{torsion} \quad (7.12)$$

V_{si}^{bond} and V_{si}^{angle} are usually cast as sums of harmonic terms in the bond lengths l and bond angles θ, respectively, of the molecule, which keep these degrees of freedom vibrating around their equilibrium positions:

$$V_l(l) = \frac{1}{2}k_l(l-l_0)^2, V_\theta(\theta) = \frac{1}{2}k_\theta(\theta-\theta_0)^2 \tag{7.13}$$

Bond length constrains are often introduced, fixing bond lengths at l_0 and dispensing with the bond stretching potential. The introduction of corresponding constraints for bond angles θ is seldom practiced, as it affects the dynamics of conformational isomerization and does not offer significant computational advantages.

The term $V_{si}^{torsion}$ is typically modeled as a sum of contributions from individual dihedral angles ϕ defined by quartets of interaction sites on the molecule, each contribution having the form of a Taylor expansion in $\cos(\phi - \phi_0)$ (equivalently, of a cosine Fourier series in $\phi - \phi_0$), with ϕ_0 corresponding to the value of ϕ at which the (periodic) torsional energy contribution from the dihedral angle in question has a global minimum. In describing the energy contribution from the skeletal torsion angles of alkanes, for example, ϕ_0 corresponds to the *trans* conformational state [6]:

$$V_\phi(\phi) = \sum_n c_n \cos^n(\phi-\phi_0) \tag{7.14}$$

Intramolecular $V_{si}^{nonbonded}$ and intermolecular $V_{sij}^{nonbonded}$ nonbonded interactions are typically modeled as sums of van der Waals and Coulombic interactions between sites on the same or different molecules:

$$V^{nonbonded} = V^{vdW} + V^{Coulomb} \tag{7.15}$$

The van der Waals term incorporates excluded volume repulsion at short distances and dispersion attraction at longer distances. It is most often modeled through the Lennard-Jones (LJ) potential:

$$V^{LJ}(r) = 4\varepsilon\left[\left(\frac{\sigma}{r}\right)^{12} - \left(\frac{\sigma}{r}\right)^6\right] \tag{7.16}$$

with r being the distance between the interacting sites. The Buckingham (exp–6) form is less commonly used, although it has a firmer theoretical basis. In the summation of van der Waals interactions a cutoff distance r_c is typically employed in connection with the minimum image convention [2]. Potential "tail" contributions to the properties may be calculated by direct integration, using pair distribution functions accumulated in the simulation, and are negligible if r_c is large enough (e.g., 13 Å).

Table 7.1 lists the parameters of two united-atom force fields that employ Eqs. (7.12)–(7.16): TraPPE [7] and NERD [8], which are used widely for linear alkanes. These force fields do an excellent job predicting fluid phase equilibria and densities, but are still not perfect (e.g., they tend to give too high populations of the *trans* conformational state relative to *gauche* and their attribution of a larger collision diameter to methylenes than to methyls is hard to justify physically).

Inclusion of Coulombic terms is necessary in cases where the interacting molecules exhibit explicit separation of electric charges. These are typically calculated after

Table 7.1 Parameters of the TraPPE (transferable potentials for phase equilibrium calculations) and NERD (Nath, Escobedo, de Pablo) force fields for linear alkanes longer than n-propane.

Type of interaction	Parameter	TraPPE	NERD
Bond stretching	Bond length	Fixed, $l = l_0 = 1.54$ Å	
Bond bending	Equilibrium bond angle	$\theta_0 = 114.0°$	
	Bending constant	$k_\theta = 62\,500$ K rad^{-2}	
Torsion	Torsion angle at *trans*	$\phi_0 = \pi$	
	Constants in Eq. (7.14)	$c_0/k_B = 919.98$ K	
		$c_1/k_B = 1748.92$ K	
		$c_2/k_B = 136.38$ K	
		$c_3/k_B = -2805.28$ K	
Nonbonded (inter- and intramolecular between pairs separated by more than three bonds)	Lennard-Jones		
	$\varepsilon_{CH3}/k_B =$	98.1 K	100.6 K
	$\sigma_{CH3} =$	3.77 Å	3.91 Å
	$\varepsilon_{CH2}/k_B =$	47.0 K	45.8 K
	$\sigma_{CH2} =$	3.93 Å	3.93 Å
	Combining rules for cross parameters: Lorentz–Berthelot	$\varepsilon_{ij} = (\varepsilon_{ii}\varepsilon_{jj})^{1/2}$ $\sigma_{ij} = (\sigma_{ii} + \sigma_{jj})/2$	

attributing partial charges to specific sites on each molecule, which may or may not coincide with atomic centers. Charges q,q' on two sites belonging to different molecules, or to the same flexible molecule, at a distance r from each other, contribute a term:

$$\mathcal{V}^{Coulomb}(r) = \frac{1}{4\pi\varepsilon_0} \frac{qq'}{r} \tag{7.17}$$

to the potential energy, with $\varepsilon_0 = 8.85 \times 10^{-12}$ C V^{-1} m^{-1} being the dielectric permittivity of free space. The values of partial charges are often based on *ab initio* electronic structure calculations on single molecules (Mulliken population analysis), with some adjustment to reproduce molecular dipole or quadrupole moments; less often, for strongly polar molecules, quantum mechanical calculations may be conducted on a molecular pair at various distances and relative orientations, and partial charges determined through regression to reproduce the energy as a function of molecular degrees of freedom. Examples of relatively simple sorbate–sorbate potentials that include partial charges for polar molecules are the simple point charge (SPC) model for water [9], representing a water molecule in terms of a LJ site and partial charges on the O and H atoms, the Murthy et al. model force field for N_2 [10], and the Harris and Yung force field for CO_2 [11]. For the summation of electrostatic interactions in the periodically continuous model systems studied in connection with zeolites, it is safest to use the Ewald summation technique (see below) [2, 3]. For sorbates of modest polarity, such as CO_2, it has been found that neglecting the electrostatic interactions when the distance between molecular centers exceeds a relatively large value (13 Å) produces indistinguishable results from the more expensive Ewald summation. Potential truncation is routinely practiced for LJ interactions between molecular centers.

Force field expressions analogous to the stretching and bending terms in Eqs. (7.11) and (7.13) and to the Coulombic terms in Eqs. (7.11) and (7.17) have been invoked to describe the zeolite part of the potential, $\mathcal{V}_z$, in the total potential energy function of Eq. (7.9). Force fields that have been invoked in such flexible models of zeolites vary very much in complexity; Reference [12] provides a review. Sophisticated force fields are needed if a faithful reproduction of the vibrational spectrum of the zeolite is required. Much simpler force fields seem to suffice for the reliable prediction of sorption and diffusion. The simple model of Demontis et al. [13] for silicalite (Chapter 18) utilizes only harmonic contributions of the type of Eq. (7.1) for the lengths between directly bonded Si and O atoms and between pairs of O atoms connected to the same Si atom. An improved parameterization for the Demontis et al. model has been proposed by Vlugt and Schenk [14]. Whenever a flexible zeolite model is used, it must be ascertained that the equilibrium structure of the pure solid resulting from $\mathcal{V}_z$ is consistent with crystallographic data. More recently, simple flexible models for zeolites and silica polymorphs have been developed and validated against vibrational normal modes, elastic constants, and thermal expansivity of the zeolite frameworks [15].

Zeolite–sorbate interactions, summarily represented by the term $\mathcal{V}_{zs}$ in Eq. (7.9), are extremely important in determining sorption thermodynamics and intracrystalline dynamics. Common practice in simulations is to use Lennard-Jones [Eq. (7.16)] and Coulombic [Eq. (7.17)] interactions between sites on the zeolite framework and counterions, and sites on the sorbate molecules. Partial charges can be attributed to the zeolite atoms on the basis of Hartree–Fock or density functional theory (DFT) electronic structure calculations, usually conducted on clusters of tetrahedral (T = Si, Al) and oxygen (O) atoms representative of the zeolite crystal structure [16]. This exercise is, of course, most straightforward in the case of fully siliceous zeolites, such as silicalite. In Al-containing zeolites the Si/Al ratio is used as input for the atomic-level reconstruction of the lattice, but the precise crystallographic positions of Al and Si atoms are seldom known with certainty. Often, Al atoms are randomly distributed among the T sites of the atomistic model framework up to the desired Al/Si ratio in such a way that Löwenstein's rule is fulfilled [17]. An alternative, but less satisfying solution, pioneered in early modeling work by Kiselev and collaborators [18], is to represent all tetrahedral atoms in the framework through a hybrid "T" atom bearing an average partial charge that depends on the particular Si/Al ratio under investigation.

LJ contributions to $\mathcal{V}_{zs}$ are often envisioned as emanating only from the O atoms of the framework and the counterions, that is, T atoms are not considered as LJ interaction sites (although they are considered as Coulomb sites), the rationale being that they are surrounded by bridging oxygens and therefore not immediately exposed to sorbate molecules occupying the channels. This representation again goes back to Kiselev and collaborators [18]. As regards the choice of ε and σ parameters for the framework oxygens, a practice introduced in early modeling work [19] was to adjust them so as to reproduce the Henry's law constant and low-occupancy isosteric heat of sorption of a simple sorbate (e.g., methane) in a fully siliceous zeolite (e.g., silicalite) at a single temperature (e.g., 300 K) and then invoke the same values in all subsequent calculations. Parameters determined in this way have been found to be reasonably portable across different hydrocarbon sorbates and different siliceous zeolites (MFI, ITQ). More refined strategies for determining LJ interaction parameters for atoms have focused on reproducing inflection points in semilogarithmic plots [$q(\ln f)$] of the sorption isotherms for specific zeolite–sorbate systems [20].

Coulombic contributions to $\mathcal{V}_z$ and $\mathcal{V}_{zs}$, if present, are typically summed using the Ewald method [2, 3], as already mentioned for $\mathcal{V}_s$.

A perfect crystalline configuration based on X-ray diffraction (XRD) is almost invariably used as a starting point for atomistic modeling investigations of sorption and diffusion in zeolites. The procedure for reconstructing a digital crystal at the atomistic level entails: (i) finding, for example, from Reference [21], the unit cell geometric characteristics that correspond to the framework type code (FTC) of the zeolite and then, from the XRD spectrum in the same reference, the position vectors of the "primary" framework atoms; (ii) applying the symmetry operations of the space group of the crystal, given, for example, in Reference [22], to generate the types and position vectors of all framework and extra-framework atoms of the unit cell; (iii) replicating the unit cell by translation along the lattice parameters to form a simulation box of the desired size. For cases where the crystallographic structure is not known, statistical mechanics-based simulation techniques may be used to solve the structure from powder diffraction data. A noteworthy effort in this direction is the work of Falcioni and Deem [23], who used the parallel tempering (replica exchange) Monte Carlo method to match geometric, density, and diffraction data to distribution functions accumulated from known zeolite structure types.

A drastic simplification of atomistic modeling calculations is achieved if an infinitely stiff zeolite model, with the positions of all framework atoms and counterions fixed in space, is invoked. With such a model, $\mathcal{V}_z$ becomes a constant and drops out of the calculations. More importantly, the contribution to $\mathcal{V}_{zs}$ from each type of interaction site present in the sorbate molecules becomes a static three-dimensional field that can be pretabulated as a function of position in space. During a simulation, contributions to $\mathcal{V}_{zs}$ and to its gradient (force) from each LJ site and each partial charge on each sorbate molecule can be readily computed by interpolation among the pretabulated values. June et al. [19] have introduced a robust and efficient method for performing the three-dimensional pretabulation and interpolation of the potential energy field on a specific site type due to the static zeolite. The pretabulation involves computing the energy, its x-, y-, and z- first derivatives, its xx-, yy-, and zz- second derivatives, and its xyz- third derivative at all points over a fine (lattice spacing 0.2 Å) cubic grid running through the accessible pore volume of the asymmetric unit of the unit cell. The energy and force at an arbitrary point in the pore volume of the model system can then be computed through a Hermite interpolation scheme, wherein the field is represented as a three-dimensional cubic spline (product of three cubic polynomials in x, y, and z). The 4^3 coefficients of the spline in each voxel of the grid are obtained by matching the 8×8 pretabulated function and derivative values at the eight apices of the voxel. This pretabulation/interpolation scheme affords a saving of two orders of magnitude in central processing unit (CPU) time relative to direct summation of LJ interactions, and an even greater saving relative to Ewald summation of Coulombic interactions.

When is it permissible to use an inflexible zeolite model? Numerous discussions of the consequences of neglecting framework flexibility on sorption thermodynamics and dynamics have appeared in the literature. Based on the accumulated experience from molecular simulations, it appears that a flexible model is necessary for tightly fitting sorbate–zeolite systems, where omission of lattice vibrations leads to an overestimation of energy barriers to intracrystalline transport. This is seen characteristically in the transition-state theory-based work of Snurr *et al.* on benzene in silicalite, where use of a rigid model that affords a good representation of sorption thermodynamics resulted in an underestimation of the diffusivity by one to two orders of magnitude relative to experiment [24]. Subsequent calculations by Forester and Smith [25], which included lattice vibrations, yielded good agreement with experiment, including state-of-the-art neutron spin echo measurements [26].

The siting of counterions in Al-containing zeolites is an important issue for atomistic modeling. To study the sorption of n-alkanes in FAU zeolites, Calero *et al.* [17] developed a model with a fixed framework with explicitly distinguished Al and Si atoms and mobile counterions. They estimated the counterion distribution by performing configurational-bias Monte Carlo simulations (see below) on a sorbate-free NaX zeolite, and then obtained the alkane–sodium, alkane–alkane, and alkane–framework interaction parameters after calibrating the force field by fitting to the entire sorption isotherms.

Imposing a cutoff r_c on Coulombic interactions is generally risky, because of their long-range nature. It is preferable, albeit more expensive computationally, to sum such interactions fully, taking advantage of the periodic structure of the model system. This can be achieved with the Ewald summation technique, which is discussed in standard simulation texts [2, 3]. To improve convergence, Ewald summation surrounds each point charge in the system by a compensating Gaussian distribution of opposite sign. The electrostatic potential and force at each point in space resulting from charges screened in this way is computed directly in real space, while the contribution due to the periodic array of Gaussian distributions introduced is summed in Fourier space. In practice, the width of the Gaussian distribution and the numbers of real-space and Fourier space terms taken into account must be chosen so as to ensure an accurate summation, while minimizing the computer time required for the summation [3]. The latter scales as $N^{3/2}$ with N being the number of point charges in the simulation box. When an inflexible zeolite model is used, the zeolite electrostatic field contributing to $\mathcal{V}_{zs}$ needs to be computed only once, at the beginning of the simulation; it can be pretabulated and interpolated for use in the simulation by the techniques discussed above. Coulomb contributions to $\mathcal{V}_s$ must, of course, be computed for each configuration visited by the simulation. For N exceeding 10^5, advanced algorithms based on the multipole expansion become more efficient than Ewald summation for calculating long-range interactions [3].

7.1.2
Ab Initio Molecular Dynamics of Zeolite-Sorbate Systems

One promising avenue to force fields for zeolite–sorbate systems, especially in cases where strong and specific interactions are involved, is to conduct electronic structure calculations on appropriately defined clusters [16], or on the entire unit cell, if feasible, and fit the results for the ground state energy as a function of configuration by a molecular mechanics-type expression for $\mathcal{V}(\mathbf{q})$ of the types discussed in Section 7.1.1. In more recent years, an approach that dispenses with the need for a force field altogether has been tried on zeolite systems: Car–Parrinello, or *ab initio*, molecular dynamics [27]. In this extended system of molecular dynamics [2, 3] method, the set of degrees of freedom of the simulation is augmented to include electronic orbitals, in addition to nuclear coordinates. Electronic degrees of freedom are treated in a density functional theory approximation. In this way, the potential energy function $\mathcal{V}(\mathbf{q})$ governing interactions among the nuclei is generated "on the fly" as the dynamical simulation proceeds, even for a chemically reactive system. Dispersion interactions cannot be captured accurately and the computational requirements of the method limit the system sizes and times that can be simulated; nevertheless, it has already been applied to some interesting problems, most often related to the catalytic activity of zeolites. In one study [28] the substitution of aluminum for silicon at various sites of a unit cell of offretite has been studied to relate proton affinity with the substitution energy and Al–O–Si bond angles. A Car–Parrinello study of sodalite [29] yielded proton stretch vibrations that were compared to infrared spectra of the zeolite, as well as interactions of methanol trapped inside a sodalite cage. More recently, the motions of acetonitrile sorbed at an isolated Brønsted site of chabazite have been simulated with *ab initio* molecular dynamics and compared against NMR experiments [30].

7.1.3
Computer Reconstruction of Meso- and Macroporous Structure

The structure of real-life zeolitic media is characterized by a range of length scales, typically exceeding the unit cell dimensions by several orders of magnitude. Transport rates and mechanisms are often significantly affected by these longer-range structural features (Sections 17.3.4 and 18.6). To capture these effects, computer models that can reproduce the precise distribution and connectivity of solid and void regions in three-dimensional space at length scales of tens of nm to tens of μm are useful.

Modeling of the structure of heterogeneous materials and of transport therein is a broad and active field of research [31]. Two general computer reconstruction methodologies that have been tried for zeolitic media are stochastic and particle-based reconstruction.

In stochastic reconstruction [32] the medium is represented as an arrangement of solid and void elements (voxels) in three-dimensional space. Input for the reconstruction is provided by experimental, typically two-dimensional, images of the material, derived from scanning or transmission electron microscopy, atomic force microscopy, fluorescence confocal optical microscopy, or high-resolution electron microscopy. These are subjected to thresholding and digitized, that is, reduced to sets of black (solid, phase $\phi = 1$) and white (void, $\phi = 0$) pixels. The digitized experimental images are then processed to derive various statistical descriptors of the structure. Such descriptors are the void fraction (fraction of pixels with $\phi = 0$), the two-point probability function (probability density function for finding two pixels at separation r, both of them belonging to the same phase ϕ), the lineal path distribution function (probability density for finding a vector r lying entirely in phase ϕ), the chord length distribution function (probability density of finding a vector r, lying entirely in phase ϕ, with its start and end on the boundary of phase ϕ), and the pore size distribution function (distribution of diameters of discretized circles that can be drawn with their center on a pixel and lying entirely in the phase of the pixel). To perform the reconstruction, one defines an objective function as a weighted sum of squared deviations in these statistical descriptors between sections of the reconstructed system and the reference images. Minimization of this objective function proceeds through a stochastic scheme that starts from a randomly generated configuration at the correct void fraction and implements random moves that swap void and filled pixels. This is in general a challenging minimization problem. Simulated annealing and parallel tempering Monte Carlo algorithms have been used to avoid the system getting trapped in the vicinity of local minima. The quality of the reconstruction can be assessed by comparing structural features not included in the objective function between the reference experimental images and the final reconstructed system. Figure 7.2b displays a model silicalite membrane that was stochastically reconstructed from fluorescent confocal microscopy images [32]. The anisotropy of the structure, resulting from preferential growth of crystallites on a substrate along the z-axis, is evident.

Particle-based reconstruction utilizes experimentally available information about the shape and size distribution of three-dimensional macroscopic objects (particles) that make up the medium. Representative samples of such particles are packed to a prescribed void fraction to generate a computer model for the medium. By defining a potential for interparticle interactions, one can cast the packing problem as an energy minimization problem. An example of this type of reconstruction is given in Figure 7.2a for a bed of NaX zeolite crystals. Electron microscopy had revealed that,

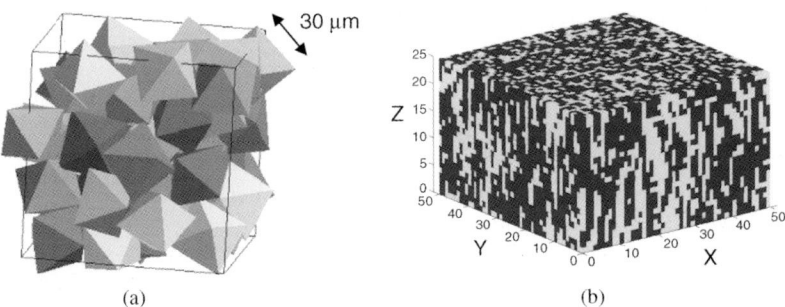

Figure 7.2 (a) Model bed of octahedral NaX zeolite crystals, as obtained by particle-based computer reconstruction [33]; the bed porosity was chosen to be consistent with the experimentally measured bed density and crystal density, while the crystal shapes and sizes were based on electron microscopy images. (b) Anisotropic silicalite membrane model, as obtained by stochastic computer reconstruction [32] based on two-dimensional fluorescence confocal microscopy images.

to an excellent approximation, the crystals have a regular octahedral shape with a mean edge length of 30 μm. In the reconstruction, the faces of the octahedra were represented as planar close-packed assemblages of soft spheres. The packing was achieved through a sequence of energy minimizations with respect to the center-of-mass positions and orientations of the crystals, under constant void fraction. Initially, a coarse representation (six spheres per tetrahedron) was employed. The resolution was progressively increased in subsequent minimizations until the graininess (sphere radius) became insignificant in relation to the dimensions of the crystals. The final configuration is a random packing of non-interpenetrating octahedra that rest on each other, satisfying the conditions of detailed mechanical equilibrium. Configurations generated in this way were used as starting points for a study of Knudsen and molecular diffusion in the intercrystalline space [33].

7.2
Monte Carlo Simulation Methods

7.2.1
Metropolis Monte Carlo Algorithm

Although this book is primarily concerned with diffusion phenomena, we will spend some time discussing ways to predict sorption equilibria and characterize the equilibrium structure and thermodynamic properties of fluids sorbed in microporous solids. These are important, because (i) an ability to describe equilibrium phenomena is a prerequisite for dealing with transport in response to driving forces that take the system away from equilibrium; (ii) checking predicted isotherms, heats of sorption, equilibrium siting, and conformation against experimental measurements is a good way for validating force fields; (iii) calculation of free energies with respect to appropriately selected collective descriptors of the configuration, or "order parameters," is an important step towards the development of coarse-grained approaches for addressing long time and length scale phenomena.

The class of stochastic simulation algorithms known as Monte Carlo (MC) plays a central role in the calculation of equilibrium properties of molecular models of materials. In 1953, Metropolis *et al.* [34] introduced an ingenious method for sampling points in a multidimensional space according to a prescribed probability density distribution defined on that space. In the applications considered here, the multidimensional space will be the configuration space, spanned by the (generalized) coordinates q of the sorbent/microporous solid system plus possibly a few macroscopic variables (e.g., number of molecules N) that are allowed to fluctuate, and the probability distribution will be defined by the equilibrium probability density ϱ^{eq} of an equilibrium ensemble of statistical mechanics [see, for example, Eq. (7.7) for the canonical or *NVT* ensemble]. Sampled configuration-space points, or "states," constitute a Markov chain, each state being formed from the previous one in a MC step. Each MC step is typically executed in two stages: One first attempts an elementary move from the current state i to a new state j with probability

$\alpha(i \rightarrow j)$. One then accepts the attempted move with probability:

$$P_{\text{accept}}(i \rightarrow j) = \min\left[1, \frac{\alpha(j \rightarrow i)\varrho^{\text{eq}}(j)J(j)}{\alpha(i \rightarrow j)\varrho^{\text{eq}}(i)J(i)}\right] \tag{7.18}$$

where J denotes a Jacobian of transformation from the set of generalized coordinates in which the move is attempted to the (Cartesian) coordinates, with respect to which ϱ^{eq} is defined. If the move is rejected, state i is retained for the current step. With these choices, transition probabilities $P(i \rightarrow j) = \alpha(i \rightarrow j) P_{\text{accept}}(i \rightarrow j)$ from state i to state j ($i \neq j$) in one step satisfy the condition of microscopic reversibility, or detailed balance:

$$\varrho^{\text{eq}}(i)J(i)P(i \rightarrow j) = \varrho^{\text{eq}}(j)J(j)P(j \rightarrow i) \tag{7.19}$$

As a consequence, the generated Markov chain of states asymptotically samples [2, 3] the probability distribution ϱ^{eq}. That is to say, in the production phase of a MC simulation, where all configuration-dependent properties fluctuate around constant average values representative of thermodynamic equilibrium, each state i is sampled with a frequency proportional to its equilibrium probability density $\varrho^{\text{eq}}(i)$. This is called "importance sampling."

The form of the stochastic matrix of attempt probabilities $\alpha(i \rightarrow j)$ is dictated by the elementary moves one chooses to implement. In the original MR^{2}T^2 algorithm [34] this matrix was symmetric, which means that αs can be dropped from the right-hand side of the acceptance criterion, Eq. (7.18), and all Js were unity. "Bias" schemes, involving non-symmetric αs and often requiring non-unity Jacobian factors, are used more often in contemporary work (see below). By designing the moves so as to take large strides in configuration space and at the same time ensure a substantial probability of acceptance, one can achieve equilibrium orders of magnitude more efficiently than with molecular dynamics. The only requirements are that the algorithm be ergodic (i.e., moves can take the system everywhere in configuration space) and microscopically reversible. Thermodynamic properties are computed as ensemble averages of microscopic configuration-dependent quantities over the equilibrated portion of an MC run.

7.2.2
Canonical Monte Carlo

Let us consider a model system consisting of components 1, 2,..., z. Component z will be the microporous solid, while components 1, 2,..., $z-1$ will be small-molecule species sorbed in its pores. The volume of the system will be denoted by V and the number of molecules of species α by N_α (for $\alpha = z$, another measure of the amount of matter, such as the number of unit cells, will be used in lieu of the number of molecules). The system is described by the set of coordinates $\mathbf{r}$ of all atoms on the microporous medium and on the sorbate molecules. Under the macroscopic constraints of constant $N_1, N_2, \ldots, N_z$, V, and temperature T, assuming classical statistical mechanics holds, the configuration-space probability density of the system is described by Eq. (7.7), with $\mathcal{V}$ being the total potential energy function. An MC algorithm designed to sample the system under these conditions (canonical ensemble) typically employs elementary moves that choose a species at random, choose a molecule of that species at random, and then impose a random displacement, reorientation, or internal reconfiguration (e.g., conformational change) on the chosen molecule [2, 3]. According to Eq. (7.18) and assuming symmetric attempt probabilities and Jacobians equal to unity, the move will be accepted with probability:

$$P_{\text{accept}}(i \to j) = \min[1, \exp(-\beta \Delta \mathcal{V})] \tag{7.20}$$

where $\Delta \mathcal{V} = \mathcal{V}_j - \mathcal{V}_i$. The move will always be accepted if it leads to lower potential energy; it will be accepted with probability $\exp(-\beta \Delta \mathcal{V})$ if it leads to an increase in potential energy.

One property that can readily be computed as an ensemble average in the canonical ensemble is the singlet probability density for the center-of-mass of a molecule of species α ($\alpha = 1, \ldots, z-1$) to be found at position $\mathbf{r}$ in three-dimensional space:

$$\varrho^\alpha(\mathbf{r}) = \left\langle \frac{1}{N_\alpha} \sum_{l=1}^{N_\alpha} \delta\left(\mathbf{r}^\alpha_{\text{cm},l} - \mathbf{r}\right) \right\rangle \tag{7.21}$$

Figure 7.3 depicts the singlet probability densities $\varrho^1(\mathbf{r})$ for n-hexane and 3-methylpentane sorbed in silicalite at 400 K at very low pressure, as obtained from an early MC calculation [19]. The three-dimensional contour plots make obvious

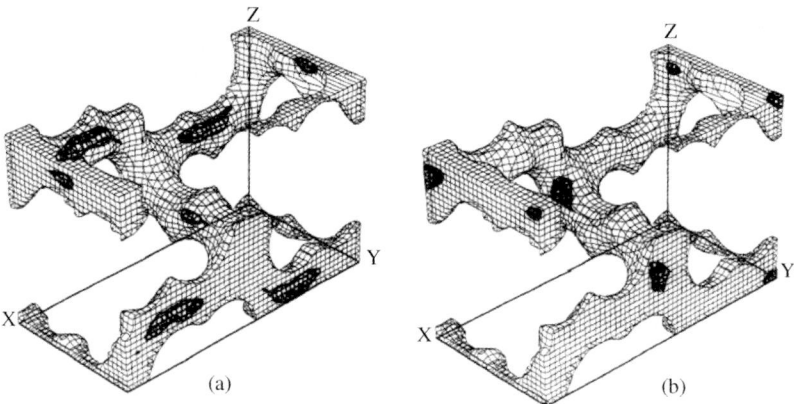

Figure 7.3 Three-dimensional contour plots of the singlet density distribution $\varrho^1(\mathbf{r})$ of the center-of-mass of n-hexane (a) and 3-methylpentane (b) sorbed in silicalite at 400 K and with very low occupancy. The lighter-gray grid outlines the interior surfaces of the pores throughout half a unit cell of the zeolite; X, Y, and Z are directed along the **a**, **b**, and **c** crystallographic axes, respectively, of the zeolite, with the other half of the unit cell being symmetric to the shown half with respect to the XZ plane. Bold lines outline the regions in which the center-of-mass of a sorbed molecule spends 50% of its time with highest probability. The distributions are dramatically different for the two isomers: n-hexane prefers to reside in the interiors of sinusoidal and straight channel segments, where it can maximize its favorable dispersive interactions with the zeolite. 3-methylpentane, in contrast, prefers the more spacious channel intersections; its methyl branch gives rise to steric exclusion from the channel interiors. With permission from [19].

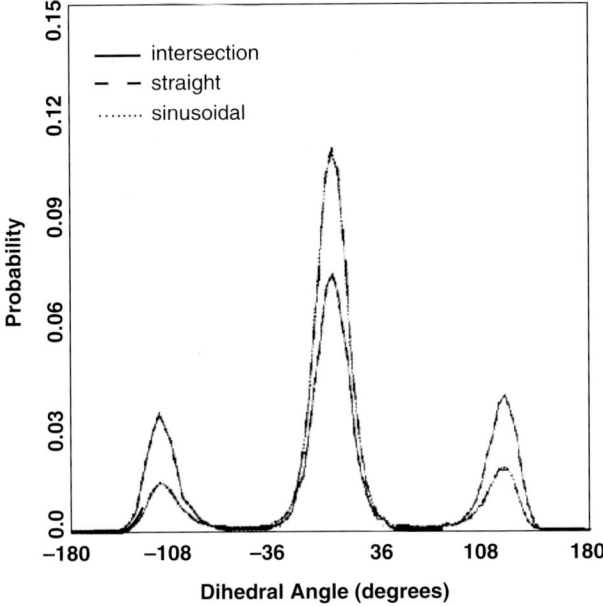

Figure 7.4 Distributions $\{\varrho^1(\phi)\}$ of the central torsion angle of n-butane sorbed in silicalite at 300 K and a loading of four molecules per unit cell. Separate distributions are shown for molecules whose center-of-mass is located in straight channel segments, in sinusoidal channel segments, and in channel intersections. The *trans* conformational state is strongly enhanced inside the straight and sinusoidal channel segments, where molecules are confined by the 5.5 Å diameter pores. In the more spacious channel intersection, the *gauche* state is populated substantially, giving rise to a distribution similar to the one encountered in the liquid state. With permission from [35].

the shape selectivity of the zeolite for these molecules. The orientational distribution of a sorbed molecule can be quantified by accumulating the distribution $\varrho^\alpha(\psi)$ of the angle between a characteristic vector affixed on the molecule (e.g., normal vector to the plane of a benzene) and a characteristic direction in the microporous medium (e.g., channel axis). The conformational changes of flexible sorbed molecules can be assessed through distributions $\varrho^\alpha(\phi)$ for specific torsion angles ϕ defined on the molecules. Figure 7.4 shows an example of the torsional distribution $\varrho^\alpha(\phi)$ of n-butane accumulated separately in the three intracrystalline environments of silicalite.

The enthalpy of sorption can readily be estimated through canonical MC simulations. Assuming a perfectly rigid model for the nanoporous solid, for a pressure P low enough for the individual sorbate components to be in the ideal gas phase:

$$\Delta_{\text{mix}} H = \langle \mathcal{V} + PV \rangle - \sum_{\alpha=1}^{z-1} N_\alpha \langle \mathcal{V}_{\text{intra}}^\alpha + PV^\alpha / N_\alpha \rangle_{\text{ig}} \qquad (7.22)$$

where

$\mathcal{V}_{\text{intra}}^{\alpha}$ denotes the intramolecular potential energy of a molecule of type α,
V^{α} is the volume of N_{α} sorbate molecules of species α in the pure fluid state at T and P,
$\langle\,\rangle$ denotes canonical ensemble averaging in the solid–sorbate system,
and $\langle\,\rangle_{\text{ig}}$ indicates canonical ensemble averaging in a pure ideal gas state of species α at T and P.

Typically, $\langle V \rangle / N_{\alpha} \approx V_z/N_{\alpha} << \langle V^{\alpha}/N_{\alpha}\rangle_{\text{ig}} = k_B T/P$, hence:

$$\Delta_{\text{mix}} H = \langle \mathcal{V} \rangle - \sum_{\alpha=1}^{z-1} N_{\alpha} \langle \mathcal{V}_{\text{intra}}^{\alpha} \rangle_{\text{ig}} - \left(\sum_{a=1}^{z-1} N_{\alpha}\right) k_B T \tag{7.23}$$

Separate contributions to the enthalpy of sorption due to interactions between each sorbate species and the solid, due to intermolecular interactions between each pair of sorbate species, and due to changes in intramolecular energy of each (flexible) sorbate species can similarly be computed.

If the compressibility of the microporous solid is not negligible, it may be more convenient to simulate the solid phase under constant pressure (isothermal–isobaric ensemble, $N_1, N_2, \ldots, N_z, P, T$ constant) or, even better, under constant stress tensor (isothermal–isostress ensemble, $N_1, N_2, \ldots, N_z, \boldsymbol{\sigma}, T$ constant). Going from one ensemble to another amounts to a Legendre transformation of the corresponding thermodynamic potential in macroscopic thermodynamics. For example, going from the canonical to the isothermal–isobaric ensemble is the microscopic equivalent of a Legendre transformation of the Helmholtz energy with respect to volume, yielding a Gibbs energy representation for the thermodynamics. The phase-space and configuration-space probability densities of these ensembles for closed systems can be derived readily from the canonical ensemble; acceptance rules for MC moves follow from these probability densities [2, 3]. In isothermal–isobaric MC one must employ volume fluctuation moves on the entire simulation box, whose acceptance depends on the imposed pressure. Isothermal–isostress simulations must allow the box shape to fluctuate as well; they are seldom practiced in an MC framework, but are convenient for studying phase transformations in solids in an MD context, using the Parrinello–Rahman extended ensemble algorithm [2].

7.2.3
Grand Canonical Monte Carlo

The grand canonical ensemble, in which the microporous solid phase is studied under constant temperature T, volume V, sorbent mass N_z, and sorbate chemical potentials $\mu_1, \mu_2, \ldots, \mu_{z-1}$, is especially convenient for studying sorption equilibria. As already mentioned, in many systems the structure and spatial extent of the solid can be assumed to be unaffected by the sorbed fluid, justifying the assumption of a constant V. The chemical potentials $\mu_1, \mu_2, \ldots, \mu_{z-1}$ or, equivalently, the fugacities $f_1, f_2, \ldots, f_{z-1}$ can readily be computed from the pressure P, temperature T, and composition (e.g., mole fractions $y_1, y_2, \ldots, y_{z-2}$) of the fluid phase, using an accurate equation of state for that phase [36]. Note that chemical potentials here are measured in units of energy per molecule; they can be converted into the units of energy per mole commonly used in the chemical engineering literature through multiplication by N_{Avo}. The following relation holds between μ_α and f_α:

$$\beta \mu_\alpha = \ln \left(\beta f_\alpha \prod_{l=1}^{s_\alpha} \Lambda_l^3 / Z^\alpha_{\text{intra,ig}} \right) \quad (7.24)$$

where

$\beta = 1/(k_B T)$, the product is taken over all s_α sites on a molecule of type α,
$\Lambda_l = [\beta h^2/(2\pi m_l)]^{\frac{1}{2}}$ is the thermal wavelength of site l with m_l being the mass of the site and h is Planck's constant,
and $Z^\alpha_{\text{intra,ig}}$ is the configurational integral of a single molecule of type α in the ideal gas state over the $(3s_\alpha - 3)$ site coordinates that specify its orientation and internal configuration.

The configuration-space probability density of the grand canonical ensemble reads:

$$\varrho^{\mu_1 \mu_2 \ldots \mu_{z-1} VT}(\mathbf{r}, N_1, N_2, \ldots, N_{z-1}) = \frac{1}{\Xi} \frac{1}{N_1! N_2! \ldots N_{z-1}!} \frac{1}{\left(\prod_{l=1}^{s_1} \Lambda_l^3\right)^{N_1} \left(\prod_{l=1}^{s_2} \Lambda_l^3\right)^{N_2} \cdots \left(\prod_{l=1}^{s_{z-1}} \Lambda_l^3\right)^{N_{z-1}}}$$

$$\times \exp[-\beta \mathcal{V}(\mathbf{r}, N_1, N_2, \ldots, N_{z-1}) + \beta(\mu_1 N_1 + \mu_2 N_2 + \ldots + \mu_{z-1} N_{z-1})]$$
$$(7.25)$$

Both the numbers of molecules of all species, $N_1, N_2, \ldots, N_{z-1}$, and the $3(N_1 s_1 + N_2 s_2 + \ldots + N_{z-1} s_{z-1})$-dimensional vector $\mathbf{r}$ of site coordinates defining the instantaneous configuration are allowed to fluctuate in this ensemble. The normalization factor Ξ is the grand partition function:

$$\Xi(\mu_1, \mu_2, \ldots, \mu_{z-1}, V, T) = \sum_{N_1=1}^{\infty} \sum_{N_2=1}^{\infty} \cdots \sum_{N_{z-1}=1}^{\infty} \frac{1}{N_1! N_2! \ldots N_{z-1}!} \frac{1}{\left(\prod_{l=1}^{s_1} \Lambda_l^3\right)^{N_1} \left(\prod_{l=1}^{s_2} \Lambda_l^3\right)^{N_2} \cdots \left(\prod_{l=1}^{s_{z-1}} \Lambda_l^3\right)^{N_{z-1}}}$$

$$\times \int d^{3(N_1 s_1 + N_2 s_2 + \ldots + N_{z-1} s_{z-1})} \mathbf{r} \, \exp[-\beta \mathcal{V}(\mathbf{r}, N_1, N_2, \ldots, N_{z-1}) + \beta(\mu_1 N_1 + \mu_2 N_2 + \ldots + \mu_{z-1} N_{z-1})]$$
$$(7.26)$$

where all spatial integrations are performed over the system volume V.

The connection with macroscopic thermodynamics is established through the grand potential, which is the Legendre transform of the Helmholtz energy with respect to the numbers of all types of molecules, through a fundamental equation of the form:

$$\Omega(\mu_1, \mu_2, \ldots, \mu_{z-1}, V, T) \equiv A - \sum_{\alpha=1}^{z-1} N_\alpha \mu_\alpha = -k_B T \ln \Xi \quad (7.27)$$

In addition to the displacement, rotation, and internal reconfiguration moves employed by canonical MC, grand canonical Monte Carlo (GCMC) employs molecule insertion and deletion moves, which change the population N_α of a sorbed species α by $+1$ or -1, respectively. For each species, insertion and deletion moves must be attempted with equal probabilities, to ensure microscopic reversibility. In an insertion attempt, the simplest scheme is to choose the position of a site (e.g., the center-of-mass) of the inserted molecule from a uniform distribution within the volume of the simulation box, to pick the overall orientation of the inserted molecule randomly, and its internal configuration with a probability proportional to the Boltzmann factor of its intramolecular potential energy $\mathcal{V}_{\text{intra}}^\alpha$. The general acceptance criterion, Eq. (7.18), in combination with Eqs. (7.24) for the chemical potential and (7.25) for the probability density, with i and j symbolizing the initial and final configurations between which the move is attempted, becomes:

$$P_{\text{accept}}(i \to j) = \min\left[1, \frac{\varrho^{\mu_1\mu_2\cdots\mu_{z-1}VT}(r_j, N_1, \ldots, N_\alpha+1, \ldots, N_{z-1})}{\varrho^{\mu_1\mu_2\cdots\mu_{z-1}VT}(r_i, N_1, \ldots, N_\alpha, \ldots, N_{z-1}) \frac{1}{V} \frac{\exp\left(-\beta \mathcal{V}_{\text{intra}}^\alpha\left(r_{N_\alpha+1}^{(j)}\right)\right)}{Z_{\text{intra,ig}}^\alpha}}\right]$$

$$= \min\left[1, \frac{\beta f_\alpha V}{N_\alpha+1} \exp\left(-\beta\left[\mathcal{V}_j - \mathcal{V}_i - \mathcal{V}_{\text{intra}}^\alpha\left(r_{N_\alpha+1}^{(j)}\right)\right]\right)\right]$$

$$= \min\left[1, \frac{\beta f_\alpha V}{N_\alpha+1} \exp(-\beta \Delta\mathcal{V}_{\text{inter}})\right]$$

(7.28)

where $\mathcal{V}_{\text{inter}}$ is the total potential energy of interaction of the sorbed molecules with the sorbent and among themselves. For an attempted deletion move, wherein a molecule of type α is selected at random and removed from the model system, causing the population of α-type molecules to change from N_α (configuration i) to $N_\alpha - 1$ (configuration j), the acceptance criterion can similarly be shown to be:

$$P_{\text{accept}}(i \to j) = \min\left[1, \frac{N_\alpha}{\beta f_\alpha V} \exp(-\beta \Delta\mathcal{V}_{\text{inter}})\right]$$

(7.29)

As output from a GCMC simulation one obtains the ensemble averaged populations $\langle N_\alpha \rangle$ for all species, under the prescribed volume V, temperature T, and fugacities $f_1, f_2, \ldots, f_{z-1}$; the latter are immediately calculable from the pressure P, temperature T, and composition of the fluid phase. For a one-component sorbate, a plot of the average molecular density $q_\alpha = \langle N_\alpha \rangle / V$ versus f_α under constant T is an adsorption isotherm.

Figure 7.5 shows sorption isotherms for N_2 and CO_2 in silicalite at room temperature, as predicted by GCMC simulation using force fields that reproduce the fluid phase equilibria of the pure fluids, along with a simple model for the zeolite as a set of LJ centers and partial charges [37].

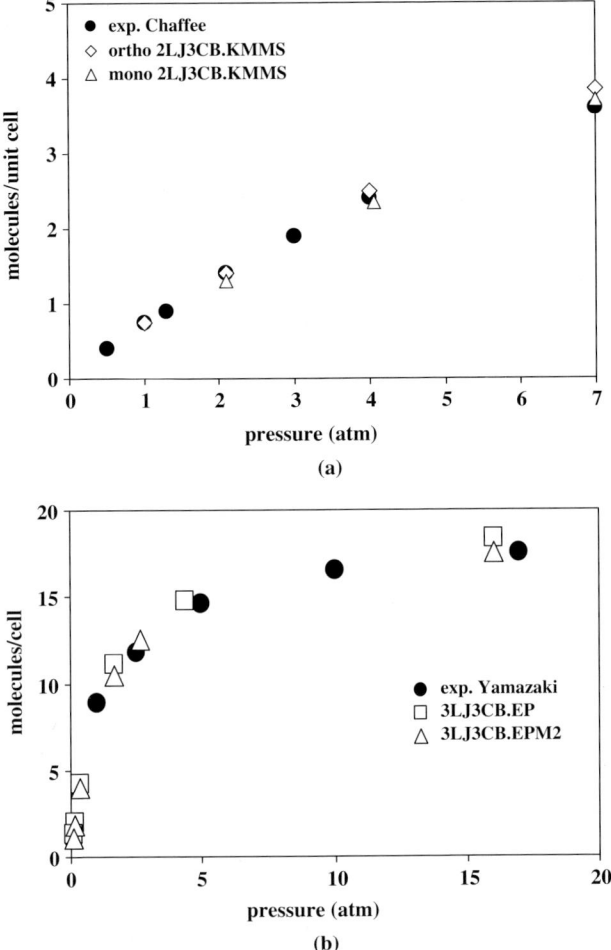

Figure 7.5 Sorption isotherms of nitrogen (a) and carbon dioxide (b) in silicalite at room temperature, as predicted by GCMC using various force fields (open symbols) and as measured experimentally (filled circles). Notice the greater affinity of the zeolite for CO_2 [37].

Measuring fluctuations in the number of sorbed particles in the course of a GCMC simulation can yield interesting thermodynamic information. From Eqs. (7.25)–(7.27) one can readily show for the variance in the number of molecules of species α that:

$$\left\langle (N_\alpha - \langle N_\alpha \rangle)^2 \right\rangle = \langle N_\alpha^2 \rangle - \langle N_\alpha \rangle^2 = -k_B T \frac{\partial^2 \Omega}{\partial \mu_\alpha^2}\bigg|_{T,V,\mu_{\gamma \neq \alpha}} = k_B T \frac{\partial \langle N_\alpha \rangle}{\partial \mu_\alpha}\bigg|_{T,V,\mu_{\gamma \neq \alpha}}$$

$$= V \frac{\partial q_\alpha}{\partial \ln f_\alpha}\bigg|_{T,\mu_{\gamma \neq \alpha}} = \langle N_\alpha \rangle \frac{\partial \ln q_\alpha}{\partial \ln f_\alpha}\bigg|_{T,\mu_{\gamma \neq \alpha}}$$

(7.30)

Equation (7.30) exemplifies the general statistical mechanical result that the fluctuation in an extensive property reduced by the average value of the property (in our case, $\langle (N_a - \langle N_a \rangle)^2 \rangle^{1/2} / \langle N_a \rangle$) scales with the inverse square root of the size of the system (in our case, $V^{-1/2}$ or $\langle N_a \rangle^{-1/2}$) and therefore vanishes in the thermodynamic limit, unless the thermodynamic susceptibility to which the fluctuation is related (in our case, $(\partial^2 \Omega / \partial \mu_a^2)_{T,V,\mu_\gamma \neq a} / \langle N_a \rangle$) diverges. We see that the fluctuation in number of molecules is immediately related to the slope of the adsorption isotherm $q_a(f_a)$. A vertical part in the isotherm, indicative of a first-order phase transition in the solid phase, is a macroscopic manifestation of divergence of the fluctuation in the number of sorbed molecules.

Useful information can be extracted from the covariance of the number of molecules and the potential energy in the grand canonical ensemble. Through standard transformations of thermodynamic derivatives, making use of the probability density of Eqs. (7.25) and (7.26) for a one-component sorbate system, one can show that:

$$\left. \frac{\partial U}{\partial \langle N_a \rangle} \right|_{T,V} = \frac{\langle N_a \mathcal{V} \rangle - \langle N_a \rangle \langle \mathcal{V} \rangle}{\langle N_a^2 \rangle - \langle N_a \rangle^2} + \langle \mathcal{K}_a \rangle \tag{7.31}$$

with $U = \langle \mathcal{V} + \mathcal{K} \rangle$ being the internal energy, the brackets denoting equilibrium averaging in the grand canonical ensemble, and $\langle \mathcal{K}_a \rangle$ being the mean kinetic energy per sorbed molecule at the prevailing temperature. If the solid phase is assumed incompressible, the left-hand side of Eq. (7.31) can be interpreted as a partial molar enthalpy $\bar{H}_a^s$ of a in the solid phase. If one subtracts the molar enthalpy of the sorbate in the fluid phase with which the solid phase is in equilibrium, one obtains an estimate of the isosteric heat of sorption, q_{st}:

$$q_{st} \equiv H_{m,a}^g - \bar{H}_a^s = N_{Avo} \langle \mathcal{V}_{intra}^a \rangle_{ig} + RT + \left(H_{m,a}^g - H_{m,a}^{ig} \right) - N_{Avo} \frac{\langle N_a \mathcal{V} \rangle - \langle N_a \rangle \langle \mathcal{V} \rangle}{\langle N_a^2 \rangle - \langle N_a \rangle^2} \tag{7.32}$$

with "g" and "ig," respectively, indicating the fluid phase and an ideal gas at the temperature of interest, and the subscript m indicating a molar property.

The assumption of a constant volume V for the solid phase can certainly be relaxed, and fluctuations in volume can be allowed by adopting a flexible model for the sorbent z. Provided the sorbent z is involatile, a convenient ensemble for studying sorption equilibria is the $f_1 f_2 f_3 \ldots f_{z-1} N_z P,T$ ensemble. This is a hybrid between grand canonical and isothermal–isobaric ensembles, where the amount of sorbent N_z, the pressure, the temperature, and the fugacities of all components in the fluid phase are constant, but the sorbent phase can fluctuate in space and swell in response to sorption. Note that only z of the quantities $f_1, f_2, \ldots, f_{z-1}, P, T$ are independent: In the fluid phase, $f_1, f_2, \ldots, f_{z-1}$ can be obtained from P, T, and $(z-2)$ mole fractions specifying the composition of the fluid phase. Although not common in simulations of sorption in zeolites, this ensemble has been used to predict sorption isotherms in polymers [38, 39].

7.2.4
Gibbs Ensemble and Gage Cell Monte Carlo

In 1987, A.Z. Panagiotopoulos [40] introduced a new ensemble for the simulation of coexisting fluid phases in equilibrium, in the absence of an interface. In the original formulation of this method, termed the Gibbs ensemble Monte Carlo (GEMC) method, which is applicable to both single- and multicomponent fluids, the total number of particles and the total volume of the two phases are constant, but the volume and the number of particles in each phase are allowed to fluctuate so as to ensure equality of pressures and chemical potentials between the phases. These fluctuations are sampled in the MC algorithm through volume and particle exchange moves between the two simulation boxes representing the coexisting phases, which are implemented along with more conventional moves that displace, rotate, or change the internal configuration of individual molecules in each box. The Gibbs ensemble method was readily adapted to confined fluids in a study of capillary phase equilibria in a structureless cylindrical wall [41].

Neimark and Vishnyakov [42] proposed a method in the spirit of GEMC, in which one of the two coexisting phases is a fluid-filled pore cell and the other is a fluid cell of limited capacity, termed the "gage cell." The total number of molecules in the two cells is fixed, mass exchange between the two cells is allowed, but the volumes of the two cells are constant, and both cells are in thermal equilibrium with a heat reservoir. If the gage cell volume is sufficiently small, density fluctuations in the confined fluid are suppressed and hysteretic phenomena observed with GCMC are circumvented. Thus, the simulation can trace metastable and unstable branches of the isotherm (Figure 7.6). The chemical potential in the gage cell can be conveniently determined in the course of a simulation via the Widom test particle insertion method [43] (see below). Thus, one can readily generate the chemical potential μ as a function of molecular density q along an isotherm through a series of gage cell simulations. The grand potential per unit volume [compare Eq. (7.27)] can then be obtained via thermodynamic integration from a state where the chemical potential is known (e.g., low-fugacity state r at point O in Figure 7.6, where the gage cell is indistinguishable from an ideal gas):

$$\begin{aligned}\frac{\Omega(\mu, V, T)}{V} &= \frac{\Omega_r(\mu_r, V, T)}{V} - \int_{\mu_r}^{\mu} q(\mu', T)\, d\mu' \\ &= \frac{\Omega_r(\mu_r, V, T)}{V} - q\mu + q_r\mu_r + \int_{q_r}^{q} \mu(q', T)\, dq'\end{aligned} \quad (7.33)$$

Thermodynamic equilibrium between two phases confined within a microporous solid (phases B, F in Figure 7.6) of constant volume dictates that the temperatures T, component chemical potentials μ, and grand potentials per unit volume Ω/V be the same between the two coexisting phases. Thus, by virtue of

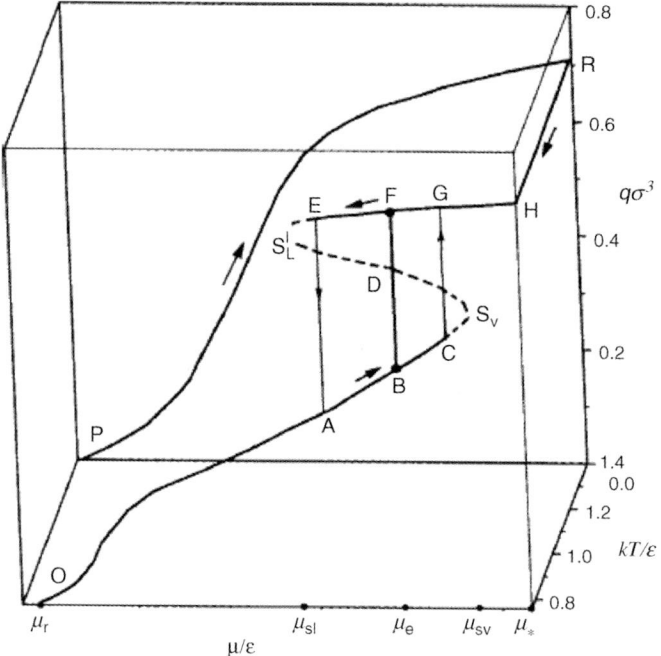

Figure 7.6 Schematic of chemical potential–temperature–molecular density surface for a pure fluid sorbed in a pore of diameter $\sim 10\sigma$, where σ is the collision diameter of the fluid molecule. Curves PR and OH trace a supercritical and a subcritical (with respect to capillary condensation) sorption isotherm, respectively. OS_V is the vapor-like and HS_L the liquid-like branch of the isotherm; the two branches terminate at the limits of stability (spinodal points) of the vapor-like phase, S_V, and of the liquid-like phase, S_L. B and F represent the two phases at equilibrium under the temperature of the subcritical isotherm. A set of GCMC simulations initiated at low chemical potentials (fugacities) at the subcritical temperature tends to by-pass point B and reach metastable vapor-like state C, where it spontaneously condenses to liquid-like state G. A set of GCMC simulations initiated at high chemical potentials tends to by-pass point F and reach metastable liquid-like state E, where it spontaneously converts into vapor-like state A. By suppressing density fluctuations, the gage-cell MC method can trace the entire van der Waals loop, including metastable (BS_V, S_LF) and unstable (S_VS_L) branches. The chemical potential at phase coexistence can then be determined by a Maxwell equal area construction. With permission from [42].

Eq. (7.33), μ at coexistence can be determined via a Maxwell equal area construction on the isotherm $\mu(q)$. Since the unstable phases are accessible, grand potential energy barriers to the nucleation of one phase inside the other can be estimated through the gage cell method [42].

Figure 7.7 shows a set of binodal curves for capillary condensation of the lower alkanes inside a single-walled carbon nanotube, as calculated by gage-cell MC [43]. Note the tremendous suppression of the coexistence region and the lowering of critical points in comparison to the corresponding bulk fluids.

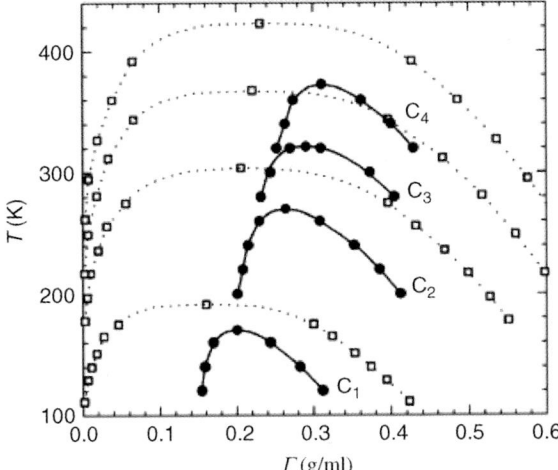

Figure 7.7 Predicted binodal curves for capillary condensation of methane (C_1), ethane (C_2), propane (C_3), and n-butane (C_4) in a (30,30) single-walled carbon nanotube of diameter 2.034 nm (filled symbols and solid lines). Calculations were performed with the gage-cell MC method, using configurational bias for the flexible molecules. The binodal curves for the bulk fluids are shown for comparison (open symbols and dotted lines). With permission from [43].

7.3
Free Energy Methods for Sorption Equilibria

The strategy followed in GCMC is to set the chemical potentials μ_α or fugacities f_α of sorbate species and extract the molecular densities q_α as an output from the simulation. An alternative strategy is to set the numbers of molecules of all species, that is, simulate the microporous solid phase as a closed system in a canonical ($N_1, N_2, \ldots, N_z, V, T$), isothermal–isobaric ($N_1, N_2, \ldots, N_z, P, T$), or isothermal–isostress ($N_1, N_2, \ldots, N_z, \boldsymbol{\sigma}, T$) ensemble, and to compute the chemical potentials or fugacities as an output of the simulation. From the chemical potentials or fugacities of the sorbate species, one can readily calculate the pressure and composition of the bulk fluid phase with which the microporous solid phase will be at equilibrium, using a reliable equation of state for the bulk fluid phase. For a microporous solid in contact with a pure fluid, calculation of the chemical potential or fugacity of the sorbate in the microporous solid phase gives one point on the sorption isotherm. The entire isotherm can be traced by repeating the calculation for different sorbate loadings in the solid phase.

The excess chemical potential of a sorbed species α is defined as the chemical potential relative to an ideal gas phase of pure α at the same molecular density and temperature:

$$\mu_\alpha^{ex}(q_1, q_2, \ldots, q_z, T) \equiv \mu_\alpha(q_1, q_2, \ldots, q_z, T) - \mu_\alpha^{ig}(q_\alpha, T)$$
$$= k_B T \ln\left(\frac{f_\alpha}{q_\alpha k_B T}\right) \quad (7.34)$$

For a microporous solid at equilibrium with a pure fluid phase of α, $\exp[-\mu_\alpha^{ex}/(k_B T)]/(k_B T)$ equals the slope q_α/f_α of a secant of the sorption isotherm at the considered state point. In the limit of very low pressure, this reduces to the slope of the sorption isotherm $q_\alpha(f_\alpha)$ at the origin and is immediately related to the Henry's law constant for sorption:

$$K_H = \frac{M_\alpha}{\varrho_z N_{\text{Avo}}} \lim_{P \to 0}\left(\frac{q_\alpha}{f_\alpha}\right) = \frac{M_\alpha}{\varrho_z RT} \lim_{P \to 0}\left[\exp\left(-\frac{\mu_\alpha^{ex}}{k_B T}\right)\right] \qquad (7.35)$$

where M_α and ϱ_z are the molar mass of the sorbate and the mass density of the microporous solid, respectively. K_H equals the mass of sorbate fluid sorbed per mass of microporous solid per unit pressure of the fluid phase in the limit of very low pressures and is a measure of the affinity of the microporous solid for the sorbate. The isosteric heat in the limit of zero loading is related to the temperature dependence of K_H:

$$q_{st,0} \equiv \lim_{q_\alpha \to 0} q_{st} = R\frac{d \ln K_H}{d(1/T)} \qquad (7.36)$$

In a classical statistical mechanical treatment, if a rigid model is invoked for the microporous solid:

$$q_{st,0} = -N_{\text{Avo}}\left[\langle \mathcal{V}^\alpha \rangle - \langle \mathcal{V}^\alpha_{\text{intra}}\rangle_{\text{ig}}\right] + RT \qquad (7.37)$$

where $\langle \mathcal{V}^\alpha \rangle$ is a canonical ensemble average of the potential energy of a single molecule of species α sorbed in the microporous solid, while $\langle \mathcal{V}^\alpha_{\text{intra}}\rangle_{\text{ig}}$ is the canonical ensemble average of the potential energy of the same molecule in the ideal gas state, where there are no intermolecular contributions. Comparison with Eq. (7.23) shows that $q_{st,0/N_{\text{Avo}}} = -\lim_{N_\alpha \to 0}(\Delta_{\text{mix}} H/N_\alpha)$ under these conditions.

How does one compute the excess chemical potential from a closed system simulation? Chemical potential, the partial molar Gibbs energy, is a "statistical" property; it has to do with entropy, which is a measure of the probability distribution of a system in phase space. Unlike "mechanical" properties (e.g., pressure, internal energy, enthalpy), which can be computed straightforwardly as averages of instantaneous quantities depending on atomic positions and interactions, statistical properties are hard to compute. A way to the calculation of the chemical potential was shown by B. Widom in 1963 [44]. Widom's "test particle insertion method" is an example of a free energy perturbation method for calculating free energy differences from simulations. These methods are very useful in connecting atomistic simulations to mesoscopic and macroscopic approaches cast in terms of fewer, coarse-grained variables in the context of multi-scale modeling and simulation approaches. This is why we will spend some time discussing them here.

As discussed in an instructive review by D. Kofke [45], methods for the calculation of the Helmholtz energy difference $A_2 - A_1$ between two systems, 1 and 2, fall into two broad categories:

1) In **density of states methods**, $A_2 - A_1$ is calculated as:

$$A_2 - A_1 = -k_B T \ln(P_2/P_1) \qquad (7.38)$$

where P_i is the probability of sampling system i in an equilibrium simulation that is given complete freedom to explore both systems. In practice, bias techniques may be used to ensure efficient sampling of both systems.

2) **Work-based methods**, on the other hand, can be envisioned as applications of the Jarzynski equality [46]:

$$A_2 - A_1 = -k_B T \ln\left[\overline{\exp(-W_{1\to 2}/k_B T)}\right] \tag{7.39}$$

where $W_{1\to 2}$ is the work associated with a (not necessarily reversible!) process that transforms system 1 into system 2, usually by modulating a parameter in the Hamiltonian of the system, and the overbar denotes averaging over an ensemble of such transformations beginning from an equilibrated system 1. When the transformation of 1 into 2 occurs reversibly, Eq. (7.39) leads to the *thermodynamic integration* method. When, on the other hand, an instantaneous transformation of individual configurations of system 1 into the corresponding configuration of system 2 is undertaken, Eq. (7.39) reduces to the *free energy perturbation* method:

$$A_2 - A_1 = -k_B T \ln\langle\exp[-(\mathcal{V}_2 - \mathcal{V}_1)/(k_B T)]\rangle_1 \tag{7.40}$$

where the ensemble average is taken over all configurations of system 1, distributed according to the canonical ensemble, and $\mathcal{V}_1$, $\mathcal{V}_2$ denote the potential energy functions of systems 1 and 2 evaluated at each sampled configuration. For free energy perturbation to work, the region of configuration space that shapes the properties of system 2 must be a subset of that of system 1. If this is not the case, staging schemes can be invoked [45].

Widom's test particle insertion method for the chemical potential [44] can be derived as a special case of Eq. (7.40), where system 2 is the real system under investigation to which an extra molecule of type α has been added, while system 1 is identical to system 2, except for the fact the extra molecule of type α is now an ideal gas molecule, capable of roaming in the entire volume of the system without experiencing any interactions. In a closed system of fixed volume, the excess chemical potential of α, relative to an ideal gas phase of pure α at the same molecular density and temperature, is computed as:

$$\mu_\alpha^{\text{ex}}(q_1, q_2, \ldots, q_Z, T) = -k_B T \ln\langle\exp(-\mathcal{V}_{\text{test}}/(k_B T))\rangle_{\text{Widom}} \tag{7.41}$$

where $\mathcal{V}_{\text{test}}$ is the energy "felt" by the extra ("test") molecule of type α inserted in the system of real molecules and the average is taken over all configurations, sampled according to the canonical ensemble, and over all points of insertion in the simulation box, sampled uniformly. If molecules of type α are flexible, Eq. (7.41) can be used by generating conformations for the inserted molecule according to the Boltzmann factor of its intramolecular energy and inserting them at random positions and orientations in the simulation box. $\mathcal{V}_{\text{test}}$ is then the potential energy experienced by the inserted molecule due to its interactions with the microporous solid plus the potential energy of intermolecular interactions between the inserted molecule and the pre-existing sorbate molecules. Typically, 10^3–10^5 insertions are used per system configuration. In "wall" regions occupied by the atoms of the microporous solid, excluded volume interactions make $\mathcal{V}_{\text{test}} \to \infty$, resulting in a zero contribution to the average of Eq. (7.41). Thus, a useful strategy, especially when inflexible models of the microporous framework are used, is to map out void regions in the solid where an insertion may lead to nonzero contribution to the ensemble average of Eq. (7.41) and confine insertions to these regions.

In the special case of a pure fluid α in equilibrium with a microporous solid z, which is treated as inflexible, Eqs. (7.35) and (7.41) lead to:

$$\frac{RT\varrho_z}{M_\alpha} K_{\text{H}} = \frac{\int d^{3s_\alpha} r_\alpha \exp[-\beta \mathcal{V}^\alpha(r_\alpha)]}{\int d^{3s_\alpha} r_\alpha \exp[-\beta \mathcal{V}^\alpha_{\text{intra}}(r_\alpha)]} = \frac{\int d^{3s_\alpha} r_\alpha \exp[-\beta \mathcal{V}^\alpha(r_\alpha)]}{V Z^\alpha_{\text{intra,ig}}} \qquad (7.42)$$

The three-dimensional volume V, over which integrations in Eq. (7.42) take place, is the volume occupied by the microporous solid. If sorption in a defect-free periodic microporous solid, such as a zeolite, is under study, it suffices to take V as the volume of the asymmetric unit of the unit cell. Equation (7.42), which was used in early calculations of low-occupancy thermodynamics in zeolites [47], makes it clear that the dimensionless quantity $RT\varrho_z K_{\text{H}}/M_\alpha$ is a partition coefficient between the microporous and the ideal gas phase, equal to the ratio of configurational integrals of a single sorbate molecule in the two phases.

Widom insertion is extremely useful in the calculation of sorption equilibria and solubilities in liquids and polymers. Unfortunately, it becomes inefficient at high concentrations and low temperatures, as most random insertions lead to overlap with $\beta \mathcal{V}_{\text{test}} \to \infty$. This insertion problem is especially severe for bulky molecules, molecules of complex shape, and molecules experiencing strong and specific interactions, such as hydrogen bonding, with their surroundings. For flexible molecules, insertions can be implemented in a configurationally biased way [48, 49], as proposed by Maginn et al. [50] for the case of long alkanes in silicalite: The flexible molecule is inserted bond by bond, adding a segment at the time. For each added segment, several possible positions are considered and one is chosen according to the Boltzmann factor of the potential energy increase resulting from addition of the segment. In this way, the added long flexible molecule is "threaded" through open spaces in the microporous material, avoiding overlaps with the surrounding walls and other sorbate molecules. The bias associated with this procedure is taken out during averaging.

To illustrate configurationally biased insertion for calculation of the excess chemical potential, assume we are inserting a linear flexible molecule in a configuration of a microporous solid that may contain other flexible molecules. The segments of the flexible molecule, $i = 1, 2, \ldots, s_\alpha$ are inserted one by one (Figure 7.8a). When we are about to insert segment i, we consider $n_{\text{trials}}(i)$ tentative positions j for this segment and pick one of them, with index $j_{\text{chosen}}(i)$, according to the probabilities:

$$p_j(i) = \frac{\exp\left[-\beta \mathcal{V}_{\text{bias}}\left(r_i^{(j)}\right)\right]}{\sum_{k=1}^{n_{\text{trials}}(i)} \exp\left[-\beta \mathcal{V}_{\text{bias}}\left(r_i^{(k)}\right)\right]} \qquad (j = 1, \ldots, n_{\text{trials}}(i)) \qquad (7.43)$$

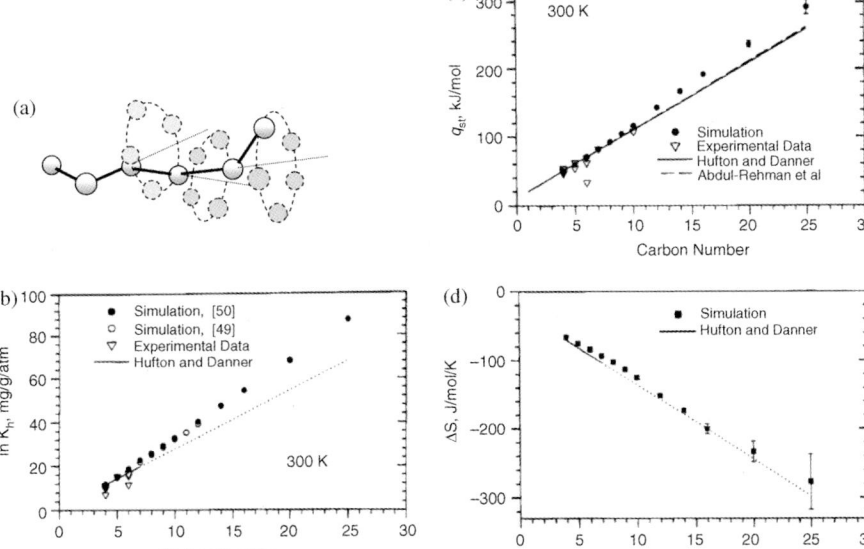

Figure 7.8 (a) Schematic of the configurationally biased insertion of a hexane molecule in a united-atom representation within a microporous medium. Candidate positions (k) for the three last inserted segments (i = 4, 5, 6) are shown with a broken outline, chosen positions (j$_{chosen}$) are shown with a solid outline. In this example, $n_{trials} = 5$. Candidate positions for each segment are chosen on a circle lying on a plane normal to the axis of the bond connecting the two previous segments, with its center on that axis. The center position and radius of the circle are chosen according to the bond angle bending and bond stretching potentials for the added bond angle and bond; these potentials constitute part of $\mathcal{V}_{bias}$ for the added segment. (b) Henry's law constants K_H for the linear alkanes C$_4$–C$_{25}$ in silicalite at 300 K, as predicted by configurationally biased insertions [49, 50] and as measured experimentally. (c) Isosteric heats in the limit of very low occupancy, $q_{st,0}$ for the linear alkanes C$_4$–C$_{25}$ in silicalite at 300 K as predicted by configurationally biased insertions and by configurationally biased canonical Monte Carlo simulations and as measured experimentally. (d) Partial molar entropy of sorption at 300 K in the limit of very low occupancy, $\Delta S = \lim_{f_\alpha \to 0} \left(\overline{S}_\alpha - S_\alpha^{ig}\right) = R\ln(RT\varrho_z K_H/M_\alpha) - q_{st,0}/T$. All three thermodynamic quantities are shown as functions of chain length. The lines are correlations based on experimental data for the lower alkanes [51, 52]. The small break in slope of the simulation results for $q_{st,0}$ around C$_{10}$ reflects a real phenomenon: beyond C$_8$, which is the longest alkane that can reside entirely within a single channel segment, sorbed molecules exhibit a preference to occupy straight channel segments, which are slightly more favorable energetically than sinusoidal channel segments. The detailed siting and conformation of the molecules in the zeolite have been quantified by simulation. With permission from [50].

where $\mathcal{V}_{bias}(\mathbf{r}_i^{(j)})$ is a bias potential for insertion of the segment i at position $\mathbf{r}_i^{(j)}$. Insertion of the entire molecule is thus associated with a bias probability:

$$W = \prod_{i=1}^{s_\alpha} p_{j_{chosen}(i)}(i) = \frac{\exp\left[-\beta \sum_{i=1}^{s_\alpha} \mathcal{V}_{bias}\left(\mathbf{r}_i^{(j_{chosen}(i))}\right)\right]}{\prod_{i=1}^{s_\alpha}\left[\sum_{k=1}^{n_{trials}(i)} \exp\left[-\beta \mathcal{V}_{bias}\left(\mathbf{r}_i^{(k)}\right)\right]\right]} \qquad (7.44)$$

The excess chemical potential of species α is computed as an average over all biased insertions in the course of the closed-system simulation:

$$\mu_\alpha^{ex} = -k_B T \ln \left\langle \frac{\exp(-\beta \mathcal{V}_{test}^{tot})}{W} \frac{1}{n_{trials}(1) \ldots n_{trials}(s_\alpha)} \right\rangle_W \quad (7.45)$$

Usually, $\mathcal{V}_{bias}(r_i^{(j)})$ is chosen as the increment in potential energy brought about by insertion of segment i at position $r_i^{(j)}$. As a consequence, $\mathcal{V}_{test}^{tot} = \sum_{i=1}^{s_\alpha} \mathcal{V}_{bias}\left(r_i^{(j_{chosen}(i))}\right)$ and Eq. (7.45) can be written as:

$$\mu_\alpha^{ex} = -k_B T \ln \left\langle \prod_{i=1}^{s_\alpha} \left[\frac{1}{n_{trials}(i)} \sum_{k=1}^{n_{trials}(i)} \exp\left(-\beta \mathcal{V}_{bias}\left(r_i^{(k)}\right)\right) \right] \right\rangle_W \quad (7.46)$$

The quantity within the angular brackets is known as the Rosenbluth weight of the inserted configuration.

Using this strategy, Smit and Siepmann [48, 49] and Maginn et al. [50] were able to predict Henry's law constants and isosteric heats of sorption of alkanes in silicalite in excellent agreement with experiment (Figure 7.8b–d). Configurational bias originally arose as a Monte Carlo move for sampling chain molecules [53–55] and has been employed very successfully in grand canonical Monte Carlo simulations of flexible sorbates in zeolites [56, 57]. When configurational bias is used as a Monte Carlo move to rearrange the configuration of a chain molecule, the reverse move has to be considered. Rosenbluth weights are computed for both the final (target) and the original configurations of the chain molecule; their ratio enters the acceptance criterion of the move [57].

7.4
Coarse-Graining and Potentials of Mean Force

In understanding the molecular mechanisms of physical phenomena (in our case, sorption and diffusion in microporous solids) and in designing computationally efficient approaches that can access long time and length scales, it is often desirable to devise descriptions cast in terms of a subset of relatively few, slowly evolving degrees of freedom (e.g., the center-of-mass coordinates of a complex molecule inside a zeolite), while integrating out the remaining degrees of freedom (e.g., vibrational degrees of freedom of the molecule and of the framework, torsional, and orientational degrees of freedom of the molecule). This procedure of deriving models in terms of fewer degrees of freedom from detailed atomistic models is generally referred to as coarse-graining. How to select the degrees of freedom that will be retained in a coarse-grained description is far from trivial; it requires good physical understanding of the system [58].

A central role in coarse-grained descriptions of material systems is played by the *potential of mean force*. Having partitioned the configuration space of a system into a subset of (slowly evolving) degrees of freedom, $\boldsymbol{R}$, on which we wish to focus, and a complementary subset of (fast) degrees of freedom, $\boldsymbol{r}$, which we wish to integrate out, we define the potential of mean force with respect to $\boldsymbol{R}$, $U(\boldsymbol{R})$, as:

$$U(\boldsymbol{R}) = -k_B T \ln \int \exp[-\beta \mathcal{V}(\boldsymbol{R}, \boldsymbol{r})] \, d\boldsymbol{r} + \text{const.} \tag{7.47}$$

with $\beta = 1/(k_B T)$. $U(\boldsymbol{R})$ is a configurational free energy that describes effective interactions among the degrees of freedom $\boldsymbol{R}$, assuming that $\boldsymbol{r}$ are always equilibrated subject to the current values of $\boldsymbol{R}$; this will be the case if the relaxation times for the motion of $\boldsymbol{r}$ are much shorter than the characteristic times governing the evolution of $\boldsymbol{R}$. The potential of mean force provides a correct and consistent thermodynamic description of the degrees of freedom $\boldsymbol{R}$, having projected out $\boldsymbol{r}$. A correct *dynamical* description cast exclusively in terms of the subset $\boldsymbol{R}$ is, in general, much more difficult to formulate. It is provided by Zwanzig and Mori's projection operation formalism [59] and commonly approximated via Brownian dynamics, Langevin dynamics, dissipative particle dynamics, and related schemes. In all these schemes, the gradient of the potential of mean force must be used as the thermodynamic driving force, to ensure consistency with the full atomistic description.

Efficient MC schemes are valuable in calculating potentials of mean force $U(\boldsymbol{R})$ with respect to selected coarse-grained degrees of freedom $\boldsymbol{R}$. An early example is provided by the work of Maginn et al. on sorption and diffusion of long linear alkanes in the zeolite silicalite [50, 60]. As shown pictorially in Figure 7.9a and b, at the coarse-grained level the zeolite was reduced to a set of intersecting curved lines representing the axes of straight and sinusoidal channels and computed directly from the force field exerted on a test particle inside the zeolite crystal. A sorbed alkane was coarse-grained into a linear wormlike object that is defined by the sequence of channel segments and intersections it goes through and by the positions x_1, x_2 of its two ends along the channel segments in which they find themselves. The potential of mean force $U(x_1, x_2)$ was computed by configurational bias MC integration. Figure 7.9c shows the typical shape of the potential of mean force. It refers to an n-octane molecule sorbed along a straight channel of silicalite in an extended conformation, in such a way as to straddle a channel intersection. A distinctive characteristic of $U(x_1, x_2)$ is a low energy trough, along which the difference $x_1 - x_2$ is more or less constant. Moving along this trough corresponds to sliding the molecule along the straight channel. The favorable energy it experiences in the trough is due to dispersive interactions with the zeolite lattice. If one tries to extend the contour of the sorbed molecule, increasing $x_1 - x_2$, one encounters a steep repulsive wall, as one has to work against intramolecular bending potentials. If one tries to compress the contour of the molecule, decreasing $x_1 - x_2$, the potential of mean force again rises due to confinement, but not as steeply as in the case of stretching the molecule. Potentials of mean force, such as that of Figure 7.9c, were used within a Brownian dynamics/transition state theory formulation to track the intracrystalline diffusion of C_4–C_{20} molecules inside silicalite [60]. Predictions were confirmed a posteriori via quasi-elastic neutron scattering measurements [61].

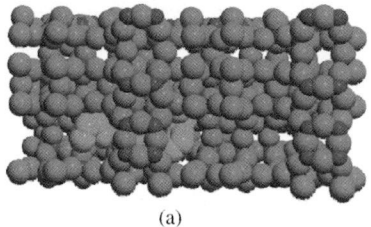
(a)

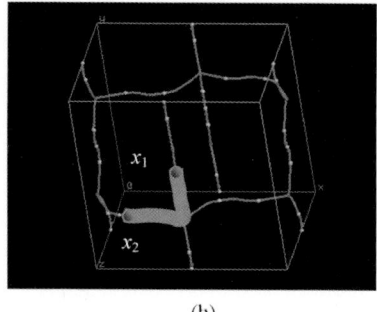
(b)

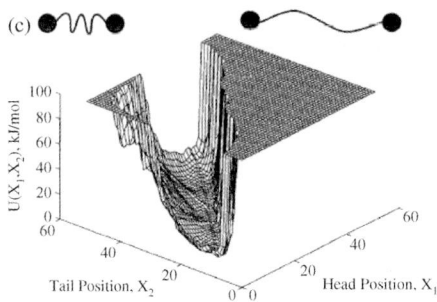
(c)

Figure 7.9 (a) Atomistic and (b) coarse-grained representation of an *n*-eicosane molecule sorbed in silicalite. At the coarse-grained level, the molecule is represented by the sequence of channel segments and intersections it goes through and by the positions x_1, x_2 of its two ends along their respective channels. (c) Potential of mean force $U(x_1, x_2)$ for an *n*-octane molecule straddling an intersection, with its two ends in straight channel segments. Large (small) values of $x_1 - x_2$ correspond to extended (compressed) conformations of the molecule inside the channels. With permission from [60].

References

1. Moore, G.E. (1965) *Electronics* **38** (8), "the experts look ahead" section.
2. Allen, M.P. and Tildesley, D.J. (1987) *Computer Simulation of Liquids*, Clarendon Press, Oxford.
3. Frenkel, D. and Smit, B. (1996) *Understanding Molecular Simulations*, Academic Press, New York.
4. Goldstein, H. (1980) *Classical Mechanics*, 2nd edn, Addison-Wesley, Reading, MA.
5. Sun, H. (1998) *J. Phys. Chem. B*, **102**, 7338.
6. Flory, P.J. (1969) *Statistical Mechanics of Chain Molecules*, Wiley Interscience, New York.
7. Martin, M.G. and Siepmann, J.I. (1998) *J. Phys. Chem. B*, **102**, 2569.
8. Nath, S.K., Escobedo, F.A., and de Pablo, J.J. (1998) *J. Chem. Phys.*, **108**, 9905.
9. Berendsen, H.J.C., Postma, J.P.M., van Gunsteren, W.F., and Hermans, J. (1981) *Intermolecular Forces* (ed. B. Pullman), Reidel, Dordrecht.
10. Murthy, C.S., Singer, K., Klein, M.L., and McDonald, I.R. (1980) *Mol. Phys.*, **41**, 1387.
11. Harris, J.G. and Yung, K.H. (1995) *J. Phys. Chem.*, **99**, 12021.
12. Demontis, P. and Suffritti, G.B. (1997) *Chem. Rev.*, **97**, 2845.
13. Demontis, P., Suffritti, G.B., Quartieri, S., Fois, E.S., and Gamba, A. (1988) *J. Phys. Chem.*, **92**, 867.
14. Vlugt, T.J.H. and Schenk, M. (2002) *J. Phys. Chem. B*, **106**, 12757.
15. Astala, R., Auerbach, S.M., and Monson, P.A. (2005) *Phys. Rev. B*, **71**, 014112.

References

16 Cook, S.J., Chakraborty, A.K., Theodorou, D.N., and Bell, A.T. (1993) *J. Phys. Chem.*, **97**, 6679.
17 Calero, S., Dubbeldam, D., Krishna, R., Smit, B., Vlugt, T.J.H., Denayer, J.F.M., Martens, J.A., and Maesen, T.L.M. (2004) *J. Am. Chem. Soc.*, **126**, 11377.
18 Bežus, A.G., Kiselev, A.V., Lopatkin, A.A., and Quang Du, P. (1978) *J. Chem. Soc., Faraday Trans 2*, **74**, 367.
19 June, R.L., Bell, A.T., and Theodorou, D.N. (1990) *J. Phys. Chem.*, **94**, 1508.
20 Dubbeldam, D., Calero, S., Vlugt, T.J.H., Krishna, R., Maesen, T.L.M., Beerdsen, E., and Smit, B. (2004) *Phys. Rev. Lett.*, **93**, 088302.
21 Baerlocher, Ch. Meier, W.M., and Olson, D.H. (2001) *Atlas of Zeolite Framework Types*, 5th edn, Elsevier, Amsterdam; the Database of Zeolite Structures can be accessed at http://www.iza-structure.org/databases (accessed 19 November 2011).
22 Hahn, T. (ed.) (2002) *International Tables for Crystallography: Volume A, Space-Group Symmetry*, 5th edn, Kluwer Academic, Norwell, MA.
23 Falcioni, M. and Deem, M.W. (1999) *J. Chem. Phys.*, **110**, 1754.
24 Snurr, R.Q., Bell, A.T., and Theodorou, D.N. (1994) *J. Phys. Chem.*, **98**, 11948.
25 Forester, T.R. and Smith, W. (1997) *J. Chem. Soc., Faraday Trans.*, **93**, 3249.
26 Jobic, H., Bée, M., and Pouget, S. (2000) *J. Phys. Chem. B*, **104**, 7130.
27 Car, R. and Parrinello, M. (1985) *Phys. Rev. Lett.*, **55**, 2471.
28 Campana, L., Selloni, A., Weber, J., Pasquarello, A., Papai, I., and Goursot, A. (1994) *Chem. Phys. Lett.*, **226**, 245.
29 Schwarz, K., Nusterer, E., and Margl, P. (1997) *Int. J. Quantum Chem.*, **61**, 369.
30 Trout, B., Suits, B.H., Gorte, R.J., and White, D. (2000) *J. Phys. Chem. B*, **104**, 11734.
31 Torquato, S. (2002) *Random Heterogeneous Materials, Microstructure and Macroscopic Properties*, Springer-Verlag, New York.
32 Makrodimitris, K., Papadopoulos, G.K., Philippopoulos, C., and Theodorou, D.N. (2002) *J. Chem. Phys.*, **117**, 5876.
33 Papadopoulos, G.K., Theodorou, D.N., Vasenkov, S., and Kärger, J. (2007) *J. Chem. Phys.*, **126**, 094702.
34 Metropolis, N., Rosenbluth, A.W., Rosenbluth, M.N., Teller, A.H., and Teller, E. (1953) *J. Chem. Phys.*, **21**, 1087.
35 June, R.L., Bell, A.T., and Theodorou, D.N. (1992) *J. Phys. Chem.*, **96**, 1051.
36 Prausnitz, J.M., Lichtenthaler, R.N., and de Azevedo, E.G. (1999) *Molecular Thermodynamics of Fluid Phase Equilibria*, 3rd edn, Prentice-Hall, Upper Saddle River, NJ.
37 Makrodimitris, K., Papadopoulos, G.K., and Theodorou, D.N. (2001) *J. Phys. Chem. B*, **105**, 777.
38 Theodorou, D.N. (1996) Molecular simulation of sorption and diffusion in amorphous polymers, in *Diffusion in Polymers* (ed. P. Neogi), Marcel Dekker, New York, pp. 67–142.
39 Theodorou, D.N. (2006) Principles of molecular simulation of gas transport in polymers, in *Materials Science of Membranes for Gas and Vapor Separation* (eds Yu. Yampolskii, I. Pinnau, and B.D. Freeman), John Wiley & Sons, Inc., Hoboken, NJ, pp. 47–92.
40 Panagiotopoulos, A.Z. (1987) *Mol. Phys.*, **61**, 813.
41 Panagiotopoulos, A.Z. (1987) *Mol. Phys.*, **62**, 701.
42 Neimark, A.V. and Vishnyakov, A. (2000) *Phys. Rev. E*, **62**, 4611.
43 Jiang, J., Sandler, S.I., and Smit, B. (2004) *Nanoletters*, **4**, 241.
44 Widom, B. (1963) *J. Chem. Phys.*, **39**, 2808.
45 Kofke, D.A. (2005) *Fluid Phase Equilib.*, **41**, 228.
46 Jarzynski, C. (1997) *Phys. Rev. Lett.*, **78**, 2690.
47 June, R.L., Bell, A.T., and Theodorou, D.N. (1990) *J. Phys. Chem.*, **94**, 1508.
48 Smit, B. and Siepmann, J.I. (1994) *Science*, **264**, 1118.
49 Smit, B. and Siepmann, J.I. (1998) *J. Phys. Chem.*, **98**, 8442.
50 Maginn, E.J., Bell, A.T., and Theodorou, D.N. (1995) *J. Phys. Chem.*, **99**, 2057.
51 Hufton, J.R., Danner, R.P. (1993) *AIChE J.* **39**, 954.
52 Abdul-Rehman, H.B., Hasanain, M.A., Loughlin, K.F. (1990) *Ind. Eng. Chem. Res.* **29**, 1525.

53 Rosenbluth, M. and Rosenbluth, A. (1955) *J. Chem. Phys.*, **23**, 356.
54 Siepmann, J.I. and Frenkel, D. (1992) *Mol. Phys.*, **75**, 59.
55 de Pablo, J.J., Laso, M., and Suter, U.W. (1992) *J. Chem. Phys.*, **96**, 2395.
56 Bates, S.P., van Well, W.J.M., van Santen, R.A., and Smit, B. (1996) *J. Am. Chem. Soc.*, **118**, 6753.
57 Smit, B. (2008) *Chem. Rev.*, **108**, 4125.
58 Öttinger, H.C. (2005) *Beyond Equilibrium Thermodynamics*, Wiley Interscience, Hoboken, NJ.
59 Hansen, J.-P. and McDonald, I.R. (1986) *Theory of Simple Liquids*, 2nd edn, Academic Press, New York.
60 Maginn, E.J., Bell, A.T., and Theodorou, D.N. (1996) *J. Phys. Chem.*, **100**, 7155.
61 Jobic, H. and Theodorou, D.N. (2007) *Microporous Mesoporous Mater.*, **102**, 21.

8
Molecular Dynamics Simulations

Having discussed how one can construct atomistic models to study zeolite/sorbate systems with molecular simulation and how one can predict equilibrium sorption thermodynamics using these models, we now address the issue of predicting the dynamics of sorbed molecules in microporous solids. We review how self-diffusivities can be computed from molecular trajectories through the Einstein and Green–Kubo equations and then show how linear response theory can serve as a basis for computing transport diffusivities from simulations carried out under equilibrium conditions. Equilibrium molecular dynamics (MD) is the method of choice for predicting diffusivities and exploring the mechanism of sorbate motion in microporous materials, provided the relevant time scales do not exceed the μs range. We briefly outline algorithms for conducting MD in the microcanonical and canonical ensembles, in the absence or presence of geometric constraints on the molecules. We present applications of equilibrium MD to pure and mixed sorbates. In the latter case, we point out how Onsager transport coefficients can be computed from equilibrium correlation functions of the molecular fluxes and, in turn, be used to extract Maxwell–Stefan diffusivities. In recent years, insights gained through equilibrium MD have spurred the development of approximate strategies for estimating Maxwell–Stefan diffusivities in multicomponent mixtures as functions of the occupancies of all species present; we briefly review these strategies. We then discuss non-equilibrium molecular dynamics (NEMD) approaches for computing the transport diffusivity.

8.1
Statistical Mechanics of Diffusion

8.1.1
Self-diffusivity

As in Chapter 7, we consider a system of fluid species (components 1, 2,..., $z-1$) sorbed within a microporous solid (component z) in volume V. The number of molecules of species i will be denoted by N_i. In the course of a dynamical simulation, the center-of-mass of each molecule l of species i follows a trajectory $\boldsymbol{r}_{li}(t)$ in three-dimensional space as a function of time t. If the motion is diffusive, the equilibrium ensemble average of the squared displacement of the molecule from its initial position will eventually grow linearly with time. The self-diffusivity, a measure of the translational mobility of the individual molecule, is defined by Einstein's Eq. (1.12), which, for our purposes, we rewrite as:

$$\mathcal{D}_i = \lim_{t \to \infty} \frac{\left\langle [r_{li}(t) - r_{li}(0)]^2 \right\rangle}{6t} \tag{8.1}$$

In simulations it is advantageous to average over all molecules of the species in question, as this reduces statistical noise:

$$\mathcal{D}_i = \lim_{t \to \infty} \frac{1}{6t} \left\langle \frac{1}{N_i} \sum_{l=1}^{N_i} [r_{li}(t) - r_{li}(0)]^2 \right\rangle \tag{8.2}$$

Note that the same strategy (i.e., averaging over many single-particle trajectories) is used to extract the self-diffusivity from experiments that track the displacement of individual molecules; see Section 11.6.2). The Einstein equation (8.1) can be expressed in the equivalent Green–Kubo form using the general equilibrium equality [1]:

$$\int_0^\infty \left\langle \dot{\mathcal{A}}_1(0) \dot{\mathcal{A}}_2(t) \right\rangle \mathrm{d}t = \lim_{t \to \infty} \frac{1}{2t} \langle [\mathcal{A}_1(t) - \mathcal{A}_1(0)][\mathcal{A}_2(t) - \mathcal{A}_2(0)] \rangle \tag{8.3}$$

where

$\mathcal{A}_1, \mathcal{A}_2$ are any scalar dynamical quantities,
$\dot{\mathcal{A}}_1, \dot{\mathcal{A}}_2$ are their time derivatives,
and the angular brackets denote equilibrium ensemble averaging.

The quantity $\left\langle \dot{\mathcal{A}}_1(0) \dot{\mathcal{A}}_2(t) \right\rangle$ in Eq. (8.3) is an equilibrium correlation function between $\dot{\mathcal{A}}_1$ and $\dot{\mathcal{A}}_2$. It can be computed as an ensemble average over a very large number of copies of the system, constituting an equilibrium ensemble and evolving between two specific times, 0 and t. Alternatively, it can be computed as a time average over a single trajectory of the system that is long enough to visit all relevant points in phase space according to the equilibrium distribution dictated by the system's Hamiltonian, by taking pairs of points along the trajectory that differ in time by t and averaging over all time origins along the trajectory. This procedure, which corresponds to "sliding" a time interval t along a very long equilibrium trajectory and averaging the product of two dynamical quantities at the beginning and end of the interval, is very often used in simulations; it is sometimes referred to as "multiple time origins." In the special case where $\mathcal{A}_1 = \mathcal{A}_2$, $\left\langle \dot{\mathcal{A}}_1(0) \dot{\mathcal{A}}_2(t) \right\rangle$ is the equilibrium time autocorrelation function of $\dot{\mathcal{A}}_1$; it measures how fast $\dot{\mathcal{A}}_1$ loses memory of its original value as a result of thermal motion. Applying Eq. (8.3) with $\mathcal{A}_1 = \mathcal{A}_2$ equal to each one of the three components of r_{li} and summing the three resulting expressions together, one transforms Eqs. (8.1) and (8.2) into:

$$\mathcal{D}_i = \frac{1}{3}\int_0^\infty \langle \boldsymbol{u}_{li}(0)\cdot \boldsymbol{u}_{li}(t)\rangle \mathrm{d}t = \frac{1}{3}\frac{1}{N_i}\sum_{i=1}^{N_i}\int_0^\infty \langle \boldsymbol{u}_{li}(0)\cdot \boldsymbol{u}_{li}(t)\rangle \mathrm{d}t \tag{8.4}$$

with $\boldsymbol{u}_{li} = \dot{\boldsymbol{r}}_{li}$ being the velocity vector of the center-of-mass of molecule l of species i. Equation (8.4) is a Green–Kubo equation for the self-diffusivity. It relates $\mathcal{D}_i$ to the time integral of the autocorrelation function of the translational velocity of a molecule.

8.1.2
Mass Fluxes in a Microporous Medium

Mass fluxes in our sorbed fluid/microporous solid system, which is assumed to be isothermal, arise in response to heterogeneities in composition, which result in nonzero gradients in the molecular chemical potentials $\nabla \mu_i$ (measured here in J m^{-1}), or in response to the imposition of external force fields $\boldsymbol{F}_i$ (measured here in N) on each species i. We will assume [2, 3] that these gradients in chemical potential or body forces are small. That is to say, one can define volume elements within the system that are small relative to macroscopic dimensions, yet large relative to molecular dimensions, in which intensive properties are uniform enough to justify considering the entire volume element as being in local thermodynamic equilibrium. Practical procedures for checking whether this is the case in NEMD simulations are discussed in Section 8.3.2. We focus on describing the mass fluxes developing in these volume elements over time scales considerably longer than the characteristic times of molecular-level relaxation processes in the system. The medium will be assumed homogeneous enough to exhibit diffusive behavior over the length scale of the considered volume elements. Anomalous or single-file diffusion of molecules may be encountered over smaller length scales, but should have crossed over to normal Einstein diffusion over the length scale of the volume elements.

We employ a laboratory-fixed frame of reference and assume that our medium satisfies the condition of mechanical equilibrium. Under constant temperature and pressure, the Gibbs–Duhem equation gives:

$$\sum_{i=1}^{z-1} N_i(\nabla \mu_i - \boldsymbol{F}_i) + N_z(\nabla \mu_z - \boldsymbol{F}_z) = 0 \tag{8.5}$$

where N_i is the number of molecules of species i in the considered volume element. Under the assumptions we have introduced, the formalism of non-equilibrium thermodynamics gives the following expression for the macroscopic flux, $\boldsymbol{J}_i$, of each species, measured in molecules m^{-2} s^{-1}, under isothermal conditions:

$$\boldsymbol{J}_i = -\sum_{j=1}^{z-1} L_{ij}(\nabla \mu_j - \boldsymbol{F}_j) - L_{iz}(\nabla \mu_z - \boldsymbol{F}_z) \tag{8.6}$$

The phenomenological transport coefficients L_{ij}, measured in molecules $J^{-1}\,m^{-1}\,s^{-1}$, satisfy Onsager's reciprocity relations $L_{ij} = L_{ji}$, Eq. (3.6). The relation between the Onsager formulation, Eq. (8.6), and the generalized Maxwell–Stefan formulation for non-equilibrium transport in microporous media has been discussed in Section 3.3.2. Note that a factor $1/T$ has been incorporated in the Onsager coefficients, as in Eq. (3.20), since isothermal conditions are considered.

A molecular interpretation for the phenomenological coefficients L_{ij} can be arrived at through linear response theory, as applied to transport processes by Mori [1, 2]. This theory establishes a very important conceptual link between the response of material systems to driving forces that take them away (but not too far) from thermodynamic equilibrium, on one hand, and spontaneous fluctuations under equilibrium conditions on the other hand. Qualitatively, in the case of our system, linear response theory states that the rate at which the considered volume element will respond to the imposition of a gradient in composition or of an external force field that causes it to depart from thermodynamic equilibrium can be estimated from the rate at which spontaneous fluctuations occurring within the same volume element in a state of thermodynamic equilibrium die out with time. This is because both spontaneous equilibrium fluctuations and the non-equilibrium response are governed by the same microscopic dynamics.

We define the microscopic (molecular) current of component i as:

$$\boldsymbol{j}_i = \sum_{l=1}^{N_i} \boldsymbol{u}_{li} \tag{8.7}$$

Linear response theory dictates:

$$L_{ij} = \frac{1}{3Vk_BT} \int_0^\infty \langle \boldsymbol{j}_i(0) \cdot \boldsymbol{j}_j(t) \rangle \, dt \tag{8.8}$$

The angular brackets in Eq. (8.8) denote ensemble averaging in an *equilibrium* system that finds itself at the same temperature, pressure, and composition as the considered volume element of the non-equilibrium system at the considered time. Equation (8.8) is of the Green–Kubo form, relating a transport coefficient to the time correlation function of microscopic currents at equilibrium.

The microscopic current–current correlation function $\langle \boldsymbol{j}_i(0) \cdot \boldsymbol{j}_j(t) \rangle$ can be analyzed as follows:

$$\langle \boldsymbol{j}_i(0) \cdot \boldsymbol{j}_j(t) \rangle = \sum_{l=1}^{N_i} \sum_{l'=1}^{N_j} \langle \boldsymbol{u}_{li}(0) \cdot \boldsymbol{u}_{l'j}(t) \rangle \quad (i \neq j) \tag{8.9}$$

$$\langle \boldsymbol{j}_i(0) \cdot \boldsymbol{j}_i(t) \rangle = \sum_{l=1}^{N_i} \langle \boldsymbol{u}_{li}(0) \cdot \boldsymbol{u}_{li}(t) \rangle + \sum_{l=1}^{N_i} \sum_{\substack{l'=1 \\ l' \neq l}}^{N_i} \langle \boldsymbol{u}_{li}(0) \cdot \boldsymbol{u}_{l'i}(t) \rangle \quad (i = j) \tag{8.10}$$

The velocity correlation and autocorrelation functions appearing in Eqs. (8.9) and (8.10) are expected to be nonzero only over a finite time t, commensurate with the "molecular relaxation time" for translational motion, and much smaller than macroscopic observation times. Furthermore, at equilibrium, velocity correlations between different molecules l and l' will be nonzero only as long as these molecules are spatially within a certain finite range of each other in the medium. Thus, the time integral appearing in Eq. (8.8) is convergent and grows linearly with the number of molecules in the considered volume element, making L_{ij} a well-defined intensive property, which depends on the local composition within the volume element V. If the time integral in Eq. (8.8) is measured in units of $m^2 s^{-1}$, L_{ij} is recovered in the units mentioned above.

Invoking the general relation, Eq. (8.3), we can convert Eq. (8.8) into the equivalent Einstein form:

$$L_{ij} = \frac{1}{6Vk_B T} \lim_{t \to \infty} \left\{ \frac{1}{t} \left\langle \left\{ \sum_{l=1}^{N_i} [r_{li}(t) - r_{li}(0)] \right\} \cdot \left\{ \sum_{l'=1}^{N_j} [r_{l'j}(t) - r_{l'j}(0)] \right\} \right\rangle \right\} \quad (8.11)$$

Equation (8.11) can be rewritten in terms of the mean square displacement of the centers-of-mass of swarms of molecules of different species in the microporous medium at equilibrium. Let:

$$r_{cm,i} = \left(\sum_{l=1}^{N_i} r_{li} \right) / N_i$$

be the center-of-mass position of a swarm of molecules of type i, consisting of N_i molecules. Equation (8.11) is equivalent to:

$$L_{ij} = \frac{N_i N_j}{6Vk_B T} \lim_{t \to \infty} \left\{ \frac{1}{t} \langle [r_{cm,i}(t) - r_{cm,i}(0)] \cdot [r_{cm,j}(t) - r_{cm,j}(0)] \rangle \right\} \quad (8.12)$$

Not all chemical potentials appearing in the flux expression, Eq. (8.6), are independent. By virtue of the Gibbs–Duhem equation [Eq. (8.5)] and symbolizing with $q_i = N_i/V$ the molecular density (concentration) of species i per unit volume of the microporous medium, we obtain for the sorbate species:

$$J_i = -\sum_{j=1}^{z-1} \left(L_{ij} - \frac{q_i}{q_z} L_{iz} \right) (\nabla \mu_j - F_j) \quad (i = 1, 2, \ldots, z-1) \quad (8.13)$$

In applying the above non-equilibrium thermodynamic formalism to microporous media, we often assume that the particles or atoms of the medium z are stationary in the laboratory frame of reference. That is to say, $u_{lz} = 0$ for every $l = 1, 2, \ldots, N_z$. Under this assumption, Eq. (8.8) gives:

$$L_{iz} = 0 \quad (i = 1, 2, \ldots, z) \quad (8.14)$$

Equation (8.13) then simplifies to:

$$J_i = -\sum_{j=1}^{z-1} L_{ij} (\nabla \mu_j - F_j) \quad (i = 1, 2, \ldots, z-1) \quad (8.15)$$

with the phenomenological coefficients given by Eq. (8.8) or its equivalent Eqs. (8.11) and (8.12). No explicit consideration of the solid component z is required.

8.1.3
Transport in the Presence of Concentration Gradients in Pure and Mixed Sorbates

In experimental diffusion work one rarely encounters the phenomenological transport coefficients L_{ij}, defined above. Instead, the concept of a transport diffusivity is used. As we have seen in Chapter 1, Eq. (1.1), the transport diffusivity D_1 of a pure sorbed species 1 is defined as the proportionality constant between the macroscopic flux $\boldsymbol{J}_1$ and the concentration gradient ∇q_1 in a system off thermodynamic equilibrium:

$$\boldsymbol{J}_1 = -D_1 \nabla q_1 \tag{8.16}$$

Note that Eq. (8.16) is strictly valid in an isotropic medium; otherwise, a more general definition of $\boldsymbol{D}_1$ as a tensorial quantity should be adopted.

On the other hand, from Eq. (8.15), specializing to the case where only components 1 and z are present, in the absence of external forces ($\boldsymbol{F}_1 = 0$), one obtains:

$$\boldsymbol{J}_1 = -L_{11} \nabla \mu_1 = -L_{11} \frac{k_B T}{q_1} \left[\frac{\partial \ln f_1}{\partial \ln q_1} \right]_{T,P} \nabla q_1 \tag{8.17}$$

with f_1 being the fugacity of 1. Note that the atoms of z have been considered stationary in deriving Eq. (8.15) and, therefore, in Eq. (8.17) as well. Combining Eqs. (8.16), (8.17), and (8.8), the transport diffusivity is expressed as:

$$D_1 = L_{11} \frac{k_B T}{q_1} \left[\frac{\partial \ln f_1}{\partial \ln q_1} \right]_T = \left[\frac{1}{3N_1} \int_0^\infty \langle \boldsymbol{j}_1(0) \cdot \boldsymbol{j}_1(t) \rangle dt \right] \left[\frac{\partial \ln f_1}{\partial \ln q_1} \right]_T \tag{8.18}$$

The transport diffusivity emerges as a product of a mobility term, incorporating the time integral of the autocorrelation function of a microscopic flux of the sorbed component in the microporous medium under equilibrium conditions, and a thermodynamic correction term, incorporating the dependence of the fugacity of the sorbed component on its concentration, that is, the slope of the sorption isotherm in logarithmic coordinates evaluated at the prevailing concentration of component 1. The subscript P has been dropped from the thermodynamic derivative in Eq. (8.18) because the microporous solid phase is considered incompressible.

As pointed out in Section 1.2.1, Eq. (1.31), Eq. (8.18) is often written as:

$$D_1 = D_{01} \left[\frac{\partial \ln f_1}{\partial \ln q_1} \right]_T \tag{8.19}$$

where:

$$D_{01} = L_{11} \frac{k_B T}{q_1} = \frac{1}{3N_1} \int_0^\infty \langle \boldsymbol{j}_1(0) \cdot \boldsymbol{j}_1(t) \rangle \, dt = N_1 \lim_{t \to \infty} \left\langle \frac{[\boldsymbol{r}_{cm,1}(t) - \boldsymbol{r}_{cm,1}(0)]^2}{6t} \right\rangle \tag{8.20}$$

is the "corrected diffusivity" (corrected, that is, for thermodynamic non-idealities and incorporating only the mobility part of D_1). The motivation for introducing D_{01} is that it exhibits weaker concentration dependence than D_1 in most systems. It should be stressed, however, that all three quantities D_1, $\mathcal{D}_1$, and D_{01} depend on q_1 and generally assume different values at finite loadings. Equation (8.19) with the substitution $D_{01} = \mathcal{D}_1$ is commonly known as the Darken equation [4]. This substitution clearly constitutes an approximation; furthermore, it should be stated clearly at which concentration the $\mathcal{D}_1$ that is used in place of D_{01} is evaluated.

In the limit of very low intracrystalline concentrations q_1 (Henry's law region) the fugacity f_1 becomes a linear function of q_1 and the thermodynamic correction term in Eqs. (8.18) and (8.19) reduces to unity. Furthermore, the microscopic current autocorrelation function in Eqs. (8.18) and (8.20) reduces to N_1 times the velocity autocorrelation function of one molecule. This is because, as seen from Eq. (8.10), the current autocorrelation function can be separated into a sum of velocity autocorrelation functions from individual molecules plus a sum of velocity cross-correlation functions between different molecules. The latter, of the form $\langle \boldsymbol{u}_{l_1}(0) \cdot \boldsymbol{u}_{l'1}(t) \rangle$, go to zero in the limit of zero occupancy, where different molecules are too far apart to influence each other's motion. In this limit, then:

$$\lim_{q_1 \to 0} \int_0^\infty \langle \boldsymbol{j}_1(0) \cdot \boldsymbol{j}_1(t) \rangle \, \mathrm{d}t = N_1 \int_0^\infty \langle \boldsymbol{u}_{l_1}(0) \cdot \boldsymbol{u}_{l_1}(t) \rangle \, \mathrm{d}t \tag{8.21}$$

The velocity autocorrelation function on the right-hand side of Eq. (8.21) is immediately related to the self-diffusion coefficient $\mathcal{D}_1$ by Eq. (8.4). In the limit of very low occupancy within a rigid microporous solid, then:

$$\lim_{q_1 \to 0} D_1 = \lim_{q_1 \to 0} D_{01} = \lim_{q_1 \to 0} \mathcal{D}_1 \tag{8.22}$$

For a binary mixture of sorbates 1 and 2 within a stationary microporous medium, the macroscopic fluxes of species 1 and 2 are, from Eq. (8.15):

$$\begin{aligned} \boldsymbol{J}_1 &= -L_{11} \nabla \mu_1 - L_{12} \nabla \mu_2 \\ \boldsymbol{J}_2 &= -L_{21} \nabla \mu_1 - L_{22} \nabla \mu_2 \end{aligned} \tag{8.23}$$

where Onsager's straight coefficients L_{11}, L_{22} and cross coefficients $L_{12} = L_{21}$ are obtainable from the self- and cross-correlation functions of molecular currents in the binary sorbate phase under *equilibrium* conditions via the Green–Kubo equation (8.8), or from molecular displacements at equilibrium via the equivalent Einstein relation, Eq. (8.11). The Onsager formulation, Eq. (8.23), is to be compared with the Maxwell–Stefan formulation (Chapter 3), which, for a binary system of sorbates in a stationary microporous medium, reads [5] [compare Eq. (3.43)]:

$$\begin{aligned} \frac{1}{k_B T} \nabla \mu_1 &= -\frac{1}{q_{1,\mathrm{sat}} \theta_1} \left(\frac{\theta_2}{\mathcal{D}_{12}} + \frac{1}{D_{01}} \right) \boldsymbol{J}_1 + \frac{1}{q_{2,\mathrm{sat}} \mathcal{D}_{12}} \boldsymbol{J}_2 \\ \frac{1}{k_B T} \nabla \mu_2 &= \frac{1}{q_{1,\mathrm{sat}} \mathcal{D}_{21}} \boldsymbol{J}_1 - \frac{1}{q_{2,\mathrm{sat}} \theta_2} \left(\frac{\theta_1}{\mathcal{D}_{21}} + \frac{1}{D_{02}} \right) \boldsymbol{J}_2 \end{aligned} \tag{8.24}$$

where $q_{i,\text{sat}}$ ($i = 1,2$) is the intracrystalline concentration of species i at saturation, that is, the maximal concentration, in molecules of species i per m^3 of microporous medium, that can be sorbed into the microporous medium from a phase of pure fluid i at the prevailing temperature in the limit of very high fugacities of i; $\theta_i = q_i/q_{i,\text{sat}}$ is the fractional occupancy of species i, defined in relation to its saturation concentration; D_{0i} ($Ð_i$ in Krishna and van Baten's original notation [5]) is a Maxwell–Stefan diffusivity incorporating the effect of interactions between the sorbed species i and the framework of the medium; and $Ð_{ij}$ ($Ð_{ij}^*$ in Krishna and van Baten's original notation [5]) is a Maxwell–Stefan exchange coefficient incorporating the effect of interactions with species j on the motion of species i (see also Chapter 3). In this formulation, the Onsager reciprocity relations lead to the condition:

$$q_{2,\text{sat}} Ð_{12} = q_{1,\text{sat}} Ð_{21} \tag{8.25}$$

Enforcing consistency between Eqs. (8.23) and (8.24) leads to the conditions:

$$Ð_{12} = \frac{k_B T}{q_{2,\text{sat}}} \frac{L_{11} L_{22} - L_{12} L_{21}}{L_{12}}, \quad Ð_{21} = \frac{k_B T}{q_{1,\text{sat}}} \frac{L_{11} L_{22} - L_{12} L_{21}}{L_{21}}$$

$$D_{01} = \frac{k_B T}{q_{1,\text{sat}} \theta_1} \frac{L_{11} L_{22} - L_{12} L_{21}}{L_{22} - \frac{q_{2,\text{sat}} \theta_2}{q_{1,\text{sat}} \theta_1} L_{12}}, \quad D_{02} = \frac{k_B T}{q_{2,\text{sat}} \theta_2} \frac{L_{11} L_{22} - L_{12} L_{21}}{L_{11} - \frac{q_{1,\text{sat}} \theta_1}{q_{2,\text{sat}} \theta_2} L_{21}} \tag{8.26}$$

Equation (8.26) can be used to extract the Maxwell–Stefan coefficients from flux correlation functions under equilibrium conditions. In general, these relations are not simple. In the limit $\theta_2 \to 0$, comparison of Eq. (8.26) with Eqs. (8.18) and (8.19) shows that D_{01} can be identified with the corrected diffusivity of pure species 1 in the microporous medium, at its prevailing concentration.

Conversely, one can express the Onsager coefficients in terms of the Maxwell–Stefan diffusivities. All one needs to do is solve the linear system of Eq. (8.24) to express J_1 and J_2 as linear combinations of $\nabla \mu_1$ and $\nabla \mu_2$ and then equate the coefficients in the resulting expressions to those of Eq. (8.23). The result is [17]:

$$L_{11} = \frac{q_1}{k_B T |B|} \left[\frac{1}{D_{02}} + \frac{\theta_1}{Ð_{21}} \right] \tag{8.27}$$

$$L_{12} = \frac{q_1}{k_B T |B|} \frac{\theta_2}{Ð_{12}} \tag{8.28}$$

$$L_{21} = \frac{q_2}{k_B T |B|} \frac{\theta_1}{Ð_{21}} \tag{8.29}$$

$$L_{22} = \frac{q_2}{k_B T |B|} \left[\frac{1}{D_{01}} + \frac{\theta_2}{Ð_{12}} \right] \tag{8.30}$$

where $|B|$ is the determinant of a matrix defined by:

$$B_{ii} = \frac{1}{D_{0i}} + \sum_{k=1, k \neq i}^{2} \frac{\theta_k}{Ð_{ik}}, \quad B_{ij(i \neq j)} = -\frac{\theta_i}{Ð_{ij}} \tag{8.31}$$

$q_{1,\text{sat}}, q_{2,\text{sat}} Ð_{12}$ and $Ð_{21}$ are the symbols appearing in Eq. (8.25).

8.2
Equilibrium Molecular Dynamics Simulations

8.2.1
Integrating the Equations of Motion: the Velocity Verlet Algorithm

Equilibrium molecular dynamics (MD) entails tracking the temporal evolution of a model system, in our case consisting of a certain amount of microporous solid and the molecules of one or more sorbate species within it (in thermodynamic equilibrium) via numerical solution of the equations of motion for all its constituent particles. Typically, the model systems we simulate have linear dimensions on the order of nm, are characterized by periodic boundary conditions [6], and are governed by a potential energy function $\mathcal{V}$ of the form discussed in Chapter 7. An initial configuration for the system is generated by atomistic reconstruction of the microporous medium and subsequent addition of the sorbate molecules. The latter can be achieved via configurationally biased insertion of the sorbed molecules, one by one, in the pore space of the medium (Chapter 7). A configuration sampled in the course of an equilibrium canonical or grand canonical MC run constitutes an excellent starting point for MD and should be used, whenever possible. Initial velocities are typically assigned from a Maxwell–Boltzmann distribution.

The integration of the equations of motion amounts to the numerical solution of an initial value problem in the coordinates and velocities. Typically, there is an "equilibration phase" of the MD simulation, in the course of which memory of the initial phase-space point is effaced, followed by a "production phase," in which the dynamical trajectory samples the probability distribution of the equilibrium ensemble dictated by the macroscopic constraints imposed on the system. Completion of the equilibration phase is evidenced by the fact that observables, such as potential and kinetic energy and their components contributed by various categories of degrees of freedom, the spatial, orientational and conformational distributions of sorbed molecules, and correlation functions describing the structure of the sorbed phase (Chapter 7 and Reference [6]), fluctuate around time-independent, equilibrium values.

Perhaps the most straightforward form of MD is one that simulates a molecular model that employs the Cartesian coordinates of all atoms, $r_1, r_2, \ldots, r_N$, as independent degrees of freedom, evolving under the influence of their mutual interactions. The total energy E and the total momentum $\sum_{i=1}^{N} p_i$, usually set to zero initially, are invariants of the motion; the MD simulation is conducted in the microcanonical, or *NVE*, ensemble.

As mentioned in Chapter 7, an approximation that is frequently invoked to save computer time is to assume that the positions of atoms constituting the microporous medium remain fixed in space. If this approximation is invoked, the index N runs over sorbate atoms only. The microporous medium is modeled merely as the source of a conservative external potential acting on interaction sites of the sorbed molecules. Sorbed molecule–wall collisions are perfectly elastic; the total energy of the sorbed molecules is preserved, making this an *NVE* simulation, but their total momentum is not; in a well-designed simulation, the total momentum of sorbate molecules fluctuates around zero. Evolution of the system to the correct equilibrium distribution of energies among the model degrees of freedom ("thermalization") relies upon sorbate–sorbate collisions. At very low loadings, short NVE MD simulations using a fixed framework may thus be problematic, as they may lead to artificial "trapping" of sorbate molecules that do not have sufficient energy to exit wells of the potential energy field of the microporous medium; such molecules would encounter other sorbate molecules too infrequently to be knocked out of the wells by them.

A simple and efficient class of algorithms for integrating the equations of motion in Cartesian coordinates was proposed by Verlet in 1967 and developed further by other investigators. Here we discuss briefly the "velocity Verlet algorithm" [7] as an example. In this algorithm, the $3N$-long vector of atomic positions r, the $3N$-long vector of atomic velocities $v = \dot{r}$, and the $3N$-long vector of atomic accelerations $a = \ddot{r}$ are stored, resulting in a total memory requirement of $9N$ bytes. Time is advanced in small steps, of length δt. At the beginning of every time step, positions at time $t + \delta t$ are calculated from positions, velocities, and accelerations at time t:

$$r(t+\delta t) = r(t) + \delta t v(t) + \frac{1}{2}(\delta t)^2 a(t) \quad (8.32)$$

The new positions $r(t+\delta t)$ are stored in place of the positions $r(t)$. Velocities at time $t + \frac{1}{2}\delta t$ (half-step) are also calculated from velocities and accelerations at time t:

$$v\left(t + \frac{1}{2}\delta t\right) = v(t) + \frac{1}{2}\delta t\, a(t) \quad (8.33)$$

Velocities $v(t+\frac{1}{2}\delta t)$ are stored in place of the velocities $v(t)$. The stored new positions at the end of the step, $r(t+\delta t)$, are then used to calculate the forces on all atoms via $F_i = -\nabla_{r_i} \mathcal{V} \equiv -\partial \mathcal{V}/\partial r_i$, and from those, though division by the respective masses, all the atomic accelerations $a(t+\delta t)$ at time $t + \delta t$. Accelerations $a(t+\delta t)$ are stored in place of the accelerations $a(t)$. Velocities at the end of the step $v(t+\delta t)$ are calculated from the half-step velocities $v(t+\frac{1}{2}\delta t)$ and end-step accelerations $a(t+\delta t)$ via:

$$v(t+\delta t) = v\left(t + \frac{1}{2}\delta t\right) + \frac{1}{2}\delta t a(t+\delta t) \quad (8.34)$$

The end-step velocities $v(t+\delta t)$ are stored in place of the mid-step velocities $v(t+\frac{1}{2}\delta t)$, completing the integration step. The calculation then loops back to perform the next step.

The prescription for advancing positions, velocities, and accelerations by a time step, Eqs. (8.32)–(8.34), can be recast in a form involving only positions and accelerations. By Eq. (8.32), applied for time $t + \delta t$ in place of t:

$$r(t+2\delta t) = r(t+\delta t) + \delta t v(t+\delta t) + \frac{1}{2}(\delta t)^2 a(t+\delta t) \quad (8.35)$$

On the other hand, by Eqs. (8.33) and (8.34):

$$v(t+\delta t) = v(t) + \frac{1}{2}\delta t[a(t) + a(t+\delta t)] \quad (8.36)$$

Substituting $\mathbf{v}(t+\delta t)$ from Eq. (8.36) into Eq. (8.35) and replacing $\delta t \mathbf{v}(t)$ in the resulting equation by its equal from Eq. (8.32), one obtains:

$$\mathbf{r}(t+2\delta t) = 2\mathbf{r}(t+\delta t) - \mathbf{r}(t) + (\delta t)^2 \mathbf{a}(t+\delta t) \tag{8.37}$$

which, when applied for time $t - \delta t$ in place of t, leads to:

$$\mathbf{r}(t+\delta t) = 2\mathbf{r}(t) - \mathbf{r}(t-\delta t) + (\delta t)^2 \mathbf{a}(t) \tag{8.38}$$

Equation (8.38) is the prescription given by the original Verlet algorithm for updating positions [8]. It can be viewed as a consequence of the following two Taylor expansions of $\mathbf{r}(t)$ around t:

$$\mathbf{r}(t+\delta t) = \mathbf{r}(t) + \delta t \mathbf{v}(t) + \frac{1}{2}(\delta t)^2 \mathbf{a}(t) + \frac{1}{6}(\delta t)^3 \dddot{\mathbf{r}}(t) + \mathcal{O}\left[(\delta t)^4\right] \tag{8.39}$$

$$\mathbf{r}(t-\delta t) = \mathbf{r}(t) - \delta t \mathbf{v}(t) + \frac{1}{2}(\delta t)^2 \mathbf{a}(t) - \frac{1}{6}(\delta t)^3 \dddot{\mathbf{r}}(t) + \mathcal{O}\left[(\delta t)^4\right] \tag{8.40}$$

Adding Eqs. (8.39) and (8.40) leads to:

$$\mathbf{r}(t+\delta t) = 2\mathbf{r}(t) - \mathbf{r}(t-\delta t) + (\delta t)^2 \mathbf{a}(t) + \mathcal{O}\left[(\delta t)^4\right] \tag{8.41}$$

upon comparison of which with Eq. (8.38) it can be concluded that the approximation error in estimating positions via the Verlet algorithms is of the order $(\delta t)^4$, even though the last term included in the update equations for the positions is of order $(\delta t)^2$. The Verlet algorithms and, especially, the velocity Verlet algorithm are simple, time-reversible, and symplectic (i.e., preserving volume in phase space). They are thus very popular in unconstrained MD simulations. Figure 8.1 shows a sample

```
........

do i = 1, N
    r(i) = r(i) + dt*v(i) + dt*dt/2*F(i)    ! update positions at t+dt using
                                             ! velocities and forces at t
    v(i) = v(i) + dt/2*F(i)                  ! update velocities at t+dt
                                             ! using forces at t
end do

call get_forces (F)                          ! calculate forces at t+dt
do i = 1, N
    F(i) = F(i)/m(i)                         ! convert forces to accelerations
end do

do i = 1, N
    v(i) = v(i) + dt/2*F(i)                  ! update velocities at t+dt
                                             ! using forces at t+dt
end do

........
```

Figure 8.1 Pseudo-code implementing a time step of the velocity Verlet algorithm.

pseudo-code [9] implementing the velocity Verlet scheme. The routine *get_forces* calculates the total force on each atom.

8.2.2
Multiple Time Step Algorithms: rRESPA

A difficulty in MD is that the spectrum of characteristic times for all the motions probed by the simulation may be extremely broad. For example, C–H bond stretching vibrations of sorbed hydrocarbon molecules, with periods around 10^{-14} s, may coexist with *gauche–trans* conformational isomerizations and shuttling translational motions within the cavities of a zeolite, with characteristic times longer than 10^{-11} s [10]. Such problems, in which there is a disparity of several orders of magnitude between the shortest and longest time constants present (formally expressed as a "condition number," that is, a ratio of largest to smallest eigenvalues of a rate constant matrix describing the dynamics) are called stiff. In the numerical integration, the time constant δt must be chosen to be considerably shorter than the shortest time constant present. In our example, the δt chosen should be smaller than 1 fs. Otherwise, the simulation will be unable to track the fast vibrational motions in the system. Calculated atomic positions will be inaccurate, the calculated trajectory diverging exponentially from the true solution of the dynamical equations of motion. Energy will not be conserved, and our simulation program will fail. Adopting a time step of a fraction of a fs is imperative to avoid these problems. On the other hand, with such a time step we will need billions of time steps to generate μs-long trajectories, as is required to track diffusional motion reliably in some systems, and our requirements in CPU time will be excessive.

One way to reduce computational burden is to resort to multiple time step MD algorithms. These employ two or more time steps. The shortest time step is used for the integration of the fastest motions, such as vibration; the longest time step is used for the integration of the slowest motions, such as translation. A long time step corresponds to many short time steps; the forces associated with the slow motions are thus updated less frequently than the forces associated with the fast motions. This is computationally advantageous, especially since calculation of the nonbonded interactions associated with slower translational motions is significantly more expensive than calculation of bond stretching potentials. The former requires invocation of neighbor lists constructed via double loops over all atoms and complex functional expressions, such as the Lennard-Jones potential, while the latter involves strictly intramolecular contributions of relatively simple (e.g., harmonic) form.

Rigorous, time-reversible multiple time step schemes have been developed by Tuckerman, Martyna, and their collaborators [11, 12] based on a mathematical technique known as Trotter factorization of the Liouville operator. For our model system of N atoms (more generally, interaction sites) evolving under the influence of their mutual interactions, the Liouville operator $\mathcal{L}$ is defined by:

$$i\mathcal{L} = \sum_{i=1}^{N} \left[\dot{r}_i \frac{\partial}{\partial r_i} + F_i \frac{\partial}{\partial p_i} \right] \tag{8.42}$$

8.2 Equilibrium Molecular Dynamics Simulations

Evolution of the representative point $\Gamma = (r,p)$ of the system in its 6 N-dimensional phase space from time 0 to time t can be described by:

$$\Gamma(t) = \exp(i\mathcal{L}t)\Gamma(0) \tag{8.43}$$

Now, if the Liouville operator is separated into two parts, $i\mathcal{L} = i\mathcal{L}_1 + i\mathcal{L}_2$ and we consider the total time t being subdivided into n intervals of length $\delta t = t/n$, the Trotter theorem states that:

$$\exp(i\mathcal{L}t) = \exp[i(\mathcal{L}_1 + \mathcal{L}_2)t/n]^n = \{\exp[i\mathcal{L}_1(\delta t/2)]\exp(i\mathcal{L}_2 \delta t)\exp[i\mathcal{L}_1(\delta t/2)]\}^n + \mathcal{O}(t^3/n^2) \tag{8.44}$$

The evolution between times 0 and t is thus broken up into n steps of duration δt, each with Liouville operator:

$$\exp[i\mathcal{L}_1(\delta t/2)]\exp(i\mathcal{L}_2 \delta t)\exp[i\mathcal{L}_1(\delta t/2)]$$

To better understand the meaning of this factorization, let us consider the example decomposition:

$$i\mathcal{L}_1 = \sum_{i=1}^{N} F_i \frac{\partial}{\partial p_i}, \quad i\mathcal{L}_2 = \sum_{i=1}^{N} \dot{r}_i \frac{\partial}{\partial r_i}$$

The Liouville operator in this case corresponds to the following update scheme:

$$r(\delta t) = r(0) + \delta t v(0) + \frac{(\delta t)^2}{2m} F(r(0))$$
$$v(\delta t) = v(0) + \frac{\delta t}{2m}[F(r(0)) + F(r(\delta t))]$$

which is seen to be identical to the velocity Verlet algorithm, Eqs. (8.32)–(8.34).

Let us now distinguish the forces acting on our model system into two categories: fast varying forces, F^f, which typically consist of short-range bonded interactions, such as bond stretching and bond-angle bending, and slowly varying forces, F^s, which typically consist of longer-range nonbonded interactions. The choice of which forces to include in each category is up to us; in applications, it is dictated by computational efficiency. The total forces on all atoms are obtained by summing fast varying and slowly varying forces, $F = F^f + F^s$. Correspondingly, we split the Liouville operator into three parts:

$$i\mathcal{L} = i\mathcal{L}_1 + i\mathcal{L}_2 + i\mathcal{L}_3 \quad \text{with} \quad i\mathcal{L}_1 = \sum_{i=1}^{N} F_i^f \frac{\partial}{\partial p_i}, \quad i\mathcal{L}_2 = \sum_{i=1}^{N} \dot{r}_i \frac{\partial}{\partial r_i}, \quad i\mathcal{L}_3 = \sum_{i=1}^{N} F_i^s \frac{\partial}{\partial p_i} \tag{8.45}$$

We also define two time steps: a long one, Δt, which will be associated with the slowly varying forces, and a short one, δt, which will be associated with the fast varying forces. The ratio of the two time steps, $n = \Delta t/\delta t$, will be an integer number.

In practical applications, n is chosen as 10 or 5. Trotter factorization of the Liouville operator describing the evolution of the system over time Δt leads to the expression:

$$\exp(i\mathcal{L}\Delta t) = \exp\left(i\mathcal{L}_3 \frac{\Delta t}{2}\right) \left[\exp\left(i\mathcal{L}_1 \frac{\delta t}{2}\right) \exp(i\mathcal{L}_2 \delta t) \exp\left(i\mathcal{L}_1 \frac{\delta t}{2}\right)\right]^n \exp\left(i\mathcal{L}_3 \frac{\Delta t}{2}\right) + \mathcal{O}\left[(\Delta t)^3\right]$$

(8.46)

which corresponds to the numerical integration scheme shown in Figure 8.2. This scheme is the essence of the reversible reference system propagator algorithm (rRESPA) proposed by Martyna and Tuckerman [11, 12].

It is seen that the typically more expensive slowly varying forces are updated by a factor of n less frequently than fast varying forces, leading to significant savings in computer time.

```
 ........

 do i = 1, N
     v(i) = v(i) + Δt/2*Fˢ(i)           ! update velocities using
                                         ! slow forces at t
 end do

 do j = 1, n
     do i = 1, N
         v(i) = v(i) + δt/2*Fᶠ(i)       ! update velocities using
                                         ! fast forces at t + (j-1)δt
         r(i) = r(i) + δt* v(i)          ! update positions at t + j δt
     end do

     call fast_forces (Fᶠ)               ! get fast forces at t + j δt
     do i = 1, N
         Fᶠ(i) = Fᶠ(i)/m(i)              ! convert forces to accelerations
     end do

     do i = 1, N
         v(i) = v(i) + δt/2*Fᶠ(i)        ! update velocities using
                                         ! fast forces at t + j δt,
     end do                              ! just calculated
 end do

 call slow_forces (Fˢ)                   ! get slow forces at t + Δt
 do i = 1, N
     Fˢ(i) = Fˢ(i)/m(i)                  ! convert forces to accelerations
 end do

 do i = 1, N
     v(i) = v(i) + Δt/2*Fˢ(i)            ! update velocities using
                                         ! slow forces at t + Δt
 end do

 ........
```

Figure 8.2 Pseudo-code implementing a long time step of the rRESPA algorithm.

8.2.3
Domain Decomposition

Present-day molecular simulations are usually carried out on computer clusters containing hundreds to thousands of processors. These offer the opportunity to conduct calculations in parallel on many processors, and therefore address model systems of large size, containing millions of interaction sites, within reasonable clock time.

We will not dwell on the details of parallel MD here. The interested reader is referred to specialized texts [13, 14]. We confine ourselves to some general remarks. A very frequently used method of parallel computation is domain decomposition [15]. In this, the simulation box is divided into smaller sub-boxes, each of which is assigned to a processor. Each processor "owns" a sub-box in three-dimensional space and is responsible for updating the positions of all atoms within its sub-box. Atoms are reassigned to new processors as they move in and out of the sub-boxes. To compute the forces on its atoms, each processor must know the positions of atoms in its neighboring sub-boxes. This requires communication of information across the sub-box boundaries. If the sub-box edge length is sufficiently large in relation to the cut-off radius for nonbonded interactions, then this communication is not rate-limiting.

Generally, to quantify the parallelizability of a computational scheme, we can study the speedup factor S_p and the parallel efficiency E_p as functions of the number of processors p, on which the calculation is conducted. The speedup factor is defined by the equation:

$$S_p = \frac{t_1}{t_p} \tag{8.47}$$

where t_1 is the execution time required for the calculation to be conducted entirely on one processor (sequential run) and t_p is the execution time of the parallel scheme on p processors [13]. The parallel efficiency, on the other hand, is defined as:

$$E_p = \frac{S_p}{p} \tag{8.48}$$

For a well-designed parallel application, S_p must be close to p (linear scaling with the number of processors) and E_p must be close to unity. A crucial aspect in any parallel algorithm is the issue of load balance. This concerns the amount of work performed by each processor during the entire simulation; ideally, this should be the same for all processors.

Multiple time step algorithms along with domain decomposition are implemented very efficiently in public-domain molecular dynamics simulation packages, such as LAMMPS (large-scale atomic molecular massively parallel simulator) (http://lammps.sandia.gov/).

8.2.4
Molecular Dynamics of Rigid Linear Molecules

Consider the problem of MD simulation of a gas composed of linear molecules, such as N_2, sorbed within a zeolite, such as silicalite. We may choose to represent each sorbate molecule with a fully flexible model, in which the positions of its atoms, r_1 and r_2, are treated as six independent degrees of freedom (compare Figure 7.1). This representation necessitates the use of a bond stretching potential to preserve the correct shape of the molecule [compare Eq. (7.13)]. The bond length in each N_2 molecule will oscillate about an equilibrium value and the integration algorithm will have to be designed so as to keep track of these fast oscillations with satisfactory energy conservation. Alternatively, we may adopt a rigid model for each sorbate molecule. In our example this will entail fixing the bond length of each molecule at its equilibrium value and dropping the bond stretching contributions from the potential energy function. This has the practical advantage that the short time scales associated with vibrational motion are removed from the problem, allowing the use of a longer integration time step. In addition, there is a fundamental motivation for freezing fast vibrational degrees of freedom. The characteristic frequencies ν associated with the motion of these degrees of freedom are very high ($h\nu/k_B T > 1$), such that a quantum mechanical partition function is required for their correct description. On the other hand, an MD simulation, as usually practiced, is completely classical. It turns out that a rigid model may provide a better approximation to the full quantum mechanical partition function of a system possessing fast vibrational degrees of freedom than a classical flexible model [16].

We can integrate the equations of motion of a model with fixed bond lengths (in general, with holonomic constraints on the Cartesian coordinates of the atoms) through one of two alternative strategies. One strategy is to perform the integration in the atomic degrees of freedom (in our example, r_1 and r_2 for each molecule). These are no longer independent, since they have to conform to the holonomic constraints at all times. Constraint dynamics strategies for integrating the equations of motion in Cartesian coordinates subject to holonomic constraints are briefly discussed in Section 8.2.6. The second alternative strategy is to integrate the equations of motion in the generalized coordinates of the constituent molecules (in our example, r_{cm}, ψ_1, ψ_2 for each molecule, see Figure 7.1). These generalized coordinates are unconstrained, that is, independent of each other. For small sorbate molecules that do not possess torsional degrees of freedom, this second strategy of integrating the equations of motion in generalized coordinates is advantageous. In the rest of this subsection we will describe it briefly for the case of linear sorbate molecules. How to integrate the equations of motion for rigid nonlinear polyatomic molecules will be discussed in the next subsection.

Envision a set of rigid linear molecules in a microporous medium. These need not be diatomic as in our example of N_2 in silicalite. The motion of polyatomic linear molecules, such as CO_2 or acetylene, can be tracked with the same algorithms we discuss here.

Each sorbate molecule will be described via its center-of-mass coordinates, r_{cm}, and a unit vector:

$$\hat{e} = \frac{r_2 - r_1}{|r_2 - r_1|}$$

directed along the axis of the molecule. The vector $\hat{e}$ corresponds to two degrees of freedom; ψ_1 and ψ_2 are the polar and azimuthal angles defining $\hat{e}$. A convenient and efficient Verlet type algorithm for tracking the temporal evolution of r_{cm} and $\hat{e}$ is the LEN algorithm proposed by Fincham [17]. The rigid body motion is analyzed into translational motion of the center-of-mass (which changes r_{cm}) and rotational motion around the center-of-mass (which changes $\hat{e}$). The center-of-mass positions r_{cm} and velocities $v_{cm} = \dot{r}_{cm}$ are updated through the equations:

$$v_{cm}\left(t + \frac{1}{2}\delta t\right) = v_{cm}\left(t - \frac{1}{2}\delta t\right) + \delta t a(t)$$
$$r_{cm}(t + \delta t) = r_{cm}(t) + \delta t v_{cm}\left(t + \frac{1}{2}\delta t\right)$$
(8.49)

in which $a = F/M$, where $F = \sum_\alpha F_\alpha$ is the total force on the considered molecule and $M = \sum_\alpha m_\alpha$ is the total mass of the molecule. Here, F_α is the force on site α due to its interactions with other sites on other molecules and with the microporous medium and m_α is the mass of site α. Equation (8.49) consists of what is known as the Verlet leapfrog scheme, which can be shown to lead to the original Verlet scheme, Eq. (8.38), and to be mathematically equivalent to the velocity Verlet scheme, Eqs. (8.32)–(8.34).

To describe rotational motion around the center-of-mass, we need the total torque relative to the center-of-mass:

$$T = \sum_\alpha (r_\alpha - r_{cm}) \times F_\alpha \tag{8.50}$$

In the case of our linear molecule, the position vectors of the atoms are of the form $r_\alpha = r_{cm} + d_\alpha \hat{e}$, where d_α is an algebraic distance of atom α from the center-of-mass along the molecular axis. The torque can then be written as:

$$T = \hat{e} \times \sum_\alpha d_\alpha F_\alpha = \hat{e} \times g \tag{8.51}$$

The vector $g = \sum_\alpha d_\alpha F_\alpha$ is sometimes called the "turning force" on the molecule. The component of T along $\hat{e}$ is clearly zero; this is understandable, since the moment of inertia of the linear molecule for rotation around its axis is zero; T can thus be rewritten as:

$$T = T - (\hat{e} \cdot T)\hat{e} = (\hat{e} \cdot \hat{e})T - (\hat{e} \cdot T)\hat{e} = \hat{e} \times (T \times \hat{e}) = \hat{e} \times [(\hat{e} \times g) \times \hat{e}]$$
$$= \hat{e} \times [(\hat{e} \cdot \hat{e})g - (\hat{e} \cdot g)\hat{e}] = \hat{e} \times [g - (\hat{e} \cdot g)\hat{e}] = \hat{e} \times g_p$$
(8.52a)

where $g_p = g - (\hat{e} \cdot g)\hat{e}$.

The torque should equal the rate of change of angular momentum with respect to the center-of-mass. In the case of our linear molecule, we can write simply:

$$T = I\dot{\omega} \tag{8.52b}$$

with ω being the angular velocity vector, $\dot{\omega}$ the angular acceleration vector, and I the moment of inertia around any axis normal to $\hat{e}$ going through the molecular center-of-mass. The vectors T, g_p, ω, $\dot{\omega}$ are normal to $\hat{e}$ at all times, with T and $\dot{\omega}$ being collinear and T and g_p being normal to each other.

From Eqs. (8.52a,b), the equations of motion for the linear molecule can be written as [14]:

$$I\frac{d\boldsymbol{\omega}}{dt} = \hat{\boldsymbol{e}} \times \boldsymbol{g}_p$$
$$\frac{d\hat{\boldsymbol{e}}}{dt} = \boldsymbol{\omega} \times \hat{\boldsymbol{e}} \tag{8.53}$$

In Fincham's LEN algorithm, the vector $\boldsymbol{u} = d\hat{\boldsymbol{e}}/dt$ is used instead of $\boldsymbol{\omega}$ in the dynamical equations. By virtue of the second part of Eq. (8.53), one can show that $\boldsymbol{\omega} = \hat{\boldsymbol{e}} \times \boldsymbol{u}$. The equations of motion are transformed into:

$$\frac{d\boldsymbol{u}}{dt} = \frac{1}{I}\boldsymbol{g}_p$$
$$\frac{d\hat{\boldsymbol{e}}}{dt} = \boldsymbol{u} \tag{8.54}$$
$$\hat{\boldsymbol{e}}^2 = 1$$

where the third equation reminds us that the length of $\hat{\boldsymbol{e}}$ must equal 1 at all times. The following calculations are performed in a time step of the LEN algorithm [17, 18]:

- An auxiliary vector $\boldsymbol{e}'(t+\delta t)$ is found from $\hat{\boldsymbol{e}}(t)$, $\boldsymbol{u}(t-\frac{1}{2}\delta t)$, and $\boldsymbol{g}_p(t)$, which would result if one temporarily ignored the unit length constraint on $\hat{\boldsymbol{e}}$:

$$\boldsymbol{e}'(t+\delta t) = \hat{\boldsymbol{e}}(t) + \delta t \boldsymbol{u}(t-\tfrac{1}{2}\delta t) + (\delta t)^2 \frac{\boldsymbol{g}_p(t)}{I} \tag{8.55}$$

- A Lagrange multiplier, λ, is computed, associated with the constraint that the length of $\hat{\boldsymbol{e}}(t+\delta t)$ remain equal to unity:

$$\lambda(t+\delta t) = -\boldsymbol{e}'(t+\delta t) \cdot \hat{\boldsymbol{e}}(t) + \left\{[\boldsymbol{e}'(t+\delta t) \cdot \hat{\boldsymbol{e}}(t)]^2 - [\boldsymbol{e}'(t+\delta t)]^2 + 1\right\}^{\frac{1}{2}} \tag{8.56}$$

- The new unit bond vector $\hat{\boldsymbol{e}}(t+\delta t)$ is computed as:

$$\hat{\boldsymbol{e}}(t+\delta t) = \boldsymbol{e}'(t+\delta t) + \lambda(t+\delta t)\hat{\boldsymbol{e}}(t) \tag{8.57}$$

- The new midstep rate of change of the axis vector, $\boldsymbol{u}(t+\tfrac{1}{2}\delta t)$, is computed as:

$$\boldsymbol{u}(t+\tfrac{1}{2}\delta t) = [\hat{\boldsymbol{e}}(t+\delta t) - \hat{\boldsymbol{e}}(t)]/(\delta t) \tag{8.58}$$

- If desired, the rate of change of the axis vector at time t, $\boldsymbol{u}(t)$, is computed as:

$$\boldsymbol{u}(t) = \left[\boldsymbol{u}(t-\tfrac{1}{2}\delta t) + \boldsymbol{u}(t+\tfrac{1}{2}\delta t)\right]/2 \tag{8.59}$$

From that the angular velocity can be estimated as $\boldsymbol{\omega}(t) = \hat{\boldsymbol{e}}(t) \times \boldsymbol{u}(t)$ and used to check for energy conservation.

The LEN algorithm has shown remarkable stability and energy conservation when applied to sorbed linear molecules in zeolites [18].

8.2.5
Molecular Dynamics of Rigid Nonlinear Molecules: Quaternions

Rigid nonlinear molecules have three principal moments of inertia, which are readily obtainable by diagonalization of their moment of inertia tensor [6]. The same diagonalization process yields the principal axes of inertia of the molecule as eigenvectors. The principal axes define a convenient body-fixed frame of reference, which we will denote as (x^b, y^b, z^b). The principal moments of inertia for rotation around these axes will be denoted as I_{xx}, I_{yy}, I_{zz}, respectively.

At any time, the molecular orientation can be described by three angles, which specify the body-fixed frame of reference with respect to the laboratory frame of reference, (x, y, z). Very often, the Euler angles (φ, θ, ψ) are used for this purpose. These are defined as shown in Figure 8.3. If v^b is the representation of a vector in the

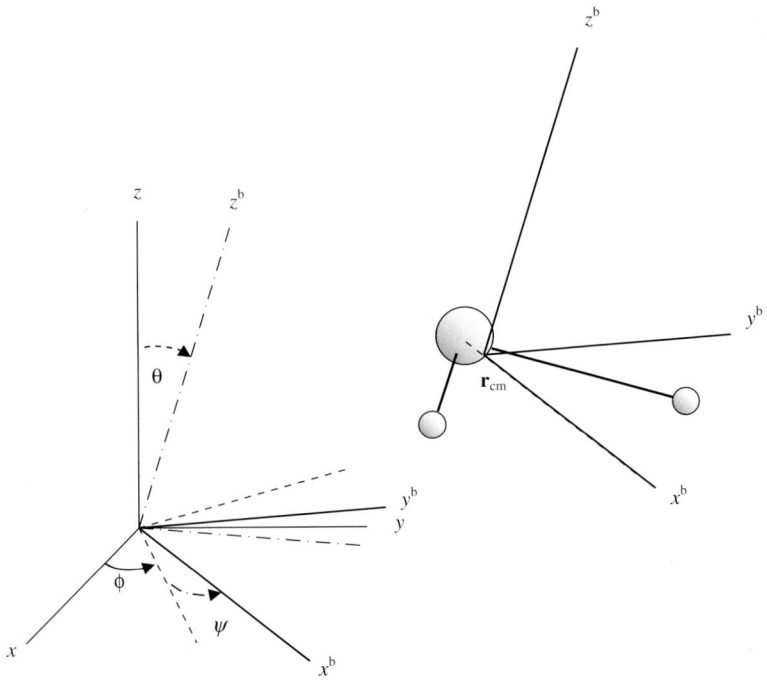

Figure 8.3 Definition of the eulerian angles for rigid body rotation of a nonlinear molecule (in this example, a water molecule). Laboratory frame of reference is denoted as (x, y, z). Body-fixed frame of reference is denoted as (x^b, y^b, z^b) and has its origin at the molecule center-of-mass, $\mathbf{r}_{cm}$. Axes x^b, y^b, z^b are chosen so as to coincide with the principal axes of rotation of the molecule. The laboratory frame (x, y, z), represented by thin solid lines, is brought onto the body-fixed frame (x^b, y^b, z^b) by (i) rotating around the z-axis by ϕ (broken lines); (ii) rotating around the new x-axis by θ (dashed-dotted lines); (iii) rotating around the new z-axis by ψ (bold solid lines).

body-fixed frame and $\mathbf{v}$ is the representation of the same vector in the laboratory frame, the two representations are related by:

$$\mathbf{v}^b = \mathbf{A}\mathbf{v} \tag{8.60}$$

Here, $\mathbf{A}$ is a transformation matrix that consists of the product of three rotation matrices:

$$\mathbf{A} = \begin{pmatrix} \cos\psi & \sin\psi & 0 \\ -\sin\psi & \cos\psi & 0 \\ 0 & 0 & 1 \end{pmatrix} \begin{pmatrix} 1 & 0 & 0 \\ 0 & \cos\theta & \sin\theta \\ 0 & -\sin\theta & \cos\theta \end{pmatrix} \begin{pmatrix} \cos\varphi & \sin\varphi & 0 \\ -\sin\varphi & \cos\varphi & 0 \\ 0 & 0 & 1 \end{pmatrix} \tag{8.61}$$

or:

$$\mathbf{A} = \begin{pmatrix} \cos\varphi\cos\psi - \sin\varphi\cos\theta\sin\psi & \sin\varphi\cos\psi + \cos\varphi\cos\theta\sin\psi & \sin\theta\sin\psi \\ -\cos\varphi\sin\psi - \sin\varphi\cos\theta\cos\psi & -\sin\varphi\sin\psi + \cos\varphi\cos\theta\cos\psi & \sin\theta\cos\psi \\ \sin\varphi\sin\theta & -\cos\varphi\sin\theta & \cos\theta \end{pmatrix}$$

$\mathbf{A}$ is orthogonal, that is, $\mathbf{A}^{-1} = \mathbf{A}^T$. Hence, the reverse transformation from the body-fixed frame to the laboratory frame representation of a vector can be accomplished via $\mathbf{v} = \mathbf{A}^T \mathbf{v}^b$.

An alternative way of describing rotations, which is particularly convenient in the numerical integration of the equations of motion, is through Hamilton's quaternions [6, 14, 19]. A quaternion is a vectorial quantity with four components, defined in terms of the Euler angles φ, θ, ψ as:

$$\begin{aligned} q_1 &= \sin(\theta/2)\cos[(\varphi-\psi)/2] \\ q_2 &= \sin(\theta/2)\sin[(\varphi-\psi)/2] \\ q_3 &= \cos(\theta/2)\sin[(\varphi+\psi)/2] \\ q_4 &= \cos(\theta/2)\cos[(\varphi+\psi)/2] \end{aligned} \tag{8.62a}$$

The four components satisfy $q_1^2 + q_2^2 + q_3^2 + q_4^2 = 1$. Only three of them are independent. The transformation matrix $\mathbf{A}$ can be written in terms of the components as:

$$\mathbf{A} = 2\begin{pmatrix} q_1^2 + q_4^2 - \frac{1}{2} & q_1 q_2 + q_3 q_4 & q_1 q_3 - q_2 q_4 \\ q_1 q_2 - q_3 q_4 & q_2^2 + q_4^2 - \frac{1}{2} & q_2 q_3 + q_1 q_4 \\ q_1 q_3 + q_2 q_4 & q_2 q_3 - q_1 q_4 & q_3^2 + q_4^2 - \frac{1}{2} \end{pmatrix} \tag{8.62b}$$

The angular velocity vector $\boldsymbol{\omega}^b$ in the body-fixed coordinate frame can be expressed simply in terms of the time derivatives of the components of the quaternions as:

$$\begin{pmatrix} \omega_x^b \\ \omega_y^b \\ \omega_z^b \\ 0 \end{pmatrix} = 2W \begin{pmatrix} \dot{q}_1 \\ \dot{q}_2 \\ \dot{q}_3 \\ \dot{q}_4 \end{pmatrix} \tag{8.63}$$

where W is the orthogonal matrix:

$$W = \begin{pmatrix} q_4 & q_3 & -q_2 & -q_1 \\ -q_3 & q_4 & q_1 & -q_2 \\ q_2 & -q_1 & q_4 & -q_3 \\ q_1 & q_2 & q_3 & q_4 \end{pmatrix} \tag{8.64}$$

The equations of motion describing rotation of the molecule dictate that the torque T, defined as in Eq. (8.50), equals the rate of change of angular momentum around the center-of-mass:

$$T = \dot{L} \tag{8.65}$$

These equations assume a particularly simple form when expressed in the body-fixed frame of principal axes of the molecule. The result is the Euler equations for rigid body rotation:

$$\begin{aligned} I_{xx}\dot{\omega}_x^b &= T_x^b + (I_{yy} - I_{zz})\omega_y^b \omega_z^b \\ I_{yy}\dot{\omega}_y^b &= T_y^b + (I_{zz} - I_{xx})\omega_z^b \omega_x^b \\ I_{zz}\dot{\omega}_z^b &= T_z^b + (I_{xx} - I_{yy})\omega_x^b \omega_y^b \end{aligned} \tag{8.66}$$

A Verlet leapfrog algorithm can be used for numerical integration of the equations of motion describing rotation of the rigid molecule around its center-of-mass, Eqs. (8.60)–(8.66), along with the Eq. (8.49) describing translation of the center-of-mass. The following calculations are involved in going from time t to time $t + dt$ [6, 20]:

- Compute the angular momentum $L(t)$ at time t from $L(t - \frac{1}{2}\delta t)$ and the torque $T(t)$, all expressed in the laboratory frame, via Eq. (8.65):

$$L(t) = L(t - \frac{1}{2}\delta t) + \frac{\delta t}{2} T(t) \tag{8.67}$$

- From the quaternion components $q = (q_1, q_2, q_3, q_4)$ at time t, form the rotation matrix $A(t) = A(q(t))$.

- Convert the angular momentum vector representation into the body-fixed frame via Eq. (8.60) and use it to compute the components of the angular velocity vector in the body-fixed frame:

$$\omega^b(t) = \begin{pmatrix} 1/I_{xx} & 0 & 0 \\ 0 & 1/I_{yy} & 0 \\ 0 & 0 & 1/I_{zz} \end{pmatrix} A(t) L(t) \qquad (8.68)$$

- Use Eq. (8.63), along with the calculated $\omega^b(t)$, to advance the quaternion components from time t to time $t + \delta t/2$:

$$q(t + \frac{1}{2}\delta t) = q(t) + \frac{\delta t}{2}\frac{1}{2} W^T(t) \begin{pmatrix} \omega^b(t) \\ 0 \end{pmatrix} \qquad (8.69)$$

- From the quaternion components $q = (q_1, q_2, q_3, q_4)$ at time $t + \delta t/2$ form the rotation matrix:

$$A(t + \frac{1}{2}\delta t) = A\left(q(t + \frac{1}{2}\delta t)\right)$$

- Advance the angular momentum in the laboratory frame from time $t - \delta t/2$ to time $t + \delta t/2$:

$$L(t + \frac{1}{2}\delta t) = L(t - \frac{1}{2}\delta t) + \delta t\, T(t) \qquad (8.70)$$

- Calculate the angular momentum vector in the body-fixed frame at time $t + \delta t/2$:

$$\omega^b(t + \frac{1}{2}\delta t) = \begin{pmatrix} 1/I_{xx} & 0 & 0 \\ 0 & 1/I_{yy} & 0 \\ 0 & 0 & 1/I_{zz} \end{pmatrix} A(t + \frac{1}{2}\delta t) L(t + \frac{1}{2}\delta t) \qquad (8.71)$$

- Use Eq. (8.63), along with the calculated $\omega^b(t + \delta t/2)$, to advance the quaternion components from time $t + \delta t/2$ to time t:

$$q(t + \delta t) = q(t + \frac{1}{2}\delta t) + \frac{\delta t}{2}\frac{1}{2} W^T(t + \frac{1}{2}\delta t) \begin{pmatrix} \omega^b(t + \frac{1}{2}\delta t) \\ 0 \end{pmatrix} \qquad (8.72)$$

- Move the linear velocity of the center-of-mass from time $t - \delta t/2$ to time $t + \delta t/2$, using the total force $F(t)$ on the molecule and its total mass M. Use the calculated center-of-mass velocity at $t + \delta t/2$ to advance the center-of-mass position from time $t - \delta t$ to time $t + \delta t$. These calculations are performed exactly as described by Eq. (8.49).
- From the quaternion components $q = (q_1, q_2, q_3, q_4)$ at time $t + \delta t$ form the rotation matrix $A(t + \delta t) = A(q(t + \delta t))$.

- From the position vectors d_α^b of all atoms α relative to the center-of-mass in the body-fixed frame of reference and the quaternion components at time $t + \delta t$, compute the position vectors of all atoms in the laboratory frame at time $t + \delta t$, using the inverse of Eq. (8.60):

$$r_\alpha(t + \delta t) = r_{cm}(t + \delta t) + A^T(t + \delta t) d_\alpha^b \tag{8.73}$$

- From the configuration at time $t + \delta t$, compute the forces $F_\alpha(t + \delta t)$ on all atoms and from those the total force $F(t + \delta t)$ and torque $T(t + \delta t)$ on the molecule.
- Advance the time by δt and loop back to perform the next step.

Center-of-mass velocities at time t are estimated from their values at half-steps obtained in the course of integrating Eq. (8.49) of translational motion for the center-of-mass as:

$$v_{cm}(t) = \frac{1}{2}\left[v_{cm}\left(t - \frac{1}{2}\delta t\right) + v_{cm}\left(t + \frac{1}{2}\delta t\right)\right]$$

An instantaneous translational temperature for a molecule can be obtained as $\frac{1}{3}M[v_{cm}(t)]^2/k_B$, and an instantaneous rotational temperature as:

$$\frac{1}{3}\left[I_{xx}\left(\omega_x^b\right)^2 + I_{yy}\left(\omega_y^b\right)^2 + I_{zz}\left(\omega_z^b\right)^2\right]/k_B$$

For a well-equilibrated simulation, these temperatures should fluctuate around a common constant value, while the total (kinetic and potential) energy should be constant.

8.2.6
Constraint Dynamics in Cartesian Coordinates

For large, flexible molecules possessing internal torsional degrees of freedom, such as the higher alkanes, a constant bond length representation is often adopted in MD simulations [10]. In the presence of torsional degrees of freedom that allow the shape of the molecule to change, however, integrating the dynamical equations in generalized coordinates becomes cumbersome. It is more straightforward and computationally more efficient to integrate the dynamical equations in the Cartesian coordinates of the interaction sites, while taking into account that these coordinates are not independent, but subject to holonomic constraints imposed by the constancy of bond lengths. The objective of constraint dynamics algorithms is to perform this integration.

To fix ideas, consider a water molecule being simulated as part of a zeolite–sorbate system with MD (Figure 8.4). For the description of the water molecules we adopt a model with constant bond lengths. This means that the Cartesian coordinate vectors

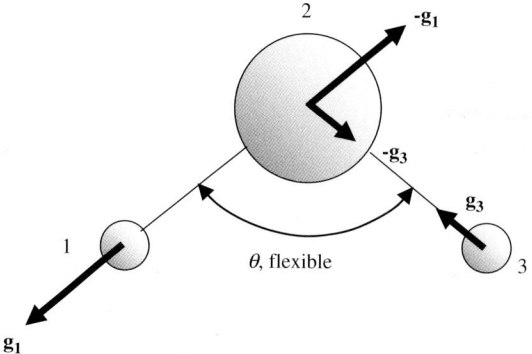

Figure 8.4 Schematic showing constraint forces (g_1, g_3) acting on the two hydrogen atoms (atoms 1, 3) and their reactions ($-g_1, -g_3$) acting on the oxygen atom (atom 2) of a water model with constant bond lengths. The bond angle θ in this model is assumed to be flexible, governed by a bending potential that is part of the potential energy function V of the system.

r_1, r_3 of the two hydrogen atoms and r_2 of the oxygen atom comprising the water molecule are subject to the following holonomic constraints at all times:

$$\chi_{12} = (r_1 - r_2)^2 - d_{12}^2 = 0$$
$$\chi_{23} = (r_2 - r_3)^2 - d_{23}^2 = 0 \tag{8.74}$$

with $d_{12} = d_{23} = 0.9572$ Å, equal to the equilibrium length of an O–H bond in water. On the other hand, the H–O–H bond angle θ is flexible in the model we invoke. This means that $V_\theta(\theta)$ of the general form of Eq. (7.13) is incorporated in the potential energy function V of the system, causing θ to fluctuate in the neighborhood of its equilibrium value $\theta_0 = 104.52$ Å. Apart from $V_\theta(\theta)$ for each water molecule present, the total potential energy function V of the system will include Lennard-Jones potentials between water oxygens and between water oxygens and oxygens on the zeolite framework, and Coulomb potentials between all pairs of partial charges in the system.

The potential energy function gives rise to "systematic" forces $F_1 = -\nabla_{r_1} V$, $F_2 = -\nabla_{r_2} V$, and $F_3 = -\nabla_{r_3} V$ on atoms 1, 2, 3, respectively. Additional forces are generated by the constraints. The holonomic constraint $\chi_{12} = 0$ will result in a constraint force g_1 directed along the bond connecting 1 with 2, acting on the hydrogen atom at r_1. The reaction of this force, $-g_1$, will be exerted on the oxygen at r_2 due to the same constraint. Similarly, the holonomic constraint $\chi_{23} = 0$ gives rise to constraint forces g_3 and $-g_3$ on the hydrogen atom at r_3 and on the oxygen atom at r_2, respectively. Each constraint can generate a force of any magnitude. The instantaneous magnitude of each constraint force is such as to ensure that atoms 1, 2, 3 continue moving without violating the constraints, Eq. (8.74). Whether the constraint forces will push or pull the atoms along their respective bonds will depend on the systematic forces experienced by the atoms at the considered instant. In the example of Figure 8.4, we have drawn force $\mathbf{g}_1$ as pushing atom 1 outwards along bond 21 and force $\mathbf{g}_3$ as pulling atom 3 inwards along bond 23. The equations of motion for the three atoms in Cartesian coordinates can be written as:

$$m_1\ddot{\mathbf{r}}_1 = \mathbf{F}_1 + \mathbf{g}_1 \tag{8.75}$$

$$m_2\ddot{\mathbf{r}}_2 = \mathbf{F}_2 - \mathbf{g}_1 - \mathbf{g}_3 \tag{8.76}$$

$$m_3\ddot{\mathbf{r}}_3 = \mathbf{F}_3 + \mathbf{g}_3 \tag{8.77}$$

with:

$$\mathbf{g}_1 = \lambda_{12}(\mathbf{r}_1 - \mathbf{r}_2) = \frac{1}{2}\lambda_{12}\nabla_{\mathbf{r}_1}\chi_{12} \tag{8.78}$$

$$\mathbf{g}_3 = \lambda_{23}(\mathbf{r}_2 - \mathbf{r}_3) = \frac{1}{2}\lambda_{23}\nabla_{\mathbf{r}_3}\chi_{23} \tag{8.79}$$

The scalar variables λ_{12} and λ_{23}, which determine the sign and magnitude of the constraint forces, are Lagrange multipliers associated with the constraints. Equations (8.75)–(8.79), along with Eq. (8.74), comprise a system of nine differential and eight algebraic equations in the unknown functions $\mathbf{r}_1(t), \mathbf{r}_2(t), \mathbf{r}_3(t), \mathbf{g}_1(t), \mathbf{g}_2(t), \lambda_{12}(t), \lambda_{23}(t)$, and also in the coordinates $\mathbf{r}_i(t)$ of all other sites of the system that interact with 1, 2, 3 and therefore enter the expressions for the systematic forces $\mathbf{F}_1, \mathbf{F}_2$, and $\mathbf{F}_3$. Constraint dynamics call for the simultaneous numerical solution of all these differential and algebraic equations for all molecules present.

A general algorithm for constraint MD has been proposed by Ryckaert *et al.* [21]. We wish to integrate the dynamical equations of motion of a system of N interaction sites:

$$m_i\ddot{\mathbf{r}}_i = \mathbf{F}_i + \mathbf{g}_i \quad (i = 1, 2, \ldots, N) \tag{8.80}$$

subject to the n_c holonomic constraints:

$$\chi_{ij} \equiv r_{ji}^2 - d_{ij}^2 = 0 \tag{8.81}$$

where we have set $\mathbf{r}_{ji} = \mathbf{r}_i - \mathbf{r}_j$ and the constraint force $\mathbf{g}_i$ on each site is obtained as:

$$\mathbf{g}_i = \frac{1}{2}\sum_j \lambda_{ij}\nabla_{\mathbf{r}_i}\chi_{ij} = \sum_j \lambda_{ij}\mathbf{r}_{ji} \tag{8.82}$$

To integrate the equations of motion between times t and $t + \delta t$, the general algorithm first estimates new positions $\mathbf{r}'_i(t + \delta t)$ for the interaction sites based on a Verlet scheme, momentarily ignoring the constraint forces:

$$\mathbf{r}'_i(t + \delta t) = 2\mathbf{r}_i(t) - \mathbf{r}_i(t - \delta t) + \frac{(\delta t)^2}{m_i}\mathbf{F}_i(t) \tag{8.83}$$

The actual positions of sites at time $t + \delta t$ will be given in terms of the still undetermined Lagrange multipliers λ_{ij} associated with the constraints as:

$$r_i(t+\delta t; \{\lambda_{ij}\}) = r'_i(t+\delta t) + \frac{(\delta t)^2}{m_i} g_i(t; \{\lambda_{ij}\}) = r'_i(t+\delta t) - \frac{(\delta t)^2}{m_i} \sum_j \lambda_{ij} r_{ij}(t) \quad (8.84)$$

Substituting expressions in Eq. (8.84) into the constraint Eq. (8.81), one obtains a system of n_c equations in the n_c unknowns $\{\lambda_{ij}\}$ of the form:

$$[r_i(t+\delta t; \{\lambda_{ij}\}) - r_j(t+\delta t; \{\lambda_{ij}\})]^2 - d_{ij}^2 = 0 \quad (8.85)$$

Given that $r_i(t+\delta t; \{\lambda_{ij}\})$, are linear in the $\{\lambda_{ij}\}$ by Eq. (8.85), the system of Eq. (8.85) is second order with respect to the $\{\lambda_{ij}\}$. The system can be solved iteratively via a Newton–Raphson method. If this strategy is followed, each iteration will require inversion of a $n_c \times n_c$ Jacobian matrix.

From the converged solution for the $\{\lambda_{ij}\}$, the new positions of all sites at time $t + \delta t$ can be obtained via Eq. (8.85).

For large molecules with many constraints, the simultaneous numerical solution of the second-order system of Eq. (8.85) at each time step becomes computationally demanding. In place of this, Ryckaert et al. [21] have proposed an iterative scheme that considers the constraints Eq. (8.84) one by one, cyclically, each time adjusting site positions to satisfy the constraint equation under consideration. This procedure continues until all constraint equations are satisfied within a pre-specified tolerance. This procedure, known as SHAKE, is widely used in public-domain simulation packages for constraint dynamics of chain molecules. A variation of SHAKE, which employs the velocity Verlet algorithm, is the RATTLE algorithm proposed by H.C. Andersen [22].

An alternative constraint dynamics method that obviates the need to solve the second-order system of Eq. (8.85) has been proposed by Edberg, Evans, and Morriss [23]. The basic idea behind this method is to differentiate each of the holonomic constraints, Eq. (8.74), twice with respect to time:

$$r_{ji}^2 - d_{ij}^2 = 0 \Rightarrow 2 r_{ji} \cdot \dot{r}_{ji} = 0 \Rightarrow r_{ji} \cdot \ddot{r}_{ji} + (\dot{r}_{ji})^2 = 0 \quad (8.86)$$

The set of constraints Eq. (8.86) must be solved along with the equations of motion obtained by combining Eqs. (8.80) and (8.82):

$$m_i \ddot{r}_i = F_i + \sum_j \lambda_{ij} r_{ji} \quad (i = 1, 2, \ldots, N) \quad (8.87)$$

Clearly, upon substituting Eq. (8.87) into Eq. (8.87), a system of n_c first-order equations in the n_c constraint variables $\{\lambda_{ij}\}$ is obtained. This system is solved directly, by matrix inversion, at each time step to obtain the constraint forces from the positions and velocities of the interaction sites and from the systematic forces on the sites. No iterations are required for determining $\{\lambda_{ij}\}$. The equations of motion,

Eq. (8.87), are then integrated, usually using a Gear predictor-corrector algorithm [6]. In practice, numerical errors cause gradual distortion of the molecules away from the geometry dictated by the constraint Eq. (8.74) and gradual deviation of the site velocities from the orthogonality condition $r_{ji} \cdot \dot{r}_{ji} = 0$ that stems from the constraints, as seen in Eq. (8.86). This problem is corrected by periodically minimizing the functions $\Phi = \sum_{ij} \left(r_{ij}^2 - d_{ij}^2\right)^2$ and $\Psi = \sum_{ij} \left(r_{ji} \cdot \dot{r}_{ji}\right)^2$ with respect to site positions and velocities, respectively. This minimization is carried out infrequently (e.g., every 100 integration time steps) and thus does not significantly augment the computational requirements of the algorithm.

8.2.7
Extended Ensemble Molecular Dynamics

So far, we have outlined algorithms for conducting molecular dynamics simulations in the microcanonical (NVE) ensemble, for microporous solid–sorbed fluid systems that evolve under the influence of interactions among their constituent particles. The temperature in such a simulation is obtained from the time averaged kinetic energy $\mathcal{K}$ at equilibrium:

$$T = \frac{2}{(3N-n_c)k_B} \langle \mathcal{K} \rangle = \frac{1}{(3N-n_c)k_B} \left\langle \sum_{i=1}^{N} \frac{p_i^2}{m_i} \right\rangle \tag{8.88}$$

with m_i, and p_i being the masses and momentum vectors, respectively, of the N atoms in the system and n_c the total number of constraints on the system. In the absence of holonomic constraints on the atomic coordinates, $n_c = 3$ when the integration algorithm maintains the total momentum $\sum_{i=1}^{N} p_i$ equal to zero. Similarly, the stress tensor on the solid–fluid system can be computed from atomic momenta, intermolecular forces, and distances through the virial theorem [6, 24].

In practice, it is often more convenient to study the dynamics of the system of interest in a statistical ensemble other than the microcanonical. For example, simulations in the canonical (NVT) ensemble may be more convenient from the point of view of comparing against experimental measurements. In response to this need, algorithms for conducting equilibrium MD simulations in statistical ensembles other than the NVE have been proposed. A variety of such algorithms are available [6]. Here we will briefly discuss one general category of such algorithms, the so-called extended ensemble methods. The basic idea in extended ensemble MD is to introduce one or more extra configurational degrees of freedom (in addition to atomic coordinates), to describe the dynamical behavior of macroscopic characteristics of the system, such as temperature or stress, which we wish to control. Physically, each extra degree of freedom represents a "reservoir," with which the original molecular system can interact. Each additional degree of freedom is represented by a "coordinate;" associated with that coordinate is a "velocity" (rate of change of the coordinate with time) and a "mass" (inertial variable, which controls fluctuations in the coordinate). Appropriate expressions are constructed for the potential and kinetic energy contributed by each additional degree of freedom. These, along with the kinetic and potential energy of the original system, are used to write a Lagrangian for the extended system. From the latter Lagrangian, equations of motion are extracted for all degrees of freedom, that is, both those of the original system and the additional ones representing the reservoir. These equations of motion are generally different from those obeyed by the original system in the microcanonical ensemble. They ensure that, at equilibrium, the degrees of freedom of the original system sample the probability distribution of the ensemble of interest.

To obtain a clearer idea of how extended system methods work, let us consider the example of simulating methane in silicalite in the canonical (NVT) ensemble. We will invoke a united-atom representation for each methane molecule i, describing it by the vector of Cartesian coordinates of its center-of-mass, r_i. An infinitely stiff model will be invoked for the zeolite framework, allowing pretabulation of the potential energy $\mathcal{V}_{zs}$ experienced by sorbate molecule i due to its interactions with the framework as a function of r_i. In addition, the energy $\mathcal{V}_s$ of sorbate–sorbate interactions will be represented as a sum of Lennard-Jones interactions between pairs of sorbate molecules. As in Chapter 7, we will use the symbol $\mathcal{V}(r_1, r_2, \ldots, r_N)$ to denote the potential energy of N methane molecules in a primary simulation box of volume V, containing a fixed number of unit cells of the zeolite crystal. The total kinetic energy of the sorbed molecules will be:

$$\mathcal{K} = \sum_{i=1}^{N} \frac{1}{2} m v_i^2 \tag{8.89}$$

with v_i and m being the velocity vector and the mass, respectively, of sorbate molecule i. Following Nosé [25] we introduce an additional degree of freedom, s, which will allow flow of energy between our original model system (sorbate molecules under the influence of the zeolite field) and a "bath" or "heat reservoir," thereby controlling the total kinetic energy of the sorbate molecules like a "thermostat." The heat reservoir will be at a fixed temperature T_{eq}.

The additional degree of freedom s is introduced as a scaling factor that can dilate or contract time, thereby modulating the molecular velocities. More specifically, in parallel with the "real" time scale, t, we introduce a "simulation time" scale, t', such that:

$$dt' = s \, dt \tag{8.90}$$

The scaling factor s is itself a function of time t. In a well-designed simulation it will fluctuate around 1. The real velocity of molecule i, v_i, can be written in terms of s and the rate of change $\dot{r}_i$ of the molecule's position vector with respect to simulation time as:

$$v_i = s\dot{r}_i \tag{8.91}$$

Dots over symbols will be used to denote derivatives with respect to simulation time.

The extra degree of freedom s is assigned a potential energy contribution:

$$\mathcal{V}_{bath}(s) = (f+1)k_B T_{eq} \ln s \tag{8.92}$$

where $f = 3N$ is the number of degrees of freedom of the original system. It is also assigned a kinetic energy contribution:

$$\mathcal{K}_{\text{bath}}(\dot{s}) = \frac{1}{2} Q \dot{s}^2 \qquad (8.93)$$

where Q is an inertial parameter controlling the rate of fluctuations in s.

The Lagrangian of the extended system (Chapter 7) is:

$$\mathcal{L}' = \mathcal{K}' - \mathcal{V}' = \mathcal{K} + \mathcal{K}_{\text{bath}} - \mathcal{V} - \mathcal{V}_{\text{bath}} = \frac{1}{2} \sum_{i=1}^{N} m s^2 \dot{r}_i^2 + \frac{1}{2} Q \dot{s}^2 - \mathcal{V} - (f+1) k_B T_{\text{eq}} \ln s \qquad (8.94)$$

The equations of motion for the extended system are derived from Eq. (7.1), with q being the column vector $(r_1, r_2, \ldots, r_N, s)$. They read:

$$\ddot{r}_i = \frac{F_i}{m s^2} - 2 \frac{\dot{s} \dot{r}_i}{s}$$

$$Q \ddot{s} = \sum_{i=1}^{3N} m s \dot{r}_i^2 - (f+1) \frac{k_B}{s} T_{\text{eq}} \qquad (8.95)$$

with $F_i = -\nabla_{r_i} \mathcal{V}$ being the systematic force felt by molecule i. Coupling with the heat reservoir results in the appearance of a friction-like term in the equations of motion of the molecules [second term on the right-hand side of the first of Eq. (8.95)], which is proportional to the velocity. The acceleration in s is driven by the difference between the instantaneous kinetic energy of the molecules and the value of the total kinetic energy expected from the equipartition theorem at temperature T_{eq}.

Generalized momenta for the molecules and for the extra degree of freedom s, with respect to simulation time, are defined based on the Lagrangian (Chapter 7) as:

$$p_i = \frac{\partial \mathcal{L}'}{\partial \dot{r}_i} = m s^2 \dot{r}_i$$

$$p_{\text{bath}} = \frac{\partial \mathcal{L}'}{\partial \dot{s}} = Q \dot{s} \qquad (8.96)$$

For the extended system, the Hamiltonian:

$$\mathcal{H}' = \sum_{i=1}^{N} \dot{r}_i \cdot p_i + \dot{s} p_{\text{bath}} - \mathcal{L}' = \frac{1}{2} \sum_{i=1}^{N} m s^2 \dot{r}_i^2 + \frac{1}{2} Q \dot{s}^2 + \mathcal{V} + (f+1) k_B T_{\text{eq}} \ln s \qquad (8.97)$$

will be an invariant of the equations of motion. The constancy of $\mathcal{H}'$ can serve as a criterion for accuracy of the numerical integration of the equations of motion.

The partition function of the extended system will be:

$$Q_{NVE'} = \frac{1}{N! h^{f+1}} \int dp_s \int d^f p \int ds \int d^f r \delta \left(\sum_i \frac{p_i^2}{2 m_i s^2} + \frac{p_s^2}{2Q} + \mathcal{V}(r) + (f+1) k_B T_{\text{eq}} \ln s - E' \right)$$

where the delta function within the integral expresses the constancy of $\mathcal{H}'$, defined by Eq. (8.97). Symbolizing by:

$$p'_i = \frac{p_i}{s} = ms\dot{r}_i = mv_i \tag{8.98}$$

the momentum of molecule i with respect to real simulation time, the partition function can be re-written as:

$$Q_{NVE'} = \frac{1}{(f+1)h}\left(\frac{2\pi Q}{k_B T_{eq}}\right)^{1/2} \exp\left(\frac{E'}{k_B T_{eq}}\right) \frac{1}{N!h^f} \int d^f p' \int d^f r \exp\left[-\mathcal{H}(p',r)\frac{1}{k_B T_{eq}}\right] \tag{8.99}$$

where $\mathcal{H}(p',r)$ is now the Hamiltonian of the original system (in our case, the set of N sorbate molecules in the zeolite). From the latter equation one can see that the Nosé MD method samples the phase space of the original system according to the probability distribution of the canonical ensemble at temperature T_{eq}. In particular, the instantaneous temperature of the system $T = \left(\sum_{i=1}^{N} mv_i^2\right)/(fk_B)$ will have a mean value $\langle T \rangle = T_{eq}$ and a variance $\left\langle (T-T_{eq})^2 \right\rangle = 2T_{eq}^2/f$, which goes to zero in the limit of large system sizes. The temporal evolution of the system in the canonical ensemble can be tracked through numerical integration of the equations of motion Eq. (8.95), taking into account Eq. (8.90) in order to convert simulation time into real time.

Care should be taken in selecting the inertial parameter Q to be used within an NVT MD simulation. For large values of Q, energy exchange between the system and the heat reservoir is slow, and in the limit $Q \to \infty$ microcanonical (NVE) dynamics is recovered. Small values of Q, on the other hand, bring about very frequent energy exchange between the system and the heat reservoir. While the model system at equilibrium adopts a correct canonical ensemble distribution, the dynamics may be perturbed by the frequent exchange with the reservoir. If the dynamics is of interest, it is always a good idea to make sure that the velocity autocorrelation functions and diffusivities obtained by NVT MD are indistinguishable from those obtained from NVE MD at the same thermodynamic state point. Nosé has given an estimate of the period of oscillations in the reservoir degree of freedom s as $t_0 = 2\pi\left[Q\langle s^2\rangle/(2fk_B T_{eq})\right]^{1/2}$ and recommends choosing Q such that t_0 is commensurate with the correlation time of the molecular velocity in the system under study.

Today, a large variety of extended ensemble techniques are available for performing simulations in ensembles other than the microcanonical. In 1980 H.C. Andersen [26] introduced a method for performing MD under constant pressure and enthalpy. Parrinello and Rahman [27] extended this idea to perform MD simulations under constant stress tensor, a method of particular value in exploring phase transformations in solids. When used in connection with reliable flexible models for zeolite frameworks, this method would be useful for predicting sorbate-induced transformations occurring in zeolite crystals (e.g., upon sorption of aromatic molecules in MFI), a calculation that does not seem to have been undertaken yet. S. Nosé [28] showed how to combine constant temperature and constant pressure extended-ensemble methods to perform isothermal-isobaric MD simulations, while W.G. Hoover reformulated the Nosé algorithm in such a way as to get rid of the distinction between simulation time and real time, allowing the use of a single "real" time clock [29]. A particularly original and useful form of extended ensemble method is the method developed by R. Car and M. Parrinello [30] (Car–Parrinello molecular dynamics, CPMD, or *ab initio* MD) for solving the ground-state electronic structure problem with density functional theory (DFT), while tracking the motion of the nuclei subject to the interactions generated by the electrons. In this algorithm, Kohn–Sham electronic orbitals are introduced as additional degrees of freedom. The method has the tremendous advantage that it does not require external specification of the potential energy as a function of the nuclear degrees of freedom, since this is computed "on the fly" by electronic DFT. CPMD can handle systems where chemical bonds or hydrogen bonds are broken and formed in the course of the simulation. Despite its heavy computational requirements, CPMD has been applied to several interesting problems involving zeolites.

8.2.8
Example Application of Equilibrium MD to Pure Sorbates

In recent years, equilibrium MD simulations have been applied to a wide variety of microporous materials and sorbates. Insights gained from MD simulations in conjunction with microscopic and macroscopic diffusion measurements are discussed in Chapters 15–20. As an illustration we present here results for the transport D and corrected D_0 diffusivities in silicalite (Chapter 18), as obtained from (i) equilibrium MD and GCMC (grand canonical Monte Carlo) simulations [Eqs. (8.19) and (8.20)]; (ii) coherent QENS and measured sorption isotherms [Section 10.2 and Eq. (8.19)]. The sorbates studied are pure N_2 at 200 K and pure CO_2 at 300 K, over a range of loadings [31]. Equilibrium MD simulations were carried out in the microcanonical (*NVE*) ensemble using Fincham's LEN algorithm (Section 8.2.4) to integrate the equations of motion of the rigid linear sorbates in generalized coordinates. The model system consisted of 27 unit cells of silicalite in a rigid framework representation. Each run encompassed an equilibration phase of 100 ps, followed by a production run of up to 20 ns. (Long production runs were necessary to ensure good statistics for D_0 at high occupancies.) D_0 values were computed from the MD simulations through Eq. (8.20). The thermodynamic correction factors $\Gamma = (\partial \ln f / \partial \ln q)_T$ needed to convert D_0 into transport diffusivities D through Eq. (8.19) were obtained from isotherms $f(q)$ computed from GCMC simulation (Section 7.2.3).

Figure 8.5 shows results from the simulations. MD predicts the loading dependence of D and D_0 in very good agreement with QENS. Small quantitative differences between predicted and measured values of the diffusivities are probably caused by inadequacies of the force field employed, which was not adjusted to experiment in any way. For both sorbates, D rises strongly with increasing loading. This is mainly a result of the thermodynamic correction factor Γ [compare Eq. (8.19)], which rises as

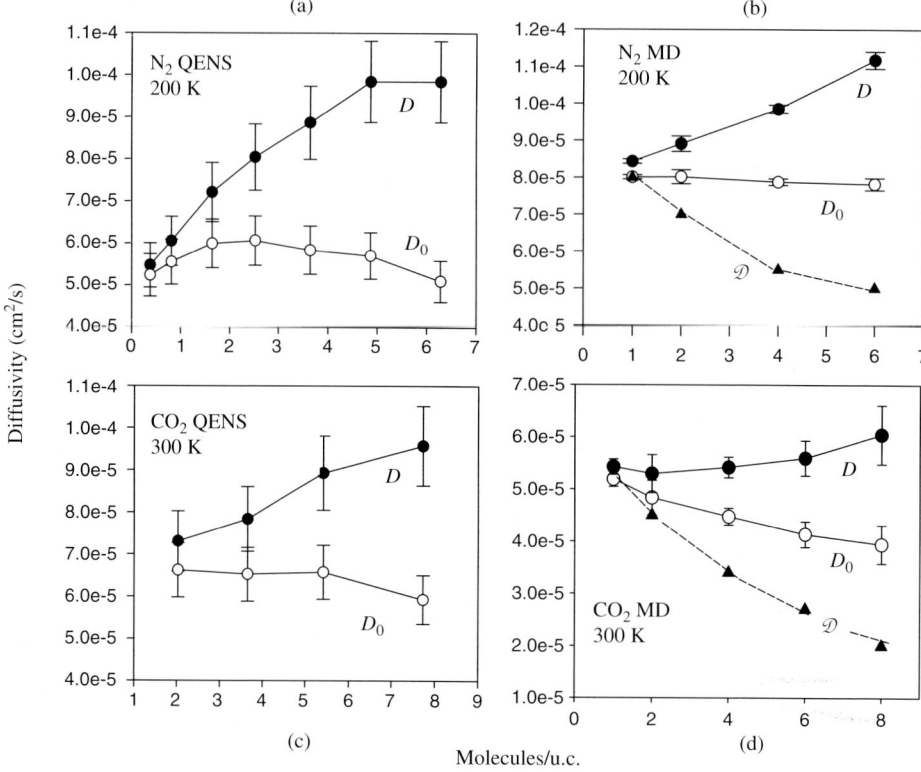

Figure 8.5 Transport D and corrected D_0 diffusivities for N_2 in silicalite at 200 K (a, b) and for CO_2 in silicalite at 300 K (c, d) as functions of loading, as obtained from coherent QENS measurements (a, c) and from equilibrium MD simulations (b, d) (With permission from [31]). Self-diffusivities $\mathcal{D}$ computed from the MD simulations are also shown.

the isotherm flattens out at high loadings. The corrected diffusivity D_0 is quite insensitive to loading in the case of N_2; it exhibits a weak maximum according to the QENS measurements. In the case of CO_2, on the other hand, D_0 exhibits a decreasing trend with increasing loading. This difference in the loading dependence of D_0 between N_2 and CO_2 can be explained in terms of the sorbate–sorbate interactions, which are considerably stronger (more attractive) in the case of CO_2. The role of sorbate–sorbate interactions in the loading dependence of D_0 can be rationalized [31] by invoking the simple model of Reed and Ehrlich (Section 4.3.2.2) [32]. For comparison, the self-diffusivity $\mathcal{D}$ is also shown as a function of loading, as obtained from the MD simulations, in Figure 8.5. It decreases with increasing loading for both sorbates. Computed results follow the inequality (3.52). In the limit of very low occupancies, D, D_0, and $\mathcal{D}$ for each sorbate converge to a common value, as theoretically expected [compare Eq. (8.22)]. Activation energies for low-occupancy diffusion, estimated by conducting the simulations at three different temperatures, were 2.7 and 5.1 kJ mol^{-1} for N_2 and CO_2, respectively, which are in excellent agreement with QENS measurements.

8.2.9
Example Applications of MD to Mixed Sorbates

Because of its obvious importance in catalytic and separation applications, the diffusion of fluid mixtures in zeolites has been investigated considerably with molecular simulations in recent years. An early MD study of self-diffusion in a binary mixture, which was complemented by QENS measurements, was performed by Gergidis *et al.* [33]. The study focused on mixtures of methane and deuterated *n*-butane in silicalite at various compositions and temperatures. Perdeuterated *n*-butane (99.7% isotopic enrichment) was used in the QENS to turn off the signal due to the *n*-butane and thus probe the self-diffusive motion of methane through the incoherent scattering from its hydrogen atoms.

As with all small linear alkanes [34], both butane and *n*-hexane prefer to reside in the interiors of straight and sinusoidal channel segments in silicalite and to avoid the more open channel intersections, where the dispersive attraction energy they can develop with the framework is weaker. At high occupancies, a competition between the two sorbates for the more favorable channel interiors will ensue. Butane, having a stronger interaction with the zeolite (heat of adsorption 47 versus 20 kJ mol^{-1} for methane), emerges as the winner from this competition. It occupies preferentially the channel interiors, forcing methane molecules into the less favorable channel intersections. This is shown characteristically in the three-dimensional contour plots of Figure 8.6, obtained from MD.

Since *n*-butane moves considerably more slowly than methane and no two molecules can bypass each other in the silicalite channels, it is expected that *n*-butane molecules will act as effective barriers to the motion of co-adsorbed methane molecules.

Figure 8.7 displays results for $\mathcal{D}$ from a series of QENS measurements and MD simulations on methane–perdeuterated *n*-butane mixtures co-adsorbed in silicalite at 200 K. The CH_4 occupancy is kept constant at 4 molecules per unit cell, while that of C_4D_{10} is varied systematically from 0 to 7.5 molecules per unit cell. Results from the measurement and from the simulation are within experimental error. MD estimates of $\mathcal{D}$ are systematically higher; this is at least partly because a perfect silicalite crystal was considered in the simulations, while experiments were conducted with a ZSM-5 sample with a Si/Al ratio of 36 [33]. A significant reduction of CH_4 self-diffusivity is observed as C_4D_{10} is increased, $\mathcal{D}_{CH_4}$ at 7.5 C_4D_{10} molecules per unit cell being by a factor of 20 lower than in pure CH_4 sorbate. The dependence of $\mathcal{D}_{CH_4}$ on $q_{C_4D_{10}}$ is concave downwards, as might be anticipated from Figure 8.6: at low $q_{C_4D_{10}}$, CH_4 molecules can find their way around co-adsorbed C_4D_{10} molecules in the maze of three-dimensionally connected channels of silicalite. As almost all channel interiors become filled with C_4D_{10}, however, a CH_4 will have to wait for a C_4D_{10} to get out of the way before it can execute significant translational motion in the zeolite (there are eight channel segments per unit cell in silicalite). At the highest loadings examined,

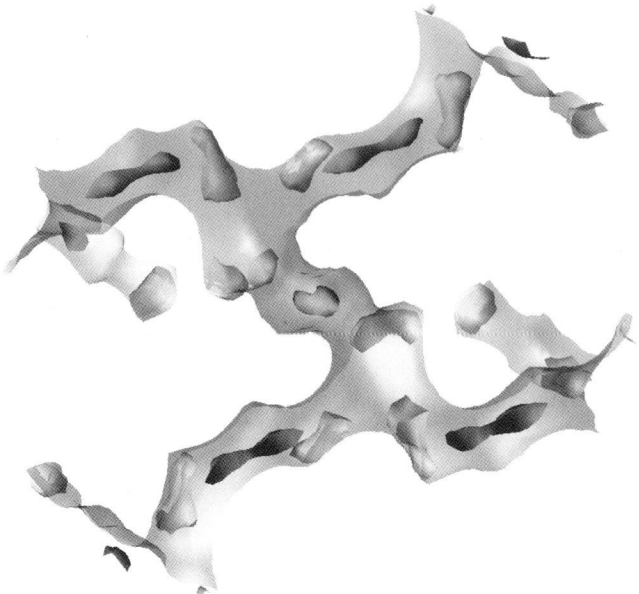

Figure 8.6 Center-of-mass distributions for methane (lighter-colored clouds) and n-butane (darker-colored clouds) co-adsorbed in silicalite, as obtained from equilibrium MD simulations at 300 K. The loading is four molecules per unit cell of methane and nine molecules per unit cell of n-butane. The distributions are shown over half a unit cell, the other half being symmetric with respect to the plane normal to the straight channels and containing the sinusoidal channel axes. The colored regions enclose the spaces where the centers of mass of the sorbed methane or butane atoms spend 50% of their time with highest probability. The interiors of the walls of the zeolite pores have been outlined with light gray, for reference.

the self-diffusivity of CH_4 becomes comparable to that of C_4D_{10}. Self-diffusivities for C_4D_{10} have been extracted from the MD simulations and found to exhibit a more modest decrease with increasing occupancies [33, 35]. At the lowest occupancies, where $\mathcal{D}$ for the C_4D_{10} can be extracted from the QENS measurements, excellent agreement between QENS and MD estimates of $\mathcal{D}_{C_4D_{10}}$ was found [33].

For practical design calculations, it is important to be able to extract the coefficients appearing in the Maxwell–Stefan formulation of multicomponent diffusion in microporous solids [compare Eq. (8.24)] from molecular simulations. Significant work in this direction has been performed in the group of R. Krishna at the University of Amsterdam.

Starting from the Maxwell–Stefan formulation of Eq. (8.24) for a binary mixture sorbed in a microporous medium, Krishna [36] developed an approximate approach for expressing the self-diffusivities in a binary mixture in terms of the diffusivities D_{0i} and $Đ_{ij}$ ($Đ_i$ and $Đ_{ij}^*$ in the notation of Krishna and van Baten [5]) appearing in the Maxwell–Stefan formulation, Eq. (8.24). D_{01} and D_{02} were assumed to vary with occupancy according to the reasonable dependence:

$$D_{0i} = D_{0i}(0)(1-\theta_1-\theta_2) \tag{8.100}$$

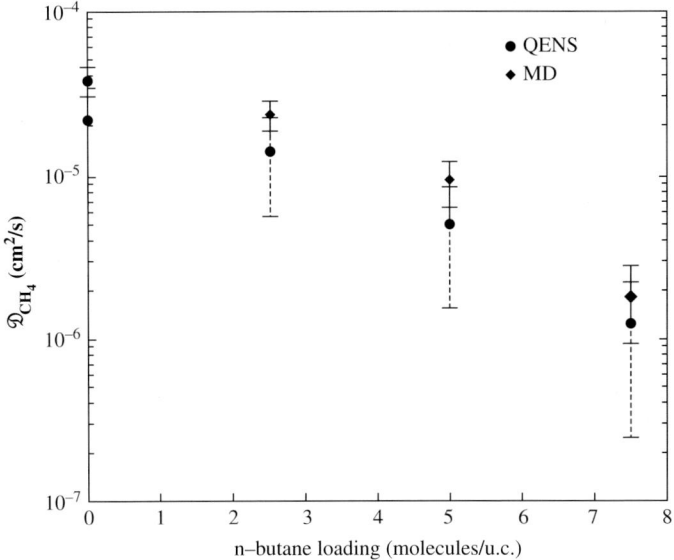

Figure 8.7 Orientationally averaged self-diffusivity of methane in mixtures of methane with perdeuterated *n*-butane co-adsorbed in silicalite at 200 K, shown as a function of the *n*-butane loading. The methane loading is constant at four molecules per unit cell. Results from MD simulations and QENS measurements shown as diamonds and circles, respectively. With permission from [33].

while $Ð_{ij}$ was expressed as a function of D_{01}, D_{02}, and the composition of the sorbed phase. The reader is reminded that $\theta_i = q_i/q_{i,\text{sat}}$ are the fractional occupancies of the two components, in relation to their respective saturation concentrations at the prevailing temperature.

Krishna [36] applied this approximate approach to the simulation data of Gergidis *et al.* [33, 35] on methane and butane in silicalite. With components 1 and 2 standing for methane and *n*-butane, respectively, he found that the simulation results could be described reasonably well using saturation loadings of 22 molecules per unit cell (methane) and 12 molecules per unit cell (*n*-butane). The parameters needed were $D_{01}(0) = 11 \times 10^{-9}\,\text{m}^2\,\text{s}^{-1}$ and $D_{02}(0) = 5 \times 10^{-9}\,\text{m}^2\,\text{s}^{-1}$ at 300 K, and $D_{01}(0) = 4 \times 10^{-9}\,\text{m}^2\,\text{s}^{-1}$ and $D_{02}(0) = 1.5 \times 10^{-9}\,\text{m}^2\,\text{s}^{-1}$ at 200 K. Figure 8.8 provides a comparison between simulation results for the self-diffusivity and Krishna's Maxwell–Stefan formulation [36].

A closer look at the Maxwell–Stefan formulation in the light of non-equilibrium molecular dynamics simulations (Section 8.3) was presented by Chempath *et al.* [37], using short alkanes and CF_4 in faujasite as a test case. For tracer diffusion of a single component *i* in a microporous medium, application of Eq. (8.24) with 1 and 2 standing, respectively, for the labeled and unlabeled molecules of type *i*, $q_{1,\text{sat}} = q_{2,\text{sat}}$, $\theta_2 = \theta - \theta_1$ and $J_1 + J_2 = 0$ (no net flux) leads to the relation (see section 3.3.3, Eq. 3.51):

$$\frac{1}{Ð_i} = \frac{1}{D_{0i}} + \frac{\theta}{Ð_{ii}} \qquad (8.101)$$

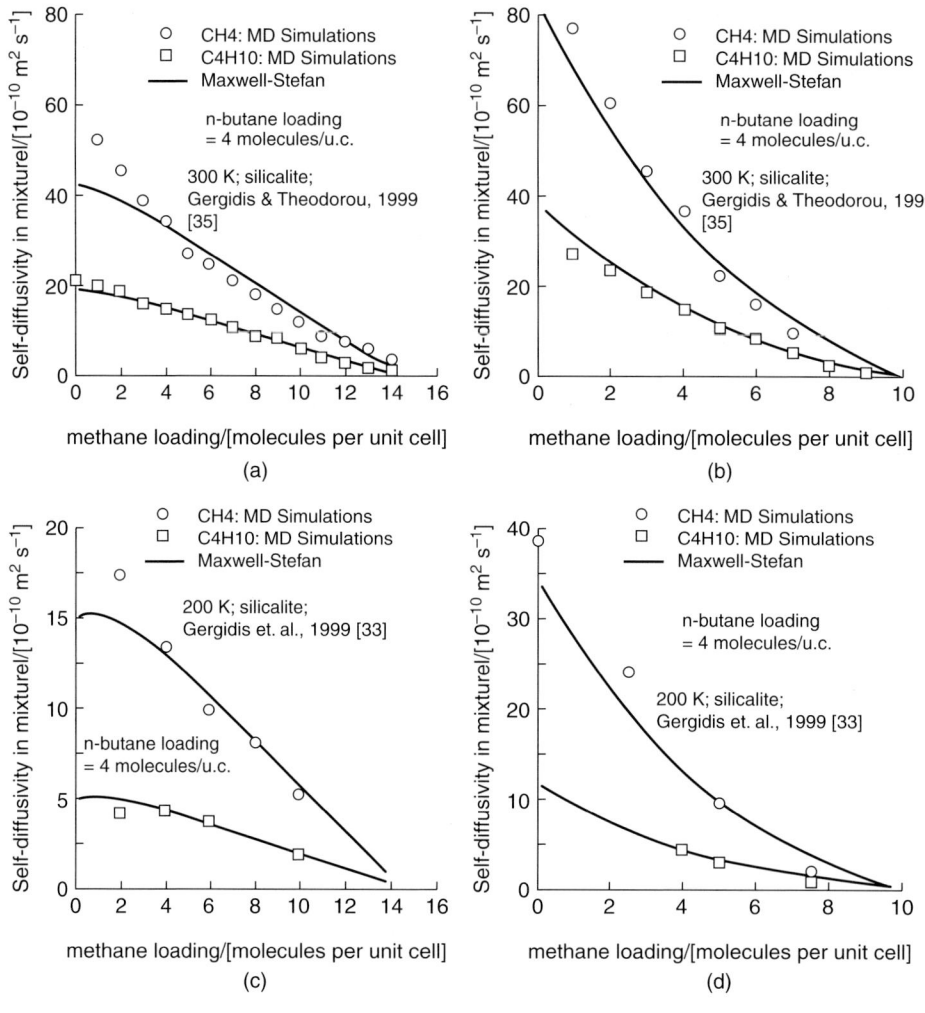

Figure 8.8 Molecular dynamics simulation predictions for the self-diffusivities of methane and n-butane in binary mixtures of various compositions sorbed within silicalite [33, 35] compared to calculations based on the Maxwell–Stefan model of Krishna [36]. Plots (a) and (b) refer to a temperature of 300 K, and (c) and (d) to 200 K. The equations and parameters of the Maxwell–Stefan model are stated in the text. With permission from [36].

which is an alternative form of Eq. (3.51). In Eq. (8.101) θ is the ratio of the concentration of the single component sorbate (labeled plus unlabeled molecules) to its saturation concentration. As usual, D_{0i} expresses the effect of interactions between the single component sorbate and the walls of the microporous medium, while $Ð_{ii}$ is an analog of the Maxwell–Stefan diffusivity $Ð_{ij}$, expressing the frictional effect of collisions between labeled and unlabeled molecules of the same species. In the systems studied by Chempath *et al.* [37], simulations indicated that the single-component D_{0i} decreased linearly with fractional occupancy:

$$D_{0i} = D_{0i}(0)(1-\theta) \tag{8.102}$$

while the ratio:

$$\zeta_i = \frac{\mathit{Ð}_{ii}}{D_{0i}} \tag{8.103}$$

decreased slightly with increasing occupancy and could safely be assumed occupancy-independent in the single component system. Thus, single-component diffusion data for species i at a given temperature can be parameterized by two constants, $D_{0i}(0)$ and ζ_i, plus the sorption isotherm of the pure component. As already pointed out (Section 8.1.3), applying Eq. (8.24) with $J_2 = 0$ and $\theta_2 = 0$, one arrives at the conclusion that the corrected diffusivity D_{0i} of the pure component i in the microporous medium, as defined in Eq. (8.20), is none else than the Maxwell–Stefan diffusivity $\mathit{Ð}_i$ at the considered occupancy:

$$D_{0i} = \mathit{Ð}_i \text{(pure sorbate } i \text{ in microporous medium, any occupancy)} \tag{8.104}$$

Therefore, the pure component transport diffusivity is readily obtained from D_{0i} and the pure component sorption isotherm via Eq. (8.19).

For a binary mixture of components 1 and 2 in the microporous medium, as is obvious from Eq. (8.24), four Maxwell–Stefan diffusivities are required: D_{01}, D_{02}, $\mathit{Ð}_{12}$, and $\mathit{Ð}_{21}$. The last two are related via the Onsager reciprocity relation, Eq. (8.25). Thus, one needs to know D_{01}, D_{02}, and $\mathit{Ð}_{12}$ as functions of q_1 and q_2 at each temperature. The simulations of Chempath et al. [37] supported the view that D_{01} and D_{02} in the binary mixture can still be safely estimated via Eq. (8.102), where $D_{0i}(0)$ are the pure-component Maxwell–Stefan diffusivities at very low occupancy and θ is now given by the following expression [compare Eq. (8.100)]:

$$\theta = \theta_1 + \theta_2 \tag{8.105}$$

For the estimation of $\mathit{Ð}_{12}$ in the mixture, Chempath et al. proposed:

$$q_{2,\text{sat}} \mathit{Ð}_{12} = \left[q_{2,\text{sat}} \mathit{Ð}_{11} \right]^{\frac{\theta_1}{\theta_1 + \theta_2}} \left[q_{1,\text{sat}} \mathit{Ð}_{22} \right]^{\frac{\theta_2}{\theta_1 + \theta_2}} \tag{8.106}$$

where now:

$$\mathit{Ð}_{11} = \zeta_1 D_{01}(\theta) \text{ and } \mathit{Ð}_{22} = \zeta_2 D_{02}(\theta) \tag{8.107}$$

Note that Eq. (8.106) and its counterpart obtained by interchanging the indices 1 and 2 satisfy the reciprocity relation, Eq. (8.25). Equations (8.102) and (8.105)–(8.107) permit the estimation of all required Maxwell–Stefan diffusivities from $D_{01}(0)$, $D_{02}(0)$, ζ_1, and ζ_2. Assuming that ζ_1 and ζ_2 remain the same in the binary mixture as they are in the pure components, a full description of diffusion in the mixture can be extracted if one computes by molecular simulation the pure-component self-diffusivities at two occupancies, along with the sorption isotherms for the mixture at the temperature of interest.

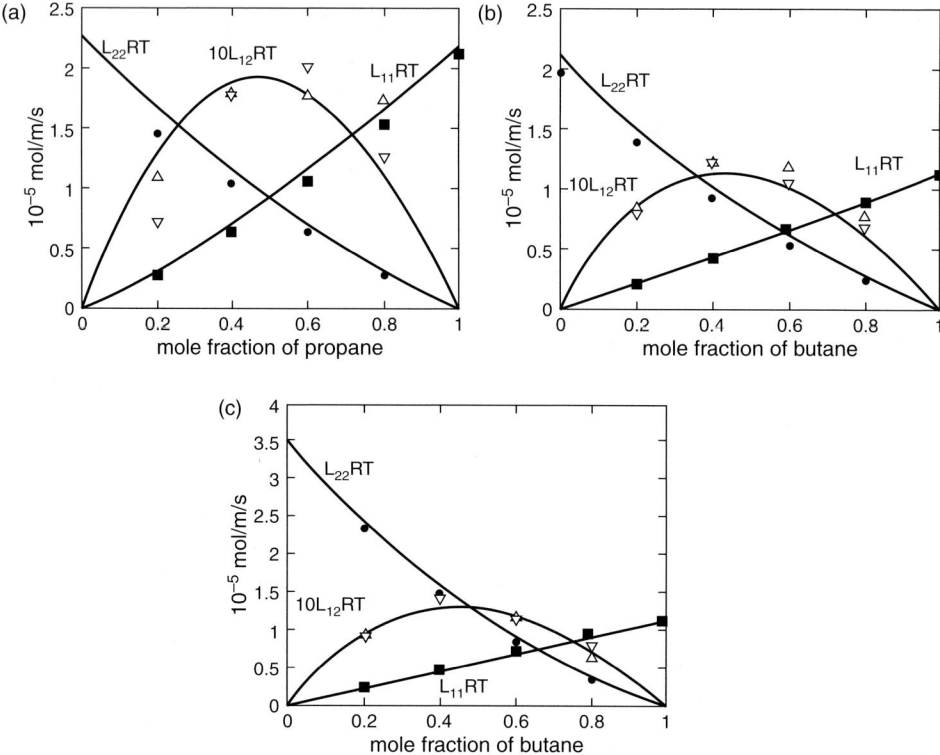

Figure 8.9 Onsager coefficients L_{11} (■), L_{22} (•), L_{12} (△), and L_{21} (▽) as functions of composition, computed via non-equilibrium MD simulations of the following binary systems sorbed in faujasite at 300 K:
(a) propane (1) and CF_4 (2) at 300 K and a total loading of 3.125 molecules per supercage;
(b) butane (1) and CF_4 (2) at 300 K and a total loading of 2.5 molecules per supercage;
(c) butane (1) and ethane (2) at a total loading of 2.5 molecules per supercage. The cross coefficients L_{12} and L_{21} have been multiplied by a factor of 10 for clarity. The lines show estimations based on the Maxwell–Stefan formulation of Reference. With permission from [37].

Once D_{01}, D_{01}, $Đ_{12}$, and $Đ_{21}$ have been determined as functions of concentration of the two species, the Onsager coefficients L_{11}, L_{22}, and $L_{12} = L_{21}$ can readily be calculated from Eqs. (8.27)–(8.31).

Figure 8.9 displays a comparison of Onsager coefficients in three binary mixtures sorbed in faujasite as computed via non-equilibrium MD simulations (points) and as calculated via the Maxwell–Stefan Eqs. (8.27)–(8.31) (lines), with Maxwell–Stefan diffusivities obtained from $D_{01}(0)$, $D_{02}(0)$, ζ_1, and ζ_2 via Eqs. (8.102) and (8.105)–(8.107), as described above. The agreement is excellent.

Krishna and van Baten [38] have extended these ideas further. They have conducted a large number of equilibrium MD and GCMC simulations on pure sorbates and binary mixtures of sorbates in various zeolites, metal organic frameworks (MOFs), covalent organic frameworks (COFs) and cylindrical silica pores with diameters from

0.6 to 30 nm. On the basis of these simulations, they have proposed a general semiempirical strategy for estimating Stefan–Maxwell diffusivities at any loading and composition of a sorbed multicomponent phase in a micro- or mesoporous medium from equilibrium MD of the pure components at various concentrations in the same medium. This strategy requires the saturation concentration of each species, which can be extracted from GCMC simulations of the pure species in the medium. A summary of this strategy is given in Section 3.4.

8.3
Non-equilibrium Molecular Dynamics Simulations

8.3.1
Gradient Relaxation Method

The MD techniques we discussed in Section 8.2 are aimed at extracting self-, transport, and corrected diffusivities by simulating a zeolite–sorbate system under conditions of thermodynamic equilibrium. The transport coefficient is extracted by quantifying the rate at which spontaneous fluctuations in sorbate intracrystalline concentration or composition relax as time elapses, based on linear response theory (Section 8.1). While equilibrium MD and use of the Einstein or equivalent Green–Kubo relations is the most straightforward and common way to study diffusion rates and mechanisms by molecular simulation, it is certainly possible to perform MD computer experiments under non-equilibrium conditions, in the presence of fluxes and driving forces. Such non-equilibrium molecular dynamics (NEMD) schemes are more problem-specific, but may yield the desired coefficients with less computational effort or provide mechanistic information that is complementary to that obtained through equilibrium MD. We will address them briefly in this section.

A simple non-equilibrium method for computing the transport diffusivity $D \equiv D_1$ of a pure sorbate in a zeolite is gradient relaxation molecular dynamics (GRMD), proposed by Maginn *et al.* in 1993 [39]. In this method, which is reminiscent of transient uptake measurements, one imposes an initial non-equilibrium sorbate concentration profile in the model system. One then integrates the equations of motion for all constituent particles under isothermal conditions to determine how the concentration profile evolves as a function of time. By fitting the relaxation of the concentration profile to the continuum solution of the diffusion equation (Fick's second law) under the initial and boundary conditions of the problem, one determines D. Several initial phase-space points must be used, all being representative of the initial non-equilibrium concentration profile and the boundary conditions imposed on the system. In the work of Maginn *et al.* [39], aimed at extracting D_y for methane along the straight channels of silicalite, the initial concentration profile had the shape of a square wave along the y-axis (Figure 8.10). The primary simulation box consisted of two sub-boxes, one with a slightly higher concentration than the other. Initial configurations for the two sub-boxes were pre-equilibrated with MC and then joined together. The dimensions of the primary box must be as large as possible

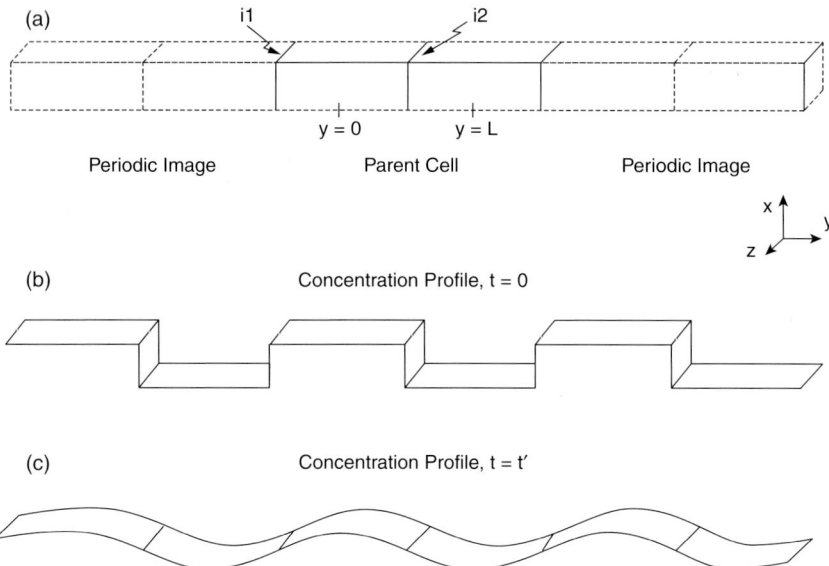

Figure 8.10 Schematic representation of the simulation box used in the gradient relaxation MD method of Maginn et al. [39]. The primary simulation box consists of two sub-boxes (a). Initially ($t=0$) the contents of each sub-box are taken from simulations equilibrated at different concentration levels. Each sub-box contains a sufficiently large integer number of unit cells. The zeolite structure is continuous across interfaces i1, i2, and their images. Thus, the initial concentration profile at the level of unit cells, averaged over all simulations conducted, has the shape of a square wave (b). An isothermal NEMD trajectory is initiated off of each starting phase-space point. The concentration profile, averaged over all trajectories at a time $t' > 0$, looks as shown in (c). An estimate of the transport diffusivity D is extracted by fitting the solution of the diffusion equation to this time-dependent periodic profile. With permission from [39].

in relation to the zeolite unit cell, to encourage the development of concentration gradients that are low enough to lie in the linear response regime during the ensuing NEMD simulation. In the simulations of Reference [39], each sub-box was eight unit cells long in the y direction and four initial state points were generated for each macroscopic initial concentration profile. Initial phase-space points were evolved for 0.5 ns by a constant kinetic energy algorithm and the spatial distributions of molecules at a unit-cell level averaged at different times to obtain an estimate of the macroscopic concentration profile $c(y,t)$ (Figure 8.10). The latter was fitted to the solution of the one-dimensional diffusion equation with the square-well initial concentration profile and zero-flux conditions at the boundaries i1 and i2. The initial 50 ps of each NEMD trajectory were excluded from the fitting, as concentration gradients prevailing there were judged to be too steep for linear response theory to hold.

8.3.2
External Field NEMD

An alternative, and computationally more advantageous, avenue for probing diffusion through NEMD simulations is offered by the "color field" method [39]. Here an external force field is introduced in the Hamiltonian, which couples with a property analogous to an electric charge that is attributed to the molecules. Following Evans and Morriss [40], this property is called a "color charge." To study diffusion of a pure fluid inside a zeolite, it is convenient to attribute a color charge of $+1$ to all sorbate molecules. The external field (color field) $\boldsymbol{F}$ induces a drift velocity on the molecules parallel to it. It does work on the system, requiring that heat be removed to maintain constant temperature. The equations of motion integrated read:

$$\dot{\boldsymbol{p}}_i = -\nabla_{r_i}\mathcal{V} + \boldsymbol{F} - \alpha(\boldsymbol{p}_i - \boldsymbol{p}_{\text{av}}) \tag{8.108}$$

where

$\boldsymbol{p}_i$ stands for the momentum vector of sorbate molecule i,
$-\nabla_{r_i}\mathcal{V}$ is the total force exerted on this molecule because of interactions with the zeolite lattice and with other sorbate molecules,
α is a thermostating multiplier,
and $\boldsymbol{p}_{\text{av}}$ is the average drift momentum of all sorbate molecules:

$$\boldsymbol{p}_{\text{av}} = \frac{1}{N}\sum_{i=1}^{N}\boldsymbol{p}_i \tag{8.109}$$

Thermostating is introduced as the requirement that the kinetic energy relative to the average drift velocity of the sorbate molecules be constant, as proposed by Evans and Morriss [40]:

$$\frac{d}{dt}\left(\frac{1}{2m}\sum_{i=1}^{N}(\boldsymbol{p}_i - \boldsymbol{p}_{\text{av}})^2\right) = 0 \tag{8.110}$$

Combining Eqs. (8.108)–(8.110) leads to the following expression for the thermostating multiplier α:

$$\alpha = \frac{\sum_{i=1}^{N}(\boldsymbol{p}_i - \boldsymbol{p}_{\text{av}})\cdot(-\nabla_{r_i}\mathcal{V} + \boldsymbol{F})}{\sum_{i=1}^{N}(\boldsymbol{p}_i - \boldsymbol{p}_{\text{av}})^2} \tag{8.111}$$

At steady state, the system is characterized by a macroscopic flux of sorbate molecules:

$$\boldsymbol{J} = \frac{1}{V}\langle \boldsymbol{j}\rangle_{\text{neq}} = \frac{1}{V}\lim_{t\to\infty}\left\langle\sum_{i=1}^{N}\boldsymbol{u}_i\right\rangle_{\text{neq}} = \frac{1}{V}\lim_{t\to\infty}\left\langle\sum_{i=1}^{N}\frac{\boldsymbol{p}_i}{m}\right\rangle_{\text{neq}} \tag{8.112}$$

By Eqs. (8.15) and (8.20), this is related to the color field F as:

$$J = L_{11} F = D_0 \frac{q}{k_B T} F = D_0 \frac{N}{V k_B T} F \tag{8.113}$$

Thus, measurement of the macroscopic flux under steady-state conditions yields the corrected diffusivity, D_0, along the direction of F. The transport diffusivity D can then be computed via Eq. (8.19), with the thermodynamic correction factor $(\partial \ln f / \partial \ln q)_T$ computed via grand canonical Monte Carlo [39].

An advantage of the color field method is that, by conducting the NEMD at progressively smaller fields F, one can directly identify the linear response regime, where D_0, as estimated via Eq. (8.113), is independent of F [39].

Figure 8.11 shows results for the transport diffusivity of methane along the straight channels of silicalite at 300 K, as obtained by the GRMD and color field NEMD methods. Both sets of results show an upward trend in D_0 with increasing intracrystalline occupancy. In this computation, GRMD underestimated the transport diffusivity, because it was partly based on the use of concentration profiles with gradients too high to be representative of the linear response regime [39]. Results from the color field method are more reliable. For the system studied in Figure 8.11 at

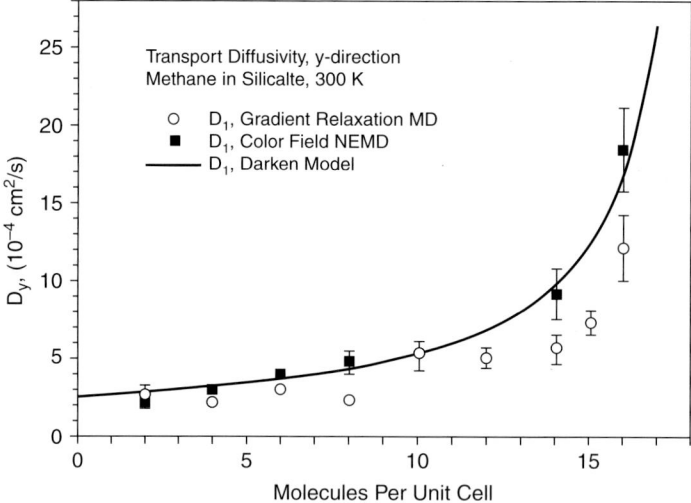

Figure 8.11 Transport diffusivity D_y for methane in silicalite at 300 K along the straight channels of the zeolite (y-axis) as a function of occupancy [39]. Estimates of D_y obtained directly by the GRMD method (○) are compared to values of $D_0 (\partial \ln f / \partial \ln q)_T$ with D_0 obtained from the color field NEMD method and the isotherm $f(q)$ computed by grand canonical Monte Carlo simulation (■). The GRMD results are less reliable, as they are subject to nonlinearities due to the large concentration gradients employed. The line labeled "Darken Model" shows $D_0 (\partial \ln f / \partial \ln q)_T$ with $D_0(0)$ being the corrected diffusivity in the limit of zero occupancy. It reproduces the color field results very well, indicating that D_0 along the y-axis is practically independent of intracrystalline occupancy in the system methane–silicalite at 300 K. With permission from [39].

300 K, D_0 turned out to be almost independent of occupancy. This, however, is far from general. More recent simulations [41] have shown that $D_0(q)$ is decreasing in CH_4–silicalite at higher temperatures. Recent coherent QENS and simulation studies in various systems have uncovered a rich variety of $D_0(q)$ dependences, which may be monotonically decreasing or display a maximum. These dependences can often be rationalized on the basis of sorbate–sorbate interactions using the simple lattice-based theory of Reed and Ehrlich [42] (Section 4.3.2.2).

8.3.3
Dual Control Volume Grand Canonical Molecular Dynamics

A non-equilibrium simulation method that can be thought of as a hybrid between grand canonical Monte Carlo (GCMC) and molecular dynamics (MD) has been developed by Heffelfinger and van Swol [43]. This method, known as dual control volume grand canonical molecular dynamics (DCV-GCMD), relies upon establishing a non-uniform steady-state chemical potential profile in the model system and measuring the flux. The transport diffusivity is then obtained through Eq. (8.16).

The original application of DCV-GCMD concerned a binary mixture of Lennard-Jones particles differing only in their "color." Pairwise interaction potentials were identical between the same and different species, making the binary system an ideal solution. Figure 8.12a shows the model system setup [43]. It was a rectangular parallelepiped of dimensions 60σ, 10σ, and 10σ along the x, y, and z directions, respectively, with σ being the characteristic length of the Lennard-Jones potential. The chemical potentials of the two species were kept at constant values within control volume A ($-20 < x/\sigma < -10$) through particle insertions and deletions, which were accepted according to the selection criteria of GCMC simulations, Eqs. (7.28) and (7.29). Likewise, the chemical potentials within control volume B ($10 < x/\sigma < 20$) were kept at constant values, different from those of control volume A, through GCMC particle insertion and deletion moves. In the application considered, the chemical potential of each species in control volume B was chosen equal to the chemical potential of the other species in control volume in A, resulting in an equimolar counter-diffusion situation at equilibrium. Every time a particle was inserted successfully, it was assigned a velocity vector whose components were drawn from Gaussian distributions corresponding to the simulation temperature. At the same time, particles everywhere in the system underwent constant temperature MD simulation. Impenetrable walls were placed at $x = -30\sigma$ and $x = 30\sigma$ and the chemical potential throughout the system was monitored through Widom test particle insertions [compare Eq. (7.41)]. The flux of particles of each species was monitored through the control surface at $x = 0$. In addition, the numbers of particles created and destroyed in each control volume were monitored, to close the mass balance.

Under these conditions, the system readily reached a steady state at which the density profiles of the two species looked as in Figure 8.12b. To a good approximation, the profiles are linear in the region $-10 < x/\sigma < 10$. From the concentration gradients in this region and the measured fluxes, the binary diffusion coefficient was readily extracted.

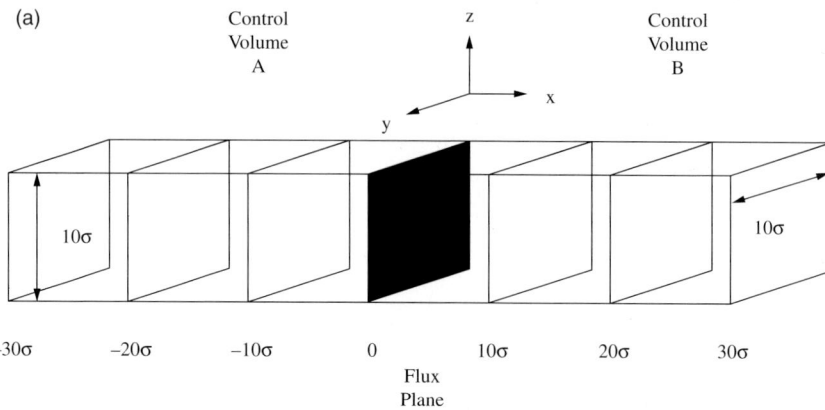

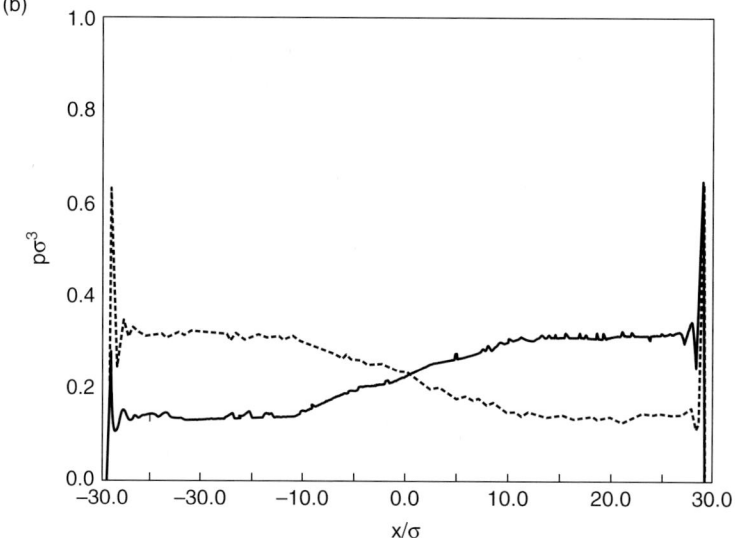

Figure 8.12 (a) Model system used in the dual control volume grand canonical dynamics (DCV-GCMD) method for simulating diffusion in a mixture of Lennard-Jones molecules with identical size and interactions. The chemical potentials of the two species are maintained at constant values through insertions and deletions of molecules in the control volumes. If the chemical potentials are different in the control volumes, nonzero fluxes develop through the plane shown in black. (b) Density profiles at steady state. The profiles are linear, to a good approximation, in the region $-10 < x/\sigma < 10$, allowing calculation of the diffusivity through measurement of the flux. Reproduced with permission from Reference [43].

An important parameter of the method is the number n_{GCMC} of attempted adjustments to the number of particles (insertions and deletions, attempted in equal numbers for each species) in each control volume per MD time step. This should be large enough to maintain the correct equilibrium densities corresponding to the set values of the chemical potentials within the control volumes, yet small enough for the MD simulation to remain stable. For the binary Lennard-Jones simulations considered by Heffelfinger and van Swol (temperature $T = \varepsilon/k_B$, MD simulation time step $\delta t = 0.01$ $(m\sigma^2/\varepsilon)^{1/2}$ using the constant temperature Verlet leapfrog algorithm of Brown and Clarke [44]) a value of $n_{GCMC} = 50$ proved optimal.

The DCV-GCMD method is conceptually attractive in that it establishes concentration or chemical potential gradients that mimic a real system and lets the flux develop as a natural consequence. Its computational requirements, however, are higher than those of alternative methods for calculating the transport diffusivity [45]: The number of molecules that need to be simulated is typically larger, because of the need to simulate the reservoirs of constant chemical potential in the two control volumes. The computational effort for keeping the chemical potential constant within each reservoir is substantial, especially for complex molecules or dense systems, where molecular insertions are subject to the problems pointed out in Section 7.3. Furthermore, momentum transfer may develop in addition to the mass transfer, complicating the interpretation of the computer experiment in terms of Onsager transport coefficients.

Arya, Chang, and Maginn [45] performed a critical comparison of three different methods for calculating the transport diffusivity D_1 for a one-component sorbate through molecular simulation. These methods were (i) equilibrium molecular dynamics (EMD), (ii) external field non-equilibrium molecular dynamics (EF-NEMD), and (iii) dual control volume grand canonical dynamics (DCV-GCMD). A sorbate consisting of spherical methane-like molecules was considered inside both single idealized pores of various diameters and within a more realistic model of an AlPO$_4$-5 molecular sieve. In the EMD approach the Onsager coefficient L_{11}, and hence D_1, was obtained through Green–Kubo integration of the time autocorrelation function of the microscopic flux, through Eq. (8.18). In the EF-NEMD, L_{11} was obtained as a proportionality constant between the macroscopic non-equilibrium flux and the externally imposed driving force from computer experiments conducted in the linear regime, through Eq. (8.113). In some DCV-GCMD simulations, a streaming velocity, in addition to the randomly oriented, Boltzmann-distributed velocity components, was imparted to the molecules inserted in the two control volumes. The magnitude of the streaming velocity was estimated as the flux out of each box (positive for one box and negative for the other) over the last 1000 integration time steps divided by the average concentration in the control volume. The number of attempted adjustments n_{GCMC} to the number of particles in each control volume per MD time step was also varied systematically. L_{11} was extracted from the DCV-GCMD simulations as the proportionality constant between the macroscopic non-equilibrium flux and the negative gradient in chemical potential established between the two control volumes under steady state conditions [see Eq. (8.17)], as described above. Good agreement between EMD and EF-NEMD results was obtained for all cases examined, which were ensured to lie in the linear response regime. The EF-NEMD afforded smaller error bars than EMD for the same computer time. Conversely, DCV-GCMD with small n_{GCMC} and no streaming velocity underestimated the flux by factors of 1.4–2.3 relative to EMD and EF-NEMD. Increasing n_{GCMC} beyond 50 while using a streaming velocity led to asymptotic values of the transport coefficient from DCV-GCMD which agreed with EMD and EF-NEMD. Matching DCV-GCMD results with those of EMD and EF-NEMD proved impossible when no streaming velocity was employed. Disparities between DCV-GCMD and the other two methods were traced to distortions in the concentration profiles at the interfaces between the control volumes and the transport region and to a non-uniform flux throughout the model system. These phenomena reflect that: (i) under conditions of low n_{GCMC}, the control volume of high chemical potential cannot supply molecules to the interface fast enough to replenish molecules lost to the transport region, or the control volume of low chemical potential cannot deplete molecules fast enough to remove molecules entering from the transport region; and (ii) when no streaming velocities are employed, insertions and deletions in the control volumes lead to artificial momentum losses. The role of control volumes in a DCV-GCMD simulation can be understood in macroscopic terms when a reaction–diffusion–convection model is invoked for the model system. Only when the reaction (creation and deletion of molecules in the two control volumes) is very fast relative to diffusion and when the convective flux in the control volumes exactly matches the diffusive flux in the transport region can DCV-GCMD give correct results for the transport diffusivity [45]. Since ensuring these conditions requires considerable expense in computer time, Arya *et al.* [45] recommend EMD or EF-NEMD as the computationally most efficient ways for computing Onsager coefficients and transport and Maxwell–Stefan diffusivities in a bulk microporous material.

References

1 Hansen, J.-P. and McDonald, I.R. (1986) *Theory of Simple Liquids*, 2nd edn, Academic Press, New York.
2 MacElroy, J.M.D. (1995) Diffusion in homogeneous media, in *Diffusion in Polymers* (ed. P. Neogi), Dekker, New York, pp. 1–66.
3 Theodorou, D.N., Snurr, R.Q., and Bell, A.T. (1996) Molecular dynamics and diffusion in microporous materials, in *Comprehensive Supramolecular Chemistry*, vol. 7 (eds G. Alberti and T. Bein), Elsevier Science, Oxford, pp. 507–548.
4 Darken, L.S. (1948) *Trans. AIME*, **175**, 184.
5 Krishna, R. and van Baten, J.M. (2009) *Chem. Eng. Sci.*, **64**, 3159.
6 Allen, M.P. and Tildesley, D.J. (1987) *Computer Simulation of Liquids*, Clarendon Press, Oxford.
7 Swope, W.C., Andersen, H.C., Berens, P.H., and Wilson, K.R. (1982) *J. Chem. Phys.*, **76**, 637.
8 Verlet, L. (1967) *Phys. Rev.*, **159**, 98.
9 Harmandaris, V. and Mavrantzas, V. (2004) Molecular dynamics simulations of polymers, in *Simulation Methods for Polymers* (eds M. Kotelyanskii and D.N. Theodorou), Marcel Dekker, pp. 177–222.
10 June, R.L., Bell, A.T., and Theodorou, D.N. (1992) *J. Phys. Chem.*, **96**, 1051.
11 Tuckerman, M., Berne, B.J., and Martyna, G.J. (1992) *J. Chem. Phys.*, **97**, 1990–2001.
12 Martyna, G.J., Tuckerman, M.E., Tobias, D.J., and Klein, M.L. (1996) *Mol. Phys.*, **87**, 1117–1157.
13 Fox, G.C., Johnson, M.A., Lyzenga, G.A., Otto, S.W., Salmon, J.K., and Walker, D.W. (1988) *Solving Problems on Concurrent Processors:*, vol. **1**, Prentice Hall, Englewood Cliffs.
14 Rapaport, D.C. (1995) *The Art of Molecular Dynamics Simulation*, Cambridge University Press, Cambridge.

15 Plimpton, S. (1995) *J. Comp. Phys.*, **117**, 1.
16 Gō, N. and Scheraga, H.A. (1976) *Macromolecules*, **9**, 535.
17 Fincham, D. (1993) *Mol. Simul.*, **11**, 79–89.
18 Makrodimitris, K., Papadopoulos, G.K., and Theodorou, D.N. (2001) *J. Phys. Chem. B*, **105**, 777.
19 Goldstein, H. (1980) *Classical Mechanics*, 2nd edn, Addison-Wesley, Reading, MA.
20 Fincham, D. (1984) More on rotational motion of linear molecules. *CCP5 Quarterly*, **12**, 47.
21 Ryckaert, J.-P., Ciccotti, G., and Berendsen, H. (1977) *J. Comput. Phys.*, **23**, 327.
22 Andersen, H.C. (1983) *J. Comput. Phys.*, **52**, 24.
23 Edberg, R., Evans, D.J., and Morriss, G.P. (1986) *J. Chem. Phys.*, **84**, 6933.
24 Theodorou, D.N., Dodd, L.R., Boone, T.D., and Mansfield, K.F. (1993) *Makromol. Chem. Theory Simul.*, **2**, 191.
25 Nosé, S. (1984) *Mol. Phys.*, **52**, 255.
26 Andersen, H.C. (1980) *J. Chem. Phys.*, **72**, 2384.
27 Parrinello, M. and Rahman, A. (1980) *Phys. Rev. Lett.*, **45**, 1196.
28 Nosé, S. (1984) *J. Chem. Phys.*, **81**, 511.
29 Hoover, W.G. (1985) *Phys. Rev. A*, **31**, 1695.
30 Car, R. and Parrinello, M. (1985) *Phys. Rev. Lett.*, **55**, 2471.
31 Papadopoulos, G.K., Jobic, H., and Theodorou, D.N. (2004) *J. Phys. Chem. B*, **108**, 12748.
32 Reed, D.A. and Ehrlich, G. (1981) *Surf. Sci.*, **102**, 588.
33 Gergidis, L.N., Theodorou, D.N., and Jobic, H. (2000) *J. Phys. Chem. B*, **104**, 5541.
34 June, R.L., Bell, A.T., and Theodorou, D.N. (1992) *J. Phys. Chem.*, **96**, 1051.
35 Gergidis, L.N. and Theodorou, D.N. (1999) *J. Phys. Chem. B*, **103**, 3380.
36 Krishna, R. (2000) *Chem. Phys. Lett.*, **326**, 477.
37 Chempath, S., Krishna, R., and Snurr, R.Q. (2004) *J. Phys. Chem. B*, **108**, 13481.
38 Krishna, R. and van Baten, J.M. (2009) *Chem. Eng. Sci.*, **64**, 3159. 40.
39 Maginn, E.J., Bell, A.T., and Theodorou, D.N. (1993) *J. Phys. Chem.*, **97**, 4173.
40 Evans, D.J. and Morriss, G.P. (1990) *Statistical Mechanics of Nonequilibrium Liquids*, Academic Press, London.
41 Skoulidas, A.I. and Sholl, D.S. (2001) *J. Phys. Chem. B*, **105**, 3151–3154.
42 Papadopoulos, G.K. and Theodorou, D.N. (2008) Computer simulation of sorption and transport in zeolites, in *Handbook of Heterogeneous Catalysis*, 2nd edn (eds G. Ertl, H. Knözinger, F. Schüth, and J. Weitkamp), Wiley-VCH Verlag GmbH, Weinheim, pp. 1662–1676, ch. 5.5.2.
43 Heffelfinger, G.H. and van Swol, F. (1994) *J. Chem. Phys.*, **100**, 7548–7552.
44 Brown, D. and Clarke, J.H.R. (1984) *Mol. Phys.*, **51**, 1243.
45 Arya, G., Chang, H.-C., and Maginn, E.J. (2001) *J. Chem. Phys.*, **115**, 8112.

9
Infrequent Event Techniques for Simulating Diffusion in Microporous Solids

As we have seen in Sections 8.2 and 8.3, molecular dynamics (MD) simulations are valuable for predicting diffusion coefficients in microporous solids from atomic-level information on structure and interactions. The Achilles' heel of MD methods is their high demand in computer time. Today, the maximum time that one can simulate with atomistic MD with conventional computational resources (e.g., a week of time on a parallel cluster of 3 GHz microprocessors) is on the order of a microsecond. Unfortunately, many dynamical processes occurring in real-life microporous materials have characteristic times that exceed the μs time scale, often by many orders of magnitude. Take, for example, diffusion of benzene in silicalite at low occupancy. As a result of the close fit experienced by a benzene molecule inside the zeolite pores, diffusion in this system is slow; a typical value based on experimental measurements is $\mathcal{D} = 2.2 \times 10^{-14}\,\mathrm{m^2\,s^{-1}}$ at 300 K (see Reference [1] and Section 18.3.1. According to Figure 18.18, experimental values of the diffusivity at 300 K range from about $1.5 \times 10^{-14}\,\mathrm{m^2/s}$ for HZSM-5 to about $5 \times 10^{-14}\,\mathrm{m^2/s}$ for silicalite). To estimate this diffusivity by MD, one would need to track the translational motion of the sorbate molecule over a time scale long enough for the root mean square displacement to be several times the unit cell size, so that a molecule would have the chance to sample all environments inside the pores and enter the Einstein regime of diffusion; let us set, rather conservatively, $\langle r^2 \rangle^{1/2} \approx 25$ nm. The time one would need to simulate in order to observe such a mean square displacement of a sorbed benzene molecule would thus be $\langle r^2 \rangle / (6\mathcal{D}) \approx (25 \times 10^{-9}\,\mathrm{m})^2 / (6 \times 2.2 \times 10^{-14}\,\mathrm{m^2\,s^{-1}})$ or 4.7 ms, which exceeds 1 μs by 3.7 orders of magnitude. Thus, the direct estimation of the diffusivity of benzene in silicalite at room temperature is impossible with "brute force" MD, and that can be said about many real-life dynamical processes that one may be interested in tracking by simulation.

Diffusion of a bulky or strongly interacting molecule inside a microporous medium is often slow because the molecule spends most of its time executing fast local motions trapped within favorable regions of the framework, and only infrequently jumps from one such region to another by overcoming a barrier in the (free) energy associated with its translational motion. In this chapter we will see how we can use the time scale separation between local motions within favorable regions in the microporous medium and infrequent transitions between such regions, to estimate

Diffusion in Nanoporous Materials. Jörg Kärger, Douglas M. Ruthven, and Doros N. Theodorou.
© 2012 Wiley-VCH Verlag GmbH & Co. KGaA. Published 2012 by Wiley-VCH Verlag GmbH & Co. KGaA.

rate constants for the transitions from atomic-level structure and interactions. We will also see how we can use these rate constants, along with the spatial arrangement of sorption sites representative of the regions, to extract the self-diffusivity. This is strategically important for overcoming the long-time limitations of MD.

Section 9.1 introduces the basic elements of the theory of infrequent events, leading to a general expression for the rate constant for transition between two states. It discusses approximations to this expression based on transition state theory and outlines a strategy for computing the rate constant from atomistic simulations. Section 9.2 introduces the master equation, whose solution provides the long-time evolution of the probability distribution of the system among its states (in our application, the evolution of a tracer concentration profile, hence the self-diffusivity), using as input the atomistically computed interstate transition rate constants. Methods for solving the master equation, including kinetic Monte Carlo simulation (KMC), are discussed in the same section. Section 9.3 presents applications of the infrequent event methodology to zeolite–sorbate systems at low and moderate occupancies.

9.1
Statistical Mechanics of Infrequent Events

9.1.1
Time Scale Separation and General Expression for the Rate Constant

A system with a rugged potential energy hypersurface may spend most of its time confined within small, favorable regions of its configuration space, and only infrequently undergo a transition between regions. We will call "states" the small regions in configuration space where the system likes to reside. The states constitute "basins" of low energy with respect to the coordinates spanning configuration space, or of low free energy with respect to a subset of coordinates providing a coarse-grained description of the system (compare Section 7.4). Each state contains one or more local minima of the (free) energy with respect to the coordinates used to describe the system. Transitions between states are infrequent events, in the sense that the mean waiting time required for a transition out of a state to occur is long in comparison to the time required for the system to establish a restricted equilibrium distribution among configurations in the state. The entire configuration space can be tessellated into states. If one represents each state by a point in configuration space and connects all pairs of states between which a transition is possible, one obtains a graph, or network of states.

If the waiting time that must elapse for the system to jump out of a state i into another state is long in comparison with the correlation time required for the system to distribute itself among the configurations of state i, then, by the time the system leaves state i, it will have lost all memory of how it entered that state. Thus, the dynamical process is reduced to a succession of independent jumps $i \rightarrow j$ between states. Under these conditions, once the system is in state i, the conditional probability per unit time that a jump $i \rightarrow j$ will occur assumes a well-defined time-independent value, which is the rate constant $k_{i \rightarrow j}$. By focusing on the energy barrier separating the states, the theory of infrequent events manages to obtain an estimate of $k_{i \to j}$ within reasonable computer time, no matter how small $k_{i \to j}$ may be. Thus, it overcomes the limitation of "brute force" molecular dynamics, which would exhaust available computer time tracking the relatively uninteresting local motions of the system within state i and hardly have the chance to sample transitions out of this state. In this Section we briefly address the question: How does one obtain an estimate of the transition rate constant $k_{i \to j}$, given the energy hypersurface and the atomic masses in the system? Section 9.2, on the other hand, addresses the question: Once one has the network of states and the rate constants for all elementary transitions between them, how can one calculate the long-time evolution of the system?

Infrequent event theory has its roots in the absolute rate theory, or transition state theory (TST), of Henry Eyring [2]. Its application to chemical reactions constitutes a cornerstone of chemical kinetics [3]. We have already seen an application of TST in the estimation of the self-diffusivity in Section 4.3.3. Ideas introduced in that Section are applied to explain pre-exponential factors of diffusivities in A-type zeolites in Section 16.5.2. Here we generalize these ideas and explain their statistical mechanical foundation. We will outline a modern development of infrequent event theory for physical processes, known as the Bennett–Chandler theory, which is particularly well-suited for the estimation of $k_{i \to j}$ from molecular simulations [4, 5]. Our treatment is inspired by the simplified derivation presented by Voter and Doll [6].

Consider a system whose dynamics are described in terms of a single degree of freedom x, associated with a mass m. To fix ideas, we may consider that x describes the position along a straight pore of an inflexible sorbate molecule that experiences a rather close fit in the pore, and m is the mass of that molecule. The dynamics in x (translational motion along the pore) will be governed by a potential of mean force, $U(x)$, obtained from the full potential energy hypersurface of the system by integrating [see Eq. (7.47)] over all other degrees of freedom (vibrational modes of the pore, orientational and vibrational degrees of freedom of the molecule), whose fluctuations are fast relative to (translational) progress along x. Assume that, locally, $U(x)$ looks as shown in Figure 9.1, allowing the definition of two states, A and B, as basins around the local minima at x_A and x_B. The states are separated by a point at $x = x^\dagger$. We will assume that the barrier in potential of mean force, $U(x^\dagger) - U(x_A)$, is high relative to $k_B T$, so that the transition from A to B is an infrequent event. The degree of freedom x will be referred to as the "reaction coordinate" and the locus of points satisfying $x = x^\dagger$ as the "dividing (hyper)surface." In this one-dimensional example, the dividing surface reduces to a single point.

We consider a canonical ensemble of systems characterized by the $U(x)$ profile of Figure 9.1 at equilibrium, at temperature T. Let N_A, N_B be the numbers of systems that find themselves in states A and B, respectively, at a particular time. We follow the dynamical trajectories of these systems. Over long times, $N_A(t)$, $N_B(t)$ will fluctuate as a result of infrequent transitions between the two basins, but the total number of copies in the ensemble, $N_A(t) + N_B(t) = N$, which is very large, will be assumed fixed.

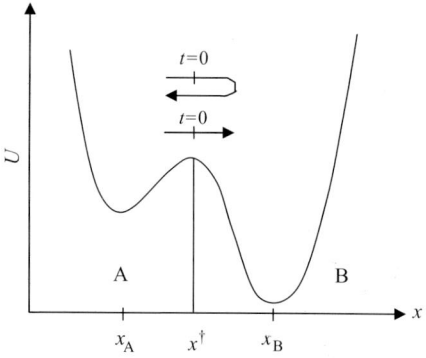

Figure 9.1 Schematic of the potential of mean force $U(x)$ experienced by the center-of-mass of a sorbate molecule along the axis of a straight pore, to explain infrequent event theory. Two states, A and B, are constructed around the local minima of $U(x)$ at x_A and x_B. The barrier (transition state) is at $x^\dagger$. In this one-dimensional case the transition state is a maximum of $U(x)$. In the more general case where more than one degree of freedom is involved in the transition, the transition state is a first-order saddle point of $U(x)$. The arrows show a successful and a recrossing trajectory. The time at which the system first passes the barrier is defined as $t=0$.

Under appropriate conditions of time scale separation, which we will try to define, $N_A(t)$ and $N_B(t)$ evolve according to the first-order kinetic laws:

$$\dot{N}_A = -k_{A \to B} N_A + k_{B \to A} N_B$$
$$\dot{N}_B = k_{A \to B} N_A - k_{B \to A} N_B \tag{9.1}$$

where dots over symbols indicate time derivatives.

The rate constants $k_{A \to B}$ and $k_{B \to A}$ have the physical meaning of a transition (conditional) probability per unit time, in the forward (A → B) and in the reverse direction. Let $\bar{N}_A$ and $\bar{N}_B$ be the time-averaged (over very long times) values of N_A and N_B, respectively. Time-averaging Eq. (9.1), assuming constancy of the rate constants with time, leads to the condition of microscopic reversibility:

$$k_{A \to B} \bar{N}_A = k_{B \to A} \bar{N}_B \tag{9.2}$$

We define the instantaneous deviations of $N_A(t)$ and $N_B(t)$ from their average values:

$$\delta N_A = N_A - \bar{N}_A$$
$$\delta N_B = N_B - \bar{N}_B \tag{9.3}$$

Clearly, the constancy of $N_A(t) + N_B(t)$ implies:

$$\delta N_A + \delta N_B = 0 \tag{9.4}$$

By virtue of Eqs. (9.1) and (9.2), δN_A satisfies the kinetic equation:

$$\delta \dot{N}_A = -k_{A \to B} \delta N_A + k_{B \to A} \delta N_B \tag{9.5}$$

Using Eq. (9.4) in Eq. (9.5):

$$\delta \dot{N}_A = -(k_{A \to B} + k_{B \to A})\delta N_A = -k_{\text{eff}} \delta N_A \tag{9.6}$$

with $k_{\text{eff}} = k_{A \to B} + k_{B \to A}$.

Multiplying Eq. (9.6) at time t with the deviation of the population of state A from its equilibrium value at time 0, $\delta N_A(0)$, and taking an equilibrium average over multiple time origins along the equilibrium ensemble trajectories, one obtains:

$$\langle \delta N_A(0) \delta \dot{N}_A(t) \rangle = -k_{\text{eff}} \langle \delta N_A(0) \delta N_A(t) \rangle \tag{9.7}$$

Note that by virtue of the time-reversibility of the motion and the equilibrium conditions prevailing:

$$\langle \delta N_A(0) \delta \dot{N}_A(t) \rangle = -\langle \delta N_A(0) \delta \dot{N}_A(-t) \rangle = -\langle \delta N_A(t) \delta \dot{N}_A(0) \rangle$$

From Eqs. (9.4) and (9.7), one obtains the exact relation:

$$k_{\text{eff}} \equiv k_{A \to B} + k_{B \to A} = \frac{\langle \delta \dot{N}_A(0) \delta N_A(t) \rangle}{\langle \delta N_A(0) \delta N_A(t) \rangle} = -\frac{\langle \delta \dot{N}_A(0) \delta N_B(t) \rangle}{\langle \delta N_A(0) \delta N_A(t) \rangle} \tag{9.8}$$

The denominator of Eq. (9.8) is a slowly varying function of time, relative to the numerator; the rate of change in δN_A will fluctuate between positive and negative values as a result of transitions occurring in the various replicas of the ensemble, and therefore decorrelate from the initial value of δN_A much more rapidly than δN_A itself. It is therefore reasonable to introduce in the denominator the approximation:

$$\langle \delta N_A(0) \delta N_A(t) \rangle \approx \langle \delta N_A(0) \delta N_A(0) \rangle \tag{9.9}$$

which leads to:

$$k_{\text{eff}} \equiv k_{A \to B} + k_{B \to A} = \frac{\langle \delta \dot{N}_A(0) \delta N_A(t) \rangle}{\langle \delta N_A(0) \delta N_A(0) \rangle} = -\frac{\langle \delta \dot{N}_A(0) \delta N_B(t) \rangle}{\langle \delta N_A(0) \delta N_A(0) \rangle} \tag{9.10}$$

Without loss of generality, now, instead of an ensemble of systems, we consider a single system evolving along its equilibrium trajectory. In other words, we set $N = 1$. Invoking the Heaviside step function $H(x)$, with $H(x) = 1$ for $x > 0$, $H(x) = 0$ for $x < 0$, and $H(x) = 1/2$ for $x = 0$, we can then write:

$$\begin{aligned} N_A(t) &= H[x^\dagger - x(t)] \\ N_B(t) &= H[x(t) - x^\dagger] \end{aligned} \tag{9.11}$$

$$\begin{aligned} \bar{N}_A &= P_A^{\text{eq}} \\ \bar{N}_B &= P_B^{\text{eq}} \end{aligned} \tag{9.12}$$

where P_A^{eq} and $P_B^{\text{eq}} = 1 - P_A^{\text{eq}}$ are the fractions of time the system spends in states A and B, respectively, at equilibrium, that is, the equilibrium probabilities of states A and B. By definition, then:

$$\delta N_A(t) = H[x^\dagger - x(t)] - P_A^{eq}$$
$$\delta N_B(t) = H[x(t) - x^\dagger] - P_B^{eq} \qquad (9.13)$$

$$\delta \dot{N}_A(0) = -\dot{x}(0)\delta[x^\dagger - x(0)] \qquad (9.14)$$

bearing in mind that the derivative of the Heaviside function is the Dirac delta function, $\delta(x)$. The autocorrelation function in the denominator of Eq. (9.10) becomes:

$$\begin{aligned}\langle \delta N_A(0)\delta N_A(0)\rangle &= \left\langle [H(x^\dagger - x(0)) - P_A^{eq}]^2 \right\rangle \\ &= \left\langle [H(x^\dagger - x(0))]^2 \right\rangle - 2P_A^{eq}\langle H(x^\dagger - x(0))\rangle + (P_A^{eq})^2 \\ &= \langle H(x^\dagger - x(0))\rangle - 2P_A^{eq}\langle H(x^\dagger - x(0))\rangle + (P_A^{eq})^2 \\ &= P_A^{eq} - 2P_A^{eq}P_A^{eq} + (P_A^{eq})^2 = P_A^{eq} - (P_A^{eq})^2 \\ &= P_A^{eq}(1 - P_A^{eq}) = P_A^{eq}P_B^{eq}\end{aligned} \qquad (9.15)$$

On the other hand, the numerator of Eq. (9.10) becomes:

$$\begin{aligned}\langle \delta \dot{N}_A(0)\delta N_B(t)\rangle &= -\langle \dot{x}(0)\delta(x^\dagger - x(0))[H(x(t) - x^\dagger) - P_B^{eq}]\rangle \\ &= -\langle \dot{x}(0)\delta(x^\dagger - x(0))H(x(t) - x^\dagger)\rangle + P_B^{eq}\langle \dot{x}(0)\delta(x^\dagger - x(0))\rangle \\ &= -\langle \dot{x}(0)\delta(x^\dagger - x(0))H(x(t) - x^\dagger)\rangle\end{aligned} \qquad (9.16)$$

where we have used the fact that $\langle \dot{x}(0)\delta(x^\dagger - x(0))\rangle$ is zero, since positive and negative values of the velocity $\dot{x}(0)$ are equally probable at equilibrium.

From Eqs. (9.10), (9.15), and (9.16) k_{eff} emerges as:

$$k_{\text{eff}} = \frac{\langle \dot{x}(0)\delta(x^\dagger - x(0))H(x(t) - x^\dagger)\rangle}{P_A^{eq}P_B^{eq}} \qquad (9.17)$$

On the other hand, the condition of detailed balance, Eq. (9.2), gives:

$$k_{A \to B}P_A^{eq} = k_{B \to A}P_B^{eq} \qquad (9.18)$$

Hence:

$$k_{\text{eff}} = k_{A \to B} + k_{B \to A} = k_{A \to B}\left(1 + \frac{P_A^{eq}}{P_B^{eq}}\right) = k_{A \to B}\frac{1}{P_B^{eq}} \qquad (9.19)$$

Combining Eqs. (9.17) and (9.19), we express the rate constant for transition from A to B as:

$$k_{A \to B}(t) = \frac{\langle \dot{x}(0)\delta(x^\dagger - x(0))H(x(t) - x^\dagger)\rangle}{P_A^{eq}} \tag{9.20}$$

where we are showing an explicit dependence of the rate constant on time.

The numerator of Eq. (9.20) can be interpreted as the flux through the dividing surface $x = x^\dagger$, modified by the step function $H(x(t) - x^\dagger)$. Trajectories that cross the dividing surface in the direction from A to B at time zero and find themselves in state B after time t contribute positively to $k_{A \to B}(t)$. Trajectories that cross the dividing surface in the direction from A to B but quickly recross it and find themselves in state A after time t do not contribute to $k_{A \to B}(t)$ (Figure 9.1). Trajectories that cross the dividing surface from B to A but quickly recross and find themselves in state B after time t contribute negatively to $k_{A \to B}(t)$. If the correlation time τ_{cor} required for the system to "thermalize," that is, establish a restricted equilibrium distribution within either state A or state B, is short relative to k_{eff}^{-1}, making the transitions A $\to$ B and B $\to$ A infrequent events, then recrossing events will quickly subside and $k_{A \to B}(t)$ will assume an asymptotic, time-independent value over a wide range of times $\tau_{cor} < t < k_{eff}^{-1}$.

9.1.2
Transition State Theory Approximation and Dynamical Correction Factor

Transition state theory (TST) relies upon a simplifying approximation to Eq. (9.20). It assumes that whenever the system crosses the dividing surface, moving from an original state to a destination state, it will continue being in the destination state at time t after the crossing. In other words, for the two-state system considered, TST ignores fast recrossing events. It assumes that, whenever a crossing of the dividing surface occurs, the system thermalizes in the state toward which it is moving. Mathematically, TST replaces the time t in the numerator of Eq. (9.20) with 0^+, a time just after the crossing at which the system is guaranteed to be in its destination state. The TST approximation to the transition rate constant is thus:

$$k_{A \to B}^{TST} = \frac{\langle \dot{x}(0)\delta(x^\dagger - x(0))H(x(0^+) - x^\dagger)\rangle}{P_A^{eq}} = \frac{\langle \dot{x}(0)\delta(x^\dagger - x(0))H(\dot{x}(0))\rangle}{P_A^{eq}}$$

In the canonical ensemble considered, positive and negative values of the velocity $\dot{x}(0)$ are equally probable. The positive and negative fluxes through $x = x^\dagger$ are equal in magnitude, and the TST expression can further be written as:

$$k_{A \to B}^{TST} = \frac{1}{2}\frac{\langle |\dot{x}(0)|\delta(x^\dagger - x(0))\rangle}{P_A^{eq}} \tag{9.21}$$

Furthermore, in a classical treatment where configurational and momentum degrees of freedom are separable:

$$k_{A \to B}^{TST} = \frac{1}{2}\langle|\dot{x}(0)|\rangle \frac{\langle\delta(x^\dagger - x(0))\rangle}{P_A^{eq}} = \left(\frac{k_B T}{2\pi m}\right)^{1/2} \frac{\exp\left(-\frac{U(x^\dagger)}{k_B T}\right)}{\int_{\text{state A}} \exp\left(-\frac{U(x)}{k_B T}\right) dx} \quad (9.22)$$

where we have used that for a one-dimensional Boltzmann distribution of velocities:

$$\langle|\dot{x}(0)|\rangle = \left(\frac{2k_B T}{\pi m}\right)^{1/2}$$

and the configuration-space probability density is proportional to:

$$\exp\left(-\frac{U(x)}{k_B T}\right)$$

Remarkably, TST has reduced the rate constant to the product of a mean molecular speed times a conditional probability per unit length along the reaction coordinate that the system will find itself on the dividing surface, provided it is allowed to sample the origin state A according to its equilibrium distribution.

When the barrier height $U(x^\dagger) - U(x_A)$ is very high relative to $k_B T$, most of the contribution to the integral in the denominator of Eq. (9.22) comes from the immediate vicinity of x_A. Under these conditions, one can invoke a "harmonic approximation", replacing $U(x)$ in the denominator by its Taylor expansion around x_A, truncated at the second order term. Setting:

$$\kappa_W = \left.\frac{\partial^2 U}{\partial x^2}\right|_{x_A}, \quad \beta = \frac{1}{k_B T}$$

one obtains:

$$\begin{aligned}
k_{A \to B}^{TST,HA} &= \left(\frac{k_B T}{2\pi m}\right)^{1/2} \frac{\exp(-\beta U(x^\dagger))}{\int_{\text{state A}} \exp\left(-\beta\left[U(x_A) + \frac{1}{2}\kappa_W(x - x_A)^2\right]\right) dx} \\
&= \left(\frac{k_B T}{2\pi m}\right)^{1/2} \frac{\exp(-\beta[U(x^\dagger) - U(x_A)])}{\int_{-\infty}^{+\infty} \exp\left(-\frac{\beta\kappa_W}{2}(x - x_A)^2\right) dx} \\
&= \left(\frac{k_B T}{2\pi m}\right)^{1/2} \left(\frac{\beta\kappa_W}{2\pi}\right)^{1/2} \exp(-\beta[U(x^\dagger) - U(x_A)]) \\
&= \frac{1}{2\pi}\left(\frac{\kappa_W}{m}\right)^{1/2} \exp(-\beta[U(x^\dagger) - U(x_A)]) \\
&= \frac{\omega_A}{2\pi} \exp(-\beta[U(x^\dagger) - U(x_A)])
\end{aligned} \quad (9.23)$$

According to TST in the harmonic approximation, Eq. (9.23), the rate constant for the transition from A to B emerges as a product of a natural frequency of oscillation in the origin state, $\omega_A/(2\pi)$, times the Boltzmann factor of the energy barrier that has to be overcome to effect the transition.

By neglecting fast recrossing events, TST typically overestimates the transition rate constant. Comparing Eq. (9.20) with the precursor to Eq. (9.21), one can define a dynamical correction factor as:

$$f_d(t) = \frac{\langle \dot{x}(0)\delta(x^\dagger - x(0))H(x(t) - x^\dagger)\rangle}{\langle \dot{x}(0)\delta(x^\dagger - x(0))H(x(0^+) - x^\dagger)\rangle} \tag{9.24}$$

The denominator of Eq. (9.24) is an average over all equilibrium trajectories crossing the dividing surface in the direction from A to B. The numerator, on the other hand, is an average over those trajectories crossing the dividing surface that find themselves in state B after time t. When the correlation time τ_{cor} and the waiting time between transitions are widely separated, the ratio $f_d(t)$ will quickly reach an asymptotic value, f_d, that does not depend on time. The transition rate constant can then be calculated as:

$$k_{A \to B} = f_d k_{A \to B}^{TST} \tag{9.25}$$

Equation (9.25) forms a basis for computing dynamically correct transition rate constants from simulation, based on transition-state theory. This will be illustrated in Section 9.3.

9.1.3
Multidimensional Transition State Theory

This formulation for calculating the rate constant of infrequent transitions from state A to state B can readily be extended to the case where more than one degree of freedom are involved in effecting a transition. In this case one works with a potential of mean force U that depends on a number of degrees of freedom, constituting a vector R [compare Eq. (7.47)]. If *all* configurational degrees of freedom of the system are included in R, then U is none else than the potential energy function $\mathcal{V}$. Multidimensional TST was first treated by Vineyard [7], in the context of self-diffusion in crystalline solids. There, the probability per unit time of observing an elementary jump of an atom from an interstitial position into an adjacent one depends not only on the coordinates of the atom itself but also on the coordinates of surrounding atoms in the lattice, which has to undergo local distortion for the elementary jump to occur. For the mathematical development of multidimensional TST it is convenient to work not with the atomic coordinates themselves but with mass-weighted atomic coordinates. These are defined as $x_i = R_i \, m_i^{1/2}$, where m_i is the mass of atom i. Thus, in place of R we will use the vector of mass-weighted atomic coordinates, x, whose dimensionality will be indicated by d. The origin state A and the destination state B are basins in the d-dimensional x-space constructed around neighboring local minima of $U(x)$. Let us denote these minima by x_A and x_B. The dividing surface is a $(d-1)$-dimensional hypersurface in x-space, which we can represent mathematically by an equation of the form $C(x) = 0$. Although there is some latitude in the definition of the dividing surface, it is advantageous to define it according to the following requirements:

1) The dividing surface passes through the first-order saddle point, $x^\dagger$, lying between x_A and x_B, which corresponds to the considered transition. By definition, the gradient $\nabla_x U$ satisfies:

$$\nabla_x U|_{x_A} = \nabla_x U|_{x_B} = \nabla_x U|_{x^\dagger} = 0 \tag{9.26}$$

while the Hessian matrix of second derivatives $\partial^2 U/\partial x \partial x^T$ is positive definite at x_A and x_B and has exactly one negative eigenvalue at $x^\dagger$. For the dividing surface we require:

$$C(x^\dagger) = 0 \tag{9.27}$$

2) At $x^\dagger$, the dividing surface is normal to the eigenvector $n^\dagger$ corresponding to the negative eigenvalue of the Hessian matrix, $\frac{\partial^2 U}{\partial x \partial x^T}\bigg|_{x^\dagger}$:

$$\frac{\nabla_x C(x)}{|\nabla_x C(x)|}\bigg|_{x=x^\dagger} \equiv n(x^\dagger) = n^\dagger \tag{9.28}$$

where we have used the symbolism $n(x)$ to denote the unit vector normal to the dividing surface at x, pointing away from state A.

3) At all points other than $x^\dagger$, the dividing surface is tangent to the gradient vector of $U(x)$:

$$\nabla_x C(x) \cdot \nabla_x U = 0, x \neq x^\dagger \tag{9.29}$$

The saddle point $x^\dagger$ is often referred to as the "transition state." The reaction path is a one-dimensional line in the d-dimensional x-space, which connects x_A and x_B and passes through $x^\dagger$. It is tangent to $n^\dagger$ at $x^\dagger$ (and therefore normal to the dividing surface at this point), and tangent to the gradient vector $\nabla_x U$ everywhere else. Note that there may be several reaction paths between two neighboring states A and B, each going through a different first-order saddle point of U. In the treatment we present here, each such path, and its dividing surface, would be treated separately and the total rate constant for transition from A to B would be obtained by summing the individual rate constants for each path. In the following discussion we are concerned with only one path between A and B.

Having defined the origin and destination states, the transition state, the reaction path, and the dividing surface based on the topography of the $U(x)$ hypersurface, we are now ready to compute the rate constant $k_{A \to B}$.

According to the TST approximation, whenever the system finds itself on the dividing surface with net momentum from state A to state B a successful transition from A to B will occur. Integrating the probability flux over the entire dividing surface one then obtains:

$$k_{A \to B}^{TST} P_A^{eq} = \int d^d x \int_{n(x) \cdot p > 0} d^d p [n(x) \cdot p] \delta[C(x)] |\nabla_x C(x)| \varrho^{NVT}(x, p) \qquad (9.30)$$

where $p = \dot{x}$ is the generalized momentum conjugate to the mass-weighted coordinate x, that is, the d-dimensional vector collecting components $m_i^{1/2} \dot{R}_i$ from each degree of freedom participating in the transition, and $\varrho^{NVT}(x, p)$ is the phase-space probability density [compare Eq. (7.3)]. The factor $\delta[C(x)]|\nabla_x C(x)|$ in Eq. (9.30) ensures that the integral over x is restricted to the dividing surface. On the other hand, the equilibrium probability of residing in state A is:

$$P_A^{eq} = \int_{\text{state A}} \varrho^{NVT}(x, p) \, d^d x \, d^d p \qquad (9.31)$$

Performing the momentum-space integrations in Eqs. (9.30) and (9.31) leads to:

$$k_{A \to B}^{TST} = \frac{1}{(2\beta\pi)^{1/2}} \frac{\int_{\text{state A}} d^d x \, \delta[C(x)] |\nabla_x C(x)| \exp[-\beta U(x)]}{\int_{\text{state A}} d^d x \, \exp[-\beta U(x)]} \qquad (9.32)$$

or, in simpler form:

$$k_{A \to B}^{TST} = \frac{1}{(2\beta\pi)^{1/2}} \frac{\int_{\text{div. surf.}} d^{d-1} x \, \exp[-\beta U(x)]}{\int_{\text{state A}} d^d x \, \exp[-\beta U(x)]} \qquad (9.33)$$

Equation (9.33) is a multidimensional analog of Eq. (9.22). Again, the transition-state estimate emerges as the product of a molecular speed times the ratio of two configurational integrals: a $(d-1)$-dimensional one, taken over the dividing surface, and a d-dimensional one, taken over the entire origin state. Recall that each of the components of x has dimensions of length times mass to the 1/2 power. As a consequence, the rate constant calculated through Eq. (9.33) has dimensions of inverse time, as it should. As in Eq. (9.22), the ratio of configurational integrals has the physical meaning of a conditional probability, per unit length along the reaction coordinate, that the system will find itself on the dividing surface, provided it is allowed to sample the origin state according to its equilibrium distribution.

An alternative expression for the TST estimate of the rate constant can be obtained if we start from Eq. (9.30), perform the integration only over the component of p that is normal to the dividing surface, divide the equation we obtain by Eq. (9.31), and recall the form of the probability density in phase space, Eq. (7.3), the definition of the partition function, Eq. (7.4), and the definition of the potential of mean force, Eq. (7.47):

$$k_{A \to B}^{TST} = \frac{k_B T}{h} \frac{Q^\dagger}{Q_A} = \frac{k_B T}{h} \exp\left(-\frac{A^\dagger - A_A}{k_B T}\right) \qquad (9.34)$$

Equation (9.34) (whence one obtains the TST relation for cage-to-cage jumps employed in Sections 4.3.3 and 16.5.2) involves the ratio of two partition functions, rather than of two configurational integrals. The partition function in the numerator, involving $(3N-1)$ configurational and $(3N-1)$ momentum degrees of freedom, is confined to the dividing surface between states A and B. The partition function in the denominator, involving $3N$ configurational and $3N$ momentum degrees of freedom, refers to the system confined in the origin state, A. The factor $k_B T$ in front of the ratio of partition functions arises from integration over the component of momentum normal to the dividing surface, while the factor h stems from the difference in dimensionality between the systems to which the two partition functions refer [compare Eq. (7.4)]. Equivalently, the ratio of partition functions is expressed as a Boltzmann factor of the Helmholtz energy difference between the dividing surface and the origin state. When changes in volume of the medium participate in the reaction coordinate for the transition, the canonical partition functions Q in Eq. (9.34) are replaced by isothermal–isobaric partition functions and the Helmholtz energies A are replaced by Gibbs energies G. Equation (9.34) is useful when some degrees of freedom playing a role in the transition (e.g., electronic, translational associated with very small mass, or vibrational of frequency $\gg k_B T/h$) have to be treated quantum mechanically; for such degrees of freedom a separation of configuration and momentum degrees of freedom is not allowed in calculating the partition function, and one has to work with the partition functions themselves, rather than with the configurational integrals. Usually, however, diffusion problems involve slow enough motions, for which a classical treatment of momentum degrees of freedom incurs little error. Thus, Eq. (9.33) for the rate constant according to TST in terms of a ratio of configurational integrals is, practically, more useful than Eq. (9.34).

Molecular simulations offer several techniques for calculating the ratio of configurational integrals in Eq. (9.33). Some of these are:

1) Identification of the dividing surface and of state A in d-dimensional space and direct evaluation of the configurational integrals by MC integration [8, 9];
2) expression of the ratio of configurational integrals as a product of ratios of configurational integrals involving a series of intermediate $(d-1)$-dimensional hypersurfaces normal to the reaction path between x and $x^\dagger$, and calculation of the individual ratios by the free energy perturbation method via MC or MD [compare Eq. (7.40)] [9];
3) calculation of density profiles within thin slices of x-space normal to the reaction path through a series of MC or MD simulations constrained to these slices, Boltzmann inversion of the profiles obtained to form configurational free energy

profiles along the reaction path, and matching of the latter profiles between adjacent slices to construct a global free energy profile between x and $x^\dagger$ (this technique is often referred to as "umbrella sampling");

4) use of the "blue moon" ensemble, which effectively drags the system from x to $x^\dagger$ along the reaction coordinate and calculates the free energy profile as an integral of the work done by the force required to perform the dragging [10].

Whatever technique is chosen to compute k^{TST} according to Eq. (9.33), it is important to realize that this computation can be accomplished with reasonable resources, no matter how small k^{TST} is. This is in sharp contrast to "brute force" MD, which, for a high barrier $U(x^\dagger) - U(x) \gg k_B T$, would get stuck within state A and never access transitions out of it. By focusing on the dividing surface, TST manages to overcome the time scale limitation of "brute force" MD. It does, however, require that one have a good idea of the variables x playing a significant role in the transition and of the states, dividing surface, and reaction path involved. In diffusion problems, the geometry of the medium (analysis of accessible volume) can be a very helpful guide for this [9, 11].

Transition state theory is not exact. As discussed in the one-dimensional treatment above, not all trajectories penetrating the dividing surface in the direction from A to B are successful in effecting a transition, because of recrossing events. A dynamical correction factor, f_d, can be computed for use in Eq. (9.25) through the following analog of Eq. (9.24):

$$f_d = \frac{\langle n(x(0)) \cdot p(0) \delta[C(x(0))] |\nabla C(x(0))| h_B(x(t)) \rangle}{\langle n(x(0)) \cdot p(0) \delta[C(x(0))] |\nabla C(x(0))| h_B(x(0^+)) \rangle} \quad (9.35)$$

where $h_B(x) = 1$ when x finds itself within state B and 0 otherwise. Equation (9.35) calls for (i) sampling a large number of initial configurations $x(0)$ on the dividing surface between A and B according to the equilibrium (Boltzmann) distribution of the system, $\varrho^{NVT}(x) \propto \exp[-\beta U(x)]$; (ii) imparting Boltzmann-distributed initial momenta, $p(0)$, to each of these configurations; and (iii) integrating the equations of motion starting at $x(0)$, $p(0)$, until the system thermalizes in one of the two states. As in the one-dimensional case, the numerator receives contribution from all trajectories that thermalize in state B, regardless of the initial direction of the momentum, while the denominator receives contribution from all trajectories for which $p(0)$ was originally directed from A to B. Thermalization will be achieved quickly (within time $\tau_{\text{cor}} \ll k_{\text{eff}}^{-1}$), yielding a time-independent value for f_d. As a consequence, the MD runs for the determination of f_d will entail only modest computational cost. By the time-reversibility of dynamical trajectories, the value of f_d will be the same for the forward (A → B) and for the backward (B → A) transitions. As a consequence, the dynamically corrected transition rate constants $k_{A \to B}$, $k_{B \to A}$ obtained from Eq. (9.25) and its analog for the inverse transition will satisfy the condition of microscopic reversibility or detailed balance, Eq. (9.18). Note that the transition state estimates $k_{A \to B}^{\text{TST}}$, $k_{B \to A}^{\text{TST}}$ do satisfy microscopic reversibility, as can be seen immediately from Eq. (9.33) and its analog for the inverse transition.

A harmonic approximation to $k_{A \to B}^{TST}$ is possible also in the case of multidimensional TST. If the pathway from A to B is highly restricted, that is, if $U(\mathbf{x})$ rises steeply relative to $k_B T$ as one moves away from $\mathbf{x}^\dagger$ on the dividing surface, then the overwhelming contribution to the integral in the numerator of Eq. (9.33) will come from the vicinity of the saddle point. Under these conditions, the dividing surface may be approximated by a hyperplane normal to the reaction path at the saddle point:

$$C(\mathbf{x}) = \mathbf{n}^\dagger \cdot (\mathbf{x} - \mathbf{x}^\dagger) = 0 \tag{9.36}$$

In the $(d-1)$-dimensional integral over this hyperplane, the potential of mean force $U(\mathbf{x})$ may be approximated through its Taylor expansion to second order around $\mathbf{x}^\dagger$. Similarly, if $U(\mathbf{x})$ presents a deep well around the minimum $\mathbf{x}_A$, then the overwhelming contribution to the integral in the denominator of Eq. (9.33) will come from the vicinity of the minimum. In the d-dimensional integral over the origin state, $U(\mathbf{x})$ may be approximated through its Taylor expansion to second order around $\mathbf{x}_A$. If these approximations are introduced, the integrals in the numerator and denominator of Eq. (9.33) reduce to multidimensional Gaussian integrals and the integrations can be performed analytically. The result is:

$$k_{A \to B}^{TST,HA} = \frac{1}{2\pi} \frac{\prod_{a=1}^{d} \omega_{a,A}}{\prod_{a=2}^{d} \omega_a^\dagger} \exp\left[-\frac{U(\mathbf{x}^\dagger) - U(\mathbf{x}_A)}{k_B T}\right] \tag{9.37}$$

The quantities $\omega_{a,A}$ ($a = 1, 2, \ldots, d$) are the normal mode angular frequencies of the system at the minimum $\mathbf{x}_A$. They are obtained readily as square roots of the eigenvalues of the Hessian matrix of second derivatives of the potential of mean force with respect to mass-weighted atomic coordinates at the minimum:

$$\left.\frac{\partial^2 U}{\partial \mathbf{x} \partial \mathbf{x}^T}\right|_{\mathbf{x}_A}$$

The quantities $\omega_a^\dagger$ ($a = 2, \ldots, d$) are the normal mode angular frequencies of the system confined to the dividing surface at the saddle point $\mathbf{x}^\dagger$. They are obtained as square roots of the positive eigenvalues of the Hessian matrix at the saddle point:

$$\left.\frac{\partial^2 U}{\partial \mathbf{x} \partial \mathbf{x}^T}\right|_{\mathbf{x}^\dagger}$$

Observe that the number of $\omega_a^\dagger$ is, by one, less than the number of $\omega_{a,A}$, as the eigendirection of negative curvature, $\mathbf{n}^\dagger$, does not contribute an angular frequency. Thus, the pre-exponential factor in Eq. (9.37) has dimensions of frequency, as it should. The exponential incorporates the barrier in potential of mean force,

$U(x^\dagger) - U(x_A)$. There is an obvious analogy between Eq. (9.37) and its one-dimensional counterpart, Eq. (9.23).

For an accurate calculation of $k_{A \to B}$, $k_{B \to A}$, it is imperative that all relevant degrees of freedom be included in x. If, for the sake of economy in computational cost, we invoke an x of lower dimensionality we may introduce serious errors in the estimation of the rate constants. This is shown characteristically in Figure 9.2, which is adapted from Bolhuis et al. [12]. A one-dimensional treatment of the transition would be satisfactory for the system of the left-hand side of the figure, but very poor for the system on the right-hand side.

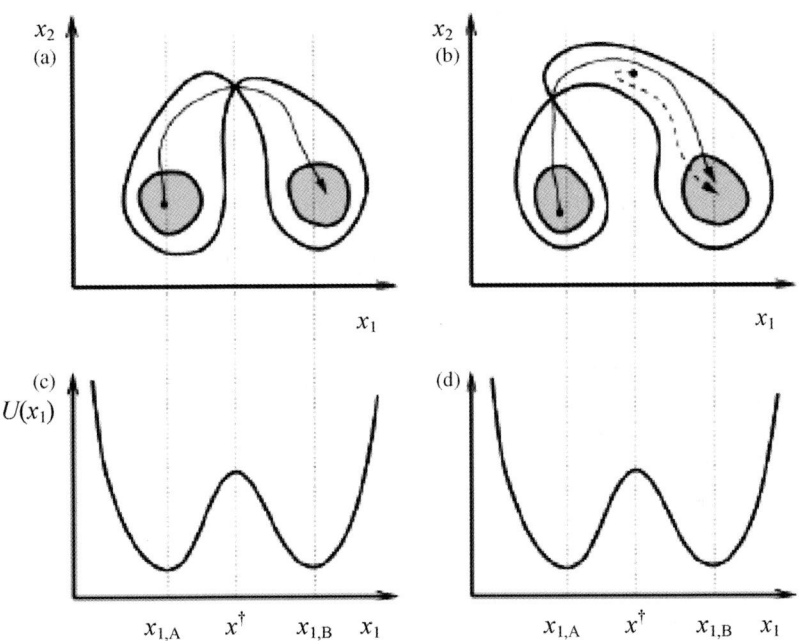

Figure 9.2 Contours of the potential energy $\mathcal{V}(x_1, x_2)$ for two two-dimensional systems (a, b) and corresponding potentials of mean force $U(x_1) = -k_B T \ln \int \exp[-\beta \mathcal{V}(x_1, x_2)] \, dx_2$ obtained by integration over x_2 (c, d). The two-dimensional potential energy $\mathcal{V}(x_1, x_2)$ possesses two minima (A, B) at roughly the same x_2 values and a saddle point at considerably higher x_2. The reaction path from A to B via the saddle point is shown as a solid line terminating in an arrow in (a) and (b). The one-dimensional potentials of mean force $U(x_1)$ are very similar, exhibiting a barrier at $x_1 = x^\dagger$, although the two-dimensional potential energy surfaces $\mathcal{V}(x_1, x_2)$ of the two systems are quite different. For the (a, c) system, x_1 is a reasonable reaction coordinate, because the dividing surface between states A and B is well approximated by the line $x_1 = x^\dagger$. Application of infrequent event analysis in one dimension, using $U(x_1)$, is expected to yield a good estimate of the rate constant $k_{A \to B}$. For the (b, d) system, x_1 is not a reasonable reaction coordinate. The orthogonal variable x_2 plays a very important role in the A $\to$ B transition, and the dividing surface for the transition is far from the maximum in $U(x_1)$, $x_1 = x^\dagger$. The dashed trajectory of the (b, d) system illustrates this point. It would be considered as contributing to the A $\to$ B transition in the one-dimensional analysis but, in reality, it is internal to state B. Adapted from Reference [12].

9.1.4
Infrequent Event Theory of Multistate Multidimensional Systems

In diffusion problems one has more than two states. There is a large number of states i between which the system can execute jumps $i \to j$, the conditional probability per unit time that the system will move to j provided it is in i being given by the rate constant $k_{i \to j}$. Infrequent event theory has been generalized to multistate systems by Voter and Doll [6]. We briefly review some basic elements of the formulation below.

Each state i in d-dimensional space is surrounded by a $(d-1)$-dimensional hypersurface that delimits it from the remaining states. In a manner analogous to the two-state case, we will assume that this boundary to state i is described by an equation $C_i(x) = 0$, where C_i is a continuous, differentiable function of the mass-weighted coordinates x. We assume:

$$C_i(x) \begin{cases} < 0, & \text{if } x \text{ is in state } i \\ 0, & \text{if } x \text{ is on the boundary of state } i \\ > 0, & \text{if } x \text{ is outside state } i \end{cases} \quad (9.38)$$

Then, $n_i(x) = \nabla C_i(x)/|\nabla C_i(x)|$ will be the unit vector normal to the boundary surface of state i at point x pointing towards the outside of state i [compare Eq. (9.28)]. We introduce the function:

$$h_i(x) = 1 - H[C_i(x)] \quad (9.39)$$

with $H(x)$ being the Heaviside step function. By definition, $h_i(x)$ equals 1 if x is in state i and 0 otherwise. The rate constant for transitions from state i to state j can then be expressed as:

$$k_{i \to j} = \frac{\langle n_i(x(0)) \cdot \dot{x}(0) \delta[C_i(x(0))] |\nabla C_i(x(0))| h_j(x(t)) \rangle}{P_i^{eq}} \quad (9.40)$$

Equation (9.40) is a multistate, multidimensional generalization of Eq. (9.20). The numerator considers all dynamical trajectories that cross the boundary surface of state i at time 0 and find themselves in state j after a (long) time t. The averaged quantity is the component of the generalized momentum normal to the surface delimiting state i.

In the multistate case, the transition state theory approximation amounts to assuming that, whenever the system exits state i through its boundary surface, it will remain outside state i at all subsequent times. As in the two-state case [compare derivation of Eqs. (9.30)–(9.33)], this leads to an expression for the rate constant for exiting state i based on transition state theory:

$$k_{i \to}^{TST} = \frac{1}{(2\beta\pi)^{1/2}} \frac{\int_{\text{bound.surf. of state } i} d^{d-1}x \exp[-\beta U(x)]}{\int_{\text{state } i} d^d x \exp[-\beta U(x)]} \quad (9.41)$$

As pointed out above, Eq. (9.41) involves a ratio of two configurational integrals and can be obtained with modest computational cost once state i and its boundary surface are known, overcoming the long-time limitation of "brute force" MD. Remember that x in Eq. (9.41) are mass-weighted coordinates, so the ratio of configurational integrals has dimensions of inverse length multiplied by the inverse square root of mass. The actual rate constant $k_{i \to j}$ for transitions from i to j is obtained from $k_{i \to}^{TST}$ through multiplication by a dynamical correction factor $f_{d,i \to j}$:

$$k_{i \to j} = k_{i \to}^{TST} f_{d,i \to j} \qquad (9.42)$$

As in Eq. (9.35), the dynamical correction factor can be obtained as:

$$f_{d,i \to j} = \frac{\langle \boldsymbol{n}_i(\boldsymbol{x}(0)) \cdot \dot{\boldsymbol{x}}(0) \delta[C_i(\boldsymbol{x}(0))] |\nabla C_i(\boldsymbol{x}(0))| h_j(\boldsymbol{x}(t)) \rangle}{\langle \boldsymbol{n}_i(\boldsymbol{x}(0)) \cdot \dot{\boldsymbol{x}}(0) \delta[C_i(\boldsymbol{x}(0))] |\nabla C_i(\boldsymbol{x}(0))| [1 - h_i(\boldsymbol{x}(0^+))] \rangle} \qquad (9.43)$$

which can be simplified to:

$$f_{d,i \to j} = \frac{\langle \boldsymbol{n}_i(\boldsymbol{x}(0)) \cdot \dot{\boldsymbol{x}}(0) \delta[C_i(\boldsymbol{x})] |\nabla C_i(\boldsymbol{x}(0))| h_j(\boldsymbol{x}(t)) \rangle}{\frac{1}{2} \langle |\boldsymbol{n}_i(\boldsymbol{x}(0)) \cdot \dot{\boldsymbol{x}}(0)| \delta[C_i(\boldsymbol{x}(0))] |\nabla C_i(\boldsymbol{x}(0))| \rangle} \qquad (9.44)$$

The numerator in $f_{d,i \to j}$ is an average over all dynamical trajectories crossing the boundary surface of state i that ultimately thermalize in state j. The denominator in Eq. (9.43) is an average over all dynamical trajectories crossing the boundary surface of state i in an outward direction. Trajectories initiated on the boundary surface thermalize in a destination state within $\tau_{cor} \ll k_{i \to}^{-1}$ and therefore their sampling entails modest computational cost. A simple sampling scheme for implementing Eq. (9.44) is discussed in Reference [6].

Interestingly, in the multistate formulation of Voter and Doll, transition state theory is applied for the total efflux from state i. The destination state j enters only through the dynamical correction factor $f_{d,i \to j}$ in Eq. (9.42). For adjacent states, sharing parts of their boundary surfaces, $f_{d,i \to j}$ starts off high (equal to the Boltzmann-weighted fraction of the boundary surface of state i that is shared with state j) and quickly decays to an asymptotic value due to dynamical recrossing and fast correlated multistate jumps. This formulation, however, accounts for transitions between nonadjacent states as well. In such a transition, the system may exit i through its boundary surface, spend a short time without thermalizing in one or more intermediate states l, then enter j, which is nonadjacent to i, and ultimately thermalize there. Such transitions are referred to as fast correlated multistate jumps. For a fast correlated multistate jump that "tunnels" from state i to a nonadjacent state j, the dynamical correction $f_{d,i \to j}$ starts off at zero at very short times and then rises to a nonzero asymptotic value [6].

9.1.5
Numerical Methods for Infrequent Event Analysis

Some words on numerical methods for identifying states, transition states, dividing surfaces, and reaction paths are in order. Given $U(\boldsymbol{x})$, local minima can be computed using a nonlinear optimization algorithm. Quasi-Newton algorithms, such as that of Broyden, Fletcher, Goldfarb, and Shanno (BFGS) [13], are particularly effective if d is not too large. For large dimensionality, when storing a numerical estimate of the Hessian becomes prohibitively expensive, one can resort to conjugate gradients algorithms [14]. Determination of the saddle point $x^\dagger$ is more challenging. Cerjan–Miller type algorithms, such as that of Baker [15], have proved useful. These algorithms are based on an iterative procedure of maximization along one mode of the Hessian matrix, with simultaneous minimization along its other modes. Saddle point calculation algorithms that do not require second derivatives are also available, such as the dimer method of Henkelman and Jónsson [16]. In both minimization and saddle point calculations, having a good initial guess, usually based on geometry and on physical understanding of the situation, is very helpful. Once $x^\dagger$ is available, construction of the entire reaction path is relatively straightforward. It entails (i) determining the Hessian at $x^\dagger$, (ii) diagonalizing it to obtain the eigenvector $n^\dagger$ corresponding to its negative eigenvalue, (iii) stepping off $x^\dagger$ by a small length in the $+n^\dagger$ and $-n^\dagger$ directions, and (iv) subsequently taking a sequence of small steps along the steepest descent direction $-\nabla_x U/|\nabla_x U|$ on either side, until a local minimum of U is reached. This construction is often referred to as "Fukui's intrinsic reaction coordinate" [17].

For complex systems, where the variables participating in the reaction coordinate are difficult to anticipate, a new method, referred to as "transition path sampling," has been developed by Chandler and collaborators for elucidating transition mechanisms and computing the corresponding rates. This method samples ensembles of dynamical trajectories connecting two states. These trajectories are generated and manipulated using importance sampling techniques based on statistical mechanics. Transition path sampling is perfectly general, but also computationally very expensive. In zeolites and other microporous solids the geometry of the medium provides considerable guidance concerning the mechanism of elementary diffusive jumps and the number of degrees of freedom involved is relatively small, so more conventional techniques based on the identification of dividing surfaces and reaction paths are usually satisfactory. The interested reader can find information on transition path sampling in some excellent reviews [12, 18].

9.2
Tracking Temporal Evolution in a Network of States

9.2.1
Master Equation

Let us assume that we have performed an exhaustive identification of states i and transition paths $i \to j$ for a penetrant inside a model zeolite and we have computed the rate constants $k_{i \to j}$ for all transitions between states. Forward and reverse transition rate constants must satisfy the condition of microscopic reversibility, Eq. (9.18), which we here rewrite as:

$$P_i^{eq} k_{i \to j} = P_j^{eq} k_{j \to i} \equiv k_{ij} \tag{9.45}$$

The equilibrium probabilities of occupancy of the states P_i^{eq} are normalized within the primary simulation box containing n states. Typically, the simulation box contains a large number of unit cells. If our system is part of a perfect infinite zeolite crystal (without defects), the probabilities P_i^{eq} within a unit cell will exhibit the space group symmetry of the zeolite, while homologous sites in different unit cells will have the same P_i^{eq}. Some links $i \to j$ will occur across the periodic boundaries of the simulation box. With each state i we will associate a position r_i in three-dimensional space that is representative of the position of the center-of-mass of the penetrant when the system resides in that state. The position r_i is best chosen as a Boltzmann-weighted average within the confines of the state. It may be approximated by the position of the center-of-mass at the energy minimum corresponding to the state. Positions r_i ($i = 1, 2, \ldots, n$) will be referred to as sorption sites. They, along with their periodic images and the connections defined by the transitions $i \to j$, form a network that constitutes a coarse-grained, projected three-dimensional rendition of the network of states.

Let $P_i(t)$ be the probability of finding the system in state i at time t. We assume here that the system starts off from an initial probability distribution that may or may not correspond to equilibrium. For example, we may start by placing the penetrant molecule in site i, which corresponds to $P_j(0) = \delta_{ij}$ with δ_{ij} being the Kronecker delta. The quantities $P_i(t)$ will evolve according to the "master equation:"

$$\frac{dP_i}{dt} = -\sum_j k_{i \to j} P_i + \sum_j k_{j \to i} P_j \tag{9.46}$$

By Eq. (9.45), the equilibrium probabilities $\{P_i^{eq}\}$ constitute a stationary solution to the master equation.

An interesting electrical analog to the master equation can be made by considering the probability ratios $p(t) \equiv P_i(t)/P_i^{eq}$. From Eqs. (9.45) and (9.46), these satisfy the equations:

$$P_i^{eq} \frac{dp_i}{dt} = -\sum_j k_{ij}(p_i - p_j) \tag{9.47}$$

If we consider an electrical network where each node i is associated with a capacitance P_i^{eq} and each link with a resistance $1/k_{ij}$, Eq. (9.47) is none other than Kirchhoff's law for the evolution of the electrostatic potential $p(t)$ in the network.

9.2.2
Kinetic Monte Carlo Simulation

A convenient stochastic method for the numerical solution of Eq. (9.46) under equilibrium conditions and the calculation of the self-diffusivity $\mathcal{D}$ is kinetic

Monte Carlo simulation (KMC). This rests on the fact that the sequence of independent interstate transitions undergone by the system constitutes a Poisson process. KMC for a diffusion problem proceeds as follows [8, 11]:

1) Consider a three-dimensional network with a large number n of sites placed at positions r_i, $i = 1, \ldots, n$, with connectivity defined by the rate constants $k_{i \to j}$. Typically, the network would encompass many (say, $100 \times 100 \times 100$) zeolite unit cells. The geometric and topological characteristics and the values of P_i^{eq} and $k_{i \to j}$ that characterize the network's nodes and links are derived from the atomistic analysis approaches described in Section 9.1.

2) Distribute a large number $\mathcal{N}_E$ of random walkers on the sites of the network according to the equilibrium probability distribution $\{P_i^{eq}\}$. To obtain good statistics, $\mathcal{N}_E$ should be much larger than the number of sites among which the walkers are initially distributed. Multiple occupancy of sites is allowed in this deployment of the random walkers. The walkers will be allowed to hop between sites without interacting with each other, that is, they behave as ghost particles towards each other. Each random walker summarily represents a system in the ensemble of microporous medium + penetrant systems whose temporal evolution we want to track with our KMC simulation. Let $\mathcal{N}_i(t)$ be the number of random walkers that find themselves in site i at time t.

3) For each site i that is occupied at the current time t, calculate the expected fluxes $R_{i \to j}(t) = \mathcal{N}_i(t) \, k_{i \to j}$ to all sites j to which it is connected. In addition, compute the overall flux $R(t) = \sum_i \sum_j R_{i \to j}(t)$ and the probabilities $q_{i \to j}(t) = R_{i \to j}(t)/R(t)$.

4) Generate a random number $\xi \in [0,1)$. Choose the time for occurrence of the next elementary jump event in the network as $\Delta t = -[\ln(1 - \xi)/R(t)]$. Choose the type of the next elementary jump event by picking one of the possible transitions $i \to j$ according to the probabilities $q_{i \to j}(t)$.

5) Of the $\mathcal{N}_i(t)$ walkers present in site i, pick one with probability $1/\mathcal{N}_i(t)$ and move it to site j.

6) Advance the simulation time by Δt. Update the array, keeping track of the current positions of all walkers to reflect the implemented hop. Update the occupancy numbers $\mathcal{N}_i(t + \Delta t) = \mathcal{N}_i(t) - 1$ and $\mathcal{N}_j(t + \Delta t) = \mathcal{N}_j(t) + 1$.

7) Return to step (3) to implement the next jump event.

The outcome from performing this stochastic simulation over a large number of steps is a set of trajectories $r(t)$ for all $\mathcal{N}_E$ random walkers. Each trajectory consists of a long sequence of jumps between sites of the network. The self-diffusivity is estimated from these trajectories through the Einstein relation, Eq. (8.2). When all rate constants $k_{i \to j}$ are small, the KMC simulation will take large strides Δt on the time axis. Thus, times on the order of milliseconds, seconds, or even hours can be accessed, which are absolutely prohibitive for "brute force" MD.

For exploration of the effects of surface resistances or intracrystalline barriers generated by twinning or other morphological defects, one may resort to a simple form of kinetic Monte Carlo simulations. In these simulations, diffusant trajectories are generated as sequences of discrete jumps in a lattice that does not necessarily represent the actual pore structure of the zeolite, with jump rate constants that are not necessarily extracted from the detailed potential energy hypersurface, but which are chosen so as to be consistent with macroscopically measured self-diffusivities in a single crystal along different directions. Such simplified kinetic Monte Carlo simulations are referred to as "dynamic Monte Carlo simulations" (see, for example, Sections 11.4.3.1, 17.2.4, 18.6.4, 18.6.5, and 18.8.1).

9.2.3
Analytical Solution of the Master Equation

When the rate constants $k_{i \to j}$ are very widely distributed, KMC may become inefficient. This is because time steps Δt must be short enough to track the fastest processes occurring in the system. In such cases of extreme dynamical heterogeneity it may be better to resort to an analytical solution of the master equation, Eq. (9.46), for the time-dependent state occupancy probabilities $\{P_i(t)\}$, given an initial condition for $\{P_i(0)\}$. We briefly outline here how such an analytical solution can be developed through spectral decomposition. Details of the formulation, which is very similar to that of Wei and Prater for the kinetics of a network of reversible first-order chemical reactions [19], are presented in Reference [20].

Again we consider a finite system consisting of a very large number of unit cells, with periodic boundary conditions. Let the total number of states of the system, and hence of the sorption sites contained in it, be n. We collect the state probabilities $P_i(t)$ to form an n-dimensional vector $\boldsymbol{P}(t)$. This time we consider an initial distribution of the penetrant that departs from equilibrium. For example, we consider an initial distribution $\boldsymbol{P}(0)$ among the sites of the system that is strongly peaked at the center. As previously, $\boldsymbol{P}(t)$ will be normalized over all sites in the system at all times. The master Eq. (9.46) can be written, in compact form, as:

$$\frac{\partial \boldsymbol{P}(t)}{\partial t} = \boldsymbol{K}\boldsymbol{P}(t) \tag{9.48}$$

with initial condition $\boldsymbol{P}(0)$. The $n \times n$ rate constant matrix $\boldsymbol{K}$ is defined by $K_{ij} = k_{j \to i}$ for $j \neq i$ and $K_{ii} = -\sum_{j \neq i} k_{i \to j}$.

To proceed with an analytical solution of Eq. (9.48), we transform the state probability vector $\boldsymbol{P}(t)$ into a reduced vector $\tilde{\boldsymbol{P}}(t)$ with elements:

$$\tilde{P}_i(t) = P_i(t)/\sqrt{P_i^{eq}}$$

This satisfies the reduced master equation:

$$\frac{\partial \tilde{\boldsymbol{P}}(t)}{\partial t} = \tilde{\boldsymbol{K}}\tilde{\boldsymbol{P}}(t)$$

with:

$$\tilde{K}_{ij} = K_{ij}\sqrt{P_j^{eq}}/\sqrt{P_i^{eq}}$$

The matrix $\tilde{\boldsymbol{K}}$ is symmetric by virtue of detailed balance, Eq. (9.45). One can readily show that it has the same eigenvalues as $\boldsymbol{K}$. Of these eigenvalues, one (corresponding to the establishment of the equilibrium distribution among states) is zero, and the remaining eigenvalues are negative. We denote these eigenvalues by $\lambda_0 = 0 \geq \lambda_1 \geq \cdots \geq \lambda_{n-1}$. We symbolize by $\tilde{u}_m = (\tilde{u}_{1,m}, \tilde{u}_{2,m}, \ldots, \tilde{u}_{i,m}, \ldots, \tilde{u}_{n,m})$ the eigenvector of $\tilde{K}$ corresponding to eigenvalue λ_m, $0 \leq m \leq n-1$. Note that:

$$\tilde{u}_{i,0} = \tilde{P}_i^{eq} = \sqrt{P_i^{eq}}$$

corresponding to the equilibrium distribution among states. The Euclidean norm of $\tilde{u}_0$ is unity by the normalization of P_i^{eq}. The solution of the master equation can be written as:

$$\tilde{P}(t) = \sum_{m=0}^{n-1} [\tilde{u}_m \cdot \tilde{P}(0)] e^{\lambda_m t} \tilde{u}_m = \tilde{P}^{eq} + \sum_{m=1}^{n-1} [\tilde{u}_m \cdot \tilde{P}(0)] e^{\lambda_m t} \tilde{u}_m \quad (9.49)$$

where the normalization condition $\sum_{j=1}^{n} P_j(0) = 1$ has been used in separating out the equilibrium contribution ($\lambda_0 = 0$). The eigenvectors $\tilde{u}_m$ form an orthonormal basis set:

$$\tilde{u}_m \cdot \tilde{u}_l = \delta_{ml}, \, 0 \leq m, l \leq n-1 \quad (9.50)$$

with δ_{ml} being the Kronecker delta. They also satisfy $\sum_{m=0}^{n-1} \tilde{u}_{i,m} \tilde{u}_{j,m} = \delta_{ij}$.

It is convenient to apply the solution, Eq. (9.49), starting from an initial distribution entirely localized in a specific state i, close to the center of the system: $P_j(0) = \delta_{ij}$. Under these conditions, Eq. (9.49) simplifies to:

$$\tilde{P}(t) - \tilde{P}^{eq} = \sum_{m=1}^{n-1} \frac{\tilde{u}_{i,m}}{\sqrt{P_i^{eq}}} \tilde{u}_m e^{\lambda_m t} \quad (9.51)$$

with all coefficients obtainable from a diagonalization of $\tilde{K}$. If we focus on i and all its homologous sites inside the network and consider times t long in relation to the time required for traversing a single unit cell, but short in comparison to the time required to reach the borders of our periodic system, the probability $P(r,t)$ of occupancy obtained from Eq. (9.51) as a function of the positions r of these sites and time t will look like a three-dimensional Gaussian, that is, like the solution to the continuum diffusion equation for a delta function initial condition. By matching Eq. (9.51) to this continuum solution, the diffusivity tensor can be extracted.

9.3
Example Applications of Infrequent Event Analysis and Kinetic Monte Carlo for the Prediction of Diffusivities in Zeolites

9.3.1
Self-diffusivity at Low Occupancies

An early application of dynamically corrected transition state theory and kinetic Monte Carlo simulation to diffusion in zeolites can be found in the work of June *et al.* on low-occupancy diffusion of xenon and SF_6 in silicalite [8]. The mobility of xenon was large enough to allow direct determination of the self diffusivity via "brute force" MD. Thus, a direct test of TST predictions against "exact" results from MD for the same molecular model was possible.

Calculations employed an infinitely stiff orthorhombic model for silicalite; thus, the vector of degrees of freedom x in which the analysis of elementary jumps was conducted was three-dimensional, corresponding to the mass-weighted position vector of a Xe molecule within the zeolite. First, an exhaustive determination of the minima of $U(x)$ [which coincide with the minima of $\mathcal{V}(r)$ for the inflexible model of the zeolite employed] was undertaken: The entire volume of the asymmetric unit of the unit cell was subdivided into small cubic elements (voxels) of edge length 0.2 Å. The potential energy felt by the Xe penetrant and appropriate derivatives were evaluated at each node of the grid formed by the voxels. Voxels of very high energy were designated as wall regions, inaccessible to the sorbate molecule. Within the remaining accessible (pore) regions, a continuous three-dimensional interpolation scheme was set up, based on the values calculated at the nodes for the fast determination of the energy, gradient, and Hessian. Next, a potential energy minimization was conducted starting from the center of each voxel lying in the accessible region of the zeolite. Energy minimizations terminated in a set of well-defined local minima of the energy. They served to define a "drainage pattern" in three-dimensional space, whereby each voxel was assigned to a local energy minimum (the one where the energy minimization initiated at the voxel terminated). In this way, states i were unequivocally determined in three-dimensional space and the entire accessible volume of the zeolite was partitioned into states. The dividing surface between states i and j was defined as the set of voxel facets (squares) that were shared between voxels assigned to state i and voxels assigned to state j. The (closed) boundary surface of each state i was also defined as the union of surfaces dividing it from other states and surfaces separating it from the wall regions.

For Xe in silicalite, a total of 24 minima were determined per unit cell: three in each sinusoidal channel segment, two in each straight channel segment, and one in each intersection (remember that, in silicalite, there are four straight channel segments, four sinusoidal channel segments, and four intersections per unit cell). The three minima in each sinusoidal channel segment were found to be separated from each other by low-lying barriers (in comparison to $k_B T$); the states of these minima were thus lumped into a single state, representative of a sinusoidal channel segment. Likewise, the barrier between the two minima in each straight channel segment was low, and the states of these minima were lumped into a single state, representative of a straight channel segment. After the lumping, a model with 12 states per silicalite unit cell was obtained. There are four straight channel states (S), four sinusoidal, or zigzag, channel states (Z), and four channel intersections (I) per unit cell. Figure 9.3a shows the boundary surfaces of the three types of states. Xenon in silicalite prefers to reside in the Z and S states and much less in the I state. The equilibrium occupancy probabilities of the states at 150 K, normalized over 1/4th of the unit cell, are $P_Z^{eq} = 0.572$, $P_S^{eq} = 0.414$, $P_I^{eq} = 0.014$. The positions of sorption sites corresponding to the 12 states within a unit cell are characterized by the symmetries of the *Pnma* space group of silicalite. All states in an infinite zeolite crystal can be obtained by

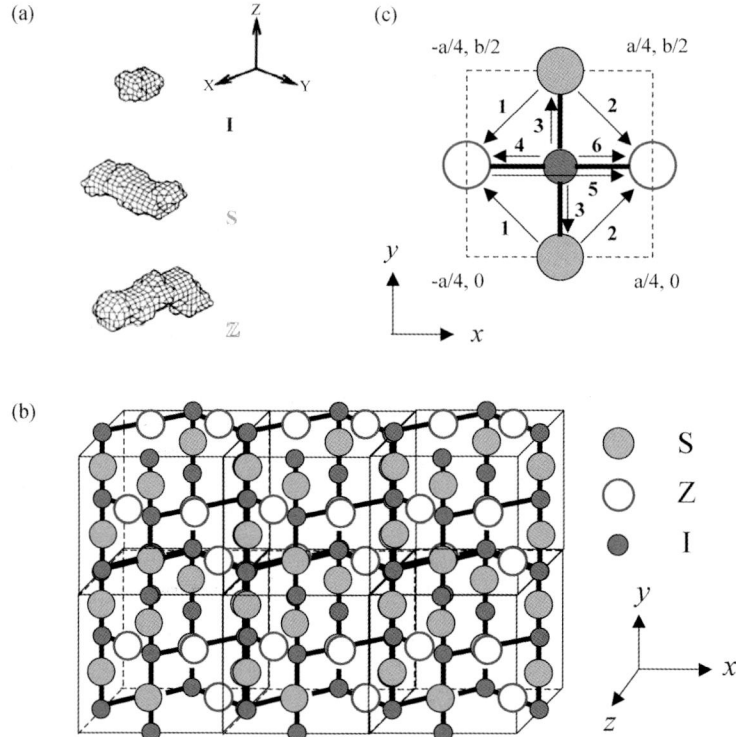

Figure 9.3 (a) Boundary surfaces of the straight channel (S), sinusoidal channel (Z), and intersection (I) states for Xe in silicalite, as determined by the analysis of June et al. [8]. (b) Periodic arrangement of the sites corresponding to these states in three-dimensional space. The bold straight lines provide a coarse rendition of the zeolite channel axes. (c) The six different transitions among states around an intersection, for which forward and reverse rate constants were estimated by the transition state theory analysis of June et al.

translating the 12 states in a unit cell along the three lattice vectors. The entire pore space is tessellated into such states. Figure 9.3b shows the spatial disposition of the periodic array of states.

June et al. [8] identified six distinct types of connections among the states for Xe in silicalite, which are shown schematically in Figure 9.3c. For each transition marked in the figure there is a reverse transition, obtained by reversing the direction of the arrow. Some transitions are marked with the same type, and therefore have the same rate constant, because of the symmetry conditions imposed by the structure of the zeolite. There are I → S (3) and I → Z (4,6) transitions, but also S → Z (1,2) and a Z → Z (5) transition.

Rate constants for all these transitions and their inverses were computed by dynamically corrected transition-state theory, following the methodology of Voter and Doll for multistate, multidimensional systems. First the total rate constant for exiting each state, $k_{i\rightarrow}^{TST}$, was estimated through transition state theory via Eq. (9.41). The two-dimensional integral over the boundary surface of the state (numerator) was computed via Monte Carlo integration over the set of voxel facets constituting the boundary surface; facets separating the state from wall regions have zero contribution in this integration. The three-dimensional integral over the state (denominator) was likewise computed by Monte Carlo integration over the set of voxels constituting the state. A total of 200 function evaluations were used for each facet or voxel in the MC integration scheme to ensure an uncertainty of approximately 1% in the integral. The reader is reminded that x in Eq. (9.41) is the vector of mass-weighted coordinates for the transition, and therefore the ratio of configurational integrals contributes an inverse square root of mass (that of a Xe atom) to $k_{i\rightarrow}^{TST}$.

Next, 1000 points on all facets of the boundary surface were sampled according to the Boltzmann weight of their potential energy. An initial velocity vector was assigned to each point from a Boltzmann distribution, and a three-dimensional trajectory of duration 10 ns was generated from that point through microcanonical MD. In practice, groups of these trajectories were generated in parallel as trajectories of noninteracting Xe molecules in the rigid zeolite framework. The trajectories quickly thermalized at a destination state j, allowing calculation of the dynamical correction factors $f_{d,i\rightarrow j}$ via Eq. (9.44). The resulting dynamically corrected rate constants $k_{i\rightarrow j}$ were used within a continuous time, discrete space kinetic Monte Carlo simulation to track the progress of 1000 ghost sorbate molecules initially distributed among the sorption sites according to the equilibrium probabilities $\{P_i^{eq}\}$. The KMC simulation was carried out for 8×10^5 moves; the total time simulated was 1.5 µs at 100 K, much beyond the capabilities of "brute force" MD at the time of the computation.

Table 9.1 shows values of the self-diffusivity at low occupancy, $\mathcal{D}$, predicted for Xe in silicalite at three different temperatures through the infrequent event calculations of June et al. [8], using an infinitely stiff model for the zeolite. The table also gives values of $\mathcal{D}$ obtained by direct MD simulation; these constitute "exact" results for the potential energy model employed and are in excellent agreement with pulsed field gradient NMR (PFG NMR) measurements by Kärger et al. [21]. No value for 100 K is provided in the MD column of Table 9.1 because, as discussed above, diffusion at this

Table 9.1 Low-temperature diffusion of Xe in silicalite as predicted by dynamically corrected TST and KMC-based infrequent event analysis and by "brute force" MD simulation.

Parameter	$10^4 \times$ Self-diffusivity $\mathcal{D}$ (cm^2 s^{-1})	
T (K)	Dynamically corrected transition state theory (TST)	Molecular dynamics (MD)
100	0.046	
200	0.44	0.51
300	1.11	1.5
Activation energy for diffusion (kJ mol^{-1})	5.2	5.5
Relative CPU time	1	10

temperature was too slow to be predicted reliably by MD. We see that the infrequent-event analysis gives estimates for the self-diffusivity and its activation energy that are in very favorable agreement with the full MD calculation, in a small fraction of the CPU time.

The rates of diffusion of aromatic hydrocarbons in silicalite cannot be predicted, unless infrequent event methods are invoked. Snurr et al. [1] conducted transition-state theory calculations on benzene at infinite dilution in silicalite in six dimensions (three translational and three orientational), using an infinitely stiff zeolite model. Their approach relied on an exhaustive determination of all saddle points of the energy in the six-dimensional configuration space of the system and subsequent construction of reaction paths through them. Rate constants were calculated by TST invoking the harmonic approximation, Eq. (9.37). A wealth of mechanistic information was extracted on translational and reorientational motions executed by the sorbed benzene. The estimated self-diffusivity in the temperature range 300–550 K was underestimated by 1–2 orders of magnitude relative to experiment, a fact attributed to the neglect of flexibility of the zeolite framework. Subsequent calculations by Forester and Smith [10], which incorporated a flexible zeolite model and invoked the blue moon ensemble MD technique to accumulate one-dimensional free energy profiles along the channels, gave self-diffusivity estimates close to experimental values.

9.3.2
Self-diffusivity at High Occupancy

The question of self-diffusion under finite occupancy was first addressed through infrequent event analysis by Beerdsen et al. and Dubbeldam et al. for methane, ethane, and propane in LTA and LTL-type zeolites [22, 23]. The approach invoked was to accumulate a potential of mean force $U(x)$ with respect to the position x of *one* molecule along the direction x involved in an elementary jump, by integrating over all other degrees of freedom of that molecule and over all positions of all *other* molecules in the system. $U(x)$ was obtained by a histogram sampling method, that is, by Boltzmann inversion of the probability of finding a molecule at x in the course of a canonical Monte Carlo simulation at finite loading [compare Eq. (7.38)]. When high free energy barriers are present, a biasing function must be used to coax a tagged molecule into accessing the barriers. For this purpose, Dubbeldam et al. [23] propose using the Widom test particle insertion profile [compare Eq. (7.41)] as a biasing function. In the cage/window structure of zeolite LTA, the free energy profile along the line connecting the centers of two adjacent large cages across an eight-ring window was shown to become more structured and to present lower barriers to the diffusing molecule as occupancy was raised. On the other hand, the dynamical correction factor f_d dropped monotonically with increasing occupancy, as collisions with other molecules enhance dynamical recrossings of a barrier. As a consequence of these two tendencies, the self-diffusivity was found to exhibit a maximum with respect to occupancy at around nine molecules per cage.

References

1. Snurr, R.Q., Bell, A.T., and Theodorou, D.N. (1994) *J. Phys. Chem.*, **98**, 11948.
2. Glasstone, S., Laidler, K.J., and Eyring, H. (1941) *The Theory of Rate Processes*, McGraw-Hill, New York.
3. Moore, J.W. and Pearson, R.G. (1981) *Kinetics and Mechanism*, 3rd edn, John Wiley & Sons, Inc., New York.
4. Chandler, D. (1978) *J. Chem. Phys.*, **68**, 2959.
5. Bennett, C.H. (1975) Exact defect calculations in model substances, in *Diffusion in Solids: Recent Developments* (eds A.S. Nowick and J.J. Burton), Academic Press, New York, pp. 73–113.
6. Voter, A.F. and Doll, J.D. (1985) *J. Chem. Phys.*, **82**, 80–92.
7. Vineyard, G.H. (1957) *J. Phys. Chem. Solids*, **3**, 121.
8. June, R.L., Bell, A.T., and Theodorou, D.N. (1991) *J. Phys. Chem.*, **95**, 8866.
9. Theodorou, D.N. (1995) Molecular simulations of sorption and diffusion in amorphous polymers, in *Diffusion in Polymers* (ed. P. Neogi), Dekker, New York, pp. 67–142.
10. Forester, T.R. and Smith, W. (1997) *J. Chem. Soc., Faraday Trans.*, **93**, 3249.
11. Theodorou, D.N. (2006) Principles of molecular simulation of gas transport in polymers, in *Materials Science of Membranes for Gas and Vapor Separation* (eds Yu. Yampolskii, I. Pinnau, and B.D. Freeman), John Wiley & Sons, Inc., Hoboken, NJ, pp. 47–92.
12. Bolhuis, P.G., Chandler, D., Dellago, C., and Geissler, P.L. (2002) *Annu. Rev. Phys. Chem.*, **53**, 291.
13. Hillstrom, K. (1976) Nonlinear Optimization Routines in AMDLIB, Technical Memorandum No 297, Argonne National Laboratory, Applied Mathematics Division (1976); Subroutine GQBFGS in AMDLIB, Argonne, IL.
14. Press, W.H., Teukolsky, S.A., Vettering, W.T., and Flannery, B.P. (2007) *Numerical Recipes: The Art of Scientific Computing*, 3rd edn, Cambridge University Press, Cambridge.
15. Baker, J. (1986) *J. Comput. Chem.*, **7**, 385.
16. Henkelman, G. and Jónsson, H. (1999) *J. Chem. Phys.*, **111**, 7010.
17. Fukui, K. (1981) *Acc. Chem. Res.*, **14**, 363.
18. Dellago, C., Bolhuis, P.G., and Geissler, P.L. (2002) *Adv. Chem. Phys.*, **123**, 1.
19. Wei, J. and Prater, C.D. (1962) The structure and analysis of complex reaction systems, in *Advances in Catalysis*, vol. **13** (ed. D.D. Eley) Academic Press, New York, pp. 204–390.
20. Boulougouris, G.C. and Theodorou, D.N. (2009) *J. Chem. Phys.*, **130**, 044905.
21. Heink, W., Kärger, J., Pfeifer, H., and Stallmach, F. (1990) *J. Am. Chem. Soc.*, **112**, 2175.
22. Beerdsen, E., Smit, B., and Dubbeldam, D. (2004) *Phys. Rev. Lett.*, **93**, 248301.
23. Dubbeldam, D., Beerdsen, E., Vlugt, T.J., and Smit, B. (2005) *J. Chem. Phys.*, **122**, 224712.

Part IV
Measurement Methods

10
Measurement of Elementary Diffusion Processes

Being among the most fundamental and omnipresent phenomena in nature, diffusion occurs in essentially all states of matter. Depending on their nature, the extension of the systems under study may range from nanometers up to intergalactic distances. In the context of this book, the length scales of interest are limited by the size of the nanoporous particles under study and the pore diameters. With respect to technological applications, the focus of interest is generally on length scales substantially greater than the pore diameter. In Section 2.1 it was pointed out that in a homogeneous medium of sufficient extension and at sufficiently long times the propagator $P(\mathbf{r}, t)$ approaches a Gaussian probability distribution function, Eq. (2.16), from which the mean square displacement is found to obey the Einstein relation $<r^2(t)> = 6\mathcal{D}t$ [Eq. (2.6)]. The measurement of $\mathcal{D}$ (the self-diffusivity) is discussed in Chapter 11.

Valuable additional information concerning the underlying elementary mechanisms of diffusion is provided by experimental techniques that are sensitive to the details of the propagation process over distances comparable with the pore sizes. In general this requires measuring the microdynamic parameters characterizing molecular motion over small intervals of time and distance at which the propagator deviates from the Gaussian form provided by Eq. (2.16). Where diffusion proceeds through a sequence of activated jumps the basic information required is the mean jump length and the average time interval between jumps. Clearly, if the diffusivity is known, an independent measurement of only one of these quantities is needed since the jump frequency $1/\tau$ and jump length l are related to the diffusivity through $\mathcal{D} = \langle l^2 \rangle / 6\tau$ [Eq. (2.7)].

In principle this feature of molecular migration may be studied by following the interaction of the diffusing molecules with each other, with their surroundings or with radiation from an external source. The methods that have been applied include nuclear magnetic resonance spectroscopy (NMR) and neutron scattering. The essential features of these techniques, as applied to the study of molecular motion in porous adsorbents, are reviewed in this chapter. In addition, because of the close relationship with neutron scattering, a brief review of the fundamentals of light scattering and some early applications of this approach are also included.

Diffusion in Nanoporous Materials. Jörg Kärger, Douglas M. Ruthven, and Doros N. Theodorou.
© 2012 Wiley-VCH Verlag GmbH & Co. KGaA. Published 2012 by Wiley-VCH Verlag GmbH & Co. KGaA.

10.1
NMR Spectroscopy

NMR spectroscopy is perhaps the most versatile of these techniques, since it can be applied to measure diffusion and self-diffusion as well as to study elementary diffusion processes. Although both types of measurement depend ultimately on the same basic principles, the methods used to study longer-range diffusion and self-diffusion are quite different from those used to study the elementary steps. Only the latter aspect is considered here; the application of NMR techniques to diffusion measurements is discussed in Chapter 11. The subject is introduced here in the simplest way; more rigorous treatments may be found in the standard texts [1–8].

10.1.1
Basic Principles: Behavior of Isolated Spins

10.1.1.1 Classical Treatment
From a strict perspective, NMR phenomena can be properly described only quantum mechanically, but most of the fundamentals relevant for the present purpose may be understood in terms of classical physics. The essential requirement is that the nucleus of the atomic species considered has a magnetic moment. Examples of such nuclei are ^{1}H, ^{2}H, ^{13}C, ^{19}F, ^{34}P, and ^{129}Xe. In a simplistic way the magnetic moment of these nuclei may be understood as arising from their rotation (the nuclear spin). The rotation of a charged object exerts the same effect on its surroundings as a current circulating within a small coil of similar dimensions. As a result, the spinning nucleus acts as an elementary magnet, the strength of which (its magnetic moment μ) is directly related to the mechanical moment (L):

$$\boldsymbol{\mu} = \gamma \boldsymbol{L} \tag{10.1}$$

The constant of proportionality (γ) is known as the gyromagnetic (or magnetogyric) ratio and is a fundamental quantity for the particular nucleus. For example, for a proton $\gamma \approx 2.67519 \times 10^8$ T^{-1}s^{-1}.

Under the influence of a constant (external) magnetic field the elementary nuclear magnets are subject to a mechanical torque, given by the vector product:

$$\boldsymbol{C} = \boldsymbol{\mu} \times \boldsymbol{B} \tag{10.2}$$

which tends to orient the magnets parallel to the applied field. The axis around which this torque acts is perpendicular to the directions of both the magnetic moment of the nuclear magnet and the magnetic field. The magnitude of the torque is given by:

$$|C| = \mu B \sin \alpha \tag{10.3}$$

where α denotes the angle between $\boldsymbol{\mu}$ and $\boldsymbol{B}$ (Figure 10.1).

The potential energy of the magnetic dipole in the magnetic field is found directly by integration:

$$E(\alpha) = -\int_{\alpha_0}^{\alpha} C(\alpha')\, d\alpha' = -\mu B(\cos \alpha - \cos \alpha_0) \tag{10.4}$$

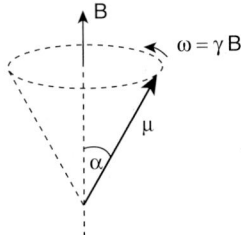

Figure 10.1 Quasi-classical representation of the orientation of a nuclear spin (dipole moment **μ**) in a constant magnetic field (**B**) showing the resulting precessional motion (Larmor frequency ω).

where α_0 denotes the angle characterized by zero potential energy. Following the usual convention α_0 is taken as $\pi/2$ so that Eq. (10.4) reduces to:

$$E(\alpha) = -\mu B \cos \alpha = -\boldsymbol{\mu} \cdot \boldsymbol{B} \tag{10.5}$$

Inserting typical values ($B = 10\,\text{T}, \mu_{\text{proton}} = 1.4 \times 10^{-26}\,\text{J T}^{-1}$) shows that, except at temperatures close to the absolute zero, this quantity is much smaller than the thermal energy $k_B T$. If one excludes cooperative effects, such as may occur for macromolecules with polar groupings, as in liquid crystals, this implies that the application of the magnetic field in NMR spectroscopy should not affect the microdynamic molecular behavior.

An NMR spectrometer is sensitive to the specific magnetization (M), which is simply the vector sum of the magnetic moments per unit volume. The probability of a magnetic dipole having an orientation α relative to the direction of the magnetic field is given by the Boltzmann factor:

$$P(\alpha) \propto \exp[\mu B \cos \alpha / k_B T] \tag{10.6}$$

The specific magnetization is obtained by integration over all possible orientations:

$$\frac{M}{N} = \frac{\int_{-1}^{+1} \exp[\mu B \cos \alpha / k_B T] \mu \cos \alpha \, d(\cos \alpha)}{\int_{-1}^{+1} \exp[\mu B \cos \alpha / k_B T] \, d \cos \alpha} \tag{10.7}$$

where N denotes the density of the nuclear spins (number per unit volume). Subject to the approximation that the interaction energy is small relative to the thermal energy, this integral yields:

$$M \approx \frac{1}{3} \frac{NB}{k_B T} \mu^2 \tag{10.8}$$

For protons at room temperature with the above values for μ and B, this leads to:

$$M \approx 10^{-5} N \mu \tag{10.9}$$

which means that the specific magnetization amounts to only about one hundred thousandth of the maximum possible value that would be realized if all spins were aligned in the same direction. This is because, except at temperatures close to absolute zero, the magnetic interaction energy is very small relative to the thermal energy. The incremental fraction of the nuclear spins that are aligned in the direction of the magnetic field is thus seen to be extremely small at ordinary temperatures, but it is only this small excess that is recorded in NMR spectroscopy. This fraction increases in proportion to the strength of the magnetic field, which has speeded up the application of superconducting magnets.

Completely novel options for enhancement of the NMR signal intensity are provided by the application of hyperpolarized nuclei [9–13]. Hyperpolarization is attained by magnetization transfer from an alkali vapor in contact with the guest molecules in the surrounding gas phase (laser-induced spin-exchange by optical pumping) [14, 15].

For an "ordinary" magnetic dipole in an external magnetic field (e.g., a compass needle in the magnetic field of the Earth) the mechanical torque exerted by the application of the field leads to an oscillation about the direction of the applied field. In the case of a nuclear magnetic dipole, however, the angular momentum of the nuclear spin leads to a precessional motion, about the direction of the applied field (as with a gyroscope in a gravitational field) rather than to an oscillation (Figure 10.1). The angular velocity of this precession can be shown to be proportional to the magnetic field, with the gyromagnetic ratio as the factor of proportionality, obeying the so-called Larmor condition [3, 4, 7, 8]:

$$\omega = \gamma B \tag{10.10}$$

10.1.1.2 Quantum Mechanical Treatment

While the relations derived so far are in fact sufficient to understand the principles of NMR self-diffusion measurements (Section 11.2), in order to understand the application of NMR to the study of the elementary steps of the diffusion process a short excursion into quantum mechanics is necessary. The angular momentum of the nuclear spin is in fact quantized:

$$\mathbf{L} = \hbar \mathbf{I} \tag{10.11}$$

where $\hbar$ denotes $h/2\pi$ (where h is Planck's constant) and $\mathbf{I}$ the spin vector. With Eq. (10.1) this means in turn that the magnetic dipole moment:

$$\boldsymbol{\mu} = \gamma \hbar \mathbf{I} \tag{10.12}$$

must also be quantized. Thus inserting Eq. (10.11) into Eq. (10.5), the Hamiltonian of a nuclear spin under the influence of an external magnetic field $\mathbf{B}_0$ is found to be:

$$H = -\gamma \hbar \mathbf{I} \mathbf{B}_0 \tag{10.13}$$

which provides the eigenvalues:

$$E_m = -\gamma \hbar B_0 I_z \tag{10.14}$$

where I_z denotes the eigenvalues of the z component of the spin vector and may assume the values:

$$I_z = m = -I, -I+1, \ldots, I-1, I \tag{10.15}$$

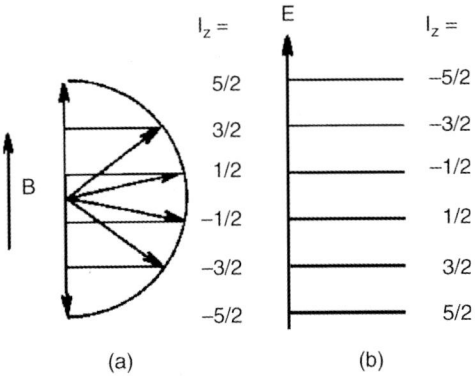

Figure 10.2 Spin orientations in a magnetic field for $I = 5/2$ (a) and resulting interaction energies (b).

in which I denotes the spin quantum number, which can assume only positive integer or half integer values. Thus, instead of being continuously distributed, the orientation of the nuclear spins may only assume certain definite values, as indicated in Figure 10.2 for spin quantum number $I = 5/2$.

For the nuclei ^{1}H, ^{13}C, ^{15}N, and ^{129}Xe, I is equal to $1/2$ so that there are only two possible orientations for the nuclear spins. In particular, with $I_z = 1/2$ and $\gamma_{proton} = 2.67 \times 10^8$ T^{-1} s^{-1}, Eq. (10.12) leads to the value used in Eq. (10.5) to estimate the relative values of the magnetic and thermal energies. In any case, irrespective of the actual value of I_z, the difference between adjacent energy levels is given by:

$$|E_{m\pm 1} - E_m| = \Delta E = |\gamma \hbar B_0| \tag{10.16}$$

It may be demonstrated by perturbation theory that the interaction between a magnetic dipole and an applied magnetic field can only lead to transitions between adjacent energy levels if the frequency of this field (ω) obeys Planck's relation:

$$\omega = \Delta E/\hbar \tag{10.17}$$

where ΔE denotes the difference in energy between the two states between which the transition occurs. Conversely, this is also the frequency of the radiation emitted from transitions between adjacent levels. Combining Eqs. (10.16) and (10.17) this may be expressed as $\omega = |\gamma B_0|$, which is seen to be identical to the Larmor condition [Eq. (10.10)]. In fact it may be shown that the behavior of an isolated spin in the classical treatment is just that of the expectation value of the corresponding operator in the rigorous quantum mechanical treatment.

10.1.2
Behavior of Nuclear Spins in Compact Material

In real macroscopic systems one is always concerned with an ensemble of spins leading to an equilibrium magnetization as given by Eq. (10.8). This equilibrium is characterized by a small excess of spins aligned in the direction of the applied

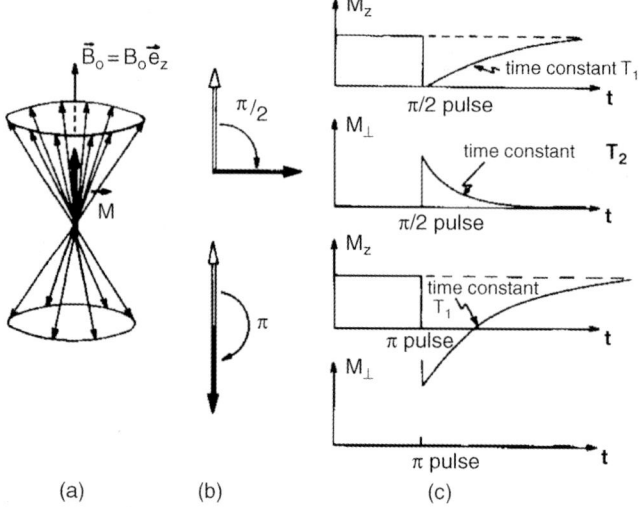

Figure 10.3 Simplified representation of the superposition of the magnetic moments of the individual spins leading to the macroscopic magnetization (a); the change in macroscopic magnetization resulting from a $\pi/2$ pulse (upper) and a π pulse (lower) (b); relaxation of the components of the magnetization parallel to ($M_{\parallel}$) and perpendicular to ($M_{\perp}$) the magnetic field (c).

magnetic field and, for reasons of symmetry, with no component in the perpendicular direction (Figure 10.3a). It can be shown within the quasi-classical approximation of the preceding section [3–5, 7, 8] that, under the influence of an appropriately chosen magnetic field (rotating in phase with the nuclear spins about the direction of the constant magnetic field, namely, by suitable radiofrequency pulses) the orientation of any individual spin is rotated through a certain angle around the direction of the applied magnetic field.

Thus the vector sum of the individual spins and hence their magnetization will also be rotated through the same angle. This is in fact identical with the result of the rigorous quantum mechanical treatment where the average (expectation) value, rather than the behavior of an individual spin, is regarded as the experimentally accessible quantity.

As an example, Figure 10.3b shows the effect of a so-called $\pi/2$ pulse and a π pulse on the transverse magnetization. Following the pulse (Figure 10.3c) the magnetization will return to the original equilibrium state. The time constants for this relaxation process are generally different for the two components and are called the longitudinal (T_1) and transverse (T_2) relaxation times, respectively.

Under the influence of a constant magnetic field, the individual isolated spins will change neither their orientation with respect to the direction of the magnetic field nor their phase relative to each other. Thus the relaxation towards the equilibrium values must be brought about by the interaction of individual spins with each other or with the surroundings. It is this interaction that is monitored in an NMR experiment and which, as illustrated later on in this section, provides, through its time dependence, information concerning the elementary steps of the diffusion process. Since, during longitudinal relaxation, the energy of the spin system changes [Eq. (10.5)] and must therefore be transferred to the environment (the "lattice") this process is also termed energy or spin–lattice relaxation. During transverse relaxation it is the entropy rather than the energy of the spin system that is changed. This process is therefore called entropy or spin–spin relaxation.

10.1.2.1 Fundamentals of the NMR Line Shape

It was shown above that the absorption of energy, which changes the orientation of the individual nuclear spins, takes place under the Larmor condition ($\omega = |\gamma B_{loc}|$). The local magnetic ($B_{loc}$) field is determined by the superposition of the constant external field (B) and the field produced by the neighboring nuclear spins (B_{add}), as illustrated in Figure 10.4 for a two-spin system. Considering the effect of spin 2 on spin 1 we notice that in arrangement (a) the additional field is parallel to the applied field while in arrangement (b) the additional field is opposite to the external field. Thus, in arrangement (a) the external field must be smaller than B_{loc}, since it is enhanced by B_{add}, while in (b) the external field must be larger by this same value in order to compensate for the effect of B_{add} being aligned in the opposite direction. In real systems there is generally a distribution of orientation angles as well as a distribution of distance between the individual spins. Consequently, energy absorption occurs over some interval of field intensities, as illustrated in Figure 10.5, rather than as a single sharp line that would be expected for an isolated spin. The NMR spectrum shown in Figure 10.5 has been obtained by measuring the dependence of the energy of absorption on the intensity of the external field at a fixed frequency ω. However, the same pattern would clearly result if the measurement were made at constant field intensity by varying the frequency.

10.1.2.2 Effect of Molecular Motion on Line Shape

We consider two spins that, at the same instant, are subjected to two different local magnetic fields, B_a and B_b. These spins will be distinctly recorded as subject to these

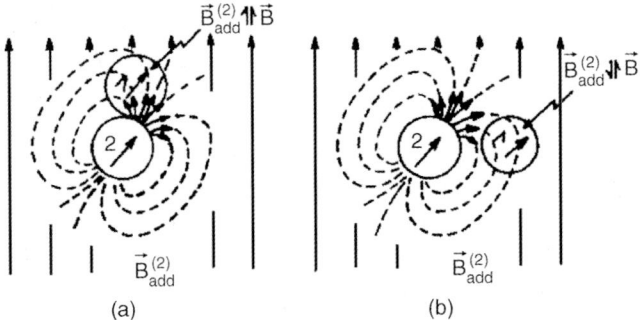

Figure 10.4 Superposition of the external magnetic field B and the field B_{add} brought about by a neighboring spin (2) on the local magnetic field B_{loc} of a spin (1), for two different relative positions.

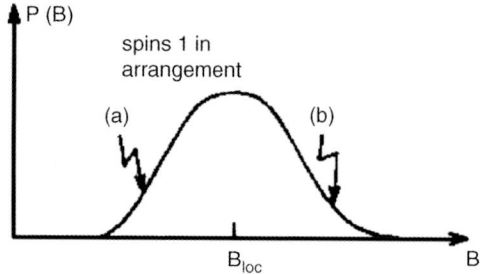

Figure 10.5 Schematic representation of the broadening of the resonance line P(B) due to magnetic dipole–dipole interaction, as illustrated in Figure 10.4.

fields provided that the local field does not change before the difference in the precessional phases is of order 1. For a qualitative estimate we introduce the "correlation time" τ_c as the time interval after which the local magnetic field acting on an individual spin will have changed significantly. If the local field is modulated by rotation [e.g., from arrangement (a) to arrangement (b) of Figure 10.4], τ_c is the mean re-orientation time for a molecule of the species considered. Since the magnetic moment of an electron paramagnetic center (e.g., iron atoms, which may be incorporated as impurities in the solid) is greater by a factor of 2000 relative to the nuclear magnetic moment, the local magnetic field experienced by a particle or molecule in that vicinity will stem almost entirely from these centers. In this situation the correlation time coincides with the mean lifetime of the particle in the vicinity of the paramagnetic center, provided that the longitudinal electron relaxation times are larger than the mean lifetime [16].

The width of the absorption line will therefore attain its maximum value corresponding to the particular geometrical arrangement when:

$$\tau_c \Delta \omega \geq 1 \tag{10.18}$$

where $\Delta \omega = |\gamma \Delta B|$ is the mean of the (absolute values of the) differences in Larmor frequency due to the difference in the local magnetic fields. Obviously, for correlation times that obey Eq. (10.18), the observed NMR spectrum should be independent of the internal dynamics of the system and, in particular, should not change with temperature. This is the limiting case of "solid state" NMR, which expresses nothing more than that the time scale of the internal processes is larger than the reciprocal NMR line width $1/\Delta \omega = 1/|\gamma \Delta B|$ (this is the situation shown on the right-hand side of Figure 10.6c).

If the correlation time τ_c is smaller than this value, the individual spins will experience some averaged local magnetic field before the difference in the accumulated phases becomes of order 1. Obviously, with decreasing magnitude of τ_c, corresponding to increasing molecular mobility, the distribution width of the local magnetic field will also decrease, leading to "motional narrowing" of the NMR line width. This situation is represented in the left-hand part of Figure 10.6c. Measurement of the temperature dependence of the NMR line width thus provides one possible way of studying local molecular motion and hence the elementary steps

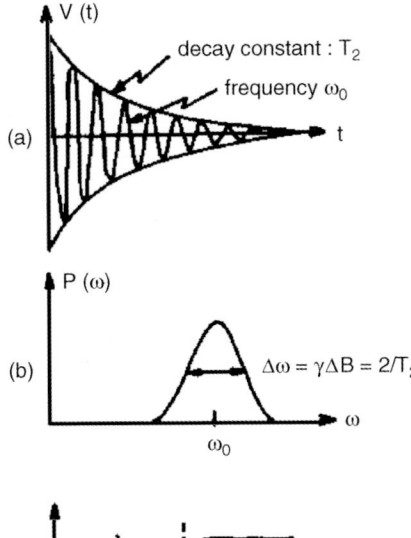

(a)

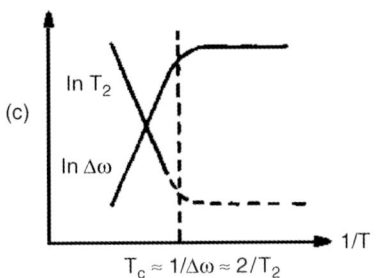

(b)

(c)

Figure 10.6 Schematic representation of (a) the free induction decay V(t) (the oscillation with the Larmor frequency ω_0 must be imagined to occur at a much larger frequency than here displayed – in fact, the indicated oscillation would correspond to the so-called offset), (b) the stationary NMR spectrum $P(\omega)$, and (c) the temperature dependence of the characteristic parameters $\Delta\omega$ and T_2. In the solid-state region [right-hand side of (c)] transverse relaxation is no longer governed by the simple exponential relation [Eq. (10.22)] and in this region T_2 is therefore only an "effective" quantity.

of diffusion. In many cases, for sufficiently low temperatures, the line width is in fact found to remain constant until, at a certain temperature, it starts to decrease continuously with increasing temperature. At this temperature the modulation of the magnetic field, which, in most cases, arises from the elementary diffusion processes, is obviously characterized by the correlation time:

$$\tau_c \approx 1/\Delta\omega \tag{10.19}$$

One may visualize, in a straightforward way, that the line shape of the NMR absorption signal (Figure 10.6b) is closely related to the transverse nuclear magnetic relaxation behavior (Figure 10.6a). For this purpose we estimate the time constant for defocusing of the transverse nuclear magnetization, that is, the transverse nuclear magnetic relaxation time T_2, by recognizing that it should be of the same order as the time interval during which the precessional phases of the spins contributing to the extreme ends of the absorption line have attained a value of order unity. Since the difference in the Larmor frequency of these two spin groups is given by $\Delta\omega$, one has:

$$T_2 \Delta\omega \approx 1.0 \tag{10.20}$$

To obtain a more precise relationship between $\Delta\omega$ and T_2 one may adopt the results of rigorous analysis [3–5, 7, 8], which shows that the shape of the NMR absorption line $P(\omega)$ (the stationary spectrum) and the NMR signal following a $\pi/2$ pulse, the "free induction decay" $V(t)$, are related as Fourier transforms:

$$P(\omega) \propto \int_0^\infty V(t) \cos \omega t \, dt, \quad V(t) \propto \int_0^\infty P(\omega) \cos \omega t \, dt \tag{10.21}$$

Thus, assuming an exponential decay of the transverse magnetization, and therefore of the free induction decay:

$$V(t) = V_0 \cos \omega_0 t \exp(-t/T_2) \tag{10.22}$$

one obtains the Lorentzian:

$$P(\omega) \propto \frac{1}{1 + (\omega - \omega_0)^2 T_2^2} \tag{10.23}$$

Defining the line width as the distance between the Larmor frequencies at which $P(\omega)$ attains half its maximum value (cf. Figure 10.6b) one has:

$$\Delta\omega = \gamma \Delta B = 2/T_2 \tag{10.24}$$

Figure 10.6a and b illustrates the relation between the free induction decay and the stationary spectrum. In addition, Figure 10.6c shows the typical form of the temperature dependence of the line width and the transverse nuclear magnetic relaxation times as expected from the motion of the spins.

10.1.2.3 Fundamentals of Pulse Measurements

Observation of the free induction decay following a $\pi/2$ pulse is the usual procedure in standard NMR measurements. Figure 10.7 presents a block diagram showing a

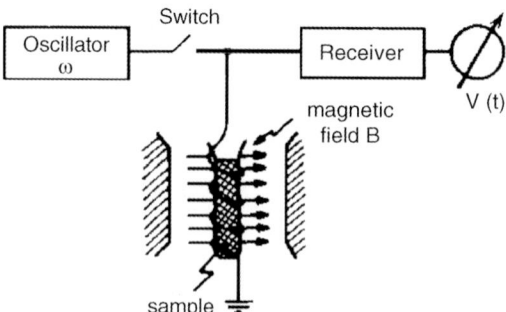

Figure 10.7 Schematic representation of an NMR spectrometer for non-stationary measurements.

typical experimental system for such measurements (a pulse NMR spectrometer). The sample is subjected to the RF magnetic field for a short period. This leads to rotation of the magnetization out of the equilibrium direction, parallel to the magnetic field and, if the magnetization is turned, for example, through an angle of $\pi/2$ by means of a $\pi/2$ pulse (Figure 10.3), there will be a macroscopic component rotating in the plane perpendicular to B_0. The oscillating magnetization in the direction of the axis of the receiver coil (Figure 10.7) induces a voltage of the same frequency (the Larmor frequency) that eventually decays to zero as the magnetization decays. The behavior of the NMR signal during this decay is adequately described by Eq. (10.22).

In addition to the effect of the internal interaction of the nuclear spins within the sample, the width of the NMR absorption line may also be increased as a result of a variation in the strength of the magnetic field within the sample. Clearly, according to Eq. (10.24) this additional broadening corresponds to a faster decay of the magnetization. This is a direct consequence of the larger spread in the Larmor frequencies, which leads to faster loss of coherence between individual nuclear spins, and thus to a faster decay of their vector sum, which is proportional to the magnetization. This behavior is illustrated in Figure 10.8a. Moreover, it is shown in Figure 10.8a that this loss of coherence may be compensated by the application of a π pulse. This effect may be understood by realizing that those spins with the highest Larmor frequencies, which are therefore ahead before the application of the compensating π pulse, will be retarded more than the slow spins. Obviously, when after the π pulse the same time (τ) has elapsed as between the $\pi/2$ pulse and the π pulse, the effects will be exactly compensated so that all spins are again in phase.

Hahn, who was the first to carry out such an experiment, visualized this effect by the "runners model" (Figure 10.8b). The effect of any inhomogeneity in the external field may be eliminated in this way. The loss of coherence resulting from the spin–spin interaction is not affected by this procedure since the directions of the spins will all be changed simultaneously in precisely the same way. The signal (voltage) induced by the magnetization at time 2τ after the $\pi/2$ pulse is called the "spin echo". Note that, in neutron scattering experiments (Section 10.2), the same mechanism of refocusing is exploited to generate the neutron spin echo. The decay of the envelope of a series of echoes with increasing values of 2τ is determined only by the transverse relaxation rate T_2 and is not affected by any inhomogeneities in the magnetic field.

To measure T_1 it is necessary to follow the re-establishment of magnetization in the direction of the external magnetic field. In contrast to the transverse magnetization, the longitudinal magnetization itself does not lead to the generation of an RF signal. The longitudinal component must therefore be monitored by converting it to a transverse magnetization by means of a $\pi/2$ pulse.

Figure 10.9 shows suitable pulse sequences for determining the time dependence of the magnetization. Also shown is the time dependence of the induced voltage and the transverse nuclear magnetization in a spin echo experiment. The times given in Figure 10.9a correspond to those given in the scheme of Figure 10.8. The relaxation curve for the transverse nuclear magnetization (broken line) may be obtained by

10 Measurement of Elementary Diffusion Processes

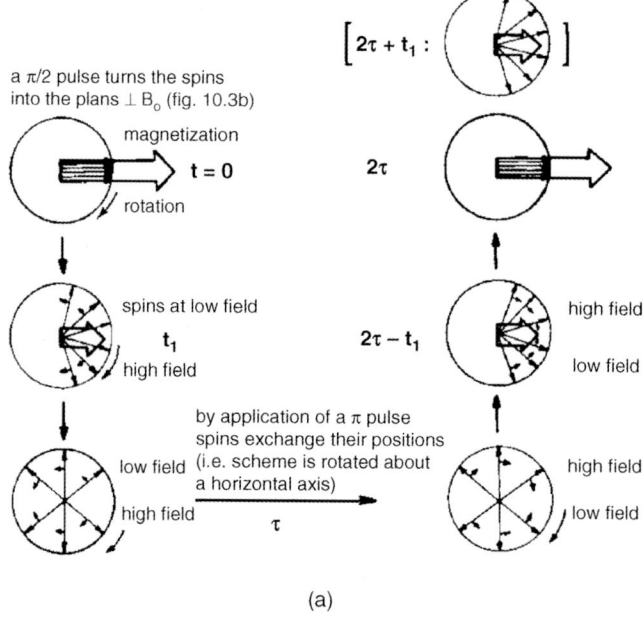

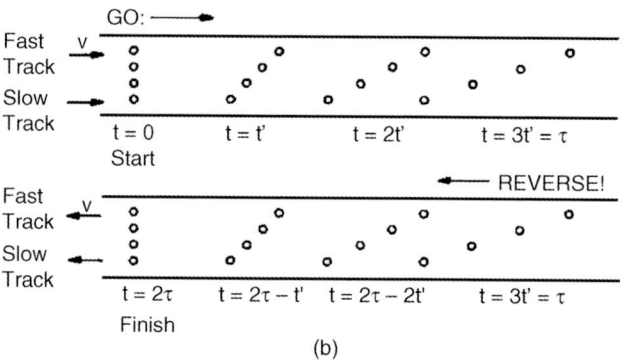

Figure 10.8 (a) Effect of an inhomogeneous field on the precessional motion of similar nuclear spins (represented in a frame rotating with the Larmor frequency) and its elimination by a π pulse; (b) illustration by the "runners model [17]." Faster runners correspond to spins at higher magnetic fields.

changing the separation between the two RF pulses (τ) and measuring the induced NMR response signal at $t = 2\tau$. Correspondingly, in Figure 10.9b, by changing the separation between the π and $\pi/2$ pulses one can monitor the longitudinal magnetization. Without a phase sensitive detector one observes only the magnitude of the induced voltage. This is indicated by the dotted line, which is proportional to

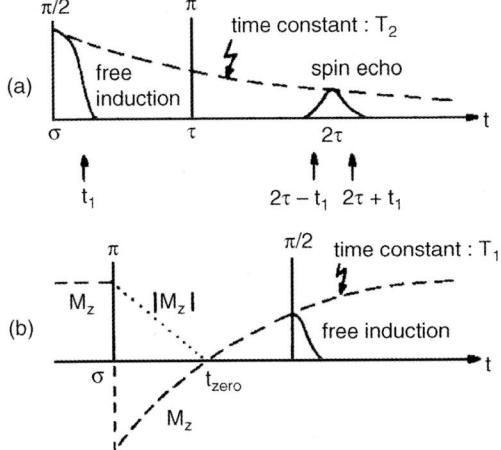

Figure 10.9 Radiofrequency pulse programs for (a) determination of transverse relaxation time (T_2) and (b) for determination of longitudinal relaxation time (T_1).

$|M_z|$ rather than to M_z. The time after which the transverse magnetization vanishes is given by:

$$M(t_{zero}) \propto 1 - 2e^{-t_{zero}/T_1} = 0 \quad \text{or} \quad t_{zero} = T_1 \ln 2 \qquad (10.25)$$

The spin-echo sequence (Figure 10.9a) is the basic pulse program for NMR self-diffusion measurements and is considered in greater detail in this context in Section 11.2. The information concerning the elementary diffusion steps, provided by T_2, has been discussed already in connection with the line width measurements. To consider the information provided by T_1 measurements we have to recall that T_1 is the time constant for transitions between different quantum states induced by the changing interaction energy between the spins and their surroundings.

The time dependence of this interaction is brought about by the elementary processes of molecular migration and/or redistribution, the time constant of which was introduced as τ_c in the discussion of line width measurements. According to perturbation theory, transitions between energy levels are only possible if the frequency of the interaction energy obeys Planck's relation [Eq. (10.17)]. This means that the greatest transition probability, that is, the minimum relaxation time, will be observed when the frequency of the interaction energy ω_{int} coincides with the Larmor frequency. Thus, by estimating the frequency of interaction using a relation of the type:

$$\omega_{int} = a/\tau_c \qquad (10.26)$$

one obtains:

$$1/\tau_c = \omega/a = \gamma B/a \qquad (10.27)$$

where a is a numerical factor of order of unity. The precise value depends on the model used.

With increasing deviation of the mean correlation time from the value given by Eq. (10.27), a continuously decreasing fraction of the spins will be subjected to motion with a correlation time that obeys Planck's relation, corresponding to continuously increasing longitudinal relaxation times. Thus a measurement of the temperature dependence of T_1 should show a distinct minimum, characterized by the condition that, at this temperature, the correlation time for nuclear magnetic interaction obeys Eq. (10.27). Section 17.2.1 illustrates how the appearance of a distinct minimum in the temperature dependence of T_1 (see, for example, Figure 17.5) may be exploited to estimate the mean jump length of saturated hydrocarbons in zeolite NaX.

10.1.3
Resonance Shifts by Different Surroundings

In the previous section, the information provided by absorption spectra has been shown to be correlated with the information provided by pulse measurements by a simple Fourier transform between time and frequency. By varying the spacing between the radiofrequency pulses in modern pulse sequences, one disposes of a whole set of time variables. This set may, correspondingly, be used for transformation into a multidimensional frequency space where the different frequencies correspond to different types of interaction. Deducing the characteristic correlation times for each of these interactions, spectra analysis is able to provide valuable information about the elementary steps of molecular reorientation and propagation [2, 4]. Since molecular jumps are generally correlated with molecular reorientation, reorientation times may be used for estimating the lower limit of the mean life time between subsequent jumps. In this way, dielectric spectroscopy has also been successfully applied for determining the translational dynamics of molecules with permanent electric dipolar moments [18–20]. We now introduce the main mechanisms of interaction that may be exploited to characterize the elementary steps of diffusion by considering the influence of the surroundings on the adsorption phenomena in nuclear magnetic resonance.

10.1.3.1 Spins in Different Chemical Surroundings
In Figure 10.4 it was shown that, due to the influence of the adjacent spins, the intensity of the local magnetic field acting on a given spin may be slightly different from that of the external field. A similar effect is produced by the electron cloud surrounding a nucleus as a result of the shielding effect. This again leads to a shift of the observed resonance line in comparison with the frequency that would be observed for a "naked" nucleus. This effect is generally discussed in terms of the so-called chemical shift [1–4]:

$$\delta_A = (B_A - B_{st})/B_{st} \tag{10.28}$$

where B_A and B_{st} denote the magnetic fields at which resonance is observed for the compound under study and for a standard. For different compounds the chemical

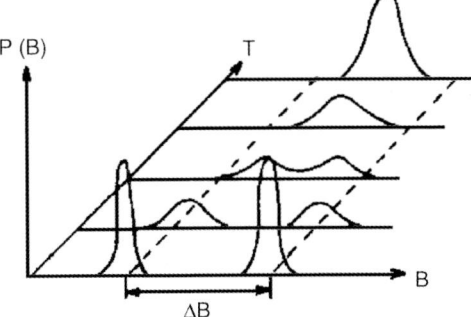

Figure 10.10 NMR spectrum for a system containing two spins in different environments at different temperatures corresponding to different exchange times.

shift of a given nucleus attains characteristic values corresponding to the chemical surroundings of the nucleus within the compound. Such shifts are of order 10^{-6} for ^{1}H and 10^{-4} for ^{13}C. Since the electron cloud may also be affected by the external environment, the chemical shift may be modified, for example, by adsorption on a solid surface.

Figure 10.10 shows a schematic representation of the NMR spectrum for a system with spins in two different environments. According to the considerations of Section 10.1.2, two distinct lines will be observed only if the mean life time (τ_1) of the spins in their chemical surroundings is larger than the reciprocal of the separation between the two lines. Thus, with increasing rates of molecular exchange the lines broaden and merge until, for $\tau_1 \ll (\Delta\omega)^{-1} = (\gamma\Delta B)^{-1}$, there is only one line determined by the weighted average of the Larmor frequencies for the two states. If the two states are in adjacent positions such observations can provide information concerning the rate of interchange that corresponds to an elementary diffusion step. Zeolites of type ZK-5 provide a good example of this type of study. They consist of two types of cages adjacent to each other so that, on their diffusion path, the guest molecules pass through continuously changing chemical surroundings. By investigating the time correlation in the chemical shifts [21] one is thus able to determine the jump rate $1/\tau$ of molecular propagation and, via Eq. (2.7), with the known cage separation l, the diffusivity. These data were found to be in good agreement with the results of MD simulations [22].

Alternatively, if the two states are further apart, such observations can provide information concerning longer-range diffusion. The interval of temperature over which the transition from a single peak to a doublet occurs depends on many variables, including the difference in the chemical shifts, the relaxation rates in the two states, the relative probabilities of the two states, and the exchange rate. Data analysis of this type has been applied for studying mass transfer phenomena in zeolites [15, 23–26] and other nanoporous materials such as Vycor® glass [27] and illustrate the advance of using hyperpolarized xenon [11, 28]. Such techniques have been used to explore the rate of mass exchange between micro- and mesopores [29].

The chemical shift of a probe molecule is known to depend on its own concentration as well as on the concentration of a second guest species [30, 31]. This makes ^{129}Xe a sensitive probe for exploring guest distribution in nanoporous host systems, in particular if this second component is much more favorably adsorbed. Measurements of this type were performed with benzene in MFI-type zeolites, yielding both intracrystalline [32] and long-range [33–35] transport diffusivities. In References [36, 37], the dependence of the chemical shift of ^{129}Xe on the xenon population in the large cages of zeolite NaA (Section 16.1.1) was exploited to determine the rate of cage-to-cage jumps and the occupation probabilities.

10.1.3.2 Anisotropy of the Chemical Shift

In many instances the electron cloud surrounding the nucleus will not be spherical and the chemical shift will therefore be different in different directions. If it were possible to fix the molecule only in two positions with respect to the external magnetic field, one would again observe two distinct lines. In reality, however, a microporous adsorbent provides a matrix in which all orientations of the sorbate molecule are possible. When the chemical shift is axially symmetric, the spectrum, for the rigid case, will appear as sketched in Figure 10.11a. The line will have its absolute maximum at $\omega_\perp = \omega_{st}(1+\delta_\perp)$, decaying towards $\omega_\parallel = \omega_{st}(1+\delta_\parallel)$ since more molecules will have the axis of symmetry oriented perpendicular ($\omega_\perp, \delta_\perp$) rather than parallel ($\omega_\parallel, \delta_\parallel$) to the magnetic field. Dipole–dipole interaction of the individual spins will lead to a broadening of the resonance line, which tends to mask the fine structure. This resonance line will coalesce to a normal single peaked response with an average chemical shift $\delta_{iso} = \frac{2}{3}\delta_\perp + \frac{1}{3}\delta_\parallel$ when complete re-orientation of the molecule occurs during the time corresponding to the reciprocal line width (Figure 10.11e).

Measurements of this type have been carried out with various nuclei (^{13}C and ^{31}P) and numerical simulations have been developed in sufficient detail to allow the measured spectra to be attributed to the corresponding reorientation times in the relevant geometry (Figure 10.11) [38–42].

10.1.3.3 Quadrupole NMR

The chemical surroundings of a nucleus may also lead to anisotropy of the local electric field. When the nucleus under consideration has an electric quadrupole moment, this anisotropy leads to an interaction energy that acts in addition to the magnetic dipole interaction given by Eq. (10.5). An electric quadrupole can occur only for spins with quantum numbers $I > {}^1/_2$. This may be intuitively understood since the spins with quantum numbers ${}^1/_2$ may only be oriented in two possible ways, which for reasons of symmetry cannot lead to different interaction energies between the nucleus and the local electric field [8]. This is no longer true for spin quantum numbers greater than ${}^1/_2$. Of special relevance in this respect is NMR with deuterium having a spin quantum number of 1. In this case interaction between the nucleus and the electric field shifts the energy level for $I_z = 0$ relative to the levels for $I_z = \pm 1$, so that, instead of a single resonance line (for the transitions between

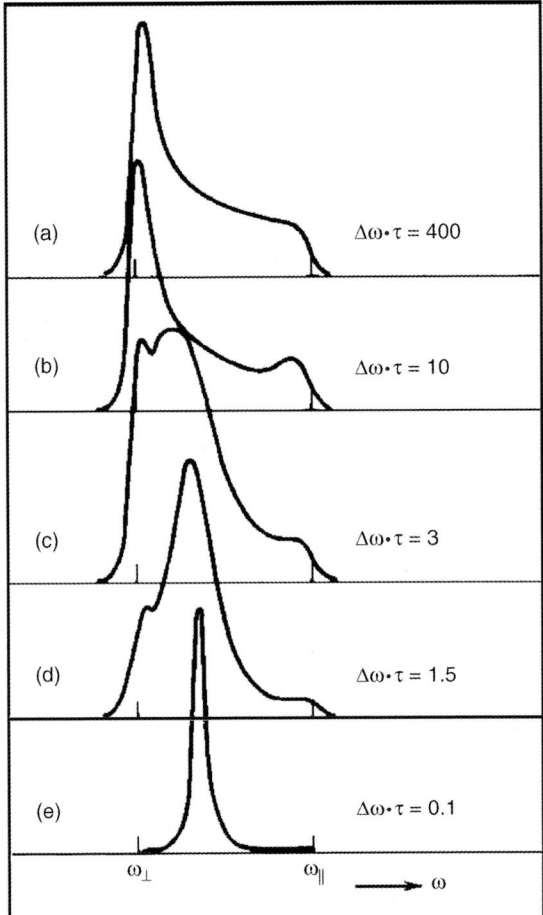

Figure 10.11 Calculated NMR line shapes due to axially symmetric shielding tensors for a jumping tetrahedron at different jump frequencies (τ^{-1}). Reprinted from Reference [38] with permission.

energy levels −1 and 0, and 0 and +1, in the unperturbed situation), one observes two resonance lines obeying the relation [8, 43, 44]:

$$\omega = \omega_0 \pm \frac{\omega_Q}{2}(3\cos^2\theta - 1 - \eta \sin^2\theta \cos 2\phi) \quad (10.29)$$

in which ω_Q is given by the relation:

$$\omega_Q = \frac{3}{4}e^2 qQ/\hbar \quad (10.30)$$

Here Q is the quadrupole moment that is a characteristic quantity for all nuclei with $I > \frac{1}{2}$, $eq = V_{zz}$ represents the zz component of the electric field gradient tensor and

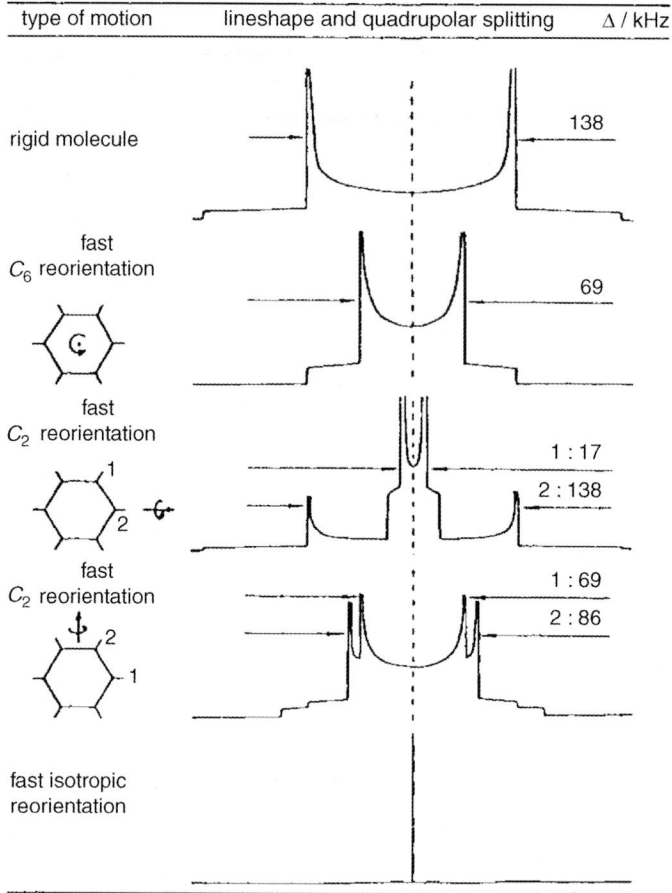

'Fast' corresponds to a correlation time which is much shorter than ω_Q^{-1}.

Figure 10.12 Theoretical ^{2}H NMR line shapes for deuterated benzene molecules undergoing different types of motion. Reprinted from Reference [44] with permission.

θ, ϕ describe the orientation of the magnetic field in the principal axes of the electric field gradient, η is the "asymmetry" parameter ($0 < \eta < 1$) defined as:

$$\eta = \frac{V_{xx} - V_{yy}}{V_{zz}} \tag{10.31}$$

where V_{xx} and V_{yy} represent the xx and yy components of the electric field gradient tensor. In deuterium NMR with hydrocarbons, the principal axis (z) is given in general by the direction of the C—D bond around which $V_{xx} = V_{yy}$ so that $\eta = 0$ and one has:

$$\omega = \omega_0 \pm \frac{\omega_Q}{2}(3\cos^2\theta - 1) \tag{10.32}$$

The corresponding line shape for statistically distributed orientations (powder spectrum) follows from averaging Eq. (10.32) over all possible values of θ. The first row of Figure 10.12 shows the result of such averaging for a rigid benzene molecule. As in the case of chemical shift anisotropy, additional interactions may lead to a slight broadening of the spectrum. When the time constant for molecular reorientation, τ_c, becomes of the same order as the reciprocal of the difference between the two extremes of the line, that is, for $\tau_c \omega_Q \approx 1$ characteristic changes of the line shape occur. The precise form of these changes depends on the molecular motion. Figure 10.12 shows an example of the different line shapes observed with deuterated benzene and the corresponding types of molecular motion. Following this principle, deuterium NMR has proved to be a most efficient tool for elucidating the elementary steps of diffusion in complex systems, including liquid crystals [45] and zeolitic host–guest systems [46–49].

10.1.4
Impact of Nuclear Magnetic Relaxation

In the given context, nuclear magnetic relaxation not only provides us with an important tool to explore the elementary steps of diffusion, as illustrated in the previous section, but also determines the conditions under which the pulsed field gradient (PFG) technique of NMR (Section 11.3) operates. This concerns both the maximum observation times (which are controlled by the longitudinal relaxation time) and the minimum molecular displacements, which can be recorded. In addition to the experimental parameters provided by the PFG NMR spectrometer, the minimum displacements depend on the time span over which the magnetic field gradients can be applied. This time span is limited by the relaxation of the transverse nuclear magnetization.

As discussed in Sections 10.1.2 and 10.1.3, spin interaction gives rise to a line broadening in the spectra of nuclear magnetic resonance that corresponds to a decrease in the transverse relaxation time. Spin interaction would only be averaged out by a sufficiently high isotropic mobility of the molecules adsorbed in the porous materials under study. Additional application of the Hahn echo (HE) [17], the Carr–Purcell–Meiboom–Gill (CPMG) [50, 51] pulse sequences, or of magic-angle spinning (MAS) [3] can reduce the effective line broadening in such systems. Figures 10.13 and 10.14 display the different pulse sequences and the operating conditions of MAS schematically. The achievable reduction of line broadening, corresponding to a prolongation of the effective transverse relaxation time T_2^*, is quintessential for PFG NMR, since the duration of the gradient pulses in PFG NMR is limited by T_2^*.

A serious problem arises if the echo decay envelope measured by the HE or CPMG pulse sequences is not mono-exponential and rather seems to represent the superposition of a fast and a slowly decaying component of the signal ([52], see also Section 11.3.3). Such a relaxation behavior affects the performance of PFG NMR in two ways. First, the signal intensity is found to be notably decreased with increasing

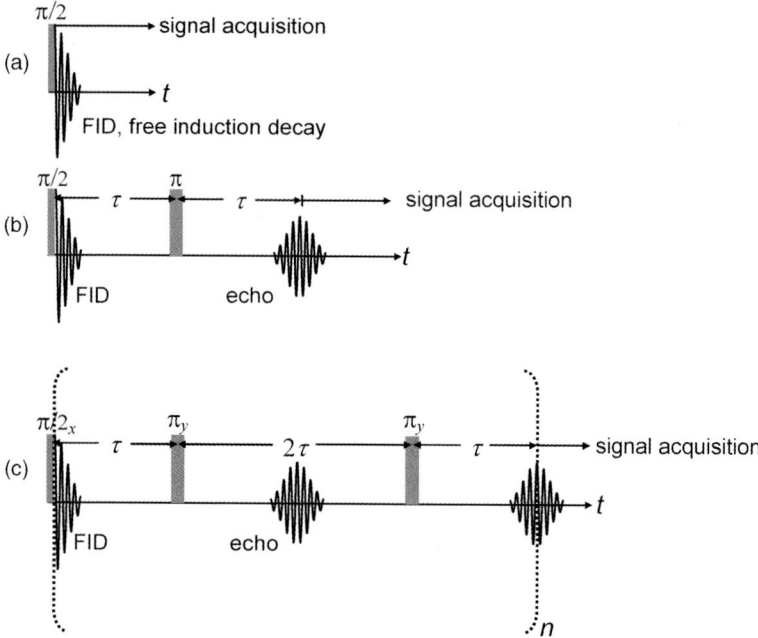

Figure 10.13 Signal acquisition after a $\pi/2$-preparation pulse for measuring the decay constant of the magnetization component perpendicular to the external magnetic field ("transverse relaxation time") following (a) immediately after the $\pi/2$ pulse (T_2^{FID}) and after use of (b) the Hahn-echo sequence (T_2^{HE}) and (c) the Carr–Purcell–Meiboom–Gill sequence (T_2^{CPMG}).

pulse spacing. Since sufficiently large spacings between the radiofrequency pulses are indispensable for the application of field gradients with large pulse widths one has to find a compromise between sufficiently large signal intensities and field gradient pulse widths. Second, and even more importantly, deviations from an exponential relaxation behavior are known to result for systems accommodating molecules of

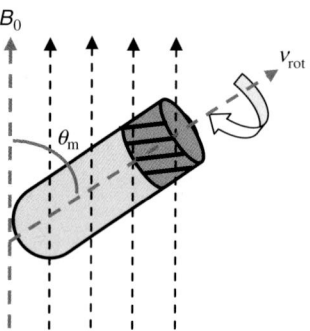

Figure 10.14 Sample tube ("rotor") in magic-angle spinning NMR. The "magic" angle between the rotation axis and the magnetic field is $\theta_m = 54.74°$ (i.e., $\cos^2\theta_m = 1/3$).

different mobility [53], in particular (i) a small fraction of highly mobile molecules with a long transverse relaxation time giving rise to a slowly decaying part at large observation times and (ii) a major fraction of less mobile molecules. Their contribution to the signal decreases rapidly with increasing observation time as a consequence of the much shorter relaxation time. In this case, application of PFG NMR would provide the diffusivity of the small fraction of highly mobile molecules rather than that of the major fraction of the molecules [54].

Figure 10.13 shows the schemes for measuring the various transverse relaxation times T_2^{FID} (a), T_2^{HE} (b), and T_2^{CPMG} (c). T_2^{FID} is obtained from the decay of the free induction, whereas T_2^{HE} and T_2^{CPMG} are obtained from the echo intensities. T_2^{MAS} experiments are similar to (a), but are performed under fast magic-angle spinning of the sample as displayed in Figure 10.14. Usually, the time constants of the decay increase in the sequence $T_2^{FID} \leq T_2^{HE} \leq T_2^{CPMG} \leq T_2^{MAS}$. This tendency is illustrated by the data of Table 10.1 [55]. They provide the decay times of transverse nuclear magnetization in ^{1}H NMR, following the above discussed pulse sequences for alkanes in zeolites silicalite and NaX and nicely illustrate the enhancement of the decay time by the pulse sequences applied. The numbers in parentheses in the HE and CPMG columns represent that fraction of magnetization that decays with the indicated, long-time constant. A first part is found to decay much faster, indicating an incomplete averaging of the intra- and intermolecular (dipolar) interaction energy of the protons. Interestingly, in the tetrahedrally coordinated pore network of NaX, this averaging is essentially complete, while an even faster movement in a zeolite of non-cubic symmetry such as MFI [$D_{CH4,silicalite} \approx 10^{-8}$ m^2 s^{-1} (Figure 18.10), in comparison with $D_{C4H10,NaX} \approx 3 \times 10^{-9}$ m^2 s^{-1} (Table 17.3, Reference [56]), data at zero loading and room temperature] cannot fully average the dipolar interaction. As a consequence, transverse nuclear magnetization as revealed by the echo attenuation notably deviates from a mono-exponential decay. This information is of particular relevance for the performance of PFG NMR diffusion experiments, since the occurrence of non-exponential magnetization attenuation would otherwise be

Table 10.1 Transverse relaxation times T_2 of adsorbed alkane molecules in the considered host–guest systems. CPMG measurements were performed with a π-pulse-distance of $2\tau = 1000\,\mu$s. The percentages in parentheses give the relative echo intensity evaluated by extrapolating the long-time part of the attenuation curves to $t = 0$. Two-exponential fitting was used for the Hahn-echo (HE) experiment, if the decay was not mono-exponential. Then, the value of T_2^{HE} in the table was obtained from the slowly decaying component [55].

	T_2^{FID} (ms)	T_2^{HE} (ms)	T_2^{CPMG} (ms)	T_2^{MAS} (ms)
Silicalite–methane	0.242	1.02 (50%)	11.7 (56%)	8.8
Silicalite–ethane	0.203	1.25 (20%)	5.0 (19%)	11.9
Silicalite–n-butane	0.137	0.91 (20%)	6.3 (17%)	6.5
Silicalite–isobutane	0.176	0.94 (20%)	4.0 (26%)	2.3
Silicalite–n-hexane	0.142	1.24 (45%)	7.5 (32%)	5.5
NaX–n-butane	0.518	1.65 (100%)	10.0 (86%)	14.4

taken as an indication of the existence of different molecules or of molecules in different states of mobility.

10.2
Diffusion Measurements by Neutron Scattering

10.2.1
Principle of the Method

Any interaction between matter and a wave of comparable scale leads to diffraction or scattering phenomena. If the scattering centers are fixed, the effect will be determined by their spatial arrangement. Information from such measurements is applied in well-established procedures for structure determination such as X-ray diffraction studies. However, when the scattering centers are mobile, exchange of energy between the incident wave and the scattering center becomes possible, leading in effect to a Doppler shift of the outgoing wave. The magnitude of any such shift may be intuitively expected to increase with the mobility of the scattering centers. Diffusion measurements by light scattering or thermal neutron scattering depend on this effect.

The application of neutrons in scattering experiments can be properly understood only by adopting the framework of quantum mechanics. A flux of neutrons (mass m and velocity v) may be described equivalently as a matter wave with wave vector:

$$k = \frac{p}{\hbar} = \frac{mv}{\hbar} \tag{10.33}$$

and frequency:

$$\omega = E/\hbar = mv^2/2\hbar \tag{10.34}$$

where p denotes the momentum of a neutron. The well-known expression for the de Broglie wavelength:

$$\lambda = h/mv \tag{10.35}$$

follows from $|k| = 2\pi/\lambda$.

Combining Eqs. (10.33) and (10.34) it follows that the wavelength and angular frequency of a matter wave are related by:

$$\lambda^2 \omega = \pi h/m \tag{10.36}$$

and cannot therefore be chosen independently. With cold neutrons the wavelength is of the order of several Ångströms, which is the required scale for investigation of molecular motion at the elementary level.

10.2.1.1 Experimental Procedure
In a neutron scattering experiment one observes the changes in both the frequency and the wave vector of a neutron beam, after passing through the sample, as a

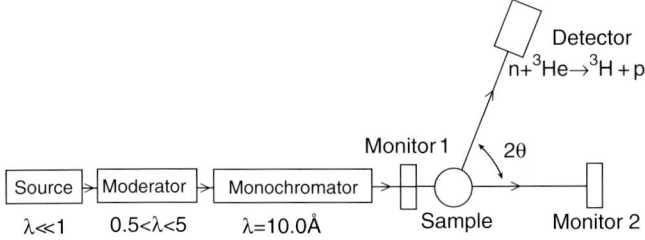

Figure 10.15 Schematic diagram showing the apparatus used for neutron scattering measurements.

function of scattering angle. A typical experimental system is shown schematically in Figure 10.15 [57]. Neutron sources are, in general, nuclear reactors in which neutrons are liberated as a result of the interaction of X-rays with matter [57–59]. The detection of neutrons is generally based on secondary processes leading to gas ionization. The easiest way to obtain "monochromatic" thermal neutrons is to moderate the neutron beam by passing it through suitable materials such as graphite or heavy water, which show relatively small absorption together with a large scattering cross section.

Neutrons with the same energy or wavelength may then be selected from the beam emerging from the moderator either by time of flight methods or by Bragg reflection from a rigid lattice. Energy analysis of the scattered beam may be achieved in the same way. In this type of analysis, information on energy exchange $\hbar\omega$ during the scattering experiment is only attainable for values exceeding the energy distribution within the beam. This limitation is overcome by the neutron spin echo technique [60–62], which is able to record relative changes in velocity down to 10^{-5} for even substantial wavelength distributions ($\Delta\lambda/\lambda \approx 0.15$) of the incident neutron flux. The name neutron spin echo has been deliberately chosen [60] since, as with the NMR spin echo (Figure 10.8), final neutron polarization (and with it the recorded signal) is established by the application of an intermediate π pulse giving rise to an eventual refocusing of the magnetization of the neutrons in the beam. Further details of the experimental procedure and analysis may be found in the relevant literature [57–59, 63–65].

10.2.2
Theory of Neutron Scattering

In the first instance we consider the neutron scatterers to be fixed in space and locate the origin of our coordinate system at the nucleus of the scattering atom. The effect of an incident neutron beam of energy $\hbar^2 k_0^2/2m$ is determined by the solution of the corresponding Schrödinger equation:

$$\left\{\nabla^2 + k_0^2 - \frac{2m}{\hbar^2} V(\mathbf{r})\right\}\psi = 0 \qquad (10.37)$$

where ψ denotes the wave function of the neutron beam and $V(r)$ is the interaction potential. In the first-order (Born's) approximation the solution of Eq. (10.37) is:

$$\psi = e^{ik_0 r} - \frac{b}{r} e^{ik_0 r} \tag{10.38}$$

where the potential has been taken as the delta function defined by:

$$V(r) = \frac{2\pi \hbar^2 b}{m} \delta(r) \tag{10.39}$$

In applying this equation it is assumed that the scattering is observed at a great distance from the scattering center so that it is sufficient to define the location (by a delta function) and the efficiency, which is given by parameter b.

A microphysical interpretation of the parameter b may be easily given by considering the scattering cross section σ. This quantity may be understood as the area which, when introduced into the beam, would remove from the beam the same number of neutrons as are scattered in the experiment under consideration. The two terms in Eq. (10.38) represent the incoming beam and the scattered beam. Since the probability of realizing a certain state is proportional to the square of the wave function ψ, one has:

$$\sigma = \frac{\text{total number of scattered particles per unit time}}{\text{total number of incident particles per unit time and area}}$$
$$= \frac{4\pi r^2 v \left| \frac{b}{r} e^{ik_0 r} \right|^2}{v \left| e^{ik_0 r} \right|^2} = 4\pi b^2 \tag{10.40}$$

Equation (10.40) shows immediately that b has the dimension of length and it is therefore termed the "scattering length."

While scattering by a single center gives an identical probability function in all directions, an ensemble of scattering centers will lead to a superposition of scattering waves with certain phase relations, thus giving rise to amplification or attenuation of the scattered beam, depending on the direction. If in addition the positions of the scattering centers are time dependent, the time dependent Schrödinger equation must be used instead of Eq. (10.37). Since, for mobile scatterers, energy exchange between the incident wave and the scattering centers becomes possible and since the spatial arrangement of the scattering centers leads to an angular variation of the intensity of the scattered beam, instead of the total scattering cross section, as defined by Eq. (10.40), the relevant quantity is the so-called differential cross section $d^2\sigma/d\Omega\, dE$ where $(d^2\sigma/dE\, d\Omega) d\Omega dE$ is the fraction of neutrons scattered into a differential solid angle $d\Omega$ having energies in the interval $E, \ldots, E + dE$. Carrying out the relevant calculations, one obtains [59]:

$$\frac{d^2\sigma}{d\Omega dE} = \frac{k_1}{k_0} \frac{1}{h} \int e^{-i\omega t} \sum_{m,n} b_m^* b_n e^{i\kappa[r_m(0) - r_n(t)]} dt \tag{10.41}$$

where the quantities κ and ω denote, respectively, the differences in the wave vectors ($\kappa = k_1 - k_0$) and the frequencies $\omega = \omega_1 - \omega_0$ between the incident and scattered neutron beams. In the literature, for the scattering vector κ the notation **Q** has also become common. The integration is to be carried out over all scattering centers whose positions at times 0 and t are, respectively, $\mathbf{r}_m(0)$ and $\mathbf{r}_n(t)$. If $\mathbf{r}_n$ is time independent this leads to a delta function, corresponding to elastic scattering from rigid scattering centers. The values b_m denote the scattering lengths of the individual nuclei. Depending on the relative orientation of the neutron spin with respect to the spin of the scattering nucleus, they may assume different values. Summing separately over like and unlike nuclei one obtains:

$$\frac{d^2\sigma}{d\Omega dE} = \frac{k_1}{k_0}\frac{1}{h}\int\left\{e^{-i\omega t}\sum_m b_m^2 e^{i\kappa[\mathbf{r}_m(0)-\mathbf{r}_m(t)]} + \sum_{m\neq n} b_m^* b_n e^{i\kappa[\mathbf{r}_m(0)-\mathbf{r}_n(t)]}\right\}dt \tag{10.42}$$

Since the actual value of b, the scattering length of a nucleus, is uncorrelated with the position of the nucleus, the summation over the individual positions may be replaced by a weighted integration using the relevant average values of the scattering length. Hence:

$$\frac{d^2\sigma}{d\Omega dE} = \frac{k_1}{k_0}\frac{1}{h}\int\int e^{-i\omega t}N[\langle b^2\rangle G_s(\mathbf{r},t) + \langle b\rangle^2 G_D(\mathbf{r},t)]e^{i\kappa\mathbf{r}}d\mathbf{r}\,dt \tag{10.43}$$

where N denotes the total number of scattering nuclei and $G_s(\mathbf{r}, t)$ and $G_D(\mathbf{r}, t)$ are the van Hove correlation functions [66]. $G_{s(D)}(\mathbf{r}, t)\,d\mathbf{r}$ denotes the probability density that, after a time t, the same (another) nucleus will be at a position shifted by the vector $\mathbf{r}$ from the original position of the nucleus under consideration. Obviously, $G_s(\mathbf{r}, t)$ coincides with the propagator (introduced in Section 2.1), which, for sufficiently large observation times, approaches a Gaussian curve. By adding $\langle b\rangle^2 G_s(\mathbf{r}, t)$ to the second term in brackets on the right-hand side of Eq. (10.43) while subtracting the same quantity from the first term, one obtains:

$$\frac{d^2\sigma}{d\Omega dE} = N\frac{\langle b^2\rangle-\langle b\rangle^2}{h}\frac{k_1}{k_0}\int e^{i(\kappa\mathbf{r}-\omega t)}G_s(\mathbf{r},t)d\mathbf{r}\,dt + N\frac{\langle b^2\rangle}{h}\frac{k_1}{k_0}\int e^{i(\kappa\mathbf{r}-\omega t)}G(\mathbf{r},t)d\mathbf{r}\,dt \tag{10.44}$$

where $G(\mathbf{r},t) = G_s(\mathbf{r},t) + G_D(\mathbf{r},t)$ denotes the probability that there is a nucleus at position $\mathbf{r}$ at time t when there was a particle (i.e., the original nucleus or another one) at the origin at time zero. This second term, referred to as the "coherent differential scattering cross section," is determined by the collective motion of the particle system, that is, by transport diffusion.

The first term is determined only by the behavior of the individual scattering centers. It is called the "incoherent differential scattering cross section" and contains the information on the propagation rate of the individual molecules, that is, on self-diffusion. The intensity of this term is determined by the mean square deviation of the scattering length from its average value:

$$\langle(b-\langle b\rangle)^2\rangle = \langle b^2\rangle - \langle b\rangle^2 \tag{10.45}$$

For hydrogen the two scattering cross sections corresponding to the two spin orientations are of almost the same magnitude but opposite sign, $\langle b \rangle^2$ is therefore much less than $\langle b^2 \rangle$ so that with hydrogen one in general observes only the incoherent scattering. For this reason, hydrogen-containing molecules are most suitable for self-diffusion measurements by neutron scattering. It turns out that hydrogen-containing molecules are also the best choice for PFG NMR measurements (Section 11.3.2) but in that case their advantage stems from the large gyromagnetic ratio.

The scattering cross section of hydrogen notably exceeds those of all other nuclei. This concerns in particular nuclei with predominantly coherent scattering cross sections, such as deuterium, carbon, and nitrogen. QENS studies of transport diffusion have therefore become amenable only quite recently [62, 67, 68], as a consequence of the notably reduced cross sections of the coherent scatterers.

10.2.3
Scattering Patterns and Molecular Motion

10.2.3.1 Fundamental Relations

According to Eq. (10.44), the quantity accessible from incoherent scattering measurements is given by:

$$S_{\text{inc}}(\boldsymbol{\kappa}, \omega) = \frac{1}{2\pi} \int \exp[i(\boldsymbol{\kappa} \cdot \boldsymbol{r} - \omega t)] G_s(\boldsymbol{r}, t) \, d\boldsymbol{r} \, dt \tag{10.46}$$

This is commonly called the "incoherent scattering function." The self-correlation function, which is the desired quantity, follows as the four-dimensional Fourier transform:

$$G_s(\boldsymbol{r}, t) = \frac{1}{(2\pi)^3} \int \exp[-i(\boldsymbol{\kappa} \cdot \boldsymbol{r} - \omega t)] S_{\text{inc}}(\boldsymbol{\kappa}, \omega) \, d\boldsymbol{\kappa} \, d\omega \tag{10.47}$$

Analysis of experimental data may be facilitated by assuming that the self-correlation function is of Gaussian form:

$$G_s(\boldsymbol{r}, t) = \frac{1}{[4\pi\gamma(t)]^{3/2}} \exp\left[\frac{-r^2}{4\gamma(t)}\right] \tag{10.48}$$

where $\gamma(t)$ is simply related to the mean square displacement of a particle during time t:

$$\langle r^2(t) \rangle = \int r^2 G_s(\boldsymbol{r}, t) \, d\boldsymbol{r} = 6\gamma(t) \tag{10.49}$$

Thus, within the limits of the Gaussian approximation, the quantity S_{inc} is completely determined by the time dependence of the mean square displacement. In particular, from Eqs. (10.46) and (10.47):

$$S_{\text{inc}}(\boldsymbol{\kappa}, \omega) = \frac{1}{2\pi} \int \exp[-\kappa^2 \gamma(t)] \exp(-i\omega t) \, dt \tag{10.50}$$

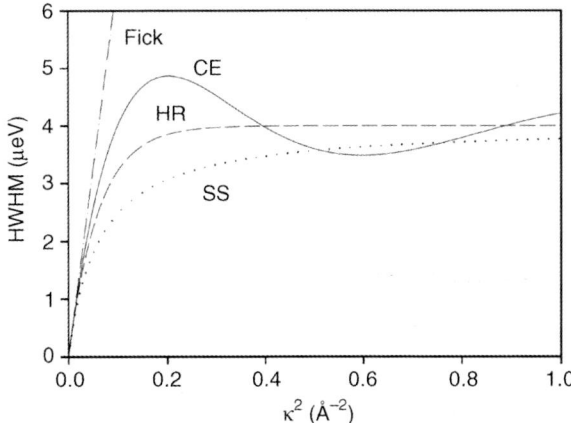

Figure 10.16 Broadening of response signal calculated as a function of κ^2 for different jump models: Chudley–Elliot (CE, solid line), Singwi–Sjölander (SS, dotted line), Hall–Ross (HR) (dashed line), and Fick (dashed-dotted line). The HWHMs are calculated for the same values of the mean squared jump length (100 Å^2) and mean time between subsequent jumps (jump rate corresponding to 4 μeV) and, therefore, for the same self-diffusivity $\mathcal{D} = 10^{-9}\,\text{m}^2\,\text{s}^{-1}$. Reprinted from Reference [64] with permission.

For normal diffusion (i.e., for observation times sufficiently large in comparison with the time scale of the elementary diffusion steps – see Section 2.1) $\gamma(t) = <r^2(t)>/6 = \mathcal{D}t$ and, with Eq. (10.50), one obtains:

$$S_{\text{inc}}^{\text{diff}} = \frac{1}{\pi} \frac{\mathcal{D}k^2}{(\mathcal{D}k^2)^2 + \omega^2} \tag{10.51}$$

that is, if plotted as a function of the energy exchange $E = \hbar\omega$ during scattering, a Lorentzian with a half-width at half-maximum (HWHM):

$$\Delta\omega(\kappa)_{\text{inc}} = \mathcal{D}\kappa^2 \tag{10.52}$$

Plotted versus κ^2, the HWFM is thus found to yield a straight line with its slope representing the self-diffusivity (Figure 10.16 [64]).

Similarly, the coherent scattering function is found to obey the relation [64, 69]:

$$S_{\text{coh}}(\kappa, \omega) = \frac{S(\kappa)}{\pi} \frac{D\kappa^2}{(D\kappa^2)^2 + \omega^2} \tag{10.53}$$

where now the HWHM:

$$\Delta\omega(\kappa)_{\text{coh}} = D\kappa^2 \tag{10.54}$$

yields the transport diffusivity rather than the self-diffusivity.

10.2.3.2 Thermodynamic Factor

Unlike its incoherent counterpart, Eq. (10.46), the coherent scattering function is proportional to the static structure factor $S(\kappa)$. In the limit of zero wave vector, $S(\kappa)$

measures the fluctuation in the number of scatterers contained in a given volume; that is [compare Eq. (7.30)]:

$$\lim_{\kappa \to 0} S(\kappa) = \frac{\langle N^2 \rangle_{fVT} - \langle N \rangle_{fVT}^2}{\langle N \rangle_{fVT}} = \frac{\partial \ln q}{\partial \ln f}\bigg|_T = \frac{1}{\Gamma} \quad (10.55)$$

with q and f being the molecular concentration and the fugacity of the sorbate, respectively; Γ, the slope of the sorption isotherm in logarithmic coordinates, is simply the thermodynamic correction factor appearing in the relation between transport and corrected diffusivity [Eq. (1.31) (implying ideal gas equation with $f = p$) and Eq. (8.19)].

Equation (10.55) is an immediate consequence of the fact that there is no scattering at all by a body where the particle densities in each volume element (with dimensions comparable with the wavelength) are identical. Thus, scattering intensity increases with the differences in particle densities, that is, with the extent of particle number fluctuations. The correlation between the intensity of particle fluctuation and the distinction between transport and corrected diffusion as provided by the thermodynamic factor may be easily rationalized in the following way: For non-interacting particles randomly distributed over the system the mean value of the squared deviation of the particle number from the mean value, $\langle (N - \langle N \rangle)^2 \rangle \equiv \langle N^2 \rangle - \langle N \rangle^2$, coincides with the mean particle number $\langle N \rangle$ and Eq. (10.55) yields the expected result that the thermodynamic factor of a system of non-interacting particles (ideal-gas approximation) equals 1. As soon as particle "competition" for the limited pore space becomes relevant (which is known to yield Langmuir-type isotherms), the mean squared deviation in the particle number is reduced. This leads, according to Eq. (10.55), to a decrease of the limiting value of $S(\kappa)$ and, hence, to an increase in the thermodynamic factor – as is required for Langmuir-type isotherms. The opposite situation occurs when intermolecular interactions significantly exceed the interaction of the guest molecules with the pore walls (e.g., for water adsorption in hydrophobic adsorbents), which is known to give rise to S-shaped (so-called type-III [70, 71]) isotherms. Facilitated by the intermolecular interactions, guest accumulations are now expected to occur more readily than in the initially considered case of non-interacting particles, leading to mean squared deviations $\langle (N - \langle N \rangle)^2 \rangle \equiv \langle N^2 \rangle - \langle N \rangle^2$ larger than $\langle N \rangle$ and, hence, to an increase of the limiting value of $S(\kappa)$ corresponding with a decrease in the thermodynamic factor. This is, once again, expected from the shape of the adsorption isotherm (see, for example, the presentation and discussion of these different possibilities in the case-study with MOF ZIF-8 in Section 19.3.1).

A detailed comparison between $\lim_{\kappa \to 0} S(\kappa)$ and $1/\Gamma$, as determined from neutron scattering measurements and configurational bias GCMC simulations, has been performed for n-hexane and n-heptane sorbed in silicalite [72]. Figure 10.17 displays the results from this comparison. Experimentally, $\lim_{\kappa \to 0} S(\kappa)$ has been estimated through the value of $S(\kappa)$ at the lowest κ studied, equal to 0.19 Å^{-1}. $S(\kappa)$ was obtained as the integral of the coherent scattering function $S_{\text{coh}}(\kappa, \omega)$ over all energy transfers.

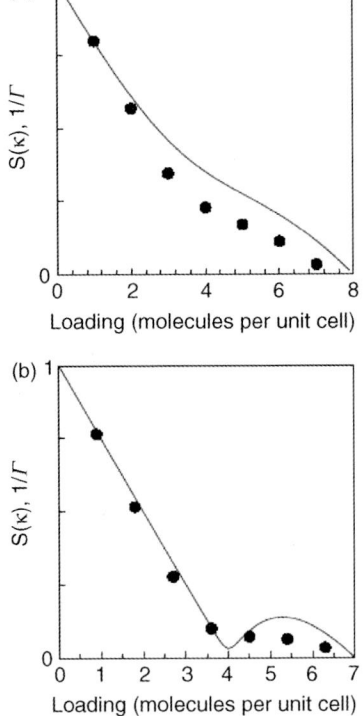
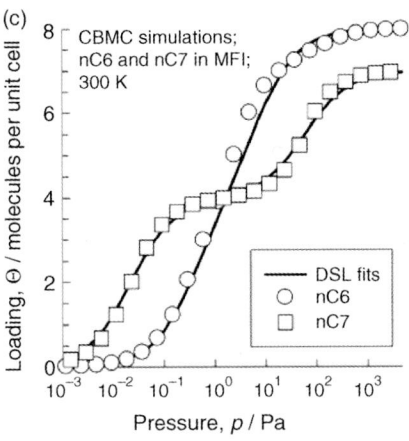

Figure 10.17 Comparison between the static structure factor, $S(\kappa)$, obtained at $\kappa = 0.19\ \text{Å}^{-1}$ from the QENS spectra (•) and the inverse of the thermodynamic factor, $1/\Gamma$ (solid line) calculated from dual site Langmuir fits of simulated adsorption isotherms for (a) n-hexane and (b) n-heptane in silicalite at 300 K. The isotherms for the two sorbates are shown in (c), where the points were determined by CBMC simulation in the grand canonical ensemble and the lines trace the dual site Langmuir fits to the points. Reprinted from Reference [72] with permission.

The agreement between static structure factors from the measurement and thermodynamic correction factors from the simulation is generally good. For n-hexane, an inflection point is seen in $1/\Gamma$ at a loading around 4 molecules per unit cell. For n-heptane, on the other hand, near-zero values of $S(\kappa)$ and $1/\Gamma$ are observed at four molecules per unit cell. At this occupancy the sorption isotherm flattens out, which is attributed to "commensurate freezing" of the n-heptane molecules within the interiors of sinusoidal channels, which are very similar in size to the n-heptane molecules [73]. At higher fugacities an inflection point is observed in the isotherm, which rises towards a saturation occupancy of approximately seven molecules per unit cell. This rapid rise of the isotherm, associated with large fluctuations in the number of sorbed molecules, causes a maximum in the simulated $1/\Gamma$ at a loading around 5.5 molecules per unit cell. These trends in $1/\Gamma$ are reflected in the loading dependence of the transport and corrected diffusivities from coherent

QENS and MD simulation, as anticipated from Eq. (8.19) (see also Section 18.2.2; notably, Figures 18.16 and 18.17).

The benefits of combining the potentials of molecular simulation and QENS have been illustrated by Figure 8.5. It provides direct comparison of transport D and corrected D_0 diffusivities in silicalite, as obtained from coherent QENS and measured sorption isotherms [Eqs. (10.53) and (8.19)] and from equilibrium MD and GCMC simulations [Eqs. (8.19) and (8.20)], revealing excellent agreement.

10.2.3.3 Evidence on the Elementary Steps of Diffusion

It must be emphasized that Eqs. (10.51)–(10.54) only hold for displacements over space and time scales that are governed by the diffusion equation, that is, Fick's second law [Eq. (1.2)], with, respectively, the self- and transport diffusivities. The deviations observed over smaller space and time scales, that is, to larger values of κ and ω as their Fourier counterparts, may provide valuable information about the elementary steps of diffusion. By considering different patterns for the elementary diffusion step, this option is already widely exploited for incoherent scattering but much less work has been performed on jump diffusion of coherent scatterers [69]. Most importantly, for isotropic systems the incoherent scattering function remains a Lorentzian [64]. However, the width (HWHM) of the energy spectra will differ from the simple $\Delta \kappa^2$ behavior. This is illustrated by Figure 10.16 [64] which displays the HWHM for different jump models.

As required for coinciding self-diffusivities, in the small-κ limit all models yield the same dependencies of Fickian (normal) diffusion. With increasing κ, details of the elementary steps become visible, appearing, for example, as different distributions and dependencies of the jump lengths or jump times. Examples are provided in Figure 10.16 [64]. Details of the jump models considered may be found in Reference [64]. The relevant model is considered to be the one that yields best agreement between the calculated dependencies and the experimental data. Until now, the limited accuracy of the primary data precludes the more rigorous treatment by fourfold Fourier transform of the scattering function as suggested by Eq. (10.47).

10.2.3.4 Intermediate Scattering Function

A quantity that is widely used in the analysis of random scattering measurements is the intermediate scattering function. It is defined according to the relation:

$$I(\kappa, t) = \int e^{i\omega t} S(\kappa, \omega) d\omega = \int e^{i\kappa r} G(r, t) dr = e^{-\kappa^2 \gamma(t)} \qquad (10.56)$$

where the last equation results by inserting Eq. (10.48).

Figure 10.18 shows the typical dependence of the intermediate scattering function I on the observation time t and on the wave vector κ. $I(\kappa,t)$ decays with time in a quasi-exponential function (note that a logarithmic scale is used in Figure 10.18 along the time axis). At larger κ (smaller length scale of observation), the decorrelation becomes more intense.

Notably, exactly this intermediate scattering function results as the primary quantity accessible in neutron spin-echo experiments [60, 75]. This is because, as

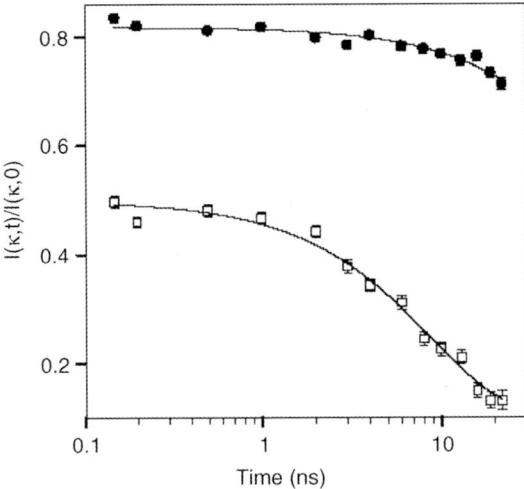

Figure 10.18 Normalized intermediate scattering functions for isobutene in silicalite at 550 K, as obtained by neutron spin echo (NSE) experiments for two different κ values: (•) 0.08 and (□) 0.2 Å^{-1}. Note that the time axis is logarithmic. The decay is clearly more expressed at the larger κ value (smaller length scale of observation). The lower wave vector value, 0.08 Å^{-1}, probes diffusion over distances of $2\pi/\kappa \approx 80$ Å [64].

with the generation of the NMR spin echo (Figure 10.8), the individual neutrons contribute to total neutron polarization only in proportion to the cosine of the deviation of their precessional angle from the mean, which is nothing other than the term ωt in the exponential [61]. We shall see that the intermediate scattering function is also the primary quantity observable in diffusion experiments by light scattering (Section 10.3 and Reference [76]). Moreover, an identical relation connects the experimentally accessible data in pulsed field gradient (PFG) NMR diffusion experiments with the underlying microdynamic processes (Section 11.1 and Reference [77]).

10.2.3.5 Range of Measurement

The main difficulty in the application of neutron scattering measurements to the study of diffusion lies in the difficulty of measuring with accuracy the energy distribution width ($\Delta\omega$) of the scattered neutrons for small scattering vectors κ since it is only this region that contains information on displacements over sufficiently large distances, that is, on the diffusional behavior. The lower limit of $\Delta\omega$, corresponding to the upper limit of the considered time scale, is imposed by the extent of deviation of the incident neutron beam from perfect monochromacy. Only in the most recent applications of the neutron spin echo is this limitation likely to be overcome.

Figure 10.19 gives an overview of the space and time scales covered by the different techniques of quasi-elastic neutron scattering [64]. The data refer to the values of momentum and energy transfer provided by the combined neutron instrumentation at the Institut Laue-Langevin, Grenoble, which presently offers worldwide the best

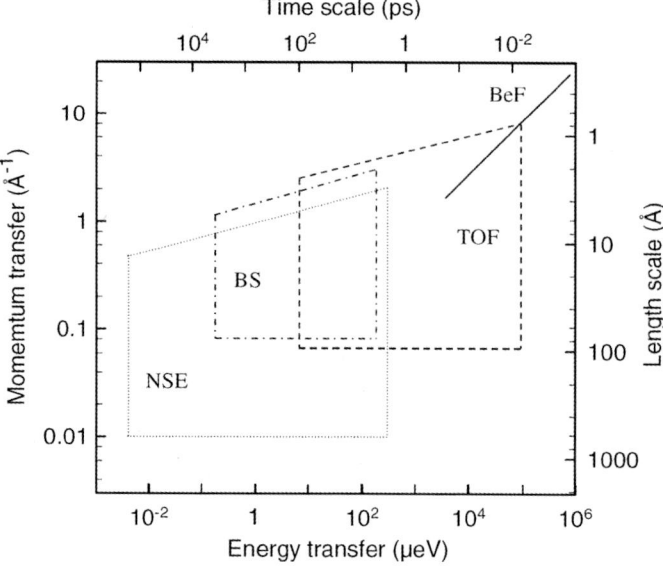

Figure 10.19 Overview of the different space and time scales following from the values of momentum and energy transfer as accessible by the different techniques of the combined neutron instrumentation at the Institut Laue-Langevin, Grenoble, namely, the neutron spin echo (NSE), back-scattering (BS), time-of-flight (TOF), and beryllium-filter (BeF) methods [74].

conditions for diffusion measurements by quasi-elastic neutron scattering. It is due to these excellent technical conditions that, over the last few years, neutron scattering has been applied to diffusion studies in nanoporous solids to a continuously increasing extent [64, 68, 75, 78–87]. Most importantly, operating at the transition between the elementary steps and long-distance diffusion, no other technique is able to monitor diffusion phenomena over such short distances.

In fact, by recording molecular displacements over a couple of nanometers, that is, over distances modestly exceeding the unit cell dimensions, neutron scattering is able to trace the features of long-range diffusion without being influenced by the effects of structural defects, which become important in real crystals at greater length scales [85]. Neutron scattering is therefore ideally suited for measuring the diffusion behavior to be expected under the influence of an ideal host structure, even in real host materials with their inevitable structural deficiencies. This makes neutron scattering a particularly useful technique for comparison with the simulation results for ideal host lattices.

Furthermore, comparison of neutron scattering measurements with the diffusivities measured over longer distances by more "macroscopic" techniques can provide useful quantitative information concerning the presence and magnitude of internal diffusion barriers (Sections 17.2.6 and 18.2.1). However, because such effects appear to be very common (perhaps even universal) in real crystals, it is also important to note that neutron scattering measurements generally do not provide useful estimates of the diffusivities encountered in real systems at practical length scales.

10.3
Diffusion Measurements by Light Scattering

The methods outlined in Section 10.2 with respect to neutron scattering can be applied, at least in principle, to obtain information concerning molecular motion from the scattering of other forms of radiation, such as light. There is, however, one serious difficulty. As a result of the low absorption cross section of most nuclei occurring in nanoporous solids for neutrons, these materials are relatively transparent to a neutron beam. This is not the case for visible light. It is for this reason that the application of light scattering to diffusion phenomena in porous materials is still in its infancy. However, promising initial results [88] and the attractive prospects offered by a successful application of light scattering to diffusion studies in nanoporous materials as a novel approach that complements existing techniques are likely to stimulate further activities in the field. We therefore include here a brief introduction to the principle of this method with some suggestions as to how the method might be further applied in practice.

In the classical theory of light scattering, an electromagnetic field from the incident radiation is considered to exert a force on the charges within the scattering volume. As a result of this force, which oscillates with the frequency of the light, the charges themselves oscillate, thus inducing an electric field. The resulting scattered field is in effect a superposition of the scattered field from all the charges or subsystems. Let us consider the subsystems as identical sites with diameters much less than the wavelength. The scattered fields will then also be identical except for a phase factor depending on the relative positions of the individual subsystems. An arbitrarily selected sub-region may therefore be paired with a second sub-region in such a way that, for any direction except forward, their scattered fields are identical in amplitude but opposite in sign, so that all side scattering is canceled. On a molecular scale, however, the fine structure of a fluid is modified by the chaotic movement of the individual molecules. This leads to a fluctuating space and time dependence of the optical properties and hence to an observable effect.

10.3.1
Theory

For an analytic treatment of this phenomenon one may start from the relation for the electric field produced as a result of scattering within the volume considered [89–91]:

$$E_s(\mathbf{r},t) = \frac{ik_f^2 E_0}{4\pi\varepsilon_0 r} \exp[i(\mathbf{k}_f \mathbf{r} - \omega_i t)] \int \delta\varepsilon(\mathbf{r}',t) e^{i\mathbf{\kappa r}'} d\mathbf{r}' \tag{10.57}$$

Here E_0 represents the electric field of the incident light, $\mathbf{k}_i$ and ω_i, represent the wave vector and frequency of the incident light and $\mathbf{k}_f$ the wave vector of the scattered light. As for neutron scattering, $\mathbf{\kappa} = \mathbf{k}_i - \mathbf{k}_f$ denotes the difference in the wave vectors between the incident and scattered beams, whence $|\mathbf{\kappa}| = (4\pi n/\lambda_0)\sin(\theta/2)$ with n, λ_0 and θ denoting, respectively, the index of refraction, the wavelength in vacuum, and the scattering angle; $\delta\varepsilon(r', t)$ denotes the deviation of the dielectric constant at position r' and time t, from the average value ε_0.

It is the time dependence of $\delta\varepsilon$ and hence of E_s that contains the information concerning the microdynamics of the system. To elucidate this effect we consider the quantity $\langle E_s^*(r', 0) E_s(r', t) \rangle$, where the brackets denote the ensemble average value, which, according to the ergodic hypothesis, should correspond with the time average value. Inserting Eq. (10.57) one obtains:

$$\langle E_s^*(r,0)E_s(r,t)\rangle = \frac{k_f^4 E_0^2}{16\pi^2\varepsilon_0^2 r^2} \left\langle \int \delta\varepsilon^*(r',v_0)e^{-i\kappa r'}dr' \int \delta\varepsilon(r'',t)e^{i\kappa r''}dr'' \right\rangle e^{-i\omega_i t}$$

$$= \frac{k_f^4 E_0^2}{16\pi^2\varepsilon_0^2 r^2} e^{-i\omega_i t} \int\int \langle \delta\varepsilon^*(r',0)\delta\varepsilon(r'+\Delta r,t)\rangle e^{i\kappa\Delta r} dr' d\Delta r$$

(10.58)

These relations contain the well-known property of electromagnetic radiation that the intensity of the scattered radiation, which is proportional to E_s^2, is proportional to the fourth power of the wave vector (or the frequency). Since, as a consequence of the averaging procedure all positions r' are equivalent, omitting all constant terms one may write:

$$\langle E_s^*(r,0)E_s(r,t)\rangle \propto e^{-i\omega_i t} \int \langle \delta\varepsilon(0,0)\delta\varepsilon(\Delta r,t)\rangle e^{i\kappa\Delta\,r} d\Delta r \quad (10.59)$$

$\langle \delta\varepsilon(0,0)\delta\varepsilon(\Delta r, t)\rangle$ is the correlation function of the deviations of the dielectric constant from the average value. Since the dielectric constant of a volume element depends on the local temperature and composition, fluctuations of temperature and composition will lead directly to fluctuations in the dielectric constant [76, 90]. The general dependency is most easily understood if the change of the dielectric constant is caused by Brownian motion of a few large molecules. Evidently then, the only terms $\langle \delta\varepsilon(0,0)\delta\varepsilon(\Delta r, t)\rangle$ in the integrand that are different from zero are those for which such a particle, located at the origin at time zero, has migrated to Δr during the time interval Δt. The probability of this is given by the propagator [Eq. (2.16)]:

$$P(\Delta r, t) = \frac{1}{(4\pi\mathcal{D}t)^{3/2}} \exp\left[-\frac{(\Delta r)^2}{4\mathcal{D}t}\right]$$

in which $\mathcal{D}$ is the self-diffusivity of the particle considered. Inserting this expression into Eq. (10.59) in place of $\langle \delta\varepsilon(0,0)\,\delta\varepsilon(\Delta r, t)\rangle$ leads to:

$$\langle E_s^*(r,0)E_s(r,t)\rangle \propto e^{-i\omega t}e^{-\kappa^2 \mathcal{D} t} \quad (10.60)$$

Equation (10.60) shows that the self-diffusion coefficient may be determined in a straightforward way from a semilogarithmic plot of the correlation function of the electric field versus the time t, which should yield a straight line of slope $\kappa^2\mathcal{D}$.

Note that the structure of the autocorrelation function of the electric field of the scattered light as given by Eq. (10.59) and, correspondingly, Eq. (10.60) coincides with that of the QENS intermediate scattering function [Eq. (10.56)], nicely reflecting the similarity in the underlying mechanisms of measurement. It is remarkable that, although based on different mechanisms (Section 11.1), the PFG NMR spin echo attenuation curve has the same form as the intermediate scattering function. With this analogy in mind, some authors refer to the PFG NMR spin-echo attenuation as a generalized scattering experiment [92–94].

Strictly speaking, the diffusivity $\mathcal{D}$ appearing in Eq. (10.60) refers to a non-equilibrium process, namely, a local (i.e., microscopic) density fluctuation, which, owing to Onsager's regression theorem [95], is controlled by transport parameters that coincide with the macroscopic transport diffusivity. Thus, light scattering reflects the situation of coherent neutron scattering [see Eqs. (10.53) and (10.54)] where the overall density fluctuation, rather than the fluctuation in the density of some labeled molecules, is recorded. With neutrons, the latter situation is provided by incoherent scattering. There is no equivalent of this situation in light scattering. For highly diluted systems, however, when any interaction of the probe molecules (i.e., the scatterers) is excluded, transport diffusion and self-diffusion are known to coincide. With this understanding, light scattering may clearly also be rationalized in relation to self-diffusion.

10.3.2
Experimental Issues for Observing Light Scattering Phenomena: Index Matching

10.3.2.1 Index Matching
The intensity of the light resulting from density fluctuations is negligibly small in comparison with that of any corruptive light due to sample imperfections, in particular due to reflections at phase boundaries between regions of different optical densities. It is for this reason that diffusion measurements in fluids by light scattering necessitate the application of fluids of extreme purity. Essentially, any contamination of either the fluid or the surroundings by (foreign) particles must be avoided [76, 89, 90, 96]. Otherwise, the light reflected by these particles would completely mask the light generated by density fluctuations within the fluid. In addition to the requirement of transparency of the porous solids under study, it is this effect of light reflection on their external surfaces that dramatically complicates the application of light scattering techniques for diffusion studies.

Will et al. [88] managed to solve this problem by considering the diffusion of polystyrene particles in "controlled porous glasses" (CPG) saturated by a mixture of water and dimethyl sulfoxide. The composition was chosen to yield a refractive index ($n = 1.465$ at the considered wavelength $\lambda_0 = 623.8$ nm), which, after "fine-tuning" by temperature variation (!), ensured best matching with the refractive index of the host and, hence, minimum reflection intensity by the host–fluid interface at the external surface of the host particle. The substantial difference between the refractive indices of the pore fluid and the polystyrene particles ($n = 1.58$) gives rise to a fluctuation-induced light scattering that may be separated from the background. In this way, the particle diffusivities could be directly determined as a function of the particle and pore diameters [88].

The application of "tracers" such as polystyrene particles in diffusion measurements by light scattering is limited to meso- and macroporous materials with sufficiently large pore diameters. However, without the presence of any dissolved tracer particles, two-component fluids give rise to light scattering, due to fluctuations in the local composition and, hence, in the refractive indices. In this case, the relevant diffusivities appearing in Eq. (10.60) are the mutual diffusivities of these two components. This type of experiment is, essentially, applicable to materials with any pore diameter, including microporous solids. The scattering intensity attainable in such experiments, however, is far smaller than that attainable by the application of extended particles so that the initial orienting experiments along these lines [97] have not been continued. For completeness, one should also mention that – under the conditions of gas phase adsorption (and in complete analogy with the situation of coherent neutron scattering) – single-component sorption at medium loadings would give rise to much larger scattering intensities, but with the transport diffusivity as the relevant quantity determining the fluctuation rates and appearing, therefore, in Eq. (10.60). Experiments of this type would exclude, however, the option of index matching provided by the application of two-component fluids at pore-space saturation so that the light scattered by density fluctuations in the interior of the host particles would most likely be obscured by light reflection from the external particle surface.

It is the correlation function of the electric field of the scattered light that, through Eqs. (10.59) and (10.60), contains all information about the underlying diffusion phenomena. There are two main ways to determine correlation functions.

10.3.2.2 Optical Mixing Techniques

Optical mixing is the analog of the beating techniques developed in radiofrequency spectroscopy. Without passing through a filter the scattered light impinges directly on the cathode of a photomultiplier. In the homodyne (or self-beat) method only the scattered light impinges on the photocathode, while in the heterodyne method a small portion of the unscattered laser beam is mixed with the scattered light on the cathode surface. Since the phototube is a square-law detector, its instantaneous current output $i(t)$ is proportional to the square of the incident field. Subsequently, the current output is usually passed into a computer (autocorrelator), which calculates its time autocorrelation function:

$$\langle i(0)i(t)\rangle \propto \left\langle |E_s(0)|^2 |E_s(t)|^2 \right\rangle \tag{10.61}$$

It may be shown that, in the case of homodyne measurement [89, 90, 98, 99] for the situation represented by Eq. (10.60):

$$\left\langle |E_s(0)|^2 |E_s(t)|^2 \right\rangle \propto 1 + \beta e^{-2\kappa^2 Dt} \tag{10.62}$$

where, in addition to the exponential decay (now with an additional factor 2 in the exponent), a DC term determining the base line for the homodyne correlation function of the scattered light is observed.

A more detailed theoretical analysis is closely related to the experimental system, especially provided by the digital (photo-count) autocorrelation techniques and would lead beyond the limits of this book. For this purpose, the more detailed introduction provided by Brown [90], Xu [89], Berne and Pecora [91], and Cummins and Pike [98] or in the original paper of Jakeman and Pike [74] is recommended.

10.3.2.3 Filter Techniques

The time correlation of a stochastic quantity $A(t)$ and its spectral density is provided by the Wiener–Khintchine relation [91, 100]:

$$\langle A^*(0)A(t) \rangle = \int_{-\infty}^{\infty} e^{i\omega t} I_A(\omega) \, d\omega \tag{10.63}$$

or vice versa:

$$I_A(\omega) = \frac{1}{2\pi} \int_{-\infty}^{\infty} e^{-i\omega t} \langle A^*(0)A(t) \rangle dt \tag{10.64}$$

where $I_A(\omega) \, d\omega$ is the total intensity of those harmonic constituents of the stochastic quantity $A(t)$, whose frequencies are in the interval $\omega, \ldots, \omega + d\omega$. Figure 10.20 shows a schematic diagram for a device to determine the intensity of the stochastic quantity (in our case of the scattering intensity) in a certain frequency range. For this purpose, the signal $A_T(t)$, measured for a period Γ, is passed through a filter that passes only the required frequency range $\omega, \ldots, \omega + d\omega$, that is, the part $A_{T0}(t)$ of the signal. Following the filter is a detector with an output proportional to $|A_{T0}(t)|^2$. Time averaging of the signal then yields $\langle |A_{T0}|^2 \rangle_T$ and the limit $T \to \infty$ of this quantity may be shown to yield:

$$\lim_{T \to \infty} \langle |A_{T0}|^2 \rangle_T = I_A(\omega_0) \Delta\omega \tag{10.65}$$

in which $I_A(\omega)$ is the function that appears in the Wiener–Khintchine relation [Eq. (10.63)].

In the following, we shall specify the stochastic function $A(t)$ to be the electric field of a scattered light wave. As a filter we may consider a grating or prism, all allowing through only a small frequency section $\Delta\omega$. Hence, according to Eq. (10.65) the output from the photomultiplier is:

$$\lim_{T \to \infty} \langle |E_{T0}|^2 \rangle_T = I_E(\omega_0) \Delta\omega \tag{10.66}$$

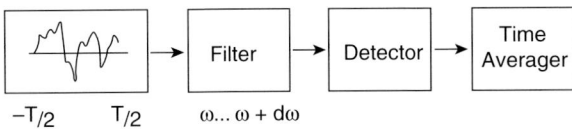

Figure 10.20 Schematic diagram showing determination of the intensity of a stochastic quantity in a certain frequency range.

where $I_E(\omega_0)$ is the spectral density of the electric field autocorrelation function, and where ω_0 and $\Delta\omega$ are determined by the filter properties. If the filter is tuned through all frequencies, the autocorrelation function of the scattered field may be determined by Fourier transformation from Eq. (10.66):

$$\langle E^*(0)E(t)\rangle = \int_{-\infty}^{\infty} e^{i\omega t} I_E(\omega) d\omega \qquad (10.67)$$

Assuming an exponentially decaying correlation function $e^{-i\omega_0 t} e^{-t/\tau}$ [see Eq. (10.70), with $1/\tau = \kappa^2 D$], by applying the inverse transform, one thus obtains for the spectrum:

$$I_E(\omega) = \frac{1}{2\pi} \int_{-\infty}^{+\infty} e^{-i(\omega-\omega_0)t} e^{-t/\tau} dt = \frac{1/\tau}{(\omega-\omega_0)^2 + (1/\tau)^2} \qquad (10.68)$$

that is, a Lorentzian with the half width at half intensity given by:

$$\Delta\omega = 1/\tau \text{ or } \Delta\nu = 1/2\pi\tau \qquad (10.69)$$

Thus, fluctuations with time constants $\tau \leq 10^{-10}$ s yield mean frequency shifts of $\nu \geq 10^9$ Hz, which are large enough to be observed by using a diffraction grating as a filter. For slower processes with 10^{-10} s $\leq \tau \leq 10^{-6}$ s corresponding to 10^5 Hz $\leq \nu \leq 10^9$ Hz, observation by Fabry–Perot interferometers is still possible. For slower processes, however, all filter techniques fail and optical mixing techniques are used. Figure 10.21 gives an example of the range of diffusion coefficients

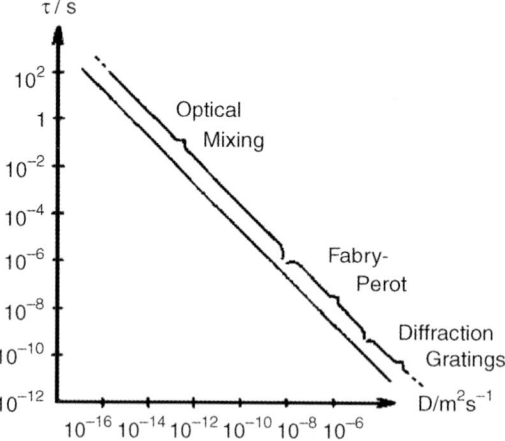

Figure 10.21 Correlation between the decay constant τ of the autocorrelation function of the electric field vector E [Eq. (10.60)] and the diffusivity D with the ranges accessible by the different observation techniques, for a scattering angle $\theta = 90°$. For a given D, τ will be shifted to higher values for smaller scattering angles and to smaller values for larger scattering angles.

accessible by light scattering using different detection methods. The representation has been determined from the relation:

$$\tau = (\kappa^2 D)^{-1} \tag{10.70}$$

for the decay constant [which follows from Eq. (10.60)] and determining the scattering vector κ for $\lambda = 500$ nm, a refraction index $n = 1.3$, and a scattering angle of $\theta = 90°$.

The best scattering properties are provided by macromolecules with diffusivities smaller than 10^{-10} m^2 s^{-1}. Evidently, from Figure 10.21, diffusivities of this magnitude can only be measured by optical mixing techniques. Furthermore, the small line width corresponding to the large decay constants requires the application of a monochromatic laser light source.

References

1. Ernst, R.R., Bodenhausen, G., and Wokaun, A. (1990) *Principles of Nuclear Magnetic Resonance in One and Two Dimensions*, Oxford University Press, Oxford.
2. Schmidt-Rohr, K. and Spiess, H.W. (1994) *Multidimensional Solid-State NMR and Polymers*, Academic Press, New York.
3. Levitt, M.H. (2001) *Spin Dynamics, Basis of Nuclear Magnetic Resonance*, John Wiley & Sons, Ltd, Chichester.
4. Blümich, B. (2005) *Essential NMR*, Springer, Berlin, Heidelberg.
5. Price, W.S. (2009) *NMR Studies of Translational Motion*, University Press, Cambridge.
6. Callaghan, P.T. (2011) *Translational Dynamics and Magnetic Resonance*, Oxford University Press, Oxford.
7. Abragam, A. (1961) *The Principles of Nuclear Magnetism*, Clarendon Press, Oxford.
8. Slichter, P. (1980) *Principles of Magnetic Resonance*, Springer, Berlin.
9. Walker, T.G. and Happer, W. (1997) *Rev. Mod. Phys.*, **69**, 629.
10. Albert, M.S., Cates, G.D., Driehuys, B., Happer, B., Saam, B., Springer, C.S., and Wishnia, A. (1994) *Nature*, **370**, 199.
11. Meersmann, T., Logan, J.W., Simonutti, R., Caldarelli, S., Comotti, A., Sozzani, P., Kaiser, L.G., and Pines, A. (2000) *J. Phys. Chem. A*, **104**, 11665.
12. Woods, J.C., Yablonskiy, D.A., and Conradi, M.S. (2009) *Diffusion Fundamentals III* (eds C. Chmelik, N.K. Kanellopoulos, J. Kärger, and D. Theodorou), Leipziger Universitätsverlag, Leipzig, p. 391.
13. Brunner, E. (1999) *Concepts Magn. Reson.*, **11**, 313.
14. LeaWoods, J.C. (2001) *Concept Magn. Reson.*, **13**, 277.
15. Moudrakovski, I.L., Ratcliffe, C.I., and Ripmeester, J.A. (1995) *Appl. Magn. Reson.*, **8**, 385.
16. Pfeifer, H. (1972) *NMR Basic Principles and Progress*, vol. 7, Springer, Berlin, p. 53.
17. Hahn, E.L. (1950) *Phys. Rev.*, **77**, 746.
18. Sangoro, J.R., Sergehei, A., Naumov, S., Galvosas, P., Kärger, J., Wespe, C., Bordusa, F., and Kremer, F. (2008) *Phys. Rev. E*, **77**, 51202.
19. Iacob, C., Sangoro, J.R., Papadopoulos, P., Schubert, T., Naumov, S., Valiullin, R., Kärger, J., and Kremer, F. (2010) *Phys. Chem. Chem. Phys.*, **12**, 13798.
20. Sangoro, J.R., Iacob, C., Naumov, S., Valiullin, R., Rexhausen, H., Hunger, J., Buchner, R., Strehmel, V., Kärger, J., and Kremer, F. (2011) *Soft Matter*, **7**, 1678.
21. Magusin, P.C.M., Schuring, D., van Oers, E.M., de Haan, J.W., and van Santen, R.A. (1999) *Magn. Reson. Chem.*, **37**, S108–S117.
22. Saengsawang, O., Schüring, A., Remsungnen, T., Loisruangsin, A., Hannongbua, S., Magusin, P.C.M., and Fritzsche, S. (2008) *J. Phys. Chem. C*, **112**, 5922.

23 Moudrakovski, I.L., Ratcliffe, C.I., and Ripmeester, J.A. (1996) *Appl. Magn. Reson.*, **10**, 559.

24 Jameson, C.J., Jameson, A.K., Gerald, R.E., and Lim, H.M. (1997) *J. Phys. Chem. B*, **101**, 8418.

25 Bonardet, J.L., Domeniconi, T., N'Gokoli-Kekele, P., Springuel-Huet, M.A., and Fraissard, J. (1999) *Langmuir.*, **15**, 5836.

26 Bonardet, J.L., Fraissard, J., Gedeon, A., and Springuel-Huet, M.A. (1999) *Catal. Rev.-Sci. Eng.*, **41**, 115.

27 Moudrakovski, I.L., Sanchez, A., Ratcliffe, C.I., and Ripmeester, J.A. (2000) *J. Phys. Chem. B*, **104**, 7306.

28 Moudrakovski, I.L., Lang, S., Ratcliffe, C.I., Simard, B., Santyr, G., and Ripmeester, J.A. (2000) *J. Magn. Reson.*, **144**, 372.

29 Liu, Y., Zhang, W., Liu, Z., Xu, S., Wang, Y., Xie, Z., Han, X., and Bao, X. (2008) *J. Phys. Chem. C*, **112**, 15375.

30 Fraissard, J., Vincent, R., Doremieux, C., Kärger, J., and Pfeifer, H. (1996) *Catalysis, Science and Technology* (eds J.R. Anderson and M. Boudart), Springer Verlag, Berlin, Heidelberg, p. 1.

31 Fraissard, J. and Ito, T. (1988) *Zeolites*, **8**, 350.

32 Kärger, J., Pfeifer, H., Wutscherk, T., Ernst, S., Weitkamp, J., and Fraissard, J. (1992) *J. Phys. Chem.*, **96**, 5059.

33 Springuel-Huet, M.A., Nosov, A., Kärger, J., and Fraissard, J. (1996) *J. Phys. Chem.*, **100**, 7200.

34 N'Gokoli-Kekele, P., Springuel-Huet, M.A., and Fraissard, J. (2002) *Adsorption*, **8**, 35.

35 Domeniconi, T., Gokoli-Kekele, P.N., Bonarded, J.L., Springuel-Huet, M.A., and Fraissard, J. (1999) *Proceedings of the 12th International Zeolite Conference* (eds M.M.J. Treacy, B.K. Marcus, M.E. Bisher, and J.B. Higgins), Material Research Society, Warrendale, p. 2991.

36 Jameson, A.K., Jameson, C.J., and Gerald, R.E. (1994) *J. Chem. Phys.*, **101**, 1775.

37 Jameson, C.J., Jameson, A.K., and Lim, H.M. (1996) *J. Chem. Phys.*, **104**, 1709.

38 Spiess, H.W. (1974) *Chem. Phys.*, **6**, 217.

39 Parker, W.O. (2000) *Comment Inorg. Chem.*, **22**, 31.

40 Becker, J., Comotti, A., Simonutti, R., Sozzani, P., and Saalwachter, K. (2005) *J. Phys. Chem. B*, **109**, 23285.

41 Stepanov, A.G. (1999) *Usp. Khim.*, **68**, 619.

42 Zibrowius, B., Bulow, M., and Pfeifer, H. (1985) *Chem. Phys. Lett.*, **120**, 420.

43 Spiess, H.W. (1978) *NMR Basic Principles and Progress*, vol. 15, Springer, Berlin.

44 Zibrowius, B., Caro, J., and Pfeifer, H. (1988) *J. Chem. Soc., Faraday Trans. 1*, **84**, 2347.

45 Xu, J. and Dong, R.Y. (2006) *J. Phys. Chem. B*, **110**, 1221.

46 Burmeister, R., Boddenberg, B., and Verfurden, M. (1989) *Zeolites*, **9**, 318.

47 Stepanov, A.G., Shubin, A.A., Luzgin, M.V., Shegai, T.O., and Jobic, H. (2003) *J. Phys. Chem. B*, **107**, 7095.

48 Favre, D.E., Schaefer, D.J., Auerbach, S.M., and Chmelka, B.F. (1998) *Phys. Rev. Lett.*, **81**, 5852.

49 Geil, B., Isfort, O., Boddenberg, B., Favre, D.E., Chmelka, B.F., and Fujara, F. (2002) *J. Chem. Phys.*, **116**, 2184.

50 Carr, H.Y. and Purcell, E.M. (1954) *Phys. Rev.*, **80**, 630.

51 Meiboom, S. and Gill, D. (1958) *Rev. Sci. Instrum.*, **29**, 688.

52 Pampel, A., Engelke, F., Galvosas, P., Krause, C., Stallmach, F., Michel, D., and Kärger, J. (2006) *Microporous Mesoporous Mater.*, **90**, 271.

53 Heink, W., Kärger, J., and Pfeifer, H. (1990) *J. Chem. Soc., Chem. Commun.*, 1454.

54 Kärger, J. and Jobic, H. (1991) *Colloids Surf.*, **58**, 203.

55 Romanova, E.E., Krause, C.B., Stepanov, A.G., Schmidt, W., van Baten, J.M., Krishna, R., Pampel, A., Kärger, J., and Freude, D. (2008) *Solid State Nucl. Magn. Reson.*, **33**, 65.

56 Kärger, J., Pfeifer, H., Rauscher, M., and Walter, A. (1980) *J. Chem. Soc., Faraday Trans. I*, **76**, 717.

57 Thomas, R.K. (1982) *Prog. Solid State Chem.*, **14**, 1.

58 Springer, T. and Lechner, R.E. (2005) *Diffusion in Condensed Matter: Methods, Materials, Models* (eds P. Heitjans and J. Kärger), Springer, Berlin, Heidelberg, p. 93.

59 Egelstaff, P.A. (1965) *Thermal Neutron Scattering*, Academic Press, London.
60 Mezei, F. (1972) *Z. Phys.*, **255**, 146.
61 Richter, D. (2005) *Diffusion in Condensed Matter: Methods, Materials, Models* (eds P. Heitjans and J. Kärger), Springer, Berlin, Heidelberg, p. 513.
62 Jobic, H., Kärger, J., and Bee, M. (1999) *Phys. Rev. Lett.*, **82**, 4260.
63 Bee, M. (1988) *Quasielastic Neutron Scattering*, Adam Hilger, Bristol.
64 Jobic, H. and Theodorou, D. (2007) *Microporous Mesoporous Mater.*, **102**, 21.
65 Springer, T. (1972) *Quasi-elastic Neutron Scattering for the Investigation of Diffusive Motion in Solids and Liquids*, Springer, Berlin.
66 van Hove, L. (1954) *Phys. Rev.*, **95**, 249.
67 Jobic, H., Makrodimitris, K., Papadopoulos, G.K., Schober, H., and Theodorou, D.N. (2004) *Stud. Surf. Sci. Catal.*, **154**, A–C, 2056.
68 Jobic, H., Methivier, A., Ehlers, G., Farago, B., and Haeussler, W. (2004) *Angew. Chem. Int. Ed.*, **43**, 364.
69 Cook, J.C., Richter, D., Scharpf, O., Benham, M.J., Ross, D.K., Hempelmann, R., Anderson, I.S., and Sinha, S.K. (1990) *J. Phys. Condens. Matter*, **2**, 79.
70 Ruthven, D.M. (2008) *Adsorption and Diffusion* (eds H.G. Karge and J. Weitkamp), Springer, Berlin, Heidelberg, p. 1.
71 Schüth, F., Sing, K.S.W., and Weitkamp, J. (2002) *Handbook of Porous Solids*, Wiley-VCH Verlag GmbH, Weinheim.
72 Jobic, H., Laloue, N., Laroche, C., van Baten, J.M., and Krishna, R. (2006) *J. Phys. Chem. B*, **110**, 2195.
73 Vlugt, T.J.H., Krishna, R., and Smit, B. (1999) *J. Phys. Chem. B*, **103**, 1102.
74 Jakeman, E. and Pike, E.R. (1968) *J. Phys. A*, **1**, 128; 2, 115, 411 (1969).
75 Jobic, H. (2008) *Adsorption and Diffusion* (eds H.G. Karge and J. Weitkamp), Springer, Berlin, Heidelberg, p. 207.
76 Leipertz, A. and Fröba, A.P. (2005) *Diffusion in Condensed Matter: Methods, Materials, Models* (eds P. Heitjans and J. Kärger), Springer, Berlin, Heidelberg, p. 579.
77 Kärger, J. and Heink, W. (1983) *J. Magn. Reson.*, **51**, 1.
78 Jobic, H., Skoulidas, A.I., and Sholl, D.S. (2004) *J. Phys. Chem. B*, **108**, 10613.
79 Chong, S.S., Jobic, H., Plazanet, M., and Sholl, D.S. (2005) *Chem. Phys. Lett.*, **408**, 157.
80 Jobic, H., Kärger, J., Krause, C., Brandani, S., Gunadi, A., Methivier, A., Ehlers, G., Farago, B., Haeussler, W., and Ruthven, D.M. (2005) *Adsorption*, **11**, 403.
81 Jobic, H., Schmidt, W., Krause, C., and Kärger, J. (2006) *Microporous Mesoporous Mater.*, **90**, 299.
82 Plant, D., Jobic, H., Llewellyn, P., and Maurin, G. (2007) *Adsorption*, **13**, 209.
83 Plant, D., Jobic, H., Llewellyn, P., and Maurin, G. (2007) *Eur. Phys. J. - Spec. Top.*, **141**, 127.
84 Kolokolov, D.I., Stepanov, A.G., Glaznev, I.S., Aristov, Y.I., Plazanet, M., and Jobic, H. (2009) *Microporous Mesoporous Mater.*, **125**, 46.
85 Feldhoff, A., Caro, J., Jobic, H., Krause, C.B., Galvosas, P., and Kärger, J. (2009) *Chem. Phys. Chem.*, **10**, 2429.
86 Salles, F., Jobic, H., Maurin, G., Koza, M.M., Llewellyn, P.L., Devic, T., Serre, C., and Férey, G. (2008) *Phys. Rev. Lett.*, **100**, 245901.
87 Rosenbach, N., Jobic, H., Ghoufi, A., Salles, F., Maurin, G., Bourrelly, S., Llewellyn, P.L., Devic, T., Serre, C., and Ferey, G. (2008) *Angew. Chem. Int. Ed.*, **47**, 6611.
88 Beschieru, V., Rathke, B., and Will, S. (2009) *Microporous Mesoporous Mater.*, **125**, 63.
89 Xu, R. (2000) *Particle Characterization: Light Scattering Methods*, Kluwer Academic Publisher, Boston.
90 Brown, W. (1996) *Light Scattering*, Clarendon Press, Oxford.
91 Berne, BJ. and Pecora, R. (1976) *Dynamic Light Scattering*, John Wiley & Sons, Inc., New York.
92 Appel, M., Fleischer, G., Geschke, D., Kärger, J., and Winkler, M. (1996) *J. Magn. Reson. A*, **122**, 248.
93 Fleischer, G. and Fujara, F. (1994) *NMR Basic Principles Prog.*, **30**, 159.
94 Valiullin, R. and Skirda, V. (2001) *J. Chem. Phys.*, **114**, 452.
95 Onsager, L. (1931) *Phys. Rev.*, **37**, 405.

96 Berne, B.J. and Pecora, R. (1976) *Dynamic Light Scattering*, John Wiley & Sons, Inc., New York.

97 Quillfeld, V. (1995) Voruntersuchungen zur Messung der intrakristallinen Diffusion an Zeolithen mit Hilfe der dynamischen Lichtstreuung, Diploma-Thesis, Leipzig University.

98 Cummins, H.Z. and Pike, E.R. (1974) *Photon Correlation and Light Beating Spectroscopy*, Plenum Press, New York.

99 Tyrell, H.J.V. and Harris, K.R. (1984) *Diffusion in Liquids*, Butterworth, London.

100 Zwanzig, R. (1965) *Annu. Rev. Phys. Chem.*, **16**, 67.

11
Diffusion Measurement by Monitoring Molecular Displacement

When considering molecular diffusion one is usually concerned with time scales that are so large that the evolution of the system under study depends exclusively on its present state and not on its history. Under such conditions, all information on system evolution is contained in the propagator (Section 2.1.1) $P(r, t; r_0, t_0)$, which denotes the probability density that a molecule, situated at position r_0 at time t_0, will have reached position r at time t. Under equilibrium conditions, all initial times t_0 are equivalent and the propagator may be denoted in the simplified form $P(r, r_0, t)$ where, now, t stands for the time elapsed during particle migration from r_0 to r.

In Section 10.2 it was shown that incoherent quasi-elastic neutron scattering (QENS) is correlated with this function, which, in the context of QENS, is generally referred to as the van Hove self-correlation function [see Eqs. (10.47) and (10.48)]. This function is determined as the mean over all starting positions r_0. The molecular displacements under study by QENS (Figure 10.19) are very small and lie within the range of the elementary steps of diffusion. The displacements studied by pulsed field gradient NMR (PFG NMR), which is the focus of this chapter, are much larger. By determining the mean propagator for displacements ranging from tens of nanometers to millimeters, PFG NMR is able to trace and discriminate between all features of mass transfer that may represent the rate-determining processes, including intracrystalline diffusion, permeation of surface resistances, and long-range diffusion in polycrystalline assemblages (zeolite pellets). This technique is also referred to as PGSE (pulsed gradient spin echo NMR).

Following an introduction to the fundamentals of this type of ensemble measurement and a discussion of the advantages and limitations of this approach the new options offered by single-particle observations and the challenge of combining the evidence from ensemble and single-molecule measurements are considered.

11.1
Pulsed Field Gradient (PFG) NMR: Principle of Measurement

11.1.1
PFG NMR Fundamentals

To understand the principles of NMR self-diffusion measurements, a detailed quantum mechanical description is not necessary. It is sufficient to start from the classical representation in which the nuclear spins are thought of as precessing with the Larmor frequency [Eq. (10.10)] about the direction of the constant magnetic field B. It was noted in Section 10.1.1 that the magnetization, that is, the vector sum of the magnetic moments of the individual nuclei per volume, will be aligned parallel to the direction of the constant magnetic field so that, on a macroscopic scale, the precession of the individual spins remains invisible. However, after application of an appropriate pulse ($\pi/2$ pulse as illustrated in Figure 10.3) magnetization is shifted into the plane perpendicular to the direction of the constant magnetic field, where it rotates with the Larmor frequency. It is this transverse component of the total magnetization that is considered in an NMR self-diffusion measurement.

The technique depends on making the magnetic field B inhomogeneous, that is, varying with position. This is achieved by superimposing a position-dependent field (gz) on the constant field B_0:

$$B = B_0 + gz \tag{11.1}$$

In general the direction of the field gradient (the z direction) coincides with the direction of the constant magnetic field (as also generally implied in our considerations) although this is not essential. The Larmor frequency for a nuclear spin in the combined field is:

$$\omega(z) = \gamma(B_0 + gz) \tag{11.2}$$

and the phase angle $\varphi(t)$ accumulated by the nuclear spins as a result of their precessional motion is given by:

$$\phi(t) = \int_0^t \omega \, dt = \int_0^t \gamma[B_0 + gz(t')] \, dt' \tag{11.3}$$

which is seen to be position dependent.

The position (z) of a nuclear spin and, hence, of the molecule hosting the atom with the nucleus under study is thus recorded by the angular phase $\varphi(t)$, which becomes accessible through the NMR signal. This correlation is the basis of magnetic resonance tomography (MRT [1–5]). As the most powerful imaging technique in medical diagnosis, MRT is today probably the most widespread application of NMR. Under its original name [1–7], NMR spin-mapping or NMR zeugmatography, this technique was first applied to the study of adsorption/desorption in 1978 [8]. Figures 6.13 and 12.3 show examples of the transient profiles recorded by this technique. Sections 12.1.5 and 14.5.4 give further details including references to more recent developments and applications.

Diffusion measurements by the PFG NMR technique are based on the application of field gradients over two short time intervals. The measurement yields the

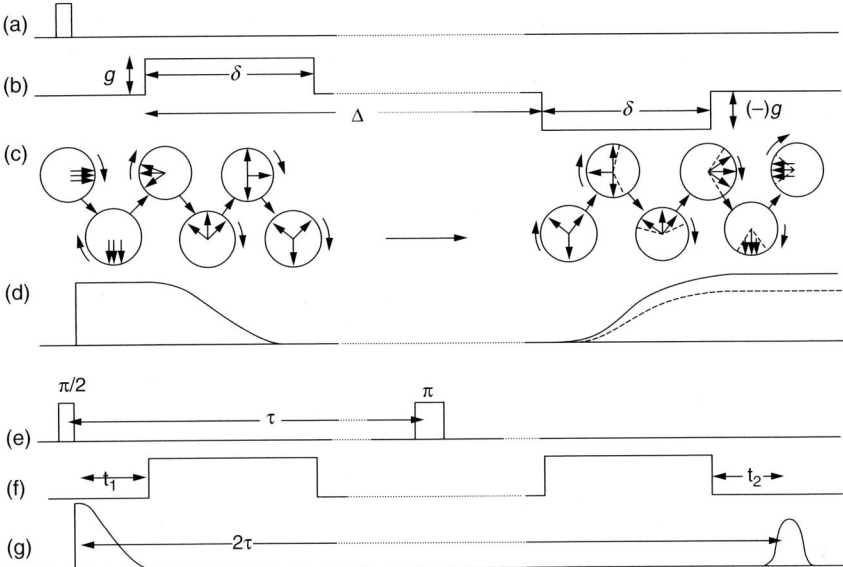

Figure 11.1 Schematic representation of the fundamentals of NMR self-diffusion measurements [(a)–(d)] and their applications in the spin-echo technique [(e)–(g)]. Broken lines in (c) and (d) indicate the behavior with molecular migration; (a) RF pulses; (b) gradient pulses; (c) transverse magnetization $M_\perp(z)$ of different regions; (d) total transverse magnetization $M_\perp$ equal to vector sum of (c); (e) RF pulses; (f) gradient pulses; and (g) magnetization. Reprinted from Kärger et al. [9] with permission.

difference in the locations of each individual molecule during these two gradient "pulses" (i.e., the molecular displacement during the time interval between these two pulses) rather than molecular positions themselves. Figure 11.1 illustrates the basic principle of this type of measurement.

To simplify the calculation we assume that the field gradient pulses are of equal magnitude and duration but of opposite sign, although in practice the experiment is not performed in precisely this way (see below). Under the influence of the field gradient pulse the precession of the transverse magnetization $M_\perp(z)$ will depend on the spatial coordinate (z). The first gradient pulse therefore gives rise to a de-phasing of the vectors of the transverse magnetization at different positions within the sample (Figure 11.1c) and as a result of this dephasing the vector sum decays (Figure 11.1d). If the individual species do not change their positions during the time interval between the two gradient pulses, the effect of the first pulse will be exactly counteracted by the second pulse (in the opposite sense) so that the phase of the spins is re-focused and the transverse magnetization vector should be restored to its original value. However, if the molecules (spins) have moved during the time interval between the gradient pulses, the re-focusing will be incomplete and the intensity of the transverse magnetization will not be fully restored to its original value. This is indicated by the broken lines in Figure 11.1c and d. The decrease in the NMR signal becomes larger as the mean square displacement increases.

For a quantitative treatment we have to calculate the vector sum of the magnetic moments $M_\perp^{(i)}$ of the individual spins in the plane perpendicular to the direction of the constant magnetic field. Denoting the difference $\phi^{(i)} - \langle\phi\rangle$ between the actual phase of spin i and the average phase by $\vartheta^{(i)}$, we obtain:

$$M_\perp = \sum_i M_\perp^{(i)} \propto \sum_i \cos\vartheta^{(i)} \tag{11.4}$$

With Eq. (11.3) and with the time program as indicated in Figure 11.1b, $\vartheta^{(i)}$ is found to be given by the relation:

$$\vartheta^{(i)} = \int \gamma(B_0 + gz^{(i)})\,dt - \int \gamma B_0\,dt = \gamma g z_1^{(i)}\delta + \gamma(-g)z_2^{(i)}\delta = \gamma\delta g\left(z_1^{(i)} - z_2^{(i)}\right) \tag{11.5}$$

$z_{1(2)}^{(i)}$ denotes the z coordinate of the i-th spin during the first (second) field gradient pulse. Introducing the particle position $z_{1(2)}^{(i)}$, we have implicitly assumed that the molecular displacements during the time interval δ of a field gradient pulse are negligibly small in comparison with those during the time interval Δ between the two field gradient pulses, that is, that $\delta \ll \Delta$. For further calculations we introduce the a priori probability $p(z_1)\,dz_1$ of finding a molecule at a position with a z coordinate between z_1 and $z_1 + dz_1$, and a reduced propagator $P(z_2, z_1, \Delta)$. In analogy to the definition of the three-dimensional propagator, $P(z_2, z_1, \Delta)dz_2$ denotes the conditional probability of finding a molecule, initially at a position with the z coordinate z_1, after a time interval Δ at a position with a z coordinate between z_2 and $z_2 + dz_2$. Therefore, the reduced propagator follows in a straightforward manner from $P(\mathbf{r}_2, \mathbf{r}_1, \Delta)$ by integration over the x and y coordinates. It was this reduced (one-dimensional) propagator with which, in Section 2.1, we started our discussion of normal diffusion. Then, by means of Eq. (11.5), Eq. (11.4) may be transformed into:

$$\psi \equiv M_\perp/M_0 = \iint \cos[\gamma\delta g(z_1 - z_2)]p(z_1)P(z_2, z_1, \Delta)\,dz_1\,dz_2 \tag{11.6}$$

where M_0 stands for the total magnetization before the field gradients are applied. On the basis of this equation, any type of translational molecular motion within the sample may be related directly to the attenuation of the NMR signal.

For molecular self-diffusion within a homogeneous region:

$$p(z_1) = \text{const} \tag{11.7}$$

and [see Eq. (2.12) with $z = z_2 - z_1$]:

$$P(z_2, z_1, \Delta) = (4\pi\mathcal{D}\Delta)^{-1/2}\exp[-(z_2 - z_1)^2/4\mathcal{D}\Delta] \tag{11.8}$$

where $\mathcal{D}$ stands for the self-diffusion coefficient. With these relations, Eq. (11.6) may be simplified to:

$$\psi(\Delta, \delta g) = \exp(-\gamma^2\delta^2 g^2 \mathcal{D}\Delta) \tag{11.9}$$

or, by use of the Einstein relation, Eq. (2.6):

$$\psi(\Delta, \delta g) = \exp[-\gamma^2\delta^2 g^2 \langle r^2(\Delta)\rangle/6] \tag{11.10}$$

where $\langle r^2(\Delta)\rangle$ stands for the molecular mean square displacement during the "observation" time Δ, that is, during the time interval between the two field gradient pulses. For anisotropic systems, Eq. (11.10) must be replaced by:

$$\psi(\Delta, \delta g) = \exp[-\gamma^2 \delta^2 g^2 \langle z^2(\Delta)\rangle/2] \qquad (11.11)$$

where $\langle z^2(\Delta)\rangle$ denotes the mean square displacement in the direction of the pulsed field gradient. According to Eqs. (11.9)–(11.11), the self-diffusion coefficients as well as the molecular mean square displacements may be determined in a straightforward manner from the slope of the semilogarithmic representation of the signal intensity (which is proportional to ψ) versus $(\delta g)^2$.

11.1.2
Basic Experiment

In the preceding discussion it was implicitly assumed that, during the absence of the field gradient pulses, the Larmor frequencies remain constant. In fact in a real system the residual heterogeneity of the constant magnetic field will give rise to some range of Larmor frequencies and thus to a rapid decay of the magnetization. It was shown in Section 10.1.2 that, by applying a second RF pulse (π pulse), after a time interval equal to the time between the two RF pulses the magnitude of the transverse magnetization can be restored, leading to the "spin echo" (Figures 10.8, 10.9a, and 11.1e and g). Thus, after the action of the π pulse, one has a situation similar to that resulting from a change in the sign of the field gradient. This is also the reason why, in the spin echo technique, two gradient pulses of the same sign (Figure 11.1f) must be applied.

The inhomogeneity of the "constant" field also leads to a signal attenuation for the diffusing molecules. However, since the inhomogeneities are small relative to the gradient pulse, this de-phasing effect is in general much smaller.

So far, we have assumed, for simplicity, that the NMR signal (i.e., the total transverse magnetization) after the second field gradient pulse is the same as after the initial $\pi/2$ pulse. In reality, however, (Section 10.1.2) the transverse magnetization decays, eventually, to zero. The time constant of this decay (T_2) is the transverse nuclear magnetic relaxation time. The de-phasing of the magnetization following the $\pi/2$ pulse does not disturb the memory of the spins concerning their phase relations with each other since the π pulse leads to re-establishment of the magnetization. In contrast, attenuations of the signal by relaxation and diffusion are irreversible processes during which the phase memory is partly lost so that full re-establishment of the initial magnetization is not possible.

Since the influences of diffusion and relaxation on the spin echo are generally independent of each other, Eqs. (11.6) and (11.9)–(11.11) are still applicable even under the influence of relaxation effects. Relaxation, however, imposes one serious limitation of PFG NMR, since a reasonable signal analysis is only possible if the attenuation of the signal from relaxation effects is not too great. Sections 11.3.2 and 11.3.3 give a detailed discussion of this problem together with some experimental techniques to minimize the adverse impact of relaxation effects.

In principle, NMR self-diffusion measurements may be carried out with any NMR pulse spectrometer (Figure 10.7) that allows production of a spin echo. In addition to the traditional facilities for providing the RF pulses, such a spectrometer must be capable of being adapted to provide the well-defined inhomogeneities in the magnetic field (the pulsed field gradients). This is generally achieved by adding an additional pulsed field gradient unit with special probe heads that transform current pulses into field gradient pulses by the use of suitably wound coils.

11.2
The Complete Evidence of PFG NMR

11.2.1
Concept of the Mean Propagator

The formulas in the last part of Section 11.1 have been determined under the simplifying assumption that the system under study is homogeneous, isotropic, and infinitely extended. In many cases this is an excellent approximation. However, application of PFG NMR is not limited to this special case and may turn out to be even more beneficial for investigating more complex systems [10–14]. To demonstrate this potential, we introduce the mean propagator [12, 15]:

$$P(z, \Delta) = \int p(z_1) P(z_1 + z, z_1, \Delta) \, dz_1 \qquad (11.12)$$

This function represents the probability density that, after time Δ, an arbitrarily selected molecule in the sample is shifted over a distance z in the direction of the z coordinate. With this notation and the abbreviation:

$$m = \gamma \delta g \qquad (11.13)$$

Equation (11.6) simplifies to:

$$\psi(m, \Delta) = \int \cos(mz) P(z, \Delta) \, dz \qquad (11.14)$$

with the inversion:

$$P(z, \Delta) = \frac{1}{2\pi} \int \cos(mz) \psi(m, \Delta) \, dm \qquad (11.15)$$

Equation (11.15) indicates that the mean propagator results directly as the Fourier transform of the PFG NMR spin echo attenuation, with the generalized pulse gradient intensity $m = \gamma \delta g$ as the Fourier conjugate of the space scale. Higher spatial resolution, that is, the measurement of smaller displacements z is thus immediately seen to be attainable with gradient pulses of larger amplitude (g) and duration (δ) and for nuclei with the largest gyromagnetic ratio γ, that is, for protons.

It should be emphasized that, as a consequence of the averaging over all starting positions z_1 [Eq. (11.12)] as part of the measuring procedure, PFG NMR is not able to provide complete information on the molecular random walk, that is, the propagator $P(r_1, r_2, \Delta)$ for any starting position r_1. For this type of information, single-particle techniques open up new perspectives. The averaging in PFG NMR is a consequence of the requirement to observe sufficiently large ensembles of nuclei (in standard equipment, with sample volumes of hundreds of microlitres this means typically more than 10^{18} protons). Clearly, combining PFG NMR measurements with NMR imaging techniques [16–20] also makes it possible to determine "local" diffusivities or, more generally, propagators that are attributed to different volume elements in the sample. However, it still remains true that these diffusivities or propagators are the averages over very large numbers of nuclei, of the order indicated above.

11.2.2
Evidence of the Mean Propagator

As an example, Figure 11.2 shows the propagator representation of molecular self-diffusion of ethane in beds of zeolite NaCaA (Section 16.1.1) with two different crystallite sizes. Being symmetric in z, for simplicity the propagator is only represented for $z > 0$. For the lowest temperature (153 K), the distribution widths of molecular displacement during the considered time intervals (5–45 ms) are found to be small in comparison with the mean radius of the larger crystallites (8 µm). This is, essentially, the situation considered in Section 11.1, where the molecules under study behave as if contained in an infinitely extended homogeneous and isotropic medium: PFG NMR monitors genuine intracrystalline self-diffusion.

In the smaller crystallites, the probability distribution of molecular displacement is found to remain unaffected by the observation time. Moreover, the mean width is of

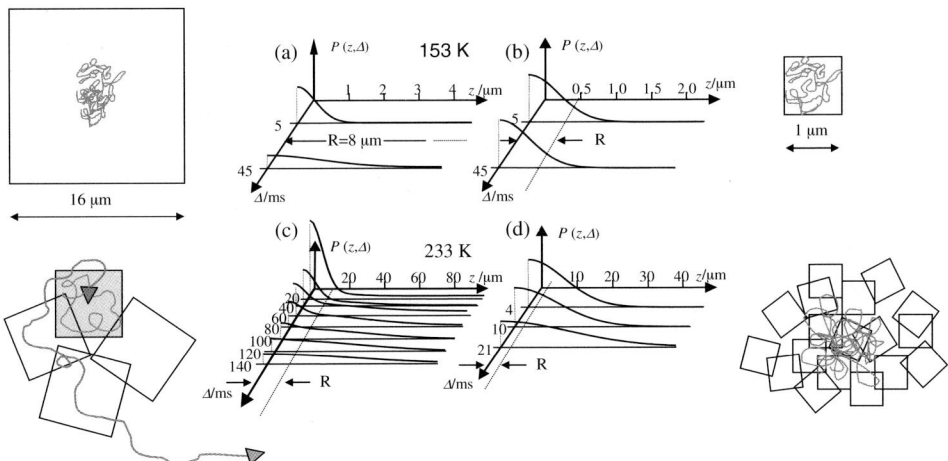

Figure 11.2 Experimental evidence provided by PFG NMR: the mean propagator of ethane in beds of zeolite NaCaA with mean crystallite radii of $R = 8$ µm (a, c) and 0.4 µm (b, d) and cartoons of typical diffusion paths under the considered situations.

the order of the crystallite radii: at the given temperature over the considered time intervals the ethane molecules are essentially confined by the individual crystallites. This confinement arises because the thermal energy of the molecules is not high enough to allow a sufficiently large number of molecules to overcome, during the "observation time" Δ, the step increase in the potential energy from the intracrystalline space into the surrounding gas phase (intercrystalline space). More specifically, one may conclude that the long-range diffusivity $\mathcal{D}_{\text{long-range}} = p_{\text{inter}} \mathcal{D}_{\text{inter}}$ is notably exceeded by the intracrystalline diffusivity $\mathcal{D}_{\text{intra}}$. Since in a crystallite of sufficiently large size the ethane molecules have been shown to be able to cover much larger diffusion paths (Figure 11.2a), molecular displacement in the small crystallites becomes a measure of the crystal size. This way of tracing the extension of microscopic regions has become popular under the name "dynamic imaging" [3, 13] and is applicable quite generally to complex systems with microscopic areas of confinement such as biological cells [14, 21, 22].

Eventually, with increasing temperature, a substantial fraction of the ethane molecules are able to leave the crystallites – corresponding to the situation in which the "long-range diffusivities" ($\mathcal{D}_{\text{long-range}} = p_{\text{inter}} \mathcal{D}_{\text{inter}}$) notably exceed the intracrystalline diffusivities $\mathcal{D}_{\text{intra}}$ (where p_{inter} and $\mathcal{D}_{\text{inter}}$ denote the relative amount of molecules in the intercrystalline space and their diffusivity – see also Sections 11.4.2 and 15.8). This may be understood as an immediate consequence of the fact that the activation energy for long-range diffusion (determined, essentially, by p_{inter}) generally exceeds the activation energy of intracrystalline diffusion. The distribution widths of molecular propagation may thus exceed the crystallite radii and, under the given conditions ($T = 233$ K), PFG NMR is able to monitor the rate of molecular propagation through the bed of crystallites and, hence, to measure "long-range" diffusivities.

In the case of the larger crystallites (Figure 11.2c) one is, at this temperature, even able to distinguish between a narrow distribution (corresponding to those molecules that remain in the interior of the individual crystallites) and a broader constituent that corresponds to those molecules which have passed several crystallites. With increasing observation time, the contribution of the broader constituent increases at the expense of the narrower one, since more and more molecules will leave the individual crystallites. A plot of the relative intensity of the broad constituent versus the observation time (i.e., the separation between the two field gradient pulses) contains information that is analogous to that of a tracer exchange experiment between a particular crystallite containing, for example, labeled molecules and the unlabeled surroundings. Therefore this way of analyzing PFG NMR data of zeolitic diffusion has been termed the "NMR tracer desorption" technique (see Section 11.4.3.3).

11.2.3
Concept of Effective Diffusivity

As in QENS [see Section 10.2.3 and Eq. (10.48)], the propagator $P(z,t)$ is often found to be closely approximated by a Gaussian. Then, with Eq. (11.14), the PFG NMR spin-echo attenuation curve is still of exactly the shape of Eq. (11.9), where now the

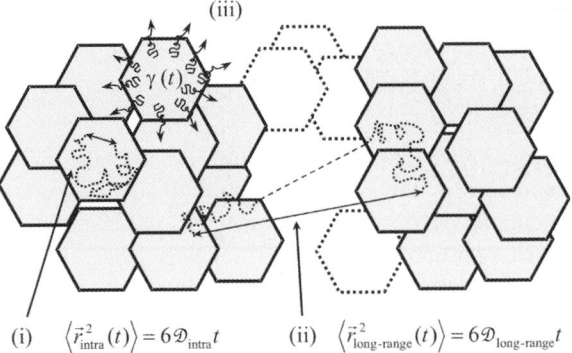

Figure 11.3 Different regimes encountered in studying molecular transport in beds of zeolite crystallites by PFG NMR.

genuine (self-)diffusivity has to be replaced by an effective diffusivity:

$$\mathcal{D}_{\text{eff}} = <z^2(t)>/2t \qquad (11.16)$$

with $<z^2(t)>$ denoting the mean square value of the molecular displacements within the sample in gradient direction. The accuracy of this approach appears in the $\ln\Psi$ versus $(g^2\delta^2)$ representation, which, according to Eq. (11.9), should be a straight line.

Thus, by approximating the $\ln\psi$ versus $(g^2\delta^2)$ dependency for $g\delta \to 0$ by a straight line, one may quite generally determine an effective diffusivity (which, in effect, is the mean over all diffusivities of the molecules under study within the sample) and, via Eq. (11.16), the mean square molecular displacement as a well-defined quantity.

Figure 11.3 summarizes the three main types of evidence that we have shown to be provided by the application of PFG NMR to beds of nanoporous particles (and, in particular) of zeolite crystallites. Before referring in Section 11.4 to the different regimes under which this type of information becomes accessible, we will discuss in more detail the conditions and limitations involved in the main fields of application of PFG NMR.

11.3 Experimental Conditions, Limitations, and Options for PFG NMR Diffusion Measurement

11.3.1 Sample Preparation

While in a transient rate measurement the sorbate is introduced into the adsorbent during the course of the experimental measurement, NMR measurements are generally made with a system in which the adsorbent has been pre-equilibrated with sorbate, under known conditions, prior to the experiment. To achieve reproducible results, the sample preparation is generally carried out as follows:

1) The sample is outgassed under vacuum at elevated temperature. Where the adsorbent has limited thermal or hydrothermal stability care must be taken not to cause structural degradation during this process. Typically, the temperature is raised very slowly while maintaining a high vacuum and the sample is spread out as a thin layer to avoid "self-steaming" effects.
2) The sorbate is then introduced, either by contacting the sample with a known pressure of sorbate vapor for a long enough period of time to permit equilibration, or by cooling the sample in liquid nitrogen while connected to a dosing volume containing a known quantity of sorbate.

The sample is usually contained in a small glass vial having a diameter in the range 5–10 mm and filled over a length of typically 10 mm. The loaded vials are then sealed and used directly for the NMR experiment. If the activation is carried out at smaller bed depths, the transfer to the measurement tube must be carried out either under vacuum or under the required vapor pressure of sorbate. The quantity of sorbate introduced into the sample may be confirmed gravimetrically, or from the intensity of the NMR signal. The first of these approaches is unambiguous but measurement of the NMR signal intensity is also desirable since this provides confirmation that all molecules admitted to the sample tube do in fact contribute to the NMR signal.

When NMR measurements are made over a range of temperatures it is important to realize that the distribution of molecules between adsorbed and vapor phases, within the sample vial, will vary with temperature. This effect is of no consequence as long as the equilibrium is sufficiently favorable that almost all of the molecules are in the adsorbed phase. However, the adsorbed phase concentration decreases as the temperature increases and so, for weakly adsorbed species at elevated temperatures, the molecules in the vapor phase may make a measurable contribution to the NMR signal. Similarly, the presence of an excess of liquid sorbate can also lead to complications of interpretation since the NMR signals of the adsorbed and liquid phases will then be superimposed.

Alternatively, the PFG NMR sample tube may be kept in continuous contact with a gas reservoir. This type of arrangement allows pressure variation in a straightforward manner and has proved to be of crucial relevance for correlating molecular uptake and diffusion under the conditions of sorption hysteresis [23–26] as discussed in Section 15.9.

11.3.2
Basic Data Analysis and Ranges of Observation

Equations (11.9) and (11.11) are the standard relations for the application of PFG NMR to "normal" diffusion, that is, to mass transfer phenomena characterized by Fick's 1st law [Eq. (1.11)] or, correspondingly, by the Einstein relation [Eq. (1.12)] in essentially infinitely extended, isotropic media (see also Section 2.1 and

Reference [27]). Plotting the signal attenuation in a logarithmic representation versus the square of the pulse width, δ, or gradient amplitude, g, yields a straight line. The mean square displacement $\langle (z^2) \rangle$ in the direction of the applied field gradient and/or the self-diffusivity $\mathcal{D}$ in this direction follow immediately from the slope of this line. Vice versa, if the PFG NMR signal attenuation is found to be given by an exponential of the type of Eqs. (11.9) or (11.11), the molecular propagator is well approximated by a Gaussian. One may easily determine the mean square displacement on the basis of Eq. (11.9) by comparison with the attenuation for a standard liquid with known diffusivity (e.g., water with $\mathcal{D} = 2.02 \times 10^{-9}\, m^2\, s^{-1}$ at 293 K [28]) by applying the same pulse program and by calculating $\langle (\Delta z)^2 \rangle$ on the basis of Eq. (2.3) from the known diffusivity. The diffusivity of the sample under study follows either by the analogous procedure from Eq. (11.9) or from the mean square displacement via Eq. (1.12).

In the derivation of Eqs. (11.6) and (11.14) it was assumed that, during the field gradient pulses, the spins assume well-defined positions. Such an assumption is clearly only acceptable if molecular displacements during the field gradient pulses are negligibly small in comparison with those in the time interval between the gradient pulses. In the case of normal diffusion it may be shown that the PFG NMR signal attenuation under the influence of field gradient pulses of finite duration becomes [10, 29, 30]:

$$\psi = \exp\left[-\gamma^2 \delta^2 g^2 \mathcal{D}(t - \delta/3)\right] \tag{11.17}$$

Equation (11.11) may be used for an estimate of the lower limit of molecular displacements accessible by PFG NMR. Under the assumption that a reliable measurement of $\langle z^2 \rangle$ is only possible if the field gradient pulses lead to a signal attenuation of $\psi = e^{-1}$, with typical maximum values for the field gradient amplitude ($g = 25\, T\, m^{-1}$) and the pulse width ($\delta = 2\, ms$), for hydrogen-containing molecules one obtains $\langle (\Delta z)^2 \rangle_{min}^{1/2} \approx 100\, nm$. With the Einstein relation, Eq. (2.3), the lower limit of the diffusivity accessible by NMR is thus found to be $\langle (\Delta z)^2 \rangle_{min}/(2 t_{max})$ where t_{max} denotes the maximum possible observation time.

If the NMR signal is generated by the $\pi/2 - \pi$ pulse sequence (the primary or Hahn spin echo sequence – see Section 10.1.2.3), t_{max} is determined by the transverse nuclear magnetic relaxation time T_2, which is typically of the order of a few milliseconds. However, the range of observation times may be significantly enhanced by applying the $\pi/2 - \pi/2 - \pi/2$ (stimulated) echo sequence, with the field gradient pulses inserted between the first and second $\pi/2$ pulses and between the third $\pi/2$ pulse and the echo, respectively. In this case, t_{max} is determined by the longitudinal relaxation time T_1, which is typically seconds. Thus, today diffusivities of less than $10^{-14}\, m^2\, s^{-1}$ [31–33] have become accessible to measurement by PFG NMR.

Equations (11.9) and (11.11) indicate that the measuring conditions deteriorate with decreasing gyromagnetic ratio. Thus, for ^{13}C ($\gamma_{13C} = 0.67 \times 10^8\, T^{-1}\, s^{-1}$, in comparison with $\gamma_{1H} = 2.67 \times 10^8\, T^{-1}\, s^{-1}$ for hydrogen) and ^{129}Xe ($\gamma_{129Xe} = 0.75 \times 10^8\, T^{-1}\, s^{-1}$) molecular displacement must be larger by a factor of about 4, and in the case of ^{15}N ($\gamma_{15N} = 0.27 \times 10^8\, T^{-1}\, s^{-1}$) by an order of magnitude greater than for hydrogen. The measurement of intracrystalline diffusion necessitates the use of crystallites with diameters significantly greater than the observed molecular displacements. Therefore, PFG NMR measurements of intracrystalline diffusion with nuclei other than hydrogen require the use of very large crystallites.

Because the amplitude (g) of the magnetic field gradient pulses is limited by the specifications of the given amplifier for the gradient pulses, sensitivity enhancement towards smaller displacements has to be based on an enhancement of the total time, δ, of gradient application. Long magnetic field gradient pulses, in turn, imply correspondingly long periods of time within the PFG NMR pulse sequences, during which nuclear magnetization is oriented in the plane perpendicular to the direction of the magnetic field. The application of magnetic field gradient pulses of long duration, δ, is limited, therefore, by the rate of the decay of transverse magnetizations. Even if applying very strong field gradients and thus reducing δ, the induction of eddy currents necessitates a delay between applications of gradient pulses and detection or subsequent RF pulses. Therefore, the rapid decay of transverse magnetization strongly hampers the application of PFG NMR experiments. However, it has recently been shown that the effects of these disturbing influences may be greatly reduced by magic angle spinning (MAS) PFG NMR, with the directions of the spinning axis and the field gradients parallel to each other (Section 10.1.4 and Figure 10.14) [34–41].

11.3.3
Pitfalls

11.3.3.1 Gradient Pulse Mismatch
The PFG NMR method assumes that the field gradient pulses acting on the NMR sample are identical. Any difference between the values of δg for the first and second field gradient pulses will lead to a signal attenuation that may be interpreted erroneously as being caused by diffusion. Hence, the correct application of PFG NMR necessitates extremely stable gradient currents (which generate the field gradient pulses within suitably structured field gradient coils [3, 10, 30, 42]) as well as a high mechanical stability of both the field gradient coils and the sample, since any movement of the sample with respect to the coils would also lead to differences in the local field at the instants of the first and second field gradient pulses. To ensure that the observed signal attenuation with increasing values of g or δ is genuinely due to diffusion (rather than to differences in the values of δg within a pair of field gradient pulses) it is, therefore, useful to apply the identical PFG NMR pulse program to a sample with a sufficiently low diffusivity (e.g., crosslinked polybutadiene with a diffusivity of $<10^{-15}$ m^2 s^{-1} and a T_2 in the ms range [43]). In this case the field gradient pulse program should not lead to any signal attenuation.

11.3.3.2 Mechanical Instabilities
With powder samples such as beds of zeolite crystallites, one must be aware of another possible pitfall [44]. Even within completely fixed sample tubes the sample particles may move under the influence of the mechanical pulses generated by the forces acting on the field gradient coils during the current pulses. This influence is

particularly stringent for short observation times (i.e., short separation between the field gradient pulses) and may be reduced by avoiding any mechanical contact between the sample tube and the probe and/or by compacting the sample material below a constriction in the sample tube.

11.3.3.3 Benefit of Extra-large Stray-Field Gradients

The develpoment of the PFG NMR method has been focused on the generation of extremely large field gradient pulses [31, 32, 42, 45–48]. The difficulties arising from the requirement of perfect matching between the two field gradient pulses may be circumvented by applying the stimulated spin echo method under the influence of a strong constant field gradient [49, 50], which is provided by the stray field of the superconducting magnet ("stray field gradient" NMR). It may be shown that the intensity of the stimulated echo is only influenced by the field gradient applied between the two first $\pi/2$ pulses and between the third $\pi/2$ pulse and the echo. Therefore, signal attenuation due to diffusion is also determined by Eqs. (11.6), (11.9)–(11.11), (11.14) or (11.17) with an effective pulse width identical to the spacing between the first two $\pi/2$ pulses (which is also the spacing between the third $\pi/2$ pulse and the maximum of the stimulated echo). It is possible by this technique to achieve very large field gradient "amplitudes" (up to 150 T m^{-1}) [51–55].

In comparison with PFG NMR, however, the signal-to-noise ratio in this technique is dramatically reduced, so that much larger acquisition times are required. Moreover, the constant magnetic field excludes the application of Fourier-transform PFG NMR for diffusion studies of multicomponent systems (Section 11.3.4.1). The measurements are additionally complicated by the fact that, by varying the "width" of the field gradient "pulses" (i.e., by changing the spacing between the first two RF pulses), the signal is affected by both diffusion and transverse nuclear magnetic relaxation.

11.3.3.4 Impedance by Internal Gradients

With the advent of superconducting magnets and the option of magnetic field strengths up to 20 T, corresponding to NMR frequencies close to the GHz range, the heterogeneity in the susceptibility of typical zeolite samples leads to significant internal field gradients that, in general, are no longer negligibly small, in comparison with the pulsed field gradient amplitudes as considered, for example, in Reference [56]. To circumvent the disturbing influences resulting from interference effects between the constant and pulsed field gradients, an extended pulse sequence, often referred to as the 13-interval pulse sequence, has been suggested [57]. This is a modification of the conventional stimulated-echo technique in which, by applying pairs of alternating field gradient pulses with a π pulse in between, rather than mere field gradient pulses, the disturbing influence of the internal field gradients arising from sample heterogeneity may be eliminated. Figure 11.4 shows the pulse sequence that is commonly used today [58], with sine-shaped (rather than rectangular) alternating gradient pulses and two further (weak) gradient pulses for the suppression of remaining eddy currents (which are anyway reduced by the application of alternating gradients).

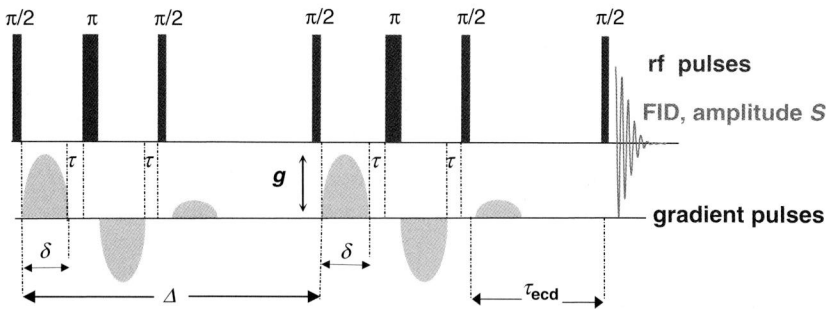

Figure 11.4 PFG NMR pulse sequence with alternating, sine-shaped gradient pulses and eddy-current quench pulses.

The decay of the amplitude S of the FID is governed by the equation [58]:

$$\psi = S/S_0 = \exp\left[-\mathcal{D}\left(\frac{4\delta g\gamma}{\pi}\right)^2\left(\Delta - \frac{\tau}{2} - \frac{2\delta}{3} - p_\pi\right)\right] \quad (11.18)$$

in which p_π denotes the duration of the π pulse and the meaning of the other symbols (in particular g and δ) is as indicated in the figure. The self-diffusivity ($\mathcal{D}$) is therefore easily obtained from a series of measurements at different field gradient intensities (g).

As a prerequisite for a successful application of this stimulated-echo sequence with two pairs of alternating gradients, the local magnetic field gradient experienced by each individual diffusant must remain constant over the whole trajectory during the observation time. However, even this constraint may be avoided by use of the so-called magic pulsed field gradient ratios [45, 59].

11.3.3.5 Impedance by Contaminants

The diffusivities resulting from analyzing the PFG NMR spin echo attenuation via Eqs. (11.6), (11.9)–(11.11), (11.14), or (11.17) refer to those molecular species that contribute to the observed signal. This is particularly important for host–guest systems where the host adsorbent contains a mixture of two different channel systems with no guest exchange during the observation time of PFG NMR. The observed signal may then result mainly from the highly mobile guest molecules within the more open channels, leading to apparent intracrystalline diffusivities that are much larger than the true values for the smaller pore system. (See Reference [60] and Section 10.1.4.)

Since the presence of highly mobile proton-containing molecules (due, for example, to the presence of traces of water) can never be absolutely excluded, the risk of corrupted measurements clearly increases with increasing accumulation numbers, unless the signal intensity is repeatedly cross checked with the molecular density of the probe molecules. Such problems obviously do not exist when operating with enriched isotopes, as, for example, in ^{13}C PFG NMR self-diffusion measurements [61–63].

11.3.4
Measurement of Self-diffusion in Multicomponent Systems

11.3.4.1 Fourier-Transform PFG NMR

Direct access to the self-diffusivities of different species within one sample is provided by combining the field gradient method with Fourier-transform NMR [64] as shown schematically in Figure 11.5. It was explained in Section 10.1.3.1 that Fourier-transform NMR allows identification of the different molecular species by the differences in the Larmor frequencies of the nuclear spins resulting from differences in the chemical surroundings (the "chemical shift"). In this way the total spin-echo magnetization may be split into its constituent components arising from the different species in the system. Their decay with increasing field gradient intensity may then be attributed to the self-diffusivities of these species. In the representation shown in Figure 11.5 it is assumed, for simplicity, that the system consists of just two species, A and B, each containing only one magnetic nucleus of the type considered. The main assumption involved in the application of Fourier-transform NMR to selective multicomponent self-diffusion measurements is that there is a distinct separation between the individual lines, that is, it is necessary that the line widths $\Delta\omega_i$ are less than the distances $|\omega_i - \omega_j|$ between the maxima of the resonance lines.

Thus, with Eqs. (10.24) and (10.28) one obtains:

$$|\omega_i - \omega_j| = \gamma B_0 |\delta_i - \delta_j| > 2/T_{2i(j)} \tag{11.19}$$

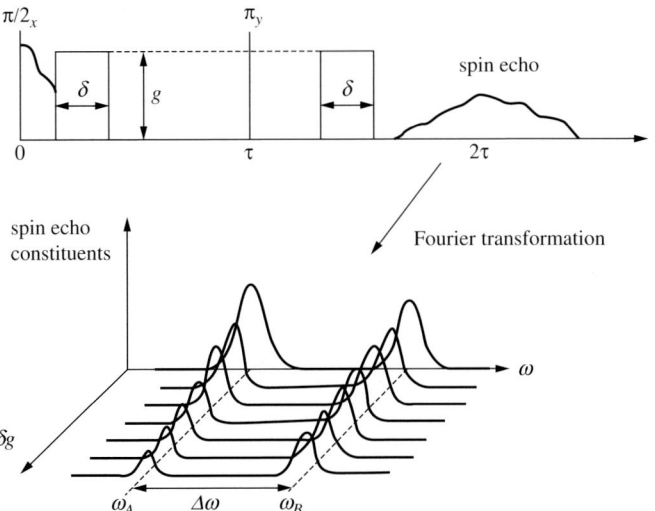

Figure 11.5 Field gradient Fourier-transform experiment. For simplicity the system is assumed to consist of nuclei having only two different resonance frequencies, ω_A and $\omega_B = \omega_A + \Delta\omega$. Reprinted from Kärger et al. [9] with permission.

It follows that the best conditions for multicomponent self-diffusion measurements are obtained for species that have large transverse relaxation times (corresponding to highly mobile molecules) and large chemical shifts (e.g., ^{13}C). Of course, the field B_0 should also be as large as possible.

11.3.4.2 Alternative Approaches

Two-component diffusion measurements are particularly easy if the molecular species under consideration contain different NMR active nuclei. Such measurements have been carried out, for example, with mixtures of C_6H_6 and C_6F_6, in both liquid and adsorbed states, by applying alternately ^{1}H and ^{19}F NMR [65]. By using alternately ^{1}H, ^{13}C, and ^{15}N PFG NMR it has even become possible to measure the long-range diffusion (Section 11.4.2) in three-component mixtures, in this case for methane, carbon dioxide, and nitrogen, in beds of zeolite NaX and NaCaA [66, 67].

An alternative approach applicable to hydrogen-containing species is to make measurements on mixtures containing only one of the components in H form (with the rest in perdeuterated form). Then the measurement of the diffusivity of the H-containing species by ^{1}H NMR is not affected by the presence of the other (perdeuterated) species. By using different mixtures, each with only one component in the hydrogen form, the diffusion characteristics of all species in the multicomponent mixture may be determined [68].

In some instances the diffusivities of different components in an adsorbed phase are very different from each other. In such a situation the plot of $\ln \psi$ versus $(\delta g)^2$, which, for a single-component system, yields a straight line of slope proportional to $\mathcal{D}$ [see Eq. (11.9)], exhibits a pronounced curvature from which – by a simple two-exponential approach – the diffusivities of the individual species may be derived. By this procedure, for example, one may also distinguish between the intracrystalline diffusivities and the diffusivity in the surrounding fluid in oversaturated samples consisting of sufficiently large particles (crystallites) [69, 70].

11.3.5
Diffusion Anisotropy

11.3.5.1 "Single-Crystal" Measurements

Subject to the assumption of sufficiently small gradient pulses, the NMR spin echo attenuation was shown in Section 11.1 [Eq. (11.11)] to be determined by the mean square displacement of the diffusing species in the direction of the applied field gradient during the time interval Δ between the gradient pulses. Thus, by a simple change of orientation of the field gradient relative to the sample, the NMR pulsed field gradient measurement can be used directly to study diffusional anisotropy.

For a quantitative discussion we introduce the angles ζ and ϑ, indicating the direction of the field relative to the principal axes of the diffusion tensor. Obviously, any displacement (z) in the direction of the field gradient may be represented by its components in the direction of the principal axes x', y', and z' by the relation:

$$z = x' \cos \zeta \sin \vartheta + y' \sin \zeta \sin \vartheta + z' \cos \vartheta \tag{11.20}$$

Taking the square and averaging, recognizing that the displacements in the different principal directions are uncorrelated, yields:

$$\langle z^2 \rangle = \langle x'^2 \rangle \cos^2\zeta \sin^2\vartheta + \langle y'^2 \rangle \sin^2\zeta \sin^2\vartheta + \langle z'^2 \rangle \cos^2\vartheta \qquad (11.21)$$

Since molecular migration in the principal directions may be represented by diffusion in one dimension the mean square displacements in these three directions are given by Eq. (2.3). Introducing the principal tensor elements $\mathcal{D}_{1(2,3)}$ corresponding to the $x'(y',z')$ directions and combining Eqs. (11.11) and (11.21) we obtain:

$$\psi(\varDelta, \delta g) = \exp[-\gamma^2 \delta^2 g^2 \varDelta (\mathcal{D}_1 \cos^2\zeta \sin^2\vartheta + \mathcal{D}_2 \sin^2\zeta \sin^2\vartheta + \mathcal{D}_3 \cos^2\vartheta)] \qquad (11.22)$$

In analogy with the case of isotropic diffusion, when the gradient pulse width becomes comparable with the observation time, $\varDelta$ must be replaced by $\varDelta - \delta/3$. Evidently, by directing the field gradient parallel to one of the principal axes Eq. (11.22) reduces to the expression for isotropic diffusion [Eq. (11.9)], thereby allowing the principal elements of the diffusion tensor to be determined ($\mathcal{D}_1$ for $\zeta = 0$ and $\vartheta = \pi/2$, $\mathcal{D}_2$ for $\zeta = \pi/2$ and $\vartheta = \pi/2$, and $\mathcal{D}_3$ for $\vartheta = 0$).

This procedure clearly implies that the orientation of the diffusion tensor as characterized by the angles ζ and ϑ is the same throughout the sample. This would require either a large single crystal or a sample in which all crystals are aligned in the same direction.

11.3.5.2 Powder Measurement

In general, however, the crystals are arranged in a random orientation so that the resulting spin-echo attenuation must be found by integration of Eq. (11.22) over the entire range [71]:

$$\psi(\varDelta, \delta g) = \frac{1}{4\pi} \int_0^{2\pi} \int_{-1}^{1} \exp\{-\gamma^2 \delta^2 g^2 \varDelta [\mathcal{D}_1 \cos^2\zeta \sin^2\vartheta + \mathcal{D}_2 \sin^2\zeta \sin^2\vartheta + \mathcal{D}_3 \cos^2\vartheta]\} \, d\zeta \, d(\cos\vartheta) \qquad (11.23)$$

where it is implied that the orientation of the diffusion tensor remains invariant for each individual molecule over the total observation time, that is, that there is no molecular exchange between different crystallites during the observation time. Under the conditions of cylindrical symmetry, with $\mathcal{D}_1 = \mathcal{D}_2 = \mathcal{D}_\perp$ and $\mathcal{D}_3 = \mathcal{D}_\parallel$, Eq. (11.23) simplifies to:

$$\psi(\varDelta, \delta g) = \exp\{-\gamma^2 \delta^2 g^2 \varDelta \mathcal{D}_\perp\} \int_0^1 \exp\{-\gamma^2 \delta^2 g^2 \varDelta (\mathcal{D}_\parallel - \mathcal{D}_\perp) x^2\} \, dx \qquad (11.24)$$

Setting, respectively, $\mathcal{D}_\parallel = 0$ and $\mathcal{D}_\perp = 0$, Eq. (11.24) is immediately found to yield the PFG NMR signal attenuation in the two limiting cases of ideal two-dimensional diffusion:

$$\psi_{2D}(\varDelta, \delta g) = \exp\{-\gamma^2 \delta^2 g^2 \varDelta \mathcal{D}_\perp\} \int_0^1 \exp\{\gamma^2 \delta^2 g^2 \varDelta \mathcal{D}_\perp x^2\} \, dx \qquad (11.25)$$

and one-dimensional diffusion:

$$\psi_{1D}(\Delta, \delta g) = \int_0^1 \exp\{-\gamma^2 \delta^2 g^2 \Delta \mathcal{D}_{\|} x^2\} \, dx \qquad (11.26)$$

11.3.5.3 Evidence on Host Structure

Complementing the information provided by adsorption and diffraction measurements or by high-resolution microscopy [72], PFG NMR can reveal structural features of nanoporous host–guest systems that determine their transport properties. Studies of this type have proved to be particularly informative when applied to ordered mesoporous materials [72, 73].

For example, MCM-41 is traversed by parallel channels with diameters of a few nanometers [74, 75], which is much smaller than the minimum displacement accessible by PFG NMR (Section 11.3.2). With perfect channels of uniform direction over the whole adsorbent particle and excluding the possibility of molecular exchange between different particles during the observation time, the PFG NMR signal attenuation is expected to follow Eq. (11.26) for ideal one-dimensional diffusion. If there is the possibility of molecular exchange between adjacent channels (e.g., through micropores in the channel walls) a dependence following Eq. (11.25), with $\mathcal{D}_{\|} \gg \mathcal{D}_{\perp}$, must be expected. In general, however, PFG NMR spin echo attenuation has been found to deviate from such patterns [76–79], indicating that the individual adsorbent particles are aggregates of domains (i.e., quasi-crystallites) rather than single bodies of ordered structure. So far, in only one case [75] were the MCM-41 particles found to reflect the behavior expected for quasi-crystallites [80]. Further details may be found in Section 15.3.

Similarly, diffusion in SBA-15 [81] also showed axial symmetry [82]. The structure of SBA-15 is comparable with MCM-41, with the option of "fine-tuning" the microporosity of the channel walls [81]. The PFG NMR studies in Reference [82] revealed a clear distinction in the rates of radial diffusion, that is, in the magnitudes of $\mathcal{D}_{\perp}$ when comparing SBA-15 samples of different texture ("fibers" or "bundles" [83]). PFG NMR measurements of diffusion anisotropy in triblock copolymers [84] (the templates in SBA-15 synthesis) revealed clearly their tendency to self-aggregation [85]. Surprisingly, axial diffusivities of the triblock copolymers in the channel direction were found to be only slightly smaller than the diffusivities in the bulk phase under the conditions of cylindrical self-organization [86]. Section 15.7.1 provides further details of this special application of PFG NMR.

11.4
Different Regimes of PFG NMR Diffusion Measurement

Since the evidence from PFG NMR studies relates directly to the mean displacements of the molecules under study [see Eqs. (11.11) and (11.16)] and since, on the other hand, the dimensions of the host particles (zeolite crystallites) are accessible by microscopic observation, data analysis may be based on a direct comparison of these two quantities. Analysis is particularly easy in the two limiting cases where the displacements are much smaller or much larger than the particle diameters. In these two cases, owing to the central limit theorem of statistics, the mean propagator [the key function for the application of PFG NMR – see Eqs. (11.12) and (11.14)] is a Gaussian [Eq. (2.13)], with the diffusivities referred to as the intracrystalline and the long-range diffusivities.

In Section 11.4.1 we consider exactly how the relation between particle size and mean square displacement affects the determination of genuine intracrystalline diffusivities (while the intracrystalline diffusivities of selected host–guest systems are discussed in greater detail in Chapters 15–19 together with the results from other measuring techniques). Section 11.4.2 gives an overview of the wealth of information provided by PFG NMR long-range diffusivity measurements. Finally, in Section 11.4.3, the intermediate case is considered in which the PFG NMR observation times are comparable with the mean life times of the guest molecules in the host particles. We shall see that, in this way, it is possible to identify additional transport resistances on the external particle boundaries, commonly referred to as "surface barriers."

11.4.1
Effect of Finite Crystal Size on the Measurement of Intracrystalline Diffusion

Application of PFG NMR to studying intracrystalline zeolitic diffusion has revealed a multitude of features of molecular transport under confinement within nanopores. This concerns, in particular, the dependence of the self-diffusivities on loading which, depending on the particular host–guest system, was found to follow various different patterns, including both increasing and decreasing trends. These dependences nicely reflect the dominant mechanisms of molecular propagation in the given host–guest system and are referred to in more detail in Section 3.5.1. Selected examples of intracrystalline diffusivities, in particular in zeolites of types LTA, FAU, and MFI, are presented in Chapters 16–18. All these measurements have profited from the progress that has been achieved in the synthesis of large zeolite crystals and in the application of field gradient pulses of high intensity (Section 11.3.2). However, in practice the range of these key parameters is limited so it is important to consider in some detail the impact of finite crystal size on the measuring procedure.

11.4.1.1 Correlating the True and Effective Diffusivities in the Short-Time Limit
A quantitative interrelation between the primary data of the measurement and the crystallite size may be based on the concept of effective diffusivities (Section 11.2.3). For intracrystalline diffusion paths that are sufficiently small in comparison with the crystallite radii, the effective diffusivity may be shown to be represented by a power series [87–89], leading to:

$$\mathcal{D}_{\mathrm{eff}}(t)/\mathcal{D} = 1 - \frac{4}{3\sqrt{\pi}} \frac{1}{R} \sqrt{\mathcal{D}t} - \frac{1}{2R^2}(\mathcal{D}t) \tag{11.27}$$

and:

$$\mathcal{D}_{\text{eff}}(t)/\mathcal{D} = 1 - \frac{2}{3\sqrt{\pi}} \frac{1}{R} \sqrt{\mathcal{D}t} - \frac{1}{R^2}(\mathcal{D}t), \quad (11.28)$$

where $\mathcal{D}$ stands for the genuine intracrystalline diffusivity. As is to be expected, the experimentally accessible quantity $\mathcal{D}_{\text{eff}}$ coincides with the true intracrystalline diffusivity in the limiting case of negligible displacements, that is, for $(\mathcal{D}t)^{1/2} \rightarrow 0$. Equation (11.27) describes the situation of ideal confinement within the intracrystalline space (Figure 11.2b). In this case, the crystallite surface acts as an (ideally) reflecting boundary for the molecules in the intracrystalline space. Equation (11.28) has been derived for absorbing boundaries. PFG NMR diffusion measurements with beds of zeolites comply with this limiting case when the long-range diffusivity is much larger than the intracrystalline diffusivity (Figure 11.2c) and one is only analyzing the intracrystalline constituent of the propagator (i.e., the narrow one).

The influence of confining boundaries on the effective diffusivity, as reflected by Eqs. (11.27) and (11.28), has been repeatedly applied to determine the pore surface in rocks or beds of sand grains [87–94] and this approach has also been successfully applied to beds of zeolite crystals [95]. Figure 11.6 shows the results of these studies, which have been performed with two different samples of NaX, one loaded with n-hexane (two molecules per supercage) and the other with n-hexane and tetrafluoromethane (one molecule of each per supercage). In both samples a temperature of 298 K was chosen for the measurements since, at this temperature, the n-hexane molecules were found to be totally confined and the data analysis could therefore be based on Eq. (11.27). The measurement of the sample containing two different

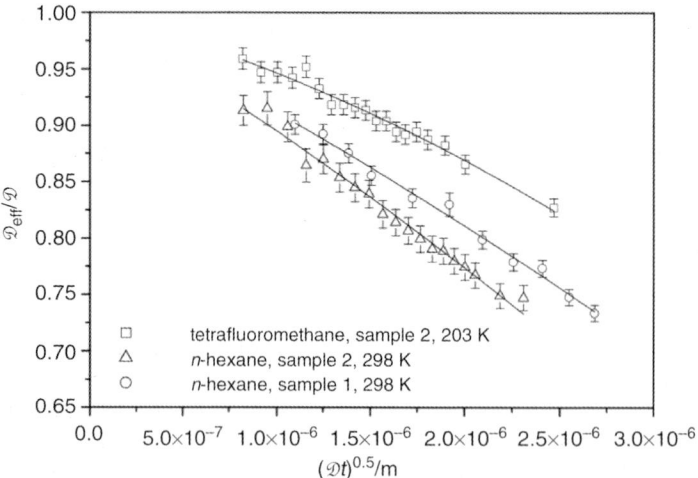

Figure 11.6 Relative effective diffusivities for n-hexane (Δ, sample 2) and tetrafluoromethane ($\square$, sample 2) under two-component adsorption and for n-hexane under single-component adsorption ($\bigcirc$, sample 1) in zeolite NaX. The lines represent the appropriate fits of Eqs. (11.27) and (11.28). Reprinted from Reference [95] with permission.

diffusants provided the option to operate with one and the same sample under the conditions of reflecting boundaries (Eq. (11.27), ^{1}H PFG NMR with n-hexane) and absorbing boundaries [Eq. (11.28), ^{19}F PFG NMR with CF$_4$]. The latter measurements had to be carried out at 203 K, since only at these low temperatures were the diffusivities small enough to allow molecular displacements sufficiently small in comparison with the crystallite radii (which is required for the series expansion approximation). Even at this low temperature the long-range diffusivity was found to be large enough to approximate the limiting case of absorbing boundaries [Eq.(11.28)]. Notably, the good fit between the experimental data and the theoretical curves was only possible by including the second-order terms in $(\mathcal{D}t)^{1/2}$.

11.4.1.2 Correlating the Effective Diffusivities with the Crystal Dimensions in the Long-Time Limit

Under conditions of complete confinement of the guest molecules within the host particle over the whole observation time, in the long-time limit the effective diffusivity in PFG NMR studies may be shown to approach well-defined values that are determined entirely by the observation time and the dimensions of the host particle. We have referred to such a situation already on discussing the propagator representation of ethane diffusion in zeolite NaCaA where, for the small crystals at the lowest temperature (Figure 11.2b), the propagator was found to remain unchanged with variation of the observation time, thus reflecting the extension of the host particle (the zeolite crystal) rather than the intracrystalline diffusivity. The quantitative treatment is based on Equation (11.6), which may be transformed into [3, 12]:

$$\lim_{t \to \infty} \psi(\gamma \delta g, t) = \iint p(z_1) P(z_2, z_1, \infty) \exp[i\gamma\delta g(z_1-z_2)] dz_1\, dz_2 \\ = \left| \int p(z) \exp(i\gamma\delta g z)\, dz \right|^2 \quad (11.29)$$

where we have written the cosine function in complex form and made use of the fact that, for sufficiently long times, the propagator becomes independent of its starting point, that is:

$$\lim_{t \to \infty} P(z_2, z_1, \infty) = p(z_2)$$

In QENS, the relation corresponding to Eq. (11.29) is referred to as the elastic incoherent structure factor (EISF) [96]. With Eq. (10.46) as the incoherent scattering function of QENS, for self-correlation functions $G_s(\mathbf{r}, t)$ remaining finite with diverging observation times, neutron scattering is found to yield a purely elastic line $\delta(\omega)$ with intensity given by the EISF.

With Eq. (11.29), the spin echo attenuation is found to be given by the square of the Fourier transform of the particle density distribution (with respect to the z coordinate, that is, the gradient direction) $p(z)$. Interestingly, this attenuation pattern coincides with the diffraction patterns of scatterers, the arrangement of which is described by the identical density function [12, 13, 22, 97–100]. Equating the expressions resulting from Eq. (11.29) for the given confining structure and Eq. (11.9) in the limit of small gradient intensity one may easily correlate the effective diffusivities in the long-time limit with the dimensions of confinement [3]. One thus obtains for diffusion confined within spheres of radius R:

$$\mathcal{D}_{\text{eff}} = R^2/(5\Delta) \tag{11.30}$$

For confinement only to the sphere surface:

$$\mathcal{D}_{\text{eff}} = R^2/(3\Delta) \tag{11.31}$$

while for confinement by cylinders of radius R in radial direction (corresponding to field gradients perpendicular to the axis direction):

$$\mathcal{D}_{\text{eff}} = R^2/(4\Delta) \tag{11.32}$$

and for confinement between parallel plates of distance L (with field gradients perpendicular to the plates):

$$\mathcal{D}_{\text{eff}} = L^2/(12\Delta) \tag{11.33}$$

11.4.2
Long-Range Diffusion

If the mean molecular displacements in the interval between the two field-gradient pulses are much larger than the crystallite diameters, the diffusivity resulting from PFG NMR measurements reflects the rate of molecular propagation through the bed of crystallites. This coefficient of long-range diffusion may be shown to be determined by [101]:

$$\mathcal{D}_{\text{l.r.}} = p_{\text{inter}} \mathcal{D}_{\text{inter}} \tag{11.34}$$

with p_{inter} and $\mathcal{D}_{\text{inter}}$ denoting, respectively, the relative amount of diffusants in the intercrystalline space and their diffusivity. Equation (11.34) may be rationalized by realizing that, at any instant of time, only the fraction p_{inter} of the total amount of molecules within the sample effectively contributes to "long-range" transportation (see also the discussions in Sections 2.4 and 15.8).

Using the simple microdynamic approach the diffusivity $\mathcal{D}_{\text{inter}}$ in the intercrystalline space may be represented by:

$$\mathcal{D}_{\text{inter}} \approx \frac{1}{3} \lambda_{\text{eff}} v / \tau_{\text{tort}} \tag{11.35}$$

with λ_{eff}, v, and τ_{tort} denoting, respectively, the effective mean free path, the mean thermal velocity, and the tortuosity factor, which takes account of the enhancement of the diffusion paths in the gas phase due to the presence of the crystallites (Sections 4.2.2 and 15.8). Typical values of the tortuosity factor are of the order 2–4 (Table 15.3).

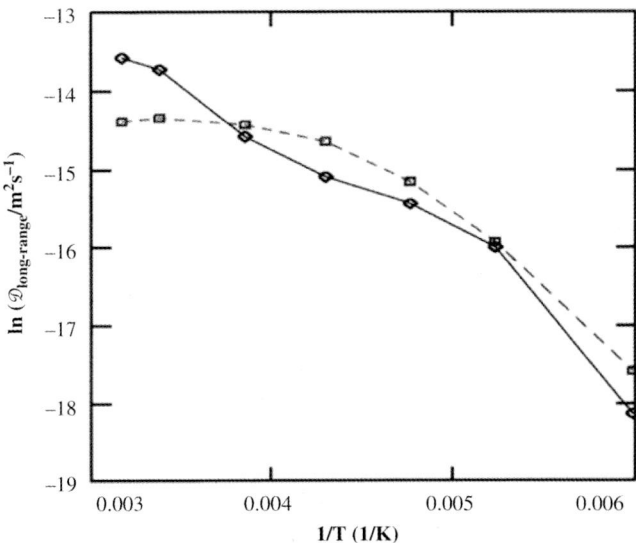

Figure 11.7 Comparison between the experimentally (diamonds) and theoretically (squares) determined long-range diffusivities for nitrogen in zeolite NaCaA. Reprinted from Reference [107] with permission.

Interestingly, in beds of nanoporous host particles, owing to the high diffusivity $\mathcal{D}_{inter}$ in the interparticle (intercrystalline) space, for sufficiently high temperatures and loadings the relative amount p_{inter} of molecules in the gas phase may attain sufficiently high values so that the long-range (bed) diffusivity of the molecules may notably exceed their diffusivity in the corresponding bulk liquid at the same temperature [102, 103]. This finding may be generalized to partially filled mesopores, where, for exactly this reason, the effective diffusivities are often found to be higher than in the bulk fluid [70, 79, 104–106].

Figure 11.7 shows the long-range diffusivity of nitrogen adsorbed on a commercial 5A type zeolite (NaCaA, with about 67% of Na^+ exchanged by Ca^{2+}) in comparison with a theoretical estimate on the basis of Eqs. (11.34) and (11.35), with p_{inter} determined from the adsorption isotherm and the packing density of the bed of crystallites [107]. For simplicity, the limiting value of the mean free path for sufficiently low temperatures has been set equal to the crystallite radius, and the tortuosity factor has been chosen to yield the best agreement with the absolute values. Despite the obvious approximations in this model, the theoretical data satisfactorily reflect the observed temperature dependence, thus providing quantitative confirmation that the decreasing slope of the Arrhenius plot with increasing temperature is a consequence of the decreasing mean free path.

11.4.2.1 Tortuosity Factor and the Mechanism of Diffusion

Having the highest value of the gyromagnetic ratio and, hence, the largest signal intensity, ^1H PFG NMR studies offer the best options for a more quantitative

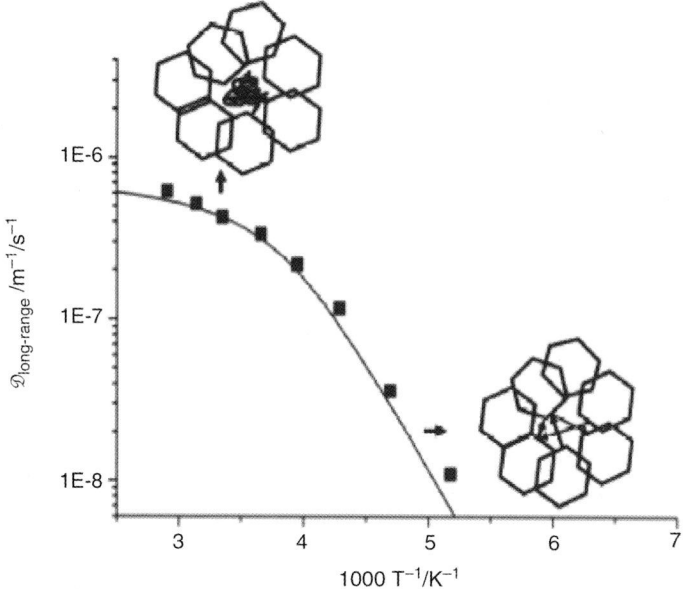

Figure 11.8 Temperature dependence of the coefficient of long-range self-diffusion of ethane measured by PFG NMR in a bed of crystallites of zeolite NaX (■) and comparison with the theoretical estimate (line). The theoretical estimate is based on the sketched models of prevailing Knudsen diffusion (low temperatures; molecular trajectories consist of straight lines connecting the points of surface encounters) and gas-phase diffusion (high temperatures, mutual collisions of the molecules lead to Brownian-type trajectories in the intercrystalline space). Reprinted from Reference [108] with permission.

comparison of experimental long-range diffusivity data with gas kinetic concepts. As an example, Figure 11.8 displays the long-range diffusivity of ethane in a bed of crystallites of zeolite NaX. The insets illustrate the two limiting situations of Knudsen diffusion (at low temperatures) and bulk diffusion (at high temperatures) in the gas phase between the crystallites. A quantitative analysis [108], based on the adsorption isotherms and the intercrystalline porosity, yielded the remarkable result that a satisfactory fit between the experimental data and the estimates of $\mathcal{D}_{l.r.} = p_{inter}\mathcal{D}_{inter}$ following Eqs. (11.34) and (11.35) could only be obtained if it is assumed that the tortuosity factors under conditions of Knudsen and molecular diffusion differ by a factor of about three.

Similar results have been obtained by dynamic Monte Carlo simulations [109–112]. In Reference [113] it is shown that the increase in the tortuosity factor under the conditions of Knudsen diffusion may be attributed to the fact that, with increasing tortuosity, subsequent jumps are more and more anti-correlated, that is, that any jump tends to reverse the displacement of the preceding jump.

11.4.3
Covering the Complete Range of Observation Times, the "NMR-Tracer Desorption" Technique

PFG NMR diffusion studies with beds of nanoporous particles are often confronted with the problem that neither of the two limiting cases of intracrystalline or long-range diffusion is really fulfilled. In addition, many zeolite samples show significant surface barriers, the effect of which should be accounted for in modeling molecular transport. However, if all these effects are incorporated, the model becomes far too complex to be useful. It is, therefore, not practical to accurately model the PFG NMR spin echo attenuation over the whole range between the limiting cases of intracrystalline and long-range diffusion. We therefore consider three different simplified models that are able to capture different features of overall mass transfer in such systems.

11.4.3.1 Two-Dimensional Monte-Carlo Simulation

The scheme displayed in Figure 11.9 [114] allows for dynamic Monte Carlo simulations taking account of the energetic conditions determining the distribution of the guest molecules and allowing for the inclusion of additional resistances at the crystal boundary (surface barrier). The guest molecules jump with uniform probability between the vertices of the structure, possibly additionally retarded by a surface barrier. The number of vertices in the crystallites and the intercrystalline space and the difference in the potential energies determine the relative populations.

Dynamic Monte Carlo simulation of a molecular random walk along the vertices yields the propagators $P(z_2, z_1, t)$ where the gradient direction (z coordinate) is chosen to be parallel to a crystal edge. The (a priori) probability is assumed to be uniform in either space, differing between them by the number of vertices and the Boltzmann factor $\exp(-\varepsilon_{\text{des}}/RT)$. Through Eq. (11.6), these probabilities allow determination of the complete PFG NMR signal attenuation curves. Here, we confine ourselves to the first decay of these attenuation curves and show, in Figure 11.10b, the effective diffusivities (Section 11.2.3) determined in this way. The complete attenuation curves may be found in Reference [114].

Both the experimental data (Figure 11.10a) and the simulation results (Figure 11.10b) reveal all limiting cases relevant for PFG NMR diffusion measurements with beds of nanoporous host particles so far considered:

1) For sufficiently low temperatures and short observation times the effective diffusivity coincides with the genuine intracrystalline diffusivity, with an activation energy $\varepsilon_{\text{diff}}$ given by the energy barrier between adjacent sites within the zeolite bulk phase (situation in Figure 11.2a).
2) Intracrystalline diffusion path lengths are confined by the crystal size (Figure 11.2b). As a consequence, the mean square displacements and hence the effective diffusivities of intracrystalline mass transfer remain restricted, approaching a limiting value that appears in the "plateaux" of the Arrhenius plots. For one-dimensional diffusion under confinement by a crystal of extension

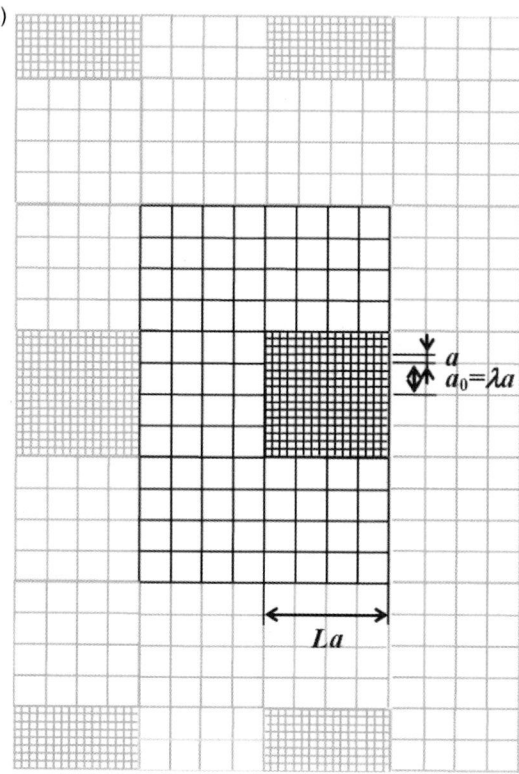

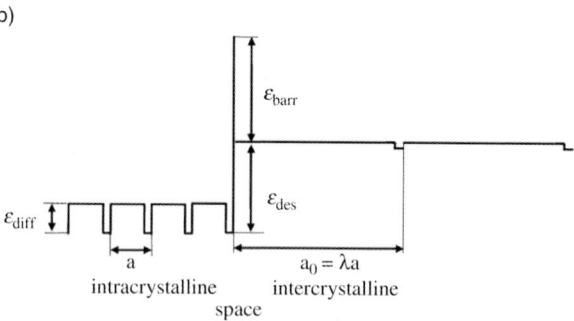

Figure 11.9 Two-dimensional model bed of nanoporous particles (a) and energies of activation (b) [114]. The periodic structure is organized by repetition of the simulation unit consisting of a square crystal with the edge length La and the outer space (indicated by bold lines).

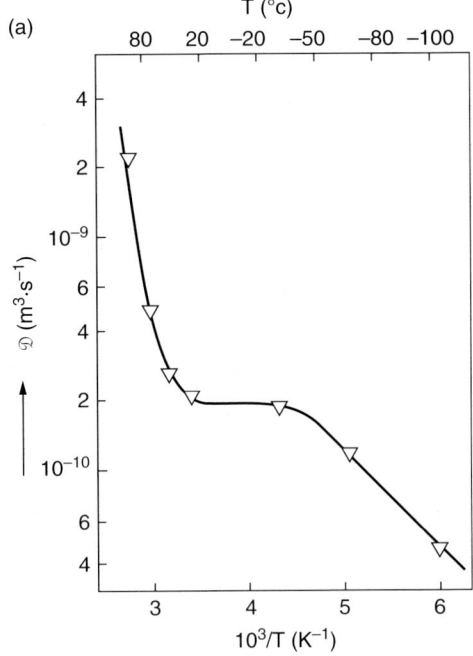

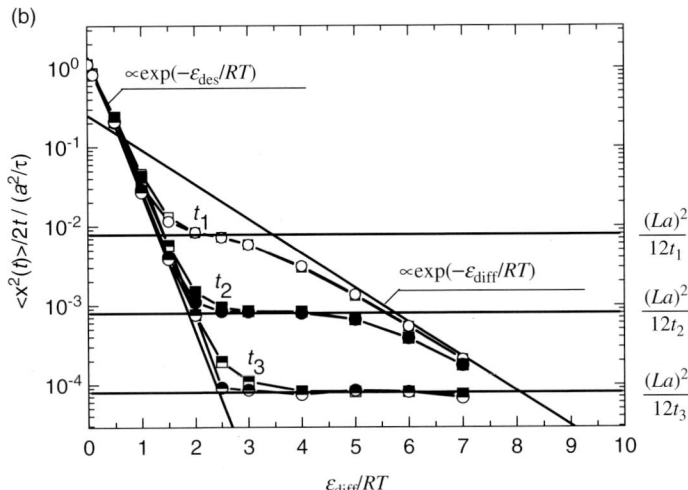

Figure 11.10 Arrhenius plot of experimental (a) and simulated (b) effective PFG NMR diffusivities of guest molecules in beds of nanoporous particles. The experimental data refer to n-hexane in a sample of NaX zeolite crystals of mean diameter 4 μm at a sorbate concentration of 20 mg g^{-1} and for an observation time of 4 ms [115]. In the simulations (b), the open, filled, and half-filled points correspond to observation times $t_1 < t_2 < t_3$, respectively, corresponding to 10^5, 10^6, and 10^7 Monte Carlo steps of duration τ. The simulations were performed with $\varepsilon_{des} = 3\varepsilon_{diff}$ and for $\varepsilon_{barr} = 0$ (squares) and $\varepsilon_{barr} = \varepsilon_{des}$ (circles). The values $\mathcal{D}_{restr} = (La)^2/(12t)$ are shown by horizontal lines.

La, Eq. (11.33) yields a limiting value of $\mathcal{D}_{restr} = (La)^2/12t$, which is in perfect agreement with the simulation results displayed in Figure 11.10b. The heights of the plateaux scale with the inverse values of the observation times as predicted.

3) With increasing temperature and, hence, with increasing thermal energy, more and more molecules are able to pass the crystallite boundaries so that, eventually, the mean square displacements may increase again (Figure 11.2c). The effective diffusivity is the product of the diffusivity of the molecules in the intercrystalline space ($\mathcal{D}_1$) and their relative amount (p_1), in agreement with Eq. (2.39). Since $\mathcal{D}_1$ has been chosen to be invariant with temperature, the activation energy of the effective diffusivity is now determined by p_1, leading to the energy of desorption, ε_{des}. In fact, in our simulations the long-range diffusivities $\mathcal{D}_{long\text{-}range} = \langle x^2(t)\rangle/2t$ for root mean square displacements $\langle x^2(t)\rangle^{\frac{1}{2}}$ much larger than the crystal dimensions La are found to agree with the product of the quantities p_1 and $\mathcal{D}_1$ within a mean error of less than 15%.

In complete agreement with the expected behavior, in each of these special cases the simulated effective diffusivities turn out to be independent of the existence and height of the surface barriers. Differences between the data simulated with and without surface barriers only appear in the transition range from restricted (intracrystalline) to long-range diffusion. We shall see in the next section that information about the existence and intensity of surface barriers is more easily accessible by considering the complete PFG NMR spin echo attenuation.

11.4.3.2 Two-Region Model

Molecular diffusion in multi-region systems may be approximated by combining Fick's 2nd law [Eq. (1.2)] with appropriate exchange terms (as provided by the Chapman–Kolmogoroff equation) [116], yielding [3, 10, 11]:

$$\frac{\partial c_i}{\partial t} = \mathcal{D}_i \, \text{div grad } c_i - \frac{1}{\tau_i} c_i + \sum_{j \neq i} \frac{p_{ji}}{\tau_j} c_j \qquad (11.36)$$

where c_i, $\mathcal{D}_i$, and τ_i denote, respectively, the concentration, the self-diffusion coefficient, and the mean life time in the i-th region, and p_{ji} is the conditional probability that if a molecule leaves region j it will transfer to region i. With Eq. (11.36), in each time interval dt an arbitrarily selected molecule in region i is considered to leave this region with the probability $dt/\tau_i \times p_{ij}$ for region j. Otherwise the molecule continues its diffusion path in region i completely unaffected. For real systems, in general, these implications mean a strong approximation. However, since in many cases it leads to simple analytical expressions for the PFG NMR spin echo attenuation, the "multi-region approach" has been adopted in numerous cases for the analytical treatment of diffusion in compartmented systems [14], in particular in biological tissues [99, 117–120].

Originally, this approach was introduced to develop an analytical expression for the PFG NMR signal attenuation in beds of zeolite crystallites, which is able to cover the complete time spectrum of the measurements, including observation times ranging from much less to much greater than the intracrystalline mean life times of the guest molecules [121–124]. The solution of Eq. (11.36) for such a "two-region" system may be written in the form:

$$\psi(\Delta) = \psi_1(\Delta) + \psi_2(\Delta) = p'_1 \exp(-\gamma^2 \delta^2 g^2 \mathcal{D}'_1 \Delta) + p'_2 \exp(-\gamma^2 \delta^2 g^2 \mathcal{D}'_2 \Delta) \tag{11.37}$$

where:

$$\mathcal{D}'_{1(2)} = \frac{1}{2}\left(\mathcal{D}_1 + \mathcal{D}_2 + \frac{1}{\gamma^2 \delta^2 g^2}\left(\frac{1}{\tau_1} + \frac{1}{\tau_2}\right)\right.$$
$$\left.\mp \left\{\left[\mathcal{D}_2 - \mathcal{D}_1 + \frac{1}{\gamma^2 \delta^2 g^2}\left(\frac{1}{\tau_2} - \frac{1}{\tau_1}\right)\right]^2 + \frac{4}{\gamma^4 \delta^4 g^4 \tau_1 \tau_2}\right\}^{\frac{1}{2}}\right) \tag{11.38}$$

$$p'_1 = 1 - p'_2, \quad p'_2 = \frac{1}{\mathcal{D}'_2 - \mathcal{D}'_1}(p_1 \mathcal{D}_1 + p_2 \mathcal{D}_2 - \mathcal{D}'_1) \tag{11.39}$$

with the indices 1 and 2 referring to the inter- and intracrystalline spaces. As a requirement of dynamic equilibrium, the relative populations are related to the mean lifetimes by the expression:

$$p_i = \tau_i / (\tau_1 + \tau_2) \tag{11.40}$$

From Eq. (11.37) the echo attenuation is seen to be governed by the superposition of two exponentials of the form of Eq. (11.9) for echo attenuation due to diffusion in a homogeneous system. The "apparent" diffusivities $\mathcal{D}'_i$ and intensities p'_i are related to the real system parameters by Eqs. (11.38)–(11.40). Interestingly, the formalism of PFG NMR for two-region diffusion as provided by Eqs. (11.37)–(11.40) also proved to be a useful tool for data analysis in single-particle tracking [124] (see also Section 11.6).

Equations (11.37)–(11.40) may be simplified by considering the inequalities $p_{1(\text{inter})} \ll p_{2(\text{intra})} \approx 1$ and $\mathcal{D}_{1(\text{inter})} \gg \mathcal{D}_{2(\text{intra})}$, which describe the situation for adsorption from the gas phase so that one obtains [121–123]:

$$\psi(\Delta) = \exp\left[-\gamma^2 \delta^2 g^2 \left(\mathcal{D}_2 + \frac{p_1 \mathcal{D}_1}{\gamma^2 \delta^2 g^2 \tau_2 p_1 \mathcal{D}_1 + 1}\right)\Delta\right] \tag{11.41}$$

Clearly, from the representation of Eq. (11.41) in Figure 11.11 this expression includes the limiting cases of both intracrystalline diffusion (for $\Delta \ll \tau_2 (\equiv \tau_{\text{intra}})$) and long-range diffusion (for $\Delta \gg \tau_2$ and $p_1 \mathcal{D}_1 \gg \mathcal{D}_2$). Moreover, with the intracrystalline mean molecular mean lifetimes $\tau_{2(\text{intra})}$ we obtain direct information about the molecular exchange rates.

The value of $\exp(-t/\tau_2)$ at the intersection of the large-gradient-intensity asymptote with the ordinate represents the fraction of molecules which, after time t, have not yet left the crystallite. For first-order reactions [which is implied by the rate expression in Eq. (11.36)] this corresponds to the fraction of molecules still in the initial state. From the general discussion in Section 6.2.1 we know that sorption kinetics follow this

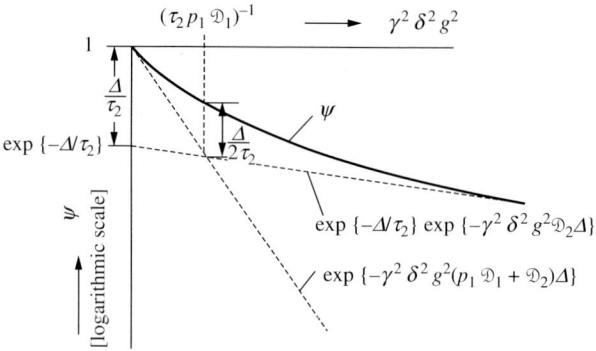

Figure 11.11 Echo attenuation for two-region diffusion with one small region of high diffusivity (region 1, corresponding to the intercrystalline space for gas-phase adsorption) and limiting results for small and high values of the gradient intensities $\gamma\delta g$. Reprinted from Kärger et al. [9] with permission.

same pattern when surface resistance is dominant. We show below that this type of analysis is not restricted to the particular model that was chosen only for mathematical convenience.

11.4.3.3 NMR Tracer Desorption (= Tracer-Exchange) Technique

In the propagator representation (Section 11.2.2, Figure 11.2c), the contribution of molecules that, after time t, have not yet left the crystallites can be easily identified as the narrow constituent (with width of the order of the crystal size) in comparison with those molecules that have larger displacements and have therefore left the individual crystallites. Since the spin-echo attenuation curves and the mean propagators are Fourier transforms [Eq. (11.15)], a distinct asymptote to the spin-echo attenuation for large gradient intensities automatically corresponds to the appearance of a distinct narrow curve in the propagator representation. The contribution of this narrow component to the total area under the propagator coincides with the relative contribution of the slowly decaying part to the total NMR signal. This correlation is illustrated by Figure 11.12. In addition, we have also plotted the relative contribution $\gamma(\Delta)$ of the broad constituent of the propagator (coinciding with the contribution of the first, steep part of the PFG NMR attenuation curve to the overall signal) as a function of the observation time. With $m_t^{(*)}$ denoting the amount of molecules within a crystal that, after time t, have been replaced by molecules from the surroundings, one may note $\gamma(\Delta) \equiv m_t^{(*)}/m_\infty^{(*)}$. This is precisely the information obtained in a traditional tracer exchange experiment (Section 13.7) so the method has also been called NMR tracer desorption or, since the measurement is made on the time scale of an NMR experiment (milliseconds to seconds), "fast tracer desorption" [125, 126].

As a prerequisite for clear distinction between these two constituents (in either case!) and, therefore, to plot the NMR tracer desorption curve, the long-range

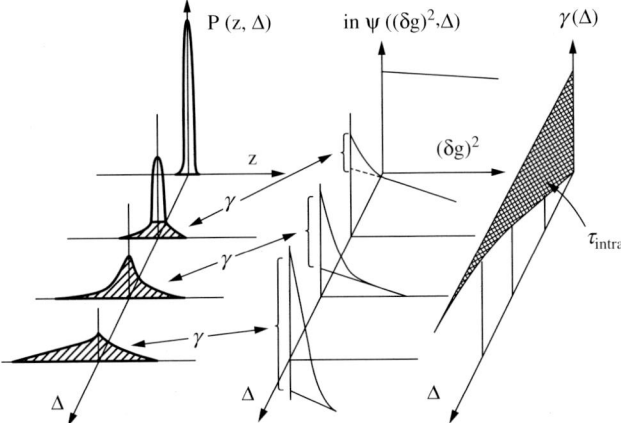

Figure 11.12 Correlation between the mean propagator $P(z,\Delta)$, the echo attenuation $\psi((\delta g)^2,\Delta)$, and the NMR tracer desorption curve $\gamma(\Delta)$.

diffusivity must be much larger than the intracrystalline diffusivity. Since the activation energy is in most cases much larger for long-range than for intracrystalline diffusion (see, for example, Figures 11.9b and 11.10) such a situation may generally be attained at sufficiently high temperatures. For quantitative analysis of the NMR tracer desorption curve it is useful to define an intracrystalline mean lifetime (τ_{intra}) by using the first statistical moment [127, 128] of the decay curve (Section 13.6):

$$\tau_{intra} = \int_{t=0}^{\infty} [1-\gamma(t)] \, dt \tag{11.42}$$

According to this definition, τ_{intra} is the area between $\gamma = \gamma(t)$ and $\gamma = 1$. In the two-region approach [coinciding with the situation for barrier-limited sorption – see Eq. (6.6)], the tracer desorption curve follows an exponential time dependence:

$$\gamma(t) = 1-\exp(-t/\tau) \tag{11.43}$$

and the first moment τ_{intra}, as defined by Eq. (11.42), coincides with the time constant τ of the approach.

For practical reasons it is sometimes necessary to carry out the pulsed field gradient measurements in an NMR tracer desorption experiment with the aid of an additional constant field gradient. Application of this procedure leads to time constants τ'_{intra} that coincide with τ_{intra} for small values (<5 ms) only and which increase more than linearly with increasing values of τ_{intra}. This leads to an enhancement of the sensitivity with which different samples can be compared with respect to their rates of molecular exchange [129].

11.4.3.4 Observation of Surface Barriers

Combining the NMR measurement of intracrystalline self-diffusion (Section 11.4.1) with an NMR tracer desorption experiment provides a direct way of detecting and measuring any resistance to mass transfer at the crystal surface. The simplest procedure is to compare the intracrystalline mean lifetime τ_{intra}, as determined directly from the tracer desorption experiment, with the corresponding value $\tau_{\text{intra}}^{\text{diff}}$, calculated from the intracrystalline diffusivity ($\mathcal{D}_{\text{intra}}$) for the appropriate geometry on the assumption that the rate of molecular exchange is controlled entirely by intracrystalline diffusion. If $\tau_{\text{intra}} \approx \tau_{\text{intra}}^{\text{diff}}$ the possibility of any significant additional resistance to mass transfer at the crystal surface can obviously be excluded. Conversely, if $\tau_{\text{intra}} > \tau_{\text{intra}}^{\text{diff}}$ the existence of such a "surface barrier" is indicated. Since the uncertainty in the values derived for τ_{intra} is typically of order 30–50% a detailed quantitative analysis is generally not justified and it is sufficient to approximate the particles as a set of uniform spheres. In this case (Section 13.6.2, Table 13.1) the relationship is simply:

$$\tau_{\text{intra}}^{\text{diff}} = R_p^2/15\mathcal{D}_{\text{intra}} \tag{11.44}$$

where R_p^2 is the mean square radius of the adsorbent particles.

The surface mass transfer resistance is usually quantified by its reciprocal value (Section 2.3.3 and Reference [130]), the surface permeance k_s (Section 6.2.1.1), defined by the relation:

$$J = k_s[q^* - q|_{R_p}] \tag{11.45}$$

The concentration driving force $q^* - q|_{R_p}$ represents the difference between the adsorbed phase concentration at equilibrium with the surrounding fluid phase and the actual adsorbed phase concentration just inside the particle [see Eq. (6.2)]. In the case of a tracer desorption measurement these quantities refer to the equilibrium concentrations of "marked" molecules, although an analogous definition may be applied to a non-equilibrium situation in which q^* and $q|_{R_p}$ then refer to total species concentrations. If the resistance is due to a surface layer of thickness δ of dramatically reduced diffusivity $\mathcal{D}_S$, comparison of Eq. (11.45) with Fick's 1st law yields:

$$k_s = D_s/\delta \tag{11.46}$$

Surface barriers may also originate from a total blockage of most of the entrances to the pore space. This option, which was considered analytically in Section 2.3.4, has been confirmed, as a special limiting case, by experimental evidence (Reference [131] and Section 19.6).

Under the combined influence of intracrystalline diffusion and a surface barrier to mass transfer, by the method of moments (Section 13.6.2) the total mean intracrystalline life time is simply the sum of the mean life times (first moments) of both processes [125, 127]:

$$\tau_{\text{intra}} = \frac{\langle R_p^2 \rangle}{15\mathcal{D}_{\text{intra}}} + \frac{R_p}{3k_s} \equiv \tau_{\text{intra}}^{\text{diff}} + \frac{R_p}{3k_s} \tag{11.47}$$

Using the values for τ_{intra} and $\mathcal{D}_{\text{intra}}$ from the NMR tracer desorption and self-diffusion experiments together with the measured value of the crystal or particle radius (R_p) yields the value of k_s.

The terms τ_{intra} and $\langle R_p^2 \rangle/(15\mathcal{D}_{\text{intra}})$ in Eq. (11.47), however, can generally be determined only with a substantial uncertainty (up to a factor of 2 or even more) so that the third term $R_p/(3k_s)$ results as the difference between two highly inaccurate values. A reasonable estimate of this term and, hence, of the surface permeability k_s is therefore only possible for barrier-limited uptake and release, that is, for $\tau_{\text{intra}} \gg \tau_{\text{intra}}^{\text{diff}}$. Such examples are presented in Sections 16.9.3 and 21.5. Better prospects for determining surface permeabilities under more general conditions are provided by considering, in addition to NMR tracer exchange, the dependence of the effective intracrystalline diffusivities on the observation time [132, 133]. However, even with this more sophisticated type of analysis, the information concerning the magnitude of surface resistances is subject to considerable uncertainty. Much more precise values can be obtained from interference microscopy (IFM) and IR micro-imaging/microscopy (IRM), as demonstrated in Section 12.3.2.

The application of NMR tracer desorption to the study of surface resistance in commercial adsorbents and coked catalysts is discussed in Sections 16.9 and 21.5, respectively.

11.5
Experimental Tests of Consistency

Before the results of PFG NMR on zeolitic diffusion were found to agree with other techniques such as quasi-elastic neutron scattering and molecular modeling, as well as with transient sorption studies, the substantial differences from many of the earlier measurements of intracrystalline diffusion called for an independent check of their validity. In this context, the versatility of PFG NMR turned out to allow for a remarkable variety of test experiments confirming the validity of these new measurements. These experiments [103, 134–137] nicely delineate the range of applicability of such methods, and so a brief summary is included here.

11.5.1
Variation of Crystal Size

The intracrystalline diffusivity is assumed to characterize molecular migration within the micropores of a particular material and as such the value should not depend on the crystal size. Repetition of diffusivity measurements with different crystal size fractions thus provides a simple means of confirming the validity of the experimental procedure and the analysis of the data. Such measurements have in fact been made for several different systems with consistent results [56, 95, 135, 136, 138].

11.5.2
Diffusion Measurements with Different Nuclei

The NMR field gradient method actually measures displacement of the resonant nuclei. This should be equivalent to the molecular displacement, provided that the measurements are made under conditions such that breakage and re-formation of chemical bonds does not occur at a significant rate. The validity of this assumption can be confirmed with a molecule such as CHF_2Cl by following the displacement according to both ^1H and ^{19}F NMR. Because the results obtained in such experiments were quite consistent, at least for this system, the possible intrusion of atomic exchange can be excluded [65].

11.5.3
Influence of External Magnetic Field

It was pointed out in Section 10.1.1.1 that the interaction energy between the magnetic field and the nuclear spins is negligibly small relative to the thermal energy and on that basis one would not anticipate any effect on the molecular mobility. This has been confirmed by repeating diffusion measurements (for n-hexane–NaX) with different magnetic field intensities (0.39 and 1.41 T). The resulting diffusivity values were identical [136].

11.5.4
Comparison of Self-diffusion and Tracer Desorption Measurements

In Section 11.4.3 we showed that a comparison between the PFG NMR self-diffusivities and NMR tracer exchange studies provides information concerning the presence or absence of surface barriers. Strong barriers have been observed for crystals of zeolite LTA subjected to severe hydrothermal pretreatment and for highly coked samples of zeolite MFI (Sections 16.9 and 21.5). In carefully dehydrated zeolite samples, however, the influence of surface barriers is found to be less pronounced, so that the intracrystalline diffusivities directly measured by PFG NMR ($\mathcal{D}_{intra}$) and those estimated from the NMR tracer desorption curves assuming diffusion-limitation ($\mathcal{D}_{des}$) are close to each other [125, 136] (Figure 16.31 and Table 7.1 of Reference [139]). Clearly, in these cases the consistency in the diffusivity values derived from two quite different experiments provides strong evidence that the underlying theory is essentially sound and that the interpretation of the measurements is correct.

11.5.5
Blocking of the Extracrystalline Space

It was shown by Figure 11.10 and explained in Section 11.4.3 that the temperature dependence of the effective diffusivity shows three distinct regions. The high temperature (high slope) region represents diffusion through the extracrystalline space. Clearly, if the external space is filled by a large molecular species that cannot penetrate the micropores, intracrystalline diffusion will be unaffected but extracrystalline diffusion will be severely retarded. This was demonstrated for diffusion of C_2H_6 in NaCaA crystals immersed in CCl_4 [135].

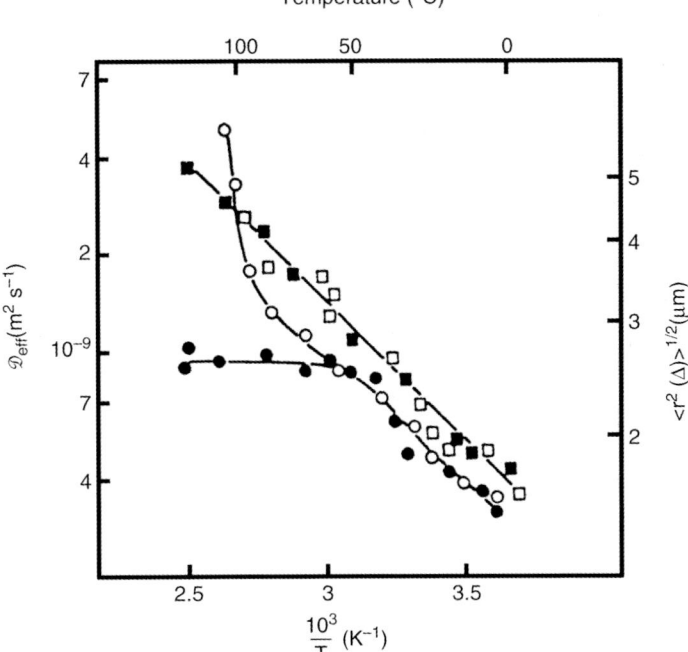

Figure 11.13 Temperature dependence of effective self-diffusivity of water in MFI-type zeolite crystals of different size: (○,●) H-ZSM-5, Si$_2$/Al ≈ 25, 7 × 4 × 3; μm^3; (□,■) NaH-ZSM-5, Si/Al ≈ 40, 16 × 12 × 8 μm^3 before (open symbols) and after (filled symbols) coating the crystals for observation time $t = 1.2$ ms. For comparison the corresponding root mean square displacements are indicated [Eq. (11.16)]. Reprinted from Caro et al. [138] with permission.

Similarly, molecules may be trapped within the interior by artificially introducing a large barrier at the external surface. Figure 11.13 shows PFG NMR diffusivity data obtained for water in crystals of MFI-type zeolites, as synthesized and after coating the external surface with silica [138]. In the range in which long-range diffusion is observed with the uncoated crystals, the coated crystals show restricted diffusion, with the mean square displacement approximately equal to the crystal size and independent of temperature. This is precisely the behavior that is to be expected for an impermeable surface barrier.

11.5.6
Determination of Crystal Size for Restricted Diffusion Measurements

The correspondence between the mean square displacement and the crystal size within the region of restricted diffusion has been referred to already in the propagator representation in Figure 11.2 and considered in detail in Section 11.4. This correspondence has been checked for several different experimental systems, thus providing additional confirmation of the underlying theory and the experimental method [103]. Notably, for the systems considered in Figure 11.6, within a range of 20%, five different ways of analyzing the size of the X-type zeolite crystals have led to coinciding results on the size of the crystallites under study [95], viz. (i) microscopic inspection, (ii) restricted diffusion in the limit of large observation times [situation shown by Figure 11.2b and Eq. (11.30)], (iii) application of Eq. (11.44) to the results of the PFG NMR tracer desorption technique, and, finally, consideration of the limit of short observation times for (iv) reflecting boundaries [Eq. (11.27)] and (v) absorbing boundaries [Eq. (11.28)].

11.5.7
Long-Range Diffusion

In the long-range diffusion regime the diffusivity is given by the relation $p_{inter}\mathcal{D}_{inter}$ [see Section 11.4.2 and Eq. (11.34)]. The distribution of molecules between adsorbed and vapor pores (p_{inter}) may be estimated from the experimental equilibrium isotherm and the voidage of the crystallite bed, while the magnitude of the extracrystalline diffusivity ($\mathcal{D}_{inter}$) can be estimated as outlined in Section 11.4.2 by using the approaches for Knudsen and gas diffusion. In all cases where such measurements have been made, the agreement with the theoretical estimates was found to be satisfactory [11, 66, 67, 69, 70, 140, 141]. Furthermore, the values have also been confirmed by direct measurement of transient uptake rates [11, 142].

11.5.8
Tracer Exchange Measurements

In addition to directly determining intracrystalline diffusivities by PFG NMR, NMR measurements have also been applied indirectly as a convenient way of following the progress of a tracer exchange experiment. A sample equilibrated with a fully deuterated hydrocarbon (e.g., C_6D_6) is exposed to an atmosphere of the undeuterated species (C_6H_6) at the same pressure. The rate of approach to equilibrium is then followed by monitoring the intensity of the proton NMR signal in the zeolite crystallites. The exchange curves are then interpreted in the same way as for any other tracer measurement (Section 13.7).

Figure 11.14 shows the experimental arrangement. The method has been applied by Förste et al. to measure tracer exchange of benzene in NaX [137, 143] and in silicalite [134]. Figure 11.15 shows representative results for the NaX system.

These measurements were carried out in a small sample tube that could be inserted directly into an NMR spectrometer to allow a direct (PFG) measurement to be carried out with the identical sample after exchange equilibrium had been reached. As may be seen, the diffusivities obtained by both techniques are quite consistent. Since the two ways of measurement are totally different and independent from each other, their validity is mutually confirmed.

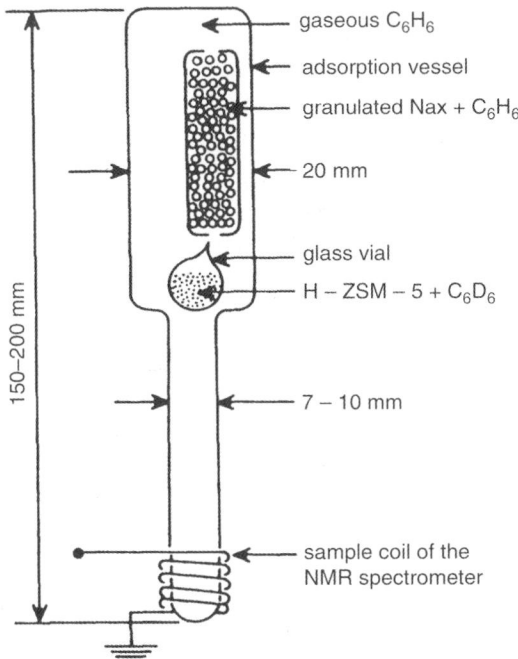

Figure 11.14 Experimental arrangement for NMR tracer exchange measurements. Reprinted from Förste et al. [143] with permission.

11.6
Single-Molecule Observation

Diffusion deals with the evolution of particle concentrations and displacement probabilities in ensembles of molecules. The accuracy of the information increases with the number of particles considered. Most of the experimental techniques presented in this book operate with very large numbers of molecules with a, correspondingly, high statistical accuracy. However, following the ground-breaking exploration of the diffusion paths of "hot" atoms on metal surfaces by scanning tunneling microscopy by Ertl and coworkers [144–146], the observation of single dye molecules via their absorption [147] or fluorescence excitation spectra [148] has also become possible [149, 150]. This type of measurement is commonly referred to as single-particle tracking (SPT). The wealth of information provided by the particle trajectories often compensates for the inherent disadvantage of this approach, which requires averaging over many trajectories to obtain meaningful diffusion data. The application of such techniques to visualization of the intracrystalline concentration profiles during reaction on a single crystal of HZSM-5 is discussed in Section 21.3.

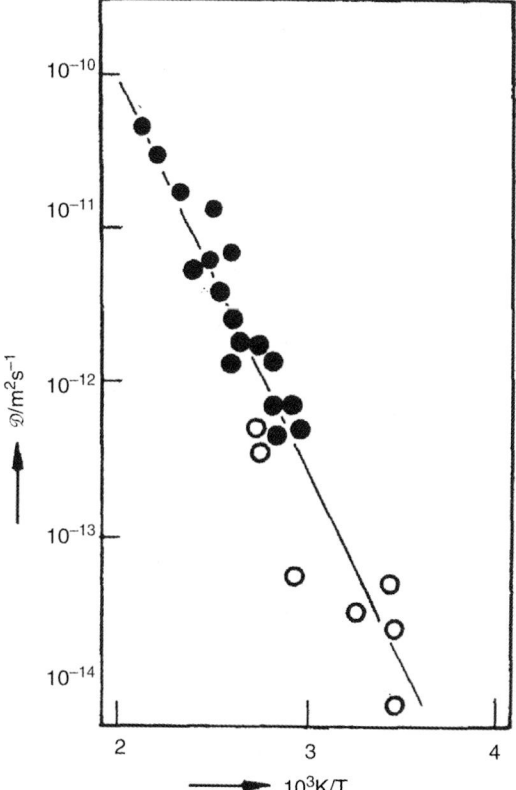

Figure 11.15 Comparison of tracer exchange self-diffusivities for benzene in NaX crystals (4.8 molecules per cage) (○) with PFG NMR diffusivities measured at equilibrium in the same sample (●). From Förste et al. [143] with permission.

11.6.1
Basic Principles of Fluorescence Microscopy

Fluorescence is the process in which molecules in an electronically excited state relax to their ground state by emitting photons of longer wavelength (i.e., of lower energy) than the photons that have caused the excitation. Application of optical filters and a beam splitter make it possible to generate fluorescence images that are not affected by the incident, exciting light. Depending on the detection geometry, particle positions may be recorded in two different ways.

Wide-field microscopy is based on the application of an array detector (e. g. a CCD camera), which allows the imaging of sample regions of several tens of micrometers in diameter at the same time [151, 152]. The single molecules appear as spots with diameters of typically 300 nm. Application of appropriate intensity fitting functions allows particle pinpointing with 5 nm precision in the focal plane, perpendicular to the direction of observation.

In confocal microscopy, the exciting light and the light emitted by the fluorescing particle are directed through pinholes, with the probe volume under study in the focus of either of them. Particle images are therefore provided if and only if a fluorescing molecule happens to be in this probe volume [153, 154]. Particle trajectories and particle distributions are attained by scanning over the whole sample volume [155–161].

In a special variant of confocal microscopy, referred to as *fluorescence correlation spectroscopy*, the fluctuation of fluorescence intensity is followed [162, 163]. This technique has been introduced for the investigation of diffusion phenomena in biological membranes. It is based on the statistical analysis of the fluctuations of fluorescence intensity (similarly to the procedure in light scattering, Section 10.3.2) in the detection area. Since it involves the observation of rather small numbers of molecules, confocal microscopy is intermediate between genuine single-particle techniques and the ensemble techniques of diffusion measurement. Special features of molecular motion, such as sub-diffusion (Section 2.6), may be deduced by an appropriate analysis of the measured decay curves [162, 164–166].

The accuracy of the diffusivities determined by single-molecule observation and, equally importantly, the evidence of possible deviations from normal diffusion depend notably on how accurately the molecular trajectories can be followed. Depending on the intensity of the emitted light and the sensitivity of the detector, the "dots" that mark the positions of the particles under study with varying time may dramatically increase in their diameters, eventually prohibiting meaningful information on particle propagation. The establishment of reliable trajectories by "connecting the dots" [167] becomes, under such conditions, a non-trivial task, requiring the development of sophisticated algorithms for data analysis [124, 168–171].

11.6.2
Time Averaging Versus Ensemble Averaging

Classical equilibrium statistics implies ergodicity, that is, coincidence of time and ensemble averages. With the advent of single-particle techniques, also particle mean square displacements have become an object to be explored with respect to ergodicity. Indeed, violation of ergodicity of the mean square displacement can be predicted to occur on the basis of certain model assumptions, including a divergence of the mean residence time between subsequent diffusion steps [164, 172, 173]. Following the discussion in Section 2.6.1, such divergences may be excluded in nanoporous host–guest systems under equilibrium conditions. Importantly, this implication holds also for hierarchical structures, as discussed in Section 2.6.2, which are known to give rise to anomalous diffusion. Under equilibrium conditions in such systems, it is of no relevance whether the mean square displacement is determined by averaging over the displacements of the same particle in many subsequent time intervals of identical duration or over the displacements of many different particles performed simultaneously for a time interval of the same duration.

Figure 11.16 shows both experimentally observed (Figure 11.16a) and computer-generated (Figure 11.16b) trajectories (generated by a simple random-walk algorithm as considered in Section 2.1) ensuring the occurrence of normal diffusion. In the

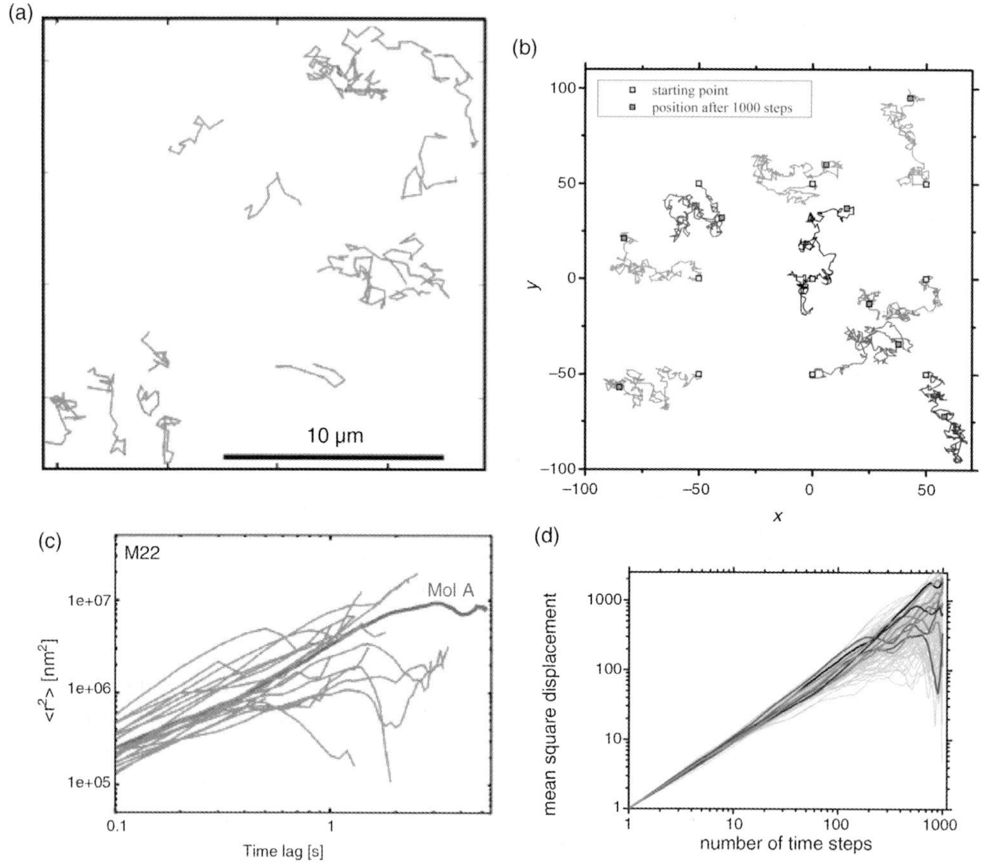

Figure 11.16 Single-particle trajectories observed experimentally in a nanoporous sol–gel glass with streptocyanine as a guest molecule ((a), reprinted from Reference [181] with permission) and computer-simulated as random walks on a two-dimensional quadratic grid (b) and the resulting mean square displacements, (c) and (d), calculated for each individual trajectory by Eq. (11.48).

representations shown, the mean square displacements resulting from each individual trajectory by time-averaging via:

$$\langle r^2(t) \rangle_T = \frac{1}{T-t} \int_0^{T-t} [r(t'+t) - r(t')]^2 \, dt' \qquad (11.48)$$

are plotted as a function of the length t of the time interval, that is, of the "time lag" considered in the experiments (Figure 11.16c) or of the number of time steps along the simulation curves (Figure 11.16d).

Evidently, reasonable data are obtained only if the total time interval (T) during which the trajectories are recorded notably exceeds the time interval t for the displacements considered (which is, of course, a prerequisite for meaningful averaging). Note that, as a consequence of the chosen way of simulation with uniform step lengths, all simulated trajectories start at a common point: 1 squared step length after 1 time step. A reasonable comparison with the experimental data has to begin, therefore, with notably larger step numbers (>10) when the superposition of the individual step displacements yields a continuous distribution curve as observed in the experiment. Interestingly, the mean square displacements (Figure 11.16c and d found following via Eq. (11.48) by time averaging over each individual trajectory (Figure 11.16a and b) are found to increase in proportion with the observation time t, that is, to obey the requirement of normal diffusion [Einstein's relation, Eq. (2.6)], long before the factors of proportionality, that is, the diffusivities (appearing in the ordinate intercepts) of the different trajectories coincide. Both the finite life time of fluorescing molecules and the restriction that, during observation, they have to remain in the focus of the monitoring device limit the time T over which particle trajectories may be recorded. The scattering in the diffusivities deduced from different trajectories is therefore a common feature of single-molecule observation, which has suggested the term "single-particle diffusivity" [174]. The true diffusivity may be obtained by averaging over a sufficiently large number of single-particle diffusivities [124]. The diffusivities in single-particle observations are therefore obtained by procedures that are intermediate between time and ensemble averaging. Notably, a similar procedure is also common in MD simulations where, for determining mean square displacements, one considers the average over many trajectories rather than over many subsequent intervals of a single trajectory [see also Section 8.1.1, Eq. (8.2)].

Their potential for exploring the conditions of ergodicity breaking in nanoporous host–guest systems make comparative single-particle and ensemble measurements a hot topic of fundamental research. Studies of this type have to contend with very large differences in the guest concentrations. Typically, in ^{1}H PFG NMR concentrations greater than one hydrogen atom per nm^3 are required to obtain sufficiently large signal intensities, whereas, in single-particle tracking concentrations of less than 1 fluorescing molecule per μm^3 are necessary to avoid the overlapping of signals from different molecules. Further problems are related to the high production costs of fluorescing molecules (which makes ensemble measurements quite expensive) and to the fact that sufficiently stable fluorescing molecules are presently only produced with sizes in the nanometer range, which, so far, excludes their application to microporous materials [149].

Investigating one and the same host–guest system by single-particle tracking (SPT) and ensemble measurement of diffusion (PFG NMR) has thus become possible under only very special conditions [175]. The chosen guest molecule, Atto532 (ATTO532-COOH, ATTO-TEC, Siegen Germany) dissolved in deuterated methanol, offered particularly favorable properties for both techniques, namely, a large enough transverse nuclear magnetic relaxation time for PFG NMR (Section 11.3.2) and sufficient photostability and quantum yield for SPT. By appropriately choosing the pore diameters of the host system (nanoporous glass, see Section 15.1.1), the guest diffusivities could be "adjusted" to those values where the sensitivity ranges of SPT

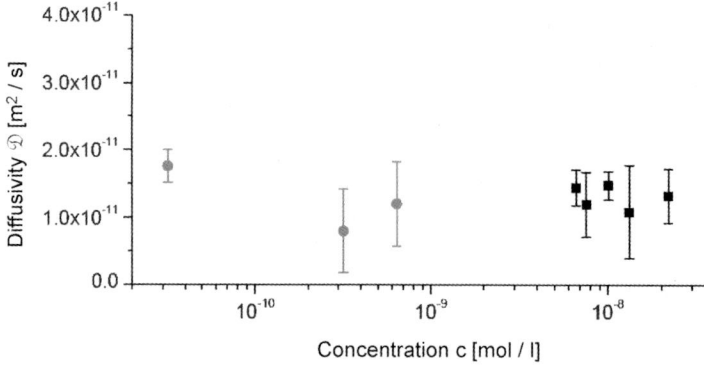

Figure 11.17 Diffusivity of Atto532, dissolved in deuterated methanol at different concentrations, in nanoporous glass (pore size 3 nm) determined by single-particle tracking (circles) and in ensemble measurements (PFG NMR, squares). Reprinted from Reference [175] with permission.

and PFG NMR overlap. Figure 11.17 shows that the diffusivities thus obtained by single-particle tracking and by measuring the ensemble average are in satisfactory agreement, representing nothing less than the experimental confirmation of ergodicity for guest diffusion in nanopores. Note that the concentrations in the PFG NMR measurements are still one order of magnitude above the concentrations considered in SPT. However, already in the PFG NMR studies the concentrations are seen to be small enough so that any mutual interference of the diffusants can be excluded, leading to essentially concentration-independent diffusivities.

11.6.3
Correlating Structure and Mass Transfer in Nanoporous Materials

It is well established that nanoporous materials generally deviate from an ideal homogeneous structure. Sections 17.3.4 and 18.2.1.1 illustrate how the measurement of ensemble diffusion by PFG NMR and QENS may reveal such deviations. With the advent of single-particle tracking in mesoporous materials, the study of such deviations may now even be localized [176–178].

Figure 11.18 illustrates the potential of such measurements. It shows the mesoporous host of type M41S, prepared as a thin film (< 100 nm) by spin-coating [177, 179], thus making it possible to keep the fluorescing molecule (a terrylendiimide, TDI, derivative [180]) in the focal plane. The trajectories reflect the highly structured manner in which the molecules travel in the hexagonal phase over distances of several µm during the acquisition time of the movie (500 s). The inset visualizes in a particularly impressive way how the guest molecules are able to explore the existence of different regions and to probe their extensions. By following the trajectory in greater detail [176], the molecule is found to travel first along the C-shaped structure on the right (1) until entering, after 65 s, side-arm (2). Then, 100 s later, it passes into the linear structure at the bottom (3). After further 144 s, it enters

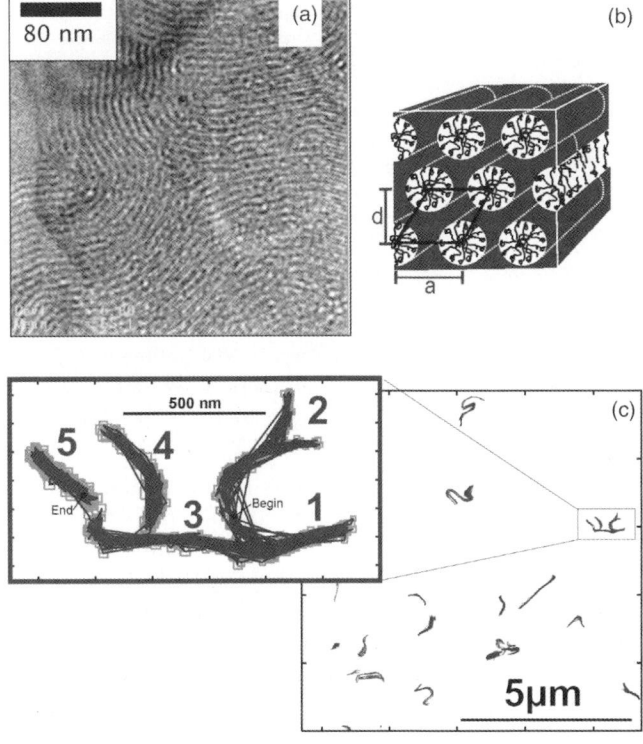

Figure 11.18 High-resolution TEM image of a mesoporous material (M41S) (a), with schematics of the hexagonal pore topology (b), and single-particle trajectories (c), with enlarged inset. Reprinted from Reference [176] with permission.

region 4 and moves around there for 69 s before coming back to region 3. Finally, it passes into region 5, where it moves back and forth for 109 s until the end of the movie (Movie S1 in the supplementary material of Reference [178]). In Reference [177] single-particle tracking is shown to yield structural details of the host system in total agreement with the information provided by high-resolution transmission electron microscopy (TEM), with the additional, unprecedented option of exploring mass transfer in the various structural domains.

References

1 Mansfield, P. and Morris, P.G. (1982) *NMR Imaging in Biomedicine*, Academic Press, New York.

2 Blümich, B. (2005) *Essential NMR*, Springer, Heidelberg, Berlin.

3 Callaghan, P.T. (2011) *Translational Dynamics and Magnetic Resonance*, University Press, Oxford.

4 Hausser, K.H. and Kalbitzer, H.R. (1991) *NMR in Medicine and Biology*, Springer, Heidelberg.

5 Kimmich, R. (1997) *NMR Tomography, Diffusometry, Relaxometry*, Springer, Berlin.

6 Garroway, A.N., Grannell, P.K., and Mansfield, P. (1974) *J. Phys. C*, 7, L457.

7 Lauterbur, P.C. (1973) *Nature*, **40**, 149.
8 Heink, W., Kärger, J., and Pfeifer, H. (1978) *Chem. Eng. Sci.*, **33**, 1019.
9 Kärger, J., Pfeifer, H., and Heink, W. (1988) *Adv. Magn. Reson.*, **12**, 1.
10 Kärger, J., Pfeifer, H., and Heink, W. (1988) *Adv. Magn. Reson.*, **12**, 2.
11 Kärger, J. (1985) *Adv. Colloid Interface*, **23**, 129.
12 Callaghan, P.T., Coy, A., MacGowan, D., Packer, K.J., and Zelaya, F.O. (1991) *Nature*, **351**, 467.
13 Cotts, R.M. (1991) *Nature*, **351**, 443.
14 Price, W.S. (2009) *NMR Studies of Translational Motion*, Cambridge University Press, Cambridge.
15 Kärger, J. and Heink, W. (1983) *J. Magn. Reson.*, **51**, 1.
16 Kärger, J., Seiffert, G., and Stallmach, F. (1993) *J. Magn. Reson. A*, **102**, 327.
17 Basser, P.J. and Pierpaoli, C. (1996) *J. Magn. Reson. B*, **111**, 209.
18 Norris, D.G. (2005) in *Diffusion Fundamentals* (eds J. Kärger, F. Grinberg, and P. Heitjans), Leipziger Universitätsverlag, Leipzig, p. 545.
19 Gladden, L.F., Mantle, M.D., and Sederman, A.J. (2008) in *Handbook of Heterogeneous Catalysis*, 2nd edn (eds G. Ertl, H. Knözinger, F. Schüth, and J. Weitkamp), Wiley-VCH Verlag GmbH, Weinheim, p. 1784.
20 Lysova, A.A. and Koptyug, I.V. (2010) *Chem. Soc. Rev.*, **39**, 4585.
21 Torres, A.M., Michniewicz, R.J., Chapman, B.E., Young, G.A.R., and Kuchel, P.W. (1998) *Magn. Reson. Imag.*, **16**, 423.
22 Price, W.S., Stilbs, P., and Soderman, O. (2003) *J. Magn. Reson.*, **160**, 139.
23 Valiullin, R., Naumov, S., Galvosas, P., Kärger, J., Woo, H.J., Porcheron, F., and Monson, P.A. (2006) *Nature*, **430**, 965.
24 Naumov, S., Valiullin, R., Galvosas, P., Karger, J., and Monson, P.A. (2007) *Eur. Phys. J.-Spec. Top.*, **141**, 107.
25 Naumov, S., Valiullin, R., Monson, P., and Kärger, J. (2008) *Langmuir*, **24**, 6429.
26 Valiullin, R., Kärger, J., and Gläser, R. (2009) *Phys. Chem. Chem. Phys.*, **11**, 2833.
27 Kärger, J. (2010) *Leipzig, Einstein, Diffusion*, Leipziger Universitätsverlag, Leipzig.
28 Holz, M., Heil, S.R., and Sacco, A. (2000) *Phys. Chem. Chem. Phys.*, **2**, 4740.
29 Stejskal, E.O. and Tanner, J.E. (1965) *J. Chem. Phys.*, **42**, 288.
30 Stilbs, P. (1987) *Prog. Nucl. Magn. Reson. Spectrosc.*, **19**, 1.
31 Komlosh, M.E. and Callaghan, P.T. (1998) *J. Chem. Phys.*, **109**, 10053.
32 Galvosas, P., Stallmach, F., Seiffert, G., Kärger, J., Kaess, U., and Majer, G. (2001) *J. Magn. Reson.*, **151**, 260.
33 Ulrich, K., Freude, D., Galvosas, P., Krause, C., Kärger, J., Caro, J., Poladli, P., and Papp, H. (2009) *Microporous Mesoporous Mater.*, **120**, 98.
34 Weybright, P., Millis, K., Campbell, N., Cory, D.G., and Singer, S. (1998) *Magn. Reson. Med.*, **39**, 337.
35 Maas, W.E., Laukien, D.G., and Cory, D.G. (1996) *J. Am. Chem. Soc.*, **118**, 13085.
36 Romanova, E.E., Krause, C.B., Stepanov, A.G., Schmidt, W., van Baten, J.M., Krishna, R., Pampel, A., Kärger, J., and Freude, D. (2008) *Solid State Nucl. Magn. Reson.*, **33**, 65.
37 Pampel, A., Kärger, J., and Michel, D. (2003) *Chem. Phys. Lett.*, **379**, 555.
38 Pampel, A., Fernandez, M., Freude, D., and Kärger, J. (2005) *Chem. Phys. Lett.*, **407**, 53.
39 Fernandez, M., Kärger, J., Freude, D., Pampel, A., van Baten, J.M., and Krishna, R. (2007) *Microporous Mesoporous Mater.*, **105**, 124.
40 Fernandez, M., Pampel, A., Takahashi, R., Sato, S., Freude, D., and Kärger, J. (2008) *Phys. Chem. Chem. Phys.*, **10**, 4165.
41 Romanova, E.E., Grinberg, F., Pampel, A., Karger, J., and Freude, D. (2009) *J. Magn. Reson.*, **196**, 110.
42 Stallmach, F. and Galvosas, P. (2007) *Ann. Rep. NMR Spectrosc.*, **61**, 51.
43 Skirda, V.D., Doroginizkij, M.M., Sundukov, V.I., Maklakov, A.I., Fleischer, G., Hausler, K.G., and Straube, E. (1988) *Makromol. Chem.-Rapid Commun.*, **9**, 603.
44 Bär, N.K., Kärger, J., Krause, C., Schmitz, W., and Seiffert, G. (1995) *J. Magn. Reson. Ser. A*, **113**, 278.

45 Galvosas, P., Stallmach, F., and Kärger, J. (2004) *J. Magn. Reson.*, **166**, 164.
46 Callaghan, P.T. (1990) *J. Magn. Reson.*, **88**, 493.
47 Callaghan, P.T. and Coy, A. (1992) *Phys. Rev. Lett.*, **68**, 3176.
48 Heink, W., Kärger, J., Seiffert, G., Fleischer, G., and Rauchfuss, J. (1995) *J. Magn. Reson. A*, **114**, 101.
49 Kimmich, R., Unrath, W., Schnur, G., and Rommel, E. (1991) *J. Magn. Reson.*, **91**, 136.
50 Fujara, F., Geil, B., Sillescu, H., and Fleischer, G. (1992) *Z. Phys. B*, **88**, 195.
51 Chang, I.Y., Fujara, F., Geil, B., Hinze, G., Sillescu, H., and Tolle, A. (1994) *J. Non-Cryst. Solids*, **172**, 674.
52 Chang, I., Hinze, G., Diezemann, G., Fujara, F., and Sillescu, H. (1996) *Phys. Rev. Lett.*, **76**, 2523.
53 Geil, B., Isfort, O., Boddenberg, B., Favre, D.E., Chmelka, B.F., and Fujara, F. (2002) *J. Chem. Phys.*, **116**, 2184.
54 Gutsze, A., Masierak, W., Geil, B., Kruk, D., Pahlke, H., and Fujara, F. (2005) *Solid State Nucl. Magn. Reson.*, **28**, 244.
55 Feiweier, T., Geil, B., Pospiech, E.-M., Fujara, F., and Winter, R. (2000) *Phys. Rev. E*, **62**, 8182.
56 Kärger, J., Pfeifer, H., and Rudtsch, S. (1989) *J. Magn. Reson.*, **85**, 381.
57 Cotts, R.M., Hoch, M.J.R., Sun, T., and Markert, J.T. (1989) *J. Magn. Reson.*, **83**, 252.
58 Wu, D.H., Chen, A.D., and Johnson, C.S. (1995) *J. Magn. Reson. A*, **115**, 260.
59 Sun, P.Z., Seland, J.G., and Cory, D. (2003) *J. Magn. Reson.*, **161**, 168.
60 Kärger, J. and Jobic, H. (1991) *Colloid Surf.*, **58**, 203.
61 Stallmach, F., Kärger, J., and Pfeife, H. (1993) *J. Magn. Reson.*, **102**, 270.
62 Schwarz, H.B., Ernst, H., Ernst, S., Kärger, J., Röser, T., Snurr, R.Q., and Weitkamp, J. (1995) *Appl. Catal. A*, **130**, 227.
63 Schwarz, H.B., Ernst, S., Kärger, J., Knorr, B., Seiffert, G., Snurr, R.Q., Staudte, B., and Weitkamp, J. (1997) *J. Catal.*, **167**, 248.
64 Stilbs, P. (1987) *Prog. Nucl. Magn. Res. Spectrosc.*, **19**, 1.
65 Kärger, J., Pfeifer, H., Rudtsch, S., Heink, W., and Gross, U. (1988) *J. Fluorine Chem.*, **39**, 349.
66 Rittig, F., Coe, C.G., and Zielinski, J.M. (2002) *J. Am. Chem. Soc.*, **124**, 5264.
67 Rittig, F., Coe, C.G., and Zielinski, J.M. (2003) *J. Phys. Chem. B*, **107**, 4560.
68 Lorenz, P., Bülow, M., and Kärger, J. (1984) *Colloid Surf.*, **11**, 353.
69 Dvoyashkin, M., Valiullin, R., Kärger, J., Einicke, W.D., and Gläser, R. (2007) *J. Am. Chem. Soc.*, **129**, 10344.
70 Dvoyashkin, M., Valiullin, M., and Kärger, J. (2007) *Phys. Rev. E*, **75**, 41202.
71 Callaghan, P.T. (1984) *Aust. J. Phys.*, **37**, 359.
72 Schüth, F., Sing, K.S.W., and Weitkamp, J. (2002) *Handbook of Porous Solids*, Wiley-VCH Verlag GmbH, Weinheim.
73 Meynen, V., Cool, P., and Vansant, E.F. (2009) *Microporous Mesoporous Mater.*, **125**, 170.
74 Beck, J.S., Vartuli, J.C., Roth, W.J., Leonowicz, M.E., Kresge, C.T., Schmitt, K.D., Chu, C.T.W., Olson, D.H., Sheppard, E.W., McCullen, S.B., Higgins, J.B., and Schlenker, J.L. (1992). *J. Am. Chem. Soc.*, **114**, 10834.
75 Oberhagemann, U., Jeschke, M., and Papp, H. (1999) *Microporous Mesoporous Mater.*, **33**, 165.
76 Hansen, E.W., Courivaud, F., Karlsson, A., Kolboe, S., and Stöcker, M. (1998) *Microporous Mesoporous Mater.*, **22**, 309.
77 Courivaud, F., Hansen, E.W., Karlsson, A., Kolboe, S., and Stöcker, M. (2000) *Microporous Mesoporous Mater.*, **6**, 327.
78 Matthae, F.P., Basler, W.D., and Lechert, H. (1998) *Stud. Surf. Sci. Catal.*, **117**, 301.
79 Stallmach, F., Graser, A., Kärger, J., Krause, C., Jeschke, M., Oberhagemann, U., and Spange, S. (2001) *Microporous Mesoporous Mater.*, **44**, 745.
80 Stallmach, F., Kärger, J., Krause, C., Jeschke, M., and Oberhagemann, U. (2000) *J. Am. Chem. Soc.*, **122**, 9237.
81 Zhao, D.Y., Huo, Q.S., Feng, J.L., Chmelka, B.F., and Stucky, G.D. (1998) *J. Am. Chem. Soc.*, **120**, 6024.
82 Naumov, S., Valiullin, R., Kärger, J., Pitchumani, R., and Coppens, M.O. (2008) *Microporous Mesoporous Mater.*, **110**, 37.

83 Pitchumani, R., Li, W., and Coppens, M.O. (2005) *Catal. Today*, **105**, 618.
84 Ulrich, K., Galvosas, P., Kärger, J., and Grinberg, F. (2009) *Phys. Rev. Lett.*, **102**, 37801.
85 Ulrich, K. (2008) Diffusionsmessungen in nanostrukturierten Systemen mit PFG NMR unter Einsatz von ultra-hohen gepulsten magnetischen Feldgradienten. PhD Thesis, Leipzig.
86 Ulrich, K., Galvosas, P., Kärger, J., Grinberg, F., Vernimmen, J., Meynen, V., and Cool, P. (2010) *J. Phys. Chem. B*, **114**, 4223.
87 Mitra, P.P., Sen, P.N., Schwartz, L.M., and Ledoussal, P. (1992) *Phys. Rev. Lett.*, **68**, 3555.
88 Mitra, P.P., Sen, P.N., and Schwartz, L.M. (1993) *Phys. Rev. B*, **47**, 8565.
89 Mitra, P.P.H.B.I. (1995) *J. Magn. Reson. Ser. A*, **113**, 94.
90 Mair, R.W., Rosen, M.S., Wang, R., Cory, D.G., and Walsworth, R.L. (2002) *Magn. Reson. Chem.*, **40**, S29–S39.
91 Mair, R.W., Sen, M.N., Hurlimann, M.D., Patz, S., Cory, D.G., and Walsworth, R.L. (2002) *J. Magn. Reson.*, **156**, 202.
92 Mair, R.W., Wang, R., Rosen, M.S., Candela, D., Cory, D.G., and Walsworth, R.L. (2003) *Magn. Reson. Imag.*, **21**, 287.
93 Stallmach, F., Vogt, C., Kärger, J., Helbig, K., and Jacobs, F. (2002) *Phys. Rev. Lett.*, **88**, 105505.
94 Stallmach, F. and Kärger, J. (2003) *Phys. Rev. Lett.*, **90**, 39602.
95 Geier, O., Snurr, R.Q., Stallmach, F., and Kärger, J. (2004) *J. Chem. Phys.*, **120**, 1.
96 Fleischer, G. and Fujara, F. (1994) *NMR Basic Principles Prog.*, **30**, 159.
97 Appel, M., Fleischer, G., Geschke, D., Kärger, J., and Winkler, M. (1996) *J. Magn. Reson. Ser. A.*, **122**, 248.
98 Price, W.S. (2005) in *Diffusion Fundamentals* (eds J. Kärger, F. Grinberg, and P. Heitjans), Leipziger Universitätsverlag, Leipzig, p. 490.
99 Price, W.S. (2005) *Diffusion Fundam.*, **2**, 112.
100 Larkin, T.J., Pages, G., Torres, A.M., and Kuchel, P.W. (2009) in *Diffusion Fundamentals III* (eds C. Chmelik, N.K. Kanellopoulos, J. Kärger, and D. Theodorou), Leipziger Universitätsverlag, Leipzig, p. 409.
101 Kärger, J., Kocirik, M., and Zikanova, A. (1981) *J. Colloid Interface Sci.*, **84**, 240.
102 Kärger, J., Pfeifer, H., Riedel, E., and Winkler, H. (1973) *J. Colloid Interface Sci.*, **44**, 187.
103 Kärger, J. and Volkmer, P. (1980) *J. Chem. Soc., Faraday Trans. I*, **76**, 1562.
104 D'Orazio, F., Bhattacharja, S., Halperin, W.P., and Gerhardt, R. (1989) *Phys. Rev. Lett.*, **63**, 43.
105 D'Orazio, F., Bhattacharja, S., Halperin, W.P., and Gerhardt, R. (1990) *Phys. Rev. B*, **42**, 6503.
106 Valiullin, R.R., Skirda, V.D., Stapf, S., and Kimmich, R. (1997) *Phys. Rev. E*, **55**, 2664.
107 McDaniel, P.L., Coe, C.G., Kärger, J., and Moyer, J.D. (1996) *J. Phys. Chem.*, **100**, 16263.
108 Geier, O., Vasenkov, S., and Kärger, J. (2002) *J. Chem. Phys.*, **117**, 1935.
109 Burganos, V.N. (1998) *J. Chem. Phys.*, **109**, 6772.
110 Malek, K. and Coppens, M.O. (2001) *Phys. Rev. Lett.*, **8712**, 125505.
111 Malek, K. and Coppens, M.O. (2002) *Colloid Surf. A*, **206**, 335.
112 Malek, K. and Coppens, M.O. (2003) *J. Chem. Phys.*, **119**, 2801.
113 Zalc, J.M., Reyes, S.C., and Iglesia, E. (2004) *Chem. Eng. Sci.*, **59**, 2947.
114 Krutyeva, M. and Kärger, J. (2008) *Langmuir*, **24**, 10474.
115 Kärger, J. and Pfeifer, H. (1987) *Zeolites*, **7**, 90.
116 Feller, W. (1970) *An Introduction to Probability Theory and its Applications*, John Wiley & Sons, Inc., New York.
117 Meier, C., Dreher, W., and Leibfritz, D. (2003) *Magnet. Reson. Med.*, **50**, 510.
118 Adalsteinsson, T., Dong, W.F., and Schönhoff, M. (2004) *J. Phys. Chem. B*, **108**, 20056.
119 Hindmarsh, J.P., Su, J.H., Flanagan, J., and Singh, H. (2005) *Langmuir*, **21**, 9076.
120 Nilsson, M., Alerstam, E., Wirestam, R., Stahlberg, F., Brockstedt, S., and Lätt, J. (2010) *J. Magn. Reson.*, **206**, 59.
121 Kärger, J. (1969) *Ann. Phys.*, **24**, 1.
122 Kärger, J. (1971) *Z. Phys. Chem., Leipzig*, **248**, 27.

123 Kärger, J. (1971) *Ann. Phys.*, **27**, 107.
124 Bauer, M., Valiullin, R., Radons, G., and Kärger, J. (2011) *J. Chem. Phys.*, **135**, 144118.
125 Kärger, J. (1982) *AICHE J.*, **28**, 417.
126 Ruthven, D.M. (1984) *Principles of Adsorption and Adsorption Processes*, New York.
127 Barrer, R.M. (1978) *Zeolites and Clay Minerals as Sorbents and Molecular Sieves*, Academic Press, London.
128 Kocirik, M. and Zikanova, A. (1975) *Ind. Eng. Chem. Fundam.*, **13**, 347.
129 Richter, R., Seidel, R., Kärger, J., Heink, H., Pfeifer, H., Fürtig, H., Höse, W., and Roscher, W. (1986) *Z Phys. Chem. - Leipzig*, **267**, 1145.
130 Crank, J. (1975) *The Mathematics of Diffusion*, Clarendon Press, Oxford.
131 Hibbe, F., Chmelik, C., Heinke, L., Li, J., Ruthven, D.M., Tzoulaki, D., and Kärger, J. (2011) *J. Am. Chem. Soc.*, **133**, 2804.
132 Krutyeva, M., Yang, X., Vasenkov, S., and Kärger, J. (2007) *J. Magn. Reson.*, **185**, 300.
133 Krutyeva, M., Vasenkov, S., Yang, X., Caro, J., and Kärger, J. (2007) *Microporous Mesoporous Mater.*, **104**, 89.
134 Förste, C., Kärger, J., Pfeifer, H., Riekert, L., Bulow, M., and Zikanova, A. (1990) *J. Chem. Soc., Faraday Trans. I*, **86**, 881.
135 Kärger, J. and Caro, J. (1977) *J. Chem. Soc., Faraday Trans. I*, **73**, 1363.
136 Kärger, J. and Ruthven, D.M. (1989) *Zeolites*, **9**, 267.
137 Förste, C., Kärger, J., and Pfeifer, H. (1990) *J. Am. Chem. Soc.*, **112**, 7.
138 Caro, J., Hocevar, S., Kärger, J., and Riekert, L. (1986) *Zeolites*, **6**, 213.
139 Kärger, J. and Ruthven, D.M. (1992) *Diffusion in Zeolites and Other Microporous Solids*, John Wiley & Sons, Inc., New York.
140 Rittig, F., Farris, T.S., and Zielinski, J.M. (2004) *AICHE J.*, **50**, 589.
141 Dvoyashkin, M., Khokhlov, A., Naumov, S., and Valiullin, R. (2009) *Microporous Mesoporous Mater.*, **125**, 58.
142 Kärger, J., Rauscher, M., and Torge, H. (1980) *J. Colloid Interface Sci.*, **76**, 525.
143 Förste, C., Kärger, J., and Pfeifer, H. (1989) in *Zeolites: Facts, Figures, Future. 8th International Zeolite Conf* (eds P.A. Jacobs and R.A. van Santen), Elsevier, Amsterdam, p. 907.
144 Wintterlin, J., Schuster, R., and Ertl, G. (1996) *Phys. Rev. Lett.*, **77**, 123.
145 Renisch, S., Schuster, R., Wintterlin, J., and Ertl, G. (1999) *Phys. Rev. Lett.*, **82**, 3839.
146 Ertl, G. (2008) *Angew. Chem. Int. Ed.*, **47**, 3524.
147 Moerner, W.E. and Kador, L. (1989) *Phys. Rev. Lett.*, **62**, 2535.
148 Orrit, M. and Bernard, J. (1990) *Phys. Rev. Lett.*, **65**, 2716.
149 Bräuchle, C., Lamb, D.C. and Michaelis, J. (eds) (2010) *Single Particle Tracking and Single Molecule Energy Transfer*, Wiley-VCH Verlag GmbH, Weinheim.
150 Michaelis, J. and Bräuchle, C. (2010) *Chem. Soc. Rev.*, **39**, 4731.
151 Hellriegel, C., Kirstein, J., Bräuchle, C., Latour, V., Pigot, T., Olivier, S., Lacombe, S., Brown, R., Guieu, V., Payrastre, C., Izquierdo, A., and Mocho, P. (2004). *J. Phys. Chem. B*, **108**, 14699.
152 Hellriegel, C., Kirstein, J., and Bräuchle, C. (2005) *New J. Phys.*, **7**, 1.
153 Song, Y., Srinivasarao, M., Tonelli, A., Balik, C.M., and McGregor, R. (2000) *Macromolecules*, **33**, 4478.
154 Ahmed, M. and Pyle, D.L. (1999) *J. Chem. Technol. Biotechnol.*, **74**, 193.
155 Cvetkovic, A., Picioreanu, C., Straathof, A.J.J., Krishna, R., and van der Wielen, L.A.M. (2005) *J. Phys. Chem. B*, **109**, 10561.
156 Cvetkovic, A., Picioreanu, C., Straathof, A.J.J., Krishna, R., and van der Wielen, L.A.M. (2005) *J. Am. Chem. Soc.*, **123**, 875.
157 Martinez, V.M., de Cremer, G., Roeffaers, M.B.J., Sliwa, M., Baruah, M., de Vos, D.E., Hofkens, J., and Sels, B.F. (2008) *J. Am. Chem. Soc.*, **130**, 13192.
158 Stavitski, E., Kox, M.H.F., and Weckhuysen, B.M. (2007) *Chem. Eur. J.*, **13**, 7057.
159 Jung, C., Schwaderer, P., Dethlefsen, M., Köhn, R., Michaelis, J., and Bräuchle, C. (2011) *Nat. Nanotechnol.*, **6**, 87.
160 de Cremer, G. (2010) In situ monitoring of dynamics in redox chemistry by

fluorescence micro- and nanoscopy. PhD thesis, Leuven.
161 de Cremer, G., Roeffaers, M.B.J., Bartholomeeusen, E., Lin, K.F., Dedecker, P., Pescarmona, P.P., Jacobs, P.A., de Vos, D.E., Hofkens, J., and Sels, B.F. (2010) *Angew. Chem. Int. Ed.*, **49**, 908.
162 Ries, J. and Schwille, P. (2008) *Phys. Chem. Chem. Phys.*, **10**, 3487.
163 Eigen, M. and Rigler, R. (1994) *Proc. Natl. Acad. Sci. USA*, **91**, 5740.
164 Lubelski, A. and Klafter, J. (2009) *Biophys. J.*, **96**, 2055.
165 Szymanski, J. and Weiss, M. (2009) *Phys. Rev. Lett.*, **103**, 38102.
166 Weiss, M., Elsner, M., Kartberg, F., and Nilsson, T. (2004) *Biophys. J.*, **87**, 3518.
167 Saxton, M.J. (2008) *Nat. Methods*, **5**, 671.
168 Forstner, M., Martin, D., Navar, A.M., and Käs, J. (2003) *Langmuir*, **19**, 4876.
169 Jaqaman, K., Loerke, D., Mettlen, M., Kuwata, H., Grinstein, S., Schmid, S.L., and Danuser, G. (2008) *Nat. Methods*, **5**, 695.
170 Serge, A., Bertaux, N., Rigneault, H., and Marguet, D. (2008) *Nat. Methods*, **5**, 687.
171 Trenkmann, I., Täuber, D., Bauer, M., Schuster, J., Bok, S., Gangopadhyay, S., and von Borczyskowski, C. (2009). *Diffusion Fundam. (online journal)*, **11**, article 108, 1–12.
172 Lubelski, A., Sokolov, I.M., and Klafter, J. (2008) *Phys. Rev. Lett.*, **100**.
173 He, Y., Burov, S., Metzler, R., and Barkai, E. (2008) *Phys. Rev. Lett.*, **101**, 58101.
174 Bauer, M., Heidernätzsch, M., Täuber, D., Schuster, J., von Borczyskowski, C., and Radons, G. (2009) in *Diffusion Fundamentals III* (eds C. Chmelik, N.K. Kanellopoulos, J. Kärger, and D. Theodorou), Leipziger Universitätsverlag, Leipzig, p. 441.
175 Feil, F., Naumov, S., Michaelis, J., Valiullin, R., Enke, D., Kärger, J., and Bräuchle, C. (2011) *Angew. Chem. Int. Ed.*, **50**, 9788–9790.
176 Jung, C., Michaelis, J., Ruthardt, N., and Bräuchle, C. (2009) *Diffusion Fundam.*, **11**, article 68, 1–18. http://www.uni-leipzig.de/diffusion/pdf/volume11/diff_fund_11(2009)108.pdf.
177 Zürner, A., Kirstein, J., Döblingern, M., Bräuchle, C., and Bein, T. (2007) *Nature*, **450**, 705.
178 Kirstein, J., Platschek, B., Jung, C., Brown, R., Bein, T., and Bräuchle, C. (2007) *Nat. Mater.*, **6**, 303.
179 Brinker, C.J., Lu, Y., Sellinger, A., and Fan, H. (1999) *Adv. Mater.*, **11**, 579.
180 Jung, C., Müller, B.K., Lamb, D.C., Nolde, F., Müllen, K., and Bräuchle, C. (2006) *J. Am. Chem. Soc.*, **128**, 5283.
181 Kirstein, J.U. (2007) Diffusion of single molecules in nanoporous mesostructured materials, PhD Thesis, Ludwig Maximilian University, Munich.

12
Imaging of Transient Concentration Profiles

As an option, intermediate between the microscopic observation of molecular propagation considered in Chapters 10 and 11 and the macroscopic measurement of sorption rates considered in Chapters 13 and 14, information on guest diffusion in nanoporous materials may also be gained by following the transient concentration profiles. Variation of the concentration profiles may be initiated by any change in the surrounding atmosphere, including an increase or decrease of pressure (or partial pressure). The simpler situations involve single-component uptake or release but one may also consider the development of concentration profiles under the conditions of multicomponent adsorption, including tracer exchange, co- and counter-diffusion and chemical reactions as important special cases among the large spectrum of phenomena that are amenable to study by such methods.

In this chapter we consider several physical phenomena and their potential application to measurements from which this type of information may be obtained. In most cases, the spatial resolution is too small to allow the determination of concentration profiles within the individual particles (crystallites) of the porous material. Therefore, the measurements can provide concentration profiles only at the scale of a macroscopic particle or within an assemblage of particles in a bed or column. Under such conditions, information about intracrystalline diffusivities and surface permeabilities can only be deduced by using appropriate mathematical models as considered in Chapters 6, 13, and 14. The reliability of the diffusivity values determined by such indirect methods obviously decreases with the number and complexity of the steps required to relate the measurement to the elementary processes of molecular transport.

Therefore (in Sections 12.2 and 12.3), we give particular emphasis to the two recent techniques, interference and IR microscopy (IFM and IRM), which have sufficient spatial resolution to allow the measurement of transient intracrystalline concentration profiles and thus to allow the direct study of intracrystalline transport phenomena in nanoporous materials. In a short time these techniques have provided an impressive array of detailed information, much of which is not accessible by other direct experimental techniques.

The present chapter is restricted to illustrating the physical principles of the measurement techniques and the main messages that they can provide. Details of the

Diffusion in Nanoporous Materials. Jörg Kärger, Douglas M. Ruthven, and Doros N. Theodorou.
© 2012 Wiley-VCH Verlag GmbH & Co. KGaA. Published 2012 by Wiley-VCH Verlag GmbH & Co. KGaA.

mathematical analysis and examples of the experimental evidence from such studies are presented elsewhere within this book.

12.1
Different Options of Observation

Monitoring guest distribution in nanoporous materials and its time variation necessitates guest-sensitive observation. Very different ranges of the electromagnetic spectrum may be applied for this purpose (Figure 12.1). Rapid progress in instrumentation over the last few years has provided us with a wealth of techniques revealing spatial heterogeneities in the guest distribution in porous solids. Benefitting from the potentials of spectroscopic analyses, these techniques are often able to provide information about the nature of the host–guest interaction, in addition to the local

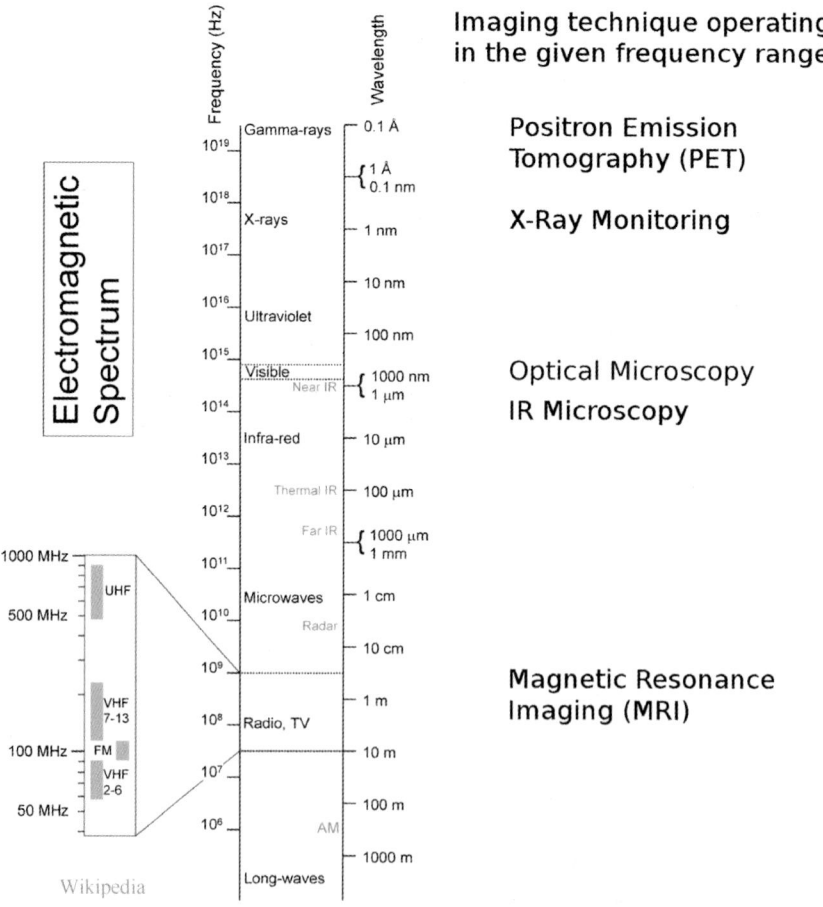

Figure 12.1 Allocation of the frequency ranges of different techniques of monitoring transient concentration profiles in the electromagnetic spectrum. Adapted from Reference [1].

concentrations. Reference [2] reviews this impressive development. Depending on the mechanisms exploited for particle monitoring, the space and time scales of observation may vary dramatically between the different techniques. For the same reason, there are also significant differences in the sensitivity of the different methods, that is, the guest densities required for observation. Within the context of this book, the techniques of greatest interest are those that are applicable for monitoring the evolution of transient intracrystalline concentration profiles. Such options are extensively provided by optical techniques, in particular by interference microscopy (IFM) and infrared microscopy (IRM). These options, which have been developed only recently, are discussed in detail in Sections 12.2 and 12.3. The present section is limited to an introduction to the principles of the different measuring techniques.

12.1.1
Positron Emission Tomography (PET)

The application of position emission tomography (PET) to monitoring concentration profiles in beds of porous particles was pioneered by R. A. van Santen [3–7]. Positrons are referred to as β^+ particles and represent the antimatter equivalent of electrons, that is, of the particles of $\beta^{(-)}$ radiation. They are emitted by nuclei of insufficient neutron content. In this way the nuclei enhance their neutron number at the expense of their protons and attain a stable state. Among the positron-emitting isotopes, ^{11}C, ^{13}N, and ^{15}O are of particular interest for studying guest dynamics in nanoporous materials. With half-life times in the range of minutes, these isotopes have to be produced on-site, normally by high-energy irradiation of an appropriate target material with protons or deuterons.

Since positrons and electrons are antiparticles, their encounters eventually lead to their annihilation, accompanied by the generation of two γ photons. Momentum conservation requires that the photons are emitted in opposite directions. Coincident photon detection by correspondingly arranged scintillators may thus be used to determine their origin and therefore for particle observation. Owing to their large energy (511 keV) the γ photons may pass through several millimeters of stainless steel thus allowing *in situ* observation even under reaction conditions.

The development in computer hard- and software and of a new breed of small self-shielding cyclotrons in the 1980s has made PET facilities widely accessible, leading to their exploitation for problems of industrial interest [3, 8, 9]. In reactors and separation columns, the information contained in 1D profiles in the axial direction is of great importance for an assessment of molecular transport and conversion. This special type of imaging has been referred to as positron emission profiling (PEP) – see Section 14.5.4. Reference [4] gives an overview of this technique and its application to studying mass transfer in beds of zeolites.

12.1.2
X-Ray Monitoring

A classical procedure for monitoring the concentration of guest molecules is based on the use of molecules containing atoms that show much stronger absorption of X-rays in comparison with the atoms of the host lattice. The guest distribution may then be followed in a straightforward way from the X-ray image of the adsorbate-adsorbent system under study. Experiments of this type have been performed with microporous carbon (MSC-5A from Takeda) [10, 11] using bromobenzene as the guest molecule. The adsorbent was in the form of cylindrical pellets 4 mm in diameter and 4–6 mm long. The cylindrical surface was sealed by an impermeable polymeric resin so that mass transfer was simplified to 1D diffusion in the axis direction. Characteristic changes in the profile evolution with increasing guest pressure during molecular uptake were found to nicely reflect the transition from macropore to micropore limitation in uptake kinetics (see also Section 12.1.5 and Figure 12.3 below).

12.1.3
Optical Microscopy

It is remarkable that the very first reported measurements of diffusion in zeolite crystals [12, 13] were carried out by using the special options offered by optical observation. The studies were performed with large crystals of natural heulandite, which, owing to the double refraction and the change in birefringence with varying concentration, allows the local concentration of the guest molecule (water) to be measured. Figure 12.2 shows schematically the essential features of

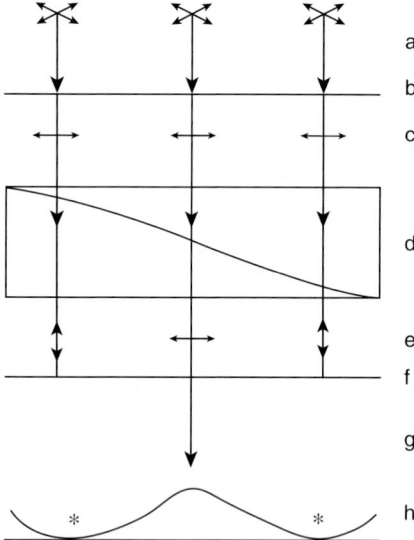

Figure 12.2 Schematic representation of the double refraction method. After passing through the polarizer (b) the light is polarized in plane (c). On passing through the crystal (d) the plane of polarization is rotated through an angle that depends on the water concentration. The second polarizer (f, the analyzer) passes only light polarized in plane (g) so the intensity distribution at (h) gives a series of dark bonds corresponding to certain concentrations of water within the crystal. Following the movement of these bands with time provides a direct way of following the propagation of the concentration profile within the crystal. (Method of Tiselius [12, 13].)

the experimental system: On passing through a non-isotropic crystal, such as heulandite, the plane of polarization of polarized light is rotated. The extent of this rotation depends on the thickness of the crystal through which the light is passed and on the water content so, in the region of the concentration front, a series of extinction bands is formed. Following the progress of the extinction bands through crossed Nicols in a polarizing microscope allows the progress of moisture concentration to be followed.

Ideally, a dry crystal would be exposed to moisture on one face only. Assuming that the guest diffusivity remains constant, the advancing concentration front would follow the solution for one-dimensional diffusion in a semi-infinite medium:

$$\frac{q(\infty)-q(z, t)}{q(\infty)-q(0)} = \mathrm{erfc}\left(\frac{z}{2\sqrt{Dt}}\right) \tag{12.1}$$

From this expression we may conclude that the distance z from the crystal boundary to a given extinction band [which is representative for a certain concentration $q(z, t)$] increases in proportion to $\sqrt{Dt}$, allowing a straightforward determination of the diffusivity. With the whole crystal being exposed to moisture, this approach is only applicable during the initial stage of uptake, since it is only in that region that there is no significant interference between the diffusion fronts from the different faces.

With the introduction of interference microscopy (Section 12.2), a further technique of optical observation has been successfully applied to diffusion studies in nanoporous material. The spectrum of information on mass transfer accessible by this novel technique greatly exceeds the options of the early optical measurements. Their application is, moreover, not restricted to birefringent host materials. In view of the new options offered for diffusion studies by interference microscopy, and of the dramatic progress in birefringence microscopy, the time appears to be ripe for a return to double refraction methods, which have been explored only in a limited way. Such efforts will surely benefit from the formalism that has been developed for analyzing and interpreting the data from interference and IR microscopy measurements (Section 12.3).

12.1.4
IR Microscopy

Since it can discriminate between different molecular species as well as giving direct information on their concentrations, IR spectroscopy provides a powerful technique for the observation of mass transfer phenomena in multicomponent systems. Following studies of pyridine uptake by Na- and H-mordenite [14], H. Karge and his coworkers extensively exploited the possibilities offered by the new options of Fourier transform IR spectroscopy (FTIR) to investigate transient sorption phenomena [15–20]. In addition to single-component studies, careful investigations of the behavior of multicomponent guest phases were also carried out, including the observation of co- and counter-diffusion (Section 13.3). The application of (FT)IR microscopy [also referred to as micro-(FT)IR spectroscopy] even allows uptake and release experiments with single zeolite crystals ("mesoscopic" diffusion measurements), in addition to the conventional "macroscopic" measurements with beds of crystals (Section 1.4 and Table 1.1) [21]. The options and limitations of these techniques are reviewed in References [18, 22, 23].

Following studies of the distribution of boron atoms in different [B]-ZSM-5 crystals [24] and of template removal in ZSM-5 [25], the first applications of IR microscopy to gain insight into the distribution (and orientation!) of guest molecules in single crystals were reported by Schüth and coworkers in the early 1990s [26–28]. Reference [29] reports the measurement of transient concentration profiles during the desorption of n-alkanes from single crystals of silicalite. The analysis and interpretation of these studies was, however, impeded by the fact that these measurements were not performed under well-defined external conditions. It was really in the PhD work of Chmelik [30] that IR microscopy was developed to become a reliable technique for microscopic diffusion measurements. IR microscopy (IRM) has been found to complement nicely the information provided by interference microscopy (IFM) and so, in Sections 12.2 and 12.3, the fundamental principles and potential applications of both these techniques are presented concurrently.

12.1.5
Magnetic Resonance Imaging (MRI)

In Section 10.1 it was shown that, under the influence of an external magnetic field B, the magnetic moment of a nuclear spin, in combination with its mechanical moment, gives rise to a precessional motion of this spin around the direction of this field (Figure 10.1), with a frequency proportional to the magnitude of the field [Eq. (10.10)]. The rotating macroscopic magnetization, being the vector sum of the magnetic moments of each individual nuclear spin, generates an NMR signal at exactly this frequency.

Both PFG NMR diffusion measurements and NMR imaging techniques (MRI) are based on the application of field gradients since, in an inhomogeneous magnetic field, the resonance frequency becomes a linear function of the space coordinate [see Eq. (11.2) and the scales in Figure 12.3]. In this way, the observed frequency dependence of the NMR signal intensity yields an image of the spatial dependence of signal intensity. Since the signal intensity is directly proportional to the number of nuclear spins, the NMR signals may thus be transferred into concentration maps. These maps provide a wealth of information concerning the molecular concentrations and this information is often further enriched by a detailed characterization of the recorded molecules with respect to their mobility and/or surroundings. This type of information is contained in the key parameters of NMR as presented in Section 10.1, including the nuclear magnetic relaxation times and chemical shifts as well as the local diffusivities.

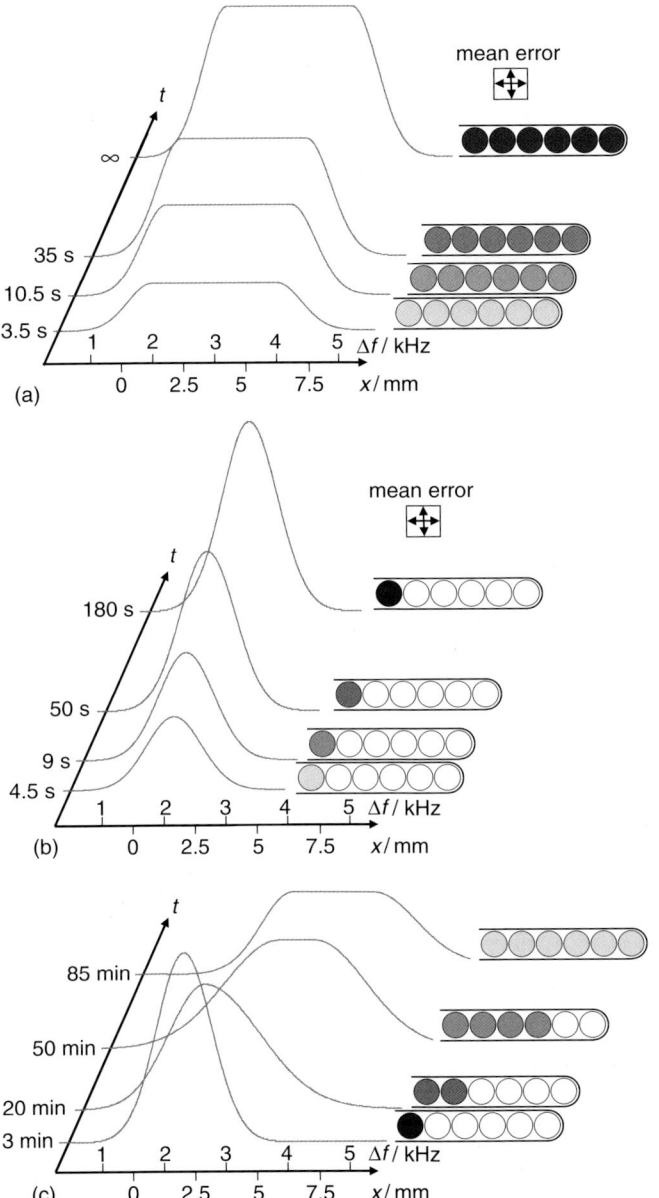

Figure 12.3 Evolution of the distribution of n-butane in a bed of crystals of zeolite NaCaA with crystal diameters of about 40 μm (a) and 4 μm [(b) and (c)] in an NMR sample tube in the axial (x-) direction as appearing in the NMR spectrum under the conditions of MRI (left-hand side) and visualized by cartoon plots (right-hand side). After activation, the sample has been brought into contact with a finite volume of gaseous n-butane (at $x=0$, the opening of the sample tube). The two scales visualize the equivalence of the spatial dimension in the gradient direction and the resonance frequency, as a common basis of the magnetic resonance imaging techniques [Eq. (11.2)]. Limitations in spatial resolution flatten ascending and descending parts in the profiles. Redrawn from Heink et al. [31].

These features have made NMR imaging the most powerful and ubiquitous imaging technique for medical diagnosis [32, 33], where it is also known as MR tomography. NMR imaging techniques are also widely applied in fluid mechanics and other areas of chemical engineering [34]. Detailed accounts of the application of MRI in heterogeneous catalysis have been given by Gladden and coworkers [35] and by Lysova and Koptyug [36].

In Section 10.1 it was noted that, in comparison with other spectroscopic techniques, NMR is relatively insensitive. This means that NMR imaging is still not able to measure the transient intracrystalline concentration profiles that are accessible by IR and interference microscopy. However, MRI can conveniently measure such profiles at the scale of an adsorbent particle, or a bed of crystals (see, for example, Figures 6.13 and 12.3). In these examples the characteristic differences in the evolution of the concentration profiles during ad- and desorption provide valuable guidelines for modeling the overall mass transfer process. It was in fact this application of NMR imaging (at that time, following Lauterbur's original terminology [37], referred to as NMR zeugmatography) that was explored during its first application in zeolite science and technology [31].

The patterns of concentration evolution are controlled by the relation between the time constants of bed uptake [$\tau_{bed} \sim L^2/D_{long\text{-}range}$, with L denoting the bed extension – see Eqs. (13.11) and (15.13)/(15.14)] and of uptake by the individual crystallites [$\tau_{intra} \sim R^2/D_{intra}$ – see Eq. (13.11)]. Thus, under the conditions leading to the patterns of Figure 12.3b and c (notably smaller crystallite radii) the ratio τ_{intra}/τ_{bed} is expected to assume much smaller values than under the conditions considered in Figure 12.3a. In complete agreement with this consideration, in Figure 12.3b the NaCaA crystals in the first layer are seen to be saturated very rapidly by the guest molecules. Only in a second, much slower process (Figure 12.3c) do the molecules, from this first layer, distribute through the whole bed of crystals. In the bed of large crystals (Figure 12.3a), however, the guest molecules are found to be distributed essentially instantaneously over the whole bed, with molecular uptake occurring essentially simultaneously in all crystals.

The evolution of the concentration profiles in Figure 12.3a may be taken as direct proof that, in this case, the overall uptake as observable by macroscopic methods is in fact limited by mass transfer phenomena within the individual crystals, including surface permeation and intracrystalline diffusion. Conversely, with the patterns of Figure 12.3b and c, one sees a situation where the assumption of uptake limitation by the individual crystallites would lead to completely erroneous conclusions.

Novel options are doubtless provided by the application of hyperpolarized guest molecules, including ^{129}Xe and ^{3}He [38–41]. As an alternative to the application of NMR, MR imaging may also be based on ESR (electron spin resonance) or EPR (electron paramagnetic resonance). The fundamentals of this technique, which is based on the spin of the electron shell rather than the nucleus, are similar to those presented in Section 10.1. However, the magnetic moments of the electron spins (and hence both their polarization and resonance frequency in a given magnetic field) exceed those of nuclear spins by three orders of magnitude. The associated dramatic gain in sensitivity allows the observation of much smaller numbers of molecules.

In this way, ESR imaging has been shown to attain spatial resolution in the range of micrometers – but still with accumulation times of hours [42, 43]. Although the instrumentation for ESR imaging is not yet as well developed as for NMR imaging, these resolutions exceed those attainable by NMR imaging over comparable accumulation times by at least one order of magnitude [42]. Since ESR imaging is subject to the requirement of non-vanishing electron magnetic moments, it is difficult to predict the extent to which, with further development of the instrumentation, ESR techniques, which have found many applications in biological fields [45], may find application in other areas, including the exploration of mass transfer in nanoporous materials.

12.2
Monitoring Intracrystalline Concentration Profiles by IR and Interference Microscopy

12.2.1
Principles of Measurement

Diffusion studies with nanoporous materials by both IFM and IR micro-imaging (IRM, also referred to as IR microscopy) are based on the measurement of concentration integrals in the direction of observation. The way in which the measurement is carried out is shown schematically in Figures 12.4 and 12.5.

IFM is based on the determination of the phase shift $\Delta\varphi^L$ between the light beams passing through the crystal and the corresponding beams that pass through the surrounding atmosphere (Figure 12.4d). Changes $\Delta(\Delta\varphi^L)$ in this difference (which appear as corresponding changes of the interference patterns [48, 49]) are caused by changes Δn of the refractive index, which are in turn a consequence of changing concentrations Δq within the crystal. To a first, linear approximation one therefore may note:

$$\Delta\left(\Delta\varphi^L(t)\right) \propto \int_0^L \Delta n(t) \mathrm{d}x \propto \int_0^L \Delta q(x, y, z, t) \mathrm{d}x \tag{12.2}$$

Spatial resolution in the plane perpendicular to observation is limited by the optical resolution of about 0.5 μm. As indicated by the signs of proportionality in Eq. (12.2), interference microscopy is unable to provide absolute concentration data. Therefore, the resulting concentration profiles are generally plotted in relative units. Information about absolute concentrations can be obtained from conventional (macroscopic) sorption measurement at the relevant guest pressures or by direct comparison with the results from IR micro-imaging (IRM). It may, in exceptional cases, also be found by comparison with the evidence of IRM that changes in $\Delta\varphi^L$ are caused by changes in the lattice and/or guest phase rather than in guest concentration.

In IRM, information about the concentration integrals in the observation direction is given by the area under a characteristic IR band of the guest molecule under study (top of Figure 12.5b). The IR microscope may operate in two modes.

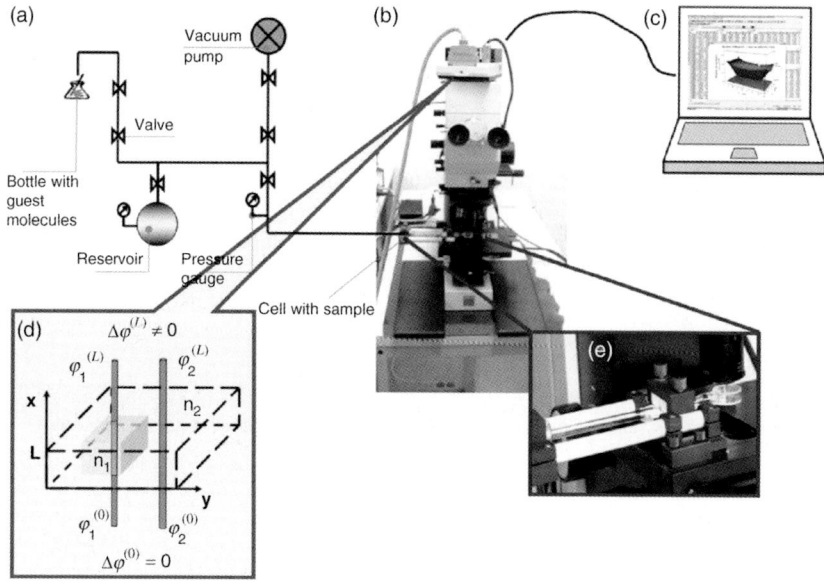

Figure 12.4 Experimental set-up and basic principle of the interference microscopy technique for diffusion measurements in nanoporous systems. (a) Schematic representation of the vacuum system (static) with optical cell containing the sample. (b) Interference microscope (Carl-Zeiss JENAPOL interference microscope with a Mach-Zehnder type interferometer [46]) with a CCD camera on top, directly connected to (c) the computer. (d) Basic principle: changes in the intracrystalline concentration during diffusion of guest molecules cause changes in the refractive index of the crystal (n_1) and, hence, in the phase difference $\Delta\varphi$ of the two beams. Having measured the difference of the optical path length one can evaluate the difference of the intracrystalline concentration by Eq. (12.2). (e) Close-up view of the optical cell containing the crystal under study. Reprinted from Reference [44] with permission.

By considering the spectrum of the whole crystal [using a single-element (SE) detector], the device is able to record the total uptake or release. This information is related to that of conventional transient sorption experiments, with the significant difference that one is now able to monitor uptake and release for a single crystal. Measurements of this type yield information intermediate between microscopic and macroscopic measurements and are, therefore, referred to as mesoscopic (Table 1.1) [50]. Thus, in the integral mode of IRM, information about the underlying diffusion phenomena can only be deduced indirectly, by applying an appropriate mathematical model to describe molecular uptake or release under the influence of the governing transport phenomena. However, even without resolving intracrystalline concentration profiles, single-crystal measurements have a great advantage in comparison with bed experiments: owing to the dramatically enlarged surface-to-volume ratio, transient sorption experiments with single crystals are essentially unaffected by the disturbing effects of heat release [51], which, for fast sorption processes in beds of zeolite crystallites [52], often becomes the rate-limiting step (Section 6.2.4).

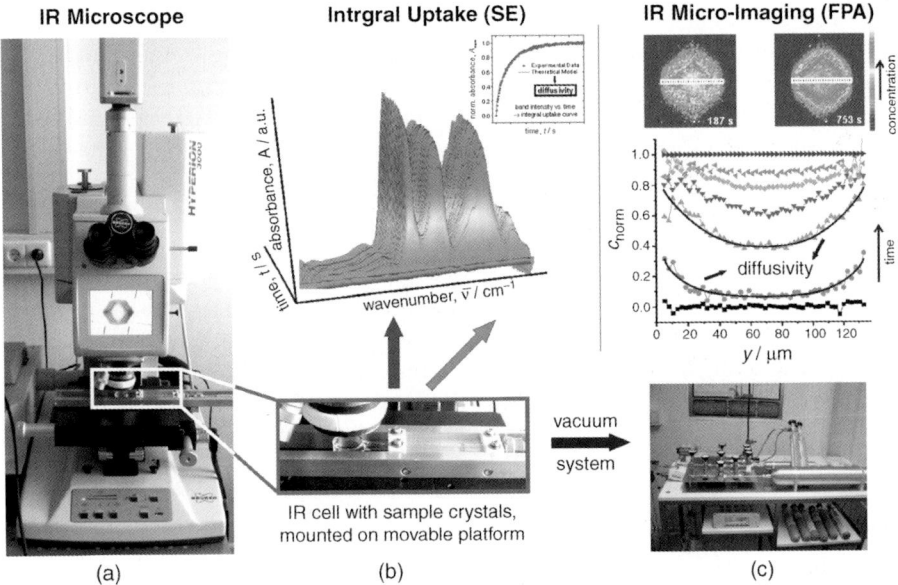

Figure 12.5 Experimental set-up of a Fourier-transform IR microscope (Bruker Hyperion 3000) composed of a spectrometer (Bruker Vertex 80v) and a microscope with a focal plane array (FPA) detector [47] as a device for micro-imaging, and basic principle of the IR microscopy technique: The optical cell is connected to a vacuum system and mounted on a movable platform under the microscope. Only one individual crystal is selected for the measurement. Changes in the area under the IR bands of the guest species are related to the guest concentration. The spectra can be recorded as signal integrated over the whole crystal (SE – single-element detector) or with a special resolution of up to 2.7 μm by the imaging detector (FPA – focal plane array). Reprinted from Reference [44] with permission.

As an alternative to the measurement of integral uptake, by the use of a focal plane array (FPA) detector, IRM also allows the spectra over each individual space element (top of Figure 12.5c) to be recorded, separately and simultaneously, with a minimum pixel size of 2.7 μm. In this way, IRM is able to follow the evolution of intracrystalline concentration profiles. In contrast to IFM, one has to take into account that the recorded data do not correspond directly to the concentration integrals over integration paths perpendicular to the plane of investigation. Rather, in an IR microscope the light is focused into the focal plane at a certain angle towards the observation direction (about 16° for the IR micro-imaging device Bruker Hyperion 3000). The resulting deviation from integration along a perpendicular integration path clearly becomes negligibly small if the extension of the crystal in the plane of observation notably exceeds the extension in the third dimension, that is, the direction of observation (see, for example, Reference [53]).

The FPA detector of the device shown in Figure 12.5 consists of an array of 128×128 single detectors, each 40 μm × 40 μm. By means of a 15× objective, in the focal plane, that is, at the position of the crystal under study, a resolution of 2.7 μm × 2.7 μm is gained. Each SE of the FPA detector records the IR signal.

The intensity of the IR light as a function of the wavelength, that is, the transmission spectrum, is determined by means of the spectrometer using Fourier transformation [54]. According to the Lambert–Beer law, the concentration of a particular molecular species is proportional to the intensity of the "absorption band," defined as the negative logarithm of the ratio between the transmission spectrum of the sample over the relevant frequency range and the corresponding background signal [23, 54]. By comparison with a standard it is therefore possible to gain information concerning the absolute number of molecules.

Both IFM and IRM operate under the conditions known from conventional macroscopic uptake and release experiments. Operating with a single crystal rather than a "macroscopic" crystal ensemble presents a particular challenge. This includes the disturbing influence of omnipresent impurity molecules, such as water, which may affect the sorption and diffusion properties of a small single crystal much more than for a macroscopic quantity of host particles.

12.2.2
Data Analysis

Both IFM and IRM are able to record the evolution of concentration profiles as exemplified in the center of Figure 12.5c. These profiles represent the integrals over concentration in the observation direction. For crystals with one- and two-dimensional pore networks extended perpendicular to the direction of observation, these integrals simply result as the product of the crystal thickness and the local concentration. In the usual case of constant thickness they therefore yield directly the local concentration. Among several possible ways to deduce, under such conditions, the relevant diffusivities from transient concentration profiles [55–57], the easiest approach is through the direct application of Fick's 1st law [Eq. (1.1)]. Assuming symmetric host crystals, the flux at position x and time t may be determined via:

$$j(x,t) \propto \frac{\int_0^x q(x',t_2)\,dx' - \int_0^x q(x',t_1)\,dx'}{t_2 - t_1} \qquad (12.3)$$

as the area (integral) between the concentration profiles $q(x,t)$ at the times t_1 and t_2 (with $t_1 < t < t_2$ and, ideally, $t_2 - t_1 \ll t$), from the crystal center to x. Following Fick's 1st law, the diffusivity is obtained by dividing the flux by the slope of the profiles.

As an alternative approach, by which asymmetric profiles may also be analyzed, the diffusivities can be determined by the microscopic application of Fick's 2nd law [Eq. (1.3)]:

$$\frac{\partial q}{\partial t} = \frac{\partial}{\partial x}\left(D\frac{\partial q}{\partial x}\right) = D\frac{\partial^2 q}{\partial x^2} + \frac{\partial D}{\partial q}\left(\frac{\partial q}{\partial x}\right)^2 \qquad (12.4)$$

In addition to the uncertainty in the determination of the second derivative of the concentration, this approach is complicated by the fact that the diffusivities appear also as their derivative $\partial D/\partial q$. However, one may eliminate this complication by considering the profiles at their minima where the disturbing term $\partial D/\partial q$ disappears (since at this point $\partial q/\partial x = 0$). The latter consideration is only necessary for transient sorption experiments (where D stands for the transport diffusivity and q denotes the total concentration). In self-diffusion experiments, that is, during tracer exchange, q denotes the concentration of the labeled molecules within a uniform overall concentration. Since the self-diffusivity is independent of the percentage of labeled molecules, the troublesome term $\partial D/\partial q$ then vanishes automatically.

The surface permeability α [in Eqs. (2.33), (2.34), and (6.2) also referred to as the film mass transfer coefficient (or permeance) k] is defined as the factor of proportionality between the particle flux through the surface, j, and the difference $q_{eq} - q_{surf}(t)$ between the intracrystalline concentration in equilibrium with the surrounding atmosphere and the actual concentration in the layer close to the crystal surface:

$$j_{surf} = \alpha \left(q_{eq} - q_{surf}(t) \right) \tag{12.5}$$

All quantities necessary for the direct experimental determination of surface permeabilities and, hence, of surface barriers (Section 2.3) are thus directly provided by the profiles [$q_{surf}(t)$ and $q_{eq} = q_{surf}(t=\infty)$] and Eq. (12.3) ($j_{surf}$, where the integration is extended from the crystal boundary to the crystal center).

The options considered so far for analysis of diffusion in one and two directions are also applicable to a three-dimensional system, provided that interference effects between the fronts propagating from the different crystal faces are negligibly small. This is always the case for sufficiently small observation times. However, with increasing time one is confronted with a typical inverse problem, namely, the determination of the "kernel" (i.e., the integrand) $q(x,y,z;t)$, which is known only through its integral $\int q(x,y,z;t) \, dx$ in the observation direction. Conventional techniques to convert such integrals into their kernels, that is, the integrands $q(x,y,z;t)$, with X-ray computed tomography (XCT) as the most prominent example, require that the integral may be taken over many orientations. IFM and IRM, however, offer only the option of observation in one direction (or in two directions if, in particular lucky cases, one succeeds in turning the crystal under study by 90° and then repeating the identical sorption experiment [56]).

In such a situation, the actual concentrations $q(x,y,z;t)$ were shown to become accessible by the application of a generic algorithm [58] starting, as a first approximation, with the solution of the diffusion equation (assuming constant diffusivities and surface permeabilities). This solution results as a series expansion (Section 6.2.2.2 and References [46, 51, 56, 57, 59]). In the subsequent iterations, the expansion parameters are varied to yield the best fit with the experimental data. This fit proved to be so perfect that, within experimental accuracy, the integrals over the simulated concentrations agreed with the results of the measurement [56, 59].

With the option to determine both the principal elements of the diffusion tensor and the surface permeabilities through each individual crystal face the separate determination of the contribution of any crystal face to the total uptake also becomes possible. Reference [57] summarizes various methods of acquiring this type of information from the primary data of the experiments and provides the results of a corresponding analysis for specimens of all-silica ferrierites.

Being able to distinguish between different molecular species, IRM is also able to record diffusion phenomena in multicomponent systems. In this way, the elements of the diffusion matrix as introduced by Eq. (3.22) also become accessible by direct measurement. This includes, in particular, the observation of the counter flux of differently labeled but otherwise identical molecules, yielding, via Eq. (1.11), the self- or tracer diffusivity.

12.3
New Options for Experimental Studies

The possibility of recording directly the transient concentration profiles by IFM and IRM has opened up a completely new field of direct experimental analysis. The impressive increase in the range of available nanoporous materials over the last few years [60] has also contributed to this development. In this section we summarize the main new possibilities for experimental studies that these techniques provide. Some of these are discussed in greater detail in Part Five of this book.

12.3.1
Monitoring Intracrystalline Concentration Profiles

The IRM and IFM techniques are unique in that they provide a clear and direct visualization of the intracrystalline transport process. In a general way these techniques follow the conventional procedure of solid-state diffusion measurements that are based on a layer-by-layer analysis of the concentration of the diffusants in the penetration direction [61–63]. However, the continuous, non-invasive measurement of the profiles over a complete crystal provides unique advantages that allow access to far more detailed information than can be obtained from more conventional solid-state diffusion measurements.

As an example of the power of the IFM technique and the new options that it presents for studying intracrystalline transport, Figures 12.6–12.8 show experimental data obtained for methanol in large crystals of all-silica ferrierite [46, 64, 65]. This system proved particularly suitable for detailed studies, since the ferrierite crystals under study [66] were extremely stable under the conditions of repeated ad- and desorption, allowing dozens of adsorption–desorption cycles to be repeated with the same crystal, without any perceptible change in the sorption and diffusion properties.

Figure 12.6 shows the host system under study and the evolution of the concentration integrals during uptake [44]. The ferrierite pore system allows diffusion in two directions (defined by the y and z coordinates, see Figure 12.6a), which are chosen to be perpendicular to the direction of observation. From structure analysis, the critical pore diameters in either direction (Figure 12.6a) are known to be determined by eight- and ten-membered silica rings. Figure 12.6b shows the measured concentration integrals at three instants of time after the start of an adsorption measurement.

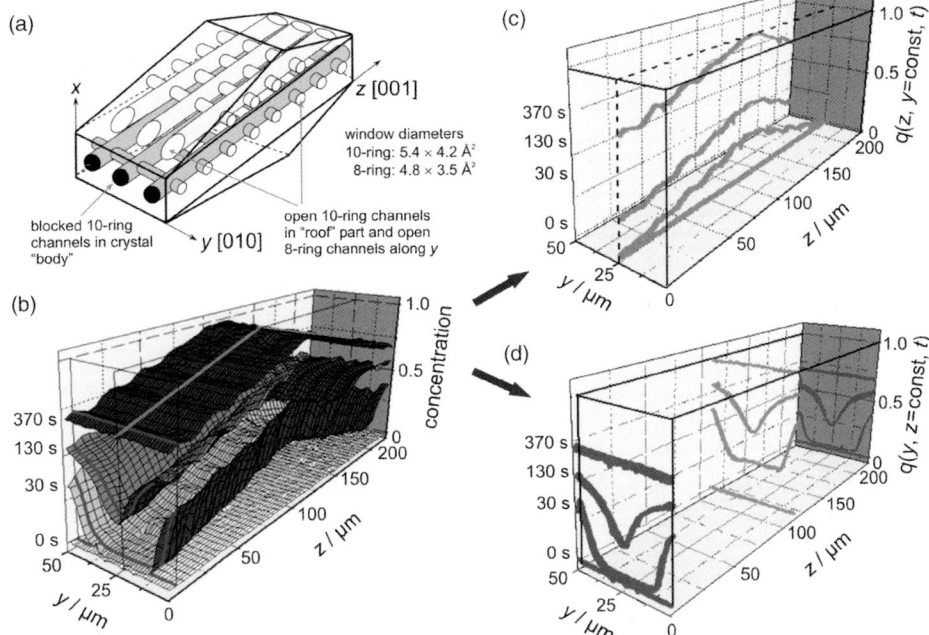

Figure 12.6 Transient profiles of methanol concentration for uptake in ferrierite (pressure step 0 → 80 mbar in the gas phase, 298 K). The 1D representations (c) and (d) are extracted from the 2D profiles (b) for a clearer view of the uptake process. The uptake process is dominated by transport along the smaller eight-ring channels (y direction). Except for the formation of a triangular "roof" at early times, we find no indications for uptake along the z direction. Obviously, the larger ten-ring channels are blocked in the crystal "body" and the molecules have to enter through the smaller eight-ring windows in the y direction [see (a)]. Reprinted from Reference [44] with permission.

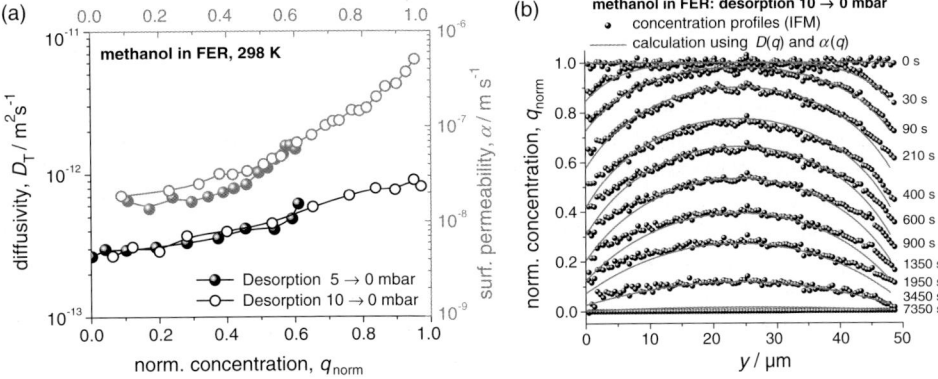

Figure 12.7 (a) Loading dependence of transport diffusivity and surface permeability as determined by local analysis of the concentration profiles for the desorption steps from 5 and 10 mbar to vacuum using Fick's second law (FER = ferrierite). (b) Comparison of the experimental and recalculated profiles for the desorption step from 10 to 0 mbar. The dots are the experimental concentration profiles; lines are calculated profiles using the concentration dependence of $D_T(q)$ and $\alpha(q)$ as shown in (a). Reprinted from Reference [64] with permission.

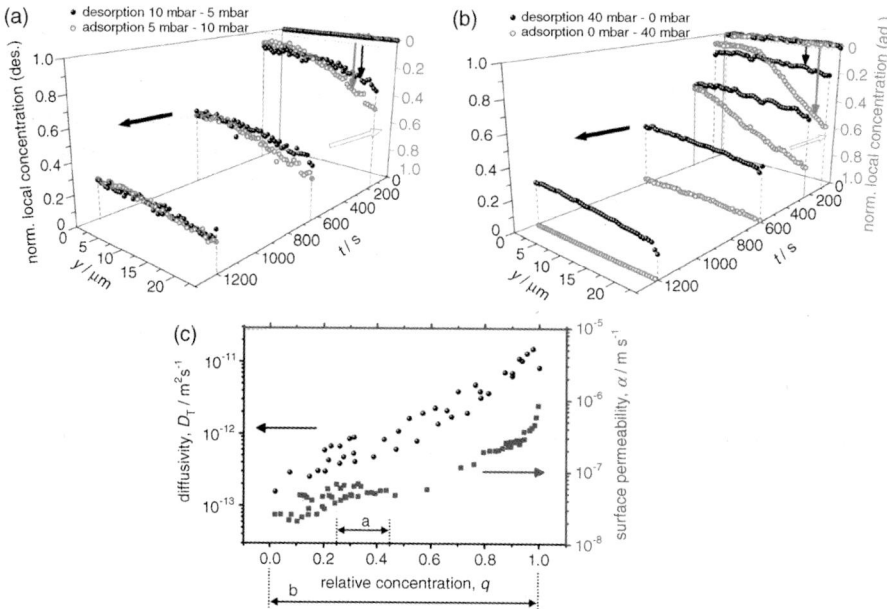

Figure 12.8 Methanol in ferrierite. The concentration dependence of the transport parameters [diffusivity and surface permeability (c)] is found to affect molecular uptake (open spheres) and release (full spheres) in strikingly different ways: For small loading steps [where the transport parameters do not dramatically change – see interval a in (c)] uptake and release proceed essentially complementary (a), for large loading steps adsorption is dramatically accelerated (b), while the desorption profiles remain similar to those for the small loading step. Note that the adsorption profiles in (a) and (b) are flipped with respect to the desorption profiles to allow their direct comparison. Reprented from Reference [67] with permission.

Selected concentration profiles in the y and z directions, taken from these profiles, are shown in Figure 12.6c and d. Clearly, from the presentation shown in Figure 12.6c a significant number of guest molecules is seen to penetrate the crystal very rapidly (essentially instantaneously). These molecules are assumed to fill the roof-like parts of the crystals that are found on top (and on the bottom) of the slab-like main body of the ferrierite crystal (Figure 12.6a). As also indicated in Figure 12.6a, this fast uptake suggests free access to the roof section through the ten-ring channels in the z direction, while there is evidently a substantial restriction to transport in this direction in the main body of the crystal. Such a restriction is necessary to explain why molecular uptake by the main crystal body occurs predominantly through the channels in the y direction, despite their smaller diameter.

The ferrierite crystals proved to be an excellent model system for considering transient sorption phenomena by diffusion through the eight-ring channels in the y direction. It is only necessary to subtract the contribution due to uptake by the roof-like sections (as immediately visible from Figure 12.6c) from the profiles of overall uptake shown in Figure 12.6d (which results in a simple shift of the base-line).

As an example, Figure 12.7b shows the transient concentration profiles during desorption, generated by a sudden decrease of the pressure in the surrounding methanol atmosphere from 10 mbar to zero. Notably, the boundary concentration (at $y = 0$ and 50 µm) does not assume the new equilibrium value $q = 0$ immediately after the onset of desorption. This indicates the existence of an additional transport resistance at the crystal surface that reduces the rate of molecular exchange between the surrounding atmosphere and the intracrystalline pore space [67]. Details of the formal kinetics for a system with internal diffusion and surface resistances may be found in Section 6.2.2.2 where, in Figure 6.4, concentration profiles as provided by IFM have been used for illustration. The limitations and possible modifications of these approaches for concentration-dependent diffusivities and surface permeabilities (as exemplified by Figure 12.7a) are discussed in Section 6.2.3.

The diffusivities and permeabilities displayed in Figure 12.7a yield the best fit between the experimentally determined concentration profiles (dots in Figure 12.7b) and the solutions of the diffusion equation [Fick's 2nd law, Eq. (12.4)], with Eq. (12.5) as the boundary condition [65]. The concentrations are normalized with respect to the equilibrium value attained with the largest pressure step. The permeabilities are plotted as a function of the given boundary concentrations. We shall return to this point in Sections 12.3.2 and 19.6, which refer in greater detail to the phenomenon of surface resistance. Notably, data analysis based on desorption with a smaller pressure step yields essentially the same values for both the diffusivities and surface permeabilities, confirming the self-consistency of the analysis.

Figure 12.8 displays the concentration profiles during desorption and adsorption in a unified representation by plotting the concentrations from bottom to top for desorption (left-hand scales in Figure 12.8a and b) and from top to bottom for adsorption (right-hand scale) [68]. For simplicity, only one half of the profiles (starting with $y = 0$ in the crystal center) is shown. During tracer exchange, to maintain overall equilibrium, at any instant of time, desorption of one component must be exactly counterbalanced by adsorption of the other. Assuming linearity in both the diffusion equation [Eq. (12.4)] and the boundary conditions [Eq. (12.5)], the same has to hold true for comparing adsorption and desorption over identical pressure steps. For sufficiently small pressure steps and, correspondingly, concentration steps, the diffusivities and surface permeabilities may be considered to be constant and the diffusion equation becomes linear. This is close to the situation shown in Figure 12.8a. The amounts adsorbed (exemplified, at $t = 200$ s, by the arrow towards open spheres) and desorbed (arrow towards full spheres) are similar to each other at any time and place. Figure 12.8c illustrates that, over the concentration range covered in the experiments, both the diffusivity and surface permeability do not vary significantly. This is totally different from the large pressure step which yields substantial differences between the transient profiles during adsorption and desorption (Figure 12.8b). Most remarkably, the rate of desorption is found to be essentially independent of the pressure step, while the rate of adsorption increases dramatically with the pressure step.

Formal analysis of the effect of the pressure step size on kinetics may be found in Section 6.2.3. The present finding may, however, be intuitively understood by recognizing that the overall process is dominated by molecular exchange between the host particle and the surroundings and, hence, by the magnitude of the diffusivities and permeabilities close to the surface. In the considered cases of desorption these values are essentially identical, leading to the observed similarity in the desorption patterns. During adsorption, however, an enhancement in the pressure step leads to a dramatic enhancement in both permeability and diffusivity close to the surface, which may easily be understood as the origin of the dramatic acceleration of the uptake process for larger pressure steps.

It is well known from classical sorption rate measurements that the measurement of adsorption and desorption over the same pressure step can yield useful insights, especially concerning the concentration dependence of the diffusivity [69, 70]. The insights provided by such measurements are even greater for IFM and IRM measurements since any differences in the form or symmetry of the transient concentration profiles can be seen directly.

12.3.1.1 New Options of Correlating Transport Diffusion and Self-diffusion

Hitherto, the comparison of self-diffusivities and transport diffusivities was complicated by the differences between the experimental arrangements applied for their determination, including the problem of comparing the results of microscopic techniques of measurement (such as NMR or QENS, see Chapters 10 and 11) with macroscopic tracer exchange (Chapter 13). Being able to determine diffusivities under equilibrium and non-equilibrium conditions within the same device (by transient sorption and tracer exchange experiments) IRM is well suited for direct comparison of self- and transport diffusivities. It is demonstrated in Section 19.3 that, with the discovery of MOF ZIF-8 [71–75], a highly stable nanoporous material with defect-free crystals of the necessary size, the comparison of the coefficients of transport diffusion and self-diffusion (i.e., parameters that refer to non-equilibrium and equilibrium conditions) under totally compatible conditions has become possible [53].

12.3.1.2 Visualizing Guest Profiles in Transient Sorption Experiments

The evolution of intracrystalline concentration profiles may be measured under conditions that are far more complex than those of the conventional uptake and release experiments such as considered in Figures 12.7 and 12.8. As an example, Figure 12.9 displays the evolution of intracrystalline concentration profiles in response to two pressure steps (up and down), with the second one applied before the first perturbation has reached equilibrium. In sorption science, this procedure is often referred to as a partial-loading experiment (Section 14.7.3 and References [70, 76]). It provides information concerning the limiting processes of molecular uptake on the basis of characteristic differences in the time dependence of desorption: If uptake is exclusively controlled by transport resistance at the crystal surface, the intracrystalline concentration should be uniform through the crystal under transient uptake conditions as well as at equilibrium. This means that desorption at "partial loading," that is, a decrease of the external pressure before attainment of equilibrium inside the crystal, leads to exactly the same time dependence of the

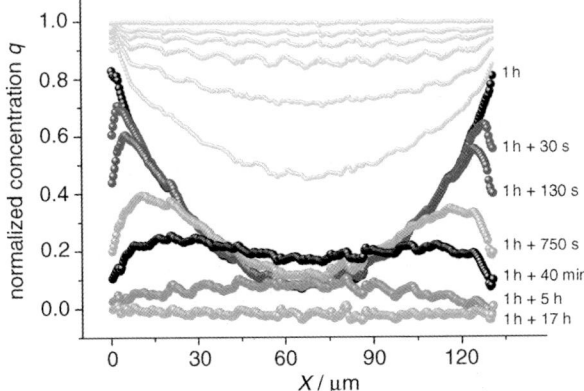

Figure 12.9 Revealing transient maxima of intracrystalline guest concentration: Guest uptake (n-butane) by a nanoporous host [Zn (tbip), 0 → 300 mbar] is followed, at time $t = 1\,h$, by desorption (pressure decreases to zero) long before equilibration. The light gray profiles show the profiles that would be recorded if the uptake process had not been interrupted and which were in fact measured in a separate run. The concentration patterns are those to be expected in "partial loading" experiments for discriminating between bulk and surface resistances in classical diffusion experiments – see Figure 14.18 and References [70, 76]. They have thus become immediately visible! Reprinted from Reference [67] with permission.

dimensionless desorption curves as under equilibrium. If the influence of diffusion is perceptible, however, desorption at partial loading would start with concentrations that are higher close to the crystal boundary, leading to a characteristic acceleration of desorption in its initial stage, as discussed in Section 14.7.4.4. Figure 12.9 displays the intracrystalline concentration profiles in exactly this situation, revealing two characteristic transient maxima close to the crystal boundary! The experimentally observed behavior in the system under study is thus found to approach the situation of diffusion control as shown in Figure 14.18a, illustrating the limiting concentration profiles during partial loading experiments.

With the concentration profiles during partial-loading experiments Figure 12.9 nicely reveals another peculiar feature of such experiments: although, in the present study, the pressure drop 1 h after onset of adsorption provokes overall desorption, increasing concentration in the center of the crystal may still be identified. This is the straightforward outcome of the two maxima in concentration giving rise to down-hill molecular fluxes both out of the crystal *and* into its center.

12.3.1.3 Testing the Relevance of Surface Resistances by Correlating Fractional Uptake with Boundary Concentration

Imaging of transient concentration profiles makes it possible to correlate the fractional uptake (m_t/m_∞), which corresponds to the area under the concentration profiles, with the ratio $q_{t,\text{surf}}/q_\infty$ at any instant of time. The respective (m_t/m_∞) versus $(q_{t,\text{surf}}/q_\infty)$ plots have been shown to provide a sensitive test for the intrusion of surface resistances on overall uptake or release [77, 78]. Details of this procedure are explained in Section 6.2.2 (Figure 6.5).

12.3.1.4 Applying Boltzmann's Integration Method for Analyzing Intracrystalline Concentration Profiles

With the application of IFM and IRM, the famous Boltzmann–Matano method for analyzing transient concentration profiles in solids (see References [61, 79, 80]) may be adopted to yield a "one-shot" determination of the complete concentration dependence of intracrystalline diffusivities. As prerequisites the following requirements must be fulfilled:

1) one-dimensional diffusion flux into or out of the host crystal,
2) instantaneous establishment of equilibrium at the boundary ($q(x=0, t=0) = q_\infty$),
3) no interference of the profile margin ("diffusion front") with the other fronts and boundaries ($q(x=\infty, t) = q_0$).

It is shown in Section 6.2.3.3 that under such conditions, by introducing the new variable $\eta = x/\sqrt{t}$ (i.e., by "Boltzmann transformation"), Fick's 2nd law [Eq. (12.4)] can be transferred into an ordinary differential equation:

$$\frac{d^2 q}{d\eta^2} + \frac{\eta}{2D}\frac{dq}{d\eta} + \frac{dD/dq}{D}\left(\frac{dq}{d\eta}\right)^2 = 0 \qquad (12.6)$$

where the concentration now appears as a function of the sole variable $\eta = x/\sqrt{t}$. Plotting the concentration profiles $q(x,t)$ as a function of this new parameter, that is, as $q(\eta = x/\sqrt{t})$, for different times t should therefore yield coinciding representations. Equation (12.6) cannot be integrated directly. However, with the above-required initial and boundary conditions, integration by parts yields [59, 81]:

$$D(q) = -\frac{1}{2}\frac{d\eta}{dq}\int_{q=q_0}^{q} \eta \, dq \qquad (12.7)$$

as a relation that yields the complete concentration dependence $D(q)$ from a single transient concentration profile (Figure 12.10 [59, 81]). The data presented refer to methanol uptake in ferrierite. It was pointed out at the beginning of Section 12.3.1 that the main uptake occurs through the small eight-ring channels, that is, in the y-direction. In addition, however, a small fraction of molecules is able to enter through the margins of the ten-ring channels. Their contribution may be estimated from the concentration increase in the middle of the eight-ring channels. Subtraction of this contribution leads to the plots shown in Figure 12.10. The sorbate profiles presented as a function of the sole variable $\eta = y/\sqrt{t}$ deviate from each other for different times t, in contrast to expectation. However, under the given circumstances, the requirements for a rigorous application of Boltzmann's integration method are, anyway, not strictly fulfilled. In Reference [82] it is shown that the deviations lead to uncertainties of less than a factor of 2 in the derived diffusivities.

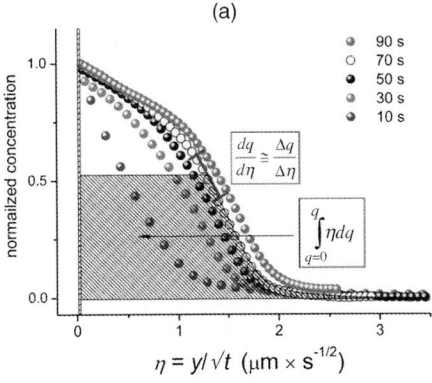 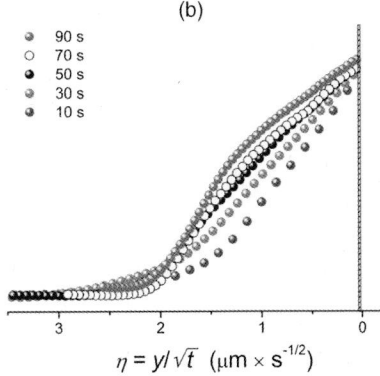

Figure 12.10 Evolution of the concentration profiles in the y-direction in the left (a) and right (b) side of the ferrierite crystal as a function of the parameter $\eta = y/\sqrt{t}$. Part (a) also illustrates the way of determining $\int_{q=q_0}^{q} \eta\, dq$ and $dq/d\eta$, yielding, via. Eq. (12.7), the diffusivity (with $q_0 = 0$). Reprinted from Reference [80] with permission.

12.3.2
Direct Measurement of Surface Barriers

The presence of a surface barrier (which implies a substantial reduction in permeance through the outer layer or layers of the crystal) can be detected by detailed analysis of the shape of the transient uptake curve, as discussed in Sections 6.2.1 and 6.2.2. The presence of significant surface barriers has been suggested as one way to explain why molecular uptake and release rates have often been found to occur more slowly than expected from the intracrystalline diffusivities provided by PFG NMR [83–85] and other microscopic measurements. For some systems a quantitative estimate of the magnitude of the surface barrier can be obtained from NMR tracer-desorption measurements (Section 11.4.3.3) [86, 87]. However, IFM and IRM techniques provide a more direct approach, which allows a much more accurate study of this phenomenon.

12.3.2.1 Concentration Dependence of Surface Permeability

Surface barriers lead to a notable reduction in the rate of molecular exchange between the bulk of the nanoporous material and the surroundings, under both equilibrium (tracer exchange) and non-equilibrium conditions (transient sorption experiments).

The surface barrier may be considered as an extended layer of dramatically reduced diffusivity (D_{barr}) at or near the surface of the crystal or particle. Alternatively, and completely equivalently to this model, transport inhibition may also be regarded as resulting from a dramatic reduction of guest solubility. Then D_{barr} is replaced by the bulk diffusivity reduced by the ratio of the concentrations in bulk and layer at equilibrium. The layer thickness (l) is assumed to be negligibly small in comparison with the crystal extension, far below the limits of spatial resolution for the imaging techniques.

In complete analogy with the self-diffusivities, permeabilities under equilibrium conditions depend only on the overall concentration, that is, on the sum of the concentrations of the labeled and unlabeled molecules. This means that the diffusivity should be constant through the surface layer so that the flux can be accurately represented by Eq. (12.5) with the permeability given by $\alpha = D_{barr}/l$. However, this is not true under transient conditions, since the concentration (and therefore the diffusivity) then varies through the surface layer. If the form of the concentration dependence of the diffusivity within the surface layer is known, this may be built into the mathematical model. A simple example of that approach is discussed in Section 6.2.3 [Eqs. (6.88)–(6.94) and Figures 6.20 and 6.21]. Clearly, the possibility that the diffusivity may vary quite strongly through the surface layer imposes some limitations to a rigorous general analysis of the transient concentration profiles. However, for some systems the approximation of a constant surface permeability appears to provide a good approximate representation of the observed behavior, since coupling Fick's diffusion equation [Fick's 2nd law, Eq. (12.4)] with the boundary conditions provided by the surface permeability [Eq. (12.5)] has been found to provide a reasonably accurate representation of the concentration profiles measured during molecular uptake and release as shown in Figure 12.7.

A different pattern of behavior has been observed for the short-chain alkanes in the MOF (metal organic framework) Zn(tbip) – see Sections 19.4 and 19.6. For these systems, the surface resistance appears to depend only on the mean concentration (the average of the actual boundary concentration and the equilibrium concentration at the crystal surface). Neither a variation of the direction of the pressure step (comparison of adsorption and desorption steps) nor variation of the magnitude of the concentration step (the difference between the actual and equilibrium conditions) leads to any significant change in the permeability, provided that the mean of the boundary and equilibrium concentrations remains constant. The finding that the surface permeability may be treated as a function of a single concentration parameter facilitates the inclusion of such resistances in the mathematical model, thus reducing the complexity and raising the prospect of developing a practically useful model for technological application.

The fundamental physics underlying this observation is not yet clear, but it clearly provides some information concerning the nature of the elementary mechanisms and processes leading to the surface resistance.

Note that the surface permeability, that is, the transport resistance at the crystal surface, is not correlated with the step in free energy (i.e., the differences in energy and entropy) between the intra- and intercrystalline spaces. It is true that the increase in potential energy from inside to outside reduces the escape rate of the molecules out of crystals, just as a decrease in entropy from outside to inside (caused by, for example, a decrease in the number of degrees of freedom due to pore confinement) decreases the entrance rate. However, these mechanisms are already accounted for in the change of the equilibrium concentrations, so that any decrease in the escape or entrance rates is exactly compensated by a corresponding increase in the respective populations and, hence, in the "attempt" rates to enter or escape from the crystals.

In complete analogy with diffusivities, which are referred to as self- (or tracer) diffusivities and transport diffusivities, one should in principle distinguish between these two types of permeabilities. However, since in the overwhelming majority of cases surface permeabilities have been determined from transient sorption rate measurements, if not stated otherwise, surface permeabilities are generally understood to refer to non-equilibrium conditions. The option of referring to "transport" and "self"- (or "tracer") permeabilities, in correspondence with the terminology used in diffusion studies, is thus left for a future decision.

12.3.2.2 Correlating the Surface Permeabilities through Different Crystal Faces

On comparing different crystallites from the same synthesis batch (including zeolites of type ferrierite [64, 65, 88–90] and SAPO STA-7 [56, 91, 92] and MOFs of type Zn(tbip) [68, 92–95]) the intracrystalline diffusivities are found to remain essentially the same, while the surface resistances may differ substantially from crystal to crystal. In view of the much greater stability of the host interior, this finding is perhaps not surprising. It is, however, remarkable that, in general, irrespective of these differences, the surface resistances observed on either side of the same crystallite prove to be identical. This suggests that the crystallization process itself does not give rise to the formation of surface barriers, since it would then be difficult to explain why the surface resistances differ from crystallite to crystallite but coincide on either side of a given crystal. The formation of surface resistances should rather be associated with the influence of the crystal surroundings, which both during crystallization and crystal storage should be much more similar on both sides of the same crystal than for different crystals.

This symmetry may be expected to be eliminated by creating areas with different "histories" on either side of a crystal. As an example, Figure 12.11 compares the evolution of the profiles of the intracrystalline concentration of propane in Zn(tbip) during tracer exchange in a typical crystal of the batch (Figure 12.11a) with a freshly broken crystal (Figure 12.11b, the freshly broken face is on the left side). By cutting, the permeability could be increased by an order of magnitude, compared to the as-synthesized face [93, 95]. The micro-image of the corresponding faces in Figure 12.12 shows that the face of lower permeability (Figure 12.12a) appears to be obstructed by layers of amorphous material that are not present on the freshly cut surfaces (Figure 12.12b).

An even more pronounced asymmetry in the surface resistances was observed with zeolites of type SAPO STA-7 [96, 97]. Figure 12.13 shows the crystal structure and habit. Most importantly, the silicon content is found to affect dramatically both the stability and transport properties of the crystals [91]. This is exemplified by the transient concentration profiles observed during methanol uptake in a low-silicon crystal of SAPO STA-7 [sample STA-7(10); Figure 12.14] and in a high-silicon crystal of the same type of material [sample STA-7(30); Figure 12.15]. The high-silicon sample deviates from STA-7(10) in two striking features. Firstly, the uptake rates are dramatically reduced which, by quantitative analysis [91], may be attributed to a decrease in the intracrystalline diffusivities by about one order of magnitude and in

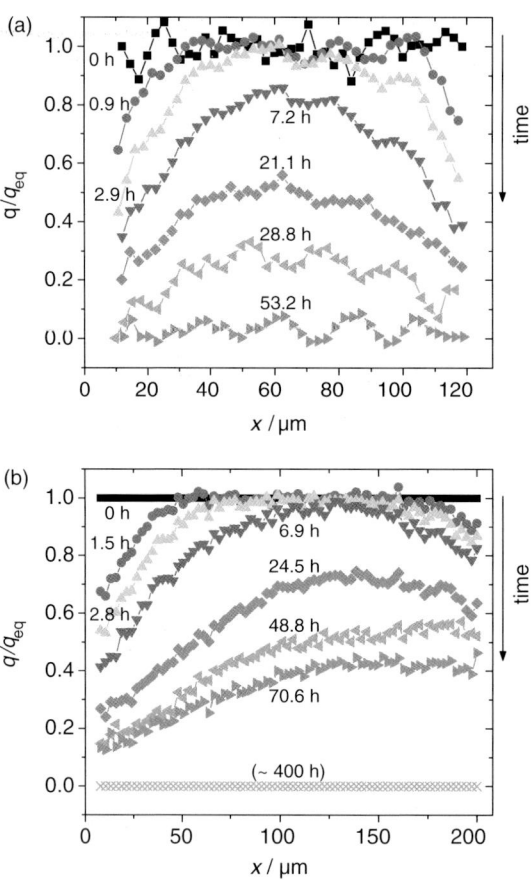

Figure 12.11 Evolution of transient concentration profiles during tracer exchange of propane (desorbing isotope) in MOF Zn(tbip) in (a) a crystal as taken from the batch and (b) a crystals with one freshly cut edge (left side). Reprinted from Reference [95] with permission.

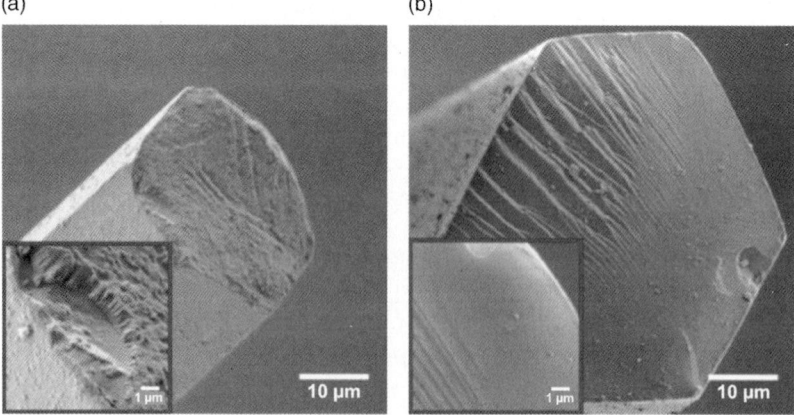

Figure 12.12 Surface of the as-synthesized Zn(tbip) crystal (a) and of the freshly broken face (b); insets: under magnification. Reprinted from Reference [93] with permission.

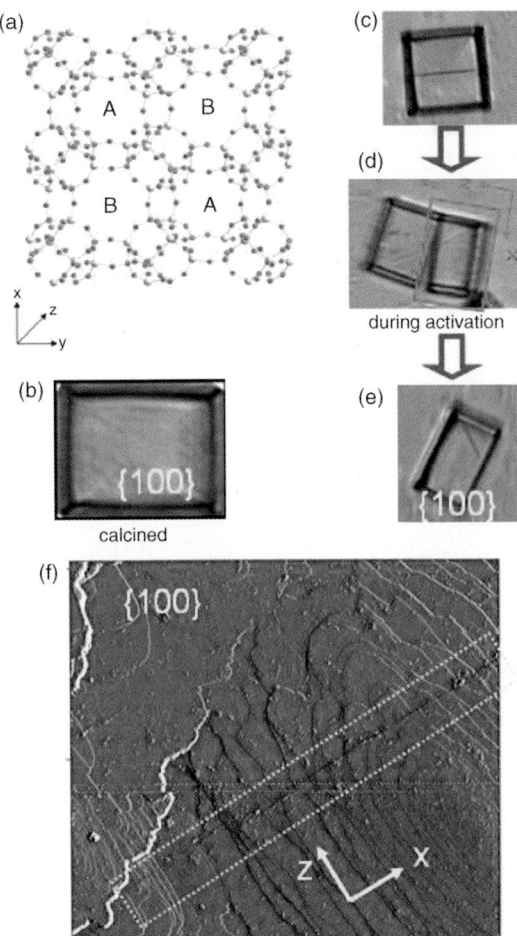

Figure 12.13 Pore structure, appearance, and surface of SAPO STA-7 crystals. (a) In the structure of SAPO STA-7 two different cages, labeled A and B, alternate, giving rise to the prismatic shape of the crystal, with a quadratic cross section ({001} plane); (b) a calcined crystal without observable defect; (c)–(e) the fracturing of a calcined STA-7(30) crystal with an observed line of weakness (and stored in the calcined state) into two parts during activation; (f) AFM error image of the {100} surface of an as-prepared STA-7(30) crystal, showing the presence of the line of weakness observed by the optical microscope (framed by dotted lines); there is, however, no fracturing at this stage, according to the AFM image. Reprinted from Reference [91] with permission.

the surface permeabilities by as much as three orders of magnitude. Secondly, the crystals are found to be highly unstable, with the superficial crack seen in Figure 12.13f becoming the source of crystal splitting as seen in Figure 12.13c–e. From the transient uptake profiles shown in Figure 12.15 it appears that the permeability through the freshly formed face notably exceeds the permeability

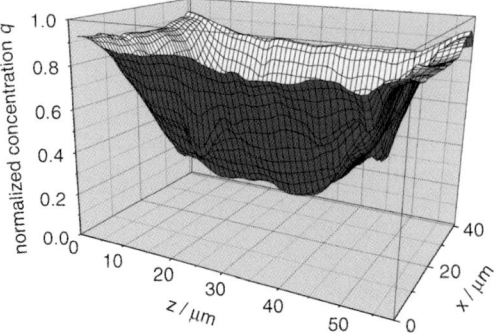

Figure 12.14 Profiles of the concentration integrals of methanol during uptake in the low-silica STA-7(10) sample; 10 s after onset of adsorption the boundary concentration approaches the equilibrium value and uptake is close to being completed. Taken from Reference [91] with permission.

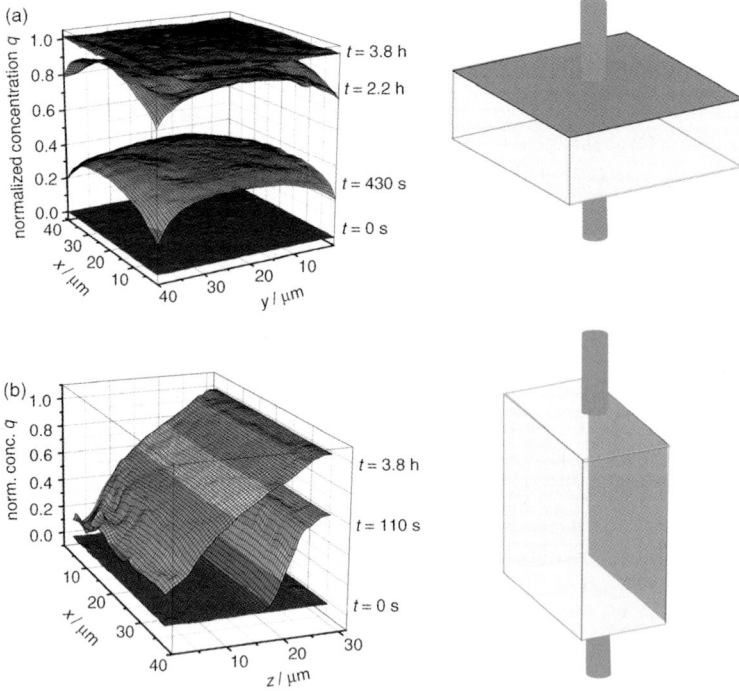

Figure 12.15 Concentration profiles of methanol during uptake in the fractured STA-7 (30) crystal. The orientation of the crystals is indicated schematically to the right: The novel (fractured) crystal face perpendicular to the [001] direction (indicated by darker shading) is perpendicular (a) and parallel (b) to the observation direction (indicated by the slim cylinder). Since the mass transfer through the fresh crystal face dominates, the recorded concentration profiles are quite flat in case (a) and strongly curved in case (b). Overall uptake is dramatically retarded in comparison with STA-7(10) shown in Figure 12.14. Taken from Reference [91] with permission.

through the regular crystal faces. Quantitative analysis suggests an enhancement by more than two orders of magnitude [91]. Interestingly, both features, the crystal instability and the high asymmetry in the surface permeabilities through the (genuine) external crystal face and the (freshly broken) face through the crystal center, are nicely supplemented by the finding from scanning electron microscopy (selected area EDX) that the silicon content decreases dramatically towards the central plane of the crystal.

12.3.3
Adapting the Concept of Sticking Probabilities to Nanoporous Particles

In catalysis, the sticking coefficient is defined as the probability that, on colliding with a plane surface, a molecule will be captured [98, 99]. Experimental studies with, for example, hydrogen, nitrogen, oxygen, carbon monoxide, and ethylene as probe molecules and a wide range of different metals and metal oxides as targets yield sticking probabilities covering a large spectrum of values from as small as 10^{-7} up to 1.

Chemical reactions involving nanoporous catalysts depend on the probability that, on encountering the surface of a catalyst particle, the reactant molecules penetrate into the particle interior and, hence, to the sites relevant for the desired conversion. Such probabilities have been estimated on the basis of both PFG NMR [100, 101] and IR [102–105] experiments, as well as by MD simulations [100]. Depending on the system under study, probabilities ranging from as small as 10^{-7} [102–104] to close to unity [100, 101] have been obtained, reproducing the range over which the sticking probabilities of gases on plane surfaces have also been observed.

Monitoring the evolution of guest profiles in nanoporous materials provides a new and direct way of measuring the sticking probabilities for nanoporous particles. It is worth referring, in this context, to the elementary mechanisms of sorption and diffusion. From the kinetic theory of gases, the flux of molecules encountering the particle surface is given by the relation $j_{Gas} = N_{Avo} p \times (2\pi RTM)^{-1/2}$, with N_{Avo} and M denoting, respectively, the Avogadro number and the molar mass [106]. On the other hand, in the short-time limit the amount adsorbed during diffusion-controlled uptake scales with $t^{1/2}$ [Eq. (6.12)], which implies an influx proportional to $t^{-1/2}$ that diverges in the limit of sufficiently short times. In the probability distribution of the particle life times in systems under diffusion exchange, this peculiarity is known to appear as an infinitely high probability density for zero life times [107]. The conflict between the finiteness of guest supply and infinitely large guest fluxes into the host system for short times exists, however, only in the mathematical formalism. In reality one has rather to remember that the diffusion equation holds only over time and space scales large enough to ensure that the superposition of the elementary steps of molecular displacement yields a Gaussian probability distribution with the variance given by the Einstein relation (Eq. (2.6), [108, 109]).

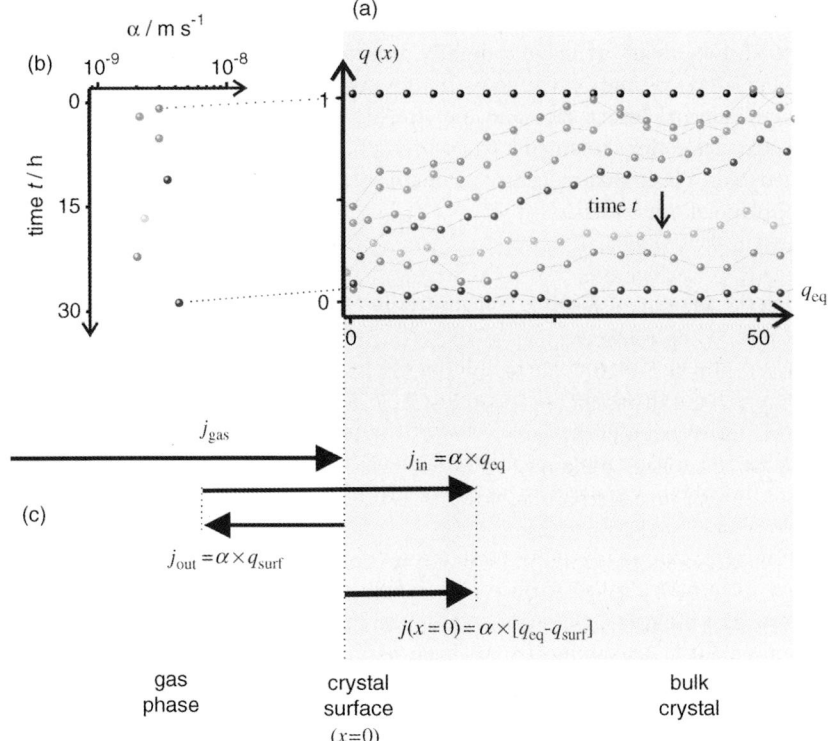

Figure 12.16 Estimating "sticking" probabilities by analyzing transient guest profiles: (a) evolution of the concentration profiles of deuterated propane in MOF Zn(tbip) during tracer exchange with non-deuterated propane; (b) the resulting surface permeabilities α; (c) with the knowledge of α, the flux j_{in} of molecules entering the crystal may be determined; with the known relation for the flux j_{gas} of molecules colliding with the external surface, the "sticking" probability j_{in}/j_{gas} that a molecule from the gas phase may continue its trajectory inside the crystal can be calculated Reprinted from Reference [68] with permission.

With the detection of surface barriers, the "bulk" phase of the nanoporous host materials is now known to be, in general, separated from the external space by a layer of dramatically reduced permeability. Whether or not this has a measurable effect on the overall kinetics depends on the ratio of the time constants for surface permeation and internal diffusion (i.e., on the dimensionless ratio $k_s r_c / D_{intra}$, see Table 13.1). Knowledge of the permeability through this layer now allows direct determination of the fraction of molecules that, after colliding with the external particle surface, are able to penetrate into the region where molecular propagation and, hence, the evolution of the guest profiles is controlled by intracrystalline diffusion.

Figure 12.16 provides an example of this type of analysis [68]. It shows the profiles of deuterated propane during tracer exchange with "normal" propane in a crystal of

MOF Zn(tbip) as considered in Section 19.4 and Figure 19.16 and the permeabilities (Figure 12.16b) resulting, via Eq. (12.5), for each individual time interval. The scattering in the permeability data with increasing time, that is, with increasing tracer exchange, reflects their uncertainty, since the overall concentration remains constant and therefore a constant permeability should be expected.

The (net) flux through the surface as accessible in our experiments is the difference between the counter-directed fluxes leaving and entering the crystal through the surface boundary (Figure 12.16c). The flux of molecules entering from the gas phase in total is thus easily found to be equal to $j_{in} = \alpha \times q_{eq}$, with q_{eq} given by the loading at equilibrium with the gas phase (since, under equilibrium, it has to be balanced by the flux of the leaving molecules). Combining this information yields a sticking probability of $j_{in}/j_{gas} \approx 5 \times 10^{-8}$. We have thus found that, out of 20 million molecules colliding with the surface of the MOF crystal under study, only a single one is able to enter into the intracrystalline bulk phase where the evolution of the guest profiles is exclusively controlled by intracrystalline diffusion in the channel direction.

It is essential to mention that this type of analysis assumes the existence of a surface barrier, that is, a thin layer covering the whole crystal surface with the effect that, after this layer, the concentration reaches exactly the value resulting from the application of Fick's 2nd law [Eq. (12.4)] with the boundary condition given by Eq. (12.5). It is the probability that, after hitting the external surface from outside, a molecule gets into this region, which may be calculated by the procedure described.

References

1 Wikipedia (2010) Electromagnetic spectrum, http://en.wikipedia.org/wiki/Electromagnetic_spectrum (accessed 27.08.2010).
2 Weckhuysen, B.M. (2009) *Angew. Chem. Int. Ed.*, **48**, 4910.
3 Jonkers, G., Vonkeman, K.A., van der Wal, S.W.A., and van Santen, R.A. (1992) *Nature*, **355**, 63.
4 Hensen, E.J.M., de Jong, A.M., and van Santen, R.A. (2008) *Adsorption and Diffusion* (eds H.G. Karge and J. Weitkamp), Springer, Berlin, p. 277.
5 Schuring, F.D., Jansen, A.P.J., and van Santen, R.A. (2000) *J. Phys. Chem. B*, **104**, 941.
6 Schumacher, R.R., Anderson, B.G., Noordhoek, N.J., de Gauw, F.J.M.M., de Jong, A.M., de Voigt, M.J.A., and van Santen, R.A. (2000) *Microporous Mesoporous Mater.*, **6**, 315.
7 Anderson, B.G., van Santen, R.A., and van Ijzendoorn, L.J. (1997) *Appl. Catal.*, **160**, 125.
8 Hawkesworth, M.R., Parker, D.J., Fowles, P., Crilly, J.E., Jefferies, N.L., and Jonkers, G. (1991) *Nucl. Instrum. Methods Phys. Res., Sect. A*, **310**, 423.
9 Ferrieri, R.A. and Wolf, A.P. (1984) *J. Phys. Chem.*, **88**, 2256.
10 Dubinin, M.M., Erashko, I.T., Kadlec, O., Ulin, V.I., Voloshchuk, A.M., and Zolotarev, P.P. (1975) *Carbon*, **13**, 193.
11 Voloshchuk, A.M. and Dubinin, M.M. (1973) *Dokl. Akad. Nauk SSSR*, **212**, 649.
12 Tiselius, A. (1934) *Z. Phys. Chem. A*, **169**, 425.
13 Tiselius, A. (1935) *Z. Phys. Chem. A*, **174**, 401.
14 Karge, H.G. and Klose, K. (1975) *Ber. Bunsen-Ges. Phys. Chem.*, **97**, 454.
15 Karge, H.G. and Niessen, W. (1991) *Cat. Today*, **8**, 451.

16 Niessen, W. and Karge, H.G. (1993) *Microporous Mater.*, **1**, 1.
17 Karge, H.G., Niessen, W., and Bludau, H. (1996) *Appl. Catal. A-Gen.*, **146**, 339.
18 Karge, H.G. (1998) *Proceedings of the Third Polish-German Zeolite Colloquium*, (ed. M. Rozwadowski), Nicholas Copernicus University Press, Torun, p. 11.
19 Schumacher, R. and Karge, H.G. (1999) *Microporous Mesoporous Mater.*, **30**, 307.
20 Schumacher, R. and Karge, H.G. (1999) *J. Phys. Chem. B*, **103**, 1477.
21 Hermann, M., Niessen, W., and Karge, H.G. (1995) *Catalysis by Microporous Materials* (eds H.K. Beyer, H.G. Karge, I. Kiricsi, and J.B. Nagy), Elsevier, Amsterdam, p. 131.
22 Karge, H.G. (2005) *Compt. Rend. Chim.*, **8**, 303.
23 Karge, H.G. and Kärger, J. (2008) *Adsorption and Diffusion* (eds H.G. Karge and J. Weitkamp), Springer, Berlin, p. 135.
24 Jansen, C.J., de Ruiter, R., and van Bekkum, H. (1989) *Stud. Surf. Sci. Cat.*, **49**, 679.
25 Nowotny, M., Lercher, J., and Kessler, H. (1991) *Zeolites*, **11**, 454.
26 Schüth, F. (1992) *J. Phys. Chem.*, **96**, 454.
27 Schüth, F. and Althoff, R. (1993) *J. Catal.*, **143**, 338.
28 Schüth, F., Demuth, D., Zibrowius, B., Kornatowksi, J.J., and Finger, G. (1994) *J. Am. Chem. Soc.*, **116**, 1090.
29 Lin, Y.S., Yamamoto, N., Choi, Y., Yamaguchi, T., Okubo, T., and Nakao, S.I. (2000) *Microporous Mesoporous Mater.*, **38**, 207.
30 Chmelik, C. (2007) FTIR microscopy as a tool for studying molecular transport in zeolites. PhD thesis, Leipzig.
31 Heink, W., Kärger, J., and Pfeifer, H. (1978) *Chem. Eng. Sci.*, **33**, 1019.
32 Vlaardingerbroek, M.T. and den Boer, J.A. (2004) *Magnetic Resonance Imaging: Theory and Practice*, Springer, Berlin.
33 Hausser, K.H. and Kalbitzer, H.R. (1991) *NMR in Medicine and Biology*, Springer, Berlin.
34 Stapf, S. and Han, S.-I. (eds) (2006) *NMR Imaging in Chemical Engineering*, Wiley-VCH Verlag GmbH, Weinheim.
35 Gladden, L.F., Mantle, M.D., and Sederman, A.J. (2008) *Handbook of Heterogeneous Catalysis*, 2nd edn (eds G. Ertl, H. Knözinger, F. Schüth, and J. Weitkamp), Wiley-VCH Verlag GmbH, Weinheim, p. 1784.
36 Lysova, A.A. and Koptyug, I.V. (2010) *Chem. Soc. Rev.*, **39**, 4585.
37 Lauterbur, P.C. (1973) *Nature*, **242**, 190.
38 Springuel-Huet, M.A., Bonardet, J.L., Gedeon, A., and Fraissard, J. (1999) *Magn. Reson. Chem.*, **37**, S1–S13.
39 Salerno, M. (2002) *Radiology*, **222**, 252.
40 LeaWoods, J.C. (2001) *Concept Magn. Reson.*, **13**, 277.
41 Chen, X.J. (1999) *Magn. Reson. Med.*, **42**, 721.
42 Blank, A., Dunnam, C.R., Borbat, P.P., and Freed, J.H. (2003) *J. Magn. Reson.*, **165**, 116.
43 Blank, A., Dunnam, C.R., Borbat, P.P., and Freed, J.H. (2004) *Rev. Sci. Instrum.*, **75**, 3050.
44 Chmelik, C., Heinke, L., Valiullin, R., and Kärger, J. (2010) *Chem. Ing. Tech.*, **82**, 779.
45 Kempe, S., Metz, H., and Mäder, K. (2010) *Eur. J. Pharm. Biopharm.*, **74**, 55.
46 Heinke, L., Chmelik, C., Kortunov, P., Ruthven, D.M., Shah, D.B., Vasenkov, S., and Kärger, J. (2007) *Chem. Eng. Technol.*, **30**, 995.
47 Roggo, Y., Edmond, A., Chalus, P., and Ulmschneider, M. (2005) *Anal. Chim. Acta*, **535**, 79.
48 Schemmert, U., Kärger, J., and Weitkamp, J. (1999) *Microporous Mesoporous Mater.*, **32**, 101.
49 Schemmert, U., Kärger, J., Krause, C., Rakoczy, R.A., and Weitkamp, J. (1999) *Europhys. Lett.*, **46**, 204.
50 Kärger, J. and Freude, D. (2002) *Chem. Eng. Technol.*, **25**, 769.
51 Heinke, L., Chmelik, C., Kortunov, P., Shah, D.B., Brandani, S., Ruthven, D.M., and Kärger, J. (2007) *Microporous Mesoporous Mater.*, **104**, 18.
52 Ruthven, D.M. and Lee, L.K. (1981) *AIChE J.*, **27**, 654.
53 Chmelik, C., Bux, H., Caro, J., Heinke, L., Hibbe, F., Titze, T., and Kärger, J. (2010) *Phys. Rev. Lett.*, **104**, 85902.

References

54 Griffiths, P.R. and de Haseth, J.A. (1986) *Fourier Transform Infrared Spectroscopy*, John Wiley & Sons, Inc., New York.
55 Heinke, L. and Kärger, J. (2008) *New J. Phys.*, **10**, 23035.
56 Heinke, L., Kortunov, P., Tzoulaki, D., Castro, M., Wright, P.A., and Kärger, J. (2008) *Europhys. Lett.*, **81**, 26002.
57 Heinke, L. and Kärger, J. (2009) *J. Chem. Phys.*, **130**, 44707.
58 Schmitt, L.M. (2001) *Theor. Comput. Sci.*, **259**, 1.
59 Heinke, L. (2008) Mass transfer in nanoporous materials: a detailed analysis of transient concentration profiles. PhD thesis, Leipzig.
60 Schüth, F., Sing, K.S.W., and Weitkamp, J. (2002) *Handbook of Porous Solids*, Wiley-VCH Verlag GmbH, Weinheim.
61 Mehrer, H. (2007) *Diffusion in Solids*, Springer, Berlin.
62 Glicksman, M.E. (2000) *Diffusion in Solids*, John Wiley & Sons, Inc., New York.
63 Heitjans, P. and Kärger, J. (2005) *Diffusion in Condensed Matter: Methods, Materials, Models*, Springer, Berlin.
64 Kärger, J., Kortunov, P., Vasenkov, S., Heinke, L., Shah, D.B., Rakoczy, R.A., Traa, Y., and Weitkamp, J. (2006) *Angew. Chem. Int. Ed.*, **45**, 7846.
65 Kortunov, P., Heinke, L., Vasenkov, S., Chmelik, C., Shah, D.B., Kärger, J., Rakoczy, R.A., Traa, Y., and Weitkamp, J. (2006) *J. Phys. Chem. B*, **110**, 23821.
66 Rakoczy, R.A., Traa, Y., Kortunov, P., Vasenkov, S., Kärger, J., and Weitkamp, J. (2007) *Microporous Mesoporous Mater.*, **104**, 1195.
67 Heinke, L., Kortunov, P., Tzoulaki, D., and Kärger, J. (2007) *Phys. Rev. Lett.*, **99**, 228301.
68 Chmelik, C., Heinke, L., Kortunov, P., Li, J., Olson, D., Tzoulaki, D., Weitkamp, J., and Kärger, J. (2009) *ChemPhysChem.*, **10**, 2623.
69 Kärger, J. and Ruthven, D.M. (1992) *Diffusion in Zeolites and Other Microporous Solids*, John Wiley & Sons, Inc., New York.
70 Ruthven, D.M., Brandani, S., and Eic, M. (2008) *Adsorption and Diffusion* (eds H.G. Karge and J. Weitkamp), Springer, Berlin, p. 45.
71 Li, H., Eddaoudi, M., O'Keeffe, M., and Yaghi, O.M. (1999) *Nature*, **402**, 276.
72 Ferey, G., Mellot-Dranznieksd, C., and Serre, C. (2005) *Science*, **309**, 2040.
73 Pan, L., Parker, B., Huang, X., Olson, D., Lee, J.-Y., and Li, J. (2006) *J. Am. Chem. Soc.*, **128**, 4180.
74 Bux, H., Chmelik, C., van Baten, J., Krishna, R., and Caro, J. (2011) *J. Membr. Sci.*, **369**, 284.
75 Bux, H., Liang, F., Li, Y., Cravillon, J., Wiebcke, M., and Caro, J. (2009) *J. Am. Chem. Soc.*, **131**, 16000.
76 Ruthven, D.M. and Brandani, S. (2000) *Recent Advances in Gas Separation by Microporous Ceramic Membranes* (ed. N.K. Kanellopoulos), Elsevier, Amsterdam, p. 187.
77 Heinke, L. (2007) *Diffusion Fundam.*, **4**, 12.
78 Heinke, L., Kortunov, P., Tzoulaki, D., and Kärger, J. (2007) *Adsorption*, **13**, 215.
79 Matano, C. (1933) *Jpn. J. Phys.*, **8**, 109.
80 Boltzmann, L. (1894) *Wied. Ann.*, **53**, 959.
81 Kortunov, P., Heinke, L., and Kärger, J. (2007) *Chem. Mater.*, **19**, 3917.
82 Heinke, L. (2007) *Diffusion Fundam*, **4**, 9.1–9.16.
83 Kärger, J. and Caro, J. (1977) *J. Chem. Soc., Faraday Trans. I*, **73**, 1363.
84 Bülow, M. (1985) *Z. Chem.*, **25**, 81.
85 Kärger, J. and Ruthven, D.M. (1989) *Zeolites*, **9**, 267.
86 Kärger, J. (1982) *AICHE J.*, **28**, 417.
87 Kärger, J., Bülow, M., Millward, B.R., and Thomas, J.M. (1986) *Zeolites*, **6**, 146.
88 Kortunov, P. (2005) Rate controlling processes of diffusion in nanoporous materials. PhD Thesis, Leipzig.
89 Kortunov, P., Chmelik, C., Kärger, J., Rakoczy, R.A., Ruthven, D.M., Traa, Y., Vasenkov, S., and Weitkamp, J. (2005) *Adsorption*, **11**, 235.
90 Chmelik, C., Kortunov, P., Vasenkov, S., and Kärger, J. (2005) *Adsorption*, **11**, 455.
91 Tzoulaki, D., Heinke, L., Castro, M., Cubillas, P., Anderson, M.W., Zhou, W., Wright, P., and Kärger, J. (2010) *J. Am. Chem. Soc.*, **132**, 11665.

92 Tzoulaki, D. (2009) *Diffusion studies on microporous materials by interference microscopy.* PhD Thesis, Leipzig.
93 Tzoulaki, D., Heinke, L., Li, J., Lim, H., Olson, D., Caro, J., Krishna, R., Chmelik, C., and Kärger, J. (2009) *Angew. Chem. Int. Ed.*, **48**, 3525.
94 Heinke, L., Tzoulaki, D., Chmelik, C., Hibbe, F., van Baten, J., Lim, H., Li, J., Krishna, R., and Kärger, J. (2009) *Phys. Rev. Lett.*, **102**, 65901.
95 Chmelik, C., Hibbe, F., Tzoulaki, D., Heinke, L., Caro, J., Li, J., and Kärger, J. (2010) *Microporous Mesoporous Mater.*, **129**, 340.
96 Wright, P.A., Maple, M.J., Slawin, A.M.Z., Patinek, V., Aitken, R.A., Welsh, S., and Cox, P.A. (2000) *J. Chem. Soc., Dalton Trans.*, 1243.
97 Castro, M., Garcia, R., Warrender, S.J., Slawin, A.M.Z., Wright, P.A., Cox, P.A., Fecant, A., Mellot-Dranznieksd, C., and Bats, N. (2007) *Chem. Commun.*, 3470.
98 Ashmore, P.G. (1963) *Catalysis and Inhibition of Chemical Reactions*, Butterworths, London.
99 Freund, H.J. (2008) *Handbook of Heterogeneous Catalysis*, 2nd edn (eds G. Ertl, H. Knözinger, F. Schüth, and J. Weitkamp), Wiley-VCH Verlag GmbH, Weinheim, p. 1375.
100 Simon, J.-M., Bellat, J.B., Vasenkov, S., and Kärger, J. (2005) *J. Phys. Chem. B*, **109**, 13523.
101 Kärger, J. and Vasenkov, S. (2006) *J. Phys. Chem. B*, **110**, 17694.
102 Jentys, A., Tanaka, H., and Lercher, J.A. (2005) *J. Phys. Chem. B*, **109**, 2254.
103 Jentys, A., Mukti, R.R., Tanaka, H., and Lercher, J.A. (2006) *Microporous Mesoporous Mater.*, **90**, 284.
104 Jentys, A., Mukti, R.R., and Lercher, J.A. (2006) *J. Phys. Chem. B*, **110**, 17691.
105 Gobin, O.C., Reitmeier, S.T., Jentys, A., and Lercher, J.A. (2009) *Microporous Mesoporous Mater.*, **125**, 3.
106 Atkins, P.W. and de Paula, J. (2006) *Physical Chemistry*, 8th edn, Oxford University Press, Oxford.
107 Rödenbeck, C., Kärger, J., and Hahn, K. (1998) *Ber. Bunsen-Ges. Phys. Chem. Chem. Phys.*, **102**, 929.
108 Kärger, J. (2010) *Leipzig, Einstein, Diffusion*, Leipziger Universitätsverlag, Leipzig.
109 Kärger, J. and Vasenkov, S. (2005) *Microporous Mesoporous Mater.*, **85**, 195.

13
Direct Macroscopic Measurement of Sorption and Tracer Exchange Rates

Perhaps the most widely used method for the determination of intraparticle or intracrystalline diffusivities involves measuring the sorption rate when a sample of adsorbent is subjected to a well-defined change in the concentration or pressure of sorbate. Measurements of this kind are commonly made in "closed" systems under conditions such that the fluid concentration is essentially uniform and each particle is therefore exposed to the same external concentration at all times. The behavior of the entire system should then replicate the behavior of an individual particle. The experimental methods differ mainly in the choice of boundary conditions and the means by which progress towards the new position of equilibrium is followed. The diffusivity or, more correctly, the diffusional time constant, is found by matching the experimental sorption curve to the dimensionless theoretical curve, calculated from the appropriate solution of the diffusion equation for the relevant boundary conditions. The solutions for some practically important situations are discussed in Chapter 6. In this chapter we consider the various experimental techniques that have been developed to obtain intraparticle diffusional time constants from uptake or tracer exchange rate measurements. The role played by heat transfer and external diffusional resistances in controlling the sorption rate was pointed out in Chapter 6. Elimination of such extraneous effects is critical to obtaining reliable intraparticle diffusion data from uptake rate measurements.

13.1
Gravimetric Methods

Gravimetric measurements offer, at least in principle, a straightforward method of following the transient adsorption/desorption curves for single-component systems. Elimination of thermal effects and extra-particle resistances to mass transfer has proved to be a much more significant challenge than originally anticipated. Elimination of extraneous heat and mass transfer resistances requires the use of a very small sample of adsorbent and, to avoid the complications introduced by nonlinearities arising from concentration dependence of the diffusivity, it is necessary to make differential measurements over small concentration steps. Reliable measurements

Diffusion in Nanoporous Materials. Jörg Kärger, Douglas M. Ruthven, and Doros N. Theodorou.
© 2012 Wiley-VCH Verlag GmbH & Co. KGaA. Published 2012 by Wiley-VCH Verlag GmbH & Co. KGaA.

therefore require a highly sensitive and rapid response balance, except for systems in which diffusion is very slow. There is also some advantage to be gained from keeping the gas volume of the system large so that the system pressure, following a sudden change in the sorbate pressure, remains essentially constant, thus approximating a perfect step change. This allows the use of simpler mathematical models to interpret the response curves.

13.1.1
Experimental System

The earliest gravimetric rate measurements were made using a McBain balance (quartz spring), which is quite accurate but suffers from the serious disadvantage of a relatively slow response (about 30 s), making it suitable only for very slow processes. The gravimetric method gained popularity during the 1970s with the advent of vacuum electrobalances (such as the Cahn balance) having a high sensitivity and a much faster response. In the Cahn balance the electrical wiring and the magnetic element are exposed to the sorbate gas, rendering the system unsuitable for use with reactive gases such as H_2S. This problem was solved by the introduction of the magnetic suspension balance [1] (Rubotherm) in which the balance elements are completely isolated from the sorbate gas. Figure 13.1 shows the essential elements of

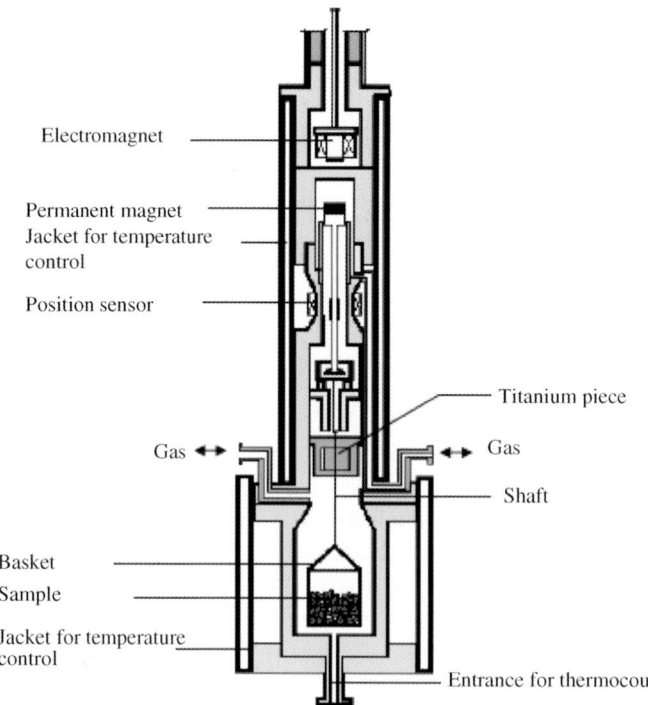

Figure 13.1 Schematic diagram showing the elements of a Rubotherm balance system.

such a system. However, the sample mass required in the Rubotherm balance is considerably greater than that required for a Cahn balance, thus making the elimination of extraneous heat and mass transfer resistances more difficult.

13.1.2
Analysis of the Response Curves

The mathematical models required for the analysis of the transient adsorption/desorption curves are discussed in detail in Chapter 6. Assuming that the measurement is made differentially and isothermally under conditions such that the step change approximation is valid, the kinetic data are conveniently analyzed using plots of the fractional approach to equilibrium against time or against the square root of time, as suggested by Eq. (6.6) (surface resistance control) and Eq. (6.13) (internal diffusion control). These plots should be linear in the short-time region. An example of such a plot, showing conformity between the adsorption and desorption curves, is shown in Figure 6.14a. A plot of $\ln(1 - m_t/m_\infty)$ versus time is also convenient, as suggested by Eq. (6.14) for diffusion control, since this should yield a linear asymptote in the long-time region. It was pointed out in Section 6.2.4 that the intrusion of heat transfer resistance is more pronounced in the later portion of the sorption curve. Therefore, if the balance response is fast enough to track the initial response, this will generally yield more reliable diffusivity values. Of course it is desirable to attain the situation in which the time constants derived from the initial and long-time regions of the curve coincide, indicating complete conformity with the isothermal diffusion model, but this is not always possible.

If the aim of the experiment is to determine the intraparticle diffusivity or surface resistance, it is desirable to adjust the experimental conditions so that only a single resistance is dominant. However, this may not be possible, especially in the case of biporous adsorbents or zeolite crystals that have both internal and surface resistances. Furthermore, it is sometimes important to understand the kinetic behavior of a particular adsorbent under specified conditions where more than one resistance may be significant. In such cases the dual resistance models discussed in Section 6.2.2 are useful.

13.1.2.1 Intrusion of Heat Effects
An example of the intrusion of heat transfer resistance in the long-time region was shown in Figure 6.25 and a further example in which the entire measurable response was controlled by heat transfer was shown in Figure 6.26. Figure 13.2 shows an example where the dominant resistance changes from essentially isothermal intracrystalline diffusion control to heat transfer control as the crystal size is reduced. As the diffusional time constant depends on the square of the crystal (or particle) radius, heat transfer resistance becomes increasingly dominant in the smaller crystals. The behavior of the larger crystals (34 and 21.5 μm) conforms closely to the isothermal diffusion model and the time constants show the expected dependence on the square of the crystal size. However, the uptake rate in the 7.3 μm crystals is only slightly faster than in the 21.5 μm crystals, since the rate in the 7.3 μm crystals is essentially

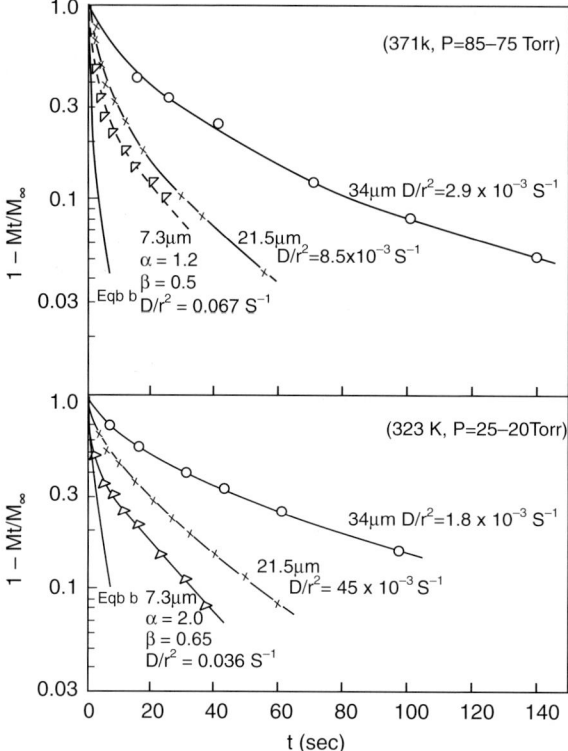

Figure 13.2 Experimental uptake curves for CO_2 in 4A zeolite crystals showing near isothermal behavior in the large (34 and 21.5 μm) crystals ($D \approx 9 \times 10^{-9}$ cm² s⁻¹ at 371 K and 5.2×10^{-9} cm² s⁻¹ at 323 K). The solid lines are the theoretical curves for isothermal diffusion from Eq. (6.10) with the appropriate value of D_c/r_c^2. The uptake curves for the smaller (7.3 μm) crystals show considerable deviation from the isothermal curves but conform well to the theoretical non-isothermal curves calculated from Eq. (6.100) with the value of D_c estimated from the data for the large crystals, the value of β' calculated from the equilibrium data, and the value of α' estimated using heat transfer parameters estimated from uptake rate measurements with a similar system under conditions of complete heat transfer control. The limiting isothermal curve [Eq. (6.10)] is also shown by a continuous line with no points. From Ruthven et al. [2] with permission.

controlled by heat transfer rather than by diffusion. It is evidently much easier to obtain reliable intracrystalline diffusivities with large crystals. However, there is some evidence that the implicit assumption that the intrinsic diffusivity should be independent of crystal size may not always be valid as a result of the role played by intracrystalline defects.

13.1.2.2 Criterion for Negligible Thermal Effects

It is possible to develop a simple criterion for the absence of significant heat effects from the non-isothermal model (Section 6.2.4). Evidently, if either α' is large (heat transfer much faster than internal diffusion) or $\beta' \to 0$ (high heat capacity or low heat of adsorption) the intrusion of heat transfer resistance will be insignificant. For a Langmuirian system it may be shown that:

$$\beta' = \left(\frac{\Delta H}{RT}\right)^2 \left(\frac{Rq_s}{C_s}\right) \theta(1-\theta) \tag{13.1}$$

Clearly, for either $\theta \to 0$ or $\theta \to 1.0$, $\beta' \to 0$, implying negligible intrusion of heat effects. Heat transfer resistance will be most important at intermediate loadings ($\theta \approx 0.5$) at which β' will have its maximum value given by:

$$\beta'_{max} = \frac{1}{4}\left(\frac{\Delta H}{RT}\right)^2 \left(\frac{Rq_s}{C_s}\right) \tag{13.2}$$

Under conditions of heat transfer control the approach to equilibrium is described by Eq. (6.103) while under isothermal conditions the corresponding expression is Eq. (6.10) [or Eq. (6.11)]. Clearly, if β' is sufficiently small heat transfer resistance will intrude only in the final stages of the approach to equilibrium, for example, the last 1% if $\beta' = 0.01$.

A second criterion may be found by comparing the exponents of Eqs. (6.103) and (6.14). This yields:

$$\frac{ha}{C_s} \cdot \frac{r_c^2}{D_c(1+\beta')} = \frac{\alpha'}{1+\beta'} \gg \pi^2 \tag{13.3}$$

We thus conclude that thermal resistance can be neglected if either $\beta' < 0.01$ or $\alpha'/(1+\beta') > 100$.

The value of β' is largely fixed by the thermodynamic parameters, although some reduction can be achieved by adding inert heat capacity to the sample. If intraparticle diffusion is slow, the second criterion may be fulfilled even for small adsorbent particles. However, when diffusion is fast, the most favorable conditions for diffusion measurements will be obtained by using the minimum possible quantity of adsorbent (to maximize the external surface area/volume ratio for the sample) and the largest available crystals (or particles) to maximize the diffusional time constant r_c^2/D_c.

13.1.2.3 Intrusion of Bed Diffusional Resistance

Since gravimetric measurements are generally made with an assemblage of particles or zeolite crystals, rather than with a single isolated particle, the adsorption/desorption kinetics may be controlled by diffusion into the sample bed, rather than by diffusion into the individual crystals (bed diffusion control). This regime was discussed in detail from the theoretical viewpoint in Section 6.2.2. Figure 13.3 shows an example where such behavior was observed experimentally. For the same pressure step the sorption rates are essentially the same for the two different crystal sizes, showing clearly that the rate-controlling process must be associated with diffusion into the crystal bed rather than into the individual crystals. Detailed analysis of this behavior has been given by Lee and Ruthven [3].

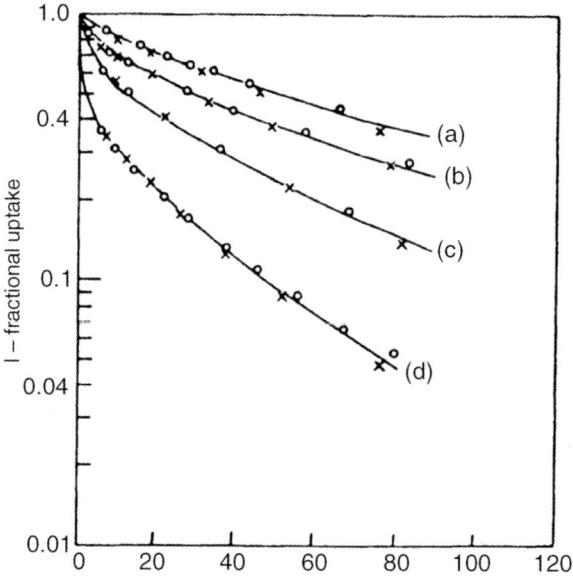

Figure 13.3 Gravimetrically measured transient sorption curves for isooctane in NaX zeolite crystals at 403 K, showing bed diffusion control. The measurements were made over similar differential pressure steps with two different crystal sizes (×, 24 μm; ○, 39 μm). Pressure steps (Torr): (a) 0.0019–0.0075; (b) 0–0.0019; (c) 0.0195–0.03; (d) 0.17–0.53 and 0.37–0.8. From Ruthven and Lee [3] with permission.

13.1.2.4 Experimental Checks

These examples show that, in any uptake rate measurements aimed at determining intraparticle diffusivities, it is essential to confirm that the conditions of the experiment conform to the basic assumptions of the model. Varying the sample quantity and the sample configuration provides a straightforward test for the intrusion of both heat transfer and bed diffusional resistances. It is not always possible to vary the size of the individual particles but if this is possible it provides a sensitive test for intraparticle diffusion control as the time constant should vary with the square of the crystal radius. Unfortunately, these tests have not always been performed in published experimental studies. In the absence of such checks the reported data must be treated with caution.

13.1.3
The TEOM and the Quartz Crystal Balance

Two novel types of microbalance that largely avoid the extraneous heat and mass transfer resistance issues associated with the use of relatively large adsorbent samples have been developed in recent years. In quartz crystal balance a few particles of adsorbent (in some cases a single particle) are attached by adhesive to a quartz crystal mounted in the gas dosing system. The resonant frequency depends sensitively on the mass and therefore varies with the sorbate loading. Such balances are very sensitive ($\approx 10^{-8}$ g), and because only a few adsorbent particles are active they are far less prone than a conventional microbalance to intrusion of extraneous resistances to mass or heat transfer. However, the use of an adhesive to attach the particle to the quartz crystal raises other issues such as the possibility that the surface may be obstructed.

The TEOM (tapered element oscillating microbalance) works in a similar way. The oscillating element consists of a tapered quartz tube containing a small sample of the adsorbent particles through which the sorbate gas flows. The gas flow rate is kept high to minimize extracrystalline mass transfer resistance. As with the quartz crystal balance the resonant frequency varies with the loading of the adsorbent, so the mass adsorbed at any time can be followed. This appears to be an excellent device that avoids most of the limitations of other types of microbalance. The earliest application of this device to the study of adsorption kinetics appears to be due to Rebo et al. [4]. A detailed description of the equipment has been given by Zhu et al. [5], who used it to study the diffusion of linear and branched alkanes in silicalite [6].

13.2
Piezometric Method

The piezometric method, which involves tracking the pressure in a constant volume system following a step addition or removal of the sorbate from the cell containing the adsorbent has been widely used as an alternative to the gravimetric method (Figure 13.4).

Essentially the same information may be obtained by either technique and many of the limitations are similar. To achieve sensitivity to relatively small changes in the amount adsorbed it is essential that the system volume be kept small. There is a temptation to increase the quantity of adsorbent to maximize the pressure change during uptake but this is not always a good idea since a larger adsorbent sample increases the probability of extracrystalline diffusional or heat transfer resistance.

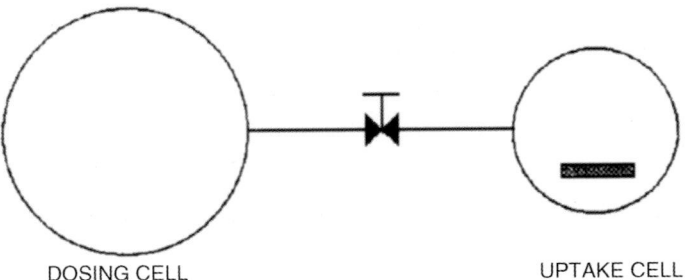

Figure 13.4 Sketch showing the essential components of a piezometric system for measuring sorption rates.

13.2.1
Mathematical Model

Another important issue that is often either ignored or not properly considered is the effect of the valve. The time delay due to the valve is strongly dependent on the quantity of gas flowing through it during the dosing step. The widely used assumption that the response time can be estimated from a blank run with no adsorbent in the uptake cell is not valid when the sorbate is strongly adsorbed. Detailed analyses of the limitations of this type of system have been presented by Schumacher et al. [7] and by Brandani [8], who derived a detailed mathematical model for the pressure response in this type of system:

$$\frac{p_d}{p_{od}} = \frac{3\delta}{1+3\delta+3\gamma} + \sum_{i=1}^{\infty} a_i \exp(-\beta_i^2 \tau) \tag{13.4}$$

$$\frac{p_u}{p_{od}} = \frac{3\delta}{1+3\delta+3\gamma} + \sum_{i=1}^{\infty} a_i (1-\beta_i^2/w) \cdot \exp(-\beta_i^2 \tau) \tag{13.5}$$

where:

$$a_i = \frac{2w^2 \delta \beta_i^2}{2w^2 \delta \beta_i^2 + (w-\beta_i^2)^2 (\beta_i^2 + z_i^2 - z_i + 2\gamma \beta_i^2)} \tag{13.6}$$

and β_i are the positive non-zero roots of the equation:

$$\beta_i \cot \beta_i = z_i \tag{13.7}$$

The dimensionless variables in these equations are defined as follows:

$$z_i = 1 + \gamma \beta_i^2 + \frac{w \delta \beta_i^2}{w - \beta_i^2}; \quad \tau = \frac{D_c t}{r_c^2}, \quad \gamma = \frac{V_u}{3KV_s}, \quad \delta = \frac{V_d}{3KV_s},$$

$$w = \frac{RT\chi D_c}{V_d r_c^2}, \quad p_d = \frac{P_d - P_{ou}}{P_\infty - P_{ou}}, \quad p_u = \frac{P_u - P_{ou}}{P_\infty - P_{ou}} \quad \chi = \text{valve constant}$$

The subscripts "u" and "d" refer to the uptake and dosing cells (volumes V_u and V_d), respectively. When w is large so that the valve delay is negligible, these equations reduce to Eqs. (16, 17, 18).

Figure 13.5 shows representative plots, calculated from these equations, showing the pressure response for both the uptake and dosing cells for several different sets of parameters. Evidently, the response depends strongly on both the adsorption equilibrium and the valve time constant. If w is small, the adsorption/desorption process is always close to equilibrium and the overall rate is controlled by flow through the valve. Obviously, no useful kinetic data can be obtained under such conditions. On the other hand, if w is large the effect of the valve is insignificant and the overall rate is controlled by the sorption kinetics. The form of the pressure response in the uptake vessel is sensitive to the nature of the kinetic regime. Under conditions of valve control the pressure in the uptake chamber increases monoton-

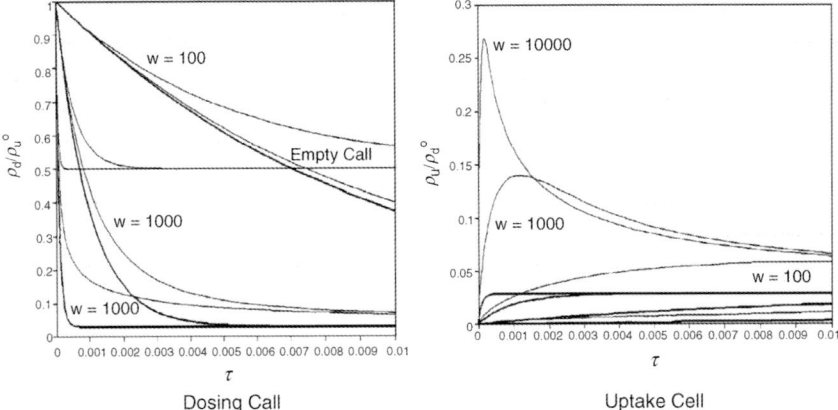

Figure 13.5 Theoretical response curves for a piezometric system calculated from Eqs. (13.4)–(13.7) with $\gamma = \delta = 0.01$, for various values of the parameter w. The curves corresponding to equilibrium control are shown by heavier lines. From Brandani [8] with permission.

ically, whereas in the kinetic regime the pressure passes through a maximum. In contrast the pressure in the dosing chamber always decreases monotonically regardless of the nature of the rate controlling process.

From a practical standpoint this means that it is important to follow the pressure in the uptake cell since this provides direct confirmation of the controlling regime. When the valve restriction is properly applied it appears that, for strongly adsorbed species, the maximum reliably measurable value of D_e/R^2 is approximately 0.005–0.01 s^{-1}, which is of the same order as for the best gravimetric systems. Thus, from the perspective of performance, piezometric and gravimetric systems are comparable. However, the piezometric system has a cost advantage, as it does not require an expensive electrobalance. Figure 13.6 presents an example showing the successful

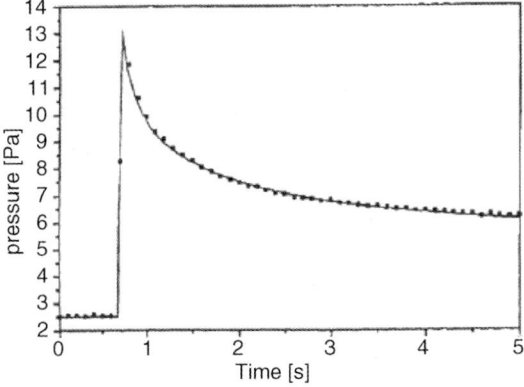

Figure 13.6 Piezometric uptake curve for ethyl benzene (2.5 molecules per unit cell) in HZSM-5 crystals ($r_c \approx 1$ μm) at 355 K ($D_c/r_c^2 \approx 0.005$ s^{-1}). From Schumacher et al. [9] with permission.

13.2.2
Single-Step Frequency Response

The "single-step frequency response" method [10, 11] is in essence simply an alternative way of carrying out a piezometric measurement. Rather than using a dosing system to change the pressure, the system volume is changed abruptly by moving a magnetically driven piston. The pressure following such a volume change is then monitored using a rapid response capacitance manometer. The pressure response is analyzed in the same way as for a piezometric system, but this approach has the important advantage that the restrictions associated with the time constant of the valve are avoided, so the mathematical model reduces to Eqs. (6.16)–(6.18). Figure 13.7 shows an example of the application of this approach.

13.2.3
Single-Step Temperature Response

The experimental technique developed by Grenier and Meunier [12, 13] is in principle similar to the "single-step frequency response" except that, rather than monitoring the pressure, the temperature response of the adsorbent resulting from a step change in the system volume is followed using a sensitive infrared detector. The sorption rate and hence the diffusional time constant are derived by application of the non-isothermal model (Section 6.2.4). An example is shown in Figure 13.8. For fast diffusing systems this approach has the advantage that the intrusion of heat effects is avoided, since these are accounted for in the model. The initial temperature rise is

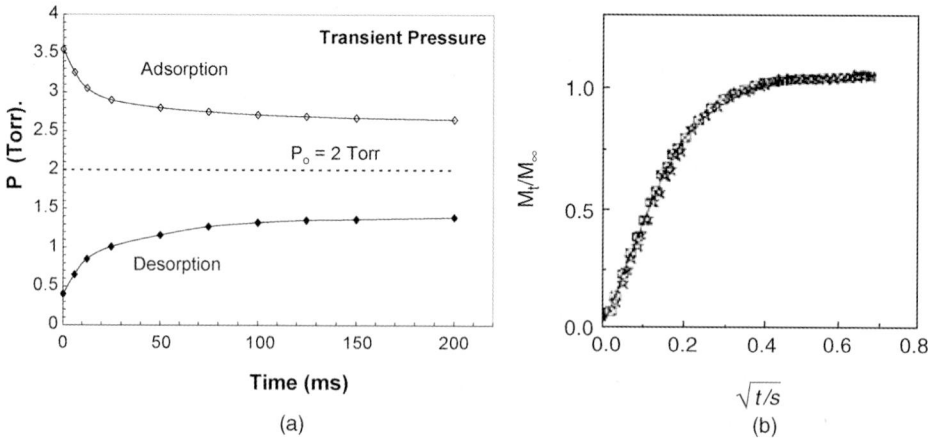

Figure 13.7 Pressure response (a) and sorption rate curve (b) for benzene in NaX zeolite (440 K, 2 Torr) derived from "single-step frequency response" measurements; (□) adsorption an (*) desorption. The line is the solution of the diffusion equation [Eq. (6.16)] with $D = 4.9 \times 10^{-10}$ m^2 s^{-1} and $\Lambda = 0.6$. From Song and Rees [10] with permission.

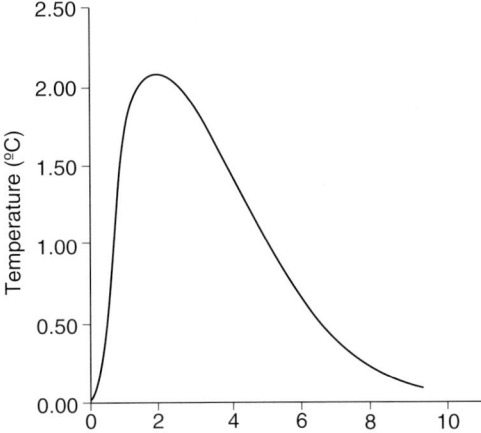

Figure 13.8 Transient temperature response for the system NaX–methanol for a pressure step of 48–80 Pa, showing conformity between the experimental temperature response and the theoretical curve calculated from the non-isothermal diffusion model [Eqs. (6.100)] with $D_o = 2.6 \times 10^{-12}$ m^2 s^{-1} and $h = 2.3$ W m^{-2} K^{-1}. From Grenier et al. [12] with permission.

controlled mainly by diffusion while the longer time decline in temperature is controlled mainly by heat transfer, so reasonably reliable values for both the heat transfer and diffusional time constants can be obtained from a single experimental curve.

13.3
Macro FTIR Sorption Rate Measurements

The IR active frequencies for an adsorbed molecule are generally significantly different from the corresponding frequencies for the same molecule in the vapor phase. As a result the absorbance (at the appropriate frequency) provides a direct measure of the adsorbed phase concentration of that particular species and, to a good approximation, the absorbance at the characteristic frequency is not affected by the presence of other species in either the gas or adsorbed phase. The "macro FTIR technique" depends on measuring the IR absorbance for an adsorbent wafer at a particular frequency in order to follow the transient adsorption or desorption in response to a step change in the ambient partial pressure of the sorbate. The prefix "macro" is intended to distinguish this technique from the single-crystal "micro FTIR" technique described in Chapter 12.

An early measurement of the equilibrium isotherm for ethene on zeolite Y by this method was reported by Liengme and Hall in 1966 [14] but the application to sorption rate measurements was pioneered by Karge and his coworkers between 1975 and 1995 [15–18]. This method can be regarded as an alternative to the gravimetric or piezometric methods and essentially similar data (the transient uptake curves) are obtained. However, the FTIR method has some notable advantages, the most

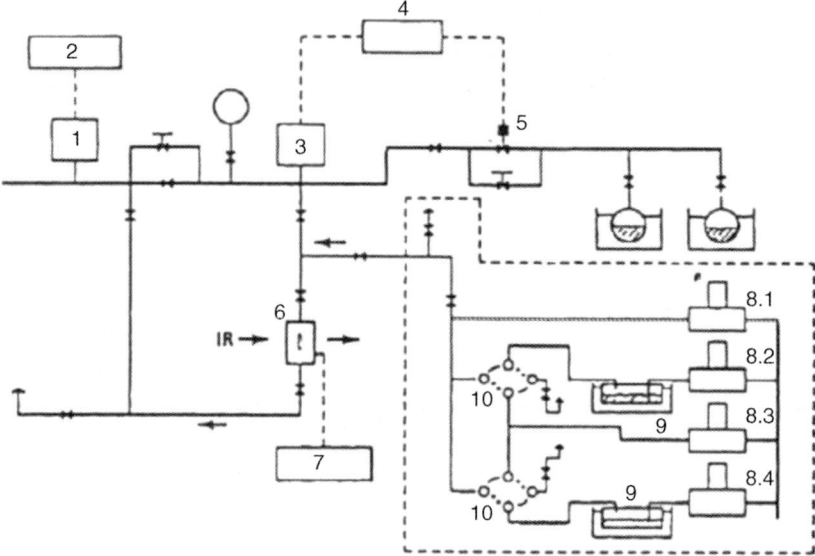

Figure 13.9 Experimental system for macro IR sorption rate measurements. Ionization gage (1), ionization gage controller (2), Baratron (3), pressure controller (4), solenoid valve (5), flow-through IR cell (6), temperature controller (7), mass flow controllers (8), saturators (9), and four-way valve (10). Reprinted from Karge and Kärger [18] with permission.

important of which is that it is not restricted to single-component measurements and can therefore be used to study co- and counter-diffusion in a binary or multicomponent system.

The essential components of the experimental system are shown schematically in Figure 13.9. The heart of the system is the IR cell in which the adsorbent wafer is mounted and through which the IR beam passes. Ancillary valves and flow controllers allow the composition of the gas flowing through the cell to be changed rapidly, for example, from pure carrier (usually He) to a well defined and constant mixture containing the sorbate, or in the reverse direction from a mixed stream to pure carrier. The transient adsorption or desorption curve is then derived directly from the time dependence of the absorbance at the selected frequency. Figure 13.10 shows an example (for p-xylene on HZSM-5 at $1516\,\text{cm}^{-1}$). Examples of binary adsorption/desorption curves, measured by this method under counter- and co-diffusion conditions, are shown in Figures 13.11 and 13.12, respectively. The co-diffusion experiments show the typical overshoot of the concentration of the less strongly adsorbed species. To determine the diffusional time constants these curves are fitted to the appropriate solution of the transient diffusion equation, just as in other uptake rate measurements.

As with other sorption rate measurements the uptake curves are, in principle, subject to the intrusion of heat transfer and extracrystalline mass transfer resistances. However, these effects can be minimized by using a small adsorbent sample (typically 10–15 mg) and designing the system to ensure that the flow rate of gas through the

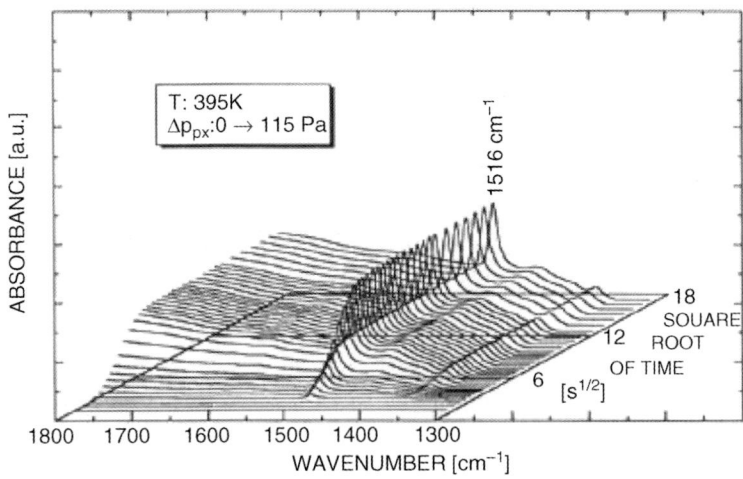

Figure 13.10 Set of p-xylene spectra at successive times showing adsorption on HZSM-5 over a period of 324 sec. Reprinted from Niessen and Karge [17] with permission.

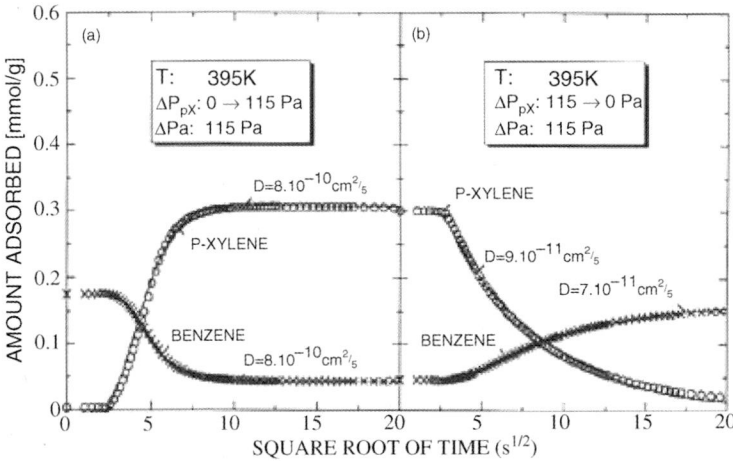

Figure 13.11 Counter-diffusion of p-xylene and benzene on HZSM-5 at 395 K, measured by FTIR, showing adsorption and desorption steps. Reprinted from Niessen and Karge [17] with permission.

cell is sufficiently high. The sensitivity of the method depends on the specific absorbance for the chosen wavelength. In general aromatics have relatively high specific absorbances and therefore high sensitivity, but for alkanes the absorbances and therefore the sensitivities are generally much lower. For that reason this technique has been widely applied to the study of aromatic sorbates and less widely to other species. The signal may be enhanced by accumulation but the signal-to-noise ratio eventually becomes limiting.

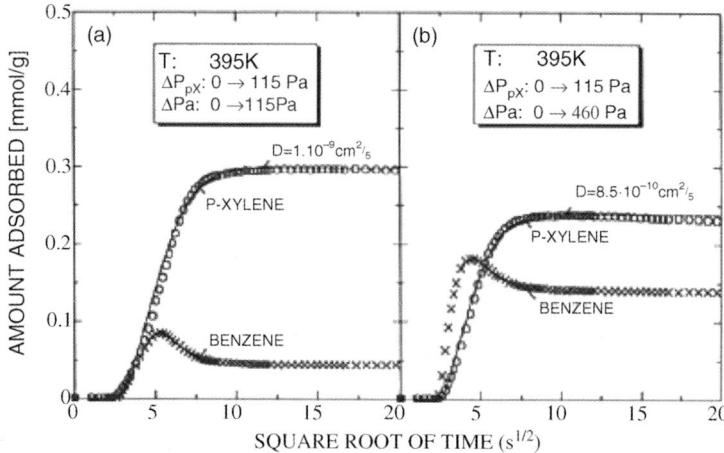

Figure 13.12 Co-diffusion of *p*-xylene and benzene in HZSM-5 at 395 K, measured by FTIR, showing adsorption and desorption steps. Reprinted from Niessen and Karge [17] with permission.

13.4
Rapid Recirculation Systems

With the exception of the FTIR technique the methods described above are only suitable for single-component studies. For the study of sorption kinetics in binary or multicomponent systems the recycle or recirculation system offers a convenient approach since, in such a system, the composition of the gas may be monitored continuously without perturbing the system. Such a device, shown schematically in Figure 13.13, was developed by Air Products to study the kinetics of tracer

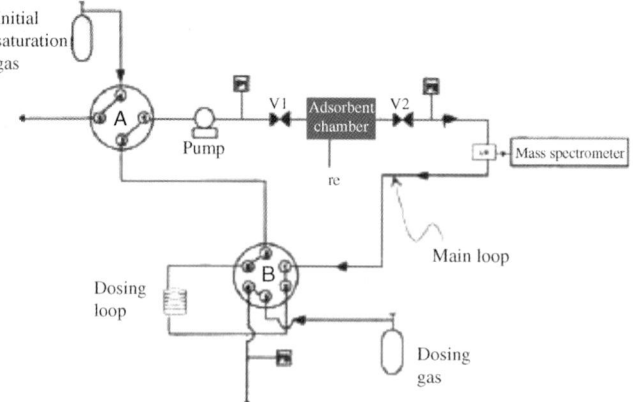

Figure 13.13 The Air Products rapid recirculation system for tracer exchange measurements. Reprinted from Rynders et al. [19] with permission.

exchange [19, 20]. The experimental data are analyzed in the same way as for a classical uptake measurement in a system of finite volume [Eqs. 6.16–6.18].

Such a system was first suggested by Denbigh [21] for the study of chemical kinetics in homogeneous systems. More recently this approach has been widely applied in the study of heterogeneous catalytic kinetics as well as for the study of sorption kinetics [22]. Various different designs have been developed. The "Berty Reactor" is a compact commercial version of such an apparatus in which the gas recirculation is internal, thus minimizing the external gas volume [23]. A major advantage of the recycle system is that by maintaining a high recirculation velocity the external fluid film resistance may be reduced to an insignificant level.

The spinning basket contactor suggested by Carberry [24] is also intended to provide good fluid–solid mass transfer characteristics. However, it is much less satisfactory than the recirculation system since, as a result of the hydrodynamic drag, the fluid phase tends to follow the spinning basket of adsorbent, so a high rate of rotation does not guarantee the high relative velocity between fluid and solid, which is the essential requirement for minimization of external mass transfer resistance. A detailed analysis of the sorption kinetic data obtained in such a system by Ma and Lee [25] in fact shows that the sorption kinetics must have been controlled by transport through the external boundary layer surrounding the spinning basket, rather than by intraparticle or intracrystalline diffusion [26].

13.4.1
Liquid Phase Systems

The measurement of sorption rates in liquid systems is formally similar to the piezometric vapor phase measurement. In general the measurement is made in the presence of a solvent, which, depending on the molecular size, may or may not penetrate the micropores of the adsorbent. The concentration is changed by adding a finite quantity of sorbate at time zero and the composition of the liquid phase is maintained constant by stirring or by using a rapid recirculation system similar to the vapor phase system shown in Figure 13.13. Progress towards equilibrium is followed by monitoring the composition of the liquid phase and the time constant is derived by matching the sorption curve to the theoretical curves for a finite volume system [Eq. (6.16)] just as for a piezometric measurement.

13.5
Differential Adsorption Bed

A simple variant of the uptake rate method that is especially useful for measuring diffusion in multicomponent systems and for systems in which external film resistance may be a problem is the differential bed method. In essence this involves contacting a small differential bed of the adsorbent with a high flow rate of the sorbate stream. The high flow insures that the composition of the fluid remains essentially constant through the bed and allows external fluid film resistance to be minimized.

After a known period of exposure the bed is isolated and purged or evacuated at elevated temperature to fully desorb all adsorbed species. By changing the contacting time, the sorption/desorption curve for any component can be determined and analyzed in the same way as for a conventional batch experiment. This approach appears to have been first applied to adsorption rate measurements by Carlsson and Dranoff [27].

The method is more time consuming than a conventional batch uptake measurement but the analysis and interpretation of the data are straightforward and unambiguous. The method has been developed and applied to liquid adsorption systems by Cantwell and coworkers [28] and to vapor phase systems by Do and coworkers [29]. One version of the experimental system is shown schematically in Figure 13.14. The differential column (C_1) is connected between two ports of a chromatographic sample injection valve. In the "saturation" position the sample solution is pumped from reservoir P_1, at a high flow rate, through the differential column and to waste. After the required contact time the valve is switched so that the column is flushed with a strong eluent from reservoir (P_2) and passed directly to the analytical column (C_2), in which the components are resolved, and thence to the LC

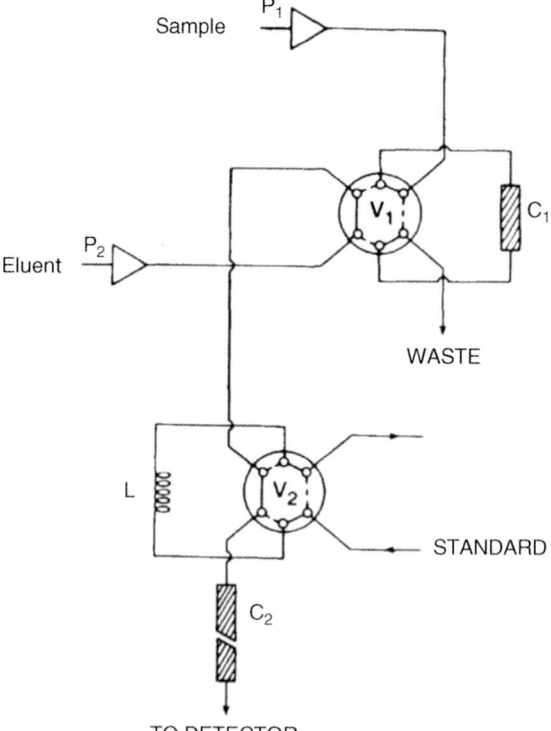

Figure 13.14 Schematic diagram showing the system for "shallow bed" kinetic measurements. After May et al. [28].

detector. The second sample valve and the loop (L) allow the system to be calibrated by injecting a known quantity of a standard solution. Blank experiments are used to correct for the liquid hold-up in the sample. In the original version this method was developed as a means of measuring adsorption equilibrium [28], but it was extended to kinetic measurements by Dubetz [30].

13.6
Analysis of Transient Uptake Rate Data

13.6.1
Time Domain Matching

The most direct way of determining the diffusional time constant from an experimental uptake curve is to match the experimental curve to the appropriate theoretical curve expressed in terms of the dimensionless time variable. The solutions for some practically important situations are summarized in Section 6.2. In general, the more complex the model the greater is the uncertainty in the derived parameters. In selecting the conditions for experimental measurements it is therefore desirable to keep the physical situation as simple as is possible within the constraints imposed by practical considerations. For example, the analysis of a linear system with constant diffusivity is in general subject to less uncertainty than the analysis of a nonlinear system with varying diffusivity. It is therefore generally desirable to make measurements differentially over small concentration steps, rather than attempting to extract differential diffusivity data from integral measurements made over large concentration steps. There are, however, some situations where integral measurements offer the only practical approach, for example, when the isotherm is highly favorable or rectangular. Some of the problems involved in the analysis and interpretation of integral uptake rate data are considered below. In a linear system, the uptake curve, expressed in terms of the fractional approach to equilibrium, is independent of the magnitude or direction of the concentration step. Varying the step size and comparing adsorption and desorption response curves thus provides a simple and straightforward experimental test for system linearity. However, invariance of the response to the magnitude of the step and conformity between adsorption and desorption provides no evidence concerning the absence of heat transfer limitations (Section 6.2.4).

Since several different transport processes may control the rates of adsorption/desorption, in the application of sorption rate measurements to the determination of intracrystalline or intraparticle diffusivities it is desirable to make the measurements under conditions such that only one of these resistances is dominant. If that is impossible then an appropriate model, taking account of more than one rate process, must be used to interpret the transient sorption curves. However, when more than one rate process is significant, the accuracy with which individual rate parameters can be determined is obviously reduced.

Verification of the rate-controlling transport process can generally be accomplished by a properly selected series of preliminary experiments in which the adsorbent particle size and/or the configuration of the adsorbent sample is varied. The application of this approach is illustrated in Figures 6.25, Figures 13.2 and 13.3. Obviously, if the sorption rate is controlled by intraparticle diffusion it should be independent of the configuration of the adsorbent sample and the time constant should vary with the square of the particle radius. In vapor phase systems the commonest intrusions are from bed diffusion and heat transfer resistance. Varying the sample configuration provides a sensitive test for both these effects.

In a composite adsorbent the macropore time constant will depend on the square of the gross particle diameter whereas the micropore time constant will depend only on the diameter of the microparticles. Under conditions of micropore control the sorption rate should therefore be independent of the overall particle size.

13.6.2
Method of Moments

Where more than one resistance is significant the time domain solutions, even for a linear system, are quite complicated and therefore somewhat inconvenient to use directly in the analysis of experimental data. In this situation one alternative is to generate solutions for the uptake curve directly by numerical solution of the governing differential equations. Another approach, which was introduced by Kocirik and Zikanova [31] and developed by Dubinin and coworkers [32], is to analyze the moments of the uptake curves. The principle of this approach is similar to the use of moments analysis in chromatography (Section 14.1.2).

The n-th moment of the uptake curve is defined by:

$$M_n = \int_{t=0}^{\infty} t^n \frac{dF}{dt} dt \equiv \int_{F=0}^{1} t^n dF \equiv n \int_0^{\infty} t^{n-1}[1-F(t)]dt \tag{13.8}$$

where $F = m_t/m_\infty$, the fractional approach to equilibrium. The experimental values of the moments can thus be determined directly from the uptake curve by integration. By the use of van der Laan's theorem [33] the expressions for the moments of the uptake curve may be derived directly from the solution of the model equations in Laplace form, thus eliminating the need to invert the transform to obtain the time domain solution:

$$M_n = (-1)^n \lim_{s \to 0} \frac{d^n}{ds^n}[\overline{F}(s)] \tag{13.9}$$

where $\overline{F}(s)$ is the Laplace transform of the uptake curve $F(t)$:

$$\overline{F}(s) = \int_0^{\infty} e^{-st} F(t) dt \tag{13.10}$$

The first moment corresponds to the time constant (τ). Table 13.1 summarizes the values for the simple cases of a parallel sided slab (thickness 2ℓ) and a spherical particle (radius r_c) with surface resistance or internal diffusion resistance control.

Table 13.1 Time constants for simple single-resistance systems.

	Slab	Sphere
Surface resistance	ℓ/k_s	$r_c/3k_s$
Internal diffusion	$\ell^2/3D$	$r_c^2/15D$

For complex systems the expressions obtained in this way are simpler and more tractable than the full expressions for the time domain solutions. For example, for an isothermal biporous adsorbent subjected to a step change in surface concentration:

$$M_1 = \frac{KR_p^2}{15\varepsilon_p D_p} + \frac{r_c^2}{15 D_c} \tag{13.11}$$

$$M_2 = 2M_1^2 + \frac{6}{1575}\left[\left(\frac{KR_p^2}{\varepsilon_p D_p}\right)^2 + \left(\frac{r_c^2}{15 D_c}\right)^2\right] \tag{13.12}$$

These expressions may be compared with the time domain solution given in Table 6.2.

In principle one may obtain both micropore and macropore diffusivities from the simultaneous solution of Eqs. (13.11) and (13.12). However, since the accuracy with which the moments can be evaluated decreases with the order, it has been more usual to use only the expression for the first moment and to vary the particle size to separate the time constants. If the model properly represents the system behavior, a plot of M_1 versus R_p^2 should yield a straight line. The values of D_p and D_c may then be found from the slope and intercept.

One advantage of the moments approach is that the contributions from time delays in the experimental system are easily accounted for. This approach has been used by Bülow and coworkers to investigate the influence of a range of extracrystalline transport resistances on piezometric response curves [34–36]. However, the method suffers from the severe disadvantage that the information contained in the *shape* of the response curve is lost. With the widespread availability of fast computers the generation of the response curves in the time domain no longer presents a serious problem and in most situations direct time-domain matching therefore appears to be preferable to the use of indirect methods such as moments analysis.

13.7
Tracer Exchange Measurements

The experimental techniques discussed in Sections 13.1–13.6 all focus on the macroscopic measurement of transient adsorption/desorption curves as a way to measure transport diffusion at the intraparticle or intracrystalline scale. Many of the techniques can be modified, by using a suitable isotopically labeled sorbate, to

Table 13.2 Some common radio-isotopes.

	Half-life	Form of radiation	Energy (keV)
^{3}H	12.5 years	β^-	19
^{14}C	5730 years	β^-	155
^{32}P	14.3 days	β^-	1710
^{60}Co	5.3 years	β^- ($+\gamma$)	474
^{133}Xe	5.25 days	β^- ($+\gamma$)	346

measure the kinetics of tracer exchange and hence to study self-diffusion. A basic assumption of the isotopic tracer technique is that the isotopic forms have identical properties. While this can never be precisely true, it is in general a reasonable approximation, particularly when the percentage difference in the atomic masses is small. Differentiation between the two isotopes, which is essential for tracer diffusion measurement, may be based on the difference in their mass (gravimetry or mass spectroscopy) or in their nuclear magnetization (NMR spectroscopy). The tracer method is particularly convenient when one of the isotopes is radioactive, since the activity then provides a sensitive and direct measure of the concentration of the labeled species. However, to avoid complications in the analysis, it is desirable that the half-time for radioactive decay should be large compared with the time scale of the diffusion measurement. In practice this is not a serious limitation, as the half-lives of most of the common radio-isotopes are relatively long. Table 13.2 gives a few examples [37].

13.7.1
Detectors for Radioisotopes

The Geiger-Müller counter is the traditional apparatus for detecting and measuring the intensity of nuclear radiation. This device consists of a tube containing an ionizable gas at low pressure. Nuclear radiation causes ionization of the gas, giving rise to a pulse of electric current that is measured. In principle a Geiger-Müller counter can detect α, β, and γ rays. However, α particles are easily absorbed by the surrounding medium, making their accurate measurement difficult. Nuclei that decay by α particle emission are therefore not convenient for use as tracers. β Particles (electrons) are also absorbed, but their range depends on their initial energy and on the thickness and density of the material traversed. For weak β emitters such as ^{3}H and ^{14}C this is a serious limitation that effectively precludes the use of a Geiger-Müller tube as a detector for such tracers.

This difficulty has been overcome by the development of the liquid scintillation counter (LSC) detector [38]. The essential component of the LSC detector is a solution containing molecules that emit a weak light flash when interacting with certain quanta of radiation. These light flashes are recorded and counted by use of a photomultiplier. The intimate mixing of the radioactive source with the detecting molecules avoids problems of absorption of radiation and thus provides a high efficiency of detection. Such solutions are called "internally counted" samples.

In principle the LSC technique may be applied to any radioactive nuclide. However, because of the costs involved in preparing the "scintillation cocktail," it is applied mainly with weak β emitters. The energy of the radiation (generally β rays) is transferred to the solvent and may appear as energy of either ionization or excitation of a solvent molecule. About 5% of the total energy absorbed eventually appears as light. In principle the light flashes might be applied directly to detect the radiation. However, since the relatively long fluorescence life times ($\sim$30 ms) may give rise to interference problems and since the emission spectrum of the solvent is often in a range that is not amenable to detection by a photomultiplier, it is normal to add a second molecular species (the solute), which is in turn stimulated by energy transfer from the solvent and emits readily detectable light quanta in the visible or near-ultraviolet regions. A detailed account of the operation of LSC detectors, including a discussion of common sources of error, is given by Dyer [38]. The LSC detection technique is particularly convenient for liquid systems and a wide range of such measurements has been reported [39–41].

13.7.2
Experimental Procedure

The basic principle of a tracer diffusion measurement is similar to the measurement of transport diffusivity by the transient uptake method, as discussed above. At time zero the sample is subjected to a change in the concentration of labeled species (in this case the radio-isotope) at the external surface and the progress towards equilibrium is followed, generally by monitoring the concentration in the fluid phase. The difference is that in the case of a tracer exchange measurement the boundary condition is chosen so as to maintain the total species concentration uniform throughout the sample; only the relative proportion of labeled and unlabeled sorbates varies. Because the total concentration is constant, the assumption of a constant diffusivity should be valid even though the change in the concentration of the labeled species may be large. Furthermore, since tracer exchange is equimolar, it is essentially isothermal. The linear isothermal models discussed in Chapter 6 are therefore used for the analysis of tracer exchange curves. However, the presence of more than one species in the external fluid phase introduces the possibility of external film resistance to mass transfer, which is not always easy to avoid.

13.8
Frequency Response Measurements

The frequency response technique, which depends on measuring the response of an adsorption system to a periodic perturbation, rather than to a single step change, was originally developed by Yasuda in the 1970s [42–45]. Since then it has been developed and applied by several different research groups and is now recognized as one of the most powerful macroscopic techniques. The basic theory, which was developed by Yasuda to interpret and analyze his experimental data, has been refined and extended to cover more complex kinetic behavior, but the original theory, which is summarized briefly here, still provides the essential theoretical framework. Further details can be found in two more recent reviews by Rees and Song [10, 11].

13.8.1
Theoretical Model

We consider an adsorption system at equilibrium so that the vapor pressure of the sorbate has its equilibrium value corresponding to the loading at the relevant temperature. Suppose that the volume of the system (V) is varied sinusoidally at angular frequency ω ($=2\pi f$) with a small perturbation (v) about its equilibrium value (V_e):

$$V = V_e[1 - v\cos(\omega t)] \tag{13.13}$$

The pressure in the system will follow a similar equation:

$$P = P_e[1 + p\cos(\omega t + \varphi)] \tag{13.14}$$

If the frequency is sufficiently low compared with the adsorption/desorption rate, the pressure will follow the volume perturbation with no time lag, so the phase difference (φ) will be zero. This will also be true if the perturbation is so fast that there is no significant adsorption on the time scale of the cycle. Between these two extremes the pressure perturbation will lag behind the imposed volume perturbation with a phase angle that depends on sorption rate.

By combining the diffusion equation, the mass balance, and the local isotherm (which is assumed to be linear) the relationship between the phase lag and the kinetic parameters may be derived. This is conveniently expressed in terms of the in-phase and out-of-phase responses. For a parallel-sided slab of half-thickness ℓ:

$$\frac{v}{p}\cos\varphi - 1 = K\delta_{in}; \quad \frac{v}{p}\sin\varphi = K\delta_{out} \tag{13.15}$$

where δ_{in} and δ_{out} are defined by:

$$\delta_{in}^1 = \frac{1}{\eta_1}\left(\frac{\sinh\eta_1 + \sin\eta_1}{\cosh\eta_1 + \cos\eta_1}\right) \quad \delta_{out}^1 = \frac{1}{\eta_1}\left(\frac{\sinh\eta_1 - \sin\eta_1}{\cosh\eta_1 + \cos\eta_1}\right) \tag{13.16}$$

where:

$$\eta_1 = \left(\frac{2\omega\ell^2}{D}\right)^{1/2}, \quad K = \frac{RT_e}{V_e}\left(\frac{dB}{dP}\right)$$

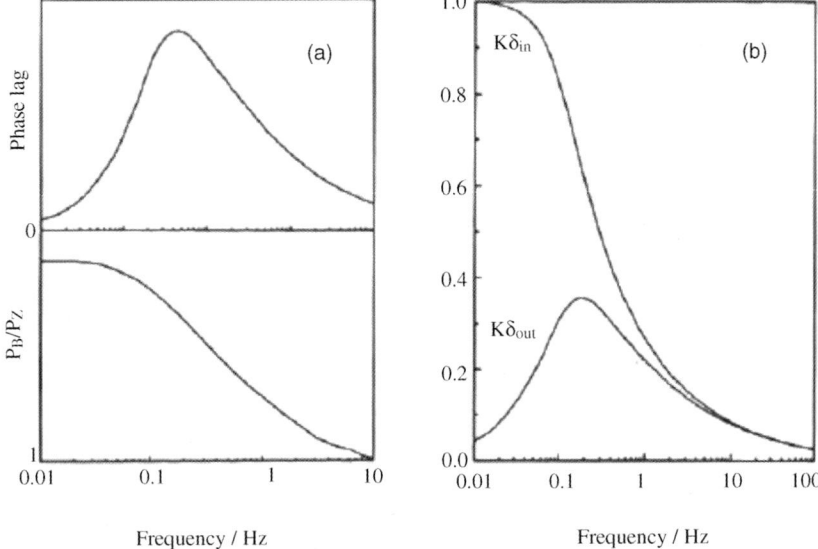

Figure 13.15 Theoretical curves calculated from Eq. (13.17) for a spherical particle of radius $a = 10$ μm, $K = 1$, and $D = 10^{-11}$ m^2 s^{-1}, showing the variation of (a) phase lag and amplitude ratio and (b) the characteristic functions with frequency. From Song and Rees [11] with permission.

and dB/dP is the local slope of the equilibrium isotherm. The corresponding expressions for a sphere of radius a are:

$$\delta_{in}^s = \frac{3}{\eta_s}\left(\frac{\sinh\eta_s - \sin\eta_s}{\cosh\eta_s - \cos\eta_s}\right) \quad \delta_{out}^s = \frac{3}{\eta_s}\left(\frac{\sinh\eta_s - \sin\eta_s}{\cosh\eta_s - \cos\eta_s} - \frac{6}{\eta_s}\right) \quad (13.17)$$

where $\eta_s = (2\omega a^2/D)^{\frac{1}{2}}$

Figure 13.15 shows the form of the phase lag and amplitude variation with frequency and the characteristic functions, for a sphere of radius $a = 10$ μm. The curves for a parallel-sided slab are qualitatively similar.

Notably, the maximum of the out-of-phase function and the inflection point of the in-phase function always occur at the frequency that is approximately the inverse of the characteristic time for the process (D/a^2 for the sphere or D/ℓ^2 for the slab). This allows a rough estimate of the time constant to be obtained directly by inspection.

For a linear system the extension to a more complex system with several mass transfer resistances is straightforward, as the overall response is simply the sum of the individual responses:

$$\delta_{in} = \sum_j \bar{K}_j \delta_{j,\,in} \quad \delta_{out} = \sum_j \bar{K}_j \delta_{j,\,out} \quad \bar{K}_j = \frac{K_j}{\sum_j K_j}$$

In this way the forms of the characteristic functions have been found for a wide range of different mass transfer mechanisms. As examples, the characteristic

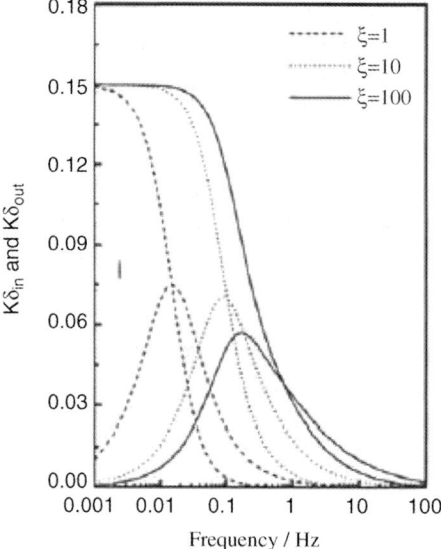

Figure 13.16 Theoretical frequency response spectra for a spherical particle with both internal diffusion and surface resistance; $K = 0.15$, $D/a^2 = 0.1\,\text{s}^{-1}$. The parameter $\xi = 3ka/D = 3L$ [Eq. (6.23)]. Reprinted from Song and Rees [11] with permission.

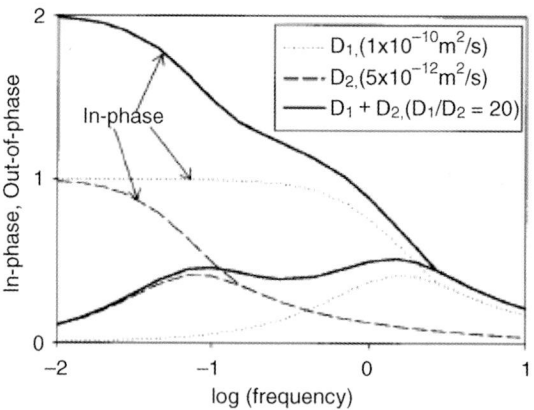

Figure 13.17 Theoretical curves showing the response for a system controlled by two independent diffusion processes with an order of magnitude difference in the two time constants. Reprinted from Turner et al. [46] with permission.

functions for a spherical particle with combined surface and internal diffusion resistance and for a particle with two independent (uncoupled) diffusion resistances are shown in Figures 13.16 and 13.17. Notably, in contrast to a single-resistance process, for combined surface and internal resistance the characteristic curves intersect at a point coinciding with or beyond the maximum. This appears to be generally true for series resistances regardless of the nature of the resistances.

It was originally thought that the frequency response technique would be insensitive to the intrusion of heat effects but Sun et al. [47, 48] have shown that this is not true. Heat transfer resistance shows up as an additional resonance in the out-of-phase response. Thus the spectrum for a non-isothermal system with a single mass transfer resistance appears very similar to that for two independent diffusion resistances. A clear differentiation between these two cases can only be obtained by extensive measurements over a wide range of conditions.

13.8.2
Temperature Frequency Response

As an alternative to the usual frequency response technique in which the pressure response to a volume perturbation is measured, Grenier, Meunier, and Sun [48–52] have developed a temperature frequency response technique in which the temperature response of the adsorbent sample is followed. This approach was made possible by the development of a very sensitive rapid response IR temperature detector capable of measuring temperature differences of a fraction of a degree ($<10^{-4}$ degree). This method has certain advantages over the pressure measurement technique. The response of the IR temperature detector is very rapid, so it imposes essentially no limitation on the frequency of operation. Since the heat balance is built into the mathematical model, the measurements are less likely to be affected by the unrecognized intrusion of heat transfer resistance. Furthermore, the phase lag between the pressure and the volume perturbation can be affected by adsorption on the cell wall, whereas the adsorbent surface temperature measurement tracks directly (through the heat balance) the sorbate loading on the adsorbent.

Meunier and his coworkers have expanded the theoretical analysis to include several more complex (non-isothermal) two- and three-resistance cases, including internal diffusion with surface resistance, two diffusional resistances (macropore/micropore) as in a commercial pelleted adsorbent and a diffusion/rearrangement model as well as anisotropic diffusion. They also showed convincingly that, contrary to early assumptions, the frequency response technique is not immune from the intrusion of heat transfer effects and that the bimodal response peaks that are commonly observed (see, for example, Figure 13.17) do not necessarily imply two independent diffusion processes but can arise also from a single mass transfer resistance coupled with finite heat transfer resistance (Section 13.8.1).

13.8.3
Measurement Limits

As with most macroscopic measurements, it is relatively easy to follow slower diffusion processes; the challenge is to extend the range of reliable measurement to faster processes, which requires measurements at higher frequencies. For reliable measurements by the frequency response method it is essential that the measure-

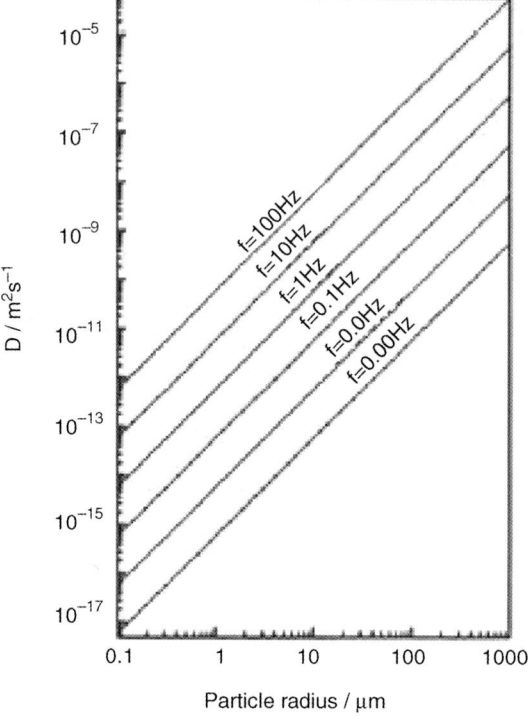

Figure 13.18 Measurable diffusivity (by frequency response) as a function of frequency and particle radius. Reprinted from Song and Rees [10] with permission.

ments extend beyond the maximum in the out-of-phase response. Using this as a criterion it is easy to calculate the relationship between the maximum measurable diffusivity, the crystal size, and the frequency. Such a plot, shown in Figure 13.18, provides rapid approximate guidance as to the system requirements.

13.8.4
Experimental Systems

There have been several different designs for experimental frequency response systems differing mainly in the way in which the volume perturbation is achieved. The main challenge has been to extend the accessible frequency range to permit measurement of faster diffusion processes. Yasuda's original device utilized a metal bellows driven through a variable speed gearbox and this had a maximum frequency of 0.25 Hz. The main limitation to the frequency arose from the use of a rather slow response pressure gage. To obtain a faster system Rees used a rapid response capacitance manometer and changed the displacement system to a magnetically driven piston, which generates a square wave rather than a sinusoidal perturbation.

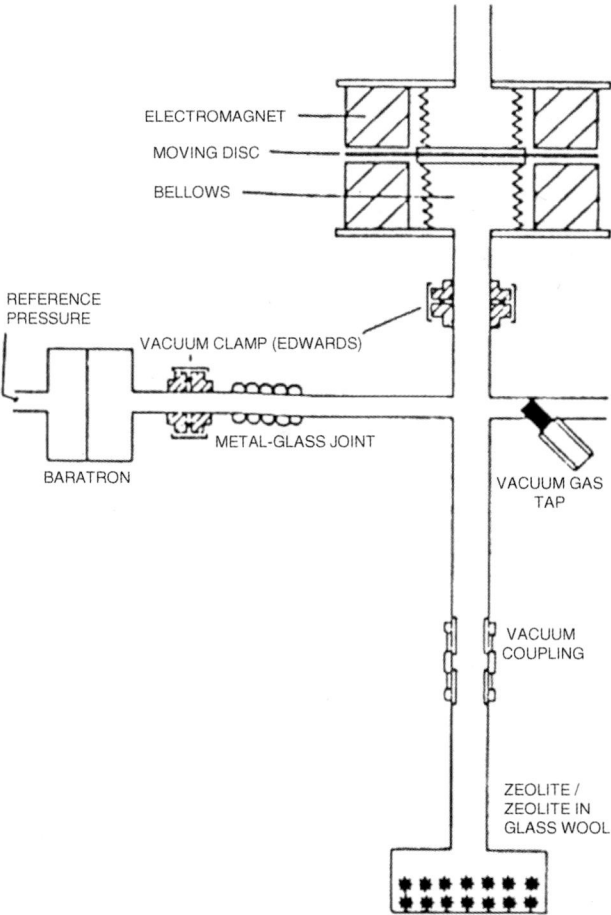

Figure 13.19 Magnetically driven frequency response system (square wave) developed by Rees.

This increases the accessible frequency range but requires somewhat more complicated data processing, since the sinusoidal components must be extracted from the experimental response by Fourier transformation. His system, which is shown schematically in Figure 13.19, was able to measure from 0.01 to about 10 Hz. More recently Turner et al. [46] designed an improved system based on a metal bellows pump that is capable of reliable sinusoidal operation up to about 30 Hz. An even faster system that utilized an acoustically coupled moving diaphragm to generate the volume perturbation was developed by Reyes [53]. Accurate sinusoidal operation up to 100 Hz is claimed for this system. The temperature frequency response system developed by Meunier and Grenier [47–52] was able to operate reliably over the range from about 10^{-4} to 30 Hz.

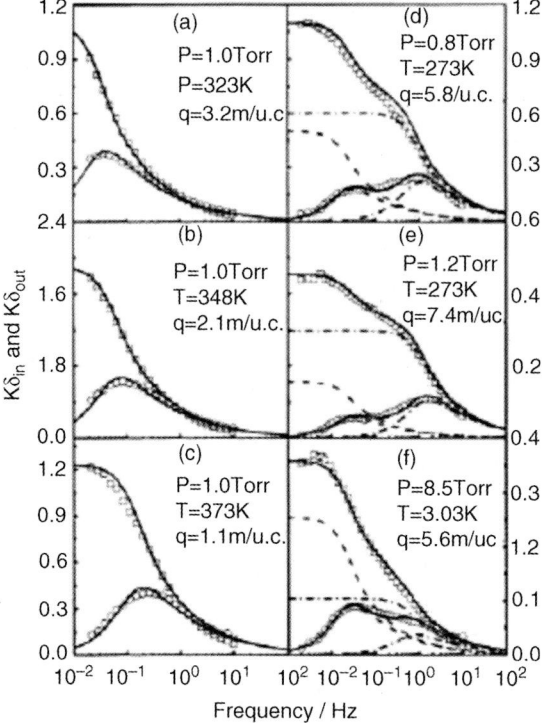

Figure 13.20 Frequency response spectra for benzene in silicalite showing the comparison between experimental response and the theoretical curves for a single (isothermal) diffusion resistance [(a)–(c)] and a dual diffusion resistance model [(d)–(f)]. Reprinted from Song and Rees [11] with permission.

13.8.5
Results

Figures 13.20 and 13.21 show several frequency response spectra selected to illustrate the variety of different patterns commonly observed.

At low loadings the behavior of benzene clearly conforms closely to the single diffusion resistance model but at higher loadings a second resonance appears. The origin of this second resonance is not completely clear and several plausible explanations have been suggested. The simplest explanation is that it is due to heat transfer resistance, which is expected to become more significant at higher loadings (Section 13.1.2).

The frequency response spectrum for CO_2 in zeolite theta, shown in Figure 13.21, shows clear evidence of two series resistances, probably surface resistance plus internal diffusion. Internal diffusion resistance appears to be dominant and the contribution from surface resistance is relatively minor ($\xi \approx$ 5–20). Clearly, the frequency response technique offers a very powerful method that can provide detailed insight into the transport behavior. By judicious application of this approach

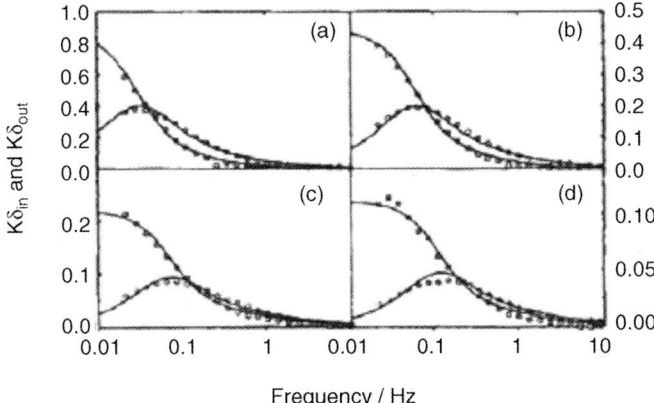

Figure 13.21 Frequency response spectra for CO_2 in zeolite theta at (a) 273, (b) 298, (c) 323, and (d) 348 K, showing conformity with the combined internal diffusion and surface resistance model. Reprinted from Song and Rees [11] with permission.

it is not only possible to determine the diffusional time constant but also to extract evidence concerning the nature of the rate limiting process.

13.8.6
Frequency Response Measurements in a Flow System

Some years ago Boniface and Ruthven [54] demonstrated the feasibility of carrying out a frequency response measurement in a flow system. Their system used a mass flow controller driven by a signal generator to impose a small sinusoidal fluctuation of the flow of an adsorbable species onto a much larger carrier flow. Like other frequency response methods, this approach has the advantage of quasi steady state operation, so a single experimental measurement is equivalent to the average of a large number of transient measurements.

The phase lag is directly related to the equilibrium hold-up in the column, while the attenuation (amplitude ratio) is related to the mass transfer resistance. Thus measurements carried out over a range of frequencies permit an unambiguous separation of the kinetic and equilibrium effects. However, as with a pulse chromatographic system, the amplitude ratio depends on the combined effects of mass transfer resistance and axial dispersion and an unambiguous resolution of these two effects is not always straightforward, except when one or other effect is dominant.

The system worked reasonably well but, due to the response time of the flow controller, the maximum frequency was limited to about 0.1 Hz. This restricts the applicability to relatively slow diffusion processes. More recently, a similar approach has been followed by LeVan and coworkers, who have developed both concentration swing and pressure swing versions of such a system [55–59]. One of the main limitations of the closed frequency response system is that heat transfer is limited, rendering the measurements susceptible to heat transfer limitations. This problem is much less severe in a flow system. However, the presence of a carrier gas renders the flow system susceptible to intrusion of gas phase mass transfer resistance, although this is generally a less severe restriction than that imposed by heat transfer limitations. The main disadvantage of this type of system is still the limitation on the maximum frequency, which, even in the more modern versions, cannot exceed about 1 Hz, thus limiting the applicability to relatively slow mass transfer processes.

References

1. Keller, J. and Staudt, R. (2004) *Gas Adsorption Equilibria*, Springer, Berlin.
2. Ruthven, D.M., Lee, L.-K., and Yucel, H. (1980) *AIChE J.*, **26**, 16.
3. Ruthven, D.M. and Lee, L.-K. (1981) *AIChE J.*, **27**, 654.
4. Rebo, H.P., Chen, D., Brownrigg, M.S.A., Moljord, K., and Holmen, A. (1997) *Collect. Czech. Chem. Commun.*, **62**, 1832.
5. Zhu, W., van de Graaf, J.M., van den Broeke, L.J.P., Kapteijn, F., and Moulijn, J.A. (1998) *Ind. Eng. Chem. Res.*, **37**, 1934.
6. Zhu, W., Kapteijn, F., and Moulijn, J. (2001) *Microporous Mesoporous Mater.*, **47**, 157–171.
7. Schumacher, R., Erhardt, K., and Karge, H.G. (1999) *Langmuir*, **15**, 3965.
8. Brandani, S. (1998) *Adsorption*, **4**, 17.
9. Schumacher, R., Lorenz, P., and Karge, H.G. (1997) *Progress in Zeolites and Microporous Materials, Studies in Surface Science and Catalysis*, vol. 105 (eds H. Chon, S.-K. Ihm, and Y.S. Uh), Elsevier Science, Amsterdam, pp. 1747–1754; see also Schumacher, R. and Karge, H.G. (1999) *Microporous Mesoporous Mater.*, **30**, 307.
10. Rees, L.V.C. and Song, L. (2000) *Recent Advances in Gas Separation by Microporous Ceramic Membranes* (ed. N.K. Kanellopoulos), Elsevier, Amsterdam, pp. 139–186.
11. Song, L. and Rees, L.V.C. (2007) *Molecular Sieves*, vol. 7 (eds H.G. Karge and J. Weitkamp), Springer-Verlag, Berlin, pp. 235–276.
12. Grenier, Ph., Meunier, F., Gray, P., Kaerger, J., Xu, Z., and Ruthven, D.M. (1994) *Zeolites*, **14**, 242.
13. Grenier, P., Bourdin, V., Sun, L.M., and Meunier, F. (1995) *AIChE J.*, **41**, 2047.
14. Liengme, B.V. and Hall, W.K. (1966) *Trans. Faraday Soc.*, **62**, 3229.
15. Karge, H.G. and Klose, K. (1975) *Ber. Bunsenges. Phys. Chem.*, **79**, 454.
16. Karge, H.G. and Weitkamp, J. (1986) *Chem.-Ing.-Tech.*, **58**, 946.
17. Niessen, W. and Karge, H.G. (1993) *Microporous Mater.*, **1**, 1–8.
18. Karge, H.G. and Kärger, J. (2007) *Molecular Sieves*, vol. 7 (eds H.G. Karge and J. Weitkamp), Springer-Verlag, Berlin, pp. 135–206.
19. Rynders, R.M., Rao, M.B., and Sircar, S. (1997) *AIChE J.*, **43**, 2456–2470.
20. Cao, D.V., Mohr, R.J., Rao, M.B., and Sircar, S. (2000) *J. Phys. Chem. B*, **104**, 10498–10501.
21. Denbigh, K.G. (1951) *J. Appl. Chem.*, **1**, 227.
22. Klemm, E. and Ehmig, G. (1997) *Chem. Eng. Sci.*, **52**, 4329–4344.
23. Berty, J.M. (1974) *Chem. Eng. Prog.*, **70**, 78.
24. Carberry, J.J. (1964) *Ind. Eng. Chem.*, **56** (11), 39; see also Carberry, J.J. (1969) *Catal. Rev.* **3**, 61.
25. Ma, Y.A. and Lee, T.Y. (1976) *AIChE J.*, **22**, 147.
26. Taylor, R.A. (1979) Ph.D. Thesis, University of New Brunswick, Fredericton.
27. Carlsson, W.W. and Dranoff, J.S. (1987) *Fundamentals of Adsorption* (ed. A.I. Liapis), (Proceeding 2nd International Conference on Adsorption, Santa Barbara, May 1986), Engineering Foundation, New York, pp. 129–144.
28. May, S., Hux, R.A., and Cantwell, F.F. (1982) *Anal. Chem.*, **54**, 1279.

29 Mayfield, P.L.J. and Do, D.D. (1990) *Gas Separation Technology* (eds E.F. Vansant and R. Dewolfs), Elsevier, Amsterdam, p. 247.
30 Dubetz, T.A. (1988) Ph.D. Thesis, University of Alberta, Edmonton.
31 Kocirik, M. and Zikanova, A. (1972) *Z. Phys. Chem. (Leipzig)*, **250**, 360;Kocirik, M. and Zikanova, A. (1974) *Ind. Eng. Chem. Fundam.*, **13**, 347.
32 Dubinin, M.M., Erashko, I.T., Kadlec, O., Ulin, V.I., Voloshchuk, A.M., and Zolotarev, P.P. (1975) *Carbon*, **13**, 193.
33 van der Laan, T. (1958) *Chem. Eng. Sci.*, **7**, 187.
34 Bülow, M., Struve, P., and Mietk, W. (1986) *Z. Phys. Chem. (Leipzig)*, **267**, 613.
35 Bülow, M., Struve, P., and Mietk, W. (1984) *J. Chem. Soc., Faraday Trans. I*, **80**, 813.
36 Struve, P., Kocirik, M., Bülow, M., Zikanova, A., and Bezus, A.G. (1983) *Z. Phys. Chem. (Leipzig)*, **264**, 49.
37 Weast, R.G. (ed.) (1974) *Handbook of Chemistry and Physics*, CRC Press, Cleveland, p. B248.
38 Dyer, A. (1970) *Liquid Scintillation Counting*, Pergamon Press, London.
39 Cundall, R.B., Dyer, A., and McHugh, J.O. (1981) *J. Chem. Soc., Faraday I*, **77**, 1039.
40 Dyer, A. and Yusof, A.M. (1987) *Zeolites*, **7**, 191.
41 Dyer, A. and Townsend, R.P. (1973) *J. Inorg. Nucl. Chem.*, **35**, 3001.
42 Yasuda, Y. (1976) *J. Phys. Chem.*, **80**, 1876.
43 Yasuda, Y. (1978) *J. Phys. Chem.*, **82**, 74.
44 Yasuda, Y. (1982) *J. Phys. Chem.*, **86**, 1913.
45 Yasuda, Y. (1994) *Heterog. Chem. Rev.*, **1**, 103.
46 Turner, M.D., Capron, L., Laurence, R.L., and Conner, W.C. (2001) *Rev. Sci. Instrum.*, **72**, 4424–4433.
47 Sun, L.M., Meunier, F., and Kärger, J. (1993) *Chem. Eng. Sci.*, **48**, 715.
48 Sun, L.M. and Bourdin, V. (1993) *Chem. Eng. Sci.*, **48**, 3783–3793.
49 Sun, L.M., Zhong, G.M., Gray, P.G., and Meunier, F. (1994) *J. Chem. Soc., Faraday Trans. 1*, **90**, 369–376.
50 Sun, L.M., Meunier, F., Grenier, Ph., and Ruthven, D.M. (1994) *Chem. Eng. Sci.*, **49**, 373–381.
51 Bourdin, V., Grenier, Ph., Meunier, F., and Sun, L.M. (1996) *AIChE J.*, **42**, 700–712.
52 Bourdin, V., Sun, L.M., Grenier, Ph., and Meunier, F. (1996) *Chem. Eng. Sci.*, **51**, 269–280.
53 Reyes, S.C., Sinfelt, J.H., DeMartin, G.J., and Ernst, R.H. (1997) *J. Phys. Chem. B*, **101**, 614–622.
54 Boniface, H. and Ruthven, D.M. (1985) *Chem. Eng. Sci.*, **40**, 1401.
55 Sward, B.K. and LeVan, M.D. (2003) *Adsorption*, **9**, 37–54.
56 Wang, Y., Sward, B.K., and LeVan, M.D. (2003) *Ind. Eng. Chem. Res.*, **42**, 4213–4222.
57 Wang, Y. and LeVan, M.D. (2005) *Ind. Eng. Chem. Res.*, **44**, 3692–3701.
58 Wang, Y. and LeVan, M.D. (2005) *Ind. Eng. Chem. Res.*, **44**, 4745–4752.
59 Wang, Y. and LeVan, M.D. (2007) *Ind. Eng. Chem. Res.*, **46**, 2141–2154.

14
Chromatographic and Permeation Methods of Measuring Intraparticle Diffusion

In the previous chapter we considered the problem of extracting intraparticle diffusivity values from experimental transient adsorption/desorption curves measured in a batch system. When diffusion is rapid the method breaks down, because intraparticle diffusion then becomes masked by other rate limiting processes such as external mass transfer resistance and/or heat dissipation. Such effects are much less severe in a flow system since, by using a sufficiently high fluid velocity, the external resistances to heat and mass transfer may be reduced, at least in principle, to an insignificant level. Various different flow methods have therefore been developed in which the diffusivity, or the diffusional time constant, is determined from measurements of the dynamic response of an adsorption column to a perturbation in the sorbate concentration at the inlet. Such methods include conventional chromatographic measurements using either a pulse or a step injection of sorbate, frequency response measurements using a sinusoidal variation of sorbate concentration, and short column measurements carried out under limiting conditions of high flow rate such that the column behaves as a differential bed [zero-length column (ZLC) method]. These techniques were originally developed for gaseous systems, but they have also been successfully extended to liquid phase sorption systems. The development of such experimental methods and the underlying theory are reviewed in this chapter. The main focus is on the ZLC technique, since this has been substantially refined and further developed since the first edition of this book [1]. In contrast, the traditional chromatographic technique, although still in widespread use, has seen relatively little further development. It is therefore reviewed only briefly and the interested reader is referred to the first edition [1] and to earlier reviews [2, 3] for a more comprehensive discussion.

In addition to transient sorption rate measurements made under flow conditions we also consider permeation rate measurements, which, in the case of the single-crystal membrane, provide a direct macroscopic measurement of diffusion at the scale of an individual crystal ("mesoscopic" measurement, see Table 1.1).

14.1
Chromatographic Method

In the usual chromatographic experiment a steady flow of an inert (non-adsorbing) carrier is passed through a small column packed with the adsorbent under study. At time zero a small pulse of sorbate is injected at the column inlet and the effluent concentration is monitored continuously. The retention volume or mean retention time is determined by the adsorption equilibrium (Henry's law constant), while the dispersion of the response peak is determined by the combined effects of mass transfer resistance and axial mixing in the column. By making measurements over a range of conditions it is possible to separate the contributions to the pulse broadening due to mass transfer resistance and axial mixing and, thus, under favorable circumstances, to determine intraparticle or intracrystalline diffusivities. To simplify the interpretation of the data, the experiment is normally carried out at low concentrations within the Henry's law region of the isotherm. Under these conditions the retention time becomes independent of the size of the sorbate pulse. Variation of the pulse size thus provides a simple and direct way of checking the validity of the linearity assumption under any given experimental conditions.

Other forms of perturbation such as a step change or a periodic variation in sorbate concentration may also be applied. For a linear system precisely the same information may be deduced from any of these measurements, so the choice is dictated by experimental convenience and practical considerations rather than by any fundamental theoretical considerations.

14.1.1
Mathematical Model for a Chromatographic Column

To analyze and interpret the data from a chromatographic experiment a mathematical model is needed. The parameters characterizing the kinetics and equilibrium of the sorption process can then be derived by matching the experimentally observed response to the model predictions. Flow of a gas or liquid through a packed column can generally be represented by the axial dispersed plug flow model. The basic continuity equation, derived from a mass balance on an element of the column (Figure 14.1), is:

$$-D_L \frac{\partial^2 c}{\partial z^2} + \frac{\partial}{\partial z}(vc) + \frac{\partial c}{\partial t} + \left(\frac{1-\varepsilon}{\varepsilon}\right) \frac{\partial \bar{q}}{\partial t} = 0 \tag{14.1}$$

If the system is isothermal with negligible pressure drop and if the concentration of the adsorbable species is small, the gas velocity may be considered as constant through the column and, under these conditions, Eq. (14.1) simplifies to:

$$-D_L \frac{\partial^2 c}{\partial z^2} + v\frac{\partial c}{\partial z} + \frac{\partial c}{\partial t} + \left(\frac{1-\varepsilon}{\varepsilon}\right) \frac{\partial \bar{q}}{\partial t} = 0 \tag{14.2}$$

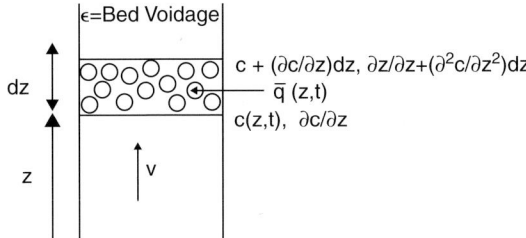

Figure 14.1 Element of a chromatographic column showing the terms involved in the derivation of Eq. (14.1).

This is the form generally used in modeling of the dynamic response of a chromatographic column.

The term $\partial \bar{q}/\partial t$ represents the local mass transfer rate averaged over an adsorbent particle. To obtain the dynamic response of the system [$c = c(z, t)$] it is necessary to solve Eq. (14.2) simultaneously with the appropriate mass transfer rate expression:

$$\frac{\partial \bar{q}}{\partial t} = f(c, q) \tag{14.3}$$

subject to the initial and boundary conditions imposed on the system. For step and pulse perturbations these are:

$$\text{step:} \quad c(z, 0) = q(z, 0) = 0, \quad c(0, t) = c_0 \tag{14.4a}$$

$$\text{pulse:} \quad c(z, 0) = q(z, 0) = 0, \quad c(0, t) = c_0 \delta(t) \tag{14.4b}$$

where $\delta(t)$ represents the Dirac delta function. In addition the "long column" condition is usually assumed:

$$c|_{z \to \infty} = 0 \tag{14.5}$$

In the application of the chromatographic method to the measurement of intraparticle diffusivities we are concerned primarily with linear systems so that the equilibrium relationship may be expressed in the form:

$$q^* = Kc \tag{14.6a}$$

or, for a perturbation from an initially uniform finite concentration level:

$$q^* = Kc + \text{constant} \tag{14.6b}$$

In practice linearity can generally be achieved by keeping the magnitude of the concentration perturbation small. In a biporous adsorbent, $K = \varepsilon_D + (1-\varepsilon_D)K_c$, where K_c is the dimensionless equilibrium constant, expressed on a solid volume basis, and K is the corresponding value on a particle volume basis.

14.1.1.1 Time Domain Solutions

The simplest choice for the rate expression [Eq. (14.3)] is the linear driving force (LDF) form:

$$\frac{\partial \bar{q}}{\partial t} = k(q^* - \bar{q}) \quad \text{or} \quad \frac{\partial \bar{q}}{\partial t} = k'(c - c^*) \quad (14.7)$$

For a linear system these forms are identical with $k' = kK$ but for nonlinear systems they are different and lead to significantly different response curves. When used as an approximate representation for an intraparticle diffusion controlled process the first of these forms is preferable, as the response curves are closer in shape to those derived from the diffusion model. This is the basis of the widely used LDF approximation, first introduced by Glueckauf [4, 5], in which Eq. (14.7) with $k = 15 D_e / R^2$ is used to represent the dynamic behavior of a diffusion-controlled system. This is an excellent approximation for linear or near-linear systems provided that the column is not too short [1]. See Section 6.2.3 and Eq. (6.72).

Solutions to Eqs. (14.2)–(14.6) have been obtained for several different rate expressions (in addition to the LDF model); some of these are listed in Table 14.1. Since the delta function is the derivative of the step function, for a linear system, the solution for a pulse input is the derivative of the solution for a step input. The solution for either boundary condition therefore yields directly both the step and pulse response.

A useful approximate expression for the step response for a plug flow system with an LDF rate expression has been given by Klinkenberg [15]:

$$\frac{c}{c_0} = \frac{1}{2} \operatorname{erfc}\left[\sqrt{\xi} - \sqrt{\tau} - \frac{1}{8\sqrt{\xi}} - \frac{1}{8\sqrt{\tau}}\right] \quad (14.8)$$

where $\xi = (kKz/v)[(1-\varepsilon)/\varepsilon]$, $\tau = k(t - z/v)$. Deviation from the exact solution is less than 0.6% for $\xi > 2.0$. The corresponding expressions for the pulse response may be found directly by differentiation. For all but the simplest forms of rate expression

Table 14.1 Analytical solutions for chromatographic response.

Flow model	Mass transfer equation	Reference
Plug flow	Linear driving force	Anzelius [11], Walter [14], Furnas [12], Nusselt [13]
Plug flow	Linear driving force	Klinkenberg [15]
Dispersed plug flow	Linear driving force	Lapidus and Amundson [6]
Plug flow	Intraparticle diffusion (+ external film resistance)	Rosen [7, 8]
Dispersed plug flow	Intraparticle diffusion (+ external film resistance)	Rasmuson and Neretnieks [9]
Plug flow	Macropore–micropore dual diffusion resistance + film resistance	Takeuchi and Kawazoe [16]
Dispersed plug flow	Macropore–micropore dual diffusion resistance + external film resistance	Rasmuson [10]

the analytic solutions are quite complicated, so nowadays it is often faster and indeed easier to calculate response curves directly by numerical solution of the model equations rather than to compute the response from the formal analytic solution [17–21].

14.1.1.2 Form of Response Curves

Figure 14.2 illustrates the general form of the theoretical response curves for a biporous diffusion-controlled adsorbent. If the time of passage through the column (L/v) is short compared with the time constant for micropore diffusion ($r_c^2/15\,KD_c$) but long compared with the time constant for macropore diffusion ($R_p^2/15\varepsilon_p D_p$) the sorbate molecules will penetrate only the macropores, giving a more or less Gaussian response peak with a mean residence time corresponding to the hold-up in the macropores and the interparticle void space ($t' \approx (L/\varepsilon v)\,[\varepsilon + (1-\varepsilon)\varepsilon_p]$). Similarly, if the time of passage is long compared with the micropore time constant, full penetration of the micropores will be achieved, leading to a nearly Gaussian response peak centered on the mean retention time (μ), which is given by:

$$\mu = \frac{L}{v}\left[1 + \left(\frac{1-\varepsilon}{\varepsilon}\right)K\right] \tag{14.9}$$

In the intermediate region where the time of passage is of the same order as the micropore time constant, we obtain a strongly tailed response peak with the maximum located between t' and $t = \mu = (L/v)\{1 + [(1-\varepsilon)/\varepsilon]K\}$. These three regimes

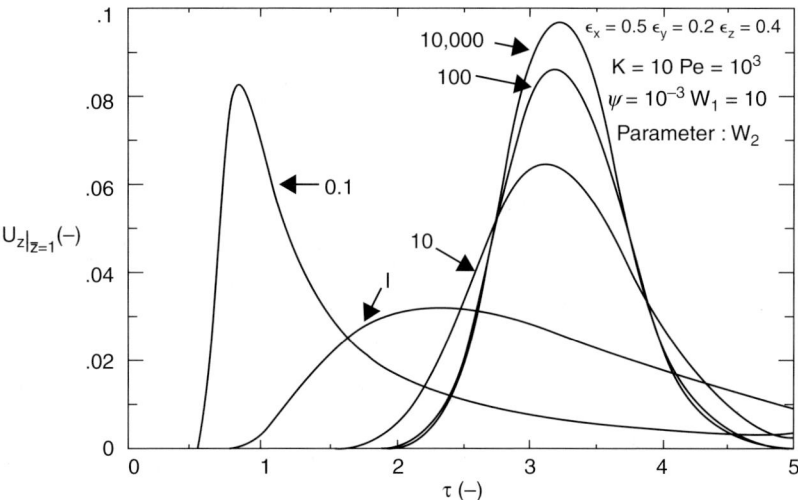

Figure 14.2 Chromatographic response for a column packed with a biporous adsorbent showing the effect of the micropore diffusional time constant on the form of the response. In the present notation: $\tau = \varepsilon v t/L, \varepsilon_z = \varepsilon, U_z = c/c_0, W_1 = (\varepsilon_n D_n/R_p^2)(L/\varepsilon v), \psi = \varepsilon_p D_p/k_f R_p, 1 + K = (K-\varepsilon_v)/\varepsilon_r(1-\varepsilon_v), W_p = (KD_p/r_p^2)\,(L/\varepsilon v)/\,(1-\varepsilon_y)$. A small value of W_2 corresponds to high micropore resistance while large W_2 corresponds to negligible micropore resistance. Reprinted from Raghavan and Ruthven [21] with permission.

are very well illustrated by the experimental data reported by Sarma and Haynes [22] for diffusion of argon in 3A, 4A, and 5A zeolites.

In the application of the chromatographic method to the measurement of micropore diffusivities a significant problem is to identify the conditions under which the dispersion of the response peak is dominated by micropore resistance. Haynes [3] has pointed out that a strongly tailed peak, under linear conditions, provides prima-facie evidence that micropore resistance is dominant and, conversely, that a near symmetric response implies mass transfer resistance is small and axial dispersion dominant. While this is generally true, it is not the complete story. It is in fact the dimensionless column length or the number of theoretical plates (Section 14.1.2) that determines whether the response is symmetric or tailed [1]. High mass transfer resistance means that the dimensionless column length is short and the peaks are strongly tailed. Conversely, if mass transfer resistance is negligible the column will contain many theoretical stages, leading to a nearly symmetric response. However, even under conditions of high mass transfer resistance, in a sufficiently long column, the response will approach a near symmetric form.

14.1.2
Moments Analysis

The time domain solutions referenced in Table 14.1 were all derived by Laplace transformation. To obtain the solutions of the model equations in the Laplace domain is straightforward; it is the inversion of the transform to obtain the solution in the time domain that is difficult. To avoid this difficulty, methods have been developed to allow the determination of model parameters directly by matching experimental response curves without recourse to the time domain solution. The most widely used of these methods depends on matching the moments of the experimental and theoretical response curves.

14.1.2.1 First and Second Moments

The expressions for the moments of the chromatographic response in terms of the model parameters may be derived directly from the solution of the model equations in Laplace form, by application of van der Laan's theorem [23]. The moments of the experimental response may be calculated directly by integration:

1) **Pulse response:**

$$\text{1st moment:} \quad \mu \equiv \bar{t} = \int_0^\infty \frac{ct\,dt}{\int_0^\infty c\,dt} = -\lim_{s \to 0} \frac{\partial \tilde{c}}{\partial s} \frac{1}{c_0}$$

$$\text{2nd moment:} \quad \sigma^2 \equiv \int_0^\infty \frac{c(t-\mu)^2\,dt}{\int_0^\infty c\,dt} = \lim_{s \to 0} \frac{\partial^2 \tilde{c}}{\partial s^2} \frac{1}{c_0} - \mu^2$$

(14.10)

2) **Step response:** [24]:

$$\text{1st moment:} \quad \mu \equiv \bar{t} = \int_0^\infty (1-c/c_0)\,dt$$

$$\text{2nd moment:} \quad \sigma^2 = 2\int_0^\infty (1-c/c_0)t\,dt - \mu^2$$

(14.11)

The model parameters may then be determined without recourse to the time domain solution. In practice the method is generally limited to the first two moments for the reasons discussed below although, in principle, additional information may be derived from the higher moments.

The expressions for the first and second moments of the pulse response for a general model of a chromatographic column, including the effects of the external film mass transfer resistance and macropore and micropore diffusional resistances, and a finite rate of adsorption at the surface (Table 14.1) were derived almost simultaneously by three independent research groups [25–27]. For physical adsorption the actual adsorption step is rapid and is therefore generally ignored. A useful general form for the second moment of the response for a biporous adsorbent is [1]:

$$\frac{\sigma^2}{2\mu^2} = \frac{D_L}{vL} + \frac{\varepsilon v}{L(1-\varepsilon)}\left[\frac{R_p}{3k_f} + \frac{R_p^2}{15\varepsilon_p D_p} + \frac{r_c^2(K-\varepsilon_p)}{15K^2 D_c}\right]\left[1 + \frac{\varepsilon}{(1-\varepsilon)K}\right]^{-2} \quad (14.12)$$

where $\mu = \bar{t}$ is given by Eq. (14.9). For gaseous systems K is often large so $\varepsilon/(1-\varepsilon)K$ is small and $(K-\varepsilon_p)/K^2 \approx 1/K$, thus simplifying Eq. (14.12) to the commonly quoted form:

$$\frac{\sigma^2}{2\mu^2} = \frac{D_L}{vL} + \frac{\varepsilon v}{L(1-\varepsilon)}\left[\frac{R_p}{3k_f} + \frac{R_p^2}{15\varepsilon_p D_p} + \frac{r_c^2}{15KD_c}\right] \quad (14.13)$$

This approximation must, however, be treated with caution since, particularly in liquid systems, the essential requirement $K \gg \varepsilon_p$ is often not fulfilled.

Clearly, from Eq. (14.12) the contributions of external film, macropore, and micropore diffusional resistances are linearly additive. If, instead of the multiple resistance model, we assume a simple LDF model [Eq. (14.7)] the corresponding expression for the reduced second moment is:

$$\frac{\sigma^2}{2\mu^2} = \frac{D_L}{vL} + \frac{\varepsilon v}{L(1-\varepsilon)}\frac{1}{kK}\left[1 + \frac{\varepsilon}{(1-\varepsilon)K}\right]^{-2}$$

$$\approx \frac{D_L}{vL} + \frac{\varepsilon v}{L(1-\varepsilon)}\frac{1}{kK} \quad \text{(for large } K\text{)} \quad (14.14)$$

To match the second moments we must therefore set:

$$\frac{1}{kK} = \frac{R_p}{3k_f} + \frac{R_p^2}{15\varepsilon_p D_p} + \frac{r_c^2}{15KD_c} \quad (14.15)$$

This relationship provides the extension of the Glueckauf approximation [4, 5] to a system in which more than one diffusional resistance is significant [21].

For gaseous systems in which the hold-up in the macropores is small compared with the micropore capacity the LDF approximation works very well except in the limiting case of a very short column (containing only a few theoretical plates). The shape of the response curve is sensitive to the magnitude of the mass transfer resistance but insensitive to the nature of this resistance. It is therefore impossible to obtain information on the nature of the rate controlling resistance simply from analysis of the shape of the chromatographic response. It should, however, be emphasized that these comments are strictly valid only for gaseous systems. In adsorption from the liquid phase the macropore capacity may be comparable with the micropore capacity and under these conditions significant differences in the response curves for macropore or micropore control are to be expected. This point has been emphasized by Weber [28].

In matching an experimental response curve, the adsorption equilibrium constant may be unambiguously determined from the first moment. Matching the second moment provides one equation containing in effect two unknown parameters; the axial dispersion coefficient and the lumped mass transfer coefficient. To determine these parameters, measurements must be conducted over a range of experimental conditions. The easiest parameter to vary experimentally is normally the fluid velocity.

14.1.2.2 Use of Higher Moments

It is in principle possible to obtain the additional equation needed to solve simultaneously for the axial dispersion and mass transfer coefficients from the third (or higher) moments. This approach has been investigated by Boniface [29, 30]. Although the equations for the second, third, and fourth moments are algebraically different, numerically they are almost degenerate. Evaluation of the higher moments therefore does not provide any additional evidence concerning the nature of the mass transfer resistance, although the higher moments can provide evidence concerning the relative importance of axial dispersion and mass transfer resistance as long as the column is not too long (in the dimensionless sense). The accuracy with which the higher moments can be calculated from experimental response peaks is severely limited since any tailing, for example, due to a small degree of nonlinearity in the equilibrium relationship, leads to a disproportionately large effect. However, with a carefully designed experimental system and a very stable baseline, the approach can provide additional information and so reduce the range of experiments needed to characterize the behavior of a given system.

14.1.2.3 HETP and the van Deemter Equation

In the preceding discussion the dynamic behavior of the column was described in terms of the differential mass balances for the adsorbed and fluid phases in an element of the column. An alternative approach that is directly analogous to the "tanks in series" model [31] for a packed-bed reactor has also been widely used. Although less physically realistic, the plate model gives results that are very similar to those obtained from the differential model and it may be shown that in a sufficiently long column the two approaches become exactly equivalent.

We consider an ideal mixing cell with mass transfer between the fluid and adsorbed phases, as illustrated in Figure 14.3. Transient mass balances for the two

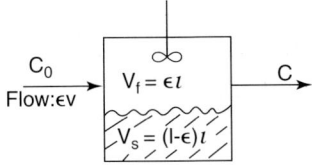

Figure 14.3 Theoretical equilibrium stage as represented in the plate theory of chromatography.

phases yield:

$$\varepsilon v c_0 = \varepsilon v c + V_f \frac{dc}{dt} + V_s \frac{d\bar{q}}{dt} \tag{14.16}$$

$$\frac{d\bar{q}}{dt} = k(q^* - \bar{q}) \tag{14.17}$$

From the Laplace transform of these two equations we obtain the expression for the transfer function:

$$\frac{\tilde{c}}{c_0} = \left\{ 1 + \frac{sl}{v}\left[1 + \frac{[(1-\varepsilon)/\varepsilon]K}{1 + s/k} \right] \right\}^{-1} \tag{14.18}$$

and for N identical stages in series:

$$\frac{\tilde{c}}{c_0} = \left\{ 1 + \frac{sl}{v}\left[1 + \frac{[(1-\varepsilon)/\varepsilon]K}{1 + s/k} \right] \right\}^{-N} \tag{14.19}$$

The expressions for the moments, derived from van der Laan's theorem, are:

$$\mu = \frac{Nl}{v}\left[1 + \left(\frac{1-\varepsilon}{\varepsilon}\right)K \right] \tag{14.20}$$

$$\frac{\sigma^2}{2\mu^2} = \frac{1}{2N} + \frac{\varepsilon v}{Nl(1-\varepsilon)} \frac{1}{kK}\left[1 + \frac{\varepsilon}{(1-\varepsilon)K} \right]^{-2} \tag{14.21}$$

Comparison shows that Eqs. (14.21) and (14.14) are identical provided that we set $N = L/l = vl/2D_L$, in which case N is then the number of theoretical plates to which the column is equivalent and the height equivalent to a theoretical plate (HETP $= H$) is seen to be given by:

$$H = \frac{\sigma^2}{\mu^2}L = \frac{2D_L}{v} + \frac{2\varepsilon v}{(1-\varepsilon)}\frac{1}{kK}\left[1 + \frac{\varepsilon}{(1-\varepsilon)K} \right]^{-2} \tag{14.22}$$

As a rough approximation for the axial dispersion coefficient one may write [32]:

$$D_L \approx 0.7 D_m + v R_p \tag{14.23}$$

so that:

$$H \approx \frac{1.4 D_m}{v} + 2R_p + \frac{2\varepsilon v}{(1-\varepsilon)}\frac{1}{kK}\left[1 + \frac{\varepsilon}{(1-\varepsilon)K} \right]^{-2} \tag{14.24}$$

which is of the same form as the classical van Deemter equation [33]:

$$H = \frac{A_1}{v} + A_2 + A_3 v \tag{14.25}$$

with: $A_1 = 1.4 D_m$, $A_2 = 2R_p$, and $A_3 \approx [2\varepsilon/(1-\varepsilon)kK]/[1+\varepsilon/(1-\varepsilon)K]^2$

The HETP provides a simple physical interpretation of the effects of axial mixing and mass transfer resistance on column performance. In an equilibrium system with no axial mixing the profiles of adsorbed and fluid phase concentrations will be coincident. However, the effect of mass transfer resistance and axial dispersion is to retard the profile in the adsorbed phase relative to that in the fluid phase. The HETP measures the distance between any concentration level in the fluid and the corresponding equilibrium concentration in the adsorbed phase.

14.2
Deviations from the Simple Theory

14.2.1
"Long-Column" Approximation

Solutions for the dynamic response are all derived on the assumption that the column is "long." The HETP provides a simple quantitative measure of the conditions under which this is a valid assumption. What matters is the number of theoretical stages or the ratio of column length to HETP. Clearly, in a short column or when mass transfer resistance is large the HETP may exceed the column length and under these conditions the basic assumption of a "long" column is obviously invalid. By solving the equations for the moments with Eqs. (14.4a) and (14.5) replaced by the proper Danckwerts boundary conditions [34]:

$$\frac{D_L}{v}\frac{\partial c}{\partial z}\bigg|_{z=0} = c|_{z=0-} - c|_{z=0+} \tag{14.26a}$$

$$\frac{\partial c}{\partial z}\bigg|_{z=L} = 0 \tag{14.26b}$$

Narayan showed that the long-column approximation is valid provided the column contains at least five theoretical plates (S. Narayan, personal communication). Deviations become severe only when the number of theoretical plates falls below about two.

14.2.1.1 Pressure Drop
Expressions for the effect of pressure drop on the moments of the response peak have been derived by three independent research groups [35–37] and the results of all three studies have been summarized and reviewed by Dixon and Ma [38]. It appears that the most useful analysis is that of Pazdernik and Schneider [37] since, by using certain approximations that appear to be generally valid within the normal range of

experimental conditions, they were able to derive the relatively simple closed form expressions:

1st moment:
$$\mu = \frac{L}{v}\left[1+\left(\frac{1-\varepsilon}{\varepsilon}\right)K\right]\left\{\frac{2}{3}\frac{(1+\gamma)^{3/2}-1}{\gamma}\right\} \quad (14.27)$$

2nd moment:
$$\sigma^2 = \left(\frac{2L}{v}\right)\left(\frac{D_L}{v^2}\right)\left[1+\left(\frac{1-\varepsilon}{\varepsilon}\right)K\right]^2\left\{1-\frac{1-e^{-Pe}}{Pe}\right\}\left\{\frac{(\gamma+1)^2-1}{2\gamma}\right\}$$
$$+\frac{2}{15}\left(\frac{L}{v}\right)\left(\frac{1-\varepsilon}{\varepsilon}\right)\left[\frac{R_p^2}{\varepsilon_p D_p}K^2+\frac{r_c^2}{D_c}(K-\varepsilon_p)\right]$$
$$\times\left\{\frac{2}{3}\frac{(1+\gamma)^{3/2}-1}{\gamma}\right\}$$
$$+\frac{2}{3}\left(\frac{L}{v}\right)\left(\frac{1-\varepsilon}{\varepsilon}\right)\left(\frac{R_p}{k_f}\right)K^2\left\{\frac{(1+\gamma)^2-1}{2\gamma}\right\} \quad (14.28)$$

where $\gamma = (P_0/P_L)^2-1$.

This is evidently a very useful contribution, since it eliminates the need to maintain isobaric conditions. The terms in braces represent the correction factors to the terms in Eq. (14.12). For moderate pressure drops ($\gamma < 0.5$) we can simplify the expression further since:

$$\frac{(\gamma+1)^2-1}{2\gamma} \approx \frac{2}{3}\frac{(\gamma+1)^{3/2}-1}{\gamma} \quad (14.29)$$

(The error is about 10% for $\gamma = 0.5$.) With this approximation we can write the expression for the reduced second moment as:

$$\frac{\sigma^2}{2\mu^2} \approx \frac{3\gamma}{2\left[(1+\gamma)^{3/2}-1\right]}\left\{\frac{D_L}{vL}+\frac{v}{L}\left(\frac{1-\varepsilon}{\varepsilon}\right)\left(\frac{R_p}{3k_f}+\frac{R_p^2}{15\varepsilon_p D_p}+\frac{r_c^2(K-\varepsilon_p)}{15K^2}\right)\times\left[1+\frac{\varepsilon}{(1-\varepsilon)K}\right]^{-2}\right\} \quad (14.30)$$

thus preserving the form of Eq. (14.12) subject to an overall correction factor.

14.2.1.2 Nonlinear Equilibrium

The theoretical analysis presented above is entirely dependent on the assumption of a linear equilibrium relationship. A linear equilibrium relationship causes neither compression nor dispersion of the concentration profile so that, in the absence of any dispersive forces (mass transfer resistance or axial mixing), a concentration disturbance will propagate without any change in shape. By contrast, if the equilibrium relationship is nonlinear, the shape of the concentration profile will change as it propagates through the column, quite apart from any dispersion arising from the effects of axial mixing or mass transfer resistance. The qualitative changes to the form of the response peak may be deduced from equilibrium theory. For a favorable (type 1) isotherm the leading edge of the concentration peak approaches a shock front, whereas the trailing profile approaches a proportional pattern form leading to a peak that spreads in proportion to the distance traveled. As a result of the isotherm curvature, the mean retention time will decrease with increasing loading. The response peak at the column outlet, plotted against time, changes from the approximately Gaussian form characteristic of a linear system to the distorted form characteristic of a nonlinear system, with a sharp rise followed by a long tail.

Varying the size of the initial pulse provides a convenient test for system linearity since, for a linear system, both the retention time and the form of the normalized response peak are independent of the size of the initial pulse.

14.2.1.3 Heat Transfer Resistance

Although heat transfer resistance between fluid and particle may be expected to be smaller in a chromatographic column than in a static batch system, there remains a possibility that the mass transfer may be significantly retarded by the finite rate at which the heat of adsorption is dissipated. This situation has been analyzed in detail by Haynes [39] who derived an expression for the heat transfer contribution to the second moment of the response. The most important conclusion from this analysis is that the test of varying the pulse size, normally used to confirm system linearity, provides also a check for isothermality. If the moments of the response are independent of the magnitude of the perturbation, then not only is the system linear but it must also be effectively isothermal.

14.3
Experimental Systems for Chromatographic Measurements

Conventional gas or liquid chromatographs can be easily adapted for measurements of this type. The best choice of detector for gaseous systems is a flame ionization detector (FID) since its high sensitivity makes it possible to work with very small concentration perturbations, thus ensuring linearity even for relatively strongly adsorbed species. However, the FID is limited to organic sorbates and so for inorganic species, notably the permanent gases, an alternative must be used. The most common choice is the thermal conductivity (hot wire) detector or katharometer. This will detect any sorbate, although it is less sensitive than the FID. However, that is generally not a serious problem, at least with the permanent gases, since these species are relatively weakly adsorbed. It is therefore not difficult to achieve linearity at concentrations that are easily measured.

For liquid phase operation the choice is generally between refractive index (RI) and ultraviolet absorption (UV) detection. The UV detector is generally the more sensitive but it is limited to sorbates that show significant absorption in the UV range. The RI detector is more versatile but generally less sensitive. It has the further disadvantage that the dead volume is generally larger than for a well-designed UV detector.

To minimize axial dispersion the column must be well packed and this can present problems if very small adsorbent particles, such as zeolite crystals, are used. Slurry packing appears to offer the best alternative for such materials.

14.3.1
Dead Volume

Minimization of any extra-column dead volume is extremely important, particularly in liquid systems. The higher molecular density of the liquid phase means that a given dead volume has a proportionately more severe effect on the response compared with a gaseous system. Careful attention to the detailed design of the system and the use of fine bore connecting tubing and low dead volume fittings are therefore essential.

In principle the magnitude of the extra-column dead space and the associated dispersion may be determined experimentally simply by removing the column and connecting the inlet directly to the outlet. However, one potential difficulty is that, with the column removed, the response time may be very short and limited by the response of the detector or recorder. This may be checked by introducing a length of small bore tubing of known volume in place of the column. To correct for the effect of dead volume is straightforward: the delay time is simply subtracted from the measured response time in calculation of the first moment, while the variance of the blank response (without the column) is subtracted directly from the variance obtained with the column to find the second moment attributable to the column itself.

14.3.2
Experimental Conditions

Insight into the conditions under which reliable diffusivity values may be derived from chromatographic experiments can be obtained from Eq. (14.13) or (14.14). Assuming $K \gg 1.0$, as is usually true for a gas phase system, the effect of mass transfer resistance will dominate over axial dispersion provided:

$$\frac{v^2}{kKD_L} \gg 1.0 \qquad (14.31)$$

where kK is given by Eq. (14.15). If measurements are made in the low Reynolds number region, which is generally desirable when small particles are used, to avoid problems of channeling and high-order dependence of axial dispersion on velocity (Section 14.4) we may assume $D_L \approx D_m$, thus reducing Eq. (14.31) to:

$$\frac{1}{kK} \gg \frac{D_m}{v^2} \qquad (14.32)$$

The advantage of using as high a velocity as possible (within the laminar regime) is immediately evident. If the objective is to measure intracrystalline, rather than intraparticle, diffusivities it is also necessary to ensure that:

$$\frac{r_c^2}{15KD_c} \gg \frac{R_p}{3k_f} + \frac{R_p^2}{15\varepsilon_p D_P} \qquad (14.33)$$

This condition can usually be achieved by reducing the gross particle size and can be confirmed experimentally by making replicate experiments with different size fractions. When intracrystalline resistance is dominant, Eq. (14.33) becomes:

$$\frac{D_c}{r_c^2} \ll \frac{v^2}{15 K D_m} \tag{14.34}$$

from which it is clear that the condition for reliable diffusivity measurement becomes more difficult to satisfy when the equilibrium constant is large. For weakly adsorbed species the chromatographic method therefore has a considerable advantage over direct measurements of the uptake rate but, for strongly adsorbed species, this advantage is largely lost.

In liquid systems the dimensionless equilibrium constant is generally much smaller, so the constraints are less severe. We may assume $D_L \approx vd$, so Eq. (14.31) becomes:

$$kK \ll v/d \tag{14.35}$$

or for an intracrystalline-controlled system:

$$D_c/r_c^2 \ll v/15 Kd \tag{14.36}$$

which is easily satisfied under most experimental conditions.

14.4
Analysis of Experimental Data

To determine the intraparticle or intracrystalline diffusivity it is necessary to match the experimental chromatograms to the theoretical response curves calculated from the appropriate dynamic model. In principle this matching may be carried out directly in the time domain, using an appropriate optimization routine to determine the best fitting values for the model parameters. However, since the time domain solutions for all but the simplest models are complicated, various alternative methods have been developed that do not require recourse to the time domain solutions [40–46]. Perhaps the most efficient procedure is to use either the moments methods or frequency domain matching to establish initial estimates of the parameters and then to refine the values, if necessary, by direct matching of the response curves in the time domain.

Regardless of the method used to determine the parameters, a direct comparison of the time domain curves is always desirable to confirm that the theoretical response actually fits the form of the experimental data. Gelbin *et al.* [47] have pointed out that quite different response curves can produce identical first and second moments. Although in principle it should be possible to determine all model parameters by analysis of a single response curve, this is not possible in practice, since the form of the response peak is only sensitive to the combined effect of the total mass transfer resistance and axial dispersion, not to the nature of the resistance (Section 14.2).

Separation of the contributions from axial dispersion and mass transfer resistance turns out to be a major problem, especially for fast diffusing species for which the effects of axial dispersion and mass transfer resistance are of similar magnitude. It is usually necessary to carry out a series of measurements with different fluid velocities and different particle sizes. It turns out that the axial dispersion problem is less severe for liquid phase systems.

14.4.1
Liquid Systems

The basic experimental method and the analysis and interpretation of the response data are in essence the same for both gas and liquid chromatographic systems, although there are some practical differences. The greater importance of dead volume in liquid systems, as a consequence of the higher molecular density, has already been noted. In general the adsorption equilibrium constant for liquid systems is much smaller than for gaseous adsorption systems so the term $\{1+[\varepsilon/(1-\varepsilon)K]\}$ in Eq. (14.14) cannot be neglected, and the approximation $K \gg \varepsilon_p$ may also be invalid.

Axial dispersion in liquid systems is dominated by eddy mixing and the role played by molecular diffusion is insignificant under all practical conditions. The axial dispersion coefficient therefore increases linearly with velocity and the term D_L/v in Eq. (14.22) should be approximately independent of velocity. A plot of HETP versus fluid velocity should therefore be approximately linear with an intercept corresponding to the axial dispersion contribution. This behavior is illustrated in Figure 14.4.

Confirmation of the magnitude of the axial dispersion term may be obtained in two simple ways by using either a rapidly diffusing adsorbate or an adsorbate that is too large to penetrate the pores of the adsorbent. In either case the contribution to the HETP from mass transfer resistance should be negligible. In aqueous systems D_2O can often be used as the "fast" tracer, while a high molecular weight (blue) dextran can be used as the non-penetrating sorbate. The application of this approach in both aqueous and non-aqueous systems is illustrated in Figures 14.4 and 14.5 using columns packed with unaggregated NaX zeolite crystals. The LC method has been applied by Ma et al. to study diffusion of alcohols and several other sorbates in pelleted silicalite and alumina adsorbents (Section 14.5) [60].

14.4.2
Vapor Phase Systems

For gas phase systems the dependence of the axial dispersion coefficient on velocity is more complex. At low Reynolds numbers axial dispersion occurs mainly by molecular diffusion and in this range the dispersion coefficient should be approximately independent of gas velocity. Equation (14.14) may be written in the alternative form:

$$\frac{\sigma^2}{2\mu^2}\frac{L}{v} \equiv \frac{H}{2v} = \frac{D_L}{v^2} + \frac{\varepsilon}{1-\varepsilon}\frac{1}{kK}\left[1+\frac{\varepsilon}{(1-\varepsilon)K}\right]^{-2} \qquad (14.37)$$

where the total mass transfer resistance $(1/kK)$ is given by Eq. (14.15). Therefore, within the low Reynolds number region a plot of $\sigma^2 L/2\mu^2 v$ versus $1/v^2$ should be linear with slope D_L and intercept equal to the total mass transfer resistance. Figure 14.6 shows examples of such plots, selected to illustrate different patterns of behavior.

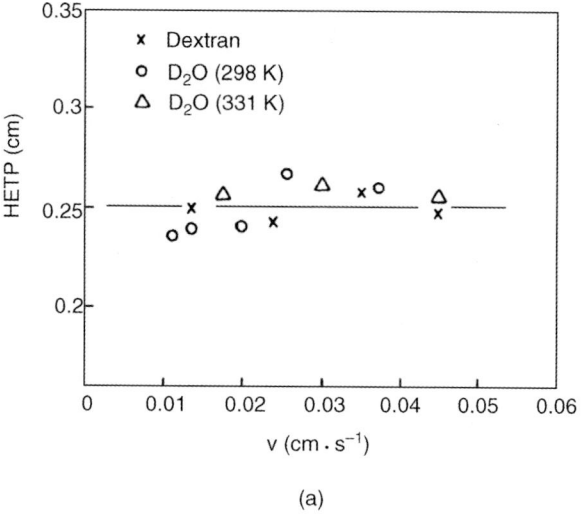

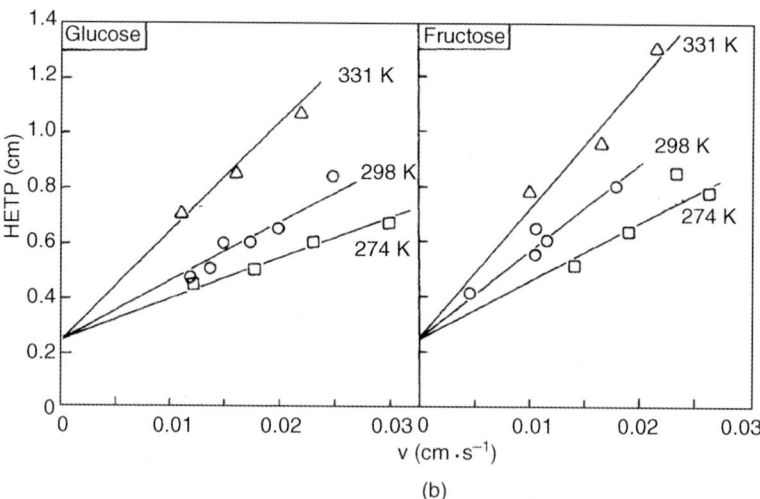

Figure 14.4 Liquid chromatographic measurements (aqueous solution) in a column ($L = 14.2$ cm) packed with 50-μm NaX crystals. (a) Measurement of axial dispersion using blue dextran (×) and D_2O (O,△). Note that the HETP is essentially constant and the same for both sorbates ($H = 2D_L/v = 0.25$ cm). (b) Variation of HETP with fluid velocity for glucose and fructose, showing a linear dependence with intercept corresponding to $2D_L/v$ [see Eq. (14.22)]. Reprinted from Ching and Ruthven [58] with permission.

14.4 Analysis of Experimental Data | 475

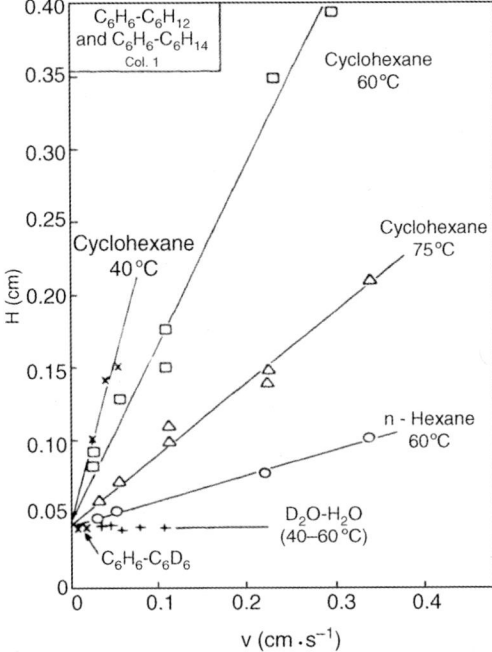

Figure 14.5 Liquid chromatographic measurements for benzene–n-hexane and benzene–cyclohexane in a column packed with 40-μm NaX zeolite crystals. The axial dispersion contribution ($2D_L/v = 0.04$ cm) is essentially the same for C_6H_6–C_6D_6 and for H_2O–D_2O. Reprinted from Awum et al. [59] with permission.

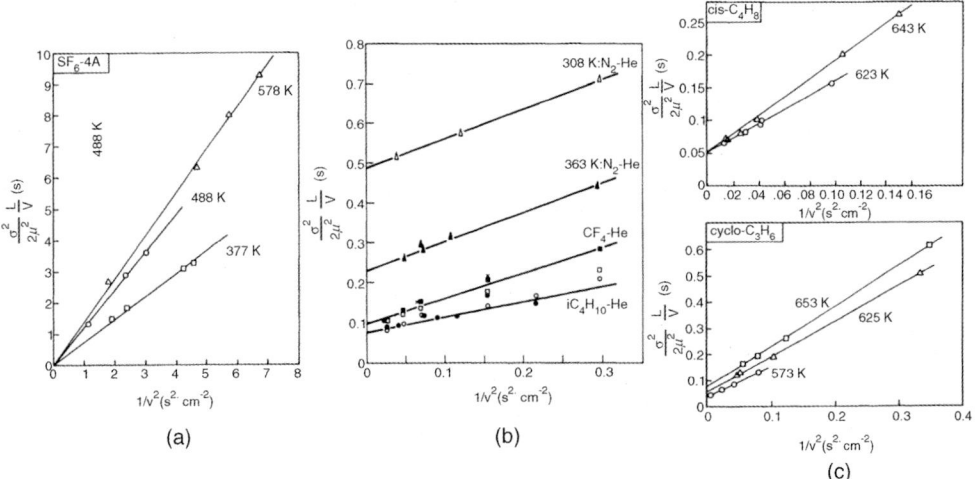

Figure 14.6 Plots of $H/2v$ versus $1/v^2$ for (a) SF_6 in 4A zeolite pellets; (b) N_2, CF_4, and i-C_4 in 4A zeolite pellets; (c) cis-butene and cyclopropane in 5A pellets. Reprinted from Haq and Ruthven [56, 57] and Kumar et al. [48] with permission.

SF$_6$ is too large to penetrate the micropores of zeolite 4A and in the small particles used in these experiments ($d \approx 0.05$ cm) macropore diffusional resistance is also negligible. As a result the dispersion of the chromatogram is determined entirely by axial dispersion. The plots of $H/2v$ versus $1/v^2$ are linear with no significant intercept and the temperature dependence of the slope (D_L) corresponds to the temperature dependence of the molecular diffusivity.

The data shown in Figure 14.6b were obtained with much larger adsorbent particles and although CF$_4$ and isobutane are too large to enter the micropores of the 4A adsorbent there is measurable macropore resistance leading to a finite intercept that is almost independent of temperature since the temperature dependence of the macropore diffusivity is modest. In contrast, nitrogen penetrates the 4A micropores, so a much larger intercept is obtained representing the sum of all three mass transfer resistances but with the largest contribution from intracrystalline diffusion. The intercept decreases markedly with increasing temperature since the activation energy for intracrystalline diffusion is, for this system, higher than the heat of adsorption [49] so that the product KD_c in Eq. (14.15) increases with temperature. The more usual situation in which the activation energy is lower than the heat of adsorption, leading to an intercept that increases with temperature, is illustrated in Figure 14.6c for cyclopropane and *cis*-butene in 5A zeolite.

As an alternative to the plot of $(\sigma^2/2\mu^2)(L/v)$ versus $1/v^2$ one may choose to plot the HETP (or the dimensionless ratio $H/d = \sigma^2 L/\mu^2 d$) directly against the gas velocity. Such a plot should have the form illustrated in Figure 14.7. At low velocities axial dispersion is dominant and is determined mainly by molecular diffusion so HETP $\propto$

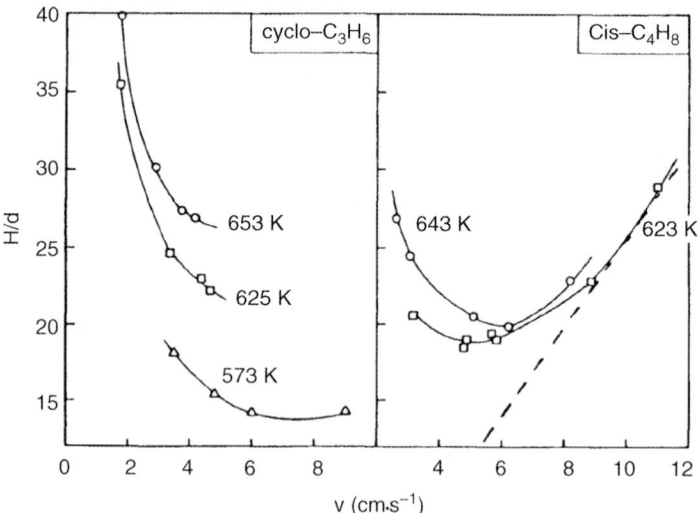

Figure 14.7 Dimensionless HETP (H/d) versus velocity for cyclopropane and *cis*-butene in 5A zeolite pellets. The high velocity asymptote corresponding to the last term of the van Deemter equation [Eq. (14.24)] is indicated by the broken line (for *cis*-butene at 623 K). Reprinted from Haq and Ruthven [57] with permission.

D_m/v and we observe a sharp rise in the curve with decreasing velocity. At high velocities mass transfer resistance becomes dominant, since this term increases with velocity while the axial dispersion term decreases. The limiting slope therefore gives the mass transfer resistance, as may be seen from Eq. (14.24).

These methods work well when mass transfer resistance is relatively large but difficulties can arise when the diffusivity is high. If the aim is to measure intracrystalline or zeolitic diffusivities, it is also desirable to use either unaggregated crystals or very small particles to minimize the intrusion of macropore resistance. The difficulty with using unaggregated crystals is that, unless the crystals are large (>50 μm), the pressure drop becomes large and the accessible range of gas velocity is then very restricted. Careful attention to the pressure correction factor (Section 14.2) is required and the accuracy of the resulting data is inevitably reduced. Furthermore, the problem of eliminating or allowing for axial dispersion becomes more serious for columns packed with very small particles. Nevertheless, the use of a column packed with un-aggregated crystals provides a useful experimental approach as long as the potential difficulties are recognized.

14.4.3
Intrusion of Axial Dispersion

In the examples presented above, the mass transfer resistance is relatively large, so the intrusion of axial dispersion does not present a major difficulty. The intrusion of axial dispersion is more serious for systems in which mass transfer resistance is smaller and seriously limits the applicability of the chromatographic method as a means of measuring intraparticle and intracrystalline diffusivities for fast diffusing systems. At high gas velocities, particularly in columns packed with small adsorbent particles, the axial dispersion contribution no longer shows a simple linear dependence on the velocity, but rather it tends to approach a second-order dependence ($D_L \propto v^2$) [50, 51]. Under these conditions the axial dispersion and mass transfer resistance terms both show similar variations with velocity, so it becomes impossible to separate the contributions from the two effects. The unwary investigator may indeed conclude that the mass transfer resistance is substantially larger than it really is. Even in the low velocity region small particles tend to stick together so that their hydrodynamic behavior resembles that of larger particles and, in particular, the axial dispersion is enhanced [53].

It is therefore important to establish the axial dispersion behavior directly under conditions as close as possible to the conditions of the experimental measurements. One way to do this is by using a gas of similar molecular weight and diffusivity that is too large to penetrate the micropore structure, in the manner noted above [48]. A somewhat less satisfactory alternative is to make measurements on a column of similar size packed with nonporous (and therefore non-adsorbing) particles of the same dimensions. This has the advantage of allowing the measurement to be made directly with the sorbate, but it suffers from the disadvantage that the axial dispersion properties may be significantly different for the two columns, since there is no guarantee that the packing will be precisely the same [19]. There is also the problem that, in the case of porous particles, some contribution to axial dispersion may arise from diffusion *through* the particle and this contribution is obviously not present with nonporous particles [52].

Another approach is to use a very narrow column, only slightly larger in diameter than the particles (the string-of-beads column). Such a system has been shown to give axial dispersion coefficients similar to those obtained in a larger-diameter bed, and significantly smaller than the values typically obtained in small-diameter packed column [51, 53]. The reason is that the major cause of enhancement of axial dispersion in a small packed column is the non-uniformities in the packing arising from wall effects. Such effects are eliminated in the "string-of-beads" system. However, this approach is generally not practical with very small adsorbent particles.

The intrusion of axial dispersion is generally not obvious from a cursory examination of experimental response peaks and a detailed and critical analysis of the experimental data is necessary. For a gaseous system the axial dispersion coefficient becomes independent of velocity only in the very low Reynolds number region. More generally the velocity dependence is given by an expression of the form [54, 55]:

$$D_L = \gamma_1 D_m + \frac{vd}{2(1+\gamma_2 D_m/vd)} \tag{14.38}$$

where γ_1 and γ_2 are constants ($\gamma \approx 0.7$, $\gamma_2 \approx 10$). For gaseous systems the Schmidt number ($\eta/\varrho D_m$) is close to 1.0 so for small particles Eq. (14.38) reduces to:

$$D_L \approx \gamma_1 D_m + \frac{1}{2} \frac{(vd)^2}{\gamma_2 D_m} \tag{14.39}$$

Combining this with Eq. (14.14), taking $k_f \approx D_m/R_p$ (Sh ≈ 2) and $D_p \approx D_m/\tau$ we obtain:

$$\frac{H}{2d} = \frac{\gamma_1 D_m}{vd} + \frac{vd}{D_m}\left\{\frac{1}{2\gamma_2} + \frac{\varepsilon}{4(1-\varepsilon)}\left(\frac{1}{3}+\frac{\tau}{15\varepsilon_p}\right)\left[1+\frac{\varepsilon}{(1-\varepsilon)K}\right]^{-2}\right\}$$
$$+ \frac{\varepsilon v}{d(1-\varepsilon)}\frac{r_c^2}{15KD_c}\left[1+\frac{\varepsilon}{(1-\varepsilon)K}\right]^{-2} \tag{14.40}$$

If intracrystalline resistance to mass transfer is negligible the last term vanishes and it is evident that H/d is a function only of (D_m/vd). Furthermore, at sufficiently low velocities the first term on the right-hand side must become dominant so that the asymptotic behavior is given by:

$$\frac{H}{2d} \rightarrow \frac{\gamma_1 D_m}{vd} \tag{14.41}$$

In the low velocity region a plot of $H/2d \equiv (\sigma^2/2\mu^2)(L/d)$ versus D_m/vd should approach a straight line through the origin with slope γ_1.

Such a plot provides a useful way of analyzing experimental data to detect the intrusion of axial dispersion, since it allows data obtained with different sorbates and at different temperatures to be compared on a common basis. An example of this

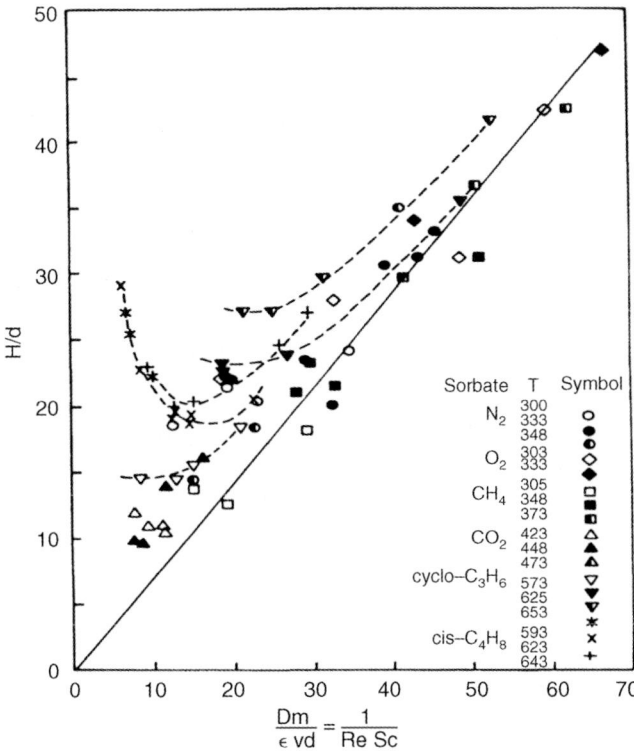

Figure 14.8 Dimensionless HETP (H/d) versus ($D_m/\varepsilon v d$) for several sorbates in 5A zeolite pellets, showing the approach to axial dispersion control for all species except cyclopropane and *cis*-butene. Reprinted from Haq and Ruthven [57] with permission.

approach is shown in Figure 14.8, which provides experimental data for several light gases in 5A zeolite pellets. Evidently, if the objective is to determine intracrystalline diffusivities, useful information can be obtained only from points that lie substantially above the asymptotic line and for which the slope is negative [indicating that the third term, rather than the second term on the right-hand side of Eq. (14.40), is dominant]. Only the data for *cis*-2-butene and, at the higher temperatures, cyclopropane meet these requirements.

14.5 Variants of the Chromatographic Method

14.5.1 Step Response

Instead of measuring the pulse response one may elect to determine the experimental breakthrough curve (the step response). Essentially the same information may be deduced from either experiment, although some practical advantages have been claimed for the step methods [61–65]. A sinusoidal input has also been employed [63–65] – see Section 13.8.

Tracer chromatography, which involves following the response to a pulse injection of an isotopically labeled tracer, usually radioactive to simplify detection, has also been developed [66]. This method has the advantage of strict linearity even when the sorbate is strongly adsorbed. In principle it measures the self-diffusivity rather than the transport diffusivity, which is the quantity determined in all other chromatographic methods.

14.5.2
Limited Penetration Regime

If mass transfer is sufficiently slow, one may reach the situation in which the sorbate does not penetrate the adsorbent particles fully on the time scale of the experiment, that is, equilibrium is never approached at any point in the chromatographic column. Under these conditions the column is equivalent to less than a single theoretical plate and the usual mathematical model is no longer applicable. It is, however, still possible to obtain kinetic information in this regime by using a different mathematical model. This is discussed in detail in the first edition of this book [1].

14.5.3
Wall-Coated Column

The standard chromatographic methods do not work well for very small particles (<5 μm), since the pressure drop through the column then becomes very high and axial dispersion effects become dominant. To circumvent these problems Delmas and Ruthven [67] used a wall-coated capillary column in which the internal wall was coated with a layer of small zeolite crystals. In such a system the flow pattern is well defined and axial dispersion can be predicted with some confidence from the Taylor–Golay model [68–70]. The form of the variation of HETP with gas velocity is similar to the van Deemter equation [Eq. (14.25)]. The parameters have similar physical significance, although their numerical form is different. As a result of the greatly reduced pressure drop, in comparison to a packed column with similar particle size, it is much easier to achieve the high-velocity regime in which the broadening of the response is controlled primarily by mass transfer resistance. Preparation of the column presents some practical problems, since it is difficult to achieve a uniform layer of crystals. Nevertheless, the viability of this approach as a way to measure the mass transfer resistance of small zeolite crystals was demonstrated and the resulting diffusivities were shown to be consistent with the values derived by other techniques for the same adsorbent sample [67].

14.5.4
Direct Measurement of the Concentration Profile in the Column

In the usual chromatographic measurement the concentration response is measured in the vapor phase at the outlet of the column. If required the (time and position dependent) profile in the adsorbed phase within the column can then be deduced from the model equations discussed in Section 14.1.1. An alternative approach involving direct measurement of the transient concentration profile within the column has also been explored. Various experimental techniques, including "positron emission profiling" (PEP) [71] and NMR imaging (MRI) (Section 12.1) [72, 73], have been successfully applied. However, there appears to be no obvious advantage over the traditional approach of following the concentration response in the fluid phase and the equipment required for both these methods is much more expensive than a standard chromatographic detector. Thus, although such measurements provide useful experimental confirmation of the theory, it seems unlikely that they will find widespread application for measurement of intraparticle diffusivities.

14.6
Chromatography with Two Adsorbable Components

Although the chromatographic method has generally been used to study diffusion in the low concentration limit with the injection of a small pulse of the adsorbable species into a non-adsorbing carrier stream, the method may be extended, by the use of a mixed carrier, to permit measurements at higher concentration levels and in a system with two adsorbable components [48, 72]. In developing the theoretical framework for these extensions it is simplest to consider first the general case of a binary system and to regard measurements at higher concentration levels with a single adsorbed species as a special case.

In a binary system, differential mass balance equations of the form of Eq. (14.2) may be written for both components. Using the constant pressure condition these two equations may be combined to yield:

$$-D_L \frac{\partial^2 y}{\partial z^2} + \frac{v \partial y}{\partial z} + \frac{\partial y}{\partial t} + \frac{1}{c}\left(\frac{1-\varepsilon}{\varepsilon}\right)\left\{(1-y)\frac{\partial q_1}{\partial t} - y\frac{\partial q_2}{\partial t}\right\} = 0 \qquad (14.42)$$

where c is the total molar concentration and y the mole fraction of component 1 in the vapor phase. The mass transfer rate expressions are written, for both components, in the form of Eq. (14.7):

$$\frac{\partial q_1}{\partial t} = k_1(q_1^* - q_1) = k_1(K_1 c y - q_1) \qquad (14.43)$$

with a similar expression for component 2. $K_1 = dq_1^*/dc_1$ represents the local slope of the equilibrium isotherm for component 1 in the binary. Provided that consideration is restricted to small differential changes in composition, the coefficients in these expressions may be considered to be constant. Equations (14.42) and (14.43) may then be solved by Laplace transformation and the expressions for the moments of the pulse response can then be extracted from the solution in Laplace form, as noted in Section 14.1.2. This procedure yields expressions formally similar to Eqs. (14.14) and (14.20) but with:

$$K = \gamma K_2 + (1-\gamma) K_1 \tag{14.44}$$

$$\frac{1}{k} = \frac{1}{K}\left[\gamma \frac{K_2}{k_2} + (1-\gamma)\frac{K_1}{k_1}\right] \tag{14.45}$$

For a system containing only one adsorbable species component 2 represents the non-adsorbing carrier, so that $K_2 = 0$. Equation (14.44) then reduces to:

$$K = (1-\gamma)K_1, \quad \frac{1}{k} = \frac{1}{k_1} \tag{14.46}$$

and in the limit for $\gamma \to 0$ the forms derived previously for a single-component system in the Henry's law region [Eqs. (14.9) and (14.14)] are recovered.

Under conditions of micropore control $k = 15\, D_c/r_c^2$ so, in a system with only one adsorbable component, the intracrystalline diffusivity may be determined directly, at any specified concentration level, from the response to a perturbation in the composition of the mixed carrier (adsorbable species plus inert). In a system with two adsorbable components the effective mass transfer coefficient depends on the diffusivities for both components [Eq. (14.45)]. These contributions cannot easily be

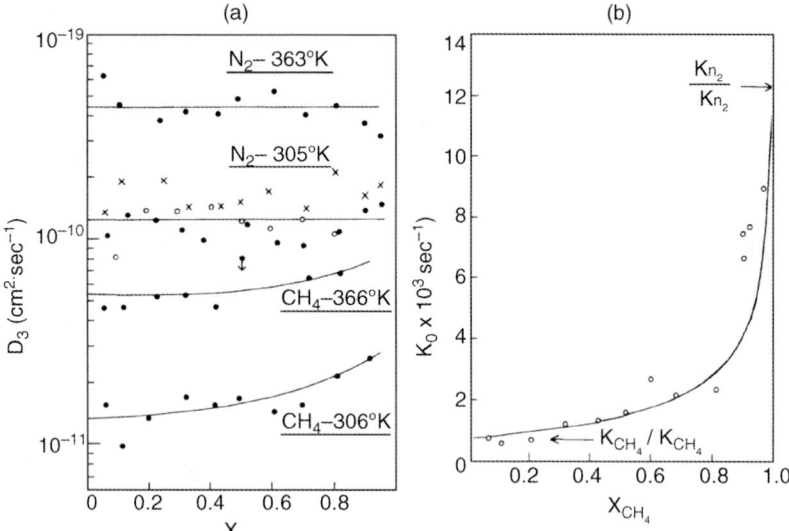

Figure 14.9 (a) Variation of intracrystalline diffusivity (D_z) with composition (x = mole fraction in carrier gas at 1 atm) for N_2 and CH_4 in 4A zeolite pellets from chromatographic measurement with a mixed carrier; (b) overall rate coefficients for the binary CH_4–N_2 mixture on 4A zeolite at 305 K, showing the comparison between experimental points, calculated from the binary response according to Eq. (14.45), and the theoretical curve calculated according to the same expression using values of k_1 and k_2 from the single-component data. Reprinted from Ruthven and Kumar [74] with permission.

separated. However, given the values of the diffusivities for the two components, for example, from a theoretical model, it is a simple matter to calculate the combined mass transfer coefficient according to Eq. (14.45), and hence to compare theory with experiment.

This approach is illustrated in Figure 14.9a, which shows experimental diffusivities for N_2 and CH_4 and 4A zeolite derived from measurements with mixed He–N_2 and He–CH_4 carriers, and in Figure 14.9b, which shows the composition dependence of the combined mass transfer coefficient in the CH_4–N_2 binary. The theoretical curves are calculated according to Eq. (14.45) on the assumption that the diffusivities in the binary are the same for the single components at comparable concentrations. Clearly, this approximation provides a good representation of the experimentally observed behavior, indicating that, at least in this system, there is no significant interaction between the two components. Since the loading is quite low ($\approx$1 molecule per cage), such a conclusion is not unexpected.

14.7
Zero-Length Column (ZLC) Method

The major advantage of the chromatographic method is that, by maintaining a relatively high carrier flow rate through the column, external mass and heat transfer resistances can be eliminated more easily than in a static system. The major disadvantage is that the dispersion of the response depends on both mass transfer resistance and axial dispersion. To determine the mass transfer resistance, and hence the intraparticle diffusivity, it is necessary to either eliminate or allow for the contribution from axial dispersion. This imposes an upper limit on the diffusional time constant (r^2/D) that can be measured since, if intraparticle diffusion is too rapid, the contribution from axial dispersion becomes dominant and it is then impossible to extract reliable kinetic data. The ZLC method was developed in an attempt to retain the basic advantages of the chromatographic method while eliminating the limitations imposed by axial dispersion [75, 76]. The method is especially useful for measuring intracrystalline diffusion in zeolites since only a very small sample of adsorbent is required and relatively high diffusivities can be measured ($D/R^2 < 0.01$ s^{-1}). It has been applied to a wide range of hydrocarbons and some other species in many different adsorbents.

14.7.1
General Principle of the Technique

The principle of the ZLC method is straightforward. A small sample of the adsorbent is equilibrated at a uniform sorbate concentration, preferably within the Henry's law range, and then desorbed by purging with an inert gas at a flow rate high enough to maintain essentially zero sorbate concentration at the external surface of the particles or crystals. The desorption rate is measured by following the composition of the effluent gas. This requires a sensitive detector, since the concentration is very low.

For organic sorbates a flame ionization detector is particularly useful, since it has the required sensitivity and can be used without interference from any inorganic purge gas. It has the further advantage that it is not sensitive to moisture, which may be present as a trace impurity or even added in controlled quantities to investigate the effect on the diffusional behavior of the system. Use of an on-line mass spectrometer as the detector extends the technique to allow tracer measurements and studies of mixture diffusion.

The method can be applied with composite pellets as well as with zeolite crystals as the adsorbent. Although originally developed for gaseous systems, the method has been extended to the study of liquid systems using an ultraviolet absorption detector to follow desorption of aromatics into a saturated hydrocarbon (n-hexane) purge stream.

Figure 14.10 shows schematically the basic experimental system for vapor phase measurements. The cell contains a very small quantity of zeolite crystals (or other adsorbent) placed between two porous sinter discs. The individual crystals are dispersed approximately as a monolayer across the area of the sinter to ensure good contact with the purge gas stream, thus minimizing external resistance to heat and mass transfer. Changing the purge gas from He to Ar or N_2 offers a convenient way of testing for the intrusion of extracrystalline resistances since, if such resistances are significant, a change in the molecular diffusivity (and thermal conductivity) of the gas should affect the mass transfer rate and thus the desorption curve.

Although the principle of the method is straightforward, to achieve reliable results detailed attention must be paid to the design of the experimental system. Since the quantity of adsorbent is very small, it is important to minimize any extraneous

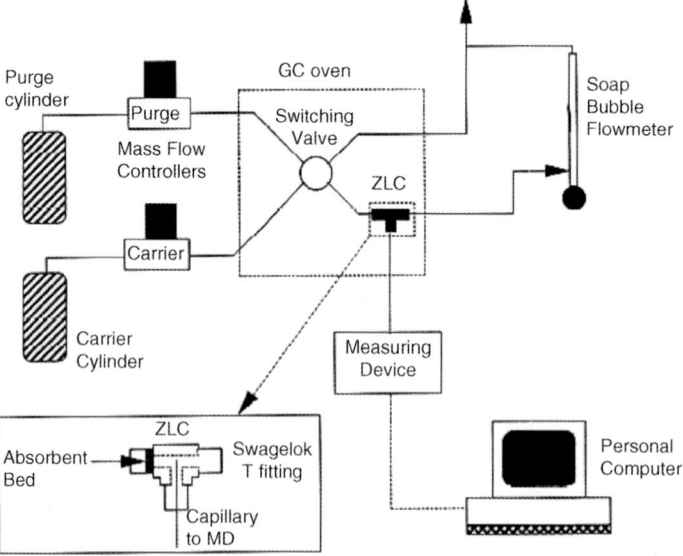

Figure 14.10 Schematic diagram of the ZLC system.

adsorption on tube walls, in valves, and so on. It is best to provide completely separate lines for saturation and purge streams and to minimize the length of the common line between the ZLC cell and the detector. Blank runs can be made with the adsorbent removed to confirm the successful elimination of extraneous adsorption and to determine the response time of the system.

14.7.2
Theory

14.7.2.1 Basic Theory for Intraparticle Diffusion Control

In the absence of significant external resistance to mass transfer, equilibrium will be maintained between the purge gas and the sorbate at the external surface of the crystal. By solving the Fickian diffusion equation together with a mass balance over the ZLC cell (neglecting hold-up in the void space in comparison with the adsorbed phase) we obtain the expression for the effluent concentration [75, 76]:

$$\frac{c}{c_0} = 2L \sum_{n=1}^{\infty} \frac{\exp(-\beta_n^2 Dt/r_c^2)}{[\beta_n^2 + L(L-1)]} \tag{14.47}$$

where β_n is given by the roots of the transcendental equation:

$$\beta_n \cot \beta_n + L - 1 = 0$$
$$L = \frac{\varepsilon v r_c^2}{3(1-\varepsilon) K D l} = \frac{1}{3} \frac{\text{purge flow rate}}{\text{crystal volume}} \frac{r_c^2}{KD} \tag{14.48}$$

The corresponding expression for a parallel-sided slab (thickness 2ℓ) is:

$$\frac{c}{c_0} = 2L \sum_{n=1}^{\infty} \frac{\exp(-\beta_n^2 Dt/\ell^2)}{[\beta_n^2 + L(L+1)]} \tag{14.49}$$

where:

$$\beta_n \tan \beta_n = L = F\ell^2/KV_s D \tag{14.50}$$

For large values of L, $\beta_n \to n\pi$ and Eq. (14.47) reduces to:

$$\frac{c}{c_0} = 2L \sum_{n=1}^{\infty} \frac{\exp(-n^2 \pi^2 Dt/r_c^2)}{[(n\pi)^2 + L(L-1)]} \tag{14.51}$$

In the long-time region only the first term of the summation is significant so Eq. (14.51) simplifies to:

$$\ln\left(\frac{c}{c_0}\right) \approx \ln\left[\frac{2L}{\beta_1^2 + L(L-1)}\right] - \beta_1^2 \frac{Dt}{R^2} \tag{14.52}$$

Evidently, a plot of $\ln(c/c_0)$ versus t should yield a linear asymptote. The relationship between the parameters β_1, L, and the intercept of such a plot, which is defined by Eq. (14.48) or (14.50), is shown, for the spherical particle case, in Figure 14.11. If L

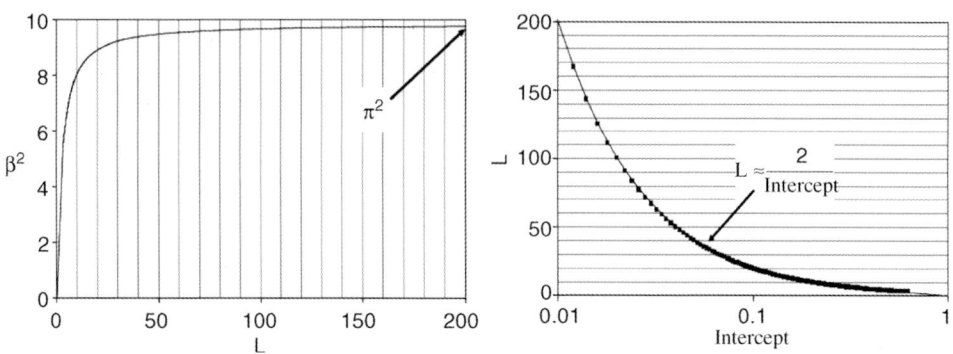

Figure 14.11 Variation of intercept with parameters β_1 and L for a spherical particle [Eqs. (14.47) and (14.48)].

is large, $\beta_1 \to \pi$, the intercept approaches $2/L$ and the slope becomes independent of L (therefore independent of flow rate) and equal to $\pi^2 D/r_c^2$. This is clearly the best regime for intracrystalline diffusion measurements.

For $L \to 0$, Eqs. (14.47) and (14.49) both approach the limiting form for equilibrium control:

$$\frac{c}{c_0} = \exp\left(-\frac{\varepsilon v t}{(1-\varepsilon)Kl}\right) = \exp\left(-\frac{Ft}{KV_s}\right) \quad (14.53)$$

Under these conditions the desorption curve contains no kinetic information. If the objective is to measure the intracrystalline diffusivity, this regime must obviously be avoided. However, measurements at low L values provide a convenient way of determining the equilibrium parameter (KV_s), which can be useful in the analysis of the desorption curves measured at higher flow rates in the kinetic regime.

Figure 14.12 presents a set of ZLC response curves for CO_2 desorbing from a fragment of a porous carbon monolith (a parallel-sided slab), showing the change in the form of the curves with increasing purge flow rate. At the lowest flow rate ($L=1$) the system is close to equilibrium control, while at the highest flow rate ($L=9.5$) the system is in the kinetic regime.

14.7.2.2 Short-Time Behavior [78]

The initial portion of the ZLC response curve is less sensitive than the long-time region to errors from baseline drift, particle size distribution, and finite rate of heat dissipation. Rather than relying entirely on analysis of the long-time asymptote, it is therefore sometimes convenient to extract parameters from the initial region of the response curve. A detailed analysis shows that the simple expression:

$$\frac{c}{c_0} \approx 1 - 2L\sqrt{\frac{D_c t}{\pi r_c^2}} \quad (14.54)$$

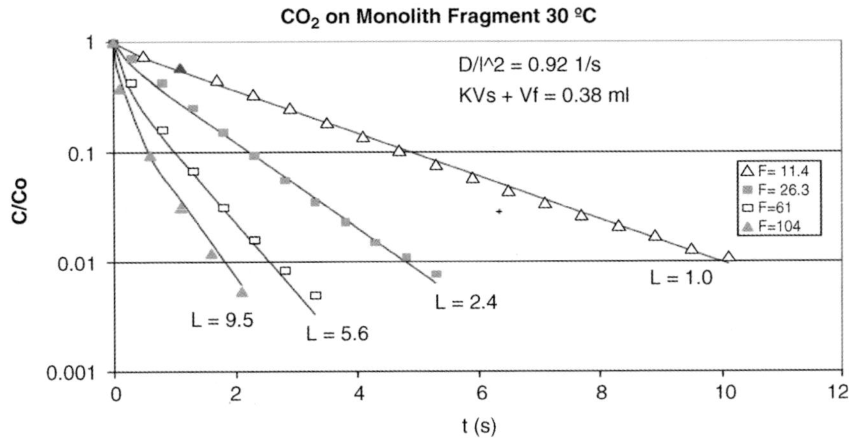

Figure 14.12 Experimental ZLC curves for desorption of CO_2 from a carbon monolith (parallel-sided slab), showing the effect of flow rate, calculated from Eqs. (14.49) and (14.50) with the given L values. Reprinted from Brandani et al. [77] with permission.

provides a good approximation to the initial region of the response curve. This equation suggests the use of a plot of $(1 - c/c_o)$ versus t as a convenient way to extract the diffusional time constant. However, such a plot is quite sensitive to any error in the zero time. A more useful approach is a plot of $(1 - c/c_o)\sqrt{t}$ versus $\sqrt{t}$, which is practically linear and may therefore be easily extrapolated to determine the time zero intercept $[2L\sqrt{(D_c/\pi r_c^2)}]$, from which the time constant (r_c^2/D_c) is found directly.

In the intermediate time region between the initial region and the long-time asymptote the response curve varies with $1/\sqrt{t}$ and is sensitive to the dimensionality of the structure:

$$\text{Spherical particle:} \quad \frac{c}{c_o} \approx \frac{1}{L}\left[\sqrt{\frac{r_c^2}{\pi D_c t}} - 1\right] \qquad (14.55)$$

$$\text{Slab:} \quad \frac{c}{c_o} \approx \frac{1}{L}\sqrt{\frac{\ell^2}{\pi D_c t}} \qquad (14.56)$$

A plot of c/c_o versus $1/\sqrt{t}$ therefore passes through the origin for a unidimensional pore system but yields a finite intercept for a spherical particle. This approach has been used to identify the main diffusion path in various intergrowths of offretite and erionite (Figure 14.13). The finite intercept for the erionite sample suggests a three-dimensional diffusion path, whereas the zero intercept for offretite suggests one-dimensional diffusion, in accordance with the difference in the structures of these materials.

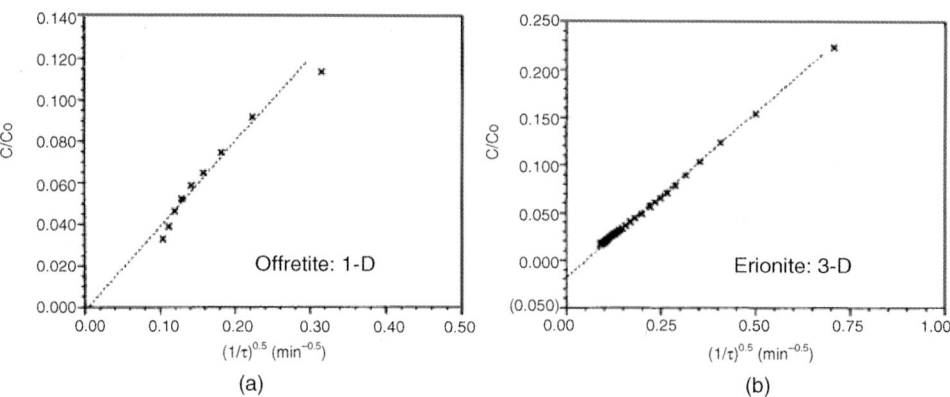

Figure 14.13 ZLC desorption curves for n-hexadecane from (a) offretite and (b) erionite crystals at 290 °C plotted in accordance with Eqs. (14.55) and (14.56) (short-time solution). Reprinted from Cavalcante et al. [79] with permission.

14.7.2.3 Diffusion in Macroporous Particles

The above equations have been written for spherical crystals but the same basic model applies equally to macroporous particles with D_c/r_c^2 replaced by D/R_p^2 where:

$$D = \frac{\varepsilon_p D_p}{\varepsilon_p + (1-\varepsilon_p)K} \qquad (14.57)$$

The solution for the ZLC response of a biporous particle has been given by Brandani [80] and by Silva and Rodrigues [81]. However, by adjusting the experimental conditions, especially the particle size, it is usually possible to approach either the macropore or micropore control case. Thus, while the biporous response is useful for understanding the system behavior, in the application of this technique to diffusion measurements it is usually desirable to work within one of the limiting regimes to avoid ambiguity in the results. Figure 14.14 presents an example of the

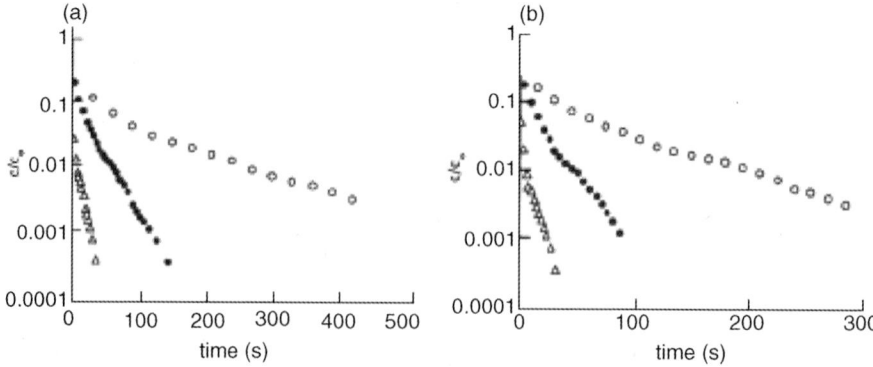

Figure 14.14 ZLC response curves for N_2 at (a) 174 and (b) 193 K in 5A zeolite beads of different diameters. R_p = (○) 1.03, (δ) 0.42, and (△) 0.24 mm. The purge rate is 20 cc STP min^{-1}. Reprinted from Ruthven and Xu [82] with permission.

application of the ZLC technique to measurement of diffusion of N_2 in pelleted 5A zeolite.

14.7.3
Deviations from the Simple Theory

14.7.3.1 Surface Resistance and/or Fluid Film Resistance
If surface resistance and/or external mass transfer resistance is significant, the analysis is basically similar except that the value of $L (= L')$ is then given by:

$$\frac{1}{L'} = \frac{1}{L} + \frac{D_c}{kr_c} = \frac{3KV_s D_c}{Fr_c^2} + \frac{D_c}{kr_c} \tag{14.58}$$

or:

$$\frac{1}{3L'} = \frac{D_c}{r_c^2}\left[\frac{KV_s}{F} + \frac{r_c}{3k}\right] \tag{14.59}$$

where $1/k = 1/k_s + K/k_f$; k_s represents the surface rate coefficient and k_f represents the gas film coefficient. In the limit when either resistance is dominant $k \to k_s$ or $k \to k_f/K$.

Equation (14.59) shows that the parameter $1/L'$ is the ratio of two reciprocal time constants; the time constant for intraparticle diffusion (r_c^2/D_c) and the combined time constant for convective washout and surface resistance. If r_c^2/D_c is large (relative to the square bracket term), $L' \to \infty$, $\beta_n \to n\pi$ and the expression for the desorption curve reduces to Eq. (14.47) for pure internal diffusion control. On the other hand, if r_c^2/D_c is small, $L' \to 0$, $(1 - \beta\cot\beta) = L' \to (\beta^2)/3$, the internal concentration profile becomes flat and the expression for the desorption curve reduces to:

$$\frac{c}{c_0} = \frac{\alpha}{1+\alpha}\exp\left(-\frac{3kt/r_c}{1+\alpha}\right) \tag{14.60}$$

where $\alpha = 3kKV_s/Fr_c$. If $\alpha \gg 1$ (i.e., $3k/r_c \gg F/KV_s$) the system is equilibrium controlled and Eq. (14.60) reduces further to Eq. (14.53). In the opposite limit ($3k/r_c \ll F/KV_s, \alpha \ll 1$) the ZLC response curve reduces to the expression for a system that is completely controlled by surface resistance [83]:

$$\frac{c}{c_0} = \frac{3kKV_s}{Fr_c}\exp\left(-\frac{3kt}{r_c}\right) \tag{14.61}$$

Interestingly, the condition for intrusion of surface resistance does not depend on the magnitude of the internal diffusional resistance but rather on the relative magnitudes of the surface (or external film) resistance and convective washout time constants.

14.7.3.2 Measurement of Surface Resistance by ZLC
The impact of moderate surface resistance is to decrease the value of the parameter L but the general form of the desorption curve remains the same. Provided that the L

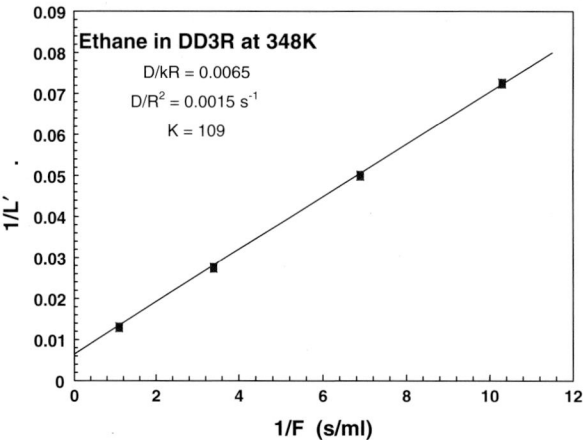

Figure 14.15 Plot of $1/L'$ versus reciprocal purge rate showing conformity with Eq. (14.58) for C_2H_6 in DD3R crystals ($r_c = 20\,\mu m$) at 348 K ([131]).

value is still large enough (>5) to yield a sufficiently small intercept, the slope of the long-time asymptote should still yield the correct value for the internal diffusional time constant. Under these conditions it is possible to determine the time constants for both internal diffusion and surface mass transfer resistance (as well as the equilibrium constant) from ZLC measurements carried out over a range of different flow rates.

It follows from Eq. (14.58) that a plot of $1/L'$ (where L' is the apparent value of L derived from the ZLC response curve) versus $1/F$ should be linear with slope $3KV_sD_c/r_c^2$ and intercept D_c/kr_c. Figure 14.15 shows an example of this approach. It should, hoever, be noted that a small degree of isotherm nonlinearity will increase the apparent value of L' so this approach will yield reliable information only if the isotherm is strictly linear under the experimental conditions.

14.7.3.3 Fluid Phase Hold-Up [84]

In the simple model discussed above, the fluid phase hold-up within the cell is considered negligible. This is generally not true for liquid phase systems and even for gas systems when adsorption is weak. A more detailed analysis taking account of the fluid phase hold-up leads to the following equations in place of Eqs. (14.47) and (14.48) (for spherical particles):

$$\frac{c}{c_o} = \sum \frac{2L\exp(-\beta_n^2\tau)}{\beta_n^2 + (L-1) + (1-L+\gamma\beta^2)^2 + \gamma\beta^2} \quad (14.62)$$

$$\beta_n \cot\beta_n + L - 1 - \gamma\beta_n^2 = 0 \quad (14.63)$$

The parameter L has its usual definition [Eq. (14.48)], $\gamma = V_f/3KV_s$ and $\tau = D_ct/r_c^2$. The parameter γ characterizes the ratio of the external to internal hold-up. For $\gamma \to 0$ these equations revert to Eqs. (14.47) and (14.48). Direct comparison of the calculated curves shows that for $\gamma < 0.1$ the effect of external hold up is negligible. This condition is usually fulfilled for adsorption from the vapor phase at atmospheric pressure, except when the Henry constant is very small. However, it is generally not fulfilled for adsorption from the liquid phase or for vapor phase adsorption at high pressure. For such systems it is important to make proper allowance for the extracrystalline hold-up by using the above equations. For example, for a very weakly adsorbed species under conditions of equilibrium control Eq. (14.53) should be replaced by:

$$\frac{c}{c_o} = \exp\left[\frac{-Ft}{V_g + KV_s}\right] \tag{14.64}$$

14.7.3.4 Effect of Isotherm Nonlinearity

The effect of isotherm nonlinearity has been investigated both theoretically [77] and experimentally [78] by considering a Langmuirian system. It was shown that the response is well approximated by the following expression:

$$\ln(c/c_o) = \ln(1-\lambda) - \lambda + \ln\left(\frac{2L_{tr}}{\beta_{tr}^2 + L_{tr}(L_{tr}-1)}\right) + \ln\left(\frac{2L}{\beta_1^2 + L(L-1)}\right) - \beta_1^2 D_o t / r_c^2 \tag{14.65}$$

where $\lambda = q_o/q_s$, $L_{tr} = L/(1-\lambda)$ and beta subscript tr is the first root of Eq. 14.48 with L replaced by Ltr. The implication of this result, verified experimentally, is that

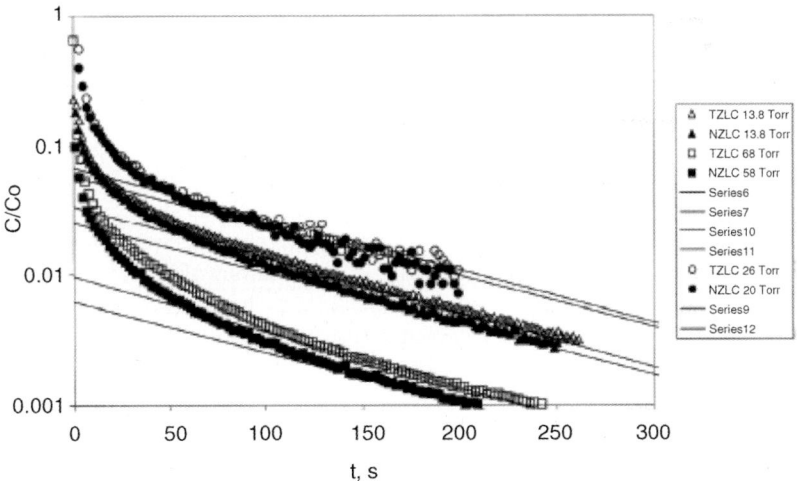

Figure 14.16 Experimental ZLC response curves for benzene–NaX at 403 K. Both the normal ZLC (NZLC) and tracer (TZLC) response curves are shown. The linear asymptotes of the NZLC curves are predicted from the tracer data in accordance with Eq. (14.65). Equilibration pressures (Torr): 58 (■), 13.8 (▲), and 2.9 (●). Open symbols = TZLC, filled symbols = NZLC. Reprinted from Brandani et al. [86] with permission.

Table 14.2 Analysis of ZLC curves for benzene–NaX at 403 K.[a]

P (Torr)	λ	$10^3 \times$ Slope (s^{-1})	Intercept	β	β_{Trac}	L	L_{Trac}
2.9	0.31	8.96	0.058	2.99	3.13	21	30.4
6.4	0.51	8.96	0.038	2.99	3.04	21	43
13.8	0.63	8.96	0.026	2.99	3.07	21	57
58	0.90	8.96	6.2×10^{-3}	2.99	3.04	21	210

a) $KV_s = 5$ cm^3, $F = 18.8$ cm^3 min^{-1}, $D_0/R^2 = 10^{-3}$ s^{-1}.

isotherm nonlinearity reduces the intercept of the long-time asymptote but has no significant effect on the slope of the plot of $\ln(c/c_0)$ versus t. Therefore, provided that the flow rate is high enough to ensure kinetic control, the diffusional time constant may be obtained from the slope of the long-time asymptote with little error even under nonlinear conditions. The form of the response curves is shown in Figure 14.16 and relevant details are given in Table 14.2. Evidently, Eq. (14.65) provides an excellent prediction of the asymptotes.

14.7.3.5 Heat Effects

A ZLC measurement is carried out with a small adsorbent sample (typically 1–3 mg) in the presence of a relatively high flow of carrier gas, which should aid heat transfer and maintain near-isothermal conditions. However, for some sorbates the heat of adsorption is large, so the possible intrusion of heat transfer effects requires consideration. A detailed analysis was carried out by Brandani et al. [87], who showed that the isothermal assumption is usually valid but can break down for highly exothermic sorbates when the adsorbent particles are large. The following criterion was developed for validity of the isothermal approximation:

$$\frac{R_p^2}{3K_o D_o} > \left(\frac{\Delta H}{R_g T}\right)^2 \frac{R_g c_o}{ha} \tag{14.66}$$

where K_0 and D_0 are the values of K and D at temperature T_0 and c_0 is the feed concentration of sorbate with which the sample was equilibrated. As a conservative approximation we may take $h \approx \lambda_g/R_g$ corresponding to $Nu = 2$ (Nu = Nusselt number) and for spherical particles $a = 3/R_p$ reducing Eq. (14.66) to:

$$\left(\frac{\Delta H}{R_g T}\right)^2 K_o D_o < \frac{\lambda_g}{R_g c_o} \tag{14.67}$$

For a He carrier at atmospheric pressure $\lambda_g \gamma / R_g c_o \approx 4.4 \, (T/300)^{1.5}$ so the criterion becomes:

$$K_o D_o < \frac{4.4}{\gamma} \left(\frac{T}{300}\right)^{1.5} \left(\frac{R_g T}{\Delta H}\right)^2 \, (\text{cm}^2 \, \text{s}^{-1}) \tag{14.68}$$

where γ is the mole fraction of sorbate in the equilibration stream. This criterion is somewhat conservative, as it does not allow for the effect of the ZLC support, which

Table 14.3 Validity of isothermal criterion in ZLC experiments.[a]

System	T (K)	(K, D)$_{mas}$	K$_a$	D (cm^2 s^{-1})	KD
Benzene–NaX	350	0.81	1.2×10^8	1.5×10^{-8}	1.8
	400	1.3	5×10^6	4×10^{-8}	0.20
	500	2.9	5.2×10^4	2×10^{-7}	0.01
p-Xylene–NaX	350	0.63	1.2×10^{10}	8×10^{-10}	9.6
	400	1.0	2.5×10^8	2.2×10^{-9}	0.55
	500	2.2	1.1×10^6	1.1×10^{-8}	0.012
N$_2$-5A (crystals)	300	5.2	35	10^{-6}	3.5×10^{-5}
N$_2$-5A (pellets)	300	5.2	22	6×10^{-4}	0.013

a) K values are dimensionless; (KD)$_{max}$ is estimated from Eq. (14.68) assuming a sorbate partial pressure of 0.01 atm. ($\gamma = 0.01$).

contributes a substantial heat capacity that tends to attenuate any temperature excursions.

Evidently, whether or not heat effects are significant depends mainly on the basic system properties (K, D, ΔH). Variables such as particle size, purge flow rate, sample volume, and so on affect both heat transfer and diffusion equally, so they have no significant influence on the relative importance of heat and mass transfer resistances. The only significant variables that can be selected arbitrarily are temperature and sorbate concentration. Varying the sorbate concentration in the feed provides a convenient experimental check to confirm the absence of a significant heat effect in a ZLC experiment. This test also confirms validity of the linearity approximation.

The isothermality criterion is easily satisfied under typical experimental conditions, but it can break down for strongly adsorbed species at low temperature and for macropore controlled systems for which very high values of $K_o D_o$ are possible (Table 14.3).

14.7.4
Practical Considerations

14.7.4.1 Choice of Operating Conditions and Extraction of the Time Constant

For diffusion measurements it is essential to operate in the kinetic regime, preferably with L value greater than about 5. Under these conditions the diffusional time constant can be conveniently determined from the slope of the long-time linear asymptote of a plot of $\ln(c/c_o)$ versus t. Other methods have been suggested but, provided that a sufficiently high L value can be attained, the use of the long-time asymptote is generally more robust and less prone to error from intrusion of nonlinearities and so on.

It is sometimes not possible to reach a high enough L value such that the slope approaches $-\pi^2 D_c/r_c^2$ but, provided that the L value is greater than about 2, it may still be possible to extract valid diffusivity data. The slope and intercept of the long-time asymptote are given by $\beta_1^2 D/r_c^2$ and $2L[\beta_1^2 + L(L-1)]$. Since L and β_1 are related through Eq. (14.63) we have three equations that can be solved for the three unknowns (β_1, L, and D/r_c^2). If this approach is adopted, an independent estimate of the L value may be obtained from measurements at very low flow rates in the equilibrium-controlled regime. Such measurements yield directly $\varepsilon/(1-\varepsilon)Kl$ [from Eq. (14.53)], so the value of LD/r_c^2 at any other velocity can be found by direct proportion.

The range of time constants that can be measured depends on several factors, so it is difficult to formulate precise limits. For slow diffusion processes the main limitation arises from the sensitivity and stability of the detector, since baseline drift over long time scales reduces the validity of the data. For fast processes the ultimate limit is imposed by the system response time (typically 1–2 s), which restricts the maximum measurable value of D/R^2 to less than about 0.01 s^{-1}.

14.7.4.2 Preliminary Checks

There are several preliminary tests that should be made to ensure that a ZLC system is generating reliable diffusivity data. It is essential to measure the blank response of the system with no adsorbent present, since this establishes the upper limit for the time scale of the measurements.

Varying the purge flow rate provides a convenient check on the rate-controlling regime. Under conditions of kinetic control the response curve, plotted as ln (c/c_o) versus t, should yield a linear asymptote in the long-time region. With increasing flow rate the intercept should decrease while the slope should remain constant. In the equilibrium-controlled regime the entire curve should remain invariant with flow rate (F) when plotted as c/c_o versus Ft.

Changing the quantity of the adsorbent sample and the nature of the carrier gas provides a convenient test for the intrusion of extracrystalline resistances to mass or heat transfer. An example is shown in Figure 14.17. For samples of less than 1 mg the desorption curve, at a fixed flow rate, is independent of either the sample size or the nature of the carrier, indicating the absence of any significant extracrystalline resistances, but for larger samples this is no longer true. Variation of the particle size can provide unequivocal confirmation of the controlling regime and the diffusivity value. This is a sensitive test, since the time constant varies with the square of the particle radius but it depends on the availability of similar particles of different size. With composite particles this is not a problem since the particles can be crushed, but for crystalline adsorbents different size fractions may not be available.

14.7.4.3 Equilibration Time [78]

During adsorption the sorbate concentration in the gas phase is governed by:

$$\frac{c}{c_o} = 1 - 2L \sum_{n=1}^{\infty} \frac{\exp(-\beta_n^2 Dt/r_c^2)}{[\beta_n^2 + L(L-1)]} \tag{14.69}$$

but the approach of the adsorbed phase to equilibrium is governed by a different expression:

$$\frac{\bar{q}}{q_o} = 1 - 6L^2 \sum_{n=1}^{\infty} \frac{\exp(-\beta_n^2 Dt/r_c^2)}{[\beta_n^2 + L(L-1)]\beta_n^2} \tag{14.70}$$

14.7 Zero-Length Column (ZLC) Method

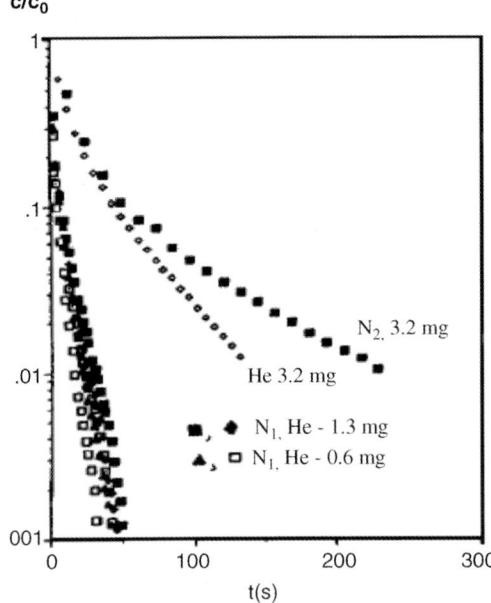

Figure 14.17 Effect of sample quantity and nature of purge gas (He or N_2) on ZLC response for benzene in 50-μm NaX crystals at 250°C. Reprinted from Brandani et al. [88] with permission.

Comparison of these two expressions shows that the fluid phase concentration approaches its final equilibrium value much more rapidly than the adsorbed phase concentration. For slowly diffusing systems the assumption that constancy of the effluent concentration means that the system is fully equilibrated may be seriously in error. The equilibration time should be at least r_c^2/D_c, which can be many minutes or even hours.

14.7.4.4 Partial Loading Experiment [70, 81]

Detection of the intrusion of surface resistance and even discrimination between surface resistance control and intraparticle diffusion control based simply on analysis of the ZLC desorption curve is not always straightforward. However, the partial loading experiment offers a simple direct experimental test for the controlling regime. The principle is illustrated in Figure 14.18. Under conditions of surface resistance control, or indeed under equilibrium control, the intraparticle concentration profile is flat, whereas under diffusion control the profile is approximately parabolic. Therefore, in a diffusion-controlled system, if the adsorbent is equilibrated for a time substantially smaller than that required to achieve equilibrium, the sorbate will be concentrated near the outer surface. If the flow is then switched to a sorbate-free purge, desorption will occur more rapidly than for a fully equilibrated particle with the same initial loading. On the other hand, for a surface-controlled system there should be no difference in the normalize curves, since the internal profile is always

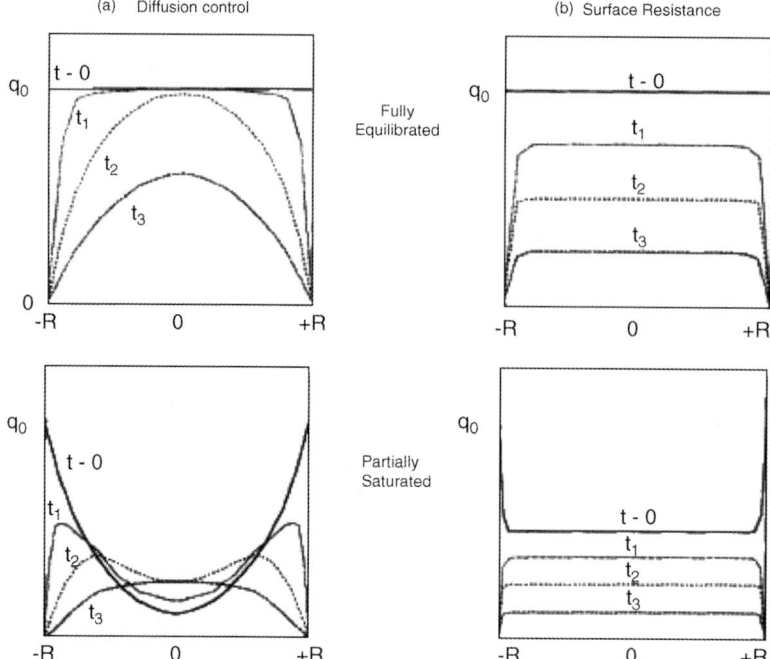

Figure 14.18 Sketch showing the forms of concentration profiles during desorption for fully equilibrated and partially equilibrated samples under (a) diffusion control and (b) surface resistance control.

flat. Concentration profiles experimentally determined under partial-loading conditions are shown in Figure 12.9.

To correct for any difference in the loading level it is simpler and more reliable to consider the response curves based on the adsorbed phase loading, which may be calculated directly by integration of the measured ZLC desorption curves:

$$\frac{\bar{q}}{q_o} = 1 - \frac{\int_0^t (c/c_o)dt}{\int_0^\infty (c/c_o)dt} \tag{14.71}$$

For surface resistance control [Eq. (14.60)] this yields a simple exponential decay:

$$\frac{\bar{q}}{q_o} = \exp\left(-\frac{3kt/r_c}{1+\alpha}\right) = \exp\left(-\frac{t}{(r_c/3k + KV_s/F)}\right) \tag{14.72}$$

Figure 14.19 shows an example showing the measured ZLC desorption curves and the calculated response curves expressed in terms of the adsorbed phase loading, for fully and partially equilibrated samples. The faster desorption of the partially equilibrated sample in (Figure 14.19a) and the deviation of the normalized adsorbed phase response curves (Figure 14.19b) from a simple exponential decay [Eq. (14.72)] provide clear evidence of intraparticle diffusion control.

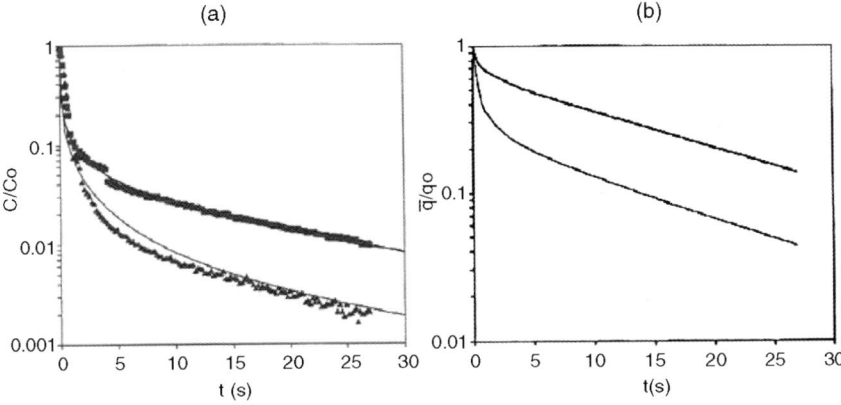

Figure 14.19 Experimental ZLC desorption curves for propane from NaX crystals at 85 °C showing (a) measured ZLC response for fully and partially equilibrated samples and (b) response curves expressed in terms of adsorbed phase loading. The curves showing faster desorption are for the partially loaded sample. Reprinted from Brandani and Ruthven [78] with permission.

14.7.5
Extensions of the ZLC Technique

14.7.5.1 Tracer ZLC [88–91]

A simple variant of the ZLC technique (tracer ZLC, TZLC) can extend the method to the measurement of self-diffusion. The sample is equilibrated with sorbate at a specified partial pressure and, at time zero, the flow is switched to a stream containing the same partial pressure of the same species in an isotopically distinguishable form. Desorption of the pre-loaded species is then followed with a species-sensitive detector such as an on-line mass spectrometer. The mathematical model is the same as for a normal ZLC experiment, except that the equilibrium constant (K) is replaced by the equilibrium concentration ratio (q^*/c). By varying the sorbate partial pressure such measurements can be made over a range of sorbate loading levels, thus yielding the concentration dependence of the self-diffusivity. Such information is not accessible through a normal ZLC experiment, which yields only the limiting value of the diffusivity at zero loading. TZLC experiments involve equimolar counter-exchange between the two labeled forms of the same sorbate, so for this type of experiment the equilibrium is always strictly linear (apart from the isotope effect, which is generally insignificant) and there can be no heat effect. Conformity between NZLC and TZLC measurements for the same sorbate at low loading levels can thus provide a useful experimental check on the validity of the experimental data.

Figure 14.20 shows representative TZLC desorption curves (for methanol–NaX). The response curves at different flow rates give the same values for the parameters D_c/r_c^2 and KV_s. Conformity of the diffusivity values obtained by TZLC, normal ZLC, and PFG NMR methods is shown in Figure 3.4.

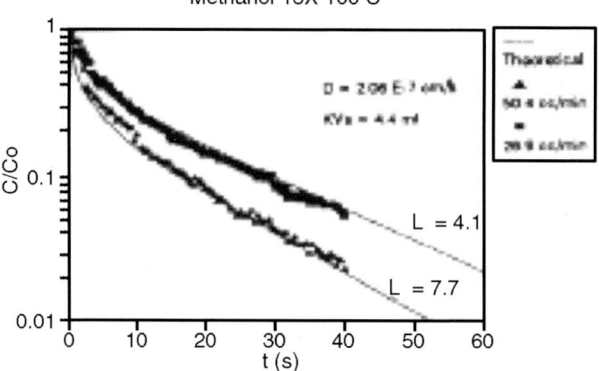

Figure 14.20 Comparison of experimental and fitted theoretical TZLC response curves for methanol–NaX (100 μm crystals at 373 K, 7.8 Torr methanol partial pressure in He). Reprinted from Brandani et al. [90] with permission.

14.7.5.2 Counter-current ZLC (CCZLC) [92, 93]

In an obvious further extension the ZLC technique can also be used to study counter-diffusion. The sample is pre-equilibrated with a certain partial pressure of species A and then at time zero the flow is switched to a stream containing an equivalent partial pressure of component B. For equimolar counter-diffusion the partial pressures are chosen to yield the same loadings of both components, but by adjusting the partial pressures it is possible to make the measurement with a net adsorption or desorption flux. In this way it is possible to investigate mutual diffusion effects by comparing the CCZLC (counter-current ZLC), NZLC, and TZLC response curves under the same conditions. Figure 14.21 shows such a comparison, for benzene (desorbing) with either p-xylene or o-xylene adsorbing, in large silicalite crystals. The NZLC and TZLC curves for benzene under similar conditions are also shown. The NZLC, TZLC, and CCZLC curves for benzene with p-xylene as the counter-diffusant all have the same asymptotic slope, showing that there is no significant counter-diffusion effect (steric hindrance) under counter-diffusion conditions. However, the asymptotic slopes for the CCZLC curves with o-xylene (which has a much lower diffusivity in silicalite), as the counter-diffusant, are much smaller, indicating substantial hindrance of benzene diffusion, especially at the higher benzene partial pressures.

14.7.5.3 Liquid Phase Measurements [84, 94]

The ZLC technique is in principle equally applicable to both gas and liquid phase systems, but there are two practical problems that make liquid phase measurements more difficult and less accurate than vapor phase measurements: limited sensitivity of liquid chromatographic detectors and the high molecular density of a liquid phase, which renders the system sensitive to any extracrystalline hold-up. For aromatic molecules the detector sensitivity is not a major issue, since an ultraviolet detector has adequate sensitivity. However, the sensitivity of a general purpose RI detector may not be sufficiently high to allow measurements to be made over a wide enough

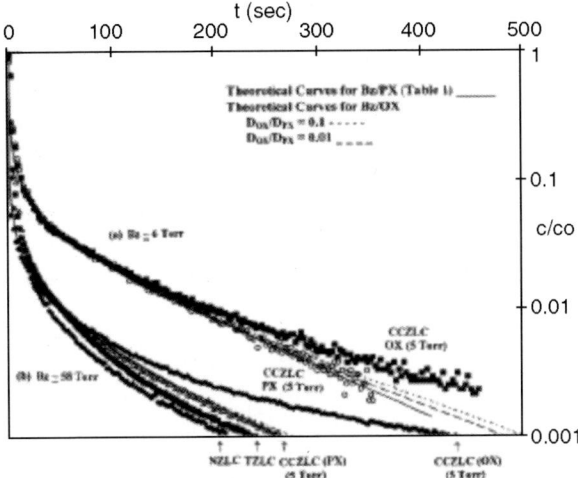

Figure 14.21 ZLC curves for benzene–silicalite at 403 K. Sample is equilibrated with benzene at (a) 6 and (b) 58 Torr. The corresponding NZLC and TZLC curves and the CCZLC curves corresponding to 5 Torr p-xylene or o-xylene are shown. The asymptotic slopes of the NZLC, TZLC, and CCZLC curves for p-xylene are all the same, showing no significant hindrance effect. However, the asymptotic slopes of the CCZLC curves with o-xylene as the counter-diffusant show substantial hindrance, especially at the higher benzene loading. Reprinted from Brandani et al. [92] with permission.

concentration range to generate reliable ZLC desorption curves. Hold-up in the detector may also be an issue.

Fluid phase hold-up is important because, as a result of the high molecular density, the hold-up in the extracrystalline liquid may well be comparable with, or even greater than, the hold-up in the adsorbed phase. The extracrystalline hold-up must be accounted for in the mathematical model and this can be accomplished by using Eq. (14.62) in place of Eq. (14.47) to represent the desorption curve, but the sensitivity of the data is reduced. The response curve for a liquid phase ZLC system has approximately the form of a two time-constant process with the faster process corresponding to the convective washout of the extracrystalline fluid (purging of the cell and detector) and the slower process corresponding to the desorption process. Clearly, the desorption rate will be measurable only if it is substantially slower than the washout. Minimization of all extracrystalline volume is therefore critical to achieving a meaningful desorption curve in a liquid ZLC system.

The limited sensitivity of the detector and the hold-up in the extracrystalline space necessitate the use of somewhat larger adsorbent samples in liquid phase measurements (typically about 10 mg compared with $\approx$1–3 mg for vapor phase measurements). Chromatography metering pumps can conveniently be used to obtain the small constant flow rates required for these measurements. Figure 14.22 shows a sketch of an experimental system and Figure 14.23 gives examples of the desorption curves measured under liquid phase conditions for benzene–n-hexane from large NaX crystals. The response curves are well represented by the theoretical model and measurements with two different crystal sizes yielded essentially the same diffusivity values.

14 Chromatographic and Permeation Methods of Measuring Intraparticle Diffusion

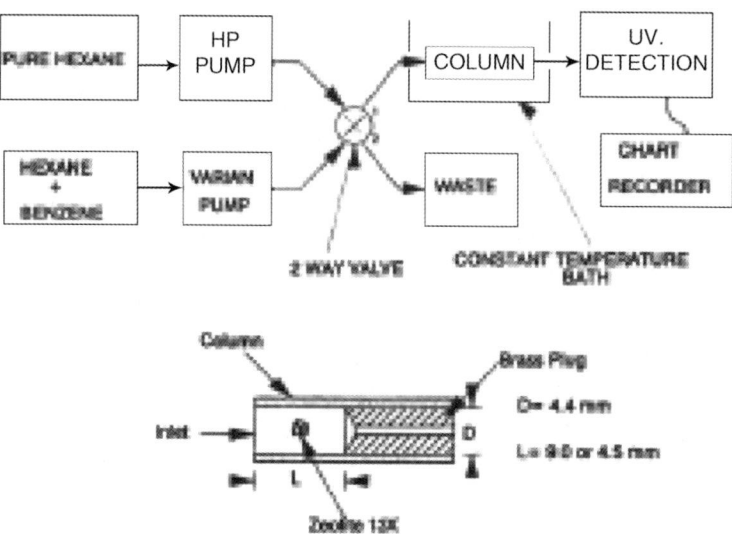

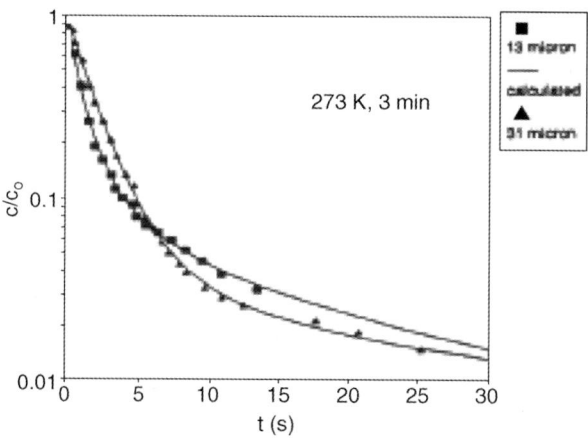

Figure 14.22 Sketch of experimental system for liquid ZLC measurements.

Figure 14.23 Experimental liquid ZLC desorption curves for benzene–n-hexane in NaX crystals (13 and 31 μm) at 273 K, 3 ml min^{-1} hexane flow, showing conformity with the theoretical curves calculated from Eqs. (14.62) and (14.63) Reprinted from Brandani and Ruthven [84] with permission.

This method has also been used by Rodrigues *et al.* to study the kinetics of cation exchange in an ion-exchange resin [95].

14.8
TAP System

The TAP (temporal analysis of products) system was originally developed to allow the rapid analysis of the products from a catalytic reaction. A small pulse of the reactant gas is injected at the inlet of a fixed bed of zeolite crystals through which it is drawn by pulling a vacuum at the bed outlet, where the reaction products are detected by an on-line mass spectrometer. The time dependence of the pulse response is a function of *inter alia* the rate of diffusion within the zeolite crystals. However, the sensitivity of the response to intracrystalline diffusion depends critically on the relative magnitudes of the other mass transfer processes involved. Application of this technique to the study of diffusion of light alkanes in silicalite has led to controversial results. Keipert and Baerns [96] concluded that reliable intracrystalline diffusion measurements could not be made for these systems, since the intracrystalline resistance is smaller than the extracrystalline resistance. Very high diffusivities consistent with PFG NMR data ($\approx 10^{-9}\,\mathrm{m^2\,s^{-1}}$) were reported by Nijhuis *et al.* [97]. However, a detailed review of their data reveals that the response peaks for all species scale accurately with the Henry's law constants, suggesting that, in conformity with the conclusion of Keipert and Baerns, the sorption kinetics must have been controlled by extracrystalline diffusion [98].

14.9
Membrane Permeation Measurements

14.9.1
Steady-State Permeability Measurements

In this type of measurement, which is closely related to the original permeation measurements of Darcy [99], the flux of a gas through a porous plug (often a single cylindrical pellet) of catalyst or adsorbent is measured under quasi-steady-state conditions such that a constant (known) pressure is maintained on one side of the plug, while the pressure on the other side of the plug is either maintained constant or allowed to increase with passage of the diffusant through the plug. If the upstream pressure is maintained at a pressure that is much higher than the downstream pressure, the pressure drop will be essentially constant and equal to the upstream pressure. Monitoring the downstream pressure, in a closed system of limited volume, thus provides a simple and convenient way of measuring the flux. The permeability (B) is then found directly since:

$$J = B \frac{\Delta p}{l} \tag{14.73}$$

where l is the thickness of the porous plug.

This method has been widely applied to porous catalyst pellets to determine, in effect, the macropore diffusivity. While the method is simple in concept and application, interpretation of the results is not always straightforward. It also suffers from the disadvantage that the permeability, rather than the diffusivity, is measured and these quantities are exactly equivalent only in the intracrystalline or Knudsen diffusion regimes. Outside these regimes Knudsen flow, molecular diffusion, Poiseuille flow, and possibly surface diffusion (Section 4.2.1) may all contribute to varying extents. As a result the relationship between the permeability and a well-defined diffusivity may be unclear and difficult to establish. This problem is particularly severe when the pore size distribution is broad or bimodal, so that different mechanisms may dominate in different ranges of pore size.

Application of this technique to measure diffusion in microporous zeolite crystals is hindered by the size of the available crystals and the difficulty of mounting a very small crystal in a permeability cell. Nevertheless, these difficulties have been successfully overcome by Wernick [100], who measured permeation rates for n-butane through a 100 µm crystal of NaX and by Hayhurst [101, 102], Shah [103], and Talu [104] and their collaborators, who measured diffusion of several light alkanes and single-ring aromatics in comparably sized crystals of silicalite (Figures 18.7 and 18.22 and Sections 18.2.1 and 18.3.2). Figure 14.24a shows a schematic diagram of their experimental system.

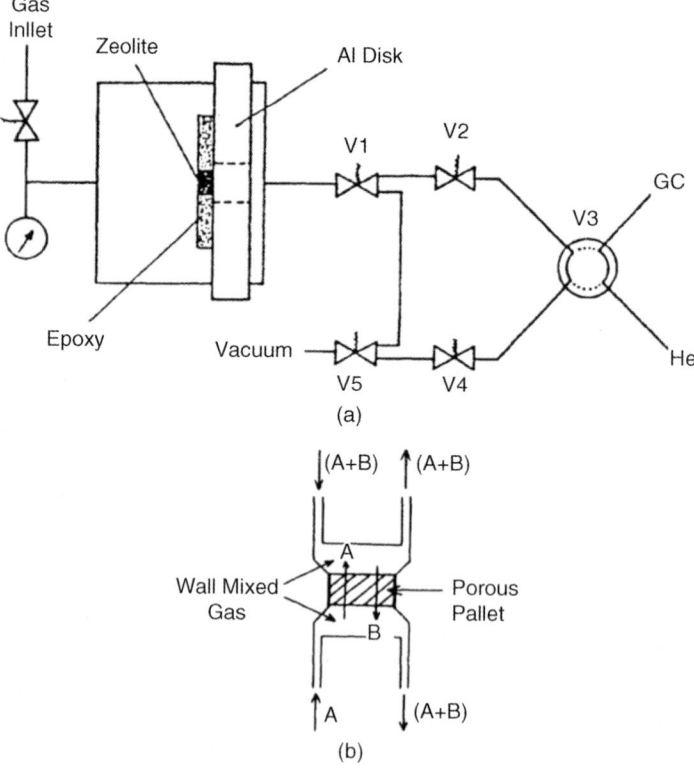

Figure 14.24 Schematic diagram showing (a) the system for permeability measurement through a single zeolite crystal [101] and (b) the "Wicke–Kallenbach" system. Reprinted from Reference [101] with permission.

14.9.2
Wicke–Kallenbach Steady-State Method

14.9.2.1 Macro/Mesopore Diffusion Measurements

Wicke and Kallenbach introduced in 1941 a variant of the permeability experiment that has become a standard method for characterization of the transport behavior of macroporous catalysts [105]. This method differs from the original permeability experiment in that the flux is measured under constant pressure conditions with a known concentration difference maintained across the membrane or pellet. One version of the apparatus is shown schematically in Figure 14.24b. A stream of carrier gas containing a small concentration of the test gas passes over one face of the pellet, while pure carrier is passed over the other face. A sensitive differential pressure cell is normally included to equalize the pressure on both faces of the pellet. The diffusion cell is designed to ensure efficient mixing at both faces and in some modifications gas stirrers are included. More commonly, mixing is achieved by maintaining a high gas velocity within the cell and keeping the dead volume as small as possible. The composition and flow rate of the gas streams leaving both sides of the cell are monitored, thus allowing the net fluxes in both directions and the effective diffusivity to be calculated directly:

$$J = \frac{dQ}{dt} = D_e \frac{\Delta c}{\ell} = D_{app} \frac{\Delta q}{\ell} \tag{14.74}$$

just as in the permeability measurement.

In the absence of Poiseuille flow and surface diffusion, transport occurs by the combined effects of Knudsen and molecular diffusion. In this regime the combined diffusivity is given by Eq. (4.11). Integrating across the membrane at constant total pressure yields:

$$D_e = \frac{\varepsilon_p D_m}{\tau \alpha \Delta y} \ln \left[\frac{(1/D_K) + (1/D_m)(1-\alpha y_A)}{(1/D_K) + (1/D_m)(1-\alpha y_B)} \right] \tag{14.75}$$

where $\alpha = 1 + N_B/N_A$. Thus both the tortuosity factor (τ) and the Knudsen diffusivity (D_K) may in principle be derived from measurements over a range of compositions.

Although constant pressure operation eliminates some of the problems of the original permeability measurement, the difficulties of interpreting the data when the pore size distribution is broad [106, 107] and when the surface diffusion contribution is significant remain.

14.9.2.2 Intracrystalline Diffusion

In the linear region of the isotherm Eq. (14.73) also applies to intracrystalline diffusion measurements but with $B = KD_c$, so knowledge of the Henry constant is required to calculate the diffusivity. Beyond the Henry's law region the analysis becomes slightly more complicated. For a Langmuirian system in which the flux is given by Eq. (1.31), integration across the membrane (from $x=0$, $q=q_1$, $\theta=\theta_1$) with a constant corrected diffusivity (D_0) yields for the loading profile (θ, x) [108]:

$$\frac{Jx}{D_0 q_s} = \ln\left(\frac{1+bp_1}{1+bp}\right) = \ln\left(\frac{1-\theta}{1-\theta_1}\right) \tag{14.76}$$

In the usual experimental arrangement $\theta \to 0$ at the permeate side ($x = \ell$) so the steady state concentration profile through the membrane is given by:

$$C = \theta/\theta_1 = \frac{1-(1-\theta_1)^{(1-X)}}{\theta_1} \tag{14.77}$$

where $X = x/\ell$. The slope is given by:

$$\frac{d\theta}{dX} = \frac{\theta_1}{q_1}\frac{dc}{dX} = (1-\theta)\ln(1-\theta_1) \tag{14.78}$$

The flow through the membrane under steady-state conditions is determined by the concentration gradient at the permeate face and is given by:

$$\frac{\Delta Q}{\ell q_1} = -\int_{\tau_1}^{\tau_2} \frac{1}{q_1} \cdot \frac{dq}{dX}\bigg|_{X=1} \cdot d\tau = -\frac{\ln(1-\theta_1)}{\theta_1} \cdot \Delta\tau \tag{14.79}$$

where $\tau = D_0 t/\ell^2$.

Comparison with Eq. (14.74) shows that for a Langmuirian system the steady-state flow (i.e., the limiting slope of a plot of $Q/\ell q_1$ versus τ) is increased by the factor $-\ln(1-\theta_1)/\theta_1$ relative to the linear case. Since this factor corresponds exactly to the ratio of the average diffusivity (over the relevant concentration range) to the limiting diffusivity, Eq. (14.79) shows that the apparent diffusivity derived from linear analysis corresponds simply to the average diffusivity:

$$\frac{D_{app}}{D_0} = \frac{1}{\theta_1}\ln\frac{1}{(1-\theta_1)} \tag{14.80}$$

Interestingly, many years ago Crank [109] suggested, on intuitive grounds, the use of the average diffusivity for this type of system.

14.9.3
Time Lag Measurements

14.9.3.1 Linear Systems (Constant D)
In the preceding discussion of membrane measurements we have focused exclusively on steady-state measurements. However, permeation measurements may also be made under transient conditions and such measurements can yield additional kinetic information. The simplest situation to consider is the behavior observed at the start of a permeability experiment with a non-adsorbing gas. Analysis of this system was first given by Daynes [110] but it was developed and popularized as a means of measuring diffusion coefficients by Barrer [111–113]. For a non-adsorbing species the mass balance equation within the porous plug is, assuming a constant effective diffusivity (defined with respect to the gas phase concentration c):

$$\frac{\partial c}{\partial t} = D_e \frac{\partial^2 c}{\partial z^2} \tag{14.81}$$

with the initial and boundary conditions for a step change in concentration at one side of the plug at time zero:

$$c(L, t) = 0, \quad c(z, 0) = 0, \quad c(0, t) = c_0 \tag{14.82}$$

The solution, expressed in terms of Q_t, the total quantity of the diffusing species that has passed through the porous plug in time t, is:

$$\frac{Q}{Lc_0} = \frac{D_e t}{L^2} - \frac{1}{6} - \frac{2}{\pi^2} \sum_{n=1}^{\infty} \frac{(-1)^n}{n^2} \exp\left(-n^2 \pi^2 \frac{D_e t}{L^2}\right) \tag{14.83}$$

which, for $t \to \infty$, approaches the asymptote:

$$Q_t = \frac{D_e c_0}{L}\left(t - \frac{L^2}{6D_e}\right) \tag{14.84}$$

which yields a straight line with intercept (time lag) $L^2/6D_e$ on the time axis.

Evidently, measurement of the time lag provides a simple and straightforward way to determine the diffusivity in such a system. The same result is obviously obtained by plotting any quantity that is directly proportional to Q_t. For example, if the diffusant is collected in a closed vessel that is initially evacuated, the pressure will be directly proportional to Q_t. If the experiment is carried out under conditions such that the upstream pressure is maintained constant and very much higher than the downstream pressure, the slope of the asymptote ($D_e c_0/L$) will correspond with the steady-state regime, so it is possible to determine the diffusivity under both steady-state and transient conditions from a single experiment. Within the Knudsen regime consistent values of diffusivity should be obtained from slope and intercept but this is not necessarily true at higher pressures when the pore size distribution is broad or bimodal.

In earlier studies of diffusion through macroporous pellets the diffusivity calculated from the time lag was commonly found to be significantly larger than the steady-state value. Such differences were generally attributed to the effect of "blind" pores [114, 115]. However, in a very detailed study it has recently been shown that such differences can arise from spatial variations in the porosity (revealed by X-ray imaging). Although the "blind pore" hypothesis appears plausible and has been widely accepted, it is not supported by the experimental data [116].

14.9.3.2 Single-Crystal Zeolite Membrane

The same analysis, with concentration expressed in terms of the adsorbed phase concentration (q), applies to permeation through a microporous zeolite crystal within the Henry's law region in which a constant (concentration independent) diffusivity is to be expected. Representative permeation curves for isobutane through a large (120 μm) single crystal of silicalite, as measured by Hayhurst and Paravar [101, 102], are shown in Figure 14.25. As for diffusion in the Knudsen regime, consistency between the transient and steady-state diffusivity values is to be expected for intracrystalline diffusion and is indeed generally observed.

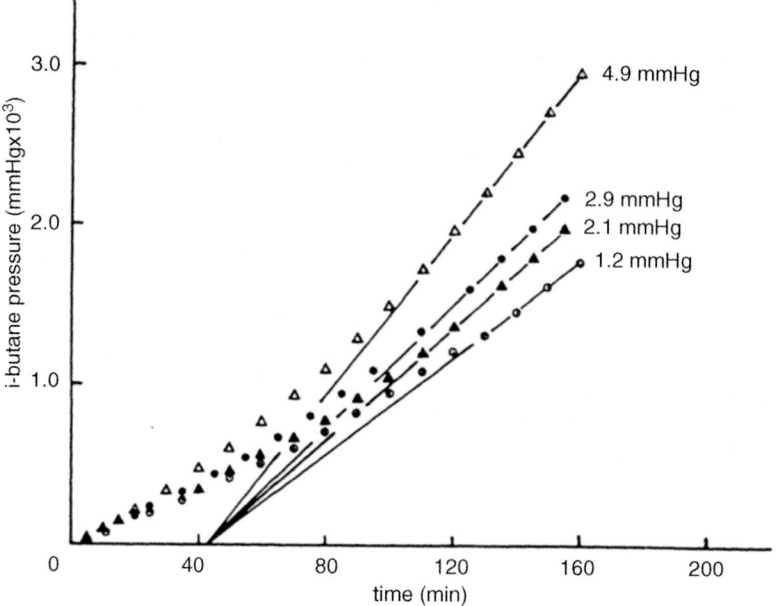

Figure 14.25 Permeability measurement through a single (100 μm) crystal of silicalite (isobutane at 334 K). Note the asymptotic approach to the same intercept [given by Eq. (14.84)] regardless of pressure. This implies that the diffusivity is independent of concentration over the relevant range. Reprinted from Hayhurst and Paravar [102] with permission.

For an absorbing gas, still within the Knudsen regime, Eq. (14.81) is replaced by:

$$(1-\varepsilon_p)\frac{\partial q}{\partial t} + \varepsilon_p \frac{\partial c}{\partial t} = \varepsilon_p D_p \frac{\partial^2 c}{\partial z^2} \qquad (14.85)$$

Within the Henry's law region, assuming local equilibrium $q = Kc$, the form of Eq. (14.81) is recovered with:

$$D_e = \frac{D_p}{(1+[(1-\varepsilon_p)/\varepsilon_p]K)} \qquad (14.86)$$

This coincides with Eq. (2.39), which was derived in Section 2.4 from the random walk model. This extension of the time lag method of measurement to adsorbing species was developed by Barrer and Grove [112]. The method was further extended by Barrer and Barrie [113] to study surface diffusion in a multicomponent system.

14.9.3.3 Nonlinear Systems

The transient behavior of a nonlinear (Langmuirian) system has also been investigated. Using the general approach suggested by Frisch [117], Rutherford and Do [118] obtained a formal general solution, but calculation of the response curves for any

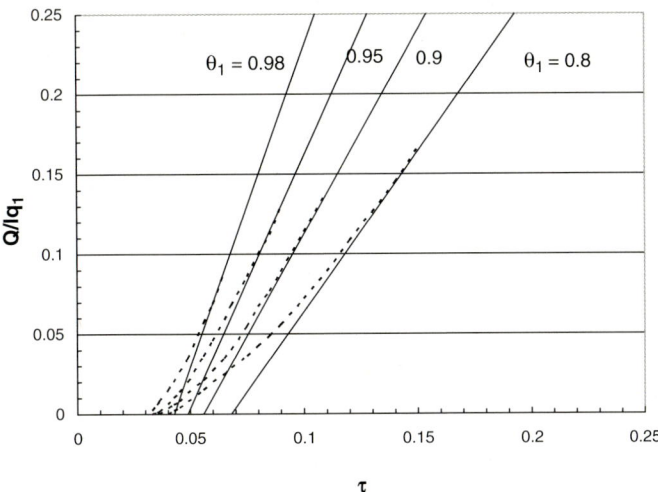

Figure 14.26 Variation of flow through membrane ($Q/\ell q_1$) with time (τ) for a Langmuirian system with different values of θ_1, showing the approach to the steady-state asymptote defined by Eq. (14.79) [108] with permission.

particular from of isotherm is quite challenging. A simpler approximate solution for a Langmuirian system with a constant value of D_o has been given by Ruthven [108]. Theoretical curves showing the time dependence of the quantity of permeate (under conditions such that the loading at the permeate side is always close to zero) calculated from this approximate solution are shown in Figure 14.26, for various degrees of nonlinearity (as measured by $\theta_1 = q/q_s$ at the high pressure side). Evidently, the form of these curves is the same as for a linear system (Figure 14.25). The slope of the long-time asymptote yields directly the apparent steady-state diffusivity which, as noted above, corresponds approximately to the average diffusivity over the loading range. At both high and low degrees of nonlinearity ($\theta_1 \rightarrow 1$ or $\theta_1 \rightarrow 0$) the apparent diffusivity calculated from the time delay [in accordance with Eq. (14.84)] is the same as the steady-state value derived from the slope of the long-time asymptote. At intermediate loadings there is some deviation between the two values but, in accordance with the detailed analysis of Frisch [117], the assumption that, for a nonlinear system, the apparent diffusivity derived from the time delay corresponds to the average diffusivity over the relevant loading range appears to be a good approximation regardless of the precise form of the concentration dependence of the diffusivity.

14.9.3.4 Evaluation of Single-Crystal Membrane Technique

This technique has the advantage of being more direct than most other experimental techniques for measurement of intracrystalline diffusion. Since heat is released (due to adsorption) at the upstream face of the crystal and then removed at the permeate face (where desorption occurs) there is, in principle, a temperature gradient through the crystal, even under steady state conditions. However, reasonable estimates of the heat conduction suggest that, under realistic conditions, any such temperature gradient will be trivial and this is especially true if heat conduction through the support is considered. This technique measures the integral diffusivity, so, for strongly adsorbed species, the extraction of the differential diffusivity is subject to error, unless the form of the concentration dependence is known. Perhaps a more serious limitation is that, like most macroscopic techniques, it is difficult, if not impossible, to distinguish between surface resistance and intracrystalline diffusional resistance, since it is the flux under a known total concentration gradient that is measured. This means that, if there is significant surface resistance, the intracrystalline diffusivity will be underestimated. A further limitation, especially for silicalite, is that it is difficult to mount the crystal in orientations other than parallel to the long axis of the crystal (the z direction). Thus, although it is in principle possible to study diffusion along different crystallographic directions, this is in practice difficult.

14.9.4
Transient Wicke–Kallenbach Experiment

In the transient version of a Wicke–Kallenbach experiment, as introduced by Grachev [119], a step change in sorbate concentration is introduced on one side of the cell at time zero and the concentration response at the other face is followed. It is obviously important to eliminate dead volume in such an experiment, but if this can be achieved the response of the system should follow Eqs. (14.83) and (14.84), just as in a transient permeability or time lag experiment. The basic advantage of a Wicke–Kallenbach system (i.e., elimination of forced flow) is therefore extended to the transient. Since it is the concentration, rather than the flux, that is measured in a Wicke–Kallenbach system, if the data are to be analyzed in accordance with Eq. (14.84), the permeate volume must be obtained by integration:

$$Q_t = F \int_0^t c_L dt \tag{14.87}$$

where $c_L(t)$ is the concentration of sorbate in the permeate stream. However, it is generally easier to obtain the time lag directly by integration of the response curve:

$$\tau = \frac{1}{c_\infty} \int_0^\infty (c_\infty - c_L) dt \tag{14.88}$$

The full solution for the transient response of a Wicke–Kallenbach system has been presented by Dudukovic [120], so the value of the effective diffusivity may be found by matching the experimental response to the dimensionless theoretical response curves. Figure 14.27 shows the form of the response. This approach has the advantage that the fit provides verification of the validity of the diffusion model.

As with a chromatographic system, a transient Wicke–Kallenbach experiment may also be made with a pulse injection of sorbate. This type of measurement was introduced by Gibilaro [121] and developed by Smith and coworkers [122–125]. The solution for the pulse response may be derived directly by differentiation from

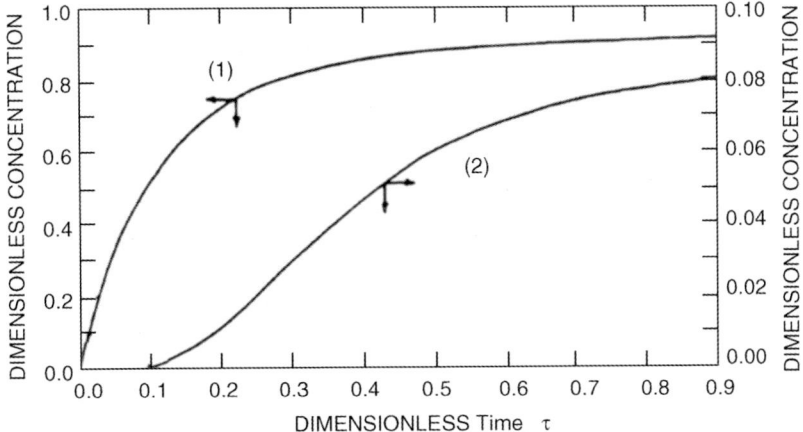

Figure 14.27 Transient response (2) of a Wicke–Kallenbach cell to a step change in feed concentration (1). Reprinted from Dudukovic [120], with permission.

Dudukovic's solution but the analysis has usually been performed using the moments of the response. The first moment (μ) yields directly the time lag:

$$\mu = \frac{\int_0^\infty tc\,dt}{\int_0^\infty c\,dt} = \frac{L^2}{6D_e} \tag{14.89}$$

This simple expression holds only when the concentration in the outlet cell is very small compared with the inlet concentration. The correction factor required when this simplification is not applicable has been given by Dogu and Ercan [126]:

$$\mu = \frac{L^2}{6D_e}\left[\frac{3+LF/AD_e}{1+LF/AD_e}\right] \tag{14.90}$$

where F is the volumetric flow rate through the outlet cell and A the cross sectional area of the pellet. Clearly, when LF/AD_e is large Eq. (14.90) reverts to Eq. (14.89). The recommended procedure is to determine μ by difference from the measurements of the concentration in the cells on both faces of the pellet. If only the outlet response is measured, a correction for the time constant of the inlet cell may be required.

The moments analysis was extended by Hashimoto et al. [124] to a biporous (micromacropore) adsorbent and further extended by Dogu et al. [126, 127] to allow for adsorption of the diffusing species. A finite rate of adsorption was also allowed for, although, in a physical adsorption system, this is normally too fast to measure. Under the conditions corresponding to the simplified boundary condition ($c_L \ll c_0$) the expressions for the first and second moments are:

$$\mu = \frac{L^2}{6D_e}\left[\varepsilon_p + (1-\varepsilon_p)K\right] \tag{14.91}$$

$$\frac{\sigma^2}{\mu^2} = \frac{2}{5} + \frac{4}{5}\frac{(D_e/L^2)}{(KD_c/r_c^2)}\left(1-\frac{\varepsilon_p}{K}\right) \tag{14.92}$$

where K is the dimensionless Henry's law constant based on total particle volume. It is remarkable that the expression for the first moment does not involve the micropore diffusivity.

14.9.5
Comparison of Transient and Steady-State Measurements

Dead end or blind pores make no contribution to the flux in a permeation or Wicke–Kallenbach type of experiment, whereas such pores do contribute to the adsorption/desorption flux under transient conditions. One might therefore expect to find the diffusivities derived from steady-state measurements to be lower than those obtained by transient methods. Comparative studies in which diffusivities in the same samples of porous solid have been measured by both transient and steady-state methods have been reported by McGreavy and Siddiqui [128], Baiker et al. [129] and Burghardt et al. [130] among others. In general, the transient diffusivities are somewhat smaller than the steady-state values, although in some instances there is good agreement. The explanation seems to be that the effect of pore size distribution, particularly in particles having a broad or bimodal distribution of pore size, is much more significant than the effect of blind pores. By far the greatest contribution to the sorption capacity comes from the smaller pores, so in a transient measurement the influence of these pores is more highly weighted than in a steady-state experiment, which measures essentially the flux through the macropores. When the pore size distribution is narrow and unimodal, the transient and steady-state methods yield reasonably consistent values. Consistent values may also be obtained for bimodal adsorbents, provided that the bimodal nature of the pore size distribution is accounted for in the model used to interpret the transient data. Where large discrepancies are seen between transient and steady state measurements, it is usually because the data for an adsorbent with a broad or bimodal distribution of pore size have been interpreted in terms of a single effective diffusivity.

References

1 Kärger, J. and Ruthven, D.M. (1992), in *Diffusion in Zeolites and Other Microporous Solids*, John Wiley & Sons, Inc., New York, ch. 10.
2 Suzuki, M. (1990), in *Adsorption Engineering*, Kodansha Elsevier, Tokyo, ch. 6.
3 Haynes, H.W. (1988) *Catal. Rev. Sci. Eng.*, **30**, 563.
4 Glueckauf, E. and Coates, J.E. (1947) *J. Chem. Soc.*, 1315.
5 Glueckauf, E. (1955) *Trans. Faraday Soc.*, **51**, 1540.
6 Lapidus, L. and Amundson, N.R. (1952) *J. Phys. Chem.*, **56**, 984.
7 Rosen, J.B. (1952) *J. Chem. Phys.*, **20**, 387.
8 Rosen, J.B. (1954) *Ind. Eng. Chem.*, **46**, 1590.
9 Rasmuson, A. and Neretnieks, I. (1980) *AIChE J.*, **26**, 686.
10 Rasmuson, A. (1982) *Chem. Eng. Sci.*, **37**, 787.
11 Anzelius, A. (1926) *Z. Angew. Math. Mech.*, **6**, 291.
12 Furnas, C.C. (1930) *Trans. AIChE*, **24**, 142.
13 Nusselt, W. (1930) *Tech. Mech. Thermodyn.*, **1**, 417.
14 Walter, J.E. (1945) *J. Chem. Phys.*, **13**, 229–332.

15 Klinkenberg, A. (1954) *Ind. Eng. Chem.*, **46**, 2285.
16 Kawazoe, K. and Takeuchi, Y. (1974) *J. Chem. Eng. Jpn.*, **7**, 431.
17 Carta, G. (1988) *Chem. Eng. Sci.*, **43**, 2877.
18 Haynes, H.W. (1975) *Chem. Eng. Sci.*, **30**, 955.
19 Hsu, L.-K.P. and Haynes, H.W. (1981) *AIChE J.*, **27**, 81.
20 Filon, L.N.G. (1928) *Proc. Roy. Soc. Edinburgh*, **XLI**, 38.
21 Raghavan, N.S. and Ruthven, D.M. (1985) *Chem. Eng. Sci.*, **40**, 699.
22 Sarma, P.N. and Haynes, H.W. (1974) *Adv. Chem.*, **133**, 205.
23 van der Laan, E.R. (1958) *Chem. Eng. Sci.*, **7**, 187.
24 Ammons, R.D., Dougharty, N.A., and Smith, J.M. (1977) *Ind. Eng. Chem. Fundam.*, **16**, 363.
25 Haynes, H.W. and Sarma, P.N. (1973) *AIChE J.*, **19**, 1043.
26 Hashimoto, N. and Smith, J.M. (1973) *Ind. Eng. Chem. Fundam.*, **12**, 353.
27 Ma, Y.H. and Mancel, C. (1973) *Adv. Chem.*, **121**, 392.
28 Weber, W.J. (1984) *Fundamentals of Adsorption* (eds A. Myers and G. Belfort), Engineering Foundation, New York, p. 679.
29 Boniface, H. (1983) Separation of argon from air using zeolites, PhD Thesis, University of New Brunswick.
30 Boniface, H. and Ruthven, D.M. (1985) *Chem. Eng. Sci.*, **40**, 1401.
31 Levenspiel, O. (1972) *Chemical Reaction Engineering*, 2nd edn, John Wiley & Sons, Inc., New York, p. 290.
32 Ruthven, D.M. (1984) *Principles of Adsorption and Adsorption Processes*, John Wiley & Sons, Inc., New York, p. 208.
33 van Deemter, J.J., Zuiderweg, F.J., and Klinkenberg, A. (1956) *Chem. Eng. Sci.*, **5**, 271.
34 van Cauwenberghe, A.R. (1966) *Chem. Eng. Sci.*, **21**, 203.
35 Kershenbaum, L.S. and Kohler, M.A. (1984) *Chem. Eng. Sci.*, **39**, 1423.
36 Chiang, A.S., Dixon, A.G., and Ma, Y.H. (1984) *Chem. Eng. Sci*, **39**, 1451.
37 Pazdernik, O. and Schneider, P. (1981) *J. Chromatogr.*, **207**, 181.
38 Dixon, A.G. and Ma, Y.H. (1988) *Chem. Eng. Sci.*, **43**, 1297.
39 Haynes, H.W. (1986) *AIChE J.*, **32**, 1750.
40 Hays, J.F., Clements, W.C., and Harris, T.R. (1967) *AIChE J.*, **13**, 373.
41 Gangwal, S.K., Hudgins, R.R., Bryson, A.W., and Silveston, P.L. (1971) *Can. J. Chem. Eng.*, **49**, 113.
42 Gangwal, S.K., Hudgins, R.R., and Silveston, P.L. (1979) *Can. J. Chem. Eng.*, **57**, 609.
43 Boersma-Klein, W. and Moulijn, J.A. (1979) *Chem. Eng. Sci.*, **34**, 959.
44 Wakao, N. and Tanska, K. (1973) *J. Chem. Eng. Jpn.*, **6**, 338.
45 Takeuchi, K. and Uraguchi, Y. (1977) *J. Chem. Eng. Jpn.*, **10**, 297.
46 Chan, T.S. and Hegedus, L.L. (1978) *AIChE J.*, **24**, 255.
47 Gelbin, D., Wolft, H.-J., and Neinass, J. (1982) *Chem. Tech.*, **34**, 559.
48 Kumar, R., Duncan, R.C., and Ruthven, D.M. (1982) *Can. J. Chem. Eng.*, **60**, 493.
49 Derrah, R.I. and Ruthven, D.M. (1972) *J. Chem. Soc., Faraday Trans. I*, **68**, 2332.
50 Aris, R. (1959) *Proc. R. Soc. London, Ser. A*, **252**, 538.
51 Hsiang, C.S. and Haynes, H.W. (1977) *Chem. Eng. Sci.*, **32**, 678.
52 Wakao, N. (1977) *Chem. Eng. Sci.*, **31**, 678.
53 Scott, D.S., Lee, W., and Papa, J. (1974) *Chem. Eng. Sci.*, **29**, 2155.
54 Langer, G., Roethe, A., Roethe, K.-P., and Gelbin, D. (1978) *Int. J. Heat Mass Trans.*, **21**, 751.
55 Hsu, L.-K.P. and Haynes, H.W. (1981) *AIChEJ.*, **27**, 81.
56 Haq, N. and Ruthven, D.M. (1986) *J. Colloid Interface Sci.*, **112**, 154.
57 Haq, N. and Ruthven, D.M. (1986) *J. Colloid Interface Sci.*, **112**, 164.
58 Ching, C.B. and Ruthven, D.M. (1988) *Zeolites*, **8**, 68.
59 Awum, F., Narayan, S., and Ruthven, D.M. (1988) *Ind. Eng. Chem. Res.*, **27**, 1510.
60 Ma, Y.H., Lin, Y.S., and Fleming, H.L. (1988) *AIChE Symp. Ser.*, **84 (264)**, 1.
61 Gibilaro, L.G. and Waldram, S.P. (1972) *Chem. Eng. J.*, **4**, 197.
62 Gibilaro, L.G. and Waldram, S.P. (1973) *Ind. Eng. Chem. Fundam.*, **12**, 472.
63 Diesler, P.F. and Wilhelm, R.H. (1953) *Ind. Eng. Chem.*, **45**, 1219.
64 Gunn, D.J. (1970) *Chem. Eng. Sci.*, **25**, 53.

65 Boniface, H.A. and Ruthven, D.M. (1985) *Chem. Eng. Sci.*, **40**, 2053.
66 Hyun, S.H. and Danner, R.P. (1985) *AIChE J.*, **31**, 1077.
67 Delmas, M.P.F., Cornu, C., and Ruthven, D.M. (1995) *Zeolites*, **15**, 45–50.
68 Taylor, G.I. (1953) *Proc. R. Soc. London, Ser. A*, **219**, 186.
69 Golay, M.J.E. (1958) *Gas Chromatography* (ed. D.H. Desty), Butterworth, London, p. 36.
70 Aris, R. (1959) *Proc. R. Soc. London, Ser. A*, **252**, 538.
71 Anderson, B.G., de Gauw, F.J.M., van Santen, R.A., and de Voigt, M.J.A. (1998) *Ind. Eng. Chem. Res.*, **37**, 815–824 and 825–834.
72 Ilg, M., Maier-Rosenkrantz, J., Mueller, W., and Bayer, E. (1990) *J. Chromatogr.*, **263**, 517.
73 Bär, N.-K., Balcom, B.J., and Ruthven, D.M. (2002) *Ind. Eng. Chem. Res.*, **41**, 2320–2329.
74 Ruthven, D.M. and Kumar, R. (1979) *Can. J. Chem. Eng.*, **57**, 342.
75 Eic, M. and Ruthven, D.M. (1988) *Zeolites*, **8**, 40.
76 Ruthven, D.M. and Eic, M. (1988) *ACS Symp. Ser.*, **368**, 362.
77 Brandani, F., Rouse, A., Brandani, S., and Ruthven, D.M. (2004) *Adsorption*, **10**, 99–109.
78 Brandani, S. and Ruthven, D.M. (1996) *Adsorption*, **2**, 133–143.
79 Cavalcante, C.L., Brandani, S., and Ruthven, D.M. (1997) *Zeolites*, **18**, 282–285.
80 Brandani, S. (1996) *Chem. Eng. Sci.*, **51**, 3283–3288.
81 Silva, J.A.C. and Rodrigues, A.E. (1996) *Gas Sep. Purif.*, **10**, 207.
82 Ruthven, D.M. and Xu, Z. (1993) *Chem. Eng. Sci.*, **48**, 3307–3312.
83 Ruthven, D.M. and Brandani, F. (2005) *Adsorption*, **11**, 31–35.
84 Brandani, S. and Ruthven, D.M. (1995) *Chem. Eng. Sci.*, **50**, 2055–2059.
85 Brandani, S. (1998) *Chem. Eng. Sci.*, **53**, 2791–2798.
86 Brandani, S., Jama, M.A., and Ruthven, D.M. (2000) *Chem. Eng. Sci.*, **55**, 1205–1212.
87 Brandani, S., Cavalcante, C., Guimaraes, A., and Ruthven, D.M. (1998) *Adsorption*, **4**, 275–285.
88 Brandani, S., Xu, Z., and Ruthven, D.M. (1996) *Microporous Mesoporous Mater.*, **7**, 323–331.
89 Brandani, S., Hufton, J., and Ruthven, D.M. (1995) *Zeolites*, **15**, 624–631.
90 Brandani, S., Ruthven, D.M., and Kärger, J. (1995) *Zeolites*, **15**, 494–496.
91 Hufton, J.R., Brandani, S., and Ruthven, D.M. (1994) *Zeolites and Related Microporous Materials: State of the Art 1994, Studies in Surface Science and Catalysis*, vol. 84 (eds J. Weitkamp, A.G. Karge, H. Pfeifer, and W. Hölderich), Elsevier, Amsterdam, pp. 1323–1330.
92 Brandani, S., Jama, M., and Ruthven, D.M. (2000) *Microporous Mesoporous Mater.*, **35**, 283–300.
93 Brandani, S., Jama, M., and Ruthven, D.M. (2000) *Ind. Eng. Chem. Res.*, **39**, 821–828.
94 Ruthven, D.M. and Stapleton, P. (1993) *Chem. Eng. Sci.*, **48**, 89–98.
95 Rodrigues, J.F., Valverde, J.L., and Rodrigues, A.E. (1998) *Ind. Eng. Chem. Res.*, **37**, 2020–2028.
96 Keipert, O.P. and Baerns, M. (1998) *Chem. Eng. Sci.*, **53**, 3623.
97 Nijhuis, T.A.J., van den Broeke, L.J.P., van de Graaf, J.M., Kapteijn, F., Makkea, M., and Moulijn, J.A. (1997) *Chem. Eng. Sci.*, **52**, 3401.
98 Brandani, S. and Ruthven, D.M. (2000) *Chem. Eng. Sci.*, **55**, 1935–1937.
99 Darcy, H. (1856) *Les Fontaines Publiques de Ie Ville de Dijon*, Balmont, Paris.
100 Wernick, D.L. and Osterhuber, E.J. (1984) (eds D. Olsen and A. Bisio), *Proceedings Sixth International Zeolite Conference, Reno, Nevada July 1983*, Butterworth, Guildford, p. 122.
101 Paravar, A. and Hayhurst, D.T. (1984) (eds D. Olson and A. Bisio), *Proceedings Sixth International Zeolite Conference, Reno, Nevada July 1983*, Butterworth, Guildford, p. 217.
102 Hayhurst, D.T. and Paravar, A. (1988) *Zeolites*, **8**, 27.
103 Shah, D.B. and Liou, H.Y. (1994) *Zeolites*, **14**, 541–548.
104 Talu, O., Sun, M.S., and Shah, D.B. (1998) *AIChE J.*, **44**, 681–694.
105 Wicke, E. and Kallenbach, R. (1941) *Kolloid Z.*, **97**, 135.

106 Wakao, N. and Smith, J.M. (1962) *Chem. Eng. Sci.*, **17**, 825.
107 Johnson, M.F.L. and Stuart, W.E. (1965) *J. Catal.*, **4**, 248.
108 Ruthven, D.M. (2007) *Chem. Eng. Sci.*, **62**, 5745–5752.
109 Crank, J. (1975) *Mathematics of Diffusion*, Oxford University Press, London.
110 Daynes, H. (1920) *Proc. R. Soc. London, Ser. A*, **97**, 296.
111 Barrer, R.M. (1951) *Diffusion in and Through Solids*, Cambridge University Press.
112 Barrer, R.M. and Grove, D.M. (1951) *Trans. Faraday Soc.*, **47**, 837.
113 Barrer, R.M. and Barrie, J.A. (1952) *Proc. R. Soc. London, Ser. A*, **213**, 250.
114 Ash, R., Baker, R.W., and Barrer, R.M. (1968) *Proc. R. Soc. London, Ser. A*, **304**, 407.
115 Pope, C.G. (1967) *Trans. Faraday Soc.*, **63**, 734.
116 Galiatsatou, P., Kanellopoulos, N., and Petropoulos, J.H. (2006) *Phys. Chem. Chem. Phys.*, **8**, 3741.
117 Frisch, H.L. (1958) *J. Phys. Chem.*, **42**, 401–404.
118 Rutherford, S.W. and Do, D.D. (1997) *Chem. Eng. Sci.*, **52**, 703–713.
119 Grachev, R.A., Ione, K.G., and Barshev, R.A. (1970) *Kinet. Catal.*, **11**, 445.
120 Dudukovic, M.P. (1982) *Chem. Eng. Sci.*, **37**, 153.
121 Gibilaro, L.G., Gioia, F., and Greco, G. (1970) *Chem. Eng. J.*, **1**, 85.
122 Dogu, G. and Smith, J.M. (1975) *AIChE J.*, **21**, 58.
123 Dogu, G. and Smith, J.M. (1976) *Chem. Eng. Sci.*, **31**, 123.
124 Hashimoto, N., Moffat, A.J., and Smith, J.M. (1976) *AIChE J.*, **22**, 944.
125 Burghardt, A. and Smith, J.M. (1979) *Chem. Eng. Sci.*, **34**, 267.
126 Dogu, G. and Ercan, C. (1983) *Can. J. Chem. Eng.*, **61**, 660.
127 Dogu, G., Keskin, A., and Dogu, T. (1987) *AIChE J.*, **33**, 322.
128 McGreavy, C. and Siddiqui, M.A. (1980) *Chem. Eng. Sci.*, **35**, 3.
129 Baiker, A., New, M., and Richarz, W. (1982) *Chem. Eng. Sci.*, **37**, 643.
130 Burghardt, A., Rogut, J., and Gotkowski, J. (1988) *Chem. Eng. Sci.*, **43**, 2463.
131 Vidoni, A. (2011) PhD Thesis, University of Maine.

Jörg Kärger,
Douglas M. Ruthven, and
Doros N. Theodorou

Diffusion in Nanoporous Materials

Related Titles

Cejka, Jiri / Corma, Avelino / Zones, Stacey (eds.)

Zeolites and Catalysis

882 Pages, 2 Volumes, Hardcover
318 Fig. (42 Colored Fig.), 75 Tab.
2010
ISBN: 978-3-527-32514-6

Kulprathipanja, Santi (ed.)

Zeolites in Industrial Separation and Catalysis

594 Pages, Hardcover
310 Fig. (10 Colored Fig.)
2010
ISBN: 978-3-527-32505-4

Hirscher, Michael (ed.)

Handbook of Hydrogen Storage

353 Pages, Hardcover
158 Fig. (6 Colored Fig.), 24 Tab.
2010
ISBN: 978-3-527-32273-2

Farrusseng, David (ed.)

Metal-Organic Frameworks

392 Pages, Hardcover
185 Fig. (26 Colored Fig.), 28 Tab.
2011
ISBN: 978-3-527-32870-3

Bruce, Duncan W. / Walton, Richard I. / O'Hare, Dermot (eds.)

Porous Materials

350 Pages, Hardcover
2010
ISBN: 978-0-470-99749-9

Bräuchle, Christoph / Lamb, Don Carroll / Michaelis, Jens (eds.)

Single particle tracking and single molecule energy transfer

343 Pages, Hardcover
112 Fig. (46 Colored Fig.), 3 Tab.
2010
ISBN: 978-3-527-32296-1

Klages, Rainer / Radons, Günter / Sokolov, Igor M. (eds.)

Anomalous Transport Foundations and Applications

584 Pages, Hardcover
163 Fig. (11 Colored Fig)
2008
ISBN: 978-3-527-40722-4

Deutschmann, Olaf (ed.)

Modeling and Simulation of Heterogeneous Catalytic Reactions From the Molecular Process to the Technical System

354 Pages, Hardcover.
155 Fig. (25 Colored Fig.), 20 Tab.
2011
ISBN: 978-3-527-32120-9

Reichl, Linda E.

A Modern Course in Statistical Physics 3rd revised and updated edition

411 Pages, Softcover
2009
ISBN: 978-3-527-40782-8

Rao, B. L. S. Prakasa

Statistical Inference for Fractional Diffusion Processes

Series: Wiley Series in Probability and Statistics
280 Pages, Hardcover
2010
ISBN: 978-0-470-66568-8

Jörg Kärger, Douglas M. Ruthven,
and Doros N. Theodorou

Diffusion in Nanoporous Materials

Volume 2

WILEY-VCH Verlag GmbH & Co. KGaA

The Authors

Prof. Dr. Jörg Kärger
Universität Leipzig
Abteilung Grenzflächenphysik
Linnestr. 5
04103 Leipzig

Prof. Dr. Douglas M. Ruthven
Dept. of Chemical Engineering
University of Maine
Orono, ME 04469-5737
USA

Prof. Doros N. Theodorou
Nat. Techn. Univ. of Athens
School of Chemical Engineering
9 Heroon Polytechniou Street
15780 Athens
Griechenland

All books published by **Wiley-VCH** are carefully produced. Nevertheless, authors, editors, and publisher do not warrant the information contained in these books, including this book, to be free of errors. Readers are advised to keep in mind that statements, data, illustrations, procedural details or other items may inadvertently be inaccurate.

Library of Congress Card No.: applied for

British Library Cataloguing-in-Publication Data
A catalogue record for this book is available from the British Library.

Bibliographic information published by the Deutsche Nationalbibliothek
The Deutsche Nationalbibliothek lists this publication in the Deutsche Nationalbibliografie; detailed bibliographic data are available on the Internet at http://dnb.d-nb.de.

© 2012 Wiley-VCH Verlag & Co. KGaA, Boschstr. 12, 69469 Weinheim, Germany

All rights reserved (including those of translation into other languages). No part of this book may be reproduced in any form – by photoprinting, microfilm, or any other means – nor transmitted or translated into a machine language without written permission from the publishers. Registered names, trademarks, etc. used in this book, even when not specifically marked as such, are not to be considered unprotected by law.

Cover Design Grafik-Design Schulz, Fußgönheim
Typesetting Thomson Digital, Noida, India
Printing and Binding betz-druck GmbH, Darmstad

Printed in the Federal Republic of Germany
Printed on acid-free paper

Print ISBN: 978-3-527-31024-1
ePDF ISBN: 978-3-527-65130-6
ePub ISBN: 978-3-527-65129-0
mobi ISBN: 978-3-527-65128-3
oBook ISBN: 978-3-527-65127-6

To Birge, Patricia and Fani

Contents

Preface *XXV*
Acknowledgments *XXIX*

Content of Volume 1

Part I	**Introduction** *1*	
1	**Elementary Principles of Diffusion** *3*	
1.1	Fundamental Definitions *4*	
1.1.1	Transfer of Matter by Diffusion *4*	
1.1.2	Random Walk *6*	
1.1.3	Transport Diffusion and Self-Diffusion *7*	
1.1.4	Frames of Reference *9*	
1.1.5	Diffusion in Anisotropic Media *10*	
1.2	Driving Force for Diffusion *12*	
1.2.1	Gradient of Chemical Potential *12*	
1.2.2	Experimental Evidence *15*	
1.2.3	Relationship between Transport and Self-diffusivities *16*	
1.3	Diffusional Resistances in Nanoporous Media *17*	
1.3.1	Internal Diffusional Resistances *17*	
1.3.2	Surface Resistance *19*	
1.3.3	External Resistance to Mass Transfer *19*	
1.4	Experimental Methods *21*	
	References *24*	
Part II	**Theory** *25*	
2	**Diffusion as a Random Walk** *27*	
2.1	Random Walk Model *27*	
2.1.1	Mean Square Displacement *27*	
2.1.2	The Propagator *29*	
2.1.3	Correspondence with Fick's Equations *32*	
2.2	Correlation Effects *33*	
2.2.1	Vacancy Correlations *33*	

2.2.2	Correlated Anisotropy 35
2.3	Boundary Conditions 35
2.3.1	Absorbing and Reflecting Boundaries 35
2.3.2	Partially Reflecting Boundary 38
2.3.3	Matching Conditions 39
2.3.4	Combined Impact of Diffusion and Permeation 40
2.4	Macroscopic and Microscopic Diffusivities 42
2.5	Correlating Self-Diffusion and Diffusion with a Simple Jump Model 44
2.6	Anomalous Diffusion 47
2.6.1	Probability Distribution Functions of Residence Time and Jump Length 47
2.6.2	Fractal Geometry 49
2.6.3	Diffusion in a Fractal System 53
2.6.4	Renormalization 54
2.6.5	Deviations from Normal Diffusion in Nanoporous Materials: A Retrospective 55
	References 57

3	**Diffusion and Non-equilibrium Thermodynamics** 59
3.1	Generalized Forces and Fluxes 59
3.1.1	Mechanical Example 59
3.1.2	Thermodynamic Forces and Fluxes 60
3.1.3	Rate of Generation of Entropy 61
3.1.4	Isothermal Approximation 63
3.1.5	Diffusion in a Binary Adsorbed Phase 64
3.2	Self-Diffusion and Diffusive Transport 65
3.3	Generalized Maxwell–Stefan Equations 67
3.3.1	General Formulation 67
3.3.2	Diffusion in an Adsorbed Phase 68
3.3.3	Relation between Self- and Transport Diffusivities 71
3.4	Application of the Maxwell–Stefan Model 72
3.4.1	Parameter Estimation 72
3.4.2	Membrane Permeation 73
3.4.3	Diffusion in Macro- and Mesopores 74
3.5	Loading Dependence of Self- and Transport Diffusivities 75
3.5.1	Self-Diffusivities 75
3.5.2	Transport Diffusivities 77
3.5.3	Molecular Simulation 78
3.5.4	Effect of Structural Defects 78
3.6	Diffusion at High Loadings and in Liquid-Filled Pores 80
	References 81

4	**Diffusion Mechanisms** 85
4.1	Diffusion Regimes 85
4.1.1	Size-Selective Molecular Sieving 85

4.2	Diffusion in Macro- and Mesopores	87
4.2.1	Diffusion in a Straight Cylindrical Pore	87
4.2.1.1	Knudsen Mechanism	88
4.2.1.2	Viscous Flow	90
4.2.1.3	Molecular Diffusion	91
4.2.1.4	Transition Region	91
4.2.1.5	Self-Diffusion/Tracer Diffusion	92
4.2.1.6	Relative Importance of Different Mechanisms	92
4.2.1.7	Surface Diffusion	92
4.2.1.8	Combination of Diffusional Resistances	93
4.2.2	Diffusion in a Pore Network	94
4.2.2.1	Dusty Gas Model	94
4.2.2.2	Effective Medium Approximation	95
4.2.2.3	Tortuosity Factor	95
4.2.2.4	Parallel Pore Model	96
4.2.2.5	Random Pore Model	97
4.2.2.6	Capillary Condensation	97
4.3	Activated Diffusion	99
4.3.1	Diffusion in Solids	99
4.3.2	Surface Diffusion Mechanisms	100
4.3.2.1	Vacancy Diffusion	100
4.3.2.2	Reed–Ehrlich Model	101
4.3.3	Diffusion by Cage-to-Cage Jumps	104
4.4	Diffusion in More Open Micropore Systems	106
4.4.1	Mobile Phase Model	106
4.4.2	Diffusion in a Sinusoidal Field	107
4.4.3	Free-Volume Approach for Estimating the Loading Dependence	108
	References	108
5	**Single-File Diffusion**	**111**
5.1	Infinitely Extended Single-File Systems	112
5.1.1	Random Walk Considerations	112
5.1.2	Molecular Dynamics	116
5.2	Finite Single-File Systems	119
5.2.1	Mean Square Displacement	119
5.2.2	Tracer Exchange	120
5.2.3	Catalytic Reactions	123
5.2.3.1	Dynamic Monte Carlo Simulations and Adapting the Analysis for Normal Diffusion	123
5.2.3.2	Rigorous Treatment	125
5.2.3.3	Molecular Traffic Control	130
5.3	Experimental Evidence	132
5.3.1	Ideal versus Real Structure of Single-File Host Systems	132
5.3.2	Experimental Findings Referred to Single-File Diffusion	135
5.3.2.1	Pulsed Field Gradient NMR	135

5.3.2.2	Quasi-elastic Neutron Scattering	136
5.3.2.3	Tracer Exchange and Transient Sorption Experiments	137
5.3.2.4	Catalysis	138
	References	139

6 Sorption Kinetics 143
6.1 Resistances to Mass and Heat Transfer 143
6.2 Mathematical Modeling of Sorption Kinetics 145
6.2.1 Isothermal Linear Single-Resistance Systems 145
6.2.1.1 External Fluid Film or Surface Resistance Control 145
6.2.1.2 Micropore Diffusion Control 146
6.2.1.3 Effect of Particle Shape 149
6.2.1.4 Macropore Diffusion Control 149
6.2.2 Isothermal, Linear Dual-Resistance Systems 151
6.2.2.1 Surface Resistance plus Internal Diffusion 151
6.2.2.2 Transient Concentration Profiles 152
6.2.2.3 Two Diffusional Resistances (Biporous Solid) 156
6.2.3 Isothermal Nonlinear Systems 160
6.2.3.1 Micropore Diffusion Control 160
6.2.3.2 Macropore Diffusion Control 160
6.2.3.3 Semi-infinite Medium 161
6.2.3.4 Adsorption Profiles 162
6.2.3.5 Desorption Profiles 165
6.2.3.6 Experimental Uptake Rate Data 167
6.2.3.7 Adsorption/Desorption Rates and Effective Diffusivities 170
6.2.3.8 Approximate Analytic Representations 172
6.2.3.9 Linear Driving Force Approximation 172
6.2.3.10 Shrinking Core Model 173
6.2.3.11 Surface Resistance Control – Nonlinear Systems 176
6.2.4 Non-isothermal Systems 179
6.2.4.1 Intraparticle Diffusion Control 181
6.2.4.2 Experimental Non-isothermal Uptake Curves 183
6.3 Sorption Kinetics for Binary Mixtures 185
6.3.1.1 Counter-Diffusion 186
6.3.1.2 Co-diffusion 187
 References 188

Part III Molecular Modeling 191

7 Constructing Molecular Models and Sampling Equilibrium Probability Distributions 193
7.1 Models and Force Fields for Zeolite–Sorbate Systems 194
7.1.1 Molecular Model and Potential Energy Function 194
7.1.2 *Ab Initio* Molecular Dynamics of Zeolite-Sorbate Systems 203

7.1.3	Computer Reconstruction of Meso- and Macroporous Structure *204*
7.2	Monte Carlo Simulation Methods *206*
7.2.1	Metropolis Monte Carlo Algorithm *206*
7.2.2	Canonical Monte Carlo *207*
7.2.3	Grand Canonical Monte Carlo *210*
7.2.4	Gibbs Ensemble and Gage Cell Monte Carlo *215*
7.3	Free Energy Methods for Sorption Equilibria *217*
7.4	Coarse-Graining and Potentials of Mean Force *222*
	References *224*

8	**Molecular Dynamics Simulations** *227*
8.1	Statistical Mechanics of Diffusion *227*
8.1.1	Self-diffusivity *227*
8.1.2	Mass Fluxes in a Microporous Medium *229*
8.1.3	Transport in the Presence of Concentration Gradients in Pure and Mixed Sorbates *232*
8.2	Equilibrium Molecular Dynamics Simulations *235*
8.2.1	Integrating the Equations of Motion: the Velocity Verlet Algorithm *235*
8.2.2	Multiple Time Step Algorithms: rRESPA *238*
8.2.3	Domain Decomposition *241*
8.2.4	Molecular Dynamics of Rigid Linear Molecules *241*
8.2.5	Molecular Dynamics of Rigid Nonlinear Molecules: Quaternions *245*
8.2.6	Constraint Dynamics in Cartesian Coordinates *249*
8.2.7	Extended Ensemble Molecular Dynamics *253*
8.2.8	Example Application of Equilibrium MD to Pure Sorbates *257*
8.2.9	Example Applications of MD to Mixed Sorbates *259*
8.3	Non-equilibrium Molecular Dynamics Simulations *265*
8.3.1	Gradient Relaxation Method *265*
8.3.2	External Field NEMD *267*
8.3.3	Dual Control Volume Grand Canonical Molecular Dynamics *269*
	References *272*

9	**Infrequent Event Techniques for Simulating Diffusion in Microporous Solids** *275*
9.1	Statistical Mechanics of Infrequent Events *276*
9.1.1	Time Scale Separation and General Expression for the Rate Constant *276*
9.1.2	Transition State Theory Approximation and Dynamical Correction Factor *281*
9.1.3	Multidimensional Transition State Theory *283*
9.1.4	Infrequent Event Theory of Multistate Multidimensional Systems *290*
9.1.5	Numerical Methods for Infrequent Event Analysis *291*

9.2	Tracking Temporal Evolution in a Network of States	292
9.2.1	Master Equation	292
9.2.2	Kinetic Monte Carlo Simulation	293
9.2.3	Analytical Solution of the Master Equation	295
9.3	Example Applications of Infrequent Event Analysis and Kinetic Monte Carlo for the Prediction of Diffusivities in Zeolites	296
9.3.1	Self-diffusivity at Low Occupancies	296
9.3.2	Self-diffusivity at High Occupancy	300
	References	301

Part IV Measurement Methods 303

10 Measurement of Elementary Diffusion Processes 305

10.1	NMR Spectroscopy	306
10.1.1	Basic Principles: Behavior of Isolated Spins	306
10.1.1.1	Classical Treatment	306
10.1.1.2	Quantum Mechanical Treatment	308
10.1.2	Behavior of Nuclear Spins in Compact Material	309
10.1.2.1	Fundamentals of the NMR Line Shape	311
10.1.2.2	Effect of Molecular Motion on Line Shape	311
10.1.2.3	Fundamentals of Pulse Measurements	314
10.1.3	Resonance Shifts by Different Surroundings	318
10.1.3.1	Spins in Different Chemical Surroundings	318
10.1.3.2	Anisotropy of the Chemical Shift	320
10.1.3.3	Quadrupole NMR	320
10.1.4	Impact of Nuclear Magnetic Relaxation	323
10.2	Diffusion Measurements by Neutron Scattering	326
10.2.1	Principle of the Method	326
10.2.1.1	Experimental Procedure	326
10.2.2	Theory of Neutron Scattering	327
10.2.3	Scattering Patterns and Molecular Motion	330
10.2.3.1	Fundamental Relations	330
10.2.3.2	Thermodynamic Factor	331
10.2.3.3	Evidence on the Elementary Steps of Diffusion	334
10.2.3.4	Intermediate Scattering Function	334
10.2.3.5	Range of Measurement	335
10.3	Diffusion Measurements by Light Scattering	337
10.3.1	Theory	337
10.3.2	Experimental Issues for Observing Light Scattering Phenomena: Index Matching	339
10.3.2.1	Index Matching	339
10.3.2.2	Optical Mixing Techniques	340
10.3.2.3	Filter Techniques	341
	References	343

11	**Diffusion Measurement by Monitoring Molecular Displacement** 347
11.1	Pulsed Field Gradient (PFG) NMR: Principle of Measurement 348
11.1.1	PFG NMR Fundamentals 348
11.1.2	Basic Experiment 351
11.2	The Complete Evidence of PFG NMR 352
11.2.1	Concept of the Mean Propagator 352
11.2.2	Evidence of the Mean Propagator 353
11.2.3	Concept of Effective Diffusivity 354
11.3	Experimental Conditions, Limitations, and Options for PFG NMR Diffusion Measurement 355
11.3.1	Sample Preparation 355
11.3.2	Basic Data Analysis and Ranges of Observation 356
11.3.3	Pitfalls 358
11.3.3.1	Gradient Pulse Mismatch 358
11.3.3.2	Mechanical Instabilities 358
11.3.3.3	Benefit of Extra-large Stray-Field Gradients 359
11.3.3.4	Impedance by Internal Gradients 359
11.3.3.5	Impedance by Contaminants 360
11.3.4	Measurement of Self-diffusion in Multicomponent Systems 361
11.3.4.1	Fourier-Transform PFG NMR 361
11.3.4.2	Alternative Approaches 362
11.3.5	Diffusion Anisotropy 362
11.3.5.1	"Single-Crystal" Measurements 362
11.3.5.2	Powder Measurement 363
11.3.5.3	Evidence on Host Structure 364
11.4	Different Regimes of PFG NMR Diffusion Measurement 364
11.4.1	Effect of Finite Crystal Size on the Measurement of Intracrystalline Diffusion 365
11.4.1.1	Correlating the True and Effective Diffusivities in the Short-Time Limit 365
11.4.1.2	Correlating the Effective Diffusivities with the Crystal Dimensions in the Long-Time Limit 367
11.4.2	Long-Range Diffusion 368
11.4.2.1	Tortuosity Factor and the Mechanism of Diffusion 369
11.4.3	Covering the Complete Range of Observation Times, the "NMR-Tracer Desorption" Technique 371
11.4.3.1	Two-Dimensional Monte-Carlo Simulation 371
11.4.3.2	Two-Region Model 374
11.4.3.3	NMR Tracer Desorption (= Tracer-Exchange) Technique 376
11.4.3.4	Observation of Surface Barriers 377
11.5	Experimental Tests of Consistency 379
11.5.1	Variation of Crystal Size 379
11.5.2	Diffusion Measurements with Different Nuclei 379
11.5.3	Influence of External Magnetic Field 380

11.5.4	Comparison of Self-diffusion and Tracer Desorption Measurements 380
11.5.5	Blocking of the Extracrystalline Space 380
11.5.6	Determination of Crystal Size for Restricted Diffusion Measurements 381
11.5.7	Long-Range Diffusion 382
11.5.8	Tracer Exchange Measurements 382
11.6	Single-Molecule Observation 383
11.6.1	Basic Principles of Fluorescence Microscopy 384
11.6.2	Time Averaging Versus Ensemble Averaging 385
11.6.3	Correlating Structure and Mass Transfer in Nanoporous Materials 388
	References 389
12	**Imaging of Transient Concentration Profiles** 395
12.1	Different Options of Observation 396
12.1.1	Positron Emission Tomography (PET) 397
12.1.2	X-Ray Monitoring 397
12.1.3	Optical Microscopy 398
12.1.4	IR Microscopy 399
12.1.5	Magnetic Resonance Imaging (MRI) 400
12.2	Monitoring Intracrystalline Concentration Profiles by IR and Interference Microscopy 403
12.2.1	Principles of Measurement 403
12.2.2	Data Analysis 406
12.3	New Options for Experimental Studies 408
12.3.1	Monitoring Intracrystalline Concentration Profiles 408
12.3.1.1	New Options of Correlating Transport Diffusion and Self-diffusion 412
12.3.1.2	Visualizing Guest Profiles in Transient Sorption Experiments 412
12.3.1.3	Testing the Relevance of Surface Resistances by Correlating Fractional Uptake with Boundary Concentration 413
12.3.1.4	Applying Boltzmann's Integration Method for Analyzing Intracrystalline Concentration Profiles 414
12.3.2	Direct Measurement of Surface Barriers 415
12.3.2.1	Concentration Dependence of Surface Permeability 415
12.3.2.2	Correlating the Surface Permeabilities through Different Crystal Faces 417
12.3.3	Adapting the Concept of Sticking Probabilities to Nanoporous Particles 421
	References 423
13	**Direct Macroscopic Measurement of Sorption and Tracer Exchange Rates** 427
13.1	Gravimetric Methods 427
13.1.1	Experimental System 428

13.1.2	Analysis of the Response Curves 429
13.1.2.1	Intrusion of Heat Effects 429
13.1.2.2	Criterion for Negligible Thermal Effects 430
13.1.2.3	Intrusion of Bed Diffusional Resistance 431
13.1.2.4	Experimental Checks 432
13.1.3	The TEOM and the Quartz Crystal Balance 432
13.2	Piezometric Method 433
13.2.1	Mathematical Model 434
13.2.2	Single-Step Frequency Response 436
13.2.3	Single-Step Temperature Response 436
13.3	Macro FTIR Sorption Rate Measurements 437
13.4	Rapid Recirculation Systems 440
13.4.1	Liquid Phase Systems 441
13.5	Differential Adsorption Bed 441
13.6	Analysis of Transient Uptake Rate Data 443
13.6.1	Time Domain Matching 443
13.6.2	Method of Moments 444
13.7	Tracer Exchange Measurements 445
13.7.1	Detectors for Radioisotopes 446
13.7.2	Experimental Procedure 447
13.8	Frequency Response Measurements 447
13.8.1	Theoretical Model 448
13.8.2	Temperature Frequency Response 451
13.8.3	Measurement Limits 451
13.8.4	Experimental Systems 452
13.8.5	Results 454
13.8.6	Frequency Response Measurements in a Flow System 455
	References 456
14	**Chromatographic and Permeation Methods of Measuring Intraparticle Diffusion** 459
14.1	Chromatographic Method 460
14.1.1	Mathematical Model for a Chromatographic Column 460
14.1.1.1	Time Domain Solutions 462
14.1.1.2	Form of Response Curves 463
14.1.2	Moments Analysis 464
14.1.2.1	First and Second Moments 464
14.1.2.2	Use of Higher Moments 466
14.1.2.3	HETP and the van Deemter Equation 466
14.2	Deviations from the Simple Theory 468
14.2.1	"Long-Column" Approximation 468
14.2.1.1	Pressure Drop 468
14.2.1.2	Nonlinear Equilibrium 469
14.2.1.3	Heat Transfer Resistance 470
14.3	Experimental Systems for Chromatographic Measurements 470

14.3.1	Dead Volume *471*	
14.3.2	Experimental Conditions *471*	
14.4	Analysis of Experimental Data *472*	
14.4.1	Liquid Systems *473*	
14.4.2	Vapor Phase Systems *473*	
14.4.3	Intrusion of Axial Dispersion *477*	
14.5	Variants of the Chromatographic Method *479*	
14.5.1	Step Response *479*	
14.5.2	Limited Penetration Regime *480*	
14.5.3	Wall-Coated Column *480*	
14.5.4	Direct Measurement of the Concentration Profile in the Column *480*	
14.6	Chromatography with Two Adsorbable Components *481*	
14.7	Zero-Length Column (ZLC) Method *483*	
14.7.1	General Principle of the Technique *483*	
14.7.2	Theory *485*	
14.7.2.1	Basic Theory for Intraparticle Diffusion Control *485*	
14.7.2.2	Short-Time Behavior *486*	
14.7.2.3	Diffusion in Macroporous Particles *488*	
14.7.3	Deviations from the Simple Theory *489*	
14.7.3.1	Surface Resistance and/or Fluid Film Resistance *489*	
14.7.3.2	Measurement of Surface Resistance by ZLC *489*	
14.7.3.3	Fluid Phase Hold-Up *490*	
14.7.3.4	Effect of Isotherm Nonlinearity *491*	
14.7.3.5	Heat Effects *492*	
14.7.4	Practical Considerations *493*	
14.7.4.1	Choice of Operating Conditions and Extraction of the Time Constant *493*	
14.7.4.2	Preliminary Checks *494*	
14.7.4.3	Equilibration Time *494*	
14.7.4.4	Partial Loading Experiment *495*	
14.7.5	Extensions of the ZLC Technique *497*	
14.7.5.1	Tracer ZLC *497*	
14.7.5.2	Counter-current ZLC (CCZLC) *498*	
14.7.5.3	Liquid Phase Measurements *498*	
14.8	TAP System *500*	
14.9	Membrane Permeation Measurements *501*	
14.9.1	Steady-State Permeability Measurements *501*	
14.9.2	Wicke–Kallenbach Steady-State Method *503*	
14.9.2.1	Macro/Mesopore Diffusion Measurements *503*	
14.9.2.2	Intracrystalline Diffusion *503*	
14.9.3	Time Lag Measurements *504*	
14.9.3.1	Linear Systems (Constant D) *504*	
14.9.3.2	Single-Crystal Zeolite Membrane *505*	
14.9.3.3	Nonlinear Systems *506*	
14.9.3.4	Evaluation of Single-Crystal Membrane Technique *507*	

14.9.4	Transient Wicke–Kallenbach Experiment	*508*
14.9.5	Comparison of Transient and Steady-State Measurements	*510*
	References	*510*

Content of Volume 2

Part V   Diffusion in Selected Systems   *515*

15   Amorphous Materials and Extracrystalline (Meso/Macro) Pores   *517*
15.1   Diffusion in Amorphous Microporous Materials   *518*
15.1.1   Diffusion in Microporous Glass   *518*
15.1.2   Diffusion in Microporous Carbon   *520*
15.2   Effective Diffusivity   *524*
15.2.1   Direct Measurement of Tortuosity in a Porous Catalyst Particle   *525*
15.2.2   Determination of Macropore Diffusivity in a Catalyst Particle under Reaction Conditions   *526*
15.3   Diffusion in Ordered Mesopores   *527*
15.4   Diffusion through Mesoporous Membranes   *530*
15.4.1   Mesoporous Vycor Glass   *530*
15.4.2   Mesoporous Silica   *531*
15.4.2.1 Permeance Measurements   *532*
15.4.2.2 Modified Mesoporous Membranes   *533*
15.5   Surface Diffusion   *534*
15.5.1   Determination of Surface Diffusivities   *534*
15.5.2   Concentration Dependence   *535*
15.6   Diffusion in Liquid-Filled Pores   *538*
15.7   Diffusion in Hierarchical Pore Systems   *539*
15.7.1   Ordered Mesoporous Material SBA-15   *539*
15.7.2   Activated Carbon with Interpenetrating Micro- and Mesopores   *541*
15.7.3   Diffusion in "Mesoporous" Zeolites   *544*
15.8   Diffusion in Beds of Particles and Composite Particles   *545*
15.8.1   Approaches by Mathematical Modeling   *546*
15.8.2   Temperature Dependence   *548*
15.8.3   Multicomponent Systems   *550*
15.9   More Complex Behavior: Presence of a Condensed Phase   *552*
15.9.1   Diffusion in Porous Vycor Glass: Hysteresis Effects   *553*
15.9.2   Supercritical Transition in an Adsorbed Phase   *555*
15.9.3   Diffusion in a Composite Particle of MCM-41   *556*
         References   *557*

16   Eight-Ring Zeolites   *561*
16.1   Eight-Ring Zeolite Structures   *561*
16.1.1   LTA Structure   *562*
16.1.2   Cation Hydration   *563*
16.1.3   Structure of CHA   *564*

16.1.3.1	Structure of DDR (ZSM-58)	*565*
16.1.4	Concentration and Temperature Dependence of Diffusivity	*566*
16.2	Diffusion in Cation-Free Eight-Ring Structures	*567*
16.2.1	Effect of Window Dimensions	*567*
16.2.2	Diffusion of CO_2 and CH_4 in DDR	*569*
16.3	Diffusion in 4A Zeolite	*571*
16.4	Diffusion in 5A Zeolite	*573*
16.4.1	Uptake Rate Measurements	*576*
16.4.2	Measurements with Different Crystal Sizes	*579*
16.4.3	PFG NMR Measurements over Different Length Scales	*580*
16.4.4	Intra-Cage Jumps	*581*
16.4.5	Comparison of Different Zeolite Samples and Different Measurement Techniques	*581*
16.5	General Patterns of Behavior in Type A Zeolites	*582*
16.5.1	Activation Energies	*584*
16.5.2	Pre-exponential Factor	*585*
16.6	Window Blocking	*587*
16.6.1	Sorption Cut-Off	*587*
16.6.2	Monte Carlo Simulations	*588*
16.6.3	Effective Medium Approximation	*590*
16.6.4	Diffusion in NaCaA Zeolites	*591*
16.6.5	3A Zeolites	*592*
16.7	Variation of Diffusivity with Carbon Number	*593*
16.7.1	Loading Dependence of Self-diffusivity	*595*
16.8	Diffusion of Water Vapor in LTA Zeolites	*595*
16.9	Deactivation, Regeneration, and Hydrothermal Effects	*596*
16.9.1	Effect of Water Vapor	*596*
16.9.2	Kinetic Behavior of Commercial 5A: Evidence from Uptake Rates	*597*
16.9.3	Effects of Pelletization and other Aspects of Technological Performance: Evidence from NMR Studies	*599*
16.10	Anisotropic Diffusion in CHA	*602*
16.11	Concluding Remarks	*603*
	References	*604*
17	**Large Pore (12-Ring) Zeolites**	*607*
17.1	Structure of X and Y Zeolites	*607*
17.2	Diffusion of Saturated Hydrocarbons	*609*
17.2.1	Evidence from NMR	*609*
17.2.2	Comparison of NMR and ZLC Data for NaX	*615*
17.2.3	Isoparaffins and Cyclohexane	*616*
17.2.4	*n*-Octane Diffusion in NaY, USY and NaX	*618*
17.2.5	Diffusion in NaCaX: Effect of Ca^{2+} Cations	*619*
17.2.6	Diffusion Measurements as Evidence of Structural Imperfection	*621*
17.3	Diffusion of Unsaturated and Aromatic Hydrocarbons In NaX	*623*
17.3.1	Light Olefins	*623*

17.3.2	C$_8$ Aromatics 624
17.3.2.1	Macroscopic Measurements 624
17.3.2.2	Comparison with Microscopic Measurements 625
17.3.3	Benzene 627
17.3.3.1	Comparison of Macroscopic and Microscopic Measurements 627
17.3.3.2	Diffusion Mechanism 628
17.3.3.3	Hysteresis 631
17.3.4	Discrepancy between Macroscopic and Microscopic Diffusivity Measurements for Aromatics in NaX 632
17.4	Other Systems 633
17.4.1	Water in NaX and NaY 633
17.4.2	Methanol 635
17.4.3	Triethylamine 635
17.4.4	Hydrogen 636
17.5	PFG NMR Diffusion Measurements with Different Probe Nuclei 636
17.5.1	Fluorine Compounds 637
17.5.2	Xenon and Carbon Monoxide and Dioxide 638
17.5.3	Nitrogen 639
17.6	Self-diffusion in Multicomponent Systems 640
17.6.1	Hydrocarbons 640
17.6.1.1	*n*-Heptane–Benzene 640
17.6.1.2	Benzene–Perfluorobenzene 641
17.6.1.3	*n*-Butane–Perfluoromethane 641
17.6.1.4	Ethane–Ethene 642
17.6.2	Self-diffusion under the Influence of Co-adsorption and Carrier Gases 643
17.6.2.1	Effect of Moisture on Self-diffusion 643
17.6.2.2	Water and Ammonia 643
17.6.2.3	Effect of an Inert Carrier Gas 644
17.6.3	Self-diffusion in Multicomponent Systems Evolving during Catalytic Conversion 645
17.6.3.1	Cyclopropane into Propene 645
17.6.3.2	Isopropanol into Acetone and Propene 647
	References 648
18	**Medium-Pore (Ten-Ring) Zeolites 653**
18.1	MFI Crystal Structure 654
18.1.1	Saturation Capacity 655
18.1.2	Molecular Sieve Behavior 659
18.2	Diffusion of Saturated Hydrocarbons 659
18.2.1	Linear Alkanes 659
18.2.1.1	Microscale Measurements 661
18.2.1.2	Macroscale Measurements 663
18.2.1.3	Molecular Simulations 664
18.2.1.4	Loading Dependence of Diffusivity 665

18.2.1.5	Microdynamic Behavior 665
18.2.2	Diffusion of Branched and Cyclic Paraffins 666
18.2.2.1	Isobutane at Low Loadings 667
18.2.2.2	Cyclohexane and Alkyl Cyclohexanes 669
18.2.2.3	2,2-Dimethylbutane 670
18.2.2.4	Comparison of Diffusivities for C_6 Isomers 670
18.2.2.5	Comparison of Diffusivities for Linear and Branched Hydrocarbons 673
18.2.2.6	Diffusion of Isobutane at High Loadings 674
18.3	Diffusion of Aromatic Hydrocarbons 676
18.3.1	Diffusion of Benzene 676
18.3.1.1	Self- and Transport Diffusion 680
18.3.1.2	Benzene Microdynamics 680
18.3.2	Diffusion of C_8 Aromatics 681
18.3.2.1	Uptake Curves 681
18.3.2.2	ZLC and TZLC Measurements 683
18.3.2.3	Frequency Response Data for p-Xylene 684
18.3.2.4	Evidence from Membrane Permeation Studies 685
18.3.2.5	Diffusion of o-Xylene and m-Xylene 686
18.4	Adsorption from the Liquid Phase 686
18.5	Microscale Studies of other Guest Molecules 688
18.5.1	Tetrafluoromethane 688
18.5.2	Water and Methanol 689
18.5.3	Ammonia 689
18.5.4	Hydrogen 691
18.6	Surface Resistance and Internal Barriers 692
18.6.1	Surface Resistance 692
18.6.1.1	Macroscopic Rate Measurements 693
18.6.1.2	Surface Etching 694
18.6.1.3	Measurement of Transient Concentration Profiles 695
18.6.1.4	Other Surface Effects 696
18.6.2	Intracrystalline Barriers 696
18.6.3	Sub-structure of MFI Crystals 697
18.6.4	Assessing Transport Resistances at Internal and External Boundaries by Micro-imaging 699
18.6.5	Evidence from PFG NMR Diffusion Studies 701
18.6.6	Differences in Diffusional Behavior of Linear and Branched Hydrocarbons 702
18.7	Diffusion Anisotropy 703
18.7.1	Correlation Rule for Structure-Directed Diffusion Anisotropy 703
18.7.2	Comparison of Measured Profiles 704
18.7.3	Evidence by PFG NMR Measurements 705
18.7.4	Limits of the Correlation Rule 707
18.7.5	Anisotropic Diffusion in a Binary Adsorbed Phase 710
18.7.6	Anisotropy of Real Crystals 710

18.8	Diffusion in a Mixed Adsorbed Phase	712
18.8.1	Blocking Effects by a Co-adsorbed Second Guest Species	712
18.8.1.1	Methane in the Presence of Benzene	712
18.8.1.2	Methane in the Presence of Pyridine and Ammonia	713
18.8.1.3	n-Butane in the Presence of Isobutane	714
18.8.2	Two-Component Diffusion	716
18.8.2.1	Methane and Tetrafluoromethane	716
18.8.2.2	Methane and Xenon	716
18.8.2.3	Methane and Ammonia	717
18.8.2.4	Permeation Properties of Nitrogen and Carbon Dioxide	717
18.8.2.5	Counter-current Desorption of p-Xylene–Benzene	717
18.8.2.6	Co- and Counter-diffusion of Benzene and Toluene	718
18.8.2.7	Counter-diffusion of Isobutane and n-Butane	720
18.9	Guest Diffusion in Ferrierite	722
	References	723
19	**Metal Organic Frameworks (MOFs)**	**729**
19.1	A New Class of Porous Solids	730
19.2	MOF-5 and HKUST-1: Diffusion in Pore Spaces with the Architecture of Zeolite LTA	732
19.2.1	Guest Diffusion in MOF-5	733
19.2.2	Guest Diffusion in CuBTC	736
19.3	Zeolitic Imidazolate Framework 8 (ZIF-8)	739
19.3.1	An Ideal Case where Experimental Self- and Transport Diffusivities are interrelated from First Principles	740
19.3.2	Membrane-Based Gas Separation Following the Predictions of Diffusion Measurements	744
19.4	Pore Segments in Single-File Arrangement: Zn(tbip)	747
19.5	Breathing Effects: Diffusion in MIL-53	751
19.6	Surface Resistance	754
19.6.1	Experimental Observations	754
19.6.2	Conceptual Model for a Surface Barrier in a One-Dimensional System: Blockage of Most of the Pore Entrances	757
19.6.2.1	Simulation Results	758
19.6.2.2	An Analytic Relationship between Surface Permeability and Intracrystalline Diffusivity	759
19.6.2.3	Surface Resistance with Three-Dimensional Pore Networks	760
19.6.2.4	Estimating the Fraction of Unblocked Entrance Windows	760
19.6.3	Generalization of the Model	761
19.6.3.1	Intracrystalline Barriers	761
19.6.3.2	Activation Energies	761
19.7	Concluding Remarks	762
	References	764

Part VI   Selected Applications   769

20   Zeolite Membranes   771
20.1 Zeolite Membrane Synthesis   772
20.2 Single-Component Permeation   773
20.2.1 Selectivity and Separation Factor   774
20.2.2 Modeling of Permeation   776
20.3 Separation of Gas Mixtures   779
20.3.1 Size-Selective Molecular Sieving   779
20.3.2 Diffusion-Controlled Permeation   781
20.3.3 Equilibrium-Controlled Permeation   783
20.4 Modeling Permeation of Binary Mixtures   784
20.4.1 Maxwell–Stefan Model   784
20.4.1.1 Concentration Profile through the Membrane   785
20.4.1.2 Importance of Mutual Diffusion   789
20.4.2 More Complex Systems   789
20.4.2.1 Membrane Thickness   791
20.4.2.2 Support Resistance   791
20.5 Membrane Characterization   792
20.5.1 Bypass Flow   792
20.5.2 Perm-Porosimetry   793
20.5.3 Isotherm Determination   795
20.5.4 Analysis of Transient Response   797
20.6 Membrane Separation Processes   797
20.6.1 Pervaporation Process for Dehydration of Alcohols   797
20.6.2 CO_2–CH_4 Separation   798
20.6.3 Separation of Butene Isomers   799
20.6.4 MOF (Metal Organic Framework) Membranes for H_2 Separation   799
20.6.5 Amorphous Silica Membranes   800
20.6.6 Stuffed Membranes   801
20.6.7 Membrane Modules   801
20.6.8 Membrane Reactors   802
20.6.9 Barriers to Commercialization   802
References   803

21   Diffusional Effects in Zeolite Catalysts   807
21.1 Diffusion and Reaction in a Catalyst Particle   807
21.1.1 The Effectiveness Factor   808
21.1.2 External Mass Transfer Resistance   811
21.1.3 Temperature Dependence   812
21.1.4 Reaction Order   814
21.1.5 Pressure Dependence   814
21.1.6 Non-isothermal Systems   815
21.1.7 Relative Importance of Internal and External Resistances   816

21.2	Determination of Intracrystalline Diffusivity from Measurements of Reaction Rate *817*	
21.2.1	Temperature Dependence of "Effective Diffusivity" *819*	
21.3	Direct Measurement of Concentration Profiles during a Diffusion-Controlled Catalytic Reaction *819*	
21.3.1	Reaction of Furfuryl Alcohol on HZSM-5 *820*	
21.3.2	Reaction in Mesoporous MCM-41 *821*	
21.4	Diffusional Restrictions in Zeolite Catalytic Processes *822*	
21.4.1	Size Exclusion *822*	
21.4.2	Catalytic Cracking over Zeolite Y *823*	
21.4.3	Catalytic Cracking Over HZSM-5 *824*	
21.4.4	Catalytic Cracking over other Zeolites *825*	
21.4.5	Activation Energies *825*	
21.4.6	Xylene Isomerization *826*	
21.4.7	Selective Disproportionation of Toluene *828*	
21.4.8	MTG Reaction *830*	
21.4.9	MTO Process *831*	
21.5	Coking of Zeolite Catalysts *833*	
21.5.1	Information from PFG NMR *834*	
21.5.2	Information from Fluorescence Microscopy *835*	
	References *835*	

Notation *839*

Index *851*

Preface

Diffusion at the atomic or molecular level is a universal phenomenon, occurring in all states of matter on time scales that vary over many orders of magnitude, and indeed controlling the overall rates of many physical, chemical, and biochemical processes. The wide variety of different systems controlled by diffusion is well illustrated by the range of the topics covered in the *Diffusion Fundamentals Conference* series (http://www.uni-leipzig.de/diffusion/). For both fundamental and practical reasons diffusion is therefore important to both scientists and engineers in several different disciplines. This book is concerned primarily with diffusion in microporous solids such as zeolites but, since the first edition was published (in 1992 under the title *Diffusion in Zeolites and Other Microporous Solids*), several important new micro- and mesoporous materials such as metal organic frameworks (MOFs) and mesoporous silicas (e.g., MCM-41 and SBA-15) have been developed. In recognition of these important developments the scope of this new edition has been broadened to include new chapters devoted to mesoporous silicas and MOFs and the title has been modified to reflect these major changes.

In addition to the important developments in the area of new materials, over the past 20 years, there have been equally important advances both in our understanding of the basic physics and in the development of new theoretical and experimental approaches for studying diffusion in micro- and mesoporous solids. Perhaps the most important of these advances is the remarkable development of molecular modeling based on numerical simulations. Building on the rapid advances in computer technology, Monte Carlo (MC) and molecular dynamic (MD) simulations of adsorption equilibrium and kinetics have become almost routine (although kinetic simulations must still be treated with caution unless confirmed by experimental data). In recognition of the importance of these developments the new edition contains three authoritative new chapters, written mainly by Doros Theodorou, dealing with the principles of molecular simulations and their application to the study of diffusion in porous materials.

With respect to experimental techniques, over the past 20 years, neutron scattering has advanced from a scientific curiosity to a viable and valuable technique for studying diffusion at short length scales. Interference microscopy (IFM) has also become a practically viable technique, providing unprecedented insights into diffusional behavior by allowing direct measurement of the internal concentration profiles during transient adsorption or desorption processes. In contrast, the early promise of light scattering techniques has not yet been fulfilled as the practical difficulties have, so far, proved insurmountable. We are, however, witnessing impressive advances in our understanding of a wide variety of systems through the application of single-molecule visualization techniques. As a highlight of this development the book includes experimental confirmation of the celebrated ergodic theorem.

As with the first edition our intention in writing this book has been to present a coherent summary and review of both the basic theory of diffusion in porous solids and the major experimental and theoretical techniques that have been developed for studying and simulating the behavior of such systems. The theoretical foundations of the subject and indeed some of the experimental approaches borrow heavily from classical theories of diffusion in solids, liquids, and gases. We have therefore attempted to include sufficient background material to allow the book to be read without frequent reference to other sources.

The book is divided into six parts, of which the first four, dealing with basic theory, molecular simulations, and experimental methods are included in Volume I. The "experimental" chapters cover both macroscopic measurements, in which adsorption/desorption rates are followed in an assemblage of adsorbent particles, and microscopic methods (mainly PFG NMR and QENS) in which the movement of the molecules themselves is followed, as well as the new imaging techniques such as IFM and IRM in which concentration profiles or fluxes within a single crystal are measured. Parts Five and Six, in Volume II, deal with diffusion in selected systems and with the practical application of zeolites as membranes and catalysts.

The first edition contained considerable discussion of the discrepancies between microscopic and macroscopic measurements. These discrepancies have now been largely resolved, but it turns out that in many zeolite crystals structural defects are much more important than was originally thought. As a result, in such systems, the measured diffusivity is indeed dependent on the length scale of the measurement and the diffusivity as a structurally perfect crystal is often approached only at the very short length scales probed by neutron scattering. Another important feature that has become apparent only through the application of detailed IFM measurements is the prevalence of surface resistance. In many zeolite and MOF crystals the resistance to transport at the crystal surface is significant and has been shown to result from the blockage of a large fraction of the pore openings. Such detailed insights, which depend on the application of new experimental techniques, have become possible only recently.

Throughout the text and in the major tables we have generally used SI units although our adherence to that system has not been slavish and, particularly with respect to pressure, we have generally retained the original units.

The selection of the material for a text of this kind inevitably reflects the biases and interests of the authors. In reviewing the literature of the subject we have made no attempt to be comprehensive but we hope that we have succeeded in covering, or at least mentioning, most of the more important developments.

Jörg Kärger, Leipzig, Germany
Douglas M. Ruthven, Orono, Maine, USA
Doros N. Theodorou, Athens, Greece

Acknowledgments

A book of this kind is inevitably a collaborative project involving not only the authors but their research students, colleagues, and associates, many of whom have contributed, both directly and indirectly, over a period of many years. Our early collaboration, in the days of the GDR, would not have been possible without the support and encouragement of two well-known pioneers of zeolite research, Professor Wolfgang Schirmer (Academy of Sciences of the GDR) and Professor Harry Pfeifer (University of Leipzig). Much of the early experimental work was carried out by Dr Jürgen Caro (now Professor of Physical Chemistry at the University of Hanover), using large zeolite crystals provided by Professor Zhdanov (University of Leningrad) and the home-made PFG NMR spectrometer that was constructed and maintained by Dr Wilfried Heink.

Since German re-unification both the formal and financial difficulties of research collaboration have been greatly reduced and the list of collaborators, many of whom are mentioned in the cited references, has become too long to name individuals. For the historical record it is, however, appropriate to mention the contributions of a few key people who were involved in the development of the new experimental and molecular modeling techniques that were used to obtain most of the information presented in this new edition. Jeffrey Hufton (now with Air Products Inc.) and Stefano Brandani (now Professor of Chemical Engineering at Edinburgh University) were mainly responsible for the development of "tracer ZLC," which allowed the first *direct* comparisons of "macroscopic" and "microscopic" measurements of self-diffusion in zeolites. The successful development of interference microscopy to allow direct visualization of the transient intracrystalline concentration profiles was largely due to the efforts of Ulf Schemmert and Sergey Vasenkov (now professors at the University of Applied Sciences in Leipzig and the University of Florida, respectively) and the parallel development of infrared microscopy to allow the visualization of the profiles of individual species in a multicomponent system was largely due to Dr Christian Chmelik (University of Leipzig). The development of molecular simulation techniques for studying sorption and diffusion in zeolites owes much to Larry June (now with Shell Oil), Randy Snurr, and Ed Maginn (now professors at Northwestern University and the University of Notre Dame, respectively), Professor Alexis Bell (University of California, Berkeley), and Dr George Papadopoulos (NTU Athens). We should also mention the work of Hervé Jobic (CNRS, Villeurbanne), who has developed neutron scattering as a viable experimental technique for studying intracrystalline diffusion over very short time and distances, comparable to those accessible by molecular dynamics simulations.

We are grateful to numerous funding agencies, especially the National Research Foundations of Germany, Canada, and the United States, the Alexander von Humboldt Foundation, DECHEMA and the Fonds der Chemischen Industrie, the European Community, and several companies, notably, ExxonMobil who have provided research support as well as valuable technical assistance over many years.

Finally, we would also like to thank Wiley-VCH and especially our editor Bernadette Gmeiner for her efficient collaboration in the preparation and editing of the manuscript and also our wives, Birge, Patricia, and Fani, for all their help and support throughout the course of this project.

Jörg Kärger
Douglas M. Ruthven
Doros N. Theodorou
5 December 2011

Part V
Diffusion in Selected Systems

15
Amorphous Materials and Extracrystalline (Meso/Macro) Pores

This book is concerned primarily with intracrystalline diffusion in zeolites and their analogs, in which the pore system is structurally regular with no distribution of pore size. This regularity facilitates our analysis and understanding of diffusional behavior and in some cases allows the diffusion mechanism to be established in detail at the molecular scale. Diffusion in amorphous pore systems is obviously more complex since the measurable diffusivities are then statistical averages over the relevant ranges of pore size and surface structure. However, apart from this complication, for a given pore size the behavior observed in random and regular pore structures is generally similar.

Depending on the method of preparation, porous glasses typically have random pore structures with pores in the size range 1–10 nm, which spans the micropore/mesopore range of the IUPAC classification (Section 1.3). Mesoporous silica materials such as MCM-41 have somewhat larger pores within the mesopore range, but the pore structure is regular with a narrow distribution of pore size. Many microporous catalysts and adsorbents have the type of biporous structure indicated in Figure 1.3b, and the sorption kinetics are commonly influenced and even controlled by diffusion in the extracrystalline pores. Similar considerations are also relevant to the determination of diffusivities from sorption kinetic studies (Chapters 6 and 13) since the transport rate within the adsorbent sample imposes an upper limit on the magnitude of the intracrystalline diffusivity that can be reliably determined from such measurements. From both practical and theoretical perspectives it is therefore important to consider diffusion in such systems.

It should also be noted that the IUPAC classification is arbitrary. The larger molecular species that are of interest in many biochemical systems have molecular diameters greater than 2 nm and cannot therefore penetrate "micropores." For such species, diffusion in pores in the "meso" range shows many of the features generally associated with micropore diffusion: significant steric effects, size exclusion, and strongly temperature-dependent pore diffusivity.

The basic mechanisms of macropore and mesopore diffusion are discussed in Section 4.2 by reference to the ideal case of a uniform cylindrical capillary and the concept of a tortuosity factor was introduced as a simple semi-empirical way to describe diffusion in a random pore network. Adsorbents that have a wide or bimodal

Diffusion in Nanoporous Materials. Jörg Kärger, Douglas M. Ruthven, and Doros N. Theodorou.
© 2012 Wiley-VCH Verlag GmbH & Co. KGaA. Published 2012 by Wiley-VCH Verlag GmbH & Co. KGaA.

distribution of macro/meso-pore sizes present a particular challenge, since different diffusion mechanisms may be dominant in the pores of different size, leading to a superposition of different effects and a complex variation of the effective diffusivity with loading, pressure, and temperature. The simplest behavior occurs when the pore size is in the middle of the mesopore range (5–10 nm), the distribution of pore size is narrow and unimodal, and the pressure is relatively low (<1 atm for light gases). Under these conditions Knudsen diffusion is generally dominant and the intrusion of viscous flow and surface diffusion are minimal. Equally simple behavior is encountered at high relative pressures when the pores are full of liquid and the effective diffusivity then corresponds to the liquid phase diffusivity, reduced by the appropriate tortuosity factor. Between these two limits surface diffusion, viscous flow, and capillary condensation are often important, leading to much more complex behavior. A detailed review of recent experimental studies of diffusion in the two-phase region has been given by Valiullin *et al.* [1].

In this chapter we present selected examples to illustrate some of the patterns of behavior that have been observed in experimental studies of diffusion in random pore systems and in structurally regular mesoporous materials. Several of the selected examples refer to diffusion in porous glass and mesoporous silica materials (such as MCM-41). These materials can be prepared relatively easily and reproducibly with controlled pore size so they have become the materials of choice for experimental studies of mesopore diffusion. The examples include so-called hierarchical porous materials where, to provide fast mass exchange with the surroundings, a microporous "bulk" phase is traversed by mesopores. We also consider surface diffusion and long-range diffusion through a bed of adsorbent particles, which is physically similar to macropore diffusion in an adsorbent particle.

15.1
Diffusion in Amorphous Microporous Materials

15.1.1
Diffusion in Microporous Glass

Detailed studies of intracrystalline diffusion in zeolite crystals have revealed that, despite the apparent regularity of the crystal structure, the measured diffusivity often decreases with the length scale of the measurement. This has been attributed to the presence of structural defects that disrupt the long-range regularity of the channel system. Thus, diffusivities measured over a few unit cells are often significantly larger than those measured over longer distances that reflect the irregularity of the actual structure (see, for example, Sections 17.3.4 and 18.6.5). Although microporous glasses do not have the advantage of crystalline uniformity, the pore structure has been shown to be statistically uniform over macroscopic distances [2–4]. As a leaching product of phase-separated sodium borosilicate glasses, porous glasses can be prepared as monolithic bodies of essentially any shape and dimensions.

By preventing uncontrolled phase separation during the cooling of the initial glass melt, pore diameters as small as 1 nm may be attained, making these materials especially useful for demonstrating the consistency of diffusion measurements with nanoporous materials carried out by different experimental techniques over different length scales [5].

Figure 15.1 shows the self-, corrected, and transport diffusivities ($\mathcal{D}$, D_0, and D, denoted in the figure by D_{self}, D_0, and D_T) for n-hexane and cyclohexane at ambient temperature, measured by several different experimental techniques, in two samples of microporous glass (mean pore diameters 1.3 and 2.0 nm) plotted as a function of fractional pore filling. Details of the preparation and characterization of these materials may be found in Reference [5]. The infrared microscopy measurements (IRM) follow the transient concentration profiles within a macroscopic sample of the porous glass adsorbent (and hence the uptake rate) in response to a

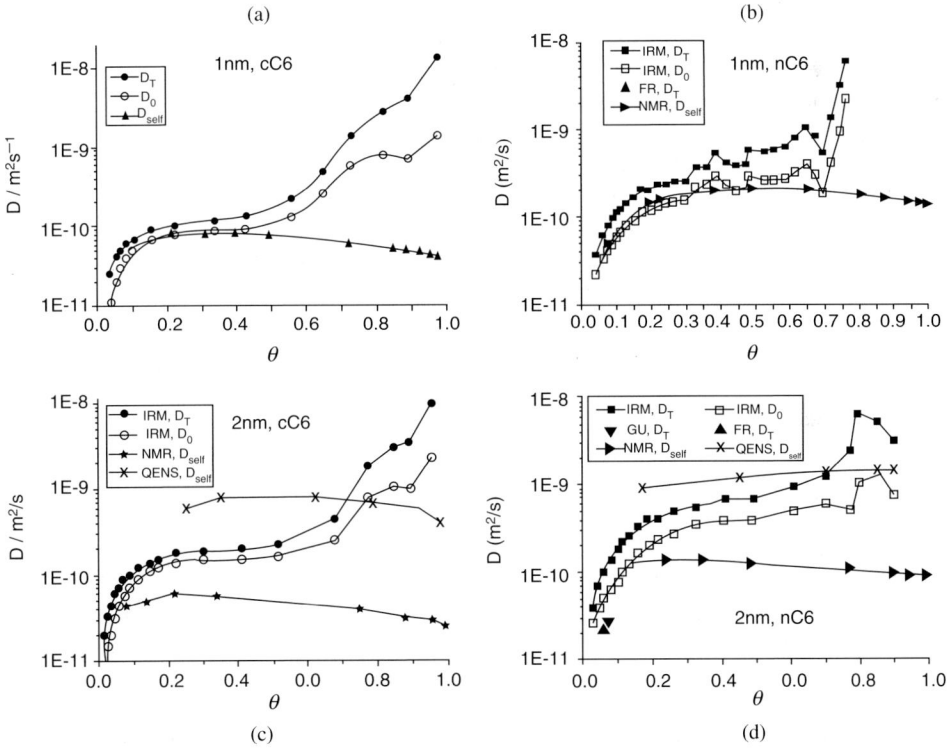

Figure 15.1 Results of diffusion measurements with cyclohexane [(a) and (c)] and n-hexane [(b) and (d)] as probe molecules in porous glasses with mean pore diameters of 1.3 nm [(a) and (b)] and 2.0 nm [(c) and (d)] at room temperature, by various experimental techniques as specified in the inserts (see text for details). Reprinted from Reference [5] with permission.

step change in the ambient concentration of the sorbate, from which the transport diffusivities may be derived directly from the diffusion equation (Section 6.2). The diffusivities so obtained are "macroscopic" values in the sense that they are measured at the length scale of the entire sample. The PFG NMR self-diffusivity measurements were carried out over a range of different time scales, corresponding to length scales from 0.1 to 10 μm. A few frequency response (FR) and gravimetric uptake (GU) measurements were also made, but only at low loadings, within or close to the Henry's law region. Corrected diffusivities (D_0) were calculated from the transport diffusivities using the thermodynamic factors ($d \ln p/d \ln q$) derived from the measured equilibrium isotherms.

The relationship between self- and transport diffusivities is discussed in Section 1.2.3 and in greater detail in Section 3.3.3 – Eq. (3.51), which may be written as $\mathcal{D}/D_0 = (1 + \theta D_0/Đ_{AA})^{-1}$. This implies that, for a system in which the equilibrium isotherm is of type 1 form (Langmuir-type isotherm, see also Section 19.3.1), $D_0 \geq \mathcal{D}$. Furthermore, if diffusion is controlled mainly by molecule–pore interactions rather than by molecule–molecule interactions (i.e., for $D_0 \ll Đ_{AA}$), $D_0 \approx \mathcal{D}$, even at higher loadings. Therefore, at low loadings the transport, corrected and self-diffusivities should all be similar while at high loadings they may be expected to diverge. The experimental data for both normal and cyclohexane in the 1.3 nm sample conform closely to this pattern, as illustrated in Figure 15.1a and b. The results of the different measurements appear quite consistent and show the expected trends. The PFG NMR self-diffusivities were independent of the time scale of the measurement and at loadings less than about 50% of saturation they approach the transport diffusivities, which also show consistency between the different macroscopic measurement techniques. At higher loadings the transport diffusivity increases rapidly with loading [in qualitative accordance with Eqs. (1.31) and (1.32)] while the self-diffusivity remains almost constant. The rapid increase in diffusivity at very low loadings suggests an energetically heterogeneous surface with progressive occupation of the strongest adsorption sites.

The data for the 2 nm sample show similar general trends but there are some significant differences. The neutron scattering measurements (QENS) yield the self-diffusivities at very small length scales, of the same order as the pore diameter (Section 10.2). Consequently, the measured diffusivities do not reflect the constraints imposed by the pore walls and are found to be close to the values for the free liquid (and substantially greater than the PFG NMR values). Furthermore, the PFG NMR diffusivities approach the corrected transport diffusivities only at very low loadings ($\theta < 0.05$ for cyclohexane and $\theta < 0.15$ for n-hexane). This is entirely consistent with Eq. (3.51) since, in the larger pores, the influence of molecule–molecule interactions will be more significant.

15.1.2
Diffusion in Microporous Carbon

"Activated" carbon adsorbents [6, 7] are produced by controlled pyrolysis and oxidation of carbonaceous materials. By careful control of the temperature, partial

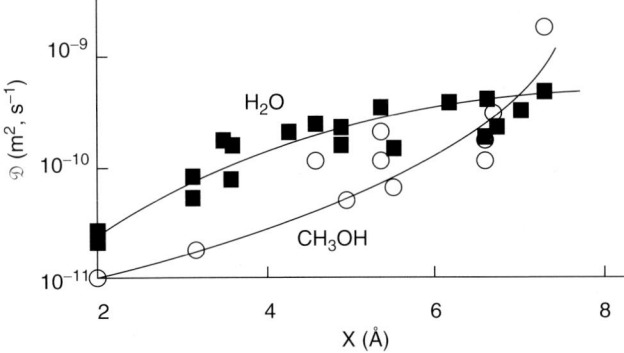

Figure 15.2 Self-diffusivities of water (■) and methanol (○) at 298 K in microporous carbons, showing the dependence on the micropore half width. Reprinted from Kärger et al. [9] with permission.

pressure, and contact time one is able to attain a remarkably uniform network of micropores with a very narrow distribution of pore size in the 3–5 Å range. Carbon molecular sieve adsorbents thus show the same kinds of size and shape selectivity as are observed with small-pore zeolites, and in commercial applications such as air separation they compete directly as an alternative to zeolites. Irrespective of their amorphous structure, in many such systems PFG NMR measurements have been found to yield normal diffusion, that is, mean square displacements of the guest molecules increasing in direct proportion to the observation time [8, 9]. This suggests that, at least on the scale of micrometers, these adsorbents can be considered as statistically uniform. Despite the differences in the nature of the surface and the shape of the micropores, which, in activated carbon, are constricted by slits formed between imperfectly packed microcrystals of graphite, the diffusional behavior is thus found to show many similarities with that of zeolites.

Figures 15.2 and 15.3 summarize some of the results of detailed PFG NMR diffusion studies with several light hydrocarbons, water, and methanol in a range of microporous carbon adsorbents [9]. In the small-pore materials (mean micropore slit width < 15 Å) the effect of steric hindrance is clearly evident. The diffusivities for both water and methanol are substantially smaller than in the free liquids and vary directly with the mean slit width (Figure 15.2). Furthermore, the diffusivity of methanol is lower than that of water, the reverse of the situation in the free liquids. This difference can be easily understood as the result of steric hindrance. The mobility of the larger methanol molecule will be hindered more severely and, furthermore, the hydrogen bonding of the water molecules, which is responsible for the anomalously low diffusivity in the free liquid, is probably to some extent inhibited in the micropores. There is, however, little difference in the activation energy for either species between the free and adsorbed states.

In the larger pores (slit width >15 Å) steric effects become insignificant and we see essentially the same ratio for both species in the free liquid and the adsorbed states.

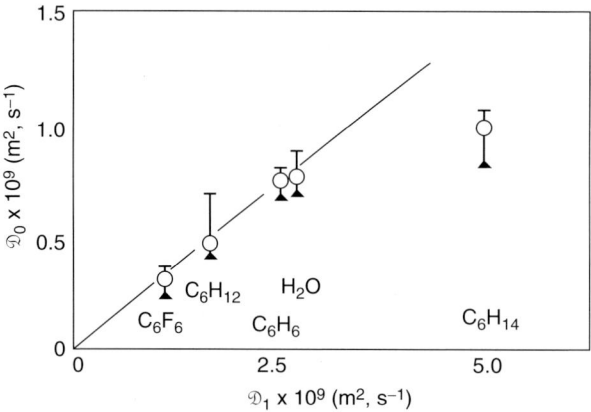

Figure 15.3 Comparison of self-diffusivities of adsorbed species $\mathcal{D}_0$ and the free liquids $\mathcal{D}_1$ (for adsorption on a large pore activated carbon at 298 K). Reprinted from Kärger et al. [9] with permision.

This is illustrated in Figure 15.3, which shows a cross plot of the diffusivities in the adsorbed and liquid phases for several C_6 hydrocarbons and water. For all species except n-hexane the ratio of the diffusivities in the free and adsorbed phases is about 3.5, which is close to the tortuosity factor expected for a random pore network (Section 4.2.2.3). The diffusivity of n-hexane in the adsorbed phase is somewhat lower than would be expected simply from the tortuosity effect, thus suggesting additional steric hindrance.

Although carbon molecular sieves are widely used as kinetically selective adsorbents in air separation and other selective adsorption processes [10, 11], diffusion in these materials has been studied less extensively than in the small-pore zeolites. Figure 15.4a and b shows uptake curves measured with different samples of molecular sieve carbon. The curves for oxygen and nitrogen in the Bergbau-Forschung sieve [12] conform well to a diffusion model, and the large difference in diffusivity between nitrogen and oxygen is clearly apparent. By contrast, it has been shown that in some other samples of carbon molecular sieve (Figure 15.4b) the uptake curves conform more closely to the surface resistance model [13, 14]. It seems likely that in these latter samples the kinetic selectivity has been enhanced by controlled cracking of hydrocarbons to reduce the free diameter of the pore mouths [15].

The most extensive experimental data are those of Chihara et al. [16], who worked with the Takeda sieve. Figure 15.5 presents representative Arrhenius plots, showing the temperature dependence of the diffusivities for several sorbates in this adsorbent. Notably, although the micropores of this adsorbent are small enough to provide good kinetic selectivity between oxygen and nitrogen, at elevated temperatures they are penetrated at a reasonably rapid rate even by benzene, which cannot penetrate the pores of the A type zeolites. This reflects the difference in the pore shapes between zeolites and carbon sieves.

15.1 Diffusion in Amorphous Microporous Materials | 523

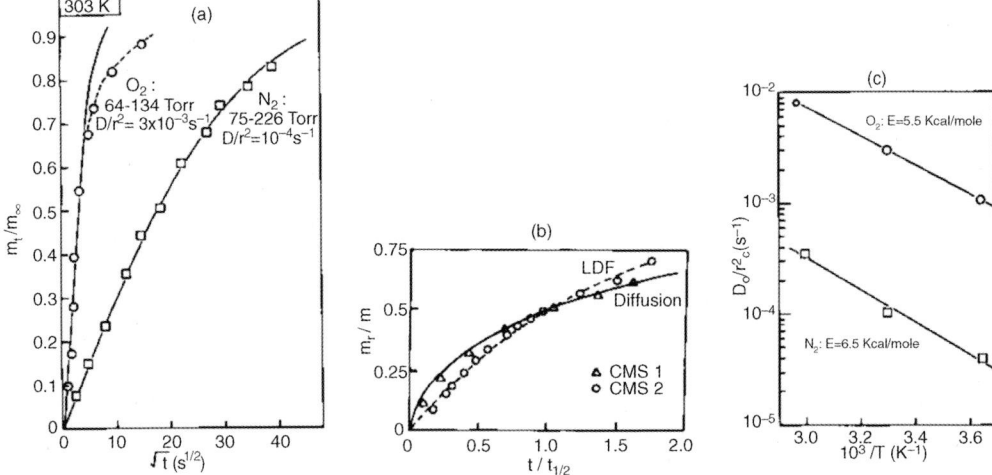

Figure 15.4 (a) Experimental uptake curves for N_2 and O_2 at 303 K in a sample of Bergbau-Forschung carbon molecular sieve, showing conformity with the diffusion model [Eq. (6.11)]. (b) Experimental uptake curves for N_2 in two different samples of carbon molecular sieve, showing conformity with the diffusion model [Eq. (6.10)] for CMS 1 and conformity with the surface resistance model [Eq. (6.6)] for sample 2. After Dominguez et al. [13]. (c) Temperature dependence of the corrected diffusivity, for the Bergbau sieve.

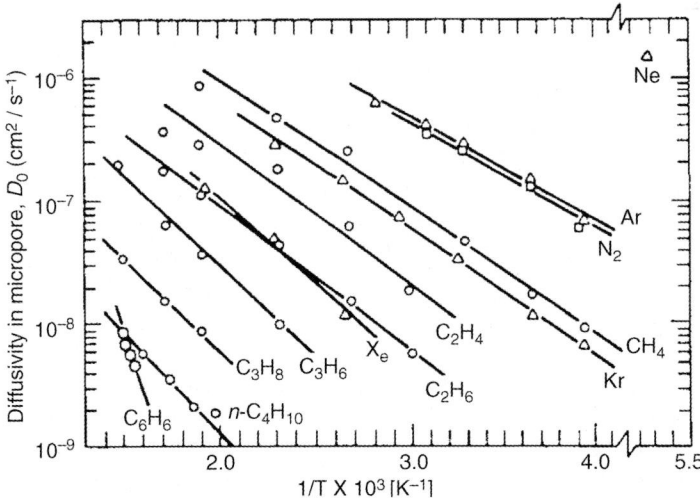

Figure 15.5 Arrhenius plot showing the temperature dependence of the corrected diffusivity (D_0) for small molecules in Takeda molecular sieve carbon "5A." Reprinted from Chihara et al. [16] with permission.

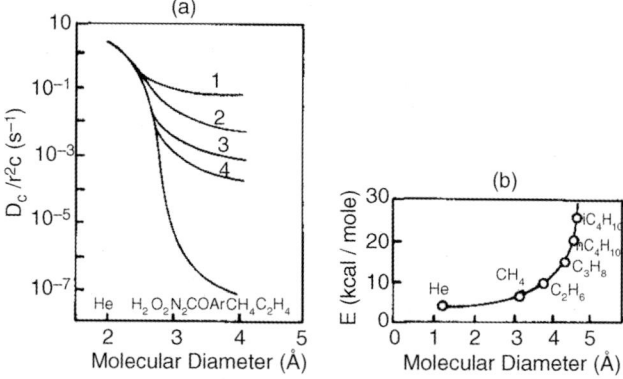

Figure 15.6 Variation of (a) diffusion time constant and (b) diffusional activation energy with mean molecular diameter. The numbers in (a) refer to differently pretreated carbon sieves Reprinted from Schröter and Jüntgen [18] with permission.

The size-selective feature of a carbon molecular sieve and the extent to which the diffusivity may be modified by different pretreatments are illustrated in Figure 15.6a. Corresponding activation energies for diffusion of several species are shown in Figure 15.6b. The direct relationship between the increase in diffusivity in the smaller pores and the decrease in the activation energy suggests that the differences in diffusivity stem largely from steric hindrance.

Chihara et al. [17] used the theoretical model presented in Section 4.4.2 to interpret their experimental diffusivity data for diffusion of small molecules in a carbon sieve. According to Eq. (4.48), with due allowance for the tortuosity of the pore structure, the relationship between the pre-exponential factor (D_∞) and activation energy (E) reduces to $D_\infty = 3.23 \times 10^{-4} (E/M)^{1/2}$ (cm^2 s^{-1}) where E is in kcal mol^{-1} and M is the molecular weight. This expression gives a reasonable fit of the experimental data for monatomic, diatomic, and small symmetric molecules but the pre-exponential factors for the polyatomic hydrocarbons are substantially lower than the predicted values, suggesting an additional entropy of activation for these species.

15.2
Effective Diffusivity

As noted previously, in microporous solids there is no clear distinction between the "adsorbed" molecules close to the pore walls and the "free" molecules in the central region of the pore. Diffusivities are therefore generally defined with respect to the total concentration of the relevant species within the pore space. However, in larger pores the distinction between "adsorbed" and "free" molecules becomes physically meaningful. Furthermore, in larger pores, it is useful to relate the diffusivity to the corresponding diffusivity in the free fluid and the ideal Knudsen diffusivity in a straight cylindrical pore (Section 4.2). This requires the introduction of porosity and tortuosity factors.

Subject to the assumptions discussed in Section 4.2.2 the diffusivity in a random pore network may be related (approximately) to the diffusivity for a straight cylindrical pore of the same average diameter, under the same conditions, by:

$$D = D_p/\tau \quad \text{or} \quad D = \varepsilon_p D_p/\tau \tag{15.1}$$

where ε_p is the porosity and τ the tortuosity factor for the pore network. The choice between these two expressions [denoted in Section 4.2.2.3 as Eq. (4.15a and b)] depends on the basis on which the fluxes are defined. Some of the simple models that have been developed to estimate and correlate tortuosity factors are discussed in Section 4.2.2.

Under steady-state conditions such an expression should be approximately valid regardless of whether or not the diffusing species is significantly adsorbed. However, under transient conditions the situation is more complex.

A common situation involves transport through the macro- or mesopores with rapid equilibration at the pore wall with an adsorbed phase (whose contribution to overall flux is assumed to be negligible). In this situation it is necessary to account for the exchange between the adsorbed phase and the fluid within the pore. For a linear system, Eq. (15.1) may be conveniently expressed in terms of the dimensionless equilibrium constant (K), yielding (for the right-hand expression):

$$D_{\text{eff}} = \frac{\varepsilon_p D_p}{\tau} \frac{n_{\text{pore}}}{n_{\text{pore}} + n_{\text{ads}}} = \frac{\varepsilon_p D_p}{\tau} \frac{1}{(\varepsilon_p + (1-\varepsilon_p)K)} \tag{15.2}$$

with n_{pore} and n_{ads} denoting, respectively, the number of molecules in the pore space and on the pore walls. When the isotherm is nonlinear K must be replaced by the isotherm slope (dq^*/dc) for transport diffusion or by the ratio q^*/c for self- or tracer diffusion. For strongly adsorbed species K is large, so the effective diffusivity in a transient situation will be much smaller than the steady state value.

15.2.1
Direct Measurement of Tortuosity in a Porous Catalyst Particle

To measure the tortuosity factor it is necessary to determine the ratio of the actual pore diffusivity to the diffusivity in a straight cylindrical pore of the same mean diameter. For a macroporous solid this can be accomplished in a straightforward way by PFG NMR since, in a liquid-filled macropore, diffusion occurs entirely by molecular diffusion. It is therefore only necessary to compare the mean square displacement in a given time interval (i.e., the effective self-diffusivity) for the free liquid and for a sample of the catalyst in which the pores are filled with the same liquid. This approach has been used by Vasenkov and Kortunov [19] and by Stallmach and Crone [20].

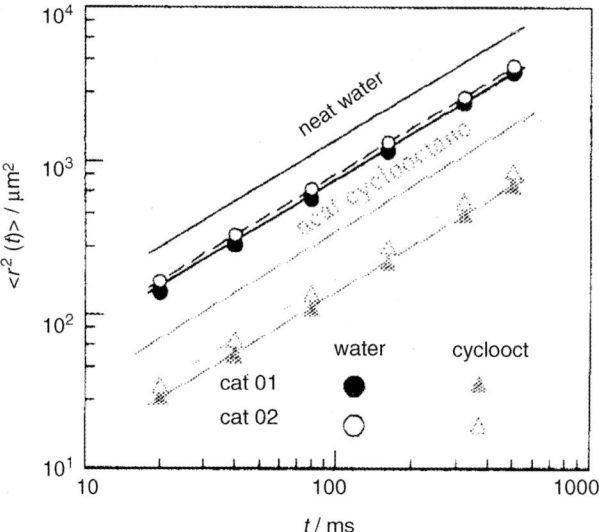

Figure 15.7 Mean square displacement as a function of time, determined by PFG NMR, for water and cyclo-octane in two samples of commercial catalyst at room temperature. Reprinted from Stallmach and Crone [20], the data for (neat)cyclooctane and the corresponding text must be presented in a better way. The figure is accessible on-line via http://www.uni-leipzig.de/diffusion/pdf/volume2/diff_fund_2(2005)105.pdf with permission.

Figure 15.7 shows representative data for two different commercial catalysts. Consistent tortuosities of 2.0 for catalyst 1 and 2.5 for catalyst 2 were obtained using both water and cyclo-octane as the test liquids.

15.2.2
Determination of Macropore Diffusivity in a Catalyst Particle under Reaction Conditions

Under transient conditions the effective macropore diffusivity is given by [see Eqs. (15.2) and (4.12)]:

$$D_{\text{eff}} = \frac{c_{\text{pore}}}{c_{\text{pore}} + c_{\text{ads}}} \cdot \frac{\varepsilon_p D_p}{\tau}; \quad \frac{1}{D_p} = \frac{1}{D_K} + \frac{1}{D_M} \tag{15.3}$$

If we assume Langmuir equilibrium the molecular densities in the adsorbed and fluid phases will be related by:

$$\frac{c_{\text{ads}}}{c_{\text{pore}}} = \frac{bq_s}{1 + bc_{\text{pore}}} \tag{15.4}$$

Thus, with the tortuosity determined as outlined above, the effective diffusivity can be estimated at any temperature provided the Langmuir parameters are

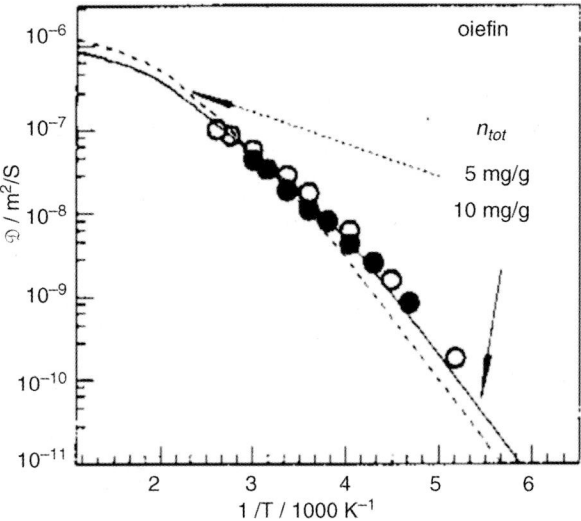

Figure 15.8 Experimental effective self-diffusivities for ethylene in a porous catalyst at two different loadings compared with the values predicted from Eqs. (15.3) and (15.4) Reprinted from Stallmach and Crone [20], with permission.

known. The application of this approach is illustrated in Figure 15.8, in which the effective self-diffusivities for ethylene in a porous catalyst, measured over a wide range of temperatures by PFG NMR, are compared with the values predicted from Eqs. (15.3) and (15.4). When extrapolating over a wide temperature range this approach offers improved accuracy over a simple linear extrapolation of the Arrhenius plot.

15.3
Diffusion in Ordered Mesopores

Ordered mesoporous materials may be obtained from materials composed of organic and inorganic mesophases with periodic structures [21]. Periodic structures are formed by the inclusion of organic molecules of "amphiphilic" nature, that is, of molecules composed of parts with differing affinity to water. This property leads, in aqueous solution, to a self-organization of these molecules and thus to the formation of liquid-crystalline domains [22, 23]. Surfactants ("surface-active agents") are the most prominent examples and have been used in the first syntheses of ordered mesoporous materials [24, 25]. Block-copolymers may also be produced with amphiphilic properties [26]. In triblock copolymers of type PEO–PPO–PEO [where PEO is (polyethylene oxide) and PPO is (polypropylene oxide)], molecular aggregation is driven by the poor solubility of the (central) PPO block. Correspondingly, only the (more) hydrophilic PEO tails are in contact with water and shield, simultaneously, the hydrophobic central (PPO) parts from water contact. This high degree of self-organization exists irrespective of the rapid movement of its constituents, that is, the amphiphilic molecules, as revealed by their diffusivities [27].

In the following, we discuss the results of PFG NMR diffusion measurements with guest molecules in the hexagonally arranged cylindrical pores of (surfactant-templated) MCM-41 [25]. Further results of diffusion measurements in hexagonally ordered mesoporous materials, comprising examples with both MCM-41 and (copolymer-templated) SBA-15 [28], are provided in Sections 15.7.1 and 15.9.3, with particular emphasis on the impact of the pore hierarchy in these materials.

Table 15.1 [29] summarizes the results of PFG NMR self-diffusion measurements with benzene in MCM-41 specimens of different origin. Irrespective of identical measuring conditions (room temperature, pore-filling factor of about 0.7, observation time 3 ms) very different diffusivities are observed. Although this result is also a consequence of differences in the particle sizes, it indicates how large the differences may be between different specimens of mesoporous materials of the same structure type. These differences may be of quite different microstructural origin, including additional transport resistances ("barriers") as well as transport-accelerating voids between the individual (micro-)domains [32, 33] of regular pore arrangements.

The diffusivity data displayed by Figure 15.9 represent the exceptional case of mesopores homogeneously arranged over the whole MCM-41 particles, which, with respect to their mesopore space, may thus be considered to represent quasi-crystals [34]. Figure 15.9 shows the results of PFG NMR diffusion measurements with water at 263 K [34] in a water-saturated specimen of MCM-41 that was synthesized with particular emphasis on attaining uniform channel orientation over the whole sample [30]. PFG NMR spin echo attenuation was found to follow the dependency required for diffusion in a powder sample of rotational symmetry [Eq. (11.24)], with the main elements of the diffusion tensor (i.e., the diffusivities parallel and perpendicular to the main channel directions) as displayed in Figure 15.9a. It is found that the diffusivity in the channel direction (D_{par}) is larger by more than one order of magnitude than in the perpendicular direction (D_{perp}) and by about one order of magnitude smaller than the diffusivity for free water. These results provide clear evidence that mass transfer can occur in the direction perpendicular to the channel, albeit at a much slower rate than in the direction of the channels. Measuring these diffusivities (and with them the corresponding displacements) as a function of time (Figure 15.9b), the agreement between the maximum displacements and the particle dimensions as illustrated by the insert nicely confirms this analysis. The situation for the largest observation times reflected by Figure 15.9b are those illustrated by Figure 11.2b. In the present case, as a result of the low temperatures (<273 K), the measurements are confined to intraparticle diffusion. Under these conditions the intercrystalline water is frozen, making the exchange extremely slow.

Table 15.1 Source and characterization of different MCM-41 material investigated by PFG NMR diffusion measurements with benzene as a probe molecule [29].[a]

Sample (source)	SEM (scanning electron microscope)	BET (Brunauer–Emmett–Teller)		PFG NMR[a]	
	Size (μm) (shape)	Specific surface (m^2 g^{-1})	Pore radius (nm)	Self-diffusivity (10^{-10} m^2 s^{-1})	RMS displacement (μm)
Si-MCM-41 [30]	0.1–10 (rod-like)	890	1.3	2	2
Si, Al-MCM-41 Si/Al = 117[b]	0.5–5	970	1.45	2	2
B, Si-MCM-41 Si/B = 18 [30]	0.5–10 (rod-like)	1000	1.5–1.7	8	4
Si-MCM + 41[c]	1–10	1000	1.5	9	4
Al-MCM-41 Si/Al = 9.8 [30]	05–10	850	0.95	27	7
MCM-41[d]	10–100 (sphere-like)	770	1.4	66	11
Si, Al-MCM-41[b]	1–50	1200	1.65	68	9
MCM-41 [31]	3 (disk-like)	1020	1.4	70	11
Si-MCM-41 [30]	<1	1050	1.5–1.7	112	16
Si-MCM-41 [30]	0.5–10 (rod-like)	1000	1.5–1.7	384	26

a) Effective self-diffusion coefficients and RMS displacements determined from the initial decays of the spin echo attenuations (pore-filling factor = 0.7; observation time t = 3 ms); samples are listed in order of increasing effective self-diffusivities.
b) H. Kosslick, 1999, personal communication, ACA, Berlin.
c) H. Gies, 1999, personal communication, Ruhr-Universität, Bochum.
d) S. Spange and A. A. Gräser, 1999, personal communication, Technische Universität Chemnitz.

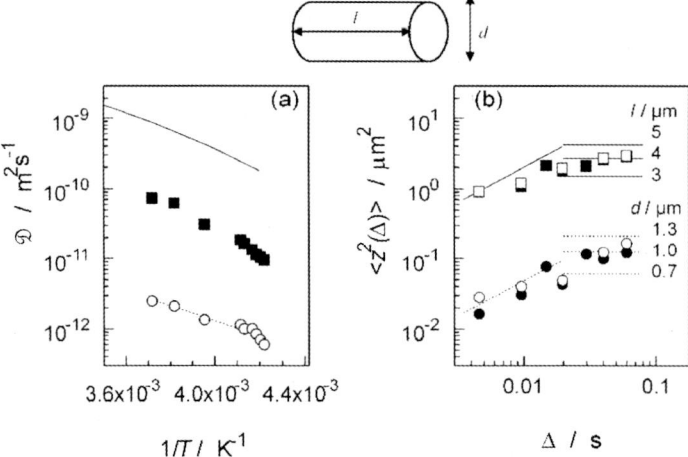

Figure 15.9 Anisotropic self-diffusion of water in MCM-41 as studied by PFG NMR. (a) Dependence of the parallel (■) and perpendicular (○) components of the axisymmetrical self-diffusion tensor on the inverse temperature at 10 ms observation time. The dotted lines are a guide for the eye. The full line represents the self-diffusion coefficients of super-cooled bulk liquid water. (b) Dependence of the parallel (■) and perpendicular (○) components of the mean square displacement on the observation time at 263 K in two different samples. The horizontal lines indicate the limiting values for the axial (full lines) and radial (dotted lines) components of the mean square displacements for restricted diffusion in cylindrical rods of lengths l and diameters d [Eqs. (11.33) and (11.32), as for Figure 11.2b]. The oblique lines, which are plotted for short observation times only (as for Figure 11.2a), represent the calculated time dependences of the mean square displacements for unrestricted (free) diffusion with $D_{par} = 1.0 \times 10^{-10}$ m² s^{-1} (full line) and $D_{perp} = 2.0 \times 10^{-12}$ m² s^{-1} (dotted line) [34] with permission.

15.4
Diffusion through Mesoporous Membranes

15.4.1
Mesoporous Vycor Glass

Vycor® porous glass can be made with a range of different pore sizes and a fairly narrow pore-size distribution. Furthermore, this material is easily formed into a self-supporting tubular membrane, thus eliminating the complications introduced by support resistance and making it very suitable for fundamental studies. For pore diameters in the 2–6 nm range at sub-atmospheric pressures viscous flow is relatively small and transport occurs primarily by the Knudsen mechanism, supplemented by surface diffusion for the more strongly adsorbed species. For such a system the diffusivity is given simply by the sum of the Knudsen and surface diffusion contributions in accordance with Eq. (4.14). This gives for the permeance (π):

$$\pi = \frac{N}{\Delta P} = \left(\frac{\varepsilon_p r}{\tau z}\right) \cdot \frac{97}{R(MT)^{\frac{1}{2}}} + \left(\frac{1-\varepsilon_p}{\tau z}\right) K D_s \qquad (15.5)$$

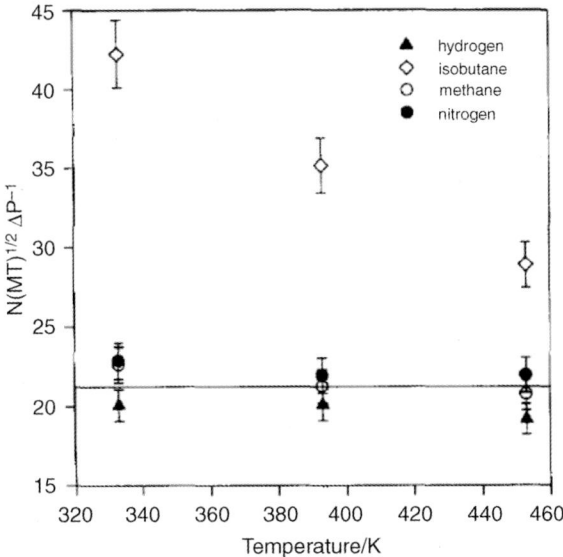

Figure 15.10 Temperature dependence of normalized permeance for H_2, N_2, CH_4, and i-C_4H_{10} in Vycor porous glass (mean pore diameter 4.5 nm) at 200 Torr. Reprinted from Fernandes and Gavalas [35] with permission.

If there is no surface diffusion the product $\pi(MT)^{1/2}$ should be independent of temperature while, if surface diffusion is significant, this product will decline with temperature since the exponential temperature decrease of K will easily outweigh the modest increase of the $T^{1/2}$ factor.

Representative results from a study by Fernandes and Gavalas [35] are shown plotted in this way in Figure 15.10. For the lighter molecules (H_2, N_2, CH_4) the normalized permeances $[N(MT)^{0.5}\Delta P^{-1}]$ are clearly almost the same and essentially independent of temperature, showing conformity with the simple Knudsen model. The normalized permeance for isobutane (a heavier and therefore more strongly adsorbed species) is clearly much larger and decreases quite strongly with temperature, suggesting a significant contribution from surface diffusion.

15.4.2
Mesoporous Silica

Since their discovery in 1992 [36], several families of ordered mesoporous silicas have been synthesized and characterized (see also Sections 15.3, 15.7, and 15.9). The potential for application of these materials as inorganic membranes with a uniform pore size was quickly recognized and a good deal of research has been directed to their preparation in membrane form. The pore diameter is quite uniform and typically in the range 4–20 nm. Without further modification such membranes can be used for ultrafiltration of colloids and large biomolecules [37, 38] but they are not useful for gas separation since, in pores of this size (at ordinary pressures), transport occurs mainly by Knudsen diffusion (Figure 4.3) [39], leading to very modest selectivities. However, because of their uniform pore size they are useful for fundamental studies.

15.4.2.1 Permeance Measurements

In modeling permeation through this type of membrane it is essential to allow for the mass transfer resistance of both the support and the active layer. The simplest situation arises at modest pressures (<1 atm for light gases) when the pores are relatively small (4–6 nm) and the pore size distribution is narrow and unimodal. In that situation Knudsen diffusion is dominant and the effects of viscous flow and molecular diffusion (within the active layer) should be minimal. Since the pores of the support are very much larger we assume viscous flow through the support with Knudsen diffusion, possibly augmented by surface diffusion, through the active layer. Using the principle of additivity of resistances the overall permeance (π) is therefore given by:

$$\frac{1}{\pi \mu RT} = \frac{z}{\mu D_e} + \left(\frac{\tau^1 z^1}{\varepsilon^1}\right) \frac{8}{\bar{P} a^2} \tag{15.6}$$

where D_e represents the "effective" Knudsen diffusivity ($D_e = \varepsilon D_K/\tau$). According to the Knudsen model $D_K = 97 r \sqrt{T/M}$ (m^2 s^{-1}), so Eq. (15.6) becomes:

$$\frac{1}{\pi \mu RT} = \left(\frac{\tau z}{\varepsilon r}\right) \cdot \frac{1}{97\mu} \cdot \sqrt{\frac{M}{T}} + \left(\frac{\tau^1 z^1}{\varepsilon^1}\right) \frac{8}{\bar{P} a^2} \tag{15.7}$$

where τ, ε, and z refer, respectively, to the active layer and τ^1, ε^1, and z^1 refer to the support.

The second term on the right-hand side of Eq. (15.7), representing the resistance of the support, is constant, and so if the Knudsen model is valid with no significant surface or viscous flow contributions, a plot of the experimental permeance data in the form $1/(\pi \mu RT)$ versus $1/\left(\mu \sqrt{T/M}\right)$ should yield a straight line with slope $\tau z/97\varepsilon r$ and intercept $8\tau^1 z^1/\varepsilon^1 \bar{P} a^2$ (the support resistance). Representative experimental data for the permeation of several light gases through a typical mesoporous silica membrane (average pore diameter $\approx$ 5.5 nm) are plotted in this way in Figure 15.11. Evidently, Eq. (15.7) provides a good representation of the data with no evidence of any systematic deviation from the simple Knudsen model. From the slope of Figure 15.11 we find $(\varepsilon r/\tau z) = 1.02 \times 10^{-5}$. With $r \approx 2.8$ nm and $\varepsilon \approx 0.12$ from porosimetry measurements and $z = 10$ μm, estimated from an electron micrograph of the membrane cross section, we find $\tau \approx 3.2$, an eminently reasonable value for the tortuosity factor.

The functional dependence of the pore diffusivity on $\sqrt{(T/M)}$, as predicted by the Knudsen model, is confirmed by the experimental data. Under the experimental conditions He is not adsorbed (to any measurable extent) whereas the other gases are significantly adsorbed, yet the permeance data for all species lie close to the same line (in Figure 15.11), implying that, in contrast to the conclusions drawn from recent theoretical modeling studies and molecular simulations [40–45], the validity of the Knudsen model is not significantly affected by at least a modest degree of adsorption.

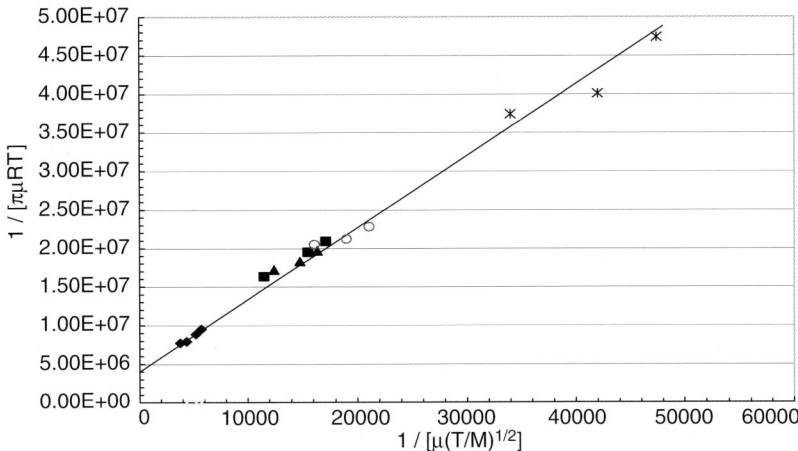

Figure 15.11 Experimental permeance data for light gases through a mesoporous silica membrane plotted as $1/(\pi\mu RT)$ versus $1/\left(\mu\sqrt{T/M}\right)$ in accordance with Eq. (15.7) (He, ◆; N$_2$, ■; Ar, ▲; CH$_4$, ○; C$_3$H$_8$, *) Reprinted from Ruthven et al. [48] with permission.

Similar behavior has also been observed in studies of diffusion in mesoporous silica adsorbents (Figure 4.3) and in porous glass membranes of similar pore size [35, 46] as well as in a single crystal of mesoporous silicon [47]. However, this simple pattern of behavior is unlikely to be replicated in materials with larger pores, at higher pressures, or with more strongly adsorbed species since the intrusion of significant contributions from viscous flow, surface diffusion, and molecular diffusion is then to be expected in addition to Knudsen diffusion.

In the Knudsen regime all components in a mixture diffuse independently with no mutual interaction. Thus the selectivity corresponds simply to the single-component permeance ratio, which in turn corresponds to the reciprocal of the ratio of the square roots of the molecular weights. Such selectivities are too small to be of much practical interest.

15.4.2.2 Modified Mesoporous Membranes

There have been numerous attempts to improve the perm-selectivity by various pretreatments that reduce the effective pore size. For example, Fernandes and Gavalas [35] pretreated their Vycor membranes with chlorosilane to reduce the pore size by silica deposition. Treatment with hexamethyldisilazane was used by Markovic et al. [46] to coat the pore surface of a mesoporous glass membrane with trimethylsilyl groups. A similar approach was also adopted by Higgins et al. [49] to modify a mesoporous silica membrane by silanation with long-chain alkyl chlorosilanes. However, the effect of such modifications has generally been found to be minimal. Some enhancement of perm-selectivity has been demonstrated, but only marginally beyond the Knudsen value. To achieve significant perm-selectivity for light gases it is evident that the pore diameter must be reduced to the "micropore" range where steric effects become dominant.

15.5
Surface Diffusion

At higher relative pressures there is increasing adsorption on the pore walls, leading to an increasing contribution from surface diffusion. This occurs at relative pressures well below the threshold for capillary condensation. Surface diffusion has been studied both experimentally and theoretically over many years. A useful review has been given by Kapoor et al. [50]. Here we focus only on selected experimental results to illustrate some of the trends that are commonly observed.

15.5.1
Determination of Surface Diffusivities

To a reasonable level of approximation the surface flux and the flux through the central region of the pore (by molecular and Knudsen diffusion and viscous flow) are additive, so the overall pore diffusivity may be expressed in the form of Eq. (4.14) with D_K replaced by the overall pore diffusivity (D_p), calculated, for example, from Eq. (4.7) or (4.11). It is generally not possible to realize conditions in which the surface contribution is completely dominant, so it is necessary to determine the surface diffusivity by difference, subtracting the known or estimated values of the pore contribution from the measured overall diffusivity.

Measurements of surface diffusivity have been made by transient methods (uptake rate and time lag) and by the steady-state (Wicke–Kallenbach) method. The earlier work has been reviewed by Barrer [51]. Several different approaches have been used to separate the pore and surface contributions. Costa et al. [52] determined the apparent diffusivities for light hydrocarbons on activated carbon from measurements of uptake rates over small incremental pressure steps. Under these conditions the equilibrium relationship can be considered linear over the small concentration step, although the slope will be a function of concentration [$K(c)$]:

$$D_e = \frac{\varepsilon_p D_p + (1-\varepsilon_p) K(c) D_s}{\varepsilon_p + (1-\varepsilon_p) K(c)} \tag{15.8}$$

Since $K(c)$ is known, one may plot $D_e \left[\varepsilon_p + (1-\varepsilon_p) K(c)\right]$ versus $K(c)$ to obtain D_s from the slope and hence D_p.

An alternative procedure is exemplified by the work of Schneider and Smith [53] who used the chromatographic method to study surface diffusion of propane on silica gel within the Henry's law region. The flux through the gas phase was measured at elevated temperature under conditions such that K was small enough to ensure a negligible contribution from the surface flux. The contribution from the flux through the gas phase (D_p) at the measurement temperature was then estimated by extrapolation and the surface diffusivity was found by difference according to Eq. (4.14). This approach is particularly straightforward if, as in this case, pore diffusion occurs mainly by the Knudsen mechanism since the temperature dependence is then modest.

Table 15.2 Surface diffusion of propane on silica gel; Data of Schneider and Smith [53].

T (°C)	$D_e \times 10^3$ (cm² s⁻¹)	$D_K \times 10^3$ (cm² s⁻¹)	$(1-\varepsilon_p/\varepsilon_p)KD_s \times 10^3$ (cm² s⁻¹)	$D_s \times 10^3$ (cm² s⁻¹)
50	1.54	0.42	1.12	0.74
75	1.45	0.44	1.01	1.31
100	1.33	0.45	0.88	1.95
125	1.22	0.47	0.75	2.8

Table 15.2 summarizes their results. It may be seen that the overall effective diffusivity decreases with temperature, but the analysis shows that this is due to the temperature dependence of K, which decreases more rapidly with temperature than D_s increases.

A third approach involving the use of a non-adsorbing calibrating gas is exemplified by the work of Rivarola and Smith [54] in their study of the surface diffusion of CO_2 on alumina (boehmite) pellets. The measurements were made by the Wicke–Kallenbach (steady-state) method. Counter-diffusion rates were first measured for a non-adsorbing gas pair (H_2/N_2). In the absence of surface diffusion the ratio of fluxes due to combined Knudsen and molecular diffusion should be equal to the ratio of the square root of the molecular weights and this was confirmed by the experimental data for H_2/N_2, both of which are not significantly adsorbed.

By contrast, the flux ratio for H_2/CO_2 was much smaller as a consequence of the enhancement of the CO_2 flux by surface flow. The effective diffusivity of CO_2 through the gas phase within the pore may be estimated from the measured diffusivity for $N_2 (D_{CO_2} = D_{N_2} \sqrt{M_{N_2}/M_{CO_2}})$ and the surface contribution may then be found directly from Eq. (4.14). In fact the procedure used was slightly more complex than this since the adsorbent had a wide bimodal distribution of pore size. The random pore model was shown to provide a good prediction of the pore diffusivity for N_2 and this model was then used to predict the pore diffusivity for CO_2 and hence to separate the pore and surface contributions. The relative importance of the surface flux varies widely depending on the conditions but contributions greater than 50% of the total are not uncommon [54].

15.5.2
Concentration Dependence

Figure 15.12 presents representative experimental data showing the concentration dependence of the surface diffusivity. Generally, although not universally, the surface diffusivity increases with concentration. Higashi et al. [55] noted that, for propane on silica glass, the form of the concentration dependence is given approximately by:

$$D_s = \frac{D_{s0}}{(1-\theta)} \qquad (15.9)$$

and he developed a theory to account for this based on Hill's "hopping model" [56].

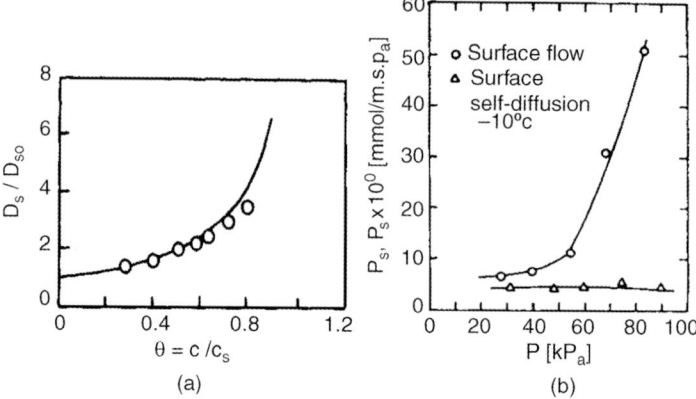

Figure 15.12 (a) Concentration dependence of surface diffusivity for propane on silica glass at 35 °C, showing the fit of the Higashi model [Eq.(15.9)] (b) concentration dependence of surface diffusivity and corrected diffusivity ("surface self-diffusivity") for SO_2 on porous Vycor glass at -10 °C. Reprinted from Okazaki et al. [57] with permission.

However, in common with most other investigators of that period, Higashi correlated his data in terms of Fickian rather than thermodynamically corrected diffusivities. Equation (15.9) is of precisely the form expected for a Langmuirian system with a constant corrected diffusivity – see Eqs. (1.31) and (1.32). It, therefore, seems likely that the observed concentration dependence arises simply from isotherm nonlinearity and complex mechanistic explanations are redundant. Similar data showing the concentration dependence of the Fickian surface diffusivity for several different sorbates on porous Vycor glass have been reported by Gilliland et al. [58] and Okazaki et al. [59]. For these systems it also appears that the concentration dependence can be accounted for entirely by the nonlinearity of the equilibrium isotherm with a constant corrected diffusivity. Representative data are shown in Figure 15.12b.

Extensive data for diffusion of light hydrocarbons on activated carbon have been reported by Costa et al. [52]. Although the data were presented as Fickian diffusivities, the equilibrium isotherms were also presented in tabular form, so the thermodynamic correction factors could be calculated. Figure 15.13 shows the results of such calculations, from which it is clear that a similar pattern of behavior applies for these systems. More complex examples showing a maximum in the surface diffusivity at intermediate loadings have also been observed [60–62]. This form of behavior can also be accounted for in the same general way [Eq. (1.31)] as the result of an inflexion in the equilibrium isotherm (e.g., if the isotherm is of BET form). Although this pattern of behavior appears to be quite common it is not universal and examples of more complex concentration dependence in which the corrected surface diffusivity is concentration dependent can certainly be found.

Figure 15.14 shows the self-diffusivities of n-hexane on the surface of mesoporous silicon as a function of loading (fractions of a monolayer) at different temperatures [63]. The mesoporous silicon samples were prepared by electrochemical etching of single-crystalline (100)-oriented p-type Si wafers with the doting and the anodization

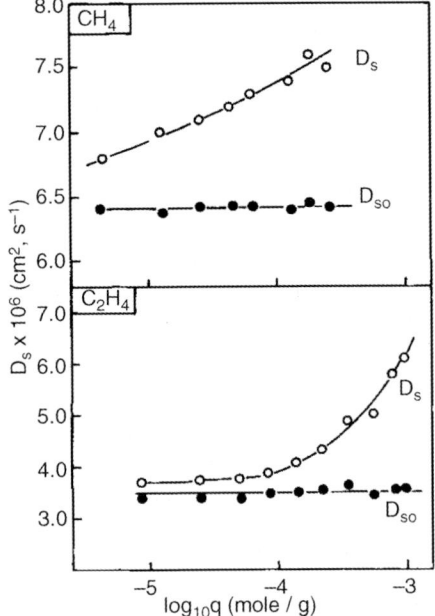

Figure 15.13 Surface diffusivity and corrected surface diffusivity for CH_4 and C_2H_4 on activated carbon at 20 °C. Data of Costa et al. [52]. Corrected diffusivities have been calculated using the tabulated equilibrium isotherm data.

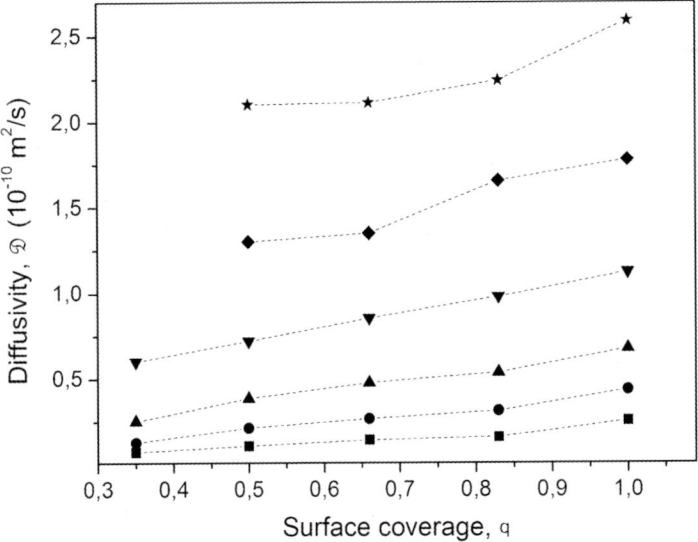

Figure 15.14 Diffusivities of n-heptane in porous silicon as a function of surface coverage for different temperatures (from bottom to top: 191, 204, 218, 237, 256, and 274 K). Reprinted from Reference [63] with permission.

current chosen to yield cylindrical pores with a mean diameter of 6 nm, perpendicular to the substrate surface [64]. Given the low vapor pressure of n-heptane, the contribution of diffusion through the vapor phase will be very small and the measured diffusivities may therefore be attributed to surface diffusion. Deviating from the behavior shown in Figure 15.13 (for light hydrocarbons on activated carbon), for heptane in mesoporous silicon the self-diffusivities show a small but significant increase with loading (Figure 15.14), which has been shown to be correlated with a decrease in the activation energy [63]. Such behavior suggests a heterogeneous surface in which the sites are occupied sequentially in accordance with their potential energies.

15.6
Diffusion in Liquid-Filled Pores

The semi-empirical approach of Eq. (15.1) has also been widely used to correlate the effective diffusivities for liquid systems. When the molecular diameter is small relative to the pore diameter, steric effects are negligible and, in the absence of surface diffusion, the situation is straightforward. Since, in the liquid phase, diffusion occurs only by the molecular mechanism, the pore diffusivity should be independent of pore diameter and Eq. (15.1), with a constant value of τ, should therefore be rigorous, even when the pore size distribution is broad. Measurements with different sorbates or at different temperatures (in the same adsorbent) should yield constant values of τ. Experimental tortuosity factors (Table 15.3) are generally of the same order as for

Table 15.3 Empirical tortuosity factors for diffusion in liquid-filled pores.

Sorbent	Sorbate–solvent	Method	$\bar{r}_p$ (Å)	ε_p	τ	Reference
Davison 5A[a] MS pellets	C_6H_{12}–C_6H_6 C_6H_6–CCl_4	Breakthrough curves	700	0.34	5.3–6.1	Lee and Ruthven [65]
Linde 5A[a] MS pellets	CCl_4–C_6H_6 C_6H_6–C_6H_{12}	Breakthrough curves	2000 (broad distribution)	0.33	1.7–2.3	Lee and Ruthven [65]
Porous Vycor glass	NaCl–H_2O	Transient uptake	25	0.23	4.5	Colton et al. [66]
			92	0.54	2.4	
			234	0.58	3.1	
			476	0.60	3.5	
			185	0.70	2.3	
Silica–alumina	NaCl–H_2O	Transient uptake	32	0.43	2.3	Satterfield et al. [67]
Alumina	Various hydrocarbons	Transient uptake	57,66	0.6	1.6	Prasher and Ma [68]
Small pore silica	Various amines in water	Chromatography	27		2.3–2.5	Ching et al. [69]

a) Intracrystalline pores are too small to admit these sorbates; only the macropore diffusivity is measured.

gaseous systems ($2 < \tau < 6$) and, where measurements have been repeated in the same adsorbent with different sorbates, constancy of the tortuosity factor is generally confirmed. In some instances rather small values of τ are found ($\tau < 1.0$) and this has generally been attributed to intrusion of surface diffusion.

When the ratio of molecular diameter to pore diameter is greater than about 0.1, steric hindrance becomes important. The extensive work on this subject, much of it involving measurements with track-etched membranes, having a well-defined and uniform pore size, has been reviewed briefly by Kärger and Ruthven [70] and by Lee [71]. Modeling of diffusion in liquid-filled micropores under conditions involving significant steric hindrance turns out to be a surprisingly challenging task [72] although a useful approximation is available [73]. This issue is considered briefly in Section 3.6.

15.7
Diffusion in Hierarchical Pore Systems

15.7.1
Ordered Mesoporous Material SBA-15

The diffusion of nitrobenzene in mesoporous material of type SBA-15 has also been studied [74] by essentially the same procedure as described in Section 15.3 for water in MCM-41 [34]. Nitrobenzene has also proved to be a versatile probe molecule for pore space exploration by NMR [75, 76], in particular via cryoporometry [77].

Corresponding to their shape, the SBA-15 particles under study are referred to as fibers and bundles (Figure 15.15a–d), with the respective diffusivities given in Figure 15.15e and f. The measurements have been carried out with overloaded samples at 253 K. This is well below the bulk freezing point ($T_{melt} = 278$ K) so that the fluid phase is confined to the mesopores by the frozen phase in the interparticle space.

As a remarkable feature, the diffusivities in the fibers (Figure 15.15e) are found to decrease with increasing observation times, while in the bundles they increase (Figure 15.15f). Diffusivities decreasing with increasing observation times may be explained by a hierarchy of transport resistances. A similar situation occurs for intracrystalline diffusion in MFI-type zeolites (Section 18.6.5). In the case of SBA-15 fibers embedded in the frozen guest phase, one may expect analogous behavior: The probability of encounters with the rigid frozen phase increases with increasing observation time, progressively inhibiting molecular propagation. The displacements covered in these experiments (a couple of micrometers in particle longitudinal extension and fractions of micrometers in the perpendicular direction) reflect the geometrical situation visualized in Figure 15.15a-d [74, 78].

Correspondingly, the absolute values of the diffusivities in the longitudinal direction (D_{par}) exceed those in the perpendicular extension by more than an

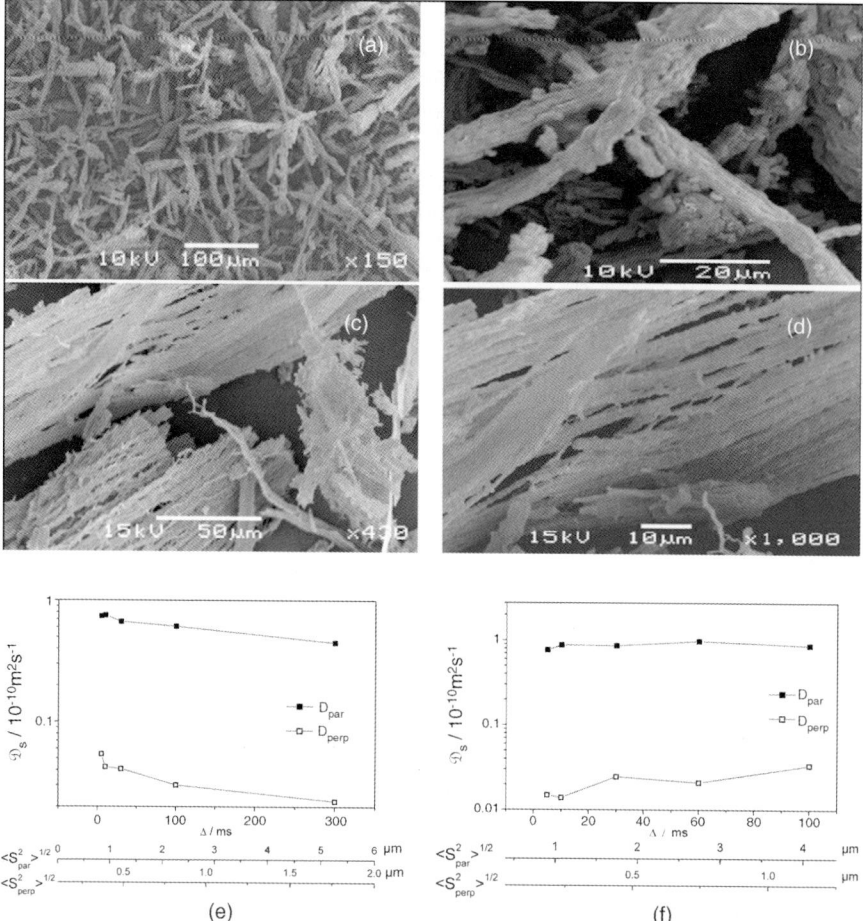

Figure 15.15 PFG NMR diffusion studies with SBA-15 SEM images of the two SBA-15 specimens under study (with "fibers" [(a) and (b)] and "bundles" [(c) and (d)], see Reference [78]) and the self-diffusivities of nitrobenzene at 253 K parallel (D_{par}) and perpendicular (D_{per}) to the main channel direction in the "fibers" (e) and "bundles" (f) as a function of the observation time. To illustrate the distances over which these diffusivities have been measured, the abscissa in (e) and (f) also shows the values of the root mean square displacements from Reference [74], with permission.

order of magnitude. The fact that diffusion is not ideally one-dimensional may be easily explained by the microporosity of SBA-15, which allows the exchange of molecules between adjacent channels. Moreover, transport perpendicular to the main channel direction may be explained quite generally by lattice defects that, although often invisible to conventional methods of structural analysis, are found, from diffusion measurements [34, 79, 80], to occur quite commonly in nanoporous materials.

Structural defects may also be considered as the origin of the unusual increase in the diffusivities with increasing observation time as observed with the SBA-15 bundles (Figure 15.15f) [74]. The bundle extensions and, hence, the space in which the guest molecules may diffuse freely exceed the mean diffusion path so that one should not expect any substantial effects of constriction as observed with the fibers. It is also possible that, on their diffusion paths, with increasing observation times, the molecules may encounter regions of higher mobility, leading to an increase in diffusivity with increasing observation time.

In comparative tracer ZLC and PFG NMR studies with toluene in SBA-15 [81], the diffusivities determined by the macroscopic TZLC technique were found to be several orders of magnitude smaller than the (microscopic) PFG NMR data. This indicates that, in addition to the barriers traced on the microscopic diffusion pathways by PFG NMR, there are much more pronounced transport resistances that become relevant only for diffusion path lengths approaching the dimensions of the SBA-15 particles.

15.7.2
Activated Carbon with Interpenetrating Micro- and Mesopores

The desire for nanoporous materials that allow fast molecular exchange between the micropore space and the surroundings has given rise to the fabrication of "hierarchical" materials in which the micropore space is traversed by mesopores. Figure 15.16 provides a visual impression of such a material, together with its pore size distribution. PFG NMR diffusion studies with cyclohexane as a probe revealed a distribution in the diffusivities [83] that must be attributed to heterogeneities in the sample structure. Since the distribution was invariant with observation time, one may conclude that the variations in sample morphology must occur with correlation lengths that exceed the maximum diffusion path lengths covered in these experiments (about 15 μm). The measurements were performed at 298 K by connecting the sample to a reservoir containing the guest molecules under a well-defined pressure. By varying the pressure a continuous variation of the sample loading could be achieved. By approaching a certain loading from either smaller or larger pressures (i.e., by adsorption or desorption) the effects of different sample "histories" could be studied.

Figure 15.17a shows the mean diffusivities resulting from the initial slope of the PFG NMR spin echo attenuations (Section 11.2.3) as a function of the pressure [84]. After each pressure step, the sample was allowed to equilibrate for 10 min. On the adsorption branch, for the last pressure step towards the saturation pressure (p_s), there was still a perceptible variation in the loading and diffusivity over a period of several hours, which is shown explicitly in Figure 15.17b and d, respectively.

One remarkable finding from these studies is the observation of hysteresis (i.e., dependence of the actual state on the past history) in both the adsorption isotherms and the diffusivities. Following investigations with purely mesoporous host systems (Vycor porous glass – see Section 15.9.1 and Reference [85]), the data

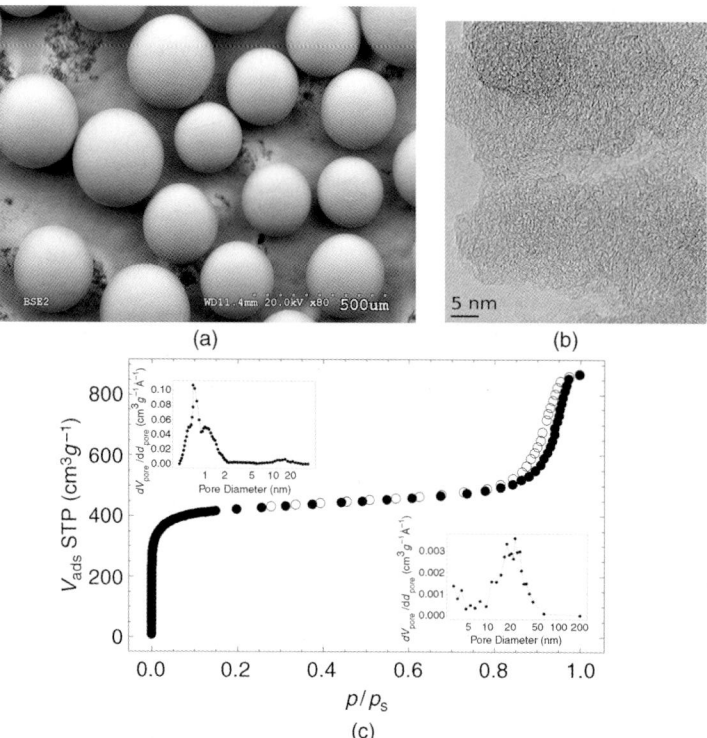

Figure 15.16 Activated carbon with hierarchical pore structure [82]: (a) SEM picture of the carbon particles; (b) TEM picture revealing both micropore and mesopore structure; (c) N_2 adsorption isotherm at 77 K with inserts showing the micropore and mesopores pore size distribution (upper left-hand side and lower right-hand side, respectively) as obtained by DFT (density functional theory) (micropore distribution) and BJH (Barrett–Joyner–Halenda) (mesopore distribution) [83, 84] with permission.

shown in Figure 15.17 represent the first observation of diffusion hysteresis in a hierarchical pore network involving both micro- and mesoporous spaces. In contrast to the observations with the pure mesoporous materials, the diffusivities on the desorption branch are now found to be larger (rather than smaller) than those on the adsorption branch. One may rationalize this difference by inspecting Figure 15.23 below and noting that the cyclohexane diffusivities in Vycor at lower loadings exceed the diffusivity at saturation over a substantial pressure range down to $p/p_s \approx 0.1$, with a pronounced maximum at about $p/p_s \approx 0.4$. Owing to the sample's "memory," the guest states on the desorption branch resemble those at saturation much more closely than those, at the identical external pressure, on the adsorption branch. This means that in Vycor the diffusivities on the desorption branch must indeed be smaller than during adsorption (corresponding to the low diffusivity at saturation), while the situation in the carbon sample considered is reversed.

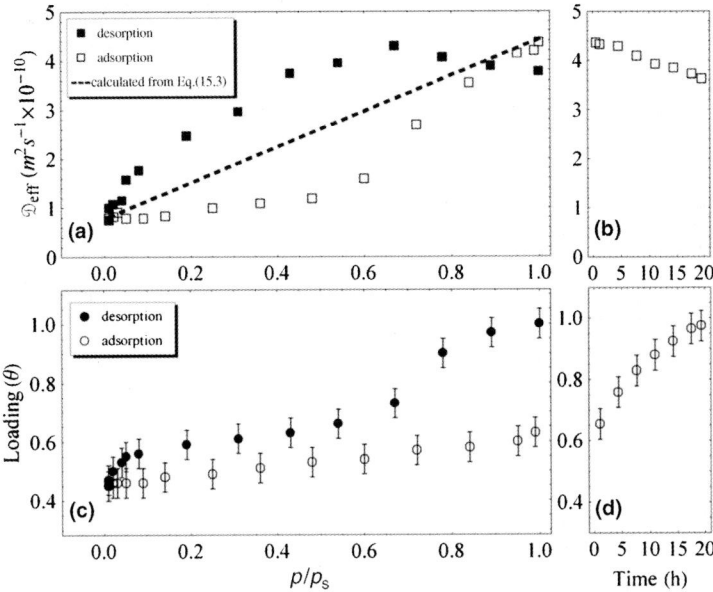

Figure 15.17 (a) Diffusivities of cyclohexane at 298 K in hierarchical activated carbon (Figure 15.16) as measured by PFG NMR as a function of the external pressure during adsorption and desorption and their estimate assuming mesopore diffusion in equilibrium with the external gas pressure (dashed line); (b) evolution of the diffusivities after the last pressure step (to $p/ps = 1$) on the adsorption branch; (c) cyclohexane loading isotherm at 298 K as a function of the external pressure during adsorption and desorption, measured 10 min after pressure variation; (d) loading evolution after the pressure step to $p/p_s = 1$ on the adsorption branch [84]. In addition to the diffusivities, the intensity of the NMR signal (free induction decay, see Section 10.1.2.3) also provides a direct measure of the relative amount of molecules adsorbed. Part (c) shows the resulting data in terms of the relative pore filling, assuming pore saturation for $p/p_s \approx 1$. Inevitable long-term instabilities cause the indicated uncertainty. The uncertainty in the measured diffusivity is smaller than the extension of the symbols, from Reference [84], with permission.

The origin of the hysteresis loop in the diffusivities on the adsorption and desorption branches (Figure 15.17a) may be visualized by estimating the contribution of the diffusion in the mesopores to overall diffusion. Since the increase of the cyclohexane content remains very small (Figure 15.17c) we have to consider only gas phase diffusion in the mesopores. Assuming fast exchange with the micropores over the NMR observation time the overall (effective) diffusivity is given by the relation:

$$\mathcal{D}_{\text{eff}} = \mathcal{D}_{\text{micro}} + p_{\text{gas}} \mathcal{D}_{\text{Knudsen}} \qquad (15.10)$$

where $\mathcal{D}_{\text{micro}}$ and $\mathcal{D}_{\text{Knudsen}}$ represent the self-diffusivities in the micro- and mesopores and where p_{gas} denotes the relative amount of molecules in the mesopore gas phase. This relation coincides with the fast-exchange limit of the two-region approach to modeling diffusion in beds of zeolites (Sections 11.4.3 and 15.7.3 and

References [86, 87]), with the important difference that now the path length of micropore diffusion is not limited by the crystal size. Moreover, with a pore size of about 20 nm, mass transfer in the mesopores follows Knudsen diffusion over the entire pressure range considered. With the relevant relations as presented in Section 15.2 and a tortuosity factor of 1.4 [84], the effective diffusivity is found to follow the dashed line in Figure 15.17a, nicely connecting the diffusivities at total micropore filling with a still negligible contribution by mesopore diffusion (p/p_s sufficiently small) with the overall diffusivity before the onset of capillary condensation. Note that the decrease in diffusivity with capillary condensation (appearing in Figure 15.17b as a decrease in the diffusivities in parallel with an increase in the loading in Figure 15.17d) indicates that, for the given system and conditions, the efficiency of liquid-phase mass transfer is considerably exceeded by mass transfer in the gas phase. The deviation of the diffusivities measured on the adsorption and desorption branches (Figure 15.17a) from the estimate by Eq. (15.10) indicates that, within most of the mesopores, the pressure deviates to smaller values during adsorption and to larger values during desorption. These deviations were found to be preserved over long intervals of time [83, 84] and correspond with the phenomenon of kinetic restriction in porous carbon as predicted theoretically [88].

15.7.3
Diffusion in "Mesoporous" Zeolites

In many technological processes the pore diameter of the host adsorbent is only slightly greater than the critical molecular diameter of the sorbate, leading to slow diffusion and reduced performance. Among the various strategies to overcome diffusion limitation, the generation of mesopores within a microporous crystal bulk phase has attracted particular attention. "Mesoporous" zeolites may be fabricated following quite different strategies, including "hard" [89] and "soft" (supramolecular) [90–92] templating methods and post-synthetic dealumination or desilication [93, 94].

Despite many efforts in the synthesis, characterization, and catalytic assessment of mesoporous zeolites, reports on their diffusion characteristics are still relatively scarce. This is probably due to the complexity of these systems and the difficulty of obtaining mesoporous zeolite crystals with sufficient structural regularity to allow reliable kinetic measurements. However, the available studies all show that the presence of a network of interconnected mesopores leads to dramatically enhanced exchange rates between the crystal bulk phase and the surroundings. Studies of this type have been performed by Groen *et al.* [95] with neopentane in MFI and by Cho *et al.* [96]. In the latter study, ^{129}Xe NMR spectroscopy provided direct evidence of accelerated guest exchange between different microporous regions. Considering xenon diffusion in a starch-templated MFI zeolite, Liu *et al.* [97] were able to demonstrate in this way that, on the NMR time scale, there is fast exchange between the mesopore and micropore spaces. The PFG NMR self-diffusivity may therefore be considered as the sum of the micropore and mesopore contributions to molecular displacement:

$$D_{\text{eff}} = p_{\text{micro}} D_{\text{micro}} + p_{\text{meso}} D_{\text{meso}} \tag{15.11}$$

in analogy to Eq. (15.10); in contrast to Eq. (15.10), where $p_{\text{gas}} \ll p_{\text{micro}} \approx 1$, there is no restriction on the relative populations of the meso- and micropores.

The first PFG NMR diffusion studies with mesoporous zeolites were reported in Reference [98]. The advantages of PFG NMR for elucidation of mass transfer in hierarchical porous materials [87] include the distinction between the contributions of mass transfer in the micro- and mesoporous spaces and the ability to distinguish between the transport in the different pore spaces and surface resistances (see also Section 11.4.3). The applicability of such methods is, however, limited by the requirement for sufficiently large particles (preferably ≥ 10 μm).

15.8
Diffusion in Beds of Particles and Composite Particles

Diffusion in a bed of microporous particles or zeolite crystals or in an aggregated adsorbent particle has often been considered to be controlled by transport through the macro/meso-pore network of the interparticle/crystalline space, with no significant contribution to long-range transport from diffusion in the micropores. Where information on macropore diffusivities is deduced from transient measurements on composite particles, the micropores are assumed to contribute to the sorption capacity but not to the effective long-range diffusivity (Section 6.2.2.3). Under practical conditions this is usually an acceptable approximation, but it should not be accepted uncritically. Mass transfer in hierarchical systems [see Section 15.7 and Eqs. (15.10) and (15.11)] provides an example where diffusion in both the micro- and meso/macro-pores may contribute to the overall transport.

In the type of pore structure sketched in Figure 1.3a the micropores are visualized as branching from macro/meso-pores and do not form an independent network over longer distances. Nevertheless it is clear that some enhancement of the overall flux may be possible since, at least over shorter distances, transport through the micropores is effectively in parallel with the flux through the larger pores. The situation in the composite type of particle sketched in Figure 1.3b is clearer. If the microparticles are non-porous, with their adsorptive capacity confined to the external surface, they can obviously make no contribution to the flux. On the other hand, if the microparticles are microporous then clearly, over certain localized regions, transport through a microparticle can provide an additional flux in parallel to the transport through the extracrystalline voids or macropores. Model considerations as presented in the next section may help to estimate the extent to which this effect may enhance the long-range diffusivity.

15.8.1
Approaches by Mathematical Modeling

We consider a composite particle or a bed of microparticles (e.g., zeolite crystals) of voidage ε through which is maintained, under steady state conditions, a gradient of tracer concentration or, equivalently, a gradient of total concentration under conditions of linear equilibrium. The flux through the composite may be represented in terms of a Fickian diffusivity defined with respect to the gradient of the average concentration in the composite:

$$J = -D_{1.\mathrm{r.}} \frac{dq'}{dz}$$
$$q' = (1-\varepsilon)q + \varepsilon c = [\varepsilon + K(1-\varepsilon)]c \tag{15.12}$$

Diffusion through the extracrystalline void spaces (or macropores) is characterized by a diffusivity D_p, defined based on the concentration in the gas phase, while diffusion through the particles or crystals is characterized by diffusivity D_c based on the adsorbed phase concentration. In Section 11.4, considering the different regimes of PFG NMR diffusion measurements with beds of zeolites, these diffusivities are referred to as D_{inter} and D_{intra}, respectively. The ratio of the diffusivities in the two phases, duly corrected for the difference in concentrations, is $\nu = KD_c/D_p$. If $\nu \to 0$ diffusion through the crystals or particles clearly makes no contribution to long-range transport:

$$J = -D_{1.\mathrm{r.}}[\varepsilon + (1-\varepsilon)K]\frac{dc}{dz} = -\varepsilon D_p \frac{dc}{dz}$$
$$\frac{D_{1.\mathrm{r.}}}{D_p} = \frac{\varepsilon}{\varepsilon + (1-\varepsilon)K} \tag{15.13}$$

If $\nu \to 1.0$ the composite will behave as a homogeneous medium and $dq'/dz = (dc/dz)[\varepsilon + K(1-\varepsilon)]$:

$$J = -D_{1.\mathrm{r.}}[\varepsilon + (1-\varepsilon)K]\frac{dc}{dz} = -D_p \frac{dc}{dz}$$
$$\frac{D_{1.\mathrm{r.}}}{D_p} = \frac{\varepsilon}{\varepsilon + (1-\varepsilon)K} \tag{15.14}$$

The behavior for large values of ν can be predicted only by more detailed modeling.

This is a classical problem of physics first formulated in terms of the electrical analog, conduction through a composite medium, by Maxwell [99] and Rayleigh [100]. Since then a range of different models have been developed for such systems; some of these are given in Table 15.4. Although these models differ in their algebraic form, the numerical predictions are very similar, as may be seen from Figure 15.18, which shows the ratio $D'_{1.\mathrm{r.}}/D_p = [\varepsilon + (1-\varepsilon)K]D_{1.\mathrm{r.}}/D_p$ plotted against ν for two different void fractions. The curves, which are all of similar form, extend over six

Table 15.4 Expressions for diffusion in a continuum with dispersed microporous particles/crystallites. Adapted from Kärger et al. [107].

Model	Expression[a]	Reference
Dilute suspension	1. $\dfrac{D'_{1.r.}}{D_p} = 1 - \dfrac{3p}{[(2+v)/(1-v)] + p}$	Maxwell [99]
	2. $\dfrac{D'_{1.r.}}{D_p} = 1 - \dfrac{3p}{[(2+v)/(1-v)] + p - 0.523[(1-v)/(4/3)-v]p^{10/3}}$	Rayleigh [100] Runge [101]
	3. $\dfrac{D'_{1.r.}}{D_p} = 1 - \dfrac{3p + 1.227(3-3v)/(4+3v)p^{10/3}}{[(2+v)(1-v) + p + 0.409[(6+3v)/(4+3v)]p^{7/3} - 0.906(3-3v)/(4+3v)p^{10/3}}$	Meredith an Tobias [102]
	4. $\dfrac{D'_{1.r.}}{D_p} = 1 - 1.21 p^{2/3} + \dfrac{p^{1/3}}{0.513(1.61 p^{-1/3} - 2 + (1 - 1/v)/[v/(v-1)\ln v - 1])}$	Jefferson et al. [103]
Series-parallel formula	5. $\dfrac{D'_{1.r.}}{D_p} = \left(\dfrac{B^{1/2}}{\{(1-\gamma)[1-B(1-v)]\}^{1/2}} \times \tan^{-1}\{[B(1-v)]/[1-B(1-v)]\}^{1/2} + 1 - B \right)^{-1}, v<1$	
	$\quad\quad = \left(\dfrac{B^{1/2}}{2\{(v-1)[1+B(v-1)]\}^{1/2}} \times \ln\dfrac{1 + \{[B(v-1)]/[1+B(v-1)]\}^{1/2}}{1 - \{[B(v-1)]/[1+B(v-1)]\}^{1/2}} + 1 - B \right)^{-1}, v>1$	
	6. $\dfrac{D'_{1.r.}}{D_p} = \dfrac{1}{(1-\Psi)/2(\Psi-\Psi_1)]\ln\{[(1-\Psi_1)(1+\Psi)/(1-\Psi)(1+\Psi_1)\} + \psi/(1-\Psi^2)}$ $\quad\quad + \dfrac{1}{(1-\psi)/\psi_1\psi + 1/v\psi}$	Cheng and Vachon [104] Jury [105, 106]

[a] $D'_{1.r.} = [\varepsilon + (1-\varepsilon)K]D_{1.r.}; v = KD_c/D_p; p = (1-\varepsilon); B = (3p/2)^{1/2}, \psi = p^{1/3}, \psi_1 = [1 + (\psi^{-2} - 1)/v]^{1/2}$.

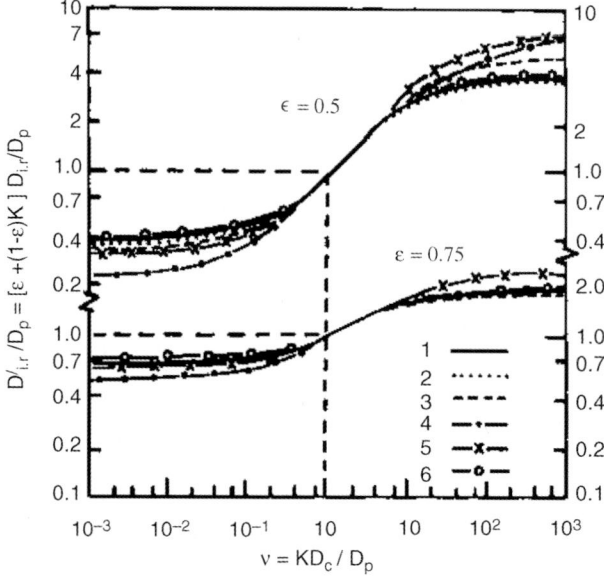

Figure 15.18 Variation of long-range diffusivity ($D'_{1,r}/D_p$) for composite pellet with the normalized ratio of diffusivities in the dispersed and continuous phases ($v = KD_c/D_p$ or, in the notation of Section 11.4, $= KD_{intra}/D_{inter}$), calculated from the models of Table 15.4. Reprinted from Kärger et al. [107] with permission.

orders of magnitude in v, yet the variation in the long-range diffusivity $D'_{1,r}$ is scarcely more than one order of magnitude. Clearly, a significant enhancement of the long-range flux is only achieved when v is substantially greater than unity and even when v is very large the limit corresponding to a monolithic crystal is never really approached. When the corrected diffusivity ratio is small, the possibility of any significant enhancement of the long-range flux as a result of micropore diffusion can be excluded.

15.8.2
Temperature Dependence

At sufficiently low pressures diffusion in the external voids or macropores occurs mainly by the Knudsen mechanism so that D_p ($=D_{inter}$) varies only slightly with temperature. Under these conditions the main contribution to the temperature dependence of the long-range diffusivity arises from the equilibrium constant (which varies according to $K = K_\infty e^{-\Delta H_0/RT}$), leading to an apparent activation energy essentially equal to the heat of adsorption, as discussed in Section 11.4.2.3 Representative experimental data, presented in Table 15.5, conform approximately to this pattern of behavior. The discrepancies can be attributed to the difficulty of making reliable diffusivity measurements over long diffusion distances and to the intrusion of some contribution from molecular diffusion within the extracrystalline pores.

15.8 Diffusion in Beds of Particles and Composite Particles

Table 15.5 Activation energies for long-range self-diffusion.

System	Reference	E (kJ mol^{-1})	$-\Delta H$ (kJ mol^{-1})^a
Propane–NaX	[61]	25	35
n-Butane–NaX	[62]	36	43
n-Pentane–NaX	[63]	53	54
n-Hexane–NaX	[63]	75	63
n-Heptane–NaX	[63]	110	71
Cyclohexane–NaX	[64]	58	56
Water–NaX	[65]	50–65	68
Ethane–5A	[66]	22–28	29

a) Isoteric heat of adsorption at low loadings.

The transition from Knudsen to molecular diffusion is illustrated in Figure 15.19 for n-butane in an assemblage of NaX crystals. At lower temperatures the gas-phase concentration is low and diffusion occurs mainly by the Knudsen mechanism. The long-range diffusivity increases with sorbate concentration as a consequence of the increase in the ratio of gas to adsorbed phase concentrations resulting from the favorable form of the equilibrium isotherm. At higher temperatures this effect is more than compensated by the increased density of the gas phase and the resulting transition from Knudsen to molecular diffusion (for which the diffusivity varies

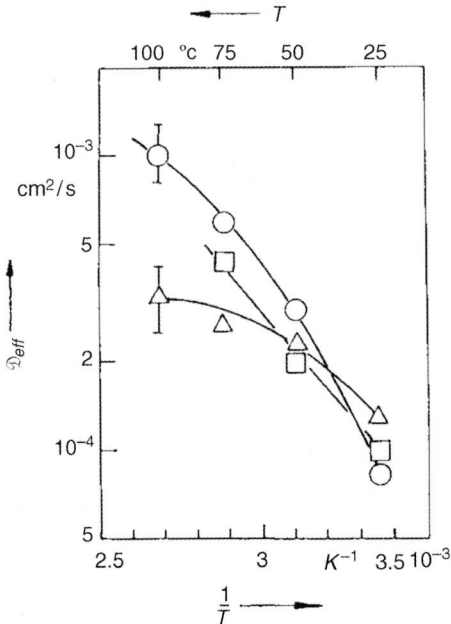

Figure 15.19 Effective self-diffusion coefficient for long-range diffusion of n-butane in a bed of NaX zeolite crystals at a loading 0.8 (○), 1.9 (□), and 2.4 (△) mmol g^{-1}. Reprinted from Kärger and Samulevich [108] with permission.

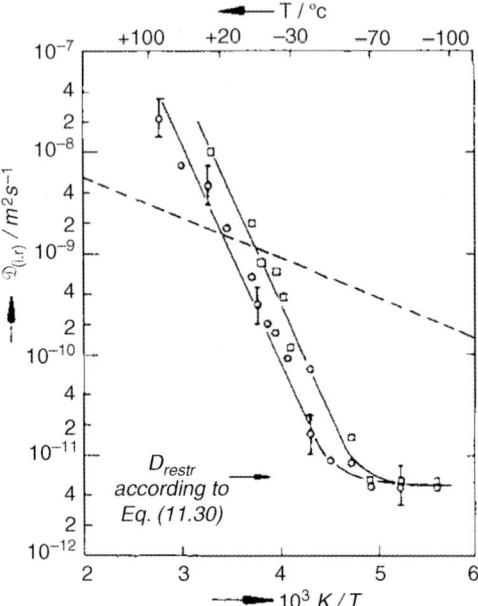

Figure 15.20 Temperature dependence of the effective self-diffusion coefficient for n-butane in a loose assemblage of NaX zeolite crystals (□) and in a pressed compact of the same crystals (○). (Mean crystal diameter ≈5 μm, loading 80 mg g^{-1}. Broken line: intracrystalline diffusivities.) Reprinted from Kärger et al. [107] with permission.

inversely with pressure or concentration). As a result, at the higher temperatures, the long-range diffusivity decreases with loading. For the sample with the highest sorbate loading the increase in the proportion of molecules in the vapor phase, at elevated temperatures, is almost exactly compensated by the decrease in the gas phase diffusivity arising from the increased gas density. As a result, the long-range diffusivity becomes almost independent of temperature, leading to pronounced curvature of the Arrhenius plot.

Figure 15.20 shows the temperature dependence of the long-range self-diffusivity for n-butane in a loose assemblage and a pressed pellet of NaX zeolite crystals. The decrease in the diffusivity by a factor of three is consistent with the decrease in the voidage from about 75% to 25%. Even at low temperatures where almost all the molecules are adsorbed within the crystals, the long-range diffusivity is controlled by diffusion through the void spaces (macropores) and the limiting behavior of a single crystal is never approached.

15.8.3
Multicomponent Systems

Figure 15.21 shows the coefficients of long-range and intracrystalline self-diffusion for cyclohexane in NaX in the single-component system and in the presence of differ-

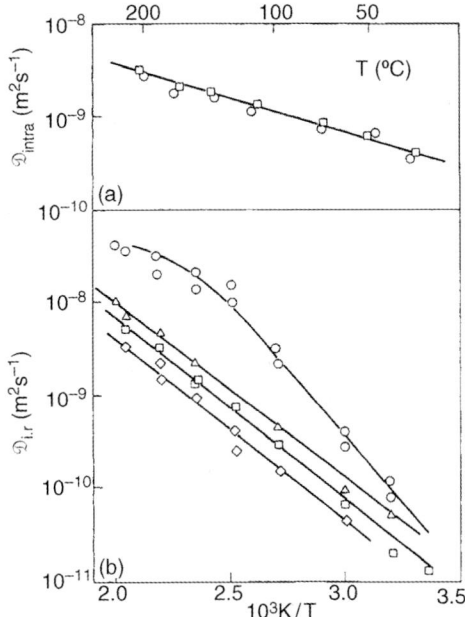

Figure 15.21 Coefficients of (a) intracrystalline and (b) long-range self-diffusion for cyclohexane in NaX zeolite crystals at a sorbate concentration of 1.9 molecules per cage for pure adsorbate (○) and in the presence of argon at 0.6 (△), 1.3 (□), and 2.0 atm (◇). Reprinted from Kärger and Pfeifer [110] with permission.

ent partial pressures of argon. Although the intracrystalline diffusion of cyclohexane in the NaX crystals is not affected, the argon atmosphere leads to a significant reduction in the long-range diffusivity [109, 110]. Evidently, in the presence of argon, diffusion in the extracrystalline void spaces occurs mainly by the molecular mechanism.

Changing the temperature alters the distribution of cyclohexane between the intracrystalline and extracrystalline phases, but does not greatly affect the molecular diffusivity in the gas phase, since the mean free path is determined by the quantity of argon, which is present in large excess. Although argon is only weakly adsorbed, the concentration in the gas phase increases somewhat with increasing temperature. As a result, the Arrhenius plot remains linear with the apparent activation energy given by the difference between the heats of adsorption of cyclohexane and argon.

Figure 15.22 [110, 111] shows the coefficients of long-range self-diffusion for a mixture of n-heptane and benzene at different compositions. From Eq. (11.34) it follows that, for sufficiently strong adsorption (i.e., $q \gg c$):

$$\left(\frac{\mathcal{D}_{1.r.}}{\mathcal{D}_{inter}}\right)_A \bigg/ \left(\frac{\mathcal{D}_{1.r.}}{\mathcal{D}_{inter}}\right)_B = \left(\frac{x}{y}\right)_B \bigg/ \left(\frac{x}{y}\right)_A = \alpha_{BA} \qquad (15.15)$$

where x and y are the mole fractions in the adsorbed and vapor phases and α is the "separation factor."

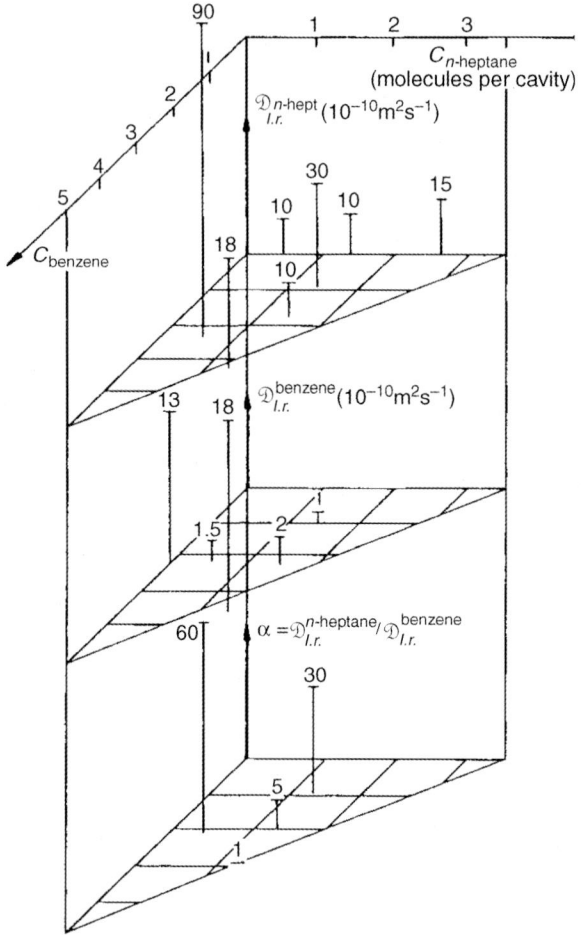

Figure 15.22 Coefficients of long-range self-diffusion of n-heptane and benzene in mixtures of these components in an assemblage of NaX crystals at 361 K, and the corresponding separation factors. Reprinted from Kärger and Pfeifer [110] with permission.

For molecules of comparable size and mass the values of $\mathcal{D}_{\text{inter}}$ are similar so that the separation factor follows directly from the ratio of the long-range diffusivities. These values, which are also indicated in Figure 15.21, are found to be in satisfactory agreement with the values determined directly from equilibrium measurements.

15.9
More Complex Behavior: Presence of a Condensed Phase

The complexity observed in the diffusional properties of mesoporous solids stems largely from the effects of capillary condensation, which occurs in mesopores at quite low relative pressures. At pressures below the two-phase region normal reversible behavior is observed and the effective diffusivity can be represented using the simple models discussed above and in Section 4.2.2. However, under the co-existence of liquid and gaseous phases the behavior becomes much more complex and a simple representation in terms of a single effective diffusivity is no longer possible. Some of the phenomena that have been observed are discussed briefly here. A more detailed summary has been given by Valiullin *et al.* [112] and complete details are given in the original papers [113–117].

15.9.1
Diffusion in Porous Vycor Glass: Hysteresis Effects

Diffusion of cyclohexane has been studied in detail by Valiullin *et al.* [113, 115–117] and by Naumov *et al.* [114] in a sample of porous Vycor glass (mean pore diameter ~6 nm) with a fairly narrow pore size distribution. Figure 15.23 shows the self-diffusivities, measured by PFG NMR, along the adsorption and desorption branches of the hysteresis isotherm.

At low relative pressures (p/p_s) the NMR signal intensity shows a simple single-exponential decay, indicating normal Fickian diffusion, and the transient sorption curves conform to the simple diffusion model. However, under the co-existence of liquid and gaseous phases the behavior becomes much more complex. The approach to equilibrium becomes much slower and is no longer governed by a simple diffusion process. There appears to be a relatively fast diffusion process followed by a slow rearrangement process as illustrated in Figure 15.24. For the same pressure step this behavior is accurately reproducible.

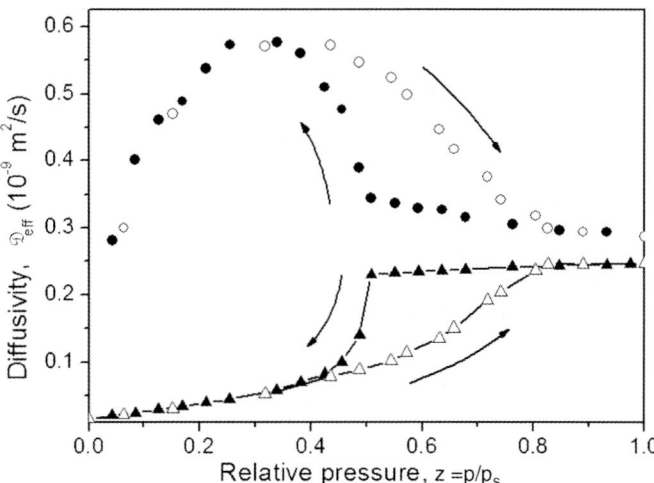

Figure 15.23 Effective self-diffusivity for cyclohexane (●, ○) and the relative amount adsorbed (▲, △) at 298 K in porous Vycor glass (pore diameter ~ 6 nm) as a function of relative pressure. Open symbols show the adsorption data and closed symbols show desorption data. Reprinted from Valiullin *et al.* [112] with permission.

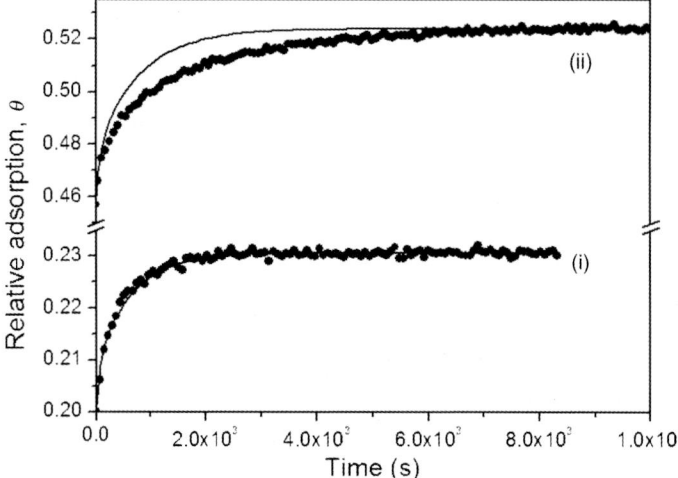

Figure 15.24 Transient uptake curves for cyclohexane in Vycor for relative pressure steps (i) 0.323–0.363 and (ii) 0.565–0.605.

The hysteresis in the kinetic data is not simply a result of the difference in the relative amounts of liquid and vapor within the pores at the same pressure during adsorption and desorption. This may be shown by re-plotting the data to show the apparent diffusivity as a function of fractional loading. As shown in Figure 15.25, the difference between the diffusivities measured in adsorption and desorption persists even when the comparison is made at the same fractional loading rather than at the same relative pressure.

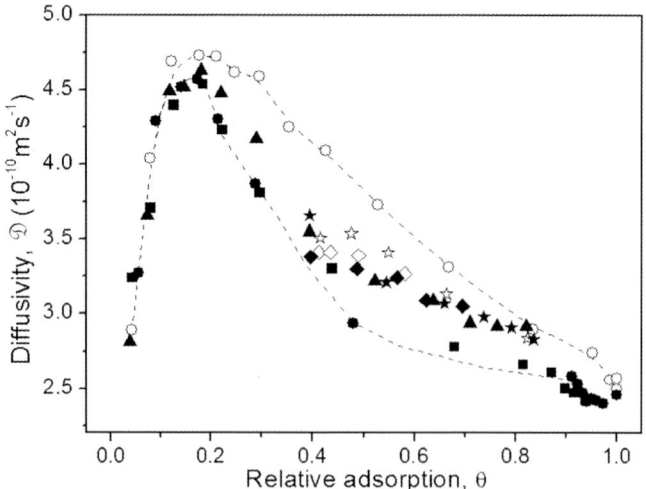

Figure 15.25 Effective self-diffusivity for cyclohexane in porous glass at 298 K as a function of loading [(○) adsorption and (●) desorption] with additional data from "scanning isotherms" (■). Reprinted from Valiullin et al. [112] with permision.

Similar results are obtained from measurements made within the hysteresis loop under non-equilibrium conditions from "scanning experiments." Representative results are included in Figure 15.25 and it is clear that the measurements within the hysteresis loop yield effective diffusivities that lie between those for the adsorption and desorption branches of the isotherm.

The different states must correspond to different local minima of free energy that have different distributions of liquid and vapor within the pores at the same overall loading and the reproducibility of the data shows that these minima must represent quasi-equilibrium states that are stable over a relatively long time scale, thus raising philosophical questions about the definition of equilibrium in this context. The rate of transition from one quasi-equilibrium state to another of lower free energy appears to occur by a relatively slow activated process involving the cooperative movement of many molecules rather than by diffusion, which occurs on a much shorter time scale.

Somewhat similar hysteresis effects have been observed with varying temperature for nitrobenzene in mesoporous silicon in the region of the freezing–melting transition [118].

15.9.2
Supercritical Transition in an Adsorbed Phase

The supercritical transition has also been studied for n-pentane in a similar sample of porous Vycor glass with mean pore diameter $\sim$6 nm. Figure 15.26 presents representative results, showing the temperature dependence of the self-diffusivity for the free fluid and the adsorbed phase.

The free fluid shows a jump in the diffusivity by more than an order of magnitude at the critical temperature ($\approx$470 K), reflecting the well-known increase in the

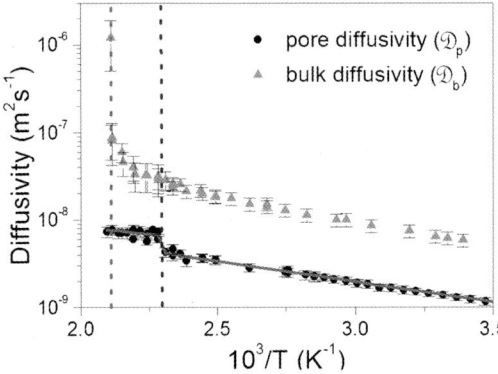

Figure 15.26 Arrhenius plot showing temperature dependence of the bulk and pore fluid self-diffusivities for n-pentane in porous Vycor glass (pore diameter $\sim$6 nm). The vertical dashed lines show the positions of the bulk- and pore-critical temperature. Reprinted from Dvoyashkin *et al.* [116] with permission.

mobility of the supercritical fluid. For the adsorbed fluid the jump in diffusivity is less dramatic and occurs at a lower temperature (438 K), showing the significant reduction in critical temperature for the confined fluid.

15.9.3
Diffusion in a Composite Particle of MCM-41

A further example of the dramatic variation in effective diffusivity that may be observed under the influence of co-existing liquid and gaseous phases is given in Figure 15.27, which shows the variation of the effective self-diffusivity with relative pressure for cyclohexane in a composite particle of MCM-41 (measured by PFG NMR). The desorption branch of the isotherm is shown in the inset and it is clear that the sharp change in diffusivity corresponds exactly with the phase transition.

The effective diffusivities are derived from the long-time slope of the spin-echo decay and therefore represent the long-range diffusivity through the composite particle. At high relative pressures the diffusivity represents mainly the contribution arising from vapor phase diffusion in the macropores (which are free from condensate) and the decrease with decreasing pressure arises from the reduction in the proportion of molecules in the vapor phase. At a relative pressure of about 0.3, evaporation of the liquid from the mesopores leads to a sharp increase in the fraction of molecules in the macropores and a consequent rise in the diffusivity. In the low pressure region there is a further rise in the diffusivity due to an increasing contribution from diffusion in the mesopores. A more detailed account is given in the original paper [119].

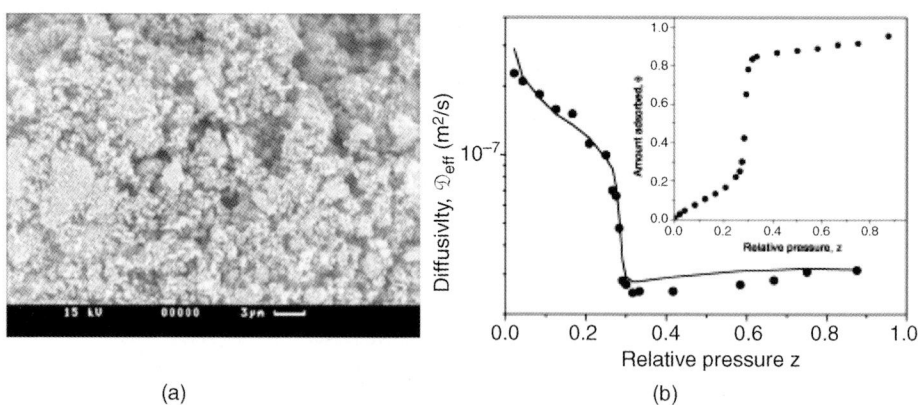

Figure 15.27 (a) SEM micrograph showing the internal texture of an MCM-41 particle formed by aggregation of individual MCM-41 crystallites. (b) Effective self-diffusivity ($\mathcal{D}_{eff}$) at $T = 297$ K for cyclohexane in MCM-41 particles as a function of the relative pressure ($z = p/p_s$); inset: desorption branch of the isotherm normalized to the volume of the mesopores. Reprinted from Valiullin et al. [112] with permission.

References

1 Valiullin, R., Kärger, J., and Gläser, R. (2009) *Phys. Chem. Chem. Phys.*, **11**, 2833.
2 Enke, D., Janowski, F., and Schwieger, W. (2003) *Microporous Mesoporous Mater.*, **60**, 19.
3 Janowski, F. and Enke, D. (2002) *Handbook of Porous Solids* (eds F. Schüth, K.S.W. Sing, and J. Weitkamp), Wiley-VCH Verlag GmbH, Weinheim, p. 1432.
4 Stoltenberg, D., Seidel-Morgenstern, A., and Enke, D. (2010) *Chem. Ing. Tech.*, **82**, 829.
5 Chmelik, C., Enke, D., Galvosas, P., Gobin, O.C., Jentys, A., Jobic, H., Kärger, J., Krause, C., Kullmann, J., Lercher, J.A., Naumov, S., Ruthven, D.M., and Titze, T. (2011). *ChemPhysChem*, **12**, 1130.
6 Marsh, H. and Rodríguez-Reinoso, F. (2006) *Activated Carbon*, Elsevier, Amsterdam.
7 Rodriguez-Reinoso, F. (2002) *Handbook of Porous Solids* (eds F. Schüth, K.S.W. Sing, and J. Weitkamp), Wiley-VCH Verlag GmbH, Weinheim, p. 1766.
8 Kärger, J., Lenzner, J., Pfeifer, H., Schwabe, H., Heyer, W., Janowski, F., Wolf, F., and Zhdanov, S.P. (1983) *J. Am. Ceram. Soc.*, **66**, 69.
9 Kärger, J., Pfeifer, H., Vartapetian, R.S., and Voloshchuk, A.M. (1989) *Pure Appl. Chem.*, **61**, 1875.
10 Yang, R.T. (1987) *Gas Separation by Adsorption Processes*, Butterworths, Stoneham, MA.
11 Ruthven, D.M. (1984) *Principles of Adsorption and Adsorption Processes*, John Wiley & Sons, Inc., New York.
12 Ruthven, D.M., Raghavan, N.S., and Hassan, M.M. (1986) *Chem. Eng. Sci.*, **41**, 1325.
13 Dominguez, J.A., Psaris, D., and LaCava, A.I. (1988) *AIChE Symp. Ser.*, **84**, 73.
14 LaCava, A.I., Dominguez, J.A., and Cardenas, J. (1989) *Adsorption Science and Technology* (eds A.E. Rodrigues, M.D. Levan, and D. Tondeur), Kluwer Academic Publisher, Amsterdam, p. 323.
15 Chihara, K. and Suzuki, M. (1979) *Carbon*, **17**, 339.
16 Chihara, K., Suzuki, M., and Kawazoe, K. (1978) *AIChE J.*, **24**, 232.
17 Chihara, K., Suzuki, M., and Kawazoe, K. (1878) *J. Colloid Interface Sci.*, **64**, 584.
18 Schröter, H.J. and Jüntgen, H. (1989) *Adsorption Science and Technology* (eds A.E. Rodrigues, M.D. Levan, and D. Tondeur), Kluwer Academic Publisher, Amsterdam, p. 269.
19 Vasenkov, S. and Kortunov, P. (2005) *Diffusion Fundam.*, **1**, 2.1–12.
20 Stallmach, F. and Crone, S. (2005) in *Diffusion Fundamentals* (eds J. Kärger, F. Grinberg, and P. Heitjans), Leipzig University Press, p. 474.
21 Di Renzo, F., Galarneau, A., Trens, P., and Fajula, F. (2002) in *Handbook of Porous Solids* (eds F. Schüth, K.S.W. Sing, and J. Weitkamp), Wiley-VCH Verlag GmbH, Weinheim, p. 1311.
22 Demus, D., Goodby, J., Gray, G.W., and Spiess, H.W. (1998) *Handbook of Liquid Crystals*, Wiley-VCH Verlag GmbH, Weinheim.
23 Huo, Q., Margolese, D.I., Ciesla, U., Feng, P., Gier, T.E., Sieger, P., Leon, R., Petroff, P.M., Schüth, F., and Stucky, G.D. (1994) *Nature*, **368**, 317.
24 Kresge, C.T., Leonowicz, M.E., Roth, M.W., Vartuli, J.C., and Beck, J.S. (1992) *Nature*, **359**, 710.
25 Beck, J.S., Vartuli, J.C., Roth, W.J., Leonowicz, M.E., Kresge, C.T., Schmitt, K.D., Chu, C.T.W., Olson, D.H., Sheppard, E.W., McCullen, S.B., Higgins, J.B., and Schlenker, J.L. (1992) *J. Am. Chem. Soc.*, **114**, 10834.
26 Hamley, I.W. (2005) *Block Copolymers in Solution*, John Wiley & Sons, Ltd, Chichester.
27 Ulrich, K., Galvosas, P., Kärger, J., and Grinberg, F. (2009) *Phys. Rev. Lett.*, **102**, 37801.

28 Zhao, D.Y., Huo, Q.S., Feng, J.L., Chmelka, B.F., and Stucky, G.D. (1998) *J. Am. Chem. Soc.*, **120**, 6024.
29 Stallmach, F., Graser, A., Kärger, J., Krause, C., Jeschke, M., Oberhagemann, U., and Spange, S. (2001) *Microporous Mesoporous Mater.*, **44**, 745.
30 Oberhagemann, U., Jeschke, M., and Papp, H. (1999) *Microporous Mesoporous Mater.*, **33**, 165.
31 Schulz-Ekloff, G., Rathousky, J., and Zukal, A. (1999) *Microporous Mesoporous Mater.*, **27**, 273.
32 Pophal, C. and Fuess, H. (1999) *Microporous Mesoporous Mater.*, **33**, 242.
33 Lin, K.F., Pescarmona, P.P., Vandepitte, H., Liang, D., van Tendeloo, G., and Jacobs, P.A. (2008) *J. Catal.*, **254**, 64.
34 Stallmach, F., Kärger, J., Krause, C., Jeschke, M., and Oberhagemann, U. (2000) *J. Am. Chem. Soc.*, **122**, 9237.
35 Fernandes, N.F. and Gavalas, G.R. (1998) *Chem. Eng. Sci.*, **53**, 1049–1058.
36 Kresge, C.T., Leonowicz, M.E., Roth, W.J., Vartuli, J.C., and Beck, J.S. (1992) *Nature*, **359**, 710–712.
37 Boissiere, C., Martines, M.U., Larbot, A., and Prouzet, E. (2005) *J. Membr. Sci.*, **251**, 17–28.
38 Boissiere, C. and Martines, M.U. (2003) *Chem. Mater.*, **15**, 460–463.
39 Reyes, S.C., Sinfelt, J.H., DeMartin, G.J., and Ernst, R.H. (1997) *J. Phys. Chem. B*, **101**, 614–622.
40 Bhatia, S.K. and Nicholson, D. (2011) *Chem. Eng. Sci.*, **66**, 284.
41 Bahtia, S., Jepps, O., and Nicholson, D. (2004) *J. Chem. Phys.*, **120**, 4472–4485.
42 Jepps, O.G., Bhatia, S.K., and Searles, D.J. (2003) *J. Chem. Phys.*, **119**, 1719–1730.
43 Jepps, O.G., Bhatia, S.K., and Searles, D.J. (2003) *Phys. Rev. Lett.*, **91**, 126102.
44 Bhatia, S.K. and Nicholson, D. (2006) *AIChE J.*, **52**, 29–38.
45 Krishna, R. and J.M. van Baten (2009). *Chem. Eng. Sci*, **64**, 870–882.
46 Markovic, A., Stoltenberg, D., Enke, D., Schluender, E.-U., and Seidel-Morgenstern, A. (2009) *J. Membr. Sci.*, **336**, 17–31.
47 Gruener, S. and Huber, P. (2008) *Phys. Rev. Lett.*, **100**, 064502.
48 Ruthven, D.M., DeSisto, W.J., and Higgins, S. (2009) *Chem. Eng. Sci.*, **64**, 3201–3203.
49 Higgins, S., DeSisto, W.J., and Ruthven, D.M. (2009) *Microporous Mesoporous Mater.*, **117**, 268–277.
50 Kapoor, A., Yang, R.T., and Wong, C. (1989) *Catal. Rev. Sci. Eng.*, **31**, 129.
51 Barrer, R.M. (1963) *Appl. Mater. Res.*, **2**, 129–143.
52 Costa, E., Calleja, G., and Comingo, F. (1985) *AIChE J.*, **31**, 982.
53 Schneider, P. and Smith, J.M. (1968) *AIChE J.*, **14**, 762.
54 Rivarola, J.B. and Smith, J.M. (1964) *Ind. Eng. Chem. Fundam.*, **3**, 308.
55 Higashi, K., Ito, H., and Oishi, J. (1963) *J. Atom. Energy Jpn.*, **5**, 846.
56 Hill, T.L. (1960) *Introduction to Statistical Thermodynamics*, Addison-Wesley, Reading MA, p. 80.
57 Okazaki, M., Tamon, H., Hyodo, T., and Toei, R. (1981) *AIChE J.*, **27**, 1035.
58 Gilliland, E.R., Baddour, R.F., Perkinson, G.P., and Sladek, K.J. (1974) *Ind. Eng. Chem. Fundam.*, **13**, 95–100.
59 Okazaki, M., Tamon, H., and Toei, R. (1981) *AIChE J.*, **27**, 262.
60 Tamon, H., Okazaki, M., and Toei, R. (1981) *AIChE J.*, **27**, 271.
61 Seyd. W. (1970) Thesis, Leipzig University.
62 Stach, H. (1977) Thesis (Promotion B) Academy of Sciences of the GDR Berlin.
63 Kärger, J. and Walter, A. (1974) **255**, 142.
64 Kärger, Zikanova, Kocirik (1984) *Z. Phys. Chem...* identical with ref [109] of the present list of references of this chapter.
65 Pfeifer, H. (1972) *NMR Basic Principles and Progress*, **7**, 53.
66 Kärger, J., Rauscher, M., and Torge, H. (1972) *J. Coll. Interf. Sci.*, **76**, 696.
67 Satterfield, C.N., Colton, C.K., and Pitcher, W.H. (1973) *AIChE J.*, **19**, 628.
68 Prasher, B.D. and Ma, Y.H. (1977) *AIChE J.*, **23**, 303.
69 Udin, M.S., Hidajat, K., and Ching, C.B. (1990) *Ind. Eng. Chem. Res.*, **29**, 647.
70 Kärger, J. and Ruthven, D.M. (1991) *Diffusion in Zeolites and Other Microporous*

Solids, 1st edn, John Wiley & Sons, Inc., New York, p. 357.
71 Lee, L. (1991) *Ind. Eng. Chem. Res.*, **30**, 29.
72 Lettat, K., Jolimaitre, E., Tayakout, M., and Tondeur, D., *AIChE J.* (2011) doi: 10.1002/aic.12679.
73 Krishna, R. and van Baten, J.M. (2010) *J. Phys. Chem. C*, **114**, 11557–11563.
74 Naumov, S., Valiullin, R., Kärger, J., Pitchumani, R., and Coppens, M.O. (2008) *Microporous Mesoporous Mater.*, **110**, 37.
75 Valiullin, R. and Furo, I. (2002) *Phys. Rev. E*, **6603**, 1508.
76 Valiullin, R. and Furo, I. (2002) *J. Chem. Phys.*, **116**, 1072.
77 Strange, J.H., Rahman, M., and Smith, E.G. (1993) *Phys. Rev. Lett.*, **71**, 3589.
78 Pitchumani, R., Li, W.J., and Coppens, M.O. (2005) *Catal. Today*, **105**, 618.
79 Vasenkov, S., Böhlmann, W., Galvosas, P., Geier, O., Liu, H., and Kärger, J. (2001) *J. Phys. Chem. B*, **105**, 5922.
80 Heinke, L., Tzoulaki, D., Chmelik, C., Hibbe, F., van Baten, J., Lim, H., Li, J., Krishna, R., and Kärger, J. (2009) *Phys. Rev. Lett.*, **102**, 65901.
81 Menjoge, A., Huang, Q., Nohair, B., Eic, M., Shen, W., Che, R., Kaliaguine, S., and Vasenkov, S. (2010) *J. Phys. Chem. C*, **114**, 16298.
82 Tennison, S.R., Kozynchenko, O.P., Strelko, V.V., and Blackburn, A.J. (2004) Porous carbons, US Pat. 024 074A1.
83 Furtado, F., Galvosas, P., Gonçalvezd, M., Kopinke, F.D., Naumov, S., Rodríguez-Reinoso, F., Roland, U., Valiullin, R., and Kärger, J. (2011) *Microporous Mesoporous Mater.*, **141**, 184.
84 Furtado, F., Galvosas, P., Gonçalvezd, M., Kopinke, F.D., Naumov, S., Rodríguez-Reinoso, F., Roland, U., Valiullin, R., and Kärger, J. (2011) *J. Am. Chem. Soc.*, **133**, 2437.
85 Valiullin, R., Naumov, S., Galvosas, P., Kärger, J., Woo, H.J., Porcheron, F., and Monson, P.A. (2006) *Nature*, **430**, 965.
86 Price, W.S. (2009) *NMR Studies of Translational Motion*, Cambridge University Press, Cambridge.
87 Valiullin, R. and Kärger, J. (2011) *Chem. Ing. Tech.*, **83**, 166.
88 Nguyen, T.X. and Bhatia, S.K. (2008) *Langmuir*, **24**, 146.
89 Fan, W., Snyder, M.A., Kumar, S., Lee, P.-S., Yoo, W.C., McCormick, A.V., Penn, R.L., Stein, A., and Tsapatsis, M. (2008) *Nat. Mater.*, **7**, 984.
90 Choi, M., Cho, H.S., Srivastava, R., Venkatesan, C., Choi, D.H., and Ryoo, R. (2006) *Nat. Mater.*, **5**, 718.
91 Wang, G. and Pinnavaia, T.J. (2006) *Angew. Chem. Int. Ed.*, **45**, 7603.
92 Na, K., Jo, C., Kim, J., Cho, K., Jung, J., Seo, Y., Messinger, R.J., Chmelka, B.F., and Ryoo, R. (2011) *Science*, **333**, 328.
93 Groen, J.C., Peffer, L.A.A., Moulijn, J.A., and Perez-Ramirez, J. (2005) *Chem. Eur. J.*, **11**, 4983.
94 Verboekend, D. and Perez-Ramirez, J. (2011) *Chem. Eur. J.*, **17**, 1137.
95 Groen, J.C., Zhu, W., Brouwer, S., Huynink, S.J., Kapteijn, F., Moulijn, J.A., and Perez-Ramirez, J. (2007) *J. Am. Chem. Soc.*, **129**, 355.
96 Cho, K., Cho, H.S., de Menorval, L.C., and Ryoo, R. (2009) *Chem. Mater.*, **21**, 5664.
97 Liu, Y., Zhang, W., Liu, Z., Xu, S., Wang, Y., Xie, Z., Han, X., and Bao, X. (2008) *J. Phys. Chem. C*, **112**, 15375.
98 Valiullin, R., Kärger, J., Cho, K., Choi, M., and Ryoo, R. (2011) *Microporous Mesoporous Mater.*, **142**, 236.
99 Maxwell, J.C. (1867) *Phil. Trans. R. Soc.*, **49**, 49.
100 Rayleigh, L. (1892). *Phil. Mag*, **34**, 481.
101 Runge, I. (1925) *Z. Tech. Phys.*, **6**, 61.
102 Meredith, R.E. and Tobias, C.W. (1960) *J. Appl. Phys.*, **31**, 1270.
103 Jefferson, T.B., Witzell, O.W., and Sibbett, W.L. (1958) *Ind. Eng. Chem. Fundam.*, **50**, 1589.
104 Cheng, S.C. and Vachon, R.I. (1969) *Int. J. Heat Mass Transf.*, **12**, 249.
105 Jury, S.H. (1977) *Can. J. Chem. Eng.*, **55**, 538.
106 Jury, S.H. (1978) *J. Frankl. Inst.-Eng. Appl. Math.*, **305**, 79.
107 Kärger, J., Kocirik, M., and Zikanova, A. (1981) *J. Colloid Interface Sci.*, **84**, 240.
108 Kärger, J. and Samulevich, N.N. (1978) *Z. Chem.*, **18**, 155.

109 Kärger, J., Zikanova, A., and Kocirik, M. (1984) *Z. Phys. Chem. (Leipzig)*, **265**, 587.

110 Kärger, J. and Pfeifer, H. (1987) *Zeolites*, **7**, 90.

111 Lorenz, P., Bülow, M., and Kärger, J. (1984) *Colloid Surf.*, **11**, 353.

112 Valiullin, R., Kärger, J., and Gläser, R. (2009) *Phys. Chem. Chem. Phys.*, **11**, 2781–2992.

113 Valiullin, R., Kortunov, P., Kärger, J., and Timoshenko, V. (2005) *J. Phys. Chem. B*, **109**, 5746–5752.

114 Naumov, S., Valiullin, R., Monson, P.A., and Kärger, J. (2008) *Langmuir*, **24**, 6429–6432.

115 Dvoyashkin, M., Khokhlov, A., Valiullin, R., and Kärger, J. (2008) *J. Chem. Phys.*, **129**, 154702.

116 Dvoyashkin, M., Valiullin, R., Kärger, J., Einicke, W.-D., and Gläser, R. (2007) *J. Am. Chem. Soc.*, **129**, 10344.

117 Valiullin, R., Naumov, S., Galvosas, P., Kärger, J., Woo, H.-J., Porcheron, F., and Monson, P.A. (2006) *Nature*, **443**, 965.

118 Dvoyashkin, M., Khokhlov, A., Valiullin, R., and Kärger, J. (2008) *J. Chem. Phys.*, **129**, 154702.

119 Valiullin, R., Dvoyashkin, M., Kortunov, P., Krause, C., and Kärger, J. (2007) *J. Chem. Phys.*, **126**, 54705.

16
Eight-Ring Zeolites

The eight-ring zeolites constitute an important class of "small-pore" zeolites, including the families LTA, CHA, ERI, and DDR. These materials, which have effective pore sizes in the 3.5–4.5 Å range, have found widespread application as size-selective adsorbents. Their importance has increased in recent years with the synthesis of a wide range of chabazite analogs that have small differences in their window dimensions, thus enabling the molecular sieve properties to be adjusted in a controlled way. For all but the smallest molecules diffusion is largely controlled by the eight-ring windows, so these materials provide a unique opportunity to study the effects of steric hindrance in pores of a well-defined and uniform size. In this chapter we focus on LTA, CHA, and DDR, since these materials have been studied in greatest detail. These structures all consist of distinct cages connected through eight-membered oxygen rings that control the jump rate between neighboring cages and hence the intracrystalline diffusivity. Propagation rates through eight-ring windows are also considered in Section 18.9, which deals with diffusion in the zeolite ferrierite, as well as in Section 12.3, where ferrierite is considered as an ideal model system for illustrating the potentials of micro-imaging. Although the pore system in ferrierite consists of a network of both eight- and ten-ring channels, the entrances to the ten-ring channels are often blocked, so that adsorption and desorption occur only through the eight-ring channels.

16.1
Eight-Ring Zeolite Structures

Figures 16.1–16.4 show schematically the structures of LTA, CHA, and DDR (ZSM-58). To translate these schematic diagrams into the actual crystal structure one has to consider that the lines represent the diameters of oxygen ions, while at each of the vertices there is either a silicon or aluminum. Since the Si and Al atoms are small, they are effectively buried in the spaces between the larger oxygen ions, so the internal surface appears to consist almost entirely of oxygen together with the exchangeable cations, the locations of which are considered below.

Diffusion in Nanoporous Materials. Jörg Kärger, Douglas M. Ruthven, and Doros N. Theodorou.
© 2012 Wiley-VCH Verlag GmbH & Co. KGaA. Published 2012 by Wiley-VCH Verlag GmbH & Co. KGaA.

16.1.1
LTA Structure [1–3]

The pore system of zeolite A consists of a cubic array of relatively large (~11.3 Å free diameter) cages (internal free volume ≈ 760 Å^3) interconnected through eight-membered oxygen windows of free diameter 4.3 Å (although this may be reduced by the presence of an exchangeable cation). Each of the large cages, therefore, has access to six similar adjacent cages. At each of the eight corners of the cage there are six-membered oxygen rings giving access to the smaller "sodalite cage." Since the six-ring can be easily penetrated only by water and other very small molecules, the sodalite cages make no contribution to the adsorption capacity for most adsorbates. The number and location of the exchangeable cations have an important effect on the diffusional behavior. There are three types of cation sites, listed in order of their energy for a small monovalent cation in Figure 16.1.

In zeolite A the Si/Al ratio is always close to unity and as a result there are 12 monovalent (or six divalent) cations per pseudo cell (i.e., per large cage). Silica-rich analogs of the A structure such as ZK-4 and ITQ-29 contain proportionately fewer exchangeable cations. If there are less than eight cations per cell (e.g., 4Na$^+$ + 4Ca^{2+}) all cations can be accommodated in the energetically favorable type I sites. In this ideal situation all windows should be free of cations and we have an "open" sieve. At the other extreme, if we have more than eleven cations per pseudo cell (e.g., 10 Na$^+$ + 1 Ca^{2+}) all windows will be obstructed, since only eight of the cations can be accommodated in the type I sites and the other three will fully occupy the type II sites. In the fully exchanged Na$^+$ form the additional Na$^+$ cation occupies a type III site. The "open" form of this structure with more than 67%

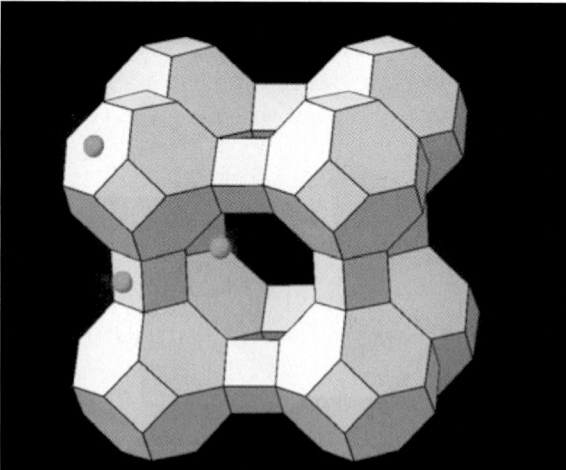

Figure 16.1 Structure of zeolite A with cation sites indicated by small spheres. I: at the center of the six rings (eight equivalent sites per pseudo cell); II: within the eight-ring windows (three equivalent sites per pseudo cell); and III: against a four-ring.

equivalent Ca^{2+} is known as "5A" while the "closed" sieve, which has a significantly smaller effective pore size as a consequence of the obstruction of the windows by Na^+ cations, is known as 4A. If Na^+ is replaced by K^+, the cation distribution remains the same but the effective pore size is reduced further (3Å) since the obstructing cation is larger.

These cation distributions are idealized and are not always attained in actual samples. The difference in energy between the type I and II sites is quite small and there is some evidence that at high temperatures the type II sites may actually become more favorable than the type I sites [4]. The relative energies can also be altered by the presence of water or other sorbates. The actual cation distribution in a zeolite A crystal may therefore depend to a considerable extent on the conditions of pretreatment. There is some evidence that an increase in the dehydration temperature leads to increasing occupation of the SII sites by Ca^{2+} cations [5]. 5A crystals dehydrated thoroughly at high temperature and cooled slowly have approximately the ideal cation distribution, whereas rapid quenching can lead to a partially obstructed sieve in which some of the type II sites are occupied and some of the type I sites are vacant.

16.1.2
Cation Hydration

Water molecules can coordinate very strongly to the exchangeable cations, particularly those in sites II and III of the A structure. Thus, in an incompletely dehydrated sample of 4A sieve, some of the windows may be blocked by Na^+-H_2O complexes. In CaA and NaCaA (5A) the situation is more complicated, as there is also the possibility of cation hydration reactions:

$$Ca^{2+} + 2H_2O \rightleftharpoons Ca(OH)^+ + H_3O^+ \tag{16.1}$$

as has been demonstrated by Coe et al. for CaX zeolite [6, 7]: This results in the replacement of one divalent cation by two monovalent cations, with the result that more of the type II (window) sites may be occupied than would be expected on the basis of the idealized cation distribution. Furthermore, the hydronium ion so created can react irreversibly with the zeolite framework, removing Al, forming a hydroxyl nest and eventually forming amorphous alumina:

$$\begin{array}{c}
| \\
Si \\
| \\
O \\
|H_3O^+ \\
-Si-O-Al-O-Si- \\
| \\
O \\
| \\
Si \\
|
\end{array} + 2H_2O \rightarrow \begin{array}{cc}
| & | \\
Si & Si \\
| & | \\
O & O \\
 & H \\
-Si-OH & HO-Si- \\
H & \downarrow \\
O & O \\
| & | \\
Si & Si \\
| & |
\end{array} + AlO_2^- \cdot H_3O^+ \tag{16.2}$$

$$\downarrow$$
$$Al(OH)_3$$
$$\downarrow$$
$$0.5Al_2O_3 + 1.5H_2O$$

The combined result of these reactions is to eliminate one of the SI sites as a hydroxyl nest and to transfer one of the Ca^{2+} cations in hydroxylated form as $Ca(OH)^+$ to an SII site (which is occupied preferentially by the larger cation). One important consequence of these effects is that, unless careful attention is paid to the conditions of dehydration, reproducibility may be difficult to achieve in experimental kinetic studies. Some of the effects of sample pretreatment can be understood on this basis. However, direct experimental proof is difficult to achieve, since measurements of cation distribution by X-ray methods are not highly precise except with large crystals, and subtle changes in the distribution are difficult to detect by powder methods.

16.1.3
Structure of CHA

The chabazite structure (Figure 16.2) [1–3] also consists of cages interconnected through eight-ring windows but the cages are substantially smaller (~380 Å^3) than the cage of zeolite A and the windows are significantly distorted from the symmetric octagonal form. In the Al-rich forms the windows are partially obstructed by the exchangeable cations, but the cation distribution has not been studied in as much detail as for zeolite A. In recent years the cation-free forms AlPO-34, SAPO-34, and SiCHA have attracted much attention. The bond lengths (Å) are: Al–O = 1.75, Si–O = 1.61, P–O = 1.53, and (Al–O + Si–O)/2 = 1.64. As a result the unit cell volume and the window dimensions can be adjusted by changing the Si/Al ratio and/or introducing phosphorus into the lattice to form the SAPO-34 or AlPO-34 analogs of the CHA structure. Figure 16.3 gives some representative examples, showing the window dimensions.

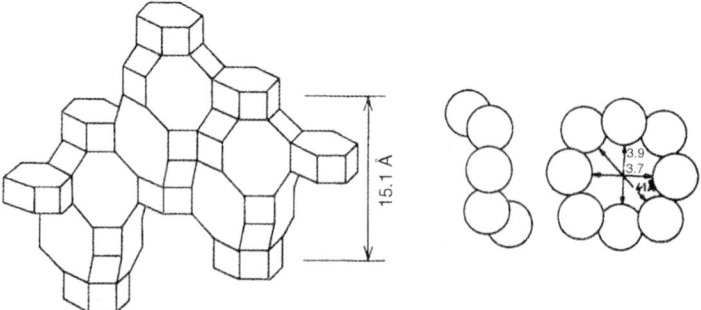

Figure 16.2 Structure of chabazite.

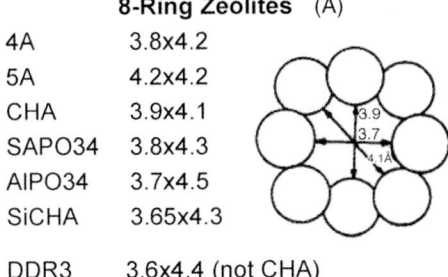

8-Ring Zeolites	(Å)
4A	3.8x4.2
5A	4.2x4.2
CHA	3.9x4.1
SAPO34	3.8x4.3
AlPO34	3.7x4.5
SiCHA	3.65x4.3
DDR3	3.6x4.4 (not CHA)

Figure 16.3 Window dimensions of some eight-ring zeolites as measured by synchrotron XRD [8].

16.1.3.1 Structure of DDR (ZSM-58) [9–11]

DDR is formally classified as a "clathrasil" rather than a zeolite. The unit cell is large, consisting of 120 tetrahedral SiO_2 units, so the structure is difficult to visualize (Figure 16.4a). The SiO_2 tetrahedra are arranged into nine pentagonal dodecahedral cages [5^{12}] and six dodecahedral cages [$4^3 5^6 6^1$] (Figure 16.4b). These secondary building units are interconnected in such a way that they form six large 19-hedral cavities [$4^3 5^{12} 6^1 8^3$] (top left of Figure 16.4b) within the unit cell. These 19-hedral cavities are interconnected through distorted eight-ring windows and form a two-dimensional pore system between the dodecahedral layers.

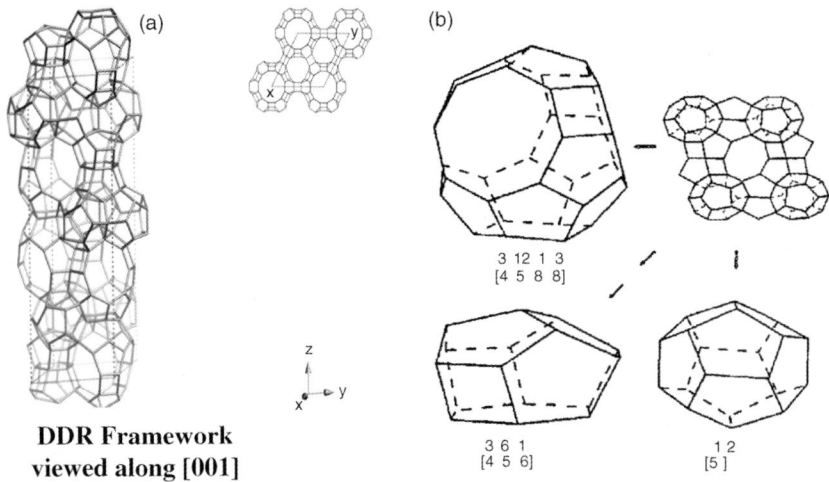

DDR Framework viewed along [001]

Figure 16.4 Structure of DDR (a) and the fundamental cages from which it is built (b). (The designation DD3R denotes the pure silica form.) Taken from the International Zeolite Association web site (http://izasc.ethz.ch/fmi/xsl/IZA-SC/ftc_fw.xsl?-db=Atlas_main&-lay=fw&-max=25&STC=DDR&-find) and Gies [9] with permission.

16.1.4
Concentration and Temperature Dependence of Diffusivity

Transport diffusivities in these structures generally increase monotonically with loading but the concentration dependence can essentially be accounted for, within experimental error, by the thermodynamic factor. Corrected diffusivities (D_0) defined in accordance with Eq. (1.31) are often found to be independent of loading, at least over the experimentally accessible range. Figure 16.5 shows an example of this pattern of behavior (for n-heptane in 5A zeolite). Such behavior is to be expected for diffusion under conditions of strong steric hindrance where molecule–pore wall interactions are dominant and molecule–molecule interactions are comparatively small. In Section 4.3.3, the origin and the limitations of this correlation are shown to be related to the application of the transition state theory to molecular propagation. It thus appears that diffusion occurs by an activated

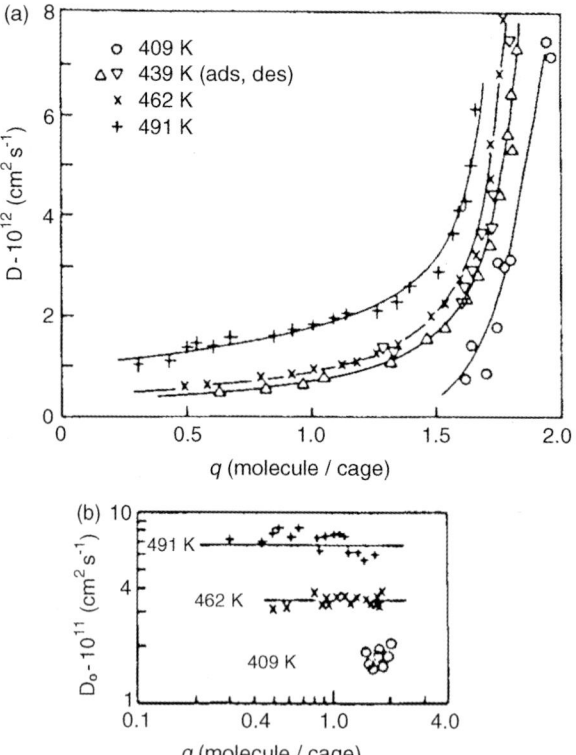

Figure 16.5 Variation of (a) diffusivity and (b) corrected diffusivity with sorbate concentration for diffusion of n-heptane in Linde 5A zeolite crystals. Reprinted from Doetsch et al. [12] with permission.

process, with the temperature dependence of the corrected diffusivity governed by an Arrhenius expression:

$$D_0 = D_\infty e^{-E/RT} \tag{16.3}$$

with activation energies typically of several kJ mol^{-1}. Both the diffusivities and the activation energies generally correlate well with the molecular diameter/window aperture, as is to be expected if the activation energy is determined mainly by the repulsive interaction energy for a molecule in the window, which corresponds to the transition state.

16.2
Diffusion in Cation-Free Eight-Ring Structures

16.2.1
Effect of Window Dimensions

The effect of window size on diffusivity in LTA structures has been clearly demonstrated by Hedin *et al.* [13], who studied the diffusion of propylene in CaA, ITQ-29 (the all-silica analog of LTA), and several partially exchanged NaCaA samples. In both 5A and ITQ-29 the eight-ring is unobstructed by cations but, due to the difference in bond lengths, the eight-ring in ITQ-29 is slightly smaller, leading to a diffusivity that is reduced by about an order of magnitude.

Diffusion in the aluminum-rich type A zeolites is complicated by the presence of the exchangeable cations. It is therefore not surprising that the diffusional behavior of the cation-free eight-ring structures shows greater reproducibility (even though some of these materials, notably pure-silica LTA [14, 15], are found to deviate substantially from ideal crystallinity). Furthermore, where measurements have been made by both microscopic and macroscopic methods, the results are generally consistent. Although these materials have been studied much less intensively than the type A zeolites, sufficient experimental data have now been accumulated to allow the pattern of variation of the diffusivities with the size of the sorbate molecules and the dimensions of the zeolite windows to be examined in detail.

Because of the small pore size the activation energy is determined mainly by the repulsive interactions in the transition state. As a result, the diffusional activation energy and the diffusivity correlate closely with the size of the diffusing molecule and the window aperture. Detailed examination of the data suggests that the minimum molecular diameter (rather than the kinetic diameter or the Lennard-Jones parameter σ) and the minimum free diameter of the window are the critical dimensions. Figure 16.6 shows the correlation of diffusivity with the minimum window diameter, for representative light molecules in several

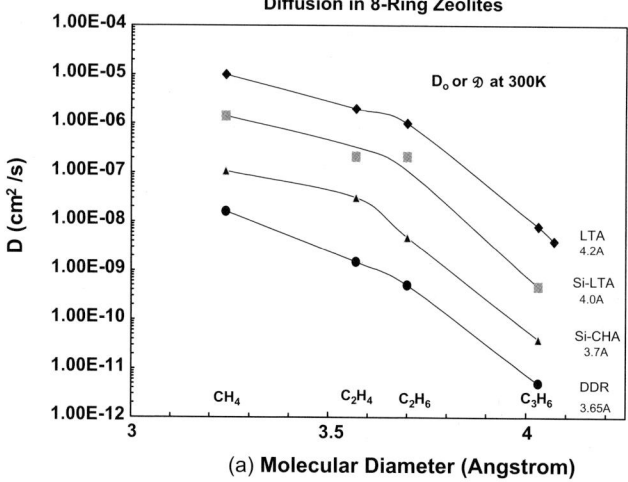

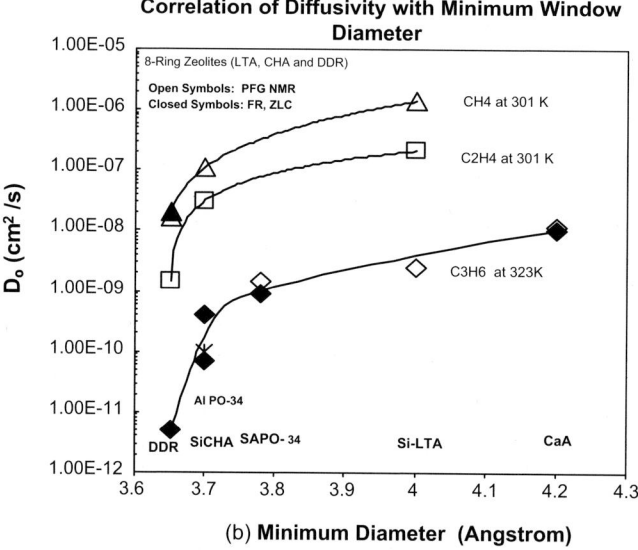

Figure 16.6 Variation of diffusivity with molecule and window dimensions for cation-free eight-ring zeolites: (a) self-diffusivity ($\mathcal{D}$) at 301 K plotted against minimum molecular diameter for several different eight-ring zeolites; (b) diffusivities ($\mathcal{D}$ and D_o) plotted against minimum window diameter for three different molecules in several different eight-ring cation-free zeolite structures. Data are from various sources [8, 13, 16–18].

different cation-free eight-ring structures while the variation of activation energy with minimum molecular diameter is shown in Figure 16.7 for several small molecules in DDR.

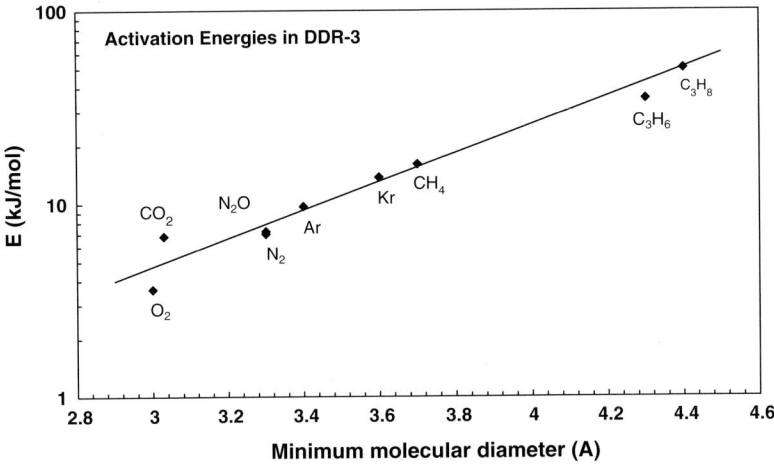

Figure 16.7 Variation of diffusional activation energy with (minimum) molecular diameter for small molecules in DDR. Data are from van den Bergh et al. [19] and Olson [17].

16.2.2
Diffusion of CO_2 and CH_4 in DDR

Because of the potential application of DDR as a selective adsorbent for CO_2–CH_4 separation the diffusion of these species in DDR has been studied in some detail, both experimentally and by molecular simulation. These studies have revealed some interesting features. Perhaps the most striking result is that it is not possible to replicate the experimental diffusivities (for both CO_2 and CH_4) by molecular simulation using the standard 6–12 force field. The simulation values so obtained over-predict the diffusivities of both species by more than an order of magnitude. It appears that the short-range repulsive interactions are significantly stronger than is predicted by the standard force field. Sholl and Jee [20] have shown that a good prediction of the experimental diffusivities can be obtained by using a 6–18 potential (Table 16.1). The equilibrium isotherm is not very sensitive to the precise form of the

Table 16.1 Comparison of experimental D_o values ($m^2 s^{-1}$) for CO_2 and CH_4 in DDR-3 at low loading at 298 K with values from molecular simulation.

	Experiment[a]	MD simulation Krishna [21, 22]	MD simulation Sholl [20]
CO_2	1.0×10^{-10}	1.6×10^{-9}	1.4×10^{-10}
CH_4	1.7×10^{-12}	8×10^{-11}	2×10^{-12}

a) Experimental values based on FR and membrane measurements were kindly provided by Dr Ron Chance (ExxonMobil).

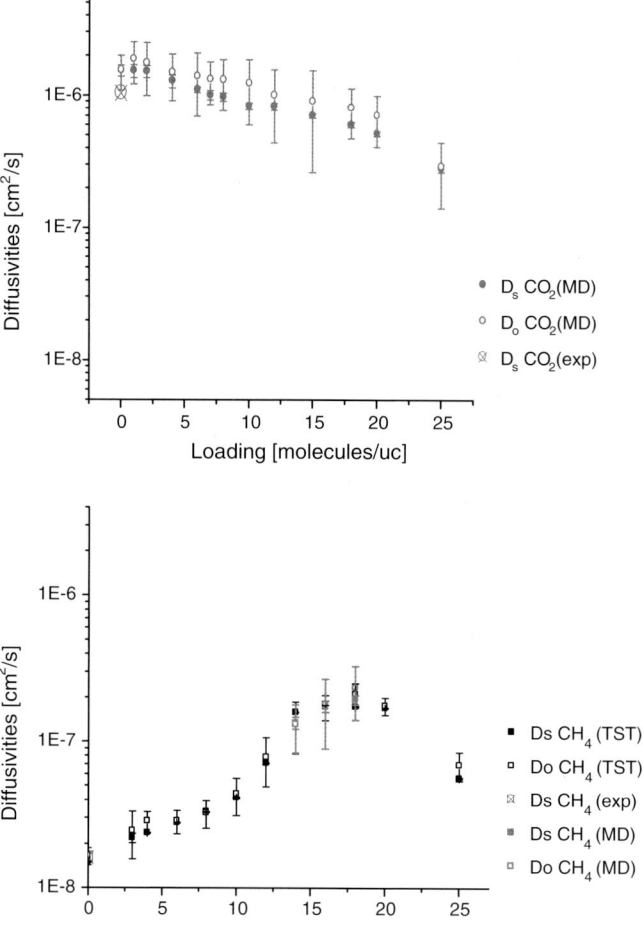

Figure 16.8 Diffusion of CO_2 and CH_4 (as single components) in DDR-3 at 298 K, showing trends with loading and comparison with experimental data at low loading. From Jee and Sholl [20] with permission.

repulsive interaction term so an equally good fit of the isotherm can be obtained with either the 6–12 or 6–18 potentials.

Figure 16.8 shows the trends of the diffusivity with loading. For both CO_2 and CH_4 the self- and corrected diffusivities are almost equal. For CO_2 the diffusivity declines slightly with loading whereas CH_4 shows the opposite trend at low loadings, passing through a maximum at high loadings.

The diffusion of CO_2–CH_4 mixtures in DDR-3 has also been studied by molecular simulation, although not experimentally. The results are shown in Figure 16.9. Even at low loadings the diffusivity of methane is significantly reduced by the presence of

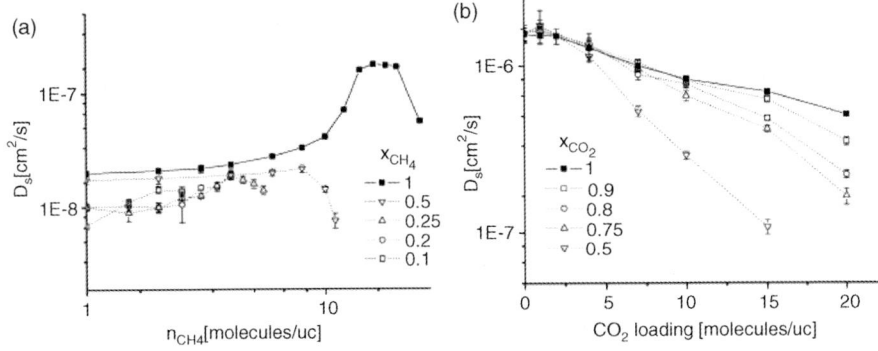

Figure 16.9 Diffusion in DDR-3 at 298 K: (a) CH_4 in presence of CO_2 (b) CO_2 in presence of CH_4. Reprinted from Jee and Sholl [20] with permission.

CO_2, at relatively small mole fractions. In contrast the effect of methane on the diffusion of CO_2 is minimal, except when the methane concentration is high. This behavior has been attributed to the preferential occupation of the eight-rings by CO_2, which leads to a reduction in the mobility of methane but has little effect on the mobility of CO_2. This has the important practical effect that it increases the perm-selectivity for CO_2 in the binary system beyond the single-component permeance ratio (Section 20.6.2).

16.3
Diffusion in 4A Zeolite

For all but the smallest molecules, diffusion in 4A zeolite is relatively slow. Sorption rates are therefore controlled mainly by intracrystalline diffusion and are relatively easily measured by classical methods. Even for methane the diffusivities in 4A are too small to allow reliable measurement by PFG NMR (pulsed field gradient nuclear magnetic resonance). Most of the available diffusivity data have, therefore, been derived from sorption rate measurements and in general there is reasonably good agreement between the data derived from macroscopic and microscopic measurements. For example, the corrected diffusivity values for methane (in unaggregated crystals) derived from sorption rate measurements are quite close to the self-diffusivity values estimated from both NMR relaxation studies [23, 24] and neutron scattering measurements [25].

Confirmation that the uptake rates in these systems are controlled by intracrystalline diffusion has been obtained from measurements with different crystal size fractions. Figure 16.10 shows representative uptake curves (for N_2–4A). The curves for the three different crystal sizes all conform to the diffusion model and the time constants show the expected variation with the square of the crystal radius so that consistent diffusivity values are obtained from all three curves.

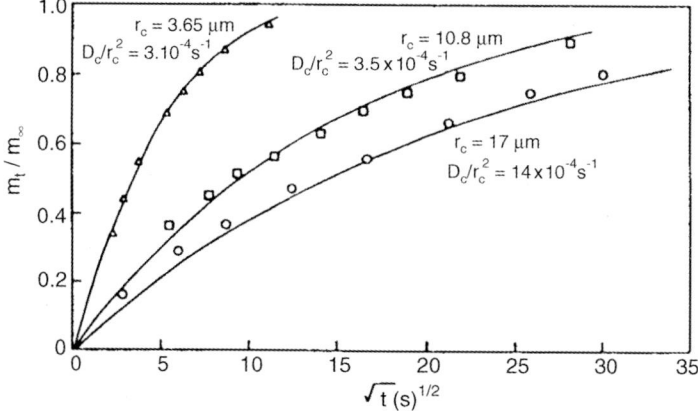

Figure 16.10 Experimental uptake curves for N_2 in three different size fractions of 4A zeolite crystals at 273 K. The different values of D_c/r_c^2 all yield essentially the same value for D_c ($= 4 \times 10^{-10}$ cm^2 s^{-1}). Reprinted from Yucel and Ruthven [26] with permission.

For less strongly adsorbed species, measurements have been made mainly at relatively low sorbate concentrations, within the Henry's law region, so no strong concentration dependence is seen. However, for more strongly adsorbed species such as C_2H_6 [26] and CO_2 [27] data were obtained over a much wider range of sorbate concentrations. These data conform to the commonly observed pattern: the diffusivity increases strongly with sorbate concentration but the corrected diffusivity is essentially constant and coincides (within the experimental uncertainty) with the tracer (or self-) diffusivity (Figure 16.11).

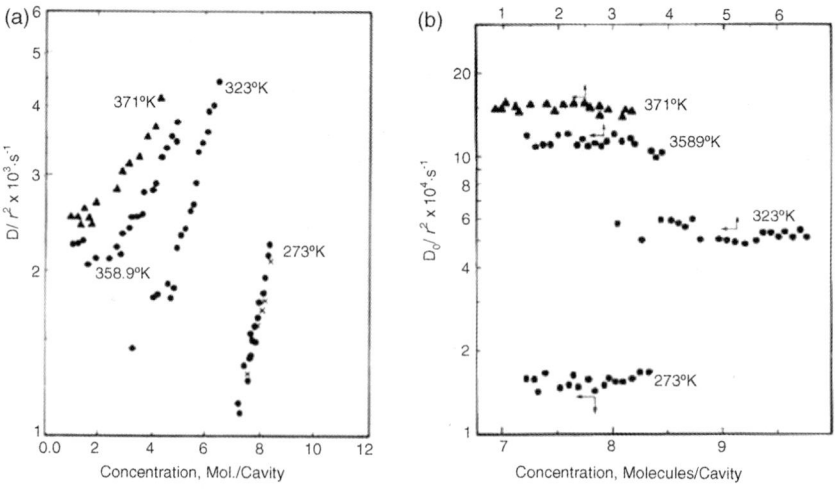

Figure 16.11 Variation of (a) diffusivity (D/r^2) and (b) corrected diffusivity (D_o/r^2) with loading for CO_2–4A at various temperatures. Reprinted from Yucel and Ruthven [27] with permission.

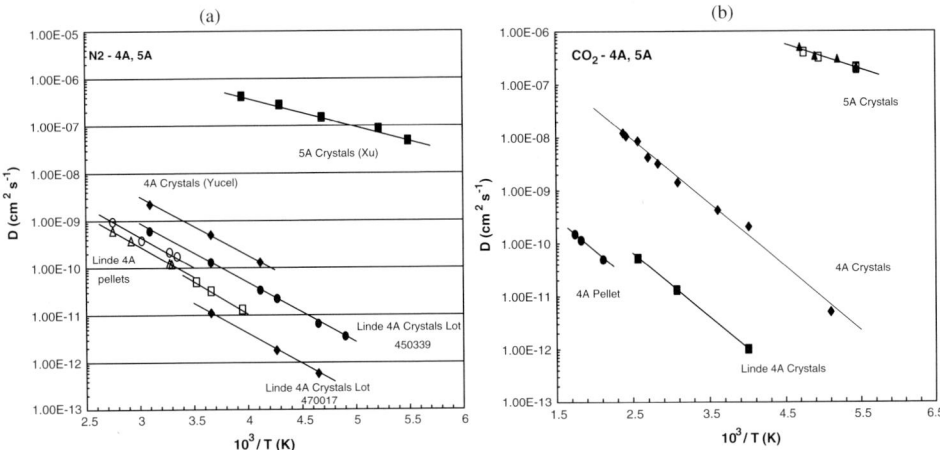

Figure 16.12 Arrhenius plots showing temperature dependence of corrected diffusivity (and self-diffusivity) for (a) N_2 and (b) CO_2 in various different samples of 4A zeolite (Data of Yucel [26, 27], Derrah [29], Kumar [33], van de Voorde [47], and Cao [46]). ZLC (zero-length column) data of Xu [48] (▲) and PFG NMR data of Kärger [49] (□) for CO_2–5A (Yucel crystals) are also shown. See also Ruthven [50].

The temperature dependence of the corrected diffusivity follows the usual Arrhenius relationship [Eq. (16.3)], as illustrated in Figure 16.12 [28]. Table 16.2 gives a more extensive summary of reported experimental data.

It is worth noting that the tracer (self-)diffusivity measurements of Cao et al. [46] agree well with earlier corrected diffusivities for similar zeolite samples. When comparing different 4A zeolite samples we often see large differences in diffusivity with very little difference in activation energy (Figure 16.12). In general, zeolite samples prepared and dehydrated in the laboratory show consistently higher diffusivity values than those obtained for commercial pellets and crystals which have been subjected to severe hydrothermal pretreatment (see also Figure 16.13). There are two plausible explanations for this behavior: blocking of a certain fraction of the windows by any of the mechanisms discussed in Section 16.1, resulting in a lower intracrystalline diffusivity, or creation of a surface barrier. Experimental evidence for and against these alternatives is considered in Section 16.5.

16.4
Diffusion in 5A Zeolite

Because of the wider window aperture (and consequently smaller activation energy) diffusion in 5A zeolite is very much faster than in 4A and, as a result, NMR methods are more widely applicable. Reliable diffusivity data for lower molecular weight sorbates can be obtained from sorption rate measurements only if relatively large crystals are used. Many of the earlier data obtained with small commercial crystals

Table 16.2 Summary of diffusivity of data for 4A zeolite.[a]

Sorbate	Sorbent	Technique	T (K)	$\mathcal{D}$ or D_o (m^2 s^{-1})	E (kJ mol^{-1})	Reference
Ar	Linde pellets	Gravimetric uptake	273	3×10^{-15}	24	Derrah and Ruthven [29]
	Linde pellets	Chromatography	273	3×10^{-15}	24	Sarma and Haynes [30] Shah and Ruthven [31]
Kr	Linde pellets	Gravimetric uptake	273	3×10^{-18}	34	Derrah and Ruthven [29]
	UOP pellets	Tracer exchange ($\mathcal{D}$)	273	5×10^{-18}	32.2	Cao [46]
O$_2$	Linde crystals	Frequency response	273	1.7×10^{-12}	—	Yasuda and Matsumoto [32]
	Linde crystals	Gravimetric uptake	273	1.6×10^{-12}	19	Derrah and Ruthven [29]
N$_2$	Linde crystals	Frequency response	273	3.7×10^{-14}	—	Yasuda and Matsumoto [32]
	Own crystals (7.3–34 μm)	Gravimetric uptake	273	4×10^{-14}	23.5	Yucel and Ruthven [26]
	Linde crystals	Gravimetric uptake	273	1.2×10^{-14}	24	Yucel and Ruthven [26]
	Linde pellets	Chromatography	273	$\sim 10^{-14}$	25	Ruthven and Kumar [33]
	Linde pellets	Chromatography	273	$\sim 10^{-14}$	19	Haq and Ruthven [34]
	Linde pellets	Gravimetric uptake	273	$\sim 10^{-15}$	25	Derrah and Ruthven [29]
	UOP pellets	Tracer exch. ($\mathcal{D}$)	273	3×10^{-15}	22.2	Cao [46]
	UOP pellets	Uptake rate	273	2.5×10^{-15}	22.3	van de Voorde [47]
CH$_4$	Linde crystals	NMR relaxation	300	5×10^{-16}	—	Freude [23]
	Lab. cynth. crystals	NMR relaxation	300	3.8×10^{-15}	21	Allonneau and Volino [35]
	Linde crystals	Neutron diffraction	350	$<<10^{-11}$	—	Cohen de Lara et al. [25]
	Own crystals (7.3–34 μm)	Gravimetric uptake	300	5×10^{-15}	24	Yucel and Ruthven [26]
	Linde pellets	Chromatography	300	3.7×10^{-15}	—	van Voorde et al. [47]
	Linde pellets	Gravimetric uptake	300	1.5×10^{-15}	—	Yang and Yeh [36]
	Linde pellets	Chromatography	300	1.2×10^{-15}	25	Kumar and Ruthven [33]
	Linde pellets	Chromatography	300	1.5×10^{-15}	19	Haq and Ruthven [34]
	UOP pellets	Tracer exch. ($\mathcal{D}$)	273	1.2×10^{-16}	26.8	Cao [46]

16.4 Diffusion in 5A Zeolite

Sorbate	Sample	Method	T (K)	D	E	Reference
CO_2	Own crystals (7.3–34 μm)	Gravimetric uptake	500	3×10^{-12}	23	Yucel and Ruthven [27]
	Linde crystals	Gravimetric uptake	500	2×10^{-14}	20	Yucel and Ruthven [27]
	Linde pellets	Chromatography	500	10^{-14}	20.5	Haq and Ruthven [34]
C_2H_6	Own crystals (7.3–40 μm)	Gravimetric uptake	323	1.4×10^{-15}	34.5	Yucel and Ruthven [26]
	Davison 4 A (0.7 μm)	Volumetric	323	1.2×10^{-15}	31	Brandt and Rudloff [37]
	Linde 4A (2.8 μm)	Gravimetric uptake	323	1.0×10^{-15}	24	Kondis and Dranoff [38]
	Linde 4A (4.1 μm)	Gravimetric uptake	323	6×10^{-16}	30	Eagan and Anderson [39] Yucel [27]
	Linde 4A (2.8 μm)	Gravimetric uptake	323	2×10^{-16}	—	Yang and Yeh [36]
	Linde 4A (pellet)	Gravimetric uptake	323	1.5×10^{-16}	22	Kondis and Dranoff [40]
	Linde 4A (pellet)	Volumetric	323	8×10^{-17}	23.5	Taylor [41]
	Linde 4A (3.2 μm)	Gravimetric uptake	323	2.2×10^{-17}	26.5	Yucel and Ruthven [26]
nC_4H_{10}	Linde 4A (~2 μm)	Gravimetric uptake	323	3×10^{-19}	40	Ruthven [42]
	Linde 4A (~1 μm)	Volumetric	323	8×10^{-19}	35.5	Walker et al. [43]
nC_6H_{14}	Linde 4A (~3.5 μm)	Chromatography	323	5×10^{-22}	50	Tobin [44]
	Linde 4A (~2 μm)	Gravimetric uptake	323	1.5×10^{-22}	49	Recalculated from data of Barrer and Clarke [45]
nC_9H_{20}	Linde 4A (~2 μm)	Gravimetric uptake	373	8×10^{-23}	68	Recalculated from data of Barrer and Clarke [45]

a) Corrected diffusivities at low loading (D_0) unless noted as $\mathcal{D}$. Reference temperatures are selected to allow comparisons between different data sets. In some cases there is a modest extrapolation beyond the experimental range.

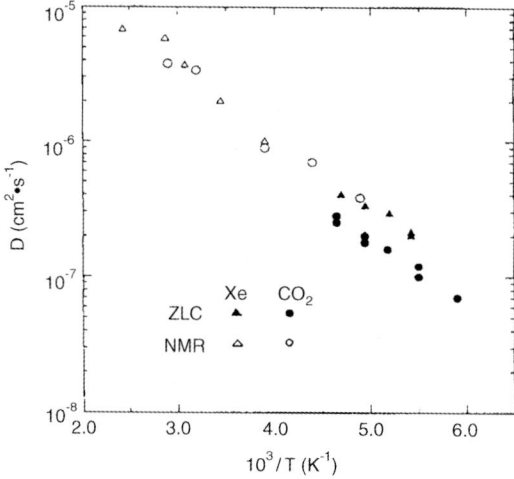

Figure 16.13 Comparison of PFG NMR self-diffusivities and limiting transport diffusivities ($\mathcal{D}_0$) measured by ZLC on 50 μm laboratory synthesized crystals of 5A. Reprinted from Ruthven [51] with permssion.

and with pelleted material are therefore probably in error as a result of unrecognized intrusion of extracrystalline resistances to heat or mass transfer.

However, if only the larger laboratory-synthesized crystals are considered, the agreement between data obtained in different laboratories and by different experimental methods becomes much closer. An example, showing the agreement between PFG NMR and ZLC measurements for CO_2 and Xe in large (50 μm) crystals of 5A zeolite is given in Figure 16.13. Further examples may also be found in Figure 16.17 below and in the data summarized in Table 16.3.

16.4.1
Uptake Rate Measurements

Figure 16.14 shows representative gravimetric uptake curves for n-butane in commercial Linde 5A (3.6 μm) crystals (curves 2–4) and in a sample of large (55 μm) crystals synthesized by Charnell's method (curve 5). All the curves were measured at 323 K over approximately the same pressure step (27–20 Torr). Despite the difference in the crystal sizes the rates are similar, showing the very large difference in diffusivities between these samples. Curve 2 is for a fresh sample of Linde crystals after initial activation at 400 °C, while curves 3 and 4 were obtained with aged samples that had been regenerated several times. In all cases the initial rate appears to be diffusion controlled (uptake directly proportional to $\sqrt{t}$) but curve 2 (the fastest) shows a significant deviation from the isothermal diffusion model at high fractional uptakes. This is probably due to the intrusion of heat transfer resistance, which becomes a serious issue in uptake rate measurements for fast systems (Section 6.2.4) [67, 68]. An uptake curve, measured under similar conditions in the same apparatus, for n-butane in NaX zeolite (for which the equilibrium is similar to 5A but the kinetics are much faster) is shown for comparison (curve 1).

Table 16.3 Summary of diffusivity data for 5A zeolite.[a]

Sorbate	Sorbent	Technique	T (K)	$\mathcal{D}$ or D_o (m^2 s^{-1})	E (kJ mol^{-1})	Reference
Xe	Own crystals (55 μm)	ZLC	200	2×10^{-11}	10	Xu [48]
	Lab. synthesized crystals	PFG NMR	293	1.5×10^{-9}	—	Heink et al. [52]
CO	Lab. synthesized crystals	PFG NMR	293	$\sim 10^{-9}$	—	Stallmach [53]
N$_2$	Own crystals (27.5 and 55 μm)	ZLC	250	5×10^{-11}	9.5	Xu [48]
CH$_4$	Linde crystals	Neutron diffraction	300	6×10^{-10}	—	Cohen de Lara et al. [25]
	Lab. synthesized crystals	PFG NMR	300	$\sim 10^{-9}$	4	Caro et al. [54]
CF$_4$	Lab. synthesized crystals	PFG NMR	473	$<3 \times 10^{-12}$	—	Kärger et al. [55]
	Own crystals (27 and 55 μm)	Gravimetric uptake	473	1.7×10^{-12}	27.5	Yucel and Ruthven [56]
	Linde crystals (lot 550045)	Gravimetric uptake	473	2×10^{-14}	38.5	Ruthven and Derrah [57]
C$_2$H$_6$	Own crystals (27.5 and 55 μm)	ZLC	250	3×10^{-11}	6.7	Xu [48]
	Lab. synthesized crystals	PFG NMR	250	6×10^{-11}	6.0	Caro et al. [54]
C$_3$H$_8$	Lab. synthesized crystals	PFG NMR	273	2×10^{-13}	—	Kärger and Ruthven [58]
	Own crystals (55 μm)	Gravimetric uptake	273	2×10^{-13}	14.5	Kärger and Ruthven [58]
	Linde crystals	Frequency response	273	2×10^{-13} and 10^{-17}	—	Yasuda and Yamaoto [59]
nC$_4$H$_{10}$	Lab. synthesized crystals	PFG NMR	400	8×10^{-13}	17	Kärger and Ruthven [58]
	Lab. synthesized crystals	NMR relaxation	400	10^{-11}	12	Michel and Rössiger [60, 61]
	Own crystals	Gravimetric uptake	400	9×10^{-13}	17	Yucel and Ruthven [56]
	Linde crystals (lot 550045)	Gravimetric uptake	400	5×10^{-15}	17	Ruthven et al. [62]
	Own crystals	ZLC	400	6×10^{-13}	17	Eic and Ruthven [63]
	Linde crystals (lot 550045)	ZLC	400	5×10^{-15}	17	Eic and Ruthven [63]
	Linde crystals (3.0 μm)	Chromatography	400	9×10^{-15}	23	Chiang et al. [64]
nC$_7$H$_{16}$	Lab. synthesized crystals (15.5 and 7.3 μm)	ZLC	473	7.4×10^{-14}	31.5	Eic and Ruthven [63]
	Linde crystals (3.7 μm)	ZLC	473	4.3×10^{-15}	33.5	Eic and Ruthven [63]

(Continued)

Table 16.3 (Continued)

Sorbate	Sorbent	Technique	T (K)	$\mathcal{D}$ or D_o (m² s⁻¹)	E (kJ mol⁻¹)	Reference
	Linde crystals (3.6 μm)	Gravimetric	473	4.3×10^{-15}	31.5	Doetsch [12], Eic and Ruthven [63]
		ZLC	473	1.1×10^{-15}	32	
	Own crystals	ZLC	400	6×10^{-13}	17	Eic and Ruthven [63]
	Linde crystals (lot 550045)	ZLC	400	5×10^{-15}	17	Eic and Ruthven [63]
	Linde crystals (3.0 μm)	Chromatography	400	9×10^{-15}	23	Chiang et al. [64]
$nC_{10}H_{22}$	Lab. synthesized crystals (15.5 μm)	ZLC	593[c]	8×10^{-13}	48	Eic and Ruthven [63]
	Lab. synthesized crystals (12–32 μm)	Piezometric	593	10^{-14}	—	Bülow et al. [65]
	Linde crystals (3.6 μm)	Gravimetric	593	1.6×10^{-14}	56	Vavlitis et al. [66]
	Linde crystals (3.6 μm)	ZLC	593	$\sim 10^{-14}$	49.5	Eic and Ruthven [63]
$nC_{13}H_{28}$	Lab. synthesized crystals (15.5 and 7.3 μm)	ZLC	573	5×10^{-14}	50	Eic and Ruthven [63]
$nC_{16}H_{34}$	Lab. synthesized crystals (15.5 μm)	ZLC	573	8×10^{-14}	54.5	Eic and Ruthven [63]

a) Values of $\mathcal{D}$ refer to low sorbate concentrations ($\Theta < 0.2$).

16.4 Diffusion in 5A Zeolite

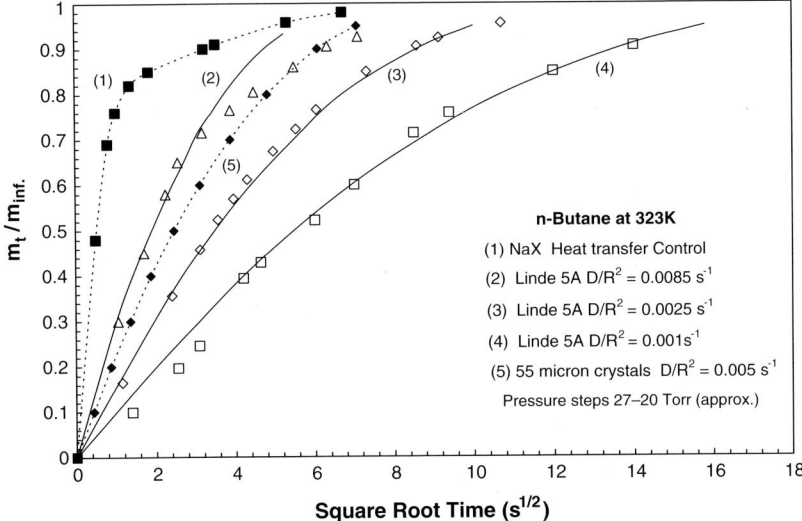

Figure 16.14 Gravimetric sorption curves for n-butane in Linde 5A crystals (lot 550045) at 323 K. Pressure step is 27–20 Torr (approximately). Theoretical lines (curves 2–4) are calculated according to Eq. (6.10) assuming isothermal diffusion control with the given time constants. Curve (1), for a comparable sample of 3.5 μm NaX for which, under similar conditions, the uptake rate is limited by heat transfer, and curve (5) for a sample of carefully dehydrated large (55 μm) crystals are also shown (as broken lines) for comparison. Reprinted from Ruthven [28] with permission.

This curve shows the rapid initial uptake followed by a slow approach to equilibrium that is typical of heat transfer control [see Eq. (6.103)]. The slower uptake curves all conform closely to the isothermal diffusion model [Eq. (6.10)] and, for the larger crystals, diffusion control was also confirmed by measurements with two different crystal sizes (Figure 16.15).

16.4.2
Measurements with Different Crystal Sizes

A detailed study of sorption rates for n-butane in several different samples of 5A crystals was carried out by Yucel [56]; some of his data are summarized in Figure 16.15.

Evidently, the measurements with 55 and 27.5 μm crystals yield consistent diffusivities, showing that the sorption rates in these experiments must have been controlled by intracrystalline diffusion. Measurements with CF_4 also yielded similar results. The variation of diffusivity with loading also generally follows a similar pattern to that observed for 4A, with the Fickian diffusivity increasing strongly at higher loadings but the corrected diffusivity (D_0) remaining essentially constant, as shown (for n-heptane) in Figure 16.5. However, as with 4A, there are considerable differences in diffusivity between different zeolite samples, often with little difference in activation energy (see Table 16.4 and Figures 16.18 and 16.19 below), and we also see that the diffusivities are generally substantially smaller for the commercial samples.

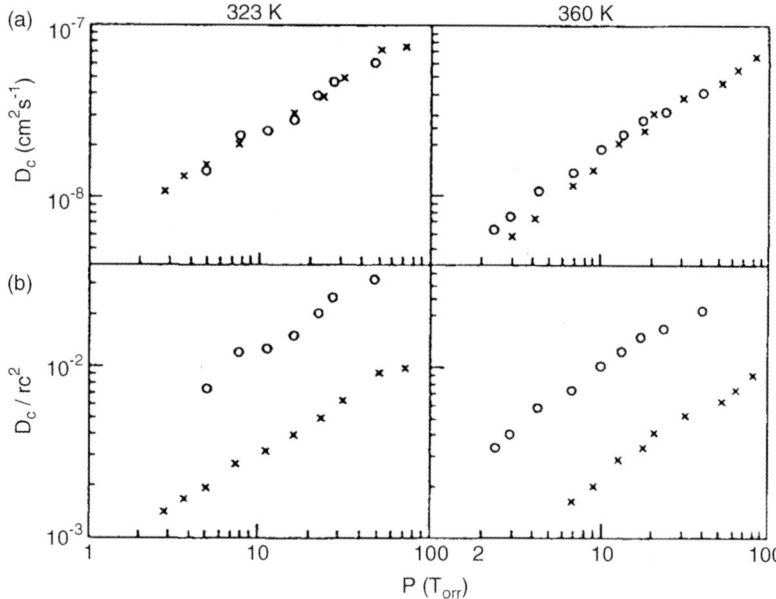

Figure 16.15 (a) Diffusivities (cm^2 s^{-1}) and (b) time constants (s^{-1}) for *n*-butane in two different size fractions of 5A zeolite crystals: (o) 27.5 and (×) 55 μm. Reprinted from Yucel and Ruthven [56] with permission.

16.4.3
PFG NMR Measurements over Different Length Scales

More recently, PFG NMR self-diffusion measurements have been carried out with 5A crystals over a range of different length scales. Figure 16.16 shows the results of such measurements (for *n*-butane). Evidently, at least for mean square displacements of up to 3 μm, the diffusivity is essentially independent of the length scale of the measurement, showing that intracrystalline (defect) barriers are not important in this system.

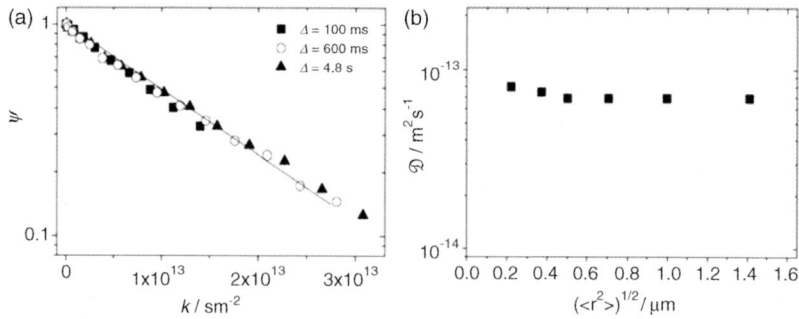

Figure 16.16 PFG NMR signal attenuation curves (a) and resulting self-diffusivities (b) of *n*-butane in 5A zeolite (at 353 K), showing constancy of measured diffusivity with varying scale of measurement [69] with permission.

16.4.4
Intra-Cage Jumps

Michel and Rössiger [60, 61] have measured NMR relaxation times for *n*-butane in 5A and thus determined the correlation time between molecular jumps. The self-diffusivity estimated on this basis, assuming a jump distance equal to the lattice parameter, is about ten times larger than the self-diffusivity derived directly from PFG NMR measurements. The simplest explanation is that most of the jumps occur within the same cage and only about one in ten jumps leads to a successful transition through the eight-ring to the next cage. Significant differences in the correlation times were observed between *n*-butane and the butenes but only small differences between the butene isomers. These were attributed to differences in the flexibility and polarizability of the molecules. However, uptake rate measurements show that the diffusion of *cis*-2-butene is much slower than diffusion of the other isomers and the diffusional activation energy is substantially greater [70]. This is, presumably, simply a reflection of the larger critical molecular diameter of the *cis*-isomer.

The large differences in intracrystalline mobility between the 4A (NaA) and 5A (CaA) forms are clearly shown by the diffusivity data summarized in Tables 16.2 and 16.3. This difference has also been demonstrated by thermal neutron scattering experiments with methane [25, 71, 72]. In particular, by application of the spin-echo technique the mean molecular lifetime of methane in a supercage of 5A, at room temperature, was found to be of the order 4×10^{-10} s, which, with Eq. (2.7), translates to a self-diffusivity 6×10^{-10} m^2 s^{-1}. In contrast, in 4A the mean molecular lifetime was shown to be greater than 10^{-8} s [25], so only an upper limit could be placed on the self-diffusivity $(\mathcal{D} \ll 10^{-11}$ m^2 s$^{-1})$.

16.4.5
Comparison of Different Zeolite Samples and Different Measurement Techniques

Outside the Henry's law region the Fickian diffusivity increases strongly with loading, reflecting the type 1 curvature of the isotherm, but the thermodynamically corrected diffusivity (D_0), defined in accordance with Eq. (1.31), is essentially independent of loading (Figure 16.5). The temperature dependence follows the usual Arrhenius form [Eq. (16.3)].

For molecular transport by cage-to-cage jumps, in Section 4.3.3 self- and transport diffusivities were shown to approach each other even at loadings beyond the Henry's law region. An example of such behavior is shown for propane–5A in Figure 16.17. This figure once again illustrates that, in comparison with the dramatic increase of the transport diffusivity with increasing loading (Figure 16.5a), the self- and corrected diffusivities are essentially constant. For the carefully dehydrated Zhdanov and Yucel crystals, corrected diffusivities and the self-diffusivities measured by TZLC (tracer zero-length column) and PFG NMR show approximate agreement. The PFG NMR self-diffusivities are in fact slightly larger than the macroscopically measured (TZLC) values and this is consistent with recent results from refined PFG NMR studies that suggest the presence of some

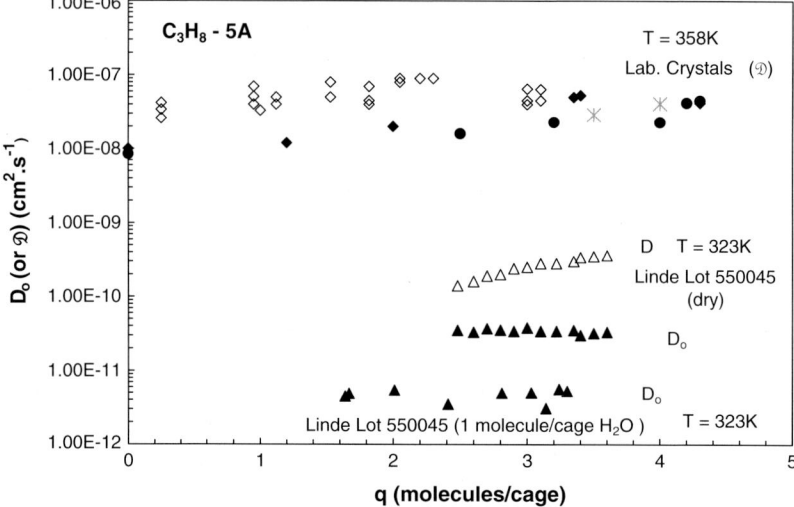

Figure 16.17 Variation of diffusivity for propane (D, D_0, or $\mathcal{D}$) with loading in various different samples of 5A zeolite: ($\diamond$) PFG NMR ($\mathcal{D}$) Zhdanov crystals; ($\blacklozenge$, $\bullet$) ZLC/TZLC (D_0, $\mathcal{D}$) Yucel crystals; ($\triangle$, $\blacktriangle$) (D, D_0) Linde 5A crystals lot 550045 (uptake rate). Data are from various sources – see Kärger and Ruthven [73], Brandani [74] and Ruthven [28].

surface resistance in such crystals [75, 76]. However, detailed analysis of the uptake curves and measurements with different crystal sizes (Figures 16.14 and 16.15) show that, in well dehydrated crystals, surface resistance is minimal and the rate is controlled mainly by intracrystalline diffusion.

The large difference in diffusivity between the commercial Linde 5A crystals and the laboratory-synthesized crystals, as well as the dramatic reduction in diffusivity for an incompletely dehydrated sample, is also apparent from Figure 16.17.

16.5
General Patterns of Behavior in Type A Zeolites

Figures 16.12–16.13, 16.18 and 16.19 summarize, in the form of Arrhenius plots, some of the extensive experimental data that have been accumulated for diffusion of several light gases in different samples of 4A and 5A zeolites. The activation energies and pre-exponential factors are summarized in Table 16.4. In several cases measurements have been made for the same sample of zeolite by both ZLC and uptake rate measurements as well as by PFG NMR, with consistent results (Figures 16.13 and 16.17–16.19). Although there is some experimental scatter, the general pattern is quite clear. As expected the diffusivities for 4A are much smaller than for 5A and the activation energies are substantially larger, reflecting the smaller effective aperture of the partially obstructed 4A windows.

16.5 General Patterns of Behavior in Type A Zeolites

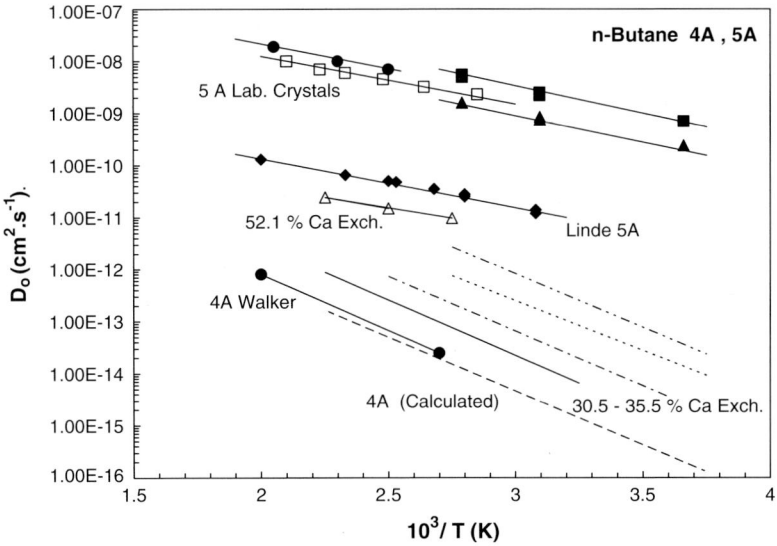

Figure 16.18 Arrhenius plot showing the temperature dependence of corrected diffusivity for n-C_4H_{10} in 5A crystals (Yucel, Zhdanov, and Linde lot 550045) and in various partially Ca exchanged samples of 4A zeolite crystals. Measurements for the 5A laboratory crystals (Yucel and Zhdanov) were made by PFG NMR (●), ZLC (□), and uptake rate (▲, ■). Data of Yucel [56], Kärger and Ruthven [73], Eic and Ruthven [63], Loughlin [77], Derrah [70], Ruthven [42], and Walker [43].

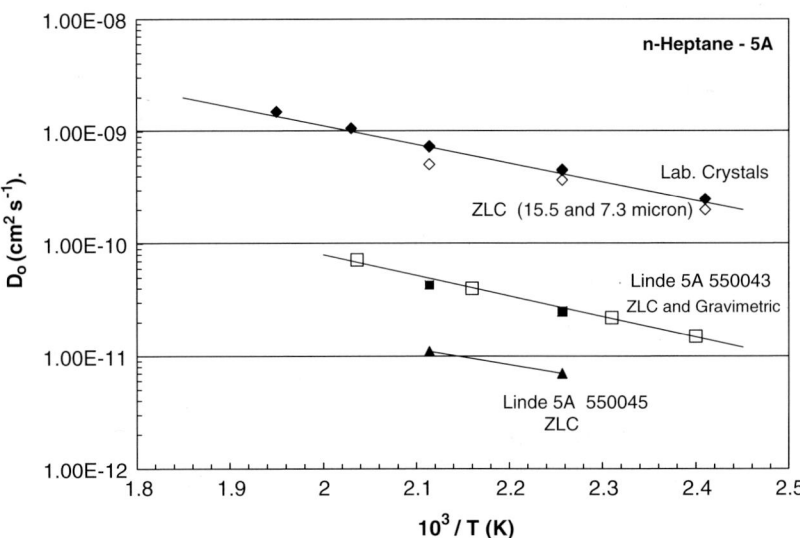

Figure 16.19 Arrhenius plot showing the temperature dependence of corrected diffusivity for n-C_7H_{16} in 5A (Yucel crystals) and in Linde 5A. Data of Doetsch [12] and Eic [63]. Note the consistency between data for different size fractions and between gravimetric (filled symbols) and ZLC (open symbols) data for Linde lot 550043.

Table 16.4 Summary of activation energies and pre-exponential factors for diffusion in NaA and CaA zeolites [28].

Sorbate	NaA crystals (Yucel)		CaA crystals (Yucel and Zhdanov)		Linde 5A
	E (kJ mol^{-1})	D_∞ (cm^2 s^{-1})	E (kJ mol^{-1})	D_∞ (cm^2 s^{-1})	D_∞ (cm^2 s^{-1})
N_2	24.2	2.2×10^{-5}	10.5	6.2×10^{-5}	—
CO_2	25.0	2.3×10^{-5}	9.4	1×10^{-4}	—
CH_4	25.0	1.4×10^{-6}	6.3	1.8×10^{-5}	—
C_2H_6	24.7	1.34×10^{-6}	6.3	6.3×10^{-6}	—
C_3H_8	36.4	1.2×10^{-8}	16.7	3.7×10^{-6}	1.1×10^{-8}
nC_4H_{10}	39.7	7.3×10^{-9}	16.7	1.1×10^{-6}	7.3×10^{-9}
nC_7H_{16}	—	—	31.7	2.4×10^{-6}	3.4×10^{-8}
$nC_{10}H_{22}$	—	—	48	5×10^{-5}	1.7×10^{-6}

Activation energies for Linde 5A are essentially the same as for Yucel and Zhdanov crystals. Reported data for smaller molecules in Linde 5A are probably unreliable as diffusion is too fast to measure accurately in such small crystals. The activation energies quoted here are average experimental values for measurements made over wide temperature ranges.

More surprisingly, for both 4A and 5A, we see large differences in diffusivity between various zeolite samples but with very little difference in the activation energy. A large reduction in diffusivity with little change in activation energy suggests that the rate-controlling energy barrier (the pore aperture) remains constant while the length and tortuosity of the diffusion path increase.

16.5.1
Activation Energies

The pattern of variation of activation energy with molecular diameter, shown in Figure 16.20, reveals significant but perhaps not unexpected differences between 4A and 5A zeolites. In 4A the activation energy increases monotonically with molecular diameter, suggesting that, even for the smaller molecules, the activation energy is mainly determined by the repulsive interactions in the transition state. However, for 5A the activation energy for the smaller molecules shows no correlation with molecular size, presumably because for these molecules the repulsive energy barrier associated with passage through the window is not fully rate controlling. A monotonic increase of activation energy is observed only for molecules larger than about 4 Å, for which the repulsive energy associated with transition through the window becomes dominant.

It has long been recognized that intracrystalline diffusivities in small-pore zeolites are strongly dependent on the size of the diffusing molecule relative to the pore diameter or, in the case of the A zeolites, the window diameter. More specifically it has been shown that the diffusional activation energies in 4A and 5A correlate directly

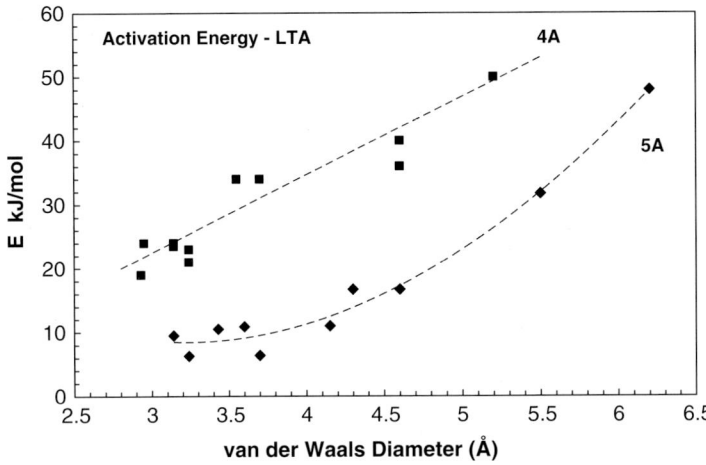

Figure 16.20 Variation of activation energy for diffusion on 4A and 5A with molecular diameter.

with the van der Waals radii of the guest molecules [62, 78]. The validity of such a correlation despite very large differences in the actual diffusivities between different samples has always seemed somewhat surprising. The data available today strongly suggest that the differences between different type A zeolite samples arise from the blocking of different fractions of the eight-ring windows, depending *inter alia* on the initial dehydration conditions, the degree of hydration, and the structural integrity of the framework, thus providing a rational explanation for the constancy of the activation energies between samples with very different diffusivities. For the larger molecules the correlation is unambiguous, but it is much less clear for the smaller molecules. In this region the energy required to leave the local site may be greater than the repulsive energy associated with passage through the eight-ring so that the correlation with molecular diameter breaks down. This appears to be the situation for N_2 and CO_2 in 5A. However, activation energies as low as this are difficult to measure with any accuracy.

For 4A zeolite the activation energies are substantially larger, reflecting the smaller window aperture. However, the activation energies for N_2, CO_2, CH_4, and C_2H_6 are all essentially the same (≈ 25 kJ mol^{-1}) despite the variation in molecular diameter. This may reflect the energy involved in displacement of the SII cation rather than the repulsive energy associated with passage of the guest molecule through the window.

16.5.2
Pre-exponential Factor

In the type A zeolite the molecule in passage through the eight-ring window is clearly identifiable as the transition state, making these systems particularly suitable for the application of transition state theory [57, 79, 80]. In Section 4.3.3 this formalism was introduced as a special case among the "infrequent event techniques." With Eq. (4.42), the pre-exponential factor in Eq. (16.3) is easily seen to be:

$$D_\infty = \frac{\ell^2}{hK_\infty} \cdot \frac{f^+}{f'_g} \qquad (16.4)$$

where we have introduced K_∞ as the pre-exponential factor in the van't Hoff expression for the temperature dependence of the dimensionless Henry constant ($K = K_\infty e^{-\Delta U/RT}$). If the molecule rotates in the transition state to the same extent as in the gas phase the ratio f^+/f'_g should approximate the reciprocal of the reduced translational partition function ($1/f'_{trans}$), which is easily calculated. For diffusion of several light molecules in the Yucel 4A and 5A crystals Table 16.5 gives a comparison of the experimental values of D_∞ with the theoretical values, estimated from Eq. (16.4) on the basis of this approximation [48]. For N_2, CO_2, CH_4, and C_2H_6 the agreement is quite close (within a factor of about 2 for 5A), but for the monatomic gases and O_2 the experimental values of D_∞ are much smaller than the predicted values. Such differences might be due to a substantial contribution to the partition function in the transition state from the degree of freedom in the plane of the window, which is neglected in the simple model.

A further prediction that follows from Eq. (16.4) is that for small nonpolar species, for which the Henry constants for 4A and 5A may be expected to be similar, the values of D_∞ for the same molecule diffusing in ideal 4A and 5A structures should also be similar, since the difference in diffusivity then arises almost entirely from the difference in the activation energy. This prediction is approximately fulfilled, as may be seen from Table 16.5, providing further evidence in support of the proposed model. In Reference [50] the similarity in diffusional activation energies for the same molecule in different samples of 4A zeolite was noted and explained on the basis of the window blocking theory, as outlined above, but with one critical difference. It was assumed that the *slowest* sample corresponds to the ideal 4A structure in which all windows contain a Na^+ cation and that the faster 4A samples contain a certain fraction of "open" windows containing no SII cation. Such a hypothesis can explain the 4A kinetic data. However, it leads to values of D_∞ for the more open 4A samples that are much larger than the corresponding values for 5A, which is inconsistent with Eq. (16.4). The assumption that the fastest of the

Table 16.5 Comparison of pre-exponential factors ($D_\infty/cm^2\ s^{-1}$) for NaA and CaA with theoretical values from Eq. (16.4).

Sorbate	D_∞ (NaA)	D_∞ (CaA)	D_∞ [Eq. (16.4)]
N_2	2.2×10^{-5}	6.2×10^{-5}	4×10^{-5}
CO_2	2.2×10^{-5}	10^{-4}	7.8×10^{-5}
CH_4	1.2×10^{-6}	1.8×10^{-5}	1.7×10^{-5}
C_2H_6	1.35×10^{-6}	6.3×10^{-6}	1.3×10^{-5}

4A samples must represent the ideal 4A structure, which yields reasonably consistent values of D_∞ for 4A and 5A, appears to be more reasonable. We have therefore to assume that in the slower samples (which have lower diffusivities but the same activation energies) a certain fraction of the windows is blocked by framework damage (probably involving de-alumination with deposition of amorphous alumina), as noted in Section 16.1.

16.6 Window Blocking

16.6.1 Sorption Cut-Off

The dramatic effect of ion exchange on the sorption properties of NaCaA zeolites was noted by Breck in his seminal paper [81]. The opening of the pores to admit larger molecules occurs progressively between about 20 and 30% Ca exchange, as shown in Figure 16.21. This behavior may be understood in terms of percolation theory, which shows that in a cubic lattice, when less than about 29% of the windows are "open," there is a sharp cut-off in the sorption capacity for any molecule that is too large to penetrate a "closed" window (i.e., a window containing a Na^+ cation) [38, 40]. This cut-off arises because, beyond this point, there are regions of the lattice that are totally enclosed by "closed" windows.

The behavior of a three-dimensional cubic pore system consisting of discrete cages interconnected through "windows" that may be either "open" or "closed" (obstructed by a cation) has been studied both by Monte Carlo simulation [42] and using the effective medium approximation [82–85].

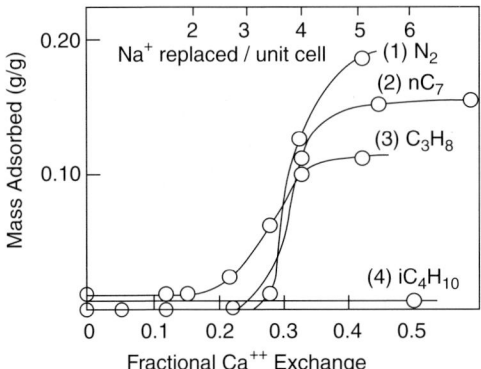

Figure 16.21 Sorption cut-off as a function of Na^+–Ca^{2+} exchange in zeolite A: (1) N_2 at 15 Torr, −195 °C; (2) n-heptane at 45 Torr, 25 °C; (3) propane at 250 Torr, 25 °C; and (4) isobutane at 400 Torr, 25 °C. Reprinted from Breck et al. [81] with permission.

Table 16.6 Variation of the fraction of open windows with the degree of ion exchange.

Cations per cell	Fraction of "open" windows
$12Na^+$	0
$10Na^+$, $1Ca^{2+}$	0
$8Na^+$, $2Ca^{2+}$	$1/3$
$6Na^+$, $3Ca^{2+}$	$2/3$
$4Na^+$, $4Ca^{2+}$	1
$2Na^+$, $5Ca^{2+}$	1
$6Ca^{2+}$	1

16.6.2
Monte Carlo Simulations

Monte Carlo simulations of a random walk have been shown to provide a good representation of the main features of the diffusional behavior in the partially exchanged forms [51]. It is assumed that the cation distribution is "ideal;" that is, all type I sites are occupied before any of the type II sites. The fraction of unobstructed (open) windows is therefore simply related to the degree of ion exchange (Table 16.6). Figure 16.22 gives the results of such simulations, showing how the diffusivity through the structure varies with the fraction of "open" windows and the diffusivity ratio $r = D_{open}/D_{closed} = D_A/D_B$.

When the diffusivity ratio for the "open" and "closed" forms is large, the activation energy changes very sharply from the value characteristic of a "closed" sieve to that for an "open" sieve at the $1/3$ threshold [42]. The data shown in Figures 16.12, 16.18 and 16.19 are consistent with this model. The large difference in diffusivity between the laboratory synthesized CaA crystals and the commercial 5A samples (Linde lots 550043 and 550045), with very little difference in activation energy, suggests that almost $2/3$ of the windows in the Linde 5A sample are "closed." Similarly, the differences between the various 4A samples (which show essentially constant activation energy) suggest that in the different samples different fractions ($<2/3$) of the windows are blocked.

Further support for this hypothesis comes from comparing the diffusivities for different sorbates (on freshly regenerated zeolite samples). If the difference in diffusivity between the different samples is indeed due to differences in the fractions of "open" and "closed" windows, the ratios of the diffusivities for the different samples should be the same for the different sorbates. Tables 16.7 and 16.8 summarize diffusivity ratios relative to the fastest samples of laboratory synthesized crystals (Yucel Sample 2). It may be seen that, despite some variations, the data are broadly consistent with this pattern. Exact conformity cannot be expected since the diffusivities are also sensitive to small differences in the pre-conditioning procedure and/or aging of the samples.

Ratios are independent of T since activation energies are essentially the same for the different 4A samples.

16.6 Window Blocking | 589

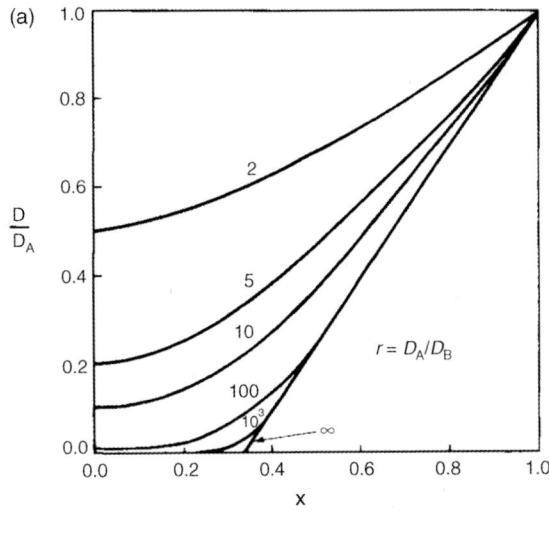

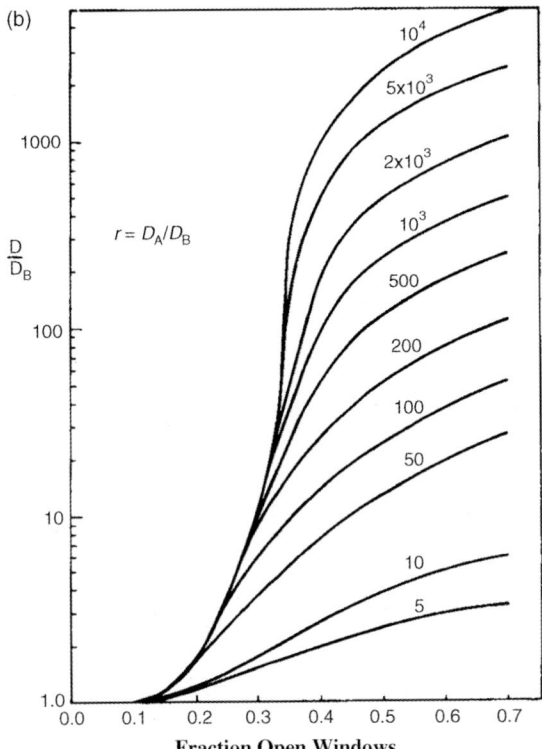

Figure 16.22 Variation in diffusivity with fraction of open windows: (a) D/D_A where D_A = diffusivity for fully open form (5A); (b) D/D_B where D_B = diffusivity for fully closed form (4A); $r = D_A/D_B$. Reprinted from Ruthven [42] with permission.

Table 16.7 Comparison of diffusivities for several different sorbates in laboratory synthesized CaA and Linde 5A crystals (Yucel lot 2 and Linde 5A lot 550045).[a]

Sorbate	D_{Yucel}/D_{Linde}
C_3H_8	230
nC_4H_{10}	190
nC_7H_{16}	200
$nC_{10}H_{22}$	150
CF_4	250

a) $D_{Yucel2}/D_{Yucel1} \approx 3$. The fastest sample (Yucel 2) is taken as the base case. The ratios are essentially independent of T as activation energies for both zeolites are the same.

Table 16.8 Comparison of diffusivities in three different 4A samples (D/D_{NaA}).[a]

	Linde 4A crystals lot 450339	Linde 4A Pellet Cao et al. [46]	Linde 4A crystals lot 470017
Ar	—	—	0.035
N_2	0.26	0.13	0.04
CH_4	0.12	0.12	0.06
C_2H_6	0.48	—	0.02
Kr	—	0.12	—
Average	0.29	0.123	0.04
ε [from Eqs. (16.5) and (16.6)]	0.47	0.58	0.64

a) D_{NaA} corresponds to the Yucel 4A crystals (7.3, 34 μm). The ratios are independent of temperature as the activation energies are essentially the same for the different 4A samles.

16.6.3
Effective Medium Approximation

According to the effective medium approximation [82–85] the diffusivity in a partially closed system is given by:

$$\frac{D}{D_{open}} = \frac{1}{4}\left[2-3\varepsilon + \frac{(3\varepsilon-1)^2}{r}\right] + \frac{1}{4}\left\{\left[2-3\varepsilon + \frac{(3\varepsilon-1)^2}{r}\right]^2 + \frac{8}{r}\right\}^{\frac{1}{2}} \quad (16.5)$$

where $r = D_{open}/D_{closed}$ and ε = fraction of "closed" windows. In the limit when r is large (negligible diffusion through the "closed" windows) this reduces to:

$$\frac{D}{D_{open}} = 1 - 1.5\varepsilon \quad \left(\text{valid for } \varepsilon < \frac{2}{3}\right) \quad (16.6)$$

Evidently, for a system in which the ratio of the diffusivities for the "open" and "closed" sieves is large, the diffusivity will drop dramatically when the fraction of "open" windows falls to about $1/3$. At this point the effective medium approximation is no longer valid but kinetic information can still be obtained from the Monte Carlo simulation. Such simulations (see Figure 16.22) show that for a "closed" sieve the diffusivity starts to increase when the fraction of "open" windows exceeds about 12%

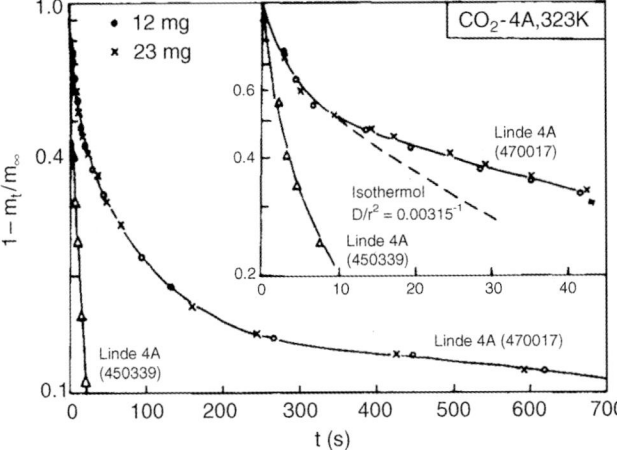

Figure 16.23 Gravimetric uptake curves for CO$_2$ in two different samples of 4A zeolite crystals (Linde lots 450339 and 470017) measured over the same pressure step (65–68 Torr) at 323 K. Essentially the same curves are obtained with two different sample weights, showing that the uptake rate is not controlled by bed diffusion or heat transfer. The curve for lot 450339 is well approximated by the diffusion equation [Eq. (6.10)] with $D/R^2 = 0.03$ s^{-1} but the curve for lot 470017 shows a distinct difference in shape with a very slow final approach to equilibrium. Reprinted from Yucel [27] with permission.

and increases very rapidly when the $1/3$ threshold is approached. This is consistent with the results of percolation measurements that show that the equilibrium capacity starts to increase well below the percolation threshold (Figure 16.21) [86, 87]. In this region the pore system consists of a network of pores controlled only by "open" windows separating regions of the structure that are accessible only through "closed" windows. The uptake curve will then show a rapid initial uptake followed by a slow exponential approach to equilibrium, qualitatively similar to the curves for heat transfer control but on a much longer time scale. Figure 16.23 shows an example of such behavior for CO$_2$ in 4A Linde (lot 470017) [27].

16.6.4
Diffusion in NaCaA Zeolites

Diffusion of n-butane has been studied in a series of partially exchanged NaCaA zeolites. The results, which are summarized in Figures 16.18 and 16.24, conform quantitatively to the window blocking model. In the Na$^+$-rich form the diffusivity increases with Ca^{2+} exchange but the activation energy remains essentially constant and equal to that for the limiting 4A form. For higher degrees of Ca^{2+} exchange the activation energy is much lower and similar to that of the 5A form. The trend of diffusivity with degree of Ca^{2+} exchange conforms quite closely to the theoretical curve with $r \approx 10^3$. However, the lower portion of this curve is quite insensitive to the actual value of r as long as it is large so that the data could be fitted just as well if one assumes a much higher diffusivity for the "open" 5A form.

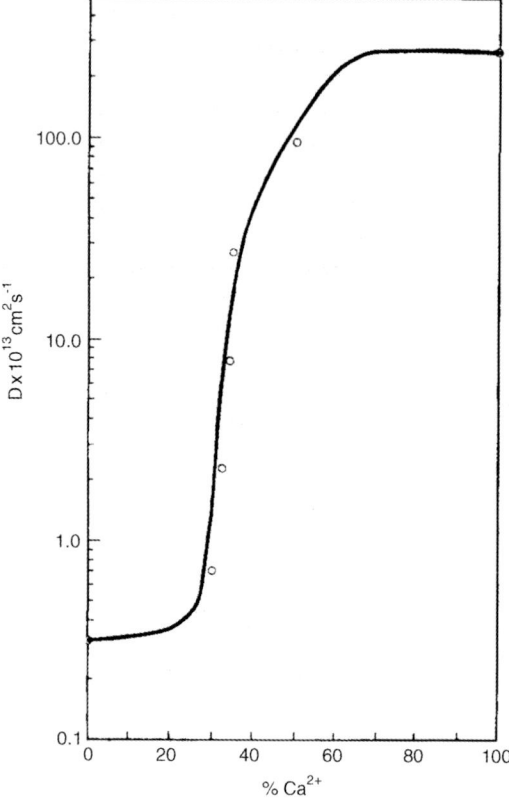

Figure 16.24 Variation of diffusivity for *n*-butane in NaCaA zeolite (at 91 °C) with ion exchange, showing comparison between the window blocking theory (line) and the experimental points. Reprinted from Ruthven [42] with permission.

16.6.5
3A Zeolites

The exchange of Na^+ for the larger K^+ cations leads to a further reduction in window size, yielding what is known commercially as 3A sieve. The smaller Na^+ cation is held preferentially at the type I sites so the K^+ cations are located primarily in the eight-ring windows. The problem of diffusion in NaKA zeolites has been studied by Yeh and Yang [36] using the "effective medium" approximation. Diffusion of CH_4 and C_2H_6 in KNaA shows a sharp cut off in diffusivity between 25 and 30% K^+ exchange. This provides strong evidence that the K^+ ions preferentially occupy the type II (window) sites; the cut-off then corresponds approximately to the value of 29% predicted from percolation theory [86, 87]. The variation of diffusivity with K^+ exchange in the region below the cut-off was found to be broadly consistent with Eq. (16.5) with a very large value for r.

16.7
Variation of Diffusivity with Carbon Number

Figures 16.25–16.27 summarize the results of a detailed study of the diffusion of linear paraffins in 5A by several different experimental techniques using the same zeolite samples [88]. For the lower carbon numbers (up to C_8) there is good

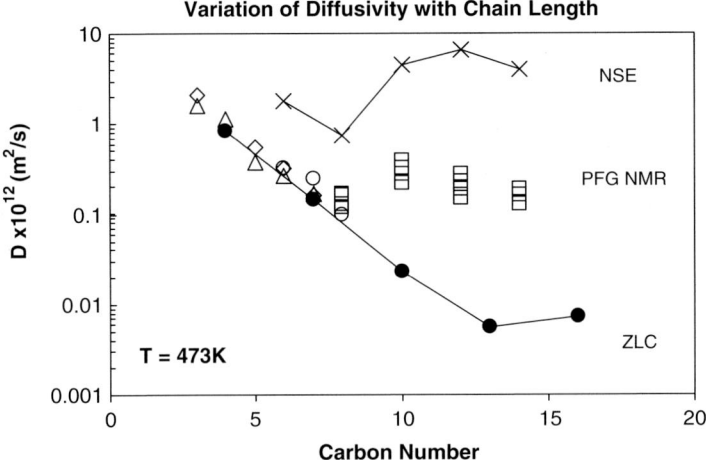

Figure 16.25 Variation of diffusivity with carbon number (at 473 K): NSE (neutron spin echo) (D, ×); ZLC (D_0, •, o); PFG NMR ($\mathcal{D}$) at 1 molecule per cage (△, □) and 2 molecules per cage, (◇). Reprinted from Jobic et al. [88] with permission.

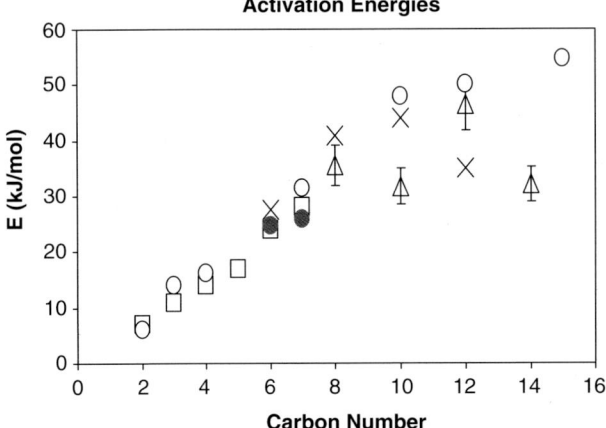

Figure 16.26 Variation of activation energy with carbon number for linear paraffins in 5A Zeolite. ZLC (D_0): old data, (o); new data, (•); PFG NMR ($\mathcal{D}$) at 1 molecule per cage, (□, △); NSE (D, ×). Reprinted from Jobic et al. [88] with permission.

agreement between the microscopic self-diffusivities (measured by PFG NMR at low loading) and the limiting D_o values measured macroscopically by ZLC and uptake rate measurements. It is worth noting that the ZLC measurements were repeated after an interval of several years by two different investigators using similar but not identical zeolite samples, with essentially consistent results [89]. At higher carbon numbers the PFG NMR and ZLC measurements diverge, with the ZLC values showing a continuous decline in diffusivity with carbon number, albeit at a reduced rate, and the PFG NMR data suggesting a local minimum and maximum at C_8 and C_{10}. The most obvious explanation for this discrepancy, that for larger molecules the desorption rate is controlled by surface resistance, can be excluded since an extensive series of partial loading ZLC experiments showed that intracrystalline diffusion rather than surface resistance is rate controlling [89].

The neutron scattering (spin echo) experiments [90] measured the Fickian diffusivity at relatively high loadings, so direct comparisons with PFG NMR self-diffusivities and ZLC values of D_0 at zero loading are not quantitatively valid. Much of the difference can probably be attributed to the thermodynamic correction factors, which, for the longer chain alkanes, are large. Nevertheless the similarity in the trends of the PFG NMR and NSE data and the divergence from the ZLC data at high carbon numbers is striking.

The variation of diffusional activation energy with carbon number is shown in Figure 16.26. The activation energy increases regularly with carbon number up to about C_8, reflecting the monotonic decline in diffusivity in this region. At higher C numbers the data become quite scattered, probably as a result of the poor measuring conditions (for PFG NMR) and the limited accessible temperature range. It may be that beyond C_8 the activation energy is almost constant, but because of the large

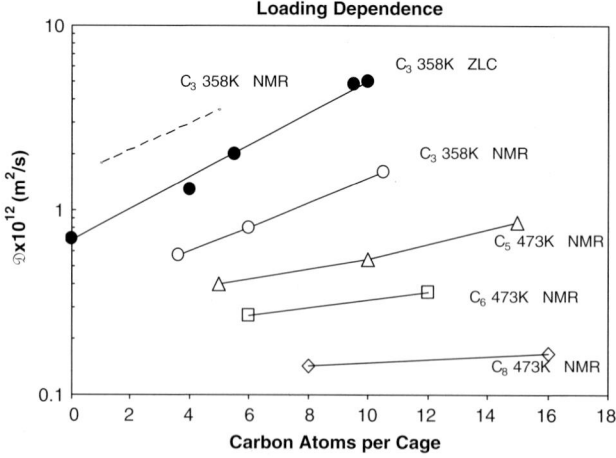

Figure 16.27 Loading dependence of self-diffusivity for linear alkanes in 5A zeolite as measured by PFG NMR and TZLC. Reprinted from Jobic et al. [88] with permssion.

scatter of the data this is not certain. The activation energies for the ZLC and NSE measurements are very similar to the PFG NMR values, especially at lower carbon numbers where the values are more reliable.

16.7.1
Loading Dependence of Self-diffusivity

Figure 16.27 shows the loading dependence of the self-diffusivity, as measured by both PFG NMR and tracer ZLC. The trends shown by the TZLC and PFG NMR data are similar. The diffusivity always increases with loading but the variation becomes less pronounced with increasing carbon number. These findings complement the message of Figure 16.5, where the values of the corrected (transport) diffusivities are found to remain essentially invariant with increasing loading. Within the model of activated jumps between adjacent supercages (Section 4.3.3), guest passage through the windows is identified as the transition state. In this case, the self-diffusivities and corrected diffusivities were shown to coincide. From Figures 16.5 and 16.27 it is clear that the relations for self-diffusion [Eq. (4.42)] and transport diffusion [Eq. (4.43)] and the relationship between them [Eq. (1.35)] are followed, at least approximately, leading to (corrected and/or self-) diffusivities that either remain constant or increase somewhat with loading.

16.8
Diffusion of Water Vapor in LTA Zeolites

Intracrystalline diffusion of water in zeolite A is too fast to measure by macroscopic methods even in the largest available crystals. Measurements can, however, be made by PFG NMR and the results of such measurements for NaA, NaCaA, and ZK4, a silica-rich analog of zeolite A, are summarized in Figure 16.28 [91]. In contrast to light hydrocarbons (C_1–C_3), which show approximately the same diffusivity in 5A and in ZK4, which also has predominantly "open" windows, the diffusivity data for water show marked differences. The diffusivities decrease in the sequence ZK4 > NaA > NaCaA. Whereas in the hydrocarbon systems diffusion is largely limited by the repulsive energy involved in passing through the windows between adjacent cages, the diffusion of water appears to be limited by interaction with the cations. The relatively low mobility in 5A reflects the strong interaction with the divalent Ca^{2+} ions while the difference between ZK4 and NaA probably reflects the reduced cation density in the ZK4 structure. The Arrhenius plots all extrapolate to about $10^{-9}\,m^2\,s^{-1}$ (for $T \to \infty$), indicating that the entropy of activation is essentially the same for all three adsorbents. At low temperatures there is a transition to a highly immobile "frozen" state in agreement with the predictions deduced from NMR absorption line measurements [92]. This is seen most clearly for ZK4 since the diffusivity is higher and the relaxation times are longer, thus permitting measurements over a wider range.

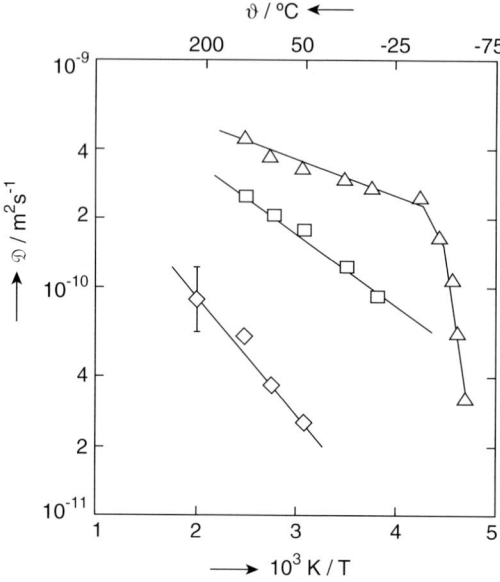

Figure 16.28 Self-diffusivity of water in ZK4 (△), 5A (◇), and 4A (□) zeolites at about 25 molecules per cage, determined by the PFG NMR method. Reprinted from Kärger et al. [91] with permission.

16.9
Deactivation, Regeneration, and Hydrothermal Effects

16.9.1
Effect of Water Vapor

The type-A zeolites, especially CaA, show a very high affinity for water vapor. The presence of even small amounts of adsorbed water can have a dramatic effect on the sorption kinetics of guest molecules, so careful control of the initial dehydration and subsequent regeneration conditions is essential for reliable and reproducible diffusion measurements. It has been established that complete dehydration of CaA requires several hours at 400 °C under a vacuum of at least 10^{-4} Torr. The rate at which the temperature is increased is also important since, although these zeolites have good thermal stability, their hydrothermal stability is poor. It is therefore important to avoid exposing the sample to a combination of high temperature and high water vapor pressure.

Using ethane as the test sorbate Kondis and Dranoff [38, 40] showed that the uptake rate in severely pretreated 4A samples was much slower than in crystals that had been carefully dehydrated under "shallow-bed" conditions, thus avoiding exposure to a high partial pressure of water vapor at elevated temperature. From these studies they concluded that the severe hydrothermal conditions encountered in the pelletization process are responsible for the generally lower diffusivities observed with pelletized commercial adsorbent samples. Similarly, it has been shown that when a sample of Charnell synthesized CaA crystals was dehydrated at 400 °C under a modest water vapor pressure ($\approx$0.1–0.01 Torr) the diffusivity for n-butane was reduced by two orders of magnitude compared with a carefully dehydrated sample of the same crystals [28, 93].

Exposure to severe hydrothermal conditions leads to "pore closure." The practical application of this phenomenon to exert a measure of control over the effective pore size of a 4A zeolite has been discussed by Breck [94]. For drying certain reactive species such as chlorofluorocarbon refrigerants it is common practice to use a "pore closed" 4A zeolite, prepared by severe hydrothermal treatment of a normal 4A sample. The reduction in the effective pore size prevents the chlorofluorocarbon from entering the micropores, thus eliminating possible hydrolysis with the formation of undesirable products such as HF and HCl. The mechanism is not completely clear but it probably involves the hydrothermal reactions (16.1) and (16.2). Such reactions are essentially irreversible so, once a sample has been hydrothermally damaged, the kinetic behavior is not restored by further vacuum dehydration at high temperature.

In addition to these irreversible effects, adsorption of relatively small traces of water at ordinary temperatures leads to a pronounced drop in sorption rates. More interestingly, the form of the uptake curve suggests that the rate controlling mechanism changes from diffusion control to surface resistance control. This effect is reversible; rapid kinetics with diffusion control are restored by careful dehydration at elevated temperature. However, the initial rate is not fully recovered and repeated regeneration leads to a slow irreversible decline. For NaA such effects can be easily explained as due to the adsorption of water by the window (SII) cations, leading to (reversible) window blocking. For CaA, however, the windows should all be "open" so this simple explanation cannot apply. It has been suggested that cation hydration may lead to migration from SI to SII sites but this has not been proved.

If the water molecules are adsorbed predominantly in the outer layers of the crystal (as is to be expected for such a high affinity sorbate) the window blocking will occur only near the surface, leading to a reversible transition from diffusion to surface resistance control, as observed. This would also explain why a very small amount of water ($\approx$1 molecule per pseudo-cell) has such a dramatic effect on the sorption kinetics and is also consistent with the observation from PFG NMR measurements that, despite the dramatic reduction in sorption rates, the intracrystalline self-diffusivity in 5A is not significantly affected, even by water loadings of up to ten molecules per cell [95, 96].

16.9.2
Kinetic Behavior of Commercial 5A: Evidence from Uptake Rates

There are, in principle, two distinct ways in which the sorption kinetics may be retarded by hydrothermal pretreatment; either the pore entrances at the crystal surface may be obstructed or the windows in the interior may be blocked. The former is equivalent to the introduction of a surface resistance to mass transfer while the latter involves a reduction in the intracrystalline diffusivity. These two possibilities are, at least in principle, distinguishable by the shape of the uptake curves. The extensive uptake rate data reported by Kondis and Dranoff for ethane in different hydrothermally treated samples of NaA [38, 40] are more consistent with the diffusion model, even though the diffusivities are much smaller than the values obtained in more recent measurements with carefully dehydrated laboratory crystals [56]. The constancy of the activation energy for different 4A samples, as illustrated in Figure 16.12, is also suggestive of diffusion control since the random blocking of a fraction of the eight-ring windows (by adsorption of water on the SII cations) will lead to exactly this pattern of behavior.

The kinetic behavior of 5A shows similar features. The kinetic data summarized in Figures 16.13–16.19 and Table 16.3 show clearly that uptake rates in commercial Linde 5A are much slower than in similarly sized crystals of carefully dehydrated laboratory 5A crystals, although the activation energies are essentially the same. The kinetic behavior of the commercial crystals could be reproduced by severe hydrothermal dehydration of similarly sized laboratory crystals [93]. Furthermore, the form of the uptake curves for both the commercial and laboratory synthesized crystals (at least for fully dehydrated samples) suggests intracrystalline diffusion control rather than surface resistance control. Both these observations are consistent with window blocking, presumably resulting from the severe hydrothermal treatment of the commercial crystals. However, studies of sorption of n-decane in various hydrothermally pretreated samples of NaCaA and NaMgA zeolite show a clear reduction in the uptake rate together with a transition from diffusion control to surface resistance control with increasing severity of hydrothermal treatment [97, 98] (Figure 16.29),

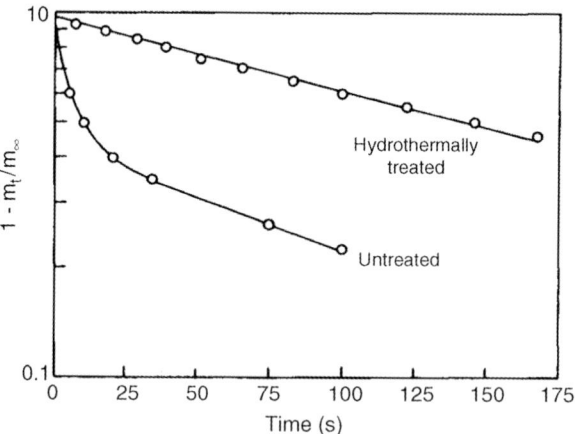

Figure 16.29 Experimental desorption curves for n-decane at 423 K ($p = 40 \rightarrow 25$ Pa) in "normally" and "hydrothermally" pretreated samples of NaMgA zeolites. Normal pretreatment: 20 h *in vacuo* at 673 K. Hydrothermal treatment: 10 h under saturated water vapor pressure plus 100 h *in vacuo* at 620 K. The data, from Bülow et al. [98], have been plotted to show conformity with Eq. (6.6) for the hydrothermally treated sample.

suggesting that a surface barrier can indeed be created by severe hydrothermal treatment.

16.9.3
Effects of Pelletization and other Aspects of Technological Performance: Evidence from NMR Studies

The experimental data for diffusion in 4A and 5A zeolites, summarized in Tables 16.2 and 16.3, show that where the same adsorbent samples are compared the agreement between the diffusivities determined by different techniques is generally quite satisfactory. However, between samples of adsorbent of different origin there are often wide variations in diffusivity. The most obvious explanation for large differences in the diffusional behavior of apparently similar zeolite samples is subtle differences in structure resulting from differences in dehydration procedure, as discussed in Section 16.9.1.

The differences between commercial 5A adsorbents and carefully dehydrated 5A crystals have also been investigated using NMR tracer desorption and NMR PFG self-diffusion measurements. The combination of these techiques provides direct evidence of the significance of surface barriers to mass transport (Section 11.4.3.3 and Figures 16.30 and 16.31). It was found that, with commercial adsorbent samples,

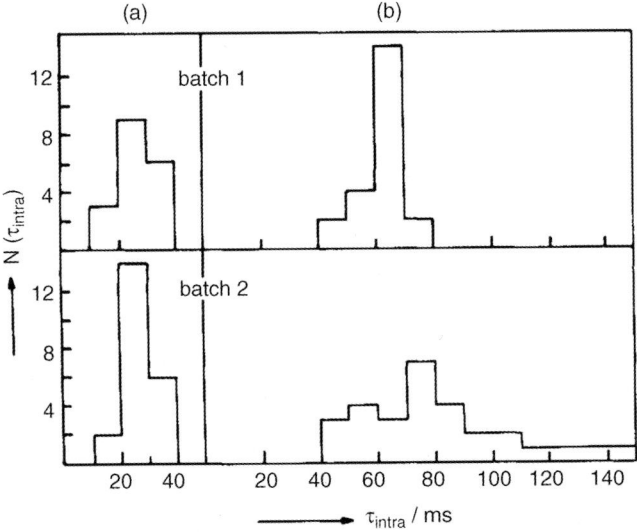

Figure 16.30 Distribution curves [$N(\tau_{intra})$ versus τ_{intra}] measured with ethane as the probe molecule at 293 K for two different samples of a commercial 5A zeolite: (a) represents the initial material and (b) the same material after hydrothermal pretreatment. Evidently, batch 1 is more resistant to hydrothermal deterioration since the lengthening of τ_{intra} is less dramatic than with batch 2. Since the intracrystalline diffusivity was the same for all samples the deterioration can be attributed to the formation of a surface barrier. Data of Seidel and Kaerger [80].

there are considerable variations in hydrothermal stability between different batches of product and even between different pellets from the same batch. The effects of various dehydration procedures were studied in detail. Using methane as the probe molecule, it was found that in carefully dehydrated crystals there is no evidence of any surface resistance, since the intracrystalline mean lifetimes directly measured and estimated from the intracrystalline diffusivities coincide [99]. In contrast, Figure 16.31 shows that in the samples of 5A zeolite that were dehydrated under severe hydrothermal conditions (open inverted triangles) there is clear evidence of the development of a surface barrier, since the apparent diffusivities $\mathcal{D}_{des}$

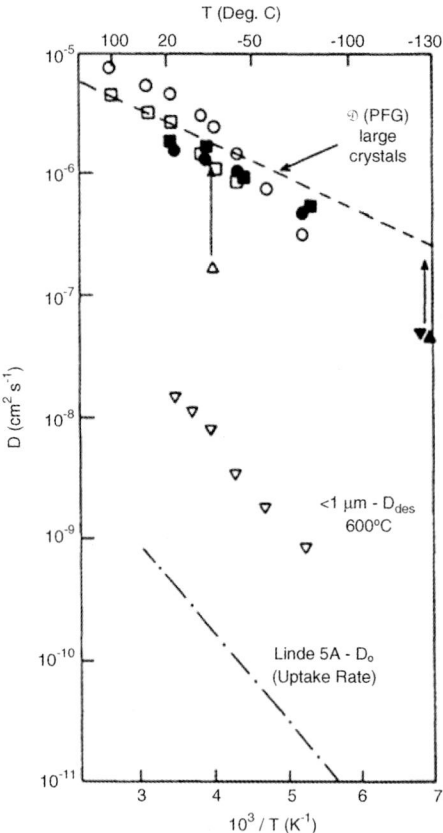

Figure 16.31 Results of tracer desorption measurements of diffusion of C_2H_6 in 5A zeolite crystals, showing the effects of crystal size and dehydration temperature. Filled symbols represent PFG NMR intracrystalline self-diffusivities. Open symbols represent tracer desorption diffusivities. Earlier self-diffusivity data obtained with large crystals (– – –) and uptake rate data (D_0) (–·—·–) for small commercial Linde crystals are shown for comparison. Treatment conditions: 400 °C (8 μm o, •) (<1 μm △, ▲); 600 °C (8 μm □, ■) (<1 μm inverted triangles). Adapted from Kärger [99].

resulting from fast tracer desorption measurements (assuming diffusion limitation) are much smaller than the values derived from intracrystalline diffusion measurements [100, 101]. The magnitude of the surface barrier increases with increasing severity of hydrothermal treatment, whereas the intracrystalline diffusivity remains almost constant [102]. This effect has been confirmed using several different probe molecules [103].

The NMR results were supplemented by X-ray photoelectron spectroscopy studies that revealed a decrease in the Al/Si ratio at the surface and a corresponding decrease in cation content, suggesting structural breakdown in the outer layers of the crystals [104, 105]. The tracer desorption measurements show that the activation energy for passage through the surface barrier is substantially greater than the activation energy for intracrystalline diffusion [102]. Clearly, therefore, the barrier

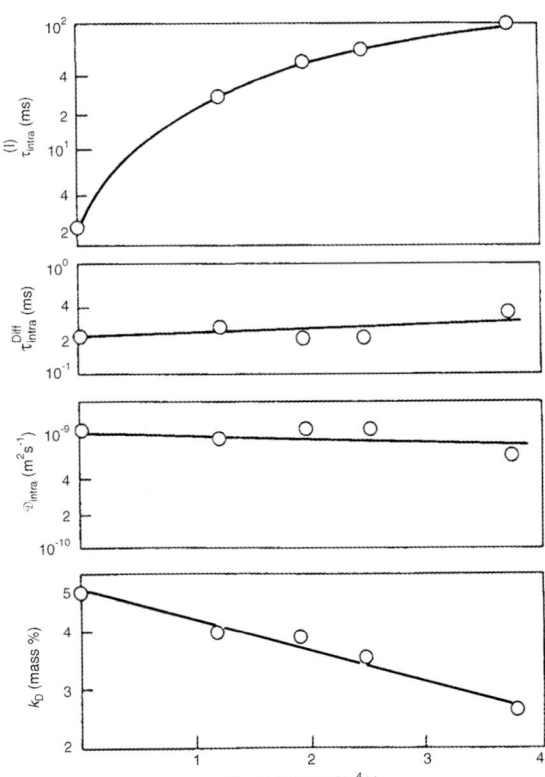

Figure 16.32 Parameters for molecular transport in granulated 5A zeolite as a function of time on stream in a petroleum refinery [102, 106]. While the intracrystalline mobility $\left(\mathcal{D}_{intra}, \tau_{intra}^{diff}\right)$ remains essentially constant, the intracrystalline mean lifetime $\left(\tau_{intra}^{(1)}\right)$ increases in parallel with a decrease in the breakthrough capacity (k_D), suggesting that the deterioration of the dynamic adsorption capacity is associated with formation of a surface barrier. Reprinted from Kärger and Pfeifer [106] with pemission.

must involve a narrowing of the pore entrances rather than complete closure of a certain fraction of the pores. The magnitude of the surface barrier is not an absolute quantity but depends on the nature of the probe molecule. Thus, samples that exhibit negligible surface resistance for methane showed a measurable surface barrier for the slightly larger xenon atom.

The effect of various other pretreatments has also been investigated by similar methods. For example, pre-adsorption of ethylbenzene was shown to create a surface barrier for the adsorption of ethane in 5A zeolite [100]. Using methane as the probe molecule the mass transfer properties of a commercial 5A adsorbent were investigated as a function of the time on stream [102, 106]. The results, some of which are shown in Figure 16.32, reveal a continuous increase in the mean intracrystalline lifetime, as measured by tracer desorption, with very little change in the intracrystalline diffusivity. It seems clear that the slow buildup of carbonaceous deposits on the crystal surface leads to an increasing surface barrier to mass transfer. The conclusion that the effect of "coking" on mass transfer rates is mainly a surface phenomenon has been confirmed by analysis of the chemical shift in ^{129}Xe NMR [107, 108].

The NMR results thus provide clear and unequivocal evidence of the development of surface resistance to mass transfer in severely hydrothermally treated samples of zeolite A. This would explain much of the kinetic variability observed between different 5A zeolite samples but it does not explain the form of the transient uptake curves or the constancy of the diffusional activation energy between the fast diffusing laboratory synthesized crystals and the much slower commercial crystals.

16.10
Anisotropic Diffusion in CHA

The chabazite structure (Figure 16.2) has trigonal symmetry, so the diffusional properties may be expected to be non-isotropic. Following the conception of structure-correlated diffusion anisotropy (Section 2.2.2), the ratio D_z/D_{xy} (where the z coordinate is in the direction of the long axis of the ellipsoidal cavity) may be estimated by a simple jump model based on the assumption that the limiting step is the passage of the molecules through the eight-rings. The resulting value of about 0.8 [109] suggests that, from the experimental perspective, the diffusional behavior should be almost isotropic. However, detailed PFG NMR self-diffusion measurements for water in two different samples of chabazite showed significant non-isotropy with $D_z/D_{xy} \approx 0.4$ [109]. This value has been confirmed by MD simulations [110, 111] and is also supported by the population density profiles derived from GCMC simulations, which show two distinct maxima within the large cavities separated by a minimum at the center [112]. It appears that, contrary to the assumptions of the simple jump model, there is a significant barrier to water transport *within* the large cavity. This provides an additional transport resistance in the z direction, thus increasing the degree of non-isotropy.

16.11
Concluding Remarks

The transport behavior of the eight-ring zeolites is now reasonably well understood. If samples with significant surface resistance are excluded we generally see good agreement between microscopically and macroscopically measured diffusivities. The dominance of repulsive interactions in determining the diffusional activation energy (and hence the diffusivity) for all but the smallest molecules has been clearly demonstrated. The Fickian diffusivities generally increase strongly with loading but this is mainly due to the effect of the thermodynamic correction factor. For many systems the corrected diffusivity appears to be approximately independent of loading, but it must be noted that it is very difficult to determine accurate values for the thermodynamic factor at high loadings and so, unless the corrected diffusivity varies quite strongly with loading, its variation may easily be concealed by uncertainties in the experimental data. For the systems in which self-diffusivities have been measured over a range of different loadings both increasing and decreasing trends are seen but the range of the variation is generally relatively modest. The strong variations predicted by many MD simulations (see, for example, Figure 4.8) are generally not observed, but this may be because the range of the experimental measurements is practically limited and seldom extends above about 70% saturation.

The cation-free CHA and DDR zeolites provide useful model systems for studying the effects of subtle differences in window dimensions on the diffusional behavior without the complications introduced by the presence of the exchangeable cations. Diffusion measurements in such materials also allow direct comparisons with the results of MD simulations. Such a comparison, for DDR, suggests that the usual MD simulations based on a 6–12 potential function overpredict the diffusivities in small pore systems as they do not adequately account for the short-range repulsion forces. A modified MD simulation based on a 6–18 potential has been found to provide a good representation of the diffusional behavior of CO_2 and CH_4 in DDR.

The large differences in diffusivity with almost constant activation energy, which are commonly seen for differently pre-treated 4A samples, are probably due to hydration of the SII cations, leading to differences in the fraction of "open" windows. The origin of the differences in the kinetic behavior of different 5A samples, however, is less clear. NMR evidence suggests that these differences arise mainly from surface resistance, induced by severe hydrothermal treatment during manufacture. However, the sorption kinetic data suggest that the differences are due to window blocking caused by occupation of a substantial fraction of the SII sites. There is evidence that this can also be caused by hydrothermal treatment. It may well be that both these structural modifications can be induced by hydrothermal treatment, and the dominant effect may therefore depend on the precise conditions of the treatment.

References

1 Breck, D.W. (1974) in *Zeolite Molecular Sieves*, John Wiley & Sons, Inc., New York, ch. 2.
2 Barrer, R.M. (1978) in *Zeolites and Clay Minerals as Sorbents and Molecular Sieves*, Academic Press, London, ch. 2.
3 Meier, W.M. and Olson, D.H. (1978) *Atlas of Zeolite Structure Types*, International Zeolites Association.
4 Röthe, K.-P., Röthe, A., and Gelbin, D. (1979) Preprints, Workshop Adsorption of hydrocarbons in zeolites, Academy of Sciences of the GDR, Berlin, November, Vol. 2, p. 108.
5 Michelena, J.A., Vansant, E.F., and de Bievre, P. (1977) *Rec. Trav. Chim. Pays-Bas*, **96** (3), 81.
6 Coe, C.G., Parris, G.E., Srinivasan, R., and Auvil, S.R. (1986) *Proceedings 7th International Zeolite Conference, Tokyo, August 1986* (eds Y. Murakami, A. Lijima, and J.W. Ward), Kodansha Elsevier, Tokyo, p. 1023.
7 Coe, C.G. (1990) *Gas Separation Technology* (eds E.F. Vansant and R. Dewolfs), Elsevier, Amsterdam, p. 149.
8 Ruthven, D.M. and Reyes, S. (2007) *Microporous Mesoporous Mater.*, **14**, 59–66.
9 Gies, H. (1986) *Z. Kristallogr.*, **175**, 93.
10 Alma-Zeestraten, N.C.M., Darrepal, J., Kejisper, J., and Gies, H. (1989) *Zeolites*, **7**, 81.
11 Meier, W.M. and Olson, D.H. (1992) *Atlas of Zeolite Structures*, 3rd edn, Butterworth-Heinemann, p. 80.
12 Doetsch, I.H., Ruthven, D.M., and Loughlin, K.F. (1974) *Can. J. Chem.*, **52**, 2717.
13 Hedin, N., DeMartin, G.J., Strohmaier, K.G., and Reyes, S.C. (2007) *Microporous Mesoporous Mater.*, **98**, 182–188.
14 Corma, A., Kärger, J., and Krause, C. (2005) *Diffusion Fundam.*, **2**, 87.
15 Kärger, J. (2008) *Microporous Mesoporous Mater.*, **116**, 715.
16 Hedin, N., DeMartin, G.J., Roth, W.J., Strohmaier, K.G., and Reyes, S.C. (2007) *Microporous Mesoporous Mater.*, **109**, 327–334.
17 Olson, D.H., Camblor, M.A., Villaescusa, L.A., and Kuehl, G.H. (2004) *Microporous Mesoporous Mater.*, **67**, 27–33.
18 U.S. Patents, Olson, D.H. 6,488,741 (2002), Strohmaier, K.G. 6,730,142 (2004), and Strohmaier, K.G. et al. 6,733,572 (2004)
19 van den Bergh, J., Zhu, W., Gascon, J., Moulijn, J.A., and Kapteijn, F. (2008) *J. Membr. Sci.*, **316**, 35–45.
20 Jee, S.E. and Sholl, D. (2009) *J. Am. Chem. Soc.*, **131**, 7896–7904.
21 Krishna, R. and van Baten, J.M. (2006) *Chem. Phys. Lett.*, **429**, 219–224.
22 Krishna, R. and van Baten, J.M. (2007) *Chem. Phys. Lett.*, **446**, 344–349.
23 Freude, D. (1986) *Zeolites*, **6**, 12.
24 Allonneau, J.M. and Volina, F. (1986) *Zeolites*, **6**, 431.
25 Cohen de Lara, E., Kahn, R., and Mezei, F. (1983) *J. Chem. Soc., Faraday Trans. I*, **79**, 1911.
26 Yucel, H. and Ruthven, D.M. (1980) *J. Chem. Soc., Faraday Trans. I*, **76**, 60.
27 Yucel, H. and Ruthven, D.M. (1980) *J. Colloid Interface Sci.*, **74**, 186.
28 Ruthven, D.M. (2010) Microporous Mesoporous Mater., in press.
29 Ruthven, D.M. and Derrah, R.I. (1975) *J. Chem. Soc., Faraday Trans. I*, **71**, 2031–2044.
30 Sarma, M. and Haynes, H.W. (1974) *Adv. Chem.*, **233**, 205.
31 Shah, D.B. and Ruthven, D.M. (1977) *AIChE J.*, **23**, 804.
32 Yasuda, Y. and Matsumoto, K. (1989) *J. Phys. Chem.*, **93**, 3195.
33 Kumar, R. and Ruthven, D.M. (1979) *Can. J. Chem. Eng.*, **57**, 342–348.
34 Haq, N. and Ruthven, D.M. (1986) *J. Colloid Interface Sci.*, **112**, 164.
35 Allonneau, J.M. and Volino, F. (1986) *Zeolites*, **6**, 431.
36 Yeh, Y.T. and Yang, R.T. (1989) *AIChE J.*, **35**, 1659.
37 Brandt, W. and Rudloff, W. (1965) *J. Phys. Chem. Solids*, **26**, 741.
38 Kondis, E.F. and Dranoff, J.S. (1971) *Adv. Chem.*, **102**, 171.

39. Eagan, J.D. and Anderson, R.B. (1975) *J. Colloid Interface Sci.*, **50**, 419.
40. Kondis, E.F. and Dranoff, J.S. (1971) *Ind. Eng. Chem. Process Design Dev.*, **10**, 108.
41. Taylor, R. (1978) PhD Thesis, Fredericton, Canada.
42. Ruthven, D.M. (1974) *Can. J. Chem.*, **52**, 3523–3528.
43. Walker, P.L., Austin, L.G., and Nandi, S.P. (1966) *Chem. Phys. Carbon*, **2**, 257.
44. Tobin, J. (1979) MSc E. Thesis, Fredericton, Canada.
45. Barrer, R.M. and Clarke, D.J. (1974) *J. Chem. Soc., Faraday Trans. I*, **70**, 535.
46. Cao, D.V., Mohr, R.J., Rao, M.B., and Sircar, S. (2000) *J. Phys. Chem. B*, **104**, 10498–11050.
47. van de Voorde, M., Tavenier, Y., Martens, J., Verleist, H., Jacobs, P., and Baron, G. (1990) *Gas Separation Technology* (eds E.V. Vansant and R. Dewolfs), Elsevier, Amsterdam, pp. 303–310.
48. Xu, Z., Eic, M., and Ruthven, D.M. (1993) *Proceedings Ninth International Zeolite Conference, Montreal, July 1992* (ed. R. von Ballmoos et al..), Butterworth-Heinemann, Boston, pp. 147–155.
49. Kärger, J., Pfeifer, H., Stallmach, F., Feoktistova, N.N., and Shdanov, S.P. (1993) *Zeolites*, **13**, 50.
50. Ruthven, D.M. (2001) *Adsorption*, **7**, 301–304.
51. Ruthven, D.M. (1993) *Zeolites*, **13**, 594.
52. Heink, W., Kärger, J., Pfeifer, H., and Stallmach, F. (1990) *J. Am. Chem. Soc.*, **112**, 2175.
53. Kärger, J., Pfeifer, H., Stallmach, F., Feoktistova, N.N., and Zhdanov, S.P. (1993) *Zeolites*, **13**, 50.
54. Caro, J., Kärger, J., Finger, G., Pfeifer, H., and Schöllner, R. (1976) *Z. Phys. Chem. Leipzig*, **257**, 903.
55. Kärger, J., Pfeifer, H., Rudtsch, S., Heink, W., and Gross, U. (1988) *J. Fluorine Chem.*, **39**, 349.
56. Yucel, H. and Ruthven, D.M. (1980) *J. Chem. Soc., Faraday Trans. I*, **76**, 71.
57. Ruthven, D.M. and Derrah, R.I. (1972) *J. Chem. Soc., Faraday Trans. I*, **68**, 2332.
58. Kärger, J. and Ruthven, D.M. (1981) *J. Chem. Soc., Faraday Trans. I*, **77**, 1485.
59. Yasuda, Y. and Yamamoto, A. (1985) *J. Catal.*, **93**, 176.
60. Michel, D. and Rössiger, V. (1976) *Surf. Sci.*, **54**, 463.
61. Läbisch, L., Schöllner, R., Michel, D., Rössiger, V., and Pfeifer, H. (1974) *Z. Phys. Chem. (Leipzig)*, **255**, 581.
62. Ruthven, D.M., Derrah, R.I., and Loughlin, K.F. (1973) *Can. J. Chem.*, **51**, 3514–3519.
63. Eic, M. and Ruthven, D.M. (1988) *Zeolites*, **8**, 472–479.
64. Chiang, A.S., Dixon, A.G., and Ma, Y.H. (1984) *Chem. Eng. Sci.*, **39**, 1461.
65. Bülow, M., Struve, P., and Rees, L.V.C. (1985) *Zeolites*, **5**, 113.
66. Vavlitis, A.P., Ruthven, D.M., and Loughling, K.F. (1981) *J. Colloid Interface Sci.*, **84**, 526.
67. Ruthven, D.M., Lee, L.K., and Yucel, H. (1980) *AIChE J.*, **26**, 16–23.
68. Ruthven, D.M. and Lee, L.-K. (1981) *AIChE J.*, **27**, 654–663.
69. Ulrich, K. (2008) Diffusionsmessungen in nanostrukturierten Systemen mit PFG NMR unter Einsatz von ultra-hohen gepulsten magnetischen Feldgradienten PhD Thesis, University of Leipzig, p.73.
70. Ruthven, D.M., Derrah, R.I., and Loughlin, K.F. (1973) *Can. J. Chem.*, **51**, 3514.
71. Cohen de Lara, E. and Kahn, R. (1981) *J. Phys. Chem.*, **42**, 1029.
72. Stockmeyer, R. (1984) *Zeolites*, **4**, 81.
73. Kärger, J. and Ruthven, D.M. (1981) *J. Chem. Soc., Faraday Trans. I*, **77**, 1485.
74. Brandani, S., Hufton, J., and Ruthven, D.M. (1995) *Zeolites*, **15**, 624–631.
75. Krutyeva, M., Yang, X., Vasenkov, S., and Kärger, J. (2007) *J. Magn. Reson.*, **185**, 300.
76. Krutyeva, M., Vasenkov, S., Yang, X., Caro, J., and Kärger, J. (2007) *Microporous Mesoporous Mater.*, **104**, 89.
77. Loughlin, K.F. (1971) PhD Thesis, University of New Brunswick, Canada.
78. Ruthven, D.M. (1983) *Principles of Adsorption and Adsorption Processes*, John Wiley & Sons, Inc., New York.

79 Kärger, J., Pfeifer, H., and Haberlandt, R. (1980) *J. Chem. Soc., Faraday Trans. I*, **76**, 1569.
80 Kärger, J. and Ruthven, D.M. (1992), *Diffusion in Zeolites and Other Microporous Solids*, John Wiley & Sons, Inc., New York, ch. 4.
81 Breck, D.W., Eversole, W.G., Milton, R.M., and Read, T.B. (1956) *J. Am. Chem. Soc.*, **78**, 5963.
82 Kirkpatrick, S. (1973) *Rev. Mod. Phys.*, **45**, 574.
83 Burgamos, V.N. and Sotirchos, S.V. (1987) *AIChE J.*, **33**, 1678.
84 Sahimi, M., Hughes, B.D., Scriven, L.E., and Davis, H.T. (1983) *J. Chem. Phys.*, **78**, 6849.
85 Sahimi, M. and Tsotsis, T.T. (1985) *J. Catal.*, **96**, 552.
86 Broadbent, S.R. and Hammersley, J.M. (1957) *Proc. Camb. Philos. Soc.*, **53**, 629.
87 Hammersley, J.M. (1963) *Methods in Computational Physics*, vol. I, Academic Press, New York, p. 281.
88 Jobic, H., Kärger, J., Krause, C., Brandani, S., Gunadi, A., Methivier, A., Ehlers, G., Farago, B., Haenssler, W., and Ruthven, D.M. (2005) *Adsorption*, **11**, 403–407.
89 Gunadi, A. and Brandani, S. (2006) *Microporous Mesoporous Mater.*, **96**, 278–283.
90 Jobic, H., Methivier, A., Ehlers, G., Farago, B., and Haeussler, W. (2004) *Angew. Chem. Int. Ed.*, **43**, 364.
91 Kärger, J., Pfeifer, H., Rosemann, M., Feoktistova, N.N., and Zhdanov, S.P. (1989) *Zeolites*, **9**, 247.
92 Pfeifer, H., Oehme, W., and Siegel, H. (1987) *Z. Phys. Chem. N. F.*, **152**, 473.
93 Tezel, O.H. (1984) PhD Thesis, University of New Brunswick, Canada.
94 Breck, D.W. (1974) *Zeolite Molecular Sieves*, John Wiley & Sons, Inc., New York, p. 490.
95 Förste, C., Germanus, A., Kärger, J., Möbius, G., Bülow, M., Zhdanov, S.P., and Feoktistova, N.N. (1989) *Isotopenpraxis*, **25**, 48.
96 Rauscher, M. (1983) Thesis, Leipzig, Karl-Marx-Universität.
97 Bülow, M., Struve, P., and Rees, L.V.C. (1985) *Zeolites*, **5**, 113.
98 Bülow, M., Struve, P., and Pikus, S. (1982) *Zeolites*, **2**, 267.
99 Kärger, J. (1982) *AIChE J.*, **28**, 417.
100 Kärger, J., Heink, W., Pfeifer, H., Rauscher, M., and Hoffman, J. (1982) *Zeolites*, **2**, 275.
101 Kärger, J., Pfeifer, H., Rauscher, M., and Bülow, M. (1981) *Z. Phys. Chem. (Leipzig)*, **262**, 567.
102 Kärger, J., Pfeifer, H., Richter, R., Fürtig, H., Roscher, W., and Seidel, R. (1988) *AIChE J.*, **34**, 1185.
103 Kärger, J., Pfeifer, H., Stallmach, F., Bülow, M., Struve, P., Entner, R., and Spindler, H. (1990) *AIChE J.*, **36**, 1500.
104 Kärger, J., Pfeifer, H., Seidel, R., Staudte, B., and Gross, Th. (1987) *Zeolites*, **7**, 282.
105 Kärger, J., Bülow, M., Millward, G.R., and Thomas, J.M. (1986) *Zeolites*, **6**, 146.
106 Kärger, J. and Pfeifer, H. (1987) *Zeolites*, **7**, 90.
107 Fraissard, J. and Ito, T. (1988) *Zeolites*, **8**, 350.
108 Fraissard, J. and Kärger, J. (1989) *Zeolites*, **9**, 351.
109 Bär, N.K., Kärger, J., Pfeifer, H., Schaefer, H., and Schmitz, W. (1998) *Microporous Mesoporous Mater.*, **22**, 289.
110 Chanajaree, R., Bopp, P.A., Fritzsche, S., and Kärger, J. (2009) *Diffusion Fundamentals III (Athens 2009)* (eds C. Chmelik, N. Kanellopoulos, J. Kärger, and D. Theodorou), Leipzig University Press, Leipzig, pp. 240–241.
111 Chanajaree, R., Bopp, P.A., Fritzsche, S., and Kärger, J. (2011) *Microporous Mesoporous Mater.*, **146**, 106.
112 Jost, S., Fritzsche, S., and Haberlandt, R. (2002), in *Impact of Zeolites and Other Porous Materials* (eds R. Aiello, G. Giordano, and F. Testa), Studies in Surface Science and Catalysis, vol. **142B**, Elsevier, Amsterdam, p. 1947.

17
Large Pore (12-Ring) Zeolites

From the practical point of view the 12-ring zeolites, particularly zeolites X and Y, are perhaps the most important of the zeolite structures. They are widely used as cracking catalysts and as selective adsorbents in several large-scale hydrocarbon separation processes [1–4]. Other 12-ring structures such as zeolite L, zeolite β, and mordenite are also used in several important industrial processes but they cannot easily be synthesized as large crystals and their diffusional properties have been studied in much less detail. We have therefore restricted this review to zeolites X and Y.

17.1
Structure of X and Y Zeolites

The synthetic zeolites X and Y and the natural mineral faujasite all have the same framework structure, shown schematically in Figure 17.1. The differences between these materials are in the Si/Al ratio (which determines the cation density) and in the nature of the cations. The framework may be thought of as a tetrahedral array of sodalite units interconnected through six-membered oxygen bridges or, equivalently, as a tetrahedral arrangement of double six-rings. The portion of the structure sketched in Figure 17.1 shows a single large "cage" formed between ten sodalite units. The "cage" is interconnected with neighboring cages through four tetrahedrally directed twelve-membered oxygen rings. The structure is very open and the distinction between "cages" and "windows" is much less clear than in the A structure.

The crystallographic unit cell consists of an array of eight "cages" containing a total of 192 AlO_2 and SiO_2 tetrahedral units. The free diameter of the 12-membered oxygen ring is about 7.5 Å. However, as a result of molecular vibration, molecules having critical diameters substantially larger than this (up to about 9.5 Å) can penetrate the pore structure at ordinary temperatures. A sorption cut-off is observed only for very large molecules such as perfluorotributylamine (critical diameter ~10.2 Å). The Si:Al ratio is in the range 1.0–1.5 for X and 1.5–3.0 for Y.

Diffusion in Nanoporous Materials. Jörg Kärger, Douglas M. Ruthven, and Doros N. Theodorou.
© 2012 Wiley-VCH Verlag GmbH & Co. KGaA. Published 2012 by Wiley-VCH Verlag GmbH & Co. KGaA.

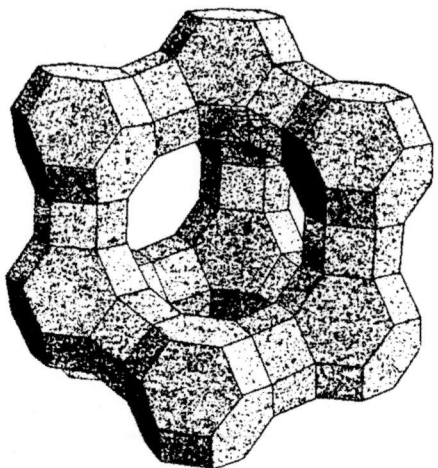

Figure 17.1 Schematic representation of the framework structure of zeolites X, Y, or natural faujasite. (There is either a Si or an Al atom at each vertex; the centers of the oxygen ions are at or near the centers of each line.)

High-silica Y zeolites with very much higher Si : Al ratios can be prepared by controlled dealumination of HY. This is sometimes followed by "desilication" [5] to generate mesopores in order to enhance the rate of mass transfer between the intracrystalline space and the surroundings of the crystal. (See Section 15.7.3). There is a corresponding difference in the number of exchangeable cations, which varies from 10 to 12 per cage for NaX to less than six for higher silica Y forms. Five different cation sites have been identified, as indicated in Figure 17.2. The cation distribution between these sites is much more complicated than in zeolite A and has been discussed by Meier, [6] Breck [7], and Smith [8]. A comprehensive summary including 69 references and giving information for many different cationic forms of both X and Y is included in Mortier's compilation of extra-framework cation sites [9].

The distribution of cations between these sites depends not only on the number and nature of the cations but also the history of the sample. The distribution can be modified by the presence of moisture and there is evidence that it may also change in the presence of organic sorbates [10–12]. Such changes in cation distribution can have a profound effect on the adsorptive properties. However, the location of a cation within the 12-membered oxygen ring is not energetically favorable so the dramatic variation of diffusivity with cation exchange, as in the NaCaA zeolites, is generally not observed in X or Y. Partial window blocking can nevertheless occur as a result of cations occupying the type II sites, which partially obstruct the 12-ring window. Thus, diffusion of critically sized molecules, such as 1-methylnaphthalene, is much slower in CaX than in NaX [13]. Since even in zeolite X the number of cation sites greatly exceeds the number of cations, the possibilities for redistribution of the cations among the energetically equivalent sites are far greater than in zeolite A. It appears that some of the complexity that is observed in the diffusional behavior can be attributed to such effects.

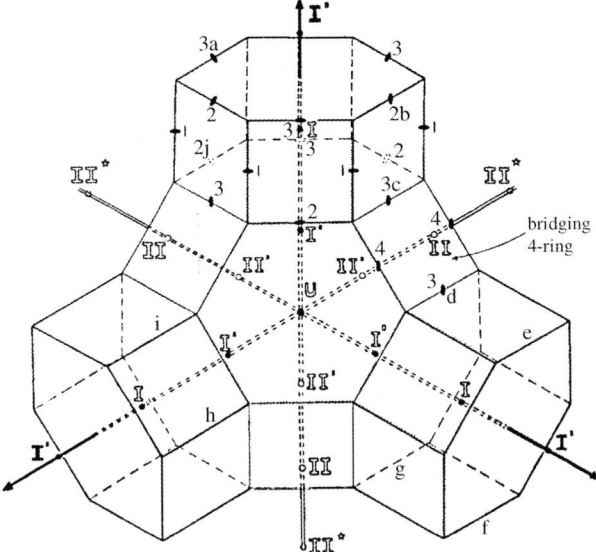

Figure 17.2 Schematic diagram showing the disposition of the cation sites in zeolites X and Y. (The numbers of each of these sites per unit cell are given in parentheses.) Site I (16) is at the center of the hexagonal prism; site I' (32) is within the sodalite cage adjacent to the double six ring (hexagonal prism); site II' (32) is within the sodalite cage adjacent to the single six ring; site II (32) is within the large cage adjacent to the single six ring (opposite site II'): site III (48) represents the positions on the wall of the supercage (adjacent to a four-ring). Site IV (16) represents the sites within the 12-ring. A unit cell contains eight supercages and 192 SiO_2/AlO_2 tetrahedra. There are 176 cation sites, of which a maximum of 96 are occupied in NaX with $Si/Al = 1.0$. The four different types of oxygen are also indicated. Reprinted from J.V. Smith [8] with permission.

Diffusion in these structures is much faster than in the smaller-pore A zeolites, so reliable diffusivity data can only be obtained by macroscopic methods if large crystals are used. Diffusion of a wide range of molecules is, however, accessible to measurement by NMR methods.

17.2
Diffusion of Saturated Hydrocarbons

17.2.1
Evidence from NMR

The linear paraffins provide a convenient example of a homologous series of structurally similar compounds and, as their relaxation times in NaX zeolite are relatively long, their diffusional behavior has been studied in considerable detail by NMR methods [14–16]. The results are summarized in Figures 17.3–17.6 and Tables 17.1–17.3. The concentration dependence of the diffusivities (Figure 17.3)

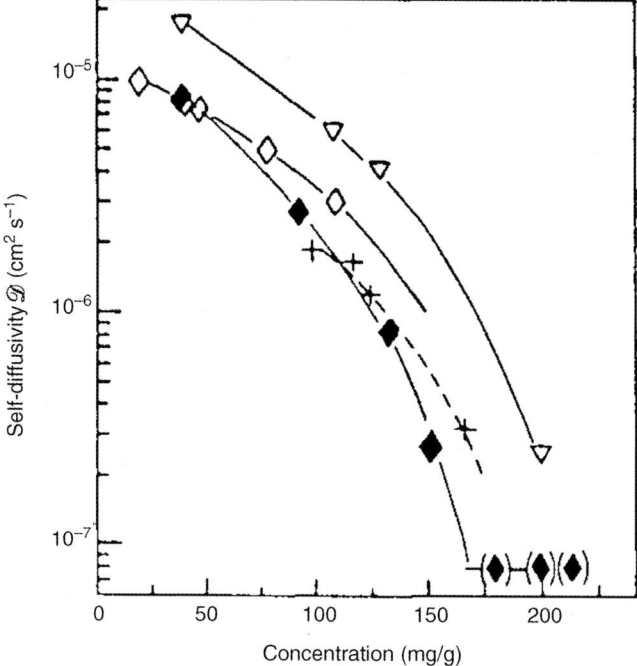

Figure 17.3 Concentration dependence of NMR self-diffusivity for paraffinic hydrocarbons in Zhdanov NaX zeolite crystals at 358 K. (Inverted open triangles) n-heptane, (◇) n-octane, (+) n-decane, and (◆) isooctane. Points in parentheses refer to pore-filling factors greater than 1.0 (i.e., free liquid present). The contribution of the free liquid has been eliminated in the data analysis. Reprinted from Kärger and Ruthven [16] with permission.

is qualitatively similar for all homologs and follows the type I pattern of zeolitic diffusion (Figure 3.3). At low loadings the self-diffusivities in NaX are about one order of magnitude lower than the corresponding values for the free liquid. This is about the magnitude of the reduction to be expected from the geometric constraints imposed by the pore structure (tortuosity) together with some contribution due to pore wall rigidity, suggesting that there is very little steric hindrance. However, this may be fortuitous, since more detailed investigation reveals that the diffusion mechanisms in the liquid and the zeolite are in fact quite different. With increasing concentration, the diffusivity of the adsorbed species decreases strongly, falling by about two orders of magnitude between zero and the saturation limit. Evidently, the mobility is severely reduced by intermolecular collisions. For methane and ethane the diffusional activation energy *decreases* with increasing sorbate concentration, but for butane and all higher homologs it was found to be almost independent of loading. The activation energy increases with carbon number up to about C_6 and then levels off at about 15 kJ mol^{-1} for the higher homologs. This is similar to the trend of activation energies for diffusion in the liquid phase. The reduction in diffusivity with increasing carbon number also follows the same trend as for diffusion in the free liquid.

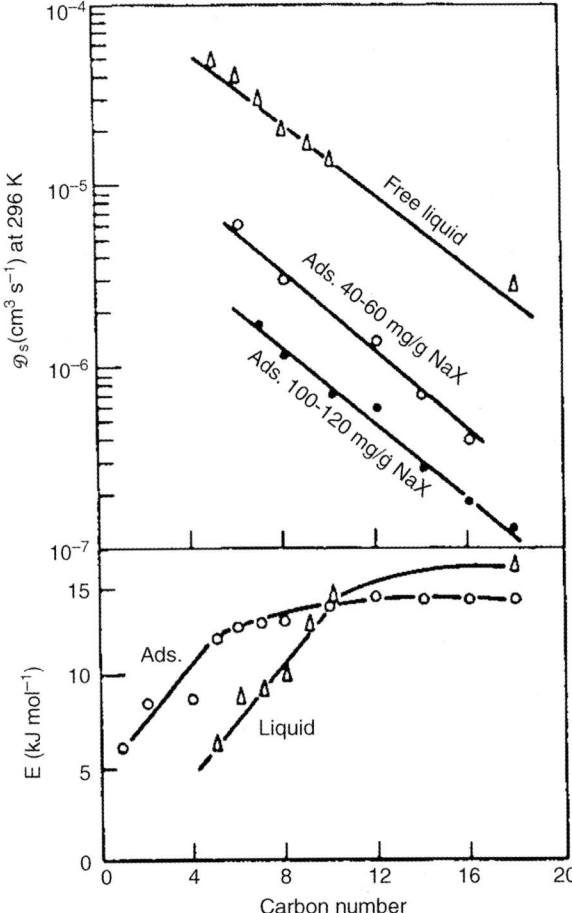

Figure 17.4 Comparison of self-diffusivities (at 298 K) and diffusional activation energies for linear paraffins as free liquids and adsorbed on NaX zeolite crystals. Data of Kärger et al. [14].

Figure 17.5 shows the variation of the longitudinal nuclear magnetic relaxation times T_1 (for n-hexane) with temperature. With increasing sorbate concentration the T_1 minimum is broadened but the temperature of the T_1 minimum remains almost constant. This is typical of a system in which the jump frequency remains constant (Section 10.1.2) but the mean jump length decreases with increasing loading. The pronounced fall in diffusivity with concentration can therefore be ascribed to a reduction in the mean jump length rather than to a change in the jump frequency. By combining the relaxation time and self-diffusion measurements the mean jump length may be calculated and the results of such calculations are shown in Figure 17.6. At low loadings the mean jump length is about 9–10 Å, suggesting that molecular jumps occur primarily between adjacent cages. However, at high loadings the mean jump length falls to about 1 Å, implying that under these conditions segmental

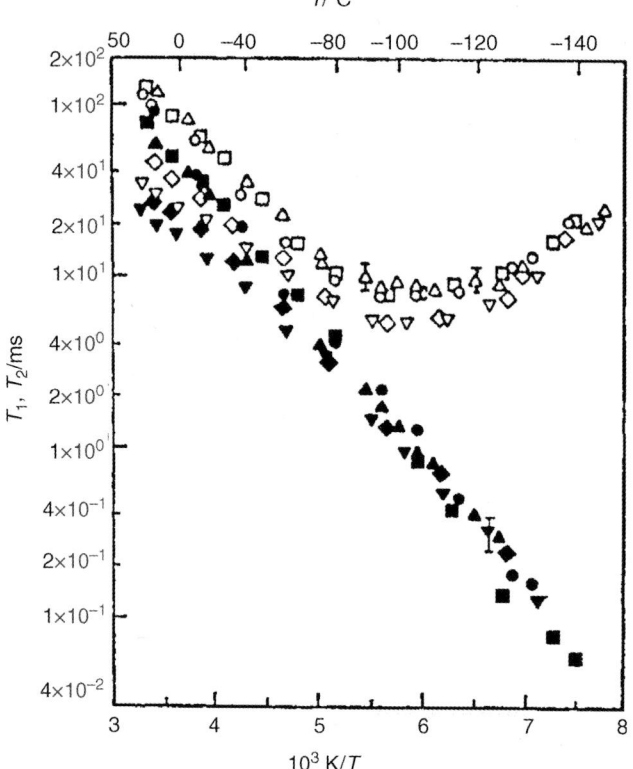

Figure 17.5 Temperature dependence of longitudinal (T_1) and transverse (T_2) relaxation times for n-hexane in NaX zeolite at different sorbate loading: T_1, open symbols; T_2, filled symbols; (△) 21, (□) 56, (○) 110, (◊) 161, and (▽) 243 mg g^{-1}. Reprinted from Kärger et al. [14] with permission.

movements are probably dominant, as for diffusion in the liquid phase. Note that this dramatic decrease of more than an order of magnitude in the jump length is associated with a corresponding decrease in the diffusivities of over two orders of magnitude with increasing loading (Figure 17.3).

The simple model [17] presented in Section 4.4.2 suggests that the pre-exponential factors for the diffusivity (at low loadings) and the Henry's law constant can both be related to the diffusional activation energy [Eqs. (4.48) and (4.49)]. The experimental data for the lower molecular weight species are in reasonable agreement with this prediction, as may be seen from Table 17.1. Furthermore, the concentration dependence is reasonably consistent with the free-volume theory (Section 4.4.3). Simple models are thus seen to provide useful approximations for the diffusivities. A more accurate treatment including a proper assessment of the guest–cation interaction [24] requires rigorous molecular dynamics simulations [25, 26].

QENS diffusion studies with n-pentane at 360 K at a loading of about 1 molecule per supercage yield a value of 3.5×10^{-9} m^2 s^{-1}, which is of the order of the above

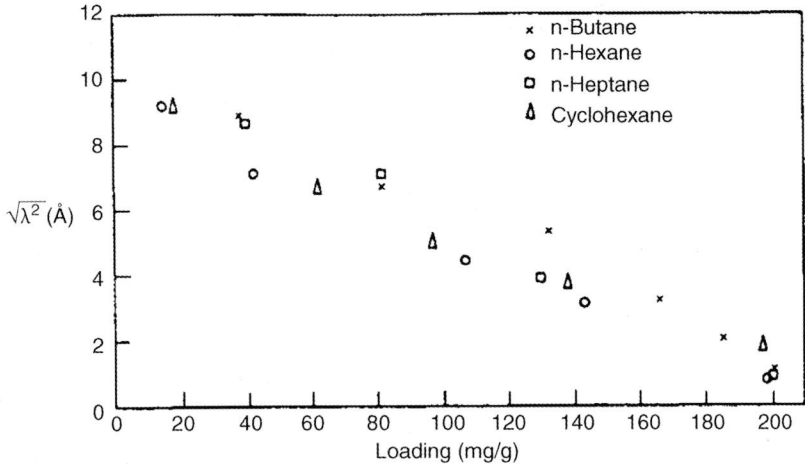

Figure 17.6 Variation of RMS jump distance with sorbate concentration for saturated hydrocarbons on NaX zeolite. Data of Kärger et al. [14].

reported PFG (pulsed field gradient) NMR diffusivities for n-butane and n-hexane [27]. With temperature variation, however, the activation energy of diffusion as obtained by QENS (6 kJ mol^{-1}) is found to be significantly smaller than the PFG NMR value. Further experimental evidence supporting the hypothesis that this difference arises from the presence of intracrystalline transport barriers acting in addition to the diffusional resistance of the genuine pore network is presented in Section 17.2.6. The spacing of these barriers must be smaller than the diffusion path length covered by PFG NMR (μm), so that the PFG NMR diffusivities are reduced, but large enough to have no influence on the QENS measurements, which operate at the nm scale.

Table 17.1 Comparison of theoretical and experimental values of K_∞ and $\mathcal{D}_\infty$ for hydrocarbons in NaX zeolites. After Kärger et al. [17].

	Experimental values			$\dfrac{(\mathcal{D}_\infty)_{\text{theory}}}{(\mathcal{D}_\infty)_{\text{expt}}}$	$\dfrac{(K_\infty)_{\text{theory}}}{(K_\infty)_{\text{expt}}}$
	V_0 (kJ mol^{-1})	$\mathcal{D}_\infty \times 10^3$ (cm^2 s^{-1})	$K_\infty \times 10^9$ (molecule per cage Pa)		
CH$_4$	6.7	3.5	7.6	1.3	1.5
C$_2$H$_6$	9.6	5.0	5.0	0.8	1.2
nC$_4$H$_{10}$	10.5	1.9	1.5	1.5	3.9
nC$_6$H$_{14}$	14.6	6.3	1.5	0.5	1.9
nC$_7$H$_{12}$	12.5	1.6	2.3	1.6	1.6
cyclo-C$_6$H$_{12}$ (sorbate concentration 15–30 mg g^{-1})	15.5	6.4	1.5	0.5	1.7

Table 17.2 Diffusion of n-butane in NaX crystals.

Crystal diameter (μm)	Expt. method	T (K)	D (m² s⁻¹)	E (kJ mol⁻¹)	Comments	Reference
1.0	Carberry mixer	308	$\sim 10^{-18}$	33.5	1.8 mmol g⁻¹	Ma and Lee [18]
2 (NaY)	Chromatography	400	2×10^{-15}	60	D_0	Fu et al. [19]
100	Membrane permeation	314	$\sim 10^{-11}$ (SDS) $\sim 10^{-8}$ (RDS)		Integral D values (0 to sat.)	Wernick and Osterhuber [20]
100	Gravimetric uptake	308	$\sim 10^{-12}$		D_0	Tezel et al. [21]
		343	1.5×10^{-11}		Effect of residual moisture noted	
80	Gravimetric uptake	298	$\geq 2 \times 10^{-11}$		D_0	Doelle and Riekert [22]
50	ZLC	353	3.5×10^{-11}			
		353	3.5×10^{-11}			
100	ZLC	373	3.9×10^{-11}	6	D_0	Eic and Ruthven [23]
		423	4.9×10^{-11}			
50	PFG NMR	223	$\sim 10^{-9}$	9.5	Extrapolated to zero concentration ($\mathcal{D}$)	Kärger et al. [14]
		358	3×10^{-9}			

Table 17.3 Comparison of NMR and ZLC diffusivity data for saturated hydrocarbons in NaX (358 K).[a]

Sorbate	ZLC		NMR	
	D_0 (m² s⁻¹)	E (kJ mol⁻¹)	$\mathcal{D}$ (m² s⁻¹)	E (kJ mol⁻¹)
nC_4H_{10}	4×10^{-11}	6	5×10^{-9}	9.5
nC_5H_{12}	2.5×10^{-11}	12.5	—	—
nC_6H_{14}	2.0×10^{-11}	14	4.5×10^{-9}	14.5
Cyclohexane	1.4×10^{-11}	15	3.5×10^{-9}	15.5
nC_7H_{16}	1.0×10^{-11}	20	2.0×10^{-9}	12.5
$nC_{10}H_{22}$	5×10^{-12}	22	8×10^{-10}	14
$nC_{14}H_{30}$	3.1×10^{-12}	20	5×10^{-10}	14.5

a) NMR diffusivity values are for low concentration limit. Data are from Kärger et al. [14] and Eic and Ruthven [23]. To facilitate comparison data are quoted at 358 K. The values at this temperature have been obtained by extrapolation where necessary.

In addition to the NMR measurements, discussed above, diffusion of n-butane in NaX has been studied by various macroscopic methods. The results, some of which are summarized in Table 17.2, provide a clear illustration of the range of reported diffusivity values and the discrepancies between the values determined by different experimental methods. It now seems clear that the very low values reported, for example, by Ma and Lee [18] who used a "Carberry mixer" can be ascribed to the unrecognized intrusion of external resistances to mass and heat transfer [28]. However, even when these extremely low values are eliminated there remains a variation of two to three orders of magnitude between the highest and lowest values reported within the temperature range 300–350 K. This is considered further in Sections 17.2.6 and 17.3.4.

17.2.2
Comparison of NMR and ZLC Data for NaX

Diffusion of several linear paraffins and cyclohexane in NaX has also been studied by the ZLC (zero-length column) method using 50 and 100 μm crystals [23]. In these crystals the desorption curves are slow enough to be easily measured (time scale of order 1–2 min) and conformity to the diffusion model was confirmed by varying the purge gas and the crystal size. Good agreement was observed between the diffusivity values derived for the two different crystal sizes and with the He and Ar as the purge gases. The data are summarized in Table 17.3 and Figure 17.7, in which the comparison with NMR values measured at low sorbate concentration is also shown. The ZLC method provides no information concerning the concentration dependence of the diffusivity, since only the limiting diffusivity value at low concentrations can be found by this technique. The trends of diffusivity and diffusional activation energy with carbon number are very similar for both the ZLC and NMR data. However, at the

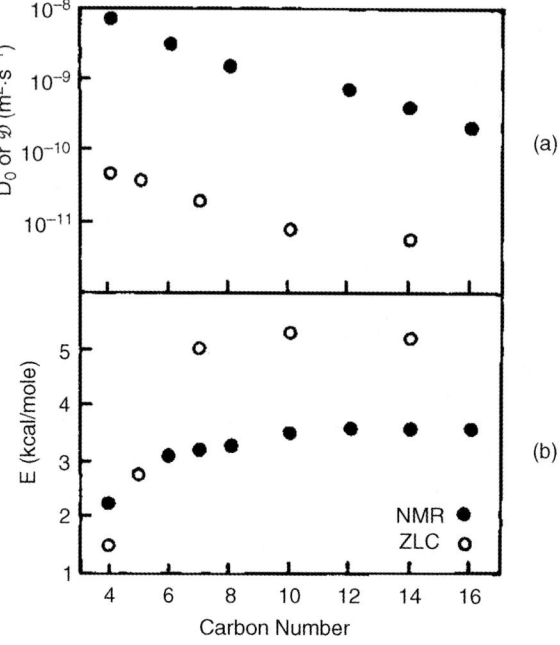

Figure 17.7 Comparison of diffusivities for linear paraffins in NaX zeolite as measured by PFG NMR (●,$\mathcal{D}$) and ZLC (○, D_0) methods: (a) diffusivities at 423 K; (b) diffusional activation energies. Reprinted from Eic and Ruthven [23] with permission.

temperature considered there is a difference of about two orders of magnitude between the ZLC and NMR diffusivities and this ratio is roughly independent of carbon number, as may be seen from Figure 17.7a.

17.2.3
Isoparaffins and Cyclohexane

PFG NMR diffusion studies with isoparaffins [16] and cyclohexane [15] in NaX showed broadly similar behavior to that of the linear paraffins [14]. Activation energies are somewhat lower than the values for the corresponding linear isomers and, for cyclohexane, the activation energy shows a significant decrease with loading (from about 3.0 to 2.1 kcal mol^{-1}). However, the diffusivity of cyclohexane is very close to that of n-heptane. In its diffusional behavior cyclohexane clearly resembles n-hexane and n-heptane more closely than it resembles benzene (Section 17.3.3). Moreover, there is a tendency that, for a constant carbon number, the diffusivities decrease with increasing bulkiness. Such behavior is exemplified by the diffusivity data shown in Figure 17.3.

Both the simulation data and the QENS diffusivities for the pentane isomers in NaY, (Figure 17.8) show a surprising trend: the largest molecule (neopentane) has

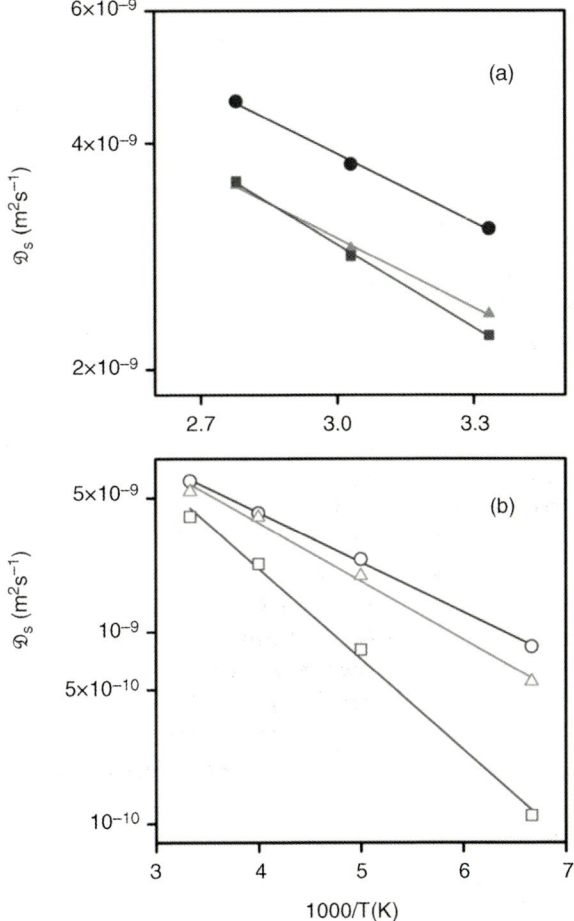

Figure 17.8 (a) QENS and (b) simulation data for the self-diffusivities of pentane isomers in NaY: (■, □) n-pentane, (▲, △) isopentane, and (●, ○) neopentane. Reprinted from Reference [26] with permission.

the highest diffusivity while the smallest molecule (n-pentane) has the lowest diffusivity. Diffusivities increasing rather than decreasing with increasing molecular size might be taken as evidence of the "levitation effect" [29–31]. In support of that idea it may be noted that the activation energy also appears to decrease with increasing molecular size. However, this evidence is certainly not unambiguous since PFG NMR measurements with pentane isomers in NaX show the opposite trend of slightly decreasing diffusivity with increasing molecular size [32], which seems intuitively more plausible. It is possible that this difference in behavior is due to differences in the cation contents of NaX and NaY and/or the greater length scale of the PFG NMR measurements, which makes them more sensitive to differences in cation position [24] and structural defects, but without consistency in the data from different techniques the interpretation of the QENS data as proof of the "levitation effect" is certainly open to question.

17.2.4
n-Octane Diffusion in NaY, USY and NaX

Diffusion measurements with NaY are rendered more difficult by the unavailability of large crystals. The crystals shown in Figure 17.9, which have mean diameters in excess of 1 μm, were prepared by Berger et al. [34]. Figure 17.9c shows the effective diffusivities from PFG NMR measurements with n-octane (at about half saturation loading) at three different temperatures, as a function of the mean molecular displacements covered in the measurements. The data provide a direct comparison with the diffusivities for the ammonium-ion exchanged zeolite Y (which, after treatment at 673 K for 40 h under high vacuum, is expected to be fully converted into HY) and the dealuminated species referred to as USY [35], which was produced by steaming the HY crystals under a water vapor pressure of 1 bar at 873 K [33]. The USY so produced contains mesopores, generally assumed to be separated from each other [36], which are formed during the process of dealumination.

The decrease in the diffusivities with increasing displacements suggests a limitation of the displacements corresponding to the crystal size (Section 11.4.1 and discussion related to Figure 11.2). Dynamic Monte Carlo simulations (Section 9.2.2) of molecular displacements in a cubic lattice with an edge length of 2.3 μm, which is close to the measured crystal size for this sample (2.7 ± 0.6 μm), shown by the solid lines in Figure 17.9, are in good agreement with the experimental measurements. Interestingly, this confinement is clearly visible only for the USY sample, while in the

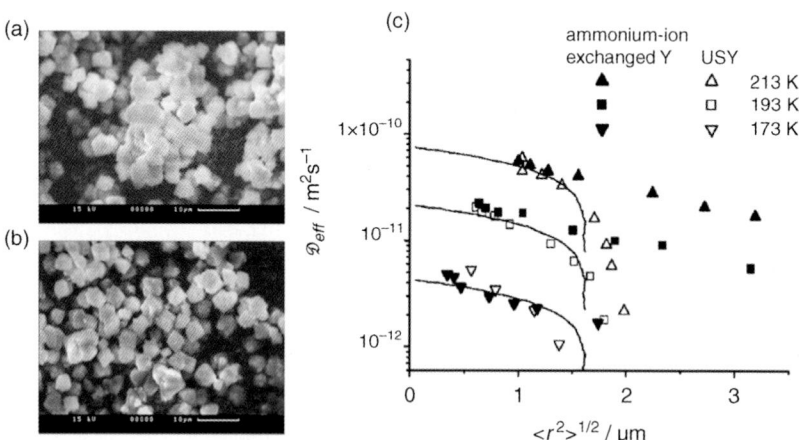

Figure 17.9 SEM images of an ammonium-ion exchanged zeolite Y (a) and of USY (b) produced by steaming of the NH_4-Y; (c) PFG NMR diffusivities (measured with n-octane as a probe molecule) in these samples (symbols), in comparison with the results of dynamic Monte Carlo calculations (lines). Reprinted from Reference [33] with permission.

ammonium-exchanged zeolite Y the observed diffusion path lengths may clearly exceed the crystal size. This finding is consistent with the direct evidence from the micrographs (Figure 17.9a and b), from which it is obvious that the original NH$_4$Y crystals show clear evidence of aggregation, whereas for the USY sample the individual crystals are separated.

Figure 17.9 also shows that the intracrystalline diffusivities of *n*-octane in the parent crystals (which do not contain mesopores) coincide with those in the dealuminated (mesoporous) USY. Such behavior might be interpreted to mean that the mesopores are not interconnected and therefore do not contribute to intracrystalline transport. However, an alternative possibility is that the fraction of molecules contained in the mesopores is too small to contribute significantly to the overall diffusivity (compare Section 15.7.3 and Reference [37]). Extrapolation of the *n*-octane diffusivities in zeolite Y (presented in Figure 17.9) to the temperature range of previous PFG NMR diffusion studies with *n*-octane in NaX [38] shows close agreement [33]. This implies not only that the effect of the mesopores on the overall diffusion rate in USY must be minimal, but also that intracrystalline diffusion of *n*-octane in the X/Y pore system is not significantly affected by the presence of Na$^+$ cations (see also Sections 17.2.5 and 17.3.3.2).

17.2.5
Diffusion in NaCaX: Effect of Ca^{2+} Cations

In view of their higher electrostatic field, calcium ions may be expected to give rise to a much stronger interaction with the saturated hydrocarbons. In zeolite NaCaA (Chapter 16) replacement of Na$^+$ by Ca^{2+} simultaneously changes the sorbate–sorbent interaction and the number of "open" and "closed" windows, thus potentially affecting the guest diffusivities in opposite directions. The more open framework of the X-type zeolites, therefore, provides better prospects for studying the influence of the cations. Figure 17.10 shows a comparison of the diffusivities of methane in zeolite Na-X and in zeolite NaCaX with 30% and 75% of Na$^+$ replaced by Ca^{++} [39]. Since the first calcium cations are known to assume positions outside the large cavities [40], they are not in direct contact with the methane molecules. Therefore, the diffusivities of methane in NaX and Na30CaX are the same (following the type I concentration dependence as classified by Figure 3.3). With higher calcium content, however, a significant change in the diffusion properties occurs: the diffusivities show exactly the opposite concentration dependence (type V concentration dependence), and at low sorbate loadings they are as much as two orders of magnitude smaller than in zeolite NaX. Obviously, the interaction with the calcium ions must dramatically reduce the mobility of methane. Interestingly, at high loadings, approaching saturation, the methane diffusivities in NaX and Na(75)CaX approach each other.

Similarly, at low loadings, the *n*-hexane diffusivity in Na75CaX is also found to be much smaller than in NaX [39]. In this case, however, the diffusivities approach each other at intermediate loadings, so that the *n*-hexane diffusivity in Na75CaX passes through a maximum, following the type III concentration pattern.

17 Large Pore (12-Ring) Zeolites

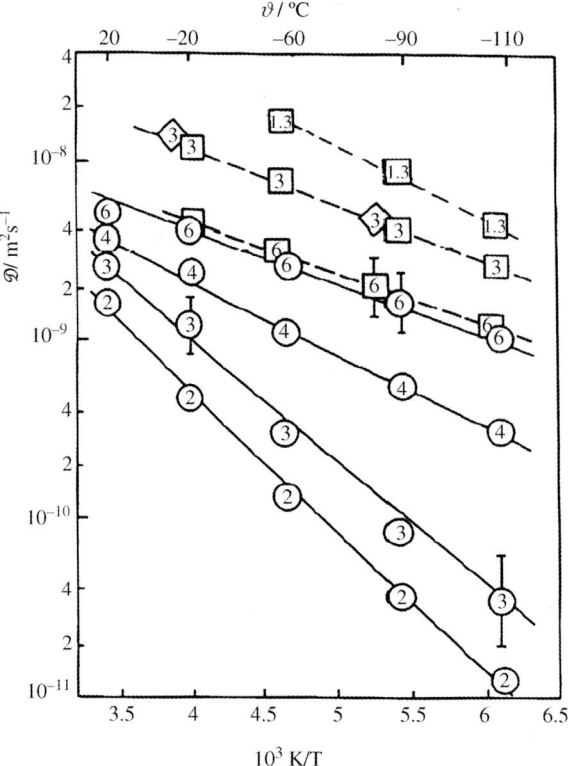

Figure 17.10 Coefficient of intracrystalline self-diffusion of methane in zeolite (○) Na75CaX, (◇) Na30CaX, and (□) NaX determined by PFG NMR; the numbers inserted in the symbols indicate the concentration in molecules per supercage. Reprinted from Reference [39] with permission.

Comparison of the diffusivities in NaCaX and NaCaA at comparable calcium exchange shows [39] that the methane diffusivities are of the same order, while the *n*-hexane diffusivity in NaCaA is much smaller than in NaCaX. One has therefore to conclude that, for diffusion of the longer *n*-alkanes in NaCaA, it is the passage through the windows that is the rate-limiting step. For methane and ethane, however, the transport properties are dominated by the interaction with the cations. MD simulations of methane diffusion in NaCaA are in satisfactory agreement with the NMR data and confirm this conclusion [41]. As a consequence of the much smaller window diameters, window blocking is more easily achieved in A than in X zeolite. However, there are examples, where the co-adsorption of a second molecular species (e.g., water in addition to C_2 or C_4 hydrocarbons) leads to a dramatic reduction in the molecular mobility, suggesting the formation of adsorption complexes in the windows between adjacent supercages (References [42, 43] and Section 17.6.2.1).

17.2.6
Diffusion Measurements as Evidence of Structural Imperfection

Although most of the diffusion studies so far presented appear to be self-consistent, the combination of the results of different measurement techniques is far from coherent. While it is probably true that some of these differences are due to shortcomings in the experimental procedure and/or in data analysis, over the last few years evidence has been accumulating that, at least in some cases, the origin of such differences must be attributed to deviations from the ideal crystal structure and the resulting differences between different specimens. Such structural irregularities may be considered to be entropic in origin and an inevitable consequence of the fact that, in a complex crystal structure, the probability that perfect regularity of the pore network will persist over millions of unit cells is negligibly small. These deviations may occur in the form of stacking faults that prohibit molecular propagation through certain planes. Since even a small number of such faults can have a dramatic effect on the intracrystalline diffusivity they are unlikely to be visible in diffraction and imaging studies. The strongest evidence of such effects comes from diffusion measurements over different length scales.

As an example, Figure 17.11 shows a comparison between the diffusivities of *n*-octane in NaX as determined by QENS (over displacements of nanometers – see Section 10.2) and by PFG NMR (with displacements over micrometers – see Section 11.2). In this study the PFG NMR diffusivity data for *n*-octane in NaX also show type-I concentration dependence, as may be seen from Figure 17.3. However, QENS measurements performed with the same zeolite sample yield notably larger

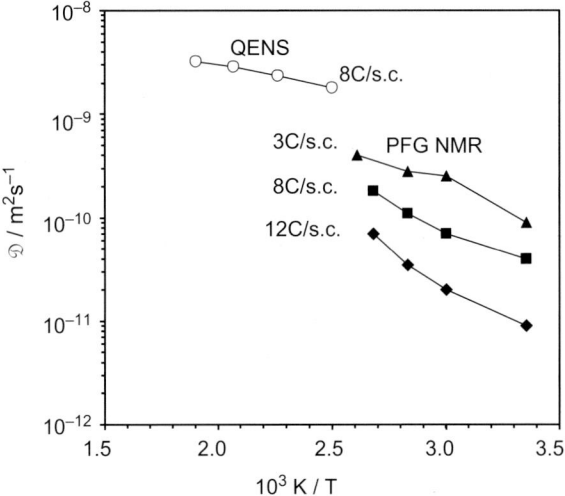

Figure 17.11 Arrhenius plots of the PFG NMR self-diffusivities of *n*-octane in a specimen of zeolite NaX at different loadings (in carbon atoms per supercage) and comparison with the diffusivity data obtained by QENS, Reprinted from Reference [44] with permission.

diffusivities. In addition, the slope of the Arrhenius plot of the QENS diffusivities is smaller, indicating a lower activation energy. Both findings suggest the existence of transport resistances with separations in the range between hundreds of nanometers and micrometers, which are large enough to be of no relevance for the QENS measurements but which clearly affect the displacements measured by PFG NMR. Moreover, if the internal barriers are formed by impenetrable layers with "holes," the effective diffusivity over length scales sufficiently exceeding the barrier separations is seen [from Eqs. (2.35)–(2.37)] to be proportional to the true intracrystalline diffusivity, even though the effect of the barriers may lead to a reduction by orders of magnitude.

If passage through the "holes" in these barriers requires a higher activation energy than diffusion through the genuine pore network, the long-distance diffusivity may, eventually, also depend on the rate of passage through theses "holes" so that the measured activation energy will become greater than the value for diffusion through the ideal pore system. The experimental findings and model considerations described in Section 19.6 (Figures 19.24 and 19.27) suggest exactly this type of behavior [for diffusion in MOF Zn(tbip)]. In zeolite X, stacking faults as revealed by transmission electron microscopy [see Figure 2.5 and Reference [44]], may well be associated with such transport barriers and this would provide a logical explanation for the observed dependence of diffusivity on the length scale of the measurement.

Figure 17.12, which shows the PFG NMR diffusivity data obtained with n-butane in two different specimens of large NaX crystals, provides another example demonstrating how diffusion studies may provide evidence of structural defects. From both XRD and the SEM images the crystals appeared to have the ideal structure [47]. However, the large difference in the diffusivities clearly shows that in one of the samples (Yang's synthesis) mass transfer is much slower. The difference in diffusivities was found to be most pronounced at small loadings, at which a broad

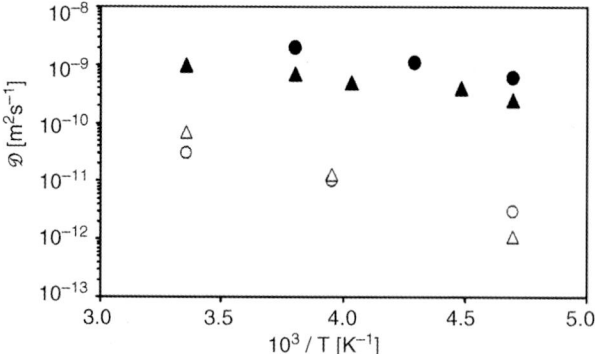

Figure 17.12 PFG NMR diffusivities of n-butane in different specimens of zeolite NaX (filled symbols): specimens synthesized by S.P. Zhdanov [38, 45]; (open symbols): specimens synthesized by X. Yang [46] at loadings of 0.75 (circles) and 3 (triangles) molecules per large cavity. Reprinted from Reference [47] with permission.

distribution of the local diffusivities was observed, suggesting a twofold heterogeneity that was not apparent from either the XRD or SEM images [47].

The presence of intraparticle transport barriers was also detected in PFG NMR studies with FAU/EMT intergrowths [48]. The intraparticle diffusivities in the intergrowths were found to be far smaller than the values estimated from the data corresponding to the pure FAU and EMT zeolites.

17.3
Diffusion of Unsaturated and Aromatic Hydrocarbons In NaX

17.3.1
Light Olefins

The PFG NMR data suggest that the self-diffusivity in NaX decreases by about one order of magnitude in the sequence n-butane > *trans*-2-butene > 1-butene > *cis*-2-butene [49]. The activation energies for all these species are similar and vary little with loading. Correlation times as deduced from nuclear magnetic relaxation measurements are also similar, suggesting that the difference in diffusivities is due to a difference in the RMS (root mean square) jump distance, which evidently varies, possibly as a result of different degrees of steric hindrance. For every jump between cages, the slowest diffusing species (*cis*-2-butene) makes on average about eight jumps within a cage compared with only about two intra-cage jumps for *trans*-2-butene. This difference is attributed to the orienting effect of the small but significant dipole of *cis*-2-butene.

Partial ion exchange of Na^+ for Ag^+ leads to a dramatic reduction in the diffusivity for olefins (more than two orders of magnitude at 20% exchange) as a result of strong interaction between the double bond and the Ag^+ cation [50]. There is a parallel increase in correlation time reflecting the dramatic increase in the mean residence time at an adsorption site, rather than any change in the RMS jump distance. Ion exchange for Tl^+ has the opposite effect, increasing the diffusivity relative to the Na^+ form. The reasons for this are less obvious.

The decrease of the translational mobility of linear olefins in AgNaX with increasing silver exchange has also been observed in ^2H NMR measurements with ethylene (Section 10.1.3.3) [51]. For the original NaX sample the diffusivities calculated by line shape analysis, assuming a jump distance of the order of the distance between the centers of adjacent supercages ($\sim$11 Å), were found to agree well with PFG NMR self-diffusion measurements [52]. For the silver-exchanged samples, however, the diffusivities estimated in the same way from the ^2H NMR relaxation measurements were about an order of magnitude smaller than the values determined from the self-diffusion measurements. The ethylene–NaX system has also been studied in some of the first diffusion measurements by neutron scattering [53]. The resulting diffusivity ($\sim 3 \times 10^{-10}\,m^2\,s^{-1}$) is smaller than the PFG NMR values by a factor of about four. More recent QENS studies with propylene adsorbed in NaY yielded diffusivities around $10^{-9}\,m^2\,s^{-1}$ [54], in closer agreement with the NMR data.

17.3.2
C$_8$ Aromatics

17.3.2.1 Macroscopic Measurements

A detailed study of diffusion of the three xylene isomers and ethylbenzene in large crystals of NaX zeolite and natural faujasite was undertaken by Goddard [55–57] and Eic [58, 59] using three different experimental techniques: sorption rate, isotopic exchange, and ZLC desorption measurements. The results are summarized in Figures 17.13–17.15 and Table 17.4.

In the larger crystals (100 μm NaX, 250 μm faujasite) uptake rates were easily measurable (half-times of the order of 200–300 s) as may be seen from Figure 17.13. Replicate measurements with different quantities of adsorbent showed no significant effect on the sorption rate, indicating no significant intrusion of external bed diffusion or heat transfer limitations. In the smaller (55 μm) crystals such effects are significant, except when a very small sample (2–3 mg) is used [59]. Time constants derived from the transient sorption curves measured with different crystal size fractions were entirely consistent (Figures 17.13 and 17.14). The time constants show

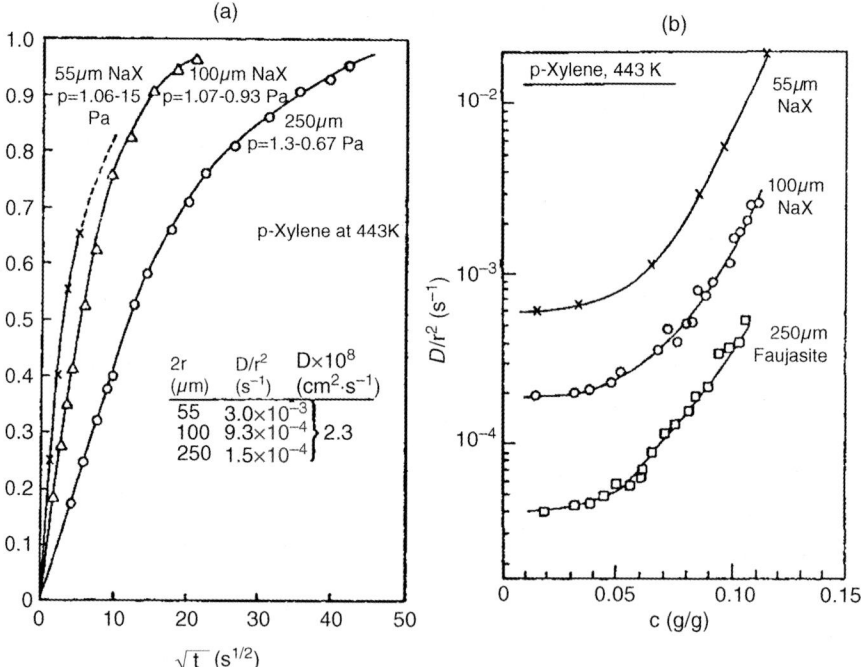

Figure 17.13 (a) Experimental uptake curves for p-xylene at 443 K, measured over comparable pressure steps in 55 μm and 100 μm NaX crystals and 250 μm natural faujasite. (b) Concentration dependence of diffusion time constants for p-xylene derived from gravimetric uptake rate measurements over small differential concentration steps [as in (a)]. The values of D/r^2 vary with $1/r^2$, confirming intracrystalline diffusion for all three samples. From Goddard and Ruthven [60], with permission.

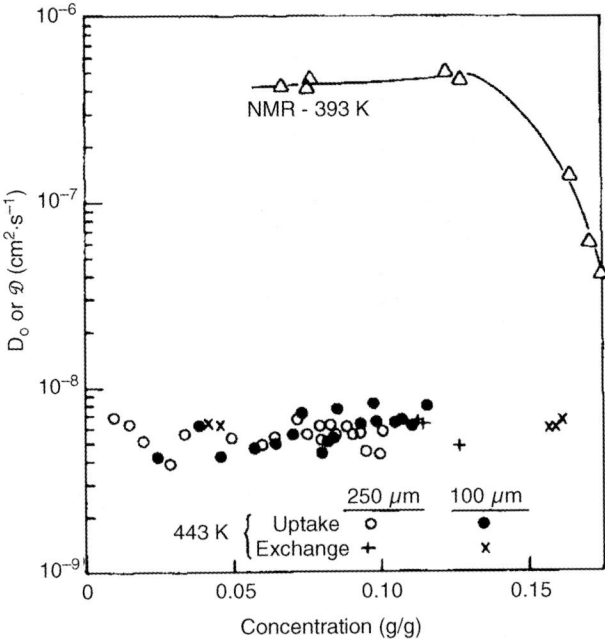

Figure 17.14 Comparison of NMR PFG self-diffusivities for xylene isomers in NaX zeolite crystals at 393 K [61] with corrected transport and tracer exchange [56, 57], diffusivities in 100 μm NaX and 250 μm natural faujasite crystals at 443 K. Adapted from Kärger and Ruthven [62] with permission.

the expected variation with the square of the crystal radius but the calculated diffusivity values are essentially constant.

The differential diffusivities show a strong increase with increasing sorbate concentration (Figure 17.13) but, as with diffusion in 5A zeolite, this trend appears to be entirely attributable to the nonlinearity of the isotherm. Corrected diffusivities, calculated according to Eq. (1.31) show a good deal of experimental scatter (Figure 17.14) but no discernible trend. Furthermore, the corrected transport diffusivities are in good agreement with self-diffusivities measured by tracer exchange using deuterated and non-deuterated sorbates. Differences in diffusivity between the four isomers are insignificant. There is, moreover, good agreement between the diffusivity values obtained with NaX and with natural faujasite, suggesting that it is the framework structure, rather than the number or nature of the cations, that controls the diffusion rate.

These diffusivity values were confirmed by ZLC measurements, which were carried out with 50 and 100 μm crystals and with He and Ar as the purge gases, with entirely consistent results. These data thus present a simple and coherent picture of diffusion, on a macroscopic time scale, in these systems.

17.3.2.2 Comparison with Microscopic Measurements

Although diffusion for C_8 aromatics in zeolite X is faster than for light paraffins in 5A, the pattern of behavior is qualitatively similar. However, in contrast to the

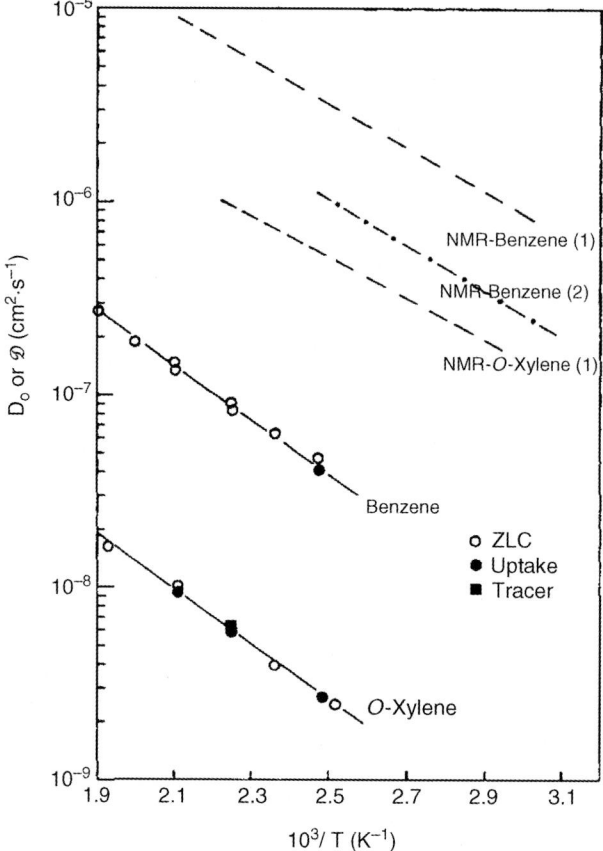

Figure 17.15 Arrhenius plot showing the temperature dependence of NMR self-diffusivity ($\mathcal{D}$) and corrected transport diffusivity (D_0) for benzene and o-xylene in NaX zeolite (and natural faujasite); crystal size range 55–250 μm. (1) Data of Germanus et al. [61] and (2) data of Kärger and Ruthven [16]. Reprinted from Kärger and Ruthven [62].

n-butane–5A and propane–5A systems, for which there was good agreement between the macroscopic and NMR diffusivity values, for the C_8 aromatics we see a large difference in diffusivity, although approximately the same activation energies (Figure 17.15).

The difference in the trends of D_o and $\mathcal{D}$ with concentration (Figure 17.14) is also striking. Only at very high sorbate concentrations is there order of magnitude agreement between the NMR and sorption rate data. Notably, QENS studies yield values close to the PFG NMR diffusivities, with, however, a smaller activation energy and diffusivities that increase with loading [63]. p-Xylene diffusivities are found to exceed those of m-xylene by a modest factor, depending on loading. This difference is enhanced to a factor of 3 in BaX [64].

Table 17.4 Comparison of PFG NMR self-diffusivities ($\mathcal{D}$) and corrected transport diffusivities (D_0) for xylene isomers in NaX zeolite at 400 K.

Sorbate	Sorbent	Technique	D_0 or $\mathcal{D}$ (m^2 s^{-1})	E (kJ mol^{-1})	Reference
o,m,p-Xylenes and ethylbenzene	NaX crystals (8–50 μm)	PFG NMR	5×10^{-11}	20–25	Germanus et al. [61]
	NaX crystals (55, 100, 250 μm)[a]	Gravimetric uptake	2.5×10^{-13}	26	Goddard and Ruthven [55, 56]
	NaX crystals (55, 100, 250 μm)[a]	Tracer exchange (C_6D_6)	2.5×10^{-13}	26	Goddard and Ruthven [57]
	NaX crystals (55, 100 μm)	ZLC	2.5×10^{-13}	26	Eic and Ruthven [58]

a) 250 μm crystals of natural faujasite.

17.3.3
Benzene

17.3.3.1 Comparison of Macroscopic and Microscopic Measurements

Diffusion of benzene in NaX has also been studied in detail by NMR, QENS, sorption rate, frequency response, and ZLC measurements. The diffusivities are compared in Table 17.5.

The diffusivity of benzene in NaX is about one order of magnitude higher than the value for the xylene isomers and as such it is close to the limit of uptake rate measurements in 100 μm crystals, although well within the range accessible by ZLC. Gravimetric measurements carried out with 250 μm crystals of natural faujasite showed a pattern of concentration dependence similar to that observed for the xylene isomers and the corrected diffusivity value agreed well with the limiting diffusivities derived from ZLC measurements with 50 μm and 100 μm NaX crystals [65]. These data were also confirmed by subsequent ZLC and TZLC (tracer zero-length column) measurements [66].

As with the xylenes, however, the values of D_0 are substantially smaller than the NMR self-diffusivities, although the activation energies are similar (Figure 17.15). The same discrepancy is seen in the patterns of the concentration dependence of D_0 and $\mathcal{D}$, where neither the corrected diffusivities nor the TZLC diffusivities (which show a moderate increase with loading) show the pronounced decrease shown by the PFG NMR diffusivities, although the data appear to converge at saturation loadings. Since in NMR tracer desorption studies [79] (Section 11.5.4) intercrystalline molecular exchange was found to occur with rates compatible with the PFG NMR data for intracrystalline diffusion [80], these discrepancies cannot be attributed to surface barriers.

Table 17.5 Diffusivity data for benzene in NaX zeolite at about 400 K and low loadings.

Sorbent	Technique	D_0 or $\mathcal{D}$ (m^2 s^{-1})	E (kJ mol^{-1})	Reference
55, 100 μm NaX and 250 μm faujasite	Gravimetric uptake and ZLC	4×10^{-12}	25	Eic et al. [65]
55 μm NaX	Gravimetric uptake	2×10^{-12}	—	Graham [67]
8–50 μm NaX	PFG NMR	$1-3 \times 10^{-10}$	25	Kärger et al. [16, 61]
	NMR tracer desorption	$1-3 \times 10^{-10}$	25	Rudtsch [68]
120 μm NaX	Piezometric	$\sim 10^{-10}$	27	Bülow et al. [69–71]
1–2 μm NaX	^{2}H NMR	1.2×10^{-11} a)	22.5	Boddenberg et al. [72]
Crystal size not given	PFG NMR	2×10^{-11} b)	17	Lechert et al. [73, 76]
50, 100 μm NaX	TZLC	1.2×10^{-11} c)	—	Brandani et al. [66]
NaX, crystal size irrelevant (and not given)	QENS	1.4×10^{-10}	17	Jobic et al. [77]
NaX, crystal size irrelevant (and not given)	QENS	7.25×10^{-11} d)	16	Jobic et al. [77]
NaY, crystal size irrelevant (and not given)	QENS	1.1×10^{-11} e)	33	Jobic et al. [77]

a) Assuming 4.0 Å jump length.
b) At 3.5 molecules per cage.
c) At 468 K.
d) At 450 K, 4 molecules per cage.
e) At 450 K, 4.5 molecules per cage, Si/Al = 2.43.

The situation becomes even more puzzling when the QENS data are considered [77]. The QENS values agree with the PFG NMR data as to the order of magnitude of the diffusivities and the trend with loading, but the activation energies are much smaller. In addition, in contrast to the uptake rate measurements, the QENS data show substantially higher diffusivities for NaX than with NaY (to which natural faujasite is assumed similar).

17.3.3.2 Diffusion Mechanism

The mechanism of benzene diffusion in zeolites NaX and NaY has been investigated in detail by a combination of PFG NMR self-diffusion and NMR relaxation measurement. The decisive role of the interaction between the double bond and the cations is revealed by a comparison of the minima in the transverse nuclear magnetic relaxation times for cyclohexane, cyclohexadiene, and benzene [74, 75], which, in this sequence, are shifted to higher temperatures, thus indicating a decreasing mobility (Section 10.1.2.2). The significant decrease of the diffusivity of benzene in CaX as observed in TZLC measurements [66] confirms the role of the cation–guest interaction in the diffusion process.

A significant shift of the T_1 minimum to higher temperatures is also observed on passing from zeolite X to Y, indicating that the rate of molecular reorientation in zeolite X is much higher than in zeolite Y [75]. This difference may be explained by

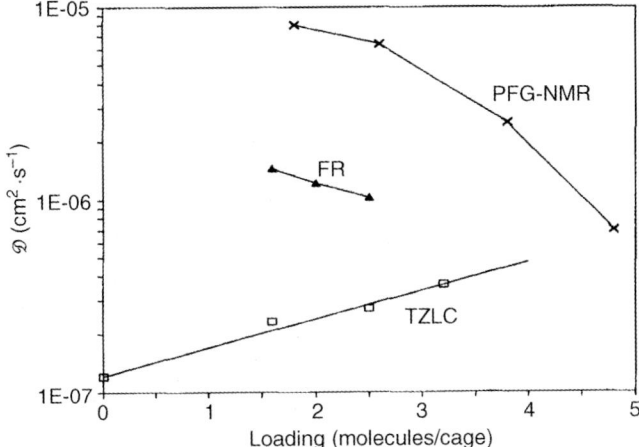

Figure 17.16 Comparison of tracer exchange (TZLC) self-diffusivities, frequency response corrected diffusivities (FR [78]), and PFG NMR self-diffusivities [53] for benzene in NaX crystals at 468 K, showing the concentration dependence of self-diffusivity. From Brandani et al. [66] with permission.

the difference in the cation densities. The lower cation density in zeolite Y leads to more pronounced potential energy minima at the adsorption sites. This difference between the X and Y zeolites is also reflected by the QENS diffusivity data shown in Table 17.5 [77] and, at least partially, in NMR self-diffusion measurements [61] where the self-diffusivities decrease slightly with increasing Si/Al ratio. However, this is not reflected in the uptake rate measurements, which show very little difference in diffusivity between NaX and natural faujasite in which the Si/Al ratio is comparable with zeolite Y [56, 57].

Application of the neutron spin-echo technique allows transport diffusion of benzene in zeolite to be studied by QENS [81] with further enhanced spatial resolution (Section 10.2.3.5). These studies, which confirmed the order of magnitude of the (self-)diffusivities from the earlier QENS studies, were accompanied by molecular dynamics simulations emphasizing the role of the cation–guest interaction in the diffusion of unsaturated hydrocarbons.

Combining the relaxation and self-diffusion measurements allows both the jump time and the RMS jump distance to be determined [75, 76, 82]. In NaY, which has no delocalized cations, the benzene molecules tend to occupy tetrahedrally disposed sites with the molecule oriented parallel to the six-ring at a distance of about 3.2 Å from the SII cation. This has been shown from detailed studies of NMR relaxation behavior [75, 76, 82] and independently by infrared spectroscopy [83]. The evidence from diffusion measurements is entirely consistent with this picture. The jump distance is found to remain constant at about 4.0 Å up to a loading of four molecules per cage. This distance corresponds approximately to the site-site distance for the tetrahedrally arranged SII cation sites, thus suggesting that the diffusion process must involve jumps between adjacent sites. This is in marked contrast to the behavior of saturated hydrocarbons, for which, at least at low loadings, the jumps occur primarily between adjacent cages.

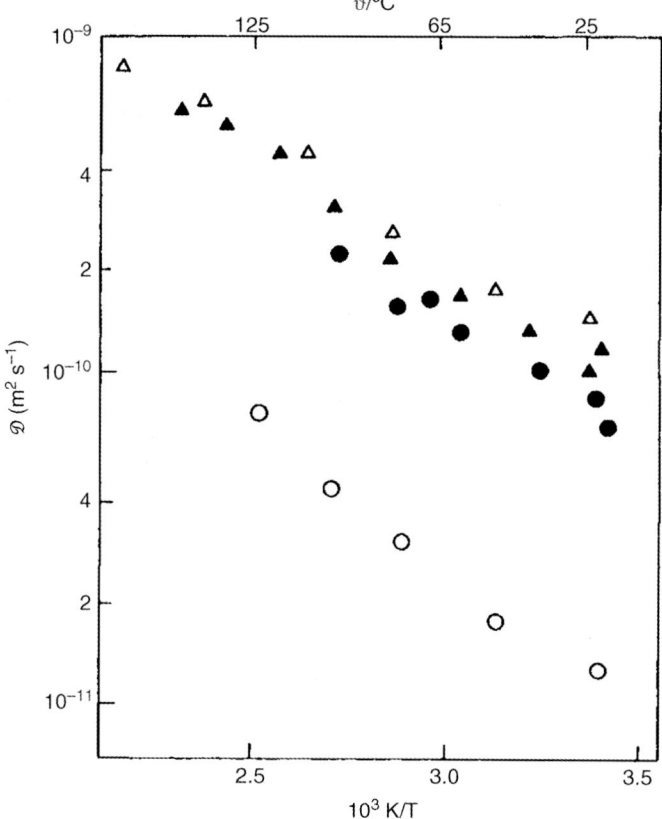

Figure 17.17 Comparison of NMR self-diffusivities for benzene (○, ●) and n-heptane (△, ▲) in NaX (open symbols) and LaX (filled symbols) at a sorbate concentration of about 2.5 molecules per cage. Reprinted from Kärger and Pfeifer [95], with permission.

Further evidence of the significance of double bond–cation interaction is provided by comparative data for the diffusion of saturated and unsaturated hydrocarbons in NaX and LaX (Figure 17.17) [84, 85]. The diffusivity of the unsaturated hydrocarbons in NaX is much lower than that of saturated hydrocarbons, but in LaX, which has no accessible cations, the diffusivities of the saturated and unsaturated species are almost the same. There is little difference in the diffusivity of n-heptane between NaX and LaX but the diffusivity of benzene is higher by more than an order of magnitude in LaX.

The elementary steps of the molecular motion of benzene in open 12-ring zeolites have been investigated in detail by ^{2}H NMR [72, 86–89] with the additional option of ^{2}H–^{13}C exchange spectroscopy [90]. In agreement with the microscopic diffusion studies, the molecular jump rates in NaY are notably below those in NaX, which is caused by a significantly larger activation energy, while the pre-exponential factors are found to remain essentially unchanged [89]. Moreover, analysis of the correlation effects brought about by site occupation of the hopping paths of the guest molecules leads [91, 92] to a similar concentration dependence of the diffusivities, as observed in the PFG NMR measurements [93]. If a mean jump length equivalent to the cage separation (~11 Å) is assumed, the correlation times yield diffusivities that are consistent with the PFG NMR data [72]. Similar agreement is obtained from neutron scattering measurements [94]. However, the mean jump lengths determined from the neutron scattering studies are in the range 3–5 Å, which is consistent with the earlier values obtained from the combination of NMR self-diffusion and relaxation measurements [73, 76] but substantially smaller than the 11 Å assumed for the jump length by Boddenberg [72]. Using the revised jump length Boddenberg's ^{2}H NMR data would suggest self-diffusivities that are about an order of magnitude lower than the values determined from the NMR PFG measurements and somewhat higher than the TZC values.

17.3.3.3 Hysteresis

An attempt to determine very accurate equilibrium data for benzene–NaX (50 μm Zhdanov crystals, Si/Al ≈ 1.2) led to the discovery of an unusual hysteresis in the isotherm [96]. The results are summarized in Figure 17.18. At low pressures (<0.02 Torr) the isotherm (curve 1) was fully reversible, adsorption/desorption rates were moderate (time scale of a few minutes), and the form of the transient uptake curves was consistent with diffusion control, with some evidence of bed diffusion resistance in the initial stages. At higher benzene pressures ($p > 1$ Torr) in the saturation region, the isotherm was again fully reversible; the sorption rates were very rapid (<1 min to equilibrium) and probably controlled by extracrystalline resistance to mass transfer. In the intermediate region the sorption rates were very slow, requiring more than a day to reach final equilibrium in some cases. The desorption

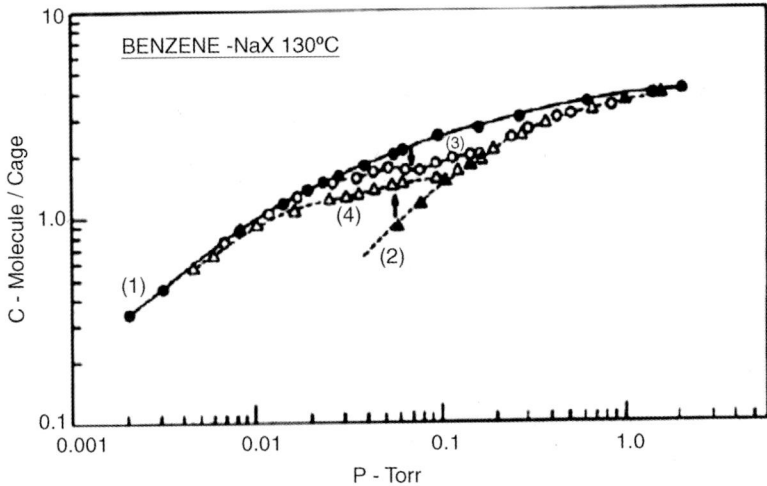

Figure 17.18 Gravimetric isotherms measured for benzene in NaX crystals (50 μm) at 403 K. (1) Adsorption measurements – 20 min endpoint (●); (2) desorption measurements – 20 min. endpoint (▲); (3) adsorption measurements – 48 h endpoint (○); and (4) desorption measurements – 48 h endpoint (Δ). Reprinted from Tezel and Ruthven [96] with permission.

isotherm in this region (curve 2) lies well below the approximate adsorption isotherm for a 20 min endpoint although, at sufficiently long times, the adsorption endpoints approach the true equilibrium corresponding to the desorption isotherm. The shapes of the transient sorption curves in this region varied from a simple exponential approach to equilibrium to a distinct two step process and in some cases *an increase in pressure* led to an initial fast uptake followed by a very slow *desorption* to a final equilibrium loading that was *below the initial level*.

Hysteresis isotherms in which the desorption branch lies above the adsorption branch are common and can have several different origins. However, in this case the desorption branch lies below the adsorption branch and that is a much rarer phenomenon that thermodynamically can be explained only by a structural change in the adsorbent [97]. The obvious conclusion is that, in the presence of benzene, the exchangeable cations adjust their positions (as a result of the strong interaction). According to this interpretation the two extreme isotherms (curves 1 and 2) would represent the equilibrium for two different cation arrangements, one of which is the stable configuration at low loading and the other at high loading. The corresponding Henry constants, estimated by extrapolation to low loading, differ by an order of magnitude.

17.3.4
Discrepancy between Macroscopic and Microscopic Diffusivity Measurements for Aromatics in NaX

The large discrepancy (approximately two orders of magnitude) between the macroscopic and microscopic diffusivity values for benzene and the C_8 aromatics in NaX is puzzling, especially as both data sets are entirely self-consistent and some of the measurements were made with the same zeolite crystals (synthesized by Zhdanov). Recent experimental evidence has shown that structural defects leading to surface and intracrystalline barriers can have a dramatic effect on sorption kinetics at the macro-scale but with only minimal influence at the micro-scale (Section 17.2.6 and Figure 17.12). The presence of internal barriers therefore appeared to offer a rational explanation for the observed differences as well as for the constancy of the diffusional activation energies measured at both macro- and micro-scales (Section 19.6.3.2). To establish whether there is any evidence of such structural defects in these particular samples a very detailed study was therefore undertaken, including repetition of the PFG NMR measurements for benzene–NaX, over a much wider range of molecular displacements, using a more modern spectrometer as well as magic-angle spinning measurements and detailed line shape analysis. The overall conclusion from these studies, which are reported in detail elsewhere [98], is that these crystals show no evidence of defects in crystal homogeneity, including indications of twinnings or other internal (or surface) barriers. Moreover, repetition of the measurements after as long as 25 years with the identical zeolite specimens yielded completely reproducible diffusivities. For these samples it is therefore difficult to believe that the discrepancy between the macroscopic and microscopic data may be ascribed to structural defects.

A possible alternative explanation is suggested by the hysteresis effect discussed above. The equilibrium arrangement of the Na^+ cations in NaX appears to be different at low and high benzene loadings and the corresponding cation rearrangement is evidently a slow process, requiring more than 24 h even at 400 K. Furthermore, the Henry constant corresponding to the freshly regenerated adsorbent (the low loading isotherm) is at least an order of magnitude larger than that obtained by extrapolation of the high-loading isotherm (Figure 17.18). Although the macro- and micro-measurements were carried out under apparently similar conditions, there is one significant and possibly critical difference: in the PFG NMR and QENS measurements the crystals and sorbate are contained in a sealed tube so the system is certainly fully equilibrated, whereas in the ZLC (and uptake rate) measurements the zeolite is freshly regenerated and only in contact with the sorbate for a short time prior to the measurements. Since the cation rearrangement requires more than a day, one might assume that the original arrangement will persist over the duration of the macro-measurements. In view of the demonstrated effect of cation interactions on diffusion rates, especially for unsaturated sorbates (Section 17.3), it seems reasonable to postulate that the diffusivities of the aromatics in NaX may be strongly affected by a rearrangement of the cations and that the slower diffusion observed in the macroscopic measurements may be attributable to the stronger interaction with the cations in the freshly regenerated adsorbent, as suggested by the higher Henry constant. It is worth mentioning that in a study of diffusion by the single-crystal membrane method, Wernick and Osterhuber [20] observed a reasonably reproducible and reversible transition, on a time scale of days, between a "fast diffusion state," favored by high loadings, and a "slow diffusion state" at low loadings (differing in permeance by more than two orders of magnitude).

By recording the rate of intracrystalline self-diffusion of n-hexane during uptake by large NaX-type crystals [99, 100], however, no indications of such a behavior could be observed: already during the process of uptake the diffusivities of the adsorbed molecules were found to be of the order of the diffusivities attained after equilibration. One must be aware of the fact, however, that these measurements were performed with relatively large pressure steps extending to concentration ranges where the differences between the corrected uptake and the PFG NMR diffusivities are much less significant than at lower loadings.

17.4
Other Systems

17.4.1
Water in NaX and NaY

Following preliminary measurements with beds of commercial zeolites NaX which, as we know today, yielded long-range diffusivities [101], the first systematic measurements of intracrystalline diffusion by PFG NMR have been performed with water in large crystals of zeolites NaX [102, 103], synthesized by S.P. Zhdanov in Leningrad [45]. They showed type IV concentration dependence, that is, a maximum in the diffusivities with increasing concentration, which, meanwhile, has also been observed for some other systems such as methanol–NaX (Figure 3.4 and Section 17.4.2). Intuitively, a concentration dependence of this type may be expected for systems with a limited number of strong adsorption sites. The first molecules are strongly adsorbed and, thus, diffuse only slowly through the pore space. Given the strong dipole moment of the guest molecules, in the present case the exchangeable cations (Figure 17.2) are easily identified as such adsorption sites. With increasing loading these sites are progressively occupied and the additional molecules are accommodated at weaker sites, leading to an overall enhancement in the diffusivity until, at sufficiently high loadings, the mutual inhibition of the diffusants may again lead to a decrease in their mobility. Notably, by carefully avoiding any self-steaming during PFG NMR sample preparation, the water diffusivities in NaX are found to approach a plateau value (rather than decrease again), following the type III pattern of concentration dependence (Figure 3.3c and Figure 17.29 below).

Exactly this scenario has been nicely confirmed by MD simulations for water in both zeolites NaX and NaY [104]. The results shown in Figure 17.19 are in good agreement with the old PFG NMR data of References [102, 103] for both the absolute values and the concentration dependence. The results of QENS measurements, which are also included, are also found to be in reasonable agreement with the trends and the orders of magnitude predicted from the simulations.

Interestingly, 15 years after the very first PFG NMR measurements of intracrystalline diffusion the identical PFG NMR sample tubes (which were stored and kept sealed over these 15 years) had already been used to confirm the old data in a twofold manner [105], namely, by (i) repeating the measurements with the identical PFG NMR sample tubes and by (ii) measurements with newly-activated and prepared

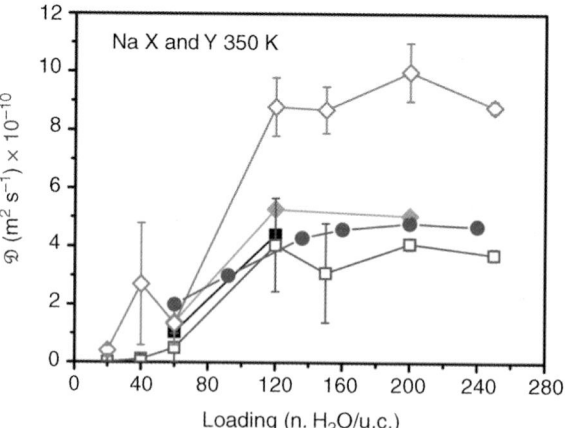

Figure 17.19 Diffusion coefficients as a function of loading, at 350 K. (■) NaX, experimental (QENS); (□) NaX, calculated; (♦) NaY, experimental (QENS); (◊) NaY, calculated; (●) NaX, experimental (PFG NMR [102, 103]). Reprinted from Reference [104] with permission.

PFG NMR samples of identical loading after having crushed the old sample tubes [106].

17.4.2
Methanol

In Figure 3.4, diffusivity data for methanol in NaX are presented as an example of the type IV concentration dependence of self-diffusion. The data include the PFG NMR self-diffusivities together with the results of both traditional ZLC, that is, for vanishing guest concentration, and tracer ZLC measurements (Section 14.7.5.1, Figure 14.20). In contrast to the behavior of saturated (Section 17.2) and unsaturated (Section 17.3) hydrocarbons, for methanol the macroscopic and microscopic diffusion measurements are found to yield consistent diffusivities over the entire concentration range.

17.4.3
Triethylamine

Triethylamine is a rather large and relatively slowly diffusing molecule and it was therefore selected as a sorbate in some of the earlier comparative studies of sorption rate [107] and NMR measurements [16]. In contrast to the hydrocarbon systems, which show large discrepancies between the corrected transport diffusivity and NMR self-diffusivities, for this system there is reasonably good agreement. Figure 17.20 shows representative experimental data. Whereas the corrected diffusivities for the hydrocarbon sorbates are almost independent of concentration, for triethylamine we see a pronounced decrease at high loadings, which is consistent with the trend of the NMR self-diffusivities. However, for practical reasons reliable uptake rate measurements for this system could be made only at relatively high concentrations. At high

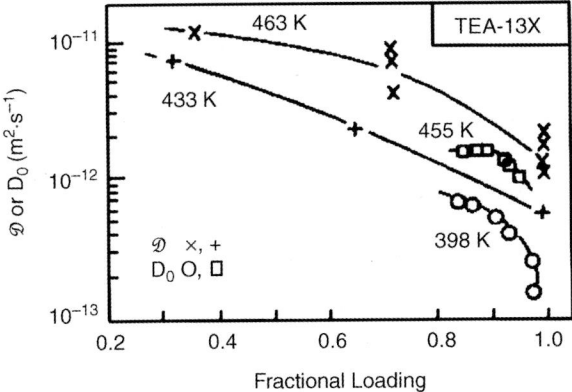

Figure 17.20 Comparison of NMR self-diffusivities ($+$, $\times$) and corrected transport diffusivities ($\bigcirc$, $\square$) for triethylamine in NaX zeolite crystals (39–55 μm) at temperatures in the range 398–463 K. Reprinted from Kärger and Ruthven [16], with permission.

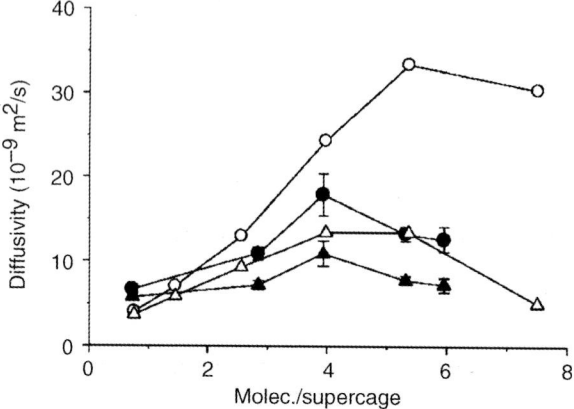

Figure 17.21 Corrected (triangles) and transport (circles) diffusivities from QENS measurements (open symbols) and MD simulations (filled symbols) of D_2 in NaX at 100 K as a function of loading. Reprinted from Reference [108] with permission.

sorbate loadings there is reasonable agreement between sorption rate and NMR measurements for many of the hydrocarbon systems. Large discrepancies become apparent only in the low concentration region as a result of the difference in the patterns of concentration dependence. It is therefore not completely clear whether the agreement between the NMR and sorption rate data for triethylamine results simply from the limited concentration range of the sorption rate measurements.

17.4.4
Hydrogen

Figure 17.21 exemplifies the good agreement generally found on comparing the QENS data with the results MD simulations for the transport and corrected diffusivities [108]. The slight maximum passed by the transport diffusivities at medium loadings is interpreted in the light of the Reed–Ehrlich model. Notably, the simulations employed a Feynman–Hibbs correction to the interaction potentials, to deal with quantum effects associated with translational motion of the light sorbates.

17.5
PFG NMR Diffusion Measurements with Different Probe Nuclei

In Section 11.3.2, the use of protons was shown to provide the best conditions for diffusion measurement by PFG NMR. For diffusion measurements with other nuclei X-type zeolites provide particularly suitable conditions, due to both their open pore network (and the resulting favorably long nuclear magnetic relaxation times) and the availability of large crystals. In addition to diffusion measurements with guest molecules that do not contain hydrogen (such as CO, CO_2 and N_2 – see Sections

17.5.2 and 17.5.3, respectively) PFG NMR diffusion measurement with different nuclei provides a useful approach to the measurement of diffusion in multicomponent systems (Sections 17.6 and 11.3.4.2).

17.5.1
Fluorine Compounds

Since ^{19}F has a gyromagnetic ratio that is only about 10% smaller than that of a proton, this species provides the next most favorable properties for PFG NMR measurements. Both relaxation [109] and self-diffusion measurements [110] have been performed for SF_6 in NaX. Figure 17.22 shows representative Arrhenius plots giving the temperature dependence of the self-diffusivity at three different loading levels. The diffusivity decreases with increasing loading but the activation energy remains almost constant. This is the same pattern of behavior as is observed for the saturated hydrocarbons (Section 17.2). Estimates from NMR relaxation studies [109] are in reasonable agreement with the self-diffusivity data.

Figure 17.23 shows a comparison of the self-diffusivities of CF_4 and CHF_2Cl in NaX, as measured by the PFG NMR method. CHF_2Cl is clearly the less mobile species, having a lower diffusivity and a higher activation energy, suggesting a stronger interaction with the lattice. This is probably due to the strongly dipolar nature of the CH_2FCl molecule (dipole moment $\approx$ 1.42 D [111]). This species is of special interest in NMR studies because it allows the application of both ^1H and ^{19}F nuclear magnetic resonances in the self-diffusion measurements. The results of both measurements are quite consistent (Section 11.5.2). Notably, the diffusivity for CF_4 in NaX is only about one order of magnitude smaller than that of CH_4 [14] while in 5A the diffusivities of these species differ by more than two orders of magnitude

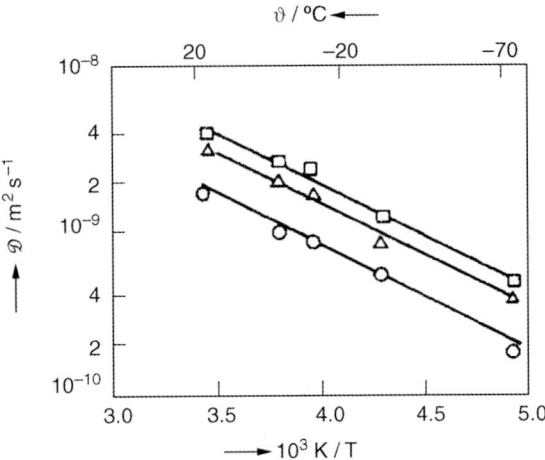

Figure 17.22 Temperature dependence of NMR self-diffusivity for SF_6 in NaX zeolite crystals at sorbate concentration levels of 1.2 (□); 3.0 (△), and 5.0 (○) molecules cage. Reprinted from Kärger et al. [110], with permission.

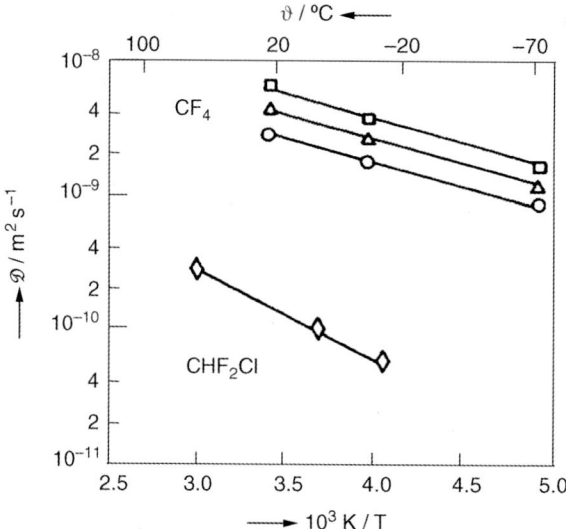

Figure 17.23 Temperature dependence of NMR self-diffusivity for CF_4 at concentrations of 2.0 (□), 3.0 (△), and 6.0 (○) molecules per cage and for CHF_2Cl at 5.0 molecules per cage (◇) in NaX zeolite. Reprinted from Kärger et al. [110], with permission.

(Table 16.3). It seems reasonable to assume that this is a reflection of the difference in the effective pore diameters. The windows of NaX are large enough to pass both CH_4 and CF_4 without significant steric hindrance whereas the diffusion of CF_4 through the smaller windows of 5A is subject to substantial hindrance.

17.5.2
Xenon and Carbon Monoxide and Dioxide

Reference [112] reports on the application of ^{129}Xe and ^{13}C PFG NMR to diffusion studies with carbon monoxide and dioxide and with xenon as guest molecules in zeolites NaX, silicalite, and Na63CaA. As an example, Figure 17.24 shows Arrhenius plots of the diffusivities in zeolite NaX for loadings of about 3 molecules (atoms) per supercage. For CO and CO_2 the activation energies of self-diffusion are found to exceed the activation energy of xenon by 3–4 kJ mol^{-1}. This finding may be attributed to the interaction between the cations and the carbon oxides, which, owing to their dipole and quadrupole moments, is much stronger than the interaction with the inert xenon atoms, even though the latter are more polarizable. In agreement with this interpretation, the activation energies of diffusion of all three molecules in the cation-free silicalite (Section 18.1) are essentially the same [112], coinciding with the activation energy for xenon in NaX. Coincidentally, in NaCaA the activation energies of all three molecules are also close to each other [112], and similar to the value for the carbon oxides in NaX. Here, the increase in the activation energy of xenon diffusion can be considered to be caused by an additional transport resistances resulting from the smaller "windows" between adjacent cavities in the A structure (Section 16.1).

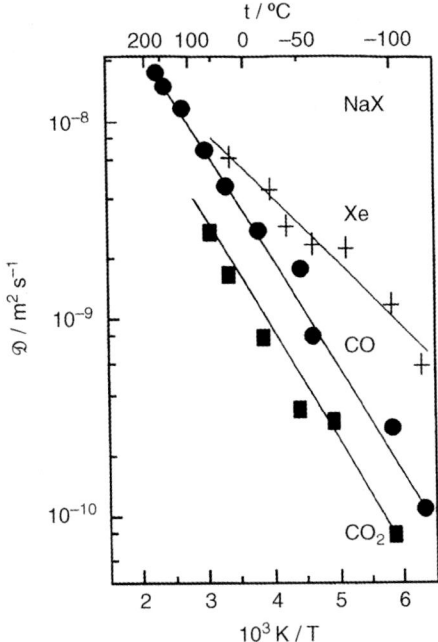

Figure 17.24 Arrhenius plot of the intracrystalline self-diffusivity of xenon, carbon dioxide, and carbon monoxide in zeolite NaX at a loading of about 3 molecules (atoms) per supercage. Reprinted from [112] with permission.

17.5.3
Nitrogen

The gyromagnetic ratio of ^{15}N ($\gamma_{15N} = 2.7106 \times 10^7\,\text{T}^{-1}\,\text{s}^{-1}$) is an order of magnitude smaller than that for hydrogen. In addition to a lower sensitivity with respect to guest concentrations, this difference shifts the range of observable displacements by an order of magnitude to larger values (Section 11.3.2). While this change might be of advantage for the observation of long-range diffusion (see, for example, Section 11.4.2), it restricts the measurement of intracrystalline diffusion. Figure 17.25 gives an example of the results of such measurements. This figure shows the effective diffusivities resulting from the slopes in the PFG NMR signal attenuations corresponding to both intracrystalline and long-range displacements. For sufficiently small temperatures (up to 190 K), intracrystalline molecular displacements during the observation time of 2 ms were found to be sufficiently small such that any significant interference with the crystal boundary and the intercrystalline space could be excluded, thus yielding true intracrystalline diffusivities. However, at higher temperatures an increasing fraction of the molecules is able to exchange positions between different crystallites. The mean square displacement of these molecules is used to calculate the effective long-range diffusivities, which are also shown in Figure 17.25. It is in this temperature range that the apparent intracrystalline

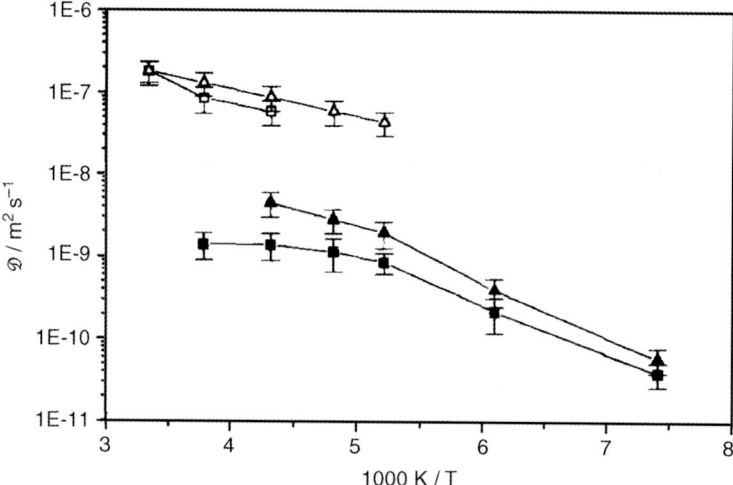

Figure 17.25 Temperature dependence of the coefficients of intracrystalline self-diffusion (filled symbols) and of long-range diffusion (open symbols) of nitrogen in zeolite NaX for 0.4 (squares) and 1.3 (triangles) molecules per supercage, from Reference [113] with permission.

diffusivities fall below their true values. Both the intracrystalline and long-range diffusivities are larger at higher loadings. This is a general observation for long-range diffusion (Section 11.4.2), resulting from the increasing fraction of sorbate molecules in the gas phase at higher temperatures. Intracrystalline diffusion obviously follows the established pattern for systems with a strong interaction between the guest molecules and the adsorption sites, that is, in the given case, the cations, as already noted for water and methanol (Sections 17.4.1 and 17.4.2, respectively).

17.6
Self-diffusion in Multicomponent Systems

17.6.1
Hydrocarbons

17.6.1.1 *n*-Heptane–Benzene

The self-diffusivity of one component in a mixed adsorbed phase may be easily determined by NMR methods by using deuterated species. Figure 17.26 shows the self-diffusivities of *n*-heptane and benzene in NaX zeolite at various sorbate concentration levels [84, 95]. In agreement with the results of single-component diffusion measurements the mobility of the *n*-heptane is reduced by an increase in the concentration of either benzene or *n*-heptane. As a result of the difference in the nature of the interaction the benzene diffusivity remains practically constant up to about half saturation. At higher loadings the mobility decreases sharply as a result of the reduction in the free volume.

17.6 Self-diffusion in Multicomponent Systems

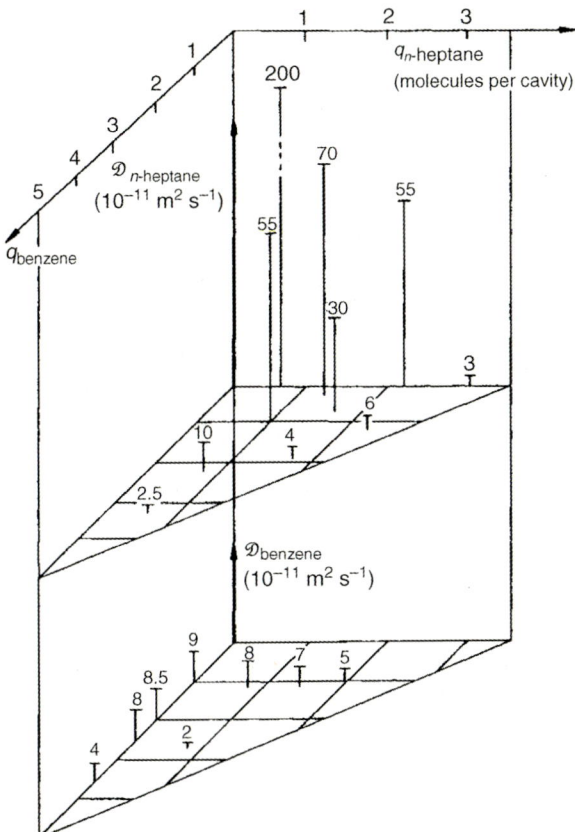

Figure 17.26 NMR self-diffusivities for the two components in heptane–benzene mixtures adsorbed on NaX at 400 K. Reprinted from Kärger and Pfeifer [95] with permission.

17.6.1.2 Benzene–Perfluorobenzene

Diffusion of two adsorbates within the same sample may be most easily followed when they contain different nuclei, both of which are accessible to measurement by the PFG NMR method. The results of such measurements for C_6H_6/C_6F_6 mixtures adsorbed on NaX have been presented by Kärger et al. [110]. The similarity in behavior of these compounds is confirmed since their diffusivities are almost identical.

17.6.1.3 n-Butane–Perfluoromethane

Following the same approach of selective diffusion measurement in two-component systems with hydrogen- and fluorine-containing compounds, Zhao and Snurr report in References [114, 115] the results of detailed PFG NMR studies with CF_4 and different hydrocarbons adsorbed in zeolite NaX. The measurements were designed to check the validity of the Maxwell–Stefan formalism for estimating the self-diffusivities in multicomponent systems from their single-component diffusivities [see Sections 3.4, 8.1.3, and 8.2.9 and Eqs. (3.51), (8.101)–(8.107)]. Figure 17.27

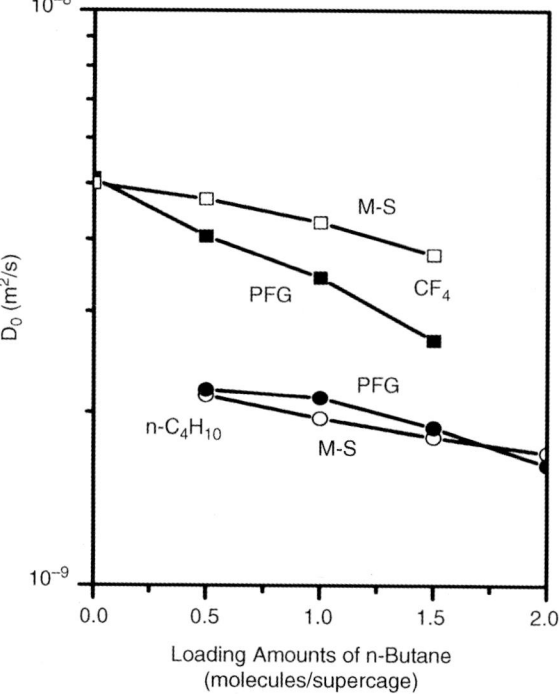

Figure 17.27 PFG NMR data (solid symbols) for the self-diffusivities of CF_4 (squares) and n-butane (circles) in NaX at 298 K under binary adsorption with n molecules of n-butane (unit of the abscissa) and $(2 - n)$ molecules CF_4 per supercage and comparison with their estimate from the single-component diffusivities using the Maxwell–Stefan formalism (open symbols), from Reference [115], with permission.

exemplifies the excellent agreement between the results of PFG NMR diffusion experiments, the theoretical estimates based on the Maxwell–Stefan model, and earlier non-equilibrium MD simulations [25].

The results show that the diffusivities of both CF_4 and n-butane decrease slowly with an increasing percentage of n-butane. n-Butane is both larger and has a lower single-component diffusivity, so as CF_4 molecules are replaced by n-butane the mixture diffusivity of both CF_4 and n-butane decreases, just as suggested already by the concentration dependence for the pure n-alkane (Figure 17.3).

17.6.1.4 Ethane–Ethene

It was pointed out in Section 11.3.4.1 that the most economic access to PFG NMR multicomponent self-diffusion measurements is provided by Fourier-transform NMR. In this method the total NMR signal may be split into the separate signals of the constituents, thus allowing simultaneous measurement of the diffusivities of all components. The application of this method (which is in common use for liquids [116, 117]) to adsorbate–adsorbent systems is complicated by the reduced mobility and a corresponding line-broadening (see Section 10.1.2.2 and the attempts to overcome these limitations by MAS NMR described in Section 10.1.4). Owing to their high mobility, ethane–ethene mixtures adsorbed on zeolite NaX provide favorable conditions for high-resolution self-diffusion measurements. In addition, the analysis of the spectra attenuation due to diffusion is substantially facilitated as the spectrum of each component is given by a single line which, for simplicity, has already been implied in Figure 11.5.

Applying ^{1}H PFG Fourier-transform NMR to an ethane–ethene (3 : 2) mixture adsorbed on NaX (at a total loading of 2.5 molecules per cavity at 293 K), the self-diffusion coefficients of the two components were found to be 4.6×10^{-9} m^2 s^{-1} and 1.25×10^{-9} m^2 s^{-1}, respectively [118]. Comparison with the single-component diffusivities at the same total loading shows that the mobility of ethene is unaffected, while the diffusivity of ethane is reduced by a factor of about two. Notably, a similar tendency is observed for benzene–n-heptane mixtures on NaX (Figure 17.26), where under the influence of the benzene molecules the heptane mobility is reduced, with very little effect on the benzene self-diffusivity.

17.6.2
Self-diffusion under the Influence of Co-adsorption and Carrier Gases

As a non-invasive technique, PFG NMR does not interfere with the internal dynamics of the nanoporous host–guest system. This makes PFG NMR ideally suited for investigating how the guest mobility may be affected by the presence of a second ubiquitous component such as a carrier gas or a common impurity such as water vapor.

17.6.2.1 Effect of Moisture on Self-diffusion
The effect of residual moisture on self-diffusion of saturated (ethane, n-butane) and unsaturated (ethylene, butylene) hydrocarbons has been studied by PFG NMR using deuterated water (Figure 17.28) [119, 220]. Except in the very low concentration region the trends are similar and it is clear that the reduction in diffusivity due to the presence of moisture is far greater than the relatively modest effect produced by the presence of additional hydrocarbon molecules. The effect is probably due to the formation of water–cation complexes in the windows [21]. The preferential localization of water molecules in the vicinity of the 12-ring windows is confirmed from Xe line shift measurements [121]. At higher moisture levels there is evidently little difference in diffusivity between the saturated and unsaturated species but at low moisture levels the difference is more than an order of magnitude. The reduced mobility of the unsaturated species evidently results from the stronger interaction with the exchangeable cations. The first few molecules of water probably screen the cations, thus increasing the diffusivity for the unsaturated species and bringing them closer to the values for the equivalent alkanes.

17.6.2.2 Water and Ammonia
Figure 17.29 shows the concentration dependence of the NMR self-diffusion coefficients of water and ammonia [119]. At low concentrations a significant decrease of the mobility of these compounds with decreasing concentration is

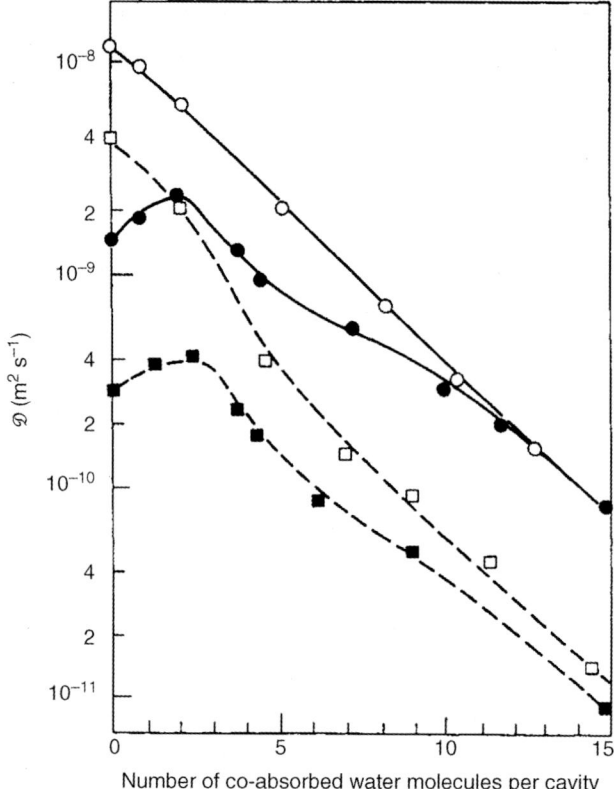

Figure 17.28 Effect of co-adsorption of moisture on the self-diffusivities of ethane (○) and ethene (●) (1.5 molecules per cage) and n-butane (□) and but-l-ene (■) (0.8 molecules per cage) in NaX at 293 K. From Kärger and Pfeifer [95] with permission.

observed. This corresponds to adsorption at a few highly favorable adsorption sites, presumably corresponding to the exchangeable cations, whose influence on the water mobility has been studied in detail by nuclear magnetic relaxation measurements [75, 122]. The dramatic increase of the molecular mobility with increasing concentration may be explained by progressive saturation of the strongest adsorption sites and the increased mobility of the cations at high sorbate loading. Small amounts of co-adsorbed hydrocarbons are seen to affect the observed concentration patterns only insignificantly.

17.6.2.3 Effect of an Inert Carrier Gas

Since a PFG NMR measurement is sensitive only to species that contain NMR active atoms, such measurements may be used to investigate the effect of an inert carrier gas on the diffusion properties. The results of a detailed study of the effect of argon on diffusion of cyclohexane in a bed of NaX crystals [123] are summarized in Section 15.8.3 (Figure 15.21). In this way it may be shown directly that the intracrystalline

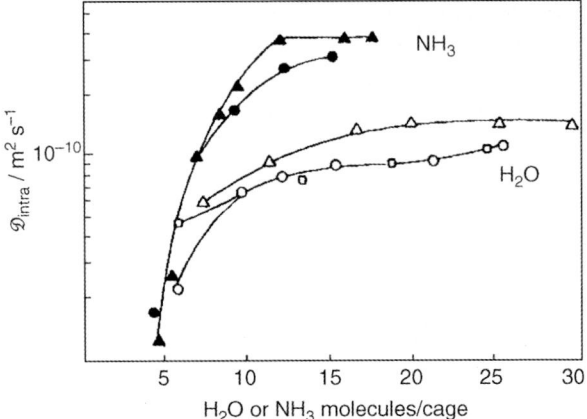

Figure 17.29 NMR self-diffusivities for H_2O (open symbols) and NH_3 (filled symbols) in NaX zeolite at 298 K as a function of loading, with and without co-adsorbed hydrocarbons: no hydrocarbon (△,▲), with 0.8 molecule per cage deuterated n-butane (○,●), and with a similar loading of deuterated butene (□). Reprinted from Förste et al. [119] with permission.

diffusivity of the guest molecules remains unaffected by the presence of an inert carrier gas, as has been generally assumed.

17.6.3
Self-diffusion in Multicomponent Systems Evolving during Catalytic Conversion

The ability of PFG NMR to monitor simultaneously the mobility of different components (Section 11.3.4 and Reference [100]) makes it a very effective tool for studying the mobility of reactant and product molecules during a chemical reaction. Since the chemical conversion leads to well-defined changes in the chemical surroundings of the nucleus of at least one of the elements, their diffusivities may be recorded in high-resolution PFG NMR measurements tuned to these particular nuclei.

17.6.3.1 Cyclopropane into Propene
Figure 17.30 shows, as an example, the results of *in situ* PFG NMR measurements during the conversion of cyclopropane into propene in zeolite NaX [124]. Following the methodology exploited in the two-component diffusion measurements with ethane and ethylene (Sections 11.3.4.1 and 17.6.1.4 and Reference [125]), the measurements were based on the difference in the proton chemical shifts in the vicinity of single and double bonds. In addition to the diffusivity of the reactant molecule (cyclopropane) and the product molecule (propene), the time dependence of the relative amounts of these molecular species is also presented. Since the conversion times are much larger than the intercrystalline exchange times determined from the diffusivities via Eq. (11.44) (Section 21.1.1), the overall rate may clearly be assumed to be reaction controlled.

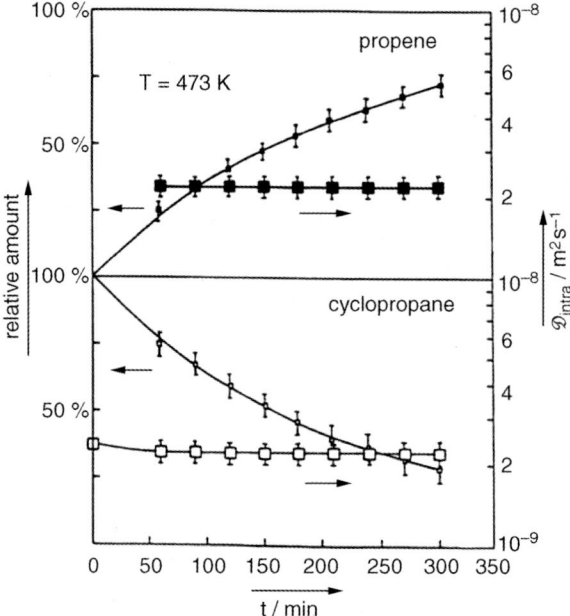

Figure 17.30 Time dependence of the relative amount of cyclopropane and propene during the conversion of cyclopropane into propene in NaX and their self-diffusion coefficients [(□) cyclopropane and (■) propene] at 473 K. Reprinted from Reference [124] with permission.

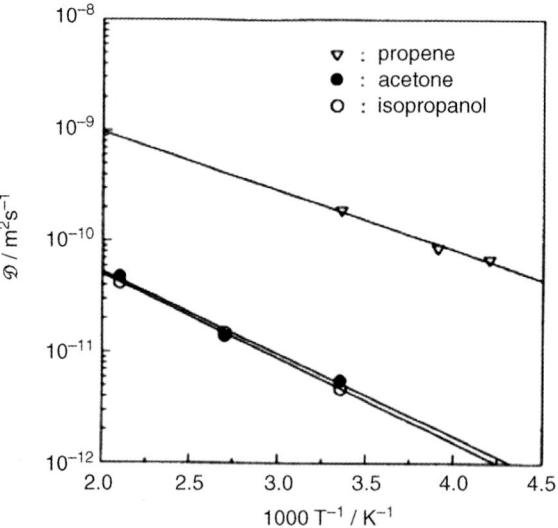

Figure 17.31 Intracrystalline diffusivities of isopropanol, acetone, and propene in NaX under the conditions of single-component adsorption at medium loadings. Reprinted from Reference [126] with permission.

The application of ^{1}H PFG NMR to the study of multicomponent diffusion in adsorbate–adsorbent systems is limited by the fact that the chemical shifts are of the order of the line width. The separation between the diffusivities of different compounds is, therefore, only possible in exceptional cases such as the rather simple reaction shown in Figure 17.30. Owing to the larger chemical shifts, ^{13}C PFG NMR provides much better conditions for such studies. On the other hand, measurements of this type are much more expensive as they require ^{13}C enriched chemical compounds. However, even with the application of ^{13}C enriched compounds accumulation times up to hours are sometimes necessary, so that *in situ* measurements are only possible for rather slow processes.

17.6.3.2 Isopropanol into Acetone and Propene

Figure 17.31 shows the diffusivities of isopropanol, acetone, and propene under the conditions of single-component adsorption on zeolite NaX [126]. All three

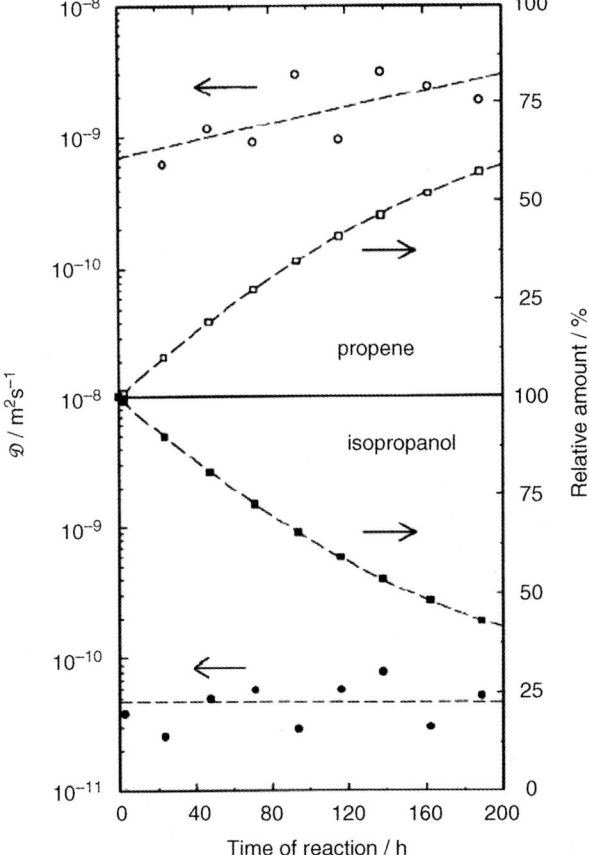

Figure 17.32 Time dependence of the relative amount of isopropanol and propene during the conversion of isopropanol in NaX and their self-diffusivities at 473 K. Reprinted from Reference [128] with permission.

compounds are involved in a well-established test reaction to discriminate between acid and basic zeolites [127]: isopropanol is dehydrated to propene on acid catalysts, while it is dehydrogenated to acetone on basic catalysts. Figure 17.31 shows that the diffusivity of propene (the product of the acid-catalyzed reaction) is more than an order of magnitude larger than the diffusivities of the reactant (isopropanol) and the product of the basic-catalyzed reaction (acetone). Therefore, if the acid- and base-catalyzed reactions were both to occur in parallel, assuming that the process is diffusion limited and the relative values of the diffusivities of the individual compounds in the mixture are similar to their single component values, the difference in the product diffusivities could lead to preferential (transport-promoted) production of the acid-catalyzed product, propene.

Figure 17.32 shows the results of the first *in situ* ^{13}C PFG NMR diffusion measurements carried out during the conversion of isopropanol in NaX [128]. In complete agreement with the single-component measurements presented in Figure 17.31, the propene diffusivities in the multicomponent system (evolving under reaction conditions) are found to be much larger than the diffusivity of isopropanol. The increase in propene diffusivity with increasing reaction time may be easily understood by realizing that the transport inhibition of propene by the less mobile isopropanol molecules becomes less significant as the amount of isopropanol decreases.

References

1 Rabo, J.A. (1976) *Zeolite Chemistry and Catalysis*, A.C.S. Monograph No. 171, American Chemical Society, Washington DC.

2 Ribeiro, F.R., Rodrigues, A.E., Rollman, L.D., and Naccache, C. (1984) *Zeolites: Science and Technology*, NATO ASI Series No. 80E, Martinus Nijhoff, The Hague.

3 Wankat, P. (1986) *Large Scale Adsorption and Chromatography*, CRC Press, Boca Raton, FL.

4 Ruthven, D.M. (1984) in *Principles of Adsorption and Adsorption Processes*, John Wiley & Sons, Inc., New York, ch 12.

5 Friedrich, H., de Jongh, P., Verkleij, A.J., and de Jong, K.P. (2009) *Chem. Rev.*, **109**, 1613.

6 Meier, W. (1968) *Molecular Sieves, Proceedings First International Conference on Zeolites*, Society of Chemical Industry, London, p. 10.

7 Breck, D.W. (1974) *Zeolite Molecular Sieves*, John Wiley & Sons, Inc., New York, p. 95.

8 Smith, J.V. (1971) *Adv. Chem.*, **101**, 171; Rabo, J. (ed.) (1976) *Zeolite Chemistry and Catalysis*, ACS Monograph No. 171, American Chemical Society, Washington, DC, p. 1.

9 Mortier, W. (1982) *Compilation of Extra Framework Cation Sites in Zeolites*, Butterworth, Guildford, UK.

10 van Dun, J.J.I., Mortier, W.J., and Uytterhoeven, J.B. (1985) *Zeolites*, **5**, 257.

11 Baker, M.D., Ozin, G.A., and Godber, J. (1985) *Catal. Rev. Sci. Eng.*, **27**, 591.

12 Gallezot, P. and Imelik, B. (1973) *J. Phys. Chem.*, **77**, 2364.

13 Culfaz, A. and Ergun, G. (1986) *Sep. Sci. Technol.*, **27**, 495.

14 Kärger, J., Pfeifer, H., Rauscher, M., and Walter, A. (1978) *Z. Phys. Chem. (Leipzig)*, **259**, 784; Kärger, J., Pfeifer, H., Rauscher, M., and Walter, A. (1980) *J. Chem. Soc., Faraday Trans. I*, **76**, 717.

15 Kärger, J., Lorenz, P., Pfeifer, H., and Bülow, M. (1976) *Z. Phys. Chem. (Leipzig)*, **257**, 209.

16 Kärger, J. and Ruthven, D.M. (1981) *J. Chem. Soc., Faraday Trans. I*, **77**, 1485.
17 Kärger, J., Bülow, M., and Haberlandt, K. (1977) *J. Colloid Interface Sci.*, **60**, 386.
18 Ma, Y.A. and Lee, T.Y. (1976) *AIChE J.*, **22**, 147.
19 Fu, C.-C., Ramesh, M.S.P., and Haynes, H.W. (1986) *AIChE J.*, **32**, 1848.
20 Wernick, D.L. and Osterhuber, E.J. (1984) *Proceedings 6th International Zeolite Conference Reno, 1983* (eds D.H. Olson and A. Bisio), Butterworth, Guildford, UK, p. 122.
21 Tezel, O.H., Ruthven, D.M., and Wernick, D.L. (1984) *Proceedings 6th International Zeolite Conference Reno, 1983* (eds D.H. Olson and A. Bisio), Butterworth, Guildford, UK, p. 232.
22 Doelle, H.J. and Riekert, L. (1977) *ACS Symp. Ser.*, **40**, 401.
23 Eic, M. and Ruthven, D.M. (1988) *Zeolites*, **8**, 472.
24 Calero, S., Dubbeldam, D., Krishna, R., Smit, B., Vlugt, T.J.H., Denayer, F.M., Martens, J., and Maesen, T.L.M. (2004) *J. Am. Chem. Soc.*, **126**, 11377.
25 Chempath, S., Krishna, R., and Snurr, R.Q. (2004) *J. Phys. Chem. B*, **108**, 13481.
26 Jobic, H., Borah, B.J., and Yashonath, S. (2009) *J. Phys. Chem. B*, **113**, 12635.
27 Jobic, H. (1999) *Phys. Chem. Chem. Phys.*, **1**, 525.
28 Taylor, R. (1979) PhD Thesis, University of New Brunswick.
29 Bhide, S.Y. and Yashonath, S. (2004) *Mol. Phys.*, **102**, 1057.
30 Yashonath, S. and Ghorai, P.K. (2008) *J. Phys. Chem. B*, **112**, 665.
31 Derycke, I., Vigneron, J.P., Lambin, P., Lucas, A.A., and Derouane, E.G. (1991) *J. Phys. Chem.*, **94**, 4620.
32 Krause, B.-C., Borah, B.J., Adem, Z., Galvosas, P., Kärger, J., and Yashonath, S. (2012) in preparation.
33 Kortunov, P., Vasenkov, S., Kärger, J., Valiullin, R., Gottschalk, P., Elia, M.F., Perez, M., Stöcker, M., Drescher, B., McElhiney, G., Berger, C., Gläser, R., and Weitkamp, J. (2005) *J. Am. Chem. Soc.*, **127**, 13055.
34 Berger, C., Gläser, R., Rakoczy, R.A., and Weitkamp, J. (2005) *Microporous Mesoporous Mater.*, **83**, 333.
35 Müller, M., Harvey, G., and Prins, R. (2000) *Microporous Mesoporous Mater.*, **34**.
36 Janssen, A.H., Koster, A.J., and de Jong, K.P. (2001) *Angew. Chem. Int. Ed.*, **40**, 1102.
37 Valiullin, R. and Kärger, J. (2011) *Chem. Ing. Tech.*, **83**, 166.
38 Kärger, J., Pfeifer, H., Rauscher, M., and Walter, A. (1980) *J. Chem. Soc., Faraday Trans. I*, **76**, 717.
39 Heink, W., Kärger, J., Ernst, S., and Weitkamp, J. (1994) *Zeolites.*, **14**, 320.
40 Breck, D.W. (1974) *Zeolite Molecular Sieves*, John Wiley & Sons, Inc., New York.
41 Fritzsche, S., Haberlandt, R., Kärger, J., Pfeifer, H., Heinzinger, K., and Wolfsberg, M. (1995) *Chem. Phys. Lett.*, **242**, 361.
42 Germanus, A., Kärger, J., and Pfeifer, H. (1984) *Zeolites*, **4**, 188.
43 Beagley, B., Dwyer, J., Evmerides, N.P., Attawa, A.I., and Ibrahim, T.K. (1982) *Zeolites*, **2**, 167.
44 Feldhoff, A., Caro, J., Jobic, H., Krause, C.B., Galvosas, P., and Kärger, J. (2009) *Chem. Phys. Chem.*, **10**, 2429.
45 Zhdanov, S.P., Khvostchov, S.S., and Feoktistova, N.N. (1990) *Synthetic Zeolites*, Gordon and Breach, New York.
46 Yang, X., Albrecht, D., and Caro, J. (2006) *Microporous Mesoporous Mater.*, **90**, 53.
47 Adem, Z., Caro, J., Futardo, F., Galvosas, P., Krause, C.B., and Kärger, J. (2011) *Langmuir*, **27**, 416.
48 Menjoge, A., Bradley, S.A., Galloway, D.B., Low, J.J., Prabhakar, S., and Vasenkov, S. (2010) *Microporous Mesoporous Mater.*, **135**, 30.
49 Kärger, J. and Michel, D. (1976) *Z. Phys. Chem. (Leipzig)*, **257**, 983.
50 Kärger, J., Michel, D., Petzold, A., Caro, J., Pfeifer, H., and Schöllner, R. (1976) *Z. Phys. Chem. (Leipzig)*, **257**, 1009.
51 Boddenberg, B. and Burmeister, R. (1988) *Zeolites*, **8**, 480.
52 Kärger, J. and Caro, J. (1977) *J. Chem. Soc., Faraday Trans. I*, **73**, 1363.
53 Wright, C.J. and Riekel, C. (1978) *Mol. Phys.*, **36**, 695.
54 Gautam, S., Tripathi, A.K., Kamble, V.S., Mitra, S., and Mukhopadhyay, R. (2008) *Pramana*, **71**, 1153–1157.

55 Goddard, M. and Ruthven, D.M. (1986) *Zeolites*, **6**, 275.
56 Goddard, M. and Ruthven, D.M. (1986) *Zeolites*, **6**, 283.
57 Goddard, M. and Ruthven, D.M. (1986) *Zeolites.*, **6**, 445.
58 Eic, M. and Ruthven, D.M. (1988) *Zeolites*, **8**, 258.
59 Eic, M., Goddard, M., and Ruthven, D.M. (1988) *Zeolites*, **8**, 258.
60 Goddard, M. and Ruthven, D.M. (1986) *Proceedings Seventh International Zeolite Conference Tokyo, 1986* (eds Y. Murakami, A. Lijima, and J.W. Ward), Kodansha, Tokyo, p. 467.
61 Pfeifer, H., Germanus, A., Schirmer, W., Bülow, M., Caro J., Concentration Dependence of Intracrystalline Self-Diffusion in Zeolites; *Ads. Sci. and Techn.* **2** (1985) 229–239.
62 Kärger, J. and Ruthven, D.M. (1989) *Zeolites*, **9**, 267.
63 Jobic, H., Bee, M., Methivier, A., and Combet, J. (2001) *Microporous Mesoporous Mater.*, **42**, 135.
64 Jobic, H., Methivier, A., and Ehlers, G. (2002) *Microporous Mesoporous Mater.*, **56**, 27.
65 Eic, M., Goddard, M., and Ruthven, D.M. (1988) *Zeolites*, **8**, 327.
66 Brandani, S., Xu, Z., and Ruthven, D.M. (1996) *Microporous Mater.*, **7**, 323.
67 Graham, A. (1980) MSc Thesis, University of New Brunswick, Fredericton.
68 Rudtsch, S. (1987) Tracerdesorptions- und Selbstdiffusionsmessungen zur Vergleichbarkeit mit Aussagen der traditionellen Sorptionstechnik" Diploma thesis, Leipzig University.
69 Bülow, M., Mietk, W., Struve, P., and Zikanova, A. (1983) *Z. Phys. Chem. (Leipzig)*, **264**, 598.
70 Bülow, M., Mietk, W., Struve, P., and Lorenz, P. (1983) *J. Chem. Soc., Faraday Trans. I*, **79**, 2457.
71 Bülow, M., Mietk, W., Struve, P., Schirmer, W., Kocirik, M., and Kärger, J. (1984) *Proceedings 6th International Conference on Zeolites Reno, 1983* (eds D. Olson and A. Bisio), Butterworth, Guildford, UK, p. 242.
72 Boddenberg, B. and Burmeister, R. (1988) *Zeolites*, **8**, 488.
73 Lechert, H., Schweitzer, W., and Kacirek, H. (1979) Preprints, Workshop, Adsorption of hydrocarbons in zeolites II, Academy of Sciences of the GDR, Berlin, p. 23.
74 Pfeifer, H., Schirmer, W., and Winkler, H. (1973) *Adv. Chem.*, **121**, 430.
75 Pfeifer, H. (1976) *Phys. Rep. Phys. Lett.*, **26**, 293.
76 Lechert, H., Wittern, K.P., and Schweitzer, W. (1978) *Acta Phys. Chem.*, **24**, 201.
77 Jobic, H. and Fitch, A.N. (2000) *J. Phys. Chem. B*, **104**, 8491.
78 Shen, D. and Rees, L.V.C. (1991) *Zeolites*, **11**, 666.
79 Rudtsch, S. (1987) "NMR-Tracerdesorptions- und Selbstdiffusionsmessungen zur Vergleichbarkeit mit Aussagen der traditionellen Sorptionstechnik" Diploma thesis, Leipzig University.
80 Kärger, J. and Ruthven, D.M. (1989) *Zeolites*, **9**, 267.
81 Jobic, H., Ramanan, H., Auerbach, S.M., Tsapatsis, M., and Fouquet, P. (2006) *Microporous Mesoporous Mater.*, **90**, 307.
82 Wittern, K.P. (1979) PhD Thesis, University of Hamburg, Germany.
83 De Mallmann, A. and Barthomeuf, D. (1986) *Proceedings 7th International Zeolite Conference, Tokyo, 1986* (eds Y. Murakami, A. Lijima, and J.W. Ward), Kodansta, Tokyo, p. 609.
84 Lorenz, P., Bülow, M., and Kärger, J. (1980) *Izv. Akad. Nauk. SSSR Ser. Khim.*, 1741.
85 Kärger, J., Pfeifer, H., and Heink, W. (1984) *Proceedings 6th International Zeolite Conference, Reno, 1983*, Butterworth, Guildford, UK, p. 184.
86 Zibrowius, B., Caro, J., and Pfeifer, H. (1988) *J. Chem. Soc., Faraday Trans.*, **84**, 2347.
87 Silbernagel, B.G., Garcia, A.R., Newsam, J.M., and Hulme, R. (1989) *J. Phys. Chem.*, **93**, 6506.
88 Hasha, D.L., Miner, V.W., Garces, J.M., and Rocke, S.C. (1985) (eds M.L. Deviney and J.L. Gland), *Catalyst Characterization Science, ACS Symposium Series*, vol. **288**, American Chemical Society, Washington DC, p. 485.
89 Auerbach, S.M., Bull, L.M., Henson, N.J., Metiu, H.I., and Cheetham, A.K. (1996) *J. Phys. Chem.*, **100**, 5923.

90 Geil, B., Isfort, O., Boddenberg, B., Favre, D.E., Chmelka, B.F., and Fujara, F. (2002) *J. Chem. Phys.*, **116**, 2184.
91 Auerbach, S.M., Henson, N.J., Cheetham, A.K., and Metiu, H.I. (1995) *J. Phys. Chem.*, **99**, 10600.
92 Auerbach, S.M. (2000) *Rev. Phys. Chem.*, **19**, 155.
93 Saravanan, C., Jousse, F., and Auerbach, S.M. (1998) *Phys. Rev. Lett.*, **80**, 5754.
94 Jobic, H., Bee, M., Kärger, J., Pfeifer, H., and Caro, J. (1990) *Chem. Commun*, 341.
95 Kärger, J. and Pfeifer, H. (1987) *Zeolites*, **7**, 90.
96 Tezel, O.H. and Ruthven, D.M. (1990) *J. Colloid Interface Sci.*, **139**, 581.
97 Riekert, L. (1969) *J. Phys. Chem.*, **73**, 2238.
98 Ulrich, K., Freude, D., Galvosas, P., Krause, C., Kärger, J., Caro, J., Poladli, P., and Papp, H. (2009) *Microporous Mesoporous Mater.*, **120**, 98.
99 Kärger, J., Seiffert, G., and Stallmach, F. (1993) *J. Magn. Reson. A*, **102**, 327.
100 Kärger, J. (2008) *Adsorption and Diffusion* (eds H.G. Karge and J. Weitkamp), Springer, Berlin, p. 85.
101 Parravano, C., Baldeschwieler, J., and Boudart, M. (1967) *Science*, **155**, 1535.
102 Kärger, J. (1971) *Z. Phys. Chem., Leipzig*, **248**, 27.
103 Pfeifer, H. (1972) *NMR Basic Principles Progr*, **7**, 53.
104 Demontis, P., Jobic, H., Gonzalez, M.A., and Suffritti, G.B. (2009) *J. Phys. Chem. C*, **113**, 12373.
105 Germanus, A. (1986) Die intrakristalline Selbstdiffusion von Molekülen in X-Zeolithen unter dem Einfluss spezifischer Wechselwirkungen, Thesis Leipzig University.
106 Ruthven, D.M., Graham, A.M., and Vavlitis, A. (1980) *Proceedings 5th International Zeolite Conference, Naples, 1980* (ed. L.V.C. Rees), Heyden, London, p. 535.
107 Pfeifer, H., Kärger, J., Germanus, A., Schirmer, W., Bülow, M., and Caro, J. (1985) Concentration Dependence of Intracrystalline Self-Diffusion in Zeolites; *Ads. Sci. and Techn.*, **2** 229–239.
108 Pantatosaki, E., Papadopoulos, G.K., Jobic, H., and Theodorou, D.N. (2008) *J. Phys. Chem. B*, **112**, 11708.
109 Thompson, J.K. and Resing, H.A. (1968) *J. Colloid Interface Sci.*, **26**, 279.
110 Kärger, J., Pfeifer, H., Rudtsch, S., Heink, W., and Groß, U. (1988) *J. Fluorine Chem.*, **39**, 349.
111 Weast, R.C. and Astle, M.J. (eds) (1983) *Handbook of Chemistry and Physics*, CRC Press, Boca Raton, FL.
112 Kärger, J., Pfeifer, H., Stallmach, F., Feoktistova, N.N., and Zhdanov, S.P. (1993) *Zeolites*, **13**, 50.
113 Bär, N.K., McDaniel, P.L., Coe, C.G., Seiffert, G., and Kärger, J. (1997) *Zeolites.*, **18**, 71.
114 Zhao, Q., Chempath, S., and Snurr, R.Q. (2005) *Diffusion Fundamentals* (eds J. Kärger, F. Grinberg, and P. Heitjans), Leipziger Universitätsverlag, Leipzig, p. 182.
115 Zhao, Q. and Snurr, R.Q. (2009) *J. Phys. Chem. A*, **113**.
116 Stilbs, P. (1987) *Prog. NMR Spectrosc.*, **19**, 1.
117 Stilbs, P. and Moseley, M.E. (1980) *J. Chem. Soc.*, **15**, 176.
118 Hong, U., Kärger, J., and Pfeifer, H. (1991) *J. Am. Chem. Soc.*, **113**, 4812.
119 Förste, C., Germanus, A., Kärger, J., Möbius, A., Bülow, M., Zhdanov, S.P., and Feoktistova, N.N. (1989) *IsotopenPraxis*, **25**, 48.
120 Germanus, A., Kärger, J., and Pfeifer, H. (1984) *Zeolites*, **4**, 188.
121 Beagley, J., Dwyer, J., Evmerides, N.P., Attava, A.I., and Ibrahim, T.K. (1982) *Zeolites* **2**, 167.
122 Pfeifer, H. (1972) *NMR Basic Principles and Progress*, vol. 7, Springer, Berlin, p. 53.
123 Kärger, J., Zikanova, A., and Kocirik, M. (1984) *Z. Phys. Chem. (Leipzig)*, **265**, 587.
124 Hong, U., Kärger, J., Hunger, B., Feoktistova, N.N., and Zhdanov, S.P. (1992) *J. Catal.*, **137**, 243.
125 Hong, U., Kärger, J., and Pfeifer, H. (1991) *J. Am. Chem. Soc.*, **113**, 4812.
126 Schwarz, H.B., Ernst, H., Ernst, S., Kärger, J., Röser, T., Snurr, R.Q., and Weitkamp, J. (1995) *Appl. Catal. A*, **130**, 227.
127 Hathaway, P.E. and Davis, M.E. (1989) *J. Catal.*, **116**, 263.
128 Schwarz, H.B., Ernst, S., Kärger, J., Knorr, B., Seiffert, G., Snurr, R.Q., Staudte, B., and Weitkamp, J. (1997) *J. Catal.*, **167**, 248.

18
Medium-Pore (Ten-Ring) Zeolites

The MFI structure type in the classification of the International Zeolite Association [1] includes ZSM-5 and its pure silica analog silicalite, as well as ZSM-11 and its pure silica analog silicalite-2. For distinction from silicalite-2, silicalite is sometimes also referred to as silicalite-1. ZSM-5 and silicalite are the most important representatives of these ten-membered oxygen ring zeolites. These materials are widely used in the petroleum and petrochemical industries both as catalysts (Chapter 21) and as selective adsorbents. Their distinctive features include high thermal/hydrothermal stability, hydrophobic/organophilic adsorptive properties, and an intermediate (ten-ring) pore size that leads to molecular sieve size selectivity for some important molecules. The Si/Al ratio ranges from 20 to 30 in a typical ZSM-5 to over 1000 in silicalite. It is the low aluminium content, rather than the framework structure, that is responsible for the hydrophobic nature of these materials; other high-silica zeolites such as highly dealuminated mordenite or zeolite Y show similar hydrophobicity [2] but the high silica form of Y is much more difficult to manufacture. In addition to its proven applications in the petrochemical industry, silicalite has attracted attention as a potential adsorbent for the selective removal of organics from dilute aqueous solutions, for example, in the concentration of alcohol produced by fermentation processes [3, 4], although it is not clear whether such processes have been developed beyond the laboratory scale.

As with the other types of zeolites, for many years researchers have been puzzled by the differences in the diffusivities in these materials, which are commonly observed, especially with light, fast diffusing species, when comparing the results obtained by different measuring techniques. We now know that, as a result of structural defects, the actual structures of most (if not all) zeolite crystals deviate from their ideal crystallographic structures and that these deviations can have a major effect on intracrystalline transport. In discussing intracrystalline diffusion it is therefore necessary to consider both those features of guest diffusion that are determined by the ideal pore structure and the effects of additional transport resistances arising from structural defects. This is particularly true for MFI since the crystals are almost always twinned and the twin ("intergrowth") structure has many different variants.

MFI-type zeolites also provide an example of a host system with structural anisotropy. The precise description of mass transfer in such systems requires the

Diffusion in Nanoporous Materials. Jörg Kärger, Douglas M. Ruthven, and Doros N. Theodorou.
© 2012 Wiley-VCH Verlag GmbH & Co. KGaA. Published 2012 by Wiley-VCH Verlag GmbH & Co. KGaA.

use of diffusion tensors rather than simple diffusion coefficients [see Eq. (1.20)]. Molecular simulations yield the diffusivities in the directions of all principal axes of the diffusion tensor. Although, in favorable situations, diffusion anisotropy is sometimes directly measurable, in real crystals the anisotropy may be substantially reduced by the substructure. Therefore, at the macroscopic scale, the use of a single mean value for the diffusivity generally provides a sufficiently accurate approximation. Examples of both types of measurement are given in this chapter.

Following a brief description of the crystal structure and the pore system the overall diffusional behavior of several different classes of guest molecules is summarized in Sections 18.2–18.5. Sections 18.6 and 18.7 deal with diffusional anisotropy and the impact of various types of structural defect (surface and intracrystalline barriers) on the transport properties while Section 18.8 summarizes the results of several studies of diffusion in a binary adsorbed phase. Zeolites of the ferrierite type are briefly considered in Section 18.9 since they are similar to MFI in that both structures consist of a network of intersecting channels. However, whereas MFI contains only ten-ring channels, the ferrierite structure contains both eight-ring and ten-ring channels. Furthermore, diffusion in ferrierite is only possible in two dimensions, making it particularly suitable for experimental studies of diffusion anisotropy.

18.1
MFI Crystal Structure

The framework structures of silicalite and ZSM-5 are essentially the same, differing only in the Si/Al ratio and the resulting cation density, which is close to zero for silicalite. These materials belong to the family of "pentasil" zeolites, so called because their frameworks can be regarded as being built up from five-membered oxygen rings. The structure of one layer of the framework is shown schematically at the top of Figure 18.1. By stacking such layers in different sequences various related structures can be obtained. The channel systems of the two end members, ZSM-5 and ZSM-11, are sketched in Figure 18.1a and b. The channels are circumscribed by ten-membered oxygen rings of free diameter ≈ 6.0 Å and their pore size is thus intermediate between the "large-pore" X and Y type zeolites and the "small-pore" A zeolites.

Figure 18.2 shows further details of ZSM-5/silicalite-1. The unit cell (Figure 18.2b and c) consists of 96 tetrahedral (SiO_2/AlO_2) units and contains two straight channel sections in the crystallographic b direction (the central one plus one quarter of each of the four corner channels) and four sinusoidal ("zigzag") channels in crystallographic a direction with four channel intersections. A three-dimensional perspective with the oxygen atoms to scale is shown in Figure 18.3. The framework density is about 1.8 g cm^{-3} and the specific micropore volume is about 0.19 cm^3 g^{-1}. More complete descriptions of the structure have been given by Kokotailo [5] and Flanigen [6]. There is no universal correlation between the crystallographic structure and crystal shape since MFI-type crystals are generally twinned and the crystal shape is determined by the twinning. A common structure is shown in Figure 18.2a [7–9]. However, more complex substructures can also occur and this topic is discussed further in Section 18.6.3.

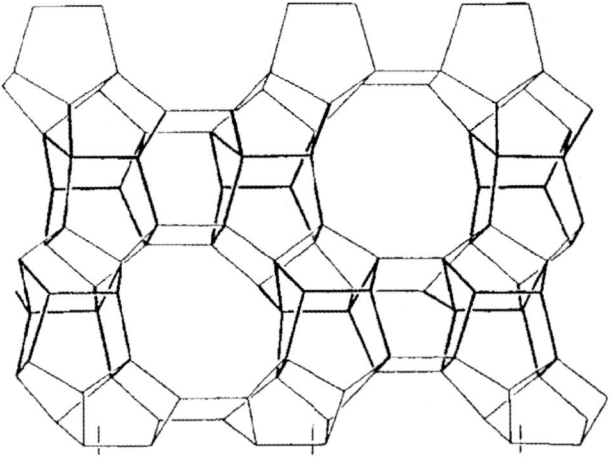

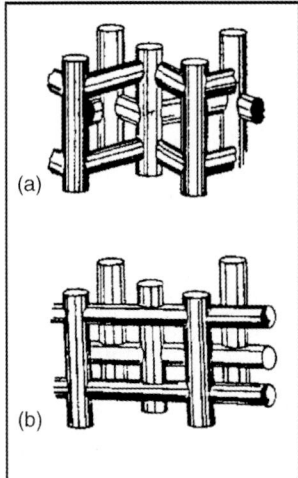

Figure 18.1 Schematic diagram of one layer of a pentasil structure, showing how the framework can be considered as being built up from double five ring units. The channel structures of (a) ZSM-5 (silicalite-1) and (b) ZSM-11 (silicalite-2) are also shown in pictorial form. Reprinted from Kokotailo *et al.* [5] with permission.

18.1.1
Saturation Capacity

Table 18.1 summarizes reported equilibrium saturation capacities for some of the more widely studied sorbates in silicalite or ZSM-5. In general there is reasonable agreement, with the notable exception of *p*-xylene, and there is evidently very little difference in capacity between silicalite and H-ZSM-5. This is to be expected since the saturation capacity is determined mainly by the specific micropore volume, which is the same for both materials.

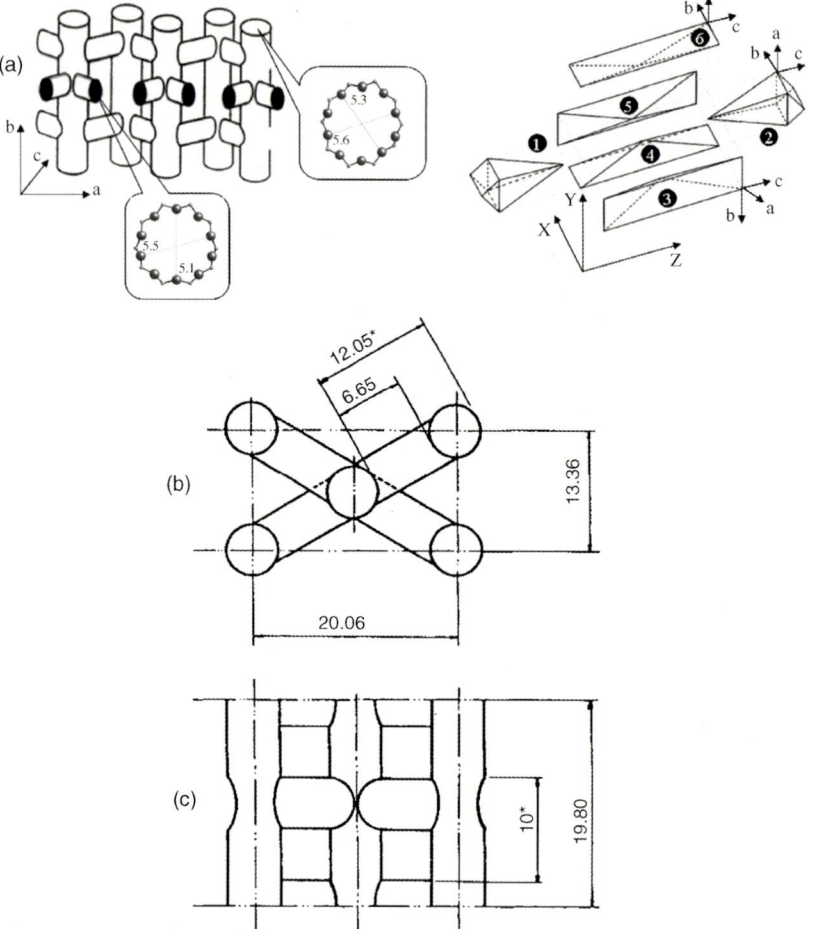

Figure 18.2 Channel system of ZSM-5/silicalite-1, showing (a) a common form of twin structure and the orientation of the straight and zigzag channels in the different sub-sections of the crystal (see Section 18.6.3 for more details); (b) a plan view looking in the b direction of the channel system/compartments shown in (a); (c) view in the c direction. Dimensions are shown in Å. Adapted from Tzoulaki et al. [15] and Wu et al. [14].

In the case of p-xylene, Olson et al. [10] and Beschmann et al. [11] report saturation capacities of about 8 molecules per unit cell, the same as for benzene, and the calorimetric data of Thamm [12, 13] also support this value. Most other authors find a somewhat lower capacity, equivalent to about 6 molecules per unit cell. Either of these limits can be rationalized in terms of the framework structure. Wu et al. [14] have put forward the plausible suggestion that the structure can accommodate one molecule of benzene or n-hexane in each section of the zigzag channels (length ≈ 6.6 Å) and two benzene molecules in each of the longer straight sections (length ≈ 19.8 Å), to

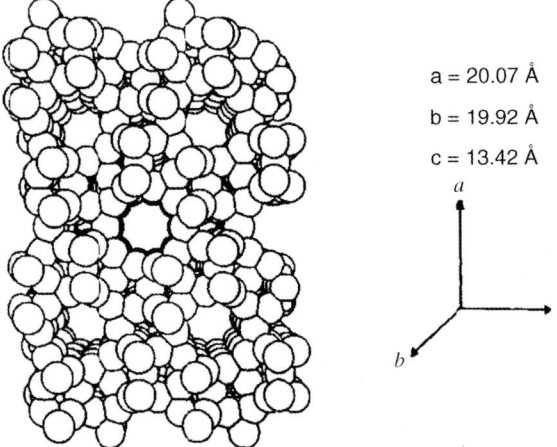

Figure 18.3 Three-dimensional view of four adjacent cells of silicalite/ZSM-5, showing the oxygen atoms to scale. Note that the c direction (also referred to as the z direction) generally corresponds to the long axis of the crystals. Reprinted from June et al. [18] with permission.

Table 18.1 Saturation capacities for H-ZSM-5 and silicalite.

Sorbate	Sorbent	Saturation capacity		Reference
		In mmol g^{-1}	Molecule per cell	
Benzene	Silicalite	1.42	8.2	Wu et al. [14], Ma [19]
	H-ZSM-5		7.6	Olson et al. [10]
Toluene	H-ZSM-5	1.07	6.6	Anderson et al. [20]
p-Xylene	H-ZSM-5	0.95	5.8	Anderson et al. [20]
	ZSM-5		5.94	Derouane [21]
	H-ZSM–5	1.42	8.2	Olson et al. [10]
	Silicalite	1.06	6.1	Wu et al. [14]
	Silicalite	1.4	8.0	Hayhurst and Lee [22]
	H-ZSM-5	1.4	8.0	Beschmann et al. [11]
Ethylbenzene	Silicalite	1.04	6.0	Wu et al. [14]
n-Hexane	Silicalite	1.42	8.2	Wu et al. [14]
	H-ZSM-5	1.23	7.6	Anderson [20]
	ZSM-5		8.14	Derouane [21]
	H-ZSM-5		7.9	Jacobs [23]
	H-ZSM-5		8.2	Olson et al. [10]
3-Methylpentane	H-ZSM-5	0.98	6.0	Anderson [20]
	ZSM-5		6.74	Derouane [21]
	H-ZSM-5		5.7	Olson [10]

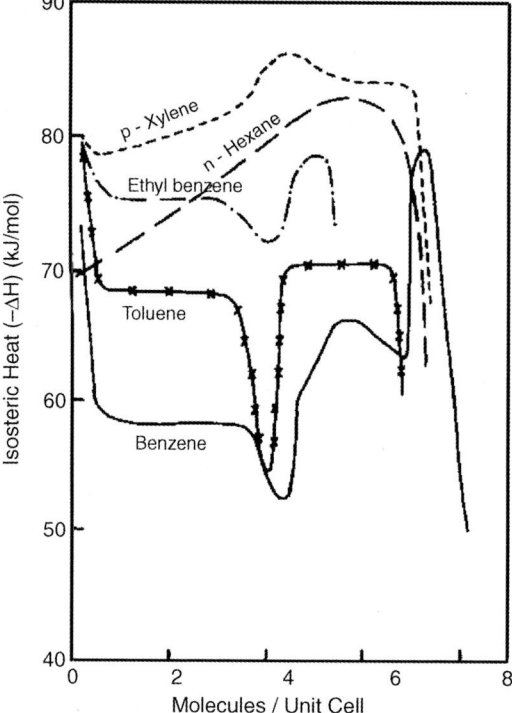

Figure 18.4 Heat of sorption versus loading for several sorbates in silicalite; calorimetric data of Thamm [12, 13]. Note, for comparison, the chromatographic values of Forni and Viscardi [16] at zero loading: benzene, −92, toluene, −94, and p-xylene, −101 kJ mol^{-1}.

give a total capacity of 8 molecules per cell. If p-xylene packs in the same way, the same capacity (in molecules per cell) will be obtained. In contrast, if one assumes that with the longer p-xylene or ethylbenzene molecules, when two such molecules are aligned in one of the straight channels, the four molecules occupying the zigzag channels will prevent occupation of the other straight channel one obtains a saturation limit of 6 molecules per cell.

Figure 18.4 shows calorimetric heats of sorption data for several sorbates taken from the work of Thamm [12, 13]. For comparison the heats of sorption for benzene, toluene, and xylene, derived by Forni and Viscardi [16] from the temperature dependence of chromatographically measured Henry constants, are given in the figure legend. These values, which refer to the low-concentration limit, are somewhat higher but quite consistent with the calorimetric data that show that the heat of sorption changes rapidly at very low loadings. The complex pattern of variation of heat of sorption with loading explains the wide variations in the reported values of the isosteric heats. The heats of sorption for ZSM-5 are generally somewhat higher than

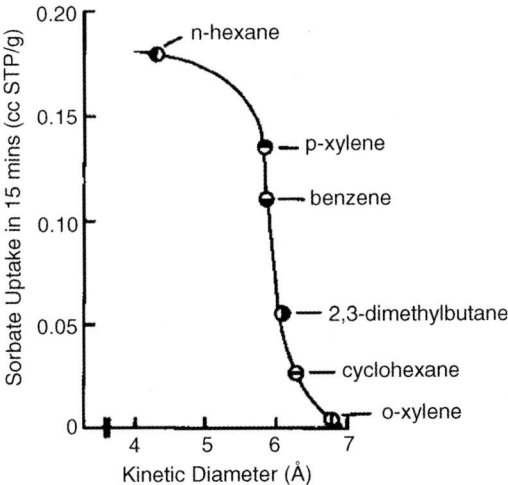

Figure 18.5 Size-selective sorption cut-off in silicalite. Uptake in cc STP per g after 15 min exposure under standardized conditions (298 K, relative pressure half saturation vapor pressure). Data of Harrison et al. [24].

for silicalite, under comparable conditions, no doubt reflecting a small contribution from polarization or electrostatic forces [17].

18.1.2
Molecular Sieve Behavior

Molecules with critical diameters less than 6.0 Å can penetrate the MFI pores freely and rapidly. The cut-off in sorption that occurs with larger molecules was demonstrated by Harrison et al. [24] and some of their data are reproduced in Figure 18.5. The sorption cut-off lies within the range of molecular size spanned by the C_8 aromatic isomers; p-xylene and ethylbenzene penetrate rapidly while m-xylene and o-xylene penetrate only slowly. This size-selective sieving effect has been suggested as a means for separating these isomers [25]. It is the basis for the selective xylene isomerization process, which, by virtue of the slow diffusion of the *meta-* and *ortho-*isomers within the catalyst micropores, allows a higher than equilibrium concentration of p-xylene to be produced. This process is discussed in greater detail in Section 21.4.6.

18.2
Diffusion of Saturated Hydrocarbons

18.2.1
Linear Alkanes

Diffusion of the lighter linear alkanes in silicalite has been studied intensively over the last two decades. This is not just because of their intrinsic interest but rather because these systems are amenable to study by a wide range of different techniques, some of which are not applicable for larger molecules. Some of the reported experimental data for diffusion of linear alkanes in MFI are summarized in Table 18.2 and in Figures 18.6 and 18.7. Silicalite is essentially cation free, so electrostatic forces are minimal. Differences in the mass, size, and shape of the diffusing molecules are therefore much more important than differences in electronic structure in determining molecular mobility. As a result, the patterns of behavior observed for the

Table 18.2 Diffusivity data for n-alkanes in silicalite/HZSM-5 at low loading.[a]

Sorbate	Sorbent	Method	T (K)	$\mathcal{D}$ or D_0 (m^2 s^{-1})	E (kJ mol^{-1})	Reference
CH$_4$	Silicalite	FR	195	7.5×10^{-9}		Rees [26]
	Silicalite/HZSM-5	PFG NMR	334	$\approx 10^{-8}$	4	Caro [27], Kärger [28]
	Silicalite	SCM	334	$\approx 10^{-10}$		Hayhurst [29]
	HZSM-5	QENS	250	3×10^{-9}		Jobic [30]
C$_2$H$_6$	Silicalite	FR	334	1.3×10^{-8}		Rees [26]
	HZSM5	PFG NMR	334	5×10^{-9}	5	Caro [27]
	Silicalite	Square wave	334	4×10^{-9}		van den Begin [31]
	Silicalite	SCM	334	2×10^{-11}		Hayhurst [29]
C$_3$H$_8$	Silicalite	FR	334	4×10^{-9}	15	Rees [32]
	Silicalite	PFG NMR	334	2×10^{-9}	7.1	Heink [33]
	Silicalite	Square wave	334	2.5×10^{-9}	6.7	van den Begin [31]
	Silicalite	ZLC	334	1.2×10^{-11}	13	Eic and Ruthven [34]
	Silicalite	SCM	334	7.3×10^{-12}		Hayhurst [29]
nC$_4$H$_{10}$	Silicalite	FR	373	6×10^{-9}		Rees [26]
	Silicalite	PFG NMR	373	1.3×10^{-9}	7.4	Heink [33]
	Silicalite	PFG NMR	334	1.5×10^{-10}	8.1	Datema [35]
	Silicalite	PCM	373	5×10^{-10}	12	Millot [36]
	Silicalite	SCM	334	$4\text{-}11 \times 10^{-12}$		Paravar [37], Hayhurst [29]
nC$_5$H$_{12}$	Silicalite	FR	373	3×10^{-9}	19	Rees [26]
	Silicalite	PFG NMR	373	6.5×10^{-10}	8.6	Heink [33]
	Silicalite	PFG NMR	334	8×10^{-11}	12.6	Datema [35]
	Silicalite	ZLC	334	2×10^{-12}	19.2	Eic and Ruthven [34]
nC$_6$H$_{14}$	Silicalite	FR	373	1.5×10^{-9}	(22)	Rees & Song [38]
	Silicalite	PCM	373	1×10^{-11}	32	Millot [36]
	Silicalite	SCM	334	5×10^{-12}	13	Talu [39]
	Silicalite	PFG NMR	373	2.5×10^{-10}	8.7	Heink [33]
	Silicalite	Square wave	379	$\approx 10^{-12}$		van den Begin [31]
nC$_8$H$_{18}$	NaZSM-5	QENS	400	1.1×10^{-10}	7.7	Jobic [40]
nC$_{10}$H$_{22}$	NaZSM-5	QENS	400	8×10^{-11}	10.3	Jobic [40]
	Silicalite	ZLC	379	5×10^{-13}	20	Eic and Ruthven [34]
nC$_{12}$H$_{26}$	NaZSM-5	QENS	400	5×10^{-11}	11.4	Jobic [40]
nC$_{14}$H$_{30}$	NaZSM-5	QENS	400	2×10^{-11}	12	Jobic [40]
	Silicalite	ZLC	379	4×10^{-13}	19.2	Eic and Ruthven [34]

a) FR = frequency response; PCM = polycrystalline membrane; SCM = single-crystal membrane. For bimodal FR signals only the faster process is quoted. Crystal sizes vary from $100 \times 100 \times 300$ to $20 \times 30 \times 50$ μm^3.

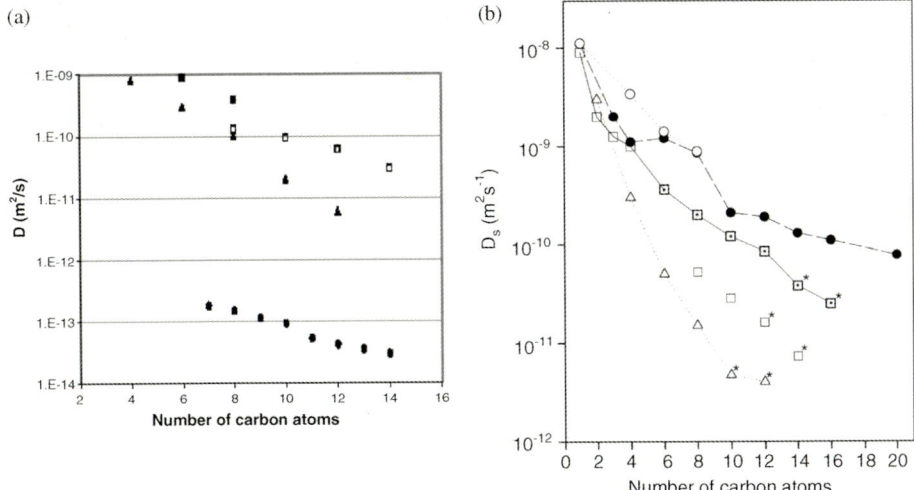

Figure 18.6 Comparison of self-diffusivities of n-alkanes at low loadings in silicalite/ZSM-5 as a function of the carbon number. (a) Comparison of macro- and microscale measurements at 423 K: QENS (□, Na ZSM-5), (■, silicalite); PFG NMR (▲); and ZLC (●). From Bourdin et al. [41, 42]. (b) Comparison of microscale measurements and molecular simulations at 300 K: QENS (□, Na ZSM-5) [40], (⊡, silicalite) [43]; PFG NMR [44] (△, silicalite); MD simulations (Section 8.2) [45] (○); and hierarchical simulation involving combined application of transition state theory and Brownian dynamics in a coarse-grained model representation tracking the positions of the two ends of each n-alkane molecule along the channels (Sections 7.4 and 9.1.3) [46] (●). Asterisked symbols correspond to extrapolation to 300 K; note that the PFG NMR diffusivity for n-hexane has been independently confirmed by Ramachandran et al. [47]. Reprinted from Jobic and Theodorou [48] with permission.

linear alkanes are likely to be replicated by other species of similar molecular shape and size such as the linear olefins.

18.2.1.1 Microscale Measurements

A key observation, illustrated in Figure 18.6, is that the intracrystalline diffusivities of the linear alkanes in MFI appear to depend on the length scale of the measurement, with substantially higher values being observed at shorter length scales. This is probably a consequence of the complex twin structure of MFI crystals, which is discussed in detail in Section 18.6.3. As a result, the "ideal" intracrystalline diffusivity for a perfect MFI crystal is approached only at very short length scales (a few unit cells), which are accessible to experimental measurement only by QENS (quasi-elastic neutron scattering). These values should replicate the values derived from molecular simulations. For the data presented in Figure 18.6b this is approximately true, but with increasing deviations for higher carbon numbers. Measurements at somewhat greater length scales (typically 10^{-7}–10^{-5} m) by PFG NMR (pulsed field gradient nuclear magnetic resonance) yield lower diffusivities. Some of these differences may be ascribed to differences in the loading levels as well as to increasing uncertainties in the simulations for the larger molecules but the general trend is clear.

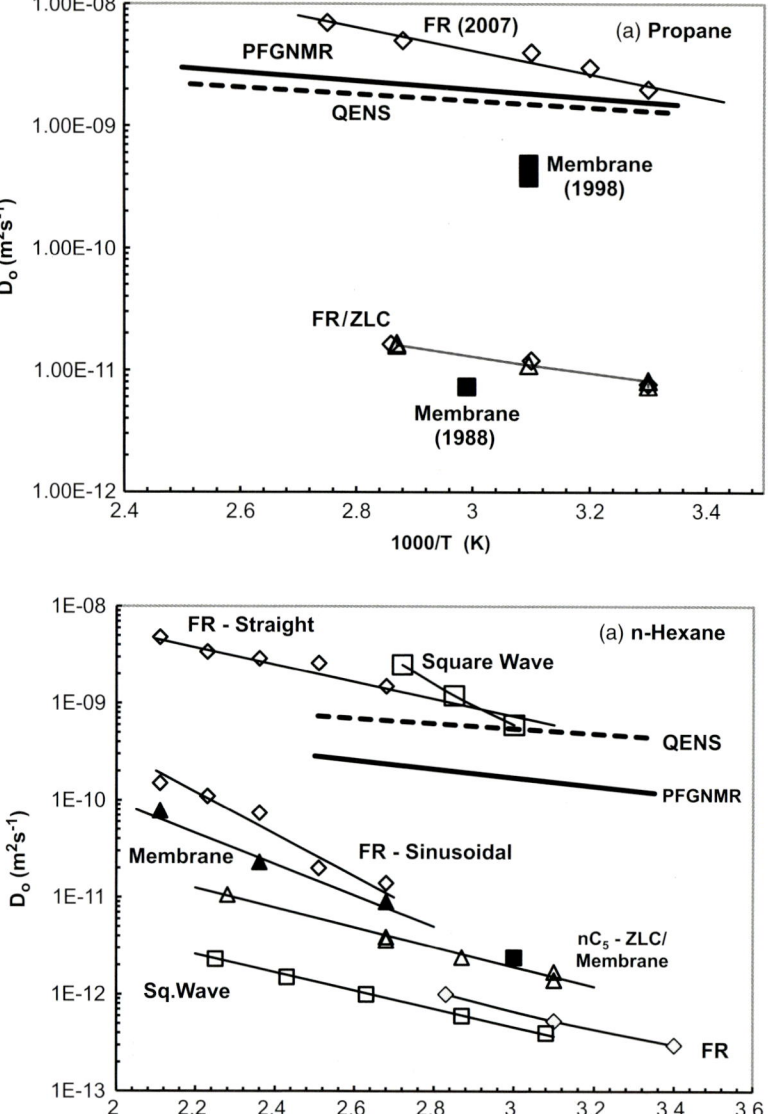

Figure 18.7 Arrhenius plots showing the temperature dependence of diffusivity for (a) propane and (b) n-hexane in different samples of silicalite, measured at low loading by different macroscopic techniques: (◇) FR [26, 32, 49], (□) square wave [31], (△) ZLC [34], (■) SCM [29, 39], and (▲) PCM [36]. The QENS and PFG NMR data (from Jobic [50] and Heink [51]) are indicated for comparison.

18.2.1.2 Macroscale Measurements

Diffusion of the lightest alkanes (methane and ethane) is too fast to measure with any confidence by macroscopic techniques (with the possible exception of frequency response) but the diffusion of propane and longer chain alkanes is easily measurable and has been studied by several different macroscopic techniques. Most macroscopic techniques yield *average* diffusivity values for the entire crystal, which are of course reduced by defects at all scales as well as by any surface resistance. Representative results are shown in Figure 18.6a and in Figure 18.7 in which the values obtained for propane and n-hexane by different measurement techniques are compared.

These results present a confusing picture. The following general observations can be made:

1) Most, but not all, of the macro-diffusivity values fall well below the values derived from the micro-scale measurements.
2) For measurements with different crystals of similar size the same measurement technique often yields widely different diffusivity values.
3) Despite large differences in diffusivity the activation energies (derived from the slopes of the Arrhenius plots) are fairly consistent for the different macro-scale techniques but generally larger than the values from the micro-scale techniques – see Table 18.2 and Figure 18.8.

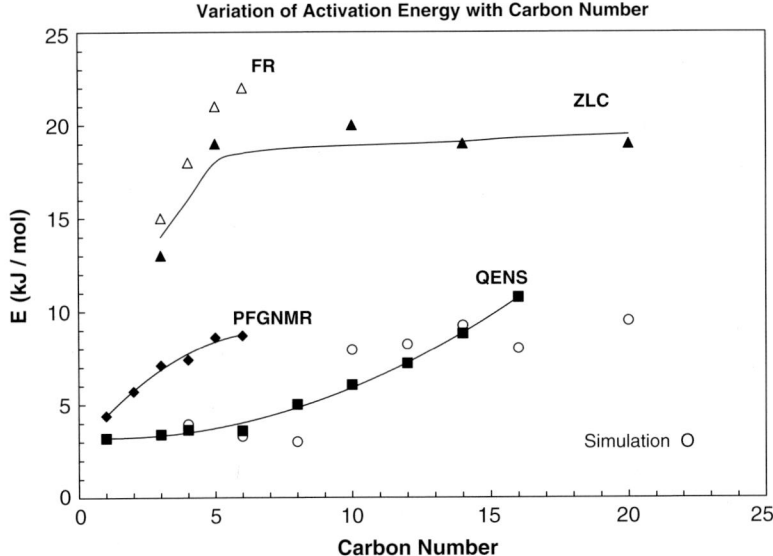

Figure 18.8 Variation of diffusional activation energy (for self- and transport diffusion at low loadings) with carbon number for linear alkanes in silicalite; comparison of experimental data with hierarchical simulation of Maginn et al. [46]. Experimental data are from Heink [51], Jobic [40], Jobic and Theodorou [43], Eic and Ruthven [34], and Rees and Song [52].

4) Some of the frequency response data show a bimodal response that has been attributed to two distinct diffusion processes corresponding to diffusion in the straight and sinusoidal channels [32, 38]. The faster of these processes (attributed to the straight channels) yields diffusivities that are comparable with or even faster than the QENS and PFG NMR values, although the activation energy is generally larger.
5) The high diffusivities detected by frequency response measurements and attributed to diffusion in the straight channels are not observed by the other macroscopic techniques, such as ZLC (zero-length column). It is unclear whether this is due to a unique feature of the frequency response technique or merely reflects the use of different crystals in the different studies.

These observations can be rationalized by considering the effect of the twin substructure of the crystals – see Section 18.6.

18.2.1.3 Molecular Simulations

Comparison between the QENS data for silicalite and NaZSM-5 (Figure 18.6b) shows that n-alkane diffusion in MFI is substantially retarded by the presence of Na^+ cations. The activation energies derived from the hierarchical simulation [46] show reasonable agreement with the QENS data [40, 43]. Evidently, the activation energy is essentially independent of carbon number up to C_6 and then increases roughly in proportion to carbon number.

The simulations by Maginn et al. [46, 53] provide a simple explanation for the observed behavior. The channel interiors are energetically more favorable for n-alkanes than the more spacious channel intersections since they allow the methylene and methyl segments to maximize their dispersive interactions with the zeolite framework. The differences in the chain-length dependence are related to the fact that the shorter n-alkanes (C_4, C_6) can reside entirely within a channel interior, while C_8 and the longer alkanes have to straddle and/or protrude into a channel intersection. The rate-limiting step for the short alkanes is thus essentially a jump across the less favorable intersection region, which is higher in energy by about 5 kJmol^{-1} relative to the channel regions. The diffusion steps for the longer alkanes are accompanied by $trans$–$gauche$ conformational changes with activation energies of about 12 kJmol^{-1}, which is again close to the observed activation energies of self-diffusion.

The agreement between the simulation results and the QENS data suggests that our understanding of diffusion of light alkanes in the ideal MFI pore system is essentially correct. However, the activation energies derived from macroscopic measurements are substantially larger and show a different pattern of variation with carbon number. The PFG NMR activation energies are intermediate, approaching the QENS values for the lower carbon numbers but with increasing deviation to higher values for higher carbon numbers.

The lower diffusivities, higher activation energies, and the different pattern of variation of activation energy with carbon number (Figure 18.8) derived from macroscopic measurements as well as from PFG NMR suggest that the controlling

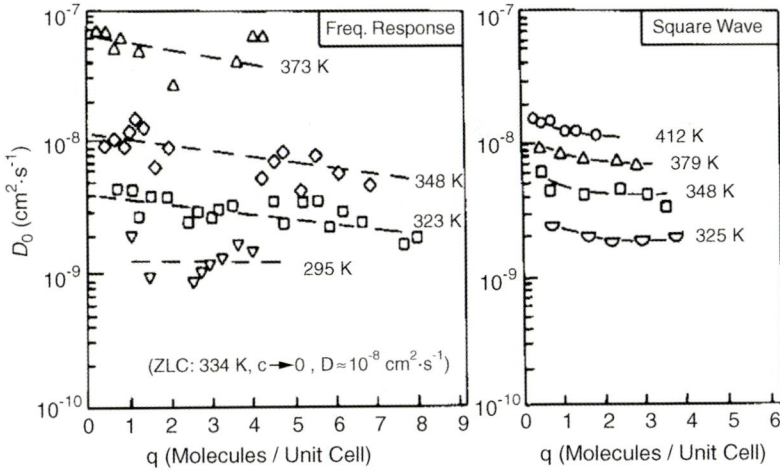

Figure 18.9 Variation of corrected diffusivity (D_0) with loading for n-C_6 in MFI, showing comparison between frequency responses [49], square wave [31], and ZLC [34] data.

mechanism at greater length scales is different, reflecting the increasing impact of the crystal sub-structure over longer distances.

18.2.1.4 Loading Dependence of Diffusivity

The ZLC and membrane measurements were performed only at low concentration within the Henry's law region and therefore provide no information concerning the concentration dependence of the diffusivity. However, the frequency response and square wave data suggest that, at least for the higher homologs, the concentration dependence of the corrected diffusivity is modest, showing a slight decreasing trend with increasing concentration (Figure 18.9). The PFG NMR data also show a decreasing trend with loading but it is much more pronounced – see Figure 18.10.

18.2.1.5 Microdynamic Behavior

The similarity between the trends observed in NaX and silicalite appears to suggest similar diffusion mechanisms. However, detailed analysis of the NMR relaxation times for propane in silicalite [27, 54] shows that the mean life time between molecular reorientations increases monotonically with concentration, as shown by a shift of the T_1 minimum to higher temperatures. As a result, the mean jump length, which, at low concentrations, is close to the distance between adjacent channel intersections, decreases only slightly with increasing concentration. Such behavior is in marked contrast to the behavior in NaX, where the mean life time between reorientations is found to be independent of concentration (constant temperature of the T_1 minimum). Both the order of magnitude of the diffusivities and the trend of the mean jump length with concentration have been confirmed, for methane, by the results of neutron scattering measurements [55, 56].

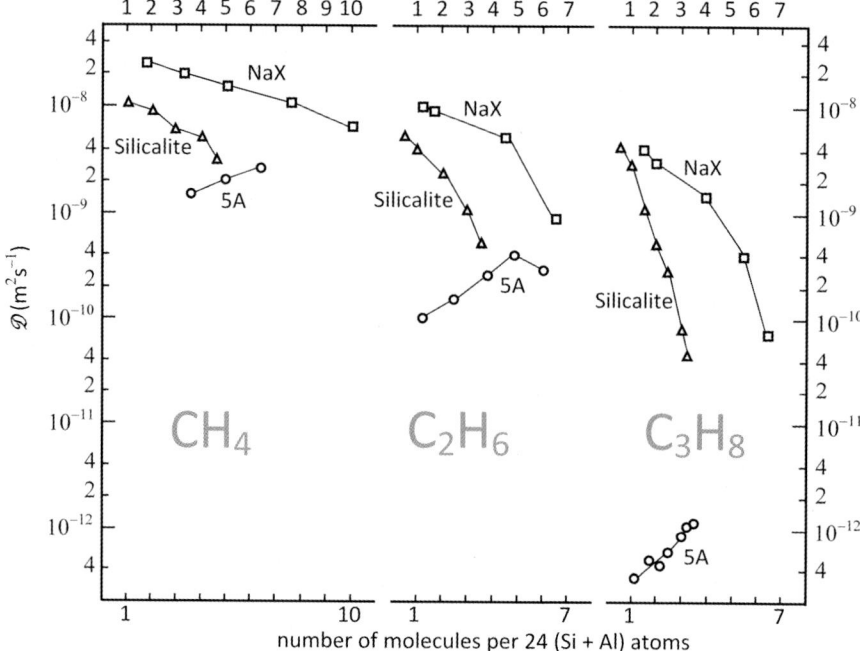

Figure 18.10 Comparison of patterns of variation of PFG NMR self-diffusivity with sorbate concentration at 300 K for CH$_4$, C$_2$H$_6$, and C$_3$H$_8$ in 5A, silicalite, and NaX zeolites; 24 (Si + Al) atoms represent one cage for the A or X structures and one channel intersection for silicalite. Adapted from Caro et al. [27] with permission.

18.2.2
Diffusion of Branched and Cyclic Paraffins

Table 18.3 summarizes experimental diffusivity data for several different branched and cyclic paraffins in various different samples of silicalite/HZSM-5. The activation energies for the branched and cyclic isomers are much higher than for the corresponding linear paraffins, leading to much lower diffusivities and quite different patterns of behavior. Most of the measurements were made by macroscopic methods since diffusion of these species is relatively slow and not easily measurable by PFG NMR and QENS as the most prominent microscopic methods. In contrast to the linear paraffins, for which the diffusivity data show wide variations between different samples and different experimental techniques, for the branched and cyclic isomers the diffusivities measured by different techniques in different samples are reasonably consistent.

A plausible explanation for this difference is that the branched and cyclic isomers, because of their larger molecular diameter, have a strong preference for the channel intersections over the channels between intersections. The diffusion process takes place by distinct jumps from one channel intersection to an adjacent intersection and

Table 18.3 Summary of diffusivity data for saturated branched and cyclic hydrocarbons in silicalite/ZSM-5.

Sorbate	Sorbent	Method	T (K)	$\mathcal{D}$ or D_o (m² s⁻¹)	E (kJ mol⁻¹)	Reference
Isobutane	Silicalite	Various[a]	400	7×10^{-12}	31	See Figure 18.11
		ZLC	400	3×10^{-11}	31	Zhu [57]
		NSE	500	2.2×10^{-11}	23	Jobic [50]
	NaZSM-5	QENS	500	3×10^{-12}	17	Millot [58]
2-Methylpentane	Silicalite	Uptake/ZLC	400	10^{-12}	46	Cavalcante [59]
3-Methylpentane	Silicalite	PCM[b]	400	2×10^{-13}	41	Millot [36]
		Uptake	296	$\sim 10^{-15}$	—	Riekert [60]
Methylcyclopentane	Silicalite	Uptake/ZLC	400	2×10^{-13}	49	Cavalcante [59]
2,3-Dimethylbutane	Silicalite	Uptake/ZLC	400	2×10^{-14}	57	Cavalcante [59]
2,2-Dimethylbutane	HZSM-5	Uptake/chromatography	400	3×10^{-18}	66	Post [61]
Cyclohexane	Silicalite	Uptake/ZLC	400	2×10^{-15}	53	Cavalcante [59]
		Uptake	400	3×10^{-15}	—	Wu et al. [14]
		Uptake	400	10^{-16}	50	Chon and Park [62]
		Uptake	400	2×10^{-16}	(65)	Xiao and Wei [63]
		Uptake/ZLC	400	2×10^{-15}	48	Magalhaes [64]
		FR	400	4×10^{-15}	—	Song and Rees [65]
Methylcyclohexane	Silicalite	Uptake/ZLC	400	3×10^{-15}	51	Magalhaes [64]
Ethylcyclohexane	Silicalite	Uptake/ZLC	400	2×10^{-15}	58	Magalhaes [64]
cis-1,4-Dimethyl cyclohexane	Silicalite	Uptake/ZLC	400	1×10^{-15}	56	Magalhaes [64]
cis 1,4-Dimethyl cyclohexane	Silicalite	FR	400	$\sim 10^{-15}$	—	Song and Rees [26]
trans-1,4-Dimethyl cyclohexane	Silicalite	Uptake/ZLC	400	$\sim 10^{-13}$	39	Magalhaes [64]
		FR	400	6×10^{-14}	—	Song and Rees [26]

a) Combined data for iC₄ as per Figure 18.11.
b) PCM = polycrystalline membrane (z aligned).

this generally occurs with almost the same probability in each of the four possible jump directions.

18.2.2.1 Isobutane at Low Loadings

Diffusion of isobutane in silicalite has been studied in a wide range of different samples by many different experimental techniques. The results from several different studies, including both microscopic and macroscopic methods, are summarized in Figure 18.11. In contrast to the complex picture presented by the data for linear alkanes (Figure 18.7), these results yield a remarkably coherent and rather simple picture of the diffusional behavior.

Despite the differences in the samples, which cover a wide range of crystal size, and the differences in the experimental techniques, which include direct uptake rate measurements, ZLC, membrane permeation, chromatography, PFG NMR, QENS,

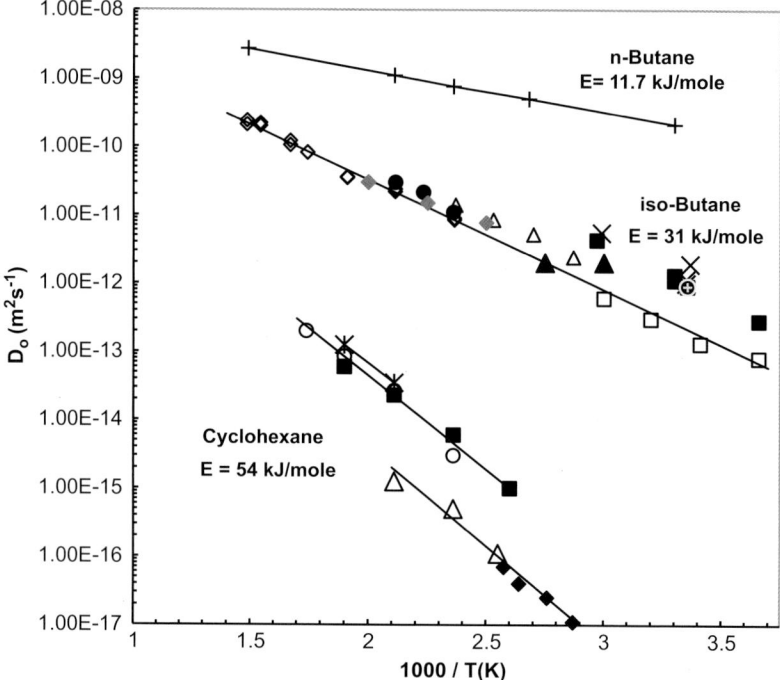

Figure 18.11 Arrhenius plot showing consistency of experimental diffusivity data (at low loadings) measured by different techniques for different silicalite samples. *Isobutane*: data of Millot [36, 58] (PCM) (◇); Jobic [40] (NSE) (◆); Hufton [66] (chromatography) (●); Banas and Brandani [67] (PFG NMR) (▲); Geier et al. [68], Tzoulaki [15], and Chmelik [69] [interference microscopy (IFM) and infrared microscopy (IRM)] (■); Hayhurst and Shah [37, 70] (single crystal membrane) (×); Zhu [57] (ZLC) (■); Jiang [71] (ZLC), (□). Data of Millot [36] for n-butane (+) are also shown for comparison. *Cyclohexane*: data of Cavalcante [59] (ZLC/gravimetric, different samples) (×, ○, △); Magalhaes [64] (■); Chon and Park [62] (◆).

IRM, and direct measurements of the transient profile by IFM, most of the data can be well represented by a simple Arrhenius expression (with $E \approx 31$ kJmol^{-1} and $D_\infty = 6.8 \times 10^{-8}$ m^2 s^{-1}). The chromatographic data of Hufton et al. [66] and the ZLC data of Zhu et al. [57] suggest slightly higher diffusivities but the same activation energy. The earlier QENS measurements, performed with NaZSM-5 [58] yielded somewhat lower diffusivities but more recent measurements carried out by the neutron spin echo (NSE) technique [50] yield values that agree well with other macroscopic techniques (Figure 18.11). This is in marked contrast to the data for linear alkanes, for which QENS and other microscopic methods yield diffusivities that are much *larger* than the macroscopic values. The consistency of the diffusivities obtained by different techniques, with different crystal sizes, and different samples, including different crystal forms, suggests that the diffusion process must be controlled by the intracrystalline pore structure, rather than by the twin sub-structure.

Even more remarkable is the close conformity between the membrane data, which measure diffusion in the z direction (along the long axis), and the values obtained by the other macroscopic methods, which measure an overall average diffusivity in all directions. This result, which is discussed in greater detail in Section 18.7.6, implies that, despite the well known non-isotropy of the ideal structure, at the crystal scale (in real crystals) the diffusional behavior is probably close to isotropic.

18.2.2.2 Cyclohexane and Alkyl Cyclohexanes

The behavior of cyclohexane appears to be generally similar to that of isobutane although the available experimental data are less extensive. Cavalcante [59] made both gravimetric and ZLC measurements with four different samples of different crystal size and shape. His results for different samples of large silicalite crystals are very consistent (Figure 18.12 and Table 18.3) and agree well with the data of Magalhaes et al. [64], which were also obtained with large crystals, as well as with the frequency response data of Song and Rees [26]. Cavalcante's data for the smaller crystals show smaller diffusivities but essentially the same activation energy (54 kJmol^{-1}). These data are in close agreement with the data of Chon and Park [62] who also used small (micron sized) crystals. The difference in apparent diffusivity between the small and large crystals can probably be attributed to surface resistance, which has a much greater impact for the smaller crystals (Section 18.6).

Magalhaes et al. [64] and Chon and Park [62] also studied diffusion of alkyl substituted cyclohexanes. Their results are summarized in Figure 18.12 and

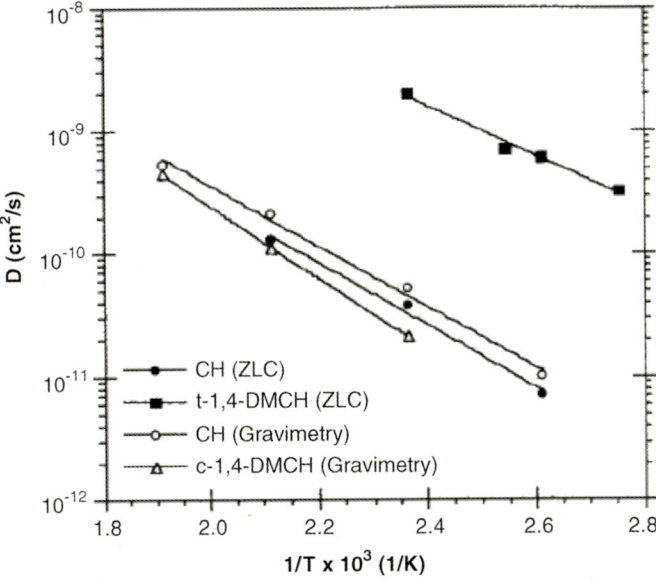

Figure 18.12 Arrhenius plots comparing the diffusivities for *cis*- and *trans*-dimethylcyclohexanes (DMCHs) and cyclohexane in silicalite. Reprinted from Magalhaes et al. [64], with permission.

Table 18.3. The diffusivities of cyclohexane, methylcyclohexane and *cis*-1,4-dimethylcyclohexane are all very similar whereas the diffusivities for *trans*-1,4-dimethylcyclohexane are larger by more than an order of magnitude, with a correspondingly lower activation energy. This behavior closely mirrors the behavior of *p*-xylene, which diffuses much faster than the other C_8 aromatics (Section 18.3.2.5). A plausible explanation is that the longer molecules (*trans*-1,4-dimethylcyclohexane or *p*-xylene) cannot fit within one channel intersection so that the variation of energy with distance (at least along the straight channels) is much less pronounced than for the shorter isomers that can fit within a single channel intersection.

18.2.2.3 2,2-Dimethylbutane

The critical molecular diameter of 2,2-dimethylbutane (6.2 Å) exceeds the nominal diameter of the silicalite pores (6 Å) so diffusion is slow, making it possible to make measurements over a wide range of temperatures with a wide range of different crystal sizes. A detailed study of this type was carried out by Post *et al.* [61], who made measurements by both gravimetric and chromatographic methods on several different samples of silicalite and H-ZSM-5 differing in crystal size and Si/Al ratio. Excellent conformity with the diffusion model was established both from the form of the uptake curves and from the variation of uptake rate with crystal size ($D/r^2 \propto 1/r^2$), and results of the gravimetric and chromatographic measurements were entirely consistent. Diffusivity values derived from measurements of the effectiveness factor in catalytic cracking reactions carried out with different crystal size fractions were also shown to be consistent with the directly measured values. Some of these results are reproduced in Figure 18.13.

From this evidence it is clear that in this system the sorption/desorption rates are controlled by intracrystalline diffusion. Furthermore, the constancy of the diffusivities obtained with many different crystal samples implies that, as for isobutane, the diffusivity must be determined by the intrinsic pore structure. The diffusional activation energy (66 kJ mol^{-1}) is almost the same as the heat of sorption, reflecting the depth of the potential well for the equilibrium state (the molecule in a channel intersection) and implying that, fortuitously, the molecule in the transition state (within the channel between two intersections) must have the same energy as in the free gas phase.

The effect of increasing Al content is shown in Figure 18.13d. Diffusivities in the Al-rich adsorbents are lower than in pure silicalite but the differences are modest, implying that the activation energy is determined mainly by the pore diameter rather than the precise nature of the pore walls.

18.2.2.4 Comparison of Diffusivities for C_6 Isomers

Figure 18.14 shows, in Arrhenius form, a comparison of the diffusivities of six different cyclic and branched C_6 paraffin isomers, measured in large silicalite crystals by gravimetric and ZLC methods. Data for 2,4-dimethylpentane are also included. The data for 2,3-dimethylbutane, discussed above, are noteworthy as they show the consistency of the diffusivities obtained with several different crystal size fractions. The Arrhenius parameters are given in Table 18.3.

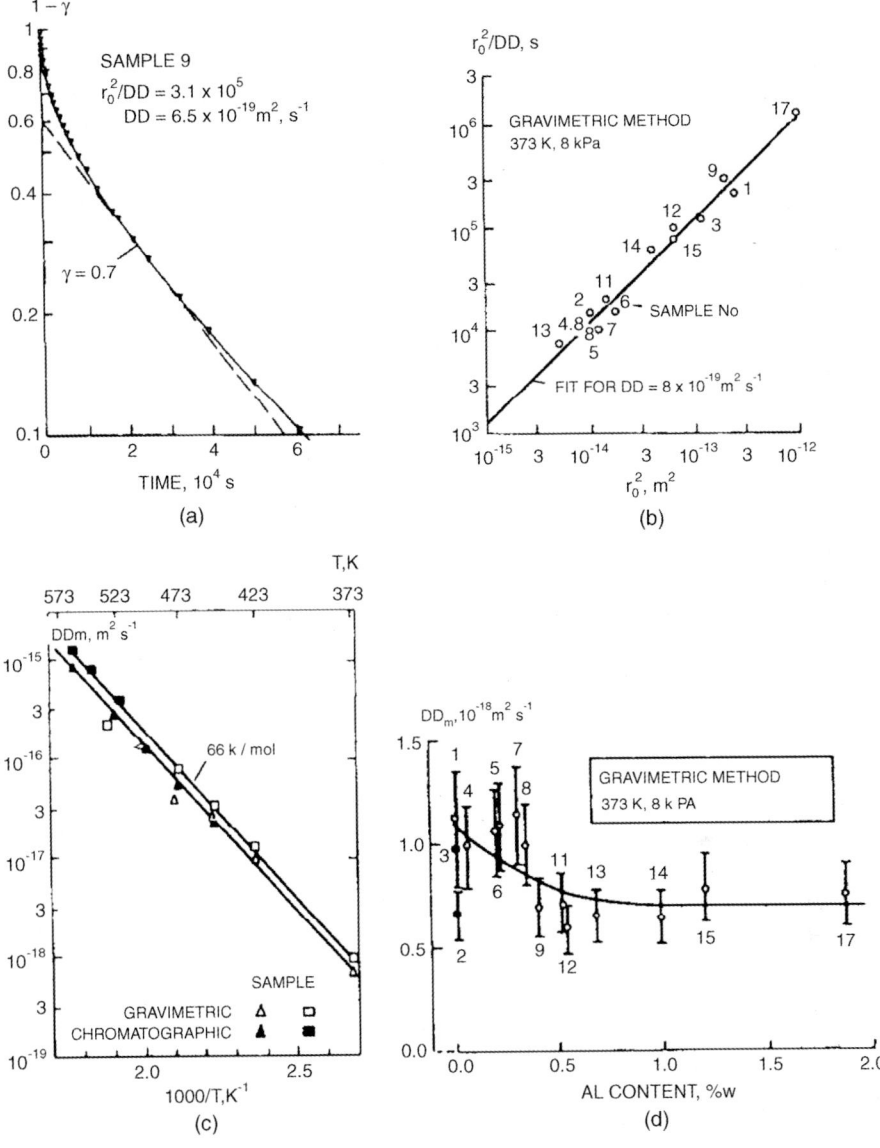

Figure 18.13 Diffusion of 2,2-dimethylbutane in silicalite and H-ZSM-5. (a) Conformity of gravimetric uptake curves to the diffusion model; (b) variation of time constant r^2/D with crystal size; (c) Arrhenius plot showing comparison between gravimetric and chromatographic data; and (d) effect of Al content on diffusivity. Reprinted from Post et al. [61] with permission.

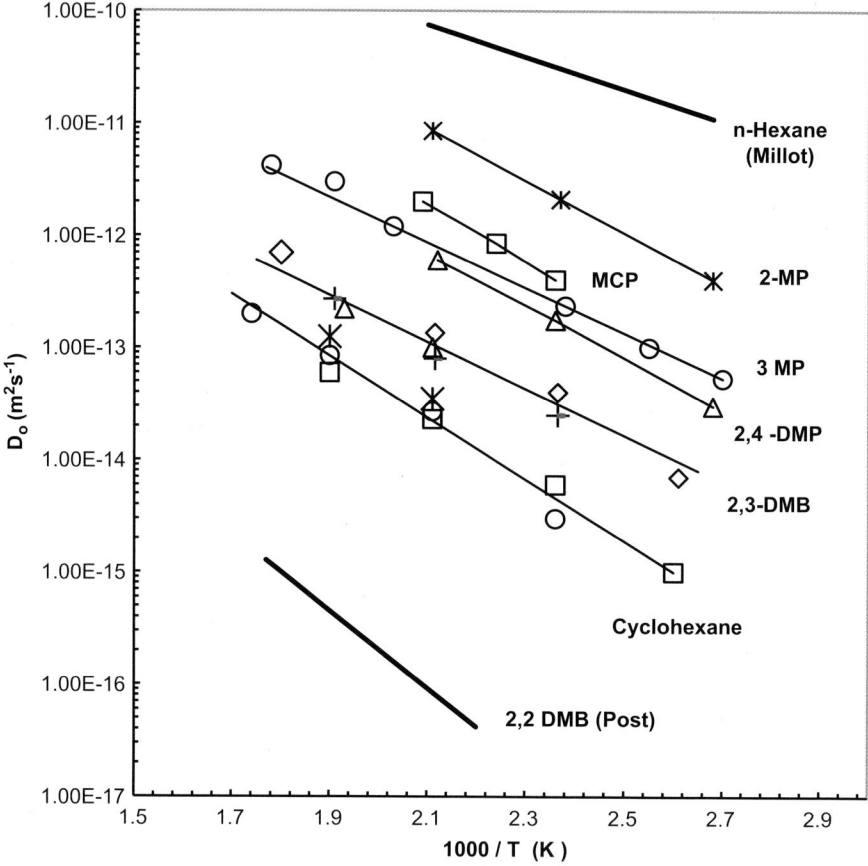

Figure 18.14 Arrhenius plot showing comparison of diffusivities for C_6 branched and cyclic paraffins. Data for n-hexane and 2,4-dimethylpentane (2,4-DMP) are included for comparison. Data are from Cavalcante [59], Millot et al. [36, 58], Post et al. [61]. Measurements for cyclohexane and 2,3-dimethylbutane were carried out with three different samples of crystals with equivalent radii from 18 to 60 μm (indicated by different symbols).

The diffusivity and activation energy depend on both the chain length and the detailed structure. For example, 2,2-dimethylbutane, in which three methyl groups and an ethyl group are connected to one carbon atom, has a larger critical molecular diameter than the other C_6 isomers, and this is reflected in a much higher activation energy and a substantially smaller diffusivity. The correlation of activation energy with critical diameter, for the C_6 isomers, shown in Figure 18.15, has the expected general form and is evidently consistent with the nominal pore diameter of about 6 Å. However, the activation energy for cyclohexane is obviously smaller than would be expected from the molecular diameter, presumably reflecting the greater flexibility of this molecule.

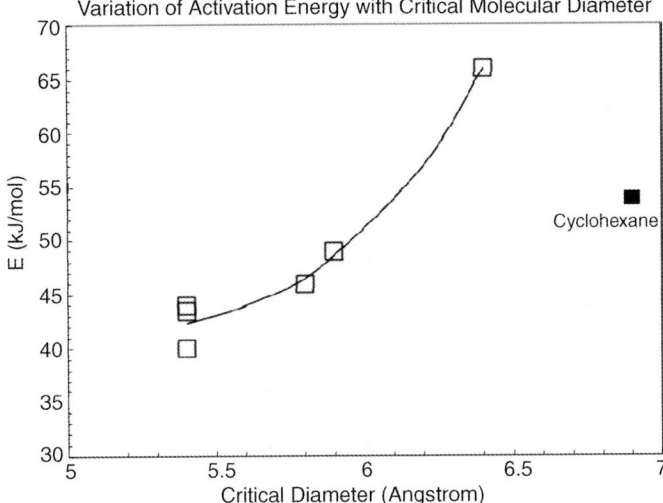

Figure 18.15 Variation of activation energy with critical molecular diameter for C_6 branched and cyclic isomers in silicalite.

18.2.2.5 Comparison of Diffusivities for Linear and Branched Hydrocarbons

Diffusion of the lighter linear alkanes in MFI is too fast to measure with confidence by most macroscopic techniques. As a result the reported diffusivities for these systems vary by more than an order of magnitude and the reliability of many of the values must be considered questionable. The most reliable values are probably those derived from frequency response or from permeance measurements with a supported membrane since, in contrast to single-crystal membrane measurements, the temperature range is not limited by the stability of the sealant, making it possible to work within the Henry's law region, at high temperatures and convenient pressures. By adjusting the thickness of the membrane the permeation rate can be reduced to an easily measurable level even for fast diffusing species.

Diffusivities for n-butane and n-hexane, derived from the high temperature permeance measurement of Millot et al. [36, 58], are included in Figures 18.11 and 18.14 to facilitate comparison with the data for the corresponding branched isomers. Evidently, the diffusivities of the linear isomers are much higher and the activation energies substantially smaller than the corresponding branched isomers. At 400 K the diffusivity ratios ($D_{linear}/D_{branched}$) are about 100 for nC_4/iC_4 and about 200 for n-hexane/2-methylpentane but, because of the differences in activation energies, these ratios vary strongly with temperature.

The large differences in diffusivity and activation energy clearly result from the substantial difference in critical molecular diameter (4.3 Å for the linear alkanes and 5.3 Å for the singly branched species). The critical diameter of the linear species (which above C_4 is independent of carbon number) is sufficiently small that they can pass easily through the 6 Å pores of the silicalite structure with minimal repulsive interaction with the pore wall, whereas the branched isomers are sufficiently large that their mobility is severely constrained by such interactions. This leads to a

18.2.2.6 Diffusion of Isobutane at High Loadings

Most of the data discussed above refer to measurements at low sorbate loadings. At higher loading levels additional complexity can arise as a result of the interplay between equilibrium and kinetic effects. As a result of the preferential location of the branched paraffins in the channel intersections, at loading levels of more than 4 molecules per unit cell (at which all guest molecules could just be accommodated in the intersections) substantial molecular rearrangement occurs and this is reflected in the shape of the adsorption isotherms. Figure 18.16, which diplays both experimental and simulation data for isobutane in silicalite at 298 K, shows this behavior. The resulting isotherms are well represented by the dual-site Langmuir model [i.e., by a functional dependence $\theta(p)$ with two different terms of the type of Eq. (1.32), corresponding to two different b values and multiplied by two different weighting factors whose sum is equal to 1] [69].

Figure 18.17a shows the (transport) diffusivities obtained by following (by IR) the isobutane uptake in a single MFI-type crystal over differential pressure steps and matching the resulting uptake curve to the appropriate solution of the diffusion equation (Table 6.1). Interestingly, the inflection at a loading of 4 molecules per unit cell (corresponding to one molecule per channel intersection) in the adsorption isotherm as presented in Figure 18.16 reappears in the corrected (Maxwell–Stefan)

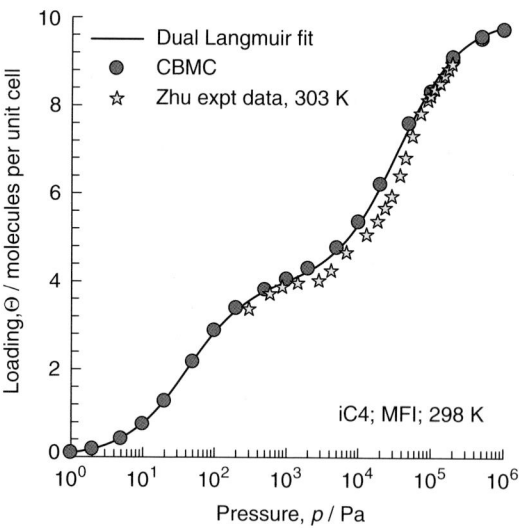

Figure 18.16 Adsorption isotherm of isobutane in silicalite-1 resulting from configurational-bias Monte Carlo (CBMC [69]) simulation (298 K) and experiment [57], showing conformity with the dual-site Langmuir model. Reprinted from Reference [69] with permission.

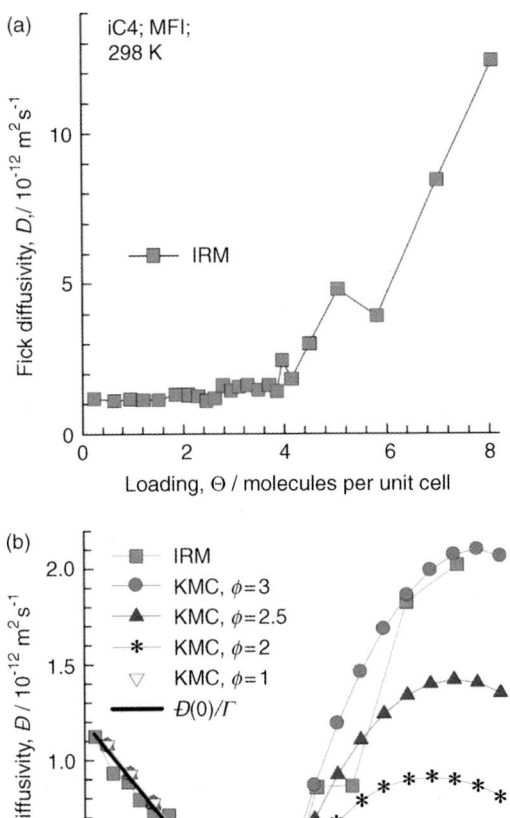

Figure 18.17 Experimentally determined transport (Fick) diffusivity of isobutane in silicalite as a function of loading (a) and comparison of the resulting corrected (M-S) diffusivities with the results of kinetic Monte Carlo (KMC) simulations following the Reed–Ehrlich model with different repulsion factors ϕ (b). Reprinted from Reference [69] with permission.

diffusivities (Figure 18.17b). Intuitively, the observed concentration dependence of the diffusivities may be attributed to an increasing mutual inhibition up to 1 molecule per intersection, that is, as long as the number of intersections is sufficient to accommodate all guest molecules. The increasing requirement for further adsorption sites with loadings larger than 4 molecules per unit cell gives rise to rearrangement of the guest molecules involving a transfer of molecules from the channel intersections to the channel segments, in which the pore blocking is much less effective. This leads to the observed dramatic increase in the diffusivities.

It is also shown in Figure 18.17b that the experimental diffusivity data are nicely approximated by kinetic Monte Carlo simulations [69, 72, 73]. Adopting the Reed–Ehrlich model (Section 4.3.2.2), the strong increase in the diffusivities with loadings above 4 molecules per intersection can be attributed to pronounced repulsive forces between the guest molecules, expressed by choosing a repulsion factor $\phi = 3$.

As a consequence of the existence of four distinct adsorption sites per unit cell, either in the channel intersections or in the channel segments, inflections in the adsorption isotherms at loadings of 4 molecules per unit cell are quite common for guest molecules in MFI-type zeolites, including benzene and the medium-chain-length n-alkanes hexane and heptane [74, 75]. The particularly pronounced occurrence with n-heptane observed by Vlugt et al. [76] is attributed to "commensurate freezing," caused by the fact that the molecular size (length) is commensurate with that of the zigzag channels [74, 75]. Again, in both MD simulations and QENS experiments [77] the inflection in the adsorption isotherm is found to reappear in the corrected diffusivities, although this is much less pronounced than for isobutane (Figure 18.17b). The reduced magnitude of this effect is understandable since the site preference of isobutane (for the intersections in preference to the straight sections) is much stronger than that of n-heptane.

18.3
Diffusion of Aromatic Hydrocarbons

18.3.1
Diffusion of Benzene

Of all the aromatic sorbates studied in silicalite, benzene has been investigated most extensively and, with a few notable exceptions, the data provide a reasonably consistent and coherent picture of the behavior. Table 18.4 summarizes representative data. The consistency of the data obtained by several different (macroscopic) techniques is shown in Figure 18.18. There are clearly some differences in diffusivity between the different samples, with the values for HZSM-5 being consistently smaller than those for silicalite, but the activation energies are essentially constant ($E \approx 27 \text{ kJ mol}^{-1}$).

The concentration dependence reported by Guo et al. [86] from differential uptake rate measurements with large crystals is shown in Figure 18.19a and b. As is commonly observed in zeolite systems, the data at 343 K show a strongly increasing trend of diffusivity with loading but the effect appears to be entirely attributable to the nonlinearity of the equilibrium isotherm. Corrected diffusivities, calculated according to Eq. (1.31), are almost independent of loading. The behavior at lower temperatures is more remarkable (Figure 18.19b). At 303 K the equilibrium isotherm is no longer of simple type I form but is sigmoid, with a well-defined inflexion at about 40% of saturation. At this loading the diffusivity shows a well-defined maximum but the corrected diffusivity remains almost constant. The consistency of the data

Table 18.4 Diffusivity data for benzene in silicalite/ZSM-5 at low concentrations.

Sorbent	Si/Al	Crystal size (μm)	Method	D_0 at 303 K (m^2 s^{-1})	D_0 at 386 K (m^2 s^{-1})	E (kJ mol^{-1})	Reference
Silicalite	1200	$r_c \approx 35$	Square wave	—	$\sim 10^{-13}$	25	van den Begin et al. [31]
Silicalite	High	Wide range (15-270)	Gravimetric	4×10^{-15}	5×10^{-14}	28	Shah et al. [78]
Silicalite	High	$r_c \approx 27.5$	ZLC	4×10^{-14}	2×10^{-13}	27	Eic and Ruthven [34]
Silicalite	$>10^3$	$30 \times 25 \times 13$	Gravimetric and piezometric	$\sim 10^{-13}$	4×10^{-13}	21	Zikanova et al. [79]
Silicalite			Tracer exch.	5×10^{-14}	—		Förste et al. [80]
Silicalite		$r_c \approx 26$	ZLC	2.4×10^{-14}	3.4×10^{-13}	30	Brandani et al. [81]
Silicalite			TZLC (⑨)	2.0×10^{-14}	10^{-13}	30	Brandani et al. [81]
Silicalite		$r_c \approx 30$	FR	—	$3-5 \times 10^{-13}$	27-29	Song and Rees [26, 52, 65]
HZSM-5	~30	$r_c \approx 8$	Gravimetric and piezometric	$\sim 10^{-14}$	9×10^{-14}	26	Zikanova et al. [79]
HZSM-5	~30	$r_c \approx 25$	Frequency response	1.5×10^{-14}	$\sim 10^{-13}$		Bülow et al. [49]
HZSM-5	~30	$r_c \approx 14$	Gravimetric	2×10^{-14}			Doelle et al. [82]
HZSM-5	~30	$r_c \approx 1.5, 4.0$	Gravimetric	8×10^{-15}			Beschmann et al. [11]
HZSM-5	110	$r_c \approx 6$	Circulating system	3×10^{-15}	2.2×10^{-14}	24	Qureshi and Wei [83]

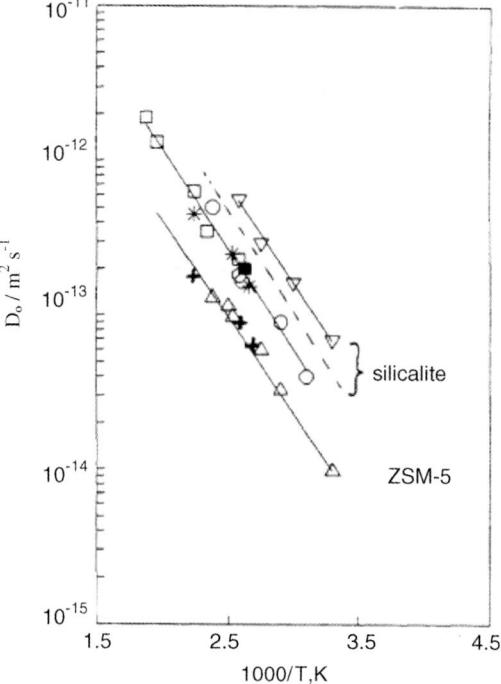

Figure 18.18 Arrhenius plot showing the temperature dependence of corrected diffusivity (D_0) for benzene in silicalite and HZSM-5 as measured by different techniques: van den Begin [31] (square wave) (□); Eic and Ruthven [34] (ZLC) (○); Zikanova et al. [79] (piezometric) (△); Shen and Rees [84] (square wave) (+); Förste et al. [80] (NMR tracer exchange) (■). The range of the ZLC/TZLC data of Brandani et al. [81] is indicated by the broken line (– – –). Reprinted from Reference Ruthven [85] with permission.

obtained with different crystal size fractions at various temperatures is shown in Figure 18.19d [78].

In contrast to the pattern of almost constant corrected diffusivities revealed by the data of Guo et al. [86] and van den Begin et al. [31], the data of Zikanova et al. [79], which were obtained by both gravimetric and piezometric methods, suggest that the corrected diffusivity decreases rather strongly with concentration. This effect was stronger in silicalite than in ZSM-5. The calculation of the corrected diffusivity depends on accurate knowledge of the equilibrium isotherm. Since, in the relevant temperature range, the isotherms are highly favorable it seems possible that significant errors in the values for the thermodynamic correction factor $[d\ln(p)/d\ln(c)]$ may have arisen from relatively small deviations from the ideal Langmuir model that was used to correlate the isotherm data.

Conformity with the diffusion model was also confirmed for HZSM-5 by Beschmann et al. using two different crystal size fractions [11], although the (integral) diffusivity values are substantially lower than those found by Zikanova et al. [79] or by Doelle et al. [82] for the same system.

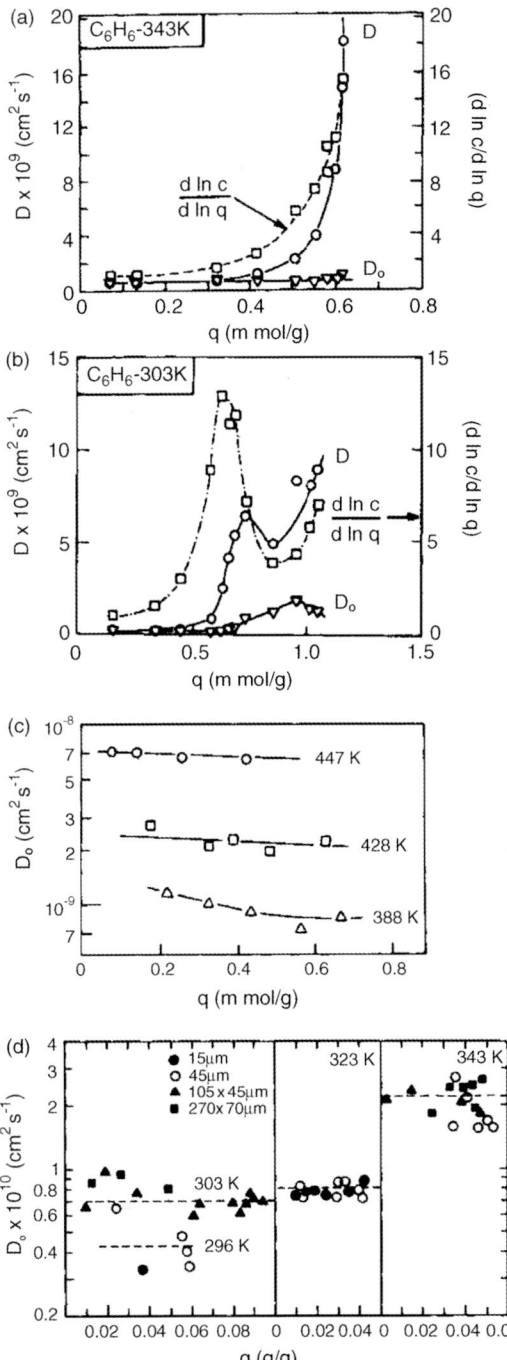

Figure 18.19 Concentration dependence of diffusivity for benzene in silicalite: (a) and (b) D, $(d \ln c/d \ln q)$ and D_0 versus concentration at 343 and 303 K, respectively; data of Guo et al. [86]; (c) D_0 versus concentration at 388, 428, and 447 K; data of van den Begin et al. [31]; (d) D_0 versus concentration, showing consistency of data for different crystal sizes; data of Shah et al. [78] with permission.

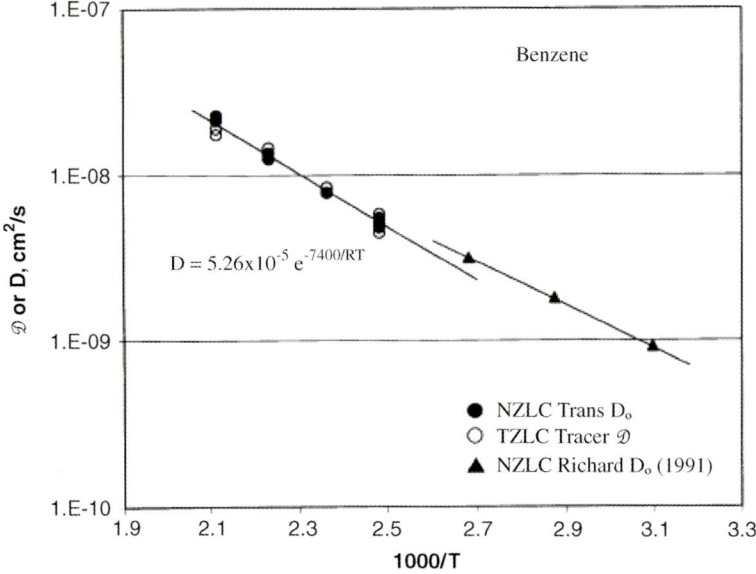

Figure 18.20 Arrhenius plot showing the temperature dependence of the limiting transport diffusivity (D_0, ●) and the tracer self-diffusivity ($\mathcal{D}$, ○) for benzene in silicalite. Also shown are the (transport) ZLC diffusivity data of Richard [88]. Reprinted from S. Brandani et al. [81] with permission.

18.3.1.1 Self- and Transport Diffusion

More recently, detailed studies of diffusion of benzene in large silicalite crystals were carried out by Shah and Liou [87] using the single-crystal membrane technique and by Brandani et al. [81] using zero-length column (ZLC) and tracer zero-length column (TZLC) methods. The results of these studies are generally consistent (Table 18.4) and they are also consistent with the earlier study of Eic and Ruthven [34, 88], as may be seen from Figure 18.20. The TZLC measurements show that the self-diffusivity is independent of loading, at least up to about 50% of saturation (0.5 molecules per intersection) and essentially equal to the corrected diffusivity (D_0), in agreement with the tracer exchange studies of Forste et al. [80]. This implies that the benzene molecules must be able to pass each other easily, which in turn implies rapid exchange between the straight and sinusoidal channels, leading to a unimodal frequency response signal [26, 52, 65] (at least at low loadings) and conformity between the diffusivities in silicalite-1 and silicalite-2 (as shown below in Figure 18.23).

18.3.1.2 Benzene Microdynamics

Diffusion of benzene in silicalite is too slow to measure by PFG NMR and incoherent quasi-elastic neutron scattering [30, 89]. However, NMR techniques (including both ^{2}H [90] and ^{13}C NMR [91]) have been successfully applied to study the elementary steps of the diffusion process. The results of these measurements show that the benzene molecules rotate rapidly about their hexad axis, just as in the liquid state or indeed in the adsorbed state in NaX. Superimposed on the hexad rotation are jumps between the sorption sites in which only restricted orientations of the benzene molecules are allowed. Semi-empirical calculations of the interaction potential between the molecule and the framework suggest that these preferred sites occur not only in the channel intersections but also in the segments between intersections [92–94]. The mean residence time between successive jumps increases with loading so, from a microscopic viewpoint, in contrast to the TZLC results discussed above, these data suggest that the self-diffusivity decreases with loading.

With the advent of the neutron spin-echo technique it is now also possible to measure diffusion of benzene in MFI by QENS. The resulting (microscopic) diffusivities are found to be in good agreement with the macroscopically measured values [95].

18.3.2
Diffusion of C_8 Aromatics

In contrast to benzene, despite intensive study, the diffusion of the C_8 aromatic isomers is still not fully understood. Table 18.5 summarizes representative diffusivity data. The diffusivities for ethylbenzene and *ortho-* and *meta-*xylenes show reasonable agreement between the various different measurements, at least for the larger crystals. The main discrepancy concerns the behavior of *p*-xylene. While most experimental studies show that *p*-xylene diffuses at a rate similar to but slightly faster than benzene, from a series of detailed frequency response measurements, Rees and coworkers [26, 52, 65, 96] concluded that the diffusivity of *p*-xylene is very much higher than that of benzene and the other C_8 isomers. This conclusion is supported by the work of Tsapatsis *et al.* [97], who showed that in specially prepared oriented silicalite membranes the permeance of *p*-xylene was at least two orders of magnitude larger than that of *o*-xylene (see Figure 20.7).

18.3.2.1 Uptake Curves
In a detailed gravimetric study in which great care was taken to eliminate any thermal effects Beschmann and Riekert [11] showed that, in contrast to benzene, the uptake curve for *p*-xylene at ambient temperature does not conform to the diffusion model. Karsli *et al.* [100] showed that, at ambient temperature, a freshly degassed sample of silicalite does not adsorb *o*-xylene, even over a period of several days. In contrast, a sample of silicalite that had previously been exposed to *p*-xylene and then evacuated did adsorb *o*-xylene slowly but at an easily measurable rate. There is evidence that the monoclinic–orthorhombic phase transition is stimulated by adsorption of *p*-xylene [101, 102], which could provide a possible explanation for the anomalous behavior observed by both Beschmann and Riekert [11] and Karsli *et al.* [100]. The substantially smaller mobility of *o*-xylene in comparison with benzene or *p*-xylene has been confirmed at the micro-scale by ^{13}C NMR line shape measurements [103].

Working at temperatures above the phase transition (330 K), using much larger silicalite crystals ($105 \times 55 \times 40 \,\mu m^3$) than had been used in earlier studies, Ruthven

Table 18.5 Diffusion of C_7 and C_8 aromatics in silicalite/ZSM-5.

Sorbate	Sorbent	r_c (μm)	D at 298 K (m^{2}s^{-1})	E (kJ mol^{-1})	Comments	Reference
Toluene	H-ZSM-5	6	2×10^{-15}	—	Circulating system, diff. step	Qureshi and Wei [83]
	Silicalite	30	1×10^{-13}	20.3	FR – extrapolated	Song and Rees [26, 96]
p-Xylene	H-ZSM-5	1.5, 4.0	10^{-15}–10^{-14}		Gravimetric method; integral step	Beschmann et al. [11]
	H-ZSM-5	1.8	6×10^{-15}		Integral step, gravimetric method	Prinz and Riekert [60]
	H-ZSM-5	0.3	3×10^{-15}		Integral step, gravimetric method	Nayak and Riekert [98]
	Silicalite	26	6×10^{-14}	26.5	ZLC (D_0)	Brandani et al. [81].
			1.8×10^{-14}	30	ZLC ($\mathcal{D}$)	
	Silicalite	30	3.5×10^{-12}	19	FR (fast response) – extrapolated	Song and Rees [26, 96]
	Silicalite	25	3×10^{-14}	26	Single crystal membrane	Shah and Liou [87] see also Ruthven [85]
Ethylbenzene	Silicalite	27.5	1.5×10^{-14}	30	ZLC, low concentration	Eic and Ruthven [88]$^{a)}$
	Silicalite	30	10^{-14}	36	FR – extrapolated	Song and Rees [26, 96]
m-Xylene	Silicalite	1.0	4×10^{-16}	—	Gravimetric diff. step, low concentration	Wu, Debebe, and Ma [14]
m-Xylene	H-ZSM-5	14	2×10^{-14}	—	Gravimetric integral step	Doelle et al. [82]
	H-ZSM-5	1.5	$\sim 10^{-19}$	—	Gravimetric integral step	Beschmann et al. [11]
o-Xylene	Silicalite	1	2×10^{-16}	—	Gravimetric diff. step, low concentration	Wu et al. [14, 99]
	Silicalite	27.5	2×10^{-15}	34	ZLC and gravimetric low concentration	Eic and Ruthven [88]$^{a)}$

a) Eic and Ruthven used the parallel-sided slab model to interpret their transient sorption rate curves. This leads to diffusivity values that are approximately double the values derived from the spherical particle approximation.

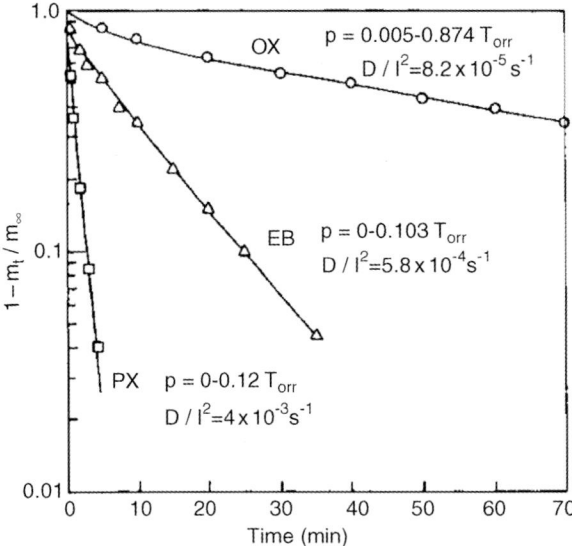

Figure 18.21 Gravimetric uptake curves for o- and p-xylenes (OX and PX) and ethylbenzene (EB) in large crystals of silicalite ($2l = 40$ μm) at 127 °C, showing conformity with the diffusion model for a parallel-sided slab [Eq. (6.26)]. Note the large difference in uptake rates between the isomers under comparable conditions. Reprinted from Ruthven et al. [88] with permission.

et al. [88] found excellent conformity with the diffusion model for a series of differential uptake rate measurements. Figure 18.21 shows representative examples of their uptake curves.

18.3.2.2 ZLC and TZLC Measurements

More recent ZLC and TZLC studies [81] show that, like benzene, the self-diffusivity of p-xylene is essentially independent of loading up to at least 3 molecules per unit cell but, in contrast to benzene, the self-diffusivity is substantially smaller than the corrected transport diffusivity ($\mathcal{D} \approx D_o/3$ – see Figure 18.22). The actual diffusivities for p-xylene in the z direction, measured by Shah and Liou [87] by the single-crystal membrane technique, lie somewhat above the extrapolated D_0 values derived from the ZLC measurements, but when the thermodynamic correction factor is included the corresponding corrected diffusivities in the z direction (D_{z0}) are entirely consistent with the TZLC self-diffusivities. This observation can be easily understood if it is assumed that (in contrast to benzene) the p-xylene molecules can jump between the straight and sinusoidal channels only with difficulty. Self-diffusion requires jumps between the two channel systems (for the diffusing molecules to pass each other) and transport diffusion in the z direction also requires such jumps so these processes may be expected to occur at similar rates. On this basis the observation that $D_{0z} \approx \mathcal{D} \approx (D_0)_{ZLC}$ is understandable.

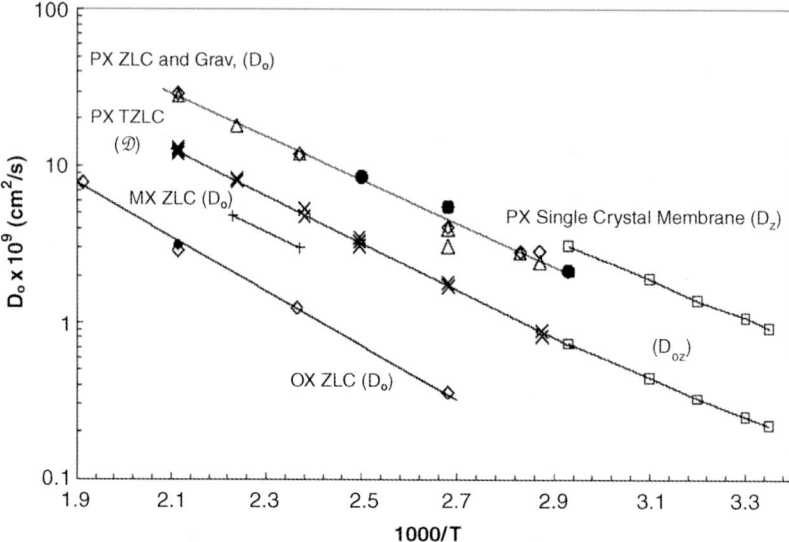

Figure 18.22 Arrhenius plot showing temperature dependence of tracer self-diffusivity ($\mathcal{D}$) and the limiting transport diffusivity (at zero loading) (D_0) for p-xylene (PX) in silicalite derived from TZLC and ZLC measurements. The corresponding values of D_0 for o-xylene and m-xylene (OX and MX) are also shown as well as the values for the diffusivity (D_z) and corrected diffusivity (D_{0z}) in the z direction, derived from the single-crystal measurements of Shah and Liou [87]. Reprinted from Ruthven [85] with permission.

18.3.2.3 Frequency Response Data for p-Xylene

Figure 18.23 shows representative results from the frequency response studies of Song and Rees [26, 52, 96]. The frequency response spectrum for p-xylene (but not for benzene) showed a bimodal response that was attributed to independent diffusion in the x (straight) and y (sinusoidal channels) directions. Such a model requires that there be negligible interchange of diffusing molecules between the two channel systems. This would require the diffusivity in the z direction to be negligible (since diffusion in that direction requires successive jumps between the two channel systems). However, the single-crystal membrane measurements (discussed above) suggest that diffusivities in the z direction are in fact comparable with the overall average diffusivities measured by other macroscopic methods.

Sun and Bourdin [104] have shown that a bimodal response may also be produced by other mechanisms, notably by the combination of diffusional resistance with a finite rate of heat transfer. However, in such a model the mass transfer peak always corresponds to the faster process. Thus, although the coupling of heat and mass transfer effects can explain the bimodal form of the response it does not explain the magnitude of the faster response peak, which suggests p-xylene diffusivities that are about two orders of magnitude greater than the values found by other experimental techniques. To resolve this issue, Shen and Rees [105] repeated the frequency response measurements with silicalite-2 (which has only straight channels – see

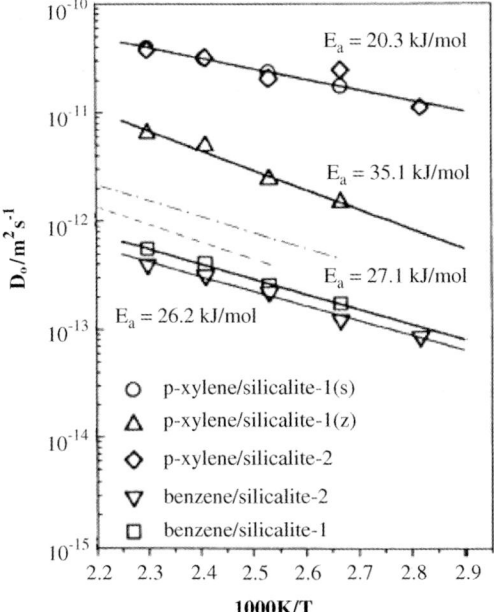

Figure 18.23 Frequency response diffusivities of Rees and coworkers [26, 52, 96] for benzene and p-xylene in silicalite-1 and silicalite-2, showing temperature dependence of the corrected diffusivity (D_0). Also shown are the ZLC data of Brandani et al. [81] for benzene (– – –) and p-xylene (–·–·–). Adapted from Rees and Song [52] with permission.

Figure 18.1). The response was unimodal and the derived diffusivities for p-xylene agreed well with the values from the faster peak for silicalite-1, thus apparently confirming the original interpretation of the bimodal response. Note that the activation energy derived from the slower response peak is larger than that derived from the faster peak, which is consistent with slower diffusion of p-xylene through the sinusoidal channels. Benzene also showed similar (but much smaller) diffusivities in both silicalite-1 and silicalite-2, with an intermediate activation energy, as may be seen from Figure 18.23.

18.3.2.4 Evidence from Membrane Permeation Studies

In a landmark paper Lai, Tsapatsis, and coworkers [97] reported the successful preparation of an oriented silicalite membrane in which the crystals were aligned in the y direction (rather than in the usual z alignment) and they showed that this membrane gave very high perm-selectivities (more than two orders of magnitude for p-xylene over o-xylene). This was initially taken as further evidence of very rapid diffusion of p-xylene in the y direction but more detailed analysis [85] suggests that the high perm-selectivity was probably due to the exclusion of o-xylene from the pores (perhaps due to surface resistance) rather than to unusually rapid diffusion of p-xylene. In fact the magnitude of the p-xylene flux corresponds to a diffusivity close to that derived from the ZLC and single-crystal permeation measurements and much smaller than the values suggested by the frequency response data.

In an ideal silicalite structure a plausible argument is that, as a result of the orienting effect of the methyl groups, the relatively long and rigid p-xylene molecule should diffuse much faster through the straight channels (in comparison with benzene or the other C_8 isomers) and that interchange between the two channel systems will be difficult. However, we know that in real crystals the channel system is disrupted by the twin structure and this probably explains why most experimental studies show only a modest difference in diffusivity between p-xylene and benzene ($D_{pX}/D_{Bz} \approx 3$) and only about one order of magnitude between p-xylene and o-xylene (Figure 18.22). More recent attempts to repeat the frequency response measurements for p-xylene in different silicalite crystals were not entirely successful. The response peaks were broad and did not show a clear bimodal structure, but that could very well reflect the quality of the crystals. It is possible that the particular crystals used by Rees were unusually free from twinning and other common structural defects.

18.3.2.5 Diffusion of *o*-Xylene and *m*-Xylene

Comparative ZLC data for diffusion of the *ortho*- and *meta*-isomers are included in Table 18.5 and Figure 18.22. The values from the gravimetric data of Richard *et al.* [88] are similar. The differences in diffusivity are relatively modest and far smaller than the very large differences (several orders of magnitude) reported by Beschmann *et al.* [11] from a detailed gravimetric study. In the light of the observations of Karsli *et al.* [100] (discussed above) it seems likely that this difference may reflect the differences in temperature and sample pre-treatment since it appears that *o*-xylene is essentially excluded from the monoclinic form of silicalite.

18.4
Adsorption from the Liquid Phase

Many important industrial processes operate in the liquid phase, so, from a practical standpoint, an understanding of sorption kinetics in liquid phase adsorption systems is important. Such systems are conveniently studied by liquid chromatographic (LC) or liquid ZLC methods – see Sections 14.4.1 and 14.7.4. The interpretation of liquid-phase sorption rate measurements is complicated by the presence of at least two components in the system. The vapor phase measurements of Qureshi and Wei [83, 106] (Section 18.8.2.6) show clearly that in a counter-diffusion experiment the diffusivity for the adsorbing species may be reduced dramatically relative to its value in a single-component system at a similar concentration level. However, the question as to whether this reduction in the Fickian diffusivity reflects merely a difference in driving force due to coupling of the equilibrium isotherms or a genuine change in mobility was not addressed (Section 6.3 and Habgood [107]).

There are several sorbates for which diffusion measurements have been carried out in both liquid and vapor phase systems at comparable temperatures. Table 18.6 compares diffusivity data for some of these systems. In general we have emphasized

Table 18.6 Diffusion in silicalite: comparison of liquid and vapor phase diffusivities.[a]

Sorbate/solvent	Critical diameter (Å)	T (K)	D_{liquid} (m^{2}s^{-1})	D_{0vapor} (m^{2}s^{-1})	D_{liq}/D_{0vapor}
Methanol–water	3.8	303	4×10^{-14}	7×10^{-14}	0.6
Ethanol–water	4.2	303	1.5×10^{-14}	1.2×10^{-14}	1.2
n-Butanol–water	4.4	303	7×10^{-14}	6×10^{-14}	1.1
Benzene–n-hexane	4.8	313	2.5×10^{-14}	5×10^{-14}	0.5
		333	4×10^{-14}	8×10^{-14}	0.5
p-Xylene–n-hexane	4.8	313	8.5×10^{-15}	8×10^{-14}	0.1
Ethyl-benzene–n-hexane	4.8	313	3.1×10^{-15}	3×10^{-14}	0.1
2-Methylpentane–n-hexane	5.5	303	2×10^{-12}	1.2×10^{-14}	170
		313	4×10^{-12}	3×10^{-14}	130
		343	10^{-11}	9×10^{-14}	110
3-Methylpentane–n-hexane	5.5	313	6×10^{-12}	1.5×10^{-15}	4000
		343	2×10^{-11}	2.3×10^{-14}	870
2,3-Dimethylbutane–n-hexane	5.8	313	2×10^{-12}	6×10^{-15}	3000
		343	2.5×10^{-12}	2×10^{-15}	1250
2,2-Dimethylbutane–n-hexane	6.3	313	2.3×10^{-12}	3×10^{-20}	10^8
		343	3×10^{-12}	10^{-19}	10^7
Cyclohexane–n-hexane	6.9	313	10^{-12}	2.0×10^{-17}	5×10^4
		343	1.3×10^{-12}	10^{-16}	1.3×10^4
Cyclohexane–toluene	6.9	303	10^{-14}	10^{-17}	10^3

a) Liquid phase measurements (LC, ZLC): Ma and Lin [108, 109], Boulicaut [113]. Vapor phase measurements (ZLC, FR, uptake rate): Cavalcante [59], Ruthven [88], Rees [26, 52], Eic [34], Post [61], Keipert [114], and Nayak and Moffat [110].

the data obtained with larger crystals since they are less prone to errors from extraneous effects such as axial dispersion. However, the data of Ma and Lin [108, 109] and Nayak and Moffat [110] have been included because, even though the measurements were made with small ($\approx 1\,\mu m$) crystals, the data appear to be quite consistent. We have not included the earlier measurements made with stirred cells since detailed analysis of these results suggests significant intrusion of external fluid phase resistance [111]. Most of the measurements were made with either water or n-hexane as the solvent. The liquid-phase self-diffusivities of these solvents are substantially larger than the diffusivities of the liquid sorbates so one may expect that the counter-diffusivity in the binary adsorbed phase will be controlled mainly by the sorbate.

The results divide clearly into two distinct categories. With the exception of p-xylene and ethylbenzene, which show significantly faster diffusion in the vapor, for those systems in which the critical molecular diameter of the sorbate is substantially smaller than the nominal diameter of the silicalite pores (≈ 6 Å) the pore diffusivities in the liquid-filled pores are similar to the corrected diffusivities for the sorbate vapor in silicalite at the same temperature. In contrast, for the sorbates that have critical molecular diameters that exceed the pore diameter (for which diffusion is significantly hindered by repulsive interaction with the pore walls) we see a dramatic difference between the liquid and vapor phase values, with the liquid values being orders of magnitude faster. This difference is most pronounced for 2,2-dimethylbutane, which, in the vapor phase, can only just penetrate the silicalite pores. As noted in Section 18.2.2.4, because of its more flexible structure, cyclohexane diffuses considerably faster than 2,2-dimethylbutane (in vapor-filled pores), even though its critical diameter is larger. Consequently the diffusivity ratio (D_{liq}/D_{vap}) for cyclohexane, although still large, is much smaller than for 2,2-dimethylbutane.

It appears that when filled with a solvent such as n-hexane the diameter of the silicalite pores increases slightly, thus reducing the steric hindrance and dramatically increasing the diffusivity for the liquid system. The swelling of silicalite crystals when fully loaded with n-hexane has been confirmed by accurate XRD measurements [112].

18.5
Microscale Studies of other Guest Molecules

18.5.1
Tetrafluoromethane

Since fluorine is a coherent scatterer, tetrafluoromethane (CF_4) has been applied in QENS measurements (Section 10.2.3) to probe the transport diffusivities in silicalite. Figure 18.24 compares the experimental data at 200 K with the results of molecular

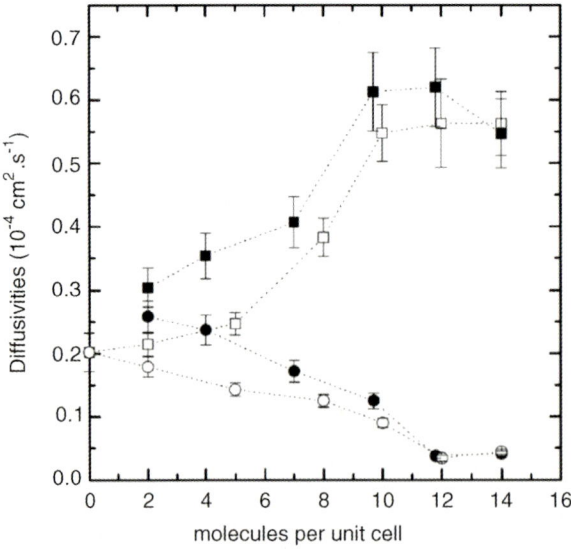

Figure 18.24 Transport diffusivities (squares) and corrected diffusivities (circles) as determined from QENS experiments (filled symbols) and atomistic simulations (open symbols) for CF_4 in silicalite at 200 K. Reprinted from Reference [115] with permission.

dynamic simulations [115]. Again, reasonably good agreement is observed between the model and the microscopic measurements. Moreover, the corrected diffusivities are found to decrease with increasing loading, following the trend observed for the saturated hydrocarbons in both the corrected diffusivities and self-diffusivities (Section 18.2.1). Similarly good agreement has been obtained at 250 K, yielding an activation energy of the order of 3 kJmol^{-1}. The results of the present study are, moreover, found to be approached by the comparative PFG NMR measurements and MD simulations of the self-diffusion in CF_4–methane mixtures in silicalite (Section 18.8.2.1 and Reference [116]).

18.5.2
Water and Methanol

In contrast to the behavior of the saturated hydrocarbons but similar to the unsaturated hydrocarbons, the self-diffusivities of water and methanol were found to decrease somewhat with decreasing Si/Al ratio [117], suggesting that the residual cations or hydroxyl groups probably act as interaction partners, thus reducing the translational mobility of the water or methanol molecules [118, 119]. The patterns of variation of the diffusivities of water and methanol with concentration are different. Whereas the diffusivity of water remains substantially constant, the diffusivity of methanol decreases with increasing concentration (Figure 18.25) [120]. This pattern is similar to the behavior of the low molecular weight alkanes and is probably simply a consequence of the self-blocking effect that occurs when the molecules have difficulty passing each other in the channels.

18.5.3
Ammonia

Comparative QENS and PFG NMR diffusion studies with ammonia in silicalite [121] revealed a remarkable difference between the results of these two techniques. Over the observation times of 1.5 and 3 ms chosen for the PFG NMR experiments, the ammonia molecules were found to undergo normal diffusion with a single average diffusivity while, over the time scale of QENS ($\approx$35 ps), the ammonia molecules were found to exist as two distinct populations. One group of molecules, corresponding to one molecule per unit cell, is found to be essentially immobile, even at the relatively high temperature of 480 K. Since within the ideal structure of silicalite there are no adsorption sites that might give rise to such large residence times the finding of the QENS studies can be considered as evidence of substantial deviations of the real structure from ideality. Figure 18.26 shows the Arrhenius plots for both the PFG NMR diffusivities and the diffusivities of the mobile guest population observed by QENS for different guest concentrations. An estimate of the overall diffusivity derived from the QENS measurements by taking account of the fraction of immobile molecules is also given. The fact that this estimate still notably exceeds the PFG NMR diffusivities may again be attributed to the influence of additional transport barriers with separations too large to affect the QENS measurements.

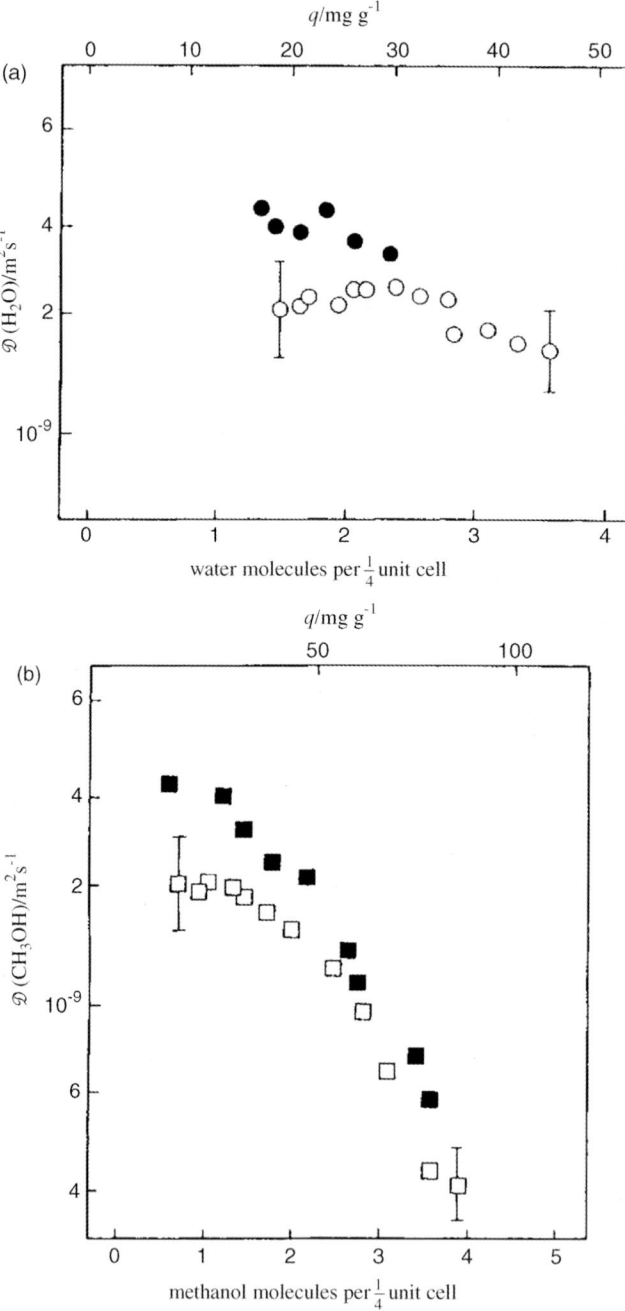

Figure 18.25 NMR self-diffusivities at 300 K for (a) water and (b) methanol in silicalite (filled symbols) and H-ZSM-5 (open symbols), showing variation with sorbate concentration. Reprinted from Caro et al. [120] with permission.

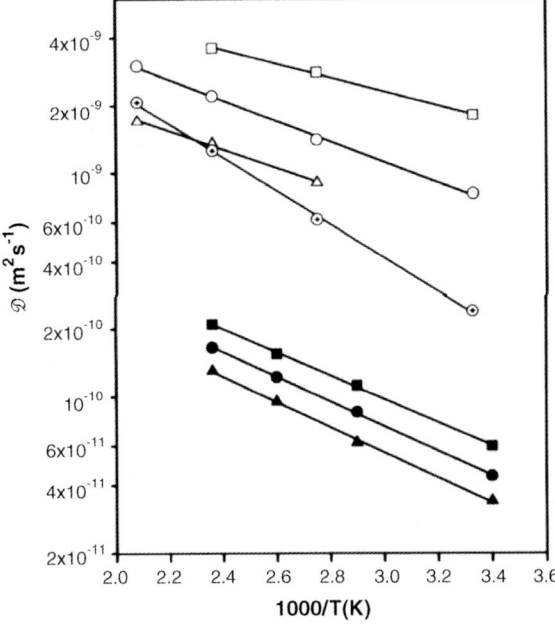

Figure 18.26 Self-diffusivities of ammonia in silicalite obtained by QENS for the mobile molecules (open symbols) and PFG NMR (filled symbols) for 1.5 (△,▲), 3.0 (○,●), and 4.3 (□,■) molecules per unit cell. The partially filled symbols indicate an estimate of the average over all molecules, as resulting from the QENS measurement for 3 molecules per unit cell. Reprinted from Jobic et al. [121] with permission.

The decrease in the diffusivities with decreasing loadings, as observed by both techniques, may be ascribed to the increasing influence of the strongest adsorption sites. As a result of the stronger sites, this trend is found to be much more pronounced in H-ZSM-5 than in silicalite. In detailed PFG NMR measurements under conditions such that the stronger sites are almost fully occupied [122] it was found that a decrease of 10% in loading, leads to a decrease of more than an order of magnitude in the diffusivity!

18.5.4
Hydrogen

Figure 18.27 shows the PFG NMR data of the self-diffusivity of hydrogen in various zeolites [123]. The diffusivities observed in NaX and NaA are found to be in good agreement with the results of QENS measurements [123, 124] and follow the expected sequence, decreasing with decreasing window size. Even smaller diffusivities are observed in chabazite (CHA), reflecting the smaller window size. A similar trend was noted for the light paraffins, with the diffusivities for silicalite lying between those for NaX and 5A (Figure 18.10).

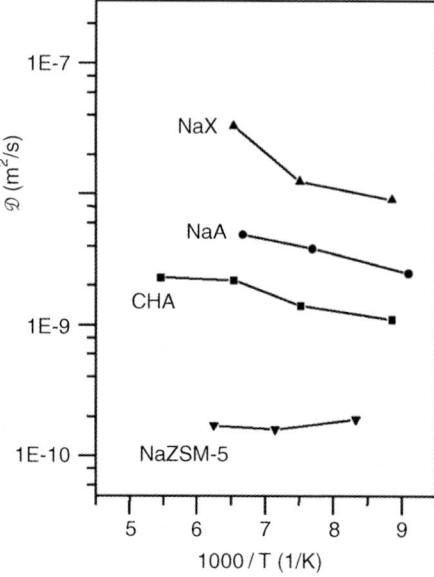

Figure 18.27 Self-diffusivities of hydrogen in different zeolites determined by PFG NMR. The loadings are about 2H$_2$ per large cavity (for NaX, NaA, and CHA) or channel intersection (ZSM-5). Reprinted from Bär et al. [123] with permission.

Surprisingly, in view of these trends, the PFG NMR diffusivities for hydrogen diffusion in the (ten-ring) MFI-type zeolites in Figure 18.27 are found to lie well below the values for the (eight-ring) zeolites NaA and chabazite. This behavior may be rationalized by the QENS measurements, which show that a large fraction of the hydrogen molecules is immobilized, most probably by being captured in the interior of the pentasil chains (top of Figure 18.1) [123]. Comparative ^{1}H MAS NMR and QENS experiments yield a mean residence time of 230 μs for the hydrogen molecules within the pentasil channels. This value is well above the QENS observation time, thus allowing separate observation of mobile and immobile molecules, and smaller than the observation time in the PFG NMR experiments, which consequently show fast exchange between all molecules.

18.6
Surface Resistance and Internal Barriers

18.6.1
Surface Resistance

Additional resistances to mass transfer at the external crystal surface (surface barriers) have been observed in many different zeolites, but they are especially common in MFI materials. In principle they can arise from several different mechanisms, including the partial or total obstruction of a fraction of the pore openings or from narrowing of the pores close to the surface. Their presence can be detected by both macroscopic and microscopic methods.

18.6.1.1 Macroscopic Rate Measurements

The forms of the transient uptake curves for surface resistance control and internal diffusion control are different (Section 6.2.1) so *prima facie* evidence of surface resistance can sometimes be obtained simply by examining the form of the uptake curve (Figure 16.29). However, since the shape of the uptake curve can also be affected by other factors such evidence is not unequivocal. A more reliable test is to compare the apparent diffusivities (or the diffusion times) derived from uptake rate measurements with several different crystal sizes (implying that both the surface permeance and the intracrystalline diffusivities remain invariant). For intracrystalline diffusion control the diffusion time is directly proportional to the square of the mean crystal radius ($t_{diff} = r_c^2/D_e$ – see Section 6.2.1.2 and Table 13.1). Such behavior was demonstrated by Post *et al.* [61] for sorption of 2,2-dimethylbutane in silicalite at 373 K (Figure 18.13b) and by Gueudré and Jolimaîte [125] for sorption of both 2,3-dimethylbutane and cyclohexane in large silicalite crystals ($r_c > 10\,\mu m$).

If surface resistance is significant the diffusion time will be given, approximately (see also the discussion of the method of moments in Section 13.6.2, Table 13.1), by:

$$t_{diff} = \frac{r_c^2}{\pi^2 D_e} = \frac{r_c^2}{\pi^2 D_c} + \frac{r_c}{3k_s} \tag{18.1}$$

This suggests that, provided that D_c and k_s can be considered constant (at a given temperature), a plot of t_{diff}/r_c versus r_c should be linear with intercept $1/3k_s$ and slope $r_c^2/(\pi^2 D_c)$. Such plots from the data of Gueudré for smaller silicalite crystals are shown in Figure 18.28 and the values of D_c and k_s (at 298 K), derived from the slopes and intercepts, are summarized in Table 18.7. Since the measurements were not all made at the same loading this analysis involves the additional approximation that the rate parameters are independent of loading. At least for cyclohexane–silicalite this is a reasonable approximation since the measurements of Cavalcante [59] and Magalhaes [64] both show that the intracrystalline diffusivity is indeed essentially independent of loading over a wide range.

Evidently, the diffusivity values are close to the values extrapolated from the large-crystal data of Cavalcante and Magalhaes (Table 18.3). It thus appears that the data for the smaller crystals studied by Gueudré *et al.* are entirely consistent with the model of combined intracrystalline diffusion and surface resistance. Since the diffusion time scales with the square of the crystal radius while the surface resistance time scales with the first power of the radius, decreasing relative importance of surface resistance with increasing crystal size is to be expected. At 298 K for the 20-micron crystals ($r_c = 10\,\mu m$) surface resistance and intracrystalline diffusion contribute almost equally to the overall mass transfer resistance while for the 4-micron crystals surface resistance accounts for almost 80% of the overall resistance.

The data at different temperatures suggest that the activation energy for the surface rate coefficient is higher than that for the intracrystalline diffusivity, leading to a

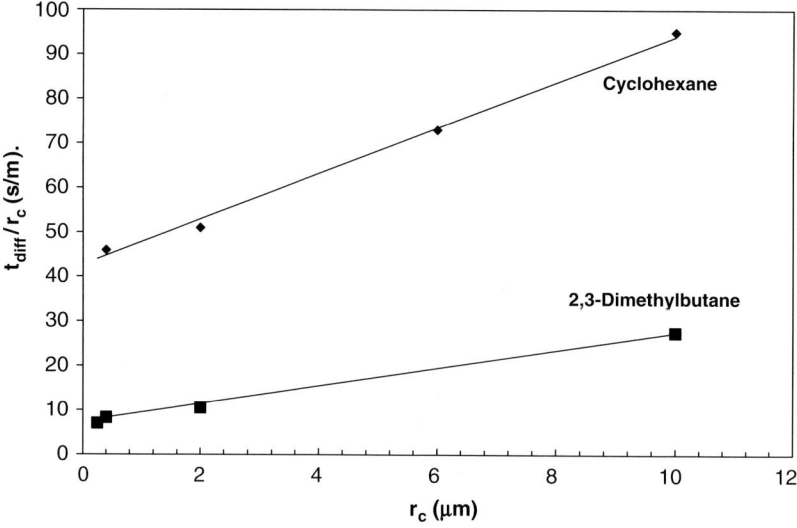

Figure 18.28 Sorption times (first statistical moments of uptake curves) for cyclohexane and 2,3-dimethylbutane in (small) silicalite-1 crystals at 298 K, showing conformity with Eq. (18.1) Data of Gueudré et al. [125].

decrease in the relative importance of surface resistance with temperature. However, in view of the approximations inherent in this analysis – especially Eq. (18.1), which implies that, even under conditions of diffusion control, the kinetics can be represented by a single time constant – any such conclusion should be treated with caution. Similar conclusions were reached by Geuedré et al. [126] from a more detailed analysis of the same experimental data, but since the model that was used assumed a strong Langmuirian dependence of the diffusivity on loading (contrary to the experimental data of Cavalcante and Magalhaes cited above) the conclusions should not be considered as definitive.

18.6.1.2 Surface Etching

An alternative approach to the demonstration of surface resistance was presented by Wloch [127], who studied the adsorption of n-hexane in silicalite crystals. He showed that the sorption rate was increased by washing the crystals in a dilute aqueous solution of HF–acetone. After several such washes the uptake rate showed no further

Table 18.7 Intracrystalline diffusivities and surface rate coefficients from Figure 18.28.

	D_c (m^2 s^{-1}) at 298 K (extrapolated from large-crystal data [59, 64])	D_c (m^2 s^{-1}) at 298 K (from Figure 18.28)	k_s (m s^{-1}) (from Figure 18.28)
Cyclohexane	1×10^{-17}	2×10^{-17}	8×10^{-12}
2,3-Dimethylbutane	6×10^{-17}	5×10^{-17}	5×10^{-11}

18.6 Surface Resistance and Internal Barriers

increase, suggesting that the surface layer, which was assumed to have been contaminated by amorphous silica, had been effectively cleaned.

18.6.1.3 Measurement of Transient Concentration Profiles

Direct measurement of the transient concentration profile in response to a step change in surface concentration by interference microscopy (Section 12.2) offers an even more direct and unambiguous way to detect surface resistance. In the absence of surface resistance, when the sorption rate is controlled by intracrystalline diffusion, the concentration just inside the surface jumps immediately to its final equilibrium value. In contrast, if surface resistance is significant there is a step change in concentration at the surface, which disappears only slowly as equilibrium is approached – see for example Figure 12.7b.

In a series of IFM measurements with various different silicalite samples it was found that most, but not all, crystals showed significant surface resistance. In particular in one specially pre-treated sample, however, the surface resistance (e.g., for n-butane in silicalite) was completely negligible (see Figures 18.29 and 18.36 below) [15]. At first sight, this seems to contradict the predictions of the existence of significant surface resistances by non-equilibrium molecular dynamics with the genuine zeolite lattice at the crystal boundary [128, 129]. However, considering that these simulations were performed with crystal sizes of only a few unit cells and that the diffusional resistance [and, hence, the diffusional time constant – see Eq. (18.1)] increases with the square of the crystal size, these MD results in fact support rather than contradict the experimental findings. The same conclusions were reached by Combariza and Sastre [130] in MD simulations with zeolite LTA.

In addition to confirming Wloch's observations by measuring the transient profiles for the fresh and washed crystals, Chmelik *et al.* [131] showed that surface resistance could be enhanced by treatment with trialkylchlorosilanes (which are too

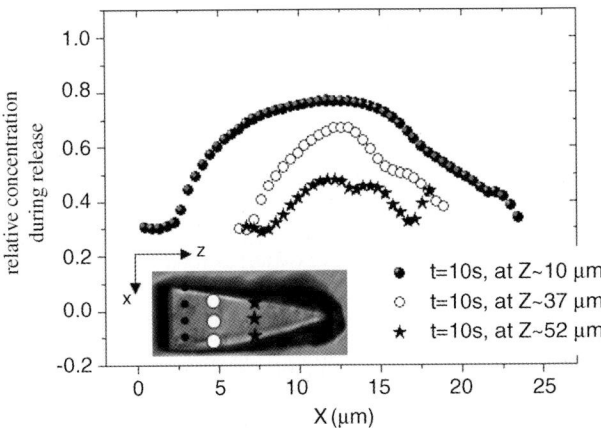

Figure 18.29 Guest profiles in a silicalite crystal fragment. Transient guest profiles (2-methylpropane) along the x-axis, 10 s after the onset of release, from a silicalite crystal fragment recorded for different cross sections (see inset). Reprinted from Reference [15] with permission.

large to enter the intracrystalline pores and presumably block the crystal surface). The permeability of the surface layer for linear alkanes showed a regular decrease with increasing chain length.

18.6.1.4 Other Surface Effects

Tzoulaki et al. [132] showed that traces of moisture enhance the surface resistance but this may be attributed to the formation of a thin water layer on the external surface and the extremely low solubility of paraffins in water.

Reitmeier et al. [133–135] made the surprising observation that the formation of an additional surface layer by chemical liquid deposition of tetraethylorthosilicate on zeolite H-ZSM-5 led to an increase rather than a decrease in the sorption rate. This effect occurs only with sufficiently small molecules (benzene), while for larger molecules, such as p-xylene, the surface layer gives rise to the expected retardation of molecular uptake and release. The possibility of exploiting this for mass separation has been suggested [136].

These observations were explained by Reitmeier et al. [133] by reference to the MD simulations of Skoulidas and Sholl [137], where the permeation rate through circular openings in an impermeable surface layer is shown to decrease dramatically with the molecular size as soon as the critical molecular diameter becomes comparable with the diameter of the openings. While this model can explain the differences in behavior between different species, it cannot explain the acceleration of the uptake that is observed for some molecules. Therefore, as an alternative explanation, one should consider the possibility that, in addition to the formation of an additional surface layer, the chemical treatment might have removed pre-existing surface resistance. Similar findings have also been reported for other systems. For example, Binder et al. [138] showed that the deposition of additional LTA layers, mediated by intermediate physisorption of polycation DADMAC [poly(diallyldimethylammonium chloride)] on the outer surface of large LTA single crystals, increases rather than decreases the surface permeability.

18.6.2
Intracrystalline Barriers

The existence of additional transport resistances on the crystal surface provides a possible explanation of why transient sorption rates are often much smaller than the rates predicted from microscopically measured intracrystalline diffusivities, leading to large discrepancies in the apparent diffusivities derived from microscopic and macroscopic diffusivity measurements. However, the intrusion of surface resistance cannot explain why, in many cases, the time dependence of molecular uptake and release still conforms to an internal diffusion control rather than surface resistance control. This is discussed in greater detail in relation to 5A zeolite in Chapter 16 as well as in Kärger and Ruthven [139]. With a very wide distribution of crystal size or a wide range of variation in surface resistance the superposition of the different surface-barrier limited sorption curves can mimic a diffusion controlled sorption rate process (see the example given with Kortunov et al. [140] and in Section 12.3.2).

However, it seems unlikely that this can offer a general explanation for such behavior since the required range of crystal size is unreasonably large.

The presence of intracrystalline barriers due to the dislocation of the intracrystalline pores arising from the twin sub-structure offers a more plausible general explanation for the diffusion-like pattern that is commonly shown by experimentally measured uptake and release curves, even when the sorption rate is much slower than that expected for ideal intracrystalline diffusion control. Only a modest number of such barriers is required to mimic intracrystalline diffusion control [141].

18.6.3
Sub-structure of MFI Crystals

Inspection of typically coffin-shaped MFI-type crystals under crossed polarizers reveals hourglass-like features that have given rise to the model sub-structure shown in Figure 18.2a. Kocirik and coworkers [142] confirmed the form of this sub-structure in an ingenious way by demonstrating that iodine atoms distribute rapidly along the interfaces between the individual sections. By heating in alkaline hydrogen peroxide solution under ultrasound treatment, Schmidt et al. [143] succeeded in breaking large crystals of silicalite and ZSM-5 into their individual segments, within which the crystal structure should be coherent. As an example, Figure 18.29 shows such a segment, which can be identified as one of the wedge-like segments (1) and (2) at the front faces in the representation of Figure 18.2a. Also shown (in Figure 18.29) are the transient guest concentration profiles (for 2-methylpropane), measured by IFM, at three different sections across the wedge, 10 s after the onset of release. These are probably the first diffusion measurements with a true single crystal of MFI structure.

Figure 18.30 provides an overview of the impressive variety of morphologies in which MFI-type zeolites may occur, depending on their synthesis procedure [144, 145]. The confocal fluorescence images shown in Figure 18.31 illustrate how

Figure 18.30 Microphotographs of specimens of MFI-type crystallites showing different morphologies. Reprinted from Reference [145] with permission.

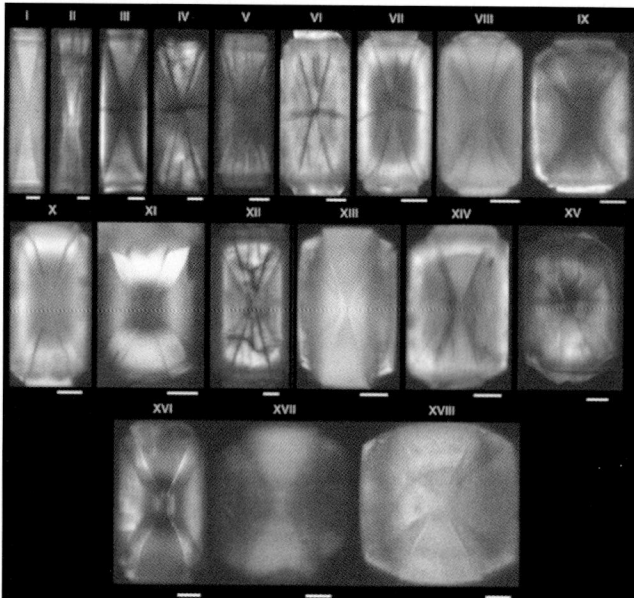

Figure 18.31 Confocal fluorescence images of the MFI-type crystallites shown in Figure 18.30, recorded halfway through the crystal. Reprinted from Reference [145] with permission.

these different morphologies reflect different sub-structures [144, 145]. Figure 18.32 shows a summary of the types of intergrowth structure to which these different patterns may be attributed [144, 145]. Evidently, the sub-structure shown in Figure 18.2a is only one among several possible variants.

Reference [146] reports that, by etching with hydrofluoric acid, MFI-type zeolites dissolved in water or acetone disintegrate into smaller particles. The patterns of disintegration were different depending on the method of synthesis.

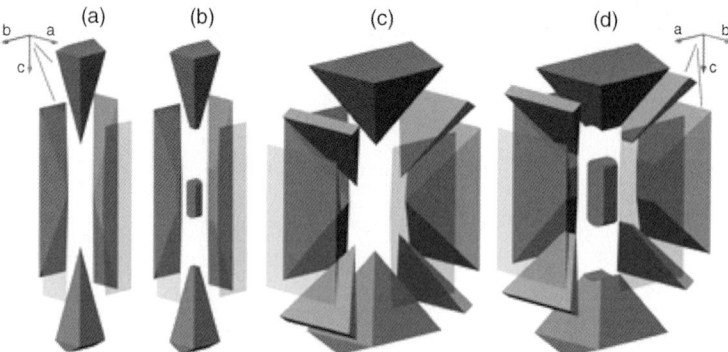

Figure 18.32 "Opened" representation of different morphologies of MFI-type crystals. Reprinted from Reference [145] with permission.

By introducing correspondingly structured dyes into the pore space fluorescence microscopy may be applied to determine the channel directions within the individual sub-sections [147, 148]. The results of such studies suggest that, at least in some crystals, the actual structure deviates from the ideal representation shown in Figure 18.2a as there is clear evidence of direct access to the straight channels from the larger side faces of the crystal.

18.6.4
Assessing Transport Resistances at Internal and External Boundaries by Micro-imaging

In the previous section (Figures 18.30–18.32) MFI-type crystallites have been shown to be traversed by intrinsic boundaries at the interfaces between the different sub-sections of these crystallites. Meso-scale and macro-scale diffusion measurements may therefore be affected by their influence.

There are two limiting cases, the occurrence of which is related to both the structure of these boundaries and the guest molecules under study. In one extreme case – as observed, for example in Reference [142] with iodine atoms – the diffusing particles distribute rapidly over the interface. In this case molecular uptake may be expected to proceed along the interfaces directly into the center of the particle, rather than through the external surface. In such a situation, as in grain boundary diffusion [149], the rates of molecular uptake would have no correlation with the rates of intracrystalline diffusion. At the other extreme, the interfaces may represent transport resistances or even impermeable barriers.

This issue has been clarified by interference microscopy in combination with dynamic Monte Carlo simulations [68], showing the potential of such an approach for structural elucidation.

For a silicalite crystal, as displayed in Figure 18.33, Figure 18.34 compares the actual measurements of transient concentration profiles during molecular uptake with the uptake simulations. The diagrams show the integrals of concentration in the x-direction as a function of z (measured along the length of the crystal) on the central line ($y = 9.8\,\mu m$) and halfway to the surface plane ($y = 4.4\,\mu m$). The simulations were performed by assuming either a moderate transport resistance along the interfaces or rapid diffusion so that the interfaces immediately equilibrate with the external gas phase after [68]. Comparison with the experimental profiles (Figure 18.34a and b) shows that the observed behavior is as expected for moderate transport resistances (Figure 18.34c and d). For this system the possibility of accelerated uptake by fast diffusion along the interfaces (which would result in internal maxima of concentration) can definitely be ruled out. Although the small iodine atoms are evidently able to slip easily along the planes of the interfaces such a process evidently does not occur for the larger alkane molecules for which, at least for the specimens under study, the internal interfaces evidently offer a small additional transport resistance. However, in another series of measurements, with another silicalite crystal and 2-methylbutane as the guest molecule, the concentration profiles (Figures 18.29 and 18.36) showed no irregularity at all, implying negligible transport resistance at the boundaries between the sub-sections.

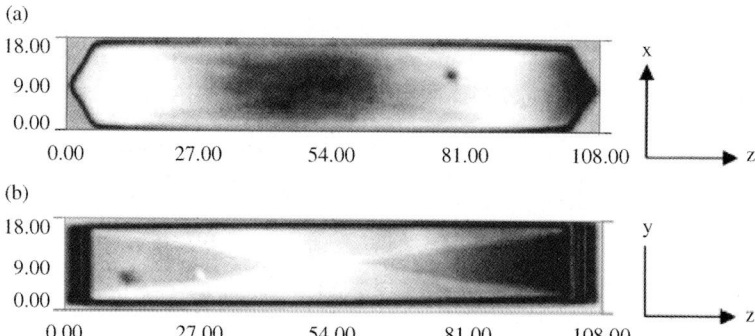

Figure 18.33 Microscopic images of a typical silicalite crystal in the two different orientations. The hourglass structure is made visible by using the shearing mechanism of the interference microscope. The length scale in the x-, y-, and z-directions is shown in micrometers, Reprinted from Reference [68] with permission.

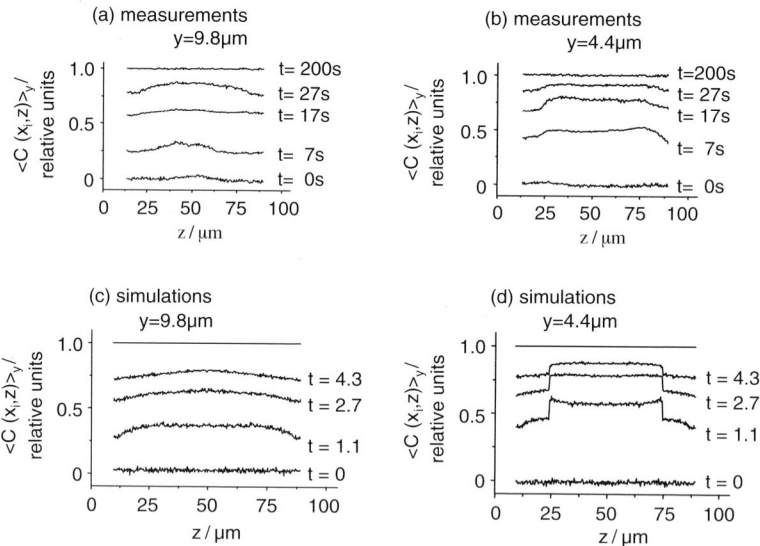

Figure 18.34 Intracrystalline concentration profiles of isobutane in the silicalite crystal shown in Figure 18.33 along the z-direction during adsorption: (a) and (b) profiles measured by interference microscopy; (c) and (d) simulated profiles, assuming that the internal interfaces serve only as transport barriers. For the simulated profiles the time unit is 10^3 elementary diffusion steps. The equilibrium values of $c(y,z)$ after the end of adsorption are equal to 1. Reprinted from Reference [68] with permission.

18.6.5
Evidence from PFG NMR Diffusion Studies

Some experimental findings that support the existence of intracrystalline barriers in X-type zeolites are summarized in Section 17.2.6. In MFI-type zeolites, the existence of intracrystalline transport resistances has been confirmed by PFG NMR self-diffusion measurements with short-chain alkanes [150, 151] with varying observation times. Figure 18.35 presents the relevant data obtained with n-butane as the probe molecule. The diffusivities are plotted as a function of the displacements over which the molecular diffusion measurements were made [estimated from the time interval of the measurements via the Einstein equation Eq. (1.12)].

The remarkable dependence of the measured diffusivities on the diffusion path lengths in a range that is much smaller than the crystal size ($100 \times 25 \times 20\,\mu m^3$) suggests the existence of extended intracrystalline transport resistances. Obviously, at the highest temperature, the thermal energy of the diffusing molecules is high enough to overcome these barriers, so that their influence becomes negligible in comparison with the transport resistance of the ideal intracrystalline pore system. The solid lines show the results of dynamic Monte Carlo simulations. These simulations have been performed with the assumption that, in addition to the energetic barrier characterizing diffusion in the ideal intracrystalline pore system, at

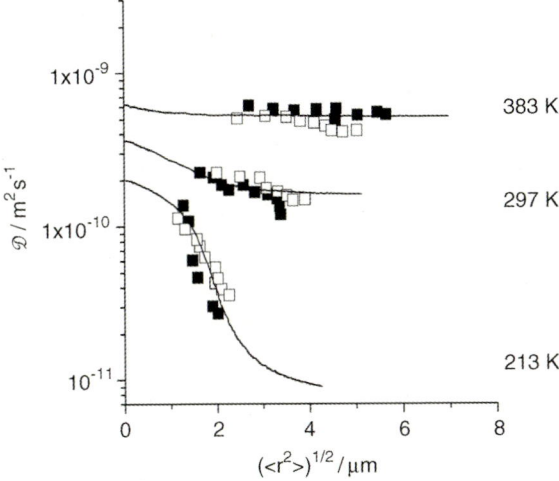

Figure 18.35 Dependencies of the diffusion coefficients of n-butane in silicalite on the root mean square displacements at different temperatures and comparison with the results of dynamic MC simulations for a barrier separation of 3 μm with the assumption that jumps across the barriers occur with an activation energy exceeding that of intracrystalline diffusion by 21.5 kJ mol^{-1}. Filled and open symbols correspond to measurements performed with two different silicalite samples. Reprinted from References [151] with permission.

intervals of 3 μm the diffusing molecules have to overcome additional potential barriers of 21.5 kJmol^{-1} [151]. It is remarkable that the main features of the PFG NMR data may be satisfactorily explained by such a simple model.

The physical nature of the intracrystalline barriers is not known but it seems likely that they may be related to the dislocation of the ideal pore geometry resulting from the complex twin sub-structure of the crystal.

18.6.6
Differences in Diffusional Behavior of Linear and Branched Hydrocarbons

Some striking differences in the diffusional behavior of linear and branched hydrocarbons were noted in Section 18.2.2.5. In addition to the much higher diffusional activation energy of the branched and cyclic species the experimental data for the branched and cyclic species show a remarkable consistency in the diffusivity values obtained by different experimental techniques and for different samples of silicalite crystals, whereas for the linear species such comparisons show wide variations. Simple energy considerations suggest that, for the branched isomers, the channel intersections are energetically favorable sites whereas for the linear species the channels between the intersections are energetically more favorable (compare Figure 7.3). The lower activation energies of the linear species also show that the variation in potential energy with position along the channel is much smaller than for the branched species. At least at low loading levels the diffusion of the cyclic and branched species almost certainly proceeds through distinct jumps between adjacent channel intersections and there is no reason to expect strong directional preference for such jumps, that is, jumps into each of the four channels leaving an intersection should occur with roughly equal probability. As a consequence of this mechanism one can expect that, for the branched and cyclic isomers, there will be little difference in the diffusivities for the straight and sinusoidal channels. The twin planes, which involve a change in the direction of the straight channels, are therefore not expected to offer any significant additional transport resistance, in accordance with the IFM results for 2-methylbutane, noted above.

The situation for the small molecules and the linear hydrocarbons is quite different. The small molecules (relative to the channel diameter) are likely to diffuse in multiple jump steps in which conservation of momentum will ensure a preference for continuing in the same direction, thus leading to a substantially higher diffusivity for the straight channels. For the longer chain alkanes the diffusion mechanism is different (probably involving reptation) but the consequence will be the same: faster diffusion along the straight channels. This difference in diffusion mechanisms means that the diffusivity of the small and linear molecules, which is dominated by fast diffusion in the straight channels, will be far more sensitive to any structural irregularities such as twin planes that disrupt the directional continuity of the straight channels. This is reflected in the wide variation in the diffusivities for different crystals (which are likely to have different substructures) and for different scales of measurement, as observed in the frequency response measurements discussed in Section 18.2.1.

18.7
Diffusion Anisotropy

Since the guest molecules are confined within the pores of the host structure, any structural anisotropy will lead to diffusional anisotropy. Examples of anisotropic host systems (and consequent anisotropic guest diffusion) are the zeolites ferrierite (Section 18.9) and SAPO STA-7 (Section 12.3.2.2). Correlating experimental diffusion data with the anisotropy of the host structure is a challenging task that is best approached through application of the various molecular modeling techniques described in Chapters 8 and 9.

There are, however, a few families of nanoporous crystallites in which the diffusivities in different directions [the principal values of the diffusion tensor – see Eq. (1.20)] may be correlated simply on the basis of the channel geometry. In Section 2.2.2 this type of interdependence of the principal elements of the diffusion tensor was referred to as structure correlation. An example of such a host system with structure-directed diffusion anisotropy, chabazite, has already been discussed briefly in Section 16.10. We now return to this topic and consider in greater detail the diffusional anisotropy of the zeolites of the pentasil family, ZSM-5/silicalite-1 and ZSM-11/silicalite-2.

18.7.1
Correlation Rule for Structure-Directed Diffusion Anisotropy

By inspecting the channel structure of ZSM-5 in the scheme provided by Figure 18.2a, guest diffusion in the crystallographic c direction (the long axis of the crystal) is seen to be possible only by interchanging displacements along channel elements in the a direction (zigzag channels) and b direction (straight channels). The elementary steps for diffusion in the c direction therefore comprise pairs of sequential steps in the a and b directions. With only the assumption that the probability of molecular displacements from a given channel intersection to a subsequent one is independent of the past history (where the molecule has come from) the mean time [τ in Eq. (2.4)] between such "diffusion-relevant" steps [l in Eq. (2.4)] in the c direction may be easily shown [152] to be the sum of the mean times between steps in the a and b directions. Representing these times via Eq. (2.4) by the respective step lengths (Figure 18.2) and diffusivities yields:

$$c^2/\mathcal{D}_z = a^2/\mathcal{D}_x + b^2/\mathcal{D}_y \tag{18.2}$$

Tetragonal pore space symmetry (Figure 18.1b) further simplifies the correlation rule of diffusion anisotropy for zeolite ZSM-11, yielding [152]:

$$c^2/\mathcal{D}_z = 2b^2/\mathcal{D}_y \tag{18.3}$$

Alternative derivations of the correlation rule, Eq. (18.2), for diffusion anisotropy in MFI-type zeolites may be found in References [153, 154].

18.7.2
Comparison of Measured Profiles

As a consequence of their microscopic size, the measurement of diffusion anisotropy in zeolite crystallites is essentially confined to microscopic techniques. Microimaging (Section 12.2) can immediately correlate the evolution of the concentration profiles and, hence, the guest fluxes with the different crystal faces and is therefore particularly useful for the study of diffusion anisotropy.

Figure 18.36 shows, as an example, the evolution of the concentration profiles of branched alkanes during desorption from a single crystal of zeolite silicalite-1, recorded over different cross-sections perpendicular to the length of the crystal [15]. We note that (i) the sequence in the desorption rates corresponds with our expectation (i.e., notably faster desorption of 2-methylpropane compared with 2-methylbutane, reflecting the smaller molecular size) and (ii) at the first point of measurement (i.e., after 10 s), for both guest molecules, the boundary concentrations are found to attain the values corresponding to equilibrium with the external gas phase (which in the present case is zero). Surprisingly, this result was found to be rather unusual; the presence of significant surface resistance, leading to significant differences between the boundary and equilibrium (i.e., final) concentrations, was far more common (Sections 12.3.2 and 18.6.1). With reference to diffusion anisotropy in real crystals, the occurrence of additional transport resistances in MFI-type zeolites is discussed further in Section 18.7.6.

The profiles are essentially the same in the X and Y directions (90° rotation around the long axis of the crystal) and independent of the position (along the long axis). Given the particular ("coffin"-like) shape of the crystal, mass transfer in the direction of the long axis should be relatively slow and is therefore not expected to contribute significantly to the molecular uptake. The similarity of the profiles recorded over different, parallel-shifted cross sections is, therefore, not unexpected. However, it is worthwhile mentioning that, under the conditions of isotropic diffusion, the profiles closer to the ends of the crystal should, nevertheless, be shifted closer to the equilibrium values. The absence of any such effect is consistent with our expectation that diffusion in the z (or c) direction (Figure 18.2a) should be substantially slower than in the transverse directions.

If perfect crystal shape could be taken as an indication of perfect crystallinity, the similarity of the profiles at right angles in any plane perpendicular to the long axis would imply equal diffusivities in the a and b directions. As mentioned in Section 18.1, however, MFI-type zeolites consist of many differently oriented regions. A similar situation occurs in apparently ideal AFI-type zeolite crystals (Section 5.3.1). In real MFI crystals the straight channels are often found to be parallel and the zigzag channels perpendicular to the longitudinally extended crystal faces (Figure 18.2a). For such a structure rotational invariance of the profiles is to be expected as a natural consequence of the overall symmetry.

18.7.3
Evidence by PFG NMR Measurements

Alternatively, diffusion anisotropy may also be studied by exploiting the potential of PFG NMR as presented in Section 11.3.5 and considering the orientation dependence of the displacements of the guest molecules. With sufficiently large single crystals, such as, for example, with natural chabazite (see Reference [155] and Section 16.10), measurements of this type may be performed quite easily by simply aligning the crystals with the desired direction oriented in the direction of the magnetic field gradient. With synthetic zeolites, the small crystal size generally precludes such a straightforward procedure, but with sufficiently large asymmetric crystals (e.g., those shown in Figure 18.36b) the required orientation can be approximated by loading the crystals in an array of parallel capillaries [156, 157]. In this way it was possible to align a sufficiently large quantity of crystallites with respect to their longitudinal extension (the crystallographic c axis). With the pulsed field gradients directed either perpendicular or parallel to the direction of the

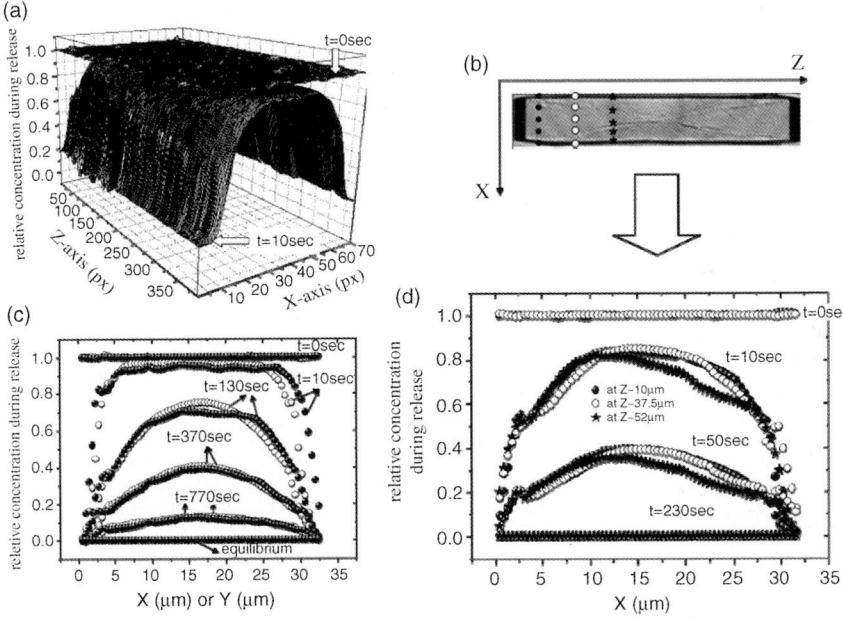

Figure 18.36 Transient profiles during release of 2-methylbutane [(a) and (c)] and 2-methylpropane [isobutane (d)] from a crystal of type silicalite (b). (a) Overview of the intracrystalline concentration profile 10 s after the onset of desorption; (b) host crystal with indication of the cross section to which the transient concentration profiles in (d) refer; (c) evolution of guest concentration profiles along x-axis (open symbols) and y-axis (filled symbols), that is, after 90° rotation around the direction of longitudinal extension, at $z \approx 10$ μm; (d) evolution of guest profiles along the x-axis during release over different cross sections as indicated in (b). Reprinted from Reference [15] with permission.

capillaries [157] the effective diffusivities provided by PFG NMR predominantly reflect molecular propagation in the xy plane or in the z direction, respectively.

Figure 18.37 shows the diffusivities of methane in a sample with ZSM-5 crystallites oriented in capillaries with the field gradients applied either parallel or perpendicular to the direction of the capillaries applied as crystal containers [156]. The diffusivities observed with field gradients perpendicular to the capillaries notably exceed those for gradients parallel to the capillaries. This is in complete agreement with the expected behavior since the straight and sinusoidal run in the a and b directions, which are perpendicular to the long axis of the crystal (Figure 18.2a).

Since the crystal alignment within the capillaries is not perfect, the measured diffusivities approximate the diffusivities in the crystallographic c direction, D_z, and the mean value $\frac{1}{2}(D_x + D_y)$ of the diffusivities in the ab plane, but the agreement is not exact. (In our notation the principal tensor elements, x, y, and z refer to the crystallographic a, b, and c directions.) Assessing the deviation from a perfect alignment on the effective diffusivities by considering the crystal geometry in correlation with the capillary diameters, one may estimate the range where the

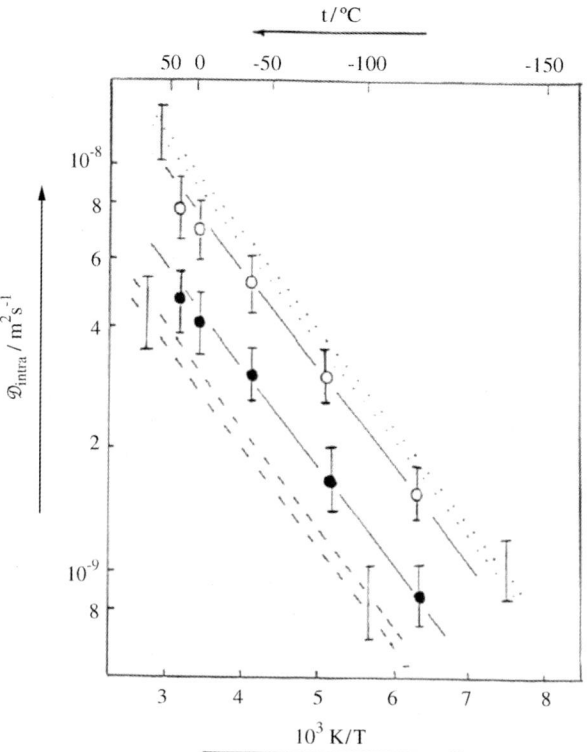

Figure 18.37 Arrhenius plot of the diffusivities of methane in ZSM-5 at a sorbate concentration of 12 molecules per unit cell with field gradients applied parallel (●) and perpendicular (○) to the capillaries of the container system. The broken (dotted) lines indicate the expected range of $\mathcal{D}_z$ and $\frac{1}{2}(\mathcal{D}_x + \mathcal{D}_y)$. Reprinted from Reference [156] with permission.

correct values of D_z and $\frac{1}{2}(D_x + D_y)$ may be expected. This range is also indicated. Since molecular alignment is only possible with respect to the c axis, a separate determination of D_x and D_y in the proposed way is not possible.

Alternatively, diffusion anisotropy may also be measured by analyzing the shape of the PFG NMR signal attenuation (in a powder sample) with increasing field gradient intensity. Since, in a powder sample, all crystal orientations with respect to the field gradient direction are equally probable, the signal attenuation is a superposition of exponentials of the type of Eq. (11.9) with diffusivities determined by the orientation of any individual crystallite. These options, which are discussed in more detail in Section 11.3.5, have been applied successfully to PFG NMR measurement of diffusion anisotropy in chabazite (Section 16.10 and Bär et al. [155]) as well as in mesoporous materials of type MCM-41 (Section 15.3) and SBA-15 (Section 15.7.1). All these systems are distinguished by their rotational symmetry so that their diffusion properties are determined by only two parameters.

The determination of all three principal elements that are necessary for a complete description of the diffusional behavior of guest molecules in MFI-type structures would be far more difficult and, indeed, essentially impossible for noisy attenuation curves. In Reference [157] this complication was circumvented by assuming the validity of the correlation rule between the different tensor elements [Eq. (18.2)]. In this way, the attenuation curves can be approximated by only two free parameters, which are conveniently chosen to be the mean diffusivity $\langle \mathcal{D} \rangle = \frac{1}{3}(\mathcal{D}_x + \mathcal{D}_y + \mathcal{D}_z)$ and the ratio $\mathcal{D}_y/\mathcal{D}_x$ of the diffusivities in the two channel systems.

Figure 18.38 compares the diffusivities determined in this way with oriented crystallites and the MD simulations by different authors. All data are found to be in reasonably good agreement. It should be clearly pointed out that, as a consequence of the fitting procedure, the accuracy of the PFG NMR data as indicated by the error bars in Figure 18.38 is only modest. The possibility that the ratio $\mathcal{D}_y/\mathcal{D}_x$ is in reality shifted to smaller values cannot therefore be excluded. Such a shift would be caused by molecular exchange between regions of different crystallographic orientation, which becomes increasingly likely when the individual crystals are composed of different regions with different crystallographic orientations (Section 18.6.3). In uptake measurements with oriented crystals [7] this effect was suggested as an explanation for the observation that the anisotropy was smaller than expected on the basis of the correlation rule [Eq. (18.2)].

18.7.4
Limits of the Correlation Rule

Section 18.6.3 illustrates in detail the variety of the habits and structural composition in which the crystallites of MFI-type zeolites may occur. The substantial deviations of real crystals from the ideal structure have prevented more detailed experimental studies of diffusion anisotropy. Tests of the validity of the correlation rule of diffusion anisotropy introduced in Section 18.7.1 have therefore been based mainly on simulations.

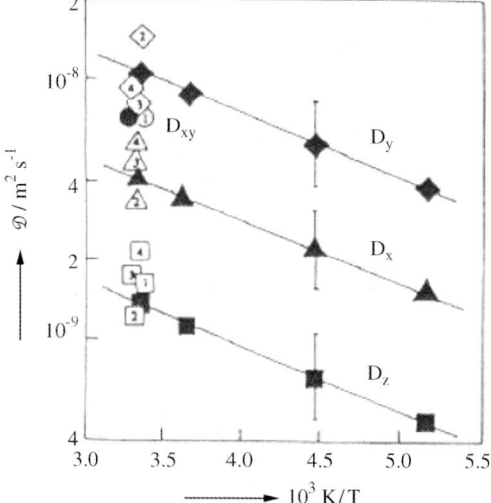

Figure 18.38 Arrhenius plot of the principal values of the diffusion tensor for methane adsorbed in ZSM-5, as determined from the best fit of the PFG NMR signal attenuation curve of a powder sample with the theoretical expression based on Eq. (11.23) by implying a correlation of the principal tensor elements via Eq. (18.2) (filled symbols), and comparison with the results of the measurement with oriented samples (1 – data from Figure 18.37) and of MD simulations presented in Demontis et al. [158] (2), Goodbody et al. [159] (3), and June et al. [160] (4) (open symbols with inserted numbers). Reprinted from Reference [157] with permission.

Deviations between the predictions of Eq. (18.2) and the simulation results led to the introduction of a memory parameter β defined by the relation:

$$\beta = \frac{c^2/D_z}{a^2/D_x + b^2/D_y} \tag{18.4}$$

With $\beta = 1$, Eq. (18.4) coincides with Eq. (18.2), indicating a complete loss of the particle's "memory;" $\beta > 1$ means a suppression of diffusion in the z direction, indicating that the guest particles prefer to continue their diffusion path along the present channel rather than switching into the other cannel, while $\beta < 1$ means exactly the opposite. Several approaches to correlate this memory parameter with the probability of the individual diffusion steps by molecular modeling have been proposed in the literature (Demontis et al. [161]; Fritzsche and Kärger [162, 163]).

A different starting point for modeling diffusion in MFI-type zeolites by steps between distinct lattice sites has been based on the finding of molecular dynamic simulations (see, for example, June et al. [164]) that, with increasing chain length, the diffusing molecules resided much more frequently in the individual channel elements rather than in the intersections. This situation is visualized in Figure 18.39, which shows the center-of-mass of a 1-butene molecule in a unit cell of silicalite on its trajectory in different projections [165]. It provides clear evidence that it is the individual channel elements rather than the intersections that may be considered as

18.7 Diffusion Anisotropy | 709

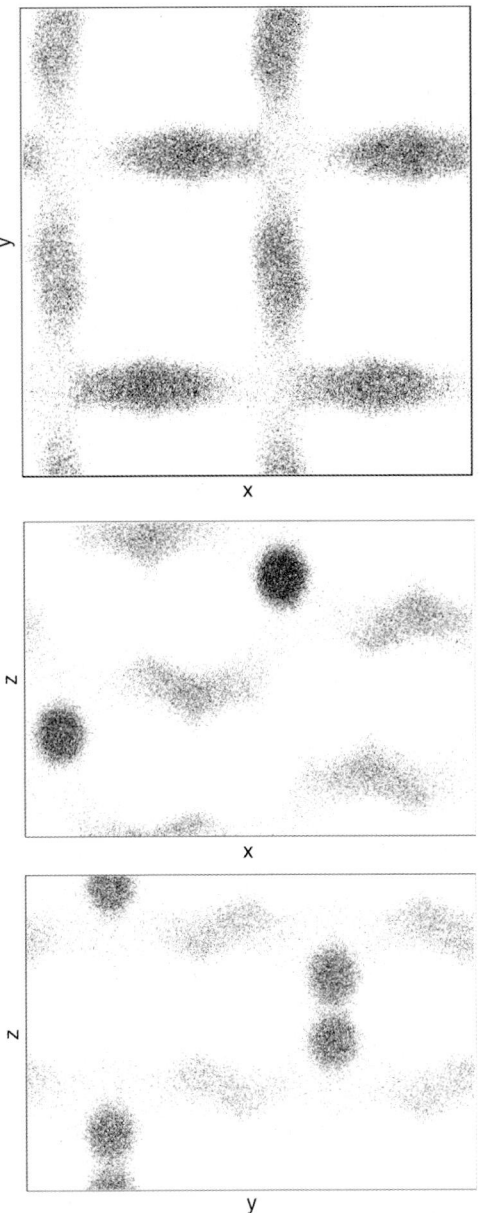

Figure 18.39 Positions registered for the center-of-mass of 1-butene molecules in the unit cell of silicalite at 300 K in different projections. Reprinted from Reference [165] with permision.

the natural positions of "rest" if the diffusion path is approached by a step sequence between distinct positions in the pore system.

It is shown in Reference [165] that the diffusion anisotropy of 1-butene in MFI may be represented completely equivalently on the basis of such a jump model or by rigorous MD simulation. It is impossible, however, to formulate a correlation rule between the different diffusivities of the type of Eq. (18.2). Since, on the other hand, the memory parameters β [see Eq. (18.4)] are found to vary between 1.2 and 1.4, Eq. (18.2) may still be considered to serve as reasonably good first approximation for diffusion anisotropy.

18.7.5
Anisotropic Diffusion in a Binary Adsorbed Phase

Figure 18.40 presents simulation results showing the mean square displacements in the (crystallographic) x, y, and z directions for diffusion of methane and butane in a binary adsorbed phase in silicalite [48, 166]. In conformity with the results from single-component diffusion measurements, the diffusivities along the straight channels (y direction) are found to be slightly greater than those along the sinusoidal channels (x direction) while the diffusivities in the z direction are notably smaller. Surprisingly, even in this two-component system, the correlation rule of diffusion anisotropy [Eq. (18.2)] is found to be accurately obeyed [48].

Evidently, only for displacements greater than a few nanometers do the mean square displacements increase in proportion to the observation time. This is because it is only after displacements of this magnitude that the diffusing molecules have moved far enough for subsequent diffusional jumps to be uncorrelated, a fundamental requirement for Fickian diffusion (Sections 2.1 and 2.6).

As observed for other two-component systems (see, for example, Figures 18.44 and 18.45 and Figures 17.27 and 17.32) the diffusivities of the two components in the mixture are found to be closer to each other than the single-component diffusivities as a result of their mutual interference (the mutual diffusion effect). The magnitude of this effect is in good agreement with the experimental results obtained in QENS studies [167].

18.7.6
Anisotropy of Real Crystals

Although the MFI crystal structure is clearly non-isotropic the experimental evidence concerning the degree of non-isotropy of real MFI crystals is somewhat contradictory. The constancy in the z direction of the transient concentration profiles for 2-methylbutane, shown in Figure 18.36, suggests that diffusion in the z direction (the direction of the long axis) must be much slower than in the transverse directions, thus providing clear evidence of the expected diffusional anisotropy at the crystal scale. Since any substantial deviation from the ideal structure might be expected to substantially reduce the degree of anisotropy, this observation also suggests that the structure of this particular crystal must have been close to ideal.

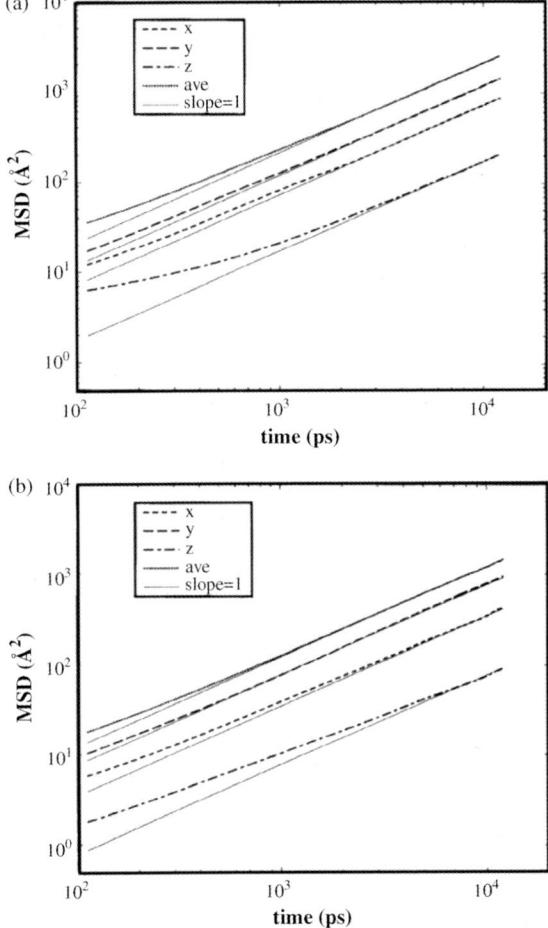

Figure 18.40 Molecular mean square displacements along the different crystallographic directions and total mean square displacement of methane (a) and n-butane (b) in silicalite at 300 K as determined from MD simulations of a guest mixture containing four molecules of methane and eight molecules n-butane per unit cell. Reprinted from References [48, 166] with permission.

Comparison of the diffusivities derived from permeation measurements with either single crystal or polycrystalline membranes oriented in the z direction with the mean diffusivity values derived from other macroscopic measurements such as ZLC can also provide evidence concerning the degree of non-isotropy. Table 18.8 summarizes comparisons of this kind for several different sorbates. The experimental data show a good deal of scatter so it is difficult to draw definite conclusions. Nevertheless, contrary to expectation, the data appear to suggest that there is little difference between the diffusivities in the z direction and the average diffusivity measured by techniques such as ZLC, suggesting approximately isotropic behavior.

Table 18.8 Comparison of corrected diffusivities (D_0) at low loading for selected hydrocarbons in silicalite.

Sorbate	T (K)	Membrane (m² s⁻¹)	ZLC (m² s⁻¹)^a	References
Propane	334	7.3×10^{-12}	1.25×10^{-11}	Hayhurst [29], Eic [34]
Isobutane	334	10^{-12}	(10^{-12})	Shah [70], Hufton [66]
	400	5.5×10^{-12}	8×10^{-12}	Millot [36], Hufton [66]
n-Pentane	334	2.4×10^{-12}	2.0×10^{-12}	Hayhurst [29], Eic [34]
3-Methylpentane	400	1.8×10^{-13}	3.0×10^{-13}	Millot [36], Cavalcante [59]
Benzene	343	7.0×10^{-14}	10^{-13}	Shah [70], Brandani [81]
	303	1.5×10^{-14}	(2.0×10^{-14})	Shah [70], Brandani [81]
p-Xylene	341	3.0×10^{-13}	2.0×10^{-13}	Shah [70], Brandani [81]

a) Values in parenthesis are extrapolated from higher temperature data.

The reasons for this behavior are not clear but deviations from the ideal pore structure resulting from the complex twin structure offer a possible explanation.

For small molecules, linear alkanes, and p-xylene in a perfect silicalite crystal the average diffusivity is determined largely by diffusion through the straight channels as this is much faster than diffusion in other directions. The twin structure destroys the continuity of the straight channels, thus making the diffusivities in the two transverse directions more nearly equal and so reducing the degree of non-isotropy. This could reduce the non-isotropy of a highly twinned crystal to a level at which it is no longer obvious from macroscopic measurements.

18.8
Diffusion in a Mixed Adsorbed Phase

Since in MFI each of the intersections in the channel network is mutually interconnected by only four channel segments, in a mixed phase, diffusion of the more mobile component is easily dominated by blocking effects due to the presence of the second, less mobile component. This limits the applicability of the Maxwell–Stefan model [169]. Effective medium approaches based on mean life and passage times offer a more useful approach for these systems.

18.8.1
Blocking Effects by a Co-adsorbed Second Guest Species

18.8.1.1 Methane in the Presence of Benzene

Since the kinetic diameter of benzene is close to the channel diameter of silicalite one may expect that the self-diffusivity of a small molecular species such as methane will be greatly reduced by the presence of co-adsorbed benzene. This trend is indeed shown by the data presented in Figure 18.41 [170, 171]. Since the mobility of benzene is small relative to that of methane it is reasonable to consider the system as a suitably distributed set of rigid obstacles within the pore structure. Results of a Monte Carlo

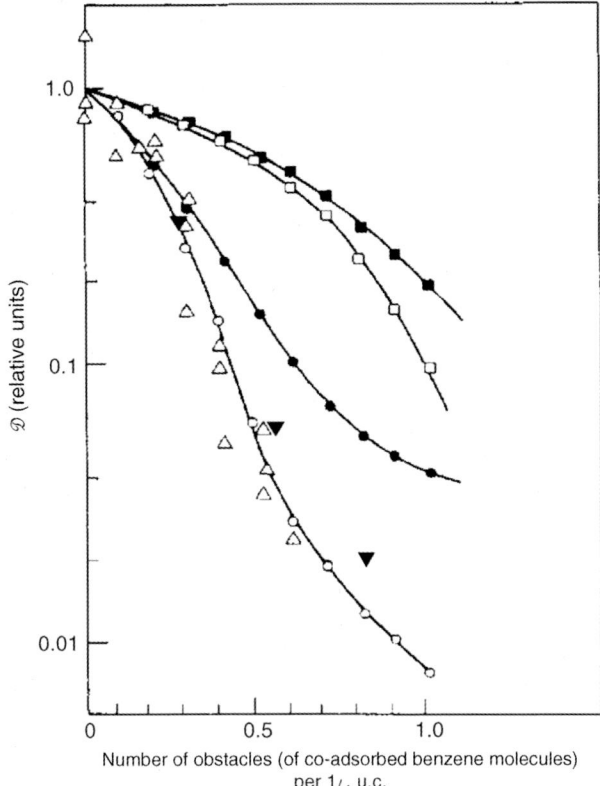

Figure 18.41 Results of dynamic Monte Carlo simulation of a random walk of methane in a two-dimensional channel network with obstacles (benzene molecules) distributed statistically in the channel segments (□) and in the channel intersections (○). Transition probabilities are 0.02 (filled symbols) and 0.004 (open symbols) [170, 171]; experimental self-diffusivity data for methane in the presence of pre-adsorbed benzene in silicalite (△) [170, 171] and the results of rigorous MD simulations with methane (three molecules per unit cell) and benzene in silicalite (▼) are also shown [172] with permission.

simulation based on such a model provide an excellent representation of the experimentally observed behavior if it is assumed that the benzene molecules are located only in the channel intersections.

In Reference [172] the methane diffusivity in silicalite under the influence of co-adsorbed benzene was considered in rigorous MD simulations. The results, which are also included in Figure 18.41, are in good agreement with both the trends observed experimentally and with the predictions of a simple two-dimensional pore network model.

18.8.1.2 Methane in the Presence of Pyridine and Ammonia

Figure 18.42 shows the self-diffusivity of methane at a loading of about 2 molecules per channel intersection in different samples of HZSM-5 (with different Si/Al) containing

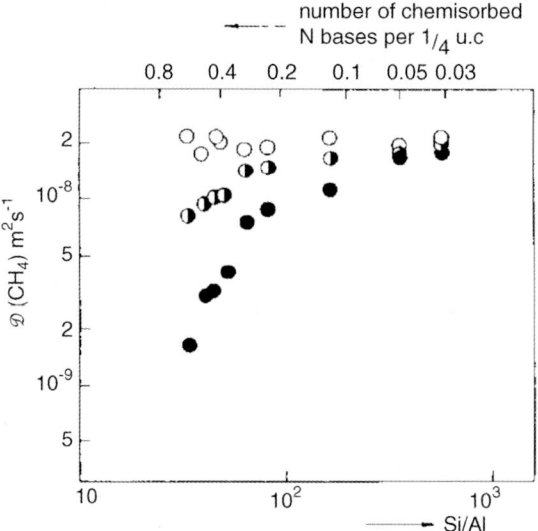

Figure 18.42 Self-diffusivities (PFG NMR) for methane in H-ZSM-5 at 300 K at a loading of about 2 molecules per channel intersection as a function of Si/Al ratio in the presence of chemisorbed N-compounds occupying about 80% of the equivalent Brønsted sites: (half-filled circle) NH_3 and (●) pyridine; (○) refer to the parent system with no pre-adsorbed N-compound. Reprinted from Caro et al. [173] with permission.

chemisorbed ammonia or pyridine [173]. The N-compounds correspond to about 80% of the Brønsted acid protons, as determined by ^{1}H magic-angle spinning (MAS – see Section 10.1.4) NMR [118, 119]. Analysis of these data in the same way as for the methane/benzene data again suggests preferential disposition of the pyridine molecules in the channel intersections. In contrast, for ammonia no such clear evidence was found. This is understandable since the small ammonia molecule will no doubt act as a much less effective obstacle to the diffusion of methane.

18.8.1.3 n-Butane in the Presence of Isobutane

The recent application of the MAS PFG NMR technique to zeolites (see Section 10.1.4 and Reference [174]) has made it possible to measure the diffusion of the individual components in a hydrocarbon guest mixture. Figure 18.43a shows the diffusivities of n-butane in a mixture with isobutane, adsorbed in silicalite, determined in such a way [175]. The data are plotted as a function of the content of the less mobile isobutane molecules. They show the expected trend: increasing isobutane loading leads to a pronounced decrease in the n-butane diffusivities. In Figure 18.43b, this experimental finding is shown to be nicely reproduced in simulations. The pronounced decrease in the n-butane diffusivities at about 2 molecules per unit cell, that is, $\frac{1}{2}$ per channel intersection, may be easily correlated with the results of dynamic Monte Carlo simulations (Section 9.2.2): whereas n-butane can locate along either straight or zigzag channels, isobutane locates preferentially at the intersections so that the percolation threshold will correspond to blockage of half the intersections.

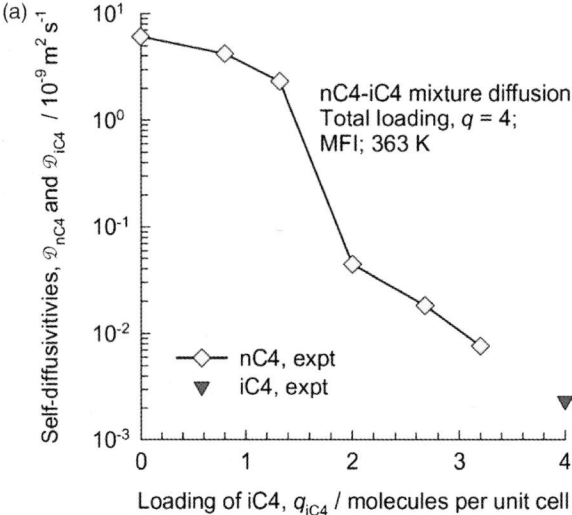

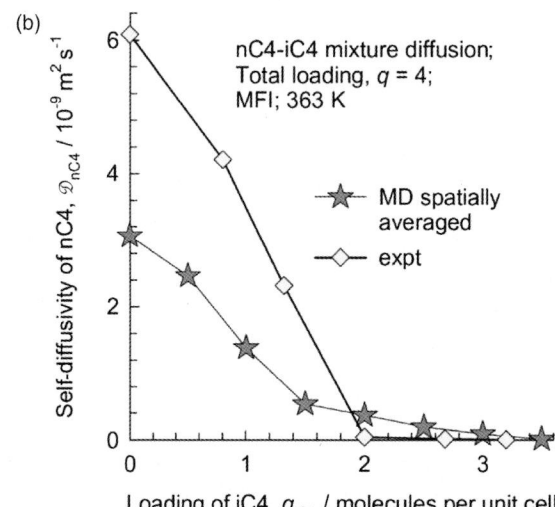

Figure 18.43 Self-diffusivities of n-butane in a mixture with isobutane at 363 K in silicalite for a total loading of 4 molecules per unit cell (corresponding to 1 molecule per channel intersection) with varying isobutane contribution (loading). (a) MAS PFG NMR data, with the isobutane diffusivity at complete isobutane loading; (b) comparison of the experimental data with the simulations. Reprinted from Reference [175] with permission.

The low mobility of isobutane in the MFI structure impedes the measuring conditions for isobutane diffusion, in particular under MAS conditions, which limit the application of extra-large field gradient intensities. Isobutane diffusion could therefore only be measured at its highest concentration, that is, in the absence of n-

butane. Comparison with the pure n-butane sample yields a difference in their diffusivities in MFI of more than three orders of magnitude, in striking contrast with NaX, in which (Section 17.2) the diffusivities of n- and isobutane are essentially the same.

18.8.2
Two-Component Diffusion

18.8.2.1 Methane and Tetrafluoromethane

Selective diffusion measurements by PFG NMR are facilitated if they can be based on different probe nuclei for the different components, as, for example, ^{1}H and ^{19}F in the two-component self-diffusion measurements of methane and tetrafluoromethane in silicalite [116]. Figure 18.44 provides an example of such measurements. It shows the diffusivities of CH_4 and CF_4 in silicalite at 200 K and a total loading of 12 molecules per unit cell. In agreement with our expectation, the diffusivities of both components increases with increasing amount of methane (the lighter and more mobile component). MD simulations are shown to be in excellent agreement with the experimental data [116].

18.8.2.2 Methane and Xenon

With the application of ^{1}H and ^{129}Xe PFG NMR, the two-component diffusion of methane and xenon in MFI also becomes easily accessible in the same way [176]. Comparison of the respective data in Figures 18.44 and 18.45 shows interesting

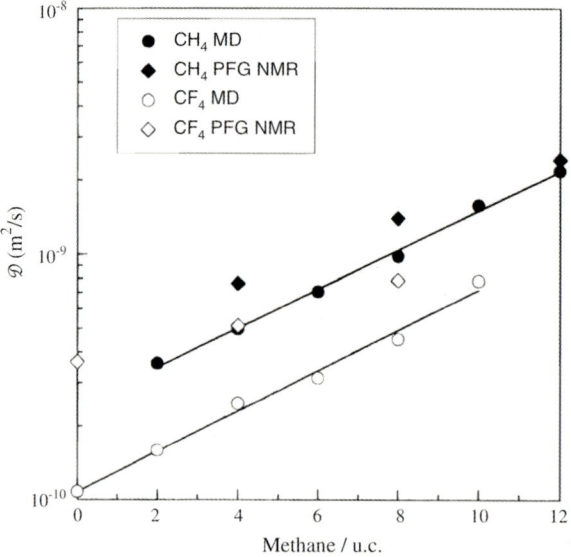

Figure 18.44 Self-diffusivities of methane and tetrafluoromethane in silicalite at 200 K and a total loading of 12 molecules per unit cell for varying composition: comparison of the results of PFG NMR measurement with MD simulations. Reprinted from Reference [116] with permission.

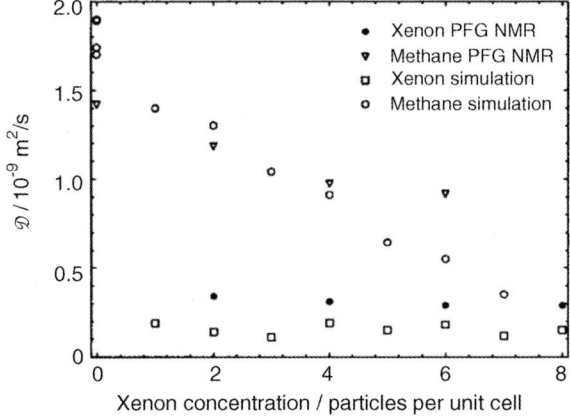

Figure 18.45 Self-diffusivities of methane and xenon in silicalite at 150 K and a total loading of 8 molecules per unit cell for varying composition: comparison of the results of PFG NMR measurement with MD simulations. Reprinted from Reference [176] with permission.

similarities and differences between the diffusional properties of CH_4–CF_4 and CH_4–Xe in MFI. While in both cases the diffusivity of the mobile component (methane) decreases with increasing content of the less mobile component (CF_4 or Xe), only the mobility of CF_4 is enhanced by an increasing content of CH_4, while the xenon mobility remains essentially unaffected. These differences also appear in the MD simulations, which, in both cases, closely replicate the experimental findings.

18.8.2.3 Methane and Ammonia

In addition to the single-component diffusivities already presented in Sections 18.2.1.1 and 18.5.3, Reference [121] also reports the results of selective PFG NMR diffusion studies with either component in methane–ammonia mixtures, at loadings of 2 molecules per unit cell in silicalite. As for single-component adsorption, the methane diffusivity in the mixture is also found to exceed the ammonia diffusivity by more than an order of magnitude. For both species the presence of the second component leads to a reduction in the mobility of the first component.

18.8.2.4 Permeation Properties of Nitrogen and Carbon Dioxide

Reference [177] reports on the computation of adsorption isotherms and self-diffusivities through molecular simulations (see Section 8.2.8 for the single-component data). Transferring the self-diffusivities into transport diffusivities by use of the corresponding thermodynamic factors, the simulation results are used to estimate the permeabilities through a perfect membrane of silicalite. The calculated permeability ratios compare reasonably well with the results of measurements reported by Yan and Gavalas [178].

18.8.2.5 Counter-current Desorption of p-Xylene–Benzene

Counter-current ZLC measurements provide a simple and straightforward way of determining the extent to which the diffusion of one component is retarded by the

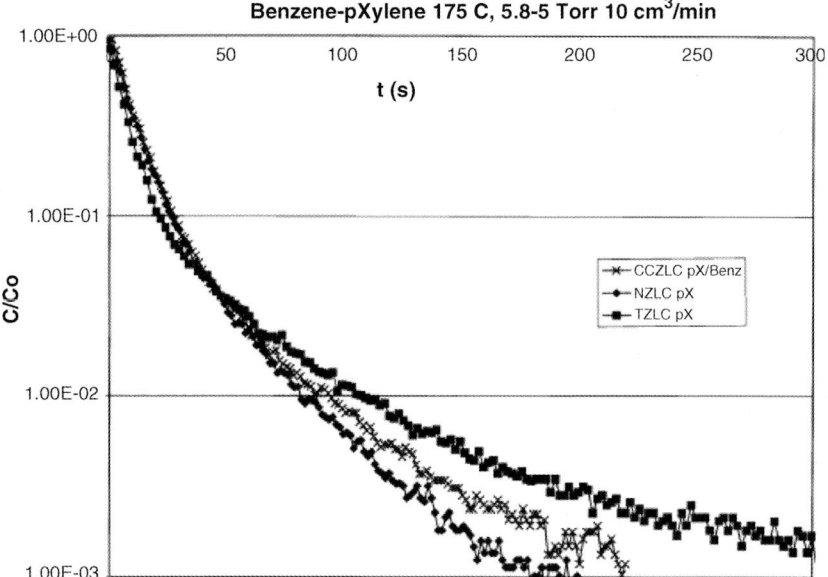

Figure 18.46 Counter-current desorption curves for benzene (5.8 Torr)–p-xylene (5.0 Torr) at 448 K (175 °C), purge flow rate 10 ml min^{-1} following p-xylene, compared with normal and tracer ZLC response curves for p-xylene under similar conditions. Reprinted from Brandani et al. [81] with permission.

counter-diffusion of a second species. As an example, Figure 18.46 shows ZLC desorption curves for p-xylene from large silicalite crystals ($r_{eq} \approx 25\,\mu m$) at 448 K. The fastest desorption curve (NZLC, normal zero-length column) corresponds to the transport diffusivity of p-xylene at this temperature ($D_0 = 1.8 \times 10^{-12}\,m^2\,s^{-1}$) while the slowest response (TZLC) corresponds to the self-diffusivity ($\mathcal{D} = 7 \times 10^{-13}\,m^2\,s^{-1}$). The intermediate curve, corresponding to an effective diffusivity of $D = 1.1 \times 10^{-12}\,m^2\,s^{-1}$, is for desorption of p-xylene with benzene counter-adsorbing. Evidently, the diffusion of p-xylene is retarded by the counter-diffusion of benzene but to a lesser extent than by p-xylene itself (under self-diffusion conditions).

18.8.2.6 Co- and Counter-diffusion of Benzene and Toluene

A detailed study of co- and counter-diffusion of benzene–toluene on H-ZSM-5 has been reported by Qureshi and Wei [83] using a recirculating gas adsorption system of the type described in Section 13.4. Measurements were made at a loading of 0.5 molecules per channel intersection. In the counter-diffusion experiments the sample was preloaded to a level of 0.5 molecules per channel intersection and the quantity of the second component added was selected to maintain the total loading essentially constant with the equilibrium adsorbed phase containing approximately equimolar quantities of each component. In the co-diffusion experiments the sample was initially loaded with both components in equimolar concentration (at a total loading

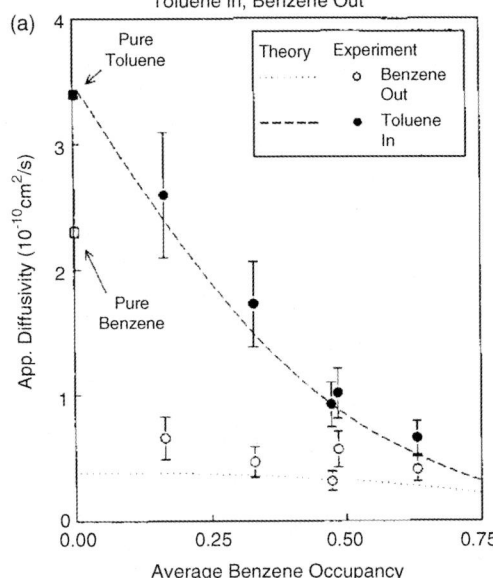

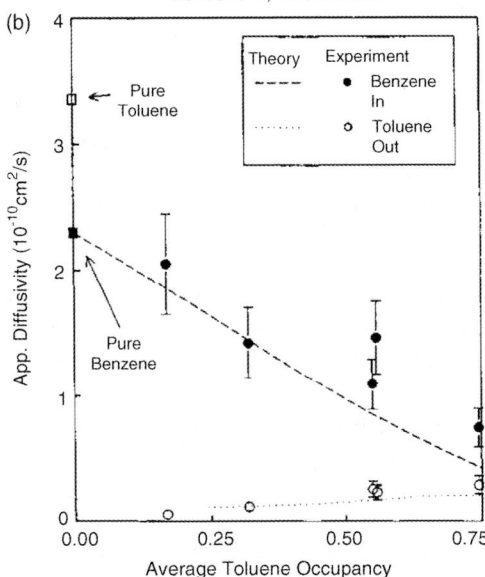

Figure 18.47 Results of counter-diffusion measurements for benzene–toluene on HZSM-5 at 65 °C and 0.05 molecules per channel intersection, showing comparison with theory: (a) toluene diffusing in, benzene out; (b) benzene diffusing in, toluene out. Reprinted from Qureshi and Wei [83] with permission.

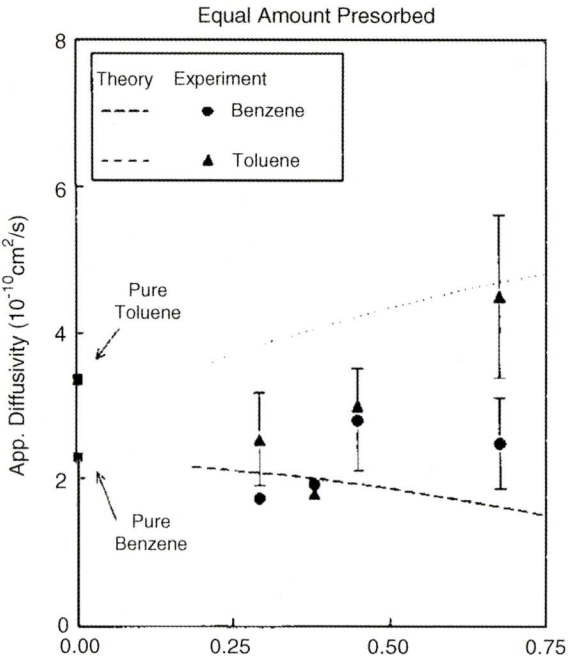

Figure 18.48 Results of co-diffusion measurements for benzene–toluene in HZSM-5 at 65 °C and 0.5 molecules per channel intersection, showing comparison with theory. Reprinted from Qureshi and Wei [83] with permission.

of 0.5 molecules per intersection). The pressures of both components were then increased by a small amount, maintaining the same ratio of partial pressures.

The results are summarized in Figures 18.47 and 18.48, which also show the theoretical curves calculated from the model outlined in Section 4.3.2.1 [see Eq. (4.27)]. In the counter-diffusion measurements the diffusivity of the adsorbing components is reduced with increasing concentration of the pre-adsorbed species but the effect is less dramatic than the reduction in the diffusivity of the desorbing species. By contrast, in the co-diffusion experiment the diffusivities for both species are close to their single-component values. The theoretical model (Eq. 4.29) provides a good representation of the behavior under all conditions.

18.8.2.7 Counter-diffusion of Isobutane and n-Butane

The interference between iso- and n-butane as guest molecules appearing in the PFG NMR self-diffusivities presented in Section 18.8.1.3 were nicely reproduced in counter-diffusion experiments by means of IRM [179]. Figure 18.49 shows the uptake of n-butane by a single crystal of silicalite with different amounts of pre-adsorbed isobutane. With the lowest amount of pre-adsorbed isobutane (Figure 18.49a), intersection blockage is still below the percolation threshold. The n-butane molecule may therefore

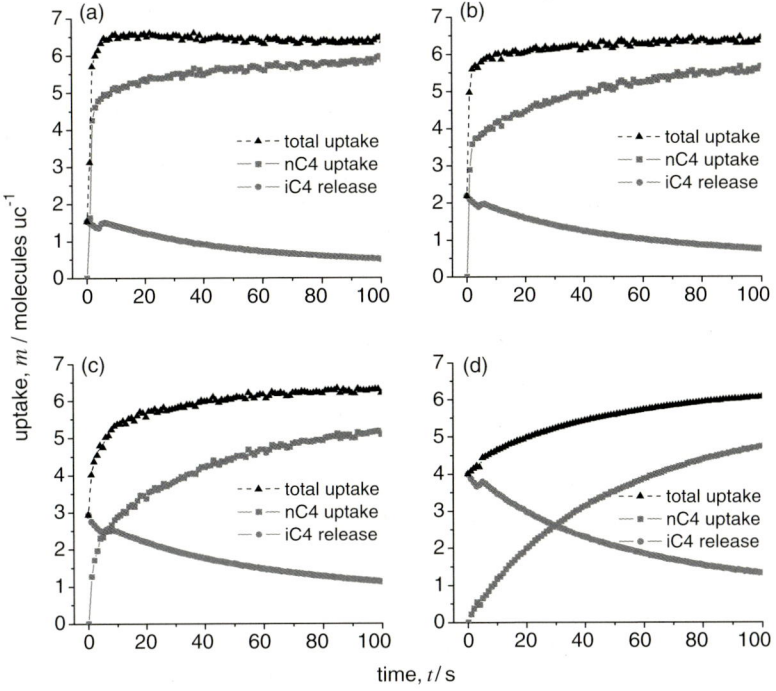

Figure 18.49 Counter-uptake of *n*-butane in silicalite in the presence of isobutane at initial loadings of (a) 1.5, (b) 2.2, (c) 2.9, and (d) 4 molecules per unit cell. The partial loadings (squares and circles) are plotted together with the total loading (triangles). Reprinted from Reference [179] with permission.

occupy all free sites and, because of the large difference in the diffusivities, this occurs almost instantaneously. The number of *n*-butane molecules adsorbing in the subsequent period of time is equal to the number of desorbing isobutane molecules. This means that there are no shielded sites and the overall concentration remains practically constant during the second step of the exchange process.

In the range of the percolation threshold, the number of sites that are directly accessible from the outer surface decreases with increasing isobutane concentration (Figure 18.49b and c). This results in an *n*-butane concentration that increases more strongly than the isobutane concentration decreases so that the overall concentration increases during the second step of the exchange process.

Finally, when all junctions are initially occupied by isobutane (Figure 18.49d) there are no adsorption sites accessible from the outer surface. Therefore, in this region, the first step of the exchange process cannot occur, with the result that the uptake of *n*-butane is limited by the release of isobutane during the entire exchange process [179]. This coupling between the diffusion of the two components greatly reduces the exchange rate, thus seriously limiting the efficiency with which MFI zeolites could be used for separating linear and branched alkanes by exploiting configurational entropy effects during sorption [180, 181].

18.9
Guest Diffusion in Ferrierite

Although ferrierite, as the other representative of the medium-pore (ten-ring) zeolites discussed in this chapter, is of some interest as a catalyst it has mainly been included here because of its favorable structural properties for diffusion studies by micro-imaging (Sections 12.2 and 12.3). Ferrierite crystals occur as thin platelets traversed by two mutually perpendicular channel systems having different diameters: eight- and ten-rings (Figure 12.6). They therefore provide an excellent model system for the study of diffusion anisotropy that, as yet, is not possible with any other system.

The ferrierite specimen considered in Section 12.3 [182] exhibited remarkable surface permeation properties, resulting in essentially complete blockage of the entrances to the larger ten-ring channels. As a consequence, molecular uptake and release were found to mainly proceed through the eight-ring channels and it was therefore only possible to provide a rough estimate of the diffusivity in the direction of the ten-ring channels [183–185].

In Reference [186] Marthala et al. showed that blockage of the ten-ring pore entrances could be removed by an appropriate pre-treatment (immersion in 0.1 M NaOH for two days) so that in the treated crystals uptake and release occurred through both channel systems. Figure 18.50 provides snapshots of the guest distribution (methanol) in both an all-silica sample (b) and an aluminium-containing ferrierite (c). The concentration profiles shown in Figure 18.50b were recorded 23 s after a step change in the surrounding atmosphere. Nevertheless the concentration front has already proceeded somewhat further into the crystal than in the case of the aluminium-containing ferrierite sample (Figure 18.50c) for which the profiles shown were recorded 400 s from the start of the uptake. Since the initial loading was essentially the same in both cases the faster uptake in the all-silica sample must be attributed to the faster diffusion through the (open) ten-ring channels.

A more detailed analysis shows that the diffusivities range from 10^{-10} to 10^{-9} m^2 s^{-1} along the larger ten-ring channels and from 10^{-13} to 10^{-11} m^2 s^{-1} along the smaller eight-ring channels, depending on the loading. The diffusional anisotropy of the Al-containing sample was much less pronounced: while the (low) diffusivities in the eight-ring channels are essentially the same for both samples, the (higher) diffusivity in the ten-ring channels was in the Al-containing sample greatly reduced.

To set the background for the interpretation of the diffusion data and their correlation with the pore system, the crystallographic orientation of the largest faces was determined in single-crystal X-ray studies. In all cases the direction [001] of the ten-membered ring channels was found to coincide with the longest extension of the crystallites, whereas the direction [010] of eight-ring channels was found to run along the second-longest crystal dimension. Hence, the channel network is always extended in the plane of the crystal with no channels running perpendicular to the [100] direction. For illustration, Figure 18.50a shows the indexing of one crystal [186].

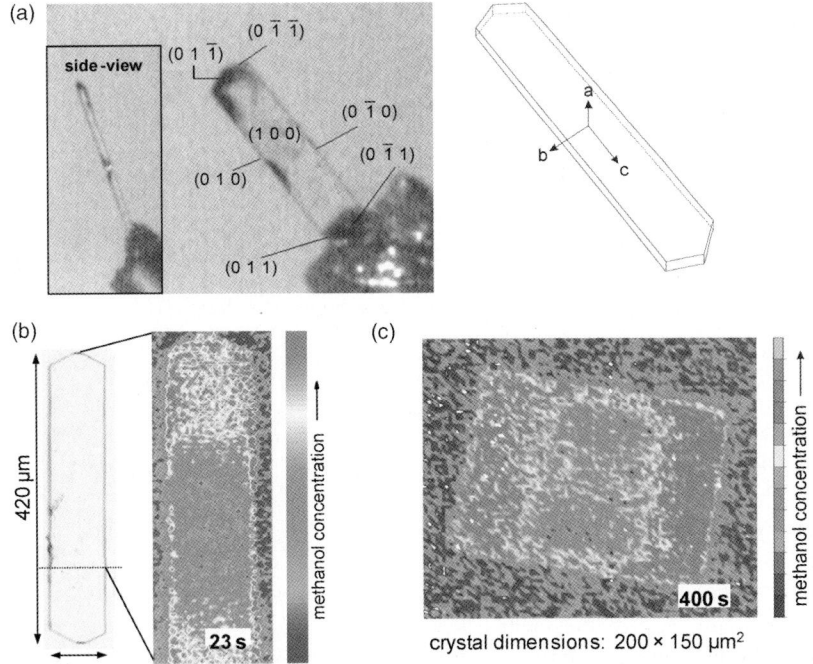

Figure 18.50 Crystallographic orientation of the largest faces of an all-silica ferrierite crystal as determined by single-crystal X-ray analysis (a). Images recorded by IR microscopy at 298 K, (b) and (c). Snapshot of the methanol distribution in an all-silica ferrierite crystal 23 s after starting the uptake; the crystal size exceeds that of the detector and is cut at the lower end (b). Uptake profile in an aluminium-containing ferrierite crystal 400 s after a step change in the surrounding gas phase (c). The different channel diameters of the two pore systems give rise to a pronounced anisotropy in their uptake rates. The uptake in the all-silica sample is faster by a factor of ca. 20. Reprinted from Reference [186] with permission.

References

1 Baerlocher, C., McCusker, L.B., and Olson, D.H. (2007) *Atlas of Zeolite Framework Types*, Elsevier, Amsterdam.
2 Chen, N.Y. (1976) *J. Phys. Chem.*, **80**, 60.
3 Milestone, N.B. and Bibby, J.M. (1981) *J. Chem. Tech. Biotechnol.*, **31**, 732.
4 Bui, S., Verykios, X., and Matharasan, R. (1985) *Ind. Eng. Chem. Process Design Dev.*, **24**, 1209.
5 Kokotailo, G.T. and Meier, W.M. (1979) *Properties and Applications of Zeolites* (ed. R.P. Townsend), Special Publication No. 33, Chemical Society, London, p. 133.
6 Flanigen, E., Bennett, J.M., Grose, W., Cohen, J.P., Patton, R.L., Kirschner, R.M., and Smith, J.V. (1978) *Nature*, **271**, 512.
7 Caro, J., Noack, M., Marlow, F., Peterson, D., Griepenstrog, M., and Kornatowksi, J.J. (1993) *J. Phys. Chem.*, **97**, 13685.
8 Weidenthaler, C., Fischer, R.X., Shannon, R.D., and Medenbach, O. (1994) *J. Phys. Chem.*, **98**, 12687.

9 Geus, E.R., Jansen, J.C., and van Bekkum, H. (1994) *Zeolites*, **14**, 82.
10 Olson, D.H., Kokotailo, G.T., Lawton, S.L., and Meier, W.M. (1981) *J. Phys. Chem.*, **85**, 2238.
11 Beschmann, K., Kokotailo, G.T., and Riekert, L. (1987) *Chem. Eng. Process*, **27**, 223.
12 Thamm, H. (1987) *J. Phys. Chem.*, **91**, 8.
13 Thamm, H., Jerschkewitz, H.-G., and Stach, H. (1988) *Zeolites*, **8**, 151.
14 Wu, P., Debebe, A., and Ma, Y.H. (1983) *Zeolites*, **3**, 118.
15 Tzoulaki, D., Heinke, L., Schmidt, W., Wilczok, U., and Kärger, J. (2008) *Angew. Chem. Int. Ed.*, **47**, 3954.
16 Forni, L. and Viscardi, C.F. (1986) *J. Catal.*, **97**, 480.
17 Pope, C.G. (1986) *J. Phys. Chem.*, **90**, 835.
18 June, R.L., Bell, A.T., and Theodorou, D.N. (1990) *J. Phys. Chem.*, **94**, 1508.
19 Ma, Y.H. (1983) *Fundamentals of Adsorption* (eds A. Myers and G. Belfort), Engineering Foundation, New York.
20 Anderson, J.R., Foger, K., Mole, T., Rajadhyaksha, R.A., and Sanders, J.V. (1979) *J. Catal.*, **58**, 114.
21 Derouane, E.G. and Gabelica, Z. (1980) *J. Catal.*, **65**, 486.
22 Hayhurst, D. and Lee, J.C. (1987) Paper 39f AIChE National Meeting, Houston, TX.
23 Jacobs, P.A., Beyer, H.K., and Valyon, J. (1981) *Zeolites*, **1**, 161.
24 Harrison, I.D., Leach, H.F., and Whan, D.A. (1984) in *Proceedings 6th International Zeolite Conference*, (eds D. Olson, A. Bisio, and D.W. Breck), Butterworths, Guildford, UK.
25 Drinkard, B.M., Allen, P.T., and Unger, E.H.(Apr 18 1972) US Patent 3, 656, 278 to Mobil Corp.
26 Song, L. and Rees, L.V.C. (2008) *Adsorption and Diffusion* (eds H.G. Karge and J. Weitkamp), Springer, Berlin, p. 235.
27 Caro, J., Bülow, M., Schirmer, W., Kärger, J., Heink, W., and Pfeifer, H. (1985) *J. Chem. Soc., Faraday Trans. I*, **81**, 2541.
28 Kärger, J., Krause, W., and Pfeifer, H. (1982) *Z. Phys. Chem. (Leipzig)*, **264**, 838.
29 Hayhurst, D.T. and Paravar, A. (1988) *Zeolites*, **8**, 27.
30 Jobic, K., Bee, M., and Dianoux, A.J. (1989) *J. Chem. Soc., Faraday Trans. I*, **85**, 2525.
31 van den Begin, N., Rees, L.V.C., Caro, J., and Bülow, M. (1989) *Zeolites*, **9**, 287.
32 Song, L. and Rees, L.V.C. (1996) *Microporous Mater.*, **6**, 363.
33 Heink, W., Kärger, J., Pfeifer, H., Datema, K.P., and Nowak, A.K. (1992) *J. Chem. Soc., Faraday Trans.*, **88**, 3505.
34 Eic, M. and Ruthven, D.M. (1989) *8th International Conference on Zeolites* (eds P.A. Jacobs and R.A. van Santen), Elsevier, Amsterdam, p. 897.
35 Datema, K.P., den Ouden, C.J.J., Ylstra, W.D., Kuipers, H.P.C.E., Post, M.F.M., and Kärger, J. (1991) *J. Chem. Soc., Faraday Trans.*, **87**, 1935.
36 Millot, B., Methivier, A., Jobic, H., Moueddeb, A., and Dalmon, A. (2000) *Microporous Mesoporous Mater.*, **38**, 85.
37 Paravar, A. and Hayhurst, D.T. (1984) *Proceedings 6th International Conference on Zeolites Reno, Nevada, 1983* (eds D.H. Olson and A. Bisio), Butterworths, Guildford, UK, p. 217.
38 Song, L. and Rees, L.V.C. (1997) *J. Chem. Soc., Faraday Trans.*, **93**, 649.
39 Talu, O., Sun, S., and Shah, D.B. (1998) *AIChE J.*, **44**, 681.
40 Jobic, H. (2000) *J. Mol. Catal. A: Chem.*, **158**, 135.
41 Bourdin, V., Brandani, S., Gunadi, A., Jobic, H., Krause, C., Kärger, J., and Schmidt, W. (2005) *Diffusion Fundamentals* (eds J. Kärger, F. Grinberg, and P. Heitjans), Leipziger Universitätsverlag, Leipzig, p. 430.
42 Bourdin, V., Brandani, S., Gunadi, A., Jobic, H., Krause, C., Kärger, J., and Schmidt, W. (2005) *Diffusion Fundamentals (Online)*, **2**, 83.
43 Jobic, H. and Theodorou, D.N. (2006) *J. Phys. Chem. B*, **110**, 1964.
44 Jobic, H., Schmidt, W., Krause, C., and Kärger, J. (2006) *Microporous Mesoporous Mater.*, **90**, 299.
45 Leroy, F., Rousseau, B., and Fuchs, A.H. (2004) *Phys. Chem. Chem. Phys.*, **6**, 775.
46 Maginn, E.J., Bell, A.T., and Theodorou, D.N. (1996) *J. Phys. Chem.*, **100**, 7155.
47 Ramachandran, C.E., Zhao, Q., Zikanova, A., Kocirik, M., Broadbelt, L.J.,

and Snurr, R.Q. (2006) *Catal. Commun.*, **7**, 936.
48 Jobic, H. and Theodorou, D. (2007) *Microporous Mesoporous Mater.*, **102**, 21.
49 Bülow, M., Schlodder, H., Rees, L.V.C., and Richards, R.E. (eds) (1986) *Proceedings of the 7th International Conference on Zeolites, Tokyo*, Elsevier.
50 Jobic, H. (2008) *Adsorption and Diffusion* (eds H.G. Karge and J. Weitkamp,), Springer, Berlin, p. 207.
51 Heink, W., Kärger, J., Pfeifer, H., Salverda, P., Datema, K.P., and Nowak, A.K. (1992) *J. Chem. Soc., Faraday Trans.*, **88**, 3505.
52 Rees, L.V.C. and Song, L. (2000) *Recent Advances in Gas Separation by Microporous Ceramic Membranes* (ed. N.K. Kanellopoulos), Elsevier, Amsterdam, p. 139.
53 Maginn, E.J., Bell, A.T., and Theodorou, D.N. (1995) *J. Phys. Chem.*, **99**, 2057.
54 Pfeifer, H., Kärger, J., Germanus, A., Schirmer, W., Bülow, M., and Caro, J. (1985) *Ads. Sci. Technol.*, **2**, 229.
55 Jobic, H., Bee, M., and Kearley, G.J. (1989) *Zeolites*, **9**, 312.
56 Jobic, H., Bee, M., Caro, J., Bülow, M., and Kärger, J. (1989) *J. Chem. Soc., Faraday Trans. I*, **85**, 4201.
57 Zhu, W., Malekian, A., Eic, M., Kapteijn, F., and Moulijn, J.A. (2004) *Chem. Eng. Sci.*, **59**, 3827.
58 Millot, B., Methivier, A., Jobic, H., Moueddeb, H., and Bee, M. (1999) *J. Phys. Chem. B*, **103**, 1096.
59 Cavalcante, C.L. and Ruthven, D.M. (1995) *Ind. Eng. Chem. Res.*, **34**, 177.
60 Prinz, D. and Riekert, L. (1986) *Ber. Bunsenges. Phys. Chem.*, **90**, 413.
61 Post, M.F.M., van Amstel, J., and Kouwenhoven, H.W. (1984) in *Proceedings 6th International Conference on Zeolites Reno, Nevada, 1983* (eds A. Bisio and D. Olson) Butterworths, Guildford, UK, p. 517.
62 Chon, H. and Park, D.H. (1988) *J. Catal.*, **114**, 1.
63 Xiao, J. and Wei, J. (1992) *Chem. Eng. Sci.*, **47**, 1143.
64 Magalhaes, F.D., Laurence, R.L., and Conner, W.C. (1998) *J. Phys. Chem. B*, **102**, 7.
65 Song, L.J., Sun, Z.L., and Rees, L.V.C. (2002) *Microporous Mesoporous Mater.*, **55**, 31.
66 Hufton, J.R., Ruthven, D.M., and Danner, R.P. (1995) *Microporous Mater.*, **5**, 39.
67 Banas, K., Brandani, F., Ruthven, D.M., Stallmach, F., and Karger, J. (2005) *Magn. Reson. Imag.*, **23**, 227.
68 Geier, O., Vasenkov, S., Lehmann, E., Kärger, J., Schemmert, U., Rakoczy, R.A., and Weitkamp, J. (2001) *J. Phys. Chem. B*, **105**, 10217.
69 Chmelik, C., Heinke, L., Kärger, J., Schmidt, W., Shah, D.B., van Baten, J.M., and Krishna, R. (2008) *Chem. Phys. Lett.*, **459**, 141.
70 Shah, D.B., Chokchai, S., and Hayhurst, D. (1991) *J. Chem. Soc., Faraday Trans.*, **89**, 3161.
71 Jiang, M., Eic, M., Meachian, S., Dalmon, J.A., and Kocirik, M. (2001) *Sep. Purif. Technol.*, **25**, 287.
72 Krishna, R. and van Baten, J.M. (2005) *J. Phys. Chem. B*, **109**, 6386.
73 Krishna, R., Paschek, D., and Baur, R. (2004) *Microporous Mesoporous Mater.*, **76**, 233.
74 Smit, B. and Maesen, T.L.M. (1995) *Nature*, **374**, 42.
75 Dubbeldam, D., Calero, S., Vlugt, T.J.H., Krishna, R., Maesen, T.L.M., and Smit, B. (2004) *J. Phys. Chem. B*, **108**, 12301.
76 Vlugt, T.J.H., Krishna, R., and Smit, B. (1999) *J. Phys. Chem. B*, **103**, 1102.
77 Jobic, H., Laloue, N., Laroche, C., van Baten, J.M., and Krishna, R. (2006) *J. Phys. Chem. B*, **110**, 2195.
78 Shah, D.B., Hayhurst, D.T., Evanina, G., and Guo, C.J. (1988) *AIChE J*, **34**, 1713.
79 Zikanova, A., Bülow, M., and Schlodder, H. (1987) *Zeolites*, **7**, 115.
80 Förste, C., Kärger, J., Pfeifer, H., Riekert, L., Bülow, M., and Zikanova, A. (1990) *J. Chem. Soc., Faraday Trans. I*, **86**, 881.
81 Brandani, S., Jama, M., and Ruthven, D. (2000) *Microporous Mesoporous Mater.*, **6**, 283.
82 Doelle, H.-J., Heering, J., and Riekert, L. (1981) *J. Catal.*, **71**, 22.
83 Qureshi, W.R. and Wei, J. (1990) *J. Catal.*, **126**, 147.

84 Shen, D. and Rees, L.V.C. (1991) *Zeolites*, **11**, 666.
85 Ruthven, D.M. (2007) *Adsorption*, **13**, 225.
86 Guo, C.J., Talu, O., and Hayhurst, D.T. (1989) 8th International Conference on Zeolites, Amsterdam, July 10–14, 1989, Recent Research Reports. *"Zeolites for the Nineties"* Jansen, J.C., Moscou, L., and Post, M.F.M., eds. No. 143, p.111.
87 Shah, D.B. and Liou, H.Y. (1994) *Zeolites*, **14**, 541.
88 Ruthven, D.M., Eic, M., and Richard, E. (1991) *Zeolites*, **11**, 647.
89 Forste, C., Heink, W., Kärger, J., Pfeifer, H., Feoktistova, N.N., and Zhdanov, S.P. (1989) *Zeolites*, **9**, 299.
90 Zibrowius, B., Caro, J., and Pfeifer, H. (1988) *J. Chem. Soc., Faraday Trans.*, **84**, 2347.
91 Zibrowius, B., Bülow, M., and Pfeifer, H. (1985) *Chem. Phys. Lett.*, **120**, 420.
92 Nowak, A.K., Cheetham, A.K., Picket, S.D., and Roundas, S. (1987) *Mol. Simul.*, **1**, 67.
93 Pickett, S.D., Nowak, A.K., Thomas, J.M., and Cheetham, A.K. (1989) *Zeolites*, **9**, 123.
94 Stach, H., Wendt, P., Fiedler, K., Grauer, B., Janchen, J., and Spindler, H. (1987) *Characterization of Porous Solids: Proceedings of the IUPAC Symposium, Bad Soden, Germany, April 26–29, 1987* (ed. K. Unger), Studies in Surface Science and Catalysis, vol. 39, Elsevier, Amsterdam.
95 Jobic, H., Bee, M., and Pouget, S. (2000) *J. Phys. Chem. B*, **104**, 7130.
96 Song, L. and Rees, L.V.C. (2000) *Microporous Mesoporous Mater.*, **35–36**, 301.
97 Lai, Z., Bouilla, G., Diaz, I., Nery, J.G., Sujaofi, K., Amat, M., Kokkoli, E., Terasaki, O., Thompson, R.W., and Tsapatsis, M. (2000) *Science*, **300**, 456.
98 Nayak, V.S. and Riekert, L. (1985) *Acta Phys. Chem.*, **31**, 157.
99 Wu, P. and Ma, Y.H. (1984) *Proceedings 6th International Conference on Zeolites Reno, Nevada, 1983* (eds D.H. Olson and A. Bisio), Butterworths, Guildford, p. 251.
100 Karsli, H., Culfaz, A., and Yucel, H. (1992) *Zeolites*, **12**, 728.
101 Fyfe, C.A., Kokotailo, G.T., Stobl, H., Gies, H., Kennedy, G.J., Pasztov, C.T., and Barlow, G.E. (1989) *Zeolites as Catalysts, Sorbents and Detergent Builders* (eds H.G. Karge and J. Weitkamp), Elsevier, Amsterdam, p. 827.
102 Kokotailo, G.T., Riekert, L., and Tissler, A. (1989) *Zeolites as Catalysts, Sorbents and Detergent Builders* (eds H.G. Karge and J. Weitkamp), Elsevier, Amsterdam, p. 843.
103 Nagy, J.B., Derouane, E.G., Resing, H.A., and Miller, G.R. (1983) *J. Phys. Chem.*, **87**, 833.
104 Sun, L.M. and Bourdin, V. (1993) *Chem. Eng. Sci.*, **48**, 3783.
105 Shen, D. and Rees, L.V.C. (1993) *J. Chem. Soc., Faraday Trans. I*, **89**, 1063.
106 Qureshi, W. and Wei, J. (1990) *J. Catal.*, **126**, 147.
107 Habgood, H.W. (1958) *Can. J. Chem.*, **36**, 1384.
108 Ma, Y.H. and Lin, Y.S. (1983) *AIChE Symp. Ser.*, **83** (259), 1.
109 Lin, Y.S. and Ma, Y.H. (1988) *Am. Chem. Soc. Symp. Ser.*, **368**, 452.
110 Nayak, V.S. and Moffat, J.B. (1989) *J. Phys. Chem.*, **92**, 7097.
111 Loughlin, K.F. (1971) Sorption and Diffusion in 5A zeolite, PhD Thesis, University of New Brunswick Fredericton, Canada.
112 Sorenson, S.G., Smyth, J.R., Kocirik, M. Zikanova, A.Noble, R.D. and Falconer, J.L. (2008) *I and E.C. Res.* **47**, 9611.
113 Boulicaut, L., Brandani, S., and Ruthven, D.M. (1998) *Microporous Mater.*, **25**, 81.
114 Keipert, O.P. and Baerns, M. (1998) *Chem. Eng. Sci.*, **53**, 3623.
115 Jobic, H., Skoulidas, A.I., and Sholl, D.S. (2004) *J. Phys. Chem. B*, **108**, 10613.
116 Snurr, R.Q. and Kärger, J. (1997) *J. Phys. Chem. B*, **101**, 6469.
117 Caro, J., Hocevar, S., Kärger, J., and Riekert, L. (1986) *Zeolites*, **6**, 213.
118 Pfeifer, H., Freude, D., and Hunger, M. (1985) *Zeolites*, **5**, 274.
119 Freude, D., Hunger, M., Pfeifer, H., and Schwieger, W. (1986) *Chem. Phys. Lett.*, **128**, 62.
120 Caro, J., Bülow, M., Richter-Mendau, J., Kärger, J., Hunger, M., Freude, D., and Rees, L.V.C. (1987) *J. Chem. Soc., Faraday Trans. I*, **83**, 1843.
121 Jobic, H., Ernst, H., Heink, W., Kärger, J., Tuel, A., and Bee, M. (1998) *Microporous Mesoporous Mater.*, **26**, 67.
122 Geier, O., Vasenkov, S., Freude, D., and Kärger, J. (2003) *J. Catal.*, **213**, 321.

123 Bär, N.K., Ernst, H., Jobic, H., and Kärger, J. (1999) *Magn. Reson. Chem.*, **37**, S79–S83.
124 Jobic, H., Kärger, J., and Bee, M. (1999) *Phys. Rev. Lett.*, **82**, 4260.
125 Gueudré, L., Jolimaîte, E., Bats, N., and Dong, W.T.(May 2009) Diffusion in Zeolites: Importance of Surface Resistance, 5th Pacific Basin Conference on Adsorption Science and Technology, May 25–27, 2009, Singapore.
126 Gueudré, L., Jolimaîte, E., Bats, N., and Dong, W. (2010) *Adsorption*, **16**, 17.
127 Wloch, J. (2003) *Microporous Mesoporous Mater.*, **62**, 81.
128 Inzoli, I., Kjelstrup, S., Bedeaux, D., and Simon, J.M. (2009) *Microporous Mesoporous Mater.*, **125**, 112.
129 Simon, J.-M., Decrette, A., Bellat, J.B., and Salazar, J.M. (2004) *Mol. Simul.*, **30**, 621.
130 Combariza, A.F. and Sastre, G. (2011) *J. Phys. Chem. C*, **115**, 13751.
131 Chmelik, C., Varma, A., Heinke, L., Shah, D.H., Kärger, J., Kremer, F., Wilczok, U., and Schmidt, W. (2007) *Chem. Mater.*, **19**, 6012.
132 Tzoulaki, D., Schmidt, W., Wilczok, U., and Kärger, J. (2008) *Microporous Mesoporous Mater.*, **110**, 72.
133 Reitmeier, S.T., Gobin, O.C., Jentys, A., and Lercher, J.A. (2009) *Angew. Chem. Int. Ed.*, **48**, 533.
134 Gobin, O.C., Reitmeier, S.T., Jentys, A., and Lercher, J.A. (2009) *Microporous Mesoporous Mater.*, **125**, 3.
135 Gobin, O.C., Reitmeier, S.J., Jentys, A., and Lercher, J.A. (2011) *J. Phys. Chem. C*, **115**, 1171.
136 Reitmeier, S.J., Gobin, O.C., Jentys, A., and Lercher, J.A. (2009) *J. Phys. Chem. C*, **113**, 15355.
137 Skoulidas, A.I. and Sholl, D.S. (2000) *J. Chem. Phys.*, **113**, 4379.
138 Binder, T., Adem, Z., Krause, B.-C., Krutyeva, M., Huang, A., and Caro, J. (2011) *Microporous Mesoporous Mater.*, **146**, 151.
139 Kärger, J. and Ruthven, D.M. (1989) *Zeolites*, **9**, 267.
140 Kortunov, P., Chmelik, C., Kärger, J., Rakoczy, R.A., Ruthven, D.M., Traa, Y., Vasenkov, S., and Weitkamp, J. (2005) *Adsorption*, **11**, 235.
141 Ruthven, D.M. (1995) in *Zeolites: a Refined Tool for Designing Catalytic Sites* p.223, Bonneviot, L. and Kaliguine, S. eds. *Studies in Surface Science and Catalysis* **97** Elsevier, Amsterdam.
142 Kocirik, M., Kornatowski, J., Masarik, V., Novak, P., Zikanova, A., and Maixner, J. (1998) *Microporous Mesoporous Mater.*, **23**, 295.
143 Schmidt, W., Wilczok, U., Weidenthaler, C., Medenbach, O., Goddard, R., Buth, G., and Cepak, A. (2007) *J. Phys. Chem. B*, **111**, 13538.
144 Karwacki, L., Kox, M.H.F., de Winter, D.A.M., Drury, M.R., Meeldijk, J.D., Stavitski, E., Schmidt, W., Mertens, M., Cubillas, P., John, N., Chan, A., Kahn, N., Bare, S.R., Anderson, M., Kornatowski, J., and Weckhuysen, B.M. (2009). *Nat. Mater.*, **8**, 959.
145 Karwacki, L. (2010) Internal architecture and molecular transport barriers of large zeolite crystals. Thesis, Utrecht.
146 Brabec, L. and Kocirik, M. (2010) *J. Phys. Chem. C*, **114**, 13685.
147 Roeffaers, M.B.J., Ameloot, R., Baruah, M., Uji-i, H., Bulut, M., de Cremer, G., Muller, U., Jacobs, P.A., Hofkens, J., Sels, B.F., and de Vos, D.E. (2008) *J. Am. Chem. Soc.*, **130**, 5763.
148 Roeffaers, M.B.J., Ameloot, R., Bons, A.J., Mortier, W., de Cremer, G., de Kloe, R., Hofkens, J., de Vos, D.E., and Sels, B.F. (2008) *J. Am. Chem. Soc.*, **130**, 13516.
149 Herzig, C. and Mishin, Y. (2005) *Diffusion in Condensed Matter: Methods, Materials, Models* (eds P. Heitjans and J. Kärger), Springer, Berlin, p. 337.
150 Vasenkov, S., Böhlmann, W., Galvosas, P., Geier, O., Liu, H., and Kärger, J. (2001) *J. Phys. Chem. B*, **105**, 5922.
151 Vasenkov, S. and Kärger, J. (2002) *Microporous Mesoporous Mater.*, **55**, 139.
152 Kärger, J. and Pfeifer, H. (1992) *Zeolites*, **12**, 872.
153 Kärger, J. (1991) *J. Phys. Chem.*, **95**, 5558.
154 Fenzke, D. and Kärger, J. (1993) *Z. Phys. D*, **25**, 345.
155 Bär, N.K., Kärger, J., Pfeifer, H., Schäfer, H., and Schmitz, W. (1998) *Microporous Mesoporous Mater.*, **22**, 289.
156 Hong, U., Kärger, J., Kramer, R., Pfeifer, H., Seiffert, G., Müller, U., Unger, K.K., Lück, H.B., and Ito, T. (1991) *Zeolites*, **11**, 816.

157 Hong, U., Kärger, J., Pfeifer, H., Müller, U., and Unger, K.K. (1991) *Z. Phys. Chem.*, **173**, 225.

158 Demontis, P., Fois, E.S., Suffritti, G., and Quartieri, S. (1990) *J. Phys. Chem.*, **94**, 4329.

159 Goodbody, S.J., Watanabe, K., MacGowan, D., Walton, J.P.B., and Quirke, N. (1991) *J. Chem. Soc. Faraday Trans.*, **87**, 1951.

160 June, R.L., Bell, A.T., and Theodorou, D.N. (1990) *J. Phys. Chem.*, **94**, 8232.

161 Demontis, P., Kärger, J., Suffritti, G.B., and Tilocca, A. (2000) *Phys. Chem. Chem. Phys.*, **2**, 1455.

162 Fritzsche, S. and Kärger, J. (2003) *Europhys. Lett.*, **63**, 465.

163 Fritzsche, S. and Kärger, J. (2003) *J. Phys. Chem. B*, **107**, 3515.

164 June, R.L., Bell, A.T., and Theodorou, D.N. (1992) *J. Phys. Chem.*, **96**, 1051.

165 Schüring, A., Fritzsche, S., Haberlandt, R., Vasenkov, S., and Kärger, J. (2004) *Phys. Chem. Chem. Phys.*, **6**, 3676.

166 Gergidis, L.N. and Theodorou, D.N. (1999) *J. Phys. Chem. B*, **103**, 3380.

167 Gergidis, L.N., Theodorou, D.N., and Jobic, H. (2000) *J. Phys. Chem. B*, **104**, 5541.

168 Eckman, R. and Vega, A.J. (1986) *J. Phys. Chem.*, **90**, 4679.

169 Liu, X., Newsome, D., and Coppens, M.-O. (2009) *Microporous Mesoporous Mater.*, **125**, 149.

170 Kärger, J. and Pfeifer, H. (1987) *Zeolites*, **7**, 90.

171 Forste, C., Germanus, A., Kärger, J., Pfeifer, H., Caro, J., Pilz, W., and Zikanova, A. (1987) *J. Chem. Soc., Faraday Trans. I*, **83**, 230.

172 Gupta, A. and Snurr, R.Q. (2005) *J. Chem. Phys. B*, **109**, 1822.

173 Caro, J., Bülow, M., Kärger, J., and Pfeifer, H. (1988) *J. Catal.*, **114**, 186.

174 Pampel, A., Engelke, F., Galvosas, P., Krause, C., Stallmach, F., Michel, D., and Kärger, J. (2006) *Microporous Mesoporous Mater.*, **90**, 271.

175 Fernandez, M., Kärger, J., Freude, D., Pampel, A., van Baten, J.M., and Krishna, R. (2007) *Microporous Mesoporous Mater.*, **105**, 124.

176 Jost, S., Bär, N.K., Fritzsche, S., Haberlandt, R., and Kärger, J. (1998) *J. Phys. Chem. B*, **102**, 6375.

177 Makrodimitris, K., Papadopoulos, G.K., and Theodorou, D.N. (2001) *J. Phys. Chem. B*, **105**, 777.

178 Yan, Y., Davis, M.E., and Gavalas, G.R. (1995) *Ind. Eng. Chem. Res.*, **34**, 1625.

179 Chmelik, C., Heinke, L., van Baten, J.M., and Krishna, R. (2009) *Microporous Mesoporous Mater.*, **125**, 11.

180 Calero, S., Dubbeldam, D., Krishna, R., Smit, B., Vlugt, T.J.H., Denayer, F.M., Martens, J., and Maesen, T.L.M. (2004) *J. Am. Chem. Soc.*, **126**, 11377.

181 Calero, S., Smit, B., and Krishna, R. (2001) *Phys. Chem. Chem. Phys.*, **3**, 4390–4298.

182 Rakoczy, R.A., Traa, Y., Kortunov, P., Vasenkov, S., Kärger, J., and Weitkamp, J. (2007) *Microporous Mesoporous Mater.*, **104**, 1195.

183 Heinke, L. and Kärger, J. (2009) *J. Chem. Phys.*, **130**, 44707.

184 Kärger, J., Kortunov, P., Vasenkov, S., Heinke, L., Shah, D.B., Rakoczy, R.A., Traa, Y., and Weitkamp, J. (2006) *Angew. Chem. Int. Ed.*, **45**, 7846.

185 Kortunov, P., Heinke, L., Vasenkov, S., Chmelik, C., Shah, D.B., Kärger, J., Rakoczy, R.A., Traa, Y., and Weitkamp, J. (2006) *J. Phys. Chem. B*, **110**, 23821.

186 Marthala, V.R.R., Hunger, M., Chmelik, C., Kettner, F., Krautscheid, H., Kärger, J., and Weitkamp, J. (2011) *Chem. Mater.*, **23**, 2521.

19
Metal Organic Frameworks (MOFs)

Metal organic frameworks (MOFs) comprise a new and important class of microporous materials that, like the zeolites, have regular and uniform intracrystalline pore systems and which therefore show many of the same characteristic features. The wide range of different "nodes" and "spacers" from which MOFs can be constructed leads to a correspondingly wide range of different pore structures and surface chemistry. Although MOF structures are usually less stable than zeolites, there are also examples with high stability. Since, moreover, they can often be grown to a larger crystal size these materials offer great potential for fundamental studies of intracrystalline transport as well as possibilities for commercial application.

The present chapter aims to provide some insight into this rapidly emerging field. Following a general introduction into this new class of nanoporous host systems Section 19.2 deals with diffusion studies carried out with two "classic" representatives in the field [MOF-5 and CuBTC (also referred to as HKUST-1)], both of which were among the earliest MOFs to be synthesized and extensively characterized. Like MOF-5 and CuBTC, the MOFs of type ZIF-8 (considered in Section 19.3; ZIF = zeolitic imidazolate framework) are also of cubic symmetry. The pore structure of ZIF-8 is similar to that of the Type A zeolites in that the pore space consists of relatively large cavities that are interconnected through smaller "windows." From well established principles such structures are expected to show a straightforward correlation between the self- and transport diffusivity and this is confirmed by the experimental data.

As potential single-file systems (Chapter 5), among the myriad of metal organic framework structures, those with one-dimensional pores are of particular interest for establishing structure–mobility correlations. Considering Zn(tbip) as a typical representative of this class of MOFs, Section 19.4 presents a detailed review of the available data for both intracrystalline diffusivities and surface permeabilities. Section 19.5 focuses on diffusion phenomena in those MOF structures, in which the channel dimensions (and hence the diffusion properties) vary dramatically depending on the nature and concentration of the guest molecules. This overview concludes in Section 19.6 with an in-depth analysis of the published diffusion and permeation data for MOF Zn(tbip) in which the relevance of these results to our

Diffusion in Nanoporous Materials. Jörg Kärger, Douglas M. Ruthven, and Doros N. Theodorou.
© 2012 Wiley-VCH Verlag GmbH & Co. KGaA. Published 2012 by Wiley-VCH Verlag GmbH & Co. KGaA.

general understanding of mass transfer phenomena in crystalline nanoporous materials is considered.

19.1
A New Class of Porous Solids

Metal organic frameworks (MOFs) represent the youngest class of nanoporous materials [1, 2]. They have emerged over the last two decades and result from the reaction between organic and inorganic species, leading to the connection of inorganic sub-units (metal ions, clusters, chains, 3D-arranged layers) by organic linkers via coordinating groups (carboxylates, phosphonates, nitrogen-containing ligands) with strong dative bonds. The structure of MOF-5, one of the most celebrated representatives of this novel class of host systems [1] (see also Section 19.2.1), shown in Figure 19.1, illustrates the principle of MOF construction in more detail. In particular, the tetrahedral Zn_4O clusters as the (inorganic) sub-units and the terephthalate (organic) linkers that are connected by carboxylated groups may be recognized.

With reference to the construction principle, the MOFs have also been referred to as porous coordination polymers (PCPs) [5, 6]. For specifying certain properties, further acronyms (such as MMOFs for microporous MOFs and GMMMOFs for

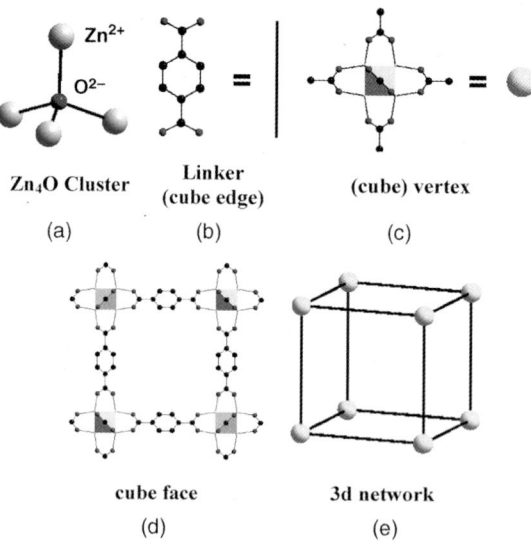

Figure 19.1 MOF-5 structure (also referred to as IRMOF-1) as an example of the construction principles of metal organic frameworks. The six edges of a tetrahedral Zn_4O cluster (a) are bridged via the carboxylate groups of the terephthalate linkers (b). The vertices thus formed (c) are connected by the organic linkers, yielding a 3D cubic network (e); (d) displays a cube face within this structure. Reprinted from Reference [3] with permission.

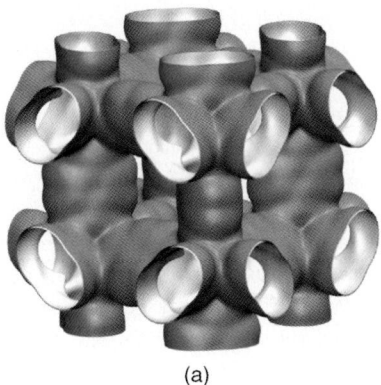
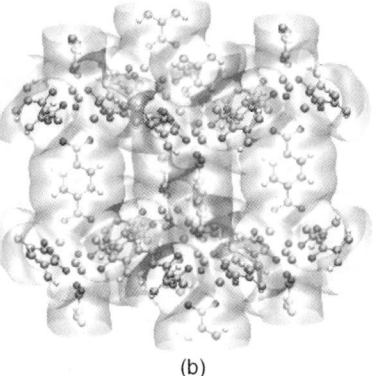

(a)   (b)

Figure 19.2 Representation of the space accessible by the guest molecules in MOF-5 (or IRMOF-1) by a suitably chosen plane of constant potential energy of the guest molecules (a). The guest molecules "perceive" the dark side of the plane while the bright side is directed towards the framework. In the representation with the framework inserted (b) both the linkers (terephthalate, Figure 19.1b) and the vertices (Zn_4O clusters with attached carboxylate groups, Figure 19.1c) may be recognized. Reprinted from Reference [4] with permission.

guest-free microporous MOFs) have also been introduced [1, 7]. The three-dimensional hybrid networks (as exemplified by Figures 19.1 and 19.2) result from the combination of inorganic sub-units with organic linkers in which almost all elements of the periodic table may be involved. The inorganic (metallic) elements include alkaline earths (Ca, Mg), transition metals (Sc, Ti,...,Zn), p metals (Al, Ga, In), and lanthanides while the organic linkers include both aliphatics and aromatics, sometimes substituted by heteroatoms (N, O, S,...) and associated with one or several coordinating functional groups, which may be either anionic or neutral. Examples include carboxylate as in MOF-5 (Figure 19.1), phosphonates, sulfonates, imidazolates, pyridines, and many more [8]. This variety exceeds by far all previous options for generating novel microporous structures. Thus, during the past decade, hundreds of different MOF structures have already been synthesized – many more than in half a century of zeolite research.

Taking into account the essentially unlimited range of inorganic units, organic linkers, and their interconnection, one may conclude that the number of different structures is infinitely large [9]. This includes, in particular, the application of linkers of increasing size so that the structure type remains invariant with continuously increasing pore diameter. The members of such families are said to follow the principles of isoreticular chemistry, with the IRMOFs (isoreticular MOFs [10]) and the MIL-88 solids [11] as particularly important representatives. The first member of the IRMOF family is discussed in Section 19.2. As a consequence of their amazing structural variability, MOFs occur in the structures of numerous minerals, including various zeotype architectures (SOD, ANA, RHO, BCT, MTN, and so on) [8, 12–14]. A representative of this class of MOFs, namely, number 8 in the series of the "imidazolate frameworks" (ZIF-8), a MOF counterpart of sodalite (SOD), is discussed in Section 19.3. Notably, the channel dimensions of some of these MOFs are in the mesopore range. In contrast to traditional mesoporous materials, such as mesoporous silica, these structures are crystalline.

Because of the wide range of different constituents and topologies (resulting in different chemistry of their internal surfaces) the MOFs offer promising prospects for technological exploitation. These options include the fabrication of crack-free films and mesoporous membranes that may be superior to inorganic membranes owing to their higher flexibility. Section 19.3 presents one such example in greater detail. Owing to their very large internal surface, MOFs are also known to provide high capacities for adsorptive storage and separation [15], which, as a consequence of the low density, appear even more impressive if referred to the sorbent mass, rather than to the sorbent volume. The potential use of MOFs as catalysts may also benefit from the option of including metals as active sites incorporated directly into the host framework. Further applications may be based on their optical and magnetic properties [8] and may also benefit from the lattice flexibility. The option of external control of the lattice structure (e.g., by irradiation at suitably chosen frequencies [16]) makes triggered drug release an obvious possibility. Diffusion phenomena under the influence of a flexible network are considered in Section 19.5.

In a chapter dealing with diffusion in MOFs it is worth citing the authors of a review on "Progress, opportunities, and challenges for applying atomically detailed modeling of molecular adsorption and transport in metal-organic framework materials" [17]. In their overview of an impressively large number of computational studies on molecular diffusion in MOFs (prior to 2009) the authors state that "one of the challenges in understanding diffusion of adsorbed molecules in MOFs is the lack of experimental data on diffusion in MOFs. Only one experimental study to date has assessed molecular diffusion in a MOF." This study is presented at the beginning of the next section. During the recent past many more such studies have been undertaken, some of them clearly groundbreaking. From the perspective of diffusion research the discovery of MOFs has already proved to be an important stimulus.

19.2
MOF-5 and HKUST-1: Diffusion in Pore Spaces with the Architecture of Zeolite LTA

Forming a simple cubic lattice of large cages, the pore space of both MOF-5 (also referred to as IRMOF-1 as the first member of the isoreticular MOF family [10]) and HKUST-1(also known as CuBTC) may be compared with those of the LTA-type zeolites (Chapter 16). In the previous section, MOF-5 has been discussed as a general introduction to the building principles of MOFs. Figure 19.2a shows a view of one large cage, circumscribed by 12 edges and eight vertices, leaving space for six "windows," connecting the cavity with the six adjacent cages. The cavities alternate in size, a cage 14.3 Å in diameter with the linker faces pointed towards the cage center and one 10.9 Å in diameter with the linker edges pointed to the cage center.

19.2 MOF-5 and HKUST-1: Diffusion in Pore Spaces with the Architecture of Zeolite LTA

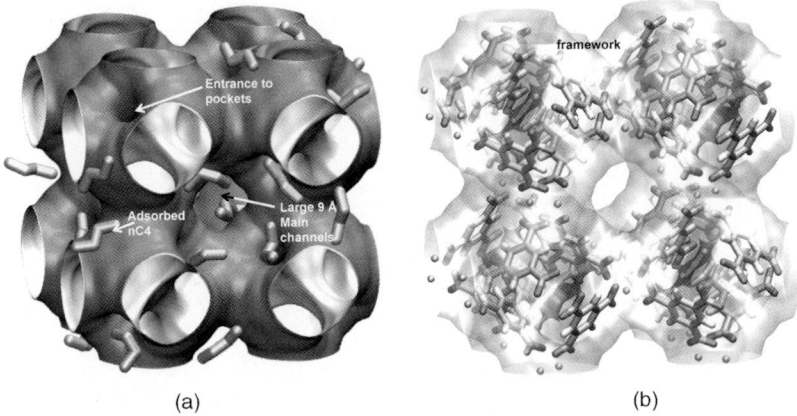

Figure 19.3 View into the pore space: (a) guest molecules "perceive" the dark side of the plane; (b) framework structure of CuBTC (or HKUST-1), from Reference [18] with permission.

Figure 19.3 presents the structure of HKUST-1 (also referred to as CuBTC: $Cu_3(BTC)_2$ with BTC = benzene-1,3,5-tricarboxylate [18]), analogous to the representation of MOF-5 by Figure 19.2 [19]. Again the eight corner elements ("vertices") surrounding the large cavity are clearly visible. The molecular clusters forming these vertices are notably larger than those in MOF-5 and contain, moreover, small pockets that can accommodate smaller molecules such as n-butane.

19.2.1
Guest Diffusion in MOF-5

MOF-5 (also named IRMOF-1, see Figure 19.2) was the first MOF material to be subjected to detailed diffusion experiments [20]. The material used in these studies was obtained from an optimized large-scale preparation that yielded crystals with sizes from 50 to 200 μm on a kilogram scale. Figure 19.4 shows the resulting material. Both X-ray diffraction and solid-state ^{1}H MAS NMR [20] show the characteristics expected for MOF-5 [10]. The compaction of the individual crystals to agglomerates would complicate the determination of diffusivities and surface permeabilities by transient sorption experiments which must be based on well-defined geometries (Chapter 6). However, this does not affect the PFG NMR diffusion measurements (Section 11.2) provided that the molecular displacements covered in the experiments are sufficiently small in comparison with the sizes of the crystallites within the agglomerates.

Figure 19.5 provides an overview of the self-diffusivities of a series of n-alkanes plus benzene in MOF-5, as measured by PFG NMR, with the material shown in Figure 19.4. They are compared with the results of MD simulations in a single rigid unit cell using the NVT ensemble (Section 8.2.7) [21]. In these simulations the authors took advantage of the result from References [21, 22] that framework

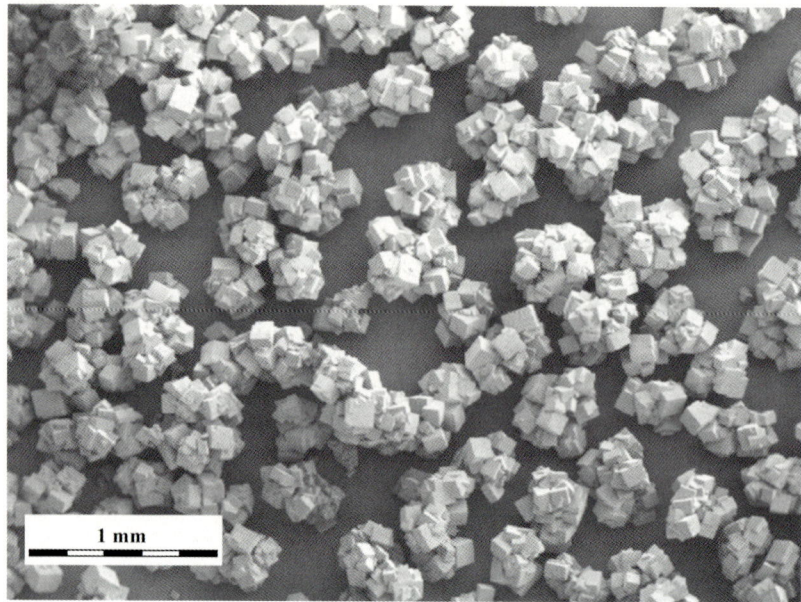

Figure 19.4 MOF-5 (IRMOF-1) material specially prepared by large-scale synthesis for PFG NMR diffusion studies, from Reference [20], with permission.

flexibility has only a minor effect on diffusion of n-alkanes and benzene in MOF-5, but it slows down the simulations dramatically.

The PFG NMR spin echo attenuations were analyzed using a bi-exponential fitting routine with a constant (non-diffusing) background [20]. The highest diffusivity, resulting from about 60% to 65% of the total NMR signal intensity, is reported as the intracrystalline self-diffusivity. The smaller diffusivities and the non-diffusing background, also appearing in the spin-echo attenuations, are assumed to be caused by residual diethylformamide (DEF) solvent molecules in parts of the MOF crystals. Such molecules will restrict the diffusion path of the n-alkanes and contribute directly to the NMR signal as a non-diffusing background.

In all cases the experimental data are found to agree well with the simulation results. For methane, the deviation of the experimental diffusivities to larger values can be attributed to the contribution of intercrystalline mass transfer. Such intrusive effects occur more easily with methane than with the longer alkanes, since here both the intracrystalline diffusivities (and, hence, the diffusion path lengths) and the intercrystalline concentrations attain their largest values.

For the chosen pore fillings, the n-alkane diffusivities are found to be close to those for the bulk liquid, slightly exceeding those in zeolite NaX (see also Section 17.2.1). This result corresponds with the larger pore size in MOF-5 and is nicely paralleled by the concentration dependences from MD simulations. In contrast to NaX, in which the self-diffusivities are found to decrease essentially monotonically with increasing concentration over the total concentration range (see, for example, Figure 17.3),

19.2 MOF-5 and HKUST-1: Diffusion in Pore Spaces with the Architecture of Zeolite LTA | 735

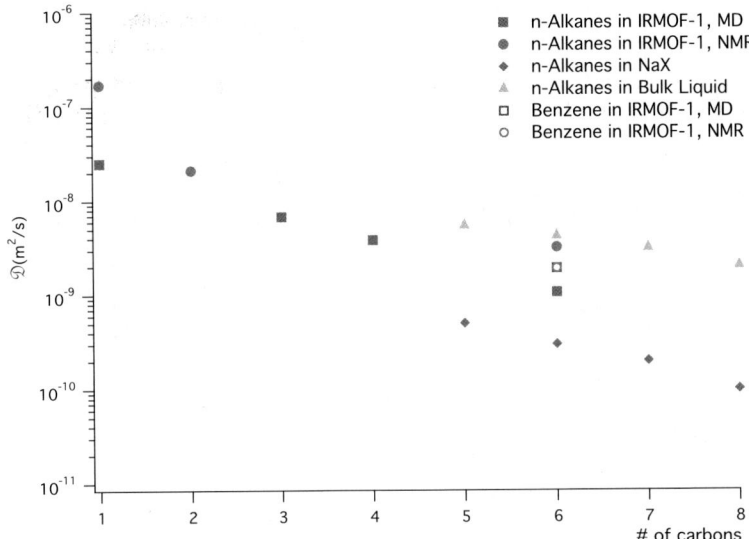

Figure 19.5 Self-diffusion coefficients of n-alkanes in MOF-5 (or IRMOF-1) measured with PFG NMR [20] and calculated from MD simulations [21] at 298 K for pore fillings of about 20–25%. MD results for benzene [21] are also shown for comparison. Self-diffusion coefficients for the bulk liquid [23, 24] and in the zeolite NaX (100–120 mg g^{-1}) [25] are also presented.

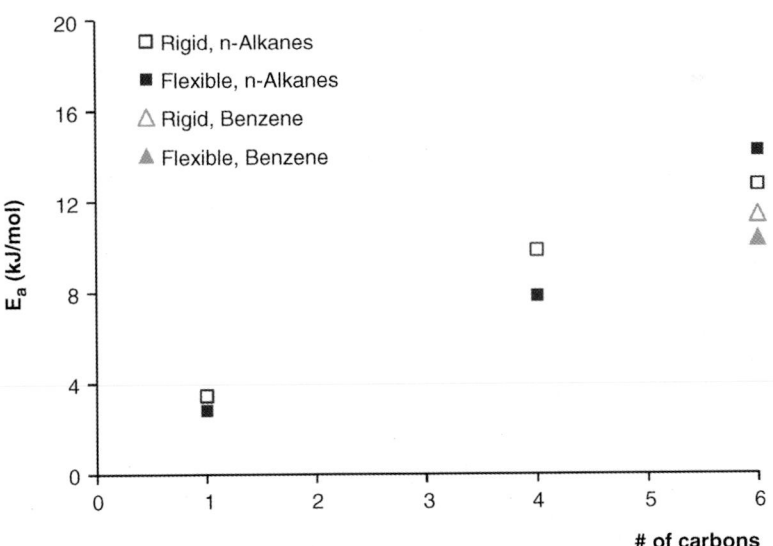

Figure 19.6 Activation energies of self-diffusion of methane, hexane, and benzene in MOF-5 (also referred to as IRMOF-1) determined by MD simulations with rigid and flexible frameworks. Reprinted from Reference [21] with permission.

following the type I pattern of concentration dependence (Figure 3.3), n-alkane diffusivities in MOF-5 are found to remain essentially constant up to medium loadings, with a steep decline at higher loadings (following the type II concentration dependence). In contrast to zeolite NaX but as known from the bulk liquid and as also reported for LaX (Figure 17.17), the benzene diffusivities in MOF-5 follow those of the alkanes of comparable size, with comparable magnitudes, concentration patterns, and activation energies as exemplified by the data shown in Figure 19.6 [21]. This difference arises because MOF-5 is free of the sodium cations that, in NaX, serve as additional adsorption sites that interact with the π bonds of the unsaturated hydrocarbons. Detailed MD simulations to explore the diffusion mechanism of benzene in MOF-5 may also be found in Reference [26]. These simulations confirm that, up to medium loadings, the diffusivity is essentially independent of loading.

In Reference [27] MD simulations were performed to determine the selectivity of MOF-5 membranes for separation of gas mixtures. Considering a model membrane of 10 μm thickness, equimolar feed, and a transmembrane pressure drop of 80%, the resulting selectivities are compared with hypothetical values, based on the corresponding ratios of the single-component permeances (Section 20.2). The dramatic deviations, observed in particular for systems including CO_2 and at high pressures, suggest the importance of the off-diagonal elements of the diffusion matrix [Section 3.1.4, Eq. (3.22)].

19.2.2
Guest Diffusion in CuBTC

As a prominent difference from MOF-5, the large cavities of CuBTC are connected with eight side pockets that are able to accommodate smaller molecules. This special feature is reflected in both their adsorption and diffusion properties (Figure 19.7) [19]. The inverse thermodynamic factors show that, when the loading corresponds to the number of side pockets, there is a pronounced inflection for the C_5 hydrocarbons, which also appears in the corrected diffusivities, but which does not appear for the smaller C_4 hydrocarbons. Moreover, there is remarkable qualitative although not quantitative agreement (since the two ordinate scales are not identical) between the corrected diffusivity and the inverse thermodynamic factor for the C_5 hydrocarbons over the whole concentration range and, for the C_4 hydrocarbons, starting for loadings above this inflection point. For a single-site Langmuir isotherm [Eq. (1.32)] the inverse thermodynamic factor $(1/\Gamma) = (1 - \Theta)$ coincides with the probability that a jump attempt between adjacent sites is directed to a vacancy and will, therefore, be successful. Considering the methane diffusivities in FAU-type zeolites [28, 29], $1/\Gamma$ was successfully used as a measure of the probability that a jump is directed to a vacancy. Adopting this interpretation in the given context points to an interesting distinction between the diffusion mechanisms of the C_4 and C_5 hydrocarbons. While the C_5 hydrocarbons follow the free-flight mechanism over the total concentration range, the C_4 hydrocarbons deviate from it for small loadings, suggesting, at these small loadings, notable differences in their microdynamics.

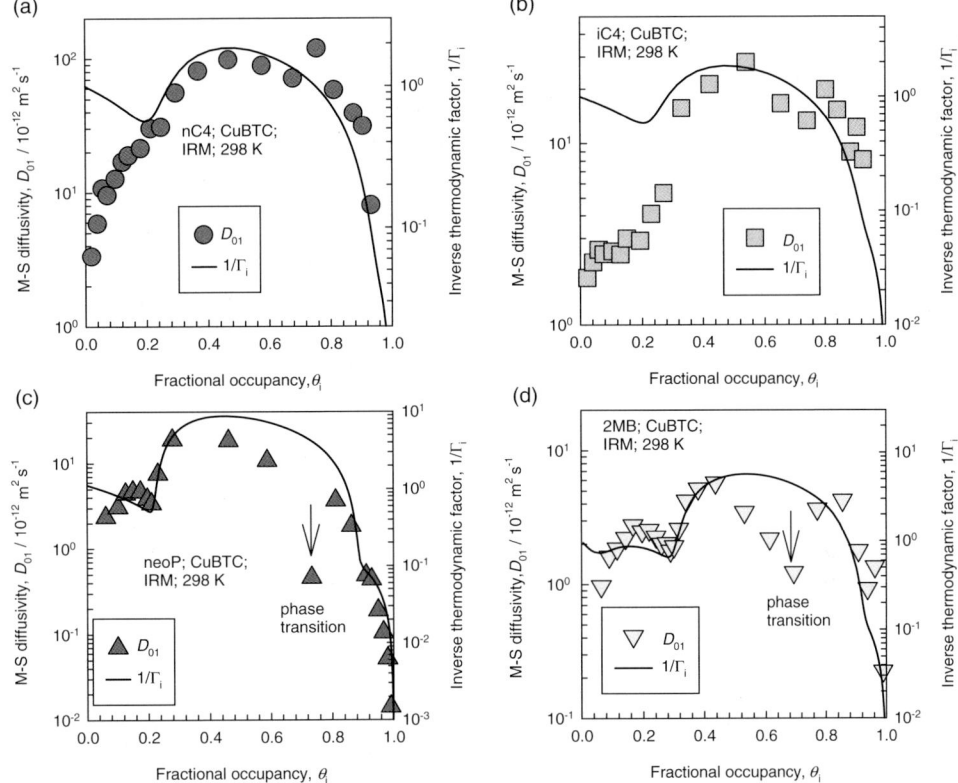

Figure 19.7 Corrected (or Maxwell–Stefan) diffusivities and inverse thermodynamic factor $1/\Gamma$ in CuBTC of (a) n-butane, (b) isobutane, (c) 2,2-dimethylpropane (neopentane), and (d) 2-methylbutane. The corrected diffusivities were calculated via Eq. (1.31) from the transport diffusivities determined by IRM (infrared microscopy) measurement in the integral mode (Section 12.2.1) under the assumption of negligible surface resistances. The thermodynamic factor is calculated according to Eq. (1.31) from the best fit of a "multi-site Sips" [30] model isotherm to experimental sorption data that were also determined by IRM, from Reference [19]. The fractional occupancies Θ refer to a loading of about 80 molecules per unit cell at saturation ($\Theta = 1$). Reprinted from Reference [19], with permission.

Though closely mimicking their concentration dependence, the MD simulations yield corrected diffusivities that exceed the experimental values by about two orders of magnitude [19]. This difference may be attributed to the extreme sensitivity of the simulated diffusivities to the choice of the Lennard-Jones σ parameters [31]. It is remarkable that, despite the numerical discrepancy, key qualitative features of adsorption and diffusion are correctly predicted from the simulations.

Figure 19.8 compares the data of Reference [19] with the self-diffusivities of n-butane in CuBTC resulting from PFG NMR measurements and MD simulations as described in Reference [32]. While the simulated self-diffusivities from the two studies are in perfect agreement, there are remarkable differences between the simulated and measured

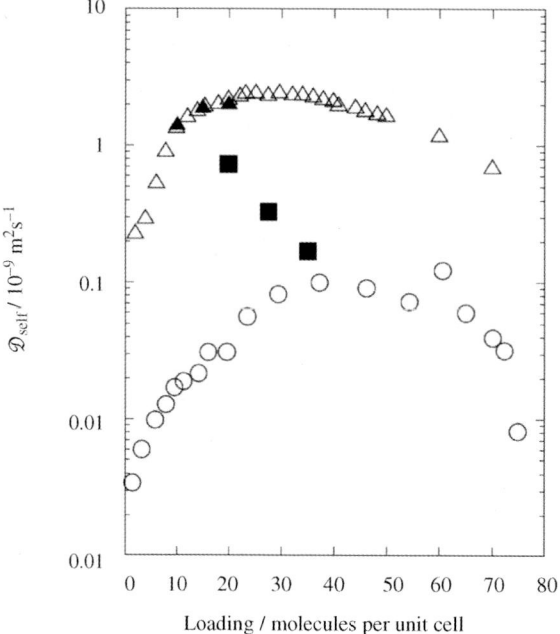

Figure 19.8 Diffusivity data of n-butane in CuBTC at room temperature: comparison of the self-diffusivities determined by PFG NMR (■ [32]) and MD simulations (▲ [32]; △ supplementary information of Reference [19]) with the corrected diffusivities measured by infrared microscopy (○ [19]). Reprinted from Reference [32] with permission.

self-diffusivities, as well as between the experimental data for the self- and corrected diffusivities, which [from Eq. (3.51)] should be expected to converge rather than to diverge in the limit of small concentrations. Similarly, as repeatedly observed with zeolite samples of different origin (see, for example, Sections 16.3 and 16.4 and 17.2.6) one clearly has to be aware that the real MOF structure may differ from the idealized structure used in the simulations and these deviations may vary from sample to sample, depending on their synthesis, storage, and pretreatment. In the given case, the sorption capacities of the CuBTC specimens of Reference [32] were found to be significantly smaller than those for the specimens considered in Reference [19], suggesting a blockage of a part of the pore space (possibly as a consequence of incomplete detemplation). Taking this into account, the PFG NMR self-diffusivity measuring points (■) in Figure 19.8 should be shifted to higher loadings, making the trends in the concentration dependence of the self- and corrected diffusivities compatible.

Simulations of multicomponent adsorption and diffusion in CuBTC membranes [33] show very clearly the potential of structure variation for optimization of the separation performance of MOF membranes. Thus, the selectivity of CuBTC membranes for CO_2/H_2 separation is predicted to exceed that of MOF-5 membranes by a factor of 20, where most of this increase in performance may be shown to be associated with the higher adsorption selectivity of CuBTC.

19.3
Zeolitic Imidazolate Framework 8 (ZIF-8)

Zeolitic imidazolate frameworks (ZIFs) form a subclass of MOFs that closely follow the topology of zeolites. Figure 19.9 illustrates the origin of this similarity. ZIFs (Figure 19.9a) are constructed by a network of transition metal ions (M) bridged by imidazolate (IM) units, which, respectively, assume the position and function of the tetrahedral silicon (or aluminum) atoms and of the oxygen bridges in zeolites (Figure 19.9b). Since the bridging angles in the metal imidazolates and in the zeolitic Si−O−Si bonds (Figure 19.9) assume similar values ($\approx 145°$) the ZIFs are able to follow the different framework patterns corresponding to the different zeolite structure types. In the ZIFs this variability is additionally enhanced by the variability in the metals and the possibility of including imidazole derivatives [34]. Examples include the zeolite structure types BCT, SOD, and RHO [35] with, among others, ZIF-1 [= $Zn(IM)_2$], ZIF-8 [= $Zn(mIM)_2$], and ZIF-71 [= $Zn(dcIM)_2$] as their ZIF equivalents (mIM = 2-methylimidazolate; dcIM = 5-chlorobenzimidazolate) [34].

In this section, we focus on ZIF-8, a ZIF counterpart of the zeolite sodalite. In comparison with other MOFs, it is distinguished by an extremely high thermal and chemical stability. It withstands temperatures up to 550 °C and survives even long-lasting exposure to boiling in water as well as alkaline conditions and various organic solvents [34], thus making this material particularly attractive for practical applications.

ZIF-8 has cubic symmetry. However, in contrast to the simple cubic arrangement of MOF-5 and CuBTC (Section 19.2) where each cage is connected with six adjacent cages through windows in the six faces of the cubic unit cell (Figures 19.2 and 19.3), the large cages of ZIF-8 are connected with eight adjacent cages via the corners of the circumscribing cubes. This arrangement is illustrated by Figure 19.10, where each of the four aligned cages is easily understood to be connected with the four cages seen in the plane "behind" and with, correspondingly, four more cages in a plane "on top" (not shown in this figure). Diffusion measurements with this host material benefit substantially from the robustness of the framework, which, obviously, prohibits substantial deviations of the actual pore structure from the idealized pattern as shown in Figure 19.10. Moreover, the small window diameters turned out to facilitate the interpretation and correlation of the measured diffusivities.

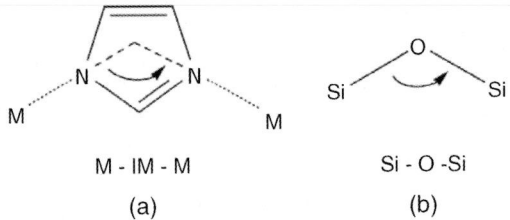

Figure 19.9 Analogies in the building blocks of zeolitic imidazolate frameworks [ZIFs (a)] and zeolites (b) and in the bridging angles.

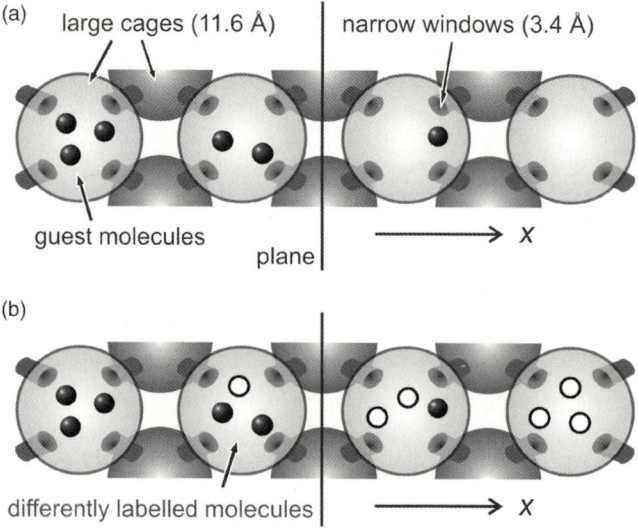

Figure 19.10 Cage arrangement in MOF ZIF-8 and illustration of the situations for diffusion measurement. In either case one determines molecular fluxes, that is, the net number of molecules passing a given plane through the crystal per time. In transport diffusion (a) the flux originates from an overall concentration gradient, while during self-diffusion (b) one considers opposing fluxes of differently labeled but otherwise identical molecules. These fluxes are identical since one considers the overall equilibrium so that the gradients of the differently labeled molecules are identical in magnitude. Reprinted from Reference [36] with permission.

19.3.1
An Ideal Case where Experimental Self- and Transport Diffusivities are interrelated from First Principles

Being able to determine diffusivities under both equilibrium and non-equilibrium conditions within the same device (by transient sorption experiments and by tracer exchange) IRM (Section 12.2.1) is ideally suited for an unbiased comparison of self- and transport diffusivities. The discovery of MOF ZIF-8 [34, 37] has provided us with a highly stable nanoporous material with essentially defect-free crystals of the necessary size, so that the evolution of observed intracrystalline guest profiles could be directly related to intracrystalline transport [36].

Figure 19.11 provides a survey of the coefficients of self-diffusion and transport diffusion in ZIF-8 resulting from such experiments [36]. We first emphasize the important finding that, for all guest species under study, self- and transport diffusion are found to converge for sufficiently small concentrations. This behavior is to be expected quite generally, as explained in Sections 3.3.3 and 8.1.3 [Eq.(8.22)].

Remarkably, with increasing loading the self-diffusivities of some guest molecules (namely, for methanol up to high loadings, see Figure 19.11a, and for ethanol up to

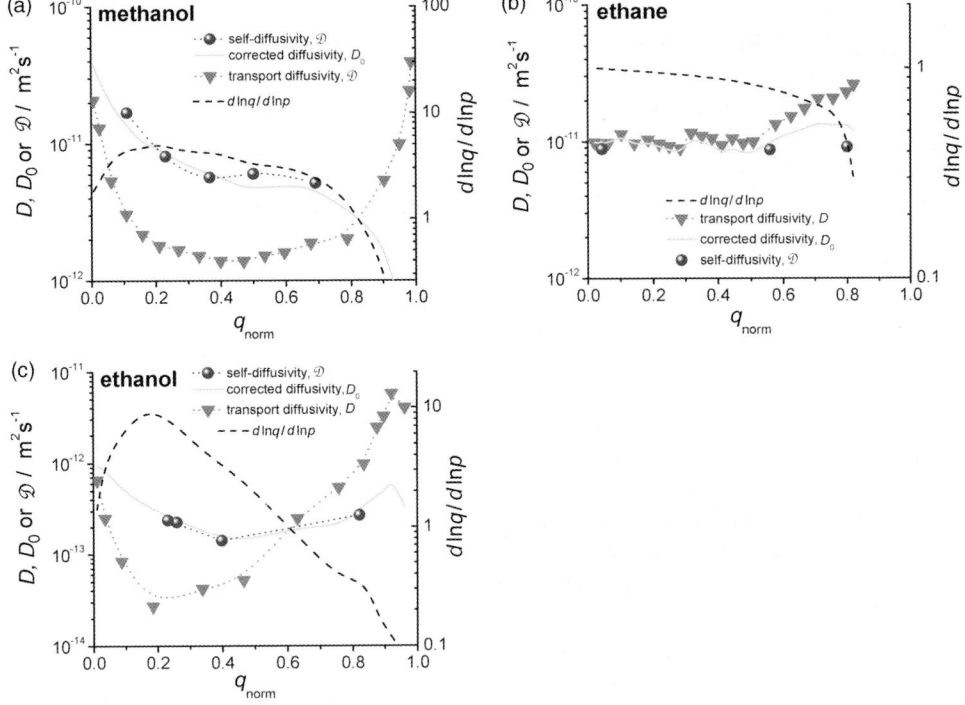

Figure 19.11 Methanol, ethane, and ethanol in ZIF-8 at 298 K. Loading dependence of transport diffusivity (D), corrected diffusivity (D_0), and self-diffusivity ($\mathcal{D}$) of (a) methanol, (b) ethane, and (c) ethanol. The inverse thermodynamic factor (obtained from Figure 19.12a) is also shown. Reprinted from Reference [36] with permission.

medium loadings, see Figure 19.11c) are found to actually exceed the transport diffusivities. In the representation shown in Figure 19.10, the overall gradient for the measurement of transport diffusion (Figure 19.10a) is chosen to coincide with the gradient of the labeled molecules (i.e., filled spheres) in the self- (or tracer) diffusion measurements (Figure 19.10b). Enhancement of self-diffusion in comparison with transport diffusion thus means an enhancement of the flux of the molecules in Figure 19.10a due to the presence of the additional (differently labeled) molecules in Figure 19.10b. Molecular propagation is evidently promoted rather than hindered under the influence of a counter flux of differently labeled but otherwise identical molecules. At first sight such behavior seems surprising, although it is in fact fully understandable, as shown below.

We refer to the considerations of Sections 1.2.3 and 3.3.3, in particular to the relation [Eq. (3.51)]:

$$1/\mathcal{D} = 1/D_0 + \theta/\DJ_{AA} \tag{19.1}$$

between the self-diffusivity $\mathcal{D}$, the "corrected" diffusivity (D_0) and the mutual Maxwell–Stefan diffusivity ($\DJ_{AA}$). The corrected diffusivity is related to the transport diffusivity D [Eq. (1.31)] via:

$$D_0 = D \frac{\partial \ln q}{\partial \ln p} \tag{19.2}$$

with $q(p)$ denoting the adsorption isotherm, that is, guest concentration q in equilibrium with the surrounding pressure p. The logarithmic derivative $\partial \ln p / \partial \ln q$ [the reciprocal value of which appears on the right-hand side of Eq. (19.2)] is known as the thermodynamic factor. Actually, given that the mutual Maxwell–Stefan diffusivity $Ð_{AA}$ is a free parameter, Eq. (19.1) is not much more than a confirmation that there is no universal correlation between the self- and transport diffusivities. It provides, however, a useful distinction between the two different mechanisms, which, intuitively, may be understood to control the flux during counter-diffusion of differently labeled molecules.

The first term on the right-hand side of Eq. (19.1) is related to the drag to which the molecules are subject under both the conditions of transport diffusion and tracer exchange, that is, to the "friction" with the host lattice. The second term, that is, the reciprocal value of the mutual Maxwell–Stefan diffusivity multiplied with the relative pore filling θ, characterizes the drag by counter-diffusion, that is, between the fluxes of the "filled" and "open" molecules in Figure 19.10b. As already indicated in the scheme of the pore space of ZIF-8 in Figure 19.10, molecular propagation in ZIF-8 is controlled by the passage through "windows" between adjacent cavities rather than by the "friction" between adjacent molecules. Therefore, the second term in Eq. (19.1) may be neglected. The self-diffusivity is thus expected to approach the corrected diffusivity, that is, [Eq. (19.2)] the product of the transport diffusivity and the inverse $\partial \ln q / \partial \ln p$ of the thermodynamic factor:

$$\mathcal{D} = D_0 = D \frac{\partial \ln q}{\partial \ln p} \tag{19.3}$$

By similar reasoning, Eq. (19.3) also follows from irreversible thermodynamics (by assuming negligibly small phenomenological cross coefficients – see Sections 3.1 and 3.2) and transition state theory (if it is assumed that molecules in the "transition state" are isolated from the remaining molecules – see Section 4.3.3).

Figure 19.11 also includes the derivatives $\partial \ln q / \partial \ln p$ determined from the adsorption isotherms shown in Figure 19.12a, as well as the resulting corrected diffusivities. Most impressively, for all molecules considered, the self-diffusivities agree well with the corrected diffusivities over the entire ranges of concentration.

Finally, we may also explain the surprising finding that, for certain guests and loadings, the self-diffusivities may exceed the transport diffusivities: with Eq. (19.3) the self-diffusivity is seen to exceed the transport diffusivities under the condition:

$$\partial \ln q / \partial \ln p \equiv \frac{\partial q / \partial p}{q/p} > 1$$

that is, for those parts of the adsorption isotherm where the slope $(\partial q / \partial p)_T$ exceeds the slope of the straight line connecting the relevant point of the adsorption isotherm with the origin (q/p). Figure 19.12b shows that this situation exists in the initial part of

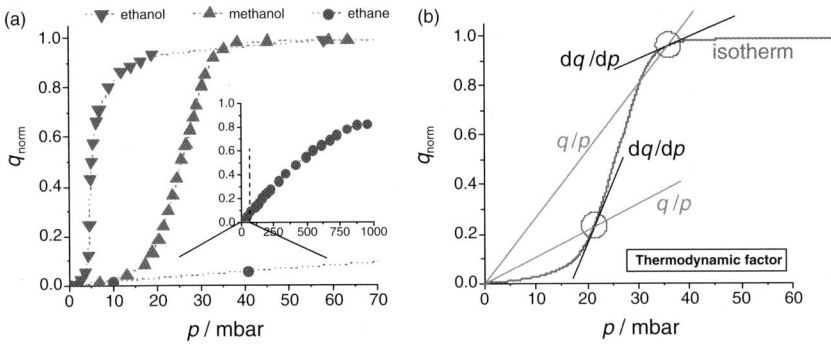

Figure 19.12 (a) Adsorption isotherms: the isotherms of methanol and ethanol exhibit a pronounced S-shape (type III or V isotherm), whereas the isotherm of ethane shows the usual ("type I") behavior. For all molecules the loading at saturated vapor pressure was normalized to reach $q_{norm} = 1$. (b) Visualization of the thermodynamic factor appearing in Eq. (19.3). For S-shaped isotherms in certain regions the slope $(\partial q/\partial p)$ exceeds the slope of the straight line connecting the relevant point of the adsorption isotherm with the origin (q/p) [36, 38].

an S-shaped isotherm (the so-called type-III or type-V isotherms), which is typical for the adsorption of polar molecules on hydrophobic surfaces [39, 40]. For the highly polar methanol molecules we may indeed expect exactly this behavior.

With increasing loading, molecular interaction must be taken into account and two opposing effects become relevant: (i) "competition" for the limited space and (ii) a tendency towards molecular aggregation, driven by the (dipolar) interaction energy of the guest molecules. Effect (i) is known to give rise to the typical Langmuir ("type-I") isotherm, that is, to $(\partial q/\partial p)_T < c/p$ (upper part of Figure 19.12b), or, in other words, to the necessity of a less than linear increase in loading with pressure. Effect (ii), by contrast, leads to a greater than linear increase in loading with increasing pressure, ensuring that $\partial q/\partial p > q/p$ (lower part of Figure 19.12b). Thus, by taking account of the prevailing mechanism, Eq. (19.3) correctly predicts the correlation between self- and transport diffusion as observed in the experiments.

On the basis of these two mechanisms, the experimentally observed behavior may be explained intuitively by comparing the situations reflected by Figure 19.10a and b: As soon as molecular interactions become relevant, with mechanism (ii) dominating, the intermolecular attractive forces will counteract the "driving force" due to the concentration gradient and reduce the flux during transport diffusion (to regions of lower overall guest content, Figure 19.10a) in comparison with self- or tracer diffusion (in which the overall concentration is uniform, as in Figure 19.10b). However, when mechanism (i) is dominant it means that the molecules will even more "willingly" move to regions of lower concentration, thus enhancing the guest flux for transport diffusion (Figure 19.10a) in comparison with self- (or tracer) diffusion (Figure 19.10b).

The differences observed between the different guest molecules nicely illustrate the conditions under which each of these mechanisms will prevail. While, over the whole loading range considered, the diffusion properties of the highly polar methanol molecules are dominated by their mutual interaction [mechanism (ii), transport diffusion exceeded by self-diffusion for all loadings], the reverse is true for non-polar ethane [mechanism (i), self-diffusion exceeded by transport diffusion]. Ethanol assumes an intermediate position, with $\mathcal{D} > D$ for small loadings [mechanism (ii) prevailing] and $\mathcal{D} < D$ for higher loadings [mechanism (i) prevailing].

The finding that the self-diffusivities may exceed the transport diffusivities has been illustrated here by referring to the remarkable situation in which molecular propagation is promoted, rather than slowed down, under the influence of a counter flux of labeled molecules. Having rationalized the origin of this finding, the situation may now more correctly be described by referring to the mere presence of the (differently labeled but otherwise identical) molecules rather than to their flux. By implying an infinitely large mutual Maxwell–Stefan diffusivity $Ð_{AA}$, that is, by excluding any intermolecular friction, the flux itself is anyway assumed to be of no influence. It is (only) the very existence of the additional molecules that enhances the propagation rate from cage to cage in comparison with the situation in transport diffusion!

19.3.2
Membrane-Based Gas Separation Following the Predictions of Diffusion Measurements

The interrelation between diffusion and permeation and its exploitation for profitable separation technologies with inorganic membranes is the focus of Chapter 20. The large diversity in structures and pore sizes, in combination with their high surface areas and adsorption affinities, make MOFs particularly attractive separation materials [41]. Their potential may be expected to exceed even that of zeolitic membranes [42] owing to their high lattice flexibility and an enhanced mechanical and thermal stability in comparison with other structures. Although there have been several attempts to synthesize MOF layers on porous supports [43–45], so far only a few [46–48] have been found to yield a dense coating [37]. Particularly favorable properties were found to be provided by MOFs with zeolitic imidazolate frameworks (ZIFs).

Interestingly, with ZIF-8 membranes particular progress in achieving dense coating has been attained by replacing dimethylformamide with pure methanol as the synthesis solvent [37, 49]. Obviously, owing to the small kinetic diameter and the weak interaction with the MOF lattice (as reflected, for example, by the type-III adsorption isotherm, see Figure 19.12a), stress on the crystal during solvent removal is strongly reduced, resulting in an essentially crack-free, dense polycrystalline layer of ZIF-8 on the porous titania support used in these studies [37].

The excellent options for membrane fabrication and for the measurement of adsorption and diffusion as exemplified in Section 19.3.1, including multicomponent adsorption, make ZIF-8 an ideal model system for correlating membrane permeabilities and selectivities with the sorption and diffusion properties of the guest molecules. Figure 19.13 shows the amount adsorbed of CO_2 and CH_4 as a function of the external pressure. Both the conditions of single-component and two-component

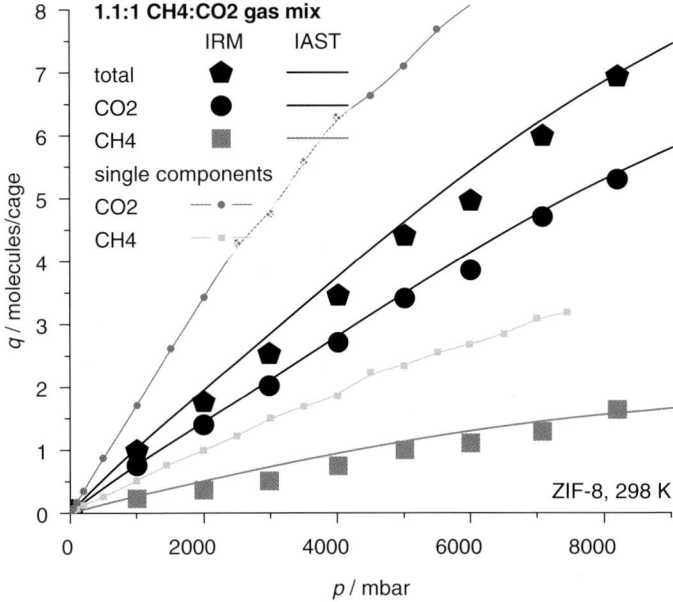

Figure 19.13 Adsorption isotherms of CO_2 and CH_4 in ZIF-8 at 298 K and comparison of the mixture data with the prediction from single-component data using the ideal adsorbed solution theory [50–52]. Reprinted from Reference [53] with permission.

adsorption (with a pressure ratio of 1 : 1.1) are considered. Figure 19.14 shows the corresponding transport diffusivities. For mixture adsorption they represent the effective transport diffusivities. They were calculated by fitting the standard relation for diffusion-limited uptake in spherical particles [Eq. (6.10)] to molecular uptake (following a step change in the total pressure), which IR micro-imaging (IRM, see Section 12.2.1) can provide separately for both components.

As a first-order estimate (Section 20.2.1), the permeation selectivity may be assumed to be the product of the adsorption selectivity with the diffusivity ratio. Figure 19.15b represents the results of such a study where grand-canonical Monte Carlo (GCMC) simulations (Section 7.2.3) have been exploited to extrapolate the experimental data into the range relevant for the permeation measurements [53].

IRM has also been applied to measure the single-component diffusivities of ethane–ethene mixtures in ZIF-8 [53, 54]. The ethene diffusivities were found to be larger by a factor of 5 in comparison with the ethane diffusivities. This may be explained by the slightly larger critical diameter of ethane. MAS PFG NMR self-diffusion studies (Sections 10.1.4 and 11.3) have been performed with the same system, considering both single-component and mixture adsorption [54]. The self-diffusivities for single-component adsorption were in perfect agreement with the results of IRM tracer exchange and confirm the validity of the IRM diffusivity data.

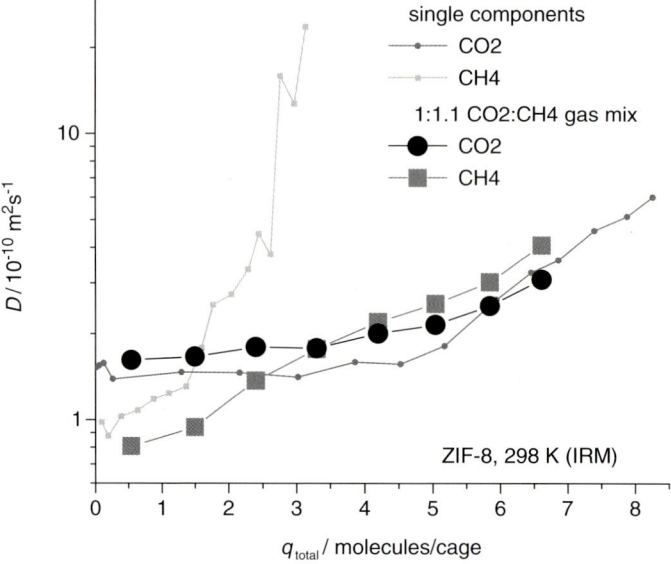

Figure 19.14 Loading dependence of the transport diffusivities of CO_2 and CH_4 in ZIF-8 under single-component adsorption and effective transport diffusivities under co-adsorption, initiated by a pressure step in the surrounding (1 : 1.1 CO_2–CH_4) gas-mixture atmosphere. Reprinted from Reference [53] with permission.

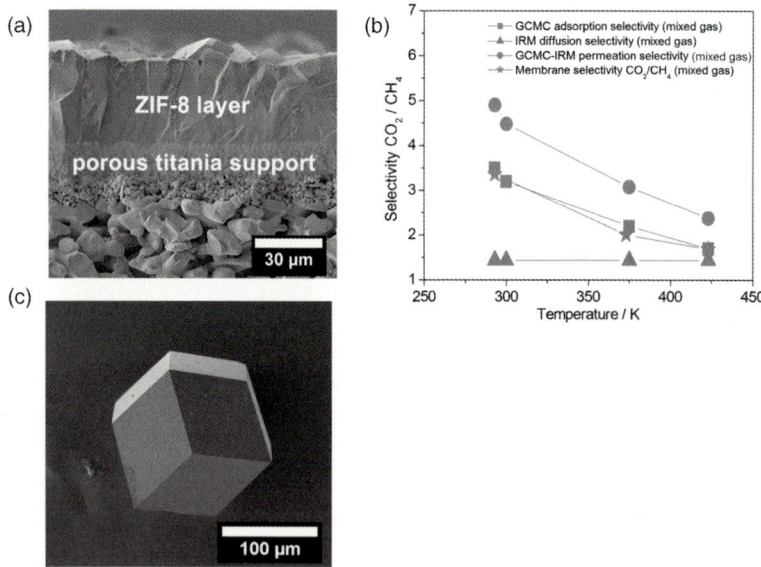

Figure 19.15 ZIF-8 membrane (a) and comparison of the membrane permeation selectivity for a CO_2–CH_4 mixture [(b) stars] with an estimate [(b) squares] based on the adsorption [(b) circles] and diffusion [(b) triangles] "selectivities" resulting from IRM measurements with a "giant" ZIF-8 crystal (c). Reprinted from Reference [53] with permission.

19.4
Pore Segments in Single-File Arrangement: Zn(tbip)

The pore space of several members of the MOF family is formed by an array of parallel chains of cavities. These examples include manganese formate [45], which, among all members of the MOF family, was the first to be considered in diffusion studies by interference microscopy (IFM) [55]. Subsequent uptake and release steps were found to give rise to structural variations, which complicated in-depth studies of mass transfer phenomena with this material. Much better prospects are provided by a MOF specimen of type Zn(tbip) (H_2tbip = 5-*tert*-butylisophthalic acid) [7]. Figure 19.16 shows the atoms of the lattice and the pore structure of MOF Zn(tbip), together with the image of a crystal considered in the diffusion studies.

In general, even after many adsorption–desorption cycles the crystals of the sample under study did not show any change in their sorption and diffusion properties, thus establishing this material as an ideal host system for exploring both intracrystalline diffusion and surface permeation [56, 57]. Altogether, more than 60 different adsorption and desorption runs with three different guest molecules (ethane, propane, and *n*-butane) and three runs of tracer exchange (between propane and deuterated propane at two different loadings) have been performed. Figures 19.17 and 19.18 show examples of the evolution of the recorded concentration profiles.

In all experiments, the boundary concentration does not immediately reach the equilibrium value. This indicates an additional transport resistance at the surface,

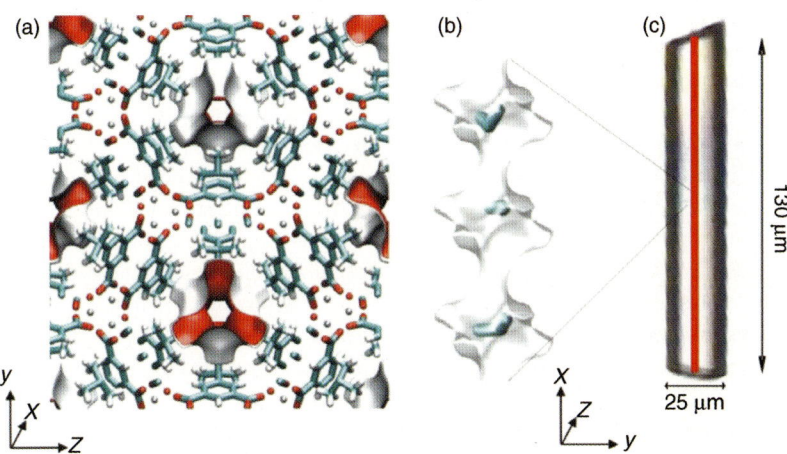

Figure 19.16 Model representations of the atoms of the crystal framework (a) and of the one-dimensional pore structure (b) and image of the MOF Zn(tbip) under study (c). The pores in the 1D arrangement are ordered like two three-leafed clovers, separated by windows of a diameter of 0.45 nm. The red planes in (a) indicate the surface of the pores as perceived by the guest molecules. The red line in (c) shows the position where the profiles are recorded. Reprinted from Reference [56] with permission.

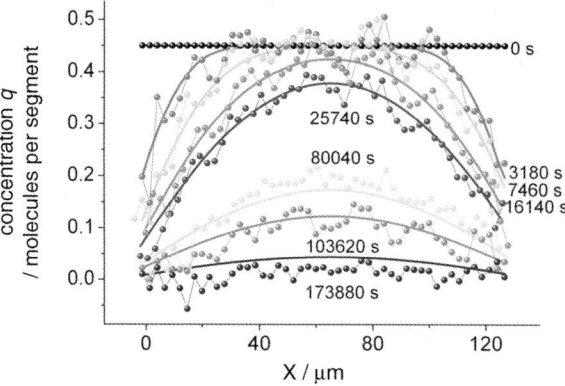

Figure 19.17 Transient concentration profiles of deuterated propane that is substituted by its undeuterated isotope. The profiles are recorded with IR micro-imaging at a pressure of 60 mbar. The thin lines represent the best fits to the solution of the corresponding diffusion equation [Eq. (6.29)]. Reprinted from Reference [56] with permission.

that is, a reduced surface permeability, to which we shall refer in more detail in Section 19.6.

During tracer exchange, molecular mobilities depend on the overall concentration rather than on the concentrations of labeled (or unlabeled) molecules. Analysis of the transient profiles has to be based, therefore, on a single value of the (tracer or self-) diffusivity and surface permeability. The data points shown in Figure 19.19b and e yield the best fits between the concentration profiles determined experimentally during tracer exchange (spheres in Figure 19.17) and the theoretical model (Eq. (6.29) [58, 59], full lines in Figure 19.17).

For transient uptake and release measurements, the possibility of concentration dependence of both the intracrystalline (transport) diffusivity and the surface permeability must be considered. In this context, the Maxwell–Stefan approach in combination with the Reed–Ehrlich model [Section 4.3.2.2 and Eqs. (4.30)–(4.37)] has been found to provide a good representation for a range of different concentration patterns. By considering only nearest-neighbor interactions on a surface of equal adsorption sites, in this approach total concentration dependence is determined by only two parameters: the corrected diffusivity $D_0(0)$ at zero loading (which coincides with the transport and self-diffusivity at zero loading) and an interaction parameter η, which is smaller than 1 for repulsion and larger than 1 for attractive interactions.

The assumption that both the intracrystalline diffusivities and the surface permeabilities follow the form of concentration dependence predicted by this formalism yields excellent agreement between theory and experiment. This is shown by the full lines in the representations of the transient concentration profiles in Figure 19.18, which have been calculated from a numerical solution of the diffusion equation with appropriately chosen values for the diffusivities and surface permeabilities at zero loading and the interaction parameters η [56]. Most importantly, this agreement is

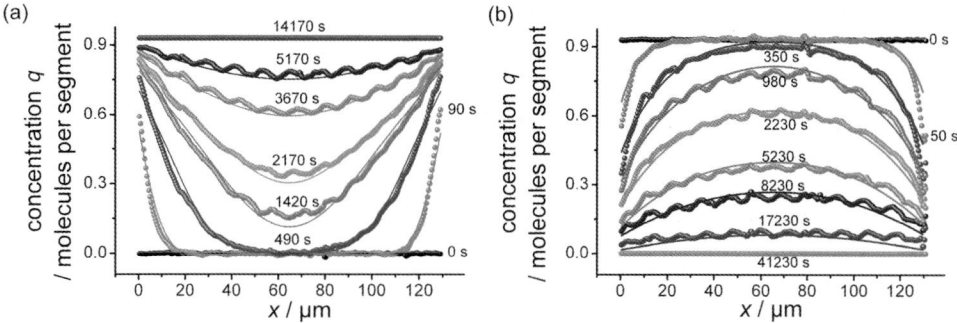

Figure 19.18 Transient intracrystalline concentration profiles of propane in a nanoporous crystal of type Zn(tbip) for a pressure step from 0 to 480 mbar (a) and from 480 mbar to vacuum (b) recorded by interference microscopy. The thin lines represent the best fits to the solution of the diffusion equation by implying the Reed–Ehrlich model [Eqs. (30)–(37)]. Reprinted from Reference [56] with permission.

found to exist for both ad- and desorption, confirming the reproducibility and reversibility of the measurements.

Figure 19.19 summarizes the concentration dependences of the transport diffusivities and surface permeabilities, together with the corresponding "corrected" quantities obtained from the respective adsorption isotherms via Eq. (1.31).

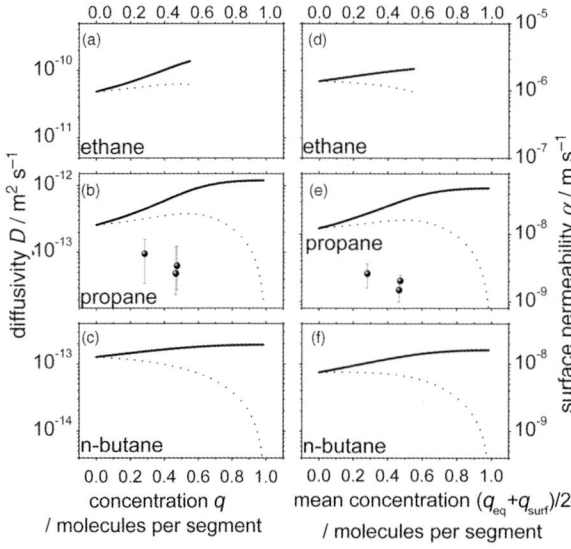

Figure 19.19 Diffusivity [(a)–(c)] and surface permeability [(d)–(f)] of ethane, propane, and n-butane in Zn(tbip) derived from the best fit of the experimentally determined transient concentration profiles. The results from (non-equilibrium) uptake/release experiments are presented by full lines and the tracer exchange data by points. The corrected (Maxwell–Stefan) parameters used in the fitting procedure are indicated by dotted lines. Reprinted from Reference [56] with permission.

With the application of micro-imaging to the study of uptake and release of light paraffins in Zn(tbip) an extensive data set of surface permeabilities could be collected and compared with the intracrystalline diffusivities for the same guest molecules. As a particular feature of these studies, the diffusivities and surface permeabilities are found to exhibit similar concentration dependences. We shall return to this point in Section 19.6 and explain why this particular finding may be of general relevance for mass transfer in nanoporous materials.

The simultaneous measurement of transport diffusion and self-diffusion in Zn(tbip) makes it possible to assess the extent to which mass transfer in the chain of pore segments is subject to single-file diffusion (Chapter 5) [60–63]. In a perfect single-file system of N sites (pore segments), with Eq. (5.14), the effective self-diffusivity for tracer exchange is known to be exceeded by $D_0(0)$ by a factor of $N\theta/(1-\theta)$. With crystal lengths $\geq 100\,\mu m$ in the present study and a site distance of $\lambda \approx 1\,nm$, the resulting factors dramatically exceed the experimental values (Figure 19.19b) by about 2 and 5 for $\theta = 0.28$ and 0.48, respectively. Hence, within the chains of pore segments in the crystals under study the propane molecules are evidently able to pass each other. The passage rate Γ for a particular pair of adjacent molecules is related to the tracer (or self-) diffusivity D by the simple random-walk expression [see, for example, Eq. (2.4) with $\Gamma = 1/(2\tau)$]:

$$D = l^2 \Gamma \tag{19.4}$$

where l ($\approx \lambda/\theta$) denotes the mean distance between adjacent molecules. With the simplifying assumptions that (i) the probability of a jump attempt towards an adjacent pore segment is unaffected by its occupation and (ii) a jump attempt towards an empty segment is always successful, while it is only successful with probability p if this segment is occupied, the passage rate may be correlated by the expression:

$$\Gamma = \frac{p(1-\theta)}{2\tau} \tag{19.5}$$

with the mean time τ between jump attempts. Inserting Eq. (19.4) into (19.5) yields:

$$D = \frac{\lambda^2}{2\tau} \frac{p(1-\theta)}{\theta^2} = D_0(0) \frac{p(1-\theta)}{\theta^2} \tag{19.6}$$

from which one obtains passage probabilities of $p = 5.5 \times 10^{-2}$ and 8.8×10^{-2} for $\theta = 0.28$ and 0.48, respectively. The increase of p with increasing loading may be attributed to the repulsive interaction of the diffusants, which, in the chosen model, has already been found to give rise to values < 1 for the fitting parameter of concentration dependence (η).

It is important to note that, with loadings up to 1 molecule per segment, the presented IFM diffusion measurements cover a range of relatively small loadings. Both in MD simulations and initial IRM measurements with ethane notably larger concentrations have been considered [64, 65]. After a monotonic decline in the self-diffusivities for loadings increasing from about two to six molecules per segment, with further increasing loadings the MD simulations are found to yield increasing diffusivities. However, this increase in the diffusivities is only observed if the lattice is assumed to be flexible. This may be because, for flexible lattices, the tighter, more constrained parts of the channels may become wider at higher concentrations, thus allowing more molecules to diffuse through the central regions of the channels.

19.5
Breathing Effects: Diffusion in MIL-53

In contrast to the usual behavior of crystalline porous solids, which show only small changes in their pore volume, lattice flexibility allows some MOFs to swell quite significantly (to "breathe" [66]) under external stimuli such as temperature, pressure, gas and solvent adsorption, or irradiation. Considering the nature of interaction and the dimensionality of the sub-network, breathing MOFs have been classified into six different types of flexibility. This was later extended to other flexible MOFs in which the swelling is even more pronounced [8]. These structures are built up from dicarboxylate linkers and either chains of corner-sharing octahedra, as in MIL-53 (Al, Cr) [67], or trimers of octahedra, as in MIL-88 [68, 69]. For MIL-53 (Figure 19.20a) reversible breathing amplitudes of up to 50%, between the hydrated and the dry forms, have been observed. These changes become immediately obvious in the adsorption isotherms (e.g., for CO_2) – see Figure 19.20b. The adsorption isotherm for methane, however, exhibits no such anomalies, indicating that, in contrast to CO_2,

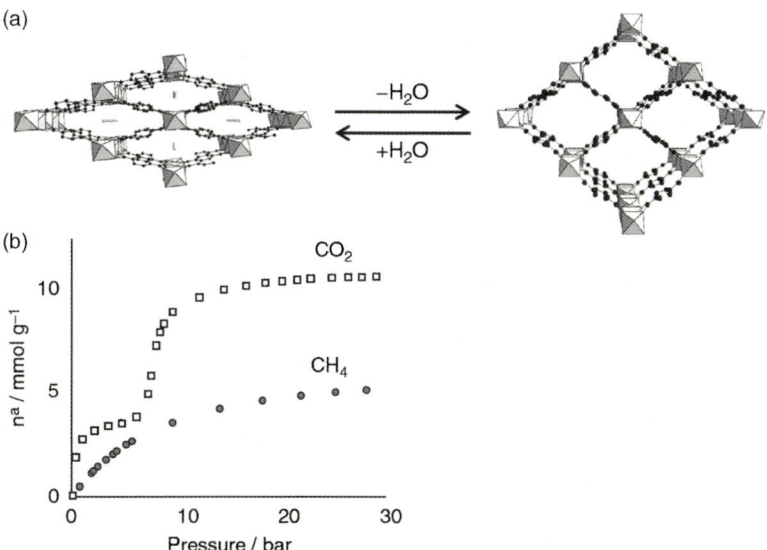

Figure 19.20 Breathing behavior of MIL-53 (Al, Cr) (a) and its appearance in the CO_2 adsorption isotherm at 303 K (b). The CH_4 isotherm at 303 K (b) does not reveal any anomaly. Reprinted from Reference [8] with permission.

the MOF lattice remains unaffected by the non-polar and almost spherical methane molecules. Guest-related changes in the lattice structure are also reported in References [70, 71] where, depending on the chosen guest molecules, the adsorption isotherms in copper-based MOFs exhibit either one or two hysteresis loops.

A spectacular example of breathing effects is observed in the isoreticular class of flexible phases (Section 19.1) denoted MIL-88A–D (each letter corresponding to a different linear dicarboxylate), where each solid swells selectively with volume variations of up to 230% [8] between the open form (filled with solvent) and the dried form.

Our understanding of the host–guest interaction in flexible MOFs has benefitted substantially from comparative MD simulations and QENS measurements of guest diffusion in MIL-47(V) [72] and MIL-53 (Cr) [67]. MIL-53 [$Cr^{(III)}$] is constructed by corner-sharing $CrO_4(OH)_2$ octahedra chains, interconnected by *trans*-benzene-dicarboxylate ligands, creating a 3D network with an array of parallel pores with a diameter of about 8.5 Å. It contains hydroxyl groups ("μ_2-OH" groups) located at the metal–oxygen–metal links. MIL-47 [$V^{(IV)}$] is isostructural with MIL-53 [$Cr^{(III)}$], with the difference that oxo-bridges ("μ_2-O" groups) assume the position of the μ_2-OH groups. Since they act as attractive sites for the guest molecules, this difference leads to differences in the diffusion and sorption mechanisms [73]. In contrast to MIL-53 (Cr), MIL-47 (V) shows no significant breathing effects. This is because the hydroxyl groups, which, in MIL-53, play a major role in the creation of symmetric hydrogen bonds with the oxygens of the occluded water (a prerequisite for the changes in the MOF lattice [66]), are missing.

Figure 19.21 compares theoretical predictions of the diffusivities of CO_2 in the flexible MIL-53 lattice with the results of extensive QENS studies by which the influence of breathing host networks on guest diffusivities has been investigated experimentally [74]. As already observed with CuBTC (Figure 19.8), there is fair agreement in the patterns of concentration dependence, while the differences in absolute values are probably explained by deficiencies in the force fields used in the simulations.

For comparison with the experimental data, the authors considered the weighted average of the simulation data for the diffusivities in the narrow-pore and large-pore lattices as shown in Figure 19.20a. The dependence of the relative contributions on the (total) CO_2 loadings was calculated on the basis of a previously validated analytical phase mixture model [75]. In the simulation shown in Figure 19.21 the transport diffusivities are estimated from the product of the corrected diffusivities (derived from the simulation) and the thermodynamic factors [Eq. (1.31)] derived from the isotherms. Similarly, as already discussed in Section 19.3.1 for methanol in MOF ZIF-8, there are parts of the adsorption isotherm for CO_2 in MOF MIL-53 (Figure 19.20b) that are convex towards the abscissa. In this concentration range the thermodynamic factor (Figure 19.12b) becomes smaller than 1 so that the corrected diffusivity exceeds the self-diffusivity. Note that here, in contrast to methanol in ZIF-8, this remarkable effect is caused by a "breathing-induced" increase in the host pore volume (occurring at medium pore fillings) rather than by a dominant guest–guest interaction.

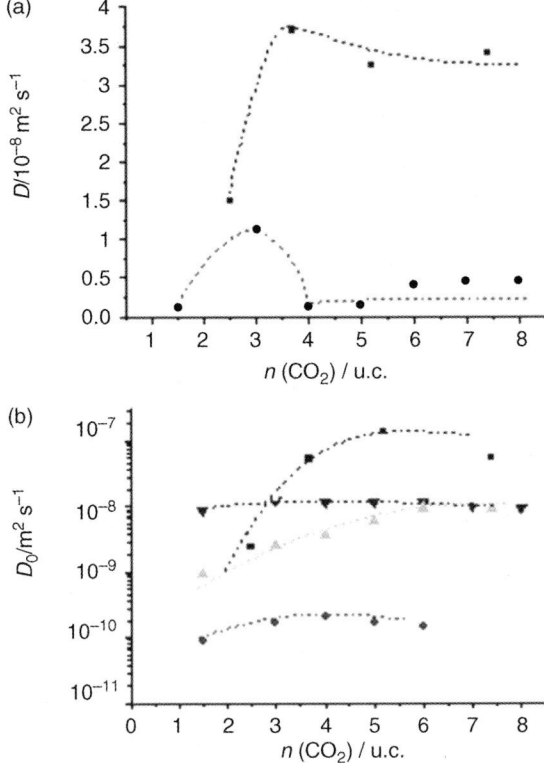

Figure 19.21 Comparison of CO_2 diffusivities in MIL-53 (Cr) determined by QENS measurements and MD simulations at 230 K as a function of loading: (a) experimental (squares) and simulation (circles) data of the transport diffusivity and (b) corrected diffusivities simulated for a rigid narrow-pore (circles) and large-pore (inverted triangles) MIL-53 lattice and their weighted average (triangles) and experimental data (squares). Reprinted from Reference [74] with permission.

Figure 19.22 shows diffusivity data analogous to those of Figure 19.21, namely, experimental and simulated corrected and transport diffusivities for CO_2 at 230 K, but this time for MIL-47 (V) [76], the non-breathing counterpart of MIL-53. Again the corrected diffusivities are found to exceed the transport diffusivities in both the experiments and the simulations and again the finding may be attributed to the influence of the thermodynamic factor, that is, to a region of the adsorption isotherm that is convex towards the abscissa, but for this system, as for ethanol in ZIF-8 (Figure 19.12a or Figure 5 in Reference [76]), only in the range of small concentrations [since there is no breathing in MIL-47 (V)]. Interestingly, contrary to the experimentally observed behavior for methanol and ethanol in ZIF-8, the MD simulations for CO_2 in MIL-47 yield self-diffusivities that deviate from the corrected diffusivities towards smaller values. This finding reflects the generally expected behavior [Eq. (3.51)] and, once again, characterizes the particular situation with

19 Metal Organic Frameworks (MOFs)

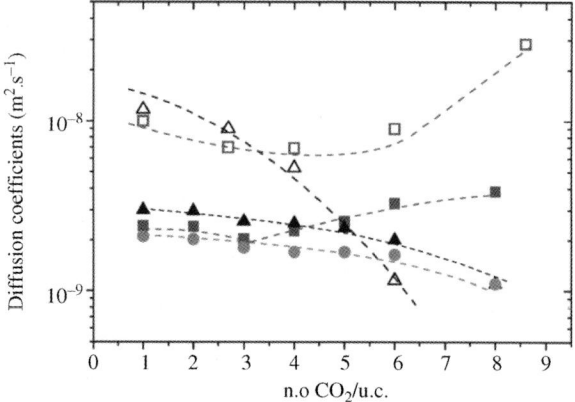

Figure 19.22 Comparison of CO_2 diffusivities in MIL-47 (V) determined by QENS measurements and MD simulations at 230 K as a function of loading: experimental data of the transport (□) and corrected (△) diffusivities and simulation data for the transport (■), corrected (▲) and self- (●) diffusivities. Reprinted from Reference [76] with permission.

methanol and ethanol in ZIF-8, where molecular propagation is dominated by narrow windows, leading to coincidence between the self- and corrected diffusivities.

In contrast to CO_2, methane does not induce any lattice breathing (Figure 19.20). Consequently, the methane diffusivities in MIL-53 (Cr) and MIL-47 (V) are found to follow similar dependencies (type I concentration pattern, see Section 3.5.1) in which the QENS self-diffusion data are found to be nicely reproduced by MD simulations [77]. The diffusivities in MIL-53 (Cr) are notably exceeded by those in MIL-47 (V), which is consistent with the calculated residence times for methane, which are longer in the presence of the μ_2-OH groups in MIL-53 (Cr) than near the μ_2-O groups in MIL-47 (V).

Qualitatively similar dependencies (type I concentration patterns) have also been observed in QENS and MD studies of hydrogen self-diffusion in MIL-53 (Cr) and MIL-47 (V) [78]. In addition, with hydrogen in MIL-53 (Cr) and MIL-47 (V) both techniques also yield coinciding data where, once again, the smaller diffusivities in MIL-53 (Cr) may be explained by the interaction with the μ_2-OH groups. Remarkably, the hydrogen diffusivities at low loadings are found to exceed those in zeolites (Figure 18.27 [79]) by two orders of magnitude.

19.6
Surface Resistance

19.6.1
Experimental Observations

It was mentioned in Section 12.3.2 that, except for tracer exchange or sorption experiments with differentially small pressure steps, surface permeabilities must

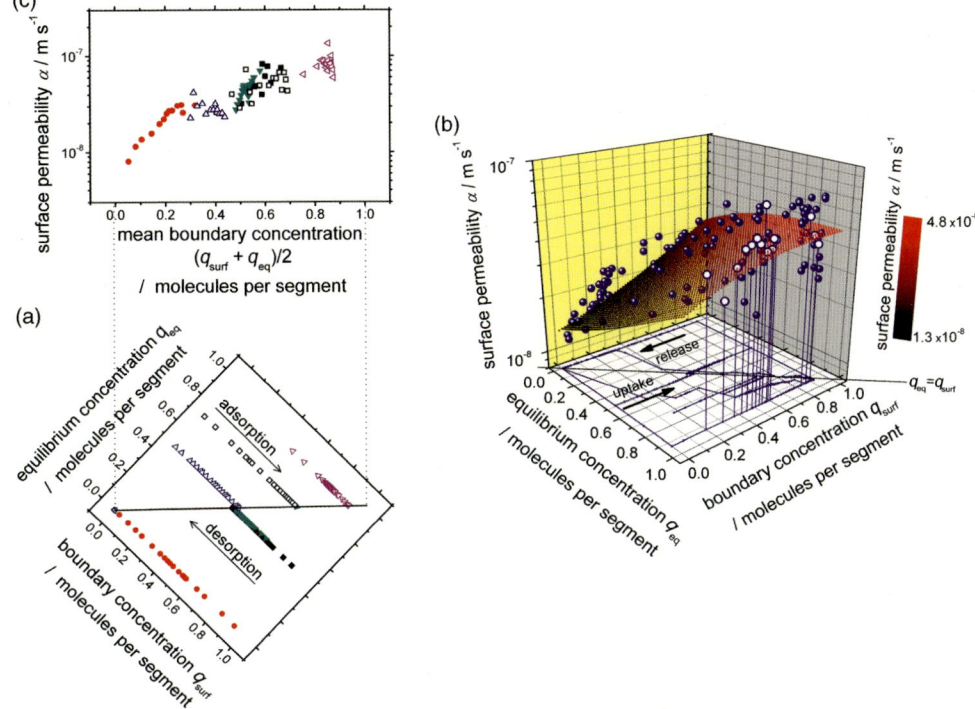

Figure 19.23 (a) The actual boundary concentration and the equilibrium concentration during the adsorption/desorption runs of propane on nanoporous crystals of type MOF Zn(tbip). (b) Surface permeability as a function of the equilibrium and the boundary concentration of the crystal in 3D representation. The blue bullets stand for surface permeabilities. Their projection on the bottom plane gives the corresponding pair of equilibrium and boundary concentration. (c) The resulting surface permeabilities, plotted [by the same symbols as in (a)] as a function of the arithmetic mean of these concentrations. Reprinted with permission from References [57, 80].

generally be considered to depend on a whole range of concentrations (namely, those between the actual boundary concentration and the concentration in equilibrium with the external atmosphere). In a systematic study, in Reference [57] the surface permeabilities of MOF crystals of type Zn(tbip) have therefore been considered as a function of the boundary and equilibrium concentrations for which they were determined. The set of permeabilities resulting from an individual adsorption run (leading, for example, to the sequence of profiles shown in Figures 19.17 and 19.18) would be considered, therefore, as a function of a constant equilibrium concentration corresponding to the externally applied pressure and of a set of boundary concentrations that increase from the initial concentration to the equilibrium concentration. In a q_{eq} versus q_{surf} plane, these concentrations would form a straight line, parallel to the q_{surf} axis and ending on the diagonal $q_{surf} = q_{eq}$, as shown in Figure 19.23a. Desorption experiments would lead to a sequence of concentrations forming analogous parallel lines, approaching the diagonal $q_{surf} = q_{eq}$ from higher surface concentrations.

Figure 19.23b shows the values of $\alpha(q_{surf}, q_{eq})$ by small bullets over all q_{surf}–q_{eq} pairs considered in the different sorption measurements. As may be seen from the "traces" in the q_{eq} versus q_{surf} plane, in one case even the external pressure (and hence the relevant equilibrium concentration) during a desorption and an adsorption experiment was varied, leading to deviations from lines strictly parallel to the q_{surf} axis. Within the uncertainty of the measurements, the resulting surface permeabilities are found to be:

1) independent of both the direction and the magnitude of the pressure step by which the sorption process is generated;
2) satisfactorily approximated by a function of a single parameter, namely, the mean value $\frac{1}{2}(q_{surf} + q_{eq})$ of the two "limiting" concentrations q_{surf} and q_{eq}.

The best fit of the surface permeabilities to a function of this parameter is indicated by the shaded area in Figure 19.23b. Figure 19.23c shows the permeability data as a function of the mean concentration $\frac{1}{2}(q_{surf} + q_{eq})$. Since a projection of the data points (q_{surf}, q_{eq}) in Figure 19.23a onto the diagonal yields exactly this mean value $\frac{1}{2}(q_{surf} + q_{eq})$, by using identical data points we are able to indicate (in Figure 19.23a) the course of the experimental runs by which the surface permeabilities (shown in Figure 19.23c) have been determined. Although there is significant experimental scatter, the permeability data from all experimental runs fit reasonably well to each other, in conformity with statement (1).

The finding that, under the given conditions, the surface permeability may be treated as a function of a single parameter, namely, the "mean" concentration, is of great practical importance: It greatly facilitates the inclusion of such resistances for modeling their influence in technological applications and allows, for example, direct comparison of experimental permeability and diffusivity data, as in Figure 19.19. Simultaneously, it is expected to contain implicitly valuable information about the nature of the elementary mechanisms and processes leading to the observed phenomenon of surface resistance.

In this context one may refer to permeation studies through a single-crystal silicalite membrane [81], where for a strongly concentration-dependent transport diffusivity $[D = D_0/(1 - q/q_s)]$ the time lags predicted from a nonlinear analysis are close to the values calculated for a constant-diffusivity system with $D = D_{av}$, in agreement with a classical, more general analysis by Frisch [82, 83]. Of course, when the diffusivity varies strongly with loading D_{av} is not the same as $D(q_{av})$. But the difference is generally small and probably within the experimental uncertainty. Moreover, in the following section a model is presented that shows that finite surface permeabilities may originate from a structure that differs substantially from the widely accepted concept of a surface layer of dramatically reduced diffusivity.

This model has been developed to rationalize the most remarkable finding contained in the data of Figure 19.19: for each of the molecules considered (ethane, propane, n-butane), and for each of the considered loadings (between, essentially, zero and 1 molecule per pore segment) and in both transient uptake/

release (non-equilibrium) experiments (with the diffusivities/permeabilities increasing with loading) and tracer-exchange experiments (with opposite concentration dependencies) the ratio α/D between the surface permeability and the diffusivity of the guest molecules in the host crystal under study is approximately constant and equal to $5 \times 10^4 \,\mathrm{m}^{-1}$!

We conclude this section by rationalizing the differences in the concentration dependencies observed under equilibrium and non-equilibrium conditions. Following the reasoning of Section 1.2.1 with Eqs. (1.31) and (1.32), with a Langmuir-type adsorption isotherm the increase in diffusivity and permeability with increasing loading under non-equilibrium may be attributed to an increase in the "driving force" due to the increasing thermodynamic factor $\partial \ln p / \partial \ln q$. This thermodynamic factor does not appear under equilibrium conditions, that is, for tracer exchange. Instead, one has to take into account the single-file character of the pore system of Zn(tbip) (Figure 19.16b) where, with increasing loading [Eq. (5.14)], the effective diffusivity is known to decrease. This holds true, irrespective of whether the file length L over which perfect single-file confinement occurs coincides with the dimensions of the crystal or whether single-file behavior persists only over shorter sections (with mutual passage of the diffusants possible between these sections). This possibility may in fact be predicted from the data presented in Section 19.4 by comparing the transport and self-diffusivities.

19.6.2
Conceptual Model for a Surface Barrier in a One-Dimensional System: Blockage of Most of the Pore Entrances

In Section 2.3.4, the existence of impermeable planes with circular holes within a 3D-pore network was shown to give rise to resistances that, within the effective medium approximation, are well described by considering planes of uniform, finite permeability. With Eq. (2.36), this permeability is found to be directly proportional to the (3D-) diffusivity, with a factor of proportionality determined by the diameter and separation of the holes. Such a proportionality would provide the desired rationalization of the experimental findings of Figure 19.19, provided that the reasoning developed for 3D-pore networks can be applied to the 1D-arrangement of pore segments in Zn(tbip). In Reference [84] this idea has been shown to provide a reasonable explanation of the experimental data.

Figure 19.24 shows the structure model used in these studies. The host system is supposed to consist of channels (1D pores) in which mass transfer occurs by random molecular jumps between adjacent (cubic) segments in the x direction. Only a small fraction p_{open} of the channel mouths to the surroundings are open. Filling of the blocked channels is ensured by defects in the channel walls which occur with the probabilities $p_y = p_z$ (making the whole system a "porous maze" [85]). Jumps are "attempted" with equal probability in all six directions from each channel segment but they are successful only if the jump attempt is in a direction in which the passage is open and the segment to which the jump is directed is empty. The equilibrium concentration is adjusted by appropriately choosing the encounter rate of molecules

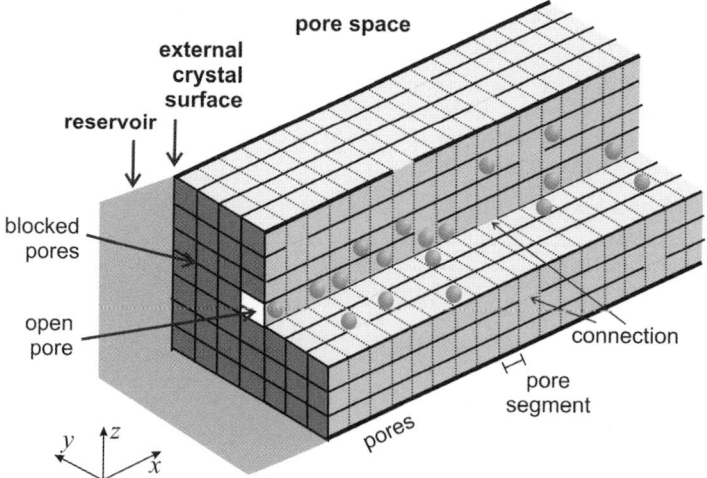

Figure 19.24 Structure model used to simulate mass transfer in Zn(tbip). The host system is supposed to consist of channels (1D pores) in which mass transfer occurs by random molecular jumps between adjacent (cubic) segments in the x direction. Only a small fraction p_{open} of the channel mouths to the surroundings are open. Filling of the blocked channels is ensured by defects in the channel walls. Reprinted from Reference [84] with permission.

from the reservoir onto the crystal surface, in comparison with the jump rate within the channels.

19.6.2.1 Simulation Results

Figure 19.25 shows the result of such a simulation: the resulting concentration profiles perpendicular to the surface (circles in Figure 19.25a) are found to be well represented by the solution of the diffusion equation, based on the intracrystalline diffusivity and surface permeability (full lines). These curves correspond with those obtained experimentally, as displayed in Figures 19.17 or 19.18. Figure 19.25b illustrates that, in the initial state of uptake or release, surface heterogeneity, that is, the variation between open and blocked entrances, gives rise to heterogeneities in guest concentrations. However, these differences soon disappear with increasing distance from the surface, which is a pre-requisite for modeling the effect of permeation through a very small number of open channels by a quasi-homogeneous layer of reduced permeability.

From these simulations it is clear that the concept discussed in Section 2.3.4 with Eq. (2.36) may also be applied to the conditions of anisotropic diffusion. Again, surface permeability and bulk diffusivity are found to be proportional to each other, since both are directly proportional to the jump rate. Hence, any variation in the jump probabilities – for example, by varying the guest molecules and the loading or even by changing from transient uptake to tracer exchange – may be expected to affect the diffusivities and permeabilities in the same way, leaving their ratio invariant. This is

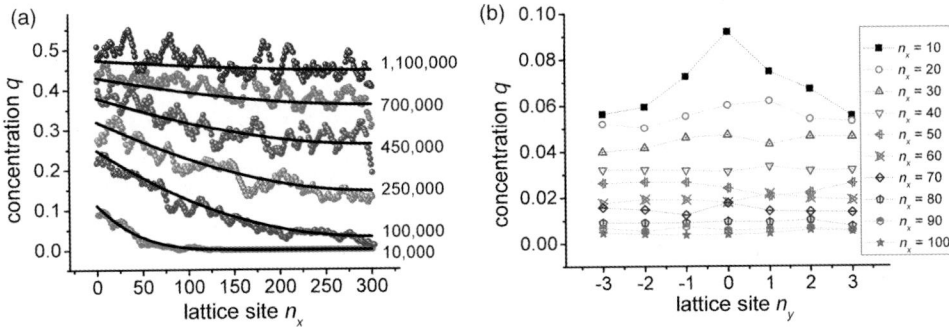

Figure 19.25 (a) Simulated concentration profiles along x (at different times) and the corresponding solutions of the diffusion equation with a finite surface permeability. (b) Dependence of the simulated profiles on the distance x from the crystal surface. With increasing distance differences in the concentration are found to vanish. Reprinted from Reference [84] with permission.

exactly the remarkable finding illustrated by Figure 19.19. The model was additionally confirmed by transient uptake and release experiments at different temperatures, in which the activation energies of intracrystalline diffusion and surface permeation were found to coincide. Mass transfer thus appears to be adequately represented by an effective-medium approach that transfers the microscopic heterogeneity into its mesoscopic correspondence [86], in complete agreement with the experimental observations and as expected from the central limit theorem of statistics (Section 2.1.2 and References [87, 88]) for displacements that significantly exceed the correlation lengths of the structural heterogeneities.

19.6.2.2 An Analytic Relationship between Surface Permeability and Intracrystalline Diffusivity

In the considered model (Figure 19.24), surface permeability and diffusivity are based on identical elementary steps, namely, jumps between adjacent cavities. Their ratio therefore depends only on the geometry of the host system, which is described by the two probabilities p_{open} and p_y ($= p_z$). Thus one has:

$$\frac{\alpha}{D} = \frac{1}{\lambda} f(p_{\text{open}}, p_y = p_z) \qquad (19.7)$$

with λ denoting the step length, that is, the distance between adjacent channel sites (length of a pore segment in Figure 19.24). Figure 19.26 displays permeability–diffusivity ratios as simulated for different geometrical conditions.

The simulation data were found to be approximated by the empirical relation:

$$\frac{\alpha \lambda}{D} \approx 0.5 \cdot p_{\text{open}} \cdot p_y \cdot \left(1 - \frac{p_y}{2 + 4p_y}\right) \cdot \frac{5}{1 + 4p_y} \qquad (19.8)$$

as illustrated by the dotted lines in the figures. Although the constituents of this approach may be correlated with probability estimates [84], the range of its applicability beyond the actual simulations has not yet been determined.

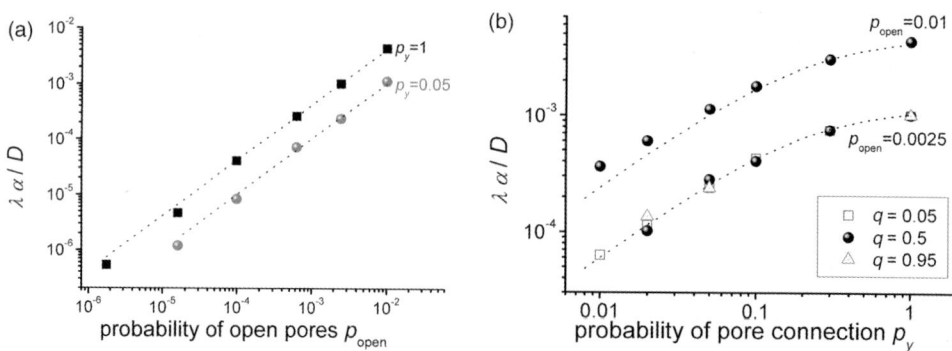

Figure 19.26 Permeability–diffusivity ratio resulting from the simulations, plotted as a function of (a) the fraction p_{open} of open pores simulated with two values p_y for the probability of channel interconnections and of (b) the probabilities of pore connections p_y ($= p_z$) simulated with two values of p_{open} and with three different concentrations (relative pore fillings) q. The dotted lines represent the analytical expression provided by Eq. (19.8). Reprinted from reference [84] with permission.

The close correlation between the surface permeabilities and the intracrystalline diffusivities as appearing in Eqs. (19.7) and (19.8) and in Figure 19.26a and b is a direct consequence of the mechanism to which the surface resistance is attributed, namely, the total blockage of most of the channel entrances. The surface permeability will obviously become smaller the larger are the "detours" that the guest molecules are forced to take to fill the whole channel system. The relative influence of these "detours" on overall uptake and release rates of guest molecules is determined by the host geometry rather than by the particular mechanism of molecular transport.

19.6.2.3 Surface Resistance with Three-Dimensional Pore Networks

With Eq. (2.36), the permeability-to-diffusivity ratio in three-dimensional pore networks, that is, with $p_y = p_z = 1$, for otherwise impenetrable surfaces with circular holes of radius a and distance L, may be noted as:

$$\frac{\alpha}{D} = \frac{2 p'_{open}}{\pi a} \tag{19.9}$$

with p'_{open} denoting the fraction of unblocked surface area ("windows"). In addition to the message that again the permeability–diffusivity ratio is found to be merely a geometry-dependent quantity, Eq. (19.9) provides the important information that, for a given fraction p'_{open} of open surface area, the permeability and, consequently, the permeability–diffusivity ratio decreases with increasing window diameter. This result indicates, not unexpectedly, a decrease of the "permeation efficiency" of the holes on clustering.

19.6.2.4 Estimating the Fraction of Unblocked Entrance Windows

In complete agreement with the simulations, for a given crystal the ratio between the experimentally determined intracrystalline diffusivities and surface permeabilities in

Zn(tbip) (Figure 19.19 and References [56, 57, 89]) is found to have a well-defined value, regardless of whether the measurements were carried out under equilibrium or non-equilibrium conditions (i.e., for both self- and transport diffusivities and the corresponding surface permeabilities) and irrespective of the chosen guest molecules, their concentration, and the temperature.

Assuming that the unblocked channel entrances are statistically distributed, Eq. (19.8) may be applied to estimate the probability p_{open} of their occurrence. To estimate the exchange probability p_y ($=p_z$) between adjacent channels we follow the reasoning of Eqs. (19.4)–(19.6) and assume that it is only by these exchanges that molecules may outrun single-file confinement. By inserting resulting typical values ($p_y = 0.05$ and $a\lambda/D \approx 5 \times 10^{-5}$ [84]) into Eq. (19.8) one obtains $p_{open} \approx 5 \times 10^{-4}$ for the probability of finding an unblocked channel entrance. This means that, on average, only one out of 45×45 channel entrances is unblocked.

19.6.3
Generalization of the Model

19.6.3.1 Intracrystalline Barriers
It is important to note that, just as they can occur close to the external surface, such planes of dramatically reduced permeability may also occur statistically distributed throughout the crystal bulk phase [90, 91]. When the separation of these planes is much smaller than the crystal diameter (or length), the uptake patterns would become indistinguishable from those under diffusion limitation [92]. The existence of such internal barriers provides a plausible explanation for the remarkable finding that, although differing by orders of magnitude, the diffusivities for many different systems deduced from "macroscopic" (e.g., uptake) and "microscopic" experiments (with displacements notably smaller than the barrier separations) often differ from each other by an approximately constant factor, leading, for example, to the same activation energies (see, for example, Figures 16.12, 16.18, and 16.19 and References [92, 93]).

19.6.3.2 Activation Energies
In the original model, for simplicity, the jump rates through the openings in the blocking plane were assumed to coincide with those in the intracrystalline space. By abandoning this condition, the relevance of this model appears to be far more general than one might suppose in view of the fact that, so far, it has been applied in detail only to the rather exceptional situation of concurrent surface permeation and bulk diffusion. The data from such simulations are shown in Figure 19.27, where, adopting physical reality, the difference in the jump rates through the "holes" in the otherwise impenetrable plane and in the pore space ($p_y = p_z = 0.05$) is attributed to the respective activation energies, which, for jumps through the "holes," are assumed to exceed those for jumps through the genuine pore space [94, 95]. The slopes of the respective Arrhenius plots differ by the ratio between these energies, which, for the simulations shown in Figure 19.27, has been set at 2. Being proportional to the jump rates, the intracrystalline diffusivity also varies with their

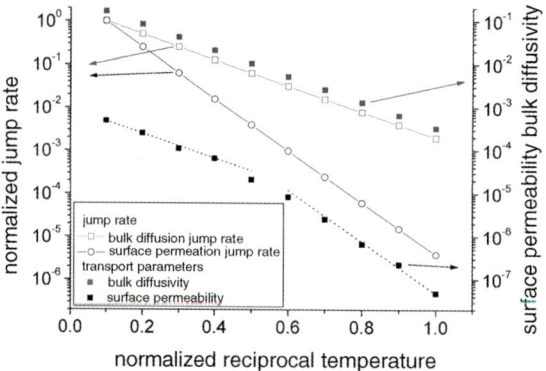

Figure 19.27 Arrhenius plots of the jump rates, surface permeability, and diffusivity. The activation energy for jumps in the surface plane is twice as large as in the bulk crystal; $p_y = p_z = 0.05$ and $p_{open} = 0.03$. Reprinted from Reference [84] with permission.

activation energy. Notably, in the high-temperature range, this is also the activation energy for surface permeation, irrespective of the fact that, except for the highest temperature considered, the jump rates in the bulk exceed those through the holes. This may be rationalized by realizing that for uptake (or pore-space filling between adjacent barriers in the crystal interior) a single step through the holes in the barrier must be followed by very many steps within the genuine pore space so that these steps may still remain rate determining. With further decreasing ratios of the jump rates (or, equivalently, as a consequence of the different activation energies) with further decreasing temperature the jumps through the barriers will eventually become increasingly influential so that the rate of (surface) permeation is no longer governed exclusively by the rate of intracrystalline diffusion. Exactly such a situation might lead to the deviations in the trends of surface permeabilities and intracrystalline diffusivities experimentally observed [55, 96–101]. At sufficiently low temperature, the activation energy of permeation will finally coincide with that of the jump rate through the holes in the barrier.

19.7
Concluding Remarks

In diffusion measurements with crystalline nanoporous materials, the guest diffusivities obtained by different techniques and with different specimens of the same host system are often found to differ in their absolute values while their dependences on temperature (i.e., the activation energies) and on the size of the guest molecule were quite similar. Examples include *n*-alkanes in zeolite LTA (Section 16.5), FAU (Section 17.2.6), and MFI (Section 18.2.1). Today, deviations from ideal crystallinity are quite generally accepted to give rise to such behavior. For example, for zeolite LTA the commonly observed large differences in diffusivity with almost constant activation energy are attributed to the hydration of the SII cations, which gives rise to differences in the fraction of "open" windows between the cavities (Section 16.11).

With MOF Zn(tbip), the nature of the mass transfer resistance at the crystal surface could be explored in greater detail than was previously possible [85]. On considering the dependence on loading, temperature, type of guest molecule, and even the observation mode (measurements under equilibrium and non-equilibrium conditions) surface permeability and intracrystalline diffusivity were shown experimentally to follow essentially identical trends (Figure 19.19 and References [56, 57, 89]). At least for this material, surface permeation and intracrystalline diffusion must therefore depend on the same elementary steps and the surface barrier evidently results from total blockage of the large majority of pore entrances.

Nanoporous crystalline materials are known to occur quite commonly as polycrystalline particles and intergrowths rather than as genuine single crystals. Transport resistances such as observed on the surface of MOF Zn(tbip) may therefore be expected to occur also in the interior of such crystals. In that situation, molecular uptake and release will proceed as if controlled by intracrystalline diffusion (Section 6.2 and Reference [92]) but with an effective diffusivity of the type given by Eq. (2.35). Therefore, eventually, the permeability (α) through the intracrystalline barriers rather than the diffusivity (D) within the genuine pore space may determine the effective diffusivity D_{eff} resulting from this type of measurement. Since, via Eqs. (19.8) and (19.9), for the type of resistances considered, permeability and diffusivity are known to be proportional to each other, it is clear that the effective diffusivities derived from different experiments may differ substantially in absolute value, although – owing to their proportionality to the "true" intracrystalline diffusivity – the effective diffusivities will all show similar dependencies with variation of the experimental conditions.

Molecular simulations (Section 19.6.2) support these conclusions. Moreover (Section 19.6.3 and Figure 19.27) simulations permit consideration of the more general (and physically more realistic) case in which the activation energy for jumps though the "holes" in the (intracrystalline or surface) barriers exceeds that for jumps within the genuine pore space. Under such conditions, the situation with equal transition rates, such as considered in Section 19.6.2, is found to occur in the high-temperature region. Now, however, the medium- and low-temperature regions are also able to reflect the situation in which the permeation rate through the holes in the barriers contributes significantly to the mass transfer resistance. The activation energy for permeation (and, with dispersed intracrystalline barriers, for the effective diffusivity derived from macroscopic uptake and release experiments) is then found to exceed the activation energy for genuine intracrystalline diffusion. This is the situation observed in numerous studies where the same system has been investigated by measuring techniques covering different diffusion path lengths (see, for example, Figure 17.11 and References [90, 93, 102]).

The model of mass transfer resistances in nanoporous materials as deduced from microscopic transient sorption and release experiments with MOF Zn(tbip) is thus able to explain a wide variety of experimental patterns that have been reported in the literature. However, it must remain the task of each individual case study to obtain independent evidence as to whether this type of additional resistance is in fact dominant or whether mass transfer is dominated by alternative mechanisms (such as percolation phenomena in the case of LTA-type zeolites, as considered in Sections 16.5 and 16.11).

References

1 Li, H., Eddaoudi, M., O'Keeffe, M., and Yaghi, O.M. (1999) *Nature*, **402**, 276.
2 Kaskel, S. (2002) *Handbook of Porous Solids* (eds F. Schüth, K.S.W. Sing, and J. Weitkamp), Wiley-VCH Verlag GmbH, Weinheim, p. 1190.
3 Bauer, S. and Stock, N. (2008) *Chem. Unserer Z.*, **42**, 12.
4 Krishna, R. (2009) *J. Phys. Chem. C*, **113**, 19756.
5 Eddaoudi, M., Moler, D.B., Li, H., Chen, B., Reineke, T.M., O'Keeffe, M., and Yaghi, O.M. (2001) *Acc. Chem. Res.*, **34**, 319.
6 Clearfield, A. (1998) *Prog. Inorg. Chem.*, **47**, 371.
7 Pan, L., Parker, B., Huang, X., Olson, D., Lee, J.-Y., and Li, J. (2006) *J. Am. Chem. Soc.*, **128**, 4180.
8 Devic, T. and Serre, C. (2008) *Ordered Porous Solids* (eds. V. Valtchev, S. Mintova, and M. Tsapatsis), Elsevier, Amsterdam, p. 77.
9 Ferey, G. (2008) *Chem. Soc. Rev.*, **37**, 191.
10 Eddaoudi, M., Kim, J., Rosi, N., Vodak, D., Wachter, J., O'Keeffe, M., and Yaghi, O.M. (2002) *Science*, **295**, 469.
11 Serre, C., Millange, F., Surble, S., and Ferey, G. (2004) *Angew. Chem. Int. Ed.*, **43**, 6285.
12 Ferey, G., Mellot-Dranznieksd, C., and Serre, C. (2005) *Science*, **309**, 2040.
13 Park, Y.K., Choi, S.B., Kim, H., Kim, K., Won, B.H., Choi, K., Choi, J.S., Ahn, W.S., Won, N., Kim, S., Jung, D.H., Choi, S.H., Kim, G.H., Cha, S.S., Jhon, Y.H., Yang, J.K., and Kim, J. (2007). *Angew. Chem. Int. Ed.*, **46**, 8230.
14 Liu, Y., Kravtsov, V.C., Larsena, R., and Eddaoudi, M. (2006) *Chem. Commun.*, 1488.
15 Li, J.R., Kuppler, R.J., and Zhou, H.C. (2009) *Chem. Soc. Rev.*, **38**, 1477.
16 Bernt, S., Feyand, M., Modrow, A., Wack, J., Senker, J., and Stock, N. (2011) *Eur. J. Inorg. Chem.*, 5378–5383. Modrow, A., Zargarani, D., Herges, R., and Stock, N. (2011) *Dalton Trans.* **40**, 4217–4222.
17 Keskin, S., Liu, J., Rankin, R.B., Johnson, J.K., and Sholl, D.S. (2009) *Ind. Eng. Chem. Res.*, **48**, 2355.
18 Chui, S.S.Y., Lo, S.M.F., Chjarmant, J.P.H., Orpen, A.G., and Williams, I.D. (1999) *Science*, **283**, 1148.
19 Chmelik, C., Kärger, J., Wiebcke, M., Caro, J., van Baten, J.M., and Krishna, R. (2009) *Microporous Mesoporous Mater.*, **117**, 22.
20 Stallmach, F., Gröger, S., Künzel, V., Kärger, J., Yaghi, O.M., Hesse, M., and Müller, U. (2006) *Angew. Chem. Int. Ed.*, **45**, 2123.
21 Ford, D.C., Dubbeldam, D., and Snurr, R.Q. (2009) in *Diffusion Fundamentals III* (eds. C. Chmelik, N. Kanellopoulos, J. Kärger, and D. Theodorou), Leipziger Universitätsverlag, p. 459.
22 Dubbeldam, D., Walton, K.S., Ellis, D.E., and Snurr, R.Q. (2007) *Angew. Chem. Int. Ed.*, **46**, 4496.
23 Tofts, P.S., Lloyd, D., Clark, C.A., Barker, G.J., Parker, G.J.M., McConville, P., Baldock, C., and Pope, J.M. (2000) *Magnet. Reson. Med.*, **43**, 368.
24 Douglass, D.C. and McCall, D.W. (1958) *J. Phys. Chem.*, **62**, 1102.
25 Kärger, J., Pfeifer, H., Rauscher, M., and Walter, A. (1980) *J. Chem. Soc., Faraday Trans. I*, **76**, 717.
26 Amirjalayer, S. and Schmid, R. (2009) *Microporous Mesoporous Mater.*, **125**, 90.
27 Keskin, S. and Sholl, D.S. (2009) *Ind. Eng. Chem. Res.*, **48**, 914.
28 Krishna, R. and van Baten, J.M. (2006) *Chem. Phys. Lett.*, **420**, 545.
29 Krishna, R. and van Baten, J.M. (2008) *Microporous Mesoporous Mater.*, **109**, 91.

30 Lee, J.W., Shim, W.G., and Moon, H. (2004) *Microporous Mesoporous Mater.*, **73**, 109.
31 Greathouse, J.A. and Allendorf, M.D. (2008) *J. Phys. Chem. C*, **112**, 5795.
32 Wehring, M., Gascon, J., Dubbeldam, D., Kapteijn, F., Snurr, R.Q., and Stallmach, F. (2010) *J. Phys. Chem. C*, **114**, 10527.
33 Keskin, S., Liu, J., Johnson, J.K., and Sholl, D.S. (2009) *Microporous Mesoporous Mater.*, **125**, 101.
34 Park, K.S., Ni, Z., Cote, A.P., Choi, J.Y., Huang, R.D., Uribe-Romo, F.J., Chae, H.K., O'Keeffe, M., and Yaghi, O.M. (2006) *Proc. Natl. Acad. Sci. USA*, **103**, 10186.
35 Baerlocher, C., McCusker, L.B., and Olson, D.H. (2007) *Atlas of Zeolite Framework Types*, Elsevier, Amsterdam.
36 Chmelik, C., Bux, H., Caro, J., Heinke, L., Hibbe, F., Titze, T., and Kärger, J. (2010) *Phys. Rev. Lett.*, **104**, 85902.
37 Bux, H., Liang, F., Li, Y., Cravillon, J., Wiebcke, M., and Caro, J. (2009) *J. Am. Chem. Soc.*, **131**, 16000.
38 Chmelik, C., Heinke, L., Valiullin, R., and Kärger, J. (2010) *Chem. Ing. Tech.*, **82**, 779.
39 Ruthven, D.M. (1984) *Principles of Adsorption and Adsorption Processes*, John Wiley & Sons, Inc. New York.
40 Schüth, F., Sing, K.S.W., and Weitkamp, J. (eds) (2002) *Handbook of Porous Solids*, Wiley-VCH Verlag GmbH, Weinheim.
41 Caro, J. (2010) *Chem. Ing. Tech.*, **82**, 837.
42 Caro, J. (2009) *Microporous Mesoporous Mater.*, **125**, 79.
43 Yoo, Y. and Jeong, H.K. (2008) *Chem. Commun.*, 2441.
44 Zacher, D., Shekhah, O., Wöll, C., and Fischer, R.A. (2009) *Chem. Soc. Rev.*, **38**, 1418.
45 Arnold, M., Kortunov, P., Jones, D.J., Nedellec, Y., Kärger, J., and Caro, J. (2007) *Eur. J. Inorg. Chem.*, 60.
46 Ranjan, R. and Tsapatsis, M. (2009) *Chem. Mater.*, **21**, 4920.
47 Gascon, J., Aguado, S., and Kapteijn, F. (2009) *Microporous Mesoporous Mater.*, **113**, 132.
48 Yoo, Y., Lai, Z., and Jeong, H.K. (2009) *Microporous Mesoporous Mater.*, **123**, 100.
49 Cravillon, J., Münzer, S., Lohmeier, S.J., Feldhoff, A., Huber, K., and Wiebcke, M. (2009) *Chem. Mater.*, **21**, 1410.
50 Krishna, R. and van Baten, J.M. (2008) *Chem. Eng. J.*, **140**, 614.
51 Murthi, M. and Snurr, R.Q. (2004) *Langmuir*, **20**, 2489.
52 Myers, A.L. and Prausnitz, J.M. (1965) *AIChE J.*, **11**, 121.
53 Bux, H., Chmelik, C., van Baten, J., Krishna, R., and Caro, J. (2010) *Adv. Mater.* **22**, 4741 and Bux, H., Chmelik, C., van Baten, J., Krishna, R., and Caro, J. (2011) *J. Membr. Sci.*, **369**, 284.
54 Chmelik, C., Freude, D., Bux, H., and Haase, J. (2012) *Microporous Mesoporous Mater.*, **147**, 135.
55 Kortunov, P., Heinke, L., Arnold, M., Nedellec, Y., Jones, D.J., Caro, J., and Kärger, J. (2007) *J. Am. Chem. Soc.*, **129**, 8041.
56 Heinke, L., Tzoulaki, D., Chmelik, C., Hibbe, F., van Baten, J., Lim, H., Li, J., Krishna, R., and Kärger, J. (2009) *Phys. Rev. Lett.*, **102**, 65901.
57 Tzoulaki, D., Heinke, L., Li, J., Lim, H., Olson, D., Caro, J., Krishna, R., Chmelik, C., and Kärger, J. (2009) *Angew. Chem. Int. Ed.*, **48**, 3525.
58 Crank, J. (1975) *The Mathematics of Diffusion*, Clarendon Press, Oxford.
59 Kärger, J. and Ruthven, D.M. (1992) *Diffusion in Zeolites and Other Microporous Solids*, John Wiley & Sons, Inc., New York.
60 Hahn, K., Kärger, J., and Kukla, V. (1996) *Phys. Rev. Lett.*, **76**, 2762.
61 Lutz, C., Kollmann, M., and Bechinger, C. (2004) *Phys. Rev. Lett.*, **93**, 26001.
62 Kollmann, M. (2003) *Phys. Rev. Lett.*, **90**, 180602.
63 Taloni, A. and Marchesoni, F. (2006) *Phys. Rev. Lett.*, **96**.
64 Seehamart, K., Nanok, T., Krishna, R., van Baten, J.M., Remsungnen, T., and Fritzsche, S. (2009) *Microporous Mesoporous Mater.*, **125**, 97.
65 Seehamart, K., Nanok, T., Kärger, J., Chmelik, C., Krishna, R., and Fritzsche, S. (2010) *Microporous Mesoporous Mater.*, **130**, 92.

66 Ferey, G. and Serre, C. (2009) *Chem. Soc. Rev.*, **38**, 1380.

67 Serre, C., Millange, F., Thouvenot, C., Nogues, M., Marsolier, G., Louer, D., and Ferey, G. (2002) *J. Am. Chem. Soc.*, **124**, 13519.

68 Serre, C., Mellot-Draznieksd, C., Surble, S., Audebrand, N., Fillinchuk, Y., and Ferey, G. (2007) *Science*, **315**, 1828.

69 Mellot-Draznieksd, C., Serre, C., Surble, S., Audebrand, N., and Ferey, G. (2005) *J. Am. Chem. Soc.*, **127**, 16273.

70 Lincke, J., Lässig, D., Moellmer, J., Reichenbach, C., Puls, A., Moeller, A., Gläser, R., Kalies, G., Staudt, R., and Krautscheid, H. (2011) *Microporous Mesoporous Mater.*, **142**, 62–69.

71 Reichenbach, C., Kalies, G., Lincke, J., Lässig, D., Krautscheid, H., and Moellmer, J. (2011) *Microporous Mesoporous Mater.*, **142**, 592.

72 Barthelet, K., Marrot, J., Riou, D., and Ferey, G. (2002) *Angew. Chem. Int. Ed.*, **41**, 281.

73 Salles, F., Kolokolov, D.I., Jobic, H., Maurin, G., Llewellyn, P.L., Devic, T., Serre, C., and Ferey, G. (2009) *J. Phys. Chem. C*, **113**, 7802.

74 Salles, F., Jobic, H., Ghoufi, A., Llewellyn, P.L., Serre, C., Bourelly, S., Ferey, G., and Maurin, G. (2009) *Angew. Chem. Int. Ed.*, **121**, 8335.

75 Ghoufi, A. and Maurin, G. (2010) *J. Phys. Chem.*, **114**, 6496.

76 Salles, F., Jobic, H., Devic, T., Llewellyn, P.L., Serre, C., Ferey, G., and Maurin, G. (2010) *ACS Nano*, **4**, 143.

77 Rosenbach, N., Jobic, H., Ghoufi, A., Salles, F., Maurin, G., Bourrelly, S., Llewellyn, P.L., Devic, T., Serre, C., and Ferey, G. (2008) *Angew. Chem. Int. Ed.*, **47**, 6611.

78 Salles, F., Jobic, H., Maurin, G., Koza, M.M., Llewellyn, P.L., Devic, T., Serre, C., and Férey, G. (2008) *Phys. Rev. Lett.*, **100**, 245901.

79 Bär, N.K., Ernst, H., Jobic, H., and Kärger, J. (1999) *Magn. Reson. Chem.*, **37**, S79–S83.

80 Chmelik, C., Heinke, L., Kortunov, P., Li, J., Olson, D., Tzoulaki, D., Weitkamp, J., and Kärger, J. (2009) *Chem. Phys. Chem.*, **10**, 2623.

81 Shah, D.B. and Liou, H.Y. (1994) *Zeolites* vol. 14 p 541.

82 Frisch, H.L. (1958) *J. Phys. Chem.*, **62**, 401.

83 Ruthven, D.M. (2007) *Chem. Eng. Sci.*, **62**, 5745.

84 Heinke, L. and Kärger, J. (2011) *Phys. Rev. Lett.*, **106**, 74501.

85 Sholl, D.S. (2011) *Nat. Chem.*, **3**, 429.

86 Tuck, C. (1999) *Effective Medium Theory*, Oxford University Press, Oxford.

87 Ben-Avraham, D. and Havlin, S. (2000) *Diffusion and Reaction in Fractals and Disordered Systems*, Cambridge University Press, Cambridge.

88 Weiss, G.H. (1994) *Aspects and Applications of Random Walks*, North-Holland, Amsterdam.

89 Hibbe, F., Chmelik, C., Heinke, L., Li, J., Ruthven, D.M., Tzoulaki, D., and Kärger, J. (2011) *J. Am. Chem. Soc.*, **133**, 2804.

90 Feldhoff, A., Caro, J., Jobic, H., Krause, C.B., Galvosas, P., and Kärger, J. (2009) *ChemPhysChem.*, **10**, 2429.

91 Agger, J.R., Hanif, N., Cundy, C.S., Wade, A.P., Dennison, S., Rawlinson, P.A., and Anderson, M.W. (2003) *J. Am. Chem. Soc.*, **125**, 830.

92 Kärger, J. and Ruthven, D.M. (1989) *Zeolites*, **9**, 267.

93 Ruthven, D.M., Brandani, S., and Eic, M. (2008) *Adsorption and Diffusion* (eds H.G. Karge and J. Weitkamp), Springer, Berlin, p. 45.

94 Dudko, O.K., Berezhkovskii, A.M., and Weiss, G.H. (2004) *J. Chem. Phys.*, **121**, 11283.

95 Dudko, O.K., Berezhkovskii, A.M., and Weiss, G.H. (2005) *J. Phys. Chem. B*, **109**, 21296.

96 Kärger, J., Kortunov, P., Vasenkov, S., Heinke, L., Shah, D.B., Rakoczy, R.A., Traa, Y., and Weitkamp, J. (2006) *Angew. Chem. Int. Ed.*, **45**, 7846.

97 Tzoulaki, D., Schmidt, W., Wilczok, U., and Kärger, J. (2008) *Microporous Mesoporous Mater.*, **110**, 72.

98 Kortunov, P., Heinke, L., Vasenkov, S., Chmelik, C., Shah, D.B., Kärger, J., Rakoczy, R.A., Traa, Y., and Weitkamp, J. (2006) *J. Phys. Chem. B*, **110**, 23821.

99 Tzoulaki, D., Heinke, L., Schmidt, W., Wilczok, U., and Kärger, J. (2008) *Angew. Chem. Int. Ed.*, **47**, 3954.

100 Tzoulaki, D., Heinke, L., Castro, M., Cubillas, P., Anderson, M.W., Zhou, W., Wright, P., and Kärger, J. (2010) *J. Am. Chem. Soc.*, **132**, 11665.

101 Heinke, L., Kortunov, P., Tzoulaki, D., Castro, M., Wright, P.A., and Kärger, J. (2008) *Europhys. Lett.*, **81**, 26002.

102 Paoli, H., Methivier, A., Jobic, H., Krause, C., Pfeifer, H., Stallmach, F., and Kärger, J. (2002) *Microporous Mesoporous Mater.*, **55**, 147.

Part VI
Selected Applications

20
Zeolite Membranes

In the context of separations technology, zeolite membranes are the "holy grail" since such membranes offer the tantalizing prospect of carrying out size-selective molecular sieve separations under steady-state continuous flow conditions, rather than under the inherently more complex unsteady-state conditions characteristic of pressure swing or thermal swing adsorption separation processes. For several important separations the performance of a zeolite membrane substantially exceeds that of the best polymeric membranes. However, the task of producing a thin coherent zeolite membrane that is substantially defect free and sufficiently robust to withstand the conditions of a commercial process is extremely challenging. A wide range of potentially interesting zeolite membranes have been produced on the laboratory scale and in many cases the fluxes and selectivities are impressive but problems of durability, scale-up, and cost have, so far, inhibited commercialization. Despite the intensive research that has been carried out over the past decade, only one zeolite membrane process has actually been developed to commercial scale – the pervaporative removal of water from alcohols using a Linde type A membrane.

The literature on this subject has expanded exponentially. An excellent review with 159 references has been given by Caro *et al.* [1] as part of a special volume on zeolite membranes [2]. The present chapter aims therefore to provide a broad overview with an emphasis on the qualitative and quantitative understanding of the observed behavior and the experimental characterization of zeolite membranes rather than a detailed review of the published work on this subject. The performance of a zeolite membrane depends on the subtle interplay of both kinetic and equilibrium phenomena. As a result, the dependence of flux and selectivity on the size and properties of the sorbate molecules and the structure of the zeolite host is complex and often counter-intuitive. Although the main focus is on zeolite membranes some discussion of MOF (metal organic framework) membranes and small-pore amorphous silica membranes is also included since, despite their amorphous nature, the behavior of these materials is similar to that of zeolite membranes of comparable pore size.

Diffusion in Nanoporous Materials. Jörg Kärger, Douglas M. Ruthven, and Doros N. Theodorou.
© 2012 Wiley-VCH Verlag GmbH & Co. KGaA. Published 2012 by Wiley-VCH Verlag GmbH & Co. KGaA.

20.1
Zeolite Membrane Synthesis

Because zeolite membranes are inherently fragile they must be grown on a suitable porous support. Various different types of support have been used but the commonest choice is porous alumina, which is readily available in various forms, including small tubes suitable for incorporation into a membrane module. A variety of different seeding strategies have been developed to grow the crystals as a thin coherent layer with uniform orientation (see, for example, References [3–10]). Careful control of all the obvious variables, especially the degree of supersaturation of the mother liquor, is essential. It has been found advantageous to grow the membrane in the inverted (upside down) configuration to avoid the incorporation of detritus from the solution. Thin coherent layers of several different zeolites, including A, X, and silicalite (MFI) and DDR, have been successfully grown by such techniques. Remarkably, even when the membrane is coherent, the individual zeolite crystals are generally easily visible in both the surface and the cross section – see Figure 20.1. The actual thickness of the active layer is difficult to determine but it is evidently much thinner than the crystal layer.

Silicalite membranes oriented along the c-axis (the longest direction of the crystal) are among the easiest to grow but other orientations, including b-axis alignment (the direction of the straight channels), have also been achieved. This alignment is potentially more interesting as it yields very high fluxes and, under the right conditions, the selectivity is also high. Oriented membranes consisting of assemblages of individual crystals of $AlPO_4$-5 and silicalite have also been produced by orienting the crystals in an electric field and depositing metal between them by either sputtering or electrolysis [11–14]. However, membranes produced in this way are obviously expensive and it is difficult to see how the technique could be simplified and scaled up for industrial production.

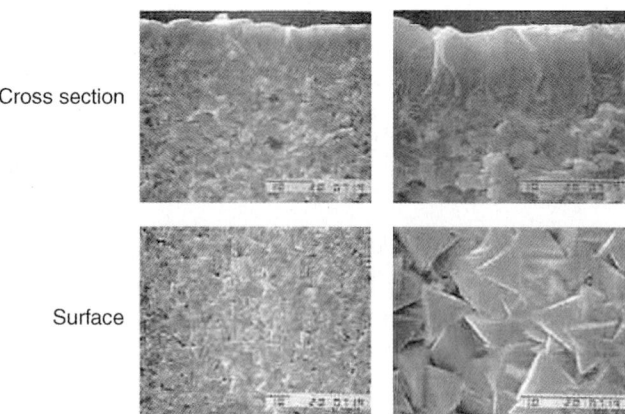

Figure 20.1 SEM photomicrographs showing the structure of a DD3R membrane. Courtesy of Professor Freek Kapteijn, Delft.

In polycrystalline membranes the active zeolite layer appears to be much thinner than the actual zeolite crystal layer (often a fraction of a micron), so fluxes can be surprisingly high. The surface of even a highly perm-selective membrane generally appears to be highly imperfect with an irregular array of cracks and grain boundaries, suggesting that the active layer must lie well below the surface. The early silicalite membranes were grown to a thickness of ten microns or more on the assumption that a thicker membrane would be more robust. It turns out that thin zeolite membranes are much more flexible and have much greater thermal stability than the thicker ones, which invariably crack on heating even to moderately elevated temperatures. This is of course advantageous since, from performance considerations, the thinner membranes are also preferable because they yield a higher flux for a given driving force.

Transport through a zeolite membrane should be directly related to the intracrystalline diffusivities for the relevant zeolite in the appropriate direction. However, although there is a general qualitative correspondence between permeation rates and measured diffusivities, the correspondence is far from straightforward. It appears that preferential transport along grain boundaries and barriers to transport due to growth planes and structural defects play an important role. The behavior of oriented $AlPO_4$-5 and b-oriented silicalite membranes is perhaps the easiest to understand since these membranes have straight pores in the direction of transport and tend to show the behavior expected for single-file diffusion systems.

20.2
Single-Component Permeation

Since the purpose of a zeolite membrane is to separate the components of gas (or liquid) mixtures the practical application of a membrane will always involve contact with a binary or multicomponent fluid phase. However, to understand the performance of a zeolite membrane it is helpful to consider first the permeation of a pure component. The simplest model, which is similar in concept to the well-known solution-diffusion model for a polymeric membrane, assumes a diffusive flux driven by the concentration gradient, in accordance with Fick's first law:

$$N = -D \frac{\partial q}{\partial z} \qquad (20.1)$$

The concentration gradient, and therefore the driving force, is provided by the pressure difference over the membrane. If the equilibrium isotherm is linear ($q^* = Kp$) Eq. (20.1) may be re-written as:

$$N = -KD \frac{dp}{dz} \qquad (20.2)$$

Integration across the membrane (from $p = p_H$ at $z = 0$ to $p = p_L$ at $z = \ell$) assuming that the equilibrium constant and diffusivity are independent of loading yields:

$$N = \frac{KD}{\ell}(p_H - p_L) \qquad (20.3)$$

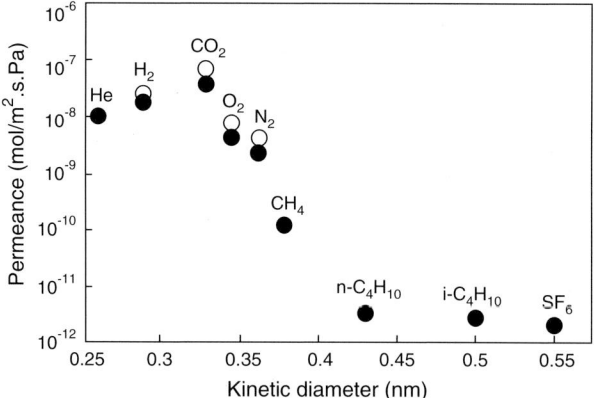

Figure 20.2 Variation of permeance with kinetic diameter for a DD3R membrane. From Tomita et al. [57] with permission.

The constant of proportionality (KD/ℓ) between the flux and the pressure difference is referred to as the *permeance* (π) while the product of the permeance and membrane thickness (KD), which is the constant of proportionality between the flux and the pressure gradient, is referred to as the *permeability*. Since the membrane thickness is often not known with any precision the permeance is generally the more useful quantity. The model outlined above is valid only in the low-concentration limit (the Henry's law region). At higher loadings the permeance (and permeability) becomes pressure dependent, thus limiting the usefulness of the simple model.

Figure 20.2 shows the typical pattern of variation of permeance with molecular size. For small molecules the permeance increases with molecular size due to the increasing strength of adsorption (increasing K values) but as the size of the molecule approaches the pore size steric hindrance becomes the dominant effect and the permeance decreases sharply.

20.2.1
Selectivity and Separation Factor

If we now consider the membrane to be exposed to a binary mixture of components A and B, with a difference in the partial pressures of both species across the membrane, at a sufficiently low loading such that the molecules of A and B diffuse independently within the membrane, we may assume that the ratio of the fluxes of A and B (normalized for any difference in partial pressures) will correspond to the ratio of the permeances:

$$S_{AB} = \frac{N_A}{N_B} \frac{(p_{HB}-p_{LB})}{(p_{HA}-p_{LA})} = \frac{K_A D_A}{K_B D_B} = \frac{\pi_A}{\pi_B} \tag{20.4}$$

This ratio is commonly called the *intrinsic selectivity* since it defines the selectivity that would be expected under ideal conditions within Henry's law region. At higher loadings the actual selectivity may be either larger or smaller than the intrinsic value but since the intrinsic value is directly related to simply measurable constants it is still a useful parameter. The practical performance of a membrane is defined by the permeance and selectivity under the relevant operating conditions and it is only within the low-concentration limit that these quantities are simply related to the Henry constants and limiting diffusivities.

In a membrane separation process the feed stream (A + B) is split into a permeate stream enriched in the more permeable component (A) and a retentate stream enriched in the less permeable component (B). If we assume perfect mixing (and therefore uniform composition) on both sides of the membrane it follows that the "separation factor" is given by:

$$\alpha \equiv \frac{\text{mole ratio A/B in permeate}}{\text{mole ratio A/B in retentate}} = \frac{y}{x}\left(\frac{1-x}{1-y}\right) \quad (20.5)$$

where x and y are the mole fractions of A in permeate and retentate, respectively. If the membrane is operated under conditions such that there is zero back pressure the flux ratio (N_A/N_B), and hence the separation factor, will approach the intrinsic selectivity, as defined by Eq. (20.4). This is the most favorable situation. Any back pressure on the permeate side will reduce the separation factor below this limiting value.

Figure 20.3a shows the variation of permeance with reciprocal temperature for N_2, CO_2, n-butane, and isobutane in a silicalite membrane. Taking account of the

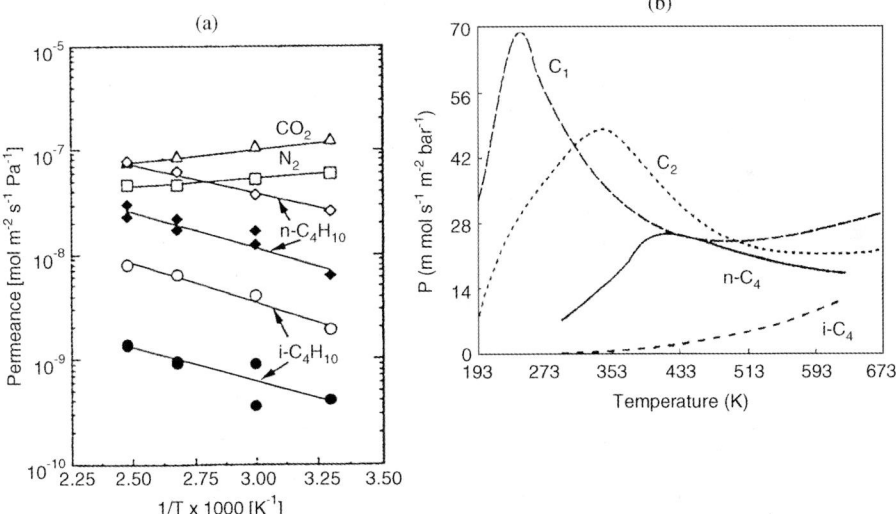

Figure 20.3 Temperature dependence of (a) permeance and (b) flux for permeation of permanent gases and light hydrocarbons through silicalite membranes. (a) Permeance data for N_2, CO_2, and n/isobutane as a function of reciprocal temperature. From Kusakabe et al. [15] with permission. Note that the data for permeation of nC_4–iC_4 mixtures (filled symbols) show a reduced flux but a higher permeability ratio (selectivity), suggesting that competitive adsorption reduces the permeance of iC_4 more than that of nC_4. (b) Permeance for CH_4, C_2H_6, C_3H_8, and n/isobutanes plotted as a function of temperature for fixed p_H, p_L values, From Bakker et al. [16] with permission.

temperature dependence of K and D [$K = K_\infty e^{-\Delta H/RT}$; $D = D_\infty e^{-E/RT}$], Eq. (20.3) predicts that the permeance should vary exponentially with reciprocal temperature, either increasing or decreasing depending on the sign of [$(-\Delta H) - E$]. This pattern of behavior is commonly observed at relatively low loadings. For CO_2 and N_2 the permeance decreases with increasing temperature, suggesting that $(-\Delta H)$ is greater than E but for the butanes the opposite trend is observed, suggesting $(-\Delta H) < E$. It can also be seen that the permeance of both species is lower in a 50:50 mixture (filled symbols) than for the pure components (open symbols), but the permeance of isobutane is reduced by a larger factor than that for n-butane. The perm-selectivity in favor of n-butane is therefore larger in the mixture than would be predicted from the single-component permeances, suggesting that isobutane is relatively less adsorbed in the mixture as a result of competitive adsorption. In general, however, if the temperature is varied over a wide enough range, the permeance passes through a maximum, as shown in Figure 20.3b.

20.2.2
Modeling of Permeation

A more satisfactory model that can account for this pattern of temperature dependence, as well as for the effects of other variables such as feed pressure, is obtained by considering the driving force to be the gradient of chemical potential rather than the pressure gradient. This leads to:

$$N = -Bq\frac{\partial \mu}{\partial z} \qquad (20.6)$$

in place of Eq. (20.1).

Assuming that the vapor phase is ideal, the chemical potential (μ) is related to the partial pressure by:

$$\mu = \mu^*(T) + RT \ln p \qquad (20.7)$$

so Eq. (20.6) becomes:

$$N = -BRTq\frac{d\ln p}{\partial q} \cdot \frac{\partial q}{\partial z} = -D_0 \frac{d\ln p}{d\ln q} \cdot \frac{\partial q}{\partial z} \qquad (20.8)$$

where $D_0 \equiv BRT$ is the "intrinsic" (or thermodynamically corrected) diffusivity (Section 1.2).

If the equilibrium isotherm can be described by the Langmuir expression:

$$\frac{q}{q_s} = \frac{bp}{1+bp} \quad \text{or} \quad bp = \frac{q/q_s}{1-q/q_s} \qquad (20.9)$$

straightforward differentiation yields:

$$\frac{d\ln p}{d\ln q} = 1 + bp \tag{20.10}$$

$$\frac{\partial q}{\partial z} = \frac{dq}{dp} \cdot \frac{\partial p}{\partial z} = \frac{bq_s}{(1+bp)^2} \cdot \frac{\partial p}{\partial z} \tag{20.11}$$

Equation (20.8) thus becomes:

$$N = \frac{-D_0 bq_s}{(1+bp)} \cdot \frac{\partial p}{\partial z} \tag{20.12}$$

and integrating across the membrane we obtain:

$$N = \frac{D_0 q_s}{\ell} \ln\left(\frac{1+bp_H}{1+bp_L}\right) \tag{20.13}$$

When the product bp_H is small ($bp_H \ll 1.0$) the logarithm may be expanded, retaining only the first term, to yield:

$$N = \frac{D_0 bq_s}{\ell}(p_H - p_L) \tag{20.14}$$

which, since $K = bq_s$, is identical with Eq. (20.3).

This formulation is formally equivalent to the model used by Talu et al. [17], which assumes that the flux through the membrane is driven by the difference in spreading pressures across the membrane.

Differentiation of Eq. (20.13), assuming the usual forms of temperature dependence for b ($b = b_\infty e^{-\Delta H/RT}$) and D_o ($D_o = D_\infty e^{-E/RT}$), shows that for given values of p_H and p_L the flux will pass through a maximum at the temperature at which the log mean value of $(1-\theta)$ corresponds to the ratio of activation energy to heat of adsorption:

$$\frac{\theta_H - \theta_L}{\ell n\left[\frac{1-\theta_L}{1-\theta_H}\right]} = \frac{E}{(-\Delta H)} \tag{20.15}$$

The commonly observed pattern of temperature dependence (Figure 20.3b) is therefore accounted for, at least qualitatively.

Figure 20.4a shows experimental permeance data for the permeation of CO_2 through an amorphous silica membrane [18–20], plotted in accordance with Eq. (20.13) [21]. The chemical potential driving force model accounts very well for the effect of varying feed and permeate pressures, and the b values derived from the permeance data agree well with values measured independently for the bulk material. The effect of increasing the feed pressure from 3 to 20 atm is also correctly predicted (Figure 20.4c).

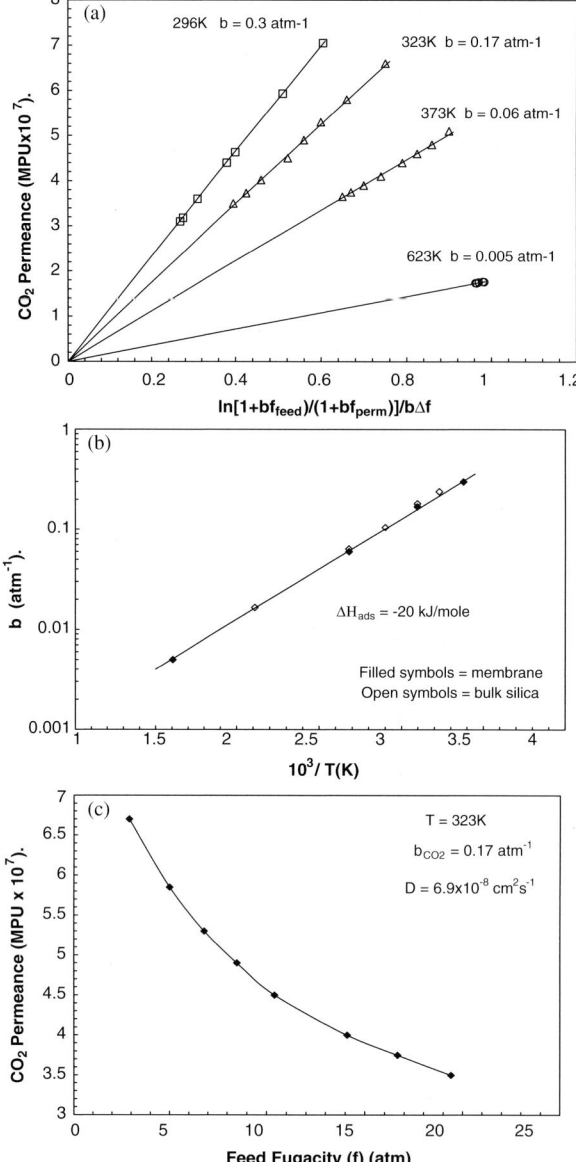

Figure 20.4 Experimental permeance data for CO_2 in a small-pore silica membrane for various temperatures and pressures showing (a) the fit of the experimental data to Eq. (20.13), from which values of b may be obtained; (b) temperature dependence of b; (c) variation of permeance with feed pressure (permeate pressure ≈ 0), showing agreement between the high-pressure data and the predictions from the low-pressure data according to Eq. (20.13). Data from Chance et al. [21].

20.3
Separation of Gas Mixtures

The behavior of zeolite membranes with respect to the separation of gas mixtures is more complex than might have been anticipated. It is useful to distinguish three different modes of separation: size-selective molecular sieving, diffusion-controlled permeation, and equilibrium-controlled permeation. However, the boundaries between these categories are not sharp.

20.3.1
Size-Selective Molecular Sieving

The concept of a size-selective molecular sieve separation is straightforward. Uniformity of pore size is a distinguishing feature of a zeolite membrane so one may expect that, in a perfect membrane, molecules that are too large to enter the pores will be completely excluded while the smaller molecules that can enter will diffuse through the pores to be recovered as the permeate. Such a process should be characterized by a very high (ideally infinite) selectivity and a high flux, since diffusion of the permeate within the pores will not be obstructed by the presence of the other component. A high-quality zeolite membrane (i.e., one with no cracks or pinholes) does indeed show a large difference in permeance between molecules that are small enough to enter the pore system and those that are too large to enter. However, the size-selective cut-off is generally not total as, in practice, there are always some non-selective paths, associated with grain boundaries or structural defects. As a result the permeance ratio between penetrating and "excluded" molecules is commonly perhaps 100 rather than infinity – see Figure 20.5a [12].

In most zeolite membranes, even those without obvious cracks or defects, the effect of non-selective paths is even more significant so, although there is still a large difference in the permeance of small and large molecules, there may be no sharp cut-off (Figure 20.5b). Even in a membrane that is sufficiently ideal to show a sharp cut-off, a high molecular sieve selectivity is observed only when there is a large difference in size between the two components. If the larger molecules are only slightly larger than the pore diameter they will penetrate slowly, leading to a catastrophic drop in both selectivity and flux since the flux of the smaller molecules is then largely controlled by the rate at which the larger molecules diffuse through the pore. This is particularly true when the pore system is unidimensional as in $AlPO_4$-5 and is illustrated very clearly in Table 20.1 [1]. Size-selective sieving with selectivities in excess of 100 is observed for n-heptane–TIPB and n-heptane–TEB but for n-heptane/m-xylene the selectivity falls to 1.7 because mesitylene is small enough to enter the pores, albeit slowly. Note that for n-heptane–toluene the selectivity is inverted. Both molecules are small enough to enter the pores quite easily and the selectivity is now determined by the preferential adsorption of toluene rather than by steric hindrance.

Table 20.1 Effect of molecular diameter on flux and selectivity for an AlPO$_4$-5 membrane.[a] From Caro et al. [1, 12, 14].

	n-Heptane (single comp.)	n-Heptane– toluene	n-Heptane– mesitylene	n-Heptane– TEB	n-Heptane– TIBP
Flux × 10^6 (mol cm^{-2} s^{-1})	3.9	0.85	0.43	1.82	0.94
Flux relative to pure n-heptane	1.0	0.22	0.11	0.47	0.24
Perm selectivity (α)	—	0.8	1.7	105	1220

a) AlPO$_4$-5 in nickel membrane at 91 °C. $\Delta P = 1$ bar. Feed is an equimolar mixture of aromatic and n-heptane; TEB = triethylbenzene and TIPB = triisopropylbenzene.

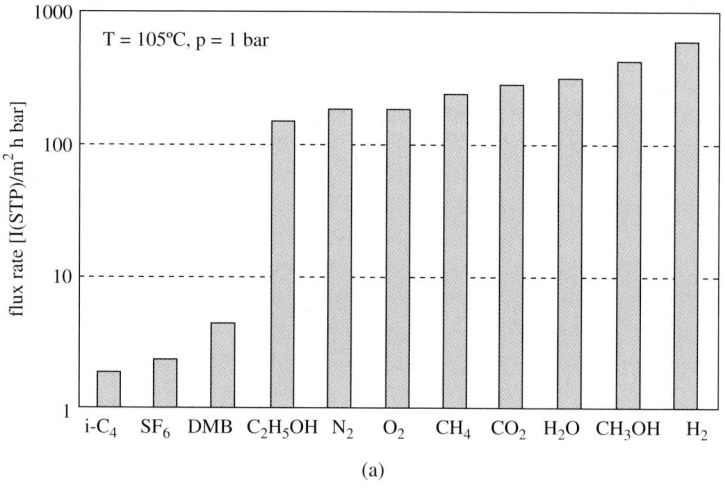

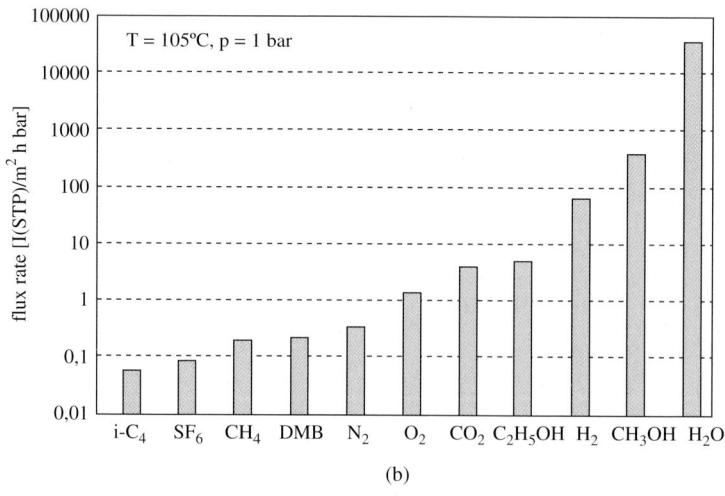

Figure 20.5 Single-component fluxes for several different sized probe molecules at 105 °C on two different silicalite membranes: (a) shows a clear evidence of size-selective sieving; (b) shows permeances that vary over five orders of magnitude but without any clear break corresponding to the nominal pore size. Reprinted from Caro et al. [1] with permission.

20.3.2
Diffusion-Controlled Permeation

Even when the conditions for size exclusion are not fulfilled, within the sterically hindered diffusion regime differences in intracrystalline diffusivity between molecules of slightly different size or shape are often large. Under these conditions it should be possible to achieve an efficient separation based on differences in permeation rates. However, in practice it is generally found that an efficient separation can be achieved only when the membrane is operated under low-loading conditions (high temperature or low sorbate pressure). At high loadings the mobility of the faster diffusing species is obstructed by the presence of the slower species (the mutual diffusion effect) so that both the flux and the selectivity are reduced. In the extreme case of a one-dimensional pore system, for large sorbate molecules, one can expect no selectivity under these conditions since single-file diffusion means that both species will diffuse at the same rate.

A good example is provided by the data shown in Figure 20.6 for permeation of n-butane–isobutane through a silicalite membrane [22]. At low sorbate pressures separation factors approaching 100 are observed. As the feed pressure increases the fluxes of both components increase but the flux of n-butane increases less rapidly than that of isobutane, leading to a decrease in the perm-selectivity.

A second example is provided by the data shown in Figure 20.7 for the separation of o- and p-xylene on a silicalite membrane [23]. For the normal orientation (c-axis) the fluxes of both components are moderately high but there is very little perm-selectivity and both fluxes and selectivity are almost independent of temperature. In this orientation the diffusing xylene molecules must jump between the straight and sinusoidal channels, a transition that has been shown to be sterically hindered for the long p-xylene molecule. As a result of this there is little difference in diffusivity between o- and p-xylene in this direction. The activation energies for both components are relatively high, approaching the heat of adsorption (≈ 17.5 kcal mol^{-1}) [24], so the fluxes show very little temperature dependence. In contrast, for the b-axis orientation, in which the straight channels are aligned with the diffusion path, we observe a very high selectivity under low-loading conditions (at high temperature). As the temper-

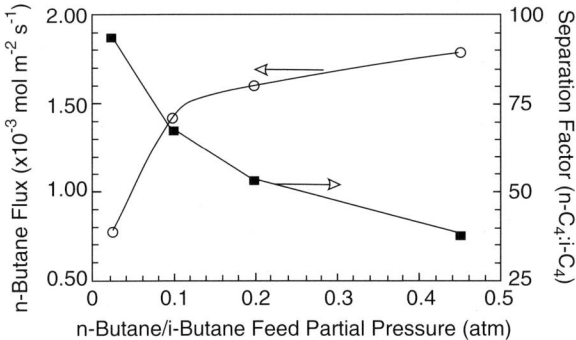

Figure 20.6 Influence of loading on the selectivity for n-butane–isobutane in a silicalite membrane. Reprinted from Tsapatsis and Gavalas [22] with permission.

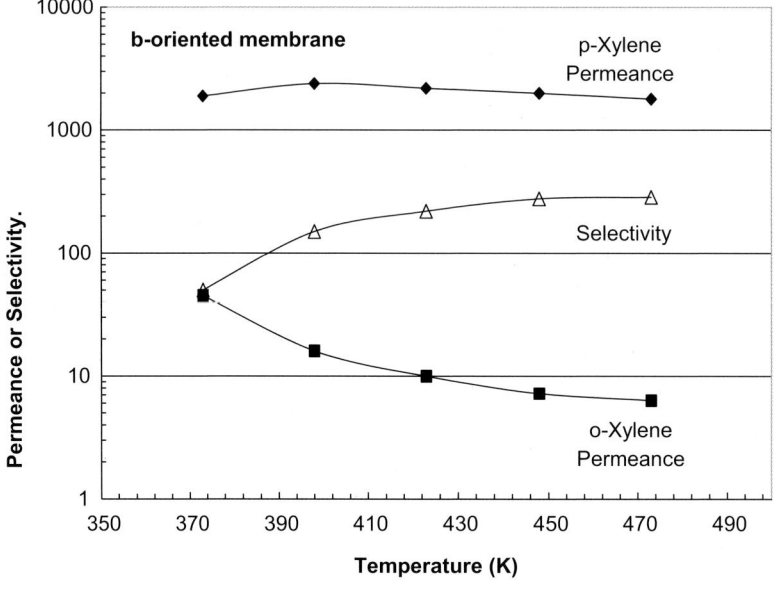

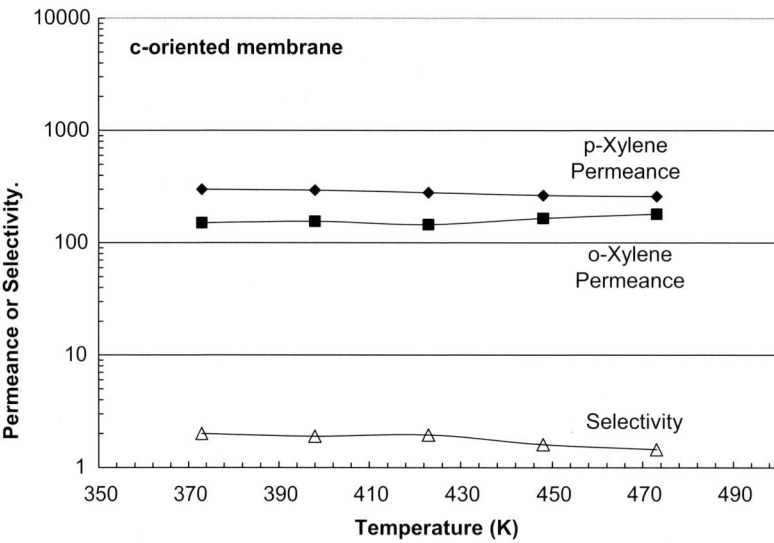

Figure 20.7 Permeance and selectivity for an equimolar mixture of o-xylene and p-xylene over c-oriented and b-oriented silicalite membranes, showing the development of high flux and high selectivity at low loadings in the b-oriented membrane (oriented in the direction of the straight channels). From data of Lai *et al.* [23].

ature is decreased (leading to increased loading) the flux of *p*-xylene remains almost constant whereas the *o*-xylene flux increases, leading to a decrease in perm-selectivity. This may be due either to a mutual diffusion effect or to the equilibrium becoming relatively more favorable for *o*-xylene at lower temperatures.

The perm-selectivity for a mixture is generally found to be lower than that estimated from the ratio of the pure-component permeances [Eq. (20.4)]. However, this is not always true. If the faster diffusing species is also the more strongly adsorbed component then under competitive adsorption conditions the adsorption of the slower component will be suppressed, leading to an increase in perm-selectivity. Such an effect has been observed for the n-hexane–DMB mixture, for which separation factors (for the mixture) greater than 1000 were observed in favor of n-hexane [25]. This is substantially larger than the measured permeance ratio for the pure components.

The equilibrium effect is particularly strong for mixtures consisting of a rapidly diffusing but weakly adsorbed species such as hydrogen with a more strongly adsorbed but slower diffusing species. For example, the pure-component permeance ratio for H_2–SF_6 in silicalite is about 136 but the binary selectivity was found to be only about 13 [26]. A similar effect was observed for H_2–n-butane, where at room temperature the binary selectivity strongly favors butane rather than hydrogen [16].

20.3.3
Equilibrium-Controlled Permeation

At high sorbate loadings the effect of adsorption equilibrium tends to become dominant. Thus for CH_4–n-butane on silicalite the pure-component permeance ratio is about 3 but in the binary mixture the selectivity is inverted, leading to preferential permeation of butane $S_{CH4-nC4} \approx 0.06$ [27].

The transient behavior of this system offers interesting insights (Figure 20.8) [28]. When a clean silicalite membrane is exposed to a 50 50 binary mixture of methane + n-butane the permeate is initially almost pure methane. The butane penetrates the membrane more slowly so that the butane appears in the permeate only after about 45 s. As the butane flux increases the methane flux declines, showing that the strongly

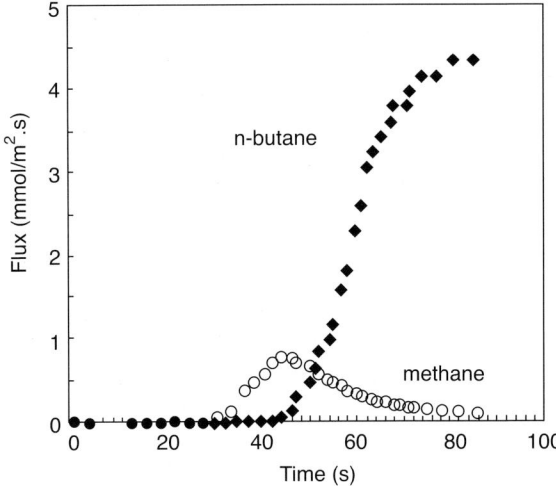

Figure 20.8 Transient permeation behavior of a 50 50 binary mixture of methane–n-butane in a silicalite membrane at 25 °C. Reprinted from Geus *et al.* [28] with permission.

adsorbing butane is preventing the methane from entering the pores. If the temperature is raised above 200 °C the loading decreases to a low level and methane again becomes the preferentially permeating species.

20.4
Modeling Permeation of Binary Mixtures

In principle the simple model developed in Eqs. (20.6)–(20.13) can be extended to a binary (or multicomponent) system simply by using the binary Langmuir isotherm to relate activity to adsorbed phase concentration. This leads to the equation:

$$N_A = \frac{D_{0A} q_s}{\delta} \ln \left\{ \frac{1 + b_A p_{AH} + b_B p_{BH}}{1 + b_A p_{AL} + b_B p_{BL}} \right\} \quad (20.16)$$

However, these expressions have been found to be of limited practical value, probably because the binary Langmuir isotherm does not provide a sufficiently accurate model for the binary equilibria.

20.4.1
Maxwell–Stefan Model

It was pointed out in Section 3.3 that in many binary systems, especially at higher loadings, mutual diffusion effects play a significant role. To describe the system behavior quantitatively it is therefore essential that the model should include such effects. The most promising approach appears to be based on the generalized Maxwell–Stefan (GMS) model [29–33], which was discussed in Chapter 3 (Sections 3.3 and 3.4). In this model the basic expression for the flux is:

$$-\frac{q_i}{RT} \nabla \mu_i = \sum_{j=i}^{n} \frac{q_j N_i - q_i N_j}{q_s \mathcal{D}_{ij}} + \frac{N_i}{q_s D_{0i}} \quad (20.17)$$

where $\mathcal{D}_{ij}$ represents the Maxwell–Stefan mutual diffusion coefficient and D_{0i} represents the thermodynamically corrected transport diffusivity for component i [see Eq. (1.31)].

The chemical potential gradient is given by a matrix of thermodynamic functions:

$$\frac{q_i}{RT} \nabla \mu_i = \sum_{j=1}^{n} \Gamma_{ij} \nabla \theta_i; \quad \cdots \Gamma_{ij} \equiv q_i \frac{\partial \ln p_i}{\partial q_j} \quad (20.18)$$

$i, j = 1, 2 \ldots n$

The adsorbed phase concentration is related to the activity (assumed equal to the partial pressure) by the adsorption equilibrium isotherm, which is often assumed to be of Langmuirian form:

$$\frac{q_i}{q_s} = \frac{b_i p_i}{1 + \sum_{j=1}^{n} b_j p_j} = \theta_i \quad (20.19)$$

This leads to the following expressions for the flux in single-component and binary systems:

$$\text{Single-component:} \quad N_i = -q_s \frac{D_{oi}}{1-\theta_i} \cdot \nabla \theta_i \tag{20.20}$$

$$\text{Binary:} \quad N_A = q_s \frac{D_{oA}}{(1-\theta_A-\theta_B)} \cdot \frac{[Đ_{AB}(1-\theta_B)+D_{oB}\theta_A]\nabla\theta_A + [Đ_{AB}\theta_A+D_{oB}\theta_A]\nabla\theta_B}{Đ_{AB}+\theta_B D_{oA}+\theta_A D_{oB}} \tag{20.21}$$

with a similar expression for N_B. If the mutual diffusivity is large ($Đ_{AB} \to \infty$) these expressions reduce to the simple Habgood formulation [see Eq. (3.49)].

The mutual diffusion coefficients are not amenable to direct measurement and must therefore be estimated. Krishna has suggested that for molecules of similar size the Vignes correlation [34] may be used:

$$Đ_{AB} = D_{0A}^{\frac{\theta_A}{\theta_A+\theta_B}} \cdot D_{0B}^{\frac{\theta_B}{\theta_A+\theta_B}} \tag{20.22}$$

whereas for differently sized molecules [33]:

$$q_{sB} Đ_{AB} = (q_{sB} D_{0A})^{\frac{\theta_A}{\theta_A+\theta_B}} \cdot (q_{sA} D_{0B})^{\frac{\theta_B}{\theta_A+\theta_B}} \tag{20.23}$$

Other empirical correlations have also been suggested – see Section 3.4.

The values of D_{oA} and D_{oB} (assumed independent of loading) are taken from single-component diffusivities or permeance data and the values of the saturation capacities are estimated from equilibrium data. Such a model has been shown to provide good predictions of the performance of a silicalite membrane for several hydrocarbon separations. Representative comparisons between the experimental permeance and selectivity and those predicted from the model with parameters derived from single-component data are shown, for methane–ethane permeation through a silicalite membrane, in Figure 20.9 [31].

The development of the GMS theory, outlined above, is based on the simple Langmuir model for adsorption equilibrium. However, the theory can be adapted to incorporate any thermodynamically consistent equilibrium model although the resulting expressions may be cumbersome. The development using the more realistic but more complex ideal adsorbed solution theory (IAS) has been presented by Kapteijn et al. [33].

20.4.1.1 Concentration Profile through the Membrane

In a typical experimental system the temperature and partial pressures of both components are known at the high pressure surface while at the low pressure surface only the total pressure is known a priori. Under stead-state conditions the fluxes of both components are constant through the membrane (i.e., they do not depend on z). The Langmuir equilibrium parameters (b_A, b_B, and q_s) and the membrane thickness (ℓ) are assumed to be known and information on the corrected diffusivities for both components (D_{0A}, D_{0B}) as a function of composition (θ_A, θ_B) is also necessary. It is usually assumed that the corrected diffusivities are constant (invariant with loading) but solutions have also been obtained for the case in which D_{0A}, and D_{0B}, decrease

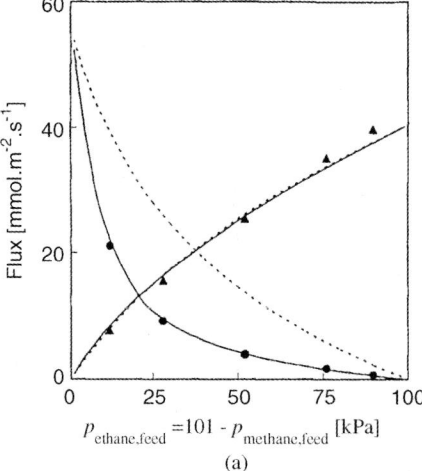
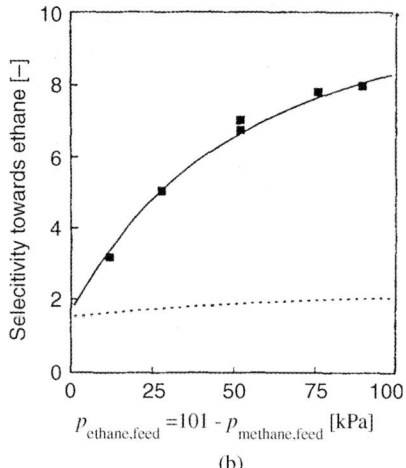

Figure 20.9 Separation of ethane–methane mixtures by permeation through a silicalite membrane: (a) flux; (b) selectivity. The continuous lines show the prediction of the Maxwell–Stefan model based on single-component diffusion and equilibrium parameters with the mutual diffusivities estimated from the Vignes correlation. The dotted lines show the predictions of the simplified (Habgood) model in which mutual diffusion effects are ignored. Reprinted from van de Graaf et al. [31] with permission.

with total loading in accordance with $D_{0A} = D_{0A}(0)(1 - \theta_A - \theta_B)$, and $D_{0B} = D_{0B}(0)(1 - \theta_A - \theta_B)$. The mutual diffusivity $Đ_{AB}$ is estimated from the Vignes correlation [Eq. (20.22)]. Because of the form of the equations and the boundary conditions the numerical integration of Eq. (20.21) is not completely straightforward. Details of one procedure have been given by Krishna and Paschek [30] and Krishna and van de Broeke [32].

When $Đ_{ij} \to \infty$ the flux equations for a binary system reduce to the Habgood [35, 36] form [Eq. (3.49)]:

$$N_A = -\frac{q_s D_{0A}}{\theta_v}\left[(1-\theta_B)\frac{d\theta_A}{dz} + \theta_A \frac{d\theta_B}{dz}\right] \tag{20.24}$$

where $\theta_v = (1 - \theta_A - \theta_B)$, with a similar expression for N_B. Krishna and Bauer [37] have shown that in this situation (with D_{0A} and D_{0B} constant and the partial pressures of both components defined at both faces of the membrane) a simple analytic solution may be obtained. The profiles of vacancies and partial pressures through the membrane are given by:

$$\frac{1-e^{-\phi\eta}}{1-e^\phi} = \frac{1/\theta_v - 1/\theta_{v0}}{1/\theta_{v\ell} - 1/\theta_{v0}} = \frac{p_i - p_{i0}}{p_{i\ell} - p_{i0}} \tag{20.25}$$

where $e^{-\phi} = \theta_{v0}/\theta_{v\ell}$ and the subscripts 0 and ℓ refer to the upstream and downstream faces of the membrane, respectively. The fluxes are given by:

$$N_i = \frac{b_i D_{0i} q_s}{\ell} \cdot \frac{\ln \theta_{ve}/\theta_{v0}}{(1/\theta_{v0} - 1/\theta_{v\ell})} \cdot (p_{i0} - p_{i\ell}) \tag{20.26}$$

where $b_i(b_A, b_B)$ and q_s are the Langmuir parameters and D_{0i} is the corrected single-component diffusivity. Since $1/\theta_v = 1 + b_A p_A + b_B p_B$ Eq. (20.26) may also be written in the equivalent explicit form:

$$N_i = \frac{D_{0i} b_i q_s}{\ell} \cdot \frac{\ln\left[\frac{1 + b_A p_{A\ell} + b_B p_{B\ell}}{1 + b_A p_{A0} + b_B p_{B0}}\right]}{b_A(p_{A0} - p_{A\ell}) - b_B(p_{B0} - p_{B\ell})} \cdot (p_{i0} - p_{i\ell}) \tag{20.27}$$

When D_{0i} decreases in proportion to the fractional occupancy $D_{0i} = D_{0i}(0)(1 - \Sigma\theta_i) = D_{0i}(0)\theta_v$ the occupancy profile decreases linearly instead of exponentially with distance:

$$\frac{\theta_v - \theta_{v0}}{\theta_{v\ell} - \theta_{v0}} = \eta = z/\ell = \frac{\theta_i - \theta_{i0}}{\theta_{i\ell} - \theta_{i0}} \tag{20.28}$$

and the fluxes are given by:

$$N_A = \frac{\theta_{v0}\theta_{v\ell}}{\ell} b_A q_s D_{0A}(p_{A0} - p_{A\ell}); \quad N_B = \frac{\theta_{v0}\theta_{v\ell}}{\ell} b_B q_s D_{B0}(p_{B0} - p_{B\ell}) \tag{20.29}$$

These solutions are exact for Langmuirian systems when mutual diffusion effects are not significant ($D_{ij} \to \infty$).

When mutual diffusion effects are important an exact analytic solution is no longer possible. However, approximate analytic solutions can still be obtained if the mutual diffusivity $Đ_{AB}$ is estimated at the upstream face of the membrane and assumed constant. For D_{oA} and D_{oB} constant the expression for the flux is then (in matrix form):

$$(N_i) = \frac{q_s}{\ell} \cdot \frac{\ln \theta_{v\ell}/\theta_{v0}}{(1/\theta_{v0} - 1/\theta_{v\ell})} \cdot [B_0]^{-1}(b_i p_{i0} - b_i p_{i\ell}) \tag{20.30}$$

$$[B_0]^{-1} = \begin{bmatrix} D_{0A} & 0 \\ 0 & D_{0B} \end{bmatrix} \begin{bmatrix} \dfrac{1 + \theta_A D_{0B}}{Đ_{AB}} & \dfrac{\theta_A D_{0B}}{Đ_{AB}} \\ \dfrac{\theta_B D_{0A}}{Đ_{AB}} & \dfrac{1 + \theta_B D_{0A}}{Đ_{AB}} \end{bmatrix} \frac{1}{1 + \dfrac{\theta_A D_{0B}}{Đ_{AB}} + \dfrac{\theta_B D_{0A}}{Đ_{AB}}} \tag{20.31}$$

The mutual diffusivity $Đ_{AB}$ (at $z = 0$) is estimated from the Vignes correlation [Eq. (20.22)].

For the situation in which D_{oA} and D_{oB} decrease in proportion to the occupancy the corresponding expression is:

$$(N_i) = \frac{q_s}{\ell} \theta_{v0} \theta_{v\ell} [B_0]^{-1}(b_i p_{i0} - b_i p_{i\ell}) \tag{20.32}$$

where $[B_0]$ has the same meaning. The validity of these approximate solutions has been confirmed by comparison with exact numerical solutions [37].

As a representative example Figure 20.10a and b shows the fluxes of 2-methylpentane and 2,2-dimethylbutane for permeation of a binary mixture of these components through a silicalite membrane. For these sorbates the corrected diffusivity decreases in proportion to the fractional occupancy, so the solutions were calculated from Eqs. (20.29) or (20.32) (for both negligible and significant mutual

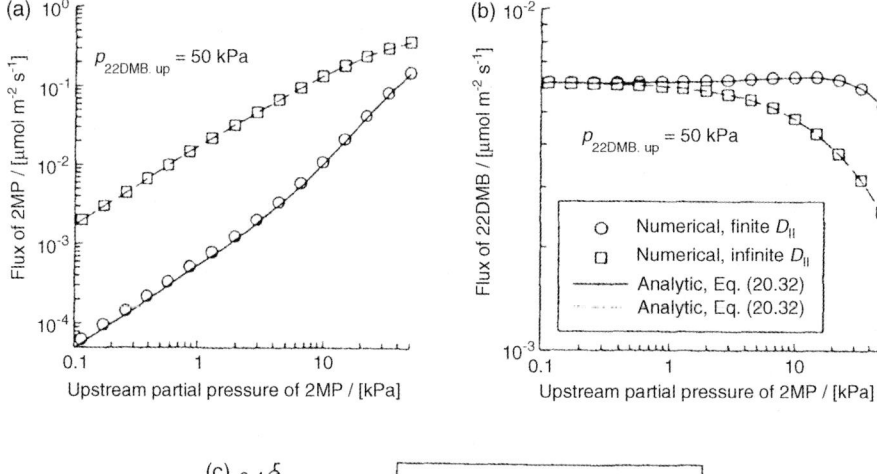
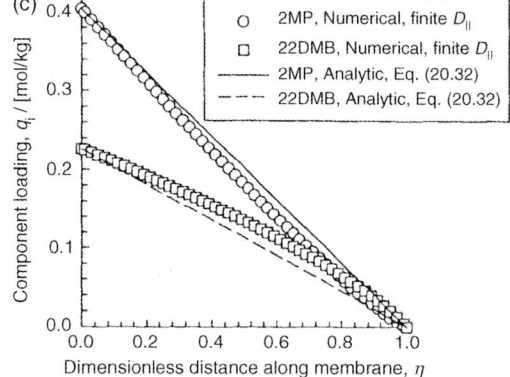

Figure 20.10 Permeation fluxes and profiles of 2-methylpentane (2-MP) and 2,2-dimethylbutane (2,2-DMB) in a silicalite membrane with the downstream side maintained under vacuum. In (a) and (b) the upstream partial pressure of 2,2-DMB is maintained at 50 kPa while the upstream partial pressure of 2-MP is varied over the range 0.1–50 kPa. Part (c) shows the loading profiles for both components when the upstream pressures are both 50 kPa. The symbols represent the results from precise numerical simulations while the lines represent the profiles calculated according to Eqs. (20.29) or (20.32) (for both $Đ_{AB}$ finite and infinite) with the corrected diffusivities varying in accordance with $D_{0A} = D_{0B}(0) (1 - \theta_A - \theta_B)$. Reprinted from Krishna and Baur [37] with permission.

diffusion). Also shown are the results of numerical calculations based on the full Stefan–Maxwell model. The agreement between the approximate analytic solution and the numerical solution is obviously excellent and, for this system, the effect of mutual diffusion is clearly important. Figure 20.10c shows the loading profiles through the membrane while Figure 20.11 shows, for ethane–methane permeation, the predicted steady-state perm-selectivity as a function of total hydrocarbon pressure. Again there is reasonably good (although not perfect) agreement between the analytic and numerical solutions. The experimental perm-selectivity data of van der Graaf et al. [31] lie close to the predicted curve allowing for mutual diffusion. The error involved in neglecting mutual diffusion is clearly substantial.

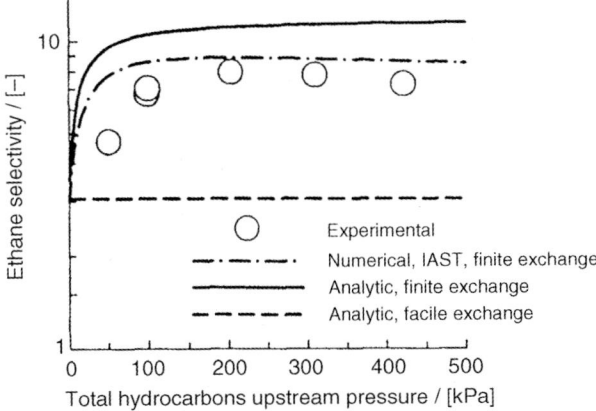

Figure 20.11 Ethane selectivity for permeation of a 50:50 mixture of ethane–methane through a silicalite membrane at 303 K. The experimental data of van de Graaf et al. [31] are compared with the analytic models [for $Đ_{AB}$ finite (—) and $Đ_{AB}$ infinite (– – –)] assuming $D_{oA} = D_{oA}(0)$ $(1 - \theta_A - \theta_B)$ with a similar expression for D_{oB}. The dash-dot line represents the results of more accurate calculation assuming $Đ_{AB}$ finite and using IAST (ideal adsorbed solution theory) rather than the Langmuir expression to represent the equilibrium isotherm. Reprinted from Krishna and Baur [37] with permission.

20.4.1.2 Importance of Mutual Diffusion

Also shown in Figure 20.9 are the permeance and selectivity predictions based on the simpler Habgood model (sometimes referred to as the non-interacting diffusion model) in which mutual diffusion effects are ignored (i.e., it is assumed that $Đ_{AB} \to \infty$). For the slower diffusing species (ethane) the flux prediction is only marginally changed by omitting the mutual diffusion terms but the effect on the faster component (methane) is considerable, leading to a major error in the predicted selectivity.

In general, mutual diffusion increases (slightly) the flux of the slower diffusing component and decreases (more substantially) the flux of the faster component. When the slower diffusing component is (because of its more favorable equilibrium) the preferentially permeating species this leads to an increased selectivity compared with the predictions of the Habgood model. A detailed theoretical analysis of mutual diffusion effects in a Langmuirian system has been carried out by Karimi and Farooq [38]. They show that the effect is generally small at low loading but it becomes important at higher loadings when the difference in the intrinsic mobilities of the components is large. Figure 20.12 shows representative results from their calculations.

20.4.2
More Complex Systems

The simple modeling approach outlined above does not work for all systems as the assumption that transport through a zeolite crystal can be represented by Eq. (20.8) with a constant value of D_o is often a poor approximation (see, for example, the data

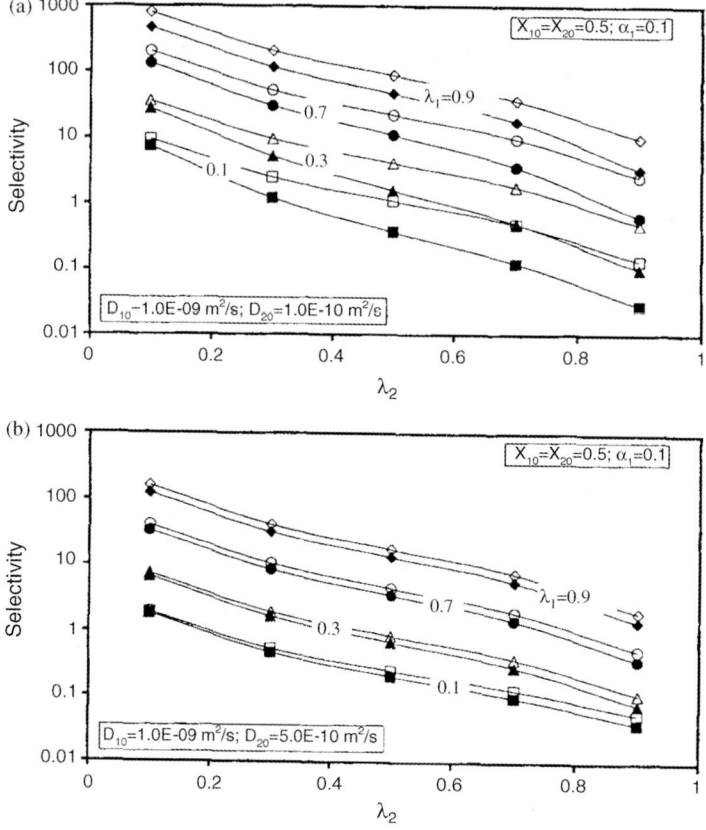

Figure 20.12 Effect of mutual diffusion on selectivity for selected conditions. The figures show the comparison between the predictions of the Maxwell–Stefan model with mutual diffusivities estimated from Eq. (20.22) (filled symbols) and the non-interacting (Habgood) model in which mutual diffusion is ignored [Eq. (20.24)] (open symbols). The parameters λ_1 and λ_2 measure the loading levels for components 1 and 2. Note that the effect of mutual diffusion becomes important only when the ratio $D_{01} : D_{02}$ is large and then only at high loadings. Reprinted from Karimi and Farooq [38] with permission.

for diffusion of hydrogen in NaX shown in Figure 3.2 and the discussion in Section 17.4.4). For CO_2–CH_4 on a SAPO-34 membrane Li et al. [39] showed that the same general approach could be used provided that the concentration dependence of the corrected diffusivities (D_{0A} and D_{0B}) is accounted for by using a suitable model such as that proposed by Reed and Ehrlich [40] (Figure 20.13). However, for CO_2–CH_4 on a DDR-3 membrane the Maxwell–Stefan approach failed. Even if the concentration dependence of the corrected diffusivities was accounted for and the mutual diffusivity used as a fitting parameter, a satisfactory fit of the binary data could not be obtained. Molecular simulations showed that, for this system, there is significant segregation in the adsorbed phase [41] with the result that the binary isotherm is not correctly predicted by the ideal adsorbed solution. The CO_2 molecules appear to sit

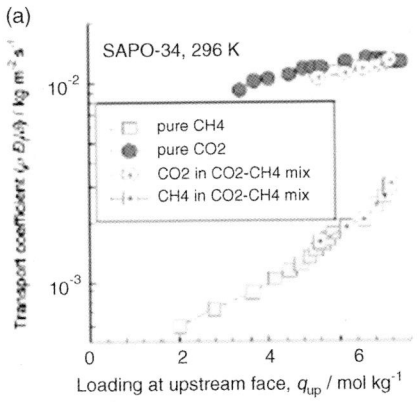 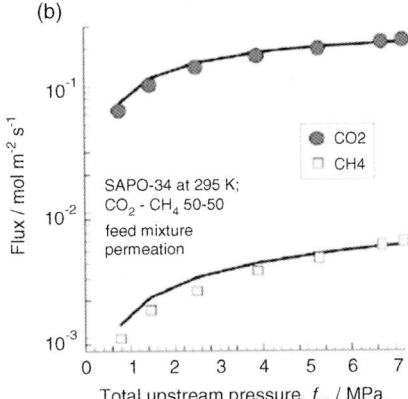

Figure 20.13 Permeation through a SAPO-34 membrane: (a) variation of corrected transport diffusivity (D_o/ℓ) for CO_2 and CH_4 in single-component and binary systems; (b) variation of flux with pressure for CO_2 and CH_4 in a binary system, showing the comparison between experimental binary data and the predictions from single-component. From Lai et al [39] with permission.

preferentially in the eight-ring windows, thus obstructing the diffusion of CH_4 and leading to a higher perm-selectivity in the binary system in comparison with the selectivity predicted from the single component permeance ratio [42] – see Section 16.2.2.

20.4.2.1 Membrane Thickness

In modeling membrane permeation it is normally assumed [as in Eq. (20.3)] that the flux varies inversely with membrane thickness. However, the available experimental data for zeolite membranes generally do not conform to this simple model, suggesting that the thickness of the active layer is probably smaller than that of the actual zeolite layer. Figure 20.14 shows a representative example [43]. For *n*-butane the flux decreases by a factor of about two for a 30-fold increase in total membrane thickness, suggesting that much of the zeolite layer is effectively by-passed. As noted in relation to Figure 20.1, in a zeolite membrane the thickness of the active layer bears little relationship to the apparent thickness.

20.4.2.2 Support Resistance

Although the support layer is generally highly permeable it is much thicker than the active zeolite layer. As a result it is generally necessary to take account of the diffusional resistance of the support layer when modeling the membrane performance. This may be accomplished by using the Maxwell–Stefan model (with an appropriate tortuosity factor) to describe transport through the support layer and hence determine the concentration profile. A less rigorous but computationally more convenient approach has been suggested by Farooq and Karimi [44] based on the use

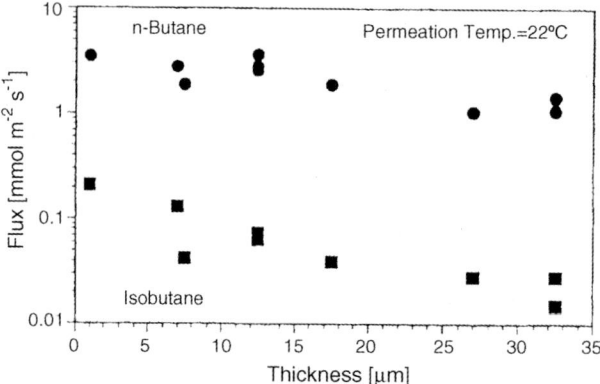

Figure 20.14 Variation of flux with the thickness of the zeolite layer for permeation of a 50:50 mixture of n-butane and isobutane through a silicalite membrane at 22 °C. Reprinted from Xomeritakis et al. [43] with permission.

of a Fickian diffusion model with a composition-dependent diffusivity. Kapteijn has pointed out that the relative importance of the support resistance depends on whether the active layer or the support is on the high-pressure side (F. Kapteijn, personal communication). As a result, reversing the flow direction can lead to a significant change in the overall permeance.

20.5
Membrane Characterization

20.5.1
Bypass Flow

It is difficult to make a perfect zeolite membrane containing no non-zeolitic transport paths. Characterization of the defect flow is therefore an essential preliminary step if the membrane is to be studied in any detail. This is usually accomplished by permeation measurements using carefully selected molecules of appropriate size. Figure 20.15 shows typical permeance measurements made with CH_4 and SF_6 over a range of pressures [21].

The defect flow is normally assumed to arise from a combination of Knudsen diffusion plus Poiseuille flow. Since the Knudsen flux is independent of pressure whereas Poiseuille flow varies linearly with pressure the two contributions can be separated by making measurements over a range of pressures and extrapolating to zero pressure to obtain the Knudsen contribution. Further confirmation may be obtained by comparing the Knudsen contributions for different components and checking that they conform to the expected inverse proportionality with the square root of the molecular mass. It is sometimes possible to reduce or even eliminate by-

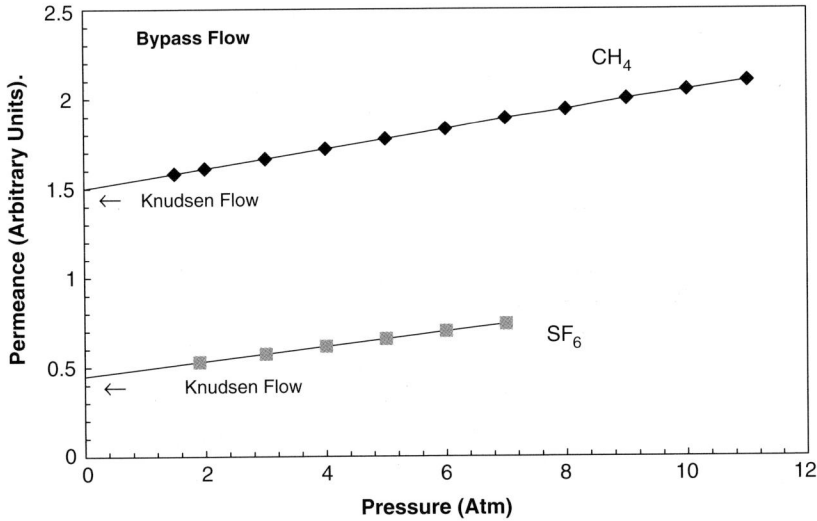

Figure 20.15 Defect flow of CH_4 and SF_6 in a small-pore silica membrane at 100 °C, showing Knudsen and Poiseuille contributions. Note that the intercepts are in the expected ratio of the square roots of the molecular weights. Data of Chance et al. [21].

pass flow by various modification procedures that block the larger pores without significantly impacting the flux through the zeolite pores.

20.5.2
Perm-Porosimetry

The pore size and pore size distribution of inorganic membranes is generally measured by *perm-porosimetry* [45]. A stream of helium containing a controlled fraction of a condensable species such as hexane is passed through the membrane, where the hexane condenses in accordance with the Kelvin equation. The idea is that the small pores are filled with *n*-hexane even at a very low activity (partial pressure), thus blocking the passage of helium (or other small test molecule), whereas the larger pores are filled with hexane only at higher activities. A plot of the helium flux as a function of hexane partial pressure is obtained, from which the pore size distribution may be estimated by application of the Kelvin equation.

This technique was originally developed for characterization of amorphous inorganic membranes but it has been successfully extended to zeolite membranes by Clark et al. [46]. It works very well for silicalite membranes since *n*-hexane is strongly and rapidly adsorbed in the intracrystalline pores. A perfect membrane will show a dramatic drop in the helium permeance (to essentially zero) at very low hexane activity. A high-quality "real" membrane will show a similar drop at low activity but a small residual helium flux corresponding to extracrystalline pores. A poor membrane will show a much smaller initial drop followed by a slow decline in helium permeance with

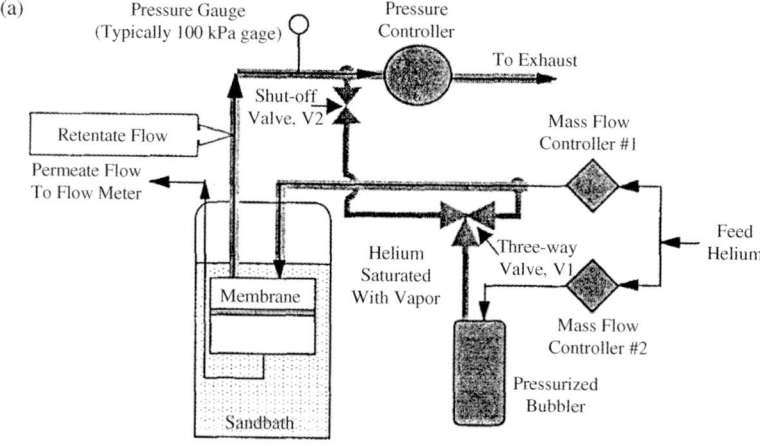

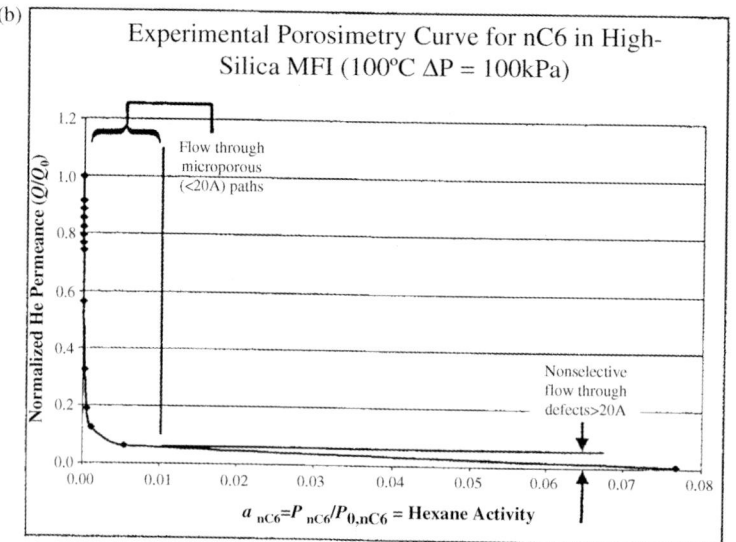

Figure 20.16 (a) Apparatus for porosimetry characterization of a zeolite membrane; (b) typical porosimetry curve for a silicalite membrane. Reprinted from Clark et al. [46] with permission.

increasing hexane activity, with the permeance falling close to zero only at relatively high hexane activity. The experimental system is shown schematically in Figure 20.16a, while Figure 20.16b shows an experimental porosimetry profile for a reasonably good silicalite membrane. The small residual helium flux at an n-hexane activity of 0.01 corresponds to the defect flow through pores larger than about 20 Å.

Although this technique has many advantages its application to silicalite membranes may be questionable since the silicalite structure is known to swell significantly when saturated with n-hexane. This swelling may seal cracks and defects with the result that an imperfect silicalite membrane may appear to be much less defective than it actually is.

20.5.3
Isotherm Determination

In addition to characterizing the defect structure perm-porosimetry provides a useful approach to determination of the equilibrium and transport properties of the membrane [46]. It is assumed that the adsorption of a condensable vapor in the membrane can be approximated by the Langmuir isotherm [Eq. (20.19)]. The fractional loading of the condensable vapor (q/q_s) is treated as uniform through the entire pore volume of the membrane and the reduction in the helium flux is assumed to vary linearly with loading:

$$\frac{N_{He}(a)}{N_{He}(o)} = 1 - q/q_s = \frac{1}{1 + bap_s} \tag{20.33}$$

where p_s is the saturation vapor pressure and $a = p/p_s$ is the "activity." In this model the shape of the porosimetry curve is therefore determined entirely by the Langmuir equilibrium constant. In the low-activity region ($a \to 0$):

$$\frac{d}{da}\left[\frac{N_{He}(a)}{N_{He}(o)}\right]_{a \to 0} = -bp_s \tag{20.34}$$

The value of b is thus easily calculated from the initial slope of the porosimetry curve. Figure 20.17 shows measurements carried out over a range of different temperatures. The temperature dependence follows the usual van't Hoff equation (Figure 20.18) and the b values agree well with the directly measured values.

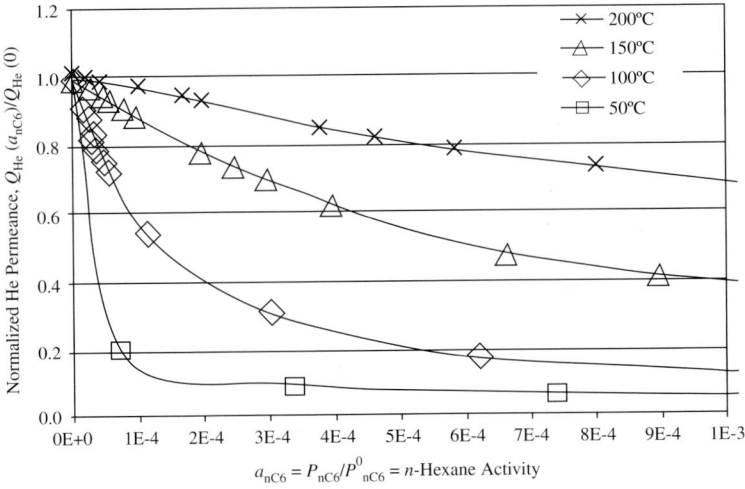

Figure 20.17 Adsorption branch perm-porosimetry curves for n-hexane in a silicalite membrane in the low-activity region at several temperatures. Reprinted from Clark et al. [46] with permission.

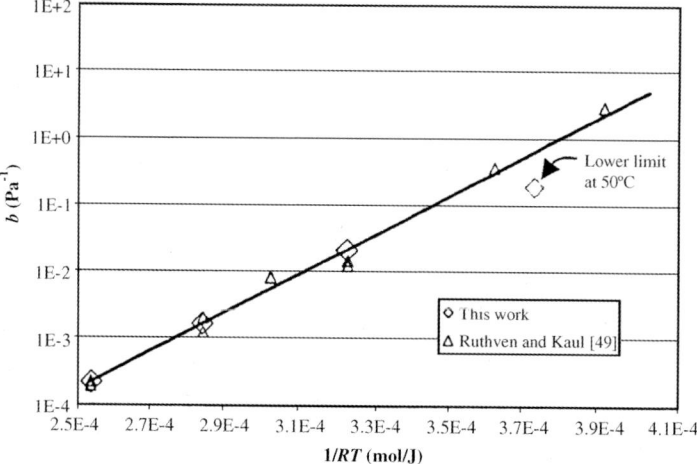

Figure 20.18 van't Hoff plot showing the temperature dependence of the Langmuir equilibrium constants for n-hexane in silicalite derived from analysis of the porosimetry curves shown in Figure 20.17. The data show excellent agreement with the values of b measured by Ruthven and Kaul for silicalite crystals [47]. Reprinted from Clark et al. [46] with permission.

If it is assumed that the same model is applicable at higher activities one may predict the entire porosimetry curve from Eq. (20.34). Figure 20.19 shows that there is good agreement between the experimental curves and the theoretical curves estimated in this way.

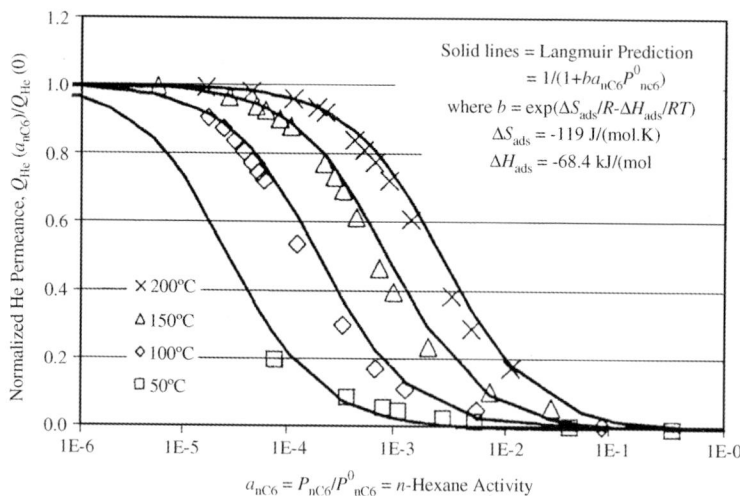

Figure 20.19 Comparison of experimental perm-porosimetry curves for n-hexane–silicalite with theoretical curves back-calculated from the Langmuir model with the derived parameters. Reprinted from Clark et al. [46] with permission.

20.5.4
Analysis of Transient Response

An alternative approach to membrane characterization, based on analysis of the transient response to a step change in feed pressure, has been developed by Gardner et al. [48–51]. The same basic expression is used for the steady-state flux [Eq. (20.33)]. The hold-up in the membrane is determined by mass balance from the transient response and this is equated to the equilibrium capacity based on the product of the membrane volume and equilibrium loading, leading to the equation:

$$q_s \left\{ 1 - \frac{b(p_H - p_L)}{(1+bp_H)(1+bp_L) \cdot \ln\left[\frac{1+bp_H}{1+bp_L}\right]} \right\} = 3 \int_0^\theta (J^1 - J) \, dt \quad (20.35)$$

where $(J^1 - J)$ represents the difference between the transient and steady state fluxes. The steady-state flux is given by Eq. (20.33). If measurements are made for a series of different values of p_H and p_L the parameters D_0, ℓ, and q_s may be determined from a regression analysis of the experimental data.

20.6
Membrane Separation Processes

Table 20.2 lists some of the commercially interesting separations that have been successfully carried out over a zeolite membrane (at laboratory scale). However, of these only the first (dehydration of ethanol by pervaporation through a 4A membrane) has actually been commercialized.

20.6.1
Pervaporation Process for Dehydration of Alcohols

A review of this process, including details of the membrane modules, has been given by Caro et al. [62]. The economic incentive for such a process is high since the

Table 20.2 Zeolite membrane separations.

System	Membrane material	Selectivity	Flux (kg m^{-2} h^{-1})	Reference
H_2O–ethanol	NaA	>10^3	5–15	Morigami et al. [52] Kondo et al. [53]
Ethanol–H_2O	Silicalite	25	10	Tuan et al. [54]
CO_2–CH_4	SAPO-34	50	2.5	Li [55, 56]
	DDR	200	1.3	Tomita [57]
CO_2–N_2	SAPO-34	16	0.6	Poshusta [58]
C_6H_6–C_6H_{12}	NaX/NaY	100	0.1	Jeong [59]
PX–MX	Oriented MFI	200	0.05	Lai [23], Hedlund et al. [60]
1-Butene–isobutene	MFI on titania	20 ($P=1$ atm)	5	Voss et al. [61]

traditional extractive or azeotropic distillation processes that are used to dehydrate alcohols beyond the azeotropic point are expensive. A pervaporation process based on using an LTA zeolite membrane in the pervaporation mode (liquid on the feed side, water vapor on the permeate side) has been developed and commercialized by Mitsui. The process operates at about 120 °C and water fluxes as high as 16 kg-water $m^{-2} h^{-1}$ have been reported. A similar process has also been demonstrated for other alcohols. For propanol and ethanol the separation factors are high (>100), suggesting that access of alcohol to the water-filled pores is severely constrained. For methanol dehydration the separation factors are lower ($\sim$10), as may be expected from both steric and equilibrium considerations.

20.6.2
CO_2–CH_4 Separation

The separation of CO_2 from methane is increasingly important in the processing of the very large reserves of low-grade natural gas. The possibility of using a zeolite membrane to achieve this separation has been demonstrated on a laboratory scale and the results, some of which are summarized in Table 20.3, appear promising. Although no such process has yet been developed to commercial scale, it is clear that a zeolite membrane is potentially a very attractive choice in comparison with other available options such as polymeric membranes or amine absorption. The permselectivity of the small-pore zeolites in favor of CO_2 is very high, especially for DDR but, even more importantly, the permeance is one to two orders of magnitude larger than for a polymeric membrane of comparable selectivity.

Although the potential economic advantage of a zeolite membrane for CO_2–CH_4 separation is clear, despite considerable research effort no such process has yet been commercialized. The practical challenges associated with scale-up, robustness, membrane life, and other practical issues have so far proved more severe than anticipated.

Table 20.3 Performance of membranes for CO_2–CH_4 separation.

Membrane type	CO_2 permeance (mol m^{-2} s^{-1} Pa^{-1})	α_{CO_2/CH_4}	Reference
Zeolite membranes			
DDR on α-alumina	3×10^{-7}	400	Himeno [63]
DDR on α-alumina	7×10^{-8}	600	van den Bergh [64]
SAPO-34 on α-alumina	10^{-7}–10^{-6}	65–170	Carreon [65], Li [39]
Polymeric membranes			
Polyimides	0.6–4×10^{-10}	10–80	Sridhar [66]
Polycarbonate	1×10^{-9}	100–150	Iqbal [67]
Polyallylamine–poly(vinyl alcohol)/polysulfone	10^{-8}	60	Cai [68]
Polyurethane/poly(vinyl alcohol) blends	10^{-10}	50	Ghalei [69]

20.6.3
Separation of Butene Isomers

The separation of isobutene (for production of butyl rubber and MTBE) from 1-butene (a co-monomer for production of high-density and linear low-density polyethylene as well as polybutylene) is technically challenging since there is very little difference in the boiling points (266 and 267 K, respectively). The critical diameter of the isobutene molecule (5 Å) is, however, substantially larger than that of 1-butene (4.3 Å), leading to a substantial difference in diffusivity in critically sized micropores such as those of silicalite, thus suggesting the possibility of a zeolite membrane separation. In laboratory-scale measurements with n-butane–isobutane mixtures (molecules with very similar dimensions) van de Graaf et al. [70] observed perm-selectivities greater than 30 at about 377 K. Similar perm-selectivities for n-butane–isobutane in a silicalite membrane were obtained by Hedlund et al. [60].

For 1-butene–isobutene, using a specially prepared titania-supported silicalite membrane, Voss et al. [61] obtained relatively high 1-butene permeances ($\approx 4\,\text{m}^3$ STP $\text{m}^{-2}\,\text{h}^{-1}\,\text{bar}^{-1}$) with perm-selectivities similar to, although slightly lower than, those for n-butane–isobutane (≈ 25 at 3 atm, 400 K). However, at higher pressures the isobutene flux increases while the 1-butene flux remains almost constant, leading to a pronounced decrease in perm-selectivity.

Single-component measurements [61] show that the isotherm for 1-butene is much more favorable than that for isobutene, approaching saturation at pressures above 5 bar. The isobutene capacity is smaller at low pressures but continues to increase up to at least 30 bar. If the mixture isotherms show similar behavior the observed variations in permeance could be explained simply by an increase in the isobutene loading at higher pressures. However, interestingly, the experimental permeance data stand in marked contrast to the predictions of molecular simulation studies [71–74], which suggests that (i) 1-butene is preferentially adsorbed at high pressures and (ii) the isobutene molecules preferentially occupy the channel intersections so that at isobutene loadings above 2 molecules per unit cell the fluxes of both components will fall to extremely low levels.

Although scientifically interesting the results of this study were technically disappointing since the economic target of a perm-selectivity greater than 20 at pressures of 20 bar with a permeance greater than $2\,\text{m}^3$ STP $\text{m}^{-2}\,\text{h}^{-1}\,\text{bar}^{-1}$ was not achieved.

20.6.4
MOF (Metal Organic Framework) Membranes for H_2 Separation

With the discovery of MOFs, many of which have framework structures similar to the zeolites, there have been numerous attempts to grow such structures as a uniform thin membrane on a macroporous support [75]. This has proved surprisingly challenging but several such membranes have now been successfully produced [76, 77]. The structures based on zeolitic imidazolate frameworks (known as ZIFs, see also Section 19.3) are of particular interest as they are very stable, even at elevated

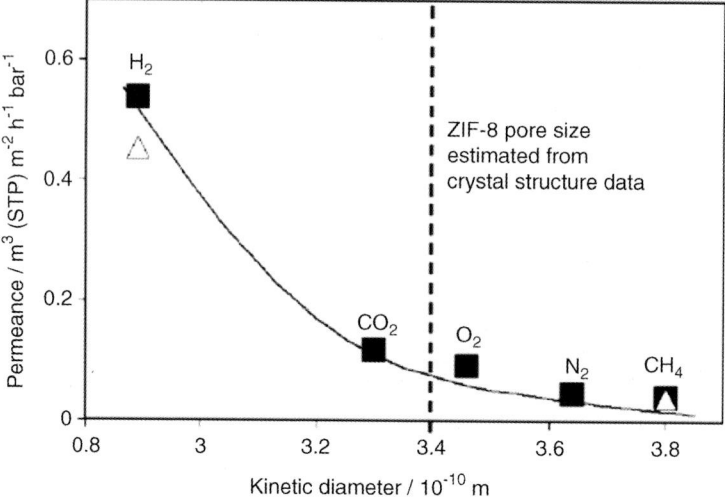

Figure 20.20 Variation of permeance with kinetic diameter for light gases in a ZIF-8 membrane at 298 K and 1 atm [single component (■) and binary mixture (Δ)]. Reprinted from Bux et al. [77] with permission.

temperatures. For example, ZIF-8 has the framework topology of sodalite. The channel dimensions are restricted by six-membered rings but with a significantly greater free diameter (about 3.4 Å compared with 2.9 Å for sodalite). Figure 20.20 shows permeation measurements for several light gases. The perm-selectivities are significantly greater than the Knudsen values, suggesting that transport is indeed occurring by diffusion through the small intracrystalline channels, rather than through structural defects. However, the sharp drop in permeance that might be expected for molecules larger than 3.4 Å is not observed, possibly as a result of the flexibility of the framework. The H_2/CH_4 perm-selectivity is about 11 (at 298 K and 1 atm), which is much larger than the corresponding Knudsen selectivity (about 2.8) but probably still not sufficiently large to be of industrial interest.

20.6.5
Amorphous Silica Membranes

Although the focus of this chapter is on zeolite membranes it should be noted that there has also been a commensurate increase in research on amorphous inorganic membranes. It has been shown that very thin (sub-micron) uniform layers of silica can be deposited by various dip-coating or spin-coating procedures [18–20]. Such membranes have effective pore sizes in the 4–5 Å range and despite their amorphous nature they show permeances and perm-selectivities that are comparable with small-pore zeolite membranes. However, traces of water are very strongly (irreversibly) adsorbed, leading to a decrease in permeance that is only partially restored by heating. This greatly limits the practical potential of such materials.

20.6.6
Stuffed Membranes

To avoid the problems involved in producing a thin coherent zeolite membrane the possibility of incorporating zeolite crystals within a polymeric material has been studied extensively. Although such "stuffed" membranes have been shown to have some interesting perm-selective properties these appear to be associated with the polymer–zeolite interface rather than with the intracrystalline diffusion properties. The lack of a coherent intracrystalline path means that the performance of such membranes generally falls far short of what might be expected for an ideal coherent zeolite membrane – see, for example, Pechar et al. [78].

20.6.7
Membrane Modules

For research and small-scale testing zeolite membranes are generally produced as small circular discs that fit conveniently into a measuring cell. However, in any practical application a membrane module is needed to achieve a large surface area within an acceptable volume. The usual design is similar to a shell and tube heat exchanger with the active zeolite layer coated on porous support tubes that are mounted between the tube plates at the ends of the module as in Figure 20.21.

From the process point of view it is desirable to operate the module under conditions of counter-current plug flow. However, the impact of deviations from ideal plug flow is more severe on the (high-pressure) feed side than on the permeate side. Since it is much easier to maintain a close approximation to plug flow through the tubes than through the shell this means that it is preferable to have the feed on the tube side. Since for high performance it is essential that the active zeolite layer be on the high pressure side of the support this means that it is desirable to grow the active zeolite layer on the inside of the tubes rather than on the outside. This can be accomplished but it becomes increasingly difficult when the diameter of the support tube is very small.

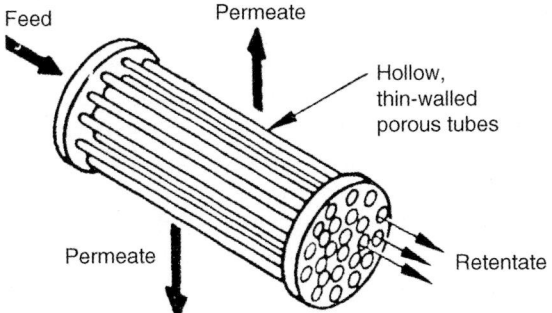

Figure 20.21 Sketch showing the construction of a tubular membrane module.

If the selectivity is sufficiently high and the purity constraints are not too severe a single membrane module may be adequate to achieve the required separation. However, by using a cascade of several stages connected in series it is possible to compensate for deviations from plug flow within an individual module and thus achieve a reasonably close approximation to an ideal counter-current flow system.

20.6.8
Membrane Reactors

The integration of a separation process with a catalytic reactor provides an attractive option for carrying out thermodynamically unfavorable reactions or for improving yields by selective removal of the desired product [79]. Dehydrogenation and partial oxidation reactions are obvious examples. Since, in contrast to a polymeric membrane, a zeolite membrane is in principle capable of operating at elevated temperatures the concept of a catalytic membrane reactor in which the catalyst is contained within a tubular zeolite membrane to allow continuous removal of product or injection of reactant along the length of the reactor is obviously attractive. Another possibility is to coat the surface of the catalyst particles with a perm-selective (zeolite) membrane to modify the selectivity by controlling the access of reactants and/or the rate of removal of products [80]. However, although such reactors have been demonstrated in the laboratory, generally with a single reactor/membrane tube, no such systems have yet been commercialized.

20.6.9
Barriers to Commercialization

In addition to the obvious issues such as the difficulty of fabrication and thermal stability there are several less obvious issues that tend to inhibit the practical development and implementation of zeolite membranes on an industrial scale. A and X zeolite membranes are unstable under acidic conditions so such membranes are unlikely to be robust enough for use in natural gas purification where acidic impurities are common. MFI membranes have better acid stability and are generally more robust. DD3R and SAPO-34 membranes appear to be promising for CO_2–CH_4 separation in the processing of low-grade natural gas. However, for many separations high selectivities and high fluxes are achieved only at low loadings. The conditions required for effective operation of the membrane therefore may not easily match the conditions required for the process. The application of such membranes in hydrogenation/dehydrogenation reactions appears practically feasible, although hydrogen perm-selectivities are not as large as might be expected. Except against very large molecules that are sterically prohibited from entering the pores the high mobility of hydrogen is largely compensated for by the stronger adsorption of the other species. Nevertheless, this seems likely to be the first area of commercialization for membrane reactors.

At this point the likely impact of MOF membranes is difficult to assess. The large range of possible organic linkers means that the pore size can be more easily tailored for a particular separation but problems of stability and the difficulties involved in the large-scale fabrication of thin coherent membranes may prove difficult to overcome.

References

1. Caro, J., Noack, M., Kolsch, P., and Schäfer, R. (2000) *Microporous Mesoporous Mater.*, **38**, 3–24; (b) See also: Noack, M. and Caro, J. (2002) *Handbook of Porous Solids*, vol. 4 (eds F. Schueth, K.S.W. Sing, and J. Weitkamp), Wiley-VCH Verlag GmbH, Weinheim, p. 2433.
2. Stöcker, M. (ed.) (2000) Microporous and mesoporous membranes and membrane systems. Special Issue: *Microporous Mesoporous Mater.*, **38**, 1–120.
3. Xomeritakis, G., Gouzinis, A., Nair, S., Okubo, T., He, M., Overney, R.M., and Tsapatsis, M. (1999) *Chem. Eng. Sci.*, **54**, 3521.
4. Lovallo, M.C., Gouzinis, A., and Tsapatsis, M. (1998) *AIChE J.*, **44**, 1903.
5. Gouzinis, A. and Tsapatsis, M. (1998) *Chem. Mater.*, **10**, 2497.
6. Mintova, S., Hedlund, J., Valtchev, V., Schoeman, B., and Sterte, J. (1997) *Chem. Commun.*, 15.
7. Mintova, S., Hedlund, J., Valtchev, V., Schoeman, B., and Sterte, J. (1998) *J. Mater. Chem.*, **10**, 2217.
8. Hedlund, J., Noack, M., Kölsch, P., Creaser, D., Caro, J., and Sterte, J. (1999) *J. Membr. Sci.*, **159**, 263.
9. Boudreau, L.C., Kuck, J.A., and Tsapatsis, M. (1999) *J. Membr. Sci.*, **152**, 41.
10. Lovallo, M.C., Tsapatsis, M., and Okubo, T. (1996) *Chem. Mater.*, **8**, 1579.
11. Caro, J., Finger, G., Jahn, E., Kornatowski, J., Marlow, F., Noack, M., Werner, L., and Zibrowius, B. (1993) *9th International Zeolite Conference, Montreal; Proceedings* (eds R. von Ballmoos, J.B. Higgins, and M.M.J. Treacy), Butterworth-Heinemann, pp. 683–691.
12. Noack, M., Kölsch, P., Venzke, D., Toussaint, P., and Caro, J. (1994) *Microporous Mater.*, **3**, 201.
13. Kölsch, P., Venske, D., Noack, M., Toussaint, P., and Caro, J. (1994) *J. Chem. Soc., Chem. Commun.*, 2491.
14. Girnus, I., Pohl, M.-M., Richter-Mendau, J., Schneider, M., Noack, M., and Caro, J. (1995) *Adv. Mater.*, **7**, 711.
15. Kusakabe, K., Murata, A., Kuroda, T., and Morooka, S. (1997) *J. Chem. Eng. Jpn*, **30**, 72–78.
16. Bakker, W.J.W., Kapteijn, F., Poppe, J., and Moulijn, J.A. (1996) *J. Membr. Sci.*, **117**, 57–78.
17. Talu, O., Sun, M.S., and Shah, D.B. (1998) *AIChE J.*, **44**, 681–694.
18. Lange, R.S.A., Heddink, J.H.A., Keizer, K., and Burggraaf, A.J. (1995) *J. Porous Mater.*, **2**, 141–149.
19. de Vos, R.M. and Verweij, H. (1998) *J. Membr. Sci.*, **143**, 37–51.
20. de Vos, R.M. (1998) High selectivity, high flux silica membranes for gas separation. PhD Thesis, Twente University, Enschede.
21. Chance, R.R., Yoon, C.J., Clark, T.E., Calabro, D.C., Deckman, H.W., Ruthven, D.M., Bex, P., and Bonekamp, B. (2004) *Adv. Nanoporous Mater.*, **68**, 71.
22. Tsapatsis, M. and Gavalas, G.R. (1999) *MRS Bull.*, **3**, 30.
23. Lai, Z.P., Bonilla, A.G., Diaz, I., Nery, J.G., Sujaoti, K., Amat, M.A., Kokkoli, E., Terasaki, O., Thompson, R.W., Tsapatsis, M., and Vlachas, D.G. (2003) *Science*, **300**, 4556.
24. Ruthven, D.M., Eic, M., and Richard, E. (1991) *Zeolites*, **11**, 647–653.
25. Coronas, J., Noble, R.D., and Falconer, J.L. (1998) *I. E.C. Res.*, **37**, 166.
26. Noble, R.D. and Falconer, J.L. (1995) *Catal. Today*, **25**, 209.
27. Vroon, Z.A.E.P., Keizer, K., Gilde, M.J., Verweij, H., and Burggraaf, J. (1998) *J. Membr. Sci.*, **144**, 65.
28. Geus, E.R., van Bekkum, H., Bakker, W.J.W., and Moulijn, J.A. (1993) *Microporous Mater.*, **1**, 131.
29. Keil, F.J., Krishna, R., and Coppens, M.-O. (2000) *Rev. Chem. Eng.*, **16**, 71–197.

30 Krishna, R. and Paschek, D. (2000) *Sep. Purif. Technol.*, **21**, 111–136.
31 van de Graaf, J.M., Kapteijn, K., and Moulijn, J.A. (1999) *AIChE J.*, **45**, 497–511.
32 Krishna, R. and van de Broeke, L.J.P. (1995) *Chem. Eng. J.*, **57**, 155–162.
33 Kapteijn, F., Moulijn, J.A., and Krishna, R. (2000) *Chem. Eng. Sci.*, **55**, 2923–2930.
34 Vignes, A. (1996) *Ind. Eng. Chem. Fundam.*, **5**, 189–199.
35 Habgood, H.W. (1958) *Can. J. Chem.*, **36**, 1384.
36 Round., G.F., Habgood, H.W., and Newton, R. (1966) *Sep. Sci.*, **1**, 219.
37 Krishna, R. and Baur, R. (2004) *Chem. Eng. J.*, **97**, 37–45.
38 Karimi, I.A. and Farooq, S. (2000) *Chem. Eng. Sci.*, **55**, 3529–3541.
39 Li, S., Falconer, J.L., Noble, R.D., and Krishna, R. (2007) *J. Phys. Chem. C*, **111**, 5075–5082.
40 Reed, D.A. and Ehrlich, G. (1981) *Surf. Sci.*, **102**, 588–609.
41 Krishna, R. and van Baten, J.M. (2007) *Chem. Phys. Lett.*, **446**, 344–349.
42 Jee, S.E. and Sholl, D. (2009) *J. Am. Chem. Soc.*, **131**, 7896.
43 Xomeritakis, G., Nair, S., and Tsapatsis, M. (2000) *Microporous Mesoporous Mater.*, **38**, 60–73.
44 Farooq, S. and Karimi, I.A. (2001) *J. Membr. Sci.*, **186**, 109–121.
45 Cao, G.Z., Meijerink, J., Brinkman, H.W., and Burggraaf, A.J. (1993) *J. Membr. Sci.*, **83**, 221–235.
46 Clark, T.E., Deckman, H.W., Cox, D.M., and Chance, R.R. (2004) *J. Membr. Sci.*, **230**, 94.
47 Ruthven, D.M. and Kaul, B.K. (1998) *Adsorption*, **4**, 269–273.
48 Gardner, T.Q., Flores, A.I., Noble, R.D., and Falconer, J.L. (2002) *AIChE J.*, **48**, 1155–1167.
49 Gardner, T.Q., Lee, J.B., Noble, R.D., and Falconer, J.L. (2002) *Ind. Eng. Chem. Res.*, **41**, 4094–4105.
50 Gardner, T.Q., Falconer, J.L., and Noble, R.D. (2002) *Desalination*, **149**, 435–440.
51 Gardner, T.Q., Falconer, J.L., Noble, R.D., and Zieverink, M.M.P. (2003) *Chem. Eng. Sci.*, **58**, 2103–2112.
52 Morigami, Y., Kondo, M., Abe, J., Kita, H., and Okamoto, K. (2001) *Sep. Purif. Technol.*, **25**, 251–260.
53 Kondo, M., Yamamura, T., Yukitake, T., Matsue, Y., Kita, K., and Okamoto, K. (2003) *Sep. Purif. Technol.*, **32**, 191–198.
54 Tuan, V.A., Li, S., Noble, R.D., and Falconer, J.L. (2002) *J. Membr. Sci.*, **196**, 111.
55 Li, S., Falconer, J.L., and Noble, R.D. (2004) *J. Membr. Sci.*, **241**, 121–135.
56 Li, S., Martinele, J., Falconer, J.L., and Noble, R.D. (2004) *I. E Res.*, **44**, 3220.
57 Tomita, T., Nakayama, K., and Sakai, H. (2004) *Microporous Mesoporous Mater.*, **68**, 71–75.
58 Poshusta, J.C., Tuan, V.A., Pope, E.A., Noble, R.D., and Falconer, J.L. (2000) *AIChE J.*, **46**, 779–789.
59 Jeong, B.H., Hasegawa, Y., Sotowa, K.I., Kusakabe, K., and Morooka, S. (2003) *J. Membr. Sci.*, **213**, 115.
60 Hedlund, J.H., Sterte, J., Athonis, M., Bons, A., Carstensen, B., Corcoran, E.W., Cox, D.M., Deckman, H.W., De Gijnst, W., de Moor, P., Lai, F., McHenry, J., Mortier, W., Reinoso, J.J., and Peters, J. (2002). *Microporous Mesoporous Mater.*, **52**, 179–189.
61 Voss, H., Diefenbacher, A., Schuch, G., Richter, H., Voigt, I., Noack, M., and Caro, J. (2009) *J. Membr. Sci.*, **329**, 11–17.
62 Caro, J., Noack, M., and Koelsch, P. (2005) *Adsorption*, **11**, 215–227.
63 Himeno, S., Tomita, T., Suzuki, K., Nakayama, K., Yajima, K., and Yoshida, S. (2007) *Ind. Eng. Chem. Res.*, **46**, 6989–6997; see also: (2007) *Kagaku Kogaku Ronbun*, **33**, 122.
64 van den Bergh, J., Zhu, W., Gascon, J., Moulijn, J.A., and Kapteijn, F. (2008) *J. Membr. Sci.*, **316**, 35–45.
65 Carreon, M.A., Li, S., Falconer, J.L., and Noble, R.D. (2008) *J. Am. Chem. Soc.*, **150**, 5412–5413.
66 Sridhar, S., Veerapur, R.S., Patil, M.B., Gudasi, K.B., and Aminabhavi, T.M. (2007) *J. Appl. Polym. Sci.*, **106**, 1585–1594.
67 Iqbal, M., Man, Z., Mukhtar, H., and Datta, B.K. (2008) *J. Membr. Sci.*, **318**, 167–175.

68 Cai, Y., Wang, Z., Yi, C., Bai, Y., Wang, J., and Wang, S. (2008) *J. Membr. Sci.*, **310**, 184–196.
69 Ghalei, B. and Semsarzadeh, M.A. (2007) *Macromol. Symp.*, 249–250 and 320–335.
70 van de Graaf, J.M., van der Bijl, E., Stol, A., Kapteijn, F., and Moulijn, J.A. (1998) *Ind. Eng. Chem. Res.*, **37**, 4071.
71 Fernandez, M., Kaerger, J., Freude, D., Pampel, A., van Baten, J.M., and Krishna, R. (2007) *Microporous Mesoporous Mater.*, **105**, 124.
72 Vlugt, T.J.H., Zhu, W., Kapteijn, F., Moulijn, J.A., Smit, B., and Krishna, R. (1998) *J. Am. Chem. Soc.*, **120**, 5599.
73 Krishna, R. and van Baten, J.M. (2008) *Chem. Eng. J.*, **140**, 614.
74 Krishna, R. and Paschek, D. (2001) *Phys. Chem. Chem. Phys.*, **3**, 453.
75 Gascon, J. and Kapteijn, F. (2010) *Angew. Chem. Int. Ed.*, **49**, 1530–1532.
76 Li, Y.-S., Liang, F.-Y., Bux, H., Feldhoff, A., Yang, W.-S., and Caro, J. (2010) *Angew. Chem. Int. Ed.*, **49**, 548–551.
77 Bux, H., Liang, F., Li, Y., Cravillon, J., Wiebcke, M., and Caro, J. (2009) *J. Am. Chem. Soc.*, **131**, 16000–16001.
78 Pechar, T.W., Kim, S., Vaughan, B., Marchand, E., Baranauskas, V., Riffle, J., Jeong, H.K., and Tsapatsis, M. (2006) *J. Membr. Sci.*, **277**, 210–218.
79 Sanchez, J. and Tsotsis, T.T. (2002) *Catalytic Membranes and Membrane Reactors*, Wiley-VCH Verlag GmbH, Weinheim.
80 van den Bergh, J. (2010) DD3R zeolite membranes in separation and catalytic processes; modelling and application. PhD Thesis, University of Delft.
81 Noack, M., Kölsch, P., Schneider, M., Toussaint, P., Sieber, I., and Caro, J. (2000) *Microporous Mesoporous Mater.*, **38**, 3–24.

21
Diffusional Effects in Zeolite Catalysts

In previous chapters we have discussed in some detail the diffusional properties of several zeolites, including ZSM-5 and zeolite Y, both of which are widely used as catalysts in the petroleum and petrochemical industries. The discussion has been focused on the fundamental aspects of the diffusional behavior of various classes of guest molecules; the practical implications have not been considered. This chapter aims, therefore, to provide a brief review of the ways in which the performance of a zeolite catalyst may be modified and indeed controlled by diffusional effects. The general subject of diffusion effects in catalytic reactions is well covered in several of the standard texts on chemical reaction engineering [1–3] so the present discussion is focussed mainly on the effects of micropore diffusional resistance. Such effects are in many ways similar to the effects of macropore diffusional resistance but there are some key differences that require emphasis. The discussion is illustrated by reference to some of the major commercial zeolite-catalyzed processes.

21.1
Diffusion and Reaction in a Catalyst Particle

The basic theory of diffusion accompanied by chemical reaction in a catalyst pore was put forward by Thiele [4] and Zeldovich [5] 70 years ago. Two excellent reviews published by Wheeler [6, 7] in the post-war years extended these basic ideas to show how diffusional resistance can modify kinetic behavior and catalyst selectivity for a range of simple model systems. Although in the intervening years there have been numerous publications extending these ideas to more complex kinetic systems, the basic theoretical framework remains very much as delineated by Wheeler. Excellent and easily readable accounts have been given by Petersen [3] and Satterfield [8] and the mathematical theory has been set forth in detail by Aris. [9] A further review of this subject would be redundant and only the briefest summary is therefore given here to serve as an introduction to those readers who may not be familiar with the basic theory.

In the prior development of the subject the focus has been directed almost entirely on conventional "macroporous" catalysts in which transport occurs mainly by the

Diffusion in Nanoporous Materials. Jörg Kärger, Douglas M. Ruthven, and Doros N. Theodorou.
© 2012 Wiley-VCH Verlag GmbH & Co. KGaA. Published 2012 by Wiley-VCH Verlag GmbH & Co. KGaA.

well-known Knudsen and molecular diffusion mechanisms. Although the same general theory can be applied to diffusion and reaction in a zeolite crystal, in such systems the differences in diffusivity between different species are very much greater and the temperature dependence of the diffusivity is generally stronger. As a result, certain of the effects are of much greater significance in micropore systems.

21.1.1
The Effectiveness Factor

Consider a porous spherical catalyst particle, as sketched in Figure 21.1, in which an irreversible first-order reaction occurs, catalyzed by the internal surface of the solid. We may characterize the reaction by an intrinsic rate constant (k_v) that is related to the surface rate constant (k_a) by $k_v = sk_a$ where s is the specific surface area per unit particle volume. A steady-state differential mass balance for a shell element of the particle yields:

$$\frac{1}{R^2}\frac{d}{dR}\left(D_e R^2 \frac{dc}{dR}\right) = k_v c \tag{21.1}$$

or, if the effective diffusivity can be regarded as independent of concentration:

$$\frac{d^2c}{dR^2} + \frac{2}{R}\frac{dc}{dR} = \frac{k_v}{D_e}c \tag{21.2}$$

which may be written in the dimensionless form:

$$\frac{d^2C}{dZ^2} + \frac{2}{Z}\cdot\frac{dC}{dZ} = \frac{R_p^2 k_v C}{D_e} = \phi_s^2 C \tag{21.3}$$

where $C = c/c_0$, $Z = R/R_p$, and $\phi_s \equiv R_p\sqrt{k_v/D_e}$ is a dimensionless parameter that measures the relative rates of reaction and diffusion within the pores. This quantity was introduced by Thiele in his 1939 paper [4] and is commonly known as the Thiele modulus. However, in a review paper presented in 1970 Gorring [10] pointed out that the same parameter had in fact been introduced by Juttner [11] in a classic paper discussing the problem of diffusion and reaction in liquid systems, which predated the work of Thiele by 30 years. Wakao [12] has indeed suggested that this quantity be referred to as the "Juttner modulus." However, in view of its widespread use in the technical literature, it seems preferable to retain the term "Thiele modulus" despite the implicit historical error.

To obtain an expression for the concentration profile through the particle it is necessary to solve Eq. (21.3) subject to the boundary conditions $R = R_p$, $Z = 1.0$, $c = c_0$, $C = 1.0$, and:

$$R = 0, Z = 0; \quad \frac{dc}{dR} = \frac{dc}{dZ} = 0 \tag{21.4}$$

If heat conduction through the particle is sufficiently rapid that the temperature can be considered as uniform through the particle, the solution assumes the very simple form:

21.1 Diffusion and Reaction in a Catalyst Particle

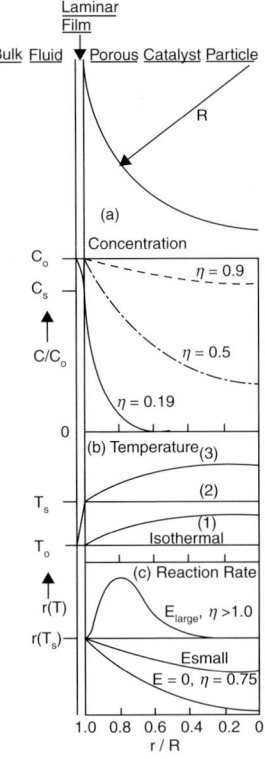

Figure 21.1 Diffusion and reaction in a porous catalyst particle: (a) Reactant concentration profiles for low, intermediate, and high degrees of diffusional limitation [expressed in terms of the effectiveness factor η – see Eqs. (21.6) and (21.7)]. The additional intrusion of external film resistance ($c_s < C_0$) is indicated for $\eta = 0.19$ (corresponding to a rapid reaction). (b) Temperature profiles (for an exothermic reaction): (1) internal heat conduction slow, heat transfer through external fluid film rapid ($T_s \approx T_0$); (2) internal heat conduction very fast, heat transfer through external fluid film relatively slow ($T_0 < T_s$, $T = T_s$ for all r); (3) both internal and external resistances to heat transfer are significant ($T_s > T_0$, $T > T_s$) (for the isothermal case $T = T_s = T_0$). (c) Profiles of local reaction rate. At the surface $r = r(T_s)$. For $E = 0$ or for an isothermal particle the rate decreases towards the particle center as a result of diffusional resistance ($\eta < 1.0$). For a non-isothermal particle this decline may be partially or fully offset by the effect of the increased temperature in the interior, leading in extreme cases to $\eta > 1.0$.

$$C = \frac{c}{c_0} = \frac{\sinh(\phi_s Z)}{Z \sinh \phi_s} \tag{21.5}$$

The form of the corresponding concentration profiles is shown in Figure 21.1a. As expected from Eq. (21.5), the shape of the profile depends only on the parameter ϕ_s and not on the individual values of the rate constant and effective diffusivity. The effect of pore diffusion on the reaction rate may be conveniently expressed in terms of the "effectiveness factor" (η), defined as:

$$\eta = \frac{\text{average reaction rate in particle}}{\text{reaction rate in absence of diffusional limitations}}$$

$$= \int_{R=0}^{R_P} \frac{4\pi R^2 k_v c(R) dR}{\left(\frac{4}{3}\pi R_P^3\right) k_v c_0} \tag{21.6}$$

With $c(R)$ given by Eq. (21.5) this integral may be easily evaluated to yield:

$$\eta = \frac{3}{\phi_s}\left[\frac{1}{\tanh \phi_s} - \frac{1}{\phi_s}\right] \tag{21.7}$$

or, for ϕ_s large, $\eta \approx 3/\phi_s$

If the same analysis is carried out for a parallel-sided slab of catalyst (thickness $2L$), rather than for a spherical particle, we have, in place of Eq. (21.2):

$$\frac{d^2c}{dz^2} = \frac{k_v}{D_e} c \tag{21.8}$$

and for the concentration profile:

$$C = \frac{\cosh(\phi_L Z)}{\cosh \phi_L} \tag{21.9}$$

where $Z = z/L$ and $\phi_L = L\sqrt{k_v/D_e}$

The corresponding expression for the effectiveness factor is:

$$\eta = \frac{\tanh \phi_L}{\phi_L} \tag{21.10}$$

or, for large ϕ_L, $\eta \approx 1/\phi_L$

The variation of effectiveness factor with Thiele modulus is shown in Figure 21.2, calculated according to both Eqs. (21.7) and (21.10) with $\phi_s = 3\phi_L$. Although these expressions are of different mathematical form, it is evident that, over the entire

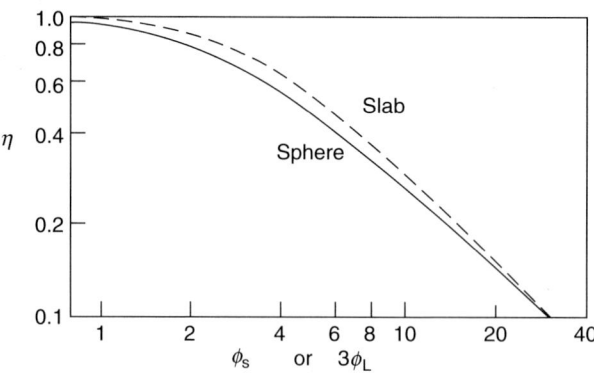

Figure 21.2 Variation of effectiveness factor with Thiele modulus for a first-order isothermal system for a sphere [Eq. (21.7)] and a parallel side slab [Eq. (21.10)].

range of ϕ_s, the two curves are very close. Setting $\phi_s = 3\phi_L$ is equivalent to defining a mean equivalent spherical radius given by $R_p = 3 \times$ particle volume/external surface area, and when compared on this basis the solutions for different particle shapes are numerically very similar. By this simple transformation, analysis of diffusion effects may therefore be carried out in the simpler geometry of a parallel sided slab and the results may then be applied, with little error, to other shapes of catalyst particle.

When the Thiele modulus is small the effectiveness factor approaches unity, there is very little variation in reactant concentration through the particle, and the rate of the reaction is controlled by the intrinsic chemical kinetics. In the other extreme where the Thiele modulus is large, the reactant concentration drops sharply near the surface and is essentially zero at the center of the particle. Under these conditions $\eta \approx 3/\phi_s$, $k_v \eta \approx (3/R_P)\sqrt{k_v D_e}$ and the reaction is said to be diffusion controlled or strongly diffusion limited. The latter phrase is preferable since, even in this region, the overall rate of reaction still depends on the chemical rate constant as well as on the effective diffusivity. Diffusional limitation is evidently favored by large particles, a high rate constant, and a low effective diffusivity while reaction control ($\phi \to 0$) is favored by small particles, a low reaction rate constant and a high effective diffusivity. Although originally developed to account for the behavior of macroporous catalyst particles this analysis applies equally to diffusion and reaction within a zeolite crystal with D_e replaced by the intracrystalline diffusivity and k_v the specific rate constant based on crystal volume.

21.1.2
External Mass Transfer Resistance

If external mass transfer resistance is significant the reactant concentration at the surface of the catalyst particle will be lower than the concentration in the bulk gas ($c_s < c_0$) and the profile through the catalyst particle will be as sketched (for $\eta = 0.19$) in Figure 21.1a. Equating the steady-state flux of reactant through the external film to the rate of reaction within the particle yields, for a first-order system:

$$4\pi R_p^2 k_f (c_0 - c_s) = \frac{4}{3}\pi R_p^3 k_v \eta c_s \tag{21.11}$$

$$\frac{c_s}{c_0} = \frac{k_f}{k_f + (R_p/3)k_v \eta} \tag{21.12}$$

$$k_{ev} = \frac{k_v \eta}{1 + k_v \eta (R_p/3k_f)} \tag{21.13}$$

or:

$$\frac{1}{k_{ev}} = \frac{1}{k_v \eta} + \frac{R_p}{3}\frac{1}{k_f} \tag{21.14}$$

where k_{ev} is the overall effective rate constant based on catalyst particle volume. The transition from internal to external control as $k_v \eta$ becomes large relative to $3k_f/R_p$ is immediately obvious.

21.1.3
Temperature Dependence

In the kinetically controlled regime ($\eta \approx 1.0$) the temperature dependence of the rate constant may be expected to follow the normal Arrhenius law:

$$k_v \eta \approx k_v = k_\infty e^{-E/RT} \qquad (21.15)$$

As the temperature increases k_v increases rapidly whereas the effective diffusivity increases more slowly, at least in a conventional catalyst where pore diffusion occurs mainly by the Knudsen or molecular mechanisms. At sufficiently high temperatures one may therefore expect that ϕ_s will become large so that there will be a gradual transition to the diffusion-limited regime in which $\eta \approx 3/\phi_s$ and:

$$k_{ev} \approx (3/R_p)\sqrt{k_v D_e} = (3/R_p)\sqrt{k_\infty D_e}\, e^{-E_d/2RT} \qquad (21.16)$$

Evidently, in this regime, if the temperature dependence of D_e is small, the apparent activation energy will be approximately half the true value.

The external film coefficient (k_f) is approximately proportional to the molecular diffusivity and therefore depends only weakly on temperature. From Eq. (21.14) it is evident that at sufficiently high temperatures $k_v \eta \gg 3k_f/R_p$ and $k_{ev} \rightarrow 3k_f/R_p$. If there is no change in mechanism one may therefore expect that, over a sufficiently wide range of temperature, the Arrhenius plot will assume the form indicated in Figure 21.3. The apparent activation energy will fall from its true value, at low temperatures in the kinetic regime, to approximately half the true value at intermediate temperatures in the pore diffusion limited regime and then to a very low value at high temperatures where the influence of external fluid film resistance becomes dominant.

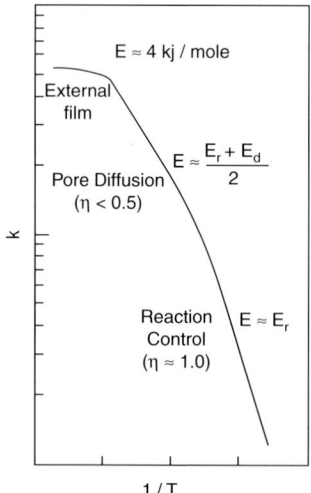

Figure 21.3 Sketch showing form of an Arrhenius plot, with a transition from intrinsic reaction rate control to combined internal diffusion and reaction control and to external mass transfer control.

This is the conventional picture that was developed for macroporous catalysts. There are two important ways in which the behavior of a zeolite catalyst subject to intracrystalline diffusion limitation may be expected to be different:

1) Intracrystalline diffusion is an activated process so that the diffusivity varies with temperature according to an equation of Arrhenius form:

$$D = D_\infty e^{-E_d/RT} \tag{21.17}$$

and the diffusional activation energy may be comparable with the activation energy for the chemical reaction.

2) Reaction occurs in the adsorbed phase so that, outside Henry's law region, the apparent reaction order (with respect to the gas phase concentrations) will differ from the true order based on the intracrystalline concentrations.

The kinetic rate expression should logically be written in terms of the adsorbed phase concentration (q). For the simplest case of a first-order reaction:

$$-\frac{1}{V}\frac{dn_A}{dt} = k'_v q_A = k_v c_A \tag{21.18}$$

If we assume Langmuir equilibrium:

$$k'_v q_A = \frac{k'_v b_A q_s}{1 + b_A c_A + b_B c_B + \cdots} c_A \tag{21.19}$$

from which it is evident that the first-order rate constant is given by:

$$k_v = \frac{k'_v b_A q_s}{1 + b_A c_A + b_B c_B + \cdots} \tag{21.20}$$

At sufficiently low concentrations (within the Henry's law region) the concentration terms in the denominator become small and:

$$k_v \approx k'_v K \tag{21.21}$$

where $K = b_A q_s$ is the Henry constant. Since the temperature dependence of the Henry constant is given by:

$$K = K_\infty e^{-\Delta U_0/RT} \tag{21.22}$$

it follows that the temperature dependence of k_v will be given by:

$$k_v = k'_\infty e^{-E/RT} K_\infty e^{-\Delta U_0/RT} = k'_\infty K_\infty e^{-(\Delta U_0 + E)/RT} \tag{21.23}$$

Since adsorption is exothermic (ΔU_0 negative) the apparent activation energy, even in the kinetic regime, may be substantially smaller than the intrinsic activation energy (of k'_v). Indeed if the energy of adsorption is larger than the activation energy the apparent activation energy may even become negative, implying a decrease in reaction rate with increasing temperature.

Taking account also of the activation energy for diffusion (E_d), for a zeolite catalyst one may expect the activation energy to change from ($E + \Delta U_0$) in the kinetic regime to $(E + \Delta U_0 + E_d)/2$ in the pore diffusion limited regime in which $k_e \propto \sqrt{k_v D_e}$. Since E_d may be of the same order as $E + \Delta U_0$, some curvature of the Arrhenius plot is to be expected but the well-defined transition to an apparent activation energy of about half the true value is unlikely to be observed in a zeolite system.

21.1.4
Reaction Order

This analysis of diffusion-limited kinetics has been presented for the simple case of a first-order irreversible reaction since the results for that case assume very simple and easily understandable forms. However, the analogous problem for a system with a more complex rate equation may be solved in a similar way by numerical methods. The results are qualitatively similar but an additional conclusion may be drawn concerning the apparent reaction order. If the true reaction order is greater than unity then, under conditions of strong pore diffusion limitation, the apparent order will be lower than the true order; if the true order is less than unity the apparent order under condition of strong diffusion limitation will be greater than the true order. This conclusion may be justified in an approximate manner in the following way. Suppose the true order is m:

$$-\frac{1}{V}\frac{dn}{dt} = k_v c^m = \left(k_v c^{m-1}\right) c \tag{21.24}$$

$$\phi_s \approx R_p \sqrt{\left(\frac{k_v c^{m-1}}{D_e}\right)} \tag{21.25}$$

Under conditions of strong pore diffusion limitation (ϕ large):

$$\eta \approx \frac{3}{\phi_s} = \frac{3}{R_p}\sqrt{\left(\frac{D_e}{k_v c^{m-1}}\right)} \tag{21.26}$$

$$-\frac{1}{V}\frac{dn}{dt} = \frac{3}{R_p}\sqrt{k_v D_e}\, c^{(m+1)/2} \tag{21.27}$$

For $m = 1$ there is evidently no change in order. For $m = 2$, the apparent order will fall to 3/2 while for $m = 1/2$ the apparent order will increase to 3/4.

21.1.5
Pressure Dependence

A further advantage of formulating the kinetics in terms of the adsorbed phase concentration is that the effect of total pressure can be visualized more easily. At elevated pressure the total adsorbed phase concentration will increase and the terms in the denominator of Eq. (21.19) become increasingly important. A decrease in apparent reaction order is therefore to be expected.

The analysis of pore diffusion effects is based on the assumption that the effective pore diffusivity in a zeolite crystal, under counter-diffusion conditions, is essentially independent of concentration. For some systems this may be a reasonable approximation, at least over a limited range of concentrations but there is evidence that, for some systems, the self-diffusivity, and therefore by extension the counter-diffusivity, decrease strongly with concentration in the saturation region (Figure 3.3a and b). At high concentrations one may therefore expect to see an increase in the Thiele modulus and a corresponding decrease in the effectiveness factor arising from the decline in diffusivity.

21.1.6
Non-isothermal Systems

The preceding analysis of diffusion effects in a catalyst particle is based on the assumption that the particle can be regarded as isothermal. Since heat is, in general, either released or absorbed in any chemical reaction this assumption will be a valid approximation only when the rate of heat conduction through the particle (or crystal) is rapid relative to the diffusion of the reactants within the pores. In the more general situation in which the diffusion of heat and matter occur at comparable rates there will be a gradient of temperature through the particle and this must be taken into account in any theoretical analysis.

To determine the steady-state concentration profiles through the particle (and hence the effectiveness factor) requires the solution of Eq. (21.2) together with the heat balance, which, for the steady state, reduces simply to:

$$-D_e \frac{dc}{dr}(-\Delta H) = \lambda_s \frac{dT}{dr} \qquad (21.28)$$

or, in integrated form:

$$T - T_s = \frac{(-\Delta H)D_e}{\lambda_s}(c_s - c) \qquad (21.29)$$

Equations (21.2) and (21.29) are coupled through the temperature dependence of k_v and D_e [Eqs. (21.17) and (21.22)] although in the classical analysis for a macroporous particle [13] the temperature dependence of D_e can be neglected. The solution gives the non-isothermal effectiveness factor, defined as the average reaction rate in the particle relative to the rate in a particle in which both temperatures and concentration are uniform at their surface values (T_s, c_s), as a function of the two parameters $\beta = c_s(-\Delta H)D_e/\lambda_s T_s$ and $\gamma = E/RT_s$.

In the endothermic case both temperature and reactant concentration decrease towards the center of the particle so the reaction rate must also decrease monotonically. The more interesting situation is the exothermic case in which the temperature increases towards the center while, as a result of diffusional limitations, the reactant concentration decreases. In this situation the reaction rate may vary in a complicated way, as indicated in Figure 21.1b and c. In the event that the increase in reaction rate due to the rise in temperature outweighs the decrease in rate due to the declining reactant concentration, the effectiveness factor may be greater than unity. This situation has attracted a great deal of theoretical attention although the ranges of parameter values for which the effectiveness factor becomes significantly greater than unity are somewhat unrealistic.

A useful asymptotic solution to the problem of non-isothermal diffusion and reaction has been given by Petersen [3] for situations in which deviations from isothermal behavior are modest. This solution leads to a very simple criterion that can be used to assess the significance of heat effects in any given situation. If:

$$|\beta\gamma| = c_s \frac{(-\Delta H)D}{\lambda_s T_s} \frac{E}{RT_s} < 0.3 \tag{21.30}$$

the effectiveness factor will differ by no more than 5% from the isothermal value. This criterion may be applied to the problem of diffusion and reaction in a zeolite crystal.

The relevant parameter values vary widely depending on the system but in general an order of magnitude analysis shows that in most zeolitic systems the assumption of uniform temperature through the *crystal* is generally a good approximation. For example, with the following typical values:

$$c_s = 2 \times 10^{-3} \text{mol cm}^{-3}; \quad -\Delta H = 50\,000 \text{ cal mol}^{-1}; \quad E = 25\,000 \text{ cal mol}^{-1};$$
$$D = 10^{-8} \text{cm}^2 \text{ s}^{-1}; \quad \lambda_s = 10^{-3} \text{cal cm}^{-1} \text{s}^{-1} \text{ deg}^{-1}; \quad T_s = 500 \text{ K}$$

$|\beta\gamma| \sim 5 \times 10^{-5}$. To exceed the Petersen criterion would evidently require an unusual combination of an extremely high diffusivity together with very high activation energy and heat of reaction.

21.1.7
Relative Importance of Internal and External Resistances

Although heat conduction through an individual crystal, and indeed through a composite catalyst particle, is generally fast enough to maintain a uniform temperature within the particle, the possibility of a temperature difference between the particle and surrounding fluid resulting from the finite rate of heat transfer at the phase boundary must also be considered. This situation, which is similar to the intrusion of external mass transfer resistance, is sketched in Figure 21.1b (curves 2 and 3). At steady state one may equate the rate of heat transfer through the external fluid film to the conductive heat flux just inside the particle:

$$h(T_s - T_0) = -\lambda_s \frac{dT}{dr}\bigg|_{R_p} \tag{21.31}$$

Following Carberry [14] we may define the parameter a_H:

$$a_H = \frac{\text{Dimensionless external temperature gradient}}{\text{Dimensionless internal temperature gradient}} = \frac{(T_g - T_s)/T_g}{\left[\frac{d}{dZ}\left(\frac{T_s}{T_g}\right)\right]_{Z=1}} = \frac{\lambda_s}{R_p h} \tag{21.32}$$

Assuming similarity between heat and mass transfer through the external film (Nu = Nusselt number and Sh = Sherwood number):

$$\mathrm{Nu} \equiv \frac{hd}{\lambda_g} = \mathrm{Sh} \equiv \frac{k_c d}{D_m} \tag{21.33}$$

In the low Reynolds number region, studies of mass and heat transfer in packed beds show Nu $\sim$ Sh $\sim$ 2.0 so Eq. (21.32) reduces to:

$$\frac{\lambda_g}{R_p h} \approx 1.0, \quad \alpha_H \approx \frac{\lambda_s}{\lambda_g} \tag{21.34}$$

Since in general the thermal conductivity of even an insulating solid is substantially higher than that of a gas $(\lambda_s \gg \lambda_g)$ this implies $\alpha_H \gg 1.0$ and we conclude that external temperature gradients will generally be more important than internal gradients. The situation is less clear at high Reynolds number since the Sherwood number may then be substantially greater than 2.0. Nevertheless, in most practical situations, the model of an isothermal particle in which the temperature may differ from the ambient gas as a result of external film heat transfer resistance is probably a reasonable representation (curve 2, Figure 21.1b).

Similar logic may be used to assess the relative importance of internal and external resistances to mass transfer (Figure 21.1a). From the steady state mass balance:

$$k_f(c_0 - c_s) = D_e \frac{dc}{dR}\bigg|_{R=R_p} \tag{21.35}$$

we obtain:

$$\alpha_M = \frac{\text{Dimensionless concentration gradient through film}}{\text{Dimensionless concentration gradient inside particle}}$$

$$= \frac{(c_0 - c_s)/c_0}{\left[\frac{d}{dR}\left(\frac{c}{c_0}\right)\right]_{R_p}} = \frac{D_e}{R_p k_f} \tag{21.36}$$

Since $\mathrm{Sh} \equiv k_f d/D_m \geq 2.0$, $D_m/R_p k_f \leq 1.0$. As the effective diffusivity (D_e) is generally substantially smaller than the molecular diffusivity it is clear that α_M should always be substantially smaller than unity, implying that, in general, external film resistance will be much less important than internal diffusional resistance. Of course this does not preclude the possibility of significant external mass transfer resistance when the catalyst is operating in the strongly pore diffusion limited regime. What it does mean is that in the absence of internal resistance, external film resistance is unlikely to be important.

21.2
Determination of Intracrystalline Diffusivity from Measurements of Reaction Rate

If the rate of a particular reaction is measured under comparable conditions with several different size fractions of zeolite crystals it is, in principle, possible to determine directly both the intrinsic rate constant and the effective diffusivity.

The simplest procedure is to make measurements first with very small crystals under conditions such that $\eta \to 1.0$. Under these conditions the apparent rate constant should be independent of crystal size so it is easy to confirm whether the desired regime has been achieved. Measurements are then repeated with larger crystals under conditions of diffusional limitation and, using the intrinsic rate constant derived for the smaller crystals, the effectiveness factor is found. The value of the Thiele modulus and hence the effective diffusivity may then be calculated from Eqs. (21.6) and (21.7). Alternatively, the same information may be derived directly from measurements with just two different crystal sizes as long as the effectiveness factor for the smaller crystals is significantly less than 1 and greater than about 0.3. If, for both crystal sizes, the effectiveness factor is small (within the asymptotic region in which $\tau \approx 3/\phi_s$) then the two equations become degenerate and cannot be solved to separate the values of k_v and D_e.

This method of measuring intracrystalline diffusivity, originally suggested by Haag et al. [16], was applied by Post et al. [17], who studied in detail the diffusion and reaction of 2,2-dimethylbutane in HZSM-5 crystals. They determined the intracrystalline diffusivity from both uptake rate and chromatographic measurements with larger crystals. The diffusivities obtained by both methods were quite consistent. Using the Henry's law constant and diffusivity extrapolated to the reaction temperature (811 K), effectiveness factors and the corresponding Thiele moduli were calculated from experimental cracking rates, measured for a range of different crystal sizes. The resulting values, as well as the earlier data of Haag et al. [16], are seen to lie quite close to the theoretical curve derived from the Thiele model (Figure 21.4). These results provide a convincing demonstration of the applicability of the Thiele model to an intracrystalline diffusion controlled system as well as an unequivocal determination of the diffusivity for 2,2-dimethylbutane in HZSM-5

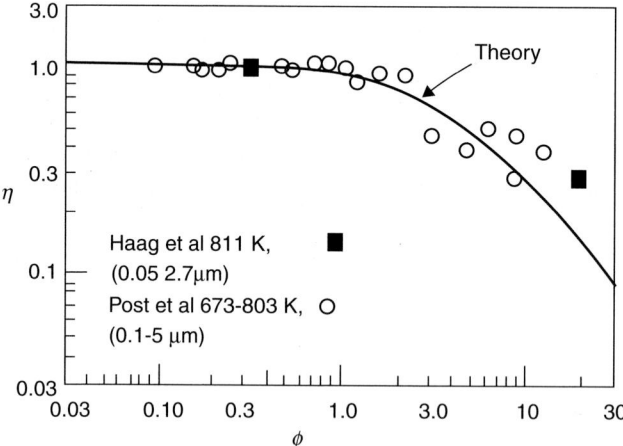

Figure 21.4 Experimental verification of the Thiele analysis as applied to intracrystalline diffusion control of the catalytic reaction of 2,2-dimethylbutane in H-ZSM-5 catalysts: (○) Post et al. [17] and (■) Haag et al. [16]. Reprinted from Post et al. [17] with permission.

Table 21.1 Intracrystalline diffusivities for hydrocarbons in H-ZSM-5 derived from effectiveness factor measurements at 811 K.

Hydrocarbon	D' (m² s⁻¹)[a]	K	D'/K (m² s⁻¹)	D_c[b] (m² s⁻¹)	E (kJ mol⁻¹)
C–C–C–C–C–C	3×10^{-8}	7	4×10^{-9}	10^{-10}	17.0
C–C–C–C–C with C branch	4×10^{-9}	7	6×10^{-10}	—	—
C–C–C–C with two C branches	2×10^{-12}	~1.0	2×10^{-12}	10^{-13}	66
C–C–C–C–C–C with C branch	3×10^{-12}	—	—	—	—

a) Values of D' derived from Thiele modulus by Haag et al. [16]. Diffusivity values (D') given by Haag et al. are based on gas-phase concentrations. To compare with the directly measured intracrystalline diffusivities these values must be divided by the dimensionless equilibrium constant.
b) Value of D_c for n-hexane estimated by extrapolation from data of Eic [18]. Value of D_c for 2-dimethylbutane estimated by extrapolation from data of Post et al. [17] (see Table 18.3).

(Table 21.1). Importantly, PFG NMR has meanwhile been successfully applied to the *in situ* measurement of the diffusivities of the different components involved in catalytic reactions (Section 17.6.3). Combining these options with the Thiele analysis for determining intracrystalline diffusivities as described in this section is among the challenges of future research.

21.2.1
Temperature Dependence of "Effective Diffusivity"

Notably, since the intracrystalline diffusivity is conventionally defined with respect to the concentration gradient in the adsorbed phase, the effective diffusivity in the Thiele modulus (under linear conditions) is the product KD_0, where K is the dimensionless Henry constant and D_0 is the corrected intracrystalline diffusivity. Since D_0 varies with temperature in accordance with an Arrhenius expression ($D_0 = D_\infty e^{-E/RT}$) while K follows a van't Hoff expression ($K = K_\infty e^{-\Delta U/RT}$) the product KD_0 will have an apparent activation energy given by the difference between the activation energy and the energy of adsorption ($E_{app} = E + \Delta U$), so if $E \approx -\Delta U$ the temperature dependence will be very weak, as noted by Garcia and Weisz [15].

21.3
Direct Measurement of Concentration Profiles during a Diffusion-Controlled Catalytic Reaction

Fluorescence microscopy (Section 11.6.1) allows the position of a single molecule to be followed in real time. By averaging over a sufficiently large number of molecules it is therefore possible to develop the average concentration profile for a reactant or product during the course of a catalytic reaction involving suitably chosen fluorescent molecules. This approach, which has attracted much research during the past decade, has yielded a great deal of important information concerning the details of catalytic reactions at the molecular level. In addition, micro-imaging by interference microscopy (IRM) (Section 12.2) offers similar prospects but the potential of this approach has not yet been explored.

A relatively simple example for the application of fluorescence microscopy is the base-catalyzed conversion of fluorescein diacetate into highly emissive fluorescein products over lactose dehydrogenase crystals, studied by Roeffaers et al. [19]. The reaction occurs only at specific active sites on the external surface of the crystal, and from the differences in the intensity of the local fluorescence the large differences in activity between different sites are clearly apparent. This study therefore provides direct experimental verification of the "active site hypothesis" and shows clearly that the characterization of catalytic activity by a single average rate constant is only a rather crude approximation.

21.3.1
Reaction of Furfuryl Alcohol on HZSM-5

A similar approach was also used to study the acid-catalyzed conversion of furfuryl alcohol into strongly fluorescent oligomers over a single HZSM-5 crystal [20]. This reactant is small enough to penetrate the intracrystalline pores but the oligomeric products can diffuse out only very slowly. The resulting profiles, reproduced in Figure 21.5, show clear evidence of diffusional limitation.

From considerations of molecular size it seems reasonable to assume that the reactant (furfuryl alcohol) can diffuse slowly into the intracrystalline pores but the oligomeric products formed within the crystal are effectively trapped as a result of their very low diffusivities. At short times we therefore see product formation only on the external surface of the crystal. At longer times we see the gradual build-up of products in the interior of the crystal, where the rate is effectively controlled by diffusion of the reactant into the pores. The form of the product profiles through the central section ("f" in Figure 21.5), which show symmetric maxima at a distance approximately midway between the external surface and the center, is consistent with this model. The higher maxima in the zones near the ends of the crystal suggest that the transport of the reactant from the external surface occurs more rapidly near the ends of the crystal, possibly as a result of facilitated transport along the twin planes, as has been observed by Kocirik et al. [21]. However, this is in contrast to the evidence presented by Geier et al. [22] (Figure 18.34) that suggests that the twin planes act as barriers to transport.

As well as providing direct experimental evidence of intracrystalline diffusion control these profiles show very clearly that the transport properties are far from uniform over the crystal. The assumption of a spatially uniform intracrystalline diffusivity that is inherent in the model generally used for the analysis of catalytic kinetics under conditions of diffusion control (Section 21.1.1) is evidently a rather gross approximation. Notably, at long times, the formation of coke deposits in the regions of high conversion is clearly apparent (Section 21.5).

21.3 Direct Measurement of Concentration Profiles during a Diffusion-Controlled Catalytic Reaction

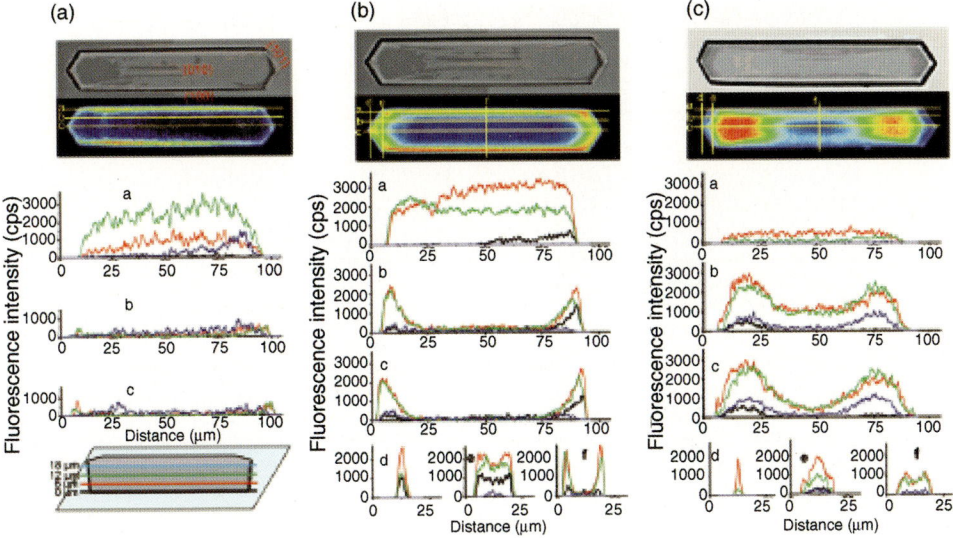

Figure 21.5 Confocal fluorescence microscopy imaging of acid-catalyzed oligomerization of furfuryl alcohol in a crystal of HZSM-5: (a) Homogeneous fluorescence intensity at the (100) face after 10 min. Intensities along selected lines a–c were measured at various depths along the crystal as indicated in the figure; (b) after 16 h the buildup of reaction products near the ends of the crystal becomes increasingly evident. Again the intensities along selected lines a–f are shown at different depths within the crystal; (c) after 50 days the accumulation of reaction products shows up as zones of high fluorescence intensity near the ends of the crystal and, at a somewhat lower intensity, in the center of the crystal. Reprinted from Roeffaers et al. [20] with permission.

21.3.2
Reaction in Mesoporous MCM-41

Similar techniques have been used to study catalytic reactions in mesoporous catalysts. For example, the epoxidation of phenylbutadienyl-substituted boron dipyrromethene difluoride in Ti-MCM-41 was studied by De Cremer et al. [23, 24] who used single-molecule fluorescence microscopy to locate each individual turnover event within a single MCM-41 particle (diameter 6 μm) during the reaction. Experiments of this type provide two distinct kinds of information: the frequency of turnover (reaction) events and the spatial variation of the reaction rate through the particle (or crystal). Accumulation of sufficient turnover events allows the reaction profile to be constructed (Figure 21.6).

In this particular reaction the rapid decline in the turnover frequency with distance from the external surface provides clear and unambiguous evidence of diffusional control. Detailed analysis of the profiles in accordance with the theory outlined in Section 21.1 yields an effectiveness factor of 0.3–0.4, a pseudo-first-order rate constant of about $10^{-3}\,s^{-1}$, and an effective diffusivity of order $10^{-16}\,m^2\,s^{-1}$.

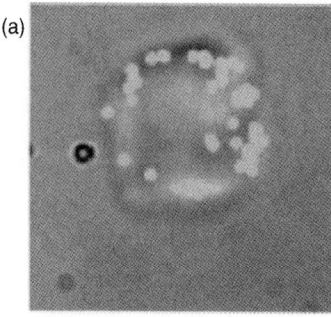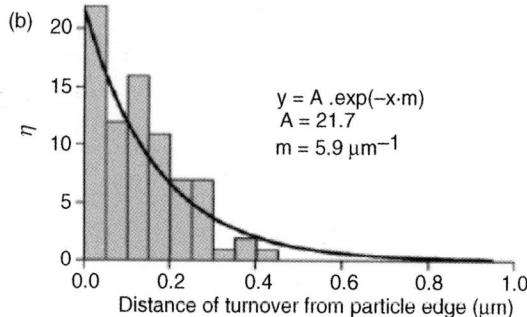

Figure 21.6 (a) High-accuracy localized reaction positions within an individual MCM-41 crystal (6 μm); (b) histogram showing location of the turnover events (distance from the external surface of the particle) together with the smoothed reaction profile derived by fitting the data to the Thiele model [Eqs. (21.8)–(21.10)]. Reprinted from De Cremer et al. [24] with permission.

21.4
Diffusional Restrictions in Zeolite Catalytic Processes

21.4.1
Size Exclusion

The most straightforward examples of diffusional restriction are to be found in situations in which one (or more) species in a mixed feed is too large to enter the zeolite pores and therefore passes through the reactor unchanged. Two examples, taken from the classic paper of Weisz et al. [25], are shown in Tables 21.2 and 21.3. In the dehydration of isomeric butanols a high level of conversion is observed for all isomers over CaX (large pore) whereas, over CaA, only n-butanol reacts to any significant extent. Similarly, when a mixture of n-hexane and 3-methylpentane is passed over NaA there is very little reaction since neither reactant can penetrate the zeolite pores at an appreciable rate. Over the larger-pore CaA we see significant cracking of n-hexane but no significant reaction of 3-methylpentane.

Table 21.2 Product distribution[a] for dehydration of primary butyl alcohols over eight-ring and 12-ring zeolites. From Weisz et al. [25] with permission.

Zeolite	Product	Temperature (°C)		
		130	230	260
CaA	Isobutanol		<2	<2
	n-Butanol		18	60
	sec-Butanol	~0	0	0
CaX	Isobutanol		46	85
	n-Butanol		9	64
	sec-Butanol	82	—	—

a) Wt% converted at $p = 1$ atm, 6 s residence time.

Table 21.3 Cracking of n-hexane and 3-methylpentane at 500 °C. From Weisz et al. [25] with permission.

Catalyst	3-Methylpentane		n-Hexane	
	Conversion (%)	$\dfrac{iC_4}{nC_4}$	Conversion (%)	$\dfrac{iC_5}{nC_5}$
96% Silica	<1		1.1	
Amorphous SiO_2/Al_2O_3	28	1.4	12.2	10
Zeolite NaA	<1		1.4	
Zeolite CaA	<1	<0.05	9.2	<0.05

21.4.2
Catalytic Cracking over Zeolite Y

Catalytic cracking of longer chain alkanes is generally carried out over zeolite Y based catalysts (commonly a mixture of H and rare earth forms). The catalyst particles are small composite pellets [typically of about 100 μm diameter formed from small (1–2 μm) zeolite crystals aggregated with a clay binder], so the ratio of particle to crystal size is typically about 50. The extent to which this important reaction is influenced by intraparticle and intracrystalline diffusion has been investigated by Kortunov et al. [26]. Figure 21.7a shows PFG NMR measurements of both intracrystalline and intraparticle diffusivities of n-octane. At 300 K these quantities are of similar magnitude but since the apparent activation energy for long-range diffusion (approximately the heat of adsorption) is higher than the activation energy for intracrystalline diffusion, at reaction temperature (≈800 K) the intraparticle diffusivity is about two orders of magnitude larger than the intracrystalline diffusivity. The relative influence of intraparticle and intracrystalline diffusion on the reaction rate depends on the ratio of the Thiele moduli for particle (ϕ_p) and crystal (ϕ_c):

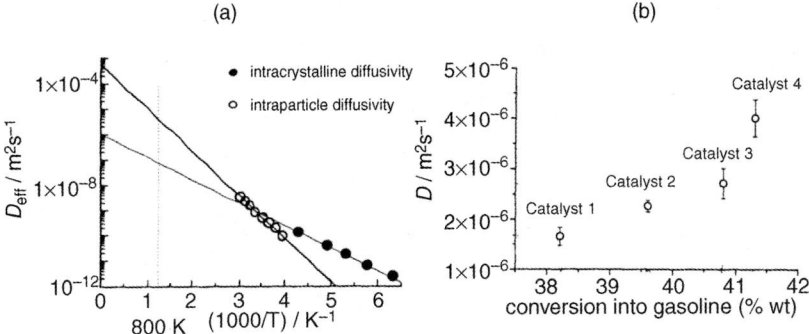

Figure 21.7 Performance of commercial zeolite Y based cracking catalysts: (a) temperature dependence of intracrystalline diffusivity and effective intraparticle diffusivity; (b) correlation of catalyst performance (at 800 K) with intraparticle diffusivity. From Kortunov et al. [26].

$$\frac{\phi_p}{\phi_c} = \frac{R_p}{r_c}\sqrt{(1-\varepsilon)\frac{D_c}{D_{l.r.}}} \qquad (21.37)$$

With $R_p/r_c \approx 50$, $D_{l.r.}/D_c \approx 100$, and $\varepsilon \approx 0.3$ this gives $\phi_p/\phi_c \approx 4$, suggesting that diffusional resistance at the particle scale is more important than intracrystalline resistance.

Figure 21.7b shows the variation in conversion with intraparticle diffusivity for four different zeolite-Y based catalysts of similar particle size operated at the same temperature (803 K), feed composition, and space velocity. The improvement in performance with increasing diffusivity is clearly apparent and provides clear evidence of intraparticle diffusion limitations under reaction conditions.

21.4.3
Catalytic Cracking Over HZSM-5

Cracking of both linear and branched hydrocarbons has been studied over various different zeolite catalysts. Unequivocal evidence that the cracking of branched isomers such as 2,2-dimethylbutane on ZSM-5 is strongly limited by intracrystalline diffusion is provided by the studies of Haag et al. [16] and Post et al. [17], which are discussed above. The evidence concerning the cracking of linear and single methyl branched species is, however, less definitive. The relative rates of cracking of different paraffin isomers over ZSM-5, taken from the work of Chen and Garwood [27], are summarized in Figure 21.8. Evidently, in general, the reaction

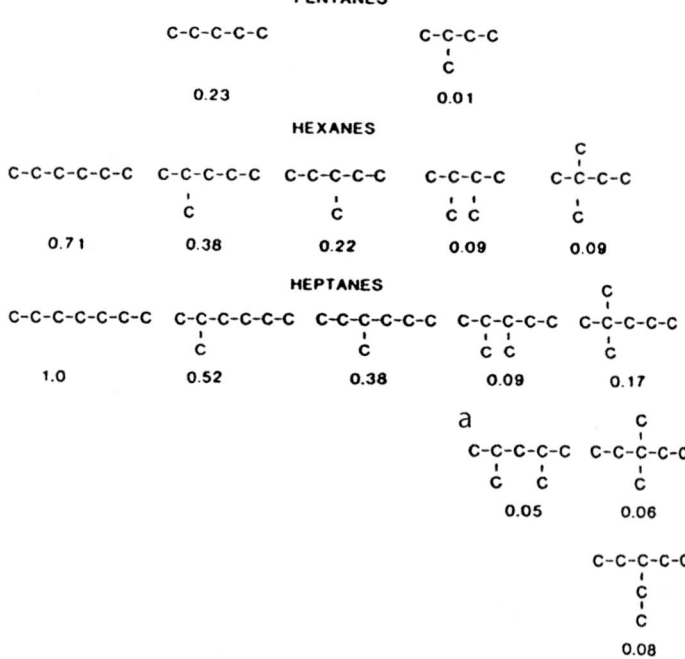

Figure 21.8 Relative rates of cracking of different paraffin hydrocarbons over H-ZSM-5 catalyst at 1.4 LHSV, 340 °C, 35 atm. Reprinted from Chen and Garwood [27] with permission.

rates are highest for the linear and singly methyl branched isomers and lowest for the bulkier doubly branched species, which appears to suggest diffusional control. However, Frillette et al. [28] showed that the relative rate of cracking of n-hexane and 3-methylpentane is independent of crystal size. It thus appears that the sequence of activities probably results from steric hindrance of the transition state rather than from diffusional limitations. A detailed study of hydrocracking of n-dodecane over HZSM-5 has also been reported by Kennedy et al. [29]. In contrast to the earlier data referred to above, these results show clear evidence of intracrystalline limitation for n-hexane and 3-methylpentane. Whether the selectivity effects observed in the cracking of linear or single methyl branched paraffins over HZSM-5 are due to intracrystalline diffusional resistance or to transition state selectivity therefore remains somewhat uncertain and may well depend on whether the catalyst is coked or coke free.

21.4.4
Catalytic Cracking over other Zeolites

In the cracking of longer chain linear paraffins ($nC_{22}H_{46}$) over zeolite T, a synthetic intergrowth of erionite and offretite, a curious bimodal product distribution was observed with maxima at C_4 and C_n [30]. This result was explained by Gorring [31] in terms of a diffusional limitation since, in a parallel study of diffusion in erionite, he observed an unusual trend of diffusivity with carbon number, showing a minimum at C_8 and a local maximum at C_{11}. These results have been widely quoted in subsequent reviews [1, 32, 33] but a detailed examination of the experimental conditions in Gorring's diffusion measurements (uptake rate method) reveals that in these experiments the rate of sorption was almost certainly controlled by extracrystalline diffusion and/or heat effects rather than by intracrystalline diffusion [34]. Two recent experimental studies [35, 36] of diffusion in zeolite T showed a monotonic decline in diffusivity with carbon number rather than the complex trend reported by Gorring.

Several selective cracking processes depending on size exclusion have been successfully commercialized. An early example is the "Selectoforming process" [37] in which the linear paraffins in reformed naphtha are cracked selectively over a (small-pore) erionite catalyst to yield lighter paraffins and propane (which is removed as LPG), thus raising the octane number. Another example is the Mobil catalytic dewaxing process [38] in which the long-chain linear paraffins in heavier distillates and lubricating oils are removed selectively by adsorption and catalytic cracking on H-ZSM-5, thus reducing the viscosity and hence the pour point and freezing point.

21.4.5
Activation Energies

Over most zeolites the activation energy for cracking of linear paraffins is about 125 kJ mol^{-1} but in their study of the cracking of n-hexane on a faujasite-based catalyst, Miale, Chen, and Weisz [39] observed an activation energy of about 60 kJ mol^{-1}, whereas n-pentane showed the "normal" value of about 125 kJ mol^{-1}. A similar difference was observed for the cracking of n-pentane and n-hexane over erionite. [40]. The low activation energy for n-hexane is consistent with diffusional limitations but this now seems unlikely in view of the evidence presented by Haag et al. [16], which shows clearly that the cracking rate for n-hexane in HZSM-5 is independent of crystal size and cannot therefore be diffusion limited.

21.4.6
Xylene Isomerization

Perhaps the most important of the zeolite-catalyzed processes in which the selectivity is determined by diffusional effects is the isomerization of the C_8 aromatics over H-ZSM-5. Figure 21.9 shows the process scheme used to refine the single-ring aromatics stream in a typical refinery. The main demand is for p-xylene, and to a lesser extent for o-xylene, while demand for m-xylene and ethylbenzene is generally minimal. o-Xylene is significantly less volatile than the other C_8 aromatic isomers and may therefore be separated by distillation (G, H in Figure 21.9). To separate p-xylene an adsorption process "Parex" (I) is generally used [41]. The raffinate stream from the "Parex" unit, which contains o-xylene, m-xylene, and ethylbenzene together with the desorbent p-diethylbenzene, is returned to the isomerization reactor and the effluent from this reactor, which contains a large fraction of p-xylene together with the other

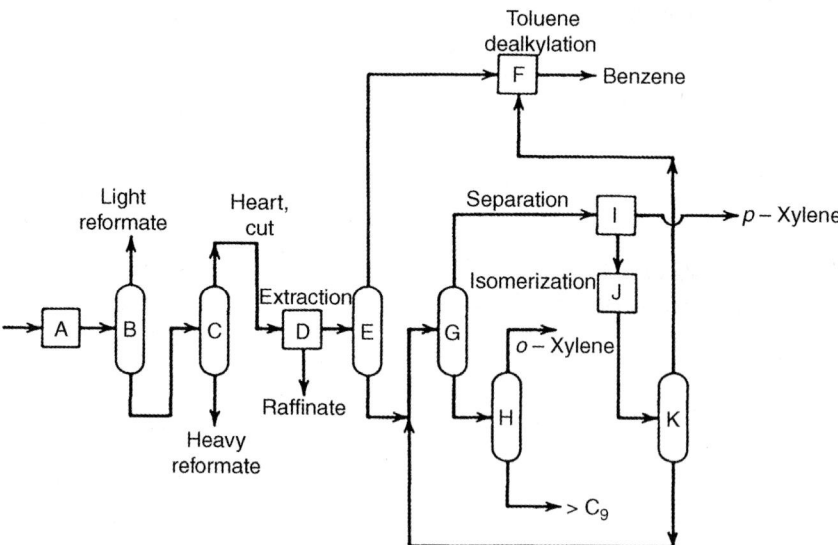

Figure 21.9 C_8 aromatics loop for a typical refinery.

C₈ isomers, is then recycled, after removal of light products (benzene and toluene) by distillation (K).

HZSM-5 shows a unique ability to isomerize xylenes with very little disproportionation. This is probably the result of transition state selectivity favoring the unimolecular isomerization path over the bimolecular disproportionation [42, 43]. However, the large difference in diffusivity between p-xylene and the ortho- and meta-isomers means that the ortho- and meta-isomers are effectively trapped within the zeolite pores whereas p-xylene (or ethylbenzene) can easily diffuse out of the crystal. As a result the ratio p-xylene/m-xylene in the product is substantially greater than the equilibrium ratio [44] (Figure 21.10).

That the para-selectivity results from diffusional effects has been shown clearly by detailed experimental studies. Small crystals of H-ZSM-5 exhibit only modest para-selectivity but the selectivity improves with increasing crystal size [45, 46] and it may also be enhanced by doping, coke deposition, or other procedures that retard intracrystalline diffusion [47].

A detailed mathematical theory to account for selective xylene isomerization has been developed by Wei [48]. Assuming a first-order reaction with rate constants chosen to satisfy the equilibrium distribution (23% o-xylene, 53% m-xylene, 24% p-xylene) and with intracrystalline diffusivities in the ratio $D_{OX} = D_{MX} = D_{PX}/1000$, reaction paths were calculated under conditions of kinetic control ($\phi \to 0$) and for strong pore diffusion control (ϕ large). As may be seen from Figure 21.10 the experimental data for nonselective catalysts conform closely to the path calculated for the kinetic regime with negligible diffusional restriction, whereas the data for the high-selectivity catalysts conform to the diffusion-limited path, giving much higher yields of p-xylene. In Wei's original calculations a diffusivity ratio (D_{PX}/D_{OX}) was assumed that is much too large. However, reducing this ratio to the more reasonable value of 20 has little effect on the results.

For the same reason a similar pattern of selectivity is found in toluene alkylations:

$$\text{methanol} + \text{toluene} = \text{xylene} + H_2O$$

Small crystals of HZSM-5 produce an equilibrium mixture of xylene isomers but with larger crystals or crystals doped with atoms such as P or Mg, which reduce the effective diameter of the pores, an enhanced yield of p-xylene is observed [49, 50].

$$\begin{array}{c} o\text{-xylene} \\ \swarrow \quad \searrow \\ m\text{-xylene} \rightleftharpoons p\text{-xylene} \end{array}$$

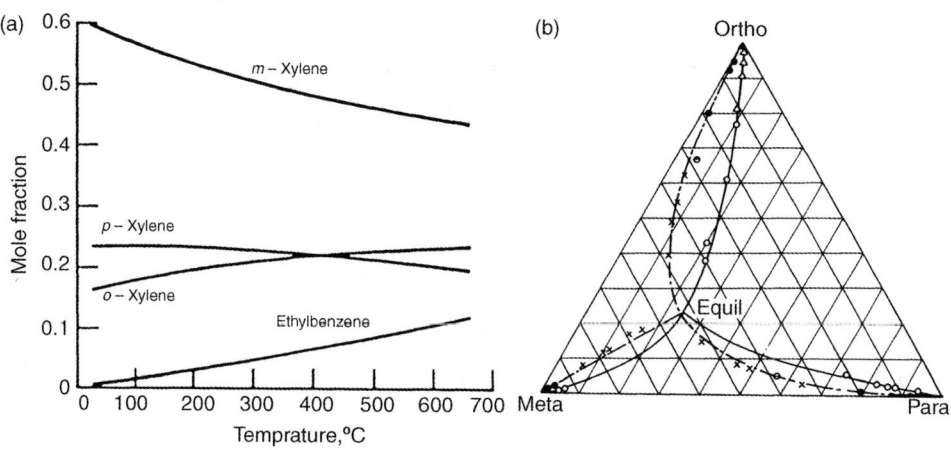

Figure 21.10 (a) Equilibrium concentration of C$_8$ aromatic isomers as function of temperature; (b) reaction path for xylene isomerization at 350 °C over nonselective (silica-alumina, zeolite Y) catalysts (×, •) and over selective (doped) ZSM-5 catalysts (Mg ZSM-5, △; P ZSM-5, ○). The broken line shows the theoretical path for a high effectiveness factor ($\eta \to 1.0$) and the full line shows the corresponding path for a low effectiveness factor ($\eta < 0.5$). Reprinted from Wei [48] with permission.

21.4.7
Selective Disproportionation of Toluene

Thermodynamic calculations show that the equilibrium mixture formed by disproportionation of toluene contains significant proportions of tri- and tetra-methyl benzenes. However, over ZSM5 catalysts this reaction proceeds almost quantitatively to benzene plus xylene and the proportion of *p*-xylene in the product commonly exceeds the equilibrium composition (Scheme 21.2).

A detailed investigation by Olson and Haag [42] reveals that this behavior can be understood as resulting from the combined effects of transition state selectivity and diffusional restriction. The 1 : 1 stoichiometry of benzene–total xylenes is found to persist regardless of crystal size, but the proportion of p-xylene–total xylenes varies with crystal size. This suggests that the formation of tri- and tetra-methylated products is prohibited by transition state selectivity while the formation of excess p-xylene arises from the slow intracrystalline diffusion of the *ortho-* and *meta-* isomers.

Figure 21.11 shows quantitative evidence in support of such a mechanism. Figure 21.11a shows the p-xylene yield, for several different catalyst samples, plotted against the diffusional time constant, as measured by the time for 30% approach to equilibrium in an uptake experiment with o-xylene at 120 °C ($at_{0.3}$). (This quantity is obviously proportional to r_c^2/D_c.) An even more general correlation is shown in

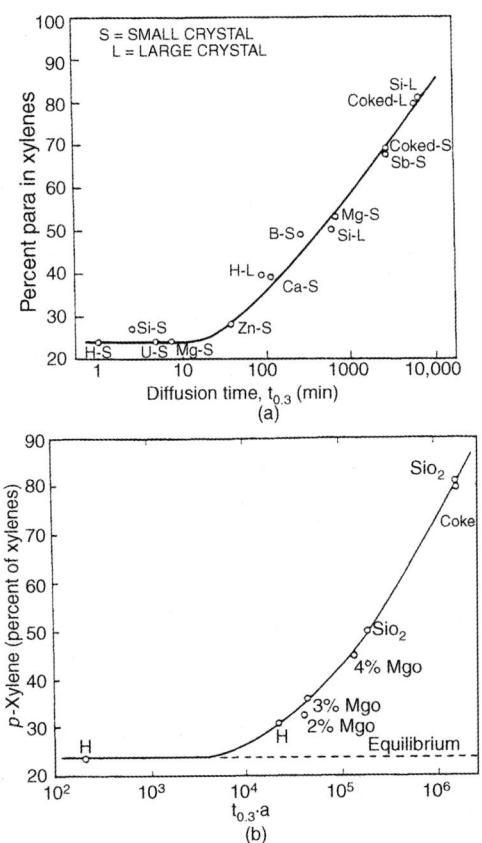

Figure 21.11 *para*-Xylene yield in toluene disproportionation, over different catalysts, plotted (a) against the diffusion time (t_{03}) as measured by the time for 30% approach to equilibrium for o-xylene at 120 °C and (b) against the product at_{03}, where a is a measure of the rate constant. (The product at_0 is therefore proportional to ϕ_s^2.) Reprinted from Olson and Haag [42] with permission.

Figure 21.11b where the p-xylene yield is plotted against the product αt_{o3}, where α is a measure of the intrinsic rate constant determined from measurements of the rate of cracking of n-hexane. Clearly, the product αt_{o3} is proportional to ϕ^2 so this plot shows that, regardless of activity, diffusivity, and crystal size, the yield is a unique function of the Thiele modulus, as is to be expected for a diffusion-controlled reaction.

21.4.8
MTG Reaction

The reaction of methanol over ZSM-5 [methanol to gasoline (MTG) reaction] yields a range of light linear paraffins together with various single-ring aromatics, including p-xylene, m-xylene, and 1,2,4-trimethylbenzene, but no o-xylene or other isomers of trimethylbenzene. This unusual product distribution has been shown by Klinowski and Anderson [51] to arise from diffusional limitations, rather than from transition state effects. Using *in situ* high resolution NMR, under reaction conditions, they showed that, within the zeolite, all possible isomers of the dimethyl- and trimethyl-benzenes are formed. The presence of only a few of these species in the reaction product must therefore reflect the difficulty with which these molecules can diffuse out of the zeolite crystal.

The overall stoichiometry follows, approximately, Scheme 21.3. Hydrocarbons in the gasoline range (C_5–C_{10}) are obtained with 90% selectivity. The resulting paraffins are highly branched and the aromatics tend to be methyl substituted, which is beneficial for the octane rating.

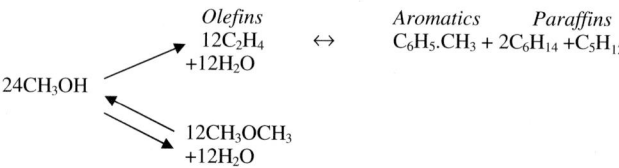

The reaction is highly exothermic, so heat removal is a major consideration. Both fixed and fluidized bed reactor systems have been developed by Chang and Silvestri [52]. The first commercial unit (14 500 barrel per day), which was built in New Zealand in 1985, used a fixed bed reactor. The fluidized bed process provides better temperature control and the reactor operates almost isothermally (at 413 °C), whereas in the fixed bed process there is a significant temperature gradient (360–415 °C) over the reactor. In addition, the fluidized bed process allows continuous recirculation and regeneration of the catalyst whereas the fixed bed system requires multiple reactors (five in the New Zealand plant) to allow for periodic regeneration. However, the advantages of the fluidized bed system are evidently not great enough to outweigh the additional cost, and the fluidized bed version of the process has so far been demonstrated only at pilot plant scale (100 barrel per day).

The process is operated at close to 100% methanol conversion to avoid the costly step of separating and recycling the methanol. Yields and product distribution are similar for both the fixed and fluidized bed processes. Gasoline yield is about 85% with about 13.5% LPG and 1–2% fuel gas containing light hydrocarbons. A more detailed account of this process has been given by Chen et al. [53, 54].

21.4.9
MTO Process

The rapidly increasing demand for ethylene and propylene for production of polyolefins has led to intensive study of the MTO (methanol to olefins) reaction, as an alternative to steam cracking of light naphtha, for production of the olefin feedstock. The earlier studies of this reaction, using HZSM-5 based catalysts, were carried out at the Mobil Laboratories in conjunction with development of the MTG process [55, 56]. The reaction of methanol over HZSM-5, at temperatures in the range 350–450 °C, yields a wide spectrum of products including light alkanes, alkanes, and single ring aromatics. The yield of $C_2 + C_3$ olefins (the desirable product for polyolefin feedstock) amounts to only 30–40%.

The introduction of SAPO-34 [57], a structural analog of chabazite (Figure 16.2), gave a dramatic improvement in both selectivity and conversion, making the process much more attractive as a route to light olefins. Under properly selected conditions light olefin yields ($C_2 + C_3$) approaching 80% can be achieved with only small amounts of higher olefins and essentially no aromatics. The $C_2 + C_3$ olefin/paraffin ratio increases with temperature, varying from about 1.0 to 2.2 over the range 648–723 K [58]. The absence of aromatic products appears to be related to constraints imposed by the small size of the chabazite cage ($\sim$350 Å^3), which is too small to allow an aromatic ring to form. However, coke formation, due to polymerization of a small fraction of the light olefins, remains a problem (Section 21.5).

The reaction mechanism has been investigated in considerable detail [54, 55]. Although the broad outline of the mechanism has been established, many important details are still not fully understood. The initial step involves the condensation of two molecules of methanol, with elimination of water, to yield dimethyl ether (DME):

$$2CH_3OH \rightarrow CH_3O\,CH_3 + H_2O \tag{21.38}$$

The second step, which appears to occur by a complex mechanism involving the reaction of DME with methanol and subsequent regeneration of the methanol, involves further elimination of water to form ethylene:

$$CH_3O\,CH_3 \rightarrow C_2H_4 + H_2O \tag{21.39}$$

The final step appears to be a polymerization to form propylene (and traces of higher olefins):

$$1.5\,C_2H_4 \Leftrightarrow C_3H_6 \tag{21.40}$$

Reaction (21.39) is reversible and exothermic so the equilibrium ratio of C_3/C_2 olefins decreases with increasing temperature. This probably accounts for the observed increase of ethylene yield with temperature [58–60].

Detailed studies of the kinetics of this reaction over different size fractions of SAPO-34 crystals together with measurements of the sorption rate and the equilibrium isotherm have been reported by Chen et al. [61–65]. These data are summarized in Figure 21.12. The dominance of intracrystalline diffusion in controlling the sorption rate was demonstrated by varying the crystal size. Values of the diffusional time constant (R^2/D_o) derived from reaction rate measurements at 698 K are close to the value extrapolated from sorption rate measurements at lower temperatures with the same batch of SAPO-34 crystals [61, 62]. The temperature dependence of the dimensionless Henry constant, also shown in Figure 21.12, yields an adsorption energy of $\Delta U \approx -7.5 \, \text{kcal mol}^{-1}$, which is almost the same as the diffusional activation energy derived from the temperature dependence of the (corrected) diffusivity ($E = 7.3 \, \text{kcal mol}^{-1}$). Consequently the product KD_0, referred to by Chen as the "steady-state diffusivity," is almost independent of temperature. A similar situation was noted by Garcia and Weisz [15] in their study of the reaction of various aromatics over HZSM-5.

As the catalyst ages, the light olefin yield and the selectivity both increase [61, 63]. This appears to be related to the build-up of coke within the intracrystalline pores, which reduces both the intrinsic rate constant and the intracrystalline diffusivity [62, 63]. Detailed measurements with different crystal sizes show that with increasing

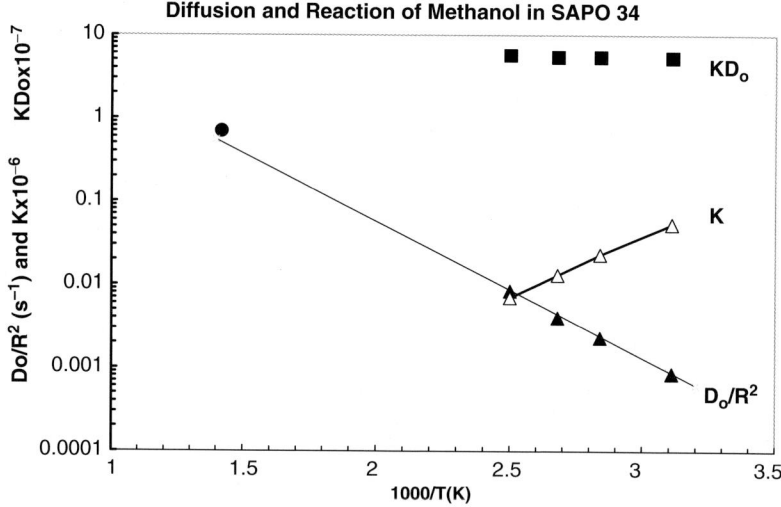

Figure 21.12 Variation of diffusional time constant (D_0/R^2), dimensionless Henry constant (K), and the product KD_0 with temperature. (From data of Chen et al. [61].) The value of D_0/R^2 derived from the reaction rate measurements (●) is also shown. Corrected diffusivities are derived from the reported integral diffusivities according to the analysis of Garg and Ruthven [66].

Table 21.4 Effect of the variation of crystal size on olefin yield. Data from Wilson and Barger [58]. Experimental conditions 673 K, 136 kPa, methanol WHSV = 1 h^{-1}.

Mean crystal size (μm)	$C_2^=+C_3^=$ (% yield)
0.6	72.9
0.7	73.2
0.9	73.9
1.2	70.6
1.4	75.3

coke levels the diffusivity declines more rapidly than the rate constant so that diffusional limitations become more pronounced as the catalyst ages. A high yield of light olefins requires that the DME formed in the first step of the reaction be retained within the crystal long enough for it to be essentially fully converted by reaction (21.39). This requires that the ratio of the Thiele moduli should be large:

$$\frac{\phi_2}{\phi_1} = \left(\frac{k_2}{k_1}\frac{D_{MeOH}}{D_{DME}}\right)^{1/2} \gg 1 \tag{21.41}$$

Since the ratio of Thiele moduli is independent of crystal size, variation of the crystal size should have no significant effect on the olefin yield. This has been confirmed experimentally (Table 21.4).

Since $k_2 < k_1$, a high ratio of D_{MeOH}/D_{DME} is necessary to achieve a high ratio ϕ_2/ϕ_1 and thus a high olefin yield. As the DME molecule is larger than the methanol molecule it is reasonable to assume that, under sterically restricted conditions, the diffusivity ratio D_{MEOH}/D_{DME} will increase as the effective pore size decreases. The observations that the olefin yield increases as the catalyst cokes and that an improvement in yield is obtained by increasing the Si/Al ratio (which decreases the unit cell size and therefore the effective window size) are consistent with this hypothesis. However, varying the Si/Al ratio also changes the strength of the acid sites, so such evidence is not entirely conclusive.

21.5
Coking of Zeolite Catalysts

In almost all catalytic reactions involving organic reactants small amounts of carbon (coke) are produced as a by-product and laid down on the catalyst surface, leading to a decline in activity as a result of both the deactivation of the actual sites and hindered access. A detailed understanding of the kinetics of coke deposition and the influence of small amounts of coke on the kinetics of the main reaction is therefore critical to the choice of operating conditions for many catalytic processes. Information concerning the spatial distribution of coke within a catalyst bed can be conveniently obtained by NMR measurements [67].

21.5.1
Information from PFG NMR

The NMR tracer exchange technique, in combination with the measurement of intracrystalline diffusion (Section 11.4.3 and Figure 11.12), has also been used to study the effect of coke deposition on the transport properties of ZSM-5 catalysts, using methane as the probe molecule [68, 69]. Figure 21.13 summarizes the results. When the sample is coked by exposure to n-hexane the intracrystalline mean lifetime (τ_{intra}) and the corresponding quantity τ_{intra}^{diff} calculated from intracrystalline diffusion

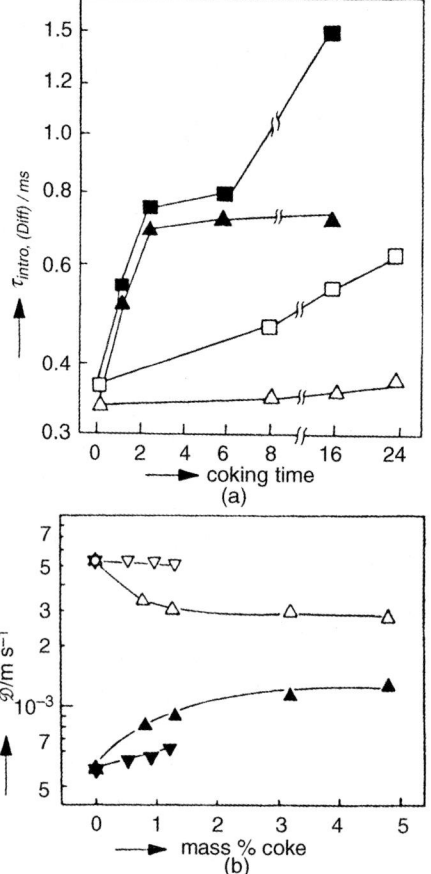

Figure 21.13 Effect of coke deposition on diffusion of methane in H-ZSM-5: (a) intracrystalline mean lifetimes τ_{intra} (squares) and τ_{intra}^{diff} (triangles) for CH_4 at 23 °C and sorbate concentration 12 molecules per unit cell as a function of coking time (indicated in hours) in samples coked by exposure to n-hexane (filled symbols) and mesitylene (which cannot penetrate the micropores, open symbols); (b) intracrystalline self-diffusion coefficients for methane in H-ZSM-5 coked by n-hexane (triangles) and mesitylene (inverted triangles) as a function of coke level in the presence of chemisorbed pyridine (filled symbols) and in the absence of pyridine (open symbols). From [68, 69] with permission.

measurements, assuming no surface resistance, both increase with coke level and the values are more or less the same, except for very severe coking when τ_{intra} increases rather strongly. This suggests that, except under very severe coking conditions, the coke is deposited mainly in the interior of the crystal with the result that the intracrystalline diffusivity falls, but without any evidence of a significant surface barrier. By contrast, when the sample was coked with mesitylene, a large molecule that cannot penetrate the micropores, we see little change in τ_{intra}^{diff} but a strong increase in τ_{intra}. This implies that under these conditions the coke is laid down almost entirely on the surface, thus introducing a surface resistance to mass transfer but with little effect on the intracrystalline diffusivity.

Figure 21.13b shows the combined effects of coke deposition and pyridine adsorption on the apparent NMR diffusivity of methane. The data for the uncoked material show a substantial reduction in diffusivity due to the effect of the chemisorbed pyridine. After coking with mesitylene this effect remains almost unchanged, but when the sample is coked with n-hexane, the blocking effect of pyridine is substantially reduced. This suggests that the internal coke is deposited at precisely those sites that act as sorption centres for the pyridine.

21.5.2
Information from Fluorescence Microscopy

Even more detailed information on coke deposition can be obtained by fluorescence microscopy since the strong fluorescence of coke molecules and their precursors allows the deposition of coke to be followed in space and time [24]. For example, Mores *et al.* [70] have used this approach to study in detail the deactivation of both ZSM-5 and SAPO-34 catalysts during the MTO process. The spatially and temporarily resolved images show clearly how the coke deposits are initially laid down at the crystal surface, moving further into the interior of the crystal as time progresses. In ZSM-5 this process is clearly hindered by the internal boundaries between different crystallographic regions, leading to clearly visible fluorescent patterns. The difference in the nature of the coke deposits in ZSM-5 and SAPO-34 could also be seen. The larger cages of SAPO-34 allow the formation of methylated polyaromatics that are less effective in splitting off the desired olefinic products.

By combining UV-visible micro-spectroscopy with fluorescence imaging it was also possible to distinguish between the fluorescent polyaromatics formed within the micropores and the light absorbing graphitic compounds on the external surface. The latter form an external shell around the crystals, thus obstructing pore openings and hindering access of the reactants to the interior.

References

1 Froment, G.F. and Bischoff, K.B. (1979) *Chemical Reactor Analysis and Design*, John Wiley & Sons, Inc., New York, ch. 3.

2 Carberry, J.J. (1976) *Chemical and Catalytic Reaction Engineering*, McGraw-Hill, New York, ch. 9.

3 Petersen, E.E. (1965) *Chemical Reaction Analysis*, Prentice Hall, Englewood Cliffs, NJ.
4 Thiele, E.W. (1939) *Ind. Eng. Chem.*, **31**, 916.
5 Zeldovich, I.B. (1939) *Zh. Fiz. Khim.*, **13**, 163.
6 Wheeler, A. (1955) *Catalysis* vol. 2, (ed. P.H. Emmett), Reinhold, New York, p. 123.
7 Wheeler, A. (1951) vol. 3, *Advances in Catalysis*, Academic Press, New York, p. 249.
8 Satterfield, C.N. (1970) *Mass Transfer in Heterogeneous Catalysis*, M.I.T. Press, Cambridge, MA.
9 Aris, R. (1975) *Mathematical Theory of Diffusion and Reaction in Permeable Catalysts*, Clarendon Press, Oxford.
10 Gorring, R.L. (March 1970) AIChE National Meeting, Houston.
11 Juttner, F. (1909) *Z. Phys. Chem.*, **65**, 595.
12 Wakao, N. and Kaguei, K. (1982) *Heat and Mass Transfer in Packed Beds*, Gordon and Breach, New York, p. 97.
13 Weisz, P.B. and Hicks, J.S. (1962) *Chem. Eng. Sci.*, **17**, 265.
14 Carberry, J.J. (1975) *Ind. Eng. Chem. Fundam.*, **14**, 129.
15 Garcia, S.F. and Weisz, P.B. (1990) *J. Catal.*, **121**, 294–311; Garcia, S.F. and Weisz, P.B. (1993) *J. Catal.* **142**, 691–696.
16 Haag, W.O., Lago, R.M., and Weisz, P.B. (1982) *Discuss. Faraday Soc.*, **72**, 317.
17 Post, M.F.M., van Amstel, J., and Kouwenhoven, H.W. (1984) in *Proceedings of the 6th International Zeolite Conference, Reno Nevada, 1983* (eds D. Olson and A. Bisio), Butterworth, Guildford, UK, p. 517.
18 Eic, M. and Ruthven, D.M. (1989) in *Proceedings of the 8th International Conference on Zeolites Amsterdam 1989* (eds P.A. Jacobs and R.A. van Santen), Elsevier, Amsterdam, p. 897.
19 Roeffaers, M.B.J., De Cramer, G., Sels, B.F., and De Vos, D.E. (2010) in *Single Particle Tracking and Single Molecule Energy Transfer* (eds C. Brauchle, D.C. Lamb, and J. Michaelis), Wiley-VCH Verlag GmbH, Weinheim, ch. 11.
20 Roeffaers, M.B.J., Sels, B.F., Uji, H., Blanpain, B., L'Hoest, P., Jacobs, P.A., De Schryver, F.C., Hofkens, J., and De Vos, D.E. (2007) *Angew. Chem., Int. Ed.*, **46**, 1706–1709.
21 Kocirik, M., Kornatowski, J., Masarik, V., Novak, P., Zikanova, A., and Maixner, J. (1998) *Microporous Mesoporous Mater.*, **23**, 295.
22 Geier, O., Vasenkov, S., Lehmann, E., Kärger, J., Schemmert, U., Rakoczy, R.A., and Weitkamp, J. (2001) *J. Phys. Chem. B*, **105**, 10217–10222.
23 De Cremer, G., Bartholomeeusen, E., Pescarmona, P.P., Lin, K., De Vos, D.E., Hofkens, J., Roeffaers, M.B.J., and Sels, B.F. (2010) *Catal. Today*, **157**, 236–242.
24 De Cremer, G., Sels, B.F., De Vos, D.E., Hofkens, J. and Roeffaers, M.B.J. (2010) *Chem. Soc. Revs.*, **39**, 4703–4715.
25 Weisz, P.B., Frillette, V.J., Maatman, R.W., and Mower, E.B. (1962) *J. Catal.*, **1**, 307.
26 Kortunov, P., Vasenkov, S., Kärger, J. *et al.* (2005) in *Diffusion Fundamentals* (eds J. Kärger, F. Grinberg, and P. Heitjans), Leipziger Universitätsverlag, Leipzig, p. 458.
27 Chen, N.Y. and Garwood, W.E. (1978) *J. Catal.*, **52**, 453.
28 Frillette, V.J., Haag, W.O., and Lago, R.M. (1981) *J. Catal.*, **67**, 218.
29 Kennedy, C.R., LaPierre, R.B., and Pereira, C.J. (1991) *Ind. Eng. Chem. Res.*, **30**, 12.
30 Chen, N.Y., Lucki, S.J., and Mower, E.B. (1969) *J. Catal.*, **13**, 329.
31 Gorring, R.L. (1973) *J. Catal.*, **31**, 13.
32 Eberly, P.E. (1976) *Zeolite Chemistry and Catalysis* (ed. J. Rabo), ACS Monograph 171, American Chemical Society, Washington, DC, p. 392.
33 Barrer, R.M. (1978) *Zeolites and Clay Minerals as Sorbents and Molecular Sieves*, Academic Press, London, p. 299.
34 Ruthven, D.M. (2006) *Microporous Mesoporous Mater.*, **96**, 262–289.
35 Cavalcante, C., Eic, M., Ruthven, D.M., and Occelli, M. (1995) *Zeolites*, **15**, 293–307.
36 Magalhaes, F.D., Laurence, R.L., and Connor, W.C. (1995) *AIChE J.*, **42**, 68–86.
37 Chen, N.Y., Maziuk, J., Schwartz, A.B., and Weisz, P.B. (1968) *Oil Gas J.*, **66** (47), 154.

38 Chen, N.Y., Gorring, R.L., Ireland, H.R., and Stein, T.R. (1968) *Oil Gas J.*, **75** (23), 154.
39 Miale, J.N., Chen, N.Y., and Weisz, P.B. (1966) *J. Catal.*, **6**, 278.
40 Robson, H.E., Hammer, G.P., and Arey, W.F. (1971) *Adv. Chem.*, **102**, 417.
41 Ruthven, D.M. (1984) in *Principles of Adsorption and Adsorption Processes*, John Wiley & Sons, Inc., New York, ch. 12.
42 Olson, D.H. and Haag, W.O. (1984) *Am. Chem. Soc. Symp. Ser.*, **248**, 275.
43 Cortes, A. and Corma, A. (1978) *J. Catal*, **51**, 338.
44 Haag, W.O. and Dwyer, F.G.(August 1979) AIChE 87th National Meeting, Boston.
45 Chutoransky, P. and Dwyer, F.G. (1970) *Adv. Chem. Ser.*, **121**, 540.
46 Le van Mao, R., Ragaini, V., Leofanti, G., and Fois, R. (1976) *J. Catal.*, **81**, 416.
47 Young, L.B., Butter, S.A., and Kaeding, W.W. (1982) *J. Catal.*, **76**, 418.
48 Wei, J. (1982) *J. Catal.*, **76**, 433.
49 Kaeding, W.W., Chu, C., Young, L.B., Weinstein, B., and Butter, S.A. (1981) *J. Catal.*, **67**, 154.
50 Kaeding, W.W., Chu, C., Young, L.B., and Butter, S.A. (1981) *J. Catal.*, **69**, 392.
51 Anderson, M. and Klinowski, J. (1989) *Nature*, **339**, 200.
52 Chang, C.D. and Silvestri, A.J. (1987) *Chemtech.*, **17**, 624.
53 Chen, N.Y., Garwood, W.E., and Dwyer, F.G. (1989) *Shape Selective Catalysis in Industrial Operations*, Marcel Dekker, New York, pp. 233–238.
54 Chen, N.Y., Degnam, T.F., and Smith, C.M. (1994) *Molecular Transport and Reaction in Zeolites*, VCH, New York.
55 Chang, C.D., Chu, C.T.W., and Socha, R.F. (1984) *J. Catal.*, **86**, 289.
56 Chang, C.D. (2011) in *Methanol Production and Use*, Marcel Dekker, New York, ch. 4, p. 133–173.
57 Kaiser, S.W. (1985) U.S. Patent 4,499,327.
58 Wilson, S. and Barger, P. (1999) *Microporous Mesoporous Mater.*, **29**, 117–126.
59 Froment, G.F., Dehertog, W.J.H., and Marchi, A.J. (1992) Chapter I, in *Catalysis*, vol. 9 (ed. J.J. Spivey), Royal Society of Chemistry, London.
60 Dahl, I.M. and Kolboe, S. (1994) *J. Catal.*, **149**, 458–464; Dahl, I.M. and Kolboe, S. (1996) *J. Catal.*, **161**, 304–309.
61 Chen, D., Rebo, H.P., Moljord, K., and Holmen, A. (1999) *Ind. Eng. Chem. Res.*, **38**, 4241–4249.
62 Chen, D., Rebo, H.P., and Holmen, A. (1999) *Chem. Eng. Sci.*, **54**, 3465–3473.
63 Chen, D., Rebo, H.P., Moljord, K., and Holmen, A. (1997) *Ind. Eng. Chem. Res.*, **36**, 3473–3479.
64 Chen, D., Rebo, H.P., Moljord, K., and Holmen, A. (1996) *Chem. Eng. Sci.*, **51**, 2687–2692.
65 Chen, D., Rebo, H.P., Moljord, K., and Holmen, A. (1998) *Stud. Surf. Sci. Catal.*, **119**, 521.
66 Garg, D.R. and Ruthven, D.M. (1972) *Chem. Eng. Sci.*, **27**, 417–423.
67 Bär, N.-K., Bauer, F., Ruthven, D.M., and Balcom, B. (2002) *J. Catal.*, **208**, 224–228.
68 Kärger, J. and Pfeifer, H. (1988) *ACS Symp. Ser.*, **368**, 376.
69 Caro, J., Bülow, M., Jobic, H., Kärger, J., and Zibrowius, B. (1993) *Adv. Catal.*, **39**, 351.
70 Mores, D., Stavitski, E., Kox, M.F.H., Kornatowski, J., Olsbye, U., and Weckhuysen, B.M. (2008) *Chem. Eur. J.*, **14**, 11320–11327.

Notation

a	scale factor (Chapter 2); absorption parameter [Eq. (2.31)]; mean molecular distance; particle radius; numerical factor (Chapter 6); specific area (external area per unit particle volume $= 3/R_p$ or $3/r_c s$); half cube edge (Table 6.1), acceleration (Chapter 8); intracrystalline step length in Monte Carlo simulation (Figure 11.9); activity (p/p_s)
a_0	intercrystalline step length in Monte Carlo simulation ($= \lambda a$; Figure 11.9)
$\mathbf{A}$	transformation matrix [Eqs. (8.60)–(8.62)]
A	arbitrary constant [Eq. (1.5)]; cross sectional area; area under Q versus γ profile (Chapter 6)
$A(N, V, T)$	Helmholtz energy [(Eq. (7.8)]
b	Langmuir equilibrium constant; van der Waals co-volume; scattering length (Chapter 10)
B	molecular mobility ($=1/f$); intensity of magnetic field; quantity of gas adsorbed (Section 13.8.1) permeability (Eq. (14.73)]
B_0	constant external magnetic field
B_{ij}	matrix elements defined by Eq. (8.31)
c	concentration; concentration in vapor phase (within a meso/macropore)
$c_{ads(pore)}$	concentration (per volume including solid and void space) of molecules on the pore walls [in the pores, Eq. (15.4)]
c^*	concentration of labeled molecules
c_0	initial value of c; value of c in bulk fluid
c_s	value of c at particle surface
c_∞	value of c as $t \to \infty$
C	torque; concentration in external fluid phase; dimensionless concentration c/c_0
C_0	initial value of C
C_∞	value of C as $t \to \infty$

Diffusion in Nanoporous Materials. Jörg Kärger, Douglas M. Ruthven, and Doros N. Theodorou.
© 2012 Wiley-VCH Verlag GmbH & Co. KGaA. Published 2012 by Wiley-VCH Verlag GmbH & Co. KGaA.

C_s	volumetric heat capacity
d	pore diameter, particle diameter
d_f	generalized dimension
$\mathcal{D}$	self-diffusivity [defined by Eqs. (1.11) or (1.12), sometimes referred to as D_s or D_{self}]
$\mathcal{D}_{1(2,3)}$ or $\mathcal{D}_{x(y,z)}$	principal elements of the self-diffusion tensor
$\mathcal{D}_\infty$	pre-exponential factor in Arrhenius expression for temperature dependence of $\mathcal{D}$ [Eq. (4.21)]
$\mathcal{D}_{eff}$	effective self-diffusion coefficient [defined by Eq. (11.16)]
$\mathcal{D}_i$	self-diffusivity in i-th region or of i-th component
$\mathcal{D}_{inter}$	self-diffusivity in intercrystalline space
$\mathcal{D}_{intra}$	intracrystalline self-diffusivity
$\mathcal{D}_{restr}$	maximum value of PFG NMR self-diffusivity under confinement [examples by Eq. (11.30) and subsequent equations]
$\mathcal{D}_{l.r}$	long-range self-diffusivity (through a bed of microporous crystals)
$\mathcal{D}_{\|(\perp)}$	self-diffusivity parallel (perpendicular) to the symmetry axis
D	(transport) diffusivity [defined by Eq. (1.1)]
D_0	corrected transport diffusivity [defined by Eq. (1.31)]
$D_{0A(B)}$	corrected transport diffusivity of component A(B) for binary mixture [Eqs. (6.107) and (8.19)]
D_c	microparticle ($\approx$intracrystalline) diffusivity
D_{eff}	effective diffusivity
D_{ij}	element of the diffusion tensor (Chapter 1) or of the diffusion matrix (Chapter 3)
D_{inter}	intercrystalline diffusivity
D_{intra}	intracrystalline diffusivity
D_K	Knudsen diffusivity
$D_{l.r.}$	long-range diffusivity
D_L	axial dispersion coefficient
D_m	molecular diffusivity
D_{macro}	macropore diffusivity
D_{micro}	micropore diffusivity
$D_{open\,(closed)}$	diffusivity through open ("closed") windows
D_p	pore diffusivity
D_s	surface diffusivity (self-diffusivity in some figures)
D_{self}	self-diffusivity (in some figures)
D_{AB}	molecular diffusivity in binary A–B mixture
D_{vis}	equivalent diffusivity for viscous (Poiseuille) flow
$\mathcal{D}_{AB}$	Stefan–Maxwell diffusivity [Eqs. (1.28) and (8.24); Chapter 3]
$\hat{e}$	unit vector along molecular axis
e	electronic charge

E	diffusional activation energy; potential energy of a magnetic dipole (Chapter 5); activation energy for reaction
E_0	energy difference between cavity and window [Eq. (4.40)]; incident electric field
E_m	energy eigenvalues of magnetic dipole in magnetic field
E_p	efficiency of parallel computing [Eq. (8.48)]
E_S	scattered electric field
f	friction factor ($=1/B$); fugacity; correlation factor (Chapter 2);
f_A	(molecular) partition function in the host cage (ground state)
f_d	dynamical correction factor to Transition-State Theory estimate of rate constant [Eq. (9.24)]
f_g	(molecular) partition function in the vapor phase per unit volume
f^+	reduced partition function for a molecule in the transition state (with the degree of freedom corresponding to passage through the transition state factored out)
$f(Q)$	concentration dependence of diffusivity [Eq. (6.40)]
$\mathbf{F}_i$	external force field acting on molecule of species i
F	mobility factor of single-file diffusion [Eqs. (5.2), (5.3)]; volumetric flow rate; fractional approach to equilibrium (Chapter 13)
F_g	gravitational force
$\mathbf{g}$	"turning force" [Eq. (8.51)]; magnetic field gradient
$G_{(s, D)}$	van Hove correlation function [for the same molecule, for different molecules; Eq. (10.43) and subsequent equations]
h	Planck's constant; heat transfer coefficient (particle–fluid)
$\hbar$	modified Planck's constant ($= h/2\pi$)
H	height equivalent to a theoretical plate (HETP)
$H(x)$	Heaviside step function [Eqs. (9.11) and (9.13)]
$\Delta H_{(0)}$	enthalpy change on adsorption (within the Henry's law region)
$\mathcal{H}(\mathbf{q},\mathbf{p})$	Hamiltonian ($= \dot{\mathbf{q}} \cdot \mathbf{p} - \mathcal{L}(\mathbf{q},\dot{\mathbf{q}})$)
I	moment of inertia
$\mathbf{I}$	spin vector
I_m	spin quantum number
$I_{xx\,(yy,zz)}$	principal moments of inertia
$I(\kappa, t)$	intermediate scattering function [Eq. (10.56)]
$\mathbf{j}_i$	microscopic (molecular) current of component $i \left(= \sum_{l=1}^{N_i} \mathbf{u}_{li}\right)$
j	molecular jump rate

$J_{(i)}$	diffusive flux (of component i) relative to plane of net molar flow – see also N
J^*	diffusive flux of labeled molecules
k	index of dimensionality in Eqs. (2.6) and (2.7); intrinsic reactivity; linear driving force mass transfer coefficient [Eq. (6.72)]
$\mathbf{k}$	wave vector (Chapter 10)
k_B	Boltzmann's constant
k_a	first order rate constant based on surface area
k_{eff}	effective transition rate
k_{ev}	effective first-order rate constant based on catalyst particle volume
k_f	external mass transfer coefficient based on fluid phase concentration difference as the driving force
$k_{i \to j}$	rate constant for transition from state i to state j
k_s	surface permeability coefficient; external mass transfer coefficient based on adsorbed phase concentration difference as the driving force
k_v	first-order rate constant based on catalyst particle volume
k_∞	pre-exponential factor [Eq. (21.15)]
$\mathbf{K}$	matrix of rate constants K_{ij}
K	dimensionless Henry's law constant based on intraparticle concentration ($=q^*/c$)
K'	dimensionless Henry's law constant based on pore volume [Eq. (4.13b)]
K_H	Henry constant based on fugacity [Eq. (7.35)]
K_{ij}	element of transition rate constant matrix ($= k_{j \to i}$ for $i \neq j$ and $-\sum_{j \neq i} k_{i \to j}$ for $i = j$)
K_∞	pre-exponential factor in expression for temperature dependence of K
$\mathcal{K}(\mathbf{p})$	kinetic energy in terms of generalized momenta $\left(= \sum_{i=1}^N \frac{p_i^2}{2m_i} \right)$
l	step size (in random walk); distance between successive barriers (Chapter 2); half-thickness of slab (Chapter 6); bond length; distance between the centers of adjacent cavities; mean distance between adjacent molecules [Eq. (19.4)]; bed depth (Table 6.3); thickness of the porous plug [Eq. (14.73)]
l_0	equilibrium bond length
L	hole distance; (crystal) length; latent heat of vaporization (Chapter 4); parameter $k_f R_p/eDp$ or $k_f r_c/KD_c$ (Chapter 6); parameter defined by Eq. (14.48); angular momentum
L'	parameter defined by Eq. (14.58)
L_{kj}	phenomenological coefficient [Eqs. (3.20), (8.8), and (8.26) and so on]

$L_{n,m}$	parameter defined in Table 6.2
L_x	cross coefficient
L_{tr}	parameter defined by Eq. (14.65);
$\mathcal{L}(\boldsymbol{q}, \dot{\boldsymbol{q}})$	Lagrangian ($= \mathcal{K} - \mathcal{V}$)
m	number of steps in positive directions (Chapter 2); mass of a molecule or particle; $\gamma \delta g$ [Eq. (11.13)]; reaction order
m_i	mass of site i
m_t	mass adsorbed at time t [$\equiv m(t)$]
m_∞	mass adsorbed at $t \to \infty$ [$\equiv m(\infty)$]
M	total quantity of diffusing substance [Eq. (1.6)], molecular weight; specific magnetization; parameter defined in Table 6.3
M_0	mass of original object
$M_{1/a}$	mass of reduced-scale replica
M_n	n-th moment
n	number of steps; number of sites; number of moles or molecules; refractive index
n_c	number of cages or cavities per unit volume
n_i	number of molecules (or moles) of species i
$n_{\text{ads(pore)}}$	number of molecules on the pore walls [in the pores, Eq. (15.2)]
N	number of sites (dimensionless file length); density of nuclear spins (Chapter 5); number of theoretical plates (Chapter 14)
$N_{(i)}$	total flux (of component i) relative to fixed frame of reference (for diffusion in pores: $N = J$)
N_A	number of molecules in state A
N_0	number of molecules in monolayer
N_{Avo}	Avogadro's number
$N_{1/a}$	number of molecules in monolayer of reduced-scale replica
N^*	fitted dimensionless file length
N_α	number of molecules of species α
$\boldsymbol{p}$	vector of generalized momenta
p	pressure; momentum of a neutron
p_H	pressure on membrane feed side
p_i	partial pressure of component i
p_{ji}	conditional probability to get from region j to i [Eq. (11.36)]
p_L	pressure at outlet of chromatographic column/membrane
p_{inter}	fraction of molecules in intercrystalline space
p_p	fraction of molecules in pore volume
p_s	fraction of molecules on pore surface; saturation pressure

p_γ	probability of free passage between adjacent channels in 1D pore systems
$p_{d(od,u)}$	relative pressure changes in piezometric measurements [Eq. (13.4) and subsequent equations]
p_{open}	fraction of unblocked entrance windows on external crystal surface
p'_{open}	fraction of unblocked surface area
p_π	duration of π pulse
$p(r)$	probability distribution function (pdf) of jump lengths
$p(t)$	probability distribution function (pdf) of residence times
$\mathbf{P}(t)$	n-dimensional vector of state probabilities $P_i(t)$
P_A^{eq}	equilibrium probability of state A
$P(\alpha)$	probability of dipole orientation α [Eq. (10.6)]
P_0	reference pressure; mean system pressure
$P_i(t)$	probability of state i in kinetic Monte Carlo simulation
$P(n, m)$	probability of event represented by n, m
$P(z, t)$	propagator function (probability density of finding a molecule or random walker at position z at time t)
$P(\mathbf{r}, t)$	propagator function (probability density of finding a molecule or random walker at position $\mathbf{r}$ at time t)
$P_{accept}(i \to j)$	probability of accepting move attempt from i to j during a Monte Carlo step [Eqs. (7.18), (7.20), (7.28) and (7.29)]
$\mathbf{q}$	generalized coordinate (for a diatomic molecule $\mathbf{r}_{cm}, \psi_1, \psi_2, l$)
q	concentration in adsorbed phase (or intracrystalline concentration in a zeolite)
q'	defined in Eq. (15.12)
$\bar{q}$	value of q averaged over a crystal or particle
q_0	initial value of q
q_∞	final value of q as $t \to \infty$
q_i	partial charge on atom i; molecular density of species i [in Eq. (8.13)]; quaternion [Eq. (8.62)]
q_s	saturation limit in Langmuir isotherm expression
q_{st}	isosteric heat of adsorption
q_{eq}	adsorbed phase concentration in equilibrium with the gas phase
$q_{surf(ace)}$	adsorbed phase concentration just inside the high resistance surface "skin"
q^*	adsorbed phase concentration in equilibrium with the gas phase ($= q_{eq}$); adsorbed phase concentration of labeled molecules
Q	heat flux (Chapter 3); normalized concentration (q/q_0); quadrupole moment
Q_t	quantity of sorbate which has diffused through a membrane in time t

$Q(N, V, T)$	canonical partition function [Eq. (7.4)]
$\mathbf{r}$	$3N$-dimensional vector of positions of all interaction sites [$\equiv (\mathbf{r}_1, \mathbf{r}_2, \ldots, \mathbf{r}_N)$]
$\mathbf{r}_{li}$	position of molecule l of species i
r	displacement (position) of a molecule; radial distance; pore radius (Chapters 2 and 4); auxiliary parameter [Eq. (6.50)]; ratio between diffusivities through open and "closed" windows [Eq. (16.5)]
r_c	radius of zeolite crystal
r_{cm}	position of center of mass
r_i	position of atom i
r_{ij}	distance between atoms i and j
R	gas constant, radial coordinate (for macroparticle); radius vector,
R_f	radius of adsorption front in irreversible adsorption model
R_g	gas constant (Chapter 14)
R_p	radius of particle
s	entropy per molecule (or per mole or unit volume); position of lone particle; Laplace transform variable; specific area of catalyst (Chapter 21)
$s(t)$	curvilinear coordinate
S	entropy; NMR signal intensity, diffusion path length
S_p	speedup factor in computation [Eq. (8.47)]
S_{AB}	intrinsic selectivity [ratio of permeabilities of A and B, Eq. (20.4)]
$S_{inc(coh)}$	incoherent (coherent) scattering function (Chapter 10)
$\mathbf{S}(\kappa)$	static structure factor, [$\lim_{\kappa \to 0} S(\kappa) = 1/\Gamma$]
t	time
t_{diff}	apparent diffusion time [= mean uptake time, Eq. (18.1)]
T	temperature; torque; total time span of single-particle tracking [Eq. (11.48)]
T_0	initial temperature
T_1	longitudinal nuclear magnetic relaxation time
T_2	transverse nuclear magnetic relaxation time
T_g	temperature of gas
T_s	temperature at particle surface; temperature of solid
$\mathbf{u}_{li}$	velocity vector of the center of mass of molecule l of species i
u	internal energy per unit volume (or per molecule); net diffusive flow velocity; variable in Eq. (6.90) and subsequent equations.
u_p	potential energy
u^*	potential energy of molecule in transition state

$\mathcal{V}(\boldsymbol{q})$	potential energy as function of the microscopic degrees of freedom
U	internal energy
U_0	difference in potential energies between cage and free gas [Eqs. (4.41) and (4.46)]
$U(\boldsymbol{R})$	potential of mean force with respect to slowly evolving degrees of freedom ($\boldsymbol{R}$), Eq. (7.47)
v	molecular velocity; interstitial fluid velocity; amplitude of (relative) volume variation in frequency response method; radii ratio in Eq. (2.38)
$\bar{v}$	mean molecular velocity
v_∞	terminal velocity
V	volume
$V_{A(B)}$	partial molar volume of component A (B)
V_0	height of energy; mean system volume
V_f	volume of fluid phase
V_s	volume of solid phase
V_I	partial molar volume
V_m	molar volume
$V(\boldsymbol{r})$	potential [Eq. (10.39)]
w	interaction energy of molecules on adjacent sites
$\boldsymbol{W}$	matrix correlating angular velocities with quaternion derivatives [Eqs. (8.63) and (8.64)]
$\hat{\boldsymbol{x}}$	unit vector along x axis
x	distance coordinate; mole fraction
$x^\dagger$	position of barrier (transition state)
X	normalized distance coordinate (x/l, Chapter 6)
X_k	generalized forces
$\hat{\boldsymbol{y}}$	unit vector along y axis
y	distance coordinate; mole fraction; time-normalized distance after Boltzmann transformation (Chapter 6)
z	distance coordinate; lattice coordination number [Eq. (2.20)]; relative pressure ($= P/P_o$); membrane thickness [Eq. (15.5) and subsequent equations]
$\hat{\boldsymbol{z}}$	unit vector along z axis
z_i	charge on cations of type i
Z	dimensionless radial coordinate R/R_p or r/r_c
$Z(N, V, T)$	configuration integral ($= \int \exp[-\beta \mathcal{V}(\boldsymbol{r})] d^{3N} r$)

Greek Letters

α	surface permeability [Eq. (12.5), synonymous with k_f]; fraction of pores filled with condensate [Eq. (4.20)]; $1 + N_B/N_A$ [Eq. (14.75)]; angle [Figure 10.1 and Eq. (10.3) and

	subsequent equations]; ratio of macropore and micropore diffusion time constants (Tables 6.2 and 6.3); separation factor; parameter defined with Eq. (14.60)
α_H	ratio of the dimensionless external and internal temperature gradients [Eq. (21.32)]
α_M	ratio of the dimensionless concentration gradients though film and inside particle [Eq. (21.36)]
α_n	roots of auxiliary equation
α'	heat transfer parameter defined in Eq. (13.3)
β	$= 1/(k_B T)$; uptake parameter defined in Tables 6.2 and 6.3; $c_s(-\Delta H)D_e/\lambda_s T_s$ (Chapter 21); memory parameter [Eq. (18.4)] of diffusion in MFI networks, parameter in Eq. (4.30)
β_n	roots of auxiliary equation
β'	heat transfer parameter defined in Eq. (13.1)
γ	activity coefficient; gyromagnetic (or magnetogyric) ratio; normalized uptake volume [Eq. (13.4) and subsequent equations]; parameter defined in Table 6.2; pressure correction factor [Eq. (14.27) and subsequent equations]; E/RT_s (Section 21.1.6)
γ_1, γ_2	constants in Eq. (14.38) and subsequent equations.
$\gamma(t)$	relative number of exchanged molecules; fraction of molecules leaving their original particle during time interval t
$\boldsymbol{\Gamma}$	point in 6N-dimensional phase space [$= (r, p)$]
Γ	thermodynamic factor [$= \partial \ln p/\partial \ln q$], passage rate between adjacent molecules in 1D arrangement
δ	film thickness; layer separation; jump distance; chemical shift; duration of magnetic field gradient pulses; normalized dosing volume [Eq. (13.4) and subsequent equations]
$\delta_{in(out)}$	in-phase (out-of-phase) characteristic functions in frequency response [Eqs. (13.16) or (13.17)]
Δ	time interval between successive gradient pulses
ε	external void fraction of adsorbent bed or membrane; Lennard-Jones force constant [Eq. (7.16)]; fraction of "closed" windows [Eq. (16.5)]
ε_p	porosity of adsorbent particle or membrane support
ε'_p	microporosity of adsorbent particle
ε_0	dielectric permittivity of free space
$\varepsilon_{\text{diff(des, barr)}}$	energies characterizing different steps during bed diffusion (Figures 11.9b and 11.10b)
ζ	angle correlating diffusion tensor with magnetic field direction in PFG NMR [Eq. (11.21)]; reduced coordinate in Eqs. (1.22) and (1.23)

η	viscosity; reduced coordinate in Eq. (1.22); interaction parameter [Eq. (4.33)]; time-normalized distance in Boltzmann transformation (Chapter 12); electric quadrupole asymmetry parameter; effectiveness factor
$\eta_{l(s)}$	parameter (square root of normalized diffusion time) in frequency response [Eqs. (13.16) and (13.17)]
η_i	probability that site i is occupied by a reactant molecule
$\eta_i^{\sigma_1...\sigma_{i-1} * \sigma_{i+1}...\sigma_N}$	single-file reactant concentration profile
ϑ	angle correlating diffusion tensor with magnetic field direction in PFG NMR [Eq. (11.21)]
$\vartheta^{(i)}$	phase difference of spin i from phased average
θ	bond angle; fractional saturation of adsorbent ($=q/q_s$); contact angle [Eq. (4.18)]; orientation angle of magnetic field with respect to electric field gradient tensor [Eq. (10.29) and subsequent equations]
θ_0	equilibrium bond angle; initial value of θ
θ_m	magic angle in MAS NMR ($= 54.74°$; i.e. $\cos^2 \theta_m = 1/3$)
θ_∞	final value of θ
θ_i	fractional saturation for component i ($=q_i/q_s$)
θ_H	fractional saturation on membrane feed side
θ_L	fractional saturation on membrane permeate side
$\theta^{\sigma_1 \sigma_2...\sigma_N}$	probability of particular occupation pattern of single-file system
κ	time exponent in Eqs. (2.57) and (2.66); reaction probability during mean jump time; normalized concentration gradient at the surface (Chapter 6); wave vector difference (neutron/light scattering vector, also referred to as Q)
λ	distance (or vector) between adjacent sites or centers of adjacent cages; parameter in the Langmuir-type presentation of the concentration dependence $f(\theta)$ of diffusivities [Eq. (6.49) and subsequent equations]; auxiliary parameter in Eq. (6.87) and subsequent equations; wavelength or de Broglie wavelength; jump length ratio in Monte Carlo simulation (Figure 11.9); $=q_o/q_s$ [in Eq. (14.65)]; thermal conductivity
λ_g	thermal conductivity of gas phase
λ_s	thermal conductivity of solid
λ_{eff}	effective mean free path
Λ	fraction of sorbate introduced to the system that is eventually taken up by adsorbent
Λ_i	thermal wavelength of site i ($= [\beta h^2/(2\pi m_i)]^{\frac{1}{2}}$)
μ	chemical potential; auxiliary parameter defined by Eqs. (6.50) and (6.51); magnetic moment; first moment of time lag or chromatographic response; viscosity [Eqs. (15.6) and (15.7)]

ν	jump frequency; vibration frequency
ξ	reduced coordinate in Eqs. (1.7) and (1.22); dimensionless bed length parameter [Eq. (14.8)]
$\Xi(\mu_1, \mu_2, \ldots, \mu_{z-1}, V, T)$	grand partition function [Eq. (7.26)]
π	permeance [Eq. (15.5)]
π_i	permeability for component i
ϱ	density; radius of curvature (Chapter 8)
$\varrho^{NVT}(\mathbf{r}, \mathbf{p})$	probability density of atomistic microstates [Eq. (7.3)]
$\varrho^{\mu_1 \mu_2 \ldots \mu_{z-1} VT}(\mathbf{r} \ldots N_{z-1})$	configuration-space probability density of the grand canonical ensemble [Eq. (7.25)]
σ	rate of entropy generation (Chapter 3); molecular diameter; Lennard-Jones force constant [Eq. (7.16)]; molecular diameter; surface tension [Eq. (4.18)]; scattering cross-section [Eq. (10.40)]
σ_i	variable indicating occupation state of site i [$\sigma_i = 0$ (1): site vacant (occupied)]
σ_2	second moment of time lag or chromatographic response
τ	time interval between molecular jumps; tortuosity (in Chapter 11: τ_{tort}); dimensionless time parameter; time interval between the $\pi/2$ and π RF pulses in the Hahn spin echo sequence; decay constant in light scattering [correlation time, Eq. (10.70)]
τ_i	mean intracrystalline life time of a molecule on site i (Chapter 5), mean life time in region i (Chapter 11)
τ_c	correlation time
τ_{intra}	(mean) intracrystalline life time
τ_{intra}^{diff}	intracrystalline/particle life time under diffusion limitation ($= R_p^2/15 \, \mathcal{D}_{intra}$)
φ	phase difference (between volume and pressure variation in frequency response)
φ^L	light phase angle
$\varphi(\tau)$	intracrystalline molecular residence time distribution function
$\varphi_i(\tau)$	intracrystalline residence time distribution of a molecule on site i
$\varphi_i^{\sigma_1 \ldots \sigma_{i-1} * \sigma_{i+1} \ldots \sigma_N}(\tau)$	single-file residence time distribution function for a given occupation pattern (configuration)
ϕ	phase angle; Thiele modulus; torsion angle; orientation angle of magnetic field with respect to electric field gradient tensor [Eq. (10.29) and subsequent equations]
ϕ_c	Thiele modulus for zeolite crystal
ϕ_L	Thiele modulus for parallel sided slab
ϕ_p	Thiele modulus for catalyst particle
ϕ_S	Thiele modulus for spherical particle

χ	phase angle; out-of-plane angle; valve constant in piezometry [Eq. (13.4) and subsequent equations]
χ_{ij}	holonomic constraint condition [$\equiv r_{ji}^2 - d_{ij}^2 = 0$, Eq. (8.81)]
ψ	wave function (Chapter 10); PFG NMR spin echo attenuation, phase lag
ψ_1	polar angle
ψ_2	azimuthal angle
ω	angular velocity or angular frequency; characteristic frequency [of radiation during state transition, Eq. (10.17)]
ω_i	frequency of (incident) light [(Eq. (10.57) and subsequent equations]
Ω	solid angle
$\Omega(\mu_1, \mu_2, \ldots, \mu_{z-1}, V, T)$	grand potential [Eq. (7.27)]

Vectors are denoted by bold italic face type

Common Dimensionless Variables

Nu	Nusselt number $= hd/\lambda_g$
Pe	Péclet number $= vd/D_L$
Re	Reynolds number $= \varepsilon \varrho v d/\eta$
Sc	Schmidt number $= \eta/\varrho D_m$
Sh	Sherwood number $= k_f d/D_m$

Index

a

absolute rate theory (See also Transition State Theory) 277
absorption line shape (NMR) 312–315
activated carbon
– adsorbents 520–524
– with hierarchical pore structure 542–543
activation energy (diffusion) 99, 521, 524, 825, 826
– variation with carbon number (5A and NaX) 593, 616
– in DDR-3 569
– for diffusional motion 131
– for intracrystalline diffusion 354, 476
– for linear paraffins 593, 616
– for long-range self-diffusion 549
– for low-occupancy diffusion 258
– for self-diffusion 108
– van der Waals diameter 585
– for xenon in NaX 638
active sites 732, 820
– for guest molecules 752
additivity of resistances, principle 146
adsorption–desorption cycles 408
adsorption equilibrium 40, 93, 434, 443, 460, 743, 783
adsorption isotherms 743
– of CO_2 and CH_4 in ZIF-8 745
adsorption/desorption kinetics 143–189
– concentration profiles for 154, 164–167, 174–175, 177–179, 400, 420, 700, 749
– effective diffusivity ratios 172
– from liquid phase 686–688
– numerical simulations 173
$AlPO_4$-5 membrane 780
aluminophosphates 3
amorphous microporous materials

– diffusion in microporous carbon 520–524
– diffusion in microporous glass 518–520
angular velocity 243, 308
anisotropic diffusion 10, 601, 703–710, 759
– in binary adsorbed phase 710
– diffusion/rearrangement model 451
anisotropy of real crystals 710–712
anomalous diffusion 27, 47
– fractal geometry 49–52
– probability distribution functions of residence time and jump length 47–49
aromatic hydrocarbons, diffusion 300, 624–631, 676–686
– benzene 627–631, 676–681
– C_8 aromatics 624–627, 681–686
Arrhenius plots 369, 573, 576, 583, 595, 600, 662, 668–672, 678, 681, 708, 761, 813
atomistic modeling 202, 203
axial diffusion 42
3A zeolite
– reduction in window size 593
4A zeolite
– activation energies 585
– diffusion in 571–573
– gravimetric uptake curves 591
– near isothermal behavior 430
– summary of diffusivity of data for 574, 575
5A zeolite
– comparison of zeolite samples 581, 582
– for cyclopropane and cis-butene in 476
– diffusion in 573–582, 595–596
– general patterns for diffusion in 582–586
– dimensionless HETP (H/d) vs. $(Dm/\varepsilon vd)$ for sorbates in 479
– experimental uptake curves 184, 572, 579
– HETP vs. cyclopropane and cis-butene in 476
– intra-cage jumpss 581

– measurement techniques 581, 582
– – with different crystal sizes 579, 580
– – uptake rate 576, 579
– PFG NMR signal attenuation curves 580

b

bed of microporous particles, diffusion 545
– expressions for diffusion in a continuum with 547
– mathematical modeling, approaches by 546
multicomponent systems 550–552
– temperature dependence 548–550
benzene, diffusion of (see also aromatic hydrocarbons) 676–686
– Arrhenius plot 678, 680
– benzene microdynamics 680, 681
– concentration dependence of 679
– diffusivity data in silicalite/ZSM-5 676, 677
– frequency response spectra, in silicalite 454
– self- and transport diffusion 680
Berty reactor 441
binary Langmuir isotherm (see also Langmuirian system) 14, 15, 46, 65–71, 74, 185, 784, 795
biporous adsorbent 143, 144, 156, 158, 429, 445, 461, 463, 465
– dual diffusion resistance model for 157
– transient concentration profile, schematic diagram 144
– uptake curves 156
boggsite (BOG) 130
Boltzmann distribution
– initial momenta 287
– of velocities 271, 282
Boltzmann factor 100, 212, 283, 307
Boltzmann–Matano method 414
Boltzmann transformation 162, 414
Born's approximation 328
boundary, absorbing 37–39
boundary conditions
– absorbing and reflecting boundaries 35–38
– diffusion and permeation, combined impact of 40–42
– matching conditions 39, 40
– partially reflecting boundary 38, 39
boundary surfaces 298, 299
branched and cyclic paraffins 666
– C_6 isomers, comparison of diffusivities for 670–673
– cyclohexane and alkyl cyclohexanes 669, 670
– diffusion
– – isobutane at high loadings 674–676
– – linear and branched hydrocarbons, comparison 673, 674
– 2,2-dimethylbutane 670
– isobutane at low loadings 667–669
– summary of diffusivity data 667
Brownian migration 7, 8
– brute force molecular dynamics 275, 277, 287, 291, 294, 297, 299
– self diffusivity 297
– time scale limitation of 287
iso-butane:
– diffusion of iso-butane in silicalite 668, 676
– MOFs 737
– NaX 610, 613, 615, 622, 623, 642
n-butane
– n-butane–perfluoromethane, selective diffusion measurement 641
– central torsion angle 209
– counter-diffusion of isobutane and 720
– diffusivity data in CuBTC 738
– diffusion in 5A 578–581
– diffusion in silicalite 660, 661, 668
– effective self-diffusion coefficient for 549
– evolution of distribution 401
– gravimetric sorption curves, in Linde 5A crystals 579
– influence of loading on the selectivity 781
– in NaX crystals 614
– and n-hexane, diffusivities for 673
– NMR relaxation times 581
– perm-selectivity 776, 799
– self-diffusivities of 715, 737
– variation of diffusivity, with ion exchange 592

c

Ca^{2+} cations 369, 562–564, 591, 619
cage-to-cage jumps 104–106
cage-type zeolites 64
Cahn balance 428
canonical partition function 197, 286
capillary condensation 97, 98, 216, 217, 518, 534, 544, 552
Carberry mixer 614, 615
carbonaceous materials, oxidation of 520
carbon molecular sieves 18, 521–524
– size-selective feature of 524
carbon nanotubes 64, 118
carbon number, variation of diffusivity with 593–595, 616, 661, 662
C_8 aromatics (see also aromatic hydrocarbons)
– diffusion of 624–627, 681, 682
– equilibrium concentration of isomers 828

– frequency response data for *p*-xylene 684, 685
– gravimetric uptake curves 683
– isomerization of 826
– loop for typical refinery 826
– macroscopic measurements 624, 625
– membrane permeation studies, evidence from 685, 686, 782
– microscopic measurements, comparison with 625, 626
– molecular sieve behavior 659
– *o*-xylene and *m*-xylene, diffusion of 624–626, 686
– in silicalite/ZSM-5 682
– uptake curves 681, 683
– ZLC and TZLC measurements 683, 684
Car–Parrinello molecular dynamics (CPMD) 257
carrier gas, effect of 644, 645
Carr–Purcell–Meiboom–Gill (CPMG) pulse sequences 323–325
Cartesian coordinates 195, 235, 236, 242, 249, 251, 254
cascades for separation processes 802
catalytic cracking 670, 823–825
cation-free eight-ring structures 567
– diffusion of CO_2 and CH_4 in DDR 569–571
– effect of window dimensions 567–569
– window dimensions, effect of 567–569
central limit theorem 30, 31, 55, 365, 759
Cerjan–Miller type algorithms 292
CHA zeolites
– diffusion in 568, 601, 791
– structure 562
chabazite 562, 564, 565
Chapman–Enskog kinetic theory 94
chemical diffusion 8
chemical potential 12–16, 215, 217
chemical shielding 318
chemical shift 318–320, 361, 362, 602, 647
chromatographic methods 459–482
– chromatographic column, element of 461
– chromatographic column, mathematical model for 460
– – form of response curves 463–464
– – time domain solutions 462–463
– inert carrier 460
– intraparticle diffusion
– – with adsorbable components 481–483
– – chromatographic column, mathematical model for 460–464
– – concentration profile, direct measurement of 480–481
– – limited penetration regime 480

– – moments analysis 464–468
– – step response 479–480
– – wall-coated column 480
– moments analysis 464–468
– – first/second moments 464–466
– – HETP/van Deemter equation 466–468
– – higher moments, use of 466
C_6 hydrocarbons 522 (see also benzene, hexane)
closed-system simulation 222
clustering effect 32
CO_2–CH_4/SAPO-34 system 74
CO_2–CH_4 separation 798
– membranes performance 798
coarse-graining 222–224
coefficients of transport diffusion 8
coherent scattering function 331, 332
coke deposition 827, 833, 834
– on diffusion of methane in H-ZSM-5, effect of 834
COMPASS force field 198
compensation effect 40, 311, 315, 383, 416, 549, 550, 802
competitive adsorption 87
composite particles 545
– approaches by mathematical modeling 546
– MCM-41 556
– multicomponent systems 550–552
– temperature dependence 548–550
– variation of long-range diffusivity, for composite pellet 548
computer reconstruction 204
concentration dependent systems, adsorption/desorption curves 172
concentration profiles, measurement of 480, 481, 819
– frequency ranges of different techniques 396
– intracrystalline, of isobutane 700
– simulated 759
– through membrane 785
– transient 152, 695
– – of deuterated propane 748
– – of propane in nanoporous crystal 749
condensation in capillaries 97, 98, 216, 217, 534, 552
condensed phase 552, 553
condition number 238
configurational integral 197
configurationally biased insertions 220
configuration-space probability density 207, 211
conservation equation 5, 62
constant-diffusivity system 148, 151, 756

constant field gradient NMR 359, 377
constant pressure (infinite volume) systems 175
constant stress tensor 210
constraint forces 250, 251
continuous time random walk (CTRW) 47
controlled porous glasses (CPG) 339
corrected diffusivities (D_0) 13, 14, 70, 73, 74, 77–79, 185, 233, 258, 265, 333, 520, 536, 537, 566, 570, 573, 581, 595, 625, 627, 629, 635, 636, 676, 683, 687, 689, 712, 736, 737, 738, 742, 753, 785, 790, 832
correlation effects 33–35
– correlated anisotropy 35
– vacancy correlations 33–35
correlation factor 34, 35
correlation function (of electric field vector) 228–230, 338, 340
correlation time 276, 281, 312, 313, 581
Coulombic interactions 199, 202, 203
counter diffusion 68, 91, 101, 119, 130, 135, 186, 187, 400, 439, 498, 718, 720, 742, 814
covalent organic frameworks (COFs) 264
crystal framework 747
crystallite radii 353, 354, 365, 367, 402
cyclohexane
– diffusion 519, 520, 672
– – in hierarchical activated carbon 543
– – measurements with 519
– PFG NMR 541
– sorption times 694
– transient uptake curves 554

d

Danckwerts boundary conditions 468
Darken equation 13, 101, 233
DDR-3 membrane 790
– diffusion of CO_2–CH_4 mixtures in 570
– SEM photomicrographs 772
– variation of permeance with kinetic diameter 774
DDR (ZSM-58),
– diffusion in 567–571
– structure of 565
de Broglie wavelength 326
density functional theory (DFT) 201, 257
– approximation 204
desorption curves
– theoretical 161, 178, 179
– experimental 168, 169, 179
desorption kinetics 160–179
– effective diffusivity ratios 172
– experimental concentration profiles 154, 409, 705, 749
– theoretical profiles 164–167
detours 760
deuterated benzene molecules, theoretical ^{2}H NMR line shapes 322
deuterated species in PFG NMR 362, 640
deuterium NMR 322, 323
diatomic molecule, description of configuration 195
diffusion
– activated 87, 99–106, 151
– – face-centered cubic crystal 99
anisotropy (See diffusion anisotropy)
– barriers 336
– bed 158, 159, 160, 179, 180, 431, 432, 444, 591, 624
– in binary adsorbed phase 64
– in carbon sieves 522, 524
– cation in zeolite 402
– of CO_2 and CH_4 in DDR 569–571
– co- and counter 395, 400, 438, 718
– in composite particles 494, 545
– concentration dependence 5, 14, 74, 101, 114, 172, 508, 535, 536, 609, 621, 629, 634, 643, 734, 737, 749, 790
– configurational 17, 85
– corrected 13, 14, 74, 77, 104, 268, 536, 566, 571, 583, 636, 684, 742, 785, 832
– deviation from ordinary 114
– driving force 5, 14, 60, 177, 230, 743, 773, 776
– elementary process 21, 45, 317, 395
– in fractals 49, 50, 53, 54, 56, 113, 114
– gas 4, 5, 43, 91, 98, 151, 160, 382
– inter, two identical species 65
– intracrystalline 10, 22, 23, 120, 150, 157, 371, 377, 378, 414, 441, 476, 505, 518, 582, 601, 627, 633, 646, 695, 734, 759, 763, 817, 820, 825
– in ion exchange resins 500
– Knudsen 96, 370, 503, 518, 533, 792
– limitation (of reaction) 811
– in liquid filled pores 80, 538, 687
– low-temperature 299
– macroscopic 42
– measurements (See diffusion measurements)
– model 172, 173, 462
– molecular dynamics (MD) simulation of 72, 117, 249, 257, 271, 319, 661, 736, 753, 754
– molecular 92, 96, 532, 548
– momentum transfer in 5, 94, 271
– Monte Carlo simulation of 44, 123, 206, 294, 295, 588, 618, 701, 713, 714

- phenomena   206, 337
- in porous glass   518, 519, 530, 554
- principles   143
- and reaction   807–809, 816, 832
- regimes, size-selective molecular sieving   85–87
- resistances   94, 156–160, 184
-- macropore and micropore   17
-- in nanoporous media   17, 18
- restricted   75, 125, 276, 381, 382, 530 (See also diffusional restrictions, zeolite catalysts)
- self and transport
-- single file   34, 49, 56, 111–121, 123, 135–138, 750
-- in sinusoidal field   107
-- in solids   99
-- Stefan-Maxwell   13, 45, 68, 69 (see also Maxwell-Stefan)
-- steric hindrance   75, 498, 539, 566, 623, 688, 779, 825
-- surface   75, 89, 93, 96, 98, 530, 531, 533–536, 539
-- tensor   11, 296, 362, 363, 708
-- thermal   63
-- tracer   21, 32, 741, 743
-- in zeolite A   45, 104, 566–567, 571–600
-- in zeolite X and Y   208, 607–648
-- in ZSM-5 (See also silicalite)   653–720
- with simple jump model   44–47
diffusional restrictions, zeolite catalysts   822
- activation energies   825, 826
- catalytic cracking over HZSM-5   824, 825
- catalytic cracking over zeolite Y   823, 824
- MTG reaction   830, 831
- MTO Process   831–833
- size exclusion   822, 823
- toluene, selective disproportionation of   828–830
- xylene isomerization   826–828
diffusion anisotropy   10, 601, 703–710, 759
- comparison of measured profiles   704
- correlation rule for structure-directed diffusion anisotropy   703, 704
- evidence by PFG NMR measurements   705–707
- host structure, evidence   364
- limits of correlation rule   707–710
- powder measurement   363, 364
- single-crystal measurements   362, 363
diffusion coefficients   13
- anomalous transport diffusion   45
- liquid phase   67
- measurement   504
- thermodynamically corrected   13

diffusion-controlled system
- concentration profiles during desorption   496
- dynamic behavior of   462
- uptake scales   421
diffusion measurements
- alternative approaches   362
- data analysis   356–358
- with different nuclei   379, 380
- different regimes of   364–379
-- intracrystalline   365
-- long-range   368, 369
-- in long-time limit   367, 368
-- in short-time limit   365–367
-- tortuosity factor and mechanism   369, 370
- experimental conditions, limitations, and options for   355
- experimental issues for observing light scattering phenomena   339–343
- extra-large stray-field gradients, benefit of   359
- filter techniques   341–343
- Fourier-transform   361, 362
- gradient pulse mismatch   358
- impedance by contaminants   360
- impedance by internal gradients   359, 360
- index matching   339, 340
- interference microscopy technique   404
- by light scattering   337–343
- mechanical instabilities   358, 359
- by monitoring molecular displacement   347
- multicomponent systems, self-diffusion measurement   361
- optical mixing techniques   340, 341
- by PFG NMR technique   348
-- pitfalls   358
-- sample preparation   355, 356
- single-molecule observation   383
-- fluorescence microscopy, basic principles of   384, 385
-- nanoporous materials, correlating structure/mass transfer   388, 389
-- time averaging vs. ensemble averaging   385–388
- theory   337–339
diffusion statistical mechanics   227–235
- mass fluxes in microporous medium   229–232
- transport in pure and mixed sorbates   232–234
- self-diffusivity   227–229
diffusivities   160, 161, 176, 411. See also diffusion
- definition of   4

– long-range 354
– permeabilities 757
diffusivity ratio 170
– variation 163
digital (photo-count) autocorrelation techniques 341
2,2-dimethylbutane (2,2-DMB), permeation fluxes and profiles 788
Dirac delta function 280, 460
discrepancy between micro and macro diffusivity values 632, 633
DDR
– diffusion of CO_2 and CH_4 in 569–571
– structure of 565
domain decomposition 241
Doppler shift 326
double refraction method, schematic representation 398
drag 45, 60, 98, 742
drift velocity 267
driving force, for diffusion 5, 12
– experimental evidence 15, 16
– gradient of chemical potential 12–15
– transport and self-diffusivities, relationship between 16, 17
dual control volume grand canonical molecular dynamics (DCV-GCMD) 269–272
– application 269
– role of control volumes 272
dual diffusion resistance model, for biporous particle 157
dusty gas model 94
dynamical correction factor 281–283, 291
dynamically corrected transition state theory 296

e

echo attenuation, for two-region diffusion 376
effective diffusivity 151, 170, 171, 524, 525
effectiveness factor 808
n-eicosane, atomistic and coarse grained representation 224
eight-ring zeolites 561–603
– anisotropic diffusion in CHA 602, 603
– 3A zeolites 592
– carbon number, variation of diffusivity with 593–595
– cation hydration 563, 564
– concentration and temperature dependence of diffusivity 566, 567
– diffusion in NaCaA zeolites 591, 592
– effective medium approximation 590, 591
– kinetic behavior of commercial 5A 597–599
– loading dependence of self-diffusivity 595
– Monte Carlo simulations 588–590
– pelletization, effects of 599–602
– self-diffusivity of water in ZK4 596
– structure of
– – CHA 564, 565
– – DDR 565
– – LTA 562, 563
– water vapor, effects of 596, 597
– window dimensions 565
Einstein relation 7, 28, 41, 294, 305, 356, 357, 421
– for ordinary diffusion 113, 114
– regime of diffusion 275
Einstein–Smoluchowski equation 28
electric field (in light scattering) 337, 338, 340
electron spin resonance (ESR) 402, 403
elementary diffusion processes measurement 305–343
– by light scattering 337–343
– by neutron scattering 326–336
– NMR spectroscopy 306–326
empirical tortuosity factors, for diffusion in liquid-filled pores 538
energy of activaton. *See* activation energy
entropy 61, 87, 103, 524
– production by internal processes 66
– rate of generation of 61–63
entropy of sorption, alkanes in silicalite 221
equations of motion 195, 196, 236, 242, 244, 247, 251, 252, 255, 257, 267, 287
equilibrium-controlled permeation 779
equilibrium isotherm 382, 437, 449, 536, 569, 776, 789
equilibrium molecular dynamics (EMD) simulations 227, 235–265, 271
– constraint dynamics in Cartesian coordinates 249–253
– domain decomposition 241
– extended ensemble molecular dynamics 253–257
– integrating the equations of motion, velocity Verlet algorithm 235–238
– to mixed sorbates, application 259–265
– multiple time step algorithms, rRESPA 238–240
– to pure sorbates, example application 257–259
– rigid linear molecules, molecular dynamics of 241–244
– rigid nonlinear molecules 245–249
equipartition theorem 255

ergodic hypothesis 338
ethane selectivity, for permeation 789
ethyl benzene, piezometric uptake curve 435
Eulerian angles 245–247
Ewald summation technique 203
excess chemical potential 218
– relation to fugacity 218
– calculation by Widom insertion 219
– calculation by configurationally biased insertion 222
experimental evidence
– findings referred to single-file diffusion 132, 135
–– catalysis 138, 139
–– pulsed field gradient NMR 135, 136
–– quasi-elastic neutron scattering 136, 137
–– tracer exchange and transient sorption experiments 137, 138
– ideal *vs.* real structure of single-file host systems 132–135
experimental methods
– classification 19–24
external field non-equilibrium molecular dynamics (EFNEMD) 266–269, 271

f

Fabry–Perot interferometers 342
fast tracer desorption 376, 601
Feynman–Hibbs correction 636
Fickian diffusivity 14, 16, 46, 90, 103
– dependencies 334
– equation 6, 416, 485
– model 792
Fick's first law 4, 12, 61, 356, 378, 406
Fick's second law 5, 265, 414
film resistance 20, 144, 179, 441, 489, 809, 812, 817
filter techniques (light scattering) 341, 342
Fincham's LEN algorithm 257
finite single-file systems 119
– catalytic reactions 123
–– adapting analysis for normal diffusion 123–125
–– dynamic Monte Carlo simulations 123
–– molecular traffic control 130–132
–– rigorous treatment 125–129
– mean square displacement 119, 120
– tracer exchange 120–122
flame ionization detector (FID) 470, 484
flexible zeolite model 201
fluctuation in number of sorbed molecules 213, 214
fluctuations in composition 229, 481
fluid–solid contactors 143

fluorescence microscopy 819
fluorine compounds (diffusion of) 637, 638
focal plane array (FPA) detector 405
force fields 197–203
Fourier transform 314, 318, 330, 334, 342, 453
– of PFG NMR spin echo attenuation 339, 352, 359, 361, 371, 374, 528, 734
– of stationary NMR spectrum 313
Fourier transform IR spectroscopy (FTIR) 399, 400, 405, 437
fractal geometry 49–52
fractal model, for system of parallel cylindrical pores 52
frames of reference 9, 10
– for binary system 10
– diffusivity for adsorbed phase 10
– interdiffusion process 9
– total volumetric flux 9
framework type code (FTC) 202
free energy methods 217–222
free induction decay 313, 314
free volume theory 108, 612
frequency response 22, 23, 436, 451, 455, 627, 681, 684, 686
– experimental systems 452–454
– in flow system 455, 456
– measurements 519
–– limits 451–452
– temperature frequency response 451
– theoretical model 448–451
– tracer exchange rates 447

g

gas diffusion 382
gaseous adsorption systems 473
gasoline range (C_5–C_{10})
– hydrocarbons 830
gasoline yield 831
gas–solid adsorption systems 43
gauche–trans conformational isomerizations 238
Gaussian approximation 330
Gaussian curve 329
Gaussian distributions 42, 203
– probability 305, 421
Gaussian response 463
Gear predictor-corrector algorithm 253
Geiger-Müller counter 446
generalized Maxwell–Stefan equations (see also Maxwell-Stefan model) 67
– application of 72
– diffusion in adsorbed phase 68–70
– general formulation 67, 68

Gibbs–Duhem equation 229, 231
Gibbs energy 98, 210, 218, 286
Gibbs ensemble Monte Carlo (GEMC) 215
Glueckauf approximation 465
Gorring's diffusion measurements 825
gradient relaxation molecular dynamics (GRMD) 265, 266, 268
Graham's law 4
grand canonical Monte Carlo (GCMC) simulations 212, 257, 264, 265, 269, 332, 745
– selection criteria 269
grand partition function 211
gravimetric methods
– gravimetric uptake (GU) measurements 427–432, 520
– tracer exchange rates 427
– – bed diffusional resistance, intrusion of 431–432
– – experimental system 428–429, 432
– – heat effects, intrusion of 429–430
– – negligible thermal effects, criterion for 430–431
– – response curves analysis 429
Green–Kubo equation 229, 233
guest diffusion
– in ferrierite 722, 723
– in nanoporous materials 35
guest uptake, nanoporous host 413
gyromagnetic ratio 308, 330, 352, 357, 369, 637, 639

h

Habgood model 785, 789
Hahn echo (HE) pulse sequences 323
Hahn's spin echo (runners' model) 323, 324
Hamilton's equations of motion 196
Hartree–Fock theory 201
H_2/CH_4 perm-selectivity 800
heat fluxes 62, 63, 816
heat transfer
– limitations 455
– resistance 182, 184, 429, 470, 815–817
Heaviside step function 279, 280, 290
He–CH_4 carriers 483
height equivalent to theoretical plate (HETP) 467, 476
Heisenberg's uncertainty principle 197
Helmholtz energy 197, 211, 218, 286
Henry constants 491, 633, 775
– of alkanes in silicalite 221
Henry's law 14, 201, 218, 233, 460, 510
– region 482, 503, 506, 774, 813

heptane–benzene mixture, experimental uptake curves 188
Hermite interpolation scheme 202
Hessian matrix 284, 288
hexamethyldisilazane 533
n-hexane 519
– activation energies for long-range self-diffusion 549
– adsorption branch perm-porosimetry curves for 795
– Arrhenius plot 672
– blocking effect 835
– cracking 823, 825, 830
– diffusivities
– – derived from high temperature 673
– – measurements 519
– – in Na75CaX 619
– – in porous silicon 537
– – in silicalite 208, 687, 795, 796
– effective self-diffusion coefficient 549
– hydroisomerization of 139
– and 3-methylpentane, over NaA 822
– relative effective diffusivities for 366
– saturation capacities for H-ZSM-5 and silicalite. 657
– self-diffusivities of 536
n-hexane–silicalite
– experimental perm-porosimetry curves, comparison of 796
hierarchical pore systems 539
– activated carbon with interpenetrating micro- and mesopores 541–544
– in mesoporous zeolites 544, 545
– ordered mesoporous material SBA-15 539–541
HKUST-1 structure 732, 733
homodyne measurement 340
van Hove correlation function 329
hydrogen bonding 521
hydrothermal treatment (of zeolite A) 597, 598
hyperpolarization 308
hysteresis effects 215, 541, 553–555, 554, 631
HZSM-5 based catalysts 824, 831

i

ideal adsorbed solution theory (IAST) 789
ideal gas 12, 209, 210, 215, 219, 220
imidazolate (IM) 739
inert carrier 460, 644, 645
inertia tensor 245
infrared microscopy measurements (IRM) 397, 519
infrequent event techniques 105, 275–301

– dynamical correction factor   281–283
– example applications   296–300
– kinetic Monte Carlo simulation   293–295
– – diffusivities in zeolites   296–300
– master equation   292, 293
– – analytical solution   295, 296
– multidimensional transition state theory   283–289
– of multistate multidimensional systems   290, 291
– numerical methods for   291, 292
– rate constant expression   276–281
– self-diffusivity
– – at high occupancy   300
– – at low occupancies   296–300
– for simulating diffusion in microporous solids   275–300
– statistical mechanics   276–292
– time scale separation   276–281
– tracking temporal evolution in network of states   292–296
– transition state theory (TST) approximation   281–283
inhomogeneous field effect   316
integral measurements   169, 443
integral sorption curves
– analysis   170
– for butylene and propylene   169
interaction energy   102, 103, 307, 317, 320, 325, 380, 567, 743
interference microscopy (IFM)   154, 379, 395, 397, 399, 400, 819, 821
– diffusion measurements   750
– monitoring intracrystalline concentration profiles   403–408, 408–415
– – Boltzmann's integration method   414–415
– – data analysis   406–408
– – principles of measurement   403–406
– – self-diffusivities and transport diffusivities   412
– – surface resistances, relevance of   413
– – transient sorption experiments, visualizing guest profiles   412–413
interferometer, Fabry-Perot   342
intermolecular interactions   219
interstitial diffusion   283
intracrystalline barriers   696, 697
– evidence from PFG NMR diffusion studies   701, 702
intracrystalline diffusion   10, 17, 85, 124, 354, 366, 367, 377, 379, 417, 482, 518, 561, 599, 602, 693, 758, 811, 817, 820, 835
– for hydrocarbons in H-ZSM-5   819
– measurements   486

– PFG NMR measurements of   358, 377
intracrystalline mean life time   124, 137–139, 375
intraparticle diffusion   459
– chromatographic measurements   470
– – dead volume   471
– – experimental conditions   470–472
– chromatographic method
– – with adsorbable components   481–483
– – chromatographic column, mathematical model for   460–464
– – concentration profile, direct measurement of   480–481
– – limited penetration regime   480
– – moments analysis   464–468
– – step response   479–480
– – wall-coated column   480
– experimental data, analysis of   472–473
– – axial dispersion, intrusion of   477–479
– – liquid systems   473
– – vapor phase systems   473–477
– long-column approximation   468
– – heat transfer resistance   470
– – nonlinear equilibrium   469–470
– – pressure drop   468–469
– membrane permeation measurements   501–510
– temporal analysis of products (TAP) system   500–501
– zero-length column (ZLC) method   483
– – counter-current ZLC (CCZLC)   498
– – deviations   489
– – extensions of   497–500
– – fluid film resistance   489
– – fluid phase hold-up   490–491
– – heat effects   492–493
– – intraparticle diffusion control, theory   485–486
– – isotherm nonlinearity, effect of   491–492
– – liquid phase measurements   498–500
– – macroporous particles, diffusion   488–489
– – practical considerations   493–497
– – principle of   483–485
– – short-time behavior   486–488
– – surface resistance, measurement   489–490
– – tracer ZLC   497–498
intrinsic diffusivity   776
intrinsic selectivity   774
IR micro-imaging/microscopy (IRM)   379, 395, 400, 403, 740
– intracrystalline concentration profiles, monitoring   403–408, 408–415
– – Boltzmann's integration method   414–415

– – data analysis 406–408
– – principles of measurement 403–406
– – self-diffusivities and transport diffusivities 412
– – surface resistances, relevance of 413
– – transient sorption experiments, visualizing guest profiles 412–413
irreversible adsorption model 176
irreversible process 351
isosteric heat of sorption 214, 218
– of alkanes in silicalite 221
isomerization of xylenes 659, 826, 827
isothermal approximation 63, 64, 182–185, 492
isothermal binary system 5
isothermal diffusion model 167
isothermal–isostress simulations 210
isothermal linear dual-resistance systems 151–160
– surface resistance plus internal diffusion 151, 152
– transient concentration profiles 152–156
– two diffusional resistances (biporous solid) 156–160
isothermal linear single-resistance systems 145–151
– external fluid film/surface resistance control 145, 146
– macropore diffusion control 149–151
– micropore diffusion control 146–149
– particle shape effect 149
isothermal MD simulation 132
isothermal nonlinear systems 160–179
– adsorption/desorption rates and effective diffusivities 170–172
– adsorption profiles 162–165
– approximate analytic representations 172
– desorption profiles 165–167
– experimental uptake rate data 167–170
– linear driving force approximation 172, 173
– macropore diffusion control 160, 161
– micropore diffusion control 160
– semi-infinite medium 161, 162
– shrinking core model 173–176
– surface resistance control–nonlinear systems 176–179
isotherm nonlinearity effect 491
isotropic diffusion 7, 363
isotropic medium 10

j

jump models 331
Juttner modulus 808

k

Kelvin equation 793
kinetic energy function 196
kinetic equation 278
kinetic Monte Carlo simulation (KMC) 276, 293–295
– for prediction of diffusivities in zeolites 296–300
kinetic theory of gases 88, 421
Kirchhoff's law 293
Knudsen diffusion 17, 49, 88, 89, 92, 150, 382, 501, 503, 504, 518, 525, 800
Knudsen flux 96, 98, 792
Koch curve 50, 51

l

Lagrange multipliers 251, 252
Lambert–Beer law 406
laminar fluid film 145
Langmuirian systems 162, 167, 176, 431, 491, 503, 776, 787, 789
– effective integral diffusivity with diffusivity ratio, variation 171
– flow through membrane and time delay 507
– theoretical adsorption and desorption curves for 178
– transient behavior of 506
Langmuir isotherm 46, 74, 101, 173, 736, 784, 795
Laplace transform 444, 467
large-scale atomic molecular massively parallel simulator (LAMMPS) 241
Larmor condition 308, 311
Larmor frequencies 314, 315
– of nuclear spins 361
lattice coordination number 34
Legendre transformation 196, 210, 211
LEN algorithm 243, 244
Lennard-Jones interactions 254
Lennard-Jones (LJ) potential 199, 202, 238
Lennard-Jones simulations 271
levitation effect 131, 132, 617
light scattering 305, 326, 337, 339, 340
linear alkanes
– Arrhenius plots 662
– comparison of self-diffusivities of n-alkanes at low loadings 661
– diffusivity data for n-alkanes 660
– loading dependence of diffusivity 665
– macroscale measurements 663, 664
– microdynamic behavior 665
– microscale measurements 661
– molecular simulations 665, 666

– variation of PFG NMR self-diffusivity 666
linear and branched hydrocarbons, differences in diffusional behavior 702, 703
linear driving force (LDF) model 462, 465
– approximation 173
linear response theory 230
linear system 165, 171
line width 312–314, 317, 320, 361, 647
liquid-filled pores
– diffusion 538, 539
– – at high loadings 80, 81
– flux expressions reduce to 81
– Habgood model 81
– mutual diffusion effects 81
liquid phase diffusion 538
liquid scintillation counting (LSC) 446, 447
liquid ZLC measurements 500
loading dependence 75
– molecular simulation 78
– self-diffusivities 75–77
– structural defects, effect of 78, 80
– transport diffusivities 77, 78
long tail behavior 470
Löwenstein's rule 201
LTA zeolites (see also Zeolite A)
– cation sites 562, 563
– diffusion in 565–567, 571–601
– structure 562–564
– diffusion of water vapor 595, 596
– membrane 797

m

macrodiffusivity 43
macro FTIR technique 437
macro IR sorption rate measurements, experimental system 438
macro/meso-pore sizes 518
macropore control 150, 169, 493
macropore diffusion 42–44, 151, 176, 179, 463
– analysis 185
– controlled system 160, 164
– determination, in catalyst particle under reaction conditions 526, 527
– diffusional resistance 476
macropores 17, 18, 143, 463, 510, 546, 550, 556
macroporous catalysts 807
macroscopic experiments 761
macroscopic flux 267
macroscopic TZLC technique 541
magic angle spinning (MAS) 323, 324
– PFG NMR 358
– – self-diffusion studies 745

magnetic dipole moment 308
magnetic field, spin orientations in 309
magnetic interaction energy 307
magnetic moments, superposition 310
magnetic resonance tomography (MRT) 348
magnetization, specific (nuclear)
– longitudinal 315
– transverse 308
Markov chain 206, 207
mass fluxes 229
mass transfer processes 9, 47, 173, 449, 456
– coefficient 41, 145, 483
– resistance at interface 40
mass transfer rates 145
mass transfer resistance 455, 460, 469, 477
master equation 292, 293, 295, 296
– analytical solution 295, 296
matching condition 39, 40, 113
mathematical modeling of sorption kinetics 145–185
matter wave 326
Maxwell–Boltzmann distribution 235
Maxwell-Stefan diffusivities 227, 234, 262, 263, 737, 741, 742, 744, 748
Maxwell–Stefan model 68–72, 784–791
– diffusion in macro- and mesopores 74, 75
– membrane permeation 73, 74, 784–791
– parameter estimation 72, 73
McBain balance 428
MCM-41 crystal
– anisotropic self-diffusion of water in 530
– high-accuracy localized reaction 822
– source and characterization 528, 529
mean free path 88, 91, 368, 369, 551
mean jump lengths 48
mean square displacement 7, 8, 27, 28, 33, 48, 56, 117, 119, 275, 349, 363, 381, 525, 701, 710, 711
membrane permeation measurements 501–510, 673, 684
– steady-state permeability measurements 501–502
– time lag measurements
– – linear systems 504–505
– – nonlinear systems 506–507
– – single-crystal membrane technique, evaluation of 507–508
– – single-crystal zeolite membrane 505–506
– transient and steady-state measurements, comparison 510
– transient Wicke–Kallenbach experiment 508–510
– Wicke–Kallenbach steady-state method 503

– – intracrystalline diffusion 503–504
– – macro/mesopore diffusion measurements 503
memory 9, 31, 47, 113, 114, 235, 276, 708, 710
Menger sponge 50, 51
mesopores 17, 75, 87, 96, 528, 544, 608, 619
mesoporous membranes, diffusion through 530
– measurement 400
mesoporous silica 531, 532
– modified mesoporous membranes 533
– permeance measurements 532, 533
mesoporous Vycor glass 530, 531
metal organic frameworks (MOFs) 4, 113, 264, 416, 729, 730
– breathing effects 751–754
– CuBTC, guest diffusion 736–738
– diffusion in MIL-53 751–754
– HKUST-1 732
– for H_2 separation 799–800
– membranes 771
– MOF-5, guest diffusion 733–736
– pore segments in single-file arrangement 747–751
– porous solids, class of 730–732
– potential use of 732
– surface resistance 754
– – activation energies 761–762
– – experimental observations 754–757
– – generalization of model 761
– – intracrystalline barriers 761
– – simulation results 758–759
– – surface barrier, conceptual model 757
– – surface permeability and intracrystalline diffusivity 759–760
– – with three-dimensional pore networks 760
– – unblocked entrance windows, fraction of 760–761
– zeolitic imidazolate framework 8 (ZIF-8) 739
– – experimental self- and transport diffusivities 740–744
– – membrane-based gas separation 744–746
– Zn(tbip) 747–751
methane
– capillary condensation, binodal curves for 217
– NMR diffusivity of 835
– self-diffusivity 261
methanol
– adsorption 155
– concentration, transient profiles 409
– deuterated 387
– diffusion and reaction 832
– to gasoline (MTG) reaction 830–831
– in NaX crystals 77
– to olefins (MTO) reaction 831–833
– over HZSM-5 831
– uptake 417
– to yield dimethyl ether (DME) 831
methanol–ferrierite, theoretical transient adsorption and desorption curves 178, 179
3-methylpentane, cracking 823
– in silicalite 208
2-methylpentane (2 MP), permeation fluxes 788
Metropolis Monte Carlo 206
MFI crystal structure 654, 655, 697–699
– molecular sieve behavior 659
– saturation capacity 655–659
– – for H-ZSM-5 and silicalite 657
– zeolite crystals, self-diffusivity of water 381
MFI-type zeolites 653
micropore-controlled systems 151
micropore diffusion 149
– diffusional resistances 465, 807
micropore/macropore diffusivities 445
– chromatographic method, application of 464
microporous carbon 398. See also carbon molecular sieves
– diffusion in 520, 521, 524
– PFG NMR 521
– properties 52
microscale studies 688
– ammonia 689–691
– hydrogen 691 692
– tetrafluoromethane 688–689
– water and methanol 689
microscopic diffusivities 42–44
microscopic techniques 412
MIL-88A–D 752
MIL-53 (Al, Cr), breathing behavior 751
MIL-88 solids 731
MIL-47 yield self-diffusivities 753
mixed adsorbed phase, diffusion in 712
– blocking effects by co-adsorbed second guest species 712
– – methane in presence of benzene 712, 713
– – methane in presence of pyridine and ammonia 713, 714
– – n-butane in presence of isobutane 714–716
mobile phase model 106
– diffusivity in self-diffusion experiment 107
– transport diffusivity 107
MOF crystal 423, 747

MOF-5 (IRMOF-1) material 734
– activation energies 735
– PFG NMR spin echo attenuations 734
– self-diffusion coefficients of *n*-alkanes in 735
MOF ZIF-8 752, 800
– cage arrangement in 740
MOF Zn(tbip) 423
molecular diffusion 17, 19, 96, 347, 501, 808
molecular displacements 114, 115
– probability distribution 353
molecular dynamics (MD) 116–118, 235–272, 275
– equilibrium molecular dynamics simulations 235–265
– extended ensemble molecular dynamics simulations 253–257
– non-equilibrium molecular dynamics simulations 265–272
– predictions 262
molecular mechanics-type expression 204
molecular migration, feature of 305
molecular models construction 193–224
– coarse-graining and mean force potentials 222–224
– Monte Carlo simulation methods 206–217
– sorption equilibria, free energy methods for 217–222
– zeolite–sorbate systems, models and force fields 194–206
molecular orientation 245
molecular relaxation time 231
molecular sieve behavior 659
molecular simulations 78
– effect of structural defects 78–80
molecular traffic control (MTC) 130
– for CH_4 and CF_4 130
– conditions, to be fulfilled by 131
molecule–molecule interactions 520
molecule–wall collisions 91
moments, method of 21, 378, 444–445
momentum transfer (in neutron scattering) 5, 271
– in gas distribution 94
Monte Carlo (MC) simulation methods 44, 134, 206–217, 370, 371
– adapting analysis for normal diffusion 123–125
– canonical Monte Carlo 207–210
– Gibbs ensemble and gage cell Monte Carlo 215–217
– grand canonical Monte Carlo 210–214
– metropolis Monte Carlo algorithm 206, 207
– of random walk 588

morphology, influence on diffusion 193, 541
Mulliken population analysis 200
multicomponent systems, self-diffusion in 67, 73, 185, 361, 441, 637, 640, 645, 784
– evolving during catalytic conversion 645
– – cyclopropane into propene 645–647
– – isopropanol into acetone and propene 647, 648
– hydrocarbons 640–643
– – benzene–perfluorobenzene 641
– – ethane–ethene 642, 643
– – *n*-butane–perfluoromethane 641, 642
– – *n*-heptane–benzene 640, 641
– under influence of co-adsorption and carrier gases 643
– – effect of inert carrier gas 644, 645
– – effect of moisture 643
– – water and ammonia 643, 644
multidimensional transition state theory 283–289
multi-region approach 374
multistate multidimensional systems 290, 291

n

NaCaA crystals 402 (see also LTA)
nanoporous host materials
– bulk phase of 422
– host–guest systems 35
nanoporous materials, deviations from normal diffusion in 55–57
nanoporous particles 305
– two-dimensional model bed of 372
nanoporous sol–gel glass, single-particle trajectories 386
Nath, Escobedo, de Pablo (NERD) force fields parameters 200
NaX, diffusion in 607–648
NaX–methanol, transient temperature response 437
– self and transport diffusivities 78
NaX zeolite crystals 165, 187, 205
– gravimetrically measured transient sorption curves 432
– model bed 205
neutron scattering 305, 326, 327, 336, 337
– diffusion measurements 326–336
– evidence on elementary steps of diffusion 334
– experimental procedure 326, 327
– fundamental relations 330, 331
– intermediate scattering function 334, 335
– measurements 520

– principle of method 326, 327
– range of measurement 335, 336
– scattering patterns and molecular motion 330–336
– theory 327–330
– thermodynamic factor 331–334
neutron spin echo
– applications 335
– experiments 334
Newton–Raphson method 252
Newton's second law of motion 196
nitrogen 639, 640
– 4A micropores 476
– curves for 522
– experimental pore diffusivities 89
– experimental studies 421
– gyromagnetic ratio 639
– kinetic selectivity 522
– long-range diffusivity of 369
– permeation properties of 717
– sorption isotherms 213
– in zeolite NaCaA 369, 572–573
non-adsorbing carrier. See inert carrier
non-equilibrium molecular dynamics (NEMD) approaches 227, 265–272
– dual control volume grand canonical molecular dynamics 269–272
– external field NEMD 266–269
– gradient relaxation method 265, 266
– simulations 267
– trajectory 266
non-isothermal systems 5, 179–185, 430, 451, 470, 815–817
– experimental non-isothermal uptake curves 183–185
– intraparticle diffusion control 181–183
nonisotropic system 10
nonlinear system, transient behavior 506
Nosé MD method 256
nuclear magnetic resonance spectroscopy (NMR) 305, 306–326, 307, 446
– anisotropy of the chemical shift 320
– basic principles, behavior of isolated spins 306–309
– behavior of nuclear spins in compact material 309–318
– classical treatment 306–308
– experiments 204, 310
– field gradient method 379
– fundamentals of line shape 311
– imaging 481
– impact 323–326
– molecular motion effect on line shape 311–314

– pulse measurements fundamentals 314–318
– quadrupole NMR 320–323
– quantum mechanical treatment 308, 309
– resonance shifts by different surroundings 318–323
– schematic representation of 314
– self-diffusion measurements 352
– signal intensity 356
– spin echo 327, 335
– spin-mapping 348
– spins in different chemical surroundings 318–320
– tracer desorption curves 376, 380
– tracer desorption experiment 377
– tracer exchange measurements 383
– zeugmatography 348, 402
nuclear spin 306, 308, 309, 348, 380, 400, 402

o

n-octane 223
– diffusion in NaY, USY and NaX 618, 619
– extrapolation of 619
– intracrystalline diffusivities of 619
– in NaX 621
– NMR self-diffusivity 610
– PFG NMR measurements 618
Ohm's law 61
olefins
– in AgNaX 623
– C_3/C_2, equilibrium ratio of 832
– crystal size on 833
– higher 831
– light 623
– methanol to 831
– reduction in diffusivity 623
– translational mobility of 623
– yield, influence of crystal size on 833
one-component sorbate system 214
one-dimensional system 153
one-dimensional transport 112
Onsager coefficients 61, 230, 234, 264
Onsager reciprocity relations 61, 234
Onsager's regression theorem 339
open micropore systems 106, 527, 528
– estimating loading dependence 108
– mobile phase model 106, 107
– sinusoidal field 107, 108
organic linkers 731
oxygen
– in Bergbau-Forschung sieve 522
– bridges in zeolites 739
– centers of 608
– zeolite crystal structure 201

p

paraffins
- HZSM-5 824
- selectoforming process 825
- solubility of 696
paramagnetic centers 312
Parrinello–Rahman extended ensemble algorithm 210
particle–particle interaction 127
partition function 197, 255
n-pentane
- cracking of 826
- diffusion in silicalite/HZSM-5 660, 661, 663
- experimental uptake curves 183
- in porous Vycor glass 555
- QENS diffusion studies 612
- supercritical transition 555
periodically continuousmodel systems 200
permeabilities 411, 774
permeability–diffusivity ratio 760
permeability measurement, through silicalite 506
permeation efficiency 760
permeation, zeolite layer 792
perm-porosimetry 793
perturbation theory 309
PFG NMR. See pulsed field gradient NMR (PFG NMR)
phase-space probability density 285
phenomenological transport coefficients 230
piezometric method 156, 433, 435
piezometric system 435
- components of 433
- theoretical response curves for 435
Planck's relation 309, 317, 318
Poiseuille flow (see also viscous flow) 90, 94, 501, 532, 793, 501, 792
Poisson process 294
polar angle 96
polycrystalline membranes 773
polyethylene oxide 527
polymer–zeolite interface 801
polypropylene oxide 527
polystyrene particles
- in controlled porous glasses (CPG) 339
- diffusion measurements by light scattering 340
pore network, diffusion in 94
- capillary condensation 97, 98
- dusty gas model 94, 95
- effective medium approximation 95
- parallel pore model 96, 97
- random pore model 97

- tortuosity factor 95, 96
pore radius 86
pore size 85, 87, 535
- distribution 96
porosity 525
porous adsorbent particles, diffusion 159
porous catalyst particle
- diffusion and reaction 809
- direct measurement of tortuosity in 525, 526
porous coordination polymers (PCPs) 730
porous glasses 518
porous Vycor glass, diffusion in 553–555
positron emission profiling (PEP) 397, 481
potential of mean force 222, 223, 278
pre-exponential factor 105, 107, 288, 584–587
probability density 214, 282
probability distribution function (PDF) 47, 48, 276
probability ratios 293
propagator function 29–31, 114
propane
- comparison of diffusion in NaX, 5A and silicalite 666
- diffusion in LTA 582
- diffusion in silicalite 662
- transient concentration profiles 748
- transient intracrystalline concentration profiles of 749
propene 645
- conversion of cyclopropane 645, 646
- conversion of isopropanol in NaX 647
- cyclopropane 645–647
- isopropanol 647, 648
- in NaX 646
- transport inhibition of 648
propylene polymerization 831
Pt/H-mordenite 139
Pt/SiO$_2$ catalysts 139
pulsed field gradient NMR (PFG NMR) 347, 348, 353, 358, 497
- application of 365
- Arrhenius plot of 373
- attenuation curve 376
- complete evidence of
-- diffusivity, concept of 354–355
-- mean propagator, concept 352–355
- consistency, experimental tests of 379
-- crystal size, determination of 381–382
-- crystal size, variation of 379
-- diffusion measurements with different nuclei 379–380
-- external magnetic field, influence of 380

–– extracrystalline space, blocking of 380–381
–– long-range diffusion 382
–– self-diffusion *vs.* tracer desorption measurements 380
–– tracer exchange measurements 382–383
– diffusion anisotropy
–– host structure, evidence 364
–– measurements of 364
–– powder measurement 363–364
–– single-crystal measurements 362–363
– diffusion measurements 151, 330, 335, 366, 371, 400
–– alternative approaches 362
–– application 325
–– data analysis 356–358
–– different regimes of 364–379
–– experimental conditions, limitations, and options for 355
–– extra-large stray-field gradients, benefit of 359
–– Fourier-transform 361–362
–– gradient pulse mismatch 358
–– impedance by contaminants 360
–– impedance by internal gradients 359–360
–– mechanical instabilities 358–359
–– multicomponent systems, self-diffusion measurement 361
–– performance 323
–– pitfalls 358
–– sample preparation 355–356
– diffusivities 381, 384
– ethane, long-range self-diffusion of
–– temperature dependence 370
– fine-tuning 364
– imaging techniques 353
– long-range diffusivity measurements 365
– measurement, principle 348
–– basic experiment 351–352
–– fundamentals 348–351
– normal diffusion 356
– pulse sequences 358
– self-diffusivities 380
– signal attenuation 357
–– in beds of zeolite crystallites 374
–– curves 371
– signal-to-noise ratio 359
– sine-shaped gradient pulses and eddy-current quench pulses 360
– spin-echo attenuation 360, 364
–– curve 354
–– Fourier transform of 352
– surface barriers, observation 377–379
– tracer desorption technique 376–377, 382
–– two-dimensional Monte-Carlo simulation 371–374
–– two-region model 374–376
– zeolite crystallites, molecular transport 355
– zeolitic diffusion 379
pyridine, blocking effect 835
pyrolysis 520

q

quadrupole moment 321
quantum mechanics/molecular mechanics (QM/MM) 194, 200
quartz crystal balance 432–433
quasi-classical approximation 310
quasi-elastic neutron scattering (QENS) 41, 347
– diffusivities for pentane isomers 616
– intermediate scattering function 339
– measurements 223, 259
– MIL-47, CO_2 diffusivities 753
– qualitatively similar dependencies 754
– self-diffusion data 754
quasi-Newton algorithms 292
quaternions 246

r

radiofrequency pulse programs 317
random walk 5
– diffusion path for 36
rapid recirculation systems 440, 441
– model
–– Fick's equations, correspondence with 32, 33
–– mean square displacement 27–29
–– propagator 29–31
RATTLE algorithm 252
Reed–Ehrlich model 104, 636, 748
– for surface diffusion 101
Rees, magnetically driven frequency response system 453
refractive index (RI) 343, 470
resistor network model 94
resonance line broadening
– schematic representation 312
reversible reference system propagator algorithm (rRESPA) 240
– pseudo-code implementing 240
Reynolds number 21
– axial dispersion 473, 474
– region 471, 478, 817
rigid zeolite framework 299
12-ring zeolites 607. *See also* saturated hydrocarbons, diffusion of
– fluorine compounds 637, 638

– hydrogen   636
– methanol   635
– PFG NMR diffusion measurements
– – with different probe nuclei   636–640
– triethylamine   635, 636
– water in NaX and NaY   633–635
– X and Y zeolites, structure of   607–609
Rosenbluth weight   222
Rubotherm balance system   428
Runners model   315

S
Saddle point calculation algorithms   292
SAPO-34 catalysts   835
SAPO-34 membranes   802
– permeation   791
SAPO STA-7 crystals, pore structure   419
saturated hydrocarbons, diffusion of   659
– cyclohexane   616–618
– diffusion in NaCaX   619, 620
– diffusion measurements as evidence of structural imperfection   621–623
– diffusion of branched and cyclic paraffins   666–676
– – summary of diffusivity data   667
– evidence from NMR   609–615
– isoparaffins   616–618
– linear alkanes   659–666
– NMR and ZLC data for NaX, comparison of   615, 616
– n-octane diffusion in NaY, USY and NaX   618, 619
SBA-15 material   528, 539
– PFG NMR diffusion studies   540, 541
scattering experiments   327
– application of neutrons   326
Schmidt number   478
Schrödinger equation   327
scintillation cocktail   447
selective surface flow   86
selectivity, mutual diffusion effect   790
selectoforming process   825
self-diffusion   7, 8, 16, 32, 65, 227–229, 258, 293, 299, 338, 360
– coefficient   45, 233, 338, 350, 351
– corrected diffusivity (D_0)   66
– cross coefficients   67
– entropy production by internal processes   66
– experiments   407
– at high occupancy   300
– loading dependence   75–77
– – MD simulation   79
– at low occupancies   296–300

– phenomenological equations   65
– relationship
– – between coefficients   65
– – between self- and corrected transport diffusivities   66, 77
– of water   521
SHAKE procedure   252
shallow bed kinetic measurements, schematic diagram   442
Sherwood number   817
shielding effect   318
shrinking core model   173–176
Si/Al ratio   653, 654
Sierpinski gasket   50, 51
– diffusion in   55
signal-to-noise ratio   439
silicalite (HZSM-5),
– anisotropy   703–706
– diffusion in   653–722
– structure   654–657
– sub-structure   696–700, 821
– surface resistance   693, 694
– surface etching   695
silicalite membranes   686, 772, 781
– methane–ethane permeation   785
– n-butane–isobutane in   781
– single-component fluxes   780
– temperature dependence   775
simple point charge (SPC) model   200
simulation box   266
simulations, of multicomponent adsorption and diffusion   738
single-component diffusion equation   186
single-file systems, infinitely extended   112
– molecular dynamics   116–118
– random walk considerations   112–115
single-particle tracking (SPT)   383, 387
single-resistance diffusion model   156
single-step frequency response method   436
singlet density distribution   208
sinusoidal channel segment   223, 297
sinusoidal field, diffusion in   107, 108
– approximation   107
– self-diffusivity   107
sinusoidal perturbation   452
size-selective molecular sieving   85–87
sodium borosilicate glasses   518
solid–sorbed fluid systems   253
sorbate–sorbate interactions   258, 269
sorbate–sorbate potentials   200
sorbate–zeolite systems   203
sorption/desorption curve   442
sorption isotherms   263
– of nitrogen and carbon dioxide   213

sorption kinetics 143–188, 441, 501
- adsorption/desorption curves 161
- for binary mixtures 185–188
- co-diffusion 187, 188
- counter-diffusion 186, 187
- isothermal linear dual-resistance systems 151–160
- isothermal linear single-resistance systems 145–151
- isothermal nonlinear systems 160–179
- mass and heat transfer, resistances to 143, 144
- mathematical modeling of 145–185
- non-isothermal systems 179–185
sorption/tracer exchange rates, direct macroscopic measurement 427–428
- differential adsorption bed 441–443
- frequency response measurements 447
-- experimental systems 452–454
-- in flow system 455–456
-- measurement limits 451–452
-- results 454–455
-- temperature frequency response 451
-- theoretical model 448–451
- gravimetric methods 427
-- bed diffusional resistance, intrusion of 431–432
-- experimental checks 432
-- experimental system 428–429
-- heat effects, intrusion of 429–430
-- negligible thermal effects, criterion for 430–431
-- response curves analysis 429
- macro FTIR sorption rate measurements 437–440
- piezometric method 433
-- mathematical model 434–436
-- single-step frequency response 436
-- single-step temperature response 436–437
- quartz crystal balance 432–433
- rapid recirculation systems 440
-- liquid phase systems 441
- tapered element oscillating microbalance (TEOM) 432–433
- tracer exchange measurements 445
-- experimental procedure 447
-- radioisotopes detectors 446–447
- transient uptake rate data, analysis
-- method of moments 444–445
-- time domain matching 443–444
spin-echo attenuation 363, 367
- curves 376
spin–lattice relaxation 311

spin quantum number 320
STA-7 (30) crystal
- methanol, concentration profiles 420
static structure factor 333
statistical mechanics
- of diffusion 227–235
- of infrequent events 276–292
statistical mechanics-based simulation techniques 202
steady-state diffusivity 832
Stefan–Maxwell diffusivity (see also Maxwell-Stefan diffusivity) 13, 45, 68–71, 781–791
Stefan–Maxwell formulation 13
steric effects 17
- in larger pores 521
steric hindrance 521, 524
stiff orthorhombic model 297
stochastic simulation algorithms 206
Stokes' law 60
straight cylindrical pore, diffusion in 87
- combination of diffusional resistances 93, 94
- different mechanisms, relative importance of 92
- Knudsen mechanism 88–90
- molecular diffusion 91
- self-diffusion/tracer diffusion 92
- surface diffusion 92, 93
- transition region 91, 92
- viscous flow 90, 91
stray field gradient 359
string-of-beads system 478
supercritical transition, in adsorbed phase 555, 556
surface diffusion 92, 506, 534
- concentration dependence 535–538
- determination of surface diffusivities 534, 535
- mechanisms 100
-- by cage-to-cage jumps 104–106
-- Reed–Ehrlich model 101–104
-- vacancy diffusion 100, 101
- of propane on silica gel 535
surface permeability 756
- bulk diffusivity 762
- diffusivity 749
- simulated concentration profiles 759
surface resistances 19, 152, 411, 692, 693
- control, concentration profiles during desorption 496
- effect 153
- external resistance to mass transfer 19–21
- macroscopic rate measurements 693, 694
- surface effects 696

- surface etching 694, 695
- theoretical frequency response 450
- transient concentration profiles, measurement of 695, 696

t

tapered element oscillating microbalance (TEOM) 433
Taylor–Golay model 480
temporal analysis of products (TAP) system 500–501
test particle insertion method 218
thermal conductivity 470, 484, 817
thermal wavelength 197, 211
thermodynamic correction factors (see also chemical potential) 257, 333
thermodynamic forces and fluxes 60, 61
Thiele concept 124, 125
Thiele moduli 123–125, 808, 810, 811, 818, 823, 833
- effectiveness factor, variation of 810
- for first-order isothermal system 810
time scale separation 275, 276–281
tortuosity factors 96, 525
total pore diffusivity 90
tracer diffusion, basic principle 447
tracer exchange measurements 445
- experimental procedure 447
- radioisotopes detectors 446–447
tracer permeabilities 417
tracking temporal evolution
- in network of states 292–296
transferable potentials for phase equilibrium calculations (TraPPE) parameters 200
transformation matrix 246
transient adsorption/desorption curves, macroscopic measurement 445
transient concentration profiles imaging 395
- observation options 396
-- IR microscopy 399–400
-- magnetic resonance imaging (MRI) 400–403
-- optical microscopy 398–399
-- positron emission tomography (PET) 397
-- X-ray monitoring 397–398
- surface barriers, direct measurement of 415
-- sticking probabilities to nanoporous particles 421–423
-- surface permeability, concentration dependence 415–417
-- surface permeability through crystal faces 417–421
transition state theory (TST) 105, 277, 281

- application 277
- approximation 281–283, 290
transport diffusivities 7, 8, 15, 16, 32, 41, 65, 232, 268, 271, 329, 339, 742
- loading dependence 77, 78, 409
-- MD simulation 79
- and self-diffusivity, relation between 71, 72
- using thermodynamic factors 520
transport resistances 111
- assessment by micro-imaging 699, 700
transverse nuclear magnetization 313, 325
transverse relaxation times 325
Trotter theorem 239
tubular membrane module, construction of 801
two-component diffusion 716
- co- and counter-diffusion of benzene and toluene 718–720
- counter-current desorption of p-xylene–benzene 717, 718
- counter-diffusion of isobutane and n-butane 720, 721
- methane and ammonia 717
- methane and n-butane 259–262
- methane and tetrafluoromethane 716
- methane and xenon 716, 717
- permeation properties of nitrogen and carbon dioxide 717
type A zeolites, general patterns of behavior in 582–584
- activation energies
-- and pre-exponential factors 584–587
-- variation, for diffusion on 4A and 5A with molecular diameter 585
- Arrhenius plot 583
TZLC desorption curves 497
- experimental $vs.$ theoretical 498

u

ultraviolet absorption (UV) detection 470
unit bond vector 244
united-atom representation 195
unsaturated, and aromatic hydrocarbons in NaX 623
- benzene 627–632
-- hysteresis 631, 632
-- macroscopic and microscopic measurements, comparison of 627, 628
-- mechanism, diffusion in zeolites NaX and NaY 628–631
- C_8 aromatics 624–627
- discrepancy in measurements 632, 633
- light olefins 623
uptake curve approaches 147–149, 152, 174

– comparison of solutions 150
– experimental 180
– – details 176
– – invariance 181
– – for moisture 175
– expression for 181, 183
– theoretical 182
– – vs. experimental 168
– transient, non-isothermal models for 180
uptake vessel, pressure response 434

v

vacancy diffusion 34, 100
– self-diffusion, in cubic lattice 34
van Deemter equation 468, 480
van der Laan's theorem 444, 464, 467
van der Waals interactions 199
van Hove correlation functions 329
– self-correlation function 347
van't Hoff equation 795, 819
Verlet algorithm 237, 251
– leapfrog algorithm 247
– pseudo-code implementing 237
vibration frequency 100
Vignes correlation 785, 787
viscous flow 90, 94, 532, 793
Vycor glass, diffusion in 319

w

water
– adsorption 332
– anisotropic self-diffusion of 530
– elimination of 831
– hydrogen bonding of 521
– in MCM-41 539
– in MFI-type zeolite crystals 381
– O-H bond 250
– oxygen atom 250
– PFG NMR self-diffusion measurements for 602
– purification 143
– self-diffusivities of 521
– solubility of paraffins 696
– as solvent 687
– zeolite–sorbate system 249
Wicke–Kallenbach method 534
Widom's test particle insertion method 219
Wiener–Khintchine relation 341, 508
window blocking 587
– sorption cut-off 587
– variation of fraction of open windows, with degree of ion exchange 588

x

X-ray computed tomography (XCT) 407
X-ray diffraction (XRD) 202
o-xylene
– CCZLC curves 498
– gravimetric uptake curves 683
– in NaX zeolite 626
– permeance/selectivity, equimolar mixture of 782
p-xylene–benzene
– with benzene counter-adsorbing 718
– counter-current desorption of 717
xylene isomers 826
– equilibrium mixture 827
– in NaX zeolite 627
– NMR PFG self-diffusivities 625
p-xylene spectra 827, 829
– co-diffusion of 440
– counter-diffusion of 439
– diffusion time constants, concentration dependence 624
– experimental uptake curves 624
– frequency response data 684
– gravimetric uptake curves 683
– permeance/selectivity, equimolar mixture of 782
– set of 439
– yield, in toluene disproportionation 829
X zeolite membranes 802

z

Zeolite A 562–601
– cation sites 562, 563
– deactivation 579, 596–601
– diffusion in 565–567, 571–596
– structure 562–564
zeolite 4A, micropores of 476
zeolite 5A, ZLC response curves for N_2 488
zeolite catalysts
– coking of 833
– – information from fluorescence microscopy 835
– – information from PFG NMR 834–835
– diffusional effects 807
– diffusional restrictions 822
– – activation energies 825–826
– – catalytic cracking over HZSM-5 824–825
– – catalytic cracking over zeolite Y 823–824
– – MTG reaction 830–831
– – MTO Process 831–833
– – size exclusion 822–823
– – toluene, selective disproportionation of 828–830
– – xylene isomerization 826–828

- diffusion and reaction 807
- diffusion-controlled catalytic reaction
-- concentration profiles, direct measurement 819–820
-- HZSM-5 crystal, furfuryl alcohol reaction 820–821
-- mesoporous MCM-41 reaction 821–822
- diffusion limitation 813, 815
- effectiveness factor 808–811
- external mass transfer resistance 811
- internal and external resistances 816–817
- intracrystalline diffusivity, determination of 817–819
-- effective diffusivity, temperature dependence of 819
- non-isothermal systems 815–816
- pressure dependence 814–815
- reaction order 814
- temperature dependence 812–814
zeolite crystals 43, 145, 181
- diffusion 398
zeolite frameworks 197, 254
- reliable flexible models for 256
zeolite membranes 771–803
- behavior of 779
- binary mixtures, modeling permeation of
-- concentration profile 785–789
-- Maxwell–Stefan model 784–785
-- membrane thickness 791
-- mutual diffusion, importance of 789
-- support resistance 791–792
- cracks/defects 779
- gas mixtures, separation of 779
-- diffusion-controlled permeation 781–783
-- equilibrium-controlled permeation 783–784
-- size-selective molecular sieving 779–780
- membrane characterization
-- bypass flow 792–793
-- isotherm determination 795–796
-- perm-porosimetry 793–794
-- transient response analysis 797
- membrane separation processes
-- alcohols dehydration, pervaporation process 797–798
-- amorphous silica membranes 800
-- barriers to commercialization 802–803
-- butene isomers, separation of 799
-- CO_2–CH_4 separation 798
-- H_2 separation, MOF membranes 799–800
-- membrane modules 801–802
-- membrane reactors 802
-- stuffed membranes 801
- polymeric membrane 773
- porosimetry characterization, apparatus 794
- separations 775, 797
- single-component permeation 773
-- permeation, modeling of 776–778
-- selectivity/separation factor 774–776
- synthesis 772–773
- transport 773
zeolite NaCaA (see also LTA)
- ethane, molecular selfdiffusion 353
- long-range diffusivities for nitrogen 369
zeolite–sorbate interactions 197, 201
zeolite-sorbate systems 193, 194, 227, 249, 265
- *ab initio* molecular dynamics 203, 204
- meso/macroporous structure, computer reconstruction 204–206
- models and force fields 194–206
- molecular model and potential energy function 194–203
zeolite theta, frequency response spectrum 454, 455
Zeolite X (and Y)
- cation sits 609
- diffusion in 609–648
- hysteresis 631
- structure 607–609
zeolite Y (see also Ch 17)
- catalytic cracking 823
- equilibrium isotherm 437
zeolitic imidazolate frameworks (ZIFs) 731, 739, 799
- analogies in building blocks 739
- ethane-ethene mixtures in 745
- membrane
-- permeance with kinetic diameter for light gases 800
- methanol, ethane, and ethanol 741
- self- and transport diffusion 740
- ZIF-8 membranes 744, 746
zero-length column (ZLC) method 459, 483
- counter-current ZLC (CCZLC) 498
- curves, for benzene–silicalite 499
- desorption curves 495, 496, 499
-- for benzene–n-hexane, experimental liquid 500
-- for propane 497
- desorption, for n-hexadecane 488
- deviations 489
- extensions of 497–500
- fluid phase hold-up 490–491
- heat effects 492–493
- intraparticle diffusion control, theory 485–486

- isotherm nonlinearity, effect of   491–492
- liquid phase measurements   498–500
- macroporous particles, diffusion   488–489
- practical considerations   493–497
- principle of   483–485
- response curves   489, 490
- – for benzene–NaX   491, 492
- – for CO_2 desorbing   486, 487
- – isothermal criterion, validity   493
- – for N_2 488
- schematic diagram   484
- short-time behavior   486–488
- surface resistance
- – fluid film resistance   489
- – measurement   489–490
- tracer ZLC   497–498

ZLC curves for benzene–silicalite   499

Zn(tbip) crystal   418, 761
- boundary and equilibrium concentrations   755
- 1D-arrangement of   757
- mass transfer, structure model   758
- MOF   755
- MOF specimen of   747
- pore system of   757
- transport diffusion and self-diffusion   750

ZSM-5 (see also silicalite), 2,2-dimethylbutane   824